METRIC (SI) MULTIPLIERS

Prefix	Abbreviation	Value
Tera	T	10^{12}
Giga	G	10^{9}
Mega	M	10^{6}
Kilo	k	10^{3}
Hecto	h	10^{2}
Deka	da	10^{1}
Deci	d	10^{-1}
Centi	c	10^{-2}
Milli	m	10^{-3}
Micro	μ	10^{-6}
Nano	n	10^{-9}
Pico	p	10^{-12}
Femto	f	10^{-15}

SI DERIVED UNITS AND THEIR ABBREVIATIONS

Quantity	Unit	Abbreviation	In terms of base units†
Force	newton	N	$kg \cdot m/s^2$
Energy and work	joule	J	$kg \cdot m^2/s^2$
Power	watt	W	$kg \cdot m^2/s^3$
Pressure	pascal	Pa	$kg/(m \cdot s^2)$
Frequency	hertz	Hz	s^{-1}
Electric charge	coulomb	C	$A \cdot s$
Electric potential	volt	V	$kg \cdot m^2/(A \cdot s^3)$
Electric resistance	ohm	Ω	$kg \cdot m^2/(A^2 \cdot s^3)$
Capacitance	farad	F	$A^2 \cdot s^4/(kg \cdot m^2)$
Magnetic field strength	tesla	T	$kg/(A \cdot s^2)$
Magnetic flux	weber	Wb	$kg \cdot m^2/A \cdot s^2$
Inductance	henry	H	$kg \cdot m^2/(s^2 \cdot A^2)$

† kg = kilogram (mass), m = meter (length), s = second (time),
A = ampere (electric current).

MATHEMATICAL SIGNS AND SYMBOLS

$\propto$	is proportional to		
$=$	is equal to	$\leqslant$	is less than or equal to
$\approx$	is approximately equal to	$\geqslant$	is greater than or equal to
$\neq$	is not equal to	Σ	sum of
$>$	is greater than	$\bar{x}$	average value of x
$\gg$	is much greater than	Δx	change in x
$<$	is less than	$\Delta x \rightarrow 0$	Δx approaches zero
$\ll$	is much less than	$n!$	$n(n-1)(n-2). \ldots (1)$

GENERAL PHYSICS

GENERAL PHYSICS

Douglas C. Giancoli

PRENTICE-HALL, INC., Englewood Cliffs, New Jersey 07632

Library of Congress Cataloging in Publication Data

Giancoli, Douglas C.
    General physics.

    Includes index.
    1. Physics.   I. Title.
QC21.2.G5   1984   530   83-17780
ISBN 0-13-350884-6

Editorial/production supervision by Linda Mihatov
Interior design and cover design by Janet Schmid
Development editor: Raymond Mullaney
Manufacturing buyer: John Hall
Acquisitions editor: Doug Humphrey
Cover Photograph by © DiMaggio/Kalish, The Image Bank

Printed in the United States of America

10 9 8 7 6 5 4 3 2 1

ISBN 0-13-350884-6

Prentice-Hall International, Inc., *London*

Prentice-Hall of Australia Pty. Limited, *Sydney*

Editora Prentice-Hall do Brasil, Ltda., *Rio de Janeiro*

Prentice-Hall Canada Inc., *Toronto*

Prentice-Hall of India Private Limited, *New Delhi*

Prentice-Hall of Japan, Inc., *Tokyo*

Prentice-Hall of Southeast Asia Pte. Ltd., *Singapore*

Whitehall Books Limited, *Wellington, New Zealand*

Contents

Preface

This introductory calculus-based physics textbook is aimed at students majoring in physics, other sciences, and engineering. It is intended to be readable, interesting, accessible to students, and yet comprehensive, with careful and detailed development of physics principles and emphasis on problem solving.

There are, of course, a number of good physics textbooks on the market. Why another one? A major reason is that the first-rate comprehensive texts tend to be dry and formal, and hence often difficult and boring for students. They lack freshness. A common approach is to treat topics formally and abstractly first, and only later (if at all) to come down to earth and relate the material to the students' own experience. This approach may be appealing (it's elegant), but it can slow down the learning process for all but the best students. My approach is to recognize that physics is a description of reality and thus to start each topic with concrete observations and experiences that students can directly relate to. Readers are then led into the more formal and abstract treatment of topics. Not only does this make the material more interesting and easier to understand, but it is closer to the way physics is actually practiced. Historically, we didn't start with the second law of thermodynamics, for example, and then derive all kinds of consequences from it; rather, the law was a generalization of all kinds of phenomena. Yet many textbooks treat topics in physics in that reversed fashion. I have avoided that dogmatic approach of stating principles and then deriving conclusions; instead I develop the principles as generalizations from specific observations.

I have tried, also, to avoid pedantic treatments by making discussions clear and concise, and to eliminate a common fault of dragging out small points to the extent of making them seem big (thus confusing students). On the other hand, I have also tried to avoid the problem of leaving certain topics "hanging," with students wondering "why did we study that?" I thus have tried to indicate why each topic is important, and to bring each topic to completion. We study static forces in structures, for example, partly because real materials are elastic and also can fracture; so I have included the latter topics in the statics chapter. After saying this, I must mention some exceptions: I have treated a very small number of topics only very briefly (such as Maxwell's equations in differential form) and have not developed them fully. The point, in these few instances, is to let the student know of their existence. When they meet them again in the future they will at least have seen them before and won't be totally ignorant. A longer treatment here could not be done at an appropriate level and/or would have made the book too long.

The order of topics is more or less traditional, but the book allows for considerable flexibility in this order. It begins with mechanics (Chapters 1–14), including fluid mechanics, followed by waves (Chapters 15–16), kinetic theory and thermodynamics (Chapters 17–21), electricity and magnetism (Chapters 22–33), and light (Chapters 34–38). Finally there are five chapters on modern physics: special relativity (Chapter 39), quantum theory and atomic physics, including material on lasers and condensed matter physics (Chapters 40–41), nuclear physics (Chapter 42), and elementary particles, including brief discussions of quarks, charm, QCD, the "standard model," and grand unified theories (Chapter 43). The topics in modern physics are treated at an appropriate level (that is, not too difficult) and necessarily briefly, but in enough detail, I hope, to whet the appetite of the students and give them a taste of what is happening in physics today. Nonetheless, the real emphasis in this book is the 38 chapters on classical physics.

The tradition of beginning with mechanics is sensible, I believe, since mechanics was developed first, historically, and since so much else in physics depends on it. Within mechanics there are various ways to order the topics. The order of the chapters here does not have to be followed precisely. Statics, for example, can be covered either before or after dynamics. One of my reasons for placing statics after dynamics is that from experience I have found that students have trouble with the concept of force without motion. Once they have understood the connection between force and motion, including Newton's third law, they seem to be better able to deal with forces without motion. Furthermore, placing statics later allows full development of the concept of torque, which is also crucial for statics and can be difficult to understand in the absence of motion. Finally, statics is really a special case of dynamics—and we study it largely so that we can prevent static structures from becoming dynamic (falling down). Nonetheless, statics (Chapter 11) has been written so that it could be covered earlier, if desired, before dynamics, after a brief introduction to vectors.

Another option is the position of the chapters on light. They are placed after the chapters on electricity and magnetism and electromagnetic waves, as is typical. However, light could be treated immediately after the chapters on waves and sound (Chapters 15 and 16), thereby keeping the various types of wave motion in one place. Another position choice involves special relativity (Chapter 39), which is treated after electromagnetic waves and light have been covered. Relativity could, however, be covered along with mechanics, say after Chapter 8, since it mainly depends (except for the optional Section 39–2) on material only through Chapter 8.

Much attention is given to problem solving. Explicit hints on how to attack problems are given in several places early in the book, notably in Sections 2–7, 4–8, and 4–10 (whose title is "Notes on Problem Solving"). The last-named section is placed after the students have had some experience wrestling with problems and hopefully will then be motivated to read and pay attention to this section. Section 4–10 can, of course, be covered much earlier if an instructor so desires.

The large selection of worked-out examples, as well as problems, covers a wide range, and I hope are more interesting than is usual; they cover not only physics, but applications to engineering, other sciences, and everyday life. The approximately 2000 problems are arranged by section and are ranked according to difficulty: level I problems are simple, usually plug-in, types designed to give students confidence, and sometimes to illustrate a simple but interesting point or application; level II are normal problems, requiring thought and often a combination of two or more concepts; level III problems are the most difficult. The arrangement by section number means only that those problems depend on material up to and including that section—earlier sections and chapters are often relied upon, particularly in level II or III problems. The ranking of the problems by difficulty (I, II, III) is necessarily subjective and is intended only as a guide. Level II problems, particularly, are of a very wide range. Level III problems will challenge even superior students. It's a good idea to check level III problems carefully before including them in regular homework assignments. Answers to odd-numbered problems are given at the back of the book. Each chapter also contains a set of questions (about 1200 in total) requiring verbal answers. SI units are used throughout; British units are defined, but not used. A limited number of problems requiring a programmable calculator (or computer) are included in a number of chapters, as is a simple discussion (in optional Section 2–10) on how to do numerical integration.

It is assumed that the readers have taken calculus or are taking it concurrently. The derivative is first introduced at the end of Chapter 2 (kinematics) in an optional section. This material can easily be covered later, say when the integral is first discussed in Chapter 6 (work and energy). Calculus is treated gently and slowly, especially at first. In fact, throughout the book, each topic is begun at a fairly low level so that understanding is accessible to a wide range of students. The rigor normally expected at this level is quickly reached; and for the most motivated students there are advanced topics (noted as optional by an asterisk), as well as a few rather difficult problems (ranked III, see above). Mathematical tools are introduced where they are first needed: the derivative and integral as mentioned above, vector addition in Chapter 3, the dot product and cross product in Chapters 6 and 10, respectively, and so on. I believe this method is preferable to putting a lot of math in Chapter 1, because it provides motivation for the student (they see immediately, for example, why the dot and cross products are defined as they are). A few topics (such as dimensional analysis and order-of-magnitude estimating) are placed in Chapter 1 to make them more visible rather than burying them at some arbitrary place in the book. These could be covered later, when the need arises.

This book contains more material than can be covered comfortably in a shorter course of, say, one year; nonetheless it can be readily adapted to such a course. Sections marked by an asterisk are considered optional. These sections contain slightly more advanced physics material, or material not usually taught at this level, or interesting applications. They contain no material needed in later chapters (except, perhaps, in later optional sections). This does not imply that all nonstarred sections must be covered; there remains considerable flexibility in the choice of material to suit the needs of students and instructors. For a short course, in addition to optional sections, much or all of Chapters 10 (except Sections 10–1 and 10–2), 11, 12, 13, 23, 31, 32, 39–43 could be omitted, as well as selected parts of Chapters 8, 16, 27, 29, 33, 35–38. The topics not covered in class can still be read by students, and the book thus provides a valuable resource as a reference book because of its wide range of coverage.

It is necessary, I feel, to pay careful attention to detail, especially when deriving an important result. Whether it is a verbal discussion, or a mathematical one, I have aimed at including all steps in a derivation so that students don't get bogged down in details and then fail to understand the concept as a whole. I have tried to make clear which equations are general, and which not, by explicitly stating the limitations of important equations in brackets next to the equation, such as

$$x = x_0 + v_0 t + \tfrac{1}{2}at^2. \qquad \text{[constant acceleration]}$$

Rotational motion is difficult for most students. As an example of attention to detail (although this is not really a "detail"), I have carefully distinguished the position vector ($\mathbf{r}$) of a point and the perpendicular distance of that point from an axis (I call this R, using a small capital). This distinction (which enters particularly for torque and angular momentum) is often not made clear in other books; some books use r for both without distinguishing—and this can be very confusing to students. Also, I have treated rotational motion by starting with the simpler situation of rotation about an axis (Chapter 9), including angular momentum and rotational kinetic energy; only in Chapter 10 is the more general rotation about a point dealt with, and this slightly more advanced material (except Sections 10–1 and 10–2 on the vector product and the torque vector) can be omitted if desired.

Among other unusual treatments is Chapter 29, Sources of Magnetic Field: here, in one chapter, are discussed the magnetic field due to currents (including Ampère's law and the law of Biot-Savart) as well as magnetic materials, ferromagnetism, paramagnetism and diamagnetism. This has resulted in a treatment that is clearer, briefer, and more of a whole, and all the content is there. Another is the treatment of conservative forces and conservation of energy in Chapter 7, which is done carefully (showing explicitly, for example, why $W_{1 \to 2} = -W_{2 \to 1}$ for a conservative force) but without the long-winded confusion that is common. There is a discussion of diffusion (Chapter 18), unusual for a book at this level, but an important topic; not only is it done more clearly and simply than in more advanced books, but "real" diffusion is discussed, not just "self-diffusion."

I wish to thank the many people who contributed in various ways to making this a better book. The professors who read the manuscript and offered many excellent comments include James B. Gerhart, Edward F. Gibson, Robert B. Hallock, Gordon E. Jones, Terrill W. Mayes, Michael A. Morrison, Edward B. Nelson, Norman Pearlman, Sheridan Simon, Gilbert H. Ward, and Thomas H. Wood. Special thanks go to John Heilbron who offered valuable suggestions for clarifying the wonderful history of our subject. Special thanks also to Professors Richard Marrus and Howard Shugart for many helpful discussions, as well as for their hospitality at the University of California, Berkeley. Finally, I wish to thank the many people at Prentice–Hall who worked on this project, particularly Logan Campbell, Doug Humphrey, Linda Mihatov, Janet Schmid, and the patient and perspicacious development editor Ray Mullaney. The responsibility for all errors lies, of course, with me. I welcome comments and corrections.

Douglas C. Giancoli

Notes to Students and Instructors on the Format

1. Sections and subsections marked with a star (*) are considered optional (see the Preface).
2. The customary conventions are used: Symbols for quantities are italicized (such as m for mass), whereas units are not italicized (m for meter); boldface (**F**) is used for vectors.
3. Important terms are italicized where they are introduced, and the most important are in boldface (such as *coefficient of friction* and **acceleration**).
4. Few equations are valid in all situations. Where practical, the limitations of important equations are stated in square brackets next to the equation.
5. Worked-out Examples and their Solutions in the text are set off with a vertical colored line in the margin.
6. Each chapter ends with a Summary, giving a brief review of important concepts and terms (the most important ones are italicized here). The Summaries are not intended to give an understanding of the material, which can only be had from a study of the chapter.
7. Following the Summary in each chapter are sets of Questions that students should attempt to answer (to themselves at least) and Problems arranged according to section and difficulty (see the Preface). Questions and Problems that relate to optional sections are starred.
8. The appendixes contain useful mathematical formulas (including derivatives and integrals), a discussion of polar coordinates, and a table of isotopes with atomic masses and other data. Tables used frequently are located inside the front and back covers.
9. The extensive Index can be a useful tool. For example, it can be used to look up concepts or words whose meanings have been forgotten.

GENERAL PHYSICS

Introduction

1

Although the earliest scientific ideas date back to early recorded history, physics as we know it today began with Galileo Galilei (1564–1642). Indeed, Galileo and his successor Isaac Newton (1642–1727) created a revolution in scientific thought. The physics that developed over the next three centuries, reaching its culmination with the electromagnetic theory of light in the latter half of the nineteenth century, is now referred to as *classical physics*. By the turn of the century, it seemed that the physical world was very well understood. But in the early years of the twentieth century, new ideas and new experiments in physics indicated that some aspects of classical physics did not work for the tiny world of the atom or for objects traveling at very high speed. This brought on a second great revolution in physics, which gave birth to what is now called *modern physics*.

1–1 Science and Creativity

The principal aim of all sciences, including physics, is generally considered to be the ordering of the complex appearances detected by our senses—that is, an ordering of what we often refer to as the "world around us." Many people think of science as a mechanical process of collecting facts and devising theories. This is not the case. Science is a creative activity that in many respects resembles other creative activities of the human mind.

Let's take some examples to see why this is true. One important aspect of science is *observation* of events. But observation requires imagination, for scientists can never include everything in a description of what they observe. Hence, scientists must make judgments about what is relevant in their observations. As an example, let us consider how two great minds, Aristotle (384–322 B.C.) and Galileo (1564–1642), interpreted motion along a horizontal surface. Aristotle noted that objects given an initial push along the ground (or on a table top) always slow down and stop. Consequently Aristotle believed that the natural state of a body is at rest. Galileo, in his reexamination of horizontal motion in the early 1600s, chose rather to study the idealized case of motion free from resistance. In fact, Galileo imagined that if friction could be eliminated, an object given an initial push allong a horizontal surface would continue to move indefinitely without stopping. He concluded that for an object to be in motion was just as natural as to be at rest. By seeing something new in the same "facts," Galileo is often given credit for founding our modern view of motion (more details in Chapter 2). This seeing of something new was surely inspired thinking.

Theories are never derived from observations—they are *created* to explain observations. They are inspirations that come from the minds of human beings. For example, the idea that matter is made up of atoms (the atomic theory) was certainly not arrived at because someone observed atoms. Rather, the idea sprang from a creative mind. The theory of relativity, the electromagnetic theory of light, and Newton's law of universal gravitation were likewise the result of inspiration.

The great theories of science may be compared, as creative achievements, with great works of art or literature. But how does science differ from these other creative activities? One important difference is that science requires *testing* of its ideas or theories to see if predictions are borne out by experiment. Indeed, careful experimentation is a crucial part of physics.

Although the testing of theories can be considered to distinguish science from other creative fields, it should not be assumed that a theory is "proved" by testing. First of all, no measuring instrument is perfect, so precise confirmation cannot be possible. Furthermore, it is not possible to test a theory in every single possible circumstance. Hence a theory can never be absolutely verified.[†] In fact theories themselves are generally not perfect—a theory rarely agrees exactly, within experimental error, in every single case in which it is tested. Indeed, the history of science tells us that theories come and go; that long-held theories are replaced by new ones. The process of one theory replacing another is an important subject in the philosophy of science today; we can discuss it here only briefly.

A new theory is accepted by scientists in some cases because its predictions are quantitatively in much better agreement with experiment than is the older theory. But in many cases, a new theory is accepted only if it explains a greater *range* of phenomena than does the older one. Copernicus's sun-centered theory of the universe, for example, was no more accurate than Ptolemy's earth-centered theory for predicting the motion of heavenly bodies. But Copernicus's theory had consequences that Ptolemy's did not: for example, it made possible a determination of the order and distance of the planets and predicted the moonlike phases in the appearances of Venus. A simpler (or no more complex) and richer theory, one which unifies and explains a greater variety of phenomena, is more useful and beautiful to a scientist. And this aspect, as well as quantitative agreement, plays a major role in the acceptance of a theory.

An important aspect of any theory is how well it can quantitatively predict phenomena; and from this point of view, a new theory may often seem to be only a minor advance over the old one. For example, Einstein's theory of relativity gives predictions that differ very little from the older theories of Galileo and Newton in nearly all everyday situations; its predictions are better mainly in the extreme case of very high speeds close to the speed of light. From this point of view, the theory of relativity might be considered as mere "fine-tuning" of the older theory. But quantitative prediction is not the only important outcome of a theory. Our view of the world is affected as well. As a result of Einstein's theory of relativity, for example, our concepts of space and time have been completely changed; and we have come to see mass and energy as a single entity (via the

---

[†] Some philosophers therefore emphasize that testing of a theory can be used only to *falsify* it, not to confirm it—and/or to put a limit on its range of validity.

famous equation $E = mc^2$). Indeed, our view of the world underwent a major change when relativity theory came to be accepted.

 ## 1-2 Models, Theories, and Laws

When scientists are trying to understand a particular set of phenomena, they often make use of a **model**. A model, in the scientists' sense, is a kind of analogy or mental image of the phenomena in terms of something we are familiar with. One example is the wave model of light. We cannot see waves of light as we can water waves; but it is valuable to think of light as if it were made up of waves because experiments on light indicate that it behaves in many respects as water waves do.

The purpose of a model is to give us a mental or visual picture—something to hold onto—when we cannot see what actually is happening. Models often give us a deeper understanding: the analogy to a known system (for instance, water waves in the above example) can suggest new experiments to perform and can provide ideas about what other related phenomena might occur.

No model is ever perfect, and scientists are constantly trying to refine their models or to think up new ones when old ones do not seem adequate. The atomic model of matter has gone through many refinements. At one time or another, atoms were imagined to be tiny spheres with hooks on them (to explain chemical bonding), or as tiny billiard balls continually bouncing against each other. More recently, the "planetary model" of the atom visualized the atom as a nucleus with electrons revolving around it, just as the planets revolve about the sun.

You may wonder what the difference is between a theory and a model. Sometimes the words are used interchangeably. Usually, however, a model is fairly simple and provides a structural similarity to the phenomena being studied, whereas a **theory** is broader, more detailed, and attempts to solve a set of problems often with mathematical precision. Often, as a model is developed and modified and corresponds more closely to experiment over a wide range of phenomena, it may come to be referred to as a theory. The atomic theory is an example, as is the wave theory of light.

Models can be very helpful, and they often lead to important theories; but it is important not to confuse a model, or a theory, with the real system or the phenomena themselves.

Scientists give the title **law** to certain concise but general statements about how nature behaves (that momentum is conserved, for example); sometimes the statement takes the form of a relationship or equation between quantities (such as Newton's law of universal gravitation, $F = G\, m_1 m_2 / r^2$).

To be called a law, a statement must be found experimentally valid over a wide range of observed phenomena; in a sense, the law brings a unity to many observations. For less general statements, the term *principle* is often used (such as Archimedes' principle). Where to draw the line between laws and principles is, of course, arbitrary, and there is not always complete consistency.

Scientific laws are different from political laws in that the latter are *prescriptive*: they tell us how we must behave. Scientific laws are *descriptive*: they do not say how nature *must* behave, but rather describe how nature *does* behave. As with theories, laws cannot be tested in the infinite variety of cases possible. So we cannot be sure that any law is absolutely true. We use the term law when its validity has been tested over a wide range of cases, and when any limitations and the range of validity are clearly understood. Even then, as new information comes in, certain laws may have to be modified or discarded.

Scientists normally do their work as if the accepted laws and theories were true; but they are obliged to keep an open mind in case new information should alter the validity of any given law or theory.

⬜ 1–3 Measurement and Uncertainty

In the quest to understand the world around us, scientists seek to find relationships among physical quantities.

We may ask, for example, in what way does the magnitude of a force on an object affect the speed or acceleration of the body. Or by how much does the pressure of gas in a closed container (such as a tire) change if the temperature is raised or lowered? Scientists normally try to express such relationships quantitatively in terms of equations whose symbols represent the quantities involved. To determine (or confirm) the form of a relationship, careful experimental measurements are required, although creative thinking also plays a role.

It is interesting to note that the emphasis on measurement and the search for quantitative relationships between physical quantities was not always the perceived goal of physical science. It became a generally accepted goal only in the eighteenth century. That scientists agreed on this as a major goal of science was, of course, a free choice; and it was not obvious then that anything so deep or important would follow from it.

Today accurate measurements are an important part of physics. But no measurement is absolutely precise; there is an uncertainty associated with every measurement. Uncertainty arises from different sources; among the most important, other than blunders, are the limited accuracy of every measuring instrument and the inability to read an instrument beyond some fraction of the smallest division shown. For example, if you were to use a centimeter ruler to measure the width of a board, the result could be claimed to be accurate to about 0.1 cm, the smallest division on the ruler (although half of this value might be a valid claim as well). The reason for this is that it is difficult for the observer to interpolate between the smallest divisions, and the ruler itself has probably not been manufactured to an accuracy much better than this.

When giving the result of a measurement, it is good practice to state the precision, or **estimated uncertainty**, in the measurement. For example, the width of a board would be written as 23.2 ± 0.1 cm. The ± 0.1 cm ("plus or minus 0.1 cm") represents the estimated uncertainty in the measurement, so that the actual width most likely lies between 23.1 and 23.3 cm. The *percent uncertainty* is simply the ratio of the uncertainty to the measured value, multiplied by 100; for example, if the measurement is 23.2 and the uncertainty about 0.1 cm, the percent uncertainty is

$$\frac{0.1}{23.2} \times 100 = 0.4\%.$$

Often the uncertainty in a measured value is not specified explicitly; in this case, it is generally accepted that the uncertainty is approximately one or two units in the last digit specified. Although this is not as precise as actually specifying the uncertainty, it is often adequate. For example if a length is given as 23.2 cm, the uncertainty is assumed to be about 0.1 cm (or perhaps 0.2 cm). It is important in this case that you not write 23.20 cm, for this implies an uncertainty of 0.01 cm; it assumes that the length is probably between 23.19 cm and 23.21 cm when actually you only believe it is between 23.1 and 23.3 cm.

The number of reliably known digits in a number is called the number of **significant figures**. Thus there are four significant figures in the number 23.21 and two in the number 0.062 cm.

When making measurements, or when doing calculations, one should not keep more digits in the final answer than the number of significant figures. For example, to calculate the area of a rectangle 11.3 cm by 6.8 cm, the result of multiplication would be 76.84 cm², but this answer is clearly not accurate to 0.01 cm², since (using the outer limits of the assumed uncertainty for each measurement) the result could be between $11.2 \times 6.7 = 75.04$ cm² and $11.4 \times 6.9 = 78.46$ cm². At best, we can quote the answer as 77 cm², which implies an uncertainty of about 1 or 2 cm^2. The other two digits (in the number 76.84 cm^2) must be dropped since they are not significant. As a general rule, *the final result of a multiplication or division should have only as many digits as the number with the least number of significant figures used in the calculation.* In our example, 6.8 cm has the least number of significant figures, namely two; thus the 76.84 cm^2 must be rounded off to 77 cm^2. Keep in mind when you use a calculator that all the digits it produces may not be significant; and they should not be quoted (or written down) unless they are truly significant figures.

It is common in physics to write numbers in "powers of ten" or "exponential" notation, such as 36,900 as 3.69×10^4, or 0.0021 as 2.1×10^{-3}. One advantage of exponential notation is that it allows the number of significant figures to be clearly expressed. For example, it is not clear whether 36,900 has three, four, or five significant figures. If the number is known to an accuracy of three significant figures we write 3.69×10^4, but if it is known to four we write 3.690×10^4.

1–4   Units, Standards, and the SI System

The measurement of any quantity is made relative to a particular standard or unit, and this unit must be specified along with the numerical value of the quantity. For example, we can measure length in such units as inches, feet, or miles, or in the metric system in centimeters, meters, or kilometers. To specify that the length of a particular object is 18.6 is meaningless. The unit *must* be given; for clearly, 18.6 meters is very different from 18.6 inches or 18.6 millimeters.

Until about 200 years ago, the units of measurement were not standardized, and that made scientific communication difficult. Different people used different units: cubits, leagues, hands; and even the length of the foot varied from place to place.

The first real international standard was the establishment of the standard *meter* by the French Academy of Sciences in 1791.[†] The meter was defined as the distance between two finely engraved marks on a particular bar of platinum-iridium alloy now kept near Paris at the International Bureau of Weights and Measures. Accurate copies of the standard meter were sent to laboratories around the world. By the end of the nineteenth century it became possible, through the work of the American physicist A. A. Michelson (see Section 36–9), to define the meter in terms of the wavelength of light. The most recent standard was chosen in 1960: the meter (abbreviated m) is now defined as 1,650,763.73 wavelengths of a particular orange light emitted by the gas krypton 86. British units (inch, foot, mile) are now defined in terms of the meter. The inch (in) is defined as precisely 2.54 centimeters (cm; 1 cm = 0.01 m). Other conversion factors are given in the table on the inside of the back cover of this book.

The standard unit of *time* is the *second* (s). For many years, the second was defined as 1/86,400 of a mean solar day. The standard second is now defined more precisely in terms of vibrations within a cesium atom. Specifically, one second is defined as the time required for 9,192,631,770 vibrations of cesium atoms when they are vibrating in a specific manner. There are, of course, 60 s in one minute (min) and 60 minutes in one hour (h).

The definitions of other standard units for other quantities will be defined as we encounter them in later chapters.

In the metric system, the larger and smaller units are defined in multiples of 10 from the standard unit, and this makes calculation particularly easy. Thus one centimeter is $\frac{1}{100}$ m, one kilometer (km) is 1000 m, and so on. The prefixes

---

[†] In a spirit of rationality, the standard meter was originally chosen to be one ten-millionth of the distance from the earth's equator to either pole. Modern measurements of the earth's circumference reveal that the intended length is off by about one-fiftieth of 1 percent.

TABLE 1–1

Metric (SI) prefixes (multipliers)

Prefix	Abbreviation	Value
Tera	T	10^{12}
Giga	G	10^{9}
Mega	M	10^{6}
Kilo	k	10^{3}
Hecto	h	10^{2}
Deka	da	10^{1}
Deci	d	10^{-1}
Centi	c	10^{-2}
Milli	m	10^{-3}
Micro	μ	10^{-6}
Nano	n	10^{-9}
Pico	p	10^{-12}
Femto	f	10^{-15}

"centi-," "kilo-," and others are listed in Table 1–1 and can be applied not only to units of length, but to units of volume, mass, or any other metric unit. For example a centiliter (cL) is $\frac{1}{100}$ liter (L) and a kilogram (kg) is 1000 grams (g).

The conversion factors between the various units of the British system (for example 12 inches in a foot) make it unwieldy for calculation. This is one reason why scientists, and nearly all countries of the world, have adopted the metric system. Indeed, the United States is slowly converting to metric units and Britain has already done so to a large extent.

When dealing with the laws and equations of physics it is very important to use a consistent set of units. To take a very simple example, suppose you want to know how far you can go in your car in 40 min at a speed of 90 km/h. As we shall see in the next chapter, the distance, x, can be written as the product of the speed, v, and the time, t: $x = vt$. But if you multiply 90 km/h by 40 min, you will get a ridiculous answer. Since the speed, v, is given in terms of hours, the time, t, must also be in hours, and in this case it is $\frac{2}{3}$ h; therefore $x = (90 \text{ km/h})(\frac{2}{3} \text{ h}) = 60$ km. (Note how the hour units cancel in this equation; the equals sign in an equation applies not only to the numerical values but also to the units.) When dealing with more complicated equations, the necessity of using a consistent set of units is even more crucial.

Several systems of units have been in use over the years. Today the most important by far is the **Système International** (French for International System), which is abbreviated SI. In SI units, the standard of length is the meter, that for time is the second, and that for mass is the kilogram. This system used to be called the MKS (meter-kilogram-second) system.

A second metric system is the cgs system in which the centimeter, gram, and second are the standard units of length, mass, and time, as abbreviated in the title.

The British engineering system takes as its standards the foot for length, the pound for force, and the second for time.

SI units are the principal ones used today in scientific work. We will therefore use SI units almost exclusively in this book, although we will give the cgs and British units for various quantities when introduced, as well as conversion factors.

 1–5 Base Vs. Derived Quantities

Physical quantities can be divided into two categories: **base quantities** and **derived quantities**. The corresponding units for these quantities are called *base units* and *derived units*. Scientists, in the interest of simplicity, want the smallest number of base quantities possible that is consistent with a full description of the physical world. This number turns out to be seven, and for the SI are given in Table 1–2. All other quantities can be defined in terms of these seven base quantities.[†] The selection of

---

[†] The only exceptions are for angle (radians—see Chapter 9) and solid angle (steradian) for which general agreement has not been reached as to whether these are base or derived quantities.

TABLE 1–2

SI base quantities and units

Quantity	Unit	Unit Abbreviation
Length	meter	m
Time	second	s
Mass	kilogram	kg
Electric current	ampere	A
Temperature	kelvin	K
Amount of substance	mole	mol
Luminous intensity	candela	cd

which quantities are to be considered base and which derived is somewhat arbitrary. For example, in the British system force is considered base, whereas mass is derived—just the opposite of SI.

Most quantities are defined in terms of the base quantities; an example is speed, which is defined as distance traveled divided by the time it takes to travel that distance (Chapter 2). The base quantities are not definable in terms of other quantities, which is why they are called base. What we can do is specify a rule (or set of rules) for measuring that base quantity; this is called an *operational definition*, and such definitions can be made for both base and derived quantities.

1-6 Dimensions and Dimensional Analysis

When we speak of the *dimensions* of a quantity, we are referring to the type of units or base quantities that make it up. The dimensions of area, for example, are always length squared (abbreviated $[L^2]$, using square brackets) and the units can be square meters, square feet, and so on. Velocity, on the other hand, can be measured in units of km/h, m/s and mi/h, but the dimensions are always a length $[L]$ divided by a time $[T]$, that is $[L/T]$. The formula for a quantity may be different in different cases, but the dimensions remain the same. For example, the area of a triangle of base b and height h is $A = \frac{1}{2} bh$, whereas the area of a circle of radius r is $A = \pi r^2$. The formulas are different in the two cases, but the dimensions in both cases are the same: $[L^2]$.

When we specify the dimensions of a quantity, we usually do so in terms of base quantities, not derived quantities. For example, force, which we will see later has the same units as mass $[M]$ times acceleration $[L/T^2]$, has dimensions of $[ML/T^2]$.

Dimensions can be used as a help in working out relationships, and such a procedure is referred to as **dimensional analysis**.[†] One useful technique is the use of dimensions to check if a relationship is *incorrect*. Two simple rules apply here. First, we can add or subtract quantities only if they have the same dimensions (we don't add centimeters and pounds); second, the quantities on each side of an equals sign must have the same dimensions.

For example, suppose you derived the equation $v = v_0 + \frac{1}{2}at^2$, where v is the speed of an object after a time t, when it starts with an initial speed v_0 and undergoes an acceleration a. Let's do a dimensional check to see if this equation is correct. We write a dimensional equation as follows, remembering that the dimensions of speed are $[L/T]$ and (as we shall see in Chapter 2) the dimensions of acceleration are $[L/T^2]$:

$$\left[\frac{L}{T}\right] \stackrel{?}{=} \left[\frac{L}{T}\right] + \left[\frac{L}{T^2}\right][T^2]$$

$$\stackrel{?}{=} \left[\frac{L}{T}\right] + [L].$$

The dimensions are incorrect: on the right side, we have the sum of quantities whose dimensions are not the same. Thus we conclude that an error was made in the derivation of the original equation.

If such a dimensional check does come out correct, it does not prove that the equation is correct; for example, a dimensionless numerical factor (such as $\frac{1}{2}$ or 2π) could be wrong; Thus a dimensional check can only tell you when a relationship is wrong; it can't tell you if it is completely right.

Dimensional analysis can also be used as a quick check on an equation you are not sure about. For example, suppose that you can't remember whether

---

[†] The techniques described in the next few paragraphs may seem more meaningful after you have studied a few chapters of this book. Reading this section now will give you an overview of the subject, and you can then return to it later as needed.

the equation for the period T (the time to make one back and forth swing) of a simple pendulum of length l is $T = 2\pi\sqrt{l/g}$ or $T = 2\pi\sqrt{g/l}$, where g is the acceleration due to gravity and, like all accelerations, has dimensions $[L/T^2]$. (Do not worry about these formulas—the correct one will be derived in Chapter 14; what we are concerned about here is a person's forgetting whether it contains l/g or g/l.) A dimensional check shows that the former is correct:

$$[T] = \sqrt{\frac{[L]}{[L/T^2]}} = \sqrt{[T^2]} = [T],$$

whereas the latter is not

$$[T] \neq \sqrt{\frac{[L/T^2]}{[L]}} = \sqrt{\frac{1}{[T^2]}} = \frac{1}{[T]}.$$

Note that the constant 2π has no dimensions and so doesn't enter here.

Finally, an important use of dimensional analysis, but one with which much care must be taken, is to obtain the *form* of an equation. That is, we may want to determine how one quantity depends on others. To take a concrete example, let us try to find an expression for the period, T, of a simple pendulum. First we try to figure out what T could depend on, and make a list of these variables. It might depend on the pendulum's length (l) on the mass (m) of the bob, on the angle (θ) of swing, and on the acceleration due to gravity (g). It might also depend on air resistance (we would use the viscosity of air), the gravitational pull of the moon, and so on; but everyday experience suggests that the earth's gravity is the major force involved, and so we ignore the others. So let's assume that T is a function of l, m, θ, and g, and that each of these factors is raised to some power:

$$T = Cl^w m^x \theta^y g^z.$$

C is a dimensionless constant, and w, x, y, and z are exponents we want to solve for. We now write down the dimensional equation for this relationship:

$$[T] = [L]^w [M]^x [L/T^2]^z;$$

because θ has no dimensions (an angle is defined as a length divided by a length—see Section 9–1) it doesn't appear. We simplify and obtain

$$[T] = [L]^{w+z} [M]^x [T]^{-2z}.$$

Since the seven base quantities (Table 1–2) are independent, in order to have dimensional consistency, we must have

$$1 = -2z$$

$$0 = w + z$$

$$0 = x.$$

We solve these equations and find $z = -\frac{1}{2}$, $w = \frac{1}{2}$, $x = 0$. Thus our desired equation must be:

$$T = C\sqrt{l/g}\, f(\theta) \tag{1–1}$$

where $f(\theta)$ is some function of θ that we cannot determine using this technique. Nor can we determine in this way the dimensionless constant C. To obtain C (it turns out to be 2π) and $f(\approx 1$ for small $\theta)$ we would have to do an analysis such as that in Chapter 14 using Newton's laws. But look what we *have* found, using only dimensional consistency. We obtained the form of the expression that relates the period of a simple pendulum to the major variables of the situation, l and g (see Eq. 14–13).

How did we do it? And how useful is this technique? Basically, we had to use our intuition as to which variables were important and which not. This is not always easy, and often requires a lot of insight. As to usefulness, the final result in our example could have been obtained from Newton's laws, as in Chapter 14. But in many physical situations, such a derivation from other laws can't be done. In those situations dimensional analysis can be a powerful tool.

In the end, any expression derived via dimensional analysis (or by any other means, for that matter) must be checked against experiment. For example, in our derivation of Eq. 1–1, we can compare the periods of two pendula of different lengths, l_1 and l_2, whose angle of swing (θ) is the same. For, using Eq. 1–1, we would have

$$\frac{T_1}{T_2} = \frac{C\sqrt{l_1/g}f(\theta)}{C\sqrt{l_2/g}f(\theta)} = \sqrt{\frac{l_1}{l_2}}.$$

Because C and $f(\theta)$ are the same for both pendula, they cancel out, so we can experimentally determine if the ratio of the periods varies as the ratio of the square roots of the lengths. This comparison to experiment checks our derivation, at least in part. C and $f(\theta)$ could be determined by further experiment.

 ## 1–7 Order of Magnitude: Rapid Estimating

We are sometimes interested only in an approximate value for a quantity. This might be because an accurate calculation would take more time than it is worth or would require additional data that are not available. In other cases, we may want to make a rough estimate in order to check an accurate calculation made on a calculator to make sure that no blunders were made when entering the numbers. Also, the correct power of 10 may be lost on a calculator or slide rule, and a rough estimate can be used to obtain it.

In general, a rough estimate is made by rounding off all numbers to one significant figure and its power of 10, and after the calculation is made, again only one significant figure is kept. Such an estimate is called an *order-of-magnitude* estimate and can be assumed to be accurate within a factor of 10, and usually better; in fact, the phrase "order of magnitude" is sometimes used to refer simply to the power of 10.

As an example, suppose that a person wants to find out how much water there is in a particular lake, which is roughly circular, about 1 km across, and has an average depth of 10 m. To find the volume, we simply multiply the average depth of the lake times its surface area (as if it were a cylinder). We approximate the surface to be a circle, so the area is πr^2, which is approximately $3 \times (5 \times 10^2 \text{ m})^2 \approx 8 \times 10^5 \text{ m}^2$, where the radius r is 500 m and π was rounded off to 3 ($\approx$ means "approximately equal to"). Then the volume is about $(8 \times 10^5 \text{ m}^2) \times (10 \text{ m}) = 8 \times 10^6 \text{ m}^3$, which is on the order of 10^7 m^3. Because of all the estimates that went into this calculation, the order-of-magnitude estimate (10^7) is probably better to quote than the 8×10^6 figure.

 ## Summary[†] ————————————————————————————

Physics, like other sciences, is a creative endeavor; it is not simply a collection of facts. Important theories are created with the idea of explaining observed facts; to be accepted they are "tested" by comparing their predictions with the results of actual experiments. Note that, in general, a theory can not really be "proved" in an absolute sense.

To understand a particular type or range of phenomena, scientists may think up a *model*, which is a kind of picture or analogy that seems to explain the phenomena and thus aids in understanding. A *theory*, often developed from a model, is usually deeper and more complex than a simple model. A scientific *law*,

[†] The summaries that appear at the end of each chapter in this book are intended to give a brief overview of the main ideas of each chapter. They *cannot* act to give an understanding of the material, which can only be had by a detailed reading of the chapter.

on the other hand, is a concise statement, often expressed in the form of an equation, which quantitatively describes a particular range of phenomena over a wide range of cases.

Measurements play a crucial role in physics, but can never be perfectly precise. Thus for any number that results from measurement, it is important to specify the *uncertainty* either by stating it directly using the ± notation and/or by keeping only the correct number of *significant figures*.

Quantities are specified relative to a particular standard or *unit*, and the unit used should always be specified. The commonly accepted set of units today is the *Système International* (SI) in which the standard units of length,

mass, and time are the meter, kilogram, and second. There are seven independent *base quantities*; all other quantities are called *derived quantities*, since they and their units can be written in terms of the base ones—for example, velocity is distance/time.

The *dimensions* of a quantity refer to the combination of base quantities that makes it up. (Velocity, for example, has dimensions of [length/time] or [L/T].) Working with only the dimensions of the various quantities in a given relationship (this technique is called *dimensional analysis*) it is possible to check a relationship for correct form, and in some cases actually to derive the general form of a relationship.

 Questions ───

1. It has been said that science is the new religion, complete with high priests and mysteries known only to the select few—the trained scientists. Do you agree? Discuss.

2. Discuss the limitations and strengths of science.

3. Discuss the distinction between science and technology.

4. It is sometimes said that science is responsible for the ills of society. Scientists may respond that their work is of a pure and intellectual nature, and it is technology—the practical applications of science—that creates the problems. Discuss.

5. According to the operationalist point of view, a quantity can have meaning in science only if we can describe a set of operations (or procedures) for determining it. What do you see as the advantages and disadvantages of this point of view?

6. It is advantageous that base standards (such as for length and time) be accessible (easy to compare to), invariable (do not change), indestructible, and reproducible. Discuss why these are advantages and whether any of these criteria can be incompatible with others.

7. What are the merits and drawbacks of using a person's foot as a standard? Discuss in terms of the criteria mentioned in question 6. Consider both (*a*) a particular person's foot and (*b*) any person's foot.

8. When traveling a highway in the mountains, elevation signs sometimes say "1220 m (4000 ft)." Critics of the metric system claim that such numbers show the metric system is more complicated. How would you alter such signs to be more consistent with a switch to the metric system?

9. The SI base units, Table 1–2, contain one quantity that uses a prefix (kilogram). Would there be an advantage to changing this to a nonprefixed unit? How might this be done? Why do you suppose the kilogram is, in fact, used?

10. Suggest a way to measure (*a*) the thickness of a sheet of paper and (*b*) the distance from earth to the sun.

11. Can you set up a complete set of base quantities, as in Table 1–2, that does not include length as one of them?

 Problems ───

[The problems at the end of each chapter are ranked I, II, or III according to estimated difficulty, with I problems being easiest. The problems are arranged by sections, meaning that the reader should have read up to and including that section, but not only that section—problems often depend on earlier material.]

SECTION 1–3

1. (I) What, approximately, is the percent uncertainty for the measurement 9.7 m?

2. (I) What is the percent uncertainty in the measurement 3.86 ± 0.17 s?

3. (II) What is the area, and its approximate uncertainty, of a circle of radius 6.7×10^4 cm?

4. (II) What is the percent uncertainty in the volume of a sphere whose radius is $r = 2.48 \pm 0.03$ m?

SECTION 1–4

5. (I) Express the following using the prefixes of Table 1–1: (*a*) 10^6 volts, (*b*) 10^{-6} meters, (*c*) 4×10^7 days (*d*) 2×10^3 bucks, (*e*) 2×10^{-9} pieces.

6. (I) Determine your height in meters.

7. (I) Determine the conversion factor between kilometers and miles, and between km/h and mi/h.

8. (I) The moon is 240,000 mi from earth. How many meters in this? Express (*a*) using powers of ten, and (*b*) using a metric prefix.

9. (I) A typical atom has a diameter of about 1.0×10^{-10} m. What is this in feet?

10. (I) (*a*) How many seconds are there in 1 year? (*b*) How many nanoseconds in 1 year? (*c*) How many years in 1 second?

11. (II) How much longer (percentage) is the 100-m dash than the 100-yd dash?

12. (II) Hold a pencil in front of your eye at a position where its tip just blocks out the moon; make appropriate measurements to estimate the diameter of the moon given that the earth-moon distance is 3.8×10^5 km.

13. (II) A *light year* is the distance light (speed = 3.00×10^8 m/s) travels in 1 year. (a) How many meters are there in 1.0 light year? (b) An astronomical unit (AU) is the average distance from the sun to earth, 1.50×10^8 km. How many AU are there in 1.0 light year? (c) What is the speed of light in AU/h?

SECTION 1–6

14. (I) The speed, v, of a body is given by the equation $v = At^3 - Bt$, where t refers to time. What are the dimensions of A and B?

15. (I) What are the SI units for the constants A and B in problem 14?

16. (II) Three students derive the following equations in which x refers to distance traveled, v the speed, a the acceleration (m/s^2), t the time, and the subscript $(_0)$ means a quantity at time $t = 0$: (a) $x = vt^2 + 2at$, (b) $x = v_0t + \frac{1}{2}at^2$, (c) $x = v_0t + 2at^2$. Which of these could possibly be correct according to a dimensional check?

17. (III) A mass m oscillates on the end of a spring with amplitude x (where $2x$ is the total distance traveled in one swing, measured in meters). Using dimensional analysis, determine the form of the period, which could depend on m, x, and the spring stiffness constant k. (k is defined as F/x where F is the force needed to stretch the spring a distance x.)

18. (III) A particle of mass m rotates in a circle of radius r with speed v. The particle has an acceleration a_c (m/s^2) called "centripetal acceleration." Use dimensional analysis to find the form of a_c.

Motion: Kinematics in One Dimension

2

The motion of objects—baseballs, automobiles, joggers, and even the sun and moon—is an obvious part of everyday life. Motion was undoubtedly the first aspect of the physical world to be thoroughly studied, and this study can be traced back to the ancient civilizations of Asia Minor. Although the ancients acquired significant insight into motion, it was not until comparatively recently, in the sixteenth and seventeenth centuries, that our modern understanding of motion was established. Many contributed to this understanding, but, as we shall soon see, two individuals stand out above the rest: Galileo Galilei (1564–1642) and Isaac Newton (1642–1727).

The study of the motion of objects, and the related concepts of force and energy, forms the field called **mechanics**. Rigid objects that move without rotating are said to undergo **translational motion**. Each part of an object undergoing pure translational motion follows the same path. An example is the diver shown in Fig. 2–1a; in Fig. 2–1b, the diver has both translational and rotational motion. For the next several chapters we will deal only with translational motion; we will take up rotational motion in Chapters 9 and 10.

In our study of motion, we will make use of the concept of an idealized **particle**. Such a particle is considered to be a mathematical point, and as such has no spatial extent (no size). A particle can undergo only translational motion. Everyday objects, of course, are not particles since they do take up space. Nonetheless, the concept of a particle is useful in many real situations where we are interested only in translational motion and the object's size is not significant. For example we might consider a billiard ball, or even a spacecraft traveling toward the moon, as a particle for many purposes. When we discuss more complicated motion in later chapters, we will find the particle concept very useful since any extended body can be considered to be made up of many tiny particles.

Mechanics is customarily divided into two parts: **kinematics**, which is the description of how objects move, and **dynamics**, which deals with why objects move as they do. This chapter and the next are concerned with the kinematics of translational motion; in Chapter 4 we begin the discussion of the dynamics of translational motion. In the present chapter we will mainly be concerned with describing an object that moves along a straight-line path (one-dimensional or

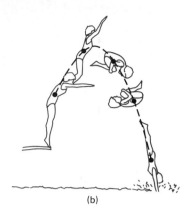

FIGURE 2–1

Motion of the diver is pure translation in (a), but is translation plus rotation in (b).

(a)   (b)

"rectilinear" motion); in Chapter 3 we will study kinematics in two and three dimensions.

2–1 Speed

"Speed" and "velocity" are words we often use interchangeably in everyday language. In physics we make a distinction between them (see Section 2–4). For now, we discuss only speed, which refers to how far an object travels in a given time interval. If a car travels 400 km in 5 hours, we say its average speed was 80 km/h. In general, the *average speed* of an object is defined as *the distance traveled divided by the time it takes to travel this distance*. If we let the symbol D stand for the distance traveled, t for the elapsed time, and v_s for the speed, then the average speed is defined as

$$\bar{v}_s = \frac{D}{t}.$$

The bar (¯) over the v_s is the standard symbol meaning "average."

EXAMPLE 2–1 How far can a cyclist travel in 4.0 h if his average speed is 11.5 km/h?

SOLUTION Since we want to find the distance traveled, we rewrite the equation above as $D = v_s t$. Since $v_s = 11.5$ km/h and $t = 4.0$ h, $D = (11.5$ km/h$)(4.0$ h$) = 46$ km.

2–2 Reference Frames

When riding in a train you may observe a bird flying by overhead and remark that it looks as if it is moving at a speed of 30 km/h. But do you mean it is traveling 30 km/h with respect to the train, or with respect to the ground?

Every measurement must be made with respect to a **frame of reference**. For example, while on a train traveling at 80 km/h, you might notice a person walk past you toward the front of the train at a speed of, say, 5 km/h. Of course this is the person's speed with respect to the train. With respect to the ground that person is moving at 85 km/h. It is always important to specify the frame of reference when stating a speed. We nearly always mean "with respect to the earth" without even thinking about it, but the reference frame should be specified whenever there might be confusion.

The values of other physical quantities also depend on the frame of reference. For example, there is no point in telling you that Yosemite National Park is 300 km away unless I specify 300 km from where. Distances are always measured in some frame of reference. Furthermore, when specifying the motion of an object, it is important to specify not only the speed but also the direction of motion. For example, if a friend leaves New York on a jet plane that travels at a speed of 1000 km/h, you would like to know in what direction this person is going—toward Washington, San Francisco, Paris, or wherever. Often we can specify a direction by using the cardinal points, north, east, south and west, and by "up" and "down." This is not always convenient; so in physics we often draw a set of coordinate axes, as shown in Fig. 2–2 to represent a frame of reference. The positive and negative directions along the axes shown are the usual conventions, although in some cases we may want to change the convention by, for instance, choosing the positive y direction to be downward instead of upward. Any point on the plane can be specified by giving its x and y coordinates. In three dimensions, a z axis is drawn perpendicular to the x and y axes.

Although most measurements are made in reference frames fixed on the earth, it is important to recognize that reference frames other than the earth are perfectly legitimate. For example, scientific measurements are often made on moving spacecraft and even on the moon.

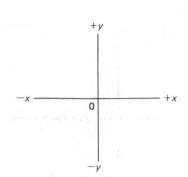

FIGURE 2–2

Standard (Cartesian) set of coordinate axes.

2–3 Changing Units

It is often necessary or useful to change from one set of units to another. For example it is often preferable to specify the speed of a car in m/s rather than in km/h, such as when measuring braking distances since the time and distances involved are usually seconds and meters rather than hours and kilometers. Also, one sometimes must change units from metric to British units, or vice versa.

To determine what a speed of 80 km/h is in m/s, we proceed as follows. There are 1000 m in 1 km, and 3600 s in 1 h. Thus:

$$80 \text{ km/h} = \left(\frac{80 \text{ km}}{1 \text{ h}}\right)\left(\frac{1000 \text{ m}}{1 \text{ km}}\right)\left(\frac{1 \text{ h}}{3600 \text{ s}}\right)$$

$$= (80)\left(\frac{1000}{3600}\right)\frac{\text{m}}{\text{s}}$$

$$= (80)(0.278) \text{ m/s} = 22 \text{ m/s}.$$

Notice that in the first line we have multiplied our original number $(80 \frac{\text{km}}{\text{h}})$ by two conversion factors $(\frac{1000 \text{ m}}{1 \text{ km}} = 1$ and $\frac{1 \text{ h}}{3600 \text{ s}} = 1)$, each of which in essence is equal to one and thus do not change the equation. The units of hours and of kilometers cancel out so we obtain m/s. (Note also that we kept only two significant figures in the final answer, 22 m/s, since the least significant number in the product contained two significant figures.)

When changing units it is often a problem to figure out whether the conversion factor is to go in the numerator or in the denominator. The easiest way to be sure is to check if the units cancel out as they did above. We would have gotten a wrong result if we had written $(80 \frac{\text{km}}{\text{h}})(\frac{1000 \text{ m}}{1 \text{ km}})(\frac{3600 \text{ s}}{1 \text{ h}})$, and this would have been obvious since the hour units do not cancel.

In the last line in the calculation above we see the factor 0.278; this is the conversion factor between m/s and km/h. This can more quickly be obtained by noting that 1 km/h is the same as 1000 m in 3600 s. That is, $1 \frac{km}{h} = \frac{1000 \, m}{3600 \, s} = 0.278$ m/s. Thus, any speed given in km/h can be changed to m/s by multiplying by the factor of 0.278.

Changes between metric and British units can be done in a similar way, and we leave these calculations as an exercise.

 ## 2–4   Average Velocity; Displacement

In physics we make a distinction between the terms speed and velocity. The term *velocity* is used to signify both the *magnitude* (numerical value) of how fast an object is moving and its *direction*. (It is therefore called a vector, as we shall see in Chapter 3.) Speed, on the other hand, is a magnitude only. There is a second difference between speed and velocity: the magnitude of the average velocity is defined in terms of "displacement," rather than total distance traveled. **Displacement** is defined as *the change in position of an object*. To see the distinction between total distance and displacement, imagine a person walking 50 m to the east and then turning around and walking back (west) a distance of 10 m. The total *distance* traveled is 60 m but the *displacement* is only 40 m since the person is now only 40 m from the starting point. At some initial time, call it t_1, suppose an object is on the x axis at point x_1 in the coordinate system shown in Fig. 2–3. At some later time, t_2, suppose it is at point x_2. Its displacement is then $x_2 - x_1$. The *magnitude* of its **average velocity** ($\bar{v}$) is defined as *the displacement divided by the elapsed time* (which is $t_2 - t_1$), and so we have

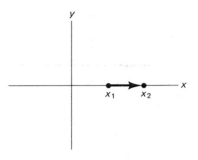

FIGURE 2–3
The arrow represents the displacement $x_2 - x_1$.

$$\bar{v} = \frac{x_2 - x_1}{t_2 - t_1}.$$

In the example above where a person walked 50 m east and then 10 m west, suppose the walk took 40 s; then the average speed was (60 m)/(40 s) = 1.5 m/s, but the average velocity in this case was only (40 m)/(40 s) = 1.0 m/s. This discrepancy between the magnitude of the velocity and the speed occurs in some cases, but only for the *average* values, and we rarely need be concerned with it. We shall see in the next section that the magnitudes of the instantaneous velocity and the instantaneous speed are always the same.

In Fig. 2–3, the displacement of our object was $x_2 - x_1$ during the time interval $t_2 - t_1$. It is convenient to write

$$\Delta x = x_2 - x_1$$

where the symbol Δ (Greek letter delta) means "change in." Then Δx means the displacement or the "change in x." Similarly, we can write the elapsed time (change in time) as $\Delta t = t_2 - t_1$. The definition of average velocity becomes

$$\bar{v} = \frac{x_2 - x_1}{t_2 - t_1} = \frac{\Delta x}{\Delta t}. \tag{2–1}$$

Notice that if x_2 is less than x_1, the object is moving to the left, so $\Delta x = x_2 - x_1$ is less than zero. The sign of the displacement, and/or velocity, indicates the direction; the average velocity for an object moving to the right along the x axis is positive, whereas it is negative when the object moves to the left.

EXAMPLE 2–2   The position of a bowling ball as a function of time is plotted as moving along the x axis of a graph. At time $t_1 = 3.00$ s, its position is $x_1 = 40.5$ m; at $t_2 = 5.50$ s, its position is $x_2 = 18.2$ m. What was its average velocity?

SOLUTION $\Delta x = x_2 - x_1 = 18.2 \text{ m} - 40.5 \text{ m} = -21.7 \text{ m}$, and $\Delta t = t_2 - t_1 = 5.50 \text{ s} - 3.00 \text{ s} = 2.50 \text{ s}$. Therefore,

$$\bar{v} = \frac{\Delta x}{\Delta t} = \frac{-21.7 \text{ m}}{2.50 \text{ s}} = -8.68 \text{ m/s}.$$

The displacement and average velocity are negative, so the ball is moving to the left along the x axis.

 2-5 Instantaneous Velocity

If you drive a car along a straight road for 150 km in 2.0 h, your average velocity is 75 km/h. It is unlikely, though, that you were moving at 75 km/h at every instant. To deal with this, we need the concept of *instantaneous velocity*, which is the velocity at any instant of time (it is what a speedometer is supposed to indicate). This is not a very precise definition of instantaneous velocity, however, since we haven't specified what is meant by an "instant."

More precisely, the **instantaneous velocity** at any moment is *the average velocity over an indefinitely short time interval*. To make this clear it is helpful to make a graph of the position of a particle versus time. An example is shown in Fig. 2-4. The particle is at position x_1 at a time t_1, and at position x_2 at time t_2; P_1 and P_2 represent these two points on the graph. A straight line drawn from point $P_1(x_1, t_1)$ to point $P_2(x_2, t_2)$ forms the hypotenuse of a right triangle whose sides are Δx and Δt. The ratio $\Delta x / \Delta t$ (the "rise" over the "run") is the *slope* of the straight line $P_1 P_2$. But $\Delta x / \Delta t$ is also the average velocity of the particle during the time interval $\Delta t = t_2 - t_1$. Therefore we conclude that the average velocity of an object during any time interval $\Delta t = t_2 - t_1$ is equal to the slope of the straight line (or *chord*) connecting the two points (x_1, t_1) and (x_2, t_2) on an x vs. t graph.

Consider now a time t_i, intermediate between t_1 and t_2, at which time the particle is at x_i (Fig. 2-5). The slope of the straight line $P_1 P_i$ is less than the slope of $P_1 P_2$ in this case. Thus the average velocity during the time interval $t_i - t_1$ is less than that during the time interval $t_2 - t_1$.

Now let us imagine that we take the point P_i in Fig. 2-5 to be closer and closer to point P_1; that is, we let the interval $t_i - t_1$, which we now call Δt, to become smaller and smaller. The slope of the line connecting the two points becomes closer and closer to the slope of a line tangent to the curve at point P_1. As we take Δt smaller and smaller, the average velocity (equal to the slope of the chord) approaches the slope of the tangent at that point. The definition of the instantaneous velocity at a given moment (say, t_1) is the limiting value of the average velocity as we let Δt approach zero; and we see that it equals the slope of the tangent to the curve at that point (which we can simply call "the slope of the curve" at that point):

$$v = \lim_{\Delta t \to 0} \frac{\Delta x}{\Delta t}.$$

This limit as $\Delta t \to 0$ is written in calculus notation as dx/dt and is called the derivative of x with respect to t:

$$v = \lim_{\Delta t \to 0} \frac{\Delta x}{\Delta t} = \frac{dx}{dt}. \qquad (2-2)$$

This equation is the definition of instantaneous velocity for one-dimensional motion.

It is important to note that we do not simply set $\Delta t = 0$, for then Δx would also be zero, and we would have an undefined number. Rather, we must consider the ratio $\Delta x / \Delta t$ as a whole; as we let Δt approach zero, Δx approaches zero

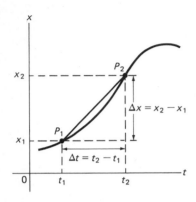

FIGURE 2-4
Graph of a particle's position x vs. time t. The slope of the straight line $P_1 P_2$ represents the average speed of the particle during the time interval $\Delta t = t_2 - t_1$.

FIGURE 2-5
Same position-vs.-time graph as in Fig. 2-4, but note that the average velocity over the time interval $t_i - t_1$ (which is the slope of $P_1 P_i$) is less than the average velocity over the time interval $t_2 - t_1$.

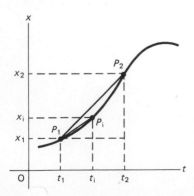

also; but the ratio $\Delta x/\Delta t$ approaches some definite value, which we call the instantaneous velocity. Note also that for instantaneous velocity we use the symbol v whereas for average velocity we use $\bar{v}$, with a bar. In the rest of this book, when we use the term "velocity" it will refer to instantaneous velocity; when we want to speak of the average velocity, we will make this clear by including the word "average."

Since the velocity at any instant equals the slope of the tangent to the x vs. t graph at that instant, we can obtain the velocity at any instant from such a graph. For example, in Fig. 2–4 as our object moves from x_1 to x_2, the slope continually increases, so the velocity is increasing. For times after t_2, however, the slope begins to decrease and in fact reaches zero (so $v = 0$) where x has its maximum value (highest point on the curve). Beyond this point, the slope is negative; the velocity is therefore negative, which makes sense since x is now decreasing (the particle is moving toward decreasing values of x, to the left on an x-y plot).

If an object moves with constant velocity over a particular time interval, then its instantaneous velocity is equal to its average velocity. Can you guess what the graph of x vs. t will be in this case of constant velocity? It will be a straight line, of course, whose slope equals the velocity. The curve of Fig. 2–4 has no straight sections, so there are no time intervals when the velocity is constant. Finally note that the instantaneous speed equals the magnitude of the instantaneous velocity under all circumstances; this is true because in the limit of $\Delta t \rightarrow 0$, the displacement also approaches zero, and so there is no way for this infinitesimal displacement to differ from the infinitesimal distance traveled.

EXAMPLE 2–3 A particle is moving along the x axis, as in Fig. 2–3. Its position as a function of time is given by the equation $x = At^2 + B$ where $A = 2.10$ m/s^2 and $B = 2.80$ m. (a) Determine the displacement of the particle during the time interval from $t_1 = 3.00$ s to $t_2 = 5.00$ s. (b) Determine the average velocity during this time interval. (c) Determine the magnitude of the instantaneous velocity at $t = 5.00$ s.

SOLUTION (a) At $t_1 = 3.00$ s, the position is

$$x_1 = At_1^2 + B = (2.10 \text{ m/s}^2)(3.00 \text{ s})^2 + 2.80 \text{ m} = 21.7 \text{ m}.$$

At $t_2 = 5.00$ s, the position is

$$x_2 = (2.10 \text{ m/s}^2)(5.00 \text{ s})^2 + 2.80 \text{ m} = 55.3 \text{ m}.$$

The displacement is thus

$$x_2 - x_1 = 55.3 \text{ m} - 21.7 \text{ m} = 33.6 \text{ m}.$$

(b) The magnitude of the average velocity is

$$\bar{v} = \frac{x_2 - x_1}{t_2 - t_1} = \frac{33.6 \text{ m}}{2.00 \text{ s}} = 16.8 \text{ m/s}.$$

(c) Let us determine the instantaneous velocity for any time t (that is, determine v as a function of t), and then set $t = 5.00$ s. At any time t, the position will be

$$x = At^2 + B.$$

At a slightly later time, $t + \Delta t$, the position will have changed by an amount Δx, and so the new position will be $x + \Delta x$ where

$$x + \Delta x = A(t + \Delta t)^2 + B$$
$$= At^2 + 2At(\Delta t) + A(\Delta t)^2 + B.$$

To find Δx, we subtract $x = At^2 + B$ from both sides of this last equation and find

$$\Delta x = 2At(\Delta t) + A(\Delta t)^2.$$

The average velocity during the time Δt is

$$\bar{v} = \frac{\Delta x}{\Delta t} = 2At + A(\Delta t).$$

To find the instantaneous velocity, v, we must let Δt become zero (see Eq. 2–2), so the last term on the right becomes zero and we have

$$v = 2At$$
$$= (4.20 \text{ m/s}^2)(t),$$

where we have substituted in $A = 2.10 \text{ m/s}^2$. Then at $t = 5.00$ s, we have

$$v = (4.20 \text{ m/s}^2)(5.00 \text{ s})$$
$$= 21.0 \text{ m/s}.$$

If you have already learned from calculus the formulas

$$\frac{d}{dt}(Ct^n) = nCt^{n-1} \quad \text{and} \quad \frac{dC}{dt} = 0$$

where C is any constant, then our final result is easily obtained:

$$\frac{dx}{dt} = \frac{d}{dt}(At^2 + B) = 2At = (4.20 \text{ m/s}^2)(t),$$

which becomes, for $t = 5.00$ s, $v = 21.0$ m/s.

2–6 Acceleration

An object whose velocity is changing in time is said to be accelerating. A car whose velocity increases from zero to 80 km/h is accelerating; if a second car can accomplish this change in velocity in less time than the first car, it is said to undergo a greater acceleration. In general, the **average acceleration**, $\bar{a}$, during a time interval $\Delta t = t_2 - t_1$ during which the velocity changes by $\Delta v = v_2 - v_1$, is defined as

$$\bar{a} = \frac{v_2 - v_1}{t_2 - t_1} = \frac{\Delta v}{\Delta t}. \tag{2–3}$$

The **instantaneous acceleration**, a, is defined as the *limiting value of the average acceleration as we let Δt approach zero*:

$$a = \lim_{\Delta t \to 0} \frac{\Delta v}{\Delta t} = \frac{dv}{dt}. \tag{2–4}$$

This limit, dv/dt, is called the derivative of v with respect to t. We will use the term "acceleration" to refer to the instantaneous value. If we want to discuss the average acceleration we will then always make this clear by including the word "average."

EXAMPLE 2–4 A car accelerates along a straight road from rest to 60 km/h in 5.0 s. What is the magnitude of its average acceleration?

SOLUTION From Eq. 2–3 we have

$$\bar{a} = \frac{60 \text{ km/h} - 0 \text{ km/h}}{5.0 \text{ s}} = 12 \text{ km/h/s}.$$

This is read as "twelve kilometers per hour per second" and means that, on the average, the velocity changed by 12 km/h each second. That is, assuming the acceleration was constant, during the first second the car's velocity increased from zero to 12 km/h; during the next second its speed increased by another 12 km/h up to 24 km/h, and so on. (Of course if the instantaneous acceleration was not constant, these numbers would be different.)

In the above example, the calculated acceleration contained two different time units: hours and seconds. We often prefer to use only seconds; to do so we can change the 60 km/h to $(60 \text{ km/h})(0.278 \frac{\text{m/s}}{\text{km/h}}) = 16 \text{ m/s}$ (see Section 2–3), and we get

$$\bar{a} = \frac{16 \text{ m/s} - 0 \text{ m/s}}{5.0 \text{ s}} = 3.2 \text{ m/s}^2.$$

We almost always write these units as m/s^2 (meters per second squared) as done here instead of m/s/s since

$$\frac{\frac{\text{m}}{\text{s}}}{\text{s}} = \frac{\text{m}}{\text{s} \cdot \text{s}} = \frac{\text{m}}{\text{s}^2}.$$

The dimensions of acceleration are always length divided by time squared, as can be seen from the definition and as illustrated in this example.

EXAMPLE 2–5 Suppose a particle is moving in a straight line so that its position is given by the relation $x = (2.10 \text{ m/s}^2)t^2 + (2.80 \text{ m})$ as already discussed in Example 2–3. Calculate (a) its average acceleration during the time interval from $t_1 = 3.00$ s to $t_2 = 5.00$ s, and (b) its instantaneous acceleration as a function of time.

SOLUTION (a) We saw in part (c) of Example 2–3 that the velocity at any time t is $v = (4.20 \text{ m/s}^2)t$. Therefore, at $t_1 = 3.00$ s, $v_1 = (4.20 \text{ m/s}^2)(3.00 \text{ s}) = 12.6 \text{ m/s}$ and at $t_2 = 5.00$ s, $v_2 = 21.0 \text{ m/s}$. Therefore

$$\bar{a} = \frac{21.0 \text{ m/s} - 12.6 \text{ m/s}}{5.00 \text{ s} - 3.00 \text{ s}} = 4.20 \text{ m/s}^2.$$

(b) We use

$$a = \lim_{\Delta t \to 0} \frac{\Delta v}{\Delta t}.$$

At any time t, $v = (4.20 \text{ m/s}^2)(t)$; a little later, at time $t + \Delta t$, the velocity will have changed by an amount Δv so it is then $v + \Delta v$ where

$$v + \Delta v = (4.20 \text{ m/s}^2)(t + \Delta t).$$

When we subtract $v = (4.20 \text{ m/s}^2)(t)$ from both sides we obtain

$$\Delta v = (4.20 \text{ m/s}^2)(\Delta t).$$

Hence:

$$a = \lim_{\Delta t \to 0} \frac{\Delta v}{\Delta t} = \lim_{\Delta t \to 0} (4.20 \text{ m/s}^2) = 4.20 \text{ m/s}^2.$$

The acceleration in this case is constant; it doesn't depend on time.

If we draw a graph of the magnitude of the velocity, v, versus time, t, as shown in Fig. 2–6, then the magnitude of the average acceleration over a time interval $\Delta t = t_2 - t_1$ is represented by the slope of the straight line connecting

FIGURE 2–6
A graph of velocity v vs. time t. The average acceleration over a time interval $\Delta t = t_2 - t_1$ is the slope of the straight line $P_1 P_2$: $\bar{a} = \Delta v / \Delta t$. The instantaneous acceleration at time t_1 is the slope of the v-vs.-t curve at that instant.

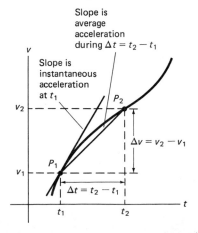

the two points P_1 and P_2 as shown.[†] The instantaneous acceleration at any time, say t_1, is the slope of the tangent to the v vs. t curve at that time, which is also shown in Fig. 2–6. Let us use this fact for the situation graphed in Fig. 2–6: as we go from time t_1 to time t_2 the velocity continually increases but the acceleration (the rate at which the velocity changes) is decreasing since the slope of the curve is decreasing.

If the velocity of an object is constant, then the acceleration is zero since $\Delta v = 0$. The x vs. t graph for constant velocity is then a straight line whose slope equals the velocity. The v vs. t graph in this case is also a straight line, but its slope is zero, so the straight line is parallel to the t axis (see Fig. 2–7).

Sometimes the acceleration of a particle moving in one dimension, as calculated by Eq. 2–3 or 2–4, comes out with a negative sign. This can happen in two ways. First, the particle may be moving with decreasing speed (decelerating) to the right along the x axis (x increasing). For example, the particle may have a velocity of 10 m/s at one instant and 2 s later have a velocity of 4 m/s; then its average acceleration was (4 m/s − 10 m/s)/2 s = −3 m/s². Second, the particle would have a negative acceleration if it is moving with increasing speed to the left along the x axis (x decreasing). In such a case the displacement Δx (or, infinitesimally, dx) will be negative. (Since x is decreasing, $x_2 < x_1$ and $\Delta x = x_2 - x_1 < 0$.) Thus the velocity (Eq. 2–1 or 2–2) must be negative. For example, if at one instant a particle has $v_1 = -6$ m/s and 2 s later it has $v_2 = -14$ m/s (it is going faster), then $\bar{a} = (-14$ m/s $+ 6$ m/s$)/2$ s $= -4$ m/s².

Note that although velocity can be positive or negative, speed is never negative since it involves the rate at which a distance is traveled, not the rate of change of displacement as velocity does.

Like velocity, acceleration is a rate. The velocity of an object is the rate at which its displacement changes with time; its acceleration, on the other hand, is the rate at which its velocity changes with time. In a sense, acceleration is a "rate of a rate." This can be expressed in equation form as follows: since $a = dv/dt$ and $v = dx/dt$, then

$$a = \frac{dv}{dt} = \frac{d}{dt}\left(\frac{dx}{dt}\right) = \frac{d^2x}{dt^2}.$$

Here, d^2x/dt^2 is called the *second derivative* of x with respect to time because we first take the derivative of x with respect to time (dx/dt) and then we again take the derivative with respect to time, $(d/dt)(dx/dt)$, to get the acceleration.

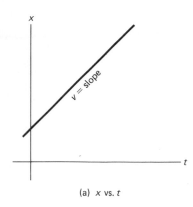

(a) x vs. t

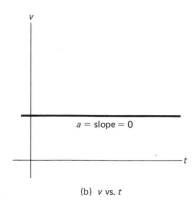

(b) v vs. t

FIGURE 2–7

For constant velocity: (a) the x-vs.-t graph, and (b) the v-vs.-t graph.

 ## 2–7 Uniformly Accelerated Motion

Many practical situations occur in which the acceleration is constant. In many other situations the variation in acceleration is sufficiently small that we are justified in assuming that it is constant. We now treat this situation of **uniformly accelerated motion**: that is, when the magnitude of the acceleration is constant and the motion is in a straight line. In this case, the instantaneous and average accelerations are equal.

To simplify our notation, let us take the initial time in any discussion to be zero: $t_1 = 0$. We can then let $t_2 = t$ be the elapsed time. The initial position (x_1) and initial velocity (v_1) of an object will now be represented by x_0 and v_0, and at time t they will be called x and v (rather than x_2 and v_2). The average velocity during the time t will be (from Eq. 2–1)

$$\bar{v} = \frac{x - x_0}{t};$$

[†] Compare this to the position vs. time graph of Fig. 2–4 for which the slope represents the magnitude of the average velocity.

$$a = \frac{v - v_0}{t}.$$

A common problem is to determine the velocity of an object after a certain time given its acceleration. We can solve such problems by solving for v in the last equation:

$$v = v_0 + at. \qquad \text{[constant acceleration]} \quad (2\text{–}5)$$

For example, it may be known that the acceleration of a particular motorcycle is 4.0 m/s^2 and we wish to determine how fast it will be going after, say, 6.0 s. Assuming it starts from rest ($v_0 = 0$), after 6.0 s the velocity will be $v = at = (4.0 \text{ m/s}^2)(6.0 \text{ s}) = 24 \text{ m/s}$.

Next, let us see how to calculate the position of an object after a time t when it is undergoing constant acceleration. From the definition of average velocity (Eq. 2–1) $\bar{v} = (x - x_0)/t$, so

$$x = x_0 + \bar{v}t.$$

Because the velocity increases at a uniform rate, the average velocity, $\bar{v}$, will be midway between the initial and final velocities:

$$\bar{v} = \frac{v + v_0}{2}. \qquad \text{[constant acceleration]} \quad (2\text{–}6)$$

(This would not necessarily be true if the acceleration were not constant.) We combine the last three equations and find

$$x = x_0 + \bar{v}t$$

$$= x_0 + \left(\frac{v + v_0}{2}\right)t$$

$$= x_0 + \left(\frac{v_0 + at + v_0}{2}\right)t$$

or

$$x = x_0 + v_0 t + \tfrac{1}{2}at^2. \qquad \text{[constant acceleration]} \quad (2\text{–}7)$$

Equations 2–5, 2–6, and 2–7 are three of the four most useful equations of uniformly accelerated motion. We now derive the fourth equation, which is useful in a situation when, for instance, the acceleration, position, and initial velocity are known, and the final velocity is desired but the time t is not known. To obtain the velocity v at time t in terms of v_0, a, x, and x_0, we begin as above:

$$x = x_0 + \bar{v}t = x_0 + \left(\frac{v + v_0}{2}\right)t$$

where we have substituted in Eq. 2–6. Next we solve Eq. 2–5 for t, obtaining $t = (v - v_0)/a$, and substituting this into the above equation we have

$$x = x_0 + \left(\frac{v + v_0}{2}\right)\left(\frac{v - v_0}{a}\right)$$

$$= x_0 + \frac{v^2 - v_0^2}{2a}.$$

We solve this for v^2 and obtain the desired relation:

$$v^2 = v_0^2 + 2a(x - x_0). \qquad \text{[constant acceleration]} \quad (2\text{–}8)$$

We now have four equations relating the various quantities important in uniformly accelerated motion (that is, a is a constant). We collect them here in one place for further reference:

$$v = v_0 + at \qquad\qquad [a = \text{constant}] \quad (2\text{--}9\text{a})$$

$$x = x_0 + v_0 t + \tfrac{1}{2}at^2 \qquad [a = \text{constant}] \quad (2\text{--}9\text{b})$$

$$v^2 = v_0^2 + 2a(x - x_0) \qquad [a = \text{constant}] \quad (2\text{--}9\text{c})$$

$$\bar{v} = \frac{v + v_0}{2}. \qquad\qquad [a = \text{constant}] \quad (2\text{--}9\text{d})$$

These equations are not valid unless a is a constant. In many cases we can set $x_0 = 0$, and this simplifies the above equation a bit. In the examples that follow, we assume $x_0 = 0$ unless stated otherwise.

EXAMPLE 2–6   Suppose a planner is designing an airport for small planes. One kind of airplane that might use this airfield must reach a speed before takeoff of 200 km/h (55.6 m/s) and can accelerate at 12.0 m/s^2. If the runway is 100 m long, can this airplane reach the proper speed to take off?

SOLUTION   We use Eq. 2–9c with $x_0 = 0$, $v_0 = 0$, $x = 100$ m, and $a = 12.0$ m/s^2. Then

$$v^2 = 0 + 2(12.0 \text{ m/s}^2)(100 \text{ m}) = 2400 \text{ m}^2/\text{s}^2$$

$$v = \sqrt{2400 \text{ m}^2/\text{s}^2} = 49.0 \text{ m/s}.$$

Unfortunately, this length runway is *not* sufficient. By solving Eq. 2–9c for $(x - x_0)$ you can determine how long a runway is needed for this plane.

One of the difficulties in doing kinematics problems (and other problems, too, for that matter) is knowing which equation to use. Perhaps the best thing to do when you're not sure is to use the following procedure: (1) write down what *is* "known" or "given," and then what you *want* to know; (2) search out an applicable[†] equation that involves only known quantities and one desired unknown, but contains no other unknowns; (3) if the desired unknown is buried within the equation—for example you may want a from Eq. 2–9c—then solve the equation for the desired unknown. In some instances a combination of equations may be needed. An important aspect of doing problems is keeping track of units. Note that an equals sign implies the units on each side must be the same just as the numbers must. By treating units carefully, you can often avoid mistakes in calculations.

EXAMPLE 2–7   How long does it take a car to travel 30 m if it accelerates from rest at a rate of 2.0 m/s^2?

SOLUTION   First we make a table:

Known	Wanted
$x_0 = 0$	t
$x = 30$ m	
$a = 2.0$ m/s^2	
$v_0 = 0$	

Since a is constant, we can use Eq. 2–9. Equation 2–9a is not helpful in this case since it contains v, an unknown, as well as the desired unknown t. Equation 2–9c

---

[†] If, for example, the acceleration is variable, you can't use Eqs. 2–9.

is worse—it contains v but not t. Equation 2–9b is perfect since the only unknown quantity is t. Before we solve for t in this equation, we simplify it by setting $v_0 = 0$ first. Then

$$x = \tfrac{1}{2}at^2$$

$$t^2 = \frac{2x}{a} = \frac{2(30 \text{ m})}{2.0 \text{ m/s}^2} = 30 \text{ s}^2$$

$$t = \sqrt{30 \text{ s}^2} = 5.5 \text{ s}.$$

EXAMPLE 2–8 We now consider the stopping distances for a car which are important not only for traffic safety but are needed for traffic design problems. This problem is best dealt with in two parts: (1) the time between the decision to apply the brakes and their actual application (the "reaction time") during which we assume $a = 0$; and (2) the actual braking period when the vehicle decelerates ($a \neq 0$). The stopping distance depends on the reaction time of the driver, the initial speed of the car (the final speed is zero), and the deceleration rate of the car. For a dry road, good brakes can decelerate a car at a rate of about 5 m/s² to 8 m/s². We make the calculation for an initial speed of 100 km/h (28 m/s) and assume the acceleration of the car is -6.0 m/s². Reaction times for normal drivers varies from perhaps 0.3 s to about 1.0 s; we take it to be 0.50 s.

SOLUTION For the first part of the problem, the car travels at a constant speed of 28 m/s during the time the driver is reacting (0.50 s). Thus:

Known	Wanted
$t = 0.50$ s	x
$v_0 = 28$ m/s	
$v = 28$ m/s	
$a = 0$	
$x_0 = 0$	

To find x we can use Eq. 2–9b (note that Eq. 2–9c won't work because x is multiplied by a which is zero):

$$x = v_0 t + 0 = (28 \text{ m/s})(0.50 \text{ s}) = 14 \text{ m}.$$

Now for the second part during which the brakes are applied and the car brought to rest:

Known	Wanted
$v_0 = 28$ m/s	x
$v = 0$	
$a = -6.0$ m/s²	

Equation 2–9a doesn't contain x; Eq. 2–9b contains x but also the unknown t. Equation 2–9c is what we want; we solve for x (after setting $x_0 = 0$):

$$v^2 - v_0^2 = 2ax$$

$$x = \frac{v^2 - v_0^2}{2a}$$

$$= \frac{0 - (28 \text{ m/s})^2}{2(-6.0 \text{ m/s}^2)} = \frac{-784 \text{ m}^2/\text{s}^2}{-12 \text{ m/s}^2}$$

$$= 65 \text{ m}.$$

The car traveled 14 m while the driver was reacting and another 65 m during the braking period before coming to a stop; the total distance traveled was then 79 m. Under wet or icy conditions, the value of *a* may be only one-third the value for a dry road since the brakes cannot be applied as hard without skidding, and hence stopping distances are much greater. Note also that stopping distance increases with the *square* of the speed, not just linearly with speed!

EXAMPLE 2–9   A baseball pitcher throws a fastball with a speed of 30.0 m/s. Calculate approximately the average acceleration of the ball during the throwing motion. It is observed that in throwing the baseball, the pitcher accelerates the ball through a total distance of about 3.50 m from behind the ear to the point where it is released (Fig. 2–8).

SOLUTION   We want to find the acceleration *a* given that $x = 3.50$ m, $v_0 = 0$, and $v = 30.0$ m/s. We use Eq. 2–9c and solve for *a*:

$$a = \frac{v^2 - v_0^2}{2x}$$

$$= \frac{(30.0 \text{ m/s})^2 - (0 \text{ m/s})^2}{2(3.50 \text{ m})} = \frac{900 \text{ m}^2/\text{s}^2}{7.00 \text{ m}} = 129 \text{ m/s}^2.$$

This is a very large acceleration!

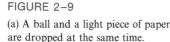

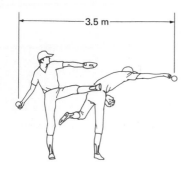

FIGURE 2–8

A baseball pitcher accelerates the ball over a distance of about 3.5 m.

2–8   Falling Bodies

One of the commonest examples of uniformly accelerated motion is that of an object allowed to fall vertically to the earth. That a falling body is accelerating may not be obvious at first. And it may seem obvious, as was widely believed until the time of Galileo, that heavier bodies fall faster than lighter bodies and that the speed of fall is proportional to how heavy the object is.

Galileo's analysis made use of his new and creative technique of abstraction and simplification—that is, of imagining what would happen in idealized (simplified) cases. For free fall, he postulated that all bodies would fall with the *same constant acceleration* in the absence of air or other resistance. He showed that this postulate predicts that for an object falling from rest, the distance traveled will be proportional to the square of the time ($D \propto t^2$). This we can see from Eq. 2–9b; but Galileo was the first to derive this mathematical relation. In fact, one of Galileo's great contributions to science was to establish such mathematical relations and to insist on their importance; another great contribution of Galileo was the proposing of a theory with specific experimental consequences that could be quantitatively checked ($D \propto t^2$).

To support his claim that the speed of falling objects increases as they fall, Galileo made use of the following argument: a heavy stone dropped from a height of 2 m will drive a stake into the ground much further than if the same stone is dropped from a height of only 10 cm; the stone must clearly be moving faster in the former case. As we saw, Galileo also claimed that *all* objects, light or heavy, fall with the *same* acceleration, at least in the absence of air. Now common sense may say that the ancients were perhaps closer to the truth; for if you hold a piece of paper horizontally in one hand and a heavier object, say, a baseball, in the other and release them at the same time (see Fig. 2–9a) surely the heavier object will reach the ground first. Repeat the experiment but this time crumple up the paper into a small wad (see Fig. 2–9b). You will find that the two objects reach the floor at nearly the same time.

Galileo was sure that air acts like a sort of friction on very light objects that have a large surface area; but in many ordinary circumstances this air resistance

FIGURE 2–9

(a) A ball and a light piece of paper are dropped at the same time. (b) Repeated, with the paper wadded up.

(a)   (b)

is negligible. In a chamber from which the air has been removed even light objects like a feather or a horizontally held piece of paper will fall with the same acceleration as any other object (see Fig. 2–10). Such a demonstration in vacuum was of course not possible in Galileo's time, which makes Galileo's achievement all the greater. Galileo is often called the "father of modern science", not only for the content of his science (astronomical discoveries, inertia, free fall), but also for his style or approach to science (idealization and simplification, mathematization of theory, prediction of testable consequences that must be checked by experiment).

Galileo's specific contribution to our understanding of the motion of falling objects can be summarized as follows: *At a given location on the earth and in the absence of air resistance, all objects fall with the same uniform acceleration.* We call this acceleration the **acceleration due to gravity** and we give it the symbol g. Its value is approximately:

$$g = 9.80 \text{ m/s}^2.$$

In British units g is about 32 ft/s^2. Actually, g varies slightly according to latitude (due to earth's rotation) and elevation (see Table 2–1) but these variations are so small that we will ignore them for most purposes. The effects of air resistance are often small and we will neglect them for the most part. However, air resistance will be noticeable even on a reasonably heavy object if the distance of fall is very large.[†]

When dealing with freely falling objects we can make use of Eqs. 2–9 where for a we use the numerical value of g given above. Also, since the motion is vertical we will substitute y in place of x and y_0 in place of x_0 (and we take $y_0 = 0$ unless otherwise specified). It is arbitrary whether we choose y to be positive in the upward direction or in the downward direction, but we must be consistent about it throughout a problem.

⬛ **EXAMPLE 2–10** Suppose that a ball is dropped from a tower 70.0 m high. How far will it have fallen after 1.00, 2.00, and 3.00 s? Assume y is positive downward.

SOLUTION Equation 2–9b is the appropriate one, with $v_0 = 0$ and $y_0 = 0$. Then, after 1.00 s,

$$y = \tfrac{1}{2} at^2 = \tfrac{1}{2}(9.80 \text{ m/s}^2)(1.00 \text{ s})^2 = 4.90 \text{ m}.$$

Similarly after 2.00 s, $y = 19.6$ m, and after 3.00 s, $y = 44.1$ m. See Fig. 2–11.

[†] The speed of an object falling in air (or other fluid) does not increase indefinitely; if it falls far enough, it will reach a maximum velocity called the *terminal velocity*. The maximum velocity is reached when the force of air resistance (which increases with speed) balances the force of gravity.

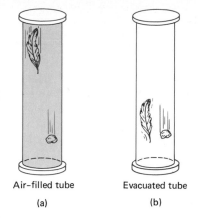

Air-filled tube   Evacuated tube
(a)   (b)

FIGURE 2–10

A rock and a feather are dropped simultaneously (a) in air, (b) in a vacuum.

FIGURE 2–11

When an object is dropped from the top of a tower, it falls with progressively greater speed and covers greater distance with every successive second.

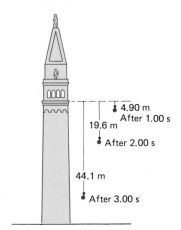

4.90 m
After 1.00 s

19.6 m
After 2.00 s

44.1 m
After 3.00 s

TABLE 2–1

Acceleration due to gravity at various locations on earth

Location	Elevation (ft)	g (m/s²)
New York	0	9.803
San Francisco	350	9.800
Denver	5400	9.796
Pikes Peak	14,100	9.789
Equator	0	9.780
North Pole (calculated)	0	9.832

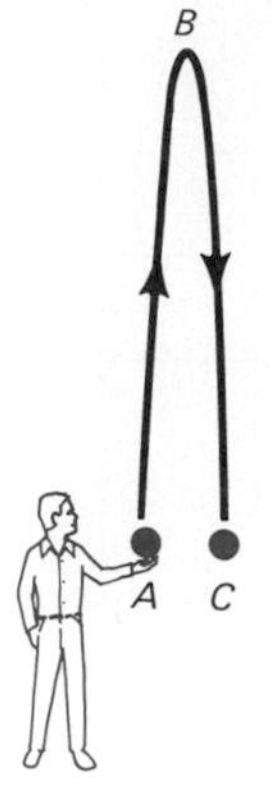

FIGURE 2–12

Object thrown
into air leaves
thrower's hand
at A, reaches
its maximum
height at B,
and returns to
original height
at C. (See
Example 2–11.)

EXAMPLE 2–11 A person throws a ball upward into the air with an initial velocity of 15.0 m/s. Calculate (*a*) how high it goes, and (*b*) how long the ball is in the air before it comes back to his hand. We are not concerned here with the throwing action, but only with the motion of the ball after it leaves the thrower's hand, Fig. 2–12.

SOLUTION Let us choose y to be positive in the upward direction and negative in the downward direction. (Note: this is a different convention from that used in the previous example.) Then the acceleration will have a negative sign (since it changes the velocity in a downward direction), $a = -g = -9.80$ m/s^2.

(*a*) To determine the maximum height we calculate the position of the ball when its velocity equals zero ($v = 0$ at the highest point). At $t = 0$ we have $y_0 = 0$, $v_0 = 15.0$ m/s; at time t (*maximum height*), $v = 0$, $a = -9.80$ m/s^2, and we wish to find y. We use Eq. 2–9c (replacing x with y) and solve for y:

$$v^2 = v_0^2 + 2ay$$

$$y = \frac{v^2 - v_0^2}{2a}$$

$$= \frac{0 - (15.0 \text{ m/s})^2}{2(-9.80 \text{ m/s}^2)} = 11.5 \text{ m.}$$

(*b*) We could do this section of the example in two parts by first calculating the time required for the ball to reach its highest point, and then calculating the time it takes to fall back down. However, it is simpler to consider the motion from A to B to C (Fig. 2–12) in one step and use Eq. 2–9b; we can do this because y (or x) represents position or displacement, and not the total distance traveled. Thus, at both points A and C, $y = 0$. We use Eq. 2–9b with $a = -9.80$ m/s^2 and find

$$y = v_0 t + \tfrac{1}{2}at^2$$

$$0 = (15.0 \text{ m/s})t + \tfrac{1}{2}(-9.80 \text{ m/s}^2)t^2.$$

There are two solutions[†]: $t = 0$ and $t = (2)(15.0 \text{ m/s})/(9.80 \text{ m/s}^2) = 3.06$ s. The first solution corresponds to the initial point, A, and the second to the return, C. Thus, the ball is in the air for 3.06 s.

The acceleration of an object, particularly rockets and fast airplanes, is often given as a multiple of $g = 9.80$ m/s^2. For example, a diving plane undergoing 3.00 g's would have an acceleration of $(3.00)(9.80 \text{ m/s}^2) = 29.4$ m/s^2.

*2–9 Variable Acceleration—Graphical Analysis and Use of Calculus

This section is optional. It assumes the reader is familiar with derivatives and simple integration. If you have not yet studied this material in your calculus course, you may want to postpone studying this section until after you have.

In Section 2–5 we saw that if we know the position of an object as a function of time, then we can determine the velocity at any instant by measuring the slope of the x vs. t graph at that instant. Alternatively, we can take the derivative of x with respect

---

† For any quadratic equation in (say) t of the form $at^2 + bt + c = 0$, where a, b, and c are constants, the quadratic formula gives the solutions:

$$t = \frac{-b \pm \sqrt{b^2 - 4ac}}{2a}.$$

to t since $v = dx/dt$. For example, as already mentioned in Example 2–3, if x can be written as a polynomial in t, then we can use the formula

$$\frac{d}{dt}(Ct^n) = nCt^{n-1};$$

if x is given by, say, $x = At^3 + Bt$ where A and B are constants, then $v = dx/dt = 3At^2 + B$. If we are given the numerical values for A and B, we can get the value of v at any time t from this equation.

In Section 2–6 we then saw that we can determine the acceleration at any time if we know the velocity as a function of time; this can be done using the fact that the acceleration equals the slope of the v vs. t graph, or by taking the derivative of v with respect to t: $a = dv/dt$. For the example above in which $v = 3At^2 + B$, $a = dv/dt = 6At$.

The reverse process is also possible. If we are given the acceleration, a, as a function of time, we can determine v as a function of time; and given v as a function of t, we can obtain the displacement, x. To see how this is done, suppose the velocity as a function of time, $v(t)$, varies as shown in Fig. 2–13a, and we are interested in the time interval from t_1 to t_2 as shown. We first divide the time axis into many small subintervals (called $\Delta t_1, \Delta t_2, \Delta t_3, \ldots$) which are indicated by the dashed vertical lines. For each subinterval, a horizontal dashed line is drawn to indicate the average velocity during that time interval. The displacement during any subinterval is given by Δx_i where the subscript i represents the particular subinterval ($i = 1, 2, 3, \ldots$). From the definition of average velocity (Eq. 2–1) we have

$$\Delta x_i = \bar{v}_i \Delta t_i.$$

Thus the displacement during each subinterval equals the product of v_i and Δt_i, and this is just the area of the rectangle shown in grey (Fig. 2–13a) for that subinterval. The total displacement between time t_1 and t_2 is the sum of the displacements over all the subintervals:

$$x_2 - x_1 = \sum_{t_1}^{t_2} \bar{v}_i \Delta t_i, \tag{2–10a}$$

(where x_1 is the position at t_1 and x_2 is the position at t_2), and this is just the area of all the rectangles shown.

It is often difficult to estimate $\bar{v}_i$ with precision for each subinterval from the graph. We can get greater accuracy in our calculation of $x_2 - x_1$ by breaking the interval $t_2 - t_1$ into more, but narrower, subintervals. Ideally, we can let each Δt_i approach zero, so we approach (in principle) an infinite number of subintervals. In this limit the area of all these infinitesimally thin rectangles becomes exactly equal to the area under the curve (Fig. 2–13b). We thus have the important result that *the total displacement between any two times is equal to the area between the velocity curve*

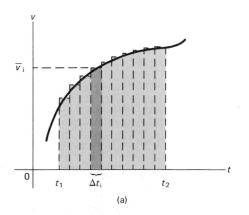

(a)

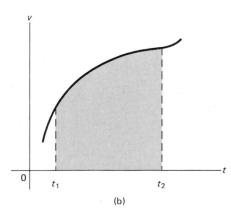

(b)

FIGURE 2–13

Graph of v vs. t for the motion of a particle. In (a), the time axis is broken into subintervals of width Δt_i; the average velocity during each Δt_i is $\bar{v}_i$, and the area of all the rectangles, $\Sigma \bar{v}_i \Delta t_i$, is numerically equal to the total displacement ($x_2 - x_1$) during the total time ($t_2 - t_1$). In (b), $\Delta t_i \rightarrow 0$ and the area under the curve is equal to ($x_2 - x_1$).

and the t *axis between the two times* t_1 *and* t_2. This limit can be written

$$x_2 - x_1 = \lim_{\Delta t \to 0} \sum_{t_1}^{t_2} \bar{v}_i \, \Delta t_i$$

or

$$x_2 - x_1 = \int_{t_1}^{t_2} v(t) \, dt. \tag{2–10b}$$

We have let $\Delta t \to 0$ and renamed it dt to indicate that it is now infinitesimally small. The average velocity, $\bar{v}$, over an infinitesimal time dt is, of course, the instantaneous velocity at that instant, which we have written $v(t)$ to remind us that v is a function of t. The symbol $\int$ is an elongated S and indicates a sum over an infinite number of infinitesimal subintervals; we say that we are taking the *integral* of $v(t)$ over dt from time t_1 to time t_2, and this is equal to the area between the $v(t)$ curve and the t axis between the times t_1 and t_2 (Fig. 2–13b). The integral in Eq. 2–10b is called a *definite integral* since definite limits (t_1 and t_2) are specified. Graphically, the integral is numerically equal to an area under a curve and that area can be accurately estimated numerically by taking small but finite subintervals, Δt_i, and summing them using a calculator or computer. In many cases integrals can be precisely evaluated using the methods of calculus.

Using Eqs. 2–10 we can obtain the total displacement if we know the velocity as a function of time by taking the integral. Similarly, if we know the acceleration as a function of time, we can obtain the velocity by the process of integration. To see this, let us use the definition of average acceleration (Eq. 2–3) and solve for Δv:

$$\Delta v = \bar{a} \, \Delta t.$$

Now if a is known as a function of t over some time interval t_1 to t_2, we can subdivide this time interval into many subintervals, Δt_i, just as we did in Fig. 2–13a. The change in velocity during each subinterval will be $\Delta v_i = a_i \Delta t_i$, where the subscript i refers to an arbitrary subinterval ($i = 1, 2, 3, \ldots$). The total change in velocity from time t_1 until time t_2 is

$$v_2 - v_1 = \sum_{t_1}^{t_2} \bar{a}_i \, \Delta t_i, \tag{2–11a}$$

where v_2 represents the velocity at t_2 and v_1 the velocity at t_1. This relation can be written as an integral by letting $\Delta t \to 0$:

$$v_2 - v_1 = \lim_{\Delta t \to 0} \sum_{t_1}^{t_2} \bar{a}_i \, \Delta t_i$$

or

$$v_2 - v_1 = \int_{t_1}^{t_2} a(t) \, dt. \tag{2–11b}$$

Equation 2–11b will allow us to determine the velocity v_2 at some time t_2 if the velocity is known at t_1 and a is known as a function of time.

Integration is the inverse process to taking the derivative. How to evaluate integrals in particular cases is discussed in textbooks and courses on calculus and we can discuss it here only briefly. An integral can always be evaluated numerically to some prescribed precision (Section 2–10). Some functional forms can be analytically evaluated using calculus; the result for some integrals will be found in Appendix B. For polynomials we can readily determine the functional form of the integral, since we know that

$$\frac{d}{dt}(Ct^n) = nCt^{n-1};$$

we multiply both sides by dt to obtain

$$nCt^{n-1} \, dt = d(Ct^n)$$

and then we integrate both sides:

$$\int nCt^{n-1}\, dt = \int d(Ct^n)$$

$$= Ct^n,$$

where the last step follows since integration is the inverse of taking the derivative. This formula is more conveniently written by dividing both sides by n and setting $n - 1 = m$:

$$\int Ct^m\, dt = \frac{1}{m+1}\, Ct^{m+1}.$$

◆ **EXAMPLE 2–12** An object starts from rest $(v_1 = 0)$ at $t_1 = 0$ and accelerates at a rate given by $a(t) = (7.00 \text{ m/s}^3)t$. What is (a) its velocity and (b) its displacement 2.00 s later?

SOLUTION (a) Let us first determine v as a function of t. We set $t_1 = 0$, $t_2 = t$, $v_1 = 0$, $v_2 = v(t)$ in Eq. 2–11b and obtain[†]

$$v(t) = \int_0^t (7.00 \text{ m/s}^3)\, t\, dt$$

$$= (7.00 \text{ m/s}^3)\left(\frac{t^2}{2}\right)\Big|_0^t = (7.00 \text{ m/s}^3)\left(\frac{t^2}{2} - 0\right)$$

$$= (3.50 \text{ m/s}^3)t^2.$$

At $t = 2.00$ s, $v = (3.50 \text{ m/s}^3)(2.00 \text{ s})^2 = 14.0$ m/s.

(b) To get the displacement, we use Eq. 2–10b with $v_1 = 0$, $v_2 = 14.0$ m/s, $t_1 = 0$, $t_2 = 2.00$ s and let us choose $x_1 = 0$:

$$x_2 = \int_0^{2.00 \text{ s}} (3.50 \text{ m/s}^3)\, t^2\, dt$$

$$= (3.50 \text{ m/s}^3)\frac{t^3}{3}\Big|_0^{2.00 \text{ s}} = 9.33 \text{ m}.$$

In sum, at $t = 2.00$ s, $v = 14.0$ m/s and $x = 9.33$ m.

◆ **EXAMPLE 2–13** Use Eqs. 2–10b and 2–11b to derive the kinematic Eqs. 2–9a, b, and c for constant acceleration.

SOLUTION We use the same notation as in Section 2–7: $t_1 = 0$, $t_2 = t$, $v_1 = v_0$, $v_2 = v$. Then from Eq. 2–11b, remembering that $a = $ constant:

$$v = v_0 + \int_0^t a\, dt = v_0 + at.$$

This is Eq. 2–9a. Now we use Eq. 2–10b:

$$x = x_0 + \int_0^t v(t)\, dt = x_0 + \int_0^t (v_0 + at)\, dt$$

$$= x_0 + v_0 t + \tfrac{1}{2}at^2.$$

This is Eq. 2–9b. To get Eq. 2–9c we use the so-called chain rule:

$$a = \frac{dv}{dt} = \frac{dv}{dx}\frac{dx}{dt};$$

[†] In the second line of the following equation the vertical bar, $\big|_0^t$, means the preceding relation has to be evaluated at the upper limit (t) and then from that is subtracted its value at the lower limit (0).

since $dx/dt = v$, we have $a = v(dv/dx)$. Then, from this last relation:

$$v \, dv = a \, dx$$

$$\int_{v_0}^{v} v \, dv = \int_{x_0}^{x} a \, dx$$

$$\frac{v^2}{2} - \frac{v_0^2}{2} = a(x - x_0)$$

or $v^2 = v_0^2 + 2a(x - x_0)$, which is Eq. 2–9c.

▥ *2–10   Variable Acceleration— Numerical Integration

We illustrate in this optional section the method of integrating numerically. We begin by taking an example that can also be evaluated analytically so we can compare the results.

▦   EXAMPLE 2–14   An object starts from rest and accelerates at a rate $a(t) = (8.00 \text{ m/s}^4)t^2$. Determine its velocity after 2.00 s using numerical methods.

SOLUTION   Let us first divide up the interval $t = 0.00$ s to $t = 2.00$ s into four subintervals each of duration $\Delta t_i = 0.50$ s (Fig. 2–14). We use Eq. 2–11a with $v_2 = v$, $v_1 = 0$, $t_2 = 2.00$ s, and $t_1 = 0$. For each of the subintervals we need to estimate $\bar{a}_i$; there are various ways to do this and we use the simple method of choosing $\bar{a}_i$ to be the acceleration $a(t)$ at the midpoint of each interval (an even simpler but usually less accurate procedure would be to use the value of a at the start of the subinterval); that is, we evaluate $a(t) = (8.00 \text{ m/s}^4)t^2$ at $t = 0.25$ s (which is midway between 0.00 s and 0.50 s), 0.75 s, 1.25 s, and 1.75 s. The results are as follows:

i	1	2	3	4
$\bar{a}_i \,(\text{m/s}^2)$	0.50	4.50	12.50	24.50

Now we use Eq. 2–11a, and note that all Δt_i equal 0.50 s (so they can be factored out):

$$v(t = 2.00 \text{ s}) = \sum_{t=0}^{t=2.00 \text{ s}} \bar{a}_i \, \Delta t_i$$

$$= (0.50 \text{ m/s}^2 + 4.50 \text{ m/s}^2 + 12.50 \text{ m/s}^2 + 24.50 \text{ m/s}^2)(0.50 \text{ s})$$

$$= 21.0 \text{ m/s}.$$

We can compare this result to the analytic solution given by Eq. 2–11b since the functional form for a is integrable analytically:

$$v = \int_{0}^{2.00 \text{ s}} (8.00 \text{ m/s}^4)t^2 \, dt = \frac{8.00 \text{ m/s}^4}{3} \, t^3 \bigg|_{0}^{2.00 \text{ s}}$$

$$= \frac{8.00 \text{ m/s}^4}{3} \left[(2.00 \text{ s})^3 - (0)^3 \right]$$

$$= 21.33 \text{ m/s}.$$

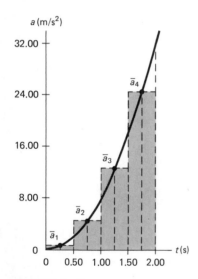

FIGURE 2–14

Example 2–14.

or 21.3 m/s to the proper number of significant figures. This analytic solution is, of course, precise, and we see that our numerical estimate is not far off even though we only used four Δt intervals. It may not, however, be close enough for purposes requiring high accuracy. If we use more and smaller subintervals, we will get a more accurate result. For example, if we use ten subintervals, each with $\Delta t = 2.00$ s/ $10 = 0.20$ s, we have to evaluate $a(t)$ at $t = 0.10$ s, 0.30 s, . . . , 1.90 s to get the $\bar{a}_i$, and these are as follows:

i	1	2	3	4	5	6	7	8	9	10
a_i (m/s^2)	0.08	0.72	2.00	3.92	6.48	9.68	13.52	18.00	23.12	28.88

Then, from Eq. 2–11a we obtain

$$v(t = 2.00 \text{ s}) = \sum \bar{a}_i \Delta t_i = \left(\sum \bar{a}_i \right)(0.200 \text{ s})$$
$$= (106.4 \text{ m/s}^2)(0.200 \text{ s}) = 21.28 \text{ m/s},$$

where we have kept an extra significant figure to show that this result is much closer to the (precise) analytic one but still is not quite identical to it. The percentage difference has dropped from 1.5 percent (0.3 m/s^2/21.0 m/s^2) for the four-subinterval computation to only 0.2 percent (.05/21.0) for the ten-subinterval one.

Note that an analytic solution (integrable in closed form) can give a precise result, whereas a numerical solution is (except for very simple cases) an approximation.

In the example above we were given an analytic function that was integrable, so we could compare the accuracy of the numerical calculation to the known precise one. (Of course, with such an easily integrable function there is no real need to bother doing a numerical calculation.) But what do we do if the function is not integrable so we can't compare our numerical result to an analytic one? That is, how do we know if we've taken enough subintervals so that we can trust our calculated estimate to be accurate to within some prescribed uncertainty, say 1 percent? Since there is no analytic value to be compared to, what we do instead is compare two successive numerical calculations: the first done with, say, n subintervals and the second with, say, twice as many subintervals ($2n$). If the two results are within the prescribed uncertainty (say 1 percent) then we can usually assume that the second calculation (with more subintervals) is within the prescribed uncertainty of the true value. If the two calculations are not that close, then a third calculation, with more subintervals (maybe double, maybe ten times as many, depending on how good the previous approximation was) must be done and compared to the previous one. For example, in Example 2–14 the difference between the calculations with four subintervals and that with ten was about 0.3 m/s^2 or 1.5 percent. If we wanted an accuracy of 1 percent, we couldn't be sure that our second calculation was quite close enough, so we would do a third calculation with, perhaps, twenty subintervals.

Such calculations can quickly become very long and tedious, particularly if the function is more complicated than our very simple example above. Even in a simple example, if we also wanted to obtain the displacement x at some time, we would have to do a second numerical integration over v, which means we would first need to calculate v for many different times—a lot of work. Programmable calculators and computers are very helpful in this regard: a program (often simple) can be put into the computer and the calculations repeated with various input variables and various numbers of subintervals until the desired accuracy is obtained.

Numerical procedures must also be used if numerical data are given instead of an analytic function. (For example, suppose the acceleration of an object is measured at a series of particular times.) The procedure is basically that outlined above, except that it may not be possible to have all the subintervals equal. A simple way to estimate the average value of the integrand, say $\bar{v}$, for each subinterval is to choose it equal to the sum of the values at the beginning and at the end of the subinterval divided by 2. [For example, in the first subinterval we would have $\bar{v} = (v_0 + v_1)/2$ where v_0 and v_1 are the values of v at the start of the first and second intervals. The value at the beginning (or end) of each interval could also be used, but then more subintervals will be required to obtain an accurate result.]

We have illustrated how to do numerical integration using a simple and direct technique. More complex techniques can also be used (such as estimating $\bar{a}_i$ more accurately for each subinterval, thus requiring fewer subintervals to obtain an accurate result), and they can be found in appropriate textbooks.

 Summary†

Kinematics deals with the description of how objects move, as compared to *dynamics*, which deals with why they move as they do. The description of the motion of any object must always be given relative to some particular *reference frame*. To describe the motion of an object, which in this chapter we restrict to only one dimension (along a line), we use the concepts of displacement, velocity, and acceleration. *Displacement* refers to the change in position of the object. Velocity is the rate of change of displacement: the *average velocity* of an object over a particular time interval is defined as the displacement Δx divided by the time interval Δt: $\bar{v} = \Delta x/\Delta t$; the *instantaneous velocity* (velocity at a given instant) is defined as the limit of the average velocity at a particular time as $\Delta t \to 0$:

$$v = \lim_{\Delta t \to 0} \frac{\Delta x}{\Delta t} = \frac{dx}{dt}$$

where dx/dt is called (in calculus) the *derivative* of x with respect to t. On a graph of position vs. time, the slope is equal to the instantaneous velocity. Acceleration is the

rate of change of velocity: the *average acceleration* over a time interval Δt is defined as $\bar{a} = \Delta v/\Delta t$ where Δv is the change of velocity during the time interval Δt; the *instantaneous acceleration* is defined as

$$a = \lim_{\Delta t \to 0} \frac{\Delta v}{\Delta t} = \frac{dv}{dt}.$$

If an object moves in a straight line with constant acceleration (*uniformly accelerated motion*), the velocity, v, and position, x, are related to the acceleration, a, the elapsed time, t, and the initial position and velocity, x_0 and v_0, by Eqs. 2–9:

$$v = v_0 + at, \qquad x = x_0 + v_0 t + \frac{1}{2} at^2,$$

$$v^2 = v_0^2 + 2a(x - x_0), \qquad \bar{v} = \frac{v + v_0}{2}.$$

(These are useful equations, but instead of memorizing them, it might be more profitable to be able to derive them from the definitions of velocity and acceleration.)

Objects that move vertically near the surface of the earth, either falling or having been projected vertically up or down, move with the constant downward *acceleration due to gravity* of about $g = 9.80$ m/s^2, if air resistance can be ignored.

---

† The summaries that appear at the end of each chapter in this book are intended to give a brief overview of the main ideas of each chapter. They *cannot* act to give an understanding of the material, which can only be had by a detailed reading of the chapter.

 Questions

1. For what purpose would it be useful to consider a moving football as a particle, and for what purposes not as a particle?

2. Does a car speedometer measure speed, velocity, or both?

3. If an accurate speedometer registers a constant value for a period of time, can you determine the average velocity over that period of time using only the speedometer? Explain.

4. Can the average velocity of a particle be nonzero over a given time interval if over a longer time interval it is zero? Explain.

5. Can the average velocity of a particle be zero over a given time interval if it is not zero over a longer time interval? Explain.

6. Can an object have a varying velocity if its speed is constant? If yes, give examples.

7. Can an object have a varying speed if its velocity is constant? If yes, give examples.

8. When an object moves with constant velocity, does its average velocity during any time interval differ from its instantaneous velocity at any instant?

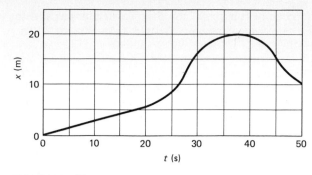

FIGURE 2–15

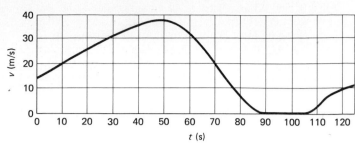

FIGURE 2–16

9. Describe in words the motion plotted in Fig. 2–15.

10. Describe in words the motion of the object graphed in Fig. 2–16.

11. Can an object have a northward velocity and a southward acceleration?

12. Can the velocity of an object be zero at the same instant its acceleration is not zero? Give an example.

13. If an object has a greater speed, does it necessarily have a greater acceleration? Explain, using examples.

14. Compare the acceleration of a motorcycle that accelerates from 80 km/h to 90 km/h with a bicycle that accelerates from rest to 10 km/h in the same time.

15. Can the velocity of an object be negative when its acceleration is positive? What about vice versa?

16. Give an example where both the velocity and acceleration are negative.

17. A rock is thrown vertically upward with speed v from the edge of a cliff. A second rock is thrown vertically downward with the same initial speed. Which rock has the greater speed when it reaches the bottom of the cliff? Ignore air resistance.

18. The acceleration due to gravity on the moon is about one-sixth what it is on earth. If an object is thrown vertically upward on the moon, how many times higher will it go as compared to earth assuming the same initial velocity.

19. A ball, thrown vertically upward, returns to the thrower's hand. Which part of the journey requires the longer time, upward or downward? Answer for (a) no air resistance, (b) in the presence of air resistance.

(Hint: the acceleration due to air resistance is always in a direction opposite to the motion.)

20. An object that is thrown vertically upward will return to its original position with the same speed as it had initially, if air resistance is neglible. If air resistance is appreciable, will this result be altered, and, if so, how?

 Problems ⎯⎯⎯⎯⎯⎯⎯⎯⎯⎯⎯⎯⎯⎯⎯⎯⎯⎯⎯⎯⎯⎯⎯⎯⎯⎯⎯⎯⎯⎯⎯⎯⎯⎯

[The problems at the end of each chapter are ranked I, II, or III according to estimated difficulty, with I problems being easiest. The problems are arranged by sections, meaning that the reader should have read up to and including that section, but not only that section—problems often depend on earlier material.]

SECTIONS 2–1 TO 2–5

1. (I) If you are driving 90 km/h and you look to the side for 2.0 s, how far do you travel during this inattentive period?

2. (I) How long does it take a bird to fly 100 km at a speed of 28 km/h?

3. (I) 55 mph is how many (a) km/h, (b) m/s, (c) ft/s?

4. (I) Determine the conversion factor between (a) km/h and mi/h, (b) m/s and ft/s, (c) mi/h and m/s.

5. (I) A person jogs eight complete laps around a quarter-mile track in a total time of 10.5 min. Calculate (a) the average speed and (b) the average velocity.

6. (I) The position of a rabbit along a straight tunnel as a function of time is plotted in Fig. 2–15. What is its instantaneous velocity (a) at $t = 10.0$ s, and (b) at $t = 30.0$ s? What is its average velocity (c) between $t = 0$ and $t = 5.0$ s, (d) between $t = 25.0$ s and $t = 30.0$ s, and (e) between $t = 40.0$ s and $t = 50.0$ s?

7. (I) In Fig. 2–15, (a) during what time periods, if any, is the rabbit's velocity constant? (b) At what time is its velocity the greatest? (c) At what time, if any, is the velocity zero? (d) Does the rabbit run in one direction or in both along its tunnel during the time shown?

8. (II) Construct the v vs. t graph for the object whose displacement as a function of time is given by Fig. 2–15.

9. (II) An airplane travels 2200 km at a speed of 1000 km/h. It then encounters a headwind that slows it to 850 km/h for the next 1750 km. What was the average speed of the plane for this trip?

10. (II) A race car driver must average 200 km/h for four laps to qualify for a race. Because of engine trouble, the car averages only 170 km/h over the first two laps. What average speed must be maintained for the last two laps?

11. (II) Calculate the carrying capacity (number of cars passing a given point per hour) on a freeway with three lanes (in one direction) using the following assumptions: the average speed is 90 km/h, the average length of a car is 6.0 m, and the average distance between cars should be 70 m.

12. (II) A rock thrown horizontally at a large bell 40 m away is heard to hit the bell 3.9 s later. If the speed of sound is 330 m/s, what was the speed of the rock? (Disregard the effect of gravity.)

13. (II) A dog runs 100 m away from its master in a straight line in 8.4 s, and then runs halfway back in one-third the time. Calculate (a) its average speed and (b) its average velocity.

14. (II) Two engines approach each other on parallel tracks. Each has a speed of 90 km/h with respect to the earth. If they are initially 8.5 km apart, how long will it be before they pass each other?

15. (II) The position of a ball rolling in a straight line is given by $x = 2.0 + 6.6t - 1.1t^2$ where x is in meters and t in seconds. (a) Determine the position of the ball at $t = 1.0$, 2.0, and 3.0 s. (b) What is the average velocity over the interval $t = 1.0$ s to $t = 3.0$ s? (c) What is its instantaneous velocity at $t = 2.0$ s and at $t = 3.0$ s?

16. (II) A car traveling 90 km/h is 100 m behind a truck traveling 50 km/h. How long will it take the car to reach the truck?

17. (II) An automobile traveling 90 km/h overtakes a 1.10-km-long train traveling in the same direction on a track parallel to the road. If the train's speed is 70 km/h, how long does it take the car to pass it and how far will the car have traveled in this time? What are the results if the car and train are traveling in opposite directions?

SECTION 2–6

18. (I) Figure 2–16 shows the velocity of a train as a function of time. (a) At what time was its velocity greatest? (b) During what periods, if any, was the velocity constant? (c) During what periods, if any, was the acceleration constant? (d) When was the magnitude of the acceleration greatest?

19. (I) A car accelerates from rest to 100 km/h in 6.6 s. What is its acceleration in m/s^2?

20. (I) At high speeds, a particular automobile is capable of an acceleration of about 3.2 m/s^2. At this rate how long does it take to accelerate from 85 km/h to 100 km/h?

21. (II) The position of a body is given by x (meters) $= At + 4Bt^3$. (a) What is its acceleration as a function of time? (b) What is its velocity and acceleration after 5.0 s? (c) What is the velocity as a function of time if $x = At + Bt^{-3}$?

22. (II) A car accelerates uniformly from rest to 15 m/s in 10 s, after which it remains at a constant speed of 15 m/s for the next 10 s, decelerates uniformly to 5.0 m/s during the following 5.0 s, and then remains at this constant speed for 5.0 s. (a) Graph the velocity as a function of time. (b) Construct a graph of displacement x vs. t.

23. (III) The position of a racing car, which starts from rest at $t = 0$ and moves in a straight line has been measured as a function of time as given in the following table. Estimate (a) its velocity and (b) its acceleration as a function of time. Display each in a table and on a graph.

t (s)	0	0.25	0.50	0.75	1.00	1.50	2.00
x (m)	0	0.11	0.46	1.06	1.94	4.62	8.55

t (s)	2.50	3.00	3.50	4.00	4.50	5.00	5.50	6.00
x (m)	13.79	20.36	28.31	37.65	48.37	60.30	73.26	87.16

SECTION 2–7

24. (I) A car accelerates from 30 km/h to 80 km/h in 5.0 s. What was its acceleration in m/s^2, and how far did it travel in this time? Assume constant acceleration.

25. (I) A car decelerates from a speed of 25 m/s to rest in a distance of 120 m. What was its acceleration, assumed constant?

26. (I) Show that $\bar{v} = (v + v_0)/2$ (see Eq. 2–9d) is not valid for the case when the acceleration $a = A + Bt$, where A and B are constants.

27. (II) A space vehicle accelerates uniformly from 55 m/s at $t = 0$ to 162 m/s at $t = 10.0$ s. How far did it move between $t = 2.0$ s and $t = 6.0$ s?

28. (II) A 90-m-long train accelerates uniformly from rest. If the front of the train passes a railway worker 130 m down the track at a speed of 25 m/s, what will be the speed of the last car as it passes the worker?

29. (II) A car traveling 90 km/h decelerates uniformly at 1.8 m/s^2. Calculate (a) the distance it goes before it stops, (b) the time it takes to stop, and (c) the distance it travels during the first and third seconds.

30. (II) In putting, the force with which a golfer strikes a ball is determined so that the ball will stop within some small distance of the cup, say 1 m long or short, even if the putt is missed. Accomplishing this from an uphill lie (that is, putting downhill) is more difficult than from a downhill lie. To see why, assume that on a particular green the ball decelerates constantly at 3.0 m/s^2 going downhill and constantly at 4.0 m/s^2 going uphill. Suppose we have an uphill lie 7.0 m from the cup. Calculate the allowable range of initial velocities we may impart to the ball so that it stops in the range 1.0 m short to 1.0 m long of the cup. Do the same for a downhill lie 7.0 m from the cup. What in your results suggests that the downhill putt is more difficult?

31. (II) A car traveling at 50 km/h strikes a tree; the front end of the car compresses and the driver comes to rest after traveling 0.70 m. What was the average deceleration of the driver during the collision? Express the answer in terms of "g's" where 1.00 $g = 9.80$ m/s^2.

32. (II) A person who is properly constrained by a shoulder harness has a good chance of surviving a car collision if the deceleration does not exceed 30 "g's" (1.00 $g = 9.80$ m/s^2). Assuming uniform deceleration at this rate, calculate the distance over which the front end of the car must be designed to collapse if a crash occurs at 100 km/h.

33. (II) Make up a table of stopping distances for an automobile with an initial speed of 60 km/h and human

reaction time of 0.80 s (a) for a deceleration $a = -4.0$ m/s²; (b) for $a = -7.0$ m/s².

34. (III) Show that the equation for the stopping distance (d_s) of a car is $d_s = v_0 t_R - v_0^2/(2a)$ where v_0 is the initial speed of the car, t_R is the driver's reaction time, and a is the rate of deceleration (and is negative).

35. (III) In designing traffic signals, it is necessary to allow the yellow light to remain on long enough that a driver can either stop or pass completely through the intersection. Thus, if a driver is less than the stopping distance, d_s (calculated in problem 34 above), from the intersection, then the light must remain on long enough for him to travel this distance plus the width of the intersection, d_I. (a) Show that the yellow light should remain on for a time $t = t_R - v_0/(2a) + d_I/v_0$ where v_0 is a typical expected speed of a car approaching the intersection, and a and t_R are as defined in problem 34. (b) A traffic planner expects cars to approach a 14.4 m wide intersection at speeds between 30.0 and 50.0 km/h. To be safe, he calculates the time for both speeds, assuming $t_R = 0.500$ s and $a = -4.00$ m/s², and chooses the longest time to be safe. What is his result?

36. (III) In the design of a rapid transit system, it is necessary to balance out the average speed of a train against the distance between stops. The more stops there are, the slower the train's average speed. To get an idea of this problem, calculate the time it takes a train to make a 30-km trip in two situations: (a) the stations at which the trains must stop are 1.00 km apart; (b) the stations are 3.00 km apart. Assume that at each station the train accelerates at a rate of 1.5 m/s² until it reaches 80 km/h, then stays at this speed until it puts on its brakes to arrive at the next station, at which time it decelerates at -3.0 m/s². Assume it stops at each station for 20 s.

37. (III) For the design of a rapid transit system as discussed in the previous problem, derive a general formula for the average speed of a train. Specify the symbols used for all quantities involved, such as the acceleration, deceleration, maximum velocity, distance between stations, and time stopped at each station.

38. (III) An unmarked police car traveling a constant 80 km/h is passed by a speeder traveling 100 km/h. Precisely 1.00 s after the speeder passes, the policeman steps on the accelerator; if the police car's acceleration is 3.00 m/s², how much time passes before the police car overtakes the speeder (assumed moving at constant speed)?

39. (III) Assume in the previous problem that the speeder's speed is not known. If the police car accelerates uniformly as given above for 6.0 s, what was the speeder's speed?

SECTION 2–8

40. (I) A stone is dropped from the top of a cliff. It is seen to hit the ground below after 4.2 s. How high is the cliff?

41. (I) (a) How long does it take a brick to reach the ground if dropped from a height of 65 m? (b) What will be its velocity just before it reaches the ground?

42. (I) A baseball is thrown vertically into the air with a speed of 18.0 m/s. (a) How high does it go? (b) How long does it take to return to the ground?

43. (I) With what minimum speed must a salmon leave the water to jump to the top of a waterfall 2.1 m high?

44. (II) A stone is dropped from the roof of a high building. A second stone is dropped 1.00 s later. How far apart are the stones when the second one has reached a speed of 23.0 m/s?

45. (II) A helicopter is ascending vertically with a speed of 8.0 m/s; at a height of 120 m above the earth, a package is dropped from a window. How much time does it take for the package to reach the ground?

46. (II) The jump of a flea can be analyzed using slow-motion photography. The motion can be separated into two parts. The first is the "push off," lasting about 10^{-3} s, during which the flea's legs push against the ground and accelerate it to a speed of about 1.0 m/s. The second part is the flight of the flea into the air subject only to gravity, assuming the jump is vertical. Calculate: (a) the acceleration of the flea during push off, expressed as a multiple of g, the acceleration of gravity, (b) the distance above the ground the flea reaches during push off, and (c) the height the flea should reach after leaving the ground when its acceleration is g. (d) The film indicates the flea reaches only two-thirds the height calculated; explain why.

47. (II) If the flea in the above problem reached a height of only 3.5 cm, calculate its actual average acceleration during its upward flight (the "second period" mentioned in problem 46).

48. (II) If an object reaches a maximum vertical height of 23.0 m when thrown vertically upward on earth, how high would it travel on the moon where the acceleration due to gravity is about one-sixth that on earth? Assume the initial velocity is the same.

49. (II) If air resistance is neglected, show that a ball thrown vertically upward with a speed v_0 will have the same speed, v_0, when it comes back down to the starting point.

50. (II) A falling stone takes 0.30 s to pass a window 2.1 m high. From what height did the stone fall?

51. (II) A stone is thrown vertically upward with a speed of 10.0 m/s from the edge of a cliff 65 m high. (a) How much later will it reach the bottom of the cliff? (b) What will be its speed just before hitting?

52. (II) A ball is thrown vertically upward with a speed of 16.0 m/s. Draw three parallel graphs for the displacement, velocity, and acceleration as a function of time.

53. (II) A stone is thrown vertically upward with a speed of 17.5 m/s. (a) How fast is it moving when it reaches a height of 12.0 m? (b) How much time is required to reach this height? (c) Why are there two answers to (b)?

54. (III) A toy rocket passes by a 2.0-m-high window whose sill is 10.0 m above the ground in 0.15 s. What was the launch speed of the rocket and how high will it go?

55. (III) A rock is dropped from a seacliff and the sound of it striking the ocean is heard 3.5 s later. If the speed of sound is 330 m/s, how high is the cliff?

56. (III) Pelicans tuck their wings and free fall straight down when diving for fish. Suppose a pelican starts its dive from a height of 25 m and cannot change its path once committed. If it takes a fish 0.15 s to perform evasive action, at what minimum height must it spot the pelican to escape? Assume the fish is at the surface of the water.

57. (III) Suppose you adjust your garden hose nozzle for a hard stream of water. You point the nozzle vertically upward at a height of 1.5 m above the ground. When you quickly move the nozzle away from the vertical, you hear the water striking the ground next to you for 2.0 s. What is the water speed as it leaves the nozzle?

*SECTION 2–9

*58. (II) In Fig. 2–16, estimate the distance the object traveled (a) during the first minute, (b) during the second minute.

*59. (II) Construct the x vs. t graph for the object whose velocity as a function of time is given by Fig. 2–16.

*60. (II) Graph $v(t) = 25 + 18t$, where v is in m/s and t is in s, from $t_1 = 1.5$ s to $t_2 = 3.5$ s. Divide this time interval into ten subintervals, estimate $\bar{v}$ for each, and estimate the total displacement by determining the area under the curve.

*61. (II) Repeat problem 60 but use calculus to determine the total displacement.

*62. (II) The acceleration of a particle is given by $a(t) = kt^{3/2}$, where k is a constant. Find the position as a function of t if $x = 0$, and $v = 0$ at $t = 0$.

*63. (II) The acceleration of a proton increases exponentially in time according to the relation $a = 6.4\,e^{2t}$ where a is in m/s² and t in s. Determine, as a function of time, the proton's (a) velocity and (b) position if it starts from rest at the origin of coordinates at $t = 0$. (c) What is the relation between $a(t)$ and $v(t)$? (d) What are the position and velocity of the proton at $t = 3.6$ s?

*64. (III) (a) Suppose the acceleration of a particle is given as a function of x. Show that

$$v_2^2 = v_1^2 + 2 \int_{x_1}^{x_2} a(x)\, dx.$$

(b) If $a = (2.0\ \mathrm{m^{-1}\,s^{-2}})x^2$, and $v_1 = 0$, find v_2 after the object has moved from $x_1 = 0.50$ m to $x_2 = 2.20$ m.

*65. (III) Air resistance acting on a falling body can be taken into account by the approximate relation for the acceleration:

$$a = \frac{dv}{dt} = g - kv,$$

where k is a constant. (a) Derive a formula for the velocity of the body as a function of time assuming it starts from rest ($v = 0$ at $t = 0$). (Hint: change variables by setting $u = g - kv$.) (b) Determine an expression for the terminal velocity, which is the maximum value the velocity reaches.

*SECTION 2–10
(NUMERICAL/PROGRAMMABLE CALCULATOR)

*66. (II) The table below gives the speed of a particular drag racer as a function of time. Estimate (a) the average acceleration (m/s²) during each time interval and (b) the total distance traveled (m) as a function of time. [Hint: for $\bar{v}$ in each interval sum the velocities at the beginning and end of the interval and divide by 2; for example, in the second interval use $\bar{v} = (6.0 + 13.2)/2 = 9.6$.] (c) Graph each of these.

t (s)	0	0.50	1.00	1.50	2.00	2.50	3.00	3.50	4.00	4.50	5.00
v (km/h)	0.0	6.0	13.2	22.3	32.2	43.0	53.5	62.6	70.6	78.4	85.1

*67. (III) [Use a calculator, preferably a programmable one.] The acceleration of an object (m/s²) is measured at 1.00-s intervals starting at $t = 0$ to be as follows: 1.25, 1.58, 1.96, 2.40, 2.66, 2.70, 2.74, 2.72, 2.60, 2.30, 2.04, 1.76, 1.41, 1.09, 0.86, 0.51, 0.28, 0.10. Use numerical methods to estimate (a) the velocity (assume $v = 0$ at $t = 0$) and (b) the displacement at $t = 8.00$ s to an accuracy of 1 percent. What is (c) the velocity and (d) the displacement at $t = 17.00$ s?

*68. (III) An object starts from rest and accelerates at a rate (m/s²) of $a = \sqrt{2 + t^2}$. Numerically integrate to determine (a) the speed and (b) the acceleration at $t = 3.00$ s. Be sure the result is accurate to at least 2 percent.

Kinematics in Two or Three Dimensions

3

We now consider the motion of a particle in two or three dimensions instead of simply along a line as we did in Chapter 2. To do so, we will need the concept of vectors. After studying the properties of vectors, we will look at motion in general and then we will examine a number of interesting types of motion including that of projectiles and of objects in circular motion.

 ## 3–1   Vectors and Scalars

We mentioned in Section 2–4 that the term *velocity* refers not only to how fast something is moving but also to its direction. A quantity such as velocity, which has *direction* as well as a *magnitude*, is called a **vector** quantity. Other quantities that are also vectors are displacement, force, and momentum. However, many quantities, such as time, temperature, and energy, have no direction associated with them; they only have a magnitude—that is, they are specified completely by giving a number (and units, if any). Such quantities are called **scalars**.

Drawing a diagram of a particular physical situation is very helpful in physics, and this is especially true when dealing with vectors. On a diagram, each vector is represented by an arrow. The arrow is always drawn so that it points in the direction of the vector it represents; the length of the arrow is drawn proportional to the magnitude of the vector. For example, in Fig. 3–1, arrows have been drawn representing the velocity of a car at various places as it rounds a curve. The magnitude of the velocity at each point can be read off this figure by measuring the length of the corresponding arrow and using the scale shown (1 cm = 90 km/h).

When we write the symbol for a vector, we will always use boldface type.[†] Thus for velocity we write **v**. If we are concerned only with the magnitude we can write either $|\mathbf{v}|$, meaning the absolute value of **v**, or more commonly we simply write v.

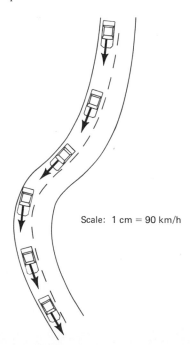

FIGURE 3–1

Car traveling on a road. The arrows represent the velocity vector at each position.

Scale: 1 cm = 90 km/h

[†] The symbol for a vector is usually written, in handwriting, with an arrow over it: as $\vec{v}$ for velocity.

3–2   Addition of Vectors— Graphical Methods

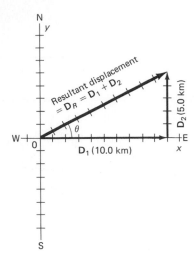

FIGURE 3–2

A person walks 10.0 km east and then 5.0 km north. These two displacements are represented by the vectors $\mathbf{D}_1$ and $\mathbf{D}_2$, which are shown as arrows. The resultant displacement vector, $\mathbf{D}_R$, which is the sum of $\mathbf{D}_1$ and $\mathbf{D}_2$, is also shown. Measurement on the graph with ruler and protractor shows that $\mathbf{D}_R$ has magnitude 11.2 km and it is at an angle $\theta = 27°$ north of east.

Because vectors are quantities that have direction as well as magnitude, they must be added in a special way. In this chapter we will deal mainly with displacement vectors (for which we use, for now, the symbol $\mathbf{D}$) and velocity vectors ($\mathbf{v}$). But the results will apply for other vectors that we encounter later.

We use simple arithmetic for adding scalars, such as time. Simple arithmetic can also be used for adding vectors if they are in the same direction. For example, if a person walks 8 km east one day, and 6 km east the next day, the person will be 8 km + 6 km = 14 km from the point of origin. We say that the net or resultant displacement is 14 km. If, on the other hand, the person walks 8 km east on the first day, and 6 km west (in the reverse direction) on the second day, then he will be 2 km from the origin after the 2 days so his resultant displacement is 2 km to the east. In this case, the resultant displacement is obtained by subtraction: 8 km − 6 km = 2 km.

But simple arithmetic cannot be used if the two vectors are not along the same line. For example, suppose a person walks 10.0 km east and then walks 5.0 km north. This motion can be represented on a graph in which the positive y axis points north and the positive x axis points east, Fig. 3–2. On this graph we draw an arrow, labeled $\mathbf{D}_1$, to represent the displacement vector of the 10.0-km displacement to the east; and a second arrow, $\mathbf{D}_2$, to represent the 5.0-km displacement to the north. Both vectors are drawn to scale.

After taking this walk, the person is now 10.0 km east and 5.0 km north of the point of origin. The resultant displacement is represented by the arrow labeled $\mathbf{D}_R$ on the diagram. If you use a ruler and a protractor, you will measure on this diagram that the person is 11.2 km from the origin at an angle of 27° north of east. In other words, the resultant displacement vector has a magnitude of 11.2 km and makes an angle (θ) with the x axis of 27°. The magnitude (length) of $\mathbf{D}_R$ can also be obtained using the theorem of Pythagoras since D_1, D_2, and D_R form a right triangle with D_R as the hypotenuse. Thus $D_R = \sqrt{D_1^2 + D_2^2} = \sqrt{(10.0 \text{ km})^2 + (5.0 \text{ km})^2} = \sqrt{125 \text{ km}^2} = 11.2$ km. You can use this theorem, of course, only when the vectors are *perpendicular* to each other.

The resultant displacement vector, $\mathbf{D}_R$, is the sum of the vectors $\mathbf{D}_1$ and $\mathbf{D}_2$. That is, $\mathbf{D}_R = \mathbf{D}_1 + \mathbf{D}_2$. This is a *vector* equation. An important feature of adding two vectors that are not along the same line is that the magnitude of the resultant vector is not equal to the sum of the magnitudes of the two separate vectors ($D_R \neq D_1 + D_2$).

Figure 3–2 illustrates the general rules for adding two vectors together, no matter what angles they make, to get their sum. Specifically, the rules are as follows: (1) on a diagram, draw one of the vectors—call it $\mathbf{V}_1$—to scale; (2) next draw the second vector, $\mathbf{V}_2$, to scale, placing its tail at the tip of the first vector and being sure its direction is correct; (3) the arrow drawn from the tail of the first vector to the tip of the second represents the *sum*, or **resultant**, of the two vectors. The length of the resultant can be measured and compared to the scale. Angles can be measured with a protractor. This method is known as the *tail-to-tip method of adding vectors*. It is, in fact, the definition of how to add vectors.

Note that it is not important in which order the vectors are taken. For example, a displacement of 5.0 km north, to which is added a displacement of 10.0 km east, yields a resultant of 11.2 km and angle $\theta = 27°$ (Fig. 3–3)—the same as when they were added in reverse order (Fig. 3–2). That is

$$\mathbf{V}_1 + \mathbf{V}_2 = \mathbf{V}_2 + \mathbf{V}_1. \qquad \text{[commutative law]} \quad (3\text{–}1a)$$

The tail-to-tip method can be extended to three or more vectors (Fig. 3–4) and, as indicated in that figure,

$$\mathbf{V}_1 + (\mathbf{V}_2 + \mathbf{V}_3) = (\mathbf{V}_1 + \mathbf{V}_2) + \mathbf{V}_3. \qquad \text{[associative law]} \quad (3\text{–}1b)$$

FIGURE 3–3

If the order of the vectors is reversed, the resultant is the same; see Fig. 3–2.

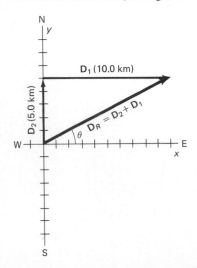

The left side of this equation means we first add $\mathbf{V}_2$ and $\mathbf{V}_3$ together and then add $\mathbf{V}_1$ to that sum to get the total sum; on the right side, $\mathbf{V}_1$ is added to $\mathbf{V}_2$, and this sum is added to $\mathbf{V}_3$. We see that the order in which two or more vectors are added doesn't affect the result.

A second way to add two vectors is the *parallelogram method*; it is equivalent to the tail-to-tip method. In this method, the two vectors are drawn from a common origin and a parallelogram is constructed using these two vectors as adjacent sides. The resultant is the diagonal drawn from the common origin. An example is shown in Fig 3–5b. In Fig. 3–5a the tail-to-tip method is shown, and it is clear that both methods yield the same result. It is a common error to draw the sum vector as the diagonal running between the tips of the two vectors, as in Fig. 3–5c; this is incorrect: it does not represent the sum of the two vectors (in fact, it represents their difference—see the next section).

Vectors, when considered as mathematical quantities, can be moved about parallel to themselves as long as their direction and magnitude are not changed. We used this above when adding vectors. (Physically, the position of a vector can be important; for example, the vector representing a force on a body must be placed at the point where the force acts if we are to understand the resulting movement of that body correctly.)

As a corollary, we define two vectors to be equal if they have the same magnitude and the same direction.

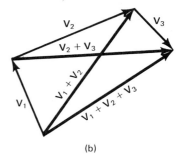

(a)

(b)

FIGURE 3–4

The three vectors in (a) can be added in any order with the same result, $\mathbf{V}_1 + \mathbf{V}_2 + \mathbf{V}_3$. In (b), it's clear that $\mathbf{V}_1 + (\mathbf{V}_2 + \mathbf{V}_3) = (\mathbf{V}_1 + \mathbf{V}_2) + \mathbf{V}_3$, which we simply write as $\mathbf{V}_1 + \mathbf{V}_2 + \mathbf{V}_3$.

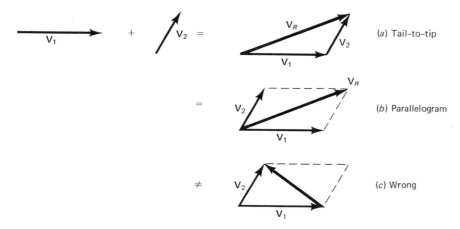

(a) Tail-to-tip

(b) Parallelogram

(c) Wrong

FIGURE 3–5

Vector addition by two different methods, (a) and (b); (c) is incorrect.

FIGURE 3–6

(a) The negative of a vector is a vector with the same length but opposite direction; (b) subtracting two vectors: $\mathbf{A} - \mathbf{B}$.

(a)

3–3 Subtraction of Vectors and Multiplication of a Vector by a Scalar

Given a vector $\mathbf{V}$, we define the *negative* of this vector $(-\mathbf{V})$ to be a vector with the same magnitude but opposite direction to $\mathbf{V}$, Fig. 3–6a. We can use this to define the subtraction of one vector from another: the difference between two vectors, $\mathbf{A} - \mathbf{B}$ is defined as

$$\mathbf{A} - \mathbf{B} = \mathbf{A} + (-\mathbf{B});$$

that is, the difference between two vectors is equal to the sum of the first plus the negative of the second. Thus our rules for addition of vectors can be applied as shown in Fig. 3–6b using the tail-to-tip method.

A vector $\mathbf{V}$ can be multiplied by a scalar c. We define this product so that $c\mathbf{V}$ has the same direction as $\mathbf{V}$ and has magnitude cV. That is, multiplication of a vector

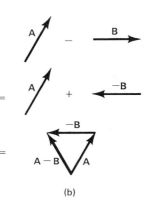

(b)

by a positive scalar c changes the magnitude of the vector by a factor c but doesn't alter the direction. If c is a negative scalar, the magnitude of the product $c\mathbf{V}$ is still cV but the direction is precisely opposite to that of $\mathbf{V}$.

3–4 Analytic Method for Adding Vectors; Components

Adding vectors graphically using a ruler and protractor is not very accurate, and is not useful for vectors in three dimensions. We discuss now a far more powerful and accurate method for adding vectors which is based on resolving vectors into their components along arbitrarily chosen coordinate axes.

Consider first a vector $\mathbf{V}$ which lies in a plane. It can be expressed as the sum of two other vectors, called the **components** of the original vector. The components are usually chosen to be along two perpendicular directions. The process of finding the components is known as *resolving the vector into its components*. An example is shown in Fig. 3–7; the vector $\mathbf{V}$ could be a displacement vector that points at an angle $\theta = 30°$ north of east, where we have chosen the x axis to be to the east and the y axis north. This vector is resolved into its x and y components by drawing lines from the tip (A) of our given vector that are perpendicular to the x and y axes (lines AB and AC). Then the lines $0B$ and $0C$ represent the x and y components of $\mathbf{V}$ respectively; these *vector components* are written $\mathbf{V}_x$ and $\mathbf{V}_y$. The magnitudes of $\mathbf{V}_x$ and $\mathbf{V}_y$, namely V_x and V_y, are referred to simply as the components, and thus are numbers (with units) that are $+$ or $-$ depending on whether they point along the positive or negative x or y axis. As can be seen in Fig. 3–7, $\mathbf{V}_x + \mathbf{V}_y = \mathbf{V}$ by the parallelogram method of adding vectors. Although the vector shown in Fig. 3–7 has its tail at the origin of the coordinates, this is not necessary in order to find the components; if the vector is placed elsewhere in the coordinate space, its components will be unchanged so long as its length and the angles it makes with the axes remain the same.

A vector in three dimensions will have three components. In a Cartesian coordinate system the vector components of a vector $\mathbf{V}$ are $\mathbf{V}_x$, $\mathbf{V}_y$, and $\mathbf{V}_z$, and

$$\mathbf{V} = \mathbf{V}_x + \mathbf{V}_y + \mathbf{V}_z.$$

Resolution of a vector in three dimensions is merely an extension of the technique described above for two dimensions. For convenience we will mainly be concerned with situations in which the vectors are in a plane and two components are all that are necessary.

The components of a given vector will be different for different choices of coordinate axes. It is therefore crucial to specify the choice of coordinate system when giving the components.

Using the definitions of the trigonometric functions, we can see from Fig. 3–7 that

$$V_x = V \cos \theta \tag{3–2a}$$

$$V_y = V \sin \theta \tag{3–2b}$$

where θ is chosen (by convention) to be the angle that $\mathbf{V}$ makes with the x axis. We can also see from the diagram that

$$V = \sqrt{V_x^2 + V_y^2} \tag{3–3a}$$

$$\tan \theta = \frac{V_y}{V_x}. \tag{3–3b}$$

We thus have two ways to specify a given vector in any coordinate system. We can give its magnitude, V, and the angle, θ, it makes with the x axis. Or we can

FIGURE 3–7

Resolving a vector $\mathbf{V}$ into its components along an arbitrarily chosen set of x and y axes.

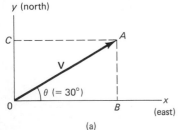

(a)

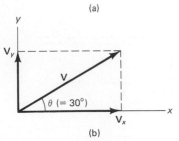

(b)

give its components, V_x and V_y. We can shift from one description to the other using Eqs. 3–2 or 3–3. Note that in two dimensions, two numbers are needed to specify a vector; in three dimensions, three numbers are needed: either V_x, V_y, and V_z; or the magnitude V and two angles.

We now discuss how to add vectors analytically by using components. The first step is to resolve each vector into its components. Next we can show (see Fig. 3–8) that the addition of two vectors $\mathbf{V}_1$ and $\mathbf{V}_2$ to give a resultant, $\mathbf{V} = \mathbf{V}_1 + \mathbf{V}_2$, implies that

$$V_x = V_{1x} + V_{2x}$$
$$V_y = V_{1y} + V_{2y} \qquad (3\text{–}4)$$
$$V_z = V_{1z} + V_{2z}$$

A careful examination of Fig. 3–8 will verify these equations for the x and y components; the result for the z component follows in a similar way. If both the magnitude and direction of the resultant vector are desired, they can be obtained using Eqs. 3–3:

$$V = \sqrt{V_x^2 + V_y^2}, \qquad \tan\theta = \frac{V_y}{V_x}.$$

The choice of coordinate axes is, of course, always arbitrary. You can often reduce the work involved in adding vectors analytically by judicious choice of axes—for example, by choosing one of the axes to be in the same direction as one of the vectors so that vector will have only one component.

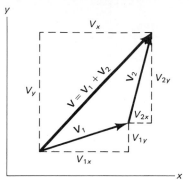

FIGURE 3–8
The components of $\mathbf{V} = \mathbf{V}_1 + \mathbf{V}_2$ are $V_x = V_{1x} + V_{2x}$ and $V_y = V_{1y} + V_{2y}$.

EXAMPLE 3–1   An explorer walks 22.0 km in a northerly direction, and then walks southeasterly (45° south of east) for 47.0 km (Fig. 3–9a). What is his resultant displacement from the origin?

SOLUTION   We choose the positive x axis to be east and the positive y axis north. We resolve each displacement vector into its components (Fig. 3–9b). Since $\mathbf{D}_1$ has magnitude 22.0 km and points north, it has only a y component

$$D_{1x} = 0, \qquad D_{1y} = 22.0 \text{ km},$$

whereas $\mathbf{D}_2$ has both x and y components:

$$D_{2x} = (47.0 \text{ km})(\cos 45°) \quad = 33.2 \text{ km}$$
$$D_{2y} = -(47.0 \text{ km})(\sin 45°) = -33.2 \text{ km},$$

since $\cos 45° = \sin 45° = 0.707$. Notice that D_{2y} is negative because this vector component points along the negative y axis. The resultant vector (**D**) has components:

$$D_x = D_{1x} + D_{2x} = 0 + 33.2 \text{ km} = 33.2 \text{ km}$$
$$D_y = D_{1y} + D_{2y} = 22.0 \text{ km} - 33.2 \text{ km} = -11.2 \text{ km}.$$

This specifies the resultant vector completely: $D_x = 33.2$ km, $D_y = -11.2$ km. We can also specify it by giving its magnitude and angle:

$$D = \sqrt{D_x^2 + D_y^2} = \sqrt{(33.2 \text{ km})^2 + (-11.2 \text{ km})^2} = 35.0 \text{ km}$$

$$\tan\theta = \frac{D_y}{D_x} = -\frac{11.2 \text{ km}}{33.2 \text{ km}} = -0.337,$$

so $\theta = 18.6°$ south of east.

FIGURE 3–9
Example 3–1.

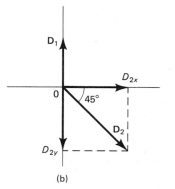

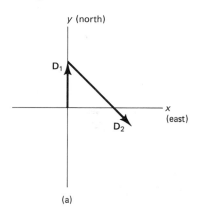

(a)

(b)

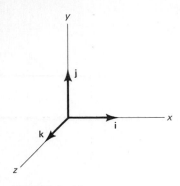

FIGURE 3–10

Unit vectors **i**, **j**, and **k** along the x, y, and z axes.

 3–5   Unit Vectors

Vectors can be conveniently written in terms of **unit vectors** (meaning vectors whose magnitude is one) which point along the chosen coordinate axes. In a rectangular coordinate system, these unit vectors are called **i**, **j**, and **k**; they point, respectively, along the positive x, y, and z axes as shown in Fig. 3–10. (Like other vectors, **i**, **j**, and **k** do not have to be placed at the origin, but can be placed elsewhere as long as their directions remain unchanged.)

Because of the definition of multiplication of a vector by a scalar (Section 3–3), the components of a vector **V** can be written $\mathbf{V}_x = V_x\mathbf{i}$, $\mathbf{V}_y = V_y\mathbf{j}$ and $\mathbf{V}_z = V_z\mathbf{k}$. Hence, any vector **V** can be written in terms of its components as

$$\mathbf{V} = V_x\mathbf{i} + V_y\mathbf{j} + V_z\mathbf{k}. \tag{3–5}$$

Unit vectors are helpful when adding vectors analytically by components. For example, Eqs. 3–4 can be seen to be true by using unit vector notation for each vector (which we write for the two-dimensional case, but the extension to three dimensions is clear):

$$\begin{aligned}
\mathbf{V} &= (V_x)\mathbf{i} + (V_y)\mathbf{j} \\
&= \mathbf{V}_1 + \mathbf{V}_2 \\
&= (V_{1x}\mathbf{i} + V_{1y}\mathbf{j}) + (V_{2x}\mathbf{i} + V_{2y}\mathbf{j}) \\
&= (V_{1x} + V_{2x})\mathbf{i} + (V_{1y} + V_{2y})\mathbf{j}.
\end{aligned}$$

Comparing the first line to the fourth line, we get Eqs. 3–4.

EXAMPLE 3–2   Write the vectors of Example 3–1 in unit vector notation, and perform the addition.

SOLUTION   In Example 3–1 we resolved $\mathbf{D}_1$ and $\mathbf{D}_2$ into components and found $D_{1x} = 0$, $D_{1y} = 22.0$ (we drop units for the moment) and $D_{2x} = 33.2$, $D_{2y} = -33.2$. Thus we have

$$\mathbf{D}_1 = 0\mathbf{i} + 22.0\mathbf{j}$$
$$\mathbf{D}_2 = 33.2\mathbf{i} - 33.2\mathbf{j}.$$

Then

$$\mathbf{D}_1 + \mathbf{D}_2 = (0 + 33.2)\mathbf{i} + (22.0 - 33.2)\mathbf{j} = 33.2\mathbf{i} - 11.2\mathbf{j}.$$

The components of the resultant displacement, **D**, are $D_x = 33.2$ km and $D_y = -11.2$ km.

3–6   Relative Velocity

If two trains are approaching one another, each with a speed of 80 km/h with respect to the ground, the speed of one train relative to the other is 160 km/h. That is, to an observer on one train, the other train seems to be approaching at 160 km/h. Similarly, when one car traveling 90 km/h passes a second car traveling 75 km/h, the first car has a speed relative to the second car of 90 km/h − 75 km/h = 15 km/h. We deal now with this subject of **relative velocity**.

When the velocities are along the same line, simple addition or subtraction is sufficient to obtain the relative velocity. But if they are not along the same line, we must make use of vector addition as the following examples show. But we note, as mentioned in Section 2–2, that when specifying a velocity, it is important to specify what the reference frame is.

EXAMPLE 3–3 An airplane whose airspeed is 200 km/h heads due north. But a 100 km/h northeast wind suddenly begins to blow. What is the resulting velocity of the plane with respect to the ground?

SOLUTION The two velocity vectors are shown resolved into components in Fig. 3–11a. They are drawn with a common origin for convenience; $\mathbf{v}_1$ represents the velocity of the airplane with respect to the air and $\mathbf{v}_2$ the velocity of the wind with respect to the ground. The resultant velocity, $\mathbf{v}_R$, is the velocity of the plane with respect to the ground. Since $\mathbf{v}_1$ is along the y-axis, it has only a y-component:

$$v_{1x} = 0 \text{ km/h}$$

$$v_{1y} = v_1 = 200 \text{ km/h}.$$

The components of $\mathbf{v}_2$ are

$$v_{2x} = -v_2 \cos 45° = -(100 \text{ km/h})(0.707) = -70.7 \text{ km/h}$$

$$v_{2y} = -v_2 \sin 45° = -(100 \text{ km/h})(0.707) = -70.7 \text{ km/h}.$$

Both v_{2x} and v_{2y} are negative because their directions are, respectively, along the negative x and y axes. The components of the resultant velocity are

$$v_{Rx} = 0 \text{ km/h} - 70.7 \text{ km/h} = -70.7 \text{ km/h},$$

$$v_{Ry} = 200 \text{ km/h} - 70.7 \text{ km/h} = 129 \text{ km/h}.$$

We find the magnitude of the resultant velocity using the Pythagorean theorem:

$$v_R = \sqrt{v_{Rx}^2 + v_{Ry}^2} = 147 \text{ km/h}.$$

To find the angle θ that $\mathbf{v}_R$ makes with the x-axis (Fig. 3–11b) we use

$$\tan \theta = \frac{v_{Ry}}{v_{Rx}} = \frac{129 \text{ km/h}}{-70.7 \text{ km/h}} = -1.82.$$

(The negative sign merely tells us that θ is with respect to the negative x axis, which we already know from the diagram.) Using trig tables we find that $\tan 61° = 1.804$ and $\tan 62° = 1.881$. So our angle θ is about 61°. A calculator (using "$\tan^{-1}$" key) gives $\theta = -61.2°$.

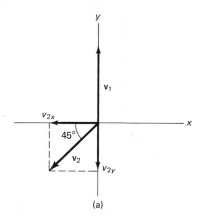

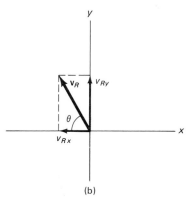

(a)

(b)

FIGURE 3–11
Example 3–3.

When determining relative velocity, it is easy to make a mistake by adding or subtracting the wrong velocities. It is useful, therefore, to use a careful labeling process that makes things clear. Each velocity is labeled by two subscripts: the first refers to the object, the second to the reference frame in which it has this velocity. For example, suppose a boat is to cross a river directly to the opposite side as shown in Fig. 3–12. We let $\mathbf{v}_{BW}$ be the velocity of the **B**oat with respect to the **W**ater. Similarly, $\mathbf{v}_{BS}$ is the velocity of the **B**oat with respect to the **S**hore, and $\mathbf{v}_{WS}$ is the velocity of the **W**ater with respect to the **S**hore (this is the river current). Note that $\mathbf{v}_{BW}$ is what the boat's motor produces (against the water), whereas $\mathbf{v}_{BS}$ is $\mathbf{v}_{BW}$ plus the effect of the current. Therefore the velocity of the boat relative to the shore is

$$\mathbf{v}_{BS} = \mathbf{v}_{BW} + \mathbf{v}_{WS}. \tag{3–6a}$$

By writing the subscripts via the convention above, we see that the inner subscripts on the right-hand side of Eq. 3–6a are the same (the two W's), and the outer subscripts on right of Eq. 3–6a (the B and the S) are the same as the two subscripts for the sum vector on the left, $\mathbf{v}_{BS}$. By following this convention, one can usually write down the correct equation relating velocities in different reference frames.[†] Our derivation of Eq. 3–6a is valid in general and can be extended to three or more velocities. For example, if a fisherman on the boat walks with a velocity $\mathbf{v}_{FB}$ relative

FIGURE 3–12

Boat must head upstream at an angle θ if it is to move directly across the river.

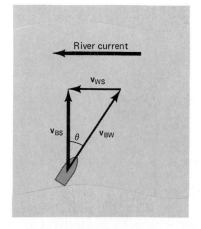

[†] We thus would know by inspection that (for example) the equation $\mathbf{v}_{BW} = \mathbf{v}_{BS} + \mathbf{v}_{WS}$ is wrong.

to the boat, his velocity relative to the shore is $\mathbf{v}_{FS} = \mathbf{v}_{FB} + \mathbf{v}_{BW} + \mathbf{v}_{WS}$. The equations involving relative velocity will be correct when adjacent inner subscripts are identical and when the outermost ones correspond exactly to the two on the variable on the left of the equation. But this works only with plus signs (on the right), not minus signs.

It is often useful to remember that for any two objects or reference frames, A and B, the velocity of A relative to B has the same magnitude but opposite direction as the velocity of B relative to A:

$$\mathbf{v}_{BA} = -\mathbf{v}_{AB}. \qquad (3\text{-}6b)$$

For example, if a train is traveling 100 km/h relative to earth in a certain direction, objects on earth (such as trees) appear to an observer on the train to be traveling 100 km/h in the opposite direction.

EXAMPLE 3–4   A boat's speed in still water is $v_{BW} = 20.0$ km/h. If the boat is to travel directly across a river whose current has speed $v_{WS} = 12.0$ km/h, at what upstream angle must the boat head (see Fig. 3–12)?

SOLUTION   Note that Fig. 3–12 has been drawn with $\mathbf{v}_{BS}$ pointing directly across the river, since this is how the boat is supposed to move. To accomplish this, the boat heads upstream, so $\mathbf{v}_{BW}$ points upstream at the angle θ. From the diagram,

$$\sin \theta = \frac{v_{WS}}{v_{BW}} = \frac{12.0 \text{ km/h}}{20.0 \text{ km/h}} = 0.600$$

so $\theta = 36.9°$.

EXAMPLE 3–5   Two automobiles approach a corner at right angles to each other with the same speed of 40.0 km/h (11.1 m/s), Fig. 3–13a. What is the relative velocity of one with respect to the other? (That is, determine the velocity of car 1 as seen by car 2.)

SOLUTION   Figure 3–13a shows the situation in a reference frame fixed to the earth. But we want to view the situation from a reference frame in which car 2 is at rest, and this is shown in Fig. 3–13b. In this reference frame (the world as seen by the driver of car 2) the earth moves toward car 2 with velocity $\mathbf{v}_{E2}$ (speed of 40.0 km/h) which is of course equal and opposite to $\mathbf{v}_{2E}$, the velocity of car 2 with respect to the earth:

$$\mathbf{v}_{2E} = -\mathbf{v}_{E2}.$$

Thus the velocity of car 1 as seen by car 2 is

$$\mathbf{v}_{12} = \mathbf{v}_{1E} + \mathbf{v}_{E2}$$

or (since $\mathbf{v}_{E2} = -\mathbf{v}_{2E}$)

$$\mathbf{v}_{12} = \mathbf{v}_{1E} - \mathbf{v}_{2E}.$$

That is, the velocity of car 1 as seen by car 2 is the difference of their velocities, $\mathbf{v}_{1E} - \mathbf{v}_{2E}$, both measured relative to the earth. See Fig. 3–13c. Since $|\mathbf{v}_{1E}| = |\mathbf{v}_{2E}| = |\mathbf{v}_{E2}|$, we see (Fig. 3–13b) that $\mathbf{v}_{12}$ points at a 45° angle toward car 2; the speed is

$$\mathbf{v}_{12} = \sqrt{(11.1 \text{ m/s})^2 + (11.1 \text{ m/s})^2} = 15.7 \text{ m/s (56.5 km/h)}.$$

FIGURE 3–13

Example 3–5.

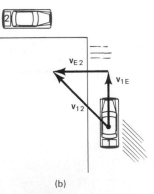

(a)

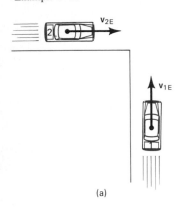

(b)

(c)

3–7   Vector Kinematics

Now that we have introduced vectors, we can extend our definitions of velocity and acceleration in a formal way to two- and three-dimensional motion.

Suppose a particle follows a path in the xy plane as shown in Fig. 3–14. At time t_1, the particle is at point P_1; and at time t_2, it is at point P_2. The vector $\mathbf{r}_1$ is the position vector of the particle at time t_1 (it represents the displacement of the particle from the origin of the coordinate system); and $\mathbf{r}_2$ is the position vector at time t_2.

In one dimension, we defined displacement as the *change in position* of the particle. In the more general case of two or three dimensions, the **displacement vector** is defined as the vector representing change in position of a particle; we call* it $\Delta\mathbf{r}$, where

$$\Delta\mathbf{r} = \mathbf{r}_2 - \mathbf{r}_1.$$

This represents the displacement during the time interval $\Delta t = t_2 - t_1$. In unit vector notation, we can write

$$\mathbf{r}_1 = x_1\mathbf{i} + y_1\mathbf{j} + z_1\mathbf{k} \qquad (3\text{–}7a)$$

where x_1, y_1, and z_1 are the coordinates of point P_1 (Fig. 3–14). Similarly,

$$\mathbf{r}_2 = x_2\mathbf{i} + y_2\mathbf{j} + z_2\mathbf{k}.$$

Hence

$$\Delta\mathbf{r} = (x_2 - x_1)\mathbf{i} + (y_2 - y_1)\mathbf{j} + (z_2 - z_1)\mathbf{k}. \qquad (3\text{–}7b)$$

If the motion is along the x axis only, then $y_2 - y_1 = 0$, $z_2 - z_1 = 0$, and the magnitude of the displacement is $\Delta r = x_2 - x_1$ which is consistent with our earlier one-dimensional equation (Section 2–4). Even in one dimension, displacement is a vector, as are velocity and acceleration.

The **average velocity vector**, $\overline{\mathbf{v}}$, over the time interval $\Delta t = t_2 - t_1$ is defined as

$$\overline{\mathbf{v}} = \frac{\Delta\mathbf{r}}{\Delta t}. \qquad (3\text{–}8)$$

Since $\mathbf{v}$ is a product of the vector $\Delta\mathbf{r}$ times a scalar $(1/\Delta t)$, the direction of $\mathbf{v}$ is the same as that of $\Delta\mathbf{r}$; and its magnitude is $\Delta r/\Delta t$.

We next consider shorter and shorter time intervals—that is, we let Δt approach zero so that the distance between points P_2 and P_1 also approaches zero. We define the **instantaneous velocity vector** as the limit of the average velocity as Δt approaches zero:

$$\mathbf{v} = \lim_{\Delta t \to 0} \frac{\Delta\mathbf{r}}{\Delta t} = \frac{d\mathbf{r}}{dt}. \qquad (3\text{–}9)$$

The direction of $\mathbf{v}$ at any moment (say at P_1 in Fig. 3–14) is along the line tangent to the path at that moment (Fig. 3–15).

Note that the magnitude of the average velocity in Fig. 3–14 is not equal to the average speed, which is the actual distance traveled, Δl, divided by Δt. In some special cases, the average speed and average velocity are equal (such as motion along a straight line in one direction), but in general they are not. However, in the limit $\Delta t \to 0$, Δr always approaches Δl so the instantaneous speed *always* equals the magnitude of the instantaneous velocity at any time.

The instantaneous velocity (Eq. 3–9) is equal to the derivative of the position vector with respect to time. Equation 3–9 can be written in terms of components[†] (see also Equation 3–7a) as

$$\mathbf{v} = \frac{d\mathbf{r}}{dt} = \frac{dx}{dt}\mathbf{i} + \frac{dy}{dt}\mathbf{j} + \frac{dz}{dt}\mathbf{k} \qquad (3\text{–}10)$$
$$= v_x\mathbf{i} + v_y\mathbf{j} + v_z\mathbf{k}$$

---

* We used **D** for the displacement vector earlier in the chapter for those simple cases. The new notation here, $\Delta\mathbf{r}$, emphasizes that it is the difference between two (position) vectors, and it will allow us more easily to discuss tiny as well as large displacements.

[†] Note that $d\mathbf{i}/dt = d\mathbf{j}/dt = d\mathbf{k}/dt = 0$ since these unit vectors are constant in magnitude and direction.

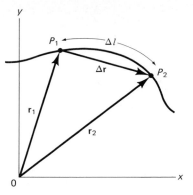

FIGURE 3–14
Path of a particle in the xy plane. At time t_1 the particle is at point P_1 given by the position vector $\mathbf{r}_1$; at t_2 the particle is at point P_2 given by the position vector $\mathbf{r}_2$. The displacement vector during the time interval $t_2 - t_1$ is $\Delta\mathbf{r} = \mathbf{r}_2 - \mathbf{r}_1$.

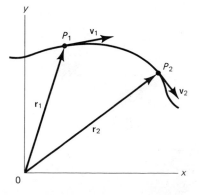

FIGURE 3–15
Velocity vectors $\mathbf{v}_1$ and $\mathbf{v}_2$ at instants t_1 and t_2 for the particle of Fig. 3–14.

where $v_x = dx/dt$, $v_y = dy/dt$, $v_z = dz/dt$ are the x, y, and z components of the velocity.

Acceleration in two or three dimensions is treated in a similar way. The **average acceleration vector**, $\bar{\mathbf{a}}$, over a time interval $\Delta t = t_2 - t_1$ is defined as

$$\bar{\mathbf{a}} = \frac{\Delta \mathbf{v}}{\Delta t} = \frac{\mathbf{v}_2 - \mathbf{v}_1}{t_2 - t_1} \tag{3–11}$$

where $\Delta \mathbf{v}$ is the change in instantaneous velocity vector during that time interval: $\Delta \mathbf{v} = \mathbf{v}_2 - \mathbf{v}_1$. Note that $\mathbf{v}_2$ in many cases (such as in Fig. 3–15) may not be in the same direction as $\mathbf{v}_1$; and $\bar{\mathbf{a}}$, then, may be in a different direction from either $\mathbf{v}_1$ or $\mathbf{v}_2$. Furthermore, $\mathbf{v}_2$ and $\mathbf{v}_1$ may have the same magnitude but different directions, and the difference of two such vectors will not be zero. Hence, acceleration can result from either a change in the magnitude of the velocity, or from a change in direction of the velocity, or from a change in both.

The **instantaneous acceleration vector** is defined as the limit of the average acceleration vector as the time interval Δt is allowed to approach zero:

$$\mathbf{a} = \lim_{\Delta t \to 0} \frac{\Delta \mathbf{v}}{\Delta t} = \frac{d\mathbf{v}}{dt}, \tag{3–12}$$

and is thus the derivative of $\mathbf{v}$ with respect to t. Using components:

$$\mathbf{a} = \frac{d\mathbf{v}}{dt} = \frac{dv_x}{dt}\mathbf{i} + \frac{dv_y}{dt}\mathbf{j} + \frac{dv_z}{dt}\mathbf{k}$$

$$= a_x\mathbf{i} + a_y\mathbf{j} + a_z\mathbf{k}. \tag{3–13}$$

The instantaneous acceleration will be nonzero not only when the magnitude of the velocity changes, but also if its direction changes. For example, a person riding in a car traveling at constant speed around a curve, or a child riding on a merry-go-round, will both experience an acceleration because of a change in the direction of the velocity, even though its magnitude may remain constant. (More on this in the following sections.)

In general, we will use the terms "velocity" and "acceleration" to mean the instantaneous values. If we want to discuss average values, we will use the word "average."

In Chapter 2 we studied the important case of one-dimensional motion for which the acceleration is a constant. We now consider motion in two or three dimensions for which the acceleration vector, $\mathbf{a}$, is constant in magnitude and direction; that is, $a_x = $ constant, $a_y = $ constant, $a_z = $ constant. The average acceleration in this case is equal to the instantaneous acceleration at any moment. The equations we derived in Chapter 2 for one dimension (Eqs. 2–9) apply separately to each component of two- or three-dimensional motion. In two dimensions we let $\mathbf{v}_0 = v_{x0}\mathbf{i} + v_{y0}\mathbf{j}$ be the initial velocity, and we apply Eqs. 3–7a, 3–10, and 3–13 for the position vector, $\mathbf{r}$, velocity, $\mathbf{v}$, and acceleration, $\mathbf{a}$. We can therefore write for two dimensions:

FOR MOTION WITH CONSTANT ACCELERATION

x Component	y Component	
$v_x = v_{x0} + a_x t$	$v_y = v_{y0} + a_y t$	(3–14a)
$x = x_0 + v_{x0}t + \frac{1}{2}a_x t^2$	$y = y_0 + v_{y0}t + \frac{1}{2}a_y t^2$	(3–14b)
$v_x^2 = v_{x0}^2 + 2a_x(x - x_0)$	$v_y^2 = v_{y0}^2 + 2a_y(y - y_0)$.	(3–14c)

The first two of these, Eqs. 3–14a and b, can be written more formally in vector notation (see Eqs. 3–7a, 3–10, and 3–13):

$$\mathbf{v} = \mathbf{v}_0 + \mathbf{a}t \tag{3–15a}$$

$$\mathbf{r} = \mathbf{r}_0 + \mathbf{v}_0 t + \frac{1}{2}\mathbf{a}t^2. \tag{3–15b}$$

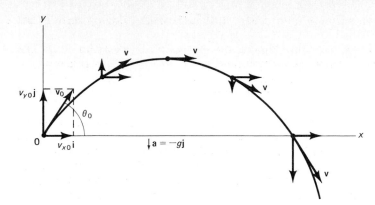

FIGURE 3–16

Motion of a projectile fired with initial velocity $\mathbf{v}_0$ at angle θ_0 to the horizontal.

Here, $\mathbf{r}$ is the position vector at any time, and $\mathbf{r}_0$ is the position vector at $t = 0$. In practical situations, we usually use the component form, Eqs. 3–14. In a sense, these component equations are separate; but in another sense they are connected since both the x equations and y equations contain the same time variable, t.

These equations, and their use, will become clearer as we use them. We next deal with several types of motion in a plane which are common in everyday life: projectile motion and circular motion.

3–8 Projectile Motion

Projectile motion refers to the motion of an object that is projected into the air at an angle. For simplicity we restrict ourselves to objects near the earth's surface. Examples are a thrown baseball, a speeding bullet, and an athlete doing the high jump. Although air resistance is often important, in many cases its effect can be ignored, and we will ignore it in the following analysis. We will not be concerned now with the process by which the object is thrown or projected; we consider only its motion after it has been projected and is moving freely through the air under the action of gravity. Thus the acceleration of the object[†] is that of gravity, $\mathbf{g}$, which acts downward with magnitude $g = 9.80 \text{ m/s}^2$.

It was Galileo who first accurately described projectile motion; he showed that it could be understood by analyzing the horizontal and vertical components of the motion separately. This was an innovative analysis, not done in this way by anyone prior to Galileo. (It was also idealized in that it did not take into account air resistance.)

Suppose a particle is projected into the air (Fig. 3–16) with an initial velocity $\mathbf{v}_0$ at an angle θ_0 to the horizontal. (If it is projected above the horizontal, θ_0 is positive; if projected downward, below the horizontal, θ_0 is negative.) We choose the coordinate axes so the motion is in the xy plane, and we choose the y axis vertical so that the acceleration is only in the y direction. Then

$$a_x = 0, \qquad a_y = -g.$$

We also choose the origin of coordinates at the point where the object begins its free flight (for example, when the baseball leaves the thrower's hand), so $x_0 = y_0 = 0$, and we choose the time $t_0 = 0$ when it begins its flight. The initial velocity $\mathbf{v}_0$ has components

$$v_{x0} = v_0 \cos \theta_0$$
$$v_{y0} = v_0 \sin \theta_0. \tag{3–16}$$

[†] We restrict ourselves to objects whose distance traveled and maximum height above the earth are small compared to the earth's radius (6.4×10^6 m) so that $\mathbf{g}$ can be considered constant.

Since $a_x = 0$, the horizontal motion occurs at constant speed, so we have (Eqs. 3–14a, b)

$$v_x = v_{x0} = v_0 \cos \theta_0 \qquad \text{[projectile motion]} \qquad (3\text{–}17)$$

$$x = v_{x0}t. \qquad \text{[projectile motion]} \qquad (3\text{–}18)$$

The vertical motion is accelerated since $a_y = -g$ so we have (Eqs. 3–14a, b, c)

$$v_y = v_{y0} - gt \qquad \text{[projectile motion]} \qquad (3\text{–}19)$$

$$y = v_{y0}t - \tfrac{1}{2}gt^2 \qquad \text{[projectile motion]} \qquad (3\text{–}20)$$

$$v_y^2 = v_{y0}^2 - 2gy. \qquad \text{[projectile motion]} \qquad (3\text{–}21)$$

For a projectile projected at an upward angle, as in Fig. 3–16, it is clear that v_y decreases in time and becomes zero at the highest point of the path; from Eq. 3–19, we see that this occurs after a time $t = v_{y0}/g$. At times greater than this, v_y becomes negative and becomes more negative as time passes, as can be seen in Fig. 3–16. Note that the particle may cross the x axis if its starting point ($x_0 = y_0 = 0$) is higher than the point where it strikes the ground.

An interesting special case is when the particle is projected horizontally so $\theta_0 = 0$. Examples are a ball rolling off the end of a table, and a bullet fired from a gun held horizontally. In this case $v_{y0} = 0$, so $v_y = -gt$ and $y = -\tfrac{1}{2}gt^2$. The vertical motion is thus the same as for a freely falling body. Thus we see—and Galileo himself predicted it on the basis of his analysis—that an object projected horizontally will hit the ground in the same time as an object dropped vertically.

EXAMPLE 3–6 A rock is thrown horizontally from a 115-m-high cliff and strikes the ground 92.5 m from the base of the cliff. At what speed was it thrown?

SOLUTION First we find how long it takes to reach the ground below. The initial velocity is horizontal, so the vertical component (v_{y0}) is zero. From Eq. 3–20, $y = -\tfrac{1}{2}gt^2$, and since $y = -115$ m we have

$$t = \sqrt{\frac{-2y}{g}} = \sqrt{\frac{230 \text{ m}}{9.80 \text{ m/s}^2}} = 4.84 \text{ s}.$$

To get the initial velocity, v_{x0}, we use Eq. 3–18:

$$v_{x0} = \frac{x}{t} = \frac{92.5 \text{ m}}{4.84 \text{ s}} = 19.1 \text{ m/s}.$$

EXAMPLE 3–7 A football is kicked at an angle $\theta_0 = 37.0°$ with a velocity of 20.0 m/s as in Fig. 3–16. Calculate (a) the maximum height, (b) the time of travel before it hits the ground, and (c) how far away it hits the ground. Assume for simplicity that the ball leaves the foot at ground level.

SOLUTION The components of the initial velocity are

$$v_{x0} = v_0 \cos 37.0° = (20.0 \text{ m/s})(0.799) = 16.0 \text{ m/s}$$

$$v_{y0} = v_0 \sin 37.0° = (20.0 \text{ m/s})(0.602) = 12.0 \text{ m/s}.$$

(a) The maximum height is attained where $v_y = 0$, and this occurs (Eq. 3–19) at time $t = v_{y0}/g = (12.0 \text{ m/s})/(9.80 \text{ m/s}^2) = 1.22$ s. From Eq. 3–20 we have

$$\begin{aligned} y &= v_{y0}t - \tfrac{1}{2}gt^2 \\ &= (12.0 \text{ m/s})(1.22 \text{ s}) - \tfrac{1}{2}(9.80 \text{ m/s}^2)(1.22 \text{ s})^2 \\ &= 7.35 \text{ m}. \end{aligned}$$

Alternatively, we could have used Eq. 3–21, solved for y, and found

$$y = \frac{v_{y0}^2 - v_y^2}{2g} = \frac{(12.0 \text{ m/s})^2 - (0 \text{ m/s})^2}{2(9.80 \text{ m/s}^2)} = 7.35 \text{ m}.$$

(b) To find the time it takes for the ball to return to the ground we use Eq. 3–20 and set $y = 0$ (ground level)

$$y = v_{y0}t - \tfrac{1}{2}gt^2$$
$$0 = (12.0 \text{ m/s})t - \tfrac{1}{2}(9.80 \text{ m/s}^2)t^2,$$

and therefore

$$t = \frac{2(12.0 \text{ m/s})}{(9.80 \text{ m/s}^2)} = 2.45 \text{ s}$$

(the solution $t = 0$ also satisfies the equation but this corresponds to the initial point when y also is zero).

(c) The total distance traveled in the x direction is found from Eq. 3–18:

$$x = v_{x0}t = (16.0 \text{ m/s})(2.45 \text{ s}) = 39.2 \text{ m}.$$

The path of a projectile through space (in the absence of air resistance) is a parabola. To show this, we solve for y as a function of x by eliminating the time, t, between Eqs. 3–18 and 3–20. From Eq. 3–18 we have $t = x/v_{x0}$, and substituting this into Eq. 3–20 we obtain

$$y = \left(\frac{v_{y0}}{v_{x0}}\right)x - \left(\frac{g}{2v_{x0}^2}\right)x^2, \tag{3-22}$$

or, using Eq. 3–16,

$$y = (\tan\theta_0)x - \left(\frac{g}{2v_0^2 \cos^2\theta_0}\right)x^2. \tag{3-23}$$

From either of these equations we see that y as a function of x has the form

$$y = ax - bx^2$$

where a and b are constants for any specific projectile motion; this is the well-known equation for a parabola.

EXAMPLE 3–8 (a) Derive a formula for the range R of a projectile which is defined as the horizontal distance it travels before returning to its original height, $y = 0$ (which is typically the ground). (b) Suppose one of Napoleon's cannons had a muzzle velocity, v_0, of 60.0 m/s. At what angle should it have been aimed (ignore air resistance) to strike a target 320 m away?

SOLUTION (a) To find a general expression for R we set $y = 0$ in Eq. 3–20 and solve for t, which gives two solutions: $t = 0$ and $t = 2v_{y0}/g$. The first corresponds to the initial instant of projection and the second ($t = 2v_{y0}/g$) is the time when the projectile returns to $y = 0$. Then the range, using the latter value of t in Eq. 3–18, is

$$R = \frac{2v_{x0}v_{y0}}{g} = \frac{2v_0^2 \sin\theta_0 \cos\theta_0}{g} = \frac{v_0^2 \sin 2\theta_0}{g},$$

where we have used the trigonometric identity $2\sin\theta\cos\theta = \sin 2\theta$. We see that the maximum range, for a given initial velocity, v_0, is obtained when the sine takes on its maximum value of 1.0; this occurs for $2\theta_0 = 90°$, so $\theta_0 = 45°$ for maximum range. (When air resistance is important, the range is less for a given v_0, and the maximum range is obtained at an angle smaller than 45°.) Note that the maximum range increases by the square of v_0, so doubling the muzzle speed of a cannon, say, increases its maximum range by a factor of four.

(b) From the equation we just derived:

$$\sin 2\theta_0 = \frac{Rg}{v_0^2} = \frac{(320 \text{ m})(9.80 \text{ m/s}^2)}{(60.0 \text{ m/s})^2} = 0.871.$$

Thus $2\theta_0 = 60.6°$, or $180° - 60.6° = 119.4°$, so that

$$\theta_0 = 30.3° \quad \text{or} \quad 59.7°.$$

Either angle gives the same range.

Often we are interested in the horizontal distance traveled when the projectile does not end up at $y = 0$. In some cases y may be greater than zero, such as when shooting an arrow at an object in a tree; or y may be less than zero, as when an airplane drops supplies to flood victims. (Note that θ_0 is negative if the projection angle is below horizontal.) In these cases, the equation derived in Example 3–8 is not useful. Instead, it is necessary to go back to the original equations and use the appropriate values of y and other variables.

 ### 3–9   Uniform Circular Motion

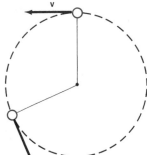

FIGURE 3–17

A particle moving in a circle, showing how the velocity changes. Note that at each point, the instantaneous velocity is in a direction tangent to the circular path.

An object that moves in a circle at constant speed v is said to undergo **uniform circular motion**. Examples are a ball on the end of a string revolved about one's head and the nearly uniform circular motion of the moon around the earth. Although the magnitude of the velocity remains constant in this case, the direction of the velocity is continually changing, Fig. 3–17. Since acceleration is defined as the rate of change of velocity, a change in direction of velocity constitutes an acceleration just as does a change in magnitude. Thus, an object undergoing uniform circular motion is accelerating. We now investigate this acceleration quantitatively.

Acceleration is defined by

$$\mathbf{a} = \lim_{\Delta t \to 0} \frac{\Delta \mathbf{v}}{\Delta t} = \frac{d\mathbf{v}}{dt}$$

where $\Delta \mathbf{v}$ is the change in velocity during the short time interval Δt. We will eventually consider the situation when Δt approaches zero and thus obtain the instantaneous acceleration; but for purposes of making a clear drawing we consider a nonzero time interval, Fig. 3–18. During the time Δt the particle in Fig. 3–18a moves from point A to point B, covering a small distance Δl which subtends a small angle $\Delta \theta$. The change in the velocity vector is $\mathbf{v} - \mathbf{v}_0 = \Delta \mathbf{v}$. If we transfer $\mathbf{v}_0$ to the right side of this equation we obtain $\mathbf{v} = \mathbf{v}_0 + \Delta \mathbf{v}$. Thus, $\Delta \mathbf{v}$ added to $\mathbf{v}_0$ is equal to $\mathbf{v}$, so $\Delta \mathbf{v}$ is the vector shown dashed in Fig. 3–18b. In this diagram we notice that when Δt is very small (approaching zero), and therefore Δl and $\Delta \theta$ are also very small, $\mathbf{v}$ will be almost parallel to $\mathbf{v}_0$ and $\Delta \mathbf{v}$ will be essentially perpendicular to them. Thus $\Delta \mathbf{v}$ points toward the center of the circle. Since $\mathbf{a}$, by its definition above, is in the same direction as $\Delta \mathbf{v}$, it too must point toward the center of the circle. Therefore, this

FIGURE 3–18

Determining the change in velocity, $\Delta \mathbf{v}$, for a particle moving in a circle.

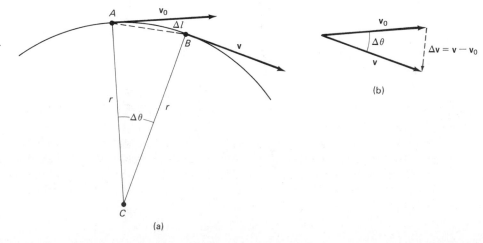

acceleration is called **centripetal acceleration** ("center seeking" acceleration) and henceforth we denote it by a_c.

Now that we have determined the direction of the acceleration, we next determine the magnitude of the acceleration, a_c. The vectors v, v_0, and Δv in Fig. 3–18b form a triangle that is geometrically similar to the triangle ABC in Fig. 3–18a. This relies on the fact that the angle between v_0 and v is equal to $\Delta\theta$, defined as the angle between CA and CB; this is true because CB is perpendicular to v, and CA is perpendicular to v_0. Thus we can write:

$$\frac{\Delta v}{v} \approx \frac{\Delta l}{r},$$

or

$$\Delta v \approx \frac{v}{r} \Delta l.$$

This is an exact equality when Δt approaches zero, for then the arc length, Δl, equals the chord AB in length. To get the magnitude of the centripetal acceleration, a_c, we use the above relation for Δv and write:

$$a_c = \lim_{\Delta t \to 0} \frac{\Delta v}{\Delta t} = \lim_{\Delta t \to 0} \frac{v}{r} \frac{\Delta l}{\Delta t};$$

and since

$$\lim_{\Delta t \to 0} \frac{\Delta l}{\Delta t}$$

is the speed, v, of the object we get

$$a_c = \frac{v^2}{r}. \tag{3-24}$$

Thus, to summarize, a particle moving in circle of radius r with constant speed v has an acceleration directed toward the center of the circle of magnitude $a_c = v^2/r$. It is not surprising that this acceleration depends on v and r. For the greater the speed v, the faster the velocity changes direction; and the larger the radius, the less rapidly the velocity changes direction. This relation was first derived by both Newton and Huygens in the latter half of the seventeenth century.

It is important to recognize that for various types of motion there is no fixed general relationship between the directions of v and a. For linear motion (such as an object falling vertically) they are, indeed, parallel. For uniform circular motion, they are perpendicular to one another (Fig. 3–19) since v is tangential to the circle, and a points toward the center; the direction of both v and a in this case is continually changing. In (nonvertical) projectile motion, a is constant in both magnitude and direction (downward, equal to g) and makes various angles with v during the course of the motion (Fig. 3–16).

Since a is constant in both direction and magnitude for free fall as well as for projectile motion, the kinematic equations for constant acceleration (Eqs. 2–9 or 3–14) can be used. But they *cannot* be used for uniform circular motion because the acceleration is not constant in direction.

FIGURE 3–19

For uniform circular motion, a is always perpendicular to v.

EXAMPLE 3–9 A satellite is put into a circular orbit 200 km above the earth's surface where the acceleration of gravity is only 9.20 m/s². Determine the satellite's speed and period (time to make one revolution).

SOLUTION The radius of the earth is about 6400 km. The radius of the satellite's orbit is therefore (6400 km + 200 km) = 6600 km = 6.6×10^6 m. The satellite is accelerated (centripetally) toward the earth's center with an acceleration $a_c = 9.20$ m/s². (If the satellite were not so accelerated, it would fly off in a straight line.) Therefore, from Eq. 3–24 with $a_c = 9.20$ m/s²:

$$v = \sqrt{ra_c} = \sqrt{(6.6 \times 10^6 \text{ m})(9.20 \text{ m/s}^2)} = 7.8 \times 10^3 \text{ m/s}.$$

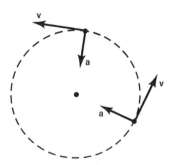

Since speed v equals distance/time, the time T for the satellite to make one revolution (a distance of $2\pi r$) is

$$T = 2\pi r/v = (6.28)(6.6 \times 10^6 \text{ m})/(7.8 \times 10^3 \text{ m/s}) = 5.3 \times 10^3 \text{ s}$$

or 88 minutes.

 EXAMPLE 3–10 The moon's nearly circular orbit about the earth has a radius of about 385,000 km and a period of 27.3 days. Determine the acceleration of the moon toward the earth.

SOLUTION The speed of the moon in its orbit about the earth is

$$v = \frac{2\pi r}{T} = \frac{(6.28)(3.85 \times 10^8 \text{ m})}{(27.3 \text{ d})(24.0 \text{ h/d})(3600 \text{ s/h})} = 1.02 \times 10^3 \text{ m/s}.$$

Therefore

$$a_c = \frac{v^2}{r} = \frac{(1.02 \times 10^3 \text{ m/s})^2}{(3.85 \times 10^8 \text{ m})} = 2.73 \times 10^{-3} \text{ m/s}^2,$$

or about $2.78 \times 10^{-4} g$, where $g = 9.80 \text{ m/s}^2$ is the acceleration of gravity at the earth's surface.[†]

3–10 Nonuniform Circular Motion

If the speed of a particle revolving in a circle is changing, then there will be a tangential acceleration, $\mathbf{a}_t$, as well as the centripetal acceleration, $\mathbf{a}_c$. The tangential acceleration arises from the change in the magnitude of the velocity:

$$a_t = \frac{dv}{dt}, \tag{3–25}$$

whereas the centripetal acceleration arises from the change in direction of the velocity (as we have already seen), and has magnitude

$$a_c = \frac{v^2}{r}.$$

The tangential acceleration always points in a direction tangent to the circle, and is in the direction of motion (parallel to $\mathbf{v}$) if the speed is increasing, as shown in Fig. 3–20 for a particle moving counterclockwise. If the speed is decreasing, $\mathbf{a}_t$ points antiparallel to $\mathbf{v}$. In either case, $\mathbf{a}_t$ and $\mathbf{a}_c$ are always perpendicular to each other, and their directions change continually as the particle moves along its circular path. The total vector acceleration, $\mathbf{a}$, is the sum of these two:

$$\mathbf{a} = \mathbf{a}_t + \mathbf{a}_c. \tag{3–26}$$

Since $\mathbf{a}_c$ and $\mathbf{a}_t$ are always perpendicular to each other, the magnitude of $\mathbf{a}$ at any moment is

$$a = \sqrt{a_t^2 + a_c^2}.$$

It is often convenient to use polar coordinates, r and θ, when describing circular motion; a brief treatment is given in Appendix C.

FIGURE 3–20

For nonuniform circular motion, the acceleration has a tangential ($\mathbf{a}_t$) as well as a centripetal ($\mathbf{a}_c$) component.

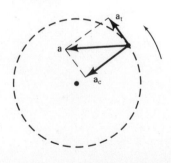

[†] Note: this acceleration $a = 2.78 \times 10^{-4} g$ is *not* the acceleration of gravity of objects at the moon's surface due to the moon's gravity. Rather it is the acceleration due to the earth's gravity of any object (such as the moon) which is 385,000 km from the earth.

Summary

A quantity that has both a magnitude and a direction is called a *vector*. A quantity that has only a magnitude is called a *scalar*.

Vectors can be added graphically by placing the tail of each successive arrow (representing a vector) at the tip of the preceding one in the sum; the sum (or resultant) vector is the arrow drawn from the tail of the first vector to the tip of the last. Two vectors can also be added using the parallelogram method. Vectors can be added more precisely using the analytic methods of adding their *components* along chosen axes with the aid of trigonometric functions. It is often helpful to express a vector in terms of its components along chosen axes using *unit vectors*, which are vectors of unit length along the chosen coordinate axes; for Cartesian coordinates the unit vectors along the x, y, and z axes are called $\mathbf{i}$, $\mathbf{j}$, and $\mathbf{k}$.

The velocity of an object relative to one reference frame can be determined by adding vectorially its velocity in a second reference frame to the velocity of the second reference frame relative to the first reference frame.

The general definitions for the instantaneous *velocity*, $\mathbf{v}$, and *acceleration*, $\mathbf{a}$, of a particle (in one, two, or three dimensions) are

$$\mathbf{v} = \frac{d\mathbf{r}}{dt} \quad \text{and} \quad \mathbf{a} = \frac{d\mathbf{v}}{dt}$$

where $\mathbf{r}$ is the position vector of the particle. The kinematic equations for motion with constant acceleration can be written for each of the x, y, and z components of the motion and have the same form as for one-dimensional motion (Eqs. 2–9 become Eqs. 3–14). Or they can be written in the more general vector form, Eq. 3–15.

Projectile motion—that of an object moving above the earth's surface—can be analyzed as two separate motions if air resistance can be ignored: the horizontal component of the motion is at constant velocity whereas the vertical component is at constant acceleration, g, just as for a vertically falling body (as long as the motion is not far above the earth's surface).

An object moving in a circle of radius r with constant speed v is said to be in *uniform circular motion*, and has a *centripetal acceleration* a_c toward the center of the circle of magnitude:

$$a_c = \frac{v^2}{r}.$$

If the speed is not constant, there will be both centripetal acceleration and tangential acceleration.

Questions

1. One car travels due east at 40 km/h, and a second car travels north at 40 km/h. Are their velocities equal? Explain.

2. Can you conclude that a car is not accelerating if its speedometer indicates a steady 60 km/h?

3. Can you give several examples of an object's motion in which a great distance is travelled but the displacement is zero?

4. Can the displacement vector for a particle moving in two dimensions ever be longer than the length of path traveled by the particle over the same time interval? Can it ever be less? Discuss.

5. During practice, a baseball player hits a very high fly ball, and then runs in a straight line and catches it. Which had the greater displacement, the player or the ball?

6. If $\mathbf{V} = \mathbf{V}_1 + \mathbf{V}_2$, is V necessarily greater than V_1 and/or V_2? Discuss.

7. Two vectors have length $V_1 = 3.5$ km and $V_2 = 4.0$ km. What are the maximum and minimum magnitudes of their vector sum?

8. Can two vectors, of unequal magnitude, add up to give the zero vector? Can *three* unequal vectors?

9. Can the magnitude of a vector ever (*a*) equal, or (*b*) be less than, one of its components?

10. Can a particle with constant speed be accelerating? What if it has constant velocity?

11. Can a vector of magnitude zero have a nonzero component?

12. What are the units of a unit vector?

13. Does the odometer of a car measure a scalar or a vector quantity? What about the speedometer?

14. Two cars with equal speed approach an intersection at right angles to each other. Will they necessarily collide? Show that when the relative velocity of approach is collinear (along the same line) with the relative displacement, we have the nautical maxim "constant bearing means collision."

15. A person sitting in an enclosed train car, moving at constant velocity, throws a ball straight up into the air in her reference frame. (*a*) Where does the ball land? What is your answer if the car (*b*) accelerates, (*c*) decelerates, (*d*) rounds a curve, (*e*) moves with constant velocity but is open to the air.

16. Two rowers who can row at the same speed set off across a river at the same time. One heads straight across and is pulled downstream somewhat by the current. The other one heads upstream at an angle so as to arrive at a point opposite the starting point. Which rower reaches the opposite side first?

17. A child wishes to determine the speed a slingshot imparts to a rock. How can this be done using only a meter stick?

18. Is it ever necessary, in projectile motion, to consider the motion in three dimensions if air resistance can be

neglected? What if air resistance can't be neglected? Discuss.

19. What factors are important for an athlete doing the broad jump? What about the high jump?

20. A projectile has the least speed at what point in its path?

21. A car rounds a curve at a steady 50 km/h. If it rounds the same curve at a steady 70 km/h, will its acceleration be any different? Explain.

22. Will the acceleration of a car be the same if it travels around a sharp curve at 60 km/h as when it travels around a gentle curve at the same speed? Explain.

 Problems _____

SECTIONS 3–1 TO 3–5

1. (I) Graphically determine the resultant of the following three vector displacements: (1) 10 m 30° north of east; (2) 6 m 37° east of north; (3) 12 m 30° west of south.

2. (I) The three vectors in Fig. 3–4 can be added in six different orders. Show on a diagram that the same resultant is obtained no matter what the order.

3. (I) Show that the vector labeled "wrong" in Fig. 3–5c is actually the difference of the two vectors. Is it $V_2 - V_1$ or $V_1 - V_2$?

4. (I) Two vectors V_1 and V_2 add to a resultant $V = V_1 + V_2$. Describe V_1 and V_2 if (a) $V = V_1 + V_2$, (b) $V^2 = V_1^2 + V_2^2$, (c) $V_1 + V_2 = V_1 - V_2$.

5. (I) If $V = 3.0i - 4.0j$; what scalar c must multiply V so that $|cV| = 7.5$?

6. (I) If $V = -2.5i + 6.0j$, what is the magnitude of cV for $c = 3.0$?

7. (I) An airplane is traveling 1000 km/h in a direction 32.5° west of north. (a) Find the components of the velocity vector in the northerly and westerly directions. (b) How far north and how far west has the plane traveled after 3.00 h?

8. (II) (a) Determine the magnitude and direction of the sum of the three vectors $V_1 = 4i - 3j$, $V_2 = i + j$, and $V_3 = -i + 4j$. (b) Determine $V_1 - V_2 + V_3$.

9. (II) The components of a vector V are often written (V_x, V_y, V_z). What are the components and length of a vector which is the sum of the two vectors V_1 and V_2 whose components are (6, 0, 2) and (1, 4, 3)?

10. (II) Given the vectors V_1 and V_2 from problem 9, determine a third vector, V_3, such that (a) $V_1 + V_2 + V_3 = 0$, (b) $V_1 - V_2 + V_3 = 0$.

11. (II) A deliveryman travels 30 m north, 25 m east, 12 m south, and then takes an elevator 36 m up into a building. What is his final displacement from the origin?

12. (II) What is the y component of a vector in the xy plane whose magnitude is 36.5 and whose x component is 25.4? What is the direction of this vector?

13. (II) Let $V_1 = 6.0i + 3.0j$ and $V_2 = -2.5i + 4.0j$. Determine the magnitude and direction of (a) V_1, (b) V_2, (c) $V_1 + V_2$, (d) $V_2 - V_1$.

14. (II) The summit of a mountain, 2150 m above a camp, is measured on a map to be 4750 m from the camp in a direction 28.2° west of north. Write down an expression, in terms of unit vectors, for the displacement vector from camp to summit. What is its length? Choose the x-axis east, y-axis north, and z-axis up.

SECTION 3–6

15. (I) A vacationer walks 4.20 km/h directly across a cruise ship whose speed relative to the earth is 9.60 km/h. What is the speed of the vacationer with respect to the earth?

16. (I) An airplane is heading due north at a speed of 425 km/h. If a wind begins blowing from the southwest at a speed of 55 km/h (average), calculate: (a) the velocity (magnitude and direction) of the plane, and (b) how far off course it will be after 15 min.

17. (II) Raindrops make an angle θ with the vertical when viewed through a moving train window. If the speed of the train is v_T, what is the speed of the raindrops in the reference frame of the earth in which they are assumed to fall vertically.

18. (II) A boat can travel 2.60 m/s in still water. (a) If the boat heads directly across a stream whose current is 0.90 m/s, what is the velocity (magnitude and direction) of the boat relative to the shore? (b) What will be the position of the boat, relative to its point of origin, after 4.0 s?

19. (II) Determine the speed of the boat with respect to the shore in Example 3–4.

20. (II) A motorboat whose speed in still water is 8.6 km/h must aim at a 65° angle upstream in order to travel directly across the stream. (a) What is the speed of the current? (b) What is the resultant speed of the boat with respect to the shore?

21. (II) A swimmer is capable of swimming 1.65 m/s in still water. (a) If she swims directly across a 180-m wide river whose current is 0.85 m/s, how far downstream (from a point opposite her starting point) will she land? (b) How long will it take her to reach the other side?

22. (II) At what upstream angle must the swimmer in the previous problem aim if she is to arrive at a point directly across the stream?

23. (II) A helicopter heads due south with an airspeed of 45 km/h. The pilot observes, however, that they have covered 25 km in the previous 50 min. in a southwesterly direction. What is the wind speed and direction?

24. (II) Two cars approach a street corner at right angles to each other. Car 1 travels at 35 km/h and car 2 at 55 km/h. What is the relative velocity of car 1 as seen by car 2? What is the velocity of car 2 relative to car 1?

25. (III) An airplane, whose speed is 550 km/h, is supposed to fly in a straight path 33.0° north of east. But a steady 120-km/h wind is blowing from the north. In what direction should the plane head?

26. (III) The speed of a boat in still water is v. The boat is to make a round trip in a river whose current travels at speed u. Derive a formula for the time needed to make a round trip of total distance D if the boat makes the round trip by moving (a) upstream and back downstream, (b) directly across the river and back. We must assume $u < v$; why?

27. (III) In hot pursuit, Agent Logan of the FBI must get directly across a 2.0-km wide river in minimum time. The river's current is 2.5 km/h, he can row a boat at 4.0 km/h, and he can run 7.0 km/h. Describe the path he should take (rowing plus running along the shore) for the minimum crossing time, and determine the minimum time.

SECTION 3–7

28. (I) The position of a particular particle as a function of time is given by $\mathbf{r} = 3.10t\mathbf{i} + 6.05\mathbf{j} - t^2\mathbf{k}$. Determine the particle's velocity and acceleration as a function of time.

29. (II) What is the shape of the path of the particle of problem 28?

30. (II) What was the average velocity of the particle in problem 28 between $t = 1.00$ s and $t = 3.00$ s? What is its instantaneous speed at $t = 2.00$ s?

31. (II) A car is moving with speed 20.0 m/s due north at one moment and 34.6 m/s due east 9.00 s later. Over this time interval, determine (a) its average velocity, (b) average acceleration (magnitude and direction for both), and (c) average speed. (Hint: Can you determine all these from the information given?)

32. (II) (a) A skier is accelerating down a 30° hill at 2.30 m/s². What is the vertical component of her acceleration? (b) How long will it take her to reach the bottom of the hill, assuming she starts from rest and accelerates uniformly, if the elevation change is 180 m?

33. (III) The position of a particle is given by $\mathbf{r} = (6.0 \cos 3.0t\mathbf{i} + 6.0 \sin 3.0t\mathbf{j})$ meters. Determine (a) the velocity vector, $\mathbf{v}$, and (b) the acceleration vector, $\mathbf{a}$. (c) What is the path of this particle? (Hint: determine $r = |\mathbf{r}|$.) (d) What is the relation between r and a (give a formula), and between $\mathbf{r}$ and $\mathbf{a}$ (give an angle). (e) Show that $a = v^2/r$.

SECTION 3–8 (Ignore air resistance in the following problems unless stated otherwise.)

34. (I) A diver running 3.2 m/s dives from the edge of a vertical cliff and reaches the water below 2.0 s later. How high was the cliff and how far from its base did the diver hit the water?

35. (I) A tiger leaps horizontally from a 16 m high rock with a speed of 7.0 m/s. How far from the base of the rock will it land?

36. (I) A fire hose held near the ground shoots water at a speed of 15.0 m/s. At what angle(s) should the nozzle point in order that the water land 18 m away? Why are there two different angles?

37. (I) An athlete executing a long jump leaves the ground at a 30° angle and travels 8.90 m. What was the takeoff speed?

38. (I) Determine how much farther a person can jump on the moon as compared to the earth if the takeoff speed and angle are the same. The acceleration due to gravity on the moon is one-sixth what it is on earth.

39. (II) A ball thrown horizontally at 22.2 m/s from the roof of a building lands 36 m from the base of the building. How high is the building?

40. (II) Show that the speed with which a projectile leaves the ground is equal to its speed just before it strikes the ground at the end of its journey, assuming the firing level equals the landing level.

41. (II) An airplane traveling 150 km/h wants to drop supplies to flood victims isolated on a patch of land 250 m below. The supplies should be dropped how many seconds before the plane is directly overhead?

42. (II) A hunter aims directly at a target (on the same level) 250 m away. (a) If the bullet leaves the gun at a speed of 550 m/s, by how much will it miss the target? (b) At what angle should the gun be aimed so the target will be hit?

43. (II) An athlete throws the shotput (mass = 7.3 kg) with an initial speed of 14.0 m/s at a 41° angle to the horizontal. Calculate the horizontal distance traveled. The shot leaves the shotputter's hand at a height of 2.2 m above the ground.

44. (II) A ball is thrown horizontally from the top of a cliff with initial speed v_0. At any moment its direction of motion makes an angle θ to the horizontal. Derive a formula for θ as a function of t before the ball reaches the ground below.

45. (II) Show that the time required for a projectile to reach its highest point is equal to the time for it to return to its original height.

46. (II) A world-class long jumper is capable of jumping 8.0 m. Assuming his horizontal speed is 9.0 m/s (top world-class sprinting speed is slightly over 10 m/s) as he leaves the ground, how long was he in the air and how high did he go? Assume that he lands standing upright, that is, the same way he left the ground.

47. (II) Derive a formula for the horizontal range, R, of a projectile when it lands at a height h above its initial point. (For $h < 0$, it lands a distance $-h$ below the starting point.) Assume it is projected at an angle θ_0 with initial speed v_0.

48. (II) At what projection angle will the range of a projectile equal its maximum height?

49. (II) 3.0 s after a projectile is fired into the air from the ground, it is observed to have a velocity $\mathbf{v} = (8.9\mathbf{i} + 3.6\mathbf{j})$ m/s where the x axis is horizontal and the y axis is positive upward. Determine (a) the horizontal range of the projectile, (b) its maximum height above the ground, and (c) its speed and angle of motion just before it strikes the ground.

50. (II) A projectile is fired into the air with an initial speed of 40.0 m/s. Plot on graph paper its trajectory for initial projection angles of $\theta = 15°, 30°, 45°, 60°, 75°, 90°$. Plot at least 10 points for each curve.

51. (III) A high diver leaves the end of a 5.0-m-high diving board and strikes the water 1.3 s later, 3.0 m beyond the end of the board. Considering the diver as a particle, determine (a) her initial velocity, $\mathbf{v}_0$, (b) the maximum height reached, and (c) the velocity $\mathbf{v}_f$ with which she enters the water.

52. (III) A hunter aims his bow and arrow directly at a monkey hanging from a high tree branch some distance away. At the instant the arrow is shot the monkey drops from the branch, hoping to avoid the arrow. Show analytically that the monkey made the wrong move. Ignore air resistance.

53. (III) A person stands at the base of a hill which is a straight incline making an angle ϕ with the horizontal. For a given initial speed v_0, at what angle θ (to the horizontal) should objects be thrown so that they land as far up the hill as possible?

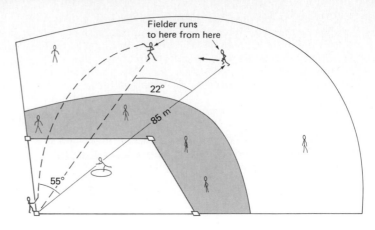

FIGURE 3–21

54. (III) A basketball leaves a player's hands at a height of 2.1 m above the floor. The basket is 2.6 m above the floor. The player likes to shoot the ball at a 35° angle. If the shot is made from a horizontal distance of 12.0 m and must be accurate to ±0.22 m (horizontally), what is the range of initial speeds allowed to make the basket?

55. (III) At $t = 0$ a batter hits a baseball with an initial speed of 35 m/s at a 55° angle to the horizontal. An outfielder is 85 m from the batter at $t = 0$, and, as seen from home plate, the line of sight to the outfielder makes a horizontal angle of 22° with the plane in which the ball moves (see Fig. 3–21). What speed and direction must the fielder choose to catch the ball at the same height from which it was struck? Give the angle with respect to the outfielder's line of sight to home plate.

56. (III) Police agents flying 180 km/h in a low-flying airplane wish to drop an explosive onto a master criminal's automobile traveling 135 km/h on a level highway 80.0 m below. At what angle (with the horizontal) should the car be in their sights when the bomb is released?

SECTION 3–9

57. (I) What is the centripetal acceleration of a child 8.2 m from the center of a merry-go-round? The child's speed is 2.1 m/s.

58. (I) A jet plane traveling 1800 km/h (500 m/s), pulls out of a dive by moving in an arc of radius 3.0 km. What is the plane's acceleration in "g's"?

59. (I) Calculate the centripetal acceleration of the earth in its orbit around the sun. Assume the earth's orbit is a circle of radius 1.5×10^{11} m.

60. (II) (a) Derive a formula for the radius of curvature of a projectile's path at the top of its arc (Fig. 3–16) in terms of θ_0 and v_0. (That is, assume the top of the arc is a tiny part of a circle.) (b) What is its "centripetal" acceleration at this point?

61. (II) Because the earth rotates once per day, the effective acceleration of gravity at the equator is slightly less than it would be if the earth didn't rotate. Estimate the magnitude of this effect. What fraction of g is this?

62. (II) What is the magnitude of the acceleration of a speck of dust on the edge of a $33\frac{1}{3}$-rpm (revolutions per minute) phonograph record whose diameter is 30 cm?

SECTION 3–10

63. (II) A particle rotates in a circle of radius 3.60 m. At a particular instant its acceleration is $0.210\,g$ in a direction that makes an angle of 28.0° to its direction of motion. Determine its speed (a) at this moment, and (b) 2.00 s later, assuming constant tangential acceleration.

64. (II) A particle starting from rest revolves with uniformly increasing speed in a clockwise circle in the xy plane. The center of the circle is at the origin of an xy coordinate system. At $t = 0$ the particle is at $x = 0.0$, $y = 2.0$ m. At $t = 2.0$ s, it is at $x = 2.0$ m, $y = 0.0$, and has speed 14.0 m/s. Determine (a) the average velocity vector and (b) the average acceleration vector during this interval.

65. (II) In problem 64, assume the tangential acceleration is constant and determine the components of the instantaneous acceleration at (a) $t = 0.0$, (b) $t = 1.0$ s, and (c) $t = 2.0$ s.

Dynamics: Newton's Laws of Motion

4

We have discussed how motion is described in terms of velocity and acceleration. Now we deal with the question of *why* objects move as they do: what makes an object at rest begin to move? What causes a body to accelerate or decelerate? What is involved when an object moves in a circle? We could answer in each case that a force is required. In this chapter we will investigate the connection between force and motion. The only restriction we make is that the velocities involved be much less than the speed of light (3.00×10^8 m/s), so we do not have to worry about relativistic effects (Chapter 39). Before we delve into this subject of *dynamics*, we first discuss the concept of *force* in a qualitative way.

 ## 4–1 Force

Intuitively, we can define **force** as any kind of a push or a pull. When you push a grocery cart, you are exerting a force on it. When children pull a wagon, they are exerting a force on the wagon. When a motor lifts an elevator, or a hammer hits a nail, or the wind blows the leaves of a tree, a force is being exerted. We say that an object falls because the force of gravity acts on it. Forces do not always give rise to motion. For example, you may push very hard on a heavy desk or refrigerator and it may not move.

Whether or not an object moves when a force is exerted on it, the object does change shape. This is obvious when you squeeze a balloon, or push a mattress. You can also see the slight deformation of the metal when you push on the side of a refrigerator or the fender of a car. There is always some deformation when a force is exerted, although it may take delicate instruments to detect it for a very rigid object like a heavy steel plate.

One way to quantitatively measure the magnitude (or strength) of a force is to make use of a spring scale (Fig. 4–1). Normally such a spring scale is used to find the weight of an object; by weight we mean the force of gravity acting on the body

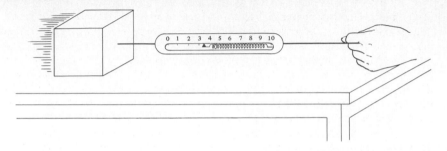

FIGURE 4–1

A spring scale used to measure a force.

(Section 4–7). The spring scale, once calibrated,[†] can be used to measure other kinds of forces as well, such as the pulling force shown in Fig. 4–1.

A force has direction as well as magnitude and is a vector that follows the rules for vector addition discussed in Chapter 3. We can represent any force on a diagram by an arrow, just as we did with velocity. The direction of the arrow is of course in the direction of the push or pull, and its length is drawn proportional to the strength or magnitude of the force. Although the definition of force as a push or a pull is adequate for the moment, we will give a more precise definition in Section 4–5.

4–2   Newton's First Law of Motion

What is the exact connection between force and motion? Aristotle believed that a force was required to keep an object moving along a horizontal plane. He would argue that to make a book move across the table, you would have to exert a force on it continuously. To Aristotle, the natural state of a body was to be at rest, and a force was believed necessary to keep a body in motion. Furthermore, Aristotle argued, the greater the force on the body, the greater its speed.

Some 2000 years later, Galileo, skeptical of these Aristotelian views just as he was of those on falling bodies, came to a radically different conclusion. Galileo claimed that it is just as natural for an object to be in horizontal motion with a constant speed as it is to be at rest. To understand Galileo's viewpoint, consider the following observations involving motion along a horizontal plane (where the effects of gravity do not enter). It will take a certain amount of force to push an object with a rough surface along a table top at constant speed. To push an equally heavy object with a very smooth surface across the table at the same speed will require less force. Finally, if a layer of oil or other lubricant is placed between the surface of the object and the table, then almost no force is required to move the object. (These observations may well be obvious to you; but if not, you should do these simple experiments for yourself.) Notice that in each step the force required was less and less. The next step is to extrapolate from these data to a situation in which the object does not rub against the table at all—or there is a perfect lubricant between them—and theorize that once started, the object would move across the table at constant speed with *no* force being applied. A steel ball bearing rolling on a hard horizontal surface approaches this situation closely.

It was Galileo's genius to imagine an idealized world—in this case, one where there is no friction—and to see it could produce a more useful view of the real world. It was this idealization that led him to his remarkable conclusion that if no

---

[†] A spring scale is calibrated by hanging from it a series of identical objects of equal weight or mass, say 1 pound or 1 kilogram. The positions of the pointer when 0, 1, 2, 3, . . . units of weight or mass are hung from it are marked. Although the amount the spring stretches is very nearly proportional to the amount of weight hung from it, as long as it isn't stretched too far, we don't have to make use of this fact in our calibration method. We only assume that the pointer rests at the same position when the same weight (force) acts. This could serve as an operational definition (Section 1–5) of force.

force is applied to a moving object, it will continue to move with constant speed in a straight line. An object slows down only if a force is exerted on it. Galileo thus interpreted friction as a force akin to ordinary pushes and pulls.

To push an object across a table at constant speed requires a force from your hand only to balance out the force of friction; in this case the pushing force is equal in magnitude to the friction force but they are in opposite directions, so the *net* force on the object is zero, Fig. 4–2. This is consistent with Galileo's viewpoint, for the object moves with constant speed when no net force is exerted on it.

The difference between Aristotle's view and Galileo's is not simply one of right or wrong. Aristotle's view was not really wrong, for our everyday experience indicates that moving objects do tend to come to a stop if not continually pushed. The real difference lies in the fact that Aristotle's view about the "natural state" of a body was almost a final statement—little further development was possible. Galileo's analysis, on the other hand, could be extended to explain a great many more phenomena. By making the creative leap of imagining the experimentally unattainable situation of no friction and by interpreting friction as a force, Galileo was able to reach his conclusion that an object will continue moving with constant velocity if no force acts to change this motion.

Upon this foundation, Newton built his great theory of motion. Newton's analysis of motion is summarized in his famous "three laws of motion." In his great work the *Principia* (which was published in 1687 and contains nearly all his work on motion) Newton readily acknowledged his debt to Galileo. In fact, **Newton's first law of motion** is very close to Galileo's conclusions[†]. It states that

Every body continues in its state of rest or of uniform speed in a straight line unless it is compelled to change that state by forces acting on it.

The tendency of a body to maintain its state of rest or of uniform motion in a straight line is called **inertia**. As a result, Newton's first law is often called the **law of inertia**.

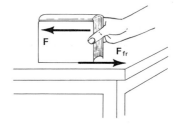

FIGURE 4–2
F represents the force applied by the person and **F**$_{fr}$ represents the force of friction.

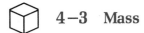

4–3   Mass

Newton's second law (which we come to in the next section) makes use of the concept of mass. Newton himself used the term *mass* as a synonym for *quantity of matter*. This intuitive notion of the mass of a body is not very precise because the concept "quantity of matter" is itself not well defined. More precisely we can say that **mass** *is a measure of the inertia of a body*. The more mass a body has, the harder it is to change its state of motion; it is harder to start it moving from rest, or to stop it when it is moving, or to change its motion sideways out of a straight-line path. A piano or a truck has much more inertia than a baseball moving at the same speed, and it is much harder to change its state. It therefore has much more mass.

To quantify the concept of mass, we must define a standard. In SI units, the unit of mass is the kilogram (kg). The actual standard is a particular platinum-iridium cylinder, kept at the International Bureau of Weights and Measures near Paris, whose mass by definition is precisely one kilogram. In cgs units, the unit of mass is the gram (g), where $1 \text{ g} = 10^{-3}$ kg; in the British system the unit of mass is called the slug (see Section 4–4). When dealing with atoms and molecules, we often make use of the *unified atomic mass unit* (u). By definition, the mass of a carbon (^{12}C) atom is assigned an atomic mass of precisely 12 u. In terms of the kilogram the best value today is

$$1 \text{ u} = (1.6605655 \pm 0.0000086) \times 10^{-27} \text{ kg}$$

[†] It is not clear in Galileo's works if he adopted "linear" inertia, or "spherical" inertia—that is, natural motion continuing along a spherical surface such as on the surface of the earth. Newton, and Descartes before him adopted the principle of inertia for motion along a straight line.

or, rounded off,

$$1 \text{ u} = 1.6606 \times 10^{-27} \text{ kg}.$$

FIGURE 4–3

An equal-arm balance.

Once we have a standard mass, there are two ways we could set up a scale of masses (that is, to define 2 kg, 3 kg, and so on). The first method corresponds roughly to Newton's "quantity of matter" definition and makes use of an equal-arm balance (Fig. 4–3). We place a 1-kg standard mass in one of the pans and claim that any mass placed in the other pan that causes the pans to balance precisely is by definition also 1 kg. We now have two 1-kg masses and when placed together we define that they make precisely 2 kg. We can obtain fractional masses by, for example, finding or making two identical masses (they balance when placed in each pan of the balance) which together balance against a single 1-kg mass. This process can be continued until we have a whole set of known masses; and any unknown mass can then be determined by balancing it against a combination of the known masses. This method depends on the fact that balance of the two pans of the equal-arm balance occurs when the gravitational pull on the two masses is equal. Because of this, mass defined in this way is often called "gravitational mass."

The second method of defining a mass scale makes use of inertia (and Newton's second law); it is described in Section 4–4 where the idea of inertial mass is put on a quantitative basis. Experiments indicate that the two methods are totally consistent with one another.

The terms *mass* and *weight* are often confused with one another, but it is important to distinguish between them. Mass is a property of a body itself (it is a measure of a body's inertia, or its "quantity of matter"). Weight, on the other hand, is a force, the force of gravity acting on a body. To see the difference, suppose we take an object to the moon. It will weigh only about one-sixth as much as it did on earth since the force of gravity is weaker. But its mass will be the same; it will have the same amount of matter and it will have just as much inertia—for in the absence of friction, it will be just as hard to start it moving or to stop it once it is moving.

 4–4   Newton's Second Law of Motion

Newton's first law states that if no net force is acting on a body it remains at rest, or if it is moving it continues moving with constant speed in a straight line. But what happens if a net force is exerted on a body? Newton perceived that the velocity will change. A net force exerted on an object may make its speed increase, or, if it is in a direction opposite to the motion, it will reduce the speed. If the net force acts sideways on a moving object, the direction as well as the magnitude of the velocity changes. Thus, a *net force gives rise to acceleration.*

What precisely is the relationship between acceleration and force? Ordinary experience can answer this question. Consider the force required to push a roller skate or cart whose friction is minimal. (If there is friction, consider the net force, which is the force you exert minus the force of friction.) Now if you push with a gentle but constant force for a certain period of time, you will make the skate or cart accelerate from rest up to some speed, say 3 km/h. If you push twice as hard, you will find that it will reach 3 km/h in half the time. That is, its acceleration will have been twice as great. If you double the force, the acceleration doubles; if you triple the force, the acceleration is tripled, and so on. Thus, the acceleration of a body is directly proportional to the net applied force. But the acceleration depends on the mass of the object as well. If you push an empty cart with the same force as you push one that is loaded, you will find that the latter accelerates more slowly. The greater the mass, the less the acceleration for a given net force. In fact, as Newton found, the acceleration of a body is inversely proportional to its mass. These relationships are found to hold in general and can be summarized as follows:

> **The acceleration of an object is directly proportional to the net force acting on it and is inversely proportional to its mass. The direction of the acceleration is in the direction of the applied net force.**

This is called **Newton's second law of motion**. In symbols, we can write it as:

$$\mathbf{a} \propto \frac{\mathbf{F}}{m}$$

where **a** stands for acceleration, m for the mass, and **F** for the net force. By **net force** we mean the vector sum of all the forces acting on the body. To change a proportion into an equation we merely need to insert a constant of proportionality. The choice of a constant is arbitrary in this case since we are relating quantities with different units; we can therefore choose the unit of force or of mass so that the proportionality constant equals one. Then $\mathbf{a} = \mathbf{F}/m$. Rearranging, we have the familiar statement of Newton's second law in equation form:

$$\mathbf{F} = m\mathbf{a}. \tag{4–1}$$

This is a vector equation; both direction and magnitude must be equal on the two sides of the equation. Newton's second law relates the description of motion to the cause of motion, force. It is one of the most fundamental relationships in physics.[†] From Eq. 4–1 we can make a more precise definition of force as an action capable of accelerating an object (more on this in the next section).

The unit of *force* is chosen so that the proportionality constant in Newton's second law, $\mathbf{F} \propto m\mathbf{a}$, will be one, and thus $\mathbf{F} = m\mathbf{a}$. With the mass in kilograms, the unit of force is called the *newton* (N). One newton, then, is the force required to impart an acceleration of 1 m/s^2 to a mass of 1 kg. Thus $1 \text{ N} = 1 \text{ kg·m/s}^2$.

In cgs units, the unit of mass is the gram (g) as mentioned earlier. The unit of force is the *dyne*, which is defined as the force required to impart an acceleration of 1 cm/s^2 to a mass of 1 g; thus $1 \text{ dyne} = 1 \text{ g·cm/s}^2$. It is easy to show that $1 \text{ dyne} = 10^{-5} \text{ N}$. (Do not confuse the abbreviation g for gram with the symbol g for the acceleration due to gravity which is always italicized.)

In the British system, the unit of force is the pound. The pound is defined as the weight (which is a force) of a body with a mass of 0.45359237 kg at a particular place on the earth where the acceleration due to gravity is $g = 32.1734 \text{ ft/s}^2$. The unit of mass is the *slug*, which is defined as that mass that will undergo an acceleration of 1 ft/s^2 when a force of 1 lb is applied to it. Thus, $1 \text{ lb} = 1 \text{ slug·ft/s}^2$. It is easy to show that $1 \text{ lb} \approx 4.45 \text{ N}$.[‡]

It is very important that only one set of units be used in a given calculation or problem, with the SI being preferred. If the force is given in, say, newtons, and the mass in grams, then before attempting to solve for the acceleration in SI units, the mass must be changed to kilograms. For example, if the force is given as 2.0 N and the mass is 500 g, we change the latter to 0.50 kg and the acceleration will then automatically come out in m/s^2 when Newton's second law is used:

$$a = \frac{F}{m} = \frac{2.0 \text{ N}}{0.50 \text{ kg}} = 4.0 \text{ m/s}^2.$$

[†] Newton originally stated his second law of motion in terms of the momentum $\mathbf{p} = m\mathbf{v}$: that is, $\mathbf{F} = d\mathbf{p}/dt$, which for constant mass reduces to $\mathbf{F} = m(d\mathbf{v}/dt) = m\mathbf{a}$. This is discussed in Chapter 8.

[‡] Other systems are occasionally used, and we mention them for reference only. In the "gravitational mKs" system, the kilogram is still the mass, but the unit of force is called the kilogram-force (kgf) and 1 kg weighs 1 kgf at a place where $g = 9.8066 \text{ m/s}^2$; in other words, 1 kgf = 9.8066 N. In the British equivalent, the pound remains the unit of force, but the unit of mass is the pound mass (lbm). A mass of 1 lbm weighs 1 lb (at a place where $g = 32.1734 \text{ ft/s}^2$), so 1 slug = 32.1734 lbm.

EXAMPLE 4–1 Calculate the average force that the pitcher in Example 2–9 exerted on the baseball, whose mass is 0.145 kg.

SOLUTION We calculated the acceleration to be 129 m/s^2 and the mass of the ball is 0.145 kg. Then:

$$F = ma$$
$$= (0.145 \text{ kg})(129 \text{ m/s}^2)$$
$$= 18.7 \text{ N}.$$

We will discuss many more examples of Newton's second law later in this chapter and, in fact, throughout the book.

As mentioned in Section 4–3, we can quantify the concept of mass using its definition as a measure of inertia. How to do this is evident from Eq. 4–1 where we see that the acceleration of a body is inversely proportional to its mass; if the same net force F acts to accelerate each of two masses, m_1 and m_2, then the ratio of their masses can be defined as the inverse ratio of their accelerations:

$$\frac{m_2}{m_1} = \frac{a_1}{a_2}.$$

If one of the masses is known (it could be the standard kilogram) and the two accelerations are precisely measured, then the unknown mass is obtained from this definition. For example, if $m_1 = 1.00$ kg, and for a particular force $a_1 = 3.00$ m/s^2 and $a_2 = 2.00$ m/s^2, then $m_2 = 1.50$ kg. With this definition of a mass scale, the inverse relationship between a and m in Newton's second law occurs because we have defined it to be so. Mass defined in this way is called *inertial mass* and is totally consistent with the method described in Section 4–3 (gravitational mass). This is a remarkable fact and is discussed more in Section 5–5.

4–5 Laws or Definitions?

According to the operationalist viewpoint (or "positivist" school of philosophy), the definition of force as a push or a pull is not a good one because it is too vague. The operationalist viewpoint argues that physical quantities should be defined in terms of one or more "operations." One way of doing this is using the spring scale as discussed in Section 4–1 (Fig. 4–1). Many thinkers believe this (perhaps primitive) definition is inadequate, in part because it depends on the spring scale being calibrated using the gravitational force. A more precise way to define force, and one which is dynamical (rather than static), is via Newton's second law itself. To determine the magnitude and direction of a certain force **F**, you allow it to act on an object of known mass m, and you measure the resulting acceleration **a**; then **F** is defined as the product m times **a**. Thus, according to this viewpoint, Newton's second law is not to be considered a law; rather it is a definition of force.

Similar remarks apply to Newton's first law. It is generally considered to define a particular kind of reference frame, called an **inertial reference frame**. An inertial reference frame is thus one in which Newton's first law is valid. (The earth for example is, for most purposes, nearly an inertial reference frame.) A **noninertial reference frame** is one in which Newton's first law does not hold. An example would be a freely falling elevator; if you were in an elevator whose cable had broken you would be pulled downward by the force of gravity with an acceleration g with respect to the ground. But if you set up a coordinate system in the elevator itself, you would be *at rest* in that coordinate system even though a net force was acting on you. Thus Newton's first law would not be valid in such a coordinate system. Because such noninertial reference frames exist, Newton's first law must be considered to be a definition rather than a law.

Note that Newton's second law ($\mathbf{F} = m\mathbf{a}$) also does not hold in a noninertial reference frame; with respect to the falling elevator, for example, a person's acceleration $\mathbf{a} = 0$ although $\mathbf{F} \neq 0$ since the force of gravity is acting.

Whether you accept the operationalist viewpoint that $\mathbf{F} = m\mathbf{a}$ is a definition of force, or whether you prefer to think of this relation as a law, has no bearing on the use of this relation in practical situations.

4–6 Newton's Third Law of Motion

The second law describes quantitatively how forces affect motion. But where, we may ask, do forces come from? Observations suggest that a force applied to any object is always applied *by another object*. A horse pulls a wagon, a person pushes a grocery cart, a hammer pushes on a nail, a magnet attracts an iron nail. In each of these examples, one object exerts the force and the other object feels it: for example the hammer exerts the force, and the nail feels it. But Newton realized things cannot be so one-sided. True, the hammer exerts a force on the nail. But the nail evidently exerts a force back on the hammer as well, for the hammer's speed is rapidly reduced to zero upon contact; only a strong force could cause such a rapid deceleration. Thus, said Newton, the two bodies must be treated on an equal basis. The hammer exerts a force on the nail, and the nail exerts a force back on the hammer. This is the essence of Newton's **third law of motion**[†]:

> **Whenever one object exerts a force on a second object, the second exerts an equal and opposite force on the first.**

This law is sometimes paraphrased as "to every action there is an equal and opposite reaction." To avoid confusion, it is very important to remember that the "action" force and the "reaction" force are acting on *different* objects.

As evidence for the validity of this third law look at your hand when you push against a grocery cart or against the edge of a desk (Fig. 4–4). Your hand's shape is distorted, clear evidence that a force is being exerted on it. You can *see* the edge of the desk pressing into your hand. You can even *feel* the desk exerting a force on your hand; it hurts. The harder you push against the desk, the harder the desk pushes back on your hand.

Consider next the ice skater in Fig. 4–5. Since there is very little friction between her skates and the ice, she will move freely if a force is exerted on her. She pushes against the wall, and then *she* starts moving backwards. Clearly there had to be a force exerted on her to make her move. The force she exerts on the wall cannot make her move, for that force pushes on the wall and can only affect the wall. Something had to exert a force on her to make her move, and that force could only have been exerted by the wall. The force with which the wall pushes on her is equal and opposite to the force she exerts on the wall.

When a person throws a package out of a boat, the boat moves (at least a little) in the opposite direction. The person exerts a force on the package; the package exerts an equal and opposite force back on the person, and this force propels the boat backward slightly. Rockets work on the same principle. A common misconception is that rockets accelerate because the gases rushing out the back of the engine push against the ground or the atmosphere. Actually, a rocket accelerates because it exerts a strong force on the gases, expelling them. The gases exert an equal and opposite force *on the rocket* and it is this force which drives

[†] Historians of science today believe that only this third law may have been original to Newton. The first two were already suggested in the works of Galileo and Descartes.

FIGURE 4–4

If you push your hand against the edge of a table, the table edge pushes back against you.

Force exerted on hand by desk.

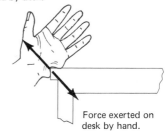

Force exerted on desk by hand.

FIGURE 4–5

When an ice skater pushes against a wall, the wall pushes back and forces her to move outward.

the rocket forward. Thus a space vehicle can maneuver in empty space just by firing its rockets in the direction opposite to that in which it wants to accelerate.

Let us consider how we walk. A person begins walking by pushing with his or her foot against the ground. The ground then exerts an equal and opposite force back on the person (Fig. 4–6), and it is this force, *on* the person, which moves him or her forward. In a similar way, a bird flies forward by exerting a force on the air, but it is the air pushing back on the bird's wings that propels the bird forward. And an automobile moves forward because of the force exerted on it by the ground, which is the reaction to the force exerted on the ground by the wheels.

From the above examples it is clear that it is quite important to remember *on* what object a given force is being exerted and *by* what object that force is exerted. The point is that a force influences the motion of an object only when it is applied *on* that object. A force exerted *by* a body does not influence that body; it only influences the other body *on* which it is exerted. Thus, to avoid confusion, the two prepositions *on* and *by* must always be used—and used with care. We will see how helpful this is in further examples in this chapter and later chapters.

We tend to associate forces with active bodies such as humans, animals, engines, or a moving object like a hammer. It is often difficult to see how an inanimate object at rest, such as a wall or a desk, can exert a force. The explanation lies in the fact that every material, no matter how hard, is elastic, at least to some degree. No one can deny that a stretched rubber band can exert a force on a wad of paper and send it flying across the room. Although other materials may not stretch as easily as rubber, they do stretch when a force is applied to them. And just as a stretched rubber band exerts a force, so does a stretched (or compressed) wall or desk.

Force exerted by person on the ground

Force exerted by ground on the person

FIGURE 4–6

We can walk forward because the ground pushes forward on our feet when we push backward against the ground.

4–7 Weight—the Force of Gravity

Galileo claimed that objects dropped near the surface of the earth will all fall with the same acceleration, g, if air resistance can be neglected. The force that gives rise to this acceleration is called the force of gravity. We now apply Newton's second law to the gravitational force and for the acceleration, **a**, we use the acceleration of gravity, **g**. Thus the force of gravity on a body, $\mathbf{F}_g$, which is also commonly called its *weight* (with symbol **w**, meaning the same as $\mathbf{F}_g$), can be written as:

$$\mathbf{F}_g = \mathbf{w} = m\mathbf{g}; \tag{4–2}$$

and the direction of this force is down toward the center of the earth.

In SI units, $g = 9.80 \text{ m/s}^2$, so the weight of a 1.00 kg mass is 1.00 kg × 9.80 m/s² = 9.80 N. The value of g varies very slightly at different places on the earth's surface as we saw in Chapter 2, but we usually won't be concerned about this. On the moon, on other planets, or in space, the weight of a given mass will be different. For example, on the moon, g is about one-sixth what it is on earth, and 1 kg weighs only 1.7 N. Although we will not have occasion to use British units, we note that for practical purposes on the earth, a mass of 1 kg weighs about 2.2 lb. (On the moon, 1 kg weighs about 0.4 lb.)

The force of gravity acts on an object when it is falling. When an object is at rest on the earth, the gravitational force does not disappear, as we know if we weigh the object on a spring scale. The same force, given by Eq. 4–2, continues to act. Why then doesn't the object move? Clearly the net force on an object at rest is zero; there must be another force on the object to balance the gravitational force. This force is supplied by the ground (see Fig. 4–7a). The ground is compressed slightly beneath the object and due to its elasticity it pushes up on the object as shown. The force exerted by the ground is often called a *contact force*, since it occurs when two objects are in contact. (The force of your hand pushing on a cart is also a contact force.) When a contact force acts normal to the common surface of contact ("normal"

FIGURE 4–7

(a) The net force on an object at rest is zero. The downward force of gravity ($\mathbf{F}_g$) is balanced by an upward force (the normal force $\mathbf{F}_N$) exerted by the ground in this case. (b) $\mathbf{F}_{GS}$ (force on the ground exerted by the statue) is the reaction to $\mathbf{F}_{SG}$, as per Newton's third law. The reaction to $\mathbf{F}_{SE}$ is not shown.

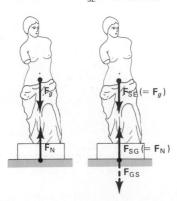

means perpendicular), it is usually referred to as the **normal force**; hence it is labeled F_N in the diagram.

The two forces shown in Fig. 4–7a are both acting on the statue, which remains at rest since the vector sum of these two forces is zero. Although these two forces have equal magnitude and opposite direction, they are *not* the equal and opposite forces spoken of in Newton's third law. This is a critical point that can cause great confusion. The action and reaction forces of Newton's third law act on *different bodies*. (The two shown in Fig. 4–7a act on the same body.) For each of the forces shown in Fig. 4–7a, we can ask, "What is the reaction force?" The upward force, F_N, on the statue is exerted by the ground. The reaction to this force is a force exerted by the statue on the ground. It is shown in Fig. 4–7b, where it is labeled F_{GS} (force on the Ground due to the Statue); the forces on the statue have also been relabeled with dual subscripts to indicate *on* what body the force is and *by* what body it is exerted: F_{SE} ($= F_g$) is the force on the Statue exerted by the Earth's gravity, and F_{SG} ($= F_N$) is the force on the Statue exerted by the Ground. F_{GS} is the reaction to F_{SG} as called for by Newton's third law. (We could equally well say the inverse: that the force F_{SG} that the ground exerts upward on the statue is the reaction to the force F_{GS} exerted by the statue on the ground.) Now, what about the other force on the statue, the force of gravity F_g? Can you guess what the reaction is to this force?[†]

4–8 Applications of Newton's Laws: Vector Forces

Newton's second law tells us that the acceleration of an object is proportional to the *net force* acting on the object; the net force, as mentioned earlier, is the *vector sum* of all forces acting on the object. The fact that forces add together as vectors comes from extensive experiments which show that forces do add together as vectors according to the rules we developed in Chapter 3. For example, in Fig. 4–8 two forces of equal magnitude are shown acting on an object at right angles to each other. Intuitively we can see that the object will move at a 45° angle and thus the net force is at a 45° angle. This is just what the rules of vector addition give. The Pythagorean theorem tells us that the resultant force must have magnitude $F_R = \sqrt{(100\ \text{N})^2 + (100\ \text{N})^2} = 141$ N. That this is the correct answer is not intuitively obvious, although clearly the net force will be less than 200 N (since the two persons are pulling at cross purposes to some extent), and it is certainly greater than zero. Carefully done experiments show that two 100-N forces acting at 90° to one another have the same effect as one force of magnitude 141N acting at 45°, in agreement with our rules for vector addition.

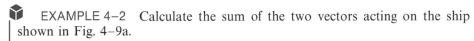

EXAMPLE 4–2 Calculate the sum of the two vectors acting on the ship shown in Fig. 4–9a.

SOLUTION These two vectors are shown resolved in Fig. 4–9b. The components of F_1 are:

$$F_{1x} = F_1 \cos 45° = (40.0\ \text{N})(0.707) = 28.3\ \text{N},$$

$$F_{1y} = F_1 \sin 45° = (40.0\ \text{N})(0.707) = 28.3\ \text{N}.$$

The components of F_2 are:

$$F_{2x} = F_2 \cos 37° = (30.0\ \text{N})(0.799) = 24.0\ \text{N},$$

$$F_{2y} = -F_2 \sin 37° = -(30.0\ \text{N})(0.602) = -18.1\ \text{N}.$$

[†] This is a hard question since we don't discuss the force of gravity in detail until Chapter 5. In fact, it is a force (F_{ES}) exerted on the earth by the statue, which is a gravitational force (like F_{SE}) and can be considered to act at the earth's center as we shall discuss in Chapter 5; we won't be concerned with it in this chapter.

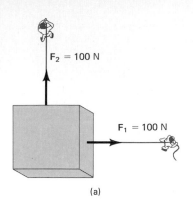

(a) $F_2 = 100$ N
$F_1 = 100$ N

(a)

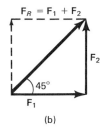

$F_R = F_1 + F_2$
F_2
45°
F_1

(b)

FIGURE 4–8

(a) Two forces, F_1 and F_2, act on an object. (b) The sum, or resultant, of F_1 and F_2 is F_R.

FIGURE 4–9

Two force vectors act on a boat (Example 4–2).

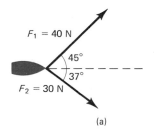

$F_1 = 40$ N
45°
37°
$F_2 = 30$ N

(a)

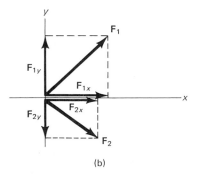

(b)

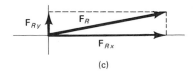

(c)

(a)

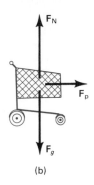

(b)

FIGURE 4–10

Forces on a cart (Example 4–3).

FIGURE 4–11

Example 4–4; (b) is the free body diagram.

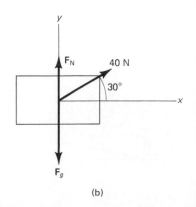

(a)

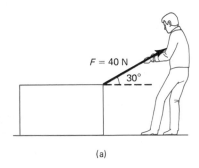

(b)

F_{2y} is negative since it points along the negative y axis. The components of the resultant force are (see Fig. 4–9c):

$$F_{Rx} = 28.3 \text{ N} + 24.0 \text{ N} = 52.3 \text{ N}$$

$$F_{Ry} = 28.3 \text{ N} - 18.1 \text{ N} = 10.2 \text{ N}.$$

To find the magnitude of the resultant force, we use the Pythagorean theorem:

$$F_R = \sqrt{F_{Rx}^2 + F_{Ry}^2} = \sqrt{(52.3)^2 + (10.2)^2} \text{ N} = 53.3 \text{ N}.$$

The only remaining question is the angle θ which the net force $\mathbf{F}_R$ makes with the x axis. We use:

$$\tan \theta = \frac{F_{Ry}}{F_{Rx}} = \frac{10.2 \text{ N}}{52.3 \text{ N}} = 0.195,$$

which corresponds to an angle of 11.0°.

EXAMPLE 4–3 Calculate the force required to accelerate the cart in Fig. 4–10a from rest to 0.50 m/s in 2.0 s. The mass of the cart is 20 kg.

SOLUTION If we ignore friction, there are three forces on the cart as shown in Fig. 4–10b: the forward pushing force exerted by the person, $\mathbf{F}_p$; the downward force of gravity, $\mathbf{F}_g$; and the upward force exerted by the floor, $\mathbf{F}_N$ (which is the reaction to the force of the cart pushing down on the floor). The vertical forces, $\mathbf{F}_g$ and $\mathbf{F}_N$, must add up to zero; if they didn't the cart would accelerate vertically. Thus the net force on the cart is simply $\mathbf{F}_p$. To calculate how large F_p must be, we first calculate the acceleration required: $a = (0.50 \text{ m/s} - 0)/2.0 \text{ s} = 0.25 \text{ m/s}^2$. Then the magnitude of the force must be $F_p = ma = (20 \text{ kg})(0.25 \text{ m/s}^2) = 5.0 \text{ N}$.

This example illustrates several important aspects in the use of Newton's laws. First we draw a sketch of the situation. Next, if there is only one body involved, we show all the forces acting on that body; if several bodies are involved, we make a diagram of each body separately, showing all the forces acting *on that body*; such a diagram is called a **free body diagram**. An example is Fig. 4–11b, which is discussed in Example 4–4. Newton's second law involves vectors, and it is usually important to resolve vectors into components. An x and a y axis should be chosen in a way that simplifies the calculation. Then Newton's second law can be applied to the x and y components separately. That is, the x component of the total force will be related to the x component of the acceleration: $F_x = ma_x$; and similarly for the y direction.

EXAMPLE 4–4 Figure 4–11a shows a 10-kg package being pulled along the floor with a force of 40 N. The force is applied at an angle of 30°. We wish to calculate (*a*) the acceleration of the package and (*b*) the magnitude of the upward force F_N exerted by the floor on the package. Assume friction can be neglected.

SOLUTION Figure 4–11b shows the free body diagram of the package. With the y axis vertical, and x axis horizontal, the pull of 40 N has components:

$$F_x = (40 \text{ N})(\cos 30°) = (40 \text{ N})(0.866) = 35 \text{ N},$$

$$F_y = (40 \text{ N})(\sin 30°) = (40 \text{ N})(0.50) = 20 \text{ N}.$$

(*a*) In the horizontal direction:

$$ma_x = F_x$$

$$a_x = (35 \text{ N})/(10 \text{ kg}) = 3.5 \text{ m/s}^2.$$

The acceleration of the package is thus 3.5 m/s².

(b) In the vertical direction we have:

$$ma_y = F_N - F_g + F_y.$$

Now $F_g = mg = (10\ \text{kg})(9.8\ \text{m/s}^2) = 98\ \text{N}$ and $F_y = 20\ \text{N}$. Furthermore, we know $a_y = 0$ since the package does not move vertically. Then

$$0 = F_N - 98 + 20$$

and

$$F_N = 78\ \text{N}.$$

Notice that in this case F_N is less than F_g; the ground does not push against the full weight of the package since part of the pull exerted by the person is in the upward direction.

We next take an example of projectile motion.

EXAMPLE 4–5 A locust jumps by extending its hind legs against the ground, Fig. 4–12a. Experiments indicate that a 3.0 g locust exerts an average force of about 0.45 N against the floor, at an angle of 57°; during this acceleration period the locust body moves a distance of about 4.0 cm. Calculate (a) the net force **F** acting on the locust, (b) its acceleration, (c) its takeoff velocity, and finally (d) the distance you expect it to travel during the jump, ignoring air resistance.

SOLUTION The first three parts deal with the acceleration period; the last part deals with projectile motion per se. (a) Figure 4–12b shows the two forces acting on the locust: the 0.45 N force exerted by the ground and the weight of the locust $mg = (0.0030\ \text{kg})(9.8\ \text{m/s}^2) = 0.030\ \text{N}$. Using the component method of adding vectors:

$$F_x = (0.45\ \text{N})(\cos 57°) = 0.25\ \text{N},$$

$$F_y = (0.45\ \text{N})(\sin 57°) - 0.030\ \text{N} = 0.35\ \text{N}.$$

Then

$$F = \sqrt{F_x^2 + F_y^2} = \sqrt{(0.25\ \text{N})^2 + (0.35\ \text{N})^2} = 0.43\ \text{N}.$$

To find the angle θ, we note that

$$\frac{F_y}{F_x} = \frac{F \sin \theta}{F \cos \theta} = \tan \theta.$$

Substituting values for F_y and F_x we find $\tan \theta = (0.35\ \text{N})/(0.25\ \text{N}) = 1.40$, so $\theta = 54°$. The net force is 0.43 N and the angle at which it acts, which will be the takeoff angle, is 54°.

(b) The acceleration leading to takeoff is

$$a = \frac{F}{m} = \frac{0.43\ \text{N}}{0.0030\ \text{kg}} = 140\ \text{m/s}^2.$$

(c) The body of the locust accelerates at 140 m/s² over a distance of 0.040 m. Therefore its speed at the end of this period, at which point the legs leave the ground and the locust becomes airborne, is (using Eq. 2–9c with D being the distance moved along the direction of the acceleration):

$$v^2 = v_0^2 + 2aD$$
$$= 0 + (2)(140\ \text{m/s}^2)(0.040\ \text{m}) = 11\ \text{m}^2/\text{s}^2$$
$$v = 3.4\ m/s.$$

(d) In this part we are dealing with projectile motion. We assume there is no air resistance (a poor assumption, as we shall see). The locust leaves the ground

FIGURE 4–12
Locust (Example 4–5).

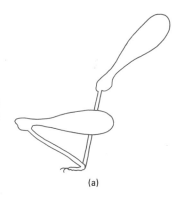

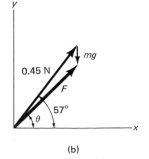

with a speed of 3.4 m/s at an angle of 54°. We proceed as in Example 3–7. First we find the time to travel the full distance (assuming the flight is over level ground):

$$y = v_{y0}t - \tfrac{1}{2}gt^2$$

$$0 = (3.4 \text{ m/s})(\sin 54°)t - (\tfrac{1}{2})(9.8 \text{ m/s}^2)t^2$$

$$t = \frac{2.8 \text{ m/s}}{4.9 \text{ m/s}^2} = 0.57 \text{ s}.$$

Then the distance $x = v_{x0}t = (v_0 \cos 54°)t = (3.4 \text{ m/s})(0.59)(0.57 \text{ s}) = 1.1$ m. In fact, no locust jumps this far (at least if it doesn't use its wings, which we have assumed). Air resistance on this light object is quite considerable and limits its range to about a third of the above value, and its actual path is not a parabola.

4–9   Applications Involving Friction, Inclines

So far we have ignored friction, but it must be taken into account in most practical situations. Friction exists between two solid surfaces because even the smoothest looking surface is quite rough on a microscopic scale. Even when a body rolls across a surface, there is still some friction, called *rolling friction*, although it is generally much less than when one body slides across the surface of another body. We will mainly be concerned with sliding friction in this section, and it is usually called *kinetic friction* (kinetic is Greek for "moving").

When a body is in motion, the force of kinetic friction always acts opposite to the direction of motion. Its magnitude depends on the nature of the two sliding surfaces. For a given surface, it is proportional to the *normal force* between the two surfaces, which is the force that either object exerts on the other, perpendicular to their common surface of contact.[†] It does not depend appreciably on the total surface area of contact; that is, the force of friction on a brick is the same whether it is being slid on its flat face or on its end. We can write the proportionality as an equation by inserting a constant of proportionality, μ_k (Greek letter "mu"):

$$F_{fr} = \mu_k F_N. \qquad \text{[kinetic friction]}$$

This is an approximate (but reasonably accurate and useful) relation. It is not a law; it is a relation between the magnitude of the friction force, F_{fr}, which acts parallel to the two surfaces, and the magnitude of the normal force, F_N, which acts perpendicular to the surfaces. It is not a vector equation since the two forces are perpendicular to one another. The term μ_k is called the *coefficient of kinetic friction* and its value depends on the two surfaces; measured values for a variety of surfaces are given in Table 4–1. These are only approximate however, since μ depends on whether the surfaces are wet or dry, on how much they have been sanded or rubbed and if any burrs remain.

What we have been discussing up to now is *kinetic friction*, when one body slides over another. There is also *static friction*, which refers to a force that resists any attempt to start a body moving. Suppose an object such as a desk is resting on a horizontal floor. If no horizontal force is exerted on the desk, there also is no friction force. But suppose, now, you try to push the desk, but it doesn't move; you are exerting a horizontal force, but the desk isn't moving, so there must be another force on the desk keeping it from moving ($\mathbf{F}_{net} = 0$). This is the force of *static friction* exerted by the floor on the desk. If you push with a greater force without moving the desk, the force of static friction also has increased. If you push hard enough, the desk will finally start to move. At this point, you have exceeded the maximum force of static friction which is given by $F_{fr} = \mu_s F_N$, where μ_s is the *coefficient of static friction* (Table 4–1). Since the force of static friction varies from zero to this

---

† Note that both the normal force and the friction force are forces exerted by one surface on the other; one is perpendicular to the contact surfaces, and the other is parallel.

TABLE 4–1

4–9 Applications Involving **69**
Friction, Inclines

Coefficients of friction[†]

Surfaces	Coefficient of Static Friction, μ_s	Coefficient of Kinetic Friction, μ_k
Wood on wood	0.4	0.2
Wood (waxed) on wet snow	0.14	0.1
Ice on ice	0.1	0.03
Metal on metal (lubricated)	0.15	0.07
Steel on steel (unlubricated)	0.6	0.3
Rubber on solids	1–4	1
Lubricated ball bearings	<0.01	<0.01
Synovial joints (in human limbs)	0.01	0.01

[†] Values are approximate and are intended only as a guide.

maximum value, we can write

$$F_{fr} \leqslant \mu_s F_N. \qquad \text{[static friction]}$$

You may have noticed that it is often easier to keep a heavy object (like a desk) moving than it is to start it moving in the first place. This is a reflection of the fact that μ_s is almost always greater than μ_k. (It can never be less.)

EXAMPLE 4–6 Recalculate the acceleration of the package in Example 4–4 assuming a coefficient of kinetic friction of 0.30. The free body diagram is shown in Fig. 4–13.

SOLUTION The force of kinetic friction always opposes the direction of motion and is parallel to the surface of contact. The calculation for the y direction is just the same as before, Example 4–4. Thus the force with which the floor pushes up on the package, the normal force, is still 78 N. For the x direction we have a new term, the force of friction which equals $\mu_k F_N = (0.30)(78 \text{ N}) = 23 \text{ N}$. Thus

$$ma_x = F_x - \mu_k F_N = 35 \text{ N} - 23 \text{ N} = 12 \text{ N}$$

$$a_x = \frac{12 \text{ N}}{10 \text{ kg}} = 1.2 \text{ m/s}^2.$$

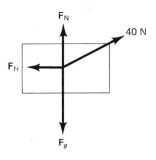

FIGURE 4–13
Free body diagram for Example 4–6.

EXAMPLE 4–7 In Fig. 4–14a two objects are connected by a rope running over a pulley. The coefficient of kinetic friction between object I and the table is 0.20. (We ignore the mass of the rope and pulley and any friction in the pulley.) We

FIGURE 4–14
Example 4–7.

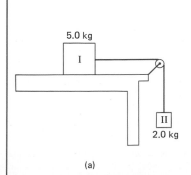

(a)

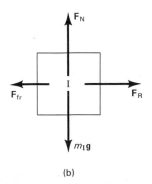

(b)

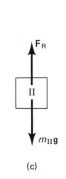

(c)

wish to find the acceleration of the system, which will be the same for both objects assuming the rope doesn't stretch.

   SOLUTION   Free body diagrams are shown for each body in Fig. 4–14b and c. Body I does not move vertically, so the normal force just balances the weight, $F_N = m_I g = (5.0 \text{ kg})(9.8 \text{ m/s}^2) = 49 \text{ N}$. In the horizontal direction there are two forces on body I: F_R, the pull of the rope (whose value we don't know), and the force of friction $= \mu_k F_N = (0.20)(49 \text{ N}) = 9.8 \text{ N}$. The horizontal acceleration is what we wish to find, and using Newton's second law, $F = ma$, we have

$$F_R - F_{fr} = m_I a.$$

Next consider block II. The force of gravity $F_g = m_{II} g = 19.6 \text{ N}$ pulls downward; and the rope pulls upward with a force F_R (which is the tension in the rope); this is the same force the rope exerts on body I since if the rope pulls on body I with a force F_R, then by Newton's third law, body I pulls back on the rope with the same force F_R; this force F_R is transmitted all along the rope to body II, just as if a person were pulling on the end of it.[†] Now we can write Newton's second law for body II:

$$m_{II} g - F_R = m_{II} a$$

We have two unknowns, a and F_R, but we now also have two equations. We solve the first for F_R:

$$F_R = F_{fr} + m_I a,$$

and substitute this into the second equation:

$$m_{II} g - F_{fr} - m_I a = m_{II} a.$$

Now we solve for a and put in numerical values:

$$a = \frac{m_{II} g - F_{fr}}{m_I + m_{II}} = \frac{19.6 \text{ N} - 9.8 \text{ N}}{5.0 \text{ kg} + 2.0 \text{ kg}} = 1.4 \text{ m/s}^2.$$

If we wish, we can calculate F_R using the first equation:

$$F_R = F_{fr} + m_I a = 9.8 \text{ N} + (5.0 \text{ kg})(1.4 \text{ m/s}^2) = 17 \text{ N}.$$

If the rope had had a mass m_R that could not be ignored, then the force at the two ends of the rope would not be the same. The rope would pull on block I with force F_{RI} which is different from the force F_{RII} it exerts at the other end on block II. By Newton's third law, block II pulls back on the rope with a force F_{RII} and block I pulls on the rope with a force F_{RI}. The net force on the rope (considered now as a third "free body") is $F_{RII} - F_{RI}$ and this must equal the mass of the rope times its acceleration:

$$F_{RII} - F_{RI} = m_R a$$

Our two earlier equations become

$$F_{RI} - F_{fr} = m_I a \quad \text{and} \quad m_{II} - F_{RII} = m_{II} a,$$

so we have three equations in three unknowns (a, F_{RI}, and F_{RII}) which can be solved.

FIGURE 4–15
Example 4–8.

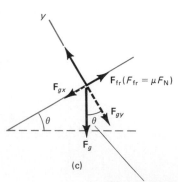

(a)

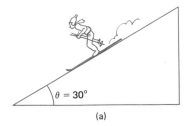

(b)

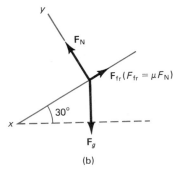

(c)

This angle is θ, the same as the slope angle, because the right sides of these two angles are perpendicular to each other and the left sides are perpendicular to each other.

We now discuss some examples of objects moving on an incline such as a hill or a ramp. Such situations commonly arise; they are interesting because gravity is the accelerating force yet the acceleration is not vertical.

   EXAMPLE 4–8   The skier in Fig. 4–15a has just begun descending the 30° slope. Assuming the coefficient of kinetic friction is 0.10, calculate (a) his acceleration and (b) the speed he will reach after 6.0 s.

   SOLUTION   (a) The free body diagram in Fig. 4–15b indicates all the forces acting on the skier: his weight ($\mathbf{F}_g = m\mathbf{g}$) downward, and the two forces exerted on

   [†] Precisely true only if the pulley is massless and frictionless, as assumed here—see discussion at end of example.

his skis by the snow—the normal force perpendicular to the snow's surface, and the friction force parallel to the surface. For convenience, we choose the x axis parallel to the snow surface with positive direction downhill, and the y axis perpendicular to the surface. Then we only have to resolve one vector into components, the weight. The components of the weight are shown as dashed lines in Fig. 4–15c; they are given by (notice which angle is θ and the reason given on the diagram):

$$F_{gx} = mg \sin \theta,$$

$$F_{gy} = mg \cos \theta.$$

First we apply Newton's second law to the y direction:

$$F_N - mg \cos \theta = ma_y$$
$$= 0,$$

since there is no motion in the y direction. Thus

$$F_N = mg \cos \theta.$$

In the x direction, $F_x = ma_x$ yields:

$$mg \sin \theta - \mu_k F_N = ma_x$$

$$mg \sin \theta - \mu_k mg \cos \theta = ma_x.$$

There is an m in each term, so they can be cancelled out. Thus:

$$a_x = g \sin 30° - \mu_k g \cos 30°$$
$$= [0.50 - (0.10)(0.866)]g$$
$$= 0.41 \, g.$$

The acceleration is 0.41 times the acceleration of gravity, which in numbers is $a = (0.41)(9.8 \text{ m/s}^2) = 4.0 \text{ m/s}^2$. It is interesting that the mass canceled out and so the acceleration doesn't depend on the mass. The fact that such a cancellation sometimes occurs, and thus saves calculation, is one advantage of working with the algebraic equations and putting in the numbers only at the end.

(b) The speed after 6.0 s is found by using Eq. 2–9a:

$$v = v_0 + at = 0 + (4.0 \text{ m/s}^2)(6.0 \text{ s}) = 24 \text{ m/s},$$

where we assumed a start from rest.

EXAMPLE 4–9   Again consider a skier descending a hill, but this time we don't know the coefficient of kinetic friction between the skis and the snow and it is this we want to determine. This can be done by observing the skier descend slopes of different angles, and noticing at what angle he descends at constant speed. The $F = ma$ equations for the x and y components will be the same as above except that $a_x = 0$ in this case, and the angle θ will not necessarily be 30°. Thus:

$$F_N - mg \cos \theta = ma_y = 0$$

$$mg \sin \theta - \mu_k F_N = ma_x = 0.$$

From the first equation we have $F_N = mg \cos \theta$; we substitute this into the second equation:

$$mg \sin \theta - \mu_k(mg \cos \theta) = 0.$$

Now we solve for μ_k:

$$\mu_k = \frac{mg \sin \theta}{mg \cos \theta} = \frac{\sin \theta}{\cos \theta} = \tan \theta,$$

where θ is the angle at which the skier moves at constant speed. For example if constant speed occurs for an angle $\theta = 5°$, then $\mu_k = \tan 5° = 0.09$.

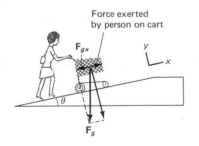

Force exerted
by person on cart

F_{gx}

y

x

θ

F_g

FIGURE 4–16
Example 4–10.

 EXAMPLE 4–10 In the design of a supermarket, there are to be several ramps connecting different parts of the store. Customers will have to push grocery carts up the ramps and it is obviously desired that this not be too difficult. The engineer has done a survey and found that almost no one complains if the force required is no more than 20 N. Ignoring friction, at what maximum angle θ should the ramps be built assuming a full 20-kg grocery cart.

SOLUTION As shown in Fig. 4–16 the upward force on the cart must just balance the x component of the weight. For the case of the maximum force of 20 N, we have

$$F_{gx} = F_g \sin \theta = 20 \text{ N}.$$

Since $F_g = mg = (20 \text{ kg})(9.8 \text{ m/s}^2) = 200$ N, we have $\sin \theta = 20/200 = 0.10$; so $\theta = 5.7°$. In a real situation friction would have to be taken into account, particularly friction in the wheels of the carts when they get old, and this would probably be done by experiment.

Friction can be a hindrance. It slows down moving objects and causes heating and binding of moving parts in machinery. Friction can be reduced by using lubricants such as oil. More effective in reducing friction between two surfaces is to maintain a layer of air or other gas between them. Devices using this concept, which is not practical for most situations, include air tracks and air tables (or games) in which the layer of air is maintained by forcing air through many tiny holes. Another technique to maintain the air layer is to suspend objects in air using magnetic fields; this is finding applications for certain types of trains. On the other hand, friction can be helpful. Our ability to walk depends on friction between the soles of our shoes (or feet) and the ground. (Does walking involve static friction or kinetic friction?) The movement of a car, and also its stability, depend on friction. When friction is low, such as on ice, safe walking or driving become difficult.

4–10   Notes on Problem Solving

Solving problems, except in the simplest cases, cannot always be done by a set procedure. Instead, it often requires creativity, for each problem is different. Nonetheless, we can outline a general approach that will be of some help in solving problems.

1. Read written problems carefully. A common error is to leave out a word or two when reading, which can completely change the sense of a problem.

2. Draw a picture or diagram of the situation. (This is probably the most overlooked, yet the most crucial, part of solving a problem.) Use arrows to represent vectors such as velocity or force; use a separate diagram for different types of vectors: say, one for force and one for velocity, if both types of vector are involved. Make sure to include all forces on a given body, and make clear what forces act on what body (or you can make an error in determining the *net force* on a particular body). A diagram showing all the forces acting on a given body (and only on that body) is called a *free body diagram* as discussed in Section 4–8.

3. Determine what the unknowns are—that is, what you are trying to determine.

4. Decide what you need in order to find the unknowns. (*a*) It may help to see if there is one or more relationships (or equations) that relate the unknowns to the knowns; but be sure the relationship(s) is applicable in the given case. Beware of formulas that are not general but apply only in a specific case. (It is dangerous for this reason to thumb through a chapter looking for an equation that will work.) It is very important to know the limitations of each formula or relationship—when it is valid and when not. In this book, the more general equations have been given numbers, but even these can have a

limited range of validity. The unnumbered equations often are valid only for very specific cases. (*b*) Also it is helpful to determine what information is relevant and what is irrelevant in a given situation or problem.

5. Solve the problem, which may include algebraic manipulation of equations and/or numerical calculations. Be sure to keep track of units, for this can serve as a check.

6. Ask yourself: *Is the answer reasonable?* Use your common sense here. Also valuable is to make an *order of magnitude estimate*, a technique described in Section 1–7; this can help to reduce decimal point errors, among other things. Also, dimensional analysis, described in Section 1–6, can help for solving certain problems.

Summary

Newton's *three laws of motion* are the basic classical laws explaining motion. Newton's *first law* states that if the net force on an object is zero, an object originally at rest remains at rest, and an object in motion remains in motion in a straight line with constant velocity. The tendency of a body to resist a change in its motion is called *inertia*. *Mass* is a measure of the inertia of a body; *weight* refers to the force of gravity on a body, and is equal to the product of the body's mass m and the acceleration of gravity g ($F_g = mg$).

Newton's *second law of motion* states that the acceleration of a body is directly proportional to the net force acting on it, and inversely proportional to its mass. In equation form:

$$\mathbf{F} = m\mathbf{a}.$$

Newton's second law is one of the most important and fundamental laws in classical physics. *Force*, which is a vector, can be considered as a push or pull; or, from Newton's second law, force can be defined as an action capable of giving rise to acceleration. *Net* force on an object refers to the vector sum of all forces acting on it.

Newton's *third law of motion* states that whenever one body exerts a force on a second body, the second body always exerts a force on the first body which is equal in magnitude but opposite in direction.

A consistent set of units must be used when making calculations. The SI is mainly used for scientific work.

When two bodies slide over one another, the force of friction which each exerts on the other can be written approximately as $F_{fr} = \mu F_N$ where F_N is the normal force (the force each body exerts on the other perpendicular to their contact surface), and μ is the coefficient of kinetic friction if there is relative motion of the bodies; if the bodies are at rest, μ is the coefficient of static friction and F_{fr} is the maximum friction force before motion starts.

Questions

1. Compare the effort (or force) needed to lift a 10-kg object when you are on the moon as compared to on earth. Compare the force needed to throw a 2-kg object horizontally with a given speed when on the moon as compared to on earth.

2. Observe the motion of the moving discs of an air hockey game. Explain how Newton's first, second, and third laws apply. The discs float on a layer of air, ejected through tiny holes, so friction is reduced to a very small amount.

3. A stone hangs by a fine thread from the ceiling and a section of the same thread dangles from the bottom of the stone. If a person gives a sharp pull on the dangling thread, where is the thread likely to break; below the stone or above it? What if the person gives a slow and steady pull? Explain your answers.

4. Why does a child in a wagon fall backward when you give the wagon a sharp pull?

5. Whiplash sometimes results from an automobile accident when the victim's car is struck violently from the rear. Explain why the head of the victim seems to be thrown backward in this situation. Is it really?

6. When a golf ball is dropped to the pavement, it bounces back up. (*a*) Is a force needed to make it bounce back up? (*b*) If so, what exerts the force?

7. Examine, in the light of Newton's first and second laws, the motion of your leg during one stride while walking.

8. A person wearing a cast on an arm or a leg experiences extra fatigue. Explain this on the basis of Newton's first and second laws.

9. If the acceleration of a body is zero, are no forces acting on it?

10. Only one force acts on an object. Can the object have zero acceleration? Can it have zero velocity?

11. Why do you push harder on the pedals of a bicycle when first starting out than when moving at constant speed?

12. What is your mass (in kg) and weight (in N)?

13. Must a force be exerted on an object to move it in a curved path? Discuss for (*a*) constant speed, (*b*) variable speed.

14. When you are running and want to stop quickly, you must decelerate quickly. (*a*) What is the origin of the force that causes you to stop? (*b*) Estimate (using your own ex-

perience) the maximum rate of deceleration of a person running at top speed to come to rest.

15. A certain horse says to himself, "There's no use in trying to pull that cart. No matter how hard I pull on the cart, it will pull back on me with exactly the same force. So I can never make it go." Explain what is wrong with the horse's reasoning. Free body diagrams, showing all forces involved, will help.

16. Why, when you walk on a log floating in water, does the log move in the opposite direction?

17. Why does it hurt your foot when you kick a football?

18. When you stand still on the ground, how large a force does the ground exert on you? Why doesn't this force make you rise up into the air?

19. The force of gravity on a 2-kg rock is twice as great as that on a 1-kg rock. Why then doesn't the heavier rock fall faster?

20. A person exerts an upward force of 40 N to hold a bag of groceries. Describe the "reaction" force (Newton's third law) by stating: (a) its magnitude, (b) its direction, (c) on what body it is exerted, and (d) by what body it is exerted.

21. According to Newton's third law, each team in a tug-of-war pulls with equal force on the other team. What, then, determines which team will win?

22. Can a coefficient of friction exceed 1.0?

23. Devise a method for measuring the coefficient of friction between two surfaces using an inclined plane.

 Problems _____

SECTION 4–4

1. (I) How much tension must a rope withstand if it is used to accelerate a 1500-kg car at 0.650 m/s^2? Ignore friction.

2. (I) What force is required to stop a 1000-kg car in 5.0 s if it is traveling at 90 km/h?

3. (I) According to a simplified model of a mammalian heart, at each pulse approximately 20 g of blood is accelerated from 0.25 m/s to 0.35 m/s during a period of 0.10 s. What is the magnitude of the force exerted by the heart muscle?

4. (I) A net force of 30.8 N accelerates an object at 8.8 m/s^2. What is the mass of the object?

5. (I) How much force is required to accelerate a 4.0-g object at 10,000 "g's" (say, in a centrifuge)?

6. (II) A 0.085-g spider is descending on a strand which supports it with a force of 4.8×10^{-4} N. What is the acceleration of the spider? Ignore air resistance.

7. (II) A person has a reasonable chance of surviving an automobile crash if the deceleration is no more than 30 "g's." Calculate the force on a 70-kg person accelerating at this rate. What distance is traveled if brought to rest at this rate from 80 km/h?

8. (II) A 0.145-kg baseball traveling 35.0 m/s strikes the catcher's mitt which, in bringing the ball to rest, recoils backward 11.0 cm. What was the average force applied by the ball on the glove?

9. (II) What is the average force exerted by a shotputter on a 7.0-kg shot it the shot is moved through a distance of 2.9 m and is released with a speed of 13 m/s?

10. (II) An elevator (mass 4750 kg) is to be designed so that the maximum acceleration is 0.0650 g. What are the maximum and minimum forces the motor should exert on the supporting cable?

11. (II) A 38-kg child wants to escape from a third-story window to avoid punishment. Unfortunately, a makeshift rope made of sheets can support a mass of only 31 kg. How can the child use this "rope" to escape? Give quantitative answer.

12. (II) What is the acceleration of a freely falling 60-kg skydiver if air resistance exerts a force of 250 N?

13. (II) A person jumps from a tower 4.7 m high. When he strikes the ground below, he bends his knees so that his torso decelerates approximately over a distance of 0.70 meters. If the mass of his torso (excluding legs) is 48 kg, find: (a) his velocity just before his feet strike the ground, and (b) the force exerted on his torso by his legs during the deceleration.

14. (II) An exceptional standing jump would raise a person 0.80 m off the ground. To do this, what force must a 75-kg person exert against the ground? Assume the person lowers himself 0.20 m prior to jumping.

15. (II) A 3.0-kg purse is dropped 55 m from the top of the Leaning Tower of Pisa and reaches the ground with a speed of 29 m/s. What was the average force of air resistance?

16. (II) A typical home-run ball will leave the bat at a speed of 80 m/s after a contact time with the bat of perhaps 5.0×10^{-4} s. If the baseball has a mass of 0.14 kg, what was the force (assume constant) exerted by the bat on the ball? Is this force large enough to lift an average person?

SECTION 4–7

17. (I) What is the weight of a 75-kg astronaut (a) on earth, (b) on the moon ($g = 1.7$ m/s^2), (c) on Venus ($g = 8.7$ m/s^2), (d) in outer space traveling with constant velocity.

SECTION 4–8

18. (I) A 500-N force acts in a northwesterly direction. A second 500-N force must be exerted in what direction so that the resultant of the two vectors points westward?

19. (II) A man pushes a lawnmower at constant speed with a force of 90 N directed along the handle, which is at an angle of 30° to the horizontal. Calculate: (a) the horizontal retarding force on the mower (whose mass is 18 kg), (b) the normal force, and (c) the force the man must exert on the lawnmower to accelerate it from rest to 4.0 km/h in 2.5 seconds.

20. (II) The 100-m dash can be run by the best sprinters in 10.0 s. (a) What is the horizontal component of force exerted on a 75 kg sprinter's feet by the ground during acceleration, which we assume is done uniformly over the first 10.0 m? (b) What is the average speed of the sprinter over the last 90 m of the race?

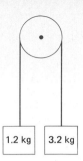

FIGURE 4–17

21. (II) The apparatus shown in Fig. 4–17 in which two masses are suspended over a pulley, is known as *Atwood's machine*. Assuming the pulley is massless and frictionless, calculate: (a) the acceleration of the system, and (b) the tension in the cord.

22. (II) At the instant a race began a 60-kg sprinter was found to exert a force of 950 N on the starting block at a 20° angle with respect to the ground. (a) What was the horizontal acceleration of the sprinter? (b) If the force was exerted for 0.32 s, with what speed did the sprinter leave the starting block?

23. (II) One paint bucket weighing 20 N is hanging by a massless rope from another paint bucket (with a handle on the bottom) also weighing 20 N, and the two are being pulled upward with an acceleration of 1.5 m/s² by another massless rope attached to the upper bucket. Calculate the tension in each rope.

24. (II) A 5000-kg helicopter accelerates upwards at 0.55 m/s² while lifting a 1500-kg car. (a) What is the lift force exerted by the air on the propellers? (b) What is the tension in the cable (ignore its mass) that connects car to helicopter?

25. (III) The two masses shown in Fig. 4–17 are each initially 1.60 m above the ground and the pulley is 4.8 m above the ground. What maximum height does the lighter object reach after the system is released?

26. (III) A train locomotive is pulling two cars of the same mass behind it. Show that the tension in the coupling between the locomotive and the first car is twice that between the first car and the second car, for any nonzero acceleration of the train.

27. (III) A heavy steel cable of length L and mass M passes over a small massless, frictionless pulley. (a) If a length y hangs on one side of the pulley (so $L - y$ hangs on the other side), calculate the acceleration of the cable as a function of y. (b) Assuming the cable starts from rest with length y_0 on one side of pulley, determine the velocity v_f at the moment the whole cable has fallen from the pulley and (c) evaluate v_f for $y_0 = \frac{2}{3} L$.

28. (III) A mass m hung by a string from the ceiling of an accelerating vehicle (train, car, airplane) can act as an accelerometer. (a) Derive the relationship between the acceleration a of the vehicle and the angle θ which the string makes with the vertical. Assume the vehicle is near the surface of the earth. (b) If $\theta = 18°$ for a car that decelerates to rest in 5.5 s, what must have been its initial speed?

SECTION 4–9

29. (I) A force of 270 N is required to start a 40-kg box moving across a concrete floor. What is the coefficient of static friction between the box and the floor?

30. (I) If the coefficient of kinetic friction between a 35-kg crate and the floor is 0.41, how much force is required to move the crate at a steady speed across the floor? How much force is required if μ_k is zero?

31. (II) An 8.0-kg box is released on a 30° incline and accelerates down the incline at 0.30 m/s². Find the friction force impeding its motion. How large is the coefficient of friction?

32. (II) Suppose you are standing on a train accelerating at 0.42 g. What minimum coefficient of friction must exist between your feet and the floor if you are not to slide?

33. (II) A car can decelerate at -5.80 m/s² without skidding when coming to rest on a level road. What would its deceleration be if the road were inclined at 12° uphill? Assume the same friction force.

34. (II) What is the maximum acceleration a car can undergo if the coefficient of static friction between the tires and the ground is 0.55?

35. (II) Referring to Fig. 4–14, how large a mass must body I have to prevent any motion from occurring? Assume $\mu_s = 0.20$.

36. (II) A wet bar of soap slides freely down a ramp 12 m long inclined at 8.8°. How long does it take to reach the bottom? Neglect friction.

37. (II) A box is given a push so that it slides across the floor. How far will it go, given that the coefficient of kinetic friction is 0.30 and the push imparts an initial speed of 3.0 m/s?

38. (II) Determine a formula for the acceleration of the system shown in Fig. 4–14 in terms of m_I, m_{II}, and the mass of the rope, m_R. Define any other variables needed.

39. (II) A roller coaster reaches the top of the steepest hill with a speed of 5.0 km/h. It then descends the hill which is at an average angle of 45° and is 50 m long. What will its speed be when it reaches the bottom? Neglect friction.

40. (II) A 1000-kg car pulls a 450-kg trailer. The car exerts a horizontal force of 3.5×10^3 N against the ground in order to accelerate. What force does the car exert on the trailer? Assume a friction coefficient of 0.45.

41. (II) (a) Show that the minimum stopping distance for an automobile traveling at speed v is equal to $(v^2)/(2\mu g)$ where μ is the coefficient of static friction between the tires and the road, and g is the acceleration of gravity. (b) What is this distance for a 1500-kg car traveling 90 km/h if $\mu = 0.85$? (c) What would it be if the car were on the moon but all else stayed the same?

42. (II) A car starts rolling down a 1-in-4 hill (1-in-4 means that for each 4 m traveled, the elevation change is 1 m). How fast is it going when it reaches the bottom after traveling 50 m? Ignore friction.

43. (II) Repeat problem 42 assuming an effective coefficient of friction equal to 0.10.

44. (II) Two crates, of mass 95 kg and 125 kg, are in contact and at rest on a horizontal surface. A 650-N force is exerted on the 95-kg crate. If the coefficient of kinetic friction is 0.25, calculate: (a) the acceleration of the system, and (b) the force that each crate exerts on the other.

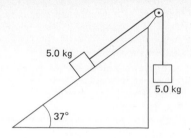

5.0 kg

5.0 kg

37°

FIGURE 4–18

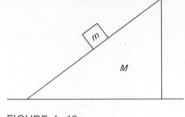

m

M

FIGURE 4–19

45. (II) A flatbed truck is carrying a 3200-kg crate of heavy machinery. If the coefficient of static friction between the crate and the bed of the truck is 0.55, what is the maximum rate the driver can decelerate when coming to a stop in order to avoid crushing the cab with the crate?

46. (II) If a bicyclist of mass 70 kg (including the bicycle) can coast down a 6.2° hill at a steady speed of 7.0 km/h, how much (average) force must be applied to climb the hill at the same speed?

47. (II) A motorcyclist is coasting with the engine off at a steady speed of 12 m/s but enters a sandy stretch where the coefficient of kinetic friction is 0.80. Will the cyclist emerge from the sandy stretch without having to start the engine if the sand lasts for 15 m? If so, what will be the speed upon emerging?

48. (II) An engineer is working on the redesign of a hilly portion of a city. An important consideration is how steep the roads can be so that even small cars can get up the hills without slowing down. It is given that a particular small car, of mass 1200 kg when fully loaded, can accelerate on the level from rest to 14 m/s (50 km/h) in 7.2 s. Using this figure, calculate the maximum steepness of a hill.

49. (II) What is the acceleration of the system shown in Fig. 4–18 if the coefficient of kinetic friction is 0.10?

50. (III) A 4.0-kg block is stacked on top of a 12.0-kg block, which is accelerating along a horizontal table at $a = 5.2 \text{ m/s}^2$. (a) What minimum coefficient of friction μ between the two blocks will prevent the 4.0-kg block from sliding off? (b) If μ is only half this minimum value, what is the acceleration of the 4.0-kg block with respect to the table, and (c) with respect to the 12.0-kg block? (d) What is the force that must be applied to the 12.0-kg block in (a) and in (b) assuming the table is frictionless.

51. (III) A bicyclist can coast down a 4.5° hill at a steady 8.5 km/h. If the force of friction (including air resistance) is proportional to the speed v so that $F_{fr} = cv$, calculate: (a) the value of the constant c, and (b) the average force that must be applied in order to descend the hill at 25 km/h. The mass of the cyclist plus bicycle is 80 kg.

52. (III) A small block of mass m rests on the sloping side of a triangular block of mass M which itself rests on a horizontal table as shown in Fig. 4–19. (a) Assuming all surfaces are frictionless, determine the acceleration of each block, and—assuming they start from rest—describe their motion. (b) What force must be applied to M so that m remains in a fixed position relative to M (that is, m doesn't move on the incline)?

53. (III) The force of air resistance, F_a, acting on a body moving with velocity v is approximately proportional to v, if v is not too large. We can therefore write $F_a = -kv$, where k is a constant. (a) Why is there a minus sign in this relation? (b) Show that Newton's second law for a body falling vertically can be written as

$$m\frac{dv}{dt} = mg - kv.$$

(c) Show that the terminal velocity is given by $v_t = mg/k$ and describe what is meant by terminal velocity. (d) Show that the velocity v at time t for a body that starts falling from rest at $t = 0$, is given by

$$v = v_t(1 - e^{-kt/m}).$$

(e) Graph a vs. t, v vs. t, y vs. t in SI units for $k = 0.30$ m and $m = 6.8$ kg. [Hint for (d): this can be shown by either integrating the equation in (b) above after changing variables to $\mu = mg/k - v$, or by substituting this relation for v into the equation in (b).]

NUMERICAL/PROGRAMMABLE CALCULATOR†

*54. (III) The force of air resistance on a rapidly falling body has the form $F = -kv^2$, so that Newton's second law applied to such an object is

$$m\frac{dv}{dt} = mg - kv^2$$

when the downward direction is considered positive. Use the numerical integration method discussed in Section 2–10 to estimate (within 2 percent) the speed and position at 1.0-s intervals, up to 15.0 s, of a 75-kg person (skydiver) who starts from rest, assuming k = 0.22 kg/m. (Hint: Use of a programmable calculator is highly advised.) Also, show that the body eventually reaches a steady speed, called the *terminal speed*, and explain why this happens.

*55. (III) Assume a net force $F = -mg - kv^2$ acts during the upward vertical motion of a 250-kg rocket, starting at the moment ($t = 0$) when the fuel has burned out and the rocket has an upward speed of 120 m/s. Let k = 0.65 kg/m. Estimate v and y at 1.0-s intervals for the upward motion only, and estimate the maximum height reached. Compare to free flight conditions without air resistance (k = 0).

† See Section 2–10.

Dynamics of Circular Motion; Gravitation and Newton's Synthesis

5

In Sections 3–9 and 3–10 we discussed the kinematics of a particle moving in a circular path. In this chapter we study the dynamics of circular motion by applying Newton's laws of motion. We will also discuss how Newton conceived of another great law by applying the concepts of circular motion to the motion of the moon and the planets. This new law was his law of universal gravitation, which was the capstone of Newton's analysis of the physical world. Indeed, Newtonian mechanics with its three laws of motion and the law of universal gravitation, was accepted for centuries (until the early 1900s) as the way the universe works.

 ## 5–1 Dynamics of Circular Motion

We saw in Section 3–9 that a particle revolving in a circle of radius r with uniform speed v is undergoing an acceleration

$$a_c = \frac{v^2}{r}.$$

This acceleration, a_c, is called centripetal acceleration because it is directed toward the center of the circle. Thus, although the magnitude of the acceleration in uniform circular motion is constant, its direction is continually changing. Hence the acceleration vector $\mathbf{a}_c$ must be considered as a variable acceleration. The direction of $\mathbf{a}_c$, in uniform circular motion, is always perpendicular to the velocity $\mathbf{v}$.

An object moving uniformly in a circle, such as a ball on the end of a string, must have a force applied to it to keep it in that circle; that is, a force is necessary to give it a centripetal acceleration. The magnitude of the required force can be calculated using Newton's second law, $F = ma$, where we use the value of the centripetal acceleration, $a_c = v^2/r$, and F must be the total (or net) force:

$$F = ma_c = m\frac{v^2}{r}.$$

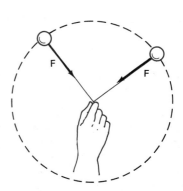

FIGURE 5–1
A force is required to keep an object moving in a circle. If the speed is constant, the force is directed toward the center of the circle.

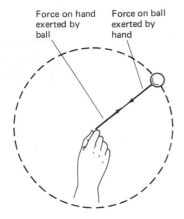

Force on hand exerted by ball

Force on ball exerted by hand

FIGURE 5–2

Swinging a ball on the end of a string. Your hand pulls in on the ball, and the ball pulls outward on your hand (Newton's third law).

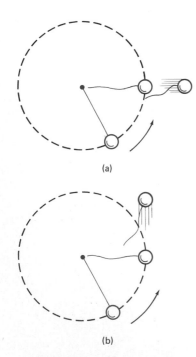

(a)

(b)

FIGURE 5–3

If centrifugal force existed, the ball would fly off as in (a) when released. In fact, it flies off as in (b).

Since a_c is directed toward the center of the circle at any moment, the force too must be directed toward the center of the circle. That a force is necessary can be seen intuitively from the fact that if no force were exerted on the object, it would not move in a circle, but in a straight line as Newton's first law tells us. To pull an object out of its "natural" straight-line path, a force to the side is necessary; in the case of circular motion, this sideways force must be toward the circle's center (see Fig. 5–1). This force is sometimes called a "centripetal force," but be aware that this term does not indicate some new kind of force. The term merely specifies that the force is directed toward the circle's center. The force must be applied by some object. For example when a person swings a ball on the end of string in a circle, the person (via the string) exerts the force on the ball.

There is a common misconception that an object moving in a circle has an outward force acting on it, a so-called centrifugal ("center-fleeing") force. Consider, for example, a person swinging a ball on the end of a string about his head (Fig. 5–2). If you have ever done this yourself, you know that you feel a force pulling outward on your hand. The misconception arises when this pull is interpreted as an outward "centrifugal" force pulling out on the ball which is transmitted along the string to the hand. But this is not what is happening at all. To keep the ball moving in a circle, the person pulls inwardly on the ball. The ball, then, exerts an equal and opposite force on the hand (Newton's third law), and *this* is the force your hand feels. The force *on the ball* is the one exerted *inwardly* on it by the person. For even more convincing evidence that a "centrifugal force" does not act on the ball, consider what happens when you let go of the string. If a centrifugal force were acting, the ball would fly outward, as shown in Fig. 5–3a. But it doesn't; it flies off tangentially (Fig. 5–3b) in the direction of the velocity it had at the moment it was released; this happens because the inward force no longer acts.

An example of centripetal acceleration occurs when a fast-moving automobile rounds a curve. In such a situation you may feel that you are thrust outward. This is not some mysterious centrifugal force pulling on you. What is happening is that you tend to move in a straight line whereas the car curves "in front of you," so to speak. To make you go in the curved path, the back of the seat, or the door of the car, exerts a force on you (Fig. 5–4). The car itself must have an inward force on it if it is to move in a curve. On a flat road this force is the friction force between the tires and the pavement. If the friction force is not great enough, as under icy conditions, sufficient force cannot be supplied and the car will skid out of a circular path into a more nearly straight path.

EXAMPLE 5–1 A 1000-kg car rounds a curve on a flat road of radius 50 m at a speed of 50 km/h (14 m/s). Will the car make the turn if (a) the pavement is dry and the coefficient of static friction is 0.60, (b) the pavement is icy and $\mu_s = 0.20$?

SOLUTION First we calculate the net force F required to accelerate the car around the curve:

$$F = m\frac{v^2}{r} = \frac{(1000\ \text{kg})(14\ \text{m/s})^2}{(50\ \text{m})} = 3900\ \text{N}.$$

The normal force, F_N, on the car is equal to the weight since the road is flat: $F_N = mg = (1000\ \text{kg})(9.8\ \text{m/s}^2) = 9800\ \text{N}$. For (a), $\mu_s = 0.60$ and the maximum friction force attainable is:

$$F_{fr} = \mu_s F_N = (0.60)(9800\ \text{N}) = 5900\ \text{N}.$$

Since a force of only 3900 N is needed, and that is how much will in fact be exerted, the car can make the turn. But in (b) the maximum friction force possible is

$$F_{fr} = \mu_s F_N = (0.20)(9800\ \text{N}) = 2000\ \text{N}.$$

The car will skid.

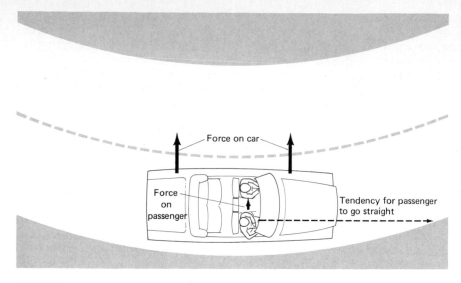

FIGURE 5–4
The road exerts an inward force on a car to make it move in a circle; and the
car exerts an inward force on the passenger.

The situation is worse if the wheels lock when the brakes are applied. When the
tires are rolling, the bottom of the tire is at rest (for that moment) against the road, so
static friction exists. But if the wheels lock, the tires slide and the friction force, which
is now kinetic friction, is less. When the road is wet or icy, locking of the wheels
occurs with less force on the brake pedal since there is less resistance from road
friction.

The banking of curves can reduce the chance of skidding because the normal
force of the road will have a component toward the center of the circle (Fig. 5–5),
thus reducing the need for friction. For a given angle of banking, there will be one
speed for which no friction is required at all. This will be the case when the horizontal
component of the normal force, $F_N \sin \theta$, is just equal to the force required to give a
vehicle its centripetal acceleration; that is

$$F_N \sin \theta = m \frac{v^2}{r}.$$

The banking angle of a road, θ, is chosen so that this condition holds for a particular
speed, called the "design speed."

EXAMPLE 5–2 (a) For a car traveling with speed v around a curve of radius r,
determine a formula for the angle at which a road should be banked so that no
friction is required. (b) What is this angle for a curve of radius 50 m at a design
speed of 50 km/h?

SOLUTION   (a) From the equation above, $F_N \sin \theta = mv^2/r$. In the vertical
direction, the forces are $F_N \cos \theta$ upward (Fig. 5–5) and the weight of the car (mg)
downward. Since there is no vertical motion, this component of the acceleration is
zero so Newton's second law tells us

$$F_N \cos \theta - mg = 0.$$

Therefore

$$F_N = \frac{mg}{\cos \theta},$$

FIGURE 5–5
Normal force on a car rounding
a banked curve, resolved into its
horizontal and vertical components.

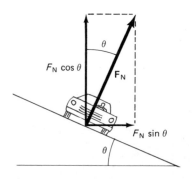

and substituting this into the equation above for the vertical motion we have

$$\frac{mv^2}{r} = F_N \sin \theta$$

$$= mg \frac{\sin \theta}{\cos \theta} = mg \tan \theta,$$

or

$$\tan \theta = \frac{v^2}{rg}.$$

(b) For $r = 50$ m and $v = 50$ km/h (or 14 m/s), $\tan \theta = (14 \text{ m/s})^2/(50 \text{ m})(9.8 \text{ m/s}^2)$ $= 0.40$, so $\theta = 22°$.

A useful device that very nicely illuminates the dynamical aspects of circular motion is the centrifuge, or the very high-speed ultracentrifuge. It is used to sediment materials quickly or to separate materials with slightly different characteristics. Test tubes or other containers are held in the centrifuge rotor which is accelerated to very high rotational speeds. This is shown in Fig. 5–6 where one test tube is shown in two different positions as the rotor turns. The small circle represents a small particle, perhaps a macromolecule, in a fluid-filled test tube. When the tube is at position A and the rotor is turning, the particle has a tendency to move in a straight line in the direction of the dashed arrow. The fluid resists[†] the motion of the particles, thus exerting the centripetal force necessary to keep the particles moving in a circle. Usually the resistance of the fluid (which could be a liquid or a gas, depending on the application) does not quite equal the required centripetal force and the particles eventually reach the bottom of the tube, and then the bottom of the tube exerts the force necessary to keep the particles moving in a circle. In fact the bottom of the tube must exert a force on the whole tube of fluid, keeping it moving in a circle. If the tube is not strong enough to exert this force, it will break.

The kinds of materials placed in a centrifuge are those which do not sediment or separate quickly under the action of gravity. The point of a centrifuge is to provide an "effective gravity" much larger than normal gravity because of the high rotational speeds, and the particles move down the tube more rapidly.

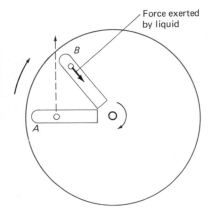

Force exerted
by liquid

B

A

FIGURE 5–6

Rotating test tube in a centrifuge.

🔳 EXAMPLE 5–3 An ultracentrifuge rotor rotates at 50,000 rpm (revolutions per minute). The top of a test-tube (perpendicular to the rotation axis) is 6.00 cm, and the bottom of the tube is 10.0 cm, from the axis of rotation. (a) Calculate the centripetal acceleration in "g's." (b) If the contents of the tube have a total mass of 12.0 g, what force must the bottom of the tube withstand?

SOLUTION (a) At the top of the tube, a particle revolves in a circle of circumference $2\pi r = 2(3.14)(0.0600 \text{ m}) = 0.377$ m. It makes 5.00×10^4 such revolutions each minute, and so its speed is

$$v = \frac{(0.377 \text{ m/rev})(5.00 \times 10^4 \text{ rev/min})}{60.0 \text{ s/min}} = 3.14 \times 10^2 \text{ m/s}.$$

The centripetal acceleration is then

$$a_c = \frac{v^2}{r} = \frac{(3.14 \times 10^2 \text{ m/s})^2}{(0.0600 \text{ m})} = 1.64 \times 10^6 \text{ m/s}^2$$

which, dividing by $g = 9.80$ m/s², is 1.67×10^5 g's. At the bottom of the tube $(r = 0.100 \text{ m})$, the speed is $v = (2\pi r)(5.00 \times 10^4)/(60.0 \text{ s}) = (0.628)(5.00 \times$

_____
[†] A type of friction, like air resistance.

$10^4)/(60.0 \text{ s}) = 5.24 \times 10^2 \text{ m/s}$ and $a_c = v^2/r = 2.74 \times 10^6 \text{ m/s}^2$ or 2.79×10^5 g's.

(b) Since the acceleration varies with distance from the axis, we estimate the force using the average acceleration $= (1.64 \times 10^6 \text{ m/s}^2 + 2.74 \times 10^6 \text{ m/s}^2)/2 = 2.19 \times 10^6 \text{ m/s}^2$. Then $F = ma = (0.0120 \text{ kg})(2.19 \times 10^6 \text{ m/s}^2) = 2.63 \times 10^4 \text{ N}$, which is equivalent to the weight of a 2680 kg mass ($m = F/g = 2.63 \times 10^4 \text{ N}/9.80 \text{ m/s}^2 = 2.68 \times 10^3 \text{ kg}$) or almost 3 tons.

We next consider an example of a different type.

EXAMPLE 5–4 A particle of mass m, suspended by a string of length L, revolves in a circle of radius $r = L \sin \theta$ where θ is the angle the string makes with the vertical (Fig. 5–7). Calculate the speed and period (time required for one revolution) of such a *conical pendulum* in terms of L, θ, and m.

SOLUTION The forces acting on the mass m are its weight $\mathbf{F}_g$ (of magnitude $F_g = mg$) and the tension $\mathbf{F}$ in the string which has horizontal and vertical components of magnitude $F \sin \theta$ and $F \cos \theta$, respectively. We apply Newton's second law to the horizontal and vertical directions. In the vertical direction, there is no motion so the acceleration is zero and we have

$$F \cos \theta - mg = 0.$$

In the horizontal direction there is only one force, of magnitude $F \sin \theta$, that acts toward the center of the circle and gives rise to the anticipated acceleration v^2/r:

$$F \sin \theta = m \frac{v^2}{r}.$$

We solve these equations for v by eliminating F between them (and using $r = L \sin \theta$):

$$v = \sqrt{\frac{rF \sin \theta}{m}} = \sqrt{\frac{r}{m}\left(\frac{mg}{\cos \theta}\right) \sin \theta} = \sqrt{\frac{Lg \sin^2 \theta}{\cos \theta}}.$$

The period T is the time required to make one revolution, a distance of $2\pi r = 2\pi L \sin \theta$. The speed v can thus be written

$$v = \frac{2\pi L \sin \theta}{T};$$

then

$$T = \frac{2\pi L \sin \theta}{v} = \frac{2\pi L \sin \theta}{\sqrt{\dfrac{Lg \sin^2 \theta}{\cos \theta}}} = 2\pi \sqrt{\frac{L \cos \theta}{g}},$$

where we again used $r = L \sin \theta$.

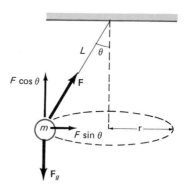

FIGURE 5–7
Conical pendulum (Example 5–4).

FIGURE 5–8
The speed of a particle moving in a circle changes if the force on it has a tangential component.

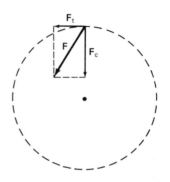

Circular motion at constant speed occurs when the force on an object is exerted toward the center of the circle. If the force is not directed toward the center, but is at an angle as shown in Fig. 5–8, then the force has two components. The component directed toward the center of the circle, $\mathbf{F}_c$, gives rise to the centripetal acceleration, $\mathbf{a}_c$, and keeps the object moving in a circle. The component $\mathbf{F}_t$, tangential to the circle, acts to increase (or decrease) the speed and thus gives rise to the tangential acceleration already discussed in Section 3–10 (Fig. 3–21).

When you first start revolving a ball on the end of string around your head, you must give it tangential acceleration. You do this by pulling on the string with your hand displaced from the center of the circle. In athletics, a hammer thrower accelerates the hammer tangentially in a similar way so that it reaches a high speed before release.

5–2 Newton's Law of Universal Gravitation

Besides developing the three laws of motion, Sir Isaac Newton also examined the motion of the heavenly bodies—the planets and the moon. In particular, he wondered about the nature of the force that must act to keep the moon in its nearly circular orbit around the earth.

Newton was also thinking about the apparently unrelated problem of gravity. Since falling bodies accelerate, Newton had concluded that they must have a force exerted on them, a force we call the force of gravity. But what, we may ask, *exerts* this force of gravity—for, as we have seen, whenever a body has a force exerted *on it*, that force is exerted *by* some other body. Every object on the surface of the earth feels this force of gravity, and no matter where the object is, the force is directed toward the center of the earth. Newton concluded that it must be the earth itself that exerts the gravitational force on objects at its surface.

According to an early account, Newton was sitting in his garden and noticed an apple drop from a tree. He is said to have been struck with a sudden inspiration: If the effect of gravity acts at the tops of trees, and even at the tops of mountains, then perhaps it acts all the way to the moon! Whether this story is true or not, and whether or not it served as inspiration we don't really know.[†] But with this idea that it is terrestrial gravity that holds the moon in its orbit, Newton developed his great theory of gravitation, with considerable help and encouragement from Robert Hooke (1635–1703).

Newton set about determining the magnitude of the gravitational effect that the earth exerts on the moon as compared to that on objects at the earth's surface. At the surface of the earth, the force of gravity accelerates objects at $9.80 \ \text{m/s}^2$. But what is the centripetal acceleration of the moon? Since the moon moves with nearly uniform circular motion, the acceleration can be calculated from $a_c = v^2/r$; we already did this calculation in Example 3–10 of Chapter 3 and found that $a_c = 2.73 \times 10^{-3} \ \text{m/s}^2$. In terms of the acceleration of gravity at the earth's surface, g:

$$a_c \approx \frac{1}{3600} \, g.$$

That is, the acceleration of the moon toward the earth is about $\frac{1}{3600}$ as great as the acceleration of objects at the earth's surface. Now the moon is 385,000 km from the earth, which is about 60 times the earth's radius of 6380 km. That is, the moon is 60 times farther from the earth's center than are objects at the earth's surface. But $60 \times 60 = 3600$—again that number 3600! Newton concluded that the gravitational force exerted by the earth on any object decreases with the square of its distance, r, from the earth's center:

$$\text{force of gravity} \propto \frac{1}{r^2}.$$

The moon, being 60 earth radii away, feels a gravitational force only $\frac{1}{60^2} = \frac{1}{3600}$ times as strong as it would if it were at the earth's surface. Any object placed

[†] Many historians of science today believe that Newton may have invented this story to push back the date of discovery to the 1660s, when in fact his correspondence and notebooks indicate he really didn't come to the law of universal gravitation until about 1685. The idea that there might be a connection between terrestrial gravity and the force on the moon, and even the $1/r^2$ dependence, had already been suggested by others, including Hooke; nonetheless, we still give most of the credit to Newton for the gravitational law since he proposed it in its fullness and gave very strong arguments to support it.

385,000 km from the earth would experience the same acceleration due to the earth's gravity as the moon experiences: 2.73×10^{-3} m/s².

Newton realized that the force of gravity on an object depends not only on distance but also on the object's mass. In fact, it is directly proportional to its mass, as we have seen. According to Newton's third law, when the earth exerts its gravitational force on an object, such as the moon, that object exerts an equal and opposite force on the earth (Fig. 5–9). Because of this symmetry, Newton reasoned, the magnitude of the force of gravity must be proportional to *both* the masses. Thus

$$F \propto \frac{m_e m_o}{r^2}$$

where m_e is the mass of the earth, m_o the mass of the other object, and r the distance from the earth's center to the center of the object.

Newton went a step further in his analysis of gravity. He had determined that the force required to hold the different planets in their orbits around the sun diminished as the inverse square of their distance from the sun. This led him to believe that it was also the gravitational force that acted between the sun and each of the planets to keep them in their orbits. And if gravity acts between these objects, why not between all objects? Thus he proposed his famous **law of universal gravitation**, which we can state as follows:

> **Every particle in the universe attracts every other particle with a force that is proportional to the product of their masses and inversely proportional to the square of the distance between them. This force acts along the line joining the two particles.**

The magnitude of this force can be written as:

$$F = G \frac{m_1 m_2}{r^2}, \tag{5–1}$$

where m_1 and m_2 are the masses of the two particles, r is the distance between them, and G is a universal constant which must be measured experimentally and has the same numerical value for all objects.

Strictly speaking, Eq. 5–1 gives the magnitude of the gravitational force that one particle exerts on a second particle that is a distance r away. For an extended (that is, not a point) object, we must consider how to measure the distance r. You might think that r would be the distance between the centers of the objects; this is not necessarily true, and to do a calculation correctly, each extended body must be considered as a collection of tiny particles and the total force is the sum of the forces due to all the particles. The sum over all these particles is often best done using integral calculus which Newton himself invented. Newton showed, however, that for two uniform spheres, Eq. 5–1 gives the correct force where r is the distance between their centers. (The proof for this is the subject of problem 23 at the end of the chapter.) Also, when extended bodies are small compared to the distance between them (as for the earth–sun system), little inaccuracy results from considering them as particles.

When we want to consider the gravitational force on one particle due to two or more other particles, such as the force on the moon due to both the earth and the sun (or when we are summing over the particles of an extended body), we must use Eq. 5–1 for each of the particles and add the forces together as vectors.

The value of G in Eq. 5–1 must be very small since we are not aware of any force existing between ordinary sized objects, such as between two heavy bowling balls. The force between two ordinary objects was first measured over 100 years after Newton's publication of his law, by Henry Cavendish in 1798. To detect and measure the incredibly small force, he used an apparatus like that shown in Fig. 5–10. Cavendish not only confirmed Newton's hypothesis that two bodies attract one another and that Eq. 5–1 accurately describes this force, but because he could measure F, m_1, m_2, and r accurately, he was then able to determine the value of the

FIGURE 5–9
The gravitational force one body exerts on a second body is directed toward the first body and is equal and opposite to the force exerted by the second body on the first.

FIGURE 5–10
Schematic diagram of Cavendish's apparatus. Two spheres are attached to a light horizontal rod, which is suspended at its center by a thin fiber. When the sphere labeled A is brought close to one of the suspended spheres, the gravitational force causes the latter to move, and this twists the fiber slightly. The tiny movement is magnified by use of a narrow light beam directed at a mirror, which is mounted on the fiber, and the beam reflects onto a scale. Previous determination of what force twisted the fiber a given amount allows one to determine the magnitude of the gravitational force between two objects.

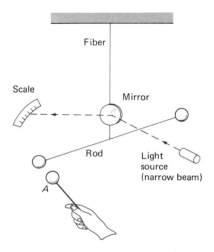

constant G as well. The accepted value today is

$$G = (6.6720 \pm 0.0041) \times 10^{-11} \text{ N}\cdot\text{m}^2/\text{kg}^2$$

or

$$G \approx 6.67 \times 10^{-11} \text{ N}\cdot\text{m}^2/\text{kg}^2.$$

EXAMPLE 5–5 Two 8.00-kg spherical lead balls are placed so their centers are 50.0 cm apart. What is the magnitude of the gravitational force each exerts on the other?

SOLUTION Assuming the spheres are uniform, Eq. 5–1 gives

$$F = \frac{(6.67 \times 10^{-11} \text{ N}\cdot\text{m}^2/\text{kg}^2)(8.00 \text{ kg})(8.00 \text{ kg})}{(0.500 \text{ m})^2} = 1.71 \times 10^{-8} \text{ N}$$

which is unnoticeably small unless very delicate instruments are used.

EXAMPLE 5–6 What is the force of gravity acting on a spacecraft when it orbits two earth radii from the earth's center (that is, 6400 km above the earth's surface)?

SOLUTION The spacecraft is twice as far from the earth's center as when at the surface of the earth; therefore, since the force of gravity decreases as the square of the distance (and $\frac{1}{2^2} = \frac{1}{4}$) the force of gravity on it will be only one fourth as great.

EXAMPLE 5–7 Determine the net force on the moon ($m_{\text{m}} = 7.36 \times 10^{22}$ kg) due to the gravitational attraction of both the earth ($m_{\text{e}} = 5.98 \times 10^{24}$ kg) and the sun ($m_{\text{s}} = 1.99 \times 10^{30}$ kg) assuming they are at right angles to each other (Fig. 5–11).

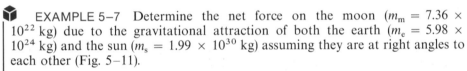

SOLUTION The earth is 3.85×10^5 km from the moon so F_{me} (force on the moon due to the earth) is

$$F_{\text{me}} = \frac{(6.67 \times 10^{-11} \text{ N}\cdot\text{m}^2/\text{kg}^2)(7.36 \times 10^{22} \text{ kg})(5.98 \times 10^{24} \text{ kg})}{(3.85 \times 10^8 \text{ m})^2}$$

$$= 1.98 \times 10^{20} \text{ N}.$$

The sun is 1.50×10^8 km from the earth, so F_{ms} (force on the moon due to the sun) is

$$F_{\text{ms}} = \frac{(6.67 \times 10^{-11} \text{ N}\cdot\text{m}^2/\text{kg}^2)(7.36 \times 10^{22} \text{ kg})(1.99 \times 10^{30} \text{ kg})}{(1.50 \times 10^{11} \text{ m})^2}$$

$$= 4.35 \times 10^{20} \text{ N}.$$

Since the two forces act at right angles in the case we are considering (Fig. 5–11), the total force is

$$F = \sqrt{(1.98)^2 + (4.35)^2} \times 10^{20} \text{ N} = 4.79 \times 10^{20} \text{ N}$$

which acts at an angle $\theta = \tan^{-1}(1.98/4.35) = 24.5°$.

The law of universal gravitation should not be confused with Newton's second law of motion, $\mathbf{F} = m\mathbf{a}$. The former describes a particular force, gravity, and how its strength varies with the distance and masses involved. Newton's second law, on the other hand, relates the force on a body—it can be any force—to the mass and acceleration of that body.

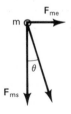

FIGURE 5–11

Orientation of sun (s), earth (e), and moon (m) for Example 5–7 (not to scale).

5-3   Vector Form of Newton's Law of Universal Gravitation

We can write Newton's law of universal gravitation in vector form as

$$\mathbf{F}_{12} = -G \frac{m_1 m_2}{r_{21}^2} \hat{\mathbf{r}}_{21} \tag{5-2}$$

where $\mathbf{F}_{12}$ is the vector force on particle 1 (of mass m_1) exerted by particle 2 (of mass m_2) which is a distance r_{21} away; $\hat{\mathbf{r}}_{21}$ is a unit vector that points from particle 2 toward particle 1 along the line joining them so that $\hat{\mathbf{r}}_{21} = \mathbf{r}_{21}/r_{21}$, where $\mathbf{r}_{21}$ is the displacement vector as shown in Fig. 5-12. The minus sign in Eq. 5-2 is necessary since the force on particle 1 due to particle 2 points in the direction opposite to $\hat{\mathbf{r}}_{21}$. The displacement vector $\mathbf{r}_{12}$ is a vector of the same magnitude as $\mathbf{r}_{21}$, but it points in the opposite direction so that

$$\mathbf{r}_{12} = -\mathbf{r}_{21}.$$

By Newton's third law, the force $\mathbf{F}_{21}$ acting on m_2 exerted by m_1 must have the same magnitude as $\mathbf{F}_{12}$ but acts in the opposite direction (Fig. 5-13) so that

$$\mathbf{F}_{21} = -\mathbf{F}_{12} = G \frac{m_1 m_2}{r_{21}^2} \hat{\mathbf{r}}_{21}$$

$$= -G \frac{m_2 m_1}{r_{12}^2} \hat{\mathbf{r}}_{12}.$$

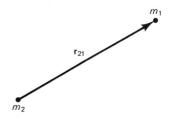

FIGURE 5-12

The displacement vector $\mathbf{r}_{21}$ points from particle of mass m_2 to particle of mass m_1.

5-4   Gravity Near the Earth's Surface

When Eq. 5-1 is applied to the gravitational force between the earth and an object at its surface, m_1 becomes the mass of the earth, m_e, and r the distance from the earth's center,[†] namely the radius of the earth, r_e. This force of gravity due to the earth is the weight of the object, which we have been writing as mg. Thus

$$mg = G \frac{m m_e}{r_e^2}.$$

Hence

$$g = G \frac{m_e}{r_e^2}. \tag{5-3}$$

Thus the acceleration of gravity at the surface of the earth, g, is determined by m_e and r_e. (Be sure not to confuse G with g; they are very different quantities, but are related by Eq. 5-3.)

In Chapter 2 we saw that the value of g varied slightly from place to place. From Eq. 5-3 we would expect g to be slightly less on the tops of mountains, for example, than at sea level since r_e, the distance from the earth's center, is slightly greater there. Indeed, this is what is found experimentally. However, Eq. 5-3 does not give precise values for g at various locations since the earth is not a perfect sphere (it not only has mountains and valleys, and bulges at the equator, but its mass is not distributed precisely uniformly; its rotation also has an effect on the measurement of g as we shall see in Example 5-8.

Until G was measured, the mass of the earth was not known. But once G was measured, Eq. 5-3 could be used to calculate the earth's mass, and Cavendish

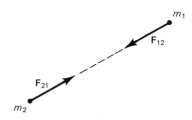

FIGURE 5-13

By Newton's third law, the gravitational force on particle 1 exerted by particle 2, $\mathbf{F}_{12}$, is equal and opposite to that on particle 2 exerted by particle 1, $\mathbf{F}_{21}$; that is $\mathbf{F}_{21} = -\mathbf{F}_{12}$.

[†] That the distance is measured from the earth's center does not imply that the force of gravity somehow emanates from that one point. Rather, all parts of the earth attract gravitationally but the net effect is a force acting toward the earth's center. (See problem 23.)

himself was the first to do so. Since $g = 9.80$ m/s^2 and the radius of the earth $r_e = 6.38 \times 10^6$ m, then from Eq. 5–3 we obtain

$$m_e = \frac{gr_e^2}{G} = \frac{(9.80 \text{ m/s}^2)(6.38 \times 10^6 \text{ m})^2}{6.67 \times 10^{-11} \text{ N·m}^2/\text{kg}^2} = 5.98 \times 10^{24} \text{ kg}$$

for the mass of the earth.

When dealing with the weight of objects at the surface of the earth, we can continue to use simply mg. If we wish to calculate the force of gravity on an object some distance from the earth, or the force due to some other heavenly body such as that exerted by the moon or a planet, we can calculate the effective value of g from Eq. 5–3, replacing r_e (and m_e) by the appropriate distance (and mass), or we can use Eq. 5–1 directly.

EXAMPLE 5–8 Assuming the earth is a perfect sphere, determine how the earth's rotation affects the value of g at the equator as compared to its value at the poles.

SOLUTION   Figure 5–14 shows a mass m hanging from a spring scale at two places on the earth. At the North Pole there are two forces acting on the mass m: the force of gravity, $\mathbf{F}_g = m\mathbf{g}$, and the force with which the spring pulls up on the mass, $\mathbf{w}$. We call this latter force $\mathbf{w}$ since it is what the scale reads as the weight of the object, for by Newton's third law it equals the force with which the mass pulls down on the spring. Since the mass is not accelerating, Newton's second law tells us

$$mg - w = 0$$

so $w = mg$; thus the weight w that the spring registers equals mg, which is no surprise. Next, at the equator, there *is* an acceleration because the earth is rotating. The same force of gravity $F_g = mg$ acts downward (we are letting g represent the acceleration of gravity in the absence of rotation and we ignore the slight bulging of the equator). The spring pulls upward with a force w'; w' is also the force with which the mass pulls on the spring (Newton's third law) and hence is the weight registered

FIGURE 5–14

Example 5–8.

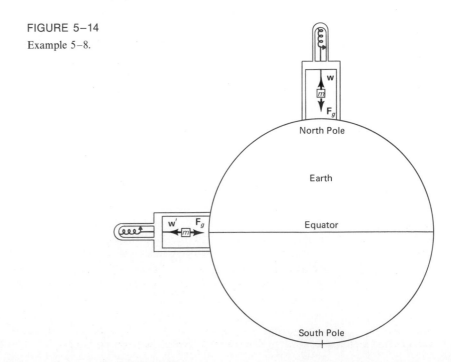

$$mg - w' = m\frac{v^2}{r_e}$$

where $r_e = 6.38 \times 10^6$ m is the earth's radius and v is the speed of m due to the earth's daily rotation and has the numerical value: $v = 2\pi r_e/1$ day $= (6.28)(6.38 \times 10^6 \text{ m})/(8.64 \times 10^4 \text{ s}) = 4.64 \times 10^2$ m/s. The effective weight is w', so the effective value of g, which we call g', is $g' = w'/m$. Solving the equation above for w' we have $w' = m(g - v^2/r_e)$ so

$$g' = \frac{w'}{m} = g - \frac{v^2}{r_e}.$$

Hence $\Delta g = g - g' = v^2/r_e = (4.64 \times 10^2 \text{ m/s})^2/(6.38 \times 10^6 \text{ m}) = 0.0337 \text{m/s}^2$. In Table 2–1 we see that the difference is actually greater than this: $(9.832 - 9.780) \text{ m/s}^2 = 0.052 \text{ m/s}^2$; this discrepancy is due mainly to the fact that the earth is slightly fatter at the equator (by 21 km) than at the poles. The calculation of the effective value of g at latitudes other than at the poles or equator is a two-dimensional problem because $\mathbf{F}_g$ acts radially toward the earth's center whereas the centripetal acceleration is directed perpendicular to the axis of rotation, parallel to the equator.

 ## * 5–5 Gravitational Versus Inertial Mass

We have dealt with two aspects of mass. In Chapter 4, we defined mass as a measure of the inertia of a body. Newton's second law relates the force acting on a body to its acceleration and its *inertial mass*, as we call it. Then in this chapter we have dealt with mass as a property related to the gravitational force—that is, mass as a quantity that determines the strength of the gravitational force between two bodies. This we call the *gravitational mass*. We mentioned this point in Section 4–4 when we discussed the two ways of defining a mass scale.

Now it is not at all obvious that the inertial mass of a body should be equal to its gravitational mass. (The force of gravity might have depended on a completely different property of a body, just as the electrical force depends on a property called electric charge as we shall see later.) Newton's and Cavendish's experiments indicated that these two types of mass are equal for a body, and modern experiments confirm it to a precision of about 1 part in 10^{11}.

If the gravitational and inertial masses were not equal, Galileo's conclusion that all bodies fall to earth with the same acceleration in the absence of air resistance would not be valid. For suppose a heavier object had the same inertia as a lighter body (and by lighter we mean the force of gravity on it is less). Then the heavier body would accelerate faster than the lighter one since $a = F/m$; that is, F would be greater for the heavy body but m would be the same as for the lighter body. Because the inertial mass is equal to the gravitational mass, the heavier body—in spite of the greater force on it—has greater inertia than the lighter one in exact proportion, so its acceleration is the same.

 ## 5–6 Satellites and Weightlessness

Artificial satellites circling the earth are now commonplace. A satellite is put into orbit by accelerating it to a sufficiently high tangential speed with the use of rockets, as shown in Fig. 5–15. If the speed is too high, the spacecraft will not be confined by the earth's gravity and will escape, never to return. If too low, it will fall back to earth. Satellites are usually put into circular, or nearly circular, orbits. This requires the least take-off speed. It is sometimes asked: "What keeps a satellite up?" The answer

FIGURE 5–15
Artificial satellites.

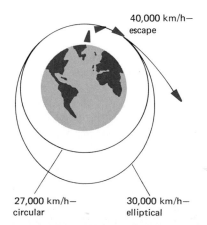

40,000 km/h—
escape

27,000 km/h—
circular

30,000 km/h—
elliptical

is its high speed. If a satellite stopped moving, it would of course fall back to earth. But at the very high speed a satellite has, it would quickly fly out into space if it weren't for the gravitational force of the earth pulling it into orbit.

For satellites that move (at least approximately) with uniform circular motion, the acceleration is v^2/r. The force that gives a satellite this acceleration is the force of gravity, and since a satellite may be at a considerable distance from the earth, we must use Eq. 5–1 for the force acting on it. When we apply Newton's second law, $F = ma$, we find

$$G\frac{mm_e}{r^2} = m\frac{v^2}{r}.\qquad(5\text{–}4)$$

This equation relates the distance of the satellite from the earth's center, r, to its speed, v. Note that only one force, gravity, is acting on the satellite.

EXAMPLE 5–9 Calculate the velocity required for a satellite moving in a circular orbit 200 km above the earth's surface.

SOLUTION We solve Eq. 5–4 for v;

$$v = \sqrt{\frac{Gm_e}{r}}.$$

The value of r is 6380 + 200 km = 6580 km, or 6.58×10^6 m. Thus

$$v = \sqrt{\frac{(6.67 \times 10^{-11}\ \text{N·m}^2/\text{kg}^2)(5.98 \times 10^{24}\ \text{kg})}{6.58 \times 10^6\ \text{m}}} = 7.79 \times 10^3\ \text{m/s}.$$

This is approximately 27,000 km/h. Note that the velocity required will decrease with increasing height.

People and other objects in a satellite circling the earth are said to experience apparent weightlessness. Before tackling the case of a satellite, however, let us first look at the simpler case of a falling elevator. In Fig. 5–16a we see an elevator at rest. Let us consider an object such as a bag hanging from a spring scale as shown. The scale indicates the downward force exerted on it by the bag. This force, exerted *on the* scale, is just equal and opposite to the force exerted *by* the scale upwards on the bag. We call this force **w**. Since the mass, m, is not accelerating, we apply $F = ma$ to the bag and obtain:

$$0 = w - mg,$$

where mg is the weight of the bag; thus $w = mg$, and since the scale indicates the force w exerted on it by the bag, it registers a force equal to the weight of the bag as we expect. If, now, the elevator has an acceleration, a, then applying $F = ma$ to the bag we have $w - mg = ma$. Solving for w we obtain

$$w = mg + ma.$$

We have chosen the positive direction up. Thus, if the acceleration is up, a is positive and the scale, which measures w, will read more than mg. If the elevator accelerates downward, a will be negative and w will be less than mg. For example, if the elevator's acceleration is $-\frac{1}{2}g$, then we find $w = mg - \frac{1}{2}mg = \frac{1}{2}mg$. That is, the scale reads one-half the actual weight. If the elevator is in *free fall* (for example if the cables break) then $a = -g$ and $w = mg - mg = 0$. The scale reads zero! See Fig. 5–16b. The bag seems weightless. If the person in the elevator let go of a pencil, say, it would not fall to the floor. True, the pencil would fall with acceleration g. But so do the floor of the elevator and the person. The pencil would hover right in front of the person. This is called *apparent weightlessness* because, in fact, gravity is still acting on the object. The objects seem weightless only because the elevator is accelerating at $-g$.

FIGURE 5–16

An object in an elevator at rest exerts a force on a spring scale equal to its weight (a); but in a freely falling elevator (b), it experiences "weightlessness."

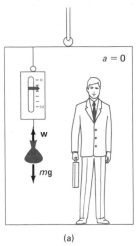

(a)

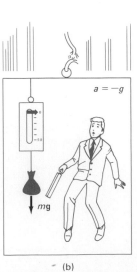

(b)

The weightlessness experienced by people in a satellite orbiting close to the earth is the same apparent weightlessness experienced in a freely falling elevator. It may seem strange, at first, to think of a satellite as freely falling. But a satellite is indeed falling toward the earth, as shown in Fig. 5–17; the force of gravity causes it to "fall" out of its natural straight-line path. The acceleration of the satellite must be the acceleration due to gravity, since the only force acting on it is gravity. (We made use of this fact in obtaining Eq. 5–4; the fact that we wrote the acceleration as v^2/r does not alter this fact.) Thus, although the force of gravity acts on objects within the satellite, the objects experience an apparent weightlessness because they, and the satellite, are accelerating as in free fall.

A completely different situation occurs when a spacecraft is out in space far from the earth and other attracting bodies such as the moon. The force of gravity due to the earth and other heavenly bodies will then be quite small due to the distances involved, and persons in such a spacecraft will experience real weightlessness.

 ## 5–7 Kepler's Laws and Newton's Synthesis

More than a half century before Newton proposed his three laws of motion and his law of universal gravitation, the German astronomer Johannes Kepler (1571–1630) had written a number of astronomical works in which we can find a detailed description of the motion of the planets about the sun. Kepler's work resulted in part from the many years he spent examining data collected by Tycho Brahe (1546–1601) on the positions of the planets in their motion through the heavens. Tucked away among these writings were three findings which we now refer to as **Kepler's laws of planetary motion**. These are summarized as follows, with additional explanation in Fig. 5–18.

Kepler's first law: The path of each planet about the sun is an ellipse with the sun at one focus (Fig. 5–18).

Kepler's second law: Each planet moves so that an imaginary line drawn from the sun to the planet sweeps out equal areas in equal periods of time (Fig. 5–18b.)

Kepler's third law: The ratio of the squares of the periods of any two planets revolving about the sun is equal to the ratio of the cubes of their average distances from the sun. That is, if T_1 and T_2 represent the periods (time needed for one revolution about the sun), and r_1 and r_2 represent their average distances from the sun, then

$$\left(\frac{T_1}{T_2}\right)^2 = \left(\frac{r_1}{r_2}\right)^3.$$

(Present-day data are given in Table 5–1; see last column.)

TABLE 5–1
Planetary data applied to Kepler's third law

Planet	Mean Distance from Sun, r (10^6 km)	Period, T (Earth Years)	$\frac{r^3/T^2}{\left(\frac{\text{km}^3}{\text{yr}^2} \times 10^{24}\right)}$
Mercury	57.9	0.241	3.34
Venus	108.2	0.615	3.35
Earth	149.6	1.0	3.35
Mars	227.9	1.88	3.35
Jupiter	778.3	11.86	3.35
Saturn	1427	29.5	3.34
Uranus	2870	84.0	3.35
Neptune	4497	165	3.34
Pluto	5900	248	3.33

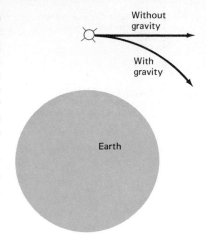

FIGURE 5–17

A moving satellite "falls," out of a straight-line path, toward the earth.

FIGURE 5–18

(a) Kepler's first law. An ellipse is a closed curve such that the sum of the distances from any point P on the curve to two fixed points (called the foci, F_1 and F_2) remains constant. That is, the sum of the distances $F_1P + F_2P$ is the same for all points P on the curve. A circle is a special case of an ellipse when the two foci coincide, at the center of the circle. (b) Kepler's second law. The two shaded regions have equal areas. It takes the planet the same time to move from point 1 to point 2 as it takes it to move from point 3 to point 4. Planets move fastest in that part of their orbit where they are closest to the sun.

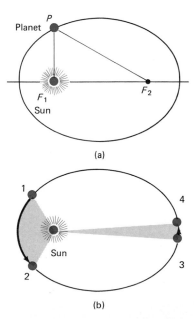

Newton was able to show that Kepler's laws could be derived mathematically from his law of universal gravitation and his laws of motion; and he showed that for any reasonable form for the gravitational force law, only one that depends on the inverse square of the distance is fully consistent with all three of Kepler's laws. He thus used Kepler's laws as evidence in favor of his law of universal gravitation, Eq. 5–1.

Kepler's third law is the easiest to derive, and we do it here for the special case of a circular orbit. (Most of the planetary orbits are fairly close to a circle, which is a special case of an ellipse.) First we write down Newton's second law of motion, $F = ma$; then for F we substitute the law of universal gravitation, Eq. 5–1, and for a the centripetal acceleration, v^2/r:

$$F = ma$$

$$G\frac{m_1 M}{r^2} = m_1 \frac{v_1^2}{r_1}.$$

Here m_1 is the mass of a particular planet, r_1 its average distance from the sun, and v_1 its average speed in orbit; M is the mass of the sun, since it is the gravitational attraction of the sun that keeps each planet in its orbit. Now the period T_1 of the planet is the time required for one complete orbit, a distance equal to $2\pi r_1$, the circumference of a circle; thus

$$v_1 = \frac{2\pi r_1}{T_1}.$$

We substitute this formula for v_1 into the equation above:

$$G\frac{m_1 M}{r_1^2} = m_1 \frac{4\pi^2 r_1}{T_1^2}.$$

We rearrange this to get

$$\frac{T_1^2}{r_1^3} = \frac{4\pi^2}{GM}.$$

We derived this for planet 1 (say Mars); the same derivation would apply for a second planet (say Saturn):

$$\frac{T_2^2}{r_2^3} = \frac{4\pi^2}{GM},$$

where T_2 and r_2 are the period and orbit radius for the second planet. Since the right sides of the two previous equations are equal, we have $T_1^2/r_1^3 = T_2^2/r_2^3$ or, rearranging,

$$\frac{T_1^2}{T_2^2} = \frac{r_1^3}{r_2^3},$$

which is Kepler's third law. Kepler's second law is also easy to derive and we do it in Chapter 10.

Accurate measurements on the orbits of the planets indicated that they did not precisely follow Kepler's laws. For example, slight deviations from perfectly elliptical orbits were observed. Newton was aware that this is to be expected from the law of universal gravitation ("every body in the universe attracts every other body . . .") because each planet exerts a gravitational force on the other planets. Since the mass of the sun is much greater than that of any planet, the force on one planet due to any other planet will be small in comparison to the force on it due to the sun. (The derivation of perfectly elliptical orbits ignores the forces due to other planets.) But because of this small force, each planetary orbit should depart from a perfect ellipse,[†] especially when a second planet is fairly close to it.

[†] Newton said thusly that the "planets neither move exactly in ellipses nor revolve twice in the same orbit." He also pointed out, for example, that the sun is not precisely at one focus of the ellipse; even in the ideal case of only two bodies (sun and one planet), the two would revolve about their common center of gravity.

Such deviations, or *perturbations*, as they are called, from perfect ellipses are indeed observed. In fact, Newton's recognition of perturbations in the orbit of Saturn was a hint that helped him formulate the law of universal gravitation, that all bodies can attract gravitationally. Observation of other perturbations later led to the discovery of Neptune and Pluto. Deviations in the orbit of Uranus, for example, could not be accounted for by perturbations due to the other known planets; careful calculation in the nineteenth century indicated that these deviations could be accounted for if there were another planet farther out in the solar system. The position of this planet was predicted from the deviations in the orbit of Uranus, and telescopes focused on that region of the sky quickly found it; the new planet was called Neptune. Similar but much smaller perturbations of Neptune's orbit led to the discovery of Pluto in 1930.

The development by Newton of the law of universal gravitation and the three laws of motion (Chapter 4) was a major achievement in Western thought. For with these laws, Newton was able to describe the motion of objects on earth and how this motion was caused. Furthermore, the motion of the planets about the sun, and the motion of the moon about the earth, were explained on the same basis. The motions of heavenly bodies and bodies on earth were seen to follow the same laws (something not previously recognized generally, although Galileo had argued strongly in its favor). For this reason (and also because Newton integrated the results of earlier workers into his system), we sometimes speak of Newton's "synthesis."

Newton's work was so encompassing that it constituted a theory of the universe, and influenced philosophy and other fields. The laws formulated by Newton are referred to as *causal laws*. By *causality* we mean the idea that one occurrence can cause another. We have repeatedly observed, for example, that when a rock strikes a window, the window almost immediately breaks. We infer that the rock *caused* the window to break. This idea of "cause and effect" took on more forceful meaning with Newton's laws. For the motion—or rather the acceleration—of any object was seen to be *caused* by the net force acting on it. As a result the universe came to be pictured by many scientists and philosophers as a big machine whose parts move in a predictable and predetermined way—according to natural laws. However, this *deterministic* view of the universe had to be rejected by scientists in the twentieth century, as we shall see in Chapters 40 and 41.

5–8 Types of Forces in Nature

We have already discussed the fact that Newton's law of universal gravitation, Eq. 5–1, describes how a particular type of force—gravity—depends on the distance between, and masses of, the objects involved. Newton's second law, $\mathbf{F} = m\mathbf{a}$, tells how a body will accelerate due to any type of force $\mathbf{F}$. But what are the types of forces that occur in nature?

Physicists today recognize only four different forces in nature. These are: (1) the gravitational force; (2) the electromagnetic force (we shall see later that electric and magnetic forces are intimately related), (3) the strong nuclear force, and (4) the weak nuclear force. We have discussed the gravitational force in detail. The nature of the electromagnetic force will be discussed in detail in later chapters. The strong and weak nuclear forces operate at the level of the atomic nucleus, and although they manifest themselves in such phenomena as radioactivity and nuclear energy, they are much less obvious in our daily lives.

But what about such forces as ordinary pushes and pulls, and friction? What category do they fit in? We call these types of force *contact forces* since they occur when one object is in contact with another object which exerts the force. According to modern quantum theory, these forces are due to the electromagnetic force. For example, the force your fingers exert on a pencil is the result of electrical repulsion between the outer electrons of the atoms of your fingers and those of the pencil.

⬡ *5–9   Gravitational Field

Most of the forces we meet in everyday life are contact forces: you push or pull on a lawn mower, a tennis racket exerts a force on a tennis ball when they make contact, or a ball exerts a force on a window when they make contact. But the gravitational force (and even the electromagnetic force, as we shall see later) acts over a distance; there is a force even when the two objects are not in contact. The earth, for example, exerts a force on a falling apple; it also exerts a force on the moon, 385,000 km away. And the sun exerts a gravitational force on the earth. The idea of a force *acting at a distance* was a difficult one for early thinkers. Newton himself felt uneasy with this concept when he published his law of universal gravitation.

The conceptual difficulties can be overcome with the concept of the *field* which was developed in the nineteenth century by Michael Faraday (1791–1867) to aid understanding of electromagnetism; only later was it applied to gravity. According to the field concept, a *gravitational field* surrounds every body that has mass, and this field permeates all of space. A second body at a particular location near the first body feels a force because of the gravitational field that exists there. Because the gravitational field at the location of the second mass is considered to act directly on this mass we are a little closer to the idea of a contact force. It must be emphasized, however, that a field should not be thought of as a kind of matter. It is a concept, invented by humans, to help us analyze the world.

The concept of gravitational field can be quantified by defining the *gravitational field intensity* as the gravitational force per unit mass at any point in space. If we want to measure the gravitational field intensity at any point, we place a small "test" mass m at that point and measure the force $\mathbf{F}$ exerted on it (making sure only gravitational forces are acting); then the gravitational field intensity, $\mathscr{G}$, at that point is defined as

$$\mathscr{G} = \frac{\mathbf{F}}{m}. \tag{5–5}$$

The units of $\mathscr{G}$ are N/kg.

If the gravitational field is due to a single body of mass M (such as when m is near the earth's surface), then

$$\mathscr{G} = \frac{GM}{r^2}$$

where r is the distance from m to M and G is the universal gravitation constant (see Eqs. 5–1 and 5–3). If several different bodies contribute to the gravitational field intensity, then $\mathscr{G}$ is the vector sum of all these contributions. In interplanetary space, for example, $\mathscr{G}$ is the vector sum of terms due to earth, sun, moon, and other bodies that contribute significantly. An important aspect of the gravitational field is that its intensity, $\mathscr{G}$, at any point in space does not depend on the value of our test mass, m, placed at that point; $\mathscr{G}$ depends only on the masses (and locations) of the bodies that create the field there.

The gravitational field intensity, $\mathscr{G}$, defined by Eq. 5–5 appears to be the same as the acceleration of gravity, g, we have discussed before. Indeed, they can have the same numerical value. But conceptually they are different. This difference is carried over in our use of units; although N/kg = m/s^2 (since 1 N is defined as 1 kg·m/s^2, Section 4–4), we use N/kg for specifying gravitational field intensity but m/s^2 for specifying the acceleration due to gravity.

It is interesting to note that the gravitational field at any point in space changes if one (or more) of the bodies creating that field changes its position. But it does not change instantaneously. It presumably takes time for the change in position of the mass to be felt at any location in space. According to Einstein's general theory of relativity, the change in field intensity propagates at the speed of light, 3.0×10^8 m/s. This propagation time has not yet been detected experimentally.

Summary

A particle revolving in a circle of radius r with constant speed v must have a force acting on it that is directed toward the center of the circle at every moment. The magnitude of this force must equal the product of the particle's mass m and its centripetal acceleration v^2/r.

Newton's *law of universal gravitation* states that every particle in the universe attracts every other particle with a force proportional to the product of their masses and inversely proportional to the square of the distance between them:

$$F = G\,\frac{m_1 m_2}{r^2}.$$

The direction of this force is along the line joining the two particles. It is this gravitational force that keeps the moon revolving around the earth and the planets revolving around the sun. The total gravitational force on any body is the vector sum of the forces exerted by all other bodies; frequently the effects of all but one or two bodies can be ignored.

Newton's three laws of motion, plus his law of universal gravitation, constituted a wide-ranging theory of the universe. With them, motion of objects on earth and in the heavens could be accurately described; and they provided a theoretical base for Kepler's experimental laws of planetary motion.

Questions

1. Sometimes it is said that water is removed from clothes in a spin dryer by centrifugal force throwing the water outward. Is this correct?

2. Technical reports often specify only the rpm for centrifuge experiments. Why is this inadequate?

3. Suppose a car moves at constant speed along a mountain road. At what places does it exert the greatest and least forces on the road: (a) at the top of a hill, (b) at a dip between two hills, (c) on a level stretch near the bottom of a hill?

4. Describe all the forces acting on a child riding a horse on a merry-go-round.

5. Does an apple exert a gravitational force on the earth? If so, how large a force? Consider an apple (a) attached to a tree, (b) falling.

6. If the earth's mass were suddenly to double, in what ways would the moon's orbit change?

7. Suppose a spacecraft travels from Earth to Mars (where $g = 3.8$ m/s² at the surface) at constant velocity. Graph the weight of (net gravitational force on) a 70-kg passenger as a function of distance between the two planets.

8. The sun's gravitational pull on the earth is much larger than the moon's. Yet the moon is mainly responsible for the tides. Explain.

9. The source of the Mississippi river is closer to the center of the earth than is its outlet in Louisiana (since the earth is fatter at the equator than at the poles). Explain how the Mississippi can flow "uphill."

10. Will an object weigh more at the equator or at the poles? What two effects are at work? Do they oppose each other?

11. Would you get more flour to the pound in a store at the equator or in one near the North Pole? Does your answer depend on whether you use a spring scale or an equal-arm balance? Explain.

12. At most locations on earth, why does a plumb bob not hang precisely in the direction of the earth's center?

13. When will your weight be the greatest, as measured by a scale in a moving elevator? When the elevator (a) accelerates downward, (b) accelerates upward, (c) is in free fall, (d) moves upward at constant speed. In which case would your weight be the least? When would it be the same as when you are on the ground?

14. If you were in a satellite orbiting the earth, how might you cope with walking, drinking, or putting a pair of scissors on a table?

15. An antenna loosens and becomes detached from a satellite in a circular orbit around the earth. Describe the antenna's motion subsequently. If it will land on earth, describe where; if not, describe how it could be made to land on earth.

16. The sun is directly below us at midnight, in line with the earth's center. Are we then heavier at midnight, due to the sun's gravitational force on us, than we are at noon? Explain.

17. The gravitational force on the moon due to the earth is only about half the force on the moon due to the sun (see Example 5–7). Why isn't the moon pulled away from the earth?

18. Astronauts who spend long periods in outer space could be adversely affected by weightlessness. One way to simulate gravity is to shape the spaceship like a bicycle wheel which rotates about an axis just like a wheel, with the astronauts walking on the inside of the "tire." Explain how this simulates gravity. Consider (a) how objects fall, (b) the force we feel on our feet, and (c) any other aspects of gravity you can think of.

19. People sometimes ask, "What keeps a satellite up in its orbit around the earth?" How would you respond?

20. Explain why a person running experiences "free fall" or "apparent weightlessness" between steps.

21. The earth moves faster in its orbit around the sun in winter than in summer. Is it closer to the sun in summer or in winter? Does this affect the seasons? Explain.

22. Discuss the conceptual differences between **g** as acceleration due to gravity and $\mathscr{G}$ as gravitational field intensity.

Problems

1. (I) Calculate the centripetal acceleration of the earth in its orbit around the sun and the net force exerted on the earth. What exerts this force on the earth? Assume the earth's orbit is a circle of radius 1.49×10^{11} m.

2. (I) What is the maximum speed with which a 1300-kg car can round a turn of radius 95 m on a flat road if the coefficient of friction between tires and road is 0.55? Is this result independent of the mass of the car?

3. (I) How large must the coefficient of friction be between the tires and the road if a car is to round a level curve of radius 62 m at a speed of 55 km/h?

4. (I) A force of 26.0 N is applied to a 0.60-kg stone to keep it rotating in a horizontal circle of radius 0.40 m. Calculate its speed.

5. (I) A child moves with a speed of 1.50 m/s when 7.8 m from the center of a merry-go-around. Calculate (a) the centripetal acceleration of the child, and (b) the net horizontal force exerted on the child (mass = 25 kg).

6. (II) How fast (in rpm) must a centrifuge rotate if a particle 9.0 cm from the axis of rotation is to experience an acceleration of 110,000 g's?

7. (II) Is is possible to whirl a bucket of water fast enough in a vertical circle so the water won't fall out? If so, what is the minimum speed?

8. (II) A coin is placed 12.0 cm from the axis of a rotating turntable of variable speed. When the speed of the turntable is slowly increased, the coin remains fixed on the turntable until a rate of 58 rpm is reached, at which point the coin slides off. What is the coefficient of static friction between the coin and the turntable?

9. (II) In a "Rotor-ride" at a carnival, riders are pressed against the inside wall of a vertical cylinder 2.9 m in radius rotating at a speed of 0.92 revolutions per second when the floor drops out. What minimum coefficient of friction is needed so a person won't slip down? Is this safe?

10. (II) What minimum speed must a roller coaster be traveling when upside down at the top of a circle (Fig. 5–19) if the passengers are not to fall out? Assume a radius of curvature of 8.0 m.

11. (II) A ball on the end of a string is revolving at a uniform rate in a vertical circle or radius 96.5 cm as shown in Fig. 5–20. If its speed is 3.15 m/s and its mass is 0.335 kg, calculate the tension in the string when the ball is (a) at the top of its path, and (b) at the bottom of its path.

12. (II) A ball of mass m is revolved in a vertical circle at the end of a cord of length L. What is the minimum speed v needed at the top of the circle if the cord is to remain taut?

13. (II) A projected space station consists of a thin circular tube which is set rotating about its center (like a bicycle wheel). The circle formed by the tube has a diameter of 1.6 km. (a) On which inner surface of the tube will people be able to walk? (b) What must be the rotation speed (revolutions per day) if an effect equal to gravity at the surface of the earth (1 g) is to be felt?

14. (III) If a curve with a radius of 60 m is properly banked for a car traveling 60 km/h, what must be the coefficient of static friction for a car not to skid when traveling at 90 km/h?

15. (III) A 1200-kg car rounds a curve of radius 65 m banked at an angle of 14°. If the car is traveling at 80 km/h, will a friction force be required? If so, how much and in what direction?

16. (III) The sides of a cone make an angle ϕ with the vertical. A small mass m is placed on the inside of the cone and the

FIGURE 5–19

The "Great American Revolution" in an amusement park in California.

FIGURE 5–20

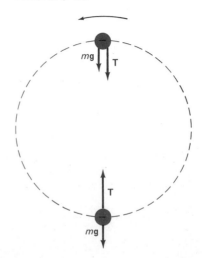

cone, with its point down, is revolved at a frequency f (revolutions per second) about its symmetry axis. If the coefficient of static friction is μ, at what positions on the cone can the mass be placed without sliding on the cone? (Give the maximum and minimum distances, r, from the axis).

SECTIONS 5–2 AND 5–3

17. (I) Calculate the force of gravity on a spacecraft 12,800 km above the earth's surface if its mass is 850 kg.

18. (I) Calculate the acceleration due to gravity at the surface of the moon. The moon's radius is about 1.7×10^6 m and its mass is 7.4×10^{22} kg.

19. (I) How far above the earth's surface will the acceleration of gravity be half what it is on the surface?

20. (II) Derive a formula for the mass of a planet in terms of its radius, r, the acceleration due to gravity at its surface, g_p, and the gravitational constant, G.

21. (II) Four 8.0-kg spheres are located at the corners of a square of side 0.50 m. Calculate the magnitude and direction of the gravitational force on one sphere due to the other three.

22. (II) At what distance from the earth will a spacecraft on the way to the moon experience zero net force because the earth and moon pull with equal and opposite forces?

23. (III) Show, using integral calculus, that the force exerted on a particle of mass m by a uniform sphere of mass M is given by Eq. 5–1 where r is the distance from m to the center of the sphere. (It is as if the mass of the sphere were concentrated at its center.) (Hint: to do this, consider a sphere as made up of very thin spherical shells; then consider each shell as made up of thin circular rings where all parts of each ring are equidistant from the particle of mass m. Determine the force on m due to each ring, sum (integrate) over the rings, and then over the shells.)

SECTION 5–4

24. (I) Calculate the effective value of g, the acceleration of gravity, (a) 3200 m, (b) 3200 km, above the earth's surface.

25. (I) Determine the mass of the sun using the known value for the period of the earth and its distance from the sun.

26. (I) Suppose the mass of the earth were doubled, but it kept the same density and spherical shape. How would the weight of objects at the earth's surface change?

27. (III) (a) Use the binomial expansion

$$(1 + x)^n = 1 + nx + \frac{n(n - 1)}{2} x^2 + \cdots$$

to show that the value of g is altered by approximately

$$\Delta g \approx -2g \frac{\Delta r}{r_e}$$

at a height Δr above the earth's surface, where r_e is the radius of the earth, as long as $\Delta r \ll r_e$. (b) What is the meaning of the minus sign in this relation? (c) Use this result to compute the effective value of g at 100 km above the earth's surface. Compare to a direct use of Eq. 5–1.

28. (III) Determine the magnitude and direction of the effective value of **g** at a latitude of $45°$ on the earth. Assume the earth is a sphere, but rotates.

29. (III) A ship steams at speed v along the equator. Show that the apparent weight w of an object as weighed on the ship is given approximately by $w = w_0(1 \pm 4\pi f v/g)$ where f is the frequency of rotation (revolutions/second) of the earth. Why is there a $\pm$ sign? Let w_0 be the measured weight of the object when the ship is at rest relative to the earth.

SECTION 5–6

30. (I) A 15.0-kg monkey hangs from a cord suspended from the ceiling of an elevator. The cord can withstand a tension of 185 N, and breaks as the elevator accelerates. What was the elevator's minimum acceleration (magnitude and direction)?

31. (I) Calculate the velocity of a satellite moving in a stable circular orbit about the earth at a height of 3200 km.

32. (II) One of the moons of Jupiter discovered by Galileo has a rotational period of 1.44×10^6 s and it is 1.9×10^9 m from Jupiter on the average. From these data, determine the mass of Jupiter.

33. (II) What will a spring scale read for the weight of a 55.0-kg woman in an elevator that moves (a) with constant speed upward of 5.0 m/s, (b) with constant speed downward of 5.0 m/s, (c) with upward acceleration of 0.33 g, (d) with downward acceleration 0.33 g, (e) in free fall?

34. (II) A 22.5-m diameter ferris wheel rotates once every 12.5 s. What is the fractional change in a person's apparent weight (a) at the top, (b) at the bottom, as compared to her weight at rest?

35. (II) What is the apparent weight of a 65-kg astronaut 4200 km from the center of the earth's moon in a space vehicle, (a) moving at constant velocity, (b) accelerating toward the moon at 3.6 m/s². State "direction" in each case.

36. (II) (a) Show that the mass of the sun is given by the formula $M_S = (4\pi^2 r^3)/(GT^2)$, where T is the period of any planet (the time to make one revolution about the sun), r is the radius of its orbit (assumed to be circular), and G is the gravitational constant. (b) Calculate the mass of the sun using the earth as the planet.

37. (II) How long would a day be if the earth were rotating so fast that objects at the equator would be weightless?

38. (II) Describe a general procedure to determine the mass of a planet from observations on the orbit of one of its satellites.

39. (II) Two stars maintain a constant distance apart of 8.0×10^{10} m and rotate about a point midway between them at a rate of one revolution every 12.6 yr. (a) Why don't the two stars crash into one another due to the gravitational force between them? (b) What must be the mass of each star (take the two masses to be the same)?

40. (III) An inclined plane, inside an elevator, makes a $30°$ angle with the floor. A mass m slides on the plane without friction. What is its acceleration if the elevator (a) accelerates upwards at 0.50 g, (b) accelerates downward at 0.50 g, (c) falls freely, (d) moves upward at constant speed?

SECTION 5–7

41. (I) Use Kepler's laws and the period of the moon (27.4 days) to determine the period of an artificial satellite orbiting near the earth's surface.

42. (I) The asteroid Icarus, though only a few hundred meters across, orbits the sun like the other planets. Its period is about 410 days. What is its average distance from the sun?

43. (I) Venus is an average distance of 1.08×10^8 km from the sun. Estimate the length of the Venusian year using the fact that the earth is 1.49×10^8 km from the sun on the average.

44. (II) Use Kepler's third law to determine at what height above the earth an artificial satellite must orbit if it is to remain at the same place relative to earth?

45. (II) (a) Use Kepler's second law to show that the ratio of the speeds of a planet at its near and far points from the sun is equal to the inverse ratio of the near and far distances: $v_N/v_F = d_F/d_N$. (b) Given that the earth's distance from the sun varies from 1.47 to 1.52×10^{11} m, determine the minimum and maximum velocities of the earth in its orbit around the sun.

*SECTION 5–9

*46. (I) What is the magnitude and direction of the gravitational field intensity midway between the earth and moon?

*47. (I) (a) What is the gravitational field intensity at the surface of the earth due to the sun? (b) Will this affect your weight significantly?

*48. (III) Two identical particles, each of mass m, are located on the x axis at $x = +x_0$ and $x = -x_0$. (a) Determine a formula for the gravitational field intensity due to these two particles for points on the y axis; that is, write $\mathscr{G}$ as a function of y, m, x_0, and so on. (b) At what point (or points) on the y axis is $\mathscr{G}$ a maximum, and what is its value there? (Hint: take the derivative $d\mathscr{G}/dy$.)

Work and Energy

6

Until now we have been studying the motion of a particle in terms of Newton's three laws of motion. In this analysis, *force* has played a central role as the quantity determining the motion. In this chapter and the two that follow, we discuss an alternative analysis of the motion of a particle in terms of the quantities *energy* and *momentum*. The importance of these quantities is that they are *conserved*; that is, in quite general circumstances they remain constant. The fact that such quantities are conserved gives us not only a deeper insight into the nature of the world but also gives us another way to attack practical problems.

The conservation laws of energy and momentum are especially valuable in dealing with systems of many objects in which a detailed consideration of the forces involved would be difficult.

This chapter is devoted to the very important concept of *energy* and the closely related concept of *work* which are scalar quantities and have no direction associated with them. Since these two quantities are scalars, they are often easier to work with than vector forces. Energy derives its importance from two sources. First, it is a conserved quantity; and second, energy is a concept that is useful not only in the study of motion, but in all areas of physics and other sciences as well. But before discussing energy itself, we first examine the concept of work.

 ## 6–1 Work Done by a Constant Force

The word *work* has a variety of meanings in everyday language. But in physics, work is given a very specific meaning to describe what is accomplished by the action of a force when it makes an object move through a distance. Specifically, the **work** done on a particle by a constant force (constant in both magnitude and direction) is defined to be *the product of the magnitude of the force times the component of the displacement parallel to the force.* In equation form we can write this as

$$W = Fd \cos \theta \qquad (6–1)$$

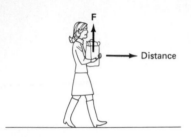

FIGURE 6–1

Work = 0 in this case since **F** is perpendicular to distance.

where F is the constant force, d is the net displacement of the particle, and θ is the angle between the directions of the force and the net displacement. (Note that $d\cos\theta$ is the component of **d** parallel to **F**.)

First let's consider the case where the motion and the force are in the same direction so $\cos\theta = 1$ and $W = Fd$. For example, if you push a loaded grocery cart a distance of 50 m by exerting a horizontal force of 30 N on the cart, you do 30 N × 50 m = 1500 N·m of work on the cart.

As this example shows, in SI units work is measured in newton-meters; for convenience, a special name is given to this unit, the *joule* (J): 1 J = 1 N·m. In the cgs system the unit of work is called the *erg* and is defined as 1 erg = 1 dyne·cm. In British units, work is measured in foot-pounds. It is easy to show that 1 J = 10^7 erg = 0.7376 ft·lb.

A force can be exerted on an object and yet do no work. For example, if you hold a heavy bag of groceries in your hands at rest, you do no work on it; you may become tired (and indeed energy is being expended by your muscles), but because the bag is not moved through a distance (the displacement is zero) the work $W = 0$. You also do no work if you carry the bag of groceries as you walk horizontally across the floor as shown in Fig. 6–1. No horizontal force is required to move the package at a constant velocity. However, you do exert an upward force **F** on the package equal to its weight. But this upward force is perpendicular to the horizontal motion of the package and thus has nothing to do with that motion; hence the upward force is doing no work. This is consistent with our definition of work, Eq. 6–1: $W = 0$ since $\theta = 90°$ and $\cos 90° = 0$. Thus when the force is perpendicular to the motion, no work is done by that force.

Consider now a boy pulling a wagon by exerting a force **F** at an angle θ to the horizontal as shown in Fig. 6–2. If he pulls the wagon a distance d along the ground, the work done is readily found using Eq. 6–1. If the force is 20 N, the angle 30°, and the wagon is pulled 100 m, the total work done is (20 N)(100 m)(0.866) = 1700 J.

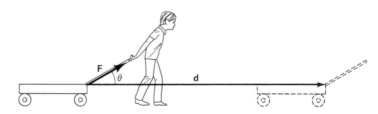

FIGURE 6–2

Work done by a force **F** acting at an angle θ to the ground is $Fd\cos\theta$.

EXAMPLE 6–1   Calculate the work done against gravity to carry a 15.0-kg backpack up a hill of height $h = 10.0$ m as shown in Fig. 6–3.

SOLUTION   Neglecting any acceleration that might occur, the person exerts a constant upward force **F** on the pack of magnitude equal to its weight: (15.0 kg)(9.80 m/s²) = 147 N. Eq. 6–1 can be written $W = F(d\cos\theta)$, and we note from Fig. 6–3 that $d\cos\theta = h$. So the work done is

$$W = Fh = (147\ \text{N})(10.0\ \text{m}) = 1470\ \text{J}.$$

Note that the work done depends only on the change in elevation and not on the steepness of the hill. The same work would be done to lift the pack vertically the same height h. To find the total work done by the person against gravity to climb the hill, we would proceed in the same way using the weight of the person plus the pack.

FIGURE 6–3

Example 6–1.

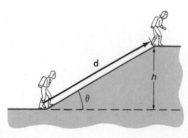

When dealing with work, as with force, it is necessary to specify whether you are talking about work done *by* a specific object, or done *on* a specific object. It is also important to specify if the work done is due to one particular force, or due to the total *net force* on the object.

The total (or net) work done on an object is the algebraic sum of the work done by each force; this total work is, of course, the work done by the resultant net force. For example, when a person gently lifts a 5.0-kg package from the ground up to a table 1.0 m high, there are two forces exerted on the package: the person pulling upward, and gravity (its weight $= mg = 49$ N) acting downward. If the package is lifted gently at constant speed, the upward force exerted by the person is equal to the weight, and the work done *by the person* is $W = (49 \text{ N})(1.0 \text{ m}) = 49$ J. If we take the upward direction as positive, then the work done *by gravity* will be $W = (-49 \text{ N})(1.0 \text{ m}) = -49$ J. (Note that when the force is in the direction opposite to the displacement, the work is negative; and in this case $\theta = 180°$ in Eq. 6–1.) Thus the *net work* done on the package is 49 J $+ (-49$ J$) = 0$. This is consistent with the fact that since the net force on the package is zero, the net work on it must be zero. This of course does not contradict the fact that the person did do work on the package equal to 49 J.

6–2 Scalar Product of Two Vectors

Although work is a scalar, it involves the product of two quantities, force and displacement, each of which are vectors. Therefore, we now investigate the multiplication of vectors.

Because vectors have direction as well as magnitude they cannot be multiplied in the same way that scalars are. Instead we must *define* what the operation of vector multiplication means. Among the many possible ways to define how to multiply vectors, there are three ways that we will find useful in physics: (1) multiplication of a vector by a scalar, which was already discussed in Section 3–3; (2) multiplication of one vector by a second vector so as to produce a scalar; (3) multiplication of one vector by a second vector so as to produce another vector. The third type, called the *vector product*, will be discussed later, in Section 10–1. We now discuss the second type, called the *scalar product*, or *dot product* (because a dot is used to indicate the multiplication). If we have two vectors, **A** and **B**, then their **scalar** (or **dot**) **product** is defined to be

$$\mathbf{A} \cdot \mathbf{B} = AB \cos \theta \qquad (6\text{–}2)$$

where A and B are the magnitudes of the vectors and θ is the smallest angle ($< 180°$) between them when their tails touch, Fig. 6–4. Since A, B, and $\cos \theta$ are scalars, then so is the scalar product $\mathbf{A} \cdot \mathbf{B}$ (read "A dot B"). This definition, Eq. 6–2, fits perfectly with our definition of the work done by a constant force, Eq. 6–1; that is, we can write the work done by a constant force as the scalar product of force and displacement:

$$W = \mathbf{F} \cdot \mathbf{d} = Fd \cos \theta. \qquad (6\text{–}3)$$

Indeed, the definition of scalar product, Eq. 6–2, is so chosen because then certain physically important quantities, such as work and others we will meet later, can be described as the scalar product of two vectors. (We could have defined the scalar product as $AB \tan \theta$ or $AB^2 \cos \theta/2$, for example, but these definitions would have been useless to us in physics.)

An equivalent definition of the scalar product is that it is the product of the magnitude of one vector (say A) and the component of the other vector along the direction of the first ($B \cos \theta$).

Since A, B, and $\cos \theta$ are scalars, it doesn't matter in what order they are multiplied. Hence the scalar product is *commutative*:

$$\mathbf{A} \cdot \mathbf{B} = \mathbf{B} \cdot \mathbf{A}.$$

FIGURE 6–4

The scalar, or dot, product of two vectors, **A** and **B**, is $\mathbf{A} \cdot \mathbf{B} = AB \cos \theta$.

It is also easy to show that it is *distributive* (see problem 23 for the proof):

$$\mathbf{A} \cdot (\mathbf{B} + \mathbf{C}) = \mathbf{A} \cdot \mathbf{B} + \mathbf{A} \cdot \mathbf{C}.$$

Using this fact, and writing each vector in terms of its rectangular components using unit vectors (Section 3–5, Eq. 3–5) it can be shown (see problem 15) that

$$\mathbf{A} \cdot \mathbf{B} = A_x B_x + A_y B_y + A_z B_z. \tag{6-4}$$

This equation is a particularly useful one.

If $\mathbf{A}$ is perpendicular to $\mathbf{B}$, then $\mathbf{A} \cdot \mathbf{B} = 0$. But the converse, given that $\mathbf{A} \cdot \mathbf{B} = 0$, can come about in three different ways: $\mathbf{A} = 0$, $\mathbf{B} = 0$, or $\mathbf{A} \perp \mathbf{B}$.

EXAMPLE 6–2　The force in Fig. 6–2 has magnitude 20 N and makes an angle of 30° to the ground. Calculate the work done using Eq. 6–4 when the wagon is dragged 100 m along the ground.

SOLUTION　We choose the x axis horizontal to the right and the y axis vertically upward. Then $\mathbf{F} = (F \cos \theta)\mathbf{i} + (F \sin \theta)\mathbf{j} = (17 \text{ N})\mathbf{i} + (10 \text{ N})\mathbf{j}$, whereas $\mathbf{d} = (100 \text{ m})\mathbf{i}$. Then, using Eq. 6–4,

$$\begin{aligned} W = \mathbf{F} \cdot \mathbf{d} &= (17 \text{ N})(100 \text{ m}) + (10 \text{ N})(0) + (0)(0) \\ &= 1700 \text{ J.} \end{aligned}$$

Note that by choosing the x axis along $\mathbf{d}$, we simplified the calculation since then $\mathbf{d}$ has only one component.

6–3　Work Done by a Non-Constant Force

In many cases the force varies in magnitude or direction during a process. For example, as a rocket moves away from the earth, work is done against the force of gravity, which varies as the inverse square of the distance from the earth's center; and the force exerted by a spring increases with the amount of stretch. How do we calculate the work done by a nonconstant force?

Figure 6–5 shows the path of a particle in the xy plane as it moves from point a to point b. The path has been divided into short intervals each of length Δl_1, $\Delta l_2, \ldots \Delta l_7$. A force $\mathbf{F}$ acts at each point on the path, and is indicated at two points as $\mathbf{F}_1$ and $\mathbf{F}_5$. During each small interval Δl, the force is approximately constant; so, for the first interval, the force does work ΔW of approximately (see Eq. 6–1)

$$\Delta W \approx F_1 \cos \theta_1 \, \Delta l_1.$$

In the second interval the work done is approximately $F_2 \cos \theta_2 \, \Delta l_2$, and so on. The total work done in moving the particle the total distance $l = \Delta l_1 + \Delta l_2 + \ldots + \Delta l_7$ is the sum of all these terms:

$$W \approx \sum_{i=1}^{7} F_i \cos \theta_i \, \Delta l_i. \tag{6-5}$$

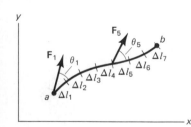

FIGURE 6–5

A particle acted on by a variable force, $\mathbf{F}$, moves along the path shown from point a to point b.

We can examine this graphically by plotting $F \cos \theta$ versus l as shown in Fig. 6–6a. The distance l has been subdivided into the same seven intervals indicated by the vertical dashed lines. The value of $F \cos \theta$ at the beginning of each interval is indicated by the horizontal dashed lines. Each of the shaded rectangles has an area $(F_i \cos \theta) (\Delta l_i)$ which is the work done during the interval. Thus the estimate of the work done given by Eq. 6–5 equals the sum of the areas of all the rectangles. If we subdivide the distance into a greater number of intervals so that each Δl_i is smaller, the estimate of the work done given by Eq. 6–5 becomes more accurate (the assumption that F is constant over each interval is more accurate). If we let each Δl_i approach zero (so we approach an infinite number of intervals) then we obtain an

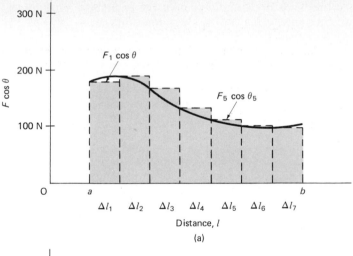

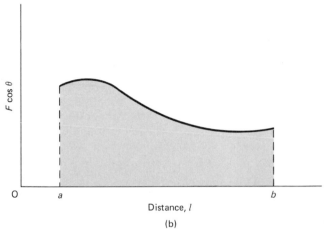

FIGURE 6–6

Work done by a force F is (a) approximately equal to the sum of the areas of the rectangles, (b) exactly equal to the area under the curve of $F \cos \theta$ vs. l.

exact result for the work done:

$$W = \lim_{\Delta l_i \to 0} \sum F_i \cos \theta_i \, \Delta l_i$$

$$W = \int_a^b F \cos \theta \, dl. \tag{6–6}$$

This limit as $\Delta l_i \to 0$ is called the *integral* of $(F \cos \theta \, dl)$ from point a to point b, and is written as shown in the second line (Eq. 6–6). The symbol for the integral, $\int$, is an elongated S to indicate an infinite *s*um; and Δl has been replaced by dl, meaning an infinitesimal distance.

In this limit as Δl approaches zero, the total area of the rectangles (Fig. 6–6a) approaches the area between the $(F \cos \theta)$ curve and the l axis from a to b as shown shaded in Fig. 6–6b. That is, *the work done by a variable force in moving a particle between two points is equal to the area under the $(F \cos \theta)$ versus (l) curve between those two points.*

In the limit as Δl approaches zero, the infinitesimal distance dl equals[†] the magnitude of the infinitesimal displacement vector $d\mathbf{l}$. The direction of the vector $d\mathbf{l}$

FIGURE 6–7

The displacement vector, $\Delta \mathbf{r}$, has a magnitude that is not necessarily equal to the distance traveled, Δl.

[†] The distance along a curve, Δl, is not generally equal to the magnitude of the displacement, $\Delta \mathbf{r}$, as shown in Fig. 6–7. But in the limit of infinitesimals they are equal: $dl = dr$, and in this limit the vector $d\mathbf{l} = d\mathbf{r}$. Note that we cannot define a vector $\Delta \mathbf{l}$ since we cannot assign it a unique direction when the path curves; we can define a direction for $d\mathbf{l}$—namely, the direction of the tangent to the curve at that point, so $d\mathbf{l} = d\mathbf{r}$ and $dl = dr$.

is along the tangent to the curve at that point, so θ is the angle between **F** and dl at any point. Thus we can write, using dot-product notation:

$$W = \int_a^b F \cos \theta \, dl = \int_a^b \mathbf{F} \cdot d\mathbf{l}. \qquad (6\text{--}7)$$

This is the *most general definition of work.* The integral in Eq. 6–7 is called a *line integral* since it is the integral of $F \cos \theta$ along the line that represents the path of the object. (Equation 6–1 for a constant force is a special case of Eq. 6–7.)

To actually use Eq. 6–6 or 6–7 to calculate the work, there are several options. If $F \cos \theta$ is known as a function of position, a graph like that of Fig. 6–6b can be made and the area determined graphically. Another possibility is to use numerical integration (numerical summing), perhaps with the aid of a computer or calculator. A third possibility is to use the analytical methods of integral calculus. To do this, we must be able to write F as a function of position: $F(x, y, z)$. As an example of the use of integral calculus, let us consider a one-dimensional situation and determine analytically the work done to stretch a coiled spring. Suppose the spring is attached at one end to a wall. To hold such a spring either stretched or compressed an amount x beyond its normal (unstretched) length requires a force that is directly proportional to x:

$$F(x) = kx,$$

where k is a constant, called the spring constant, and is a measure of the stiffness of the particular spring. This *spring equation*, sometimes referred to as *Hooke's law,*[†] is accurate for springs as long as x is not too great. Let us now calculate the work needed to stretch (or compress) the spring from its normal (unstretched) length, $x_a = 0$, to an extra length, $x_b = x$. We assume the stretching is done slowly, so that the acceleration is essentially zero. The force is exerted parallel to the axis of the spring (along the x axis) so **F** and dl are parallel. Hence the work done is (since $dl = dx \, \mathbf{i}$ in this case)[‡]

$$W = \int_{x_a = 0}^{x_b = x} \left[F(x)\mathbf{i} \right] \cdot \left[dx \, \mathbf{i} \right] = \int_0^x F(x) \, dx = \int_0^x kx \, dx = \tfrac{1}{2}kx^2 \Big|_0^x = \tfrac{1}{2}kx^2.$$

(Notice that we have used x to represent both the variable of integration, and the particular value of x at the end of the interval $x_a = 0$ to $x_b = x$.) Thus we see that the work needed is proportional to the square of the distance stretched (or compressed), x. This same result can be obtained by computing the area under the F versus x graph (with $\cos \theta = 1$ in this case) as shown in Fig. 6–8. Since the area is a triangle of altitude kx and base x, the work done, equal to the area, is

$$W = \tfrac{1}{2}(x)(kx) = \tfrac{1}{2}kx^2,$$

which is the same result as before.

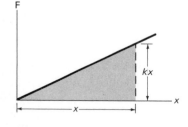

FIGURE 6–8

Work done to stretch a spring a distance x equals the triangular area under the curve $F = kx$; the area of a triangle is $\tfrac{1}{2} \times$ base $\times$ altitude, so $W = \tfrac{1}{2}(x)(kx) = \tfrac{1}{2}kx^2$.

■ **EXAMPLE 6–3** A simple pendulum consists of a small object of mass m (the "bob") suspended by a massless string of length L (Fig. 6–9a). A force **F** is applied in the horizontal direction (so $\mathbf{F} = F\mathbf{i}$) which moves the bob very slowly so the acceleration is essentially zero. (Note that the magnitude of **F** will vary with the angle θ that the string makes to the vertical at any moment.) (a) Determine the work done by this force, **F**, to move the pendulum from $\theta = 0$ to $\theta = \theta_0$. (b) Determine the work done by the gravitational force on the bob, $\mathbf{F}_g = m\mathbf{g}$, and the work done by the force $\mathbf{F}_s$ which the string exerts on the bob.

SOLUTION (a) We choose the coordinates of the bob to be $x = y = 0$ at its lowest point ($\theta = 0$). Since the acceleration is zero, the net force on the pendulum bob must be zero at all positions. Hence the x and y components of the total force

---

[†] The term "law" applied to this relation is not really appropriate since, first of all, it is only an approximation, and secondly because it refers only to a limited set of phenomena. Most physicists prefer to reserve the word "law" to those relations which are deeper and more encompassing and precise, such as Newton's laws of motion or the law of conservation of energy (Chapter 7).

[‡] See the Table of Integrals, Appendix B.

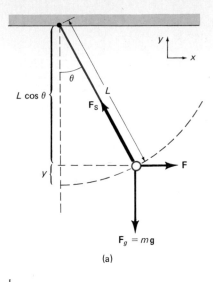

FIGURE 6–9

(a) The simple pendulum (Example 6–3). (b) Enlargement of the bob, showing infinitesimal displacement.

(a)

(b)

must be zero (Fig. 6–9a):

$$F - F_s \sin \theta = 0$$

$$F_s \cos \theta - mg = 0.$$

If we solve for F_s in the second equation, and substitute it into the first, we find

$$F = mg \tan \theta \quad \text{or} \quad \mathbf{F} = mg \tan \theta \, \mathbf{i}.$$

This tells us how F varies with position of the bob, where position is given in terms of θ. We see from Fig. 6–9b that at any point,

$$d\mathbf{l} = dx \, \mathbf{i} + dy \, \mathbf{j}$$

where

$$dx = \cos \theta \, dl$$

$$dy = \sin \theta \, dl.$$

The work done by the force $\mathbf{F}$ to move the bob from $\theta = 0$ to $\theta = \theta_0$ is

$$W_F = \int_{\theta = 0}^{\theta_0} \mathbf{F} \cdot d\mathbf{l} = \int_{\theta = 0}^{\theta_0} (mg \tan \theta \mathbf{i}) \cdot (\cos \theta \, dl \mathbf{i} + \sin \theta \, dl \mathbf{j})$$

$$= \int_{\theta = 0}^{\theta_0} mg \tan \theta \cos \theta \, dl$$

$$= \int_{\theta = 0}^{\theta_0} mg \sin \theta \, dl$$

$$= \int_{y = 0}^{y_0} mg \, dy = mgy \Big|_0^{y_0}$$

$$= mgy_0.$$

From Fig. 6–9a, $y = L - L \cos \theta$ so $y_0 = L(1 - \cos \theta_0)$ and in terms of θ_0, $W_F = mgL(1 - \cos \theta_0)$.

(b) The force $\mathbf{F}_s$ does no work since its direction is perpendicular to the displacement of the bob at every moment so $\mathbf{F}_s \cdot d\mathbf{l} = 0$ at all points. The work done by gravity is

$$W_g = \int_{y = 0}^{y_0} m\mathbf{g} \cdot d\mathbf{l} = \int_{y = 0}^{y_0} (-mg \, \mathbf{j}) \cdot (dx \, \mathbf{i} + dy \, \mathbf{j}) = -\int_0^{y_0} mg \, dy = -mgy_0.$$

Note that the total work done by all the forces is $mgy_0 + 0 - mgy_0 = 0$, consistent with the fact that the net force is zero.

6−4 Kinetic Energy and the Work−Energy Theorem

Energy is one of the most important concepts in science. Yet we cannot give a simple but accurate and general definition of energy in only a few words. Energy of various types can be defined fairly simply; and in this chapter we will define (translational) kinetic energy and (mechanical) potential energy. In later chapters we will define other types of energy, such as that related to heat (Chapters 19 to 21). The crucial aspect of all the types of energy is that they can be defined consistently with one another with the units of work (force × distance) and in such a way that the sum of all types, the *total energy*, is the same after any process occurs as it was before: that is, the quantity "energy" can be defined so that it is a conserved quantity. But more on this later.

For the purposes of this chapter, we could define energy in the usual way as "the ability to do work." This simple definition is not very precise, nor is it really valid for all types of energy.[†] It is not incorrect, however, for mechanical energy, which we discuss in the present chapter; and it can serve as a motivation for defining energy in purely mechanical systems since the connection between work and energy is fundamental. We now define and discuss one of the basic types of energy: kinetic energy.

A moving object can do work on another object it strikes. A flying cannonball does work on a brick wall it knocks down; a moving hammer does work on a nail it strikes. In either case, a moving object exerts a force on a second object and moves it through a distance. An object in motion has the ability to do work and thus can be said to have energy. The energy of motion is called **kinetic energy**, from the Greek word *kinetikos*, meaning "motion."

In order to obtain a quantitative definition for kinetic energy, let us calculate how much work a moving object actually can do in a particular case of motion in one dimension. For now we assume the object is a particle, or in any case is undergoing only translational motion. Suppose the moving object has mass m and velocity v, and it strikes a second object on which it does work and then comes to rest. To be concrete, we might think of a hammer striking a nail. Let us assume that, upon striking the nail, the hammer exerts a constant force $\mathbf{F}$ on the nail over a distance d parallel to $\mathbf{F}$, Fig. 6−10, until the hammer comes to rest. This force thus does an amount of work on the nail, $W = Fd$. We really want to focus our attention on the hammer, and by Newton's third law, the nail exerts a force equal to $-F$ on the hammer. Let us assume that $-F$ represents the net force on the hammer (imagine the person's hand as merely supporting the hammer after it has reached speed v). Then, by Newton's second law, the force that the nail exerts on the hammer decelerates it from its initial speed v to rest: $-F = ma$. The acceleration a of the hammer (which is constant since we assumed F is constant during this process) can be related to the initial speed v and the distance d using Eq. 2−9c; since the final speed is zero, $v^2 = -2ad$. Combining these equations, we find the work done by the hammer is

$$W = Fd = -mad = m\left(\frac{v^2}{2d}\right)d = \tfrac{1}{2}mv^2.$$

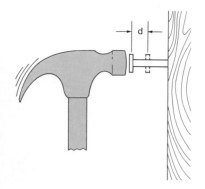

FIGURE 6−10

A hammer does work when striking a nail.

Thus, a body of mass m moving with a velocity v can do an amount of work equal to $\tfrac{1}{2}mv^2$ when it is brought to rest. We therefore *define* the quantity $\tfrac{1}{2}mv^2$ as the **translational kinetic energy** (KE) of the body:

$$\text{KE} = \tfrac{1}{2}mv^2. \tag{6−8}$$

This definition of KE makes quantitative the idea of energy as the ability to do work. (We shall see shortly that this definition applies also in three dimensions.) And, very

---

[†] Energy associated with heat is often not available to do work, as we will discuss in detail in Chapter 21.

importantly, as we shall see soon, with this definition of KE we will be able to deal with a more general quantity, energy, that is conserved in all processes.

We have seen that a moving object can do work. The opposite is true as well: work must be done on an object to give it KE. To find the precise relationship, we reverse the above argument. Suppose an object of mass m is moving in a straight line with an initial speed v_1, and to accelerate it uniformly to a speed v_2 a constant net force F is exerted on it parallel to its motion over a distance d; then the work done on it is $W = Fd$. Using Newton's second law, $F = ma$, and Eq. 2–9c, now written as $v_2^2 = v_1^2 + 2ad$, where v_1 is the initial speed and v_2 the final speed, we find

$$W = Fd = mad = m\left(\frac{v_2^2 - v_1^2}{2d}\right)d$$

or

$$W = \tfrac{1}{2}mv_2^2 - \tfrac{1}{2}mv_1^2 = \Delta\text{KE}. \tag{6–9}$$

That is,

The net work done on an object is equal to its change in kinetic energy.

This is known as the **work-energy theorem**. Notice, however, that since we made use of Newton's second law, $F = ma$, F must be the *net* force—that is, the sum of all forces acting on the object. Thus, the work-energy theorem is valid only if W is the *net work* done on the object.

The connection between work and kinetic energy operates in both directions. First, if work is done on an object, its kinetic energy increases. Second, if an object has kinetic energy, it can do work on something else; and if it does, its own kinetic energy decreases. Another way of saying this is that if the net work W done on an object is positive, the KE increases, whereas if W is negative, the KE decreases. If the net work done is zero, the KE remains constant.

We have proved the work-energy theorem, Eq. 6–9, for motion in one dimension with a constant force. It is valid even if the force is variable and the motion is in two or three dimensions; this can be proved as follows. Suppose the net force **F** on an object varies in both magnitude and direction and the path of the object is a curve as in Fig. 6–5. The force may be considered to be a function of l, the distance along the curve. The work done by this force is (Eq. 6–6):

$$W = \int F \cos\theta\, dl = F_{||}\, dl$$

where $F_{||}$ represents the component of F parallel[†] to the curve at any point. By Newton's second law,

$$F_{||} = ma_{||} = m\frac{dv}{dt}$$

where $a_{||}$, the component of a parallel to the curve at any point, is equal to the rate of change of speed, dv/dt. We can think of v as being a function of l and, using the chain rule for derivatives, we have

$$\frac{dv}{dt} = \frac{dv}{dl}\frac{dl}{dt} = v\frac{dv}{dl}$$

since dl/dt is the speed v. Thus (letting 1 and 2 refer to the initial and final quantities,

---

† Notice that a force (or component of a force) acting perpendicular to the velocity vector does no work. Such a force changes only the direction of the velocity; it does not affect the magnitude of the velocity. One example of this is uniform circular motion in which an object moving with constant speed in a circle has a ("centripetal") force acting on it toward the center of the circle; this force does no work on the object.

respectively):

$$W = \int_1^2 F_{||}\, dl = \int_1^2 m\frac{dv}{dt}\, dl = \int_1^2 mv\frac{dv}{dl}\, dl = \int_1^2 mv\, dv$$

which integrates to

$$W = \tfrac{1}{2}mv_2^2 - \tfrac{1}{2}mv_1^2 = \Delta\text{KE}.$$

This is, of course, the work-energy theorem which we have now proved for motion in three dimensions with a variable net force. Note, incidentally, that the work-energy theorem is not a new independent law. Rather, it has been derived from the definitions of work and kinetic energy using Newton's second law.

Because of the direct connection between work and kinetic energy, energy must be measured in the same units as work: joules in SI units, ergs in the cgs, and foot-pounds in the British system. Like work, kinetic energy is a scalar quantity. The kinetic energy of a set of particles is the (scalar) sum of the kinetic energies of the individual particles.

EXAMPLE 6–4 How much work is required to accelerate a 1000-kg car from 20 m/s to 30 m/s?

SOLUTION The work needed is equal to the increase in the kinetic energy:

$$W = \tfrac{1}{2}mv_2^2 - \tfrac{1}{2}mv_1^2$$
$$= \tfrac{1}{2}(1000\ \text{kg})(30\ \text{m/s})^2 - \tfrac{1}{2}(1000\ \text{kg})(20\ \text{m/s})^2$$
$$= 2.5 \times 10^5\ \text{J}.$$

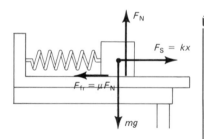

FIGURE 6–11
Example 6–5.

EXAMPLE 6–5 A horizontal spring has spring constant $k = 180$ N/m. (a) How much work is required to compress it from its unstretched length ($x = 0$) to $x = 11.0$ cm? (b) If a 1.85-kg block is placed against the spring and the spring is released, what will be the speed of the block when it separates from the spring at $x = 0$? Ignore friction. (c) Repeat part (b) but assume the block is moving on a table, Fig. 6–11, and the coefficient of kinetic friction is $\mu_k = 0.38$.

SOLUTION (a) We saw in Section 6–3 that the work, W, needed to compress a spring by a distance x is $W = \tfrac{1}{2}kx^2$. Therefore the work required is $W = \tfrac{1}{2}(180\ \text{N/m})(0.110\ \text{m})^2 = 2.16$ J, where we have converted all units to SI.

(b) In returning to its unstretched length, the spring does 2.16 J of work on the block [same calculation as in part (a), only in reverse]. According to the work-energy theorem the block requires KE = 2.16 J; since KE $= \tfrac{1}{2}mv^2$, the block's speed must be

$$v = \sqrt{\frac{2\text{KE}}{m}} = \sqrt{\frac{2(2.16\ \text{J})}{1.85\ \text{kg}}} = 2.34\ \text{m/s}.$$

(c) There are two forces on the block: that exerted by the spring and that exerted by friction. The spring does 2.16 J of work on the block. The work done by friction on the block is, since the normal force F_N equals the weight mg,

$$W_{fr} = (-\mu_k mg)(x) = -(0.38)(1.85\ \text{kg})(9.8\ \text{m/s}^2)(0.110\ \text{m}) = -0.76\ \text{J}.$$

This work is negative since the force of friction is in the direction opposite to the displacement X. The net work done on the block is $W = 2.16$ J $- 0.76$ J $= 1.40$ J. From the work-energy theorem, Eq. 6–9 (with $v_2 = v$ and $v_1 = 0$), we have

$$v = \sqrt{\frac{2W}{m}} = \sqrt{\frac{2(1.40\ \text{J})}{1.85\ \text{kg}}} = 1.26\ \text{m/s}.$$

6-5 Potential Energy

An object can be said to have energy not only by virtue of its motion, but also by virtue of its position or configuration (shape). This is called **potential energy** (PE).

A wound-up watch spring, for example, has potential energy. For, as it unwinds, it does work to move the watch hands around. The watch spring acquired its potential energy because work was done *on* it by the person winding the watch.

Perhaps the most common example of potential energy is *gravitational potential energy*. A heavy brick held high in the air has potential energy because of its position. It has the ability to do work, for if it is released it will fall to the ground and can do work on, say, a stake, driving it into the ground. Let us determine quantitatively the gravitational PE of an object near the surface of the earth. In order to lift an object of mass m vertically, a force at least equal to its weight, mg, must be exerted on it, say, by a person's hand. In order to lift it without acceleration to a height h above the ground, Fig. 6-12a, the work done on it by the person will be the product of the force, mg, and the vertical distance h; that is, $W = mgh$. If we instead allow the object to fall freely under the action of gravity and drive a stake into the ground, the object will do an amount of work equal to mgh on the stake, a fact which can be verified by using the kinematical equations as we did in Section 6-4 for kinetic energy. Thus, to raise an object of mass m to a height h *requires* an amount of work equal to mgh; and once at height h the object has the *ability* to do an amount of work equal to mgh.

Suppose, instead of moving vertically upward, the object follows some arbitrary path in the xy plane as shown in Fig. 6-12b. The object starts at a vertical height y_1 and reaches a height y_2, where $y_2 - y_1 = h$. To calculate the work done by gravity, W_g, during this process we use Eq. 6-7:

$$W_g = \int_1^2 \mathbf{F}_g \cdot d\mathbf{l} = \int_1^2 mg \cos \theta \, dl.$$

We now let $\phi = 180° - \theta$ be the angle between $d\mathbf{l}$ and its vertical component dy as shown in Fig. 6-12b; then, since $\cos \theta = -\cos \phi$ and $dy = dl \cos \phi$ we have

$$W_g = -\int_{y_1}^{y_2} mg \, dy$$
$$= -mg(y_2 - y_1). \tag{6-10}$$

Since $(y_2 - y_1)$ is the vertical height h, we see that the work done depends only on the vertical height and does not depend on the particular path taken! In the case shown in Fig. 6-12b, $y_2 > y_1$ and therefore the work done by gravity is negative. If $y_2 < y_1$, so that the object is falling, then W_g is positive.

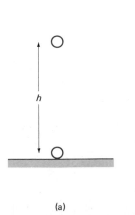

(a)

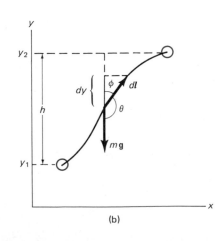

(b)

FIGURE 6-12

Object of mass m raised a height $h = y_2 - y_1$ (a) vertically, (b) along an arbitrary two-dimensional path.

If an object is allowed to fall a vertical distance $h = y_2 - y_1$, it can do an amount of work (on some other object or system) equal to $mgh = mg(y_2 - y_1)$. In accordance with the definition of energy as the ability to do work, we therefore define the *change in gravitational potential energy*, U, when an object moves from a height y_1 to a height y_2 as

$$\Delta U = U_2 - U_1 = mg(y_2 - y_1). \qquad \text{[gravitational PE]} \qquad (6\text{--}11)$$

Equation 6–11 defines the change in potential energy between two points. The potential energy, U, at any point a vertical height y above some reference point (the origin of the coordinate system) can be defined as

$$U = mgy. \qquad \text{[gravitational PE]} \qquad (6\text{--}12)$$

We could also define the gravitational PE at a point to be

$$U = mgy + C$$

where C is a constant, since this is also consistent with Eq. 6–11 (the constants C would cancel each other when we subtracted U_1 from U_2). We usually choose C to be zero, for convenience, since U depends on the choice of coordinate system anyway (that is, where we choose y to be zero). The gravitational PE of a book held high above a table, for example, depends on whether we measure y from the top of the table, from the floor, or from some other reference point.[†] What is physically important in any situation is the *change* in potential energy because that is what is related to the work done. We can thus choose the potential energy to be zero at any point that is convenient, but we must be consistent throughout any given problem. The change in PE between any two points does not depend on this choice.

EXAMPLE 6–6 A 1000-kg car moves from point A, Fig. 6–13, to point B and then point C. (a) What is its PE at B and C relative to point A? (b) What is its change in potential energy when it goes from B to C? (c) Repeat parts (a) and (b) but take the reference point ($y = 0$) to be at point C.

SOLUTION (a) Let us measure the heights from point A, so initially the car has zero PE. At point B, where $y = 10$ m,

$$PE_B = mgy = (1000 \text{ kg})(9.8 \text{ m/s}^2)(10 \text{ m}) = 1.0 \times 10^5 \text{ J}.$$

At point C, $y = -15$ m since C is below A. Therefore,

$$PE_C = mgy = (1000 \text{ kg})(9.8 \text{ m/s}^2)(-15 \text{ m}) = -1.5 \times 10^5 \text{ J}.$$

(b) In going from B to C, the potential energy change is:

$$PE_C - PE_B = (-1.5 \times 10^5 \text{ J}) - (1.0 \times 10^5 \text{ J}) = -2.5 \times 10^5 \text{ J}.$$

That is, the car loses 2.5×10^5 J.

(c) In this case, the car has a potential energy initially (at A) equal to $(1000 \text{ kg})(9.8 \text{ m/s}^2)(15 \text{ m}) = 1.5 \times 10^5$ J since $y = 15$ m. At B, its PE is 2.5×10^5 J, and at C it is zero. But the change in PE going from B to C (or from A to B) gives the same result as in part (b).

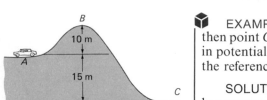

FIGURE 6–13
Example 6–6.

By comparing Eqs. 6–10 and 6–11, we see that we have defined the change in gravitational potential energy so it is equal to the negative of the work done by gravity when the object moves from height y_1 to y_2:

$$\Delta U = -W_g = -\int_1^2 \mathbf{F}_g \cdot d\mathbf{l}.$$

[†] A similar thing is true of kinetic energy: the KE of an object depends on the choice of reference frame; for example, a person sitting on a bus moving at speed v has zero KE when measured in the reference frame of the bus, but has $KE = \frac{1}{2}mv^2$ in a reference frame fixed to the ground. However, the *change* in PE does not depend on the reference frame, whereas for KE, even the change in KE depends on the reference frame.

(This is equivalent to saying that the change in potential energy, ΔU, equals the (positive of the) work done by a second force (of equal magnitude)—say, exerted by a person—to raise the object against the force of gravity.)

There are other types of potential energy besides gravitational. In general we define the *change in potential energy associated with a particular force* **F** *as the negative of the work done by that force*:

$$\Delta U = U_2 - U_1 = -\int_1^2 \mathbf{F} \cdot d\mathbf{l}. \qquad (6\text{-}13)$$

We will see in the next chapter that this definition does not have meaning for all possible forces, but makes sense only for so-called conservative forces such as gravity.

We consider now one other type of potential energy, that associated with elastic materials. This includes a great variety of practical applications. To take a simple example, consider a spring such as that in a watch, or better yet the coil spring shown in Fig. 6-14. The spring has potential energy when compressed (or stretched); for when released it can do work on a ball as shown. Like other elastic materials, a spring is described by Hooke's equation (as we discussed earlier in Section 6-3) as long as the displacement x is not too great. Let us take our coordinate system so the end of the unstretched spring is at $x = 0$ (Fig. 6-14a) and x is positive to the right. To hold the spring compressed (or stretched) a distance x, a person must exert a force $F = kx$. By Newton's third law, the spring pushes back with a force

$$F_s = -kx.$$

The negative sign appears because the force F_s is in the direction opposite to the displacement x (see Fig. 6-14b). From Eq. 6-13, the change in potential energy of the spring between $x_1 = 0$ (its uncompressed position) and $x_2 = x$, is

$$\Delta U = U(x) - U(0) = -\int_0^x (-kx)\,dx = \tfrac{1}{2}kx^2.$$

Here, $U(x)$ means the potential energy at x, and $U(0)$ means U at $x = 0$. It is convenient to choose the potential energy at $x = 0$ to be zero: $U(0) = 0$, so we have that the PE of a spring compressed or stretched an amount x from equilibrium is

$$U(x) = \tfrac{1}{2}kx^2. \qquad [\text{elastic PE}] \qquad (6\text{-}14)$$

In each of the preceding examples of potential energy—gravitational or elastic—an object has the capacity or potential to do work even though it is not yet actually doing it. That is why we use the term "potential" energy. From these examples we can also see that energy can be stored, for later use, in the form of potential energy. It is also worth noting that although there is a single, universal formula for the kinetic energy of a particle, $\tfrac{1}{2}mv^2$, there is no single formula for potential energy; instead, the mathematical form of the PE depends on the force or forces involved.

Potential energy is always associated with a force; and that force is always exerted on one body by another body (the earth exerts a gravitational force on a falling stone; a compressed spring exerts a force on a ball, and so on). Thus potential energy is not something a body "has" by itself, but rather it is associated with the interaction of two (or more) bodies.

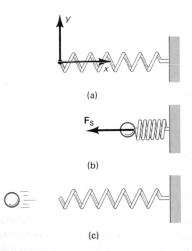

FIGURE 6-14

A spring (a) can store energy (elastic *PE*) when compressed (b), which can be used to do work when released (c).

6-6 Other Forms of Energy

Besides the KE and PE of ordinary objects, other forms of energy can be defined as well. These include electric energy, nuclear energy, thermal energy, and the chemical energy stored in food and fuels. With the advent of the atomic theory, these other forms of energy have come to be considered as kinetic or potential energy at the atomic or molecular level. For example, according to the atomic theory, thermal energy is interpreted as the kinetic and/or potential energy of rapidly moving molecules—when an object is heated, the molecules that make up the object move faster. On the other hand, the energy stored in food and fuel, such as gasoline, can be regarded as potential energy stored by virtue of the relative positions of the atoms within a molecule. For this energy to be used to do work, it must be released, usually through a chemical reaction. This is something like a compressed spring which, when released, can do work.

Electric, magnetic, and nuclear energy can also be considered examples of kinetic and potential energy. We will deal with these other forms of energy in later chapters.

 6–7   Transformation of Energy

Energy can be transformed from one form to another. A stone held high in the air has potential energy; as it falls, it loses potential energy since its height above the ground decreases; at the same time it gains in kinetic energy since its velocity is increasing. Potential energy is being transformed into kinetic energy.

Often the transformation of energy involves a transfer of energy from one body to another. The potential energy stored in the spring of Fig. 6–14b is transformed into kinetic energy of the ball, Fig. 6–14c. Water at the top of a dam has potential energy which is transformed into kinetic energy as the water falls; at the base of the dam, the kinetic energy of the water can be transferred to turbine blades and further transformed into electric energy as we shall see in Chapter 30. The potential energy stored in a bent bow can be transformed into the kinetic energy of an arrow.

In each of these examples, the transfer of energy is accompanied by the performance of work. The spring does work on the ball; water does work on the turbine's blades; the bow does work on the arrow. This observation gives us a further insight into the relation between work and energy:

Work is done whenever energy is transferred from one object to another.

Another example is a person throwing a ball or pushing a grocery cart. The work done is a manifestation of energy being transferred from the person (ultimately derived from the chemical potential energy stored in food) to the ball or cart.

The importance of the energy concept is that energy is a conserved quantity. That is, in any process, energy can be transformed from one type to another, but the total energy neither increases nor decreases. This is the law of conservation of energy. We discuss this great law, one of the most important in physics, in the next chapter.

 Summary _____

Work is done on a body by a force when that force moves the object through a distance. The work W done by a constant force $\mathbf{F}$ on an object whose position changes by a displacement $\mathbf{d}$ is given by

$$W = Fd \cos \theta = \mathbf{F} \cdot \mathbf{d}$$

where θ is the angle between $\mathbf{F}$ and $\mathbf{d}$.

The last expression is called the scalar product of $\mathbf{F}$ and $\mathbf{d}$. In general, the *scalar product* of any two vectors $\mathbf{A}$ and $\mathbf{B}$ is defined as $\mathbf{A} \cdot \mathbf{B} = AB \cos \theta$ where θ is the angle between $\mathbf{A}$ and $\mathbf{B}$.

The work W done by a variable force $\mathbf{F}$ on a particle that moves from point a to point b is

$$W = \int_a^b \mathbf{F} \cdot d\boldsymbol{l} = \int_a^b F \cos \theta \, dl$$

when dl represents an infinitesimal displacement along the path of the particle and θ is the angle between dl and $\mathbf{F}$ at each point of the particle's path.

The *kinetic energy* of a particle of mass m moving with speed v is defined to be

$$\mathrm{KE} = \tfrac{1}{2}mv^2.$$

The *work-energy theorem* states that the net work done on a body (by the net resultant force) is equal to the change in kinetic energy of the body:

$$W = \tfrac{1}{2}mv_2^2 - \tfrac{1}{2}mv_1^2.$$

An object can have *potential energy* because of its position or configuration. Examples are: gravitational potential energy (equal to mgy for a particle of mass m near the earth's surface that is a height y above some reference point); elastic potential energy (equal to $\tfrac{1}{2}kx^2$ for a spring with stiffness constant k stretched or compressed a distance x from equilibrium); and chemical, electrical, and nuclear energy. Potential energy is always specified relative to some reference point, since only changes in potential energy are physically meaningful. The change in potential energy, ΔU, of a body that moves from point 1 to point 2 under the action of a force $\mathbf{F}$ is defined as the negative of the work done by the force:

$$\Delta U = U_2 - U_1 = -\int_1^2 \mathbf{F} \cdot d\mathbf{l}.$$

(Potential energy can be defined only for certain forces—see Chapter 7.)

Questions

1. In what ways is the word "work" as used in everyday language the same as defined in physics? In what ways is it different?

2. A lever such as that shown in Fig. 6–15 can be used to lift objects we might not otherwise be able to lift. It is easy to show that the ratio of output force, F_o, to input force, F_i, is related to the lengths l_i and l_o from the pivot by $F_o/F_i = l_i/l_o$ (ignoring friction and the mass of the lever). Show that although a lever can be used to increase a force, it does not save us work.

3. Why is the work done by kinetic friction forces always negative?

4. A woman swimming upstream is not moving with respect to the shore. Is she doing any work? If she stops swimming and merely floats, is work done on her?

5. Show that $\mathbf{A} \cdot (-\mathbf{B}) = -\mathbf{A} \cdot \mathbf{B}$.

6. Does the scalar product of two vectors depend on the choice of coordinate system?

7. Can a dot product ever be negative? If yes, describe under what conditions this can be so.

8. If $\mathbf{A} \cdot \mathbf{C} = \mathbf{B} \cdot \mathbf{C}$, is it necessarily true that $\mathbf{A} = \mathbf{B}$?

9. Can a centripetal force ever do work? Explain.

10. Can the normal force on an object ever do work? Explain.

11. You have two springs that are identical except that spring 1 is stiffer than spring 2 ($k_1 > k_2$). On which spring is more work done (a) if they are stretched using the same force, (b) if they are stretched the same distance?

12. Can kinetic energy ever be negative? Explain.

13. Does the net work done on a particle depend on the choice of reference frame? How does this affect the work-energy theorem?

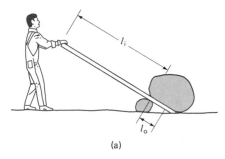

(a)

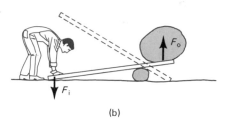

(b)

FIGURE 6–15
A simple lever.

14. By approximately how much does your gravitational potential energy change when you jump as high as you can?

15. Describe the energy transformations when a child hops around on a pogo stick.

16. Describe the energy transformations that take place when a skier starts skiing down a hill, but after a time is brought to rest by striking a snowdrift.

Problems

SECTION 6–1

1. (I) A 48-kg woman climbs a flight of stairs 4.5 m high. How much work is required?

2. (I) How far must a 37.8-kg pile driver fall if it is to do 5.60×10^4 J of work?

3. (I) A car does 5.8×10^4 J of work in traveling 1.25 km at constant speed. What was the average force of friction (from all sources) acting on the car?

4. (I) How much work did a horse do that pulled a 300-kg wagon 50 km along a level road if the effective coefficient of

friction was 0.060? Assume the horse exerts a force that is constant and horizontal.

5. (I) A 59-kg crate rests on the floor. How much work is required to move it at constant speed (a) 12.0 m along the floor against a friction force of 150 N, and (b) 12.0 m vertically?

6. (I) Determine the conversion factor between joules and ergs and between joules and foot-pounds.

7. (II) How much work is needed to push a 1250-kg car 115 m up a 13.5° incline at constant velocity? (a) Ignore friction. (b) Assume the coefficient of friction is 0.090.

8. (II) Ten bricks, each 6.0 cm thick and with mass 1.5 kg, lie flat on a table. How much work is required to stack them one on top of another?

9. (II) Calculate the net work done on a helicopter of mass M that is accelerating upward at $0.10\,g$ for a distance h.

10. (II) or (III) A 300-kg piano slides 4.5 m down a 25° incline and is kept from accelerating by a man who is pushing back on it. The effective coefficient of friction is 0.39. Calculate (a) the net work done on the piano, (b) the work done by the man on the piano, (c) the work done by gravity on the piano.

SECTION 6–2

11. (I) Show that

$$i \cdot i = j \cdot j = k \cdot k = 1$$

and

$$i \cdot j = i \cdot k = j \cdot k = 0,$$

where i, j, and k are unit vectors along the x, y, and z axes of a rectangular coordinate system.

12. (I) For any vector $V = V_x i + V_y j + V_z k$ show that

$$V_x = i \cdot V, \quad V_y = j \cdot V, \quad V_z = k \cdot V$$

13. (I) Vector V_1 points along the z axis and has magnitude $V_1 = 36$. Vector V_2 lies in the xz plane, has magnitude $V_2 = 55$ and makes a 25° angle with the x axis. What is the scalar product $V_1 \cdot V_2$?

14. (I) Calculate the angle between the vectors: $A = 6i + 4j - 3k$ and $B = -2i + 12j - 7k$.

15. (I) Prove that $A \cdot B = A_x B_x + A_y B_y + A_z B_z$ starting from Eq. 6–2 and using the distributive law (proved in problem 23).

16. (II) If $A = 4.0i - 3.5j$, $B = -1.7i + 8.1j + 4.6k$, and $C = 6.3i - 2.2j$, determine (a) $A \cdot (B + C)$, (b) $(A + C) \cdot B$, and (c) $(B + A) \cdot C$.

17. (II) Suppose that in Fig. 6–9 the force F is always exerted along the arc so that it makes an angle θ with the horizontal at any moment. Calculate the work done by this force to move the bob slowly at constant speed from its lowest point to a height y_0.

18. (II) Given vectors $A = 5.2i - 3.8j$ and $B = 3.6i + 5.1j$ determine the vector C which lies in the xy plane perpendicular to B and whose dot product with A is 16.4.

19. (II) Show that if two vectors have the same magnitude their sum must be perpendicular to their difference.

20. (II) Let $V = 6.8i + 12.0j - 4.7k$. What angles does this vector make with the x, y, and z axes?

21. (II) Use the scalar product to prove the law of cosines for a triangle:

$$c^2 = a^2 + b^2 - 2ab \cos \theta,$$

where a, b, and c are the lengths of the sides of a triangle and θ is the angle opposite side c.

22. (II) Vectors A and B are in the xy plane and their scalar product is 24.6 units. If A makes a 30° angle with the x axis and has magnitude $A = 11.7$ units, what can you say about B?

23. (III) Show that the scalar product of two vectors is distributive: $A \cdot (B + C) = A \cdot B + A \cdot C$. (Hint: use a diagram showing all three vectors in a plane and indicate dot products on the diagram.)

SECTION 6–3

24. (I) A spring has $k = 74$ N/m. Use a graph (such as that in Fig. 6–8) to determine the work needed to stretch it from $x = 3.0$ cm to $x = 5.5$ cm where $x = 0$ refers to its unstretched length.

25. (I) In pedaling a bicycle, a particular cyclist exerts a downward force of 85 N during each stroke. If the diameter of the circle traced by each pedal is 36 cm, calculate how much work is done in each stroke.

26. (II) The force exerted on a particle increases linearly from zero at $x = 0$ to 24.0 N at $x = 3.0$ m. It remains constant at 24.0 N from $x = 3.0$ m to $x = 8.0$ m and then decreases linearly to zero at $x = 11.0$ m. Determine the work done to move the particle from $x = 0$ to $x = 11.0$ m by graphically determining the area under F versus x graph.

27. (II) If the hill in Example 6–1 were not a slope but rather irregular as in Fig. 6–16, show that the same result would be obtained as in Example 6–1: namely, that the work done by gravity depends only on the height of the hill and not on its shape or the path taken.

28. (II) The force needed to hold a spring compressed an amount x from its normal length is given by $F = kx + ax^3 + bx^4$. How much work must be done to compress it by an amount X?

29. (II) In Fig. 6–6a assume the distance axis is linear and that $l_a = 4.0$ m and $l_b = 28.5$ m. Estimate the work done by this force in moving a 25-kg object from l_a to l_b.

30. (III) A 1400-kg space vehicle falls vertically from a height of 2500 km above the earth's surface. (a) Determine how

FIGURE 6–16

much work is done by the force of gravity in bringing the vehicle to the earth's surface by first constructing an F versus r graph (using Eq. 5–1), where r is the distance from the earth's center; then determine the work graphically to an accuracy of 3 percent. (b) Repeat using integration.

SECTION 6–4

31. (I) A carbon atom of mass 1.99×10^{-26} kg has 4.64×10^{-19} J of kinetic energy. How fast is it moving?

32. (I) How much work must be done to slow down a 1000-kg car from 90 km/h to 30 km/h?

33. (I) How much work does it take to accelerate an electron ($m = 9.11 \times 10^{-31}$ kg) from rest to 2.10×10^6 m/s?

34. (I) (a) A 3.0-g locust reaches a speed of 3.40 m/s during its jump. What is its KE at this speed? (b) If the locust transforms energy with 40 percent efficiency, how much energy was required for the jump?

35. (I) A 1250-kg car rolling on a horizontal surface has speed $v = 40$ km/h when it strikes a horizontal coiled spring and is brought to rest in a distance of 2.5 m. What is the spring constant of the spring?

36. (I) A baseball ($m = 140$ g) traveling 30 m/s moves a fielder's glove backward 35 cm when the ball is caught. What was the average force exerted by the ball on the glove?

37. (II) The force required to compress a horizontal spring an amount x is given by $F = 230x + 2.7x^3$ where x is in meters and F in newtons. If the spring is compressed 2.0 m, what speed will it give to a 3.0-kg ball held against it and released?

38. (II) If the speed of a car is doubled, by what factor will its minimum braking distance be increased, assuming all else is the same?

39. (II) One car has twice the mass of a second car, but only half as much KE. When both cars increase their speed by 3.0 m/s, they then have the same KE. What were the original speeds of the two cars?

40. (II) A 130-kg load is lifted 30 m vertically by a single cable with an acceleration $a = 0.15\, g$. Determine (a) the tension in the cable, (b) the net work done on the load, (c) the work done by the cable on the load, (d) the work done by gravity on the load, (e) the final speed of the load assuming it started from rest.

41. (II) An elevator cable breaks when a 750-kg elevator is 25 m above a huge spring ($k = 4.0 \times 10^4$ N/m) at the bottom of the shaft. Calculate (a) the work done by gravity on the elevator before it hits the spring, (b) the speed of the elevator just before striking the spring, (c) the amount the spring compresses (note that work is done by both the spring and gravity in this part).

42. (III) What should be the spring constant, k, of a spring designed to bring a 1500-kg car to rest from a speed of 90 km/h so that the occupants undergo a maximum acceleration of 15 g?

SECTION 6–5

43. (I) A spring has a spring constant, k, of 320 N/m. How much must this spring be compressed to store 50 J?

44. (I) A 4.2-kg monkey swings from one branch to another 1.7 m higher. What is the change in its potential energy?

45. (I) A 1.80-m tall person lifts a 230-g book so it is 2.15 m off the ground. What is the potential energy of the book relative to (a) the ground, (b) the top of the person's head? (c) How is the work done related to the answers in parts (a) and (b)?

46. (I) A 65-kg hiker starts at an elevation of 1500 m and climbs to the top of a 2600-m peak. (a) What is the hiker's change in potential energy? (b) What is the minimum work required? (c) Can the actual work done be more than this? Explain.

47. (II) (a) A spring of spring constant k is initially compressed a distance x_0 from its equilibrium length. What is the change in potential energy if it is now compressed to an amount x from equilibrium? (b) The spring is now *stretched* a distance x_0 from the equilibrium length. What is the change in potential energy as compared to when it is compressed by an amount x_0 from the equilibrium length?

48. (II) The force of electrical attraction between two charged particles varies as the inverse square of the distance r between them, $F = C/r^2$, where we take C as a constant. Determine (a) work required to increase the distance between the two particles from r_1 to $r_2 = r_1 + \Delta r$, and (b) the potential energy as a function of r, taking $U = 0$ at $r = \infty$.

49. (III) Assume a cyclist of weight mg can exert a force on the pedals equal to 0.90 mg on the average. If the pedals rotate in a circle of radius 18 cm, the wheels have a radius of 34 cm, and the front and back sprockets on which the chain runs have 42 and 27 points, respectively, determine the maximum steepness of hill the cyclist can climb at constant speed. Assume the mass of the bike is 12 kg and that of the rider is 60 kg. Ignore friction. Assume the cyclist's average force is always (a) downward, (b) tangential to pedal motion.

NUMERICAL/PROGRAMMABLE CALCULATOR[†]

*50. (II) The net force along the linear path of a particle of mass 480 g has been measured at 10.0-cm intervals, starting at $x = 0.0$, to be 26.0, 28.5, 35.6, 29.6, 32.8, 40.1, 46.6, 42.2, 48.8, 52.6, 55.8, 60.2, 60.6, 58.2, 53.7, 50.3, 45.6, 45.2, 43.2, 38.9, 35.1, 30.8, 27.2, 21.0, 22.2, 18.6. Determine the total work done on the particle over this entire range.

*51. (III) The force on a 150-g particle is given by the relation

$$F = \frac{848}{x^2 + 14.0x}$$

where F is in newtons and x in meters. Use numerical integration to estimate (to 2 percent accuracy) the work done by this force to move the particle from $x = 2.00$ m to $x = 6.50$ m. (Hint: Use of a programmable calculator or computer is advised.)

*52. (III) Refer to problem 54, Chapter 4. In that problem estimate (within 2 percent) the net work done on the skydiver over the 15.0-s interval.

[†] See Section 2–10.

Conservation of Energy

7

This chapter continues the discussion of the concepts of work and energy begun in the previous chapter. Now we will see why these concepts, particularly energy, are so important. The reason, ultimately, is because energy is conserved—the total energy remains constant in any process. That such a quantity can be defined which remains constant, as far as our best experiments can tell, is a remarkable statement about nature. The law of conservation of energy is, in fact, one of the great unifying principles of science.

This law also gives us another tool, another approach, for solving problems. There are many situations for which an analysis based on Newton's laws would be difficult or impossible (the forces may not be known or accessible to measurement), but which can be dealt with using the law of conservation of energy, and in some cases other conservation laws (such as the conservation of momentum) which we will deal with in later chapters.

 ## 7–1 Conservative Forces and the Work–Energy Theorem

We will find it useful to categorize forces into two types: conservative and non-conservative. By definition, we call any force a **conservative force** if (*a*) the *force depends only on position* and (*b*) the *work done by the force on a particle moving between any two positions depends only on the initial and final positions* and so *is independent of the particular path taken.*

One example of a conservative force is the force of gravity. We saw in Section 6–5 that the work done by the gravitational force $\mathbf{F}_g = m\mathbf{g}$ on an object (near the earth's surface), when it moves from one point to another, does not depend on the path taken. For instance the work done by gravity when an object moves vertically from an elevation y_1 to an elevation y_2 is the same as when the object takes a curved path. Hence, gravity is a conservative force.

An exactly equivalent definition of a *conservative force* is this: *a force is a conservative force if the net work done by the force is zero whenever a particle moves along any closed path that returns it to its original position.* (We can call such a path a "round-trip".) To see why this is equivalent to our earlier definition, consider a particle that moves from point 1 to point 2 via either of two paths labeled A and B in Fig. 7–1a. If we assume a conservative force acts on the particle, the work done by this force is the same whether the particle takes path A or path B, by our first definition. This work to get from point 1 to point 2 we will call W. Now consider the round trip shown in Fig. 7–1b. The particle moves from 1 to 2 via path A and our force does work W. Our particle then returns to point 1 via path B. How much work is done during the return? In going from 1 to 2 via path B the work done is W which by definition equals $\int_1^2 \mathbf{F} \cdot d\mathbf{l}$. In doing the reverse, going from 2 to 1, the force $\mathbf{F}$ at each point is the same but $d\mathbf{l}$ is directed in precisely the opposite direction; hence $\mathbf{F} \cdot d\mathbf{l}$ has the opposite sign at each point so the total work done in making the return trip from 2 to 1 must be $-W$. Hence the total work done in going from 1 to 2 and back to 1 is $W + (-W) = 0$, which proves the equivalence of the two above definitions for conservative force.

The second definition of a conservative force illuminates an important aspect of such a force: the *work done by a conservative force is recoverable* in the sense that if work is done *by* a particle on something else on one part of its path, an equivalent amount of work will be done *on* our particle on its return path.

As mentioned above, the force of gravity is conservative, and it is easy to show (see the problems) that the elastic force ($F = kx$) is also conservative. But not all forces are conservative. The force of friction, for example, is a *nonconservative* force. For example, the work done in moving a heavy crate across a level floor is equal to the product of the friction force and the total *distance traveled*, since the friction force is always directed precisely opposite to the direction of motion. Hence the work done to move the object between two points along a straight line is less than if the path between the two points is curved, such as a semicircle.

Note also in the case of kinetic friction that since the force of friction always opposes the motion, the work done on a body by friction is negative. Thus when an object is moved in a round trip, say from some point 1 to point 2 and back to 1, the total work done by friction is never zero—it is always negative. Thus the work done by a nonconservative force is not recoverable, as it is for a conservative force.

An important fact, and a main reason for distinguishing conservative from nonconservative forces, is that *potential energy can be defined only for a conservative force.* This can be seen by looking at the definition of potential energy (Eq. 6–13):

$$\Delta U = U_2 - U_1 = -\int_1^2 \mathbf{F} \cdot d\mathbf{l}.$$

The integral can be evaluated only if $\mathbf{F}$ is a function of position and the integral has a unique value only if

$$\int_1^2 \mathbf{F} \cdot d\mathbf{l}$$

(which equals the work done) depends only on the end points (1 and 2) and not on the path taken. If this integral were evaluated for a nonconservative force, its value would depend not only on the positions of points 1 and 2 but also on the path taken between these two points; ΔU would depend on path and we could not say that the U had a particular value at each point in space; thus the concept of potential energy is meaningless for a nonconservative force.

In the one-dimensional case, where a conservative force can be written as a function, say, of x, the potential energy becomes

$$U(x) = -\int F(x)\, dx.$$

This relation tells us how to obtain $U(x)$ when given $F(x)$. If, instead, we are given $U(x)$, we can obtain $F(x)$ by inverting the above equation: that is, we take the derivative of both sides, remembering that integration and taking the derivative are inverse operations

$$\frac{d}{dx} \int F(x)\, dx = F(x).$$

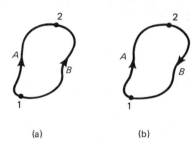

FIGURE 7–1

(a) A particle moves between points 1 and 2 via two different paths, A and B. (b) The particle makes a round trip from point 1 to point 2 via path A and back to point 1 via path B.

Thus

$$F(x) = -\frac{dU(x)}{dx}.$$ (7-1)

In three dimensions, this relation becomes

$$\mathbf{F}(x, y, z) = -\mathbf{i}\frac{\partial U}{\partial x} - \mathbf{j}\frac{\partial U}{\partial y} - \mathbf{k}\frac{\partial U}{\partial z}.$$

Here, $\partial/\partial x$, and so on, are called partial derivatives; $\partial/\partial x$, for example, means that although U may be a function of $x, y,$ and z [written $U(x, y, z)$], we take the derivative only with respect to x with the other variables held constant. The components of $\mathbf{F}$ are

$$F_x = -\frac{\partial U}{\partial x}, \quad F_y = -\frac{\partial U}{\partial y}, \quad F_z = -\frac{\partial U}{\partial z}.$$

EXAMPLE 7–1 Suppose $U(x) = -ax/(b^2 + x^2)$ where a and b are constants; what is F as a function of x?

SOLUTION Since $F(x) = -dU/dx$,

$$F(x) = -\frac{d}{dx}\left[-\frac{ax}{b^2 + x^2}\right] = \frac{a}{b^2 + x^2} - \frac{ax}{(b^2 + x^2)^2}2x = \frac{a(b^2 - x^2)}{(b^2 + x^2)^2}.$$

We can now state the **work-energy theorem** (discussed earlier in Section 6–4) to include potential energy. Suppose several forces act on a particle, some of which are conservative and we can write a potential energy function U for these conservative forces. Then *the work, W', done by all the other forces acting on a particle is equal to the total change in kinetic and potential energy of the particle*:

$$W' = \Delta\text{KE} + \Delta U.$$ (7-2)

We can now show this is true by recalling Eqs. 6–13 and 6–7:

$$\Delta U = -\int_1^2 \mathbf{F} \cdot d\mathbf{l} = -W_\text{c}$$

where $-W_\text{c}$ is the negative of the work done by the *c*onservative forces; thus, if we bring the ΔU term in Eq. 7–2 to the left side of the equation we have

$$W' + W_\text{c} = \Delta\text{KE}$$

where $W' + W_\text{c}$ is the work done by *all* the forces acting, and hence is the *net work* done; this is just the work-energy theorem in its original form, Eq. 6–9. This shows the validity of Eq. 7–2.

It must be emphasized that all the forces acting on a body must be included in Eq. 7–2, either in the potential energy term on the right, or in the work term, W', on the left (but not in both!). It is arbitrary whether a given conservative force is considered as doing work (and therefore included in W') or as involved in a change in potential energy, whereas nonconservative forces (such as friction) must be included in the work term, W'.

 7–2   Mechanical Energy and Its Conservation

Let us consider now a particle or rather a system (that is, a group) of particles in which only conservative forces act and in which energy is transformed from kinetic to potential, or vice versa.

We must consider a system, and not a single particle, since potential energy does not exist for an isolated body: potential energy is associated with a force, and a force on one object is always exerted by some other object. The potential energy is thus a property of the system as a whole. For the system of a mass m on the end of a spring the potential energy increases by $\frac{1}{2}kx^2$ when the spring is stretched a distance x from equilibrium, and this is related to the force $F = -kx$ exerted by the spring on the mass. Likewise, when a particle is raised to a height y above the earth's surface, the potential energy change is mgy; the system here is the particle plus the earth.

Suppose we have a conservative system (meaning only conservative forces act), such as a mass m either on the end of a spring or in the earth's gravitational field. Since only conservative forces act, when we write down the work-energy theorem, Eq. 7-2, we can include all the forces in the potential energy term. Hence $W' = 0$ and we have

$$\Delta \text{KE} + \Delta U = 0. \qquad \text{[conservative forces only]} \qquad (7\text{-}3)$$

What this equation tells us is that in any process, if the kinetic energy of the system increases, the potential energy decreases by an equal amount, and vice versa. We now define a quantity E, called the **total mechanical energy** of our system, as the sum of the kinetic energy[†] of our particle of mass m and the potential energy U:

$$E = \tfrac{1}{2}mv^2 + U.$$

For a conservative system we then have, from Eq. 7-3, that

$$E = \tfrac{1}{2}mv^2 + U = \text{constant}. \qquad \text{[conservative forces only]} \qquad (7\text{-}4)$$

That is, since any increase or decrease in KE is accompanied by a corresponding decrease or increase in U, the *total mechanical energy of a conservative system remains constant*. We also say that the total mechanical energy is *conserved*, meaning it neither increases nor decreases. This is called the **principle of conservation of mechanical energy for conservative forces**. We now see the reason for the name "conservative force": because for such forces the mechanical energy is conserved. The conservation of mechanical energy is an example of a more general law, the law of conservation of energy (to be discussed in the next section) that includes all forms of energy.

If we let v_1 and U_1 represent the velocity and potential energy at one instant, and v_2 and U_2 represent them at a second instant, then we can rewrite Eq. 7-4 as

$$\tfrac{1}{2}mv_1^2 + U_1 = \tfrac{1}{2}mv_2^2 + U_2. \qquad \text{[conservative system]} \qquad (7\text{-}5)$$

This is just another way to write the fact that the sum of kinetic and potential energies remains constant for conservative systems. From this equation we can see again that it doesn't make any difference where we choose the potential energy to be zero: adding a constant to U (as discussed in Section 6-5) merely adds a constant to both sides of the above equation, and these cancel. A constant also doesn't affect the force obtained from Eq. 7-1, $F = -dU/dx$, since the derivative of a constant is zero. Because we deal only with changes in the potential energy, the absolute value of U doesn't matter.

Let us now consider some examples of the use of the conservation of mechanical energy for conservative systems. We thus neglect friction and other nonconservative forces. First consider a rock falling under gravity (Fig. 7-2); the rock initially has only potential energy. As it falls, its PE decreases, but its KE increases to compensate so that the sum of the two remains constant. At any point along the path, the total mechanical energy is given by $\frac{1}{2}mv^2 + mgy$ where y is its height above the ground at that point and v is its velocity at that point. If we let the subscript 1 represent the rock at one point along its path (say, the initial point), and 2 represent it at some other point, then according to Eq. 7-5, we have:

$$\tfrac{1}{2}mv_1^2 + mgy_1 = \tfrac{1}{2}mv_2^2 + mgy_2. \qquad \text{[only gravity acting]} \qquad (7\text{-}6)$$

[†] We assume there is no other significant KE in the system. If, for example, m is on the end of a spring and the spring's mass cannot be neglected, then we would need an additional KE term.

FIGURE 7-2
The stone's PE changes to KE as it falls.

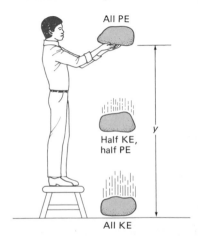

All PE

Half KE, half PE

y

All KE

To see the practical value of this relation, suppose the original height of the stone in Fig. 7–2 is $y_1 = 3.0$ m, and we wish to calculate its speed when it has fallen to 1.0 m above the ground. Then, since $v_1 = 0$ (the moment of release), $y_2 = 1.0$ m, and $g = 9.8$ m/s², Eq. 7–6 tells us

$$0 + (m)(9.8 \text{ m/s}^2)(3.0 \text{ m}) = \tfrac{1}{2}mv_2^2 + (m)(9.8 \text{ m/s}^2)(1.0 \text{ m}).$$

The m's cancel out, and solving for v_2^2 we find $v_2^2 = 2[(9.8 \text{ m/s}^2)(3.0 \text{ m}) - (9.8 \text{ m/s}^2)(1.0 \text{ m})] = 39.2$ m²/s², and $v_2 = \sqrt{39.2}$ m/s $= 6.3$ m/s.

Equation 7–6 can be applied to any object moving without friction under the action of gravity since, as we have seen, the PE depends only on position (in this case the height y) and not on the path taken. For example, Fig. 7–3 shows a car starting from rest at the top of a hill, and coasting without friction to the bottom and up the hill on the other side. Initially, the car has only potential energy. As it coasts down the hill, it loses PE and gains in KE, but the sum of the two remains constant. At the bottom of the hill, it has its maximum KE and as it climbs up the other side, the KE changes back to PE. When the car comes to rest again, all of its energy will be PE. Since PE is proportional to height and because energy is conserved, the car comes to rest at a height equal to its original height. If the two hills are the same height, the car just barely reaches the top of the second hill when it stops. If the second hill is lower than the first, not all of the car's KE will be transformed to PE and the car continues over the top and down the other side. If the second hill is higher, the car will only reach a height on it equal to its original height on the first hill. This is true no matter how steep the hill is, since PE depends only on the vertical height.

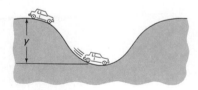

FIGURE 7–3

Car coasting down a hill illustrates conservation of energy.

◼ **EXAMPLE 7–2**   Assuming the height of the hill in Fig. 7–3 is 40 m, calculate (a) the velocity of the car at the bottom of the hill and (b) the height at which it will have half this speed.

SOLUTION   (a) Applying conservation of mechanical energy, we use Eq. 7–6 with $v_1 = 0$, $y_1 = 40$ m, and $y_2 = 0$. Then

$$0 + (m)(9.8 \text{ m/s}^2)(40 \text{ m}) = \tfrac{1}{2}mv_2^2 + 0.$$

The m's cancel out and we find $v_2 = 28$ m/s.
(b) We use the same equation but now $v_2 = 14$ m/s and y_2 is unknown;

$$0 + (m)(9.8 \text{ m/s}^2)(40 \text{ m}) = \tfrac{1}{2}(m)(14 \text{ m/s})^2 + (m)(9.8 \text{ m/s}^2)(y_2).$$

We cancel the m's and solve for y_2 and find $y_2 = 30$ m. That is, the car has a speed of 14 m/s when it is 30 *vertical* meters above the lowest point, both when descending the left-hand hill and when ascending the right-hand hill.

There are many interesting examples of the conservation of energy in athletics, one of which is the pole vault illustrated in Fig. 7–4. In terms of energy, the sequence of events is as follows: KE of the running athlete is transformed into elastic PE of the

FIGURE 7–4

Transformation of energy during a pole vault.

bending pole, and as he leaves the ground, into gravitational PE; then, as he reaches the top and the pole straightens out again, it has all been transformed to gravitational PE. (We ignored his low speed as he travels over the bar.) The pole does not supply any energy, but it acts as a very convenient device to *store* energy and thus aid in the transformation of KE into gravitational PE, which is the net result. The energy required to pass over the bar depends on how high the center of mass[†] (cm) of the vaulter must be raised. By bending their bodies, pole vaulters can keep their cm so low that it actually passes slightly beneath the bar, thus enabling them to cross over a higher bar than would otherwise be possible.

EXAMPLE 7–3   Calculate the kinetic energy and the velocity required for a 70-kg pole vaulter to pass over a 5.0-m high bar. Assume the vaulter's center of mass is initially 0.90 m off the ground and reaches its maximum height at the level of the bar itself.

SOLUTION   We equate the total energy just before the vaulter places the end of the pole into the ground (and the pole begins to bend and store PE) with his total energy as he passes over the bar (we ignore the small amount of KE at this point). Thus

$$\tfrac{1}{2}mv^2 + 0 = 0 + mgy$$

and

$$\text{KE} = \tfrac{1}{2}mv^2 = mgy$$
$$= (70 \text{ kg})(9.8 \text{ m/s}^2)(4.1 \text{ m}) = 2.8 \times 10^3 \text{ J}.$$

The velocity $v = \sqrt{\dfrac{2\text{KE}}{m}} = \sqrt{\dfrac{2(2800 \text{ J})}{70 \text{ kg}}} = 8.9 \text{ m/s}.$

Notice that the kinematic equations for constant acceleration, Eqs. 2–9, could have been used to treat the simple case of a falling rock (discussed immediately after Eq. 7–6), but they could not be used to solve either Example 7–2 or 7–3 since the acceleration is not constant in these cases. Here we have seen the great power of using the conservation of energy principle. Even though this principle is equivalent to Newton's second law, $\mathbf{F} = m\mathbf{a}$, and can be derived from it[‡], we have used the conservation of energy to solve problems without having to deal with force and acceleration which vary in a complicated way for these cases.

As another example of the conservation of mechanical energy let us consider a mass m connected to a spring whose own mass can be neglected and whose stiffness constant is k. The mass m has speed v at any moment and the potential energy of the system is $\tfrac{1}{2}kx^2$ where x is the displacement of the spring from its unstretched length. If neither friction nor any other force is acting, the conservation of energy principle tells us that

$$\tfrac{1}{2}mv_1^2 + \tfrac{1}{2}kx_1^2 = \tfrac{1}{2}mv_2^2 + \tfrac{1}{2}kx_2^2$$

where the subscripts 1 and 2 refer to the velocity and displacement at two different times.

FIGURE 7–5

Example 7–4.

EXAMPLE 7–4   A ball of mass $m = 2.6$ kg, starting from rest, falls a vertical distance $h = 55$ cm before striking a vertical coiled spring, which it compresses (see Fig. 7–5). If the spring has stiffness constant $k = 72$ N/m, what is the maximum compression of the spring? Let us measure all distances from the point where the ball first touches the unstretched spring.

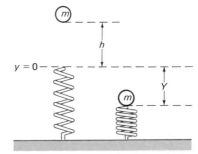

[†] The center of mass (cm) of a body is that point at which the entire mass of the body can be considered as concentrated for purposes of describing its translational motion—that is, so we can consider it as a particle. (This is discussed in Chapter 8.)

[‡] This is done via the work-energy theorem. (See Sections 6–4, 7–1, and 7–2.)

SOLUTION   Since the motion is vertical, we use y instead of x (y positive upward). Let the maximum compression of the spring be called Y. Then initially, the total energy of the system is

$$E_1 = mgh.$$

When the spring has been compressed to its maximum, the total energy is

$$E_2 = \tfrac{1}{2}kY^2 - mgY.$$

The first term on the right is the elastic PE, the second term is the gravitational PE since the ball fell a distance h before striking the spring ($y = 0$ at the top of the unstretched spring) and then drops a further distance Y as the spring compresses ($Y > 0$ so $y_2 = -Y$). The gravitational PE is $-mgY$ at this point. At both points 1 and 2 the kinetic energy is zero, although at intermediate points it is nonzero since the ball is moving. Because energy is conserved, $E_1 = E_2$, so

$$mgh = \tfrac{1}{2}kY^2 - mgY$$

or

$$\tfrac{1}{2}kY^2 - mgY - mgh = 0.$$

We use the quadratic formula to find Y:

$$Y = \frac{+mg \pm \sqrt{m^2g^2 + 2mghk}}{k} = 1.1 \text{ m}.$$

We used the root with the plus sign since this gives $Y > 0$, as assumed. The root with the negative sign yields $Y = -0.36$ m which would correspond to the ball and spring, connected to each other, having sprung back up a distance of 0.36 m above the unstretched ($y = 0$) length.

 ## 7–3   The Law of Conservation of Energy

We now take into account nonconservative forces such as friction, since they are important in real situations. For example, consider again the car in Fig. 7–3, but this time let us include friction. The car will not in this case reach the same height on the second hill as it had on the first hill because of friction.

In this and in other natural processes the sum of the kinetic and potential energies does not remain constant but decreases. Because of this, the friction force is called a *dissipative force*. Historically this was one factor that prevented the formulation of a comprehensive conservation of energy law until well into the nineteenth century. It was not until then that heat, which is always produced when friction is present (try rubbing your hands together), was seen to be related to energy. Quantitative studies by nineteenth-century scientists (Chapter 19) showed that the work done to produce a given amount of heat is always the same. They saw that it is possible to interpret the heat produced by friction as a new form of energy which we call *thermal energy*. A block sliding freely across a table, for example, comes to rest because of friction. Its initial kinetic energy is transformed into thermal energy. The block and table are a little warmer as a result of this process. A more obvious example of the transformation of kinetic energy into thermal energy can be observed by vigorously striking a nail several times with a hammer and then gently touching the nail with your finger.

According to the atomic theory, thermal energy represents kinetic energy of rapidly moving molecules. We shall see in Chapter 18 that a rise in temperature corresponds to an increase in the average kinetic energy of the molecules. Because thermal energy represents the energy of atoms and molecules that make up a body, it is often called *internal energy*. Internal energy[†], from the atomic point of view, can include not only kinetic energy of molecules but also potential energy (electrical in nature) because of the relative positions of atoms within molecules. On a macroscopic level, internal energy corresponds to nonconservative forces such as friction; but at the atomic level, the energy is kinetic and potential and the corresponding forces are mainly conservative.

To establish the more general law of conservation of energy, it required nineteenth-century physicists to recognize electrical, chemical, and other forms of energy in addition to heat and to explore if in fact they could fit into a conservation law. For each type of force, conservative or nonconservative, it has always been found possible to define a type of energy to correspond to the work done by such forces. And it has been found experimentally that the total energy E always remains constant. That is, the change in the total energy, kinetic plus potential plus all other forms of energy, equals zero:

$$\Delta\text{KE} + \Delta U + [\text{change in all other forms of energy}] = 0. \qquad (7\text{–}8)$$

This is one of the most important principles in physics. It is called the **law of conservation of energy** and can be stated as follows:

> **The total energy is neither increased nor decreased in any process. Energy can be transformed from one form to another, and transferred from one body to another, but the total amount remains constant.**

For conservative mechanical systems, this law can be derived from Newton's laws (Section 7–2) and thus is equivalent to them. But in its full generality, the validity of the law of conservation of energy rests on experimental observation. And even though Newton's laws are found to fail in the submicroscopic world of the atom, the law of conservation of energy has been found to hold there and in every experimental situation so far tested.

In Section 7–2 we discussed several examples of the law of conservation of energy for conservative systems. Now let us consider some examples that involve nonconservative forces as well.

Suppose, for example, that the car rolling on the hills of Fig. 7–3 is subject to frictional forces. In going from some point 1 to a second point 2, the work done by the friction force, $\mathbf{F}_{fr}$, acting in the car is $W' = \int_1^2 \mathbf{F}_{fr} \cdot d\mathbf{l}$. If $\mathbf{F}_{fr}$ is constant in magnitude, $W' = -F_{fr}l$ where l is the actual distance along the path traveled by the object in going from point 1 to point 2. (The minus sign comes from the fact that $\mathbf{F}_{fr}$ opposes the motion and thus is in the opposite direction to $d\mathbf{l}$.) From the work-energy theorem, Eq. 7–2, we have

$$W' = -F_{fr}l = \Delta\text{KE} + \Delta U = (\tfrac{1}{2}mv_2^2 - \tfrac{1}{2}mv_1^2) + (mgy_2 - mgy_1)$$

or

$$\tfrac{1}{2}mv_1^2 + mgy_1 = \tfrac{1}{2}mv_2^2 + mgy_2 + F_{fr}l.$$

The quantity $F_{fr}l$ represents the internal energy, U_{int}, produced by the friction force and is equal to the work done against this force; that is, $U_{int} = -W' = F_{fr}l$, and we can write

$$\text{KE}_1 + U_1 = \text{KE}_2 + U_2 + U_{int}.$$

In other words, the initial total energy of the car ($\text{KE}_1 + U_1$) is equal to the kinetic plus potential energy of the car at any subsequent point along its path plus the amount of internal (or thermal) energy produced in the process.

[†] The term *internal energy* can also be used to refer to kinetic and potential energy of the internal parts of a body, such as vibration, when we are primarily interested in the body's motion as a whole.

EXAMPLE 7–5 The car in Example 7–2 is found to reach a vertical height of only 25 m on the second hill before coming to a stop. It traveled a total distance of 400 m. Calculate the average friction force on the car. Assume the car has a mass of 1000 kg.

SOLUTION We take point 1 to be the instant when the car started moving and point 2 when it stopped. Then $v_1 = 0$, $y_1 = 40$ m, $v_2 = 0$, $y_2 = 25$ m, and $l = 400$ m. Thus

$$0 + (1000 \text{ kg})(9.8 \text{ m/s}^2)(40 \text{ m}) = 0 + (1000 \text{ kg})(9.8 \text{ m/s}^2)(25 \text{ m}) + F_{fr}(400 \text{ m}).$$

We solve this for F_{fr} to find $F_{fr} = 370$ N.

Next we consider another nonconservative force, that of a person slowly pulling on a pendulum bob as in Fig. 6–9.

 EXAMPLE 7–6 Calculate the work done by the force **F** which acts horizontally to slowly raise a pendulum bob as shown in Fig. 6–9 from $\theta = 0$ to $\theta = \theta_0$. (This was previously done in Example 6–3.)

SOLUTION Since the process is carried out slowly, the KE is essentially zero at all points. By the law of conservation of energy (or by the work-energy theorem, Eq. 7–2) the work done by the force **F** must be equal to the increase in potential energy of the bob. (There is no other energy or work involved since the tension, **T**, acts perpendicular to the motion and hence does no work.) Thus, if the pendulum is raised (see Fig. 6–9) to a height $y_0 = L(1 - \cos \theta_0)$, then

$$W = \Delta\text{PE} = mgy_0 = mgL(1 - \cos \theta_0).$$

Notice how much more easily we obtained this answer using energy than we did in Example 6–3.

7–4   Significance of the Conservation of Energy

The law of conservation of energy is important for two reasons. First, because it allows us to solve problems that would otherwise be difficult or impossible to deal with. Second, it is a unifying concept, an ordering principle—it shows us an order in the world around us and thus tells us something rather remarkable.

The concept of energy and its conservation permeates not only all of physics, but the other sciences as well. It is one thread that holds the various sciences together.

But note that the law of conservation of energy was not something that was discovered, in the same sense that gold was discovered in California in 1848. Rather, it is a concept invented by human minds. Energy is not a substance nor matter, but rather a concept, an abstraction. Some scientists might say it is a mathematical function. This of course doesn't lessen its importance. It took creativity on the part of many people to develop (create) this concept of energy—a quantity (mathematical, if you will) that when applied to any system of the physical world remains constant during any process.

It is important to distinguish the use of the word "conservation" in everyday life from its specific use in physics. In everyday usage, conservation means "saving" or "using wisely"—as when it is said that we should "conserve energy". (Actually, this statement really means that we should conserve *fuel*; energy *is* conserved.) In physics, the word conservation is applied to a quantity that remains strictly *constant*. To be clear, we can say that energy is conserved (we don't have to try to conserve it), but we should try to conserve our fuel resources so they are not used up too quickly.

7–5 Gravitational Potential Energy and Escape Velocity; Central Forces

We have been dealing with gravitational potential energy so far in this chapter assuming the force of gravity is constant, $\mathbf{F} = m\mathbf{g}$. This is an accurate assumption for ordinary objects near the surface of the earth. But to deal with gravity more generally we must use the fact that the gravitational force exerted by the earth on a particle of mass m decreases inversely as the square of the distance r from the earth's center. The precise relationship is given by Newton's universal law of gravitation (Sections 5–2 and 5–3):

$$\mathbf{F} = -G\frac{mM_e}{r^2}\hat{\mathbf{r}}$$

where M_e is the mass of the earth and $\hat{\mathbf{r}}$ is a unit vector (at the position of m) directed radially away from the earth's center; the minus sign indicates that the force on m is directed toward the earth's center, in the direction opposite to $\hat{\mathbf{r}}$. This equation can also be used to describe the gravitational force on a mass m in the vicinity of other heavenly bodies, such as the moon, planets, or sun, in which case M_e must be replaced by that body's mass.

Suppose an object of mass m moves from one position to another along an arbitrary path (Fig. 7–6) so that its distance from the earth's center changes from r_1 to r_2. The work done by the gravitational force is

$$W = \int_1^2 \mathbf{F} \cdot d\boldsymbol{l} = -GmM_e\int_1^2 \frac{\hat{\mathbf{r}} \cdot d\boldsymbol{l}}{r^2}$$

where $d\boldsymbol{l}$ represents an infinitesimal displacement. Since $\hat{\mathbf{r}} \cdot d\boldsymbol{l} = dr$ is the increment of $d\boldsymbol{l}$ along $\hat{\mathbf{r}}$ (see Fig. 7–6), then

$$W = -GmM_e\int_{r_1}^{r_2} \frac{dr}{r^2} = GmM_e\left(\frac{1}{r_2} - \frac{1}{r_1}\right)$$

or

$$W = \frac{GmM_e}{r_2} - \frac{GmM_e}{r_1}. \qquad \text{[gravity]}$$

Since the value of the integral depends only on the position of the end points (r_1 and r_2) and not on the path taken, the gravitational force must be a conservative force. We can therefore use the concept of potential energy for the gravitational force. Since the change in potential energy is always defined (Section 6–5) as the negative of the work done by the force, we have

$$\Delta U = U_2 - U_1 = -\frac{GmM_e}{r_2} + \frac{GmM_e}{r_1}. \qquad (7\text{–}9)$$

From Eq. 7–9 the potential energy at any distance r from the earth's center can be written:

$$U(r) = -\frac{GmM_e}{r} + C$$

where C is a constant. It is usual to choose $C = 0$, so that

$$U(r) = -\frac{GmM_e}{r}. \qquad \text{[gravity]} \qquad (7\text{–}10)$$

With this choice for C, $U = 0$ at $r = \infty$. As an object approaches the earth, its potential energy decreases and is always negative.

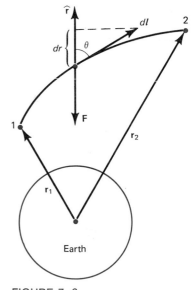

FIGURE 7–6
Arbitrary path of particle of mass m moving from point 1 to point 2.

The total energy of a particle of mass m which feels only the force of the earth's gravity is conserved since gravity is a conservative force. Therefore we can write

$$\tfrac{1}{2}mv_1^2 - G\frac{mM_e}{r_1} = \tfrac{1}{2}mv_2^2 - G\frac{mM_e}{r_2} = \text{constant.} \qquad \text{[only gravity acting]}$$

$$(7\text{–}11)$$

EXAMPLE 7–7 A package falls from a rocket traveling at a speed of 1800 m/s when 1600 km above the earth's surface. The package eventually falls to the earth. What is its speed just before impact? Ignore air resistance.

SOLUTION The package initially has a speed relative to earth equal to the speed of the rocket from which it falls. Thus we have $v_1 = 1.80 \times 10^3$ m/s, $r_1 = 1.60 \times 10^6$ m + 6.40×10^6 m = 8.00×10^6 m, and $r_2 = 6.40 \times 10^6$ m (the radius of the earth). We want to find v_2, which we do by using the conservation of energy—that is, by solving for v_2 in Eq. 7–11:

$$v_2 = \sqrt{v_1^2 - 2GM_e\left(\frac{1}{r_1} - \frac{1}{r_2}\right)}$$

$$= \left[(1.80 \times 10^3 \text{ m/s})^2 - 2(6.67 \times 10^{-11} \text{ N·m}^2/\text{kg}^2)(5.98 \times 10^{24} \text{ kg}) \right.$$
$$\left. \times \left(\frac{1}{8.00 \times 10^6 \text{ m}} - \frac{1}{6.40 \times 10^6 \text{ m}}\right)\right]^{1/2}$$

$$= 5310 \text{ m/s.}$$

In reality, the speed will be somewhat less than this because of air resistance. Note, incidentally, that the direction of the velocity never entered the problem. This is one of the advantages of the energy method.

When a body is projected into the air from the earth, it will return to earth unless its speed is very high. But if the speed *is* high enough it will continue out into space never to return to earth (barring other forces or collisions). The minimum initial speed needed so that an object never returns to earth is called the *escape velocity* from earth, v_{esc}. To determine v_{esc} from the earth's surface (ignoring air resistance), we use Eq. 7–11 with $v_1 = v_{esc}$ and $r_1 = R_e = 6.40 \times 10^6$ m, the radius of the earth. Since we want the minimum speed for escape, we need the object to reach $r_2 = \infty$ with merely zero speed, $v_2 = 0$. Applying Eq. 7–11 we have

$$\tfrac{1}{2}mv_{esc}^2 - G\frac{mM_e}{R_e} = 0 + 0$$

or

$$v_{esc} = \sqrt{\frac{2GM_e}{R_e}} = 1.12 \times 10^4 \text{ m/s}$$

or 11.2 km/s. It is important to note that although a mass can escape from the earth (or solar system) never to return, the force on it due to the earth's gravitational field is never zero for a finite value of r, although the force does become very small and usually ignorable at great distances.

Near the surface of the earth, it is possible to show that the change in gravitational potential energy, Eq. 7–9, reduces to our familiar form $mg(y_2 - y_1)$. (This is done in problem 29.)

The gravitational force belongs to a category of force called *central forces*. A *central force* is defined as any force whose magnitude depends only on the distance r from some single point, called the origin (which typically is a second particle, or the center of another object), and whose direction is either toward or away from this origin. Such a force can be written as

$$\mathbf{F} = F(r)\hat{\mathbf{r}}.$$

If we use the same argument we did in reaching Eq. 7–9, and notice again from Fig. 7–6 that $\hat{\mathbf{r}} \cdot d\mathbf{l} = dr$, then we can define a meaningful potential energy function

$$U_2 - U_1 = -\int_1^2 \mathbf{F} \cdot d\mathbf{l} = -\int_1^2 F(r)\hat{\mathbf{r}} \cdot d\mathbf{l} = -\int_{r_1}^{r_2} F(r) \, dr$$

since the value of the last integral depends only on the end points, r_1 and r_2. We have the conclusion, then, that any central force is a conservative force, no matter what the form of $F(r)$. The potential energy of a central force is a function only of r: $U = U(r)$. Another example of a central force is the electrostatic force, which we will meet later.

*7–6 Potential Energy Diagrams; Stable and Unstable Equilibrium

Suppose we are given the potential energy of a particle of mass m as a function of position, and we wish to find the motion of the particle as a function of time. Let us assume the motion is in one dimension—let us call it x—so we are given $U(x)$ and want to find $x(t)$. We consider a case where all the forces acting are conservative and are included in the given potential energy; hence the total mechanical energy E is constant, and we can write

$$\tfrac{1}{2}mv^2 + U(x) = E = \text{constant}.$$

We now solve for v:

$$v = \frac{dx}{dt} = \sqrt{\frac{2}{m}[E - U(x)]}. \tag{7–12}$$

Next we separate the variables (x and t) and integrate:

$$\int_{x_0}^x \frac{dx}{\sqrt{\dfrac{2}{m}[E - U(x)]}} = \int_0^t dt = t$$

where we have taken the particle to be at x_0 when $t = 0$. If the integration can be performed,[†] then we have obtained the desired relation between x and t.

In many cases this integral is too difficult to perform. In such cases a great deal can still be learned about the motion by making and examining a potential energy diagram—that is, by making a graph of $U(x)$ versus x. Even when the integration can be done, a PE diagram is still often useful. An example of a PE diagram is shown in Fig. 7–7. The rather complex curve represents some complicated potential energy $U(x)$. The total energy E is constant and thus can be represented as a horizontal line on this graph. Actually four different possible values for E are shown, labeled E_0, E_1, E_2,

---

[†] According to Newton's second law,

$$F = ma = m\frac{dv}{dt} = m\frac{d^2x}{dt^2}.$$

To get $x(t)$ when F is known requires *two* integrations because of the second derivative. The first integral of Newton's second law:

$$\int dv = \int \frac{F}{m} \, dt$$

corresponds to Eq. 7–12, and one more integral must be done to get x.

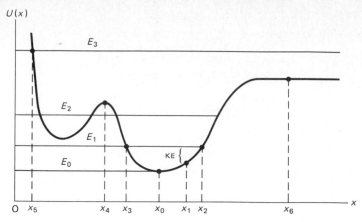

FIGURE 7–7

Potential energy curve.

and E_3. What the actual value of E will be for a given system depends on the initial conditions. (For example, the total energy E of a particle oscillating on the end of a spring depends on the amount the spring is initially compressed or stretched.) The important thing to remember is that E stays constant as long as no additional forces act. From Eq. 7–12, it is clear that $U(x)$ must be less than or equal to E for all situations:

$$U(x) \leq E.$$

If this were not true, then the velocity v would be imaginary (the square root of a negative number) which makes no sense physically. Thus the minimum value which the total energy can take for the PE shown in Fig. 7–7 is that labeled E_0. For this value of E, the particle can only be at rest at $x = x_0$. It has potential energy but no kinetic energy.

If the particle's total energy E is greater than E_0, say it is E_1 on our plot, the particle can have both kinetic and potential energy. Since energy is conserved, KE + PE = E or

$$\text{KE} = E - U(x).$$

Since the curve represents $U(x)$ at each x, the kinetic energy at any value of x is represented by the distance between the E line and the curve $U(x)$ at that value. In the diagram, the KE for a particle at x_1, when its total energy is E_1, is indicated by the notation "KE". A particle with energy E_1 can oscillate only between the points x_2 and x_3; for if $x > x_2$ or $x < x_3$, the potential energy would be greater than E, meaning KE < 0 and v would be imaginary, and so would be impossible. At x_2 and x_3 the velocity is zero since $E = U$ at these points. Hence x_2 and x_3 are called the *turning points* of the motion since, if the particle is at x_0, say, moving to the right, its KE (and speed) decreases until it reaches zero at $x = x_2$. The particle then reverses direction, proceeding to the left and increasing in speed until it passes x_0 again; it continues to move, now decreasing in speed until it reaches $x = x_3$ where again $v = 0$, and the particle again reverses direction.

If the particle has energy $E = E_2$ in Fig. 7–7, there are four turning points. The particle can move in only one of the two potential energy valleys, depending on where it is initially; it cannot get from one valley to the other because of the barrier between them—for at points such as x_4, $U > E_2$ which again means v would be imaginary.[†] For energy E_3, there is only one turning point since $U(x) < E_3$ for all $x > x_5$; then our particle, if moving initially to the left, varies in speed as it passes the potential valleys but eventually stops and turns around at $x = x_5$; it then proceeds to the right indefinitely, never to return.

[†] Although this is true according to Newtonian physics, modern quantum mechanics predicts that particles can tunnel through such a barrier, and such processes have been observed at the atomic and subatomic level.

How do we know the particle reverses direction at the turning points? Because of the force exerted on it. The force F is related to the potential energy U by Eq. 7–1:

$$F = -\frac{dU}{dx}.$$

That is, the force F is equal to the negative of the slope of the U versus x curve at any point x. At $x = x_2$, for example, the slope is positive so the force is negative which means it acts to the left (toward decreasing values of x); at $x = x_3$, the slope is negative so F is positive, meaning it acts toward the right. In both cases the force acts to reverse the direction of the particle's motion.

At $x = x_0$ the slope is zero so $F = 0$. At such a point the particle is said to be in *equilibrium*. This term means simply that the net force on the particle is zero. Hence, its acceleration is zero, and so if it is initially at rest, it remains at rest. If the particle at rest at $x = x_0$ were moved slightly to the left or right, a nonzero force would then act on it and the direction of the force would be such as to move the particle back toward x_0. If the particle were to move a little to the left of x_0, for example, the force would be to the right (since $dU/dx < 0$), and vice versa. A particle which returns toward its equilibrium point when displaced slightly is said to be in *stable equilibrium*. Any *minimum* in the potential energy curve represents a point of stable equilibrium.

A particle at $x = x_4$ would also be in equilibrium since $F = -dU/dx = 0$. If the particle were displaced a bit to either side of x_4, a force would act to pull the particle *away* from the equilibrium point. Points like x_4, where the potential energy curve has a maximum, are points of *unstable equilibrium*. The particle will *not* return to equilibrium if displaced slightly, but moves further away.

When a particle is in a region over which U is constant, such as at $x = x_6$ in Fig. 7–7, the force is zero over the whole region. The particle is in equilibrium and if displaced slightly to one side the force is still zero. The particle is said to be in *neutral equilibrium* in this region.

It is worth noting that the motion described in the above analysis resembles (though is not precisely the same as) a ball sliding on a track of the same shape as $U(x)$. You might reread the above with this in mind.

It is also clear from this analysis (and in fact follows from Eq. 7–1) that a force always acts in a direction to lower the potential energy of a particle.

EXAMPLE 7–8 Draw a potential energy diagram and analyze the motion of a mass m resting on a frictionless horizontal table and connected to a horizontal spring with stiffness constant k (Fig. 7–8a). The mass is pulled a distance to the

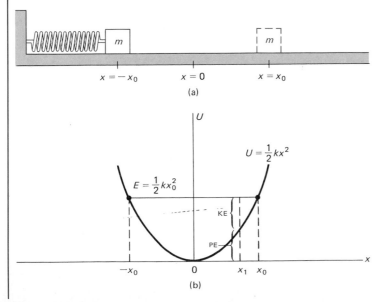

FIGURE 7–8

(a) Mass oscillating on end of spring, for which (b) is the potential energy diagram.

right so the spring is stretched a distance x_0 initially, and then the mass is released from rest.

SOLUTION   The potential energy is

$$U(x) = \tfrac{1}{2}kx^2,$$

and this is plotted in Fig. 7–8b. Initially, $x = x_0$ and $v = 0$ so the total energy $E = \tfrac{1}{2}kx_0^2$. The total E is conserved, so at any point x

$$\tfrac{1}{2}mv^2 + \tfrac{1}{2}kx^2 = E = \tfrac{1}{2}kx_0^2.$$

From this equation we can obtain v at any position x. At the point $x = x_1$, the magnitudes of the KE and PE are indicated and can be read off the graph. After the mass is released at $x = x_0$, the force $F = -dU/dx = -kx$ acts to the left; the mass accelerates to the left, increasing in speed until it passes $x = 0$; then it begins to decelerate (since F is now positive, to the right) and eventually comes to rest at $x = -x_0$. The force $F = -kx$ is still acting, and accelerates the mass to the right. Thus the mass oscillates back and forth between x_0 and $-x_0$. It has its greatest speed, and greatest KE, where U is a minimum: at $x = 0$.

7–7   Power

Power is defined as the rate at which work is done. The *average power*, $\bar{P}$, when an amount of work W is done in a time t, is

$$\bar{P} = W/t.$$

The *instantaneous power*, P, is

$$P = \frac{dW}{dt}.$$

The work done in a process is equal to the energy transformed from one form to another. Hence we can also say that power is the rate at which energy is transformed:

$$P = \frac{dE}{dt}.$$

The power of a horse refers to how much work it can do per unit time. The power rating of an engine refers to how much chemical or electrical energy can be transformed into mechanical energy per unit time. In SI units, power is measured in joules per second and this is given a special name, the *watt* (W): 1 W = 1 J/s. We are most familiar with the watt for measuring the rate at which an electric light bulb or heater changes electric energy into light or heat energy, but it is used for other types of energy transformation as well. In the British system, the unit of work is the foot-pound per second (ft · lb/s). For practical purposes a larger unit is often used, the *horsepower*. One (British) horsepower[†] (hp) is defined as 550 ft · lb/s, which equals 746 watts. In the metric system, 1 hp is defined as 750 W. The metric and British horsepower units differ by less than 1 percent.

To see the distinction between energy and power, consider the following example. A person is limited in the work he can do not only by the total energy required, but also by the rate this energy is used; that is, by power. For example a

---

[†] The unit was first chosen by James Watt (1736–1819) who needed a way to specify the power of his newly developed steam engines. He found by experiment that a good horse can work all day at an average rate of about 360 ft · lb/s. So as not to be accused of exaggeration in the sale of his steam engines, he multiplied this by roughly $1\tfrac{1}{2}$ when he defined the hp.

person may be able to walk a long distance or climb many flights of stairs before having to stop because he has used up so much energy. On the other hand, if a person runs very quickly up stairs, he may fall exhausted after only a flight or two. He is limited in this case by power, the rate at which his body can transform chemical energy into mechanical energy.

EXAMPLE 7–9 A 70-kg man runs up a long flight of stairs in 4.0 s. The vertical height of the stairs is 4.5 m. Calculate his power output in watts and horsepower.

SOLUTION The work done is against gravity. Hence the average power $\bar{P}$ is

$$\bar{P} = \frac{mgh}{t} = \frac{(70 \text{ kg})(9.8 \text{ m/s}^2)(4.5 \text{ m})}{4.0 \text{ s}} = 770 \text{ W}.$$

Since there are 750 W in 1 hp, the man was doing work at a rate of just over 1 hp. It is worth noting that a human cannot do work at this rate for very long.

Automobiles do work to overcome the force of friction (and air resistance), to climb hills, and to accelerate. A car is limited by the rate at which it can do work, which is why automobile engines are rated in horsepower. A car needs power most when it is climbing hills and when accelerating (as for passing). In the next example we will calculate how much power is needed in these situations for a car of reasonable size. Even when a car travels on the level, it needs some power for it must do work to overcome friction and air resistance; these forces depend on the conditions but are usually in the region of 400–1000 N.

It is often convenient to write the power in terms of the net force **F** applied to an object and its velocity **v**. Since $P = dW/dt$ and from Eq. 6–7, $dW = \mathbf{F} \cdot d\mathbf{l}$, then

$$P = \frac{dW}{dt} = \mathbf{F} \cdot \frac{d\mathbf{l}}{dt}$$

$$P = \mathbf{F} \cdot \mathbf{v}. \tag{7–13}$$

EXAMPLE 7–10 Calculate the power required of a 1400-kg car under the following circumstances: (a) the car climbs a 10° hill at a steady 80 km/h and (b) the car accelerates from 90 to 110 km/h in 6.0 s to pass another car. Assume the force of friction on the car is 700 N. See Fig. 7–9.

SOLUTION (a) To move at a steady speed up the hill, the car must exert a force equal to the sum of the friction force, 700 N, and the component of gravity parallel to the hill, $mg \sin 10° = (1400 \text{ kg})(9.8 \text{ m/s}^2)(0.174) = 2400$ N. Since $v = 80$ km/h $= 22$ m/s, and is parallel to **F**,

$$P = Fv$$
$$= (2400 \text{ N} + 700 \text{ N})(22 \text{ m/s}) = 6.8 \times 10^4 \text{ W}$$
$$= 91 \text{ hp}.$$

(b) The car accelerates from 25.0 m/s to 30.6 m/s (90 to 110 km/h). Thus the car must exert a force that overcomes the 700 N friction force plus that required to give it the acceleration $a = (30.6 \text{ m/s} - 25.0 \text{ m/s})/6.0 \text{ s} = 0.93$ m/s². Since the mass of the car is 1400 kg, the force required for acceleration is $F = ma = (1400 \text{ kg})(0.93 \text{ m/s}^2) = 1300$ N, and the total force required is then 2000 N. The average speed $v = 27.8$ m/s. Thus, using Eq. 7–13, the average power needed is

$$\bar{P} = (2000 \text{ N})(27.8 \text{ m/s})$$
$$= 5.6 \times 10^4 \text{ W}$$
$$= 75 \text{ hp}.$$

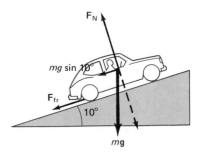

FIGURE 7–9

Calculation of power needed for car to climb a hill (Example 7–10).

Even taking into account the fact that only 60 to 80 percent of the engine's power reaches the wheels, it is clear from these calculations that an engine of 100 to 150 hp is quite adequate from a practical point of view even in a reasonably heavy car. In view of fuel requirements and environmental damage, larger engines are becoming an extravagance that society is questioning.[†]

† But note that horsepower ratings given by manufacturers are often higher than the real available horsepower.

Summary

A *conservative force* is a force that is a function only of position and for which the work done by the force in moving a particle from one position to another depends only on the two positions and not on the path taken. The work done by a conservative force is recoverable, which is not true for nonconservative forces such as friction. A potential energy, U, can be defined only for conservative forces and is related to the force in the one-dimensional case by the equation $F = -dU/dx$.

When only conservative forces are acting, the total mechanical energy, E, defined as the sum of kinetic and potential energies, is conserved:

$$E = \text{KE} + U = \text{constant}.$$

If nonconservative forces also act, additional types of energy are involved, such as thermal energy. It has been found experimentally that when all forms of energy are included the total energy is conserved. This is the *law of conservation of energy*.

The gravitational force, as described by Newton's law of universal gravitation, is a conservative force; the potential energy of a particle of mass m due to the gravitational force exerted on it by the earth is given by $U(r) = -GmM_e/r^2$ where M_e is the mass of the earth, and r is the distance of the particle from the earth's center (r ≥ radius of earth).

Power is defined as the rate at which work is done or the rate at which energy is transformed from one form to another.

Questions

1. List some everyday forces that are not conservative and explain why they aren't.

2. You lift a heavy book from a table to a high shelf. List the forces on the book during this process and state whether each is conservative or nonconservative.

3. The net force acting on a particle is conservative and increases its KE by 300 J. What is the change in (a) the potential energy, (b) the total energy, of the particle?

4. Does the following statement make sense? "Fiberglass poles led to higher pole vaults because the additional potential energy of bending was converted to gravitational potential energy." Explain.

5. An inclined plane has a height h. A body of mass m is released from rest at the top. Does its velocity at the bottom depend on the angle of the plane if (a) there is no friction, (b) there is friction?

6. Why is it tiring to push hard against a solid wall even though no work is done?

7. Analyze the motion of a simple swinging pendulum in terms of energy, (a) ignoring friction, and (b) taking it into account. Explain why a grandfather clock has to be wound up.

8. When a "superball" is dropped, can it rebound to a greater height than its original height?

9. Is it true that a rock thrown with a certain speed from the top of a cliff will enter the water below with the same speed whether the rock is thrown horizontally or at any angle? Explain.

10. A coil spring of mass m rests upright on a table. If you compress the spring by pressing down with your hand, and then release it, can the spring actually leave the table? Explain using the law of conservation of energy.

11. What happens to the potential energy of the water at the top of a waterfall after it reaches the pool below?

12. Experienced hikers prefer to step over a fallen log in their path rather than stepping on top and jumping down on the other side. Explain.

13. Consider two observers who are in different reference frames that move with velocity v with respect to each other. They both observe an object being pulled on a rough horizontal surface. Do they agree as to the value of (a) the object's kinetic energy, (b) the total work done on the object, (c) the amount of energy transformed from mechanical to thermal due to friction? Does your answer to (c) contradict (a) and (b)? Explain.

14. (a) Where does the kinetic energy come from when a car accelerates uniformly starting from rest? (b) How is the increase in KE related to the friction force the road exerts on the tires?

15. The earth is closest to the sun in winter. When is its gravitational PE the greatest?

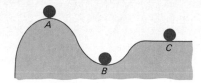

FIGURE 7-10

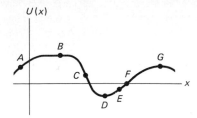

FIGURE 7-11

16. Name the type of equilibrium for each position of the balls in Fig. 7–10.

17. Can the total mechanical energy $E = \text{KE} + U$ ever be negative? Explain.

18. (a) Describe in detail the velocity changes of a particle that has energy E_3 in Fig. 7–7 as it moves from x_6 to x_5 and back to x_6. (b) Where is its kinetic energy the greatest and the least?

19. Give several examples of stable, unstable, and neutral equilibrium.

20. In what state of equilibrium is a cube (a) when resting on its face. (b) when on its edge?

21. Figure 7–11 shows a potential energy curve, $U(x)$. (a) At which point does the force have greatest magnitude? (b) For each labeled point, state whether the force acts to the left, the right, or is zero. (c) Where is there equilibrium and of what type is it?

22. Why does an overloaded car have trouble climbing a steep hill?

23. Suppose you lift a suitcase from the floor to a table. Does the work you do on the suitcase depend on (a) whether you lift it straight up or along a more complicated path, (b) the time it takes, (c) the height of the table, (d) the weight of the suitcase?

24. Repeat the previous question for the *power* needed rather than the work.

25. Why is it easier to climb a mountain via a zigzag trail rather than to climb straight up?

 Problems _____

SECTION 7-1

1. (I) If $U = x^2 + 2xy + 4y^2z$, what is the force, $\mathbf{F}$?

2. (II) Show that the elastic force of an ideal spring, $F = -kx$, is a conservative force.

3. (II) A particular spring obeys the force law $\mathbf{F} = (-kx + ax^3 + bx^4)\mathbf{i}$. (a) Is this force conservative? Explain why or why not. (b) If it is conservative, determine the form of the potential energy function.

4. (II) The nuclear force between two neutrons in a nucleus is described roughly by the Yukawa potential

$$U(r) = -U_0 \frac{r_0}{r} e^{-r/r_0},$$

where r is the distance between the neutrons, and U_0 and $r_0 \approx 10^{-15}$ m are constants. (a) Determine the force $F(r)$. (b) What is the ratio $F(3r_0)/F(r_0)$? (c) Calculate this same ratio for the force between two electrically charged particles where $U(r) = -C/r$, with C being a constant. Why is the Yukawa force referred to as being a "short-range" force?

5. (II) Air resistance can be represented by a force proportional to the speed v of an object: $F = -kv$. Is this force conservative? Explain.

6. (III) Show that the force $\mathbf{F} = F(r)\hat{\boldsymbol{\theta}}$ is nonconservative; $\hat{\boldsymbol{\theta}}$ is the unit vector in polar coordinates (Appendix C) perpendicular to $\hat{\mathbf{r}}$. (Hint: choose a particular path.) In Section 7–5 it was shown that $\mathbf{F} = F(r)\hat{\mathbf{r}}$ *is* conservative.

SECTION 7-2

7. (I) With what speed must a salmon leave the water if it is to jump a waterfall 2.8 m high? (Use conservation of energy.)

8. (I) In the high jump, the KE of an athlete is transformed into gravitational potential energy without the aid of a pole. If the athlete leaves the ground with a speed of 5.0 m/s, and crosses the bar with a speed of 1.2 m/s, how high was his center of mass raised?

9. (I) Tarzan is running at top speed (8.0 m/s) and grabs a vine hanging vertically from a tall tree in the jungle. How high can he swing upward? Does the length of the vine (or rope) affect your answer?

10. (I) A mass m is attached to the end of a spring of spring constant k. The mass is given an initial displacement x_0, after which it oscillates back and forth. Write down a formula for the total mechanical energy, E (ignore friction), at any position x between $-x_0$ and x_0. Express in terms of the given variables.

11. (II) A vertical spring whose spring constant is 850 N/m stands on a table and is compressed 0.400 m. (a) What speed can it give to a 0.300-kg ball when released? (b) How high above its original position (spring compressed) will the ball fly?

12. (II) A ball is attached to a horizontal cord of length L whose other end is fixed, Fig. 7–12. (a) If the ball is released, what will be its speed at the lowest point of its path? (b) A peg is located a distance h directly below the point of attachment of the cord. If $h = 0.75 L$, what will be the speed of the ball when it reaches the top of its circular path about the peg?

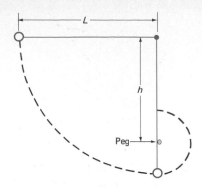

FIGURE 7–12

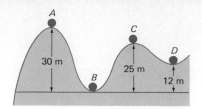

FIGURE 7–13

13. (II) A roller coaster is shown in Fig. 7–13. Assuming no friction, calculate the speed at points B, C, D, assuming the coaster has a speed of 2.30 m/s at point A.

14. (II) A football is thrown with a speed of 12.5 m/s at an angle of 30°. (a) What is its speed at its highest point, and (b) how high does it go? (Use conservation of energy.)

15. (II) A mass m is attached to a vertical spring (spring constant k). (a) Find the distance the mass will move down, after it is attached, if it is lowered gently and comes to rest in equilibrium. (b) If the mass is allowed to fall freely, what will be the maximum extension of the spring?

16. (III) A small mass m slides without friction along the looped apparatus shown in Fig. 7–14. (a) If the object is to remain on the track, even at the top of the circle (whose radius is r), from what minimum height h_m must it be released (from rest)? (b) If it is released at a height $h = 0.80h_m$, at what height will it lose contact with the track?

17. (II) A cyclist intends to cycle up a 10° hill 100 m high. Assuming the mass of bicycle plus person is 80 kg, (a) how much work must be done against gravity? (b) If each complete revolution of the pedals moves the bike 5.1 m along its path, calculate the average force that must be exerted on the pedals tangent to their circular path. Neglect friction and other losses. The pedals turn in a circle of diameter 36 cm.

18. (III) A particle of mass m starts from rest at the top of a solid sphere of radius r and slides down its frictionless surface. (a) At what angle θ (Fig. 7–15) will the mass leave the sphere? (b) If friction were present, would the mass fly off at a greater or lesser angle?

19. (III) In the Bohr model of the hydrogen atom, an electron is held in its circular orbit of radius r around the nucleus by a force $F = -C/r^2$ where C is constant. If the electron moves from a circular orbit with radius $r = r_0$ to a circular orbit with $r = n^2 r_0$ where n is an integer, determine the change in total mechanical energy.

20. (III) An engineer is designing a spring to be placed at the bottom of an elevator shaft; if the elevator cable should break at a height h above the top of the spring, calculate

the required value of the spring constant, k, so that passengers undergo an acceleration of no more than 10 g when brought to rest. Let M be the total mass of the elevator and passengers.

21. (III) Show that $h \geq 0.60\,L$ if the ball in Fig. 7–12 is to make a complete circle around the peg.

SECTION 7–3

22. (I) A 17-kg child descends a slide 4.6 m high and reaches the bottom with a speed of 2.2 m/s. How much heat was generated in this process?

23. (I) A 145-g baseball is dropped from a tree 22 m above the ground. (a) With what speed would it hit the ground if air resistance could be ignored? (b) If it actually hits the ground with a speed of 9.0 m/s, what was the average force of air resistance exerted on it?

24. (II) A skier traveling 12.6 m/s reaches the foot of a steady upward 20° incline and glides 11.4 m up (along) the incline before coming to rest. What was the average coefficient of friction?

25. (II) A ski starts from rest and slides down a 30 m long 18° incline. (a) If the coefficient of friction is 0.080, what is the ski's speed at the base of the incline? (b) If the snow is level at the foot of the incline, and has the same coefficient of friction, how far will the ski travel along the level? Use energy methods.

26. (II) The roller coaster in Fig. 7–13 passes point A with a speed of 1.2 m/s. If the average force of friction is equal to one-fifth its weight, with what speed will it reach point B? The distance traveled is 70 m.

27. (III) A 200-g wood block is firmly attached to a horizontal spring, Fig. 7–16. The block can slide along a table where the coefficient of friction is 0.40. A force of 10 N compresses the spring 18 cm. (a) If the spring is released from this position, how far beyond its equilibrium position will it stretch on its first swing? (b) What total distance will it travel

FIGURE 7–15

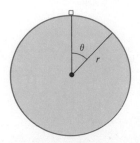

FIGURE 7–14

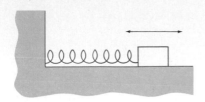

FIGURE 7–16

before coming to rest? (c) How much thermal energy will be produced by the time the block comes to rest?

SECTION 7–5

28. (II) Determine the escape velocity from the sun for an object (a) at its surface ($R = 7.0 \times 10^5$ km, $M = 2.0 \times 10^{30}$ kg) and (b) at the average distance of the earth (1.49×10^8 km). Compare to the speed of the earth in its orbit.

29. (II) Show that Eq. 7–9 for gravitational potential energy reduces to Eq. 6–11, $\Delta U = mg(y_2 - y_1)$ for objects near the surface of the earth.

30. (II) Show that the change in potential energy of an object between the earth's surface and a height h above is

$$\Delta U = mgh \bigg/ \left(1 + \frac{h}{r_e}\right)$$

where r_e is the radius of the earth and h is not necessarily small.

31. (II) (a) Show that the total mechanical energy of a satellite (mass m) orbiting at a distance r from the center of the earth (mass M_e) is

$$E = -\frac{1}{2}\frac{GmM_e}{r}.$$

(b) Show that although friction causes the value of E to slowly decrease in time, if the orbit remains a circle the kinetic energy must actually increase. (c) Show that $v = \sqrt{GM/r}$.

32. (II) Show that the escape velocity for any satellite in a circular orbit is $\sqrt{2}$ times its velocity.

33. (II) The earth's distance from the sun varies from 1.471×10^8 km to 1.521×10^8 km during the year. Determine the difference in (a) the potential energy, (b) kinetic energy, (c) total energy, between these extreme points.

34. (II) Take into account the earth's rotational speed (1 rev/day) and determine the necessary speed, with respect to earth, for a rocket to escape if fired from the earth at the equator in a direction (a) eastward, (b) westward, (c) vertically upward.

35. (II) (a) Determine a formula for the maximum height h a rocket will reach if launched vertically from the earth's surface with speed $v_0(<v_{esc})$. Express in terms of v_0, r_e, M_e, and G. (b) How high does a rocket go if $v_0 = 8.2$ km/s? Ignore air resistance and the earth's rotation.

36. (II) (a) Determine the rate at which the escape velocity from earth changes with height above the earth's surface, dv_{esc}/dr. (b) Use the approximation $\Delta v \approx (dv/dr) \, \Delta r$ to determine the escape velocity for a spacecraft orbiting the earth at a height of 300 km.

37. (II) A meteor has a speed of 85.1 m/s when 720 km above the earth. It is falling vertically (ignore air resistance) and

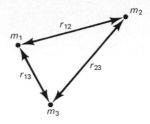

FIGURE 7–17

strikes a bed of sand in which it is brought to rest in 3.25 m. (a) What was its speed just before striking the sand? (b) How much work did the sand do to stop the meteor (mass = 575 kg) (c) What was the average force exerted by the sand on the meteor? (d) How much thermal energy was produced?

38. (II) To escape the solar system, an interstellar spacecraft must overcome the gravitational attraction of both the earth and sun. Ignoring the effects of other bodies in the solar system, determine (a) the escape velocity, and (b) the total energy required, for a 5.0×10^5-kg spacecraft to escape the solar system if launched from the earth's surface.

39. (II) (a) Suppose we have three masses, m_1, m_2, and m_3, that initially are infinitely far apart from each other. Show that the work needed to bring them to the positions shown in Fig. 7–17 is

$$W = -G\left(\frac{m_1 m_2}{r_{12}} + \frac{m_1 m_3}{r_{13}} + \frac{m_2 m_3}{r_{23}}\right).$$

(b) Can we say that this formula also gives the potential energy of the system? Or the PE of one or two of the bodies? (c) Is W equal to the binding energy of the system (equal to the energy required to separate the components to an infinite distance)? Explain.

40. (III) A sphere of radius r_1 has a concentric spherical cavity of radius r_2. Assume this spherical shell of thickness $r_1 - r_2$ is uniform and has a total mass M. Show that the gravitational PE of a mass m a distance r from the center of the shell ($r > r_1$) is given by

$$U = -\frac{GmM}{r}.$$

*SECTION 7–6

*41. (II) (a) Show that the potential energy diagram for a particle in the vicinity of the earth, acted on by gravity, is given by a curve like that of Fig. 7–18. In particular, why is the

FIGURE 7–18

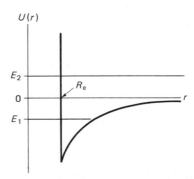

curve nearly vertical at $r = R_e$, the radius of the earth. (b) Describe the motion for a particle with energy E_1, and (c) with energy E_2, as shown. (d) What energy corresponds to escape velocity?

42. (III) The potential energy of the two atoms in a diatomic molecule can be written

$$U(r) = -\frac{a}{r^6} + \frac{b}{r^{12}},$$

where r is the distance between the two atoms and a and b are positive constants. (a) At what values of r is $U(r)$ a minimum? A maximum? (b) At what values of r is $U(r) = 0$? (c) Plot $U(r)$ as a function of r from $r = 0$ to r at a sufficiently large value that all the above features are shown. (d) Describe the motion of one atom with respect to the second when $E < 0$, and for $E > 0$. (e) Let F be the force one atom exerts on the other. For what values of r is $F > 0$, $F < 0$, $F = 0$? (f) Determine F as a function of r.

*43. (III) The *binding energy* of a two-particle system is defined as the energy required to separate the two particles from their state of lowest energy to $r = \infty$. Determine the binding energy for the molecule discussed in problem 42.

SECTION 7–7

44. (I) How long will it take a 1.25-hp motor to lift a 325-kg piano to a sixth-story window 16.0 m above?

45. (I) Show that a British horsepower is equal to 746 W.

46. (I) Electric energy is often stated in kilowatt-hours. Show that the kilowatt-hour (kWh) is a unit of energy and is equal to 3.6×10^6 J.

47. (I) An 80-kg football player traveling 5.0 m/s is stopped in 1.0 s by a tackler. (a) What was the original KE of the player? (b) What average power was required to stop him?

48. (I) If a car generates 15 hp when traveling at a steady 90 km/h, what must be the average force exerted on the car due to friction and air resistance?

49. (I) A shotputter accelerates a 7.3-kg shot from rest to 15 m/s. If this motion takes 2.0 s, what average power was developed?

50. (II) What minimum horsepower must a motor have to be able to drag a 200-kg box along a level floor at a speed of 1.22 m/s if the coefficient of friction is 0.45?

51. (II) A 70-kg hiker climbs to the top of a 3120-m-high mountain. The climb is made in 4.0 h starting at an elevation of 1850 m. Calculate (a) the work done against gravity, (b) the average power output in watts and in horsepower, and (c) the rate of energy input required, assuming the body is 15 percent efficient.

52. (II) How fast must a cyclist climb a 12° hill to maintain a power output of 0.20 hp? Ignore friction and assume the mass of cyclist plus bicycle is 85 kg.

53. (II) A pump is to lift 5.0 kg of water per minute through a height of 4.2 m. What output rating should the pump motor have?

54. (II) A 1200-kg car slows down from 90 km/h to 70 km/h in about 5.0 s on the level when it is in neutral. Approximately what power (hp) is needed to keep the car traveling at a constant 80 km/h?

55. (II) A bicyclist coasts down a 6.0° hill at a steady speed of 7.0 km/h. Assuming a total mass of 75 kg (bicycle plus rider), what must be the cyclist's power output to climb the same hill at the same speed?

56. (II) A 1250-kg car has a maximum power output of 90 hp. How steep a hill can it climb at a constant speed of 60 km/h if the frictional forces add up to 450 N?

57. (II) Water flows over a dam at the rate of 750 kg/s and falls vertically 130 m before striking the blades of a turbine. Calculate (a) the velocity of the water just before striking the turbine blades, and (b) the rate at which mechanical energy is transferred to the turbine blades. Assume the water loses 80 percent of its speed when striking the blades, and that 12 percent of the initial energy is transformed to heat.

58. (III) A bicyclist of mass 80 kg (including the bicycle) can coast down a 4.0° hill at a steady speed of 6.0 km/h. Pumping hard, the cyclist can descend the hill at a constant speed of 40 km/h. Using the same power, at what speed can the cyclist climb the same hill? Assume the force of friction is directly proportional to the speed v; that is, $F_{fr} = bv$, where b is constant.

59. (III) The position of a 280-g particle is given (in meters) by $x = 6.2t^3 - 3.0t^2 - 88t$, where t is in seconds. Determine the net rate of work done on this particle (a) at $t = 2.0$ s and (b) at $t = 4.0$ s. (c) What was the average net power input during this time interval?

Conservation of Linear Momentum; Many Bodies and Collisions

8

The law of conservation of energy, which we discussed in the previous chapter, is one of several great conservation laws in physics. Among the other quantities found to be conserved are linear momentum, angular momentum, and electric charge. We will eventually discuss all of these since the conservation laws are among the most important in all of science. In this chapter we discuss linear momentum and its conservation; we will also make use of the laws of conservation of linear momentum and of energy to analyze and solve an important class of phenomena known as collisions.

The law of conservation of momentum is particularly useful when dealing with two or more bodies. Our focus up to now has been mainly on the motion of a single particle. But now we deal with the problem of two or more particles, and with extended bodies which can be considered as collections of particles. An important idea for this study is that of center of mass, which we now introduce.

 ## 8–1 Center of Mass

As just mentioned, we have until now been mainly concerned with the motion of a single particle; when we have dealt with an extended body (that is, a body that has size) we have assumed that it approximated an ideal particle or that it underwent only translational motion. Real "extended" bodies, however, can undergo rotational and other types of motion as well. For example, the diver in Fig. 8–1a undergoes only translational motion (all parts of the body follow the same path), whereas the diver in Fig. 8–1b undergoes both translational and rotational motion. An extended body can also vibrate and its various parts may move internally in complicated ways as well (such as the arms of a diver). We will refer to motion that is not pure translation as *general motion*.

We will be able to show that even if a body rotates, or there are several bodies that move relative to one another, there is one point in the body (or group of bodies) that moves in the same path that a particle would if subjected to the same net force.

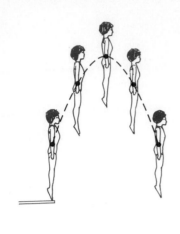

(a)

(b)

FIGURE 8–1

Motion of the diver is pure translation in (a), but is translation plus rotation in (b).

FIGURE 8–2

The center of mass of a two-particle system lies on the line joining the two masses.

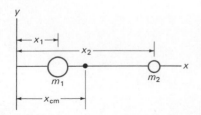

This point is called the **center of mass** (cm). Because of this, the general motion of an extended body, or system of bodies, can be considered as the sum of the translational motion of its cm plus rotational, vibrational, or other type of motion about its cm.

As an example, consider the motion of the center of mass of the diver in Fig. 8–1: the cm follows a parabolic path even when the diver rotates as shown in Fig. 8–1b. This is the same parabolic path that a projected particle follows when acted on only by the force of gravity (that is, projectile motion). Other points in the rotating diver's body follow more complicated paths.

Now let us define center of mass. We can consider any extended body as being made up of many tiny particles. But let us first consider a system made up of only two particles, of mass m_1 and m_2; we choose a coordinate system so that both particles lie on the x axis at positions x_1 and x_2, Fig. 8–2. The center of mass of this system is defined to be at the position, x_{cm}, given by

$$x_{cm} = \frac{m_1 x_1 + m_2 x_2}{m_1 + m_2} = \frac{m_1 x_1 + m_2 x_2}{M},$$

where $M = m_1 + m_2$ is the total mass of the system. The center of mass lies on the line joining m_1 and m_2. If the two masses are equal ($m_1 = m_2 = m$), x_{cm} is midway between them, since then $x_{cm} = m(x_1 + x_2)/2m = (x_1 + x_2)/2$. If one mass is greater than the other, say $m_1 > m_2$, then the cm is closer to the larger mass. If all the mass is concentrated at x_2, say, so $m_1 = 0$, then $x_{cm} = (0x_1 + m_2 x_2)/(0 + m_2) = x_2$, as we would expect.

Now let us consider a system consisting of n particles, where n could be very large; this system could be an extended body which we consider as being made up of n tiny particles. If these n particles are all along a straight line (call it the x axis), we define the cm of the system to be located at

$$x_{cm} = \frac{m_1 x_1 + m_2 x_2 + \cdots + m_n x_n}{m_1 + m_2 + \cdots + m_n} = \frac{\sum_{i=1}^{n} m_i x_i}{M},$$

where $m_1, m_2, \ldots m_n$ are the masses of each particle and $x_1, x_2, \ldots x_n$ are their positions; the symbol $\sum_{i=1}^{n}$ is the summation sign meaning to sum over all the particles, where i takes on integer values from 1 to n. (Often we simply write $\sum m_i x_i$, leaving out the $i = 1$ to n.) The total mass of the system is $M = \sum m_i$. If the particles are spread out in two or three dimensions (as for a typical extended body) then we define the coordinates of the cm as

$$x_{cm} = \frac{\sum m_i x_i}{M}, \quad y_{cm} = \frac{\sum m_i y_i}{M}, \quad z_{cm} = \frac{\sum m_i z_i}{M}, \tag{8–1}$$

where x_i, y_i, z_i are the coordinates of the particle of mass m_i and again $M = \sum m_i$ is the total mass.

Although from a practical point of view we usually calculate the components of the cm (Eq. 8–1), it is sometimes convenient (for example, for derivations) to write Eq. 8–1 in vector form. If $\mathbf{r}_i = x_i \mathbf{i} + y_i \mathbf{j} + z_i \mathbf{k}$ is the position vector of the ith particle, and $\mathbf{r}_{cm} = x_{cm} \mathbf{i} + y_{cm} \mathbf{j} + z_{cm} \mathbf{k}$ is the position vector of the center of mass, then

$$\mathbf{r}_{cm} = \frac{\sum m_i \mathbf{r}_i}{M}. \tag{8–2}$$

For an extended body, it is often convenient to think of it as made up of a continuous distribution of matter. In other words, we consider it as made up of n particles where n is allowed to approach infinity. Then the summations in Eqs. 8–1 and 8–2 become integrals:

$$x_{cm} = \frac{1}{M} \int x \, dm, \quad y_{cm} = \frac{1}{M} \int y \, dm, \quad z_{cm} = \frac{1}{M} \int z \, dm; \tag{8–3}$$

in vector notation, this becomes

$$\mathbf{r}_{cm} = \frac{1}{M} \int \mathbf{r} \, dm. \qquad (8\text{--}4)$$

The quantity dm is a formal way of representing an infinitesimal element of mass; in actual calculations it normally must be written in terms of geometric (space) variables as we shall see in the next section.

8–2 Locating the Center of Mass

Let us now see how to locate the cm for a few situations.

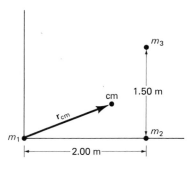

EXAMPLE 8–1   Three particles, each of mass 2.50 kg, are located at the corners of a right triangle whose sides are 2.00 m and 1.50 m long as shown in Fig. 8–3. Locate the center of mass.

SOLUTION   We choose our coordinate system as shown (to simplify calculations) with m_1 at the origin, and m_2 on the x axis; then m_1 has coordinates $x_1 = y_1 = 0$; m_2 has coordinates $x_2 = 2.0$ m, $y_2 = 0$; and m_3 has coordinates $x_3 = 2.0$ m, $y_3 = 1.5$ m. Then, from Eq. 8–1,

$$x_{cm} = \frac{(2.50 \text{ kg})(0) + (2.50 \text{ kg})(2.00 \text{ m}) + (2.50 \text{ kg})(2.00 \text{ m})}{3(2.50 \text{ kg})} = 1.33 \text{ m}$$

$$y_{cm} = \frac{(2.50 \text{ kg})(0) + (2.50 \text{ kg})(0) + (2.50 \text{ kg})(1.50 \text{ m})}{7.50 \text{ kg}} = 0.50 \text{ m}.$$

The cm and the position vector $\mathbf{r}_{cm}$ are shown in Fig. 8–3.

FIGURE 8–3
Example 8–1.

For symmetrically shaped bodies of uniform composition, such as spheres, cylinders, and rectangular solids, the cm is located at the geometric center of the body. To see why this is true, consider a uniform circular cylinder, such as a solid wheel. We expect the cm to be at the center of the circle. To show that it is, we first choose a coordinate system whose origin is at the center of the circle with the z axis perpendicular to the cylinder (Fig. 8–4). When we take the sum $\sum m_i x_i$ in Eq. 8–1, there is as much mass at any $+x_i$ as there is at $-x_i$; so all terms cancel out in pairs and $x_{cm} = 0$; the same is true for y_{cm}. In the vertical (z) direction, the cm must lie halfway between the circular faces, since if we choose our origin of coordinates at that point there is as much mass at any $+z_i$ as at $-z_i$, so $z_{cm} = 0$. For other uniform symmetrically shaped bodies, we can make similar arguments to show that the cm must lie on a line of symmetry. If a symmetric body is *not* uniform, then these arguments do not hold. For example, the cm of a wheel weighted on one side is not at the geometric center, but is closer to the weighted side.

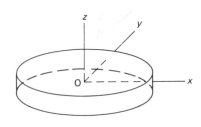

FIGURE 8–4

Cylinder with origin of coordinates at geometric center.

FIGURE 8–5

Locating the cm of a uniform cone.

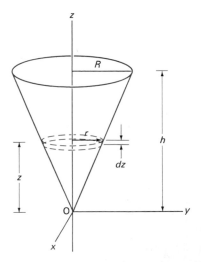

EXAMPLE 8–2   Determine the cm of a uniform cone of height h and radius R.

SOLUTION   We choose our coordinate system so the origin is at the tip of the cone and the z axis is along the line of symmetry, Fig. 8–5. Then $x_{cm} = y_{cm} = 0$ since the cm must lie on the line of symmetry by the arguments used above. To find z_{cm}, we divide the cone up into an infinite number of cylinders of thickness dz, one of which is shown. The mass of each infinitesimal cylinder is $dm = \rho \, dV = \rho \pi r^2 \, dz$ where ρ is the density (mass per unit volume), which is constant since the cone is uniform, and $dV = \pi r^2 \, dz$ is the volume of this very thin cylinder. We now integrate using Eq. 8–3.

$$z_{cm} = \frac{1}{M} \int z \, dm = \frac{1}{M} \int_0^h z \rho \pi r^2 \, dz.$$

Now r, the radius of each infinitesimal cylinder, is related to z by the ratio $r/z = R/h$, so $r = Rz/h$; we then have

$$z_{cm} = \frac{1}{M} \int_0^h \frac{\rho\pi R^2}{h^2} z^3\, dz = \frac{\rho\pi R^2}{Mh^2} \frac{z^4}{4}\Big|_0^h = \frac{\rho\pi R^2 h^2}{4M}.$$

The total mass M of the cone is equal to the density, ρ, times the total volume of the cone, $\pi R^2 h/3$; that is, $M = \rho\pi R^2 h/3$; thus we have

$$z_{cm} = \tfrac{3}{4}h.$$

So the cm is $\tfrac{3}{4}h$ from the tip of the cone, or $\tfrac{1}{4}h$ from its base.

To locate the center of mass of a group of extended bodies, we can use Eq. 8–1 where the m_i are the masses of these bodies, and x_i, y_i, and z_i are the coordinates of the cm of each of the bodies. To see why this is true, consider two extended bodies, call them A and B. Let body A have mass M_A and be made up of n_A particles, and body B have mass M_B and be made up of n_B particles; the total number of particles is $n = n_A + n_B$. Then, from Eq. 8–1 for the x coordinate, we have

$$Mx_{cm} = \sum_{i=1}^n m_i x_i = \sum_{i=1}^{n_A} m_i x_i + \sum_{i=1}^{n_B} m_i x_i,$$

where $M = M_A + M_B$ is the total mass of the system. The two terms on the right represent the sum over the particles of body A and body B separately. But

$$\sum_{i=1}^{n_A} m_i x_i = M_A x_{Acm} \quad \text{and} \quad \sum_{i=1}^{n_B} m_i x_i = M_B x_{Bcm}$$

by Eq. 8–1 for these separate bodies alone. We combine these with the previous equation to get

$$Mx_{cm} = M_A x_{Acm} + M_B x_{Bcm}$$

or

$$x_{cm} = \frac{M_A x_{Acm} + M_B x_{Bcm}}{M}.$$

This is just Eq. 8–1 for two bodies; the same argument can be applied to the y and z coordinates and can easily be extended to include more than two bodies. Thus we see that the cm of a system of several extended bodies can be found by treating each body as a particle all of whose mass is concentrated at its center of mass.

EXAMPLE 8–3 Determine the cm of the uniform thin L-shaped object shown in Fig. 8–6.

SOLUTION Consider the object as two rectangles: rectangle A which is 2.06 m × 0.20 m, and rectangle B which is 1.48 m × 0.20 m. With the origin at 0 as shown, the cm of A is at $x_A = 1.03$ m, $y_A = 0.10$ m; the cm of B is at $x_B = 1.96$ m, $y_B = -0.74$ m. The mass of A is $M_A = (2.06$ m$)(0.20$ m$)(t)(\rho) = (0.412$ m$^2)(\rho t)$ where ρ and t are the density and thickness respectively. The mass of B is $(1.48$ m$)(0.20$ m$)(\rho t) = (0.296$ m$^2)(\rho t)$ and the total mass is $M = (0.708$ m$^2)(\rho t)$. Thus

$$x_{cm} = \frac{M_A x_A + M_B x_B}{M} = \frac{(0.412\text{ m}^2)(1.03\text{ m}) + (0.296\text{ m}^2)(1.96\text{ m})}{(0.708\text{ m}^2)} = 1.42 \text{ m},$$

where ρt was canceled out in numerator and denominator. Similarly

$$y_{cm} = \frac{(0.412\text{ m}^2)(0.10\text{ m}) + (0.296\text{ m}^2)(-0.74\text{ m})}{(0.708\text{ m}^2)} = -0.25 \text{ m},$$

which puts the cm approximately at the point so labeled in Fig. 8–6.

FIGURE 8–6

Example 8–3.

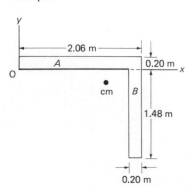

Note in this last example that the cm can actually lie *outside* the body. Another example is a doughnut whose cm is at the center of the hole.

For irregularly shaped objects, it is often difficult to perform the sums of Eq. 8–1 or the integrals of Eq. 8–3. The cm can always be found experimentally, however, using the following technique. Consider a flat uniform body whose cm is known, such as the oval in Fig. 8–7. The oval object is suspended from a point near its edge (you could think of it hanging from a nail); in the position shown, its cm is displaced to one side. We can think of the whole mass of the object as being concentrated at the cm, so we have the equivalent of a simple pendulum (Fig. 6–9); and just as the bob of a simple pendulum, when released, will remain at rest only when the mass is directly below the point of support, so an extended body suspended freely at one point will remain at rest only when the cm is directly below the point of support.

To determine the center of mass of any odd-shaped object, we make use of this fact—that its center of mass will lie on a vertical line directly below any point from which it is suspended. If the object is two dimensional or has a plane of symmetry, it need only be hung from two different pivot points and the respective vertical lines drawn; then the center of mass will be at the intersection of the two lines, as in Fig. 8–8. If the object is suspended from a third point, a vertical line will also pass through the cm. If the object doesn't have a plane of symmetry, the cm with respect to the third dimension is found by suspending the object from at least three points that are not in a plane.

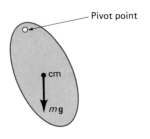

FIGURE 8–7

Determining the cm of an irregularly shaped object.

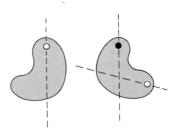

FIGURE 8–8

Finding the cm of an object.

 ## 8–3   Center of Mass and Translational Motion

As mentioned at the start of this chapter, a major reason for the importance of the concept of center of mass is that the motion of the cm for a system of particles (or an extended body) can often be described simply since it is related to the net force on the system. We now show this. Let us examine the motion of a system of n particles of total mass M, which we assume remains constant. From Eq. 8–2 we have

$$M\mathbf{r}_{cm} = \sum m_i \mathbf{r}_i.$$

We now differentiate this equation with respect to time:

$$M \frac{d\mathbf{r}_{cm}}{dt} = \sum m_i \frac{d\mathbf{r}_i}{dt}$$

or

$$M\mathbf{v}_{cm} = \sum m_i \mathbf{v}_i \qquad (8\text{–}5)$$

where $\mathbf{v}_i = d\mathbf{r}_i/dt$ is the velocity of the ith particle of mass m_i, and $\mathbf{v}_{cm}$ is the velocity of the cm. We take the derivative with respect to time again and obtain

$$M \frac{d\mathbf{v}_{cm}}{dt} = \sum m_i \mathbf{a}_i$$

where $\mathbf{a}_i = d\mathbf{v}_i/dt$ is the acceleration of the ith particle. Now $d\mathbf{v}_{cm}/dt$ is the acceleration of the cm, $\mathbf{a}_{cm}$; and by Newton's second law, $m_i \mathbf{a}_i = \mathbf{F}_i$ where $\mathbf{F}_i$ is the net force on the ith particle. Therefore

$$M\mathbf{a}_{cm} = \mathbf{F}_1 + \mathbf{F}_2 + \ldots + \mathbf{F}_n = \sum \mathbf{F}_i. \qquad (8\text{–}6)$$

That is, *the vector sum of all the forces acting on the system is equal to the total mass of the system times the acceleration of its center of mass.* Note that our system of n particles could be the n particles that make up one or more extended bodies.

The force $\mathbf{F}_i$ exerted on the particles of the system can be divided into two types: (1) forces exerted by objects outside our system (such as the gravitational pull of the earth), which we call *external forces*; and (2) forces that particles within the

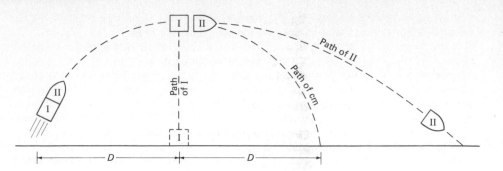

FIGURE 8-9
Example 8-4.

system exert on other particles of the system (such as electrical and gravitational forces, or forces due to springs connecting the different particles), which are referred to as *internal forces*. By Newton's third law, the internal forces occur in pairs: if one particle exerts a force on a second particle in our system, the second exerts a force back on the first with the same magnitude but opposite direction. Thus, in the sum over all the forces in Eq. 8-6, these internal forces cancel each other in pairs. We are left, then, with only the external forces on the right side of Eq. 8-6; so we have

$$M\mathbf{a}_{cm} = \mathbf{F}_{ext} \qquad (8-7)$$

where $\mathbf{F}_{ext}$ is the sum of all the external forces acting on our system, which is the *net force* acting on it. Since Eq. 8-7 is the same as Newton's second law for a particle, we conclude that the *center of mass of a system of particles (or of an extended body of total mass M) moves like a single particle of mass M which is acted on by the same net external force.* This is a useful result. It tells us that the system moves as if all its mass were concentrated at the cm and all the external forces acted at that point. We can thus treat the translational notion of any body or system of bodies as the motion of a particle. (We have, in fact, done just that in earlier chapters.) This theorem clearly simplifies our analysis of the motion of complex systems or extended bodies (such as that shown in Fig. 8-1b). Although the motion of various parts of the system may be complicated, we may often be satisfied with knowing the motion of the cm. This theorem also allows us to solve certain types of problems very easily, as illustrated by the next example.

EXAMPLE 8-4   A rocket is fired into the air as shown in Fig. 8-9. At the moment it reaches its highest point, a horizontal distance D from its starting point, it separates into two parts of equal mass. Part I falls vertically to earth. Where does part II land? Assume $\mathbf{g}$ = constant.

SOLUTION   The path of the cm of the system continues to follow the parabolic trajectory of a projectile acted on by a constant gravitational force. The cm will thus arrive at a point $2D$ from the origin. Since the masses of I and II are equal, the cm must be midway between them. Therefore II lands a distance $3D$ from the origin. (If part I had been given a kick up or down, instead of merely falling, the answer would have been somewhat more complicated.)

 8-4 **Linear Momentum and Its Relation to Force**

The **momentum**, $\mathbf{p}$, of a particle is defined as the product of its mass, m, and its velocity, $\mathbf{v}$:

$$\mathbf{p} = m\mathbf{v}. \qquad (8-8)$$

Momentum is a vector, since it is the product of a scalar and a vector; its direction is the direction of the velocity, **v**, and its magnitude is $p = mv$. Since **v** depends on the reference frame, this frame must be specified. The unit of momentum is that of mass × velocity, which in SI units is kg·m/s. The momentum, **p**, is often referred to as **linear momentum** to distinguish it from angular momentum (which is associated with rotational motion, Chapter 9). The term *momentum* alone always refers to the linear momentum, **p**; when we wish to speak of angular momentum we will always specify "angular" so there will be no confusion.

Everyday usage of the term *momentum* is in accord with the definition above. For according to Eq. 8–8, a fast-moving car has more momentum than a slow-moving car of the same mass; and a heavy truck has more momentum than a small car moving with the same speed. The more momentum an object has, the harder it is to stop it, and the greater effect it will have if it is brought to rest by impact or collision. A football player is more likely to be stunned if tackled by a heavy opponent running at top speed than by a lighter or slower-moving tackler. And a heavy fast-moving truck can do more damage than a slow-moving car.

A force is required to change the momentum of an object whether it is to increase the momentum, decrease it (such as to bring a moving object to rest), or to change its direction. Newton originally stated his second law in terms of momentum (although he called the product mv the "quantity of motion"). Today, translated into modern language, Newton's second law may be stated as, "the rate of change of momentum of a particle is proportional to the net force applied to it." In symbols

$$\mathbf{F} = \frac{d\mathbf{p}}{dt} \tag{8-9}$$

where **F** is the net force applied to the particle. Our earlier statement of this law, $\mathbf{F} = m\mathbf{a}$, can readily be derived from Eq. 8–9 as follows. We assume the mass m is constant, and using the definition of acceleration, $\mathbf{a} = -d\mathbf{v}/dt$, we have

$$\mathbf{F} = \frac{d(m\mathbf{v})}{dt} = m\frac{d\mathbf{v}}{dt} = m\mathbf{a}.$$

Newton's statement, Eq. 8–9, is more general than the more familiar one ($\mathbf{F} = m\mathbf{a}$) since it includes the situation in which the mass may change. Examples involving changing mass, which are not very common, are discussed in Section 8–12 (rockets losing mass as they burn fuel) and Chapter 39 (relativity).

Equation 8–9 applies to a single particle. Let us now consider a system of n particles of total mass $M = m_1 + m_2 + \cdots + m_n$. Assume the particles have momentum $\mathbf{p}_1 = m_1\mathbf{v}_1$, $\mathbf{p}_2 = m_2\mathbf{v}_2, \ldots \mathbf{p}_n = m_n\mathbf{v}_n$, where $\mathbf{v}_1$, $\mathbf{v}_2, \ldots \mathbf{v}_n$, are the velocities of the individual particles. Then the *total momentum*, **P**, of the system is defined as

$$\mathbf{P} = m_1\mathbf{v}_1 + m_2\mathbf{v}_2 + \cdots + m_n\mathbf{v}_n = \sum \mathbf{p}_i. \tag{8-10}$$

From Eq. 8–5 ($M\mathbf{v}_{cm} = \sum m_i\mathbf{v}_i$) we have

$$\mathbf{P} = M\mathbf{v}_{cm}. \tag{8-11}$$

Thus, *the total linear momentum of a system of particles is equal to the product of the total mass M and the velocity of the center of mass of the system. Or, the linear momentum of an extended body is the product of the body's mass and the velocity of its cm.*

If we differentiate Eq. 8–11 with respect to time, we obtain, assuming the total mass M is constant,

$$\frac{d\mathbf{P}}{dt} = M\frac{d\mathbf{v}_{cm}}{dt} = M\mathbf{a}_{cm}$$

$$= \mathbf{F}_{ext} \tag{8-12}$$

where $\mathbf{F}_{ext}$ is the net external force on the system and we have used Eq. 8–7. Equation 8–12 is **Newton's second law for a system of particles.** It is the equivalent of

Eq. 8–9 for a single particle. It is valid for any definite fixed system of particles. Equation 8–12 can also be applied to systems whose mass varies, although care must be taken (this is discussed later in this chapter).

EXAMPLE 8–5 Water leaves a hose at a rate of 5.0 kg/s with a speed of 50 m/s. It strikes a wall, which stops it. (That is, we ignore any splashing back.) What is the force exerted by the water on the wall?

SOLUTION In each second, water with a momentum of (5.0 kg)(50 m/s) = 250 kg·m/s is brought to rest. The magnitude of the force (assumed constant) required to change the momentum by this amount is

$$F = \frac{dP}{dt} = \frac{0 - 250 \text{ kg·m/s}}{1.0 \text{ s}} = -250 \text{ N}.$$

The minus sign indicates that the force on the water is opposite to its original velocity. The wall exerts a force of 250 N to stop the water; so, by Newton's third law, the water exerts a force of 250 N on the wall.

8–5 Conservation of Linear Momentum

If the net external force on a system of particles is zero, then Eq. 8–12 becomes

$$\frac{d\mathbf{P}}{dt} = 0, \quad \text{or} \quad \mathbf{P} = \text{constant.} \qquad [\mathbf{F}_{\text{ext}} = 0] \qquad (8\text{–}13)$$

Thus:

When the net external force on a system is zero, the total momentum remains constant.

This is the **law of conservation of momentum**. It can also be stated as: *the total momentum of an isolated system of bodies remains constant.* By an **isolated system**, we mean one on which no external forces act—the only forces acting are those between particles of the system.

Like energy, the importance of the momentum concept is that momentum is conserved under quite general conditions. Although the law of conservation of momentum follows from Newton's second law, as we have seen, it is in fact more general than Newton's laws. In the tiny world of the atom, Newton's laws fail, but the great conservation laws—those of energy, momentum, angular momentum, and electric charge—have been found to hold in every experimental situation tested. It is for this reason that the conservation laws are considered more basic than Newton's laws.

As an example, consider the head-on collision of the two billiard balls shown in Fig. 8–10. Although the momentum of each of the two balls changes as a result of the collision, the *sum* of their momenta is found to be the same before and after the collision. If $m_1\mathbf{v}_1$ is the momentum of ball 1 and $m_2\mathbf{v}_2$ the momentum of ball 2, both measured before the collision, then the total momentum of the two balls before the collision is $m_1\mathbf{v}_1 + m_2\mathbf{v}_2$. After the collision, the balls each have a different velocity and momentum which we will designate by a prime on the velocity: $m_1\mathbf{v}_1'$ and $m_2\mathbf{v}_2'$. The total momentum after the collision is $m_1\mathbf{v}_1' + m_2\mathbf{v}_2'$. No matter what the velocities and masses involved are, it is found that total momentum before the collision is the same as after whether the collision is head-on or not:

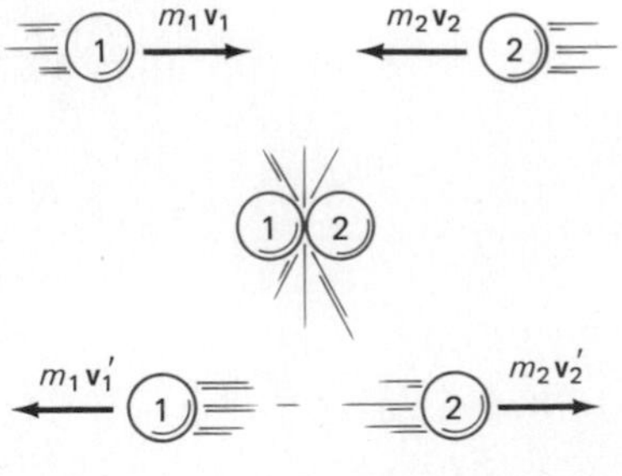

$$m_1\mathbf{v}_1 + m_2\mathbf{v}_2 = m_1\mathbf{v}_1' + m_2\mathbf{v}_2'. \qquad [\text{collision}] \qquad (8\text{–}14)$$

That is, the total momenta of the two balls is conserved. We can rewrite this as

$$m_1\mathbf{v}_1' - m_1\mathbf{v}_1 = -(m_2\mathbf{v}_2' - m_2\mathbf{v}_2),$$

FIGURE 8–10

Momentum is conserved in collision of two balls.

which tells us that any momentum lost by one ball is gained by the other. The total momentum thus stays constant.

EXAMPLE 8–6 A 10,000-kg railroad car traveling at a speed of 24.0 m/s strikes an identical car at rest. If the cars lock together as a result of the collision, what is their common speed afterward?

SOLUTION The initial total momentum is

$$m_1 v_1 + m_2 v_2 = (10,000 \text{ kg})(24.0 \text{ m/s}) + (10,000 \text{ kg})(0 \text{ m/s})$$
$$= 2.40 \times 10^5 \text{ kg·m/s}.$$

After the collision, the total momentum will be the same but it will be shared by both cars; since the two cars become attached, they will have the same velocity, call it v'. Then:

$$(m_1 + m_2)v' = 2.40 \times 10^5 \text{ kg·m/s}$$

$$v' = \frac{2.40 \times 10^5 \text{ kg·m/s}}{2.00 \times 10^4 \text{ kg}} = 12 \text{ m/s}.$$

The law of conservation of momentum is particularly useful when we are dealing with fairly simple systems such as collisions and certain types of explosions. For example rocket propulsion, which we saw in Chapter 4 can be understood on the basis of action and reaction, can also be explained on the basis of the conservation of momentum: before a rocket is fired, the total momentum of rocket plus fuel is zero; as the fuel burns, the total momentum remains unchanged, but the backward momentum of the expelled gases is just balanced by the forward momentum of the rocket itself. Similar examples are the recoil of a gun and the throwing of a package from a boat.

EXAMPLE 8–7 Calculate the recoil velocity of a 4.0-kg rifle which shoots a 0.050-kg bullet at a speed of 280 m/s.

SOLUTION The total momentum of the system is conserved. We let the subscripts B represent the bullet and R the rifle; the final velocities are indicated by primes. Then

$$m_B v_B + m_R v_R = m_B v'_B + m_R v'_R$$

$$0 + 0 = (0.050 \text{ kg})(280 \text{ m/s}) + (4.0 \text{ kg})(v'_R)$$

$$v'_R = -3.5 \text{ m/s}.$$

Since the rifle has a much larger mass, its velocity is much less than that of the bullet. The minus sign indicates that the velocity (and momentum) of the rifle is in the opposite direction to that of the bullet. Notice that it is the *vector sum* of the momenta that is conserved.

If a net external force acts on a system, the law of conservation of momentum will not be valid. However, in some cases the system can be redefined so as to include other objects, so that the conservation of momentum principle then will apply. For example, if we take as our system a falling rock, it does not conserve momentum—its momentum increases as it falls toward the earth. Its momentum is not conserved because an external force, the force of gravity exerted by the earth, is acting on it. However, if we include the earth in the system, the total momentum of rock plus earth, as seen from an inertial reference frame, is conserved. This of course means that the earth comes up to meet the ball. However, since the earth's mass is so great, its upward velocity is very tiny.

8-6 Collisions and Impulse

Conservation of momentum is a very useful tool for dealing with collision processes, as we already saw in the examples of the previous section. Collisions are a common occurrence in everyday life: a tennis racket striking a tennis ball, a baseball bat or golf club striking a ball, two billiard balls colliding, one railroad car striking another, a hammer hitting a nail. At the atomic and subatomic level, scientists learn about the structure of atoms and nuclei, and about the nature of the forces involved, by careful study of collisions between atoms and nuclei.

What exactly do we mean by a collision? The interaction between two bodies is called a collision if the interaction occurs over a short time interval and is so strong that other forces acting are insignificant compared to the forces each body exerts on the other during the collision. Normally the collision time is very short compared to the time we have to observe it, so we can clearly distinguish the periods before the collision and after the collision during which the collision forces are not acting. In a game of tennis, for example, there is a distinct period before and after each stroke of the ball when the ball moves as a projectile under the forces of gravity and air resistance. When the racket strikes the ball, however, these other forces are insignificant; the radical change in motion of the ball is due almost entirely to the very brief but very strong force exerted by the racket. An abrupt change in the motion of at least one of the bodies (and often both) is characteristic of most collisions. Both objects in a collision are deformed, often considerably, because of the strong forces involved, Fig. 8–11.

Let us now look at the collision process in more detail. We assume for now that the masses of particles remain constant, and that none of the speeds is close to the speed of light, so we can ignore relativistic effects (Chapter 39). We consider the motion of the cm of each body and represent the momentum of its cm by $\mathbf{p}$.

When a collision between ordinary objects occurs, the force normally jumps from zero at the moment of contact to a very large value within a very short time, and then abruptly returns to zero again. A graph of the magnitude of the force one object exerts on the other during a collision, as a function of time, is typically like that shown in Fig. 8–12. The time interval $\Delta t = t_f - t_i$, where t_i is the "initial" time (when the force starts acting) and t_f is the "final" time (when the force stops acting), is usually very distinct and usually very small.

From Newton's second law the *net* force on an object is equal to the rate of change of its momentum:

$$\frac{d\mathbf{p}}{dt} = \mathbf{F}.$$

(This equation, of course, applies to *each* of the objects in a collision.) During the infinitesimal time interval dt, the momentum changes by

$$d\mathbf{p} = \mathbf{F}\,dt.$$

If we integrate this over the period during a collision, we have

$$\mathbf{p}_f - \mathbf{p}_i = \int_{p_i}^{p_f} d\mathbf{p} = \int_{t_i}^{t_f} \mathbf{F}\,dt,$$

where $\mathbf{p}_i$ and $\mathbf{p}_f$ are the momenta of the object just before and just after the collision. The integral of the force over the time interval during which it acts is called the **impulse, J**:

$$\mathbf{J} = \int_{t_i}^{t_f} \mathbf{F}\,dt,$$

Thus we have that the change in momentum of an object, $\Delta\mathbf{p} = \mathbf{p}_f - \mathbf{p}_i$ is equal to the impulse acting on it:

$$\Delta\mathbf{p} = \mathbf{p}_f - \mathbf{p}_i = \int_{t_i}^{t_f} \mathbf{F}\,dt = \mathbf{J}. \qquad (8\text{–}15)$$

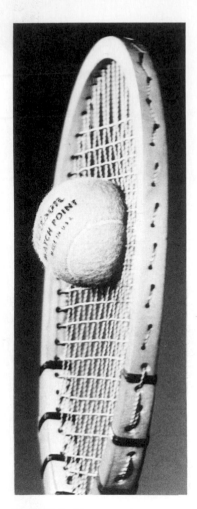

FIGURE 8–11

Tennis racket striking a ball. Note deformation of both ball and racket due to the large force each exerts on the other. (Photograph by Russ Kinne, Photo Researchers, Inc.)

FIGURE 8–12

Force as a function of time during a typical collision.

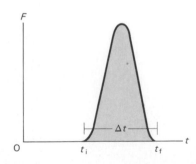

The units for impulse are the same as for momentum, kg · m/s (or N · s) in SI. Since $\mathbf{J} = \int \mathbf{F}\,dt$, we have that the impulse $\mathbf{J}$ of a force is equal to the area under the F versus t curve, as indicated by the shading in Fig. 8–12.

Equation 8–15 is true only if $\mathbf{F}$ is the *net* force on the object. It is valid for any net force $\mathbf{F}$ where $\mathbf{p}_i$ and $\mathbf{p}_f$ correspond precisely to the times t_i and t_f. But the impulse concept is really most useful for so-called *impulsive forces*; that is, for a force like that shown in Fig. 8–12 which has a very large magnitude over a very short time interval, and is essentially zero outside this time interval. For most collision processes, the impulsive force is much larger than any other force acting, and the others can be neglected; so the impulsive force is essentially the net force, and the change in momentum of an object during a collision is due almost entirely to the impulsive force. For such an impulsive force, the time interval over which we take the integral in Eq. 8–15 is not critical as long as we start before t_i and end after t_f, since $\mathbf{F}$ is essentially zero outside this time interval $\Delta t = t_f - t_i$. (Of course, if the chosen time interval is too large, the effect of the other forces becomes significant—such as the flight of a tennis ball which slowly falls under gravity after being struck by a racket.)

It is sometimes useful to speak of the average force, $\bar{\mathbf{F}}$, during a collision. It is defined as that constant force which, when acting over the same time interval $\Delta t = t_f - t_i$ as the actual force, produces the same impulse and change in momentum. Thus

$$\bar{\mathbf{F}}\,\Delta t = \int_{t_i}^{t_f} \mathbf{F}\,dt.$$

Figure 8–13 shows the magnitude of the average force, $\bar{F}$, for the impulsive force of Fig. 8–12. The rectangular area $\bar{F}\Delta t$ equals the area under the impulsive force curve.

◆ EXAMPLE 8–8 (a) Calculate the impulse suffered when a 70-kg person lands on firm ground after jumping from a height of 5.0 m. Then estimate the average force exerted on the person's leg if the landing is (b) stiff-legged, and (c) with bent legs. In the former case, assume the cm of the body moves 1.0 cm during impact and in the second case, when the legs are bent, about 50 cm.

SOLUTION (a) Although we don't know F and thus can't calculate the impulse from

$$\mathbf{J} = \int \mathbf{F}\,dt,$$

we can use the fact that the impulse equals the change in momentum of the object. After a fall of 5.0 m, the velocity of the person will be (using the appropriate kinematics equation for constant acceleration, Eq. 2–9c, in the vertical $[y]$ direction, with $a = g = 9.8$ m/s², and initial velocity $v_0 = 0$):

$$v = \sqrt{2a(y - y_0)} = \sqrt{2(9.8 \text{ m/s}^2)(5.0 \text{ m})} = 9.9 \text{ m/s}.$$

After striking the ground, the momentum is quickly brought to zero, Fig. 8–14. The impulse on the person is then

$$J = \bar{F}\,\Delta t = \Delta p = p_f - p_i$$
$$= 0 - (70 \text{ kg})(9.9 \text{ m/s}) = -690 \text{ N·s}.$$

The negative sign tells us the force must be opposed to the original momentum.

(b) In coming to rest, the cm decelerates from 9.9 m/s to zero in a distance $D = 1.0 \times 10^{-2}$ m. The average speed during this period was $(9.9 \text{ m/s} + 0 \text{ m/s})/2 = 5.0$ m/s, so the time of the collision was $\Delta t = D/\bar{v} = (1.0 \times 10^{-2} \text{ m})/(5.0 \text{ m/s}) = 2.0 \times 10^{-3}$ s. The average force was then

$$\bar{F} = \frac{J}{\Delta t} = \frac{690 \text{ N · s}}{2.0 \times 10^{-3} \text{ s}} = 3.5 \times 10^5 \text{ N}.$$

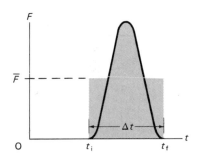

FIGURE 8–13
The average force $\bar{F}$ over Δt gives the same impulse ($\bar{F}\,\Delta t$) as the actual force.

FIGURE 8–14
Period during which impulse acts (Example 8–8).

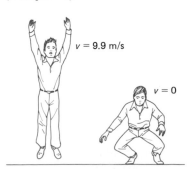

$v = 9.9$ m/s

$v = 0$

(c) This is just like part (b) except $d = 0.50$ m so $\Delta t = (0.50 \text{ m})/(5.0 \text{ m/s}) = 0.10$ s and

$$\bar{F} = \frac{690 \text{ N} \cdot \text{s}}{0.10 \text{ s}} = 6.9 \times 10^3 \text{ N}.$$

Clearly the force on the feet and legs is much less when the knees are bent. In fact the ultimate strength of the leg bone (see Chapter 11, Table 11–2) is not high enough to support the force calculated in part (b), so the leg would undoubtedly break in such a landing, whereas it probably would not in part (c).

8–7 Conservation of Momentum and Energy in Collisions

During most collisions, we usually don't know how the collision force varies as a function of time. Nonetheless, we can still determine some of the details of the motion after a collision, given the initial motion, by making use of the conservation laws for momentum and energy. We continue to make the reasonable assumption that the impulsive forces acting during a collision are much greater than any other forces acting.

Let us consider two bodies, of mass m_1 and m_2, that have momenta $\mathbf{p}_1$ and $\mathbf{p}_2$ before they collide, and $\mathbf{p}_1'$ and $\mathbf{p}_2'$ after they collide, as in Fig. 8–15. The primes (') denote quantities *after* the collision. During the collision, suppose the force exerted by body 1 on body 2 at any moment is $\mathbf{F}$; then, by Newton's third law, the force exerted by body 2 on body 1 is $-\mathbf{F}$. During the very short collision time, the impulsive force $\mathbf{F}$ is assumed to be much greater than any other (external) forces acting, so $\mathbf{F}$ represents the net force to a very good approximation; therefore the change in momentum of body 2 is

$$\Delta\mathbf{p}_2 = \mathbf{p}_2' - \mathbf{p}_2 = \int_{t_i}^{t_f} \mathbf{F} \, dt$$

and for body 1 it is

$$\Delta\mathbf{p}_1 = \mathbf{p}_1' - \mathbf{p}_1 = -\int_{t_i}^{t_f} \mathbf{F} \, dt.$$

We compare these two equations and see that

$$\Delta\mathbf{p}_1 = -\Delta\mathbf{p}_2$$

$$\mathbf{p}_1' - \mathbf{p}_1 = -(\mathbf{p}_2' - \mathbf{p}_2).$$

Rearranging, we have

$$\mathbf{p}_1 + \mathbf{p}_2 = \mathbf{p}_1' + \mathbf{p}_2'. \tag{8–16}$$

The total momentum before the collision is the same as the total momentum after the collision. *Momentum is conserved.* This is, of course, the law of conservation of momentum, which we now see holds to an excellent approximation even when external forces do act, as long as they are much smaller than the impulsive force during the time of collision. The change in momentum due to the collision force is so large compared to that due to the external forces that the latter can be ignored. If significant external forces do act, then to be most accurate, Eq. 8–16 should be applied just before and just after the collision; if $\mathbf{p}_1$, $\mathbf{p}_2$, $\mathbf{p}_1'$, and $\mathbf{p}_2'$ are measured long before and long after the collision, the external forces may alter the momenta significantly since the integral $\int \mathbf{F} \, dt = \Delta\mathbf{p}$ over longer times changes the value of $\Delta\mathbf{p}$; an example (already mentioned) is the effect of gravity on a tennis ball in its flight, after it has been struck. Thus we can apply conservation of momentum to a collision only if the impulsive forces are much greater than the external forces. We assume this to be true for collisions we consider here (indeed, this is how we

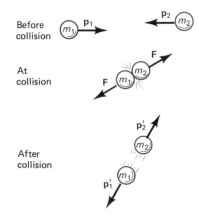

FIGURE 8–15

Collision of two objects. Their momenta before collision are $\mathbf{p}_1$ and $\mathbf{p}_2$, and after collision are $\mathbf{p}_1'$ and $\mathbf{p}_2'$. At any moment during the collision each exerts forces on the other of equal magnitude but opposite direction.

define what we mean by a collision). Note that we must not only determine the momenta at times close enough to the collision time that external forces do not change the momenta significantly, but we must also be sure to measure the velocities only after the impulsive forces between the bodies have ceased to act.

The law of conservation of energy tells us that the total energy of our two objects will also be conserved during a collision. But because energy can take many forms, this law is not always useful. There are, however, collisions in which the total *kinetic* energy of the two particles is conserved; such a collision is called an **elastic collision**, and for them we can write

$$\tfrac{1}{2}m_1 v_1^2 + \tfrac{1}{2}m_2 v_2^2 = \tfrac{1}{2}m_1 v_1'^2 + \tfrac{1}{2}m_2 v_2'^2. \qquad \text{[elastic collision]} \quad (8\text{-}17)$$

Of course during the brief moment when the two objects are in contact, some (or all) of the energy is stored momentarily in the form of potential energy (elastic, electrical, or some other type); but the total KE before the collision is the same as the total KE after the collision.

At the atomic level, the collisions of atoms, nuclei, and elementary particles are often elastic. But in the macroscopic world of everyday objects, an elastic collision is an ideal that is never quite reached since at least a little thermal energy (and perhaps sound and other forms of energy) is always produced during a collision. The collision of two hard elastic balls, however, such as billiard balls, is close to being elastic and we sometimes treat it as such.

8-8   Elastic Collisions in One Dimension

We now apply the conservation laws of momentum and kinetic energy to an elastic collision between two particles that is head-on, so all the motion is along a line. To be general, let us assume both particles are moving initially, with velocities v_1 and v_2, along the x axis, Fig. 8-16a. After the collision, their velocities are v_1' and v_2', Fig. 8-16b. For any $v > 0$, the particle is moving to the right (increasing x); whereas for $v < 0$, the particle is moving to the left, toward decreasing values of x.

From conservation of momentum we have

$$m_1 v_1 + m_2 v_2 = m_1 v_1' + m_2 v_2'$$

and because the collision is assumed to be elastic, kinetic energy is also conserved:

$$\tfrac{1}{2}m_1 v_1^2 + \tfrac{1}{2}m_2 v_2^2 = \tfrac{1}{2}m_1 v_1'^2 + \tfrac{1}{2}m_2 v_2'^2.$$

We have two equations; therefore we can solve for two unknowns. If we know the masses and initial velocities, then we can solve these two equations for the velocities after the collision, v_1' and v_2'. We will do this in a moment for some special cases, but first we derive a useful theorem. To do so we rewrite the momentum equation as

$$m_1(v_1 - v_1') = m_2(v_2' - v_2); \qquad (8\text{-}18a)$$

and we rewrite the KE equation as

$$m_1(v_1^2 - v_1'^2) = m_2(v_2'^2 - v_2^2)$$

or

$$m_1(v_1 - v_1')(v_1 + v_1') = m_2(v_2' - v_2)(v_2' + v_2). \qquad (8\text{-}18b)$$

We divide Eq. 8-18b by 8-18a (assuming $v_1 \neq v_1'$ and $v_2 \neq v_2'$)[†] and obtain

$$v_1 + v_1' = v_2' + v_2. \qquad (8\text{-}18c)$$

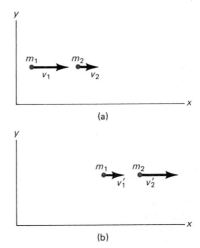

FIGURE 8-16

Two particles, of mass m_1 and m_2, (a) before collision, (b) after collision.

[†] Note that Eqs. 8-18a and 8-18b, which are the conservation laws for momentum and kinetic energy, are both satisfied by the solution $v_1' = v_1$ and $v_2' = v_2$. This is a valid solution, but not very interesting. It corresponds to no collision at all—when the two particles miss each other.

We can rewrite this equation as

$$v_1 - v_2 = v_2' - v_1'. \tag{8-19}$$

This is the theorem we sought. It is an interesting result: it tells us that the relative speed of the two particles after the collision is the same as before, for any elastic head-on collision, no matter what the masses are.

Now let us look at some special cases of elastic head-on collisions. We assume $v_1, v_2, m_1,$ and m_2 are known, and we wish to solve for v_1' and v_2', the velocities of the two particles after the collision.

1. Equal masses, $m_1 = m_2$. From conservation of momentum we have:

$$v_1 + v_2 = v_1' + v_2'.$$

We need a second equation since there are two unknowns. We could use the conservation of kinetic energy equation, but it is easier to use Eq. 8-19, which tells us that the relative velocity before and after the collision is the same:

$$v_1 - v_2 = v_2' - v_1'.$$

Now we add these two equations and get

$$v_2' = v_1$$

and then subtract these two equations to get

$$v_1' = v_2.$$

That is, the particles exchange velocities as a result of the collision; particle 2 acquires the velocity that particle 1 had before the collision, and vice versa. If particle 2 is at rest initially, so $v_2 = 0$, we have

$$v_2' = v_1 \quad \text{and} \quad v_1' = 0.$$

That is, particle 1 is stopped dead in its tracks and particle 2 takes off with a velocity equal to what particle 1 had before the collision. This effect is well known to billiard and pool players.

2. Particle 2 at rest initially, $v_2 = 0$. This is a very common practical situation, for a moving object to strike one at rest. Combining the momentum equation (which now is we have, using the relations for v_2' and v_1' above,

$$v_2' = v_1 \left(\frac{2m_1}{m_1 + m_2} \right)$$

$$v_1' = v_1 \left(\frac{m_1 - m_2}{m_1 + m_2} \right).$$

Some special cases of this are:

a. $v_2 = 0$, and $m_1 = m_2$. In this case we have

$$v_2' = v_1 \text{ and } v_1' = 0.$$

This is the same case treated in item 1 above, and we get the same result: for particles of equal mass, one of which is initially at rest, the velocity of the particle moving initially is completely transferred to the particle originally at rest.

b. $v_2 = 0$ and $m_1 \gg m_2$. A very heavy moving object strikes a light object at rest, and we have, using the relations for v_2' and v_1' above,

$$v_2' \approx 2v_1$$

$$v_1' \approx v_1.$$

Thus the velocity of the heavy incoming particle is practically unchanged, whereas the light particle, originally at rest, takes off with twice the velocity of the heavy one. The velocity of a heavy bowling ball, for example, is hardly affected by striking the much lighter bowling pins.

c. $v_2 = 0$ and $m_1 \ll m_2$. A moving light particle strikes a very massive particle at rest.
In this case

8-8   Elastic Collisions   **149**
in One Dimension

$$v_2' \approx 0$$

$$v_1' \approx -v_1.$$

The massive object remains essentially at rest and the very light incoming particle rebounds with essentially its same speed but in the opposite direction. In other words, each rebounds with approximately the same speed (but opposite direction) it had before the collision. For example, a tennis ball colliding head-on with a stationary bowling ball will hardly affect the bowling ball, but will rebound with nearly the same speed it had initially, just as if it had struck a hard wall.

It can readily be shown (it is given as problem 50) for any elastic head-on collision that

$$v_2' = v_1 \left(\frac{2m_1}{m_1 + m_2} \right) + v_2 \left(\frac{m_2 - m_1}{m_1 + m_2} \right) \qquad (8\text{-}20\text{a})$$

and

$$v_1' = v_1 \left(\frac{m_1 - m_2}{m_1 + m_2} \right) + v_2 \left(\frac{2m_2}{m_1 + m_2} \right). \qquad (8\text{-}20\text{b})$$

These general equations, however, should not be memorized. If need be, they can be derived quickly from the conservation laws. For many problems, it is simplest just to start from scratch as we did in the special cases above and as exemplified by the next example.

EXAMPLE 8-9   A proton of mass 1.01 u (unified atomic mass units) traveling with a speed of 3.60×10^4 m/s has an elastic head-on collision with a helium (He) nucleus ($m_{He} = 4.00$ u) initially at rest. What are the velocities of the proton and helium nucleus after the collision?

SOLUTION   We saw in Section 4-3 that $1 \text{ u} = 1.6606 \times 10^{-27}$ kg, but we won't really need this fact. We have $v_2 = v_{He} = 0$ and $v_1 = v_p = 3.60 \times 10^4$ m/s. We want to find the velocities v_p' and v_{He}' after the collision. From conservation of momentum we have

$$m_p v_p + 0 = m_p v_p' + m_{He} v_{He}'.$$

From conservation of kinetic energy we have

$$\tfrac{1}{2} m_p v_p^2 + 0 = \tfrac{1}{2} m_p v_p'^2 + \tfrac{1}{2} m_{He} v_{He}'^2.$$

From the first equation

$$v_p' = v_p - \left(\frac{m_{He}}{m_p} \right) v_{He}'.$$

We substitute this into the second equation and obtain

$$(v_{He}')^2 - (v_{He}') \left(\frac{2 m_p v_p}{m_p + m_{He}} \right) = 0.$$

There are two solutions,

$$v_{He}' = 0 \quad \text{and} \quad v_{He}' = \frac{2 m_p v_p}{m_p + m_{He}} = 1.45 \times 10^4 \text{ m/s}.$$

These correspond to

$$v_p' = v_p - \left(\frac{m_{He}}{m_p} \right) v_{He}' = v_p = 3.60 \times 10^4 \text{ m/s},$$

and

$$v_p' = v_p - \left(\frac{m_{He}}{m_p}\right)\left(\frac{2m_p v_p}{m_p + m_{He}}\right) = -0.597 v_p = -2.15 \times 10^4 \text{ m/s}.$$

The first solution ($v_{He}' = 0$, $v_p' = v_p$) corresponds to no collision. The second is the interesting one, $v_{He}' = 1.45 \times 10^4$ m/s, $v_p' = -2.15 \times 10^4$ m/s. Note that the proton reverses direction upon collision, although its speed is less than its incident speed.

◈ EXAMPLE 8–10　In a nuclear reactor, neutrons (and energy) are released when uranium ($^{235}_{92}$U) nuclei[†] undergo fission (breaking apart), and this occurs at a reasonable rate only when the $^{235}_{92}$U nuclei are struck by very slowly moving neutrons. The neutrons emitted by previous fissions are fast-moving; so to sustain a chain reaction, the neutrons must be rapidly slowed down so they can cause more fissions before they escape from the uranium fuel rods. The material used to slow down the neutrons is called the *moderator*. What type of material would best serve this purpose?

SOLUTION　The atoms of the moderator can be considered to be essentially at rest. This corresponds to $v_2 = 0$, where v_1 represents the velocity of the neutron. This is our special case 2 discussed above. In particular, we saw that if m_2 (mass of atoms in moderator) is either much greater or much less than that of m_1 (the neutron), the neutron's speed will not be greatly reduced. But if the moderator is made of material whose atomic mass is close to the neutron mass ($m_1 = m_2$) the neutron's velocity after a head-on collision will be essentially zero, $v_1' \approx 0$. The lightest atom, hydrogen ($^1_1$H), has a mass almost precisely equal to that of the neutron, and so would be the ideal moderator; unfortunately, ordinary hydrogen has a strong tendency to absorb the neutrons, so is useless. But the isotope of hydrogen that has twice this mass ($^2_1$H, called *deuterium* or *heavy hydrogen*) does not absorb many neutrons; it is therefore the best practical moderator. It is usually incorporated into water molecules (H_2O), which is then called "heavy water." Carbon ($^{12}_6$C) is also often used as a moderator since its mass is not too large, and it is very practical for other reasons. This is pursued more fully in problem 51.

▱ *8–9　Elastic Collisions in Two or Three Dimensions

Conservation of momentum and energy can also be applied to collisions in two or three dimensions, and in this case the vector nature of momentum comes into play. The two most commonly analyzed collisions are (1) those for which one particle (called the "projectile") strikes a second particle initially at rest (the "target" particle), and (2) those for which the initial momenta of the two particles are along the same line (usually directed toward each other). In either case, if the collision is not head-on, the final momenta after the collision will be in different directions than the initial momenta, and the collision is essentially two dimensional since the paths of both particles will be in the plane defined by the initial and final momenta (see question 29).

Let us consider now elastic collisions of the first type: a projectile colliding with a stationary target particle. (The assumption of a stationary target is not really restrictive since we can choose any reference frame we want, and we can

[†] The subscript 92, is the number of protons in this nucleus; the superscript, 235, is the total number of protons and neutrons, and is approximately equal to the mass in atomic mass units (see Chap. 42.)

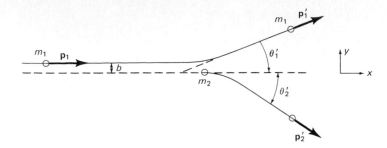

FIGURE 8–17

Particle 1, the projectile, collides with particle 2, the target; the impact parameter is b. They move off, after the collision, with momenta $\mathbf{p}'_1$ and $\mathbf{p}'_2$ at angles θ'_1 and θ'_2.

always choose one where $v_2 = 0$.) This is the common situation for experiments in atomic and nuclear physics. The projectiles, from radioactive decay or accelerated by a high-energy accelerator, strike a stationary target nucleus.

Figure 8–17 shows particle 1 (the projectile, m_1) heading along the x axis toward particle 2 (the target, m_2) which is initially at rest. If these are, say, billiard balls, m_1 strikes m_2 and they go off at the angles θ'_1 and θ'_2, which are measured relative to m_1's initial direction (the x axis). If they are electrically charged particles, or nuclear particles, they may begin to deflect even before they touch because of the force (electric or nuclear) between them. (You might think, for example, of two magnets oriented so they repel each other: when one moves toward the other, the second moves away before the first one touches it.)

The distance b in Fig. 8–17 is called the *impact parameter*. This is the perpendicular distance between the initial line of motion of the projectile and a parallel line through the center of the target particle. It is a measure of the directness of the collision. For $b = 0$, the collision is head-on.

Let us now apply the laws of conservation of momentum and kinetic energy for an elastic collision like that of Fig. 8–17. From conservation of kinetic energy, since $v_2 = 0$, we have

$$\tfrac{1}{2}m_1 v_1^2 = \tfrac{1}{2}m_1 v_1'^2 + \tfrac{1}{2}m_2 v_2'^2. \tag{8–21a}$$

We choose the xy plane to be the plane in which lie the initial and final momenta. Since momentum is a vector, and is conserved, its components in the x and y directions remain constant. In the x direction

$$m_1 v_1 = m_1 v_1' \cos \theta'_1 + m_2 v_2' \cos \theta'_2. \tag{8–21b}$$

Since there is no motion in the y direction initially, the y component of the total momentum is zero:

$$0 = m_1 v_1' \sin \theta'_1 + m_2 v_2' \sin \theta'_2. \tag{8–21c}$$

We have three independent equations.[†] This means we can solve for at most three unknowns. If we are given m_1, m_2, v_1 (and v_2, if not zero), we cannot uniquely predict the final variables v_1', v_2', θ'_1, and θ'_2 since there are four of them; θ'_2, for example, can be anything. However, if we measure one of these variables, say θ'_1, then the other three variables (v_1', v_2', and θ'_2) are uniquely determined and we can calculate them using the above three equations. We must be sure, however, that all velocities and angles are measured sufficiently long before and after the collision that the force between the particles is no longer acting.

EXAMPLE 8–11 A proton traveling with speed 8.2×10^5 m/s collides elastically with a stationary proton in a hydrogen target. One of the protons is observed to be scattered at a 60° angle. At what angle will the second proton be observed, and what will be the velocities of the two protons after the collision?

[†] Note that Eqs. 8–21b and c are true even if the collision is inelastic and KE is not conserved.

SOLUTION Since $m_1 = m_2$, Eqs. 8–21a, b, and c become

$$v_1^2 = v_1'^2 + v_2'^2$$

$$v_1 = v_1' \cos \theta_1' + v_2' \cos \theta_2'$$

$$0 = v_1' \sin \theta_1' + v_2' \sin \theta_2'$$

where $v_1 = 8.2 \times 10^5$ m/s and $\theta_1' = 60°$ are known. In the second and third equations, we take the v_1' terms to the left side and square both sides of these two equations:

$$v_1^2 - 2v_1 v_1' \cos \theta_1' + v_1'^2 \cos^2 \theta_1' = v_2'^2 \cos^2 \theta_2'$$

$$v_1'^2 \sin^2 \theta_1' = v_2'^2 \sin^2 \theta_2'.$$

We add these two equations and use the fact that $\sin^2 \theta + \cos^2 \theta = 1$ to get:

$$v_1^2 - 2v_1 v_1' \cos \theta_1' + v_1'^2 = v_2'^2.$$

Into this equation we substitute, from the first equation above, $v_2'^2 = v_1^2 - v_1'^2$, and obtain

$$2v_1'^2 = 2v_1 v_1' \cos \theta_1'$$

or

$$v_1' = v_1 \cos \theta_1' = (8.2 \times 10^5 \text{ m/s})(\cos 60°) = 4.1 \times 10^5 \text{ m/s}.$$

To obtain v_2', we use the first equation (conservation of KE):

$$v_2' = \sqrt{v_1^2 - v_1'^2} = 7.1 \times 10^5 \text{ m/s}.$$

Finally, from the third equation:

$$\sin \theta_2' = -\frac{v_1'}{v_2'} \sin \theta_1' = -\left(\frac{4.1 \times 10^5 \text{ m/s}}{7.1 \times 10^5 \text{ m/s}}\right)(0.866) = -0.50,$$

so $\theta_2' = -30°$. (The minus sign merely means particle 2 moves at an angle below the x axis if particle 1 is above the axis (as in Fig. 8–17). An example of such a collision is shown in the bubble chamber photo of Fig. 8–18. Notice that the two trajectories are at right angles to each other after the collision. This can be shown to be true in general for non–head-on collisions of two particles of equal mass, one of which was at rest initially (see the problems).

Since the analysis of the speeds and directions of colliding particles does not deal with the forces themselves, this analysis is sometimes referred to as "kinematics."

FIGURE 8–18

Photo of a proton-proton collision in a hydrogen bubble chamber (a device that makes visible the paths of elementary particles). The many lines represent incoming protons which can strike the protons of the hydrogen in the chamber. (Courtesy of Brookhaven National Laboratory.)

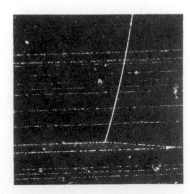

✳ 8–10   Center-of-Momentum (CM) Reference Frame

In any collision, be it elastic or inelastic, when external forces can be neglected the total momentum $\mathbf{P} = m_1\mathbf{v}_1 + m_2\mathbf{v}_2$ is conserved; $\mathbf{P}$ is the same before the collision as after it. This will be true in any coordinate system or reference frame we choose to use; although $\mathbf{P}$ will not change as a result of the collision, the numerical value of $\mathbf{P}$ will depend on the choice of reference frame.

Collisions can be viewed or analyzed from a variety of reference frames. But there is one reference frame that is unique and simpler than any other. This is the reference frame in which the total momentum $\mathbf{P}$ is zero. It is called the *center-of-momentum (CM) reference frame*. It is also called the *center-of-mass reference frame* since the center of mass of the two colliding particles, given by

$$\mathbf{r}_{cm} = \frac{m_1\mathbf{r}_1 + m_2\mathbf{r}_2}{m_1 + m_2}$$

must be at rest since

$$\mathbf{P}_{cm} = (m_1 + m_2)\frac{d\mathbf{r}_{cm}}{dt} = \left(m_1\frac{d\mathbf{r}_1}{dt} + m_2\frac{d\mathbf{r}_2}{dt} \right) = m_1\mathbf{v}_1 + m_2\mathbf{v}_2,$$

which is zero by definition of the center-of-momentum reference frame:[†]

$$m_1\mathbf{v}_1^* + m_2\mathbf{v}_2^* = 0. \qquad \text{[CM frame]}$$

We use an asterisk (✳) to represent variables that are measured in the CM reference frame.

In the CM system, which we discuss here only briefly, the two particles move toward each other with equal and opposite momenta, $\mathbf{p}_1^* = -\mathbf{p}_2^*$ since $\mathbf{p}_1^* + \mathbf{p}_2^* = \mathbf{P} = 0$. If the collision is elastic and head-on, the two particles rebound in the directions from which they came with the same speed. If it is not head-on, they have the same speeds as before the collision, but rebound at some angle θ'^*. As shown in Fig. 8–19, they move in opposite directions (since $\mathbf{p}_1'^* + \mathbf{p}_2'^* = \mathbf{P}'^* = 0$), so only one angle θ'^* specifies the directions after the collision. Even if the collision is inelastic we still have $\mathbf{p}_1'^* = -\mathbf{p}_2'^*$: the momenta have equal magnitude and opposite directions. If the collision is elastic, the magnitudes of the momenta don't change so $p_1^* = p_2^* = p_1'^* = p_2'^*$. For both elastic and inelastic collisions, the momenta do not depend on the angle, as they do in other coordinate systems (such as that in Fig. 8–17). The final state is essentially in one dimension, though rotated by an angle θ'^* from the initial line of motion. Collisions viewed in the center-of-momentum system are thus highly symmetric, and are more simply analyzed than in any other system.

[†] Although center-of-momentum and center-of-mass reference frames are identical for classical mechanics, they are not quite the same at high speeds where, as the theory of relativity tells us, the mass of particles depends on velocity. See Chapter 39.

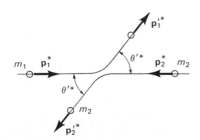

FIGURE 8–19

Elastic collision in the CM system:
$p_1^* = p_2^* = p_1'^* = p_2'^*$.

Not only is the kinematics (description) of collisions simplest in the CM reference frame but so are the theories developed by physicists for trying to understand nuclear forces. For these reasons, and because laboratory systems differ, it is customary to analyze nuclear and elementary particle collisions in the CM system. Although some physics experiments involving elementary particles are actually carried out in the CM system (with the use of so-called storage rings, so that particles are made to move toward each other with equal but opposite momenta), many experiments are done with one of the particles at rest; that is, a high speed (projectile) particle is made to strike a stationary target particle, as in Fig. 8–17. This is the usual laboratory setup, and this reference frame where one particle (the target) is at rest is generally called the laboratory frame or *lab frame*. Experimental measurements carried out in the lab frame are normally transferred, on paper, into the CM frame. We won't go into this in more detail here, but will leave it for the problems.

*8–11   Inelastic Collisions

Collisions in which kinetic energy is not conserved are called *inelastic collisions*. Some of the initial kinetic energy in such collisions is transformed into other types of energy such as thermal or potential energy, so the total final KE is less than the total initial KE; the inverse can also happen when potential energy (such as chemical or nuclear) is released, and then the total final KE can be greater than the initial KE. If two objects stick together as a result of a collision, the collision is said to be *completely inelastic*. Two colliding balls of putty that stick together, or two railroad cars that couple together when they collide (Example 8–6) are examples. The kinetic energy is not necessarily all transformed to other forms of energy in an inelastic collision. In Example 8–6, for instance, we saw that when a traveling railroad car collided with a stationary one, the coupled cars traveled off with some KE. Even though KE is not conserved in inelastic collisions, the total energy is conserved, and the total vector momentum is also always conserved.

EXAMPLE 8–12   For the totally inelastic collision we considered in Example 8–6, calculate how much of the initial kinetic energy is transferred to thermal or other forms of energy.

SOLUTION   Initially, the total kinetic energy is $\frac{1}{2}m_1v_1^2 = \frac{1}{2}(10{,}000 \text{ kg})(24.0 \text{ m/s})^2 = 2.88 \times 10^6$ J. After the collision, the total kinetic energy is $\frac{1}{2}(20{,}000 \text{ kg})(12.0 \text{ m/s})^2 = 1.44 \times 10^6$ J. Hence the energy transformed to other forms is 2.88×10^6 J $- 1.44 \times 10^6$ J $= 1.44 \times 10^6$ J.

EXAMPLE 8–13   The *ballistic pendulum* is a device used to measure the speed of a projectile, such as a bullet. The projectile, of mass m, is fired into a large block (of wood or other material) of mass M which is suspended like a pendulum. (Usually M is somewhat greater than m.) As a result of the collision, the center of mass of the pendulum (and embedded projectile) swings up to a maximum height h, Fig. 8–20. Determine the relationship between the initial speed of the projectile, v, and the height h.

SOLUTION   We analyze this process by dividing it into two parts: (1) the collision itself, and (2) the subsequent motion of the pendulum from the vertical position to height h. In part (1) we assume the collision time is very short, and so the projectile comes to rest before the block has moved significantly from its position directly below its support. Thus there is no net external force and momentum is conserved:

$$mv_1 = (m + M)v'$$

FIGURE 8–20

Ballistic pendulum.

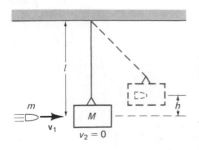

where v' is the speed of block and embedded projectile just after the collision, before they have moved significantly. Once the pendulum begins to move, there will be a net external force (gravity, tending to pull it back to the vertical position). So, for part (2), we cannot use conservation of momentum. But we can use conservation of mechanical energy since the kinetic energy immediately after the collision is changed entirely to gravitational potential energy when the pendulum reaches its maximum height, h. Therefore

$$\tfrac{1}{2}(m + M)v'^2 = (m + M)gh,$$

so $v' = \sqrt{2gh}$. (Why is the work done by the tension force in the supporting cord equal to zero?) We combine these two equations to obtain

$$v_1 = \frac{m + M}{m}\, v' = \frac{m + M}{m}\sqrt{2gh},$$

which is the final result. To obtain this result, we had to be opportunists in that we used what conservation laws we could: in (1) we could use only conservation of momentum since the collision is inelastic and conservation of mechanical energy is not valid; and in (2) conservation of mechanical energy is valid, but not conservation of momentum. If there is significant motion of the pendulum during the deceleration of the projectile in the block, then there *is* an external force during (1)—so conservation of momentum is not valid, and this would have to be taken into account.

*8–12 Systems of Variable Mass

We now treat systems whose mass varies. Such systems could be considered a type of inelastic collision, but it is simpler to return to Eq. 8–12, $d\mathbf{P}/dt = \mathbf{F}_{ext}$, where $\mathbf{P}$ is the total momentum of the system and $\mathbf{F}_{ext}$ is the net external force exerted on it. Great care must be taken to define the system, and to include all changes in momentum. An important application is to rockets, which propel themselves forward by the ejection of burnt gases; the force exerted by the gases on the rocket accelerates the rocket. In this case, the mass M of the rocket decreases during this process, so $dM/dt < 0$. Another application is the dropping of material (gravel, packaged goods) onto a conveyor belt. In this situation, the mass M of the loaded conveyor belt increases, so $dM/dt > 0$.

To treat the general case of a system of variable mass, let us consider the system shown in Fig. 8–21. At some time, t, we have a system of mass M and momentum $M\mathbf{v}$; we also have a tiny (infinitesimal) mass dM traveling with velocity $\mathbf{u}$ which is about to enter our system. An infinitesimal time dt later, the mass dM combines with the system. For simplicity we will refer to this as a "collision." So our system has changed in mass from M to $M + dM$ in the time dt. (Note that dM can be less than zero, as for a rocket propelled by ejected gases.)

In order to apply Eq. 8–12, $d\mathbf{P}/dt = \mathbf{F}_{ext}$, we must consider a definite fixed system of particles. That is, in considering the change in momentum, $d\mathbf{P}$, we must consider the momentum of the same particles initially and finally. We will define our *total system* as including M plus dM. Then initially, at time t, the total momentum is $M\mathbf{v} + \mathbf{u}\,dM$. At time $t + dt$, after dM has combined with M, the velocity of the whole is now $\mathbf{v} + d\mathbf{v}$ and the total momentum is $(M + dM)(\mathbf{v} + d\mathbf{v})$. So the change in momentum $d\mathbf{P}$ is

$$d\mathbf{P} = (M + dM)(\mathbf{v} + d\mathbf{v}) - (M\mathbf{v} + \mathbf{u}\,dM)$$
$$= M\,d\mathbf{v} + \mathbf{v}\,dM + dM\,d\mathbf{v} - \mathbf{u}\,dM.$$

FIGURE 8–21

(a) At time t, a mass dM is about to be added to our system M. (b) At time $t + dt$, the mass dM has been added to our system.

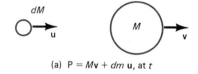

(a) P = $Mv + dm$ **u**, at t

(b) P = $(M + dM)$ (**v** + dv), at $t + dt$

Then, by Eq. 8–12, we have

$$\mathbf{F}_{ext} = \frac{d\mathbf{P}}{dt} = \frac{M\,d\mathbf{v} + \mathbf{v}\,dM - \mathbf{u}\,dM}{dt},$$

or

$$\mathbf{F}_{ext} = M\frac{d\mathbf{v}}{dt} - (\mathbf{u} - \mathbf{v})\frac{dM}{dt}, \tag{8–22a}$$

where we have dropped the term $dM\,d\mathbf{v}/dt$ since in the limit of infinitesimals it is zero. Note that the quantity $(\mathbf{u} - \mathbf{v})$ is the relative velocity, $\mathbf{v}_{rel}$, of dM with respect to M; that is

$$\mathbf{v}_{rel} = \mathbf{u} - \mathbf{v}$$

is the velocity of the entering mass dM as seen by an observer on M. So we can rearrange Eq. 8–22a:

$$M\frac{d\mathbf{v}}{dt} = \mathbf{F}_{ext} + \mathbf{v}_{rel}\frac{dM}{dt}. \tag{8–22b}$$

The first term on the right, $\mathbf{F}_{ext}$, refers to the external force on the system as a whole (for a rocket, it would include the force of gravity and air resistance). It does *not* include the force that dM exerts on M as a result of their collision since this is an internal force within the total system. The second term on the right, $\mathbf{v}_{rel}(dM/dt)$, represents the rate at which momentum is being transferred into or out of the system because of the mass that leaves or is added to it. It thus can be interpreted as the force exerted on the system of mass M due to the addition (or ejection) of mass. For a rocket this term is called the *thrust* since it represents the force exerted on the rocket by the expelled gases.

EXAMPLE 8–14 A fully fueled rocket has a mass of 21,000 kg, of which 15,000 kg is fuel. The burned fuel is exhausted at a rate of 190 kg/s at a speed of 2800 m/s relative to the rocket. If the rocket is fired vertically upward, calculate: (*a*) the thrust of the rocket; (*b*) the net force on the rocket at blastoff and just before burnout (when all the fuel has been used up); (*c*) the rocket's velocity at burnout. Ignore air resistance and assume the acceleration due to gravity is constant at $g = 9.80$ m/s^2.

SOLUTION (*a*) The thrust is

$$v_{rel}\frac{dM}{dt} = (2800 \text{ m/s})(190 \text{ kg/s}) = 5.3 \times 10^5 \text{ N}.$$

(*b*) $F_{ext} = Mg = (2.1 \times 10^4 \text{ kg})(9.80 \text{ m/s}^2) = 2.1 \times 10^5$ N initially, and $(6.0 \times 10^3 \text{ kg})(9.80 \text{ m/s}^2) = 5.9 \times 10^4$ N at burnout. Hence, the net force at blastoff is

$$5.3 \times 10^5 \text{ N} - 2.1 \times 10^5 \text{ N} = 3.2 \times 10^5 \text{ N} \quad \text{(blast off)}$$

and just before burnout it is

$$5.3 \times 10^5 \text{ N} - 5.9 \times 10^4 \text{ N} = 4.7 \times 10^5 \text{ N} \quad \text{(burnout)}.$$

After burnout, of course, the net force is that of gravity, -5.9×10^4 N.

(*c*) From Eq. 8–22b we have

$$dv = \frac{F_{ext}}{M}\,dt + v_{rel}\frac{dM}{M}$$

where $F_{ext} = -Mg$, and M is the mass of the rocket and is a function of time. Since

v_{rel} is constant, we can integrate this easily:

$$\int_{v_0}^{v} dv = -\int_{0}^{t} g\, dt + v_{rel} \int_{M_0}^{M} \frac{dM}{M}$$

or

$$v = v_0 - gt + v_{rel} \ln \frac{M}{M_0},$$

where v is the rocket's velocity and M its mass at any time t. Note that v_{rel} is negative (-2800 m/s in our case) since it is opposite to the motion, and that $\ln(M/M_0)$ is also negative since $M_0 > M$; hence, the last term—which represents the thrust—is positive and acts to increase the velocity. The time required to reach burnout is the time needed to use up all the fuel (15,000 kg) at a rate of 190 kg/s; so at burnout,

$$t = \frac{1.50 \times 10^4 \text{ kg}}{190 \text{ kg/s}} = 79.0 \text{ s}.$$

If we take $v_0 = 0$, then

$$v = -(9.80 \text{ m/s}^2)(79 \text{ s}) + (-2800 \text{ m/s})\left(\ln \frac{6000 \text{ kg}}{21{,}000 \text{ kg}} \right)$$

$$= 2830 \text{ m/s}$$

since $\ln(6/21) = -\ln 3.50 = -1.25$.

 EXAMPLE 8–15 A hopper drops gravel onto a conveyor belt at a rate of 75.0 kg/s. If the belt moves at a constant speed $v = 2.20$ m/s, what force is needed to keep the conveyor belt moving? Ignore friction. See Fig. 8–22.

SOLUTION We assume the hopper is at rest so $u = 0$, and that the belt has just started moving, so $dM/dt = 75.0$ kg/s. Since the belt moves at a constant speed ($dv/dt = 0$), we have from Eq. 8–22a

$$F_{ext} = M\frac{dv}{dt} - (u - v)\frac{dM}{dt}$$

$$= 0 - (0 - 2.20 \text{ m/s})(75.0 \text{ kg/s}) = 165 \text{ N}.$$

If friction were not neglected, we would have had to write $F_{ext} = F_{motor} - F_{friction} = 165$ N where F_{motor} is the force exerted by the motor and we placed a minus sign in front of the magnitude of $F_{friction}$ since friction always opposes the motion.

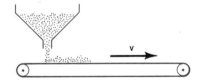

FIGURE 8–22
Gravel dropped from hopper onto conveyor belt.

Summary

For a system of particles, or for an extended body that can be considered as having a continuous distribution of matter, the *center of mass* (cm) is defined as

$$x_{cm} = \frac{\sum m_i x_i}{M}, \quad y_{cm} = \frac{\sum m_i y_i}{M}, \quad z_{cm} = \frac{\sum m_i z_i}{M}$$

or

$$x_{cm} = \frac{1}{M}\int x\, dm, \quad y_{cm} = \frac{1}{M}\int y\, dm, \quad z_{cm} = \frac{1}{M}\int z\, dm,$$

where M is the total mass of the system. The center of mass of a system is important because this point moves like a single particle of mass M which is acted on by the same net external force, $\mathbf{F}_{ext}$. In equation form, this is just Newton's second law for systems of particles (or extended bodies):

$$M\mathbf{a}_{cm} = \mathbf{F}_{ext}$$

where M is the total mass of the system (or body), $\mathbf{a}_{cm}$ is the acceleration of the cm of the system, and $\mathbf{F}_{ext}$ is the total (net) external force acting on all parts of the system. In terms of the total linear momentum, $\mathbf{P} = \sum m_i \mathbf{v}_i = M\mathbf{v}_{cm}$ of the system, this can be written

$$\frac{d\mathbf{P}}{dt} = \mathbf{F}_{ext}.$$

When the net external force on a system is zero, the total momentum remains constant. This is the *law of conservation of momentum*. Stated another way, the total momentum of an isolated system of particles remains constant.

The law of conservation of momentum is very useful in dealing with the class of events known as *collisions*. In a collision two (or more) bodies interact with each other for a very short time and the force between them during this time is very large compared to any other forces acting. The *impulse* of such a force on a body is defined as

$$\mathbf{J} = \int \mathbf{F}\, dt,$$

and is equal to the change in momentum of the body:

$$\Delta\mathbf{p} = \mathbf{p}_f - \mathbf{p}_i = \int_{t_i}^{t_f} \mathbf{F}\, dt = \mathbf{J}.$$

Momentum is conserved in any collision. The total energy is also conserved, but this may not be useful unless the only type of energy transformation involves kinetic energy; in this case, kinetic energy is conserved and the collision is called an *elastic collision*. If kinetic energy is not conserved, the collision is called *inelastic*.

 Questions _____

1. Why is the cm of a 1-m length of pipe at its midpoint whereas this is not true for your arm or leg?

2. Show on a diagram how your cm shifts when you change from a lying position to a sitting position.

3. Describe an analytic way of determining the cm of any triangular-shaped, thin uniform plate.

4. Place yourself facing the edge of an open door. Position your feet astride the door with your nose and abdomen touching the door's edge. Try to rise on your tiptoes. Why can't this be done?

5. Why is it more difficult to do situps when your knees are bent than when your legs are stretched out?

6. Why is it not possible to sit upright in a chair and rise to one's feet without first leaning forward?

7. Explain why a uniform rectangular brick can be placed so that slightly less than half its length can be suspended over the edge of a table, but no more.

8. Why do you tend to lean backward when carrying a heavy load in your arms?

9. Analyze the motion of the spool in Fig. 8–23 if you pull the thread (*a*) directly upward, (*b*) horizontally. Consider the effect on the motion of how hard you pull.

10. If only an external force can change the momentum of the center of mass of an object, how can the internal force of the engine accelerate a car?

11. How can a rocket change direction when it is far out in space and is essentially in a vacuum?

12. Is it possible for an object to have momentum without having energy? Can it have energy but no momentum? Explain.

13. We claim that momentum is conserved. Yet most moving objects eventually slow down and stop. Explain.

14. Two blocks of mass m_1 and m_2 rest on a frictionless table and are connected by a spring. The blocks are pulled apart, stretching the spring, and then released. Describe the subsequent motion of the two blocks.

15. A rocket following a parabolic path through the air suddenly explodes into many pieces. What can you say about the motion of this system of pieces?

16. A light body and a heavy body have the same kinetic energy. Which has the greater momentum?

17. When a person jumps from a tree to the ground, what happens to the momentum of the person upon striking the ground?

18. Explain, on the basis of conservation of momentum, how a fish propels itself forward by swishing its tail back and forth.

19. Why, when you release an inflated balloon, does it fly across the room?

20. It is said that in ancient times a rich man with a bag of gold coins was frozen to death stranded on the surface of a frozen lake. Because the ice was frictionless, he could not push himself to shore. What could he have done to save himself had he not been so miserly?

21. If a falling ball were to make a perfectly elastic collision with the floor, would it rebound to its original height? Explain.

22. According to Eq. 8–15, the shorter the impact time of an impulse, the greater the force must be and hence the greater the deformation of the object on which the force acts. Explain on this basis the value of "air bags" which are intended to inflate during an automobile collision and reduce the possibility of fracture or death.

23. Is it possible for a body to receive a larger impulse from a small force than from a large force?

24. How could a force give zero impulse over a nonzero time interval even though it is not zero for at least a part of that time interval?

FIGURE 8–23
A spool of thread.

25. In a collision between two cars, which would you expect to be more damaging to the occupants: if the cars collide and remain together, or if the two rebound backward? Explain.

26. A superball is dropped from a height h onto a hard steel plate (fixed to the earth), from which it rebounds at very nearly its original speed. (a) Is the momentum of the ball conserved during any part of this process? (b) If we consider the ball and earth as our system, during what parts of the process is momentum conserved? (c) Answer part (b) for a piece of putty that falls and sticks to the steel plate.

27. At a hydroelectric power plant, water is directed at high speed against turbine blades which are connected to an axle that turns an electric generator. Do you think the turbine blades should be designed so that the water is brought to a dead stop, or so that the water rebounds?

*28. In observations of nuclear β-decay, the electron and recoil nucleus often do not separate along the same line. Use conservation of momentum in two dimensions to explain why this implies the emission of at least one other particle in the disintegration.

*29. Use conservation of (vector) momentum to show that a two-particle collision occurs in a plane if (a) one particle is initially at rest; (b) the two particles have momenta in the same or opposite directions.

*30. For what initial conditions (if any) could a two-particle collision require a description in three dimensions? Explain. What if you analyze a collision in the CM system.

*31. Explain how a CM reference frame can be at rest when the two colliding bodies are moving.

32. Can you choose a reference frame in which the cm of two colliding bodies moves at a greater speed than either of the two bodies?

 Problems _____

SECTIONS 8–1 AND 8–2

1. (I) The distance between a carbon atom ($m = 12$ u) and an oxygen atom ($m = 16$ u) in the CO molecule is 1.13×10^{-10} m. How far from the oxygen atom is the center of mass of the molecule?

2. (I) Determine the cm of (a) an outstretched arm and (b) an arm bent at right angles, using Table 8–1.

3. (I) From the data in Table 8–1 determine the height above the ground of the cm of the entire body for a 1.75-m-tall person.

4. (I) An empty 1500-kg car has its cm 3.10 m from the front axle. How far from the front of the car will the cm be if two people sit in the front seat 2.60 m from the front of the car and three in the back seat 3.85 m from the front. Assume each person has a mass of 65 kg.

5. (II) Calculate how far below the torso's median line the cm will be when a jumper is in a position such that his arms and legs are hanging vertically, and the trunk and head are horizontal. Will this be outside the body? Use Table 8–1.

6. (II) A square uniform raft, 30 m by 30 m of mass 7400 kg, is used as a ferryboat carrying cars northward across a river. (a) If three cars, each of mass 1400 kg, occupy the NE, SE, and SW corners, determine the cm of the loaded ferryboat. (b) If the car in the SW corner accelerates northward at 1.80 m/s², where will the cm be after 4.0 seconds?

7. (II) A uniform thin semicircular wire has radius r. Determine the coordinates of its center of mass with respect to an origin of coordinates at the center of the "full" circle.

8. (II) Three spheres, of radius r_0, $2r_0$, and $3r_0$, are placed next to one another (in contact) with their centers along a straight line and the $r = 2r_0$ sphere in the center. Where is the cm of this system?

9. (III) Determine the cm of a uniform thin semicircular plate.

10. (III) Determine the cm of a uniform pyramid which has four triangular faces with equal sides of length s.

TABLE 8–1

Center of mass of parts of typical male human body[†]

Distance of Hinge Points from Floor (%)	Hinge Points (•) (Joints)		Center of Mass (×) (% Height Above Floor)		Percent Mass
91.2	Base of skull on spine		Head	93.5	6.9
81.2	Shoulder joint		Trunk and neck	71.1	46.1
		elbow 62.2	Upper arms	71.7	6.6
			Lower arms	55.3	4.2
		wrist 46.2	Hands	43.1	1.7
52.1	Hip		Upper legs (thighs)	42.5	21.5
28.5	Knee			18.2	9.6
			Lower legs	1.8	3.4
4.0	Ankle		Feet	58.0	100.0

[†] From *Bioastronautics Data Book*, NASA, Washington, D.C.

11. (I) The masses of the earth and moon are 5.98×10^{24} kg and 7.36×10^{22} kg, respectively, and they are separated by about 3.80×10^8 m. (a) Where is the cm of this system located? (b) What can you say about the motion of the earth-moon system about the sun, and of the earth and moon separately about the sun?

12. (I) Two 35-kg masses have velocities (in m/s) of $\mathbf{v}_1 = 12\mathbf{i} - 16\mathbf{j}$ and $\mathbf{v}_2 = -20\mathbf{i} + 14\mathbf{j}$. Determine the momentum of the center of mass of the system.

13. (II) (a) Suppose that in Example 8–4 (Fig. 8–9) $m_{\text{II}} = 3m_{\text{I}}$, where then would m_{II} land? (b) What if $m_{\text{I}} = 3m_{\text{II}}$?

14. (II) A 50-kg girl and a 70-kg boy stand 8.0 m apart on frictionless ice. (a) How far from the girl is their cm? (b) If they hold on to the two ends of a rope, and the boy pulls on the rope so that he moves 2.2 m, how far from the girl will he be now? (c) How far will the boy have moved when he collides with the girl?

15. (II) A helium balloon and its carriage, of mass M, are stationary with respect to the ground. A passenger, of mass m, then climbs out and slides down a rope with speed v, measured with respect to the balloon. With what speed and direction (relative to earth) does the balloon then move? What happens if the passenger stops?

16. (III) A 20-m long, 200-kg flatcar is moving with a speed of 6.0 m/s along horizontal frictionless rails. A 75-kg man then starts walking from one end of the car to the other in the direction of motion, with speed 2.5 m/s with respect to the car. In the time it takes for him to reach the other end, how far has the flatcar moved?

SECTION 8–4

17. (I) The momentum of a particle, in SI units, is given by $\mathbf{p} = 4.0t^2\mathbf{i} - 2.6\mathbf{j} - 3.9t\mathbf{k}$. What is the force as a function of time?

18. (II) The air in a 200 km/h wind strikes a 30 m by 20 m face of a building at a rate of 5.4×10^4 kg/s. Calculate the net force on the building, assuming the air is brought to rest on impact.

19. (II) (a) What is the momentum of a 50-g sparrow with a speed of 15 m/s? (b) What will be its momentum 12 s later if a 2.0×10^{-2} N force due to air resistance acts on it.

20. (II) The force on a particle of mass m is given by $\mathbf{F} = 26\mathbf{i} - 12t^2\mathbf{j}$ (newtons). What will be the change in momentum of the particle between $t = 1.0$ s and $t = 3.0$ s?

21. (II) A 145-g baseball, moving along the x axis with speed 30 m/s, strikes a fence at a 45° angle and rebounds along the y axis. Give its change in momentum using unit vector notation.

22. (II) A 7200-kg rocket is traveling in outer space with a velocity of 150 m/s toward the sun. It wishes to alter its course by 30°, and can do this by shooting its rockets briefly in a direction perpendicular to its motion. If the rocket gases are expelled at a speed of 2400 m/s, what mass of gas must be expelled?

SECTION 8–5

23. (I) A 140-kg tackler moving at 3.0 m/s meets head-on (and tackles) a 90-kg halfback moving at 7.5 m/s. What will be their mutual speed immediately after the collision?

24. (I) An atomic nucleus at rest decays radioactively into an alpha particle and a smaller nucleus. What will be the speed of this recoiling nucleus if the speed of the alpha particle is 6.2×10^5 m/s? Assume the nucleus has a mass 57 times greater than that of the alpha particle.

25. (I) A 12,000-kg open railroad car moves with a constant speed of 15.0 m/s on a level frictionless track. Snow begins to fall vertically, and fills the car at a rate of 3.00 kg/min. What is the speed of the car after 60.0 min?

26. (I) A 10,000-kg railroad car travels alone on a level frictionless track with a constant speed of 22 m/s. An additional 5000 kg load is dropped onto the car. What will its speed be now?

27. (II) A 44-g bullet strikes a 15.4-kg block of wood placed on a horizontal surface just in front of the gun. If the coefficient of kinetic friction between the block and the surface is 0.28, and the impact drives the block a distance of 18.0 m before it comes to rest, what was the muzzle speed of the bullet? Assume the bullet becomes embedded in the block.

28. (II) An explosion breaks an object, originally at rest, into two fragments. One fragment acquires twice the kinetic energy of the other. What is the ratio of their masses? Which is greater?

29. (II) A particle of mass m traveling with speed v_0 along the x axis suddenly shoots out one-third its mass parallel to the y axis with speed $2v_0$. Express the velocity of the remainder of the particle in $\mathbf{i}, \mathbf{j}, \mathbf{k}$ notation.

30. (II) The decay of a neutron into a proton, an electron, and a neutrino is an example of a three-particle decay process. Use the vector nature of momentum to show that if the neutron is initially at rest, the velocity vectors of the three must be coplanar (that is, all in the same plane). The result is not true for numbers greater than three.

31. (II) Two blocks, of mass m_1 and m_2, resting on a frictionless table, are connected by a stretched spring and then released. (a) Is there a net external force on the system? (b) Determine the ratio of their velocities, v_1/v_2. (c) What is the ratio of their kinetic energies? (d) Describe the motion of the cm of this system. (e) How would the presence of friction alter the above results?

32. (II) A radioactive nucleus at rest decays into a second nucleus, an electron, and a neutrino. The electron and neutrino are emitted at right angles and have momenta of 8.6×10^{-23} kg·m/s and 6.2×10^{-23} kg·m/s respectively. What is the magnitude and direction of the recoiling nucleus?

33. (II) A 940-kg two-stage rocket is traveling at a speed of 6.2×10^3 m/s with respect to earth when a predesigned explosion separates the rocket into two sections of equal mass that then move with a relative speed of 2.4×10^3 m/s along the original line of motion. (a) What is the speed and direction of each section after the explosion? (b) How much energy was supplied by the explosion? (Hint: what is the change in KE as a result of the explosion?)

34. (III) A 300-kg projectile, fired with a speed of 130 m/s at a 45° angle, breaks into three pieces of equal mass at the highest point of its arc. Two of the fragments move with the same speed right after the explosion, one vertically downward and the other horizontally. If the explosion provides 5.0×10^5 J of additional KE to the fragments, determine the velocity of the third fragment immediately after the explosion.

35. (I) The tennis ball may leave the racket of a top player on the serve with a speed of 65 m/s. If the ball's mass is 0.060 kg and is in contact with the racket for 0.030 s, what is the average force on the ball? Would this force be large enough to lift an average size person?

36. (I) A 0.145-kg baseball pitched at 35 m/s is hit on a horizontal line drive straight back at the pitcher at 50 m/s. If the contact time between bat and ball is 5.0×10^{-4} s, calculate the force (assumed to be constant) between the ball and bat.

37. (II) A 150-kg astronaut (including space suit) acquires a speed of 2.5 m/s by pushing off with his legs from a 2200-kg space capsule. (a) What is the change in speed of the space capsule? (b) If the push lasts 0.40 s, what was the average force exerted by each on the other? Use the reference frame of the capsule before the push. (c) What is the KE of each after the push?

38. (II) A tennis ball of mass m and speed v strikes a wall at a 45° angle and rebounds with the same speed at 45°. What is the impulse given the wall?

39. (II) Water strikes the turbine blades of a generator so that its rebounding velocity is reversed in direction but undiminished in magnitude. If the flow rate is 60 kg/s and the water speed is 16 m/s, what is the average force on the blades?

40. (II) (a) A molecule of mass m at speed v strikes a wall at right angles and rebounds with the same speed; if the collision time is Δt, what is the average force on the wall during the collision? (b) If molecules, all of this type, strike the wall on the average a time t apart, what is the average force on the wall averaged over a long time?

41. (II) Suppose the force acting on a tennis ball (mass 0.060 kg) as a function of time is given by the graph of Fig. 8–24. Use graphical methods to (a) estimate the total impulse given the ball; (b) estimate the speed of the ball after being struck, assuming the ball is being served so it is nearly at rest initially.

42. (II) The force on a bullet is given by the formula $F = 480 - 1.6 \times 10^5 t$ over the time interval $t = 0$ to $t = 3.0 \times 10^{-3}$ s; in this formula t is in seconds and F is in newtons. (a) Plot a graph of F versus t for $t = 0$ to $t = 3.0 \times 10^{-3}$ s. (b) Estimate, using graphical methods, the impulse given the bullet. (c) Determine the impulse by integration. (d) If the bullet achieves a speed of 320 m/s as a result of this impulse, given to it in the barrel of a gun, what must its mass be?

43. (III) From what maximum height can a 60-kg person jump without breaking the lower leg bone whose cross-sectional area is 3.0×10^{-4} m^2? Ignore air resistance and assume the cm of the person moves a distance of 0.60 m from the standing to the seated position (that is, in breaking the fall). Assume the breaking strength (force per unit area—see Section 11–4) of bone is 170×10^6 N/m^2.

44. (III) A scale is adjusted so that when a large, flat pan is placed on it, it reads zero. A water faucet at height $h = 3.0$ m above is turned on and water falls into the pan at a rate $R = 0.20$ kg/s. Determine (a) a formula for the scale reading as a function of time t, and (b) the reading for $t = 15$ s. (c) Repeat, but replace the flat pan with a tall, narrow cylindrical container of area $A = 20$ cm^2.

SECTION 8–8

45. (I) Two billiard balls of equal mass undergo a perfectly elastic head-on collision. If the speed of one ball was initially 2.0 m/s, and of the other 3.0 m/s in the opposite direction, what will be their speeds after the collision?

46. (II) A 0.60 kg ball makes an elastic head-on collision with a second ball initially at rest. The second ball moves off with half the original speed of the first ball. (a) What is the mass of the second ball? (b) What fraction of the original kinetic energy gets transferred to the second ball?

FIGURE 8–24

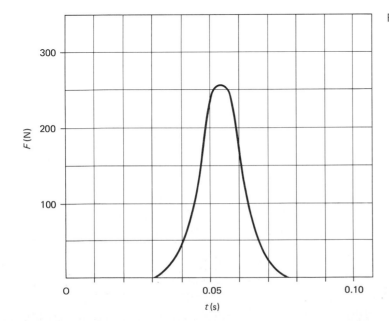

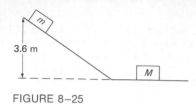

FIGURE 8–25

FIGURE 8–26

47. (II) A ball of mass m makes a head-on elastic collision with a second ball (at rest) and rebounds with a speed equal to one-third its original speed. What is the mass of the second ball?

48. (II) A particle of mass m_2 at rest is struck head-on by a particle of mass m_1, moving with speed v_1. For a given value of v_1 what value would you choose for m_1 (give in terms of m_2) if you wanted m_2 to have the greatest possible (a) speed, (b) kinetic energy, or (c) momentum. What if you wanted m_2 to have the least possible (d) speed, (e) kinetic energy, or (f) momentum?

49. (II) A block of mass $m = 2.0$ kg slides down a $30°$ incline which is 3.6 m high. At the bottom, it strikes a block of mass $M = 6.0$ kg which is at rest on a horizontal surface, Fig. 8–25. If the collision is elastic, and friction can be ignored, determine (a) the speeds of the two blocks after the collision and (b) how far back up the incline the smaller mass will go.

50. (II) Show that, in general, for any head-on one-dimensional elastic collision the speeds after collision are:

$$v_2' = v_1\left(\frac{2m_1}{m_1 + m_2}\right) + v_2\left(\frac{m_2 - m_1}{m_1 + m_2}\right)$$

and

$$v_1' = v_1\left(\frac{m_1 - m_2}{m_1 + m_2}\right) + v_2\left(\frac{2m_2}{m_1 + m_2}\right).$$

51. (II) Refer to Example 8–10. Determine the fraction of kinetic energy lost by a neutron ($m_1 = 1.01$ u) when it collides head-on and elastically with (a) $^1_1$H ($m = 1.01$ u); (b) $^2_1$H (heavy hydrogen, $m = 2.01$ u); (c) $^{12}_6$C ($m = 12.00$ u); (d) $^{208}_{82}$Pb (lead, $m = 208$ u). Discuss the usefulness of each of these as a moderator in a nuclear reactor (Example 8–10).

52. (III) In problem 49, Fig. 8–25, what maximum value can the mass m have if it is to rebound from M and return down the incline and collide with M again?

53. (III) A 3.5-kg block slides along a frictionless table top at 8.0 m/s toward a second block (at rest) of mass 6.0 kg. A coil spring, which obeys Hooke's law and has spring constant $k = 750$ N/m, is attached to the second block in such a way that it will be compressed when struck by the moving block, Fig. 8–26. (a) What will be the maximum compression of the spring? (b) What will be the final velocities of the blocks after the collision? (c) Was the collision elastic?

*SECTION 8–9

*54. (II) A particle of mass m traveling with speed v collides elastically with a target particle of mass $2m$ (initially at rest) and is scattered at $90°$. (a) At what angle does the target particle move after the collision? (b) What are the final speeds of the two particles? (c) What fraction of the initial KE is transferred to the target particle?

*55. (II) Two billiard balls of equal mass move at right angles and meet at the origin of an xy coordinate system. One is moving upward along the y axis at 3.0 m/s, the other is moving to the right along the x axis with speed 4.8 m/s. After the collision, the second ball is moving along the positive y axis. What is the final direction of the first ball, and what are their two speeds?

*56. (II) A billiard ball of mass $M_A = 0.40$ kg strikes a second ball, initially at rest, of mass $M_B = 0.60$ kg. As a result of this elastic collision, ball A is deflected at an angle of $30°$ and ball B at $53°$. What is the ratio of their speeds after the collision?

*57. (II) Show that for an elastic collision between two equal-mass particles of the type discussed in Example 8–11 (where one particle is at rest initially) the sum of the final angles is always $90°$: that is, $\theta_1' + \theta_2' = 90°$. Assume the collision is not head-on.

*58. (III) A neutron collides elastically with a helium nucleus (at rest initially) whose mass is four times that of the neutron; the helium is observed at an angle $\theta_2 = 36°$. Determine the angle of the neutron, θ_1, and the speeds of the two particles, v_n', v_{He}', after the collision. The neutron's initial speed was 6.8×10^5 m/s.

*59. (III) For an elastic collision between a projectile particle of mass m_1 and a target particle (at rest) of mass m_2, show that the scattering angle, θ_1', of the projectile (a) can take any value, 0 to $180°$, for $m_1 < m_2$, but (b) has a maximum angle ϕ given by $\cos^2 \phi = 1 - (m_2/m_1)^2$ for $m_1 > m_2$.

*SECTION 8–10

*60. (I) Two particles, of mass m_1 and m_2, are moving in the same direction with speeds v_1 and v_2 before they collide. What will be the speed of their center of mass?

*61. (II) Determine the CM velocity of particle 2 ($v_2'^*$, $\theta_2'^*$) after the collision described in Example 8–11.

*62. (II) In the laboratory, a particle of momentum $m_1 v_1$ strikes a particle of mass m_2 initially at rest ($v_2 = 0$). (a) Show that the velocity of the cm reference frame relative to the lab is $\mathbf{v}_{cm} = m_1 \mathbf{v}_1/(m_1 + m_2)$; (b) show that the velocity of any particle in the cm system is given by $\mathbf{v}_i^* = \mathbf{v}_i - \mathbf{v}_{cm}$ where v_i is its velocity in the lab and i can be either 1 or 2;

*63. (II) In a proton-proton collision in the CM system, the total KE is 3.2×10^{-10} J. What KE does the projectile proton have in the lab (one proton at rest)?

*64. (III) A projectile is fired with a speed of 450 m/s at a $45°$ angle. After 60 s it breaks into three fragments of equal mass, all of which have the same speed relative to the original projectile at the moment it broke up. An observer on earth sees that one fragment falls vertically to earth and a second fragment moves horizontally immediately after the collision. How far from the original point of projection does the third fragment land?

*SECTION 8–11

*65. (II) A meteor whose mass was about 10^8 kg struck the earth ($m = 6.0 \times 10^{24}$ kg) with a speed of about 15 m/s and came to rest in the earth. (a) What was the earth's recoil speed? (b) What fraction of the meteor's kinetic energy was transformed to KE of the earth? (c) By how much did the earth's KE change as a result of this collision?

*66. (II) An eagle ($m_1 = 6.6$ kg) moving with speed $v_1 = 12.4$ m/s is on a collision course with a second eagle ($m_2 = 8.3$ kg) moving at $v_2 = 9.1$ m/s in a direction at right angles to the first. After they collide, they hold onto one another. In what direction, and with what speed, are they moving after the collision?

*67. (II) An explosion breaks an object into two pieces, one of which has 1.5 times the mass of the other. If 4500 J were released in the explosion, how much kinetic energy did each piece acquire?

*68. (II) (a) Derive a formula for the fraction of kinetic energy lost, $\Delta\text{KE}/\text{KE}$, for the ballistic pendulum collision of Example 8–13. (b) Evaluate for $m = 10.0$ g and $M = 105$ g.

*69. (II) After a completely inelastic collision between two objects of equal mass, each having initial speed v, the two move off with speed $v/3$. What was the angle between their initial directions?

*70. (III) For the ballistic pendulum, let us determine the accuracy of our approximation that the block (mass M) doesn't move during the collision. Use the symbols of Example 8–13 and Fig. 8–20, and assume the projectile is decelerated uniformly within the block in a distance d. Determine: (a) the collision time Δt; (b) by how much is momentum not conserved, $\Delta\mathbf{p}$, because there is a net external force acting during this time; and (c) by what fraction the calculated speed of the bullet will be in error if we use the equation given in Example 8–13.

*SECTION 8–12

*71. (II) Suppose the conveyor belt of Example 8–15 is retarded by a friction force of 140 N. Determine the required output power (horsepower) of the motor as a function of time from the moment gravel first starts falling ($t = 0$) until 3.0 s after the gravel has started being dumped off the end of the conveyor belt. The end of the conveyor belt is 20 m from the hopper.

*72. (II) (a) Show that the power provided by the force exerted on a conveyor belt, such as that of problem 71, is $P = v^2(dM/dt) = 2(d/dt)(\text{KE})$; in other words, show that the energy is put in at a rate that is *twice* the rate at which the kinetic energy of the system is increasing. (b) Where is the other half of this energy going?

*73. (II) Consider the railroad car of problem 25, which is slowly filling with snow. (a) Determine the speed of the car as a function of time. (b) Determine, using Eq. 8–22, the speed of the car after 60 min. Does this agree with the simpler calculation (problem 25) based on Section 8–5?

*74. (II) The jet engine of an airplane takes in 100 kg of air per second, which is burned with 4.2 kg of fuel per second. The burned gases leave the plane at a speed of 550 m/s. If the plane is traveling 270 m/s (600mi/h), what is (a) the net thrust of the engine; and (b) the power (hp) delivered?

*75. (II) A rocket traveling 1850 m/s away from the earth at an altitude of 6400 km fires its rockets which eject gas at a speed of 1200 m/s (relative to rocket). If the mass of the rocket at this moment is 25,000 kg, and an acceleration of 1.7 m/s^2 is desired, at what rate must the gases be ejected?

*76. (II) A 2500-kg rocket is to be accelerated at 3.0 g at take-off. If the gases can be ejected at a rate of 30 kg/s, what must be their exhaust speed?

*77. (II) A sled filled with sand slides without friction down a 30° slope. Sand leaks out a hole in the sled at a rate of 2.0 kg/s. If the sled starts from rest with an initial total mass of 40.0 kg, how long does it take the sled to travel 120 m along the slope?

Rotational Motion About an Axis

9

We have, until now, been concerned with translational motion of particles or systems of particles. In this chapter and the next, we will deal with rotational motion. Although we will deal with systems of many particles, we will mainly be concerned with rigid bodies; by a *rigid body* we mean a body that has a definite shape which doesn't change, so that the particles that make it up stay in fixed positions relative to one another. Of course any real body is capable of vibrating or deforming when a force is exerted on it; these effects are often very small, so the concept of an ideal rigid body is very useful as a good approximation in many cases.

The motion of a rigid body (as mentioned earlier and which we will prove in Chapter 10) can be analyzed as the translational motion of its center of mass, plus rotational motion about its center of mass. We have already discussed translational motion in detail; so now, in this chapter, we focus our attention on purely rotational motion. By *purely rotational motion* we mean that all points in the body move in circles, such as the point P in the rotating wheel of Fig. 9–1a, and that the centers of these circles all lie on a line called the **axis of rotation**. We assume the axis is fixed in an inertial reference frame, but we will not insist that the axis pass through the center of mass.

FIGURE 9–1

(a) Looking down on a wheel that is rotating counter-clockwise about an axis (perpendicular to the page), through the wheel's center at O. (b) Showing the distinction between **r** and **R** for a point P on the edge of a cylinder rotating about the z axis.

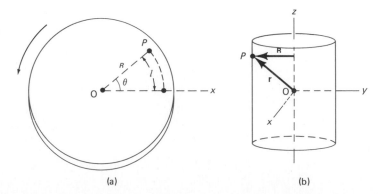

(a)   (b)

9–1 Angular Quantities

Since we are going to consider three-dimensional rigid bodies rotating about a fixed axis, we will find it useful to use the symbol R to represent the *perpendicular distance of a point or particle from the axis of rotation*. We do this to distinguish R from r, which will continue to represent the position of a particle with reference to the origin (a point) of some coordinate system.* This distinction is illustrated in Fig. 9–1b. For a flat, very thin body, like a wheel, with origin in the plane of the body (at the center of the wheel, for example), R and r will be the same.

Every particle in a body rotating about a fixed axis moves in a circle whose center is on the axis and the radius of that circle is R. A line drawn perpendicular from the axis to any particle sweeps out the same angle θ in the same time. To indicate the position of the body, or how far it has rotated, we specify the angle θ of some particular line in the body with respect to some reference line, such as the x axis (see Fig. 9–1a). A particle of the body (such as P in Fig. 9–1a) moves through an angle θ when it travels the distance l measured along the circumference of its circular path. Angles are commonly measured in degrees, but the mathematics of circular motion is much simpler if we use the *radian* for angular measure. One *radian* (rad) is defined as the angle subtended by an arc whose length is equal to the radius. (For example, in Fig. 9–1a, particle P is a distance R from the axis of rotation and it has moved a distance l along the arc of a circle; if $l = R$, then θ is exactly equal to 1 rad.) In general, any angle θ is given by

$$\theta = \frac{l}{R} \qquad\qquad (9-1)$$

where R is the radius of the circle, and l is the arc length subtended by the angle θ; θ itself will then be in radians. Radians can be related to degrees in the following way. In a complete circle there are 360° and this of course must correspond to an arc length equal to the circumference of the circle, $l = 2\pi R$; thus $\theta = l/R = 2\pi R/R = 2\pi$ rad in a complete circle, so

$$360° = 2\pi \text{ rad.}$$

One radian is then $360°/2\pi = 360°/6.28 = 57.3°$.

EXAMPLE 9–1 A particular bird's eye can just distinguish objects which subtend an angle no smaller than about 3×10^{-4} rad (how many degrees is this?). How small an object can the bird just distinguish when flying at a height of 100 m?

SOLUTION From Eq. 9–1, $l = R\theta$. Strictly speaking, l is the arc length, but for small angles the linear distance subtended is approximately the same.† Since $R = 100$ m and $\theta = 3 \times 10^{-4}$ rad we find

$$l = (100 \text{ m})(3 \times 10^{-4} \text{ rad}) = 3 \times 10^{-2} \text{ m} = 3 \text{ cm.}$$

Had the angle been given in degrees, we would first have had to change it to radians to make this calculation.

Note in this example that we used the fact that the radian is dimensionless (has no units) since it is the ratio of two lengths.

---

* Some books use the symbol ρ instead of R. Since we use ρ to represent density (mass per unit volume), we avoid confusion by using R rather than ρ to represent the perpendicular distance of a point from an axis.

† Even for an angle as large as 15°, the error in making this estimate is only 1 percent; but for larger angles the error increases rapidly.

Angular velocity is defined in analogy with ordinary linear velocity. Instead of linear displacement, we use the angular displacement. Let θ_1 and θ_2 represent the angular positions of the body at times t_1 and t_2, respectively. Then we define the magnitude of the **average angular velocity** (denoted by ω, the Greek lowercase letter omega) as:

$$\bar{\omega} = \frac{\theta_2 - \theta_1}{t_2 - t_1} = \frac{\Delta\theta}{\Delta t}$$

where $\Delta\theta = \theta_2 - \theta_1$, is the angular displacement of the body during the time interval $\Delta t = t_2 - t_1$. We define the magnitude of the **instantaneous angular velocity** as the limit of this ratio as Δt approaches zero:

$$\omega = \lim_{\Delta t \to 0} \frac{\Delta\theta}{\Delta t} = \frac{d\theta}{dt}. \tag{9-2}$$

Angular velocity is generally specified in radians per second. Note that all points in the body rotate with the same angular velocity. This follows since every position in the body moves through the same angle in the same time interval.

Angular acceleration, in analogy with ordinary linear acceleration, is defined as the change in angular velocity divided by the time required to make this change. We let ω_1 and ω_2 be the magnitude of the instantaneous angular acceleration at times t_1 and t_2, respectively. Then the magnitude of the **average angular acceleration** (denoted by α, the Greek lowercase letter alpha) is defined as

$$\bar{\alpha} = \frac{\omega_2 - \omega_1}{t_2 - t_1} = \frac{\Delta\omega}{\Delta t}.$$

We define the magnitude of the **instantaneous angular acceleration** as the limit of this ratio as Δt approaches zero:

$$\alpha = \lim_{\Delta t \to 0} \frac{\Delta\omega}{\Delta t} = \frac{d\omega}{dt}. \tag{9-3}$$

Since ω is the same for all points of a rotating body, Eq. 9-3 tells us that α also will be the same for all points. With ω measured in radians per second and t in seconds, α will be expressed as radians per second squared (rad/s²).

Each particle or point of a rotating rigid body has, at any moment, a linear velocity v, and linear acceleration a. We can relate these linear quantities, v and a, of each particle to the angular quantities, ω and α, of the rotating body as a whole. Consider a particle located a distance R from the axis of rotation. If the body rotates with angular velocity ω, any particle will have a linear velocity whose direction is tangent to its circular path; the magnitude of its linear velocity, v, is found by writing Eq. 9-1 as $l = R\theta$ and differentiating it, using the fact that the linear velocity is $v = dl/dt$; thus

$$v = \frac{dl}{dt} = R\frac{d\theta}{dt}$$

$$v = R\omega \tag{9-4}$$

where we have used the fact that $R =$ constant for any given particle since each particle rotates in a circle of constant radius. Thus, although ω is the same for every point in the rotating body at any instant, the linear velocity v is greater for points farther from the axis.

The angular acceleration α is related to the tangential linear acceleration a_T of a particle by

$$a_T = \frac{dv}{dt} = R\frac{d\omega}{dt}$$

$$a_T = R\alpha \tag{9-5}$$

where we used Eq. 9–4. In this equation R is the radius of the circle in which the particle is moving, and the subscript T in a_T stands for "tangential" since the acceleration considered here is along the circle, that is, tangent to it. The total linear acceleration of a particle is

$$\mathbf{a} = \mathbf{a}_T + \mathbf{a}_c$$

where the radial component, $\mathbf{a}_c$, the "centripetal acceleration," points toward the center of the particle's circular path; as we saw in Chapter 3, $a_c = v^2/R$, and we rewrite this in terms of ω using Eq. 9–4:

$$a_c = \frac{v^2}{R} = \omega^2 R. \tag{9–6}$$

Equations 9–4, 9–5 and 9–6 relate the angular quantities describing the rotation of a body to the linear quantities for each particle of the body.

It is sometimes useful to relate the angular velocity ω to the frequency of rotation, f. By frequency, we mean the number of complete revolutions (rev) per second. One revolution (of, say, a wheel) corresponds to an angle of 2π radians, and thus 1 rev/s = 2π rad/s; hence, in general the frequency f is given by:

$$f = \frac{\omega}{2\pi}$$

or

$$\omega = 2\pi f.$$

EXAMPLE 9–2   What is the linear speed of a point on the edge of a steadily rotating 33-rpm (revolutions per minute) phonograph record whose diameter is 30 cm?

SOLUTION   First we find the angular velocity in radians per second: the frequency f = 33 rev/min = 33 rev/60 s = 0.55 rev/s; then ω = $2\pi f$ = 3.5 rad/s. The radius R is 0.15 m, so the speed v at the edge is

$$v = R\omega = (0.15\ \text{m})(3.5\ \text{rad/s}) = 0.52\ \text{m/s}.$$

EXAMPLE 9–3   What is the magnitude of the acceleration of a speck of dust on the edge of the record in Example 9–2?

SOLUTION   From Example 9–2, ω = 3.5 rad/s and v = 0.52 m/s. Since ω = constant, α = 0 so a_T = 0. From Eq. 9–6, $a_c = \omega^2 R = (3.5\ \text{rad/s})^2(0.15\ \text{m})$ = 1.8 m/s²; or $a_c = v^2/R = (0.52\ \text{m/s})^2/(0.15\ \text{m})$ = 1.8 m/s² which is the same result.

EXAMPLE 9–4   A centrifuge rotor is accelerated from rest to 20,000 rpm in 5.0 min. What is its average angular acceleration?

SOLUTION

$$\omega = (20{,}000\ \text{rev/min})\left(\frac{2\pi\ \text{rad/rev}}{60\ \text{s/min}}\right) = 2100\ \text{rad/s};$$

then, since the initial angular velocity is 0, and the final angular velocity is 2100 rad/s, we have

$$\bar{\alpha} = \frac{2100\ \text{rad/s} - 0}{300\ \text{s}} = 7.0\ \text{rad/s}^2.$$

9–2 Kinematic Equations for Uniformly Accelerated Rotational Motion

In Chapter 2, we derived the important equations (2–9) that relate acceleration, velocity, and distance for the situation of uniform linear acceleration. Those equations were derived from the definitions of linear velocity and acceleration assuming constant acceleration. The definitions of angular velocity and angular acceleration are the same as for their linear counterparts, except that θ has replaced the displacement x, ω has replaced v; and α has replaced a. Therefore, the angular equations for constant angular acceleration will be analogous to Eqs. 2–9 with x replaced by θ, v by ω, and a by α, and they can be derived in exactly the same way. We summarize them here, opposite their linear equivalents:

Angular	Linear		
$\omega = \omega_0 + \alpha t$	$v = v_0 + at$	[constant α, a]	(9–7a)
$\theta = \omega_0 t + \frac{1}{2}\alpha t^2$	$x = v_0 t + \frac{1}{2}at^2$	[constant α, a]	(9–7b)
$\omega^2 = \omega_0^2 + 2\alpha\theta$	$v^2 = v_0^2 + 2ax$	[constant α, a]	(9–7c)

Note that we have chosen the initial position $\theta_0 = 0$ (also $x_0 = 0$) for simplicity; ω_0 represents the angular velocity at $t = 0$, whereas θ and ω represent the angular position and velocity at time t. Since the angular acceleration is constant, $\alpha = \bar{\alpha}$. These equations are of course valid for constant angular velocity, in which case $\alpha = 0$ and we have $\omega = \omega_0$, $\theta = \omega_0 t$, and $\bar{\omega} = \omega$.

EXAMPLE 9–5 Through how many turns has the centrifuge rotor of Example 9–4 turned during its acceleration period? Assume constant angular acceleration.

SOLUTION   We know that $\omega_0 = 0$, $\omega = 2100$ rad/s, $\alpha = \bar{\alpha} = 7.0$ rad/s², and $t = 300$ s. We could use either Eq. 9–7b or 9–7c; the former gives

$$\theta = 0 + \tfrac{1}{2}(7.0 \text{ rad/s}^2)(300 \text{ s})^2 = 3.2 \times 10^5 \text{ rad}.$$

To find the total number of revolutions we divide by 2π and obtain 5.0×10^4 revolutions. (To decide whether to multiply or to divide by 2π, it helps to remember that there are more radians than revolutions since 2π rad = 1 rev.)

EXAMPLE 9–6   Derive Eqs. 9–7a and b using calculus.

SOLUTION   Since $\alpha = d\omega/dt$ where $\alpha = $ constant:

$$\int_{\omega_0}^{\omega} d\omega = \int_0^t \alpha \, dt$$

$$\omega - \omega_0 = \alpha t$$

or $\omega = \omega_0 + \alpha t$ which is Eq. 9–7a. Next, since $\omega = d\theta/dt$, we have $d\theta = \omega dt = (\omega_0 + \alpha t) \, dt$ and integrating we get

$$\int_0^\theta d\theta = \int_0^t (\omega_0 + \alpha t) dt$$

$$\theta = \omega_0 t + \tfrac{1}{2}\alpha t^2$$

which is Eq. 9–7b.

9-3 Vector Nature of Angular Quantities

The linear quantities displacement, velocity, and acceleration are vectors. We will see in this section that angular velocity and angular acceleration can also be treated as vectors, although angular displacement, θ, is not a vector.

First let us see why the angular displacement cannot be a vector. One property of a vector is that when two vectors are added, you get the same result no matter in what order you add them. That is, if the two vectors are called $\mathbf{V}_1$ and $\mathbf{V}_2$, then

$$\mathbf{V}_1 + \mathbf{V}_2 = \mathbf{V}_2 + \mathbf{V}_1.$$

(This is called the commutative law for addition—see Eq. 3-1a.) Suppose, now, we rotate a book by $\theta_1 = 90°$ around the x axis followed by a rotation $\theta_2 = 90°$ around the y axis, as shown in Fig. 9-2a. If, instead, we first rotate the book by $\theta_2 = 90°$ around the y axis followed by $\theta_1 = 90°$ around the x axis, Fig. 9-2b, we do *not* get the same result! In other words, $\boldsymbol{\theta}_1 + \boldsymbol{\theta}_2 \neq \boldsymbol{\theta}_2 + \boldsymbol{\theta}_1$. Hence θ cannot be a vector.

But now consider a rotation $\theta_1 = 15°$ about the x axis and $\theta_2 = 15°$ about the y axis. In this case, $\boldsymbol{\theta}_1 + \boldsymbol{\theta}_2$ is nearly(but not quite) equal to $\boldsymbol{\theta}_2 + \boldsymbol{\theta}_1$ as shown in Fig. 9-3. However, in the limit of infinitesimal angles of rotation, the equality $d\boldsymbol{\theta}_1 + d\boldsymbol{\theta}_2 = d\boldsymbol{\theta}_2 + d\boldsymbol{\theta}_1$ is exact. We get the same result when we add two infinitesimal rotation angles in either order. Hence, an infinitesimal angular displacement, $d\boldsymbol{\theta}$ *is* a vector although a finite angular displacement is not.

The angular velocity, $\boldsymbol{\omega}$, must also be a vector since it is the product of a vector $(d\boldsymbol{\theta})$ and a scalar $(1/dt)$;

$$\boldsymbol{\omega} = \frac{d\boldsymbol{\theta}}{dt}.$$

Similarly, since $\boldsymbol{\omega}$ is a vector, the angular velocity

$$\boldsymbol{\alpha} = \frac{d\boldsymbol{\omega}}{dt}$$

is also a vector.

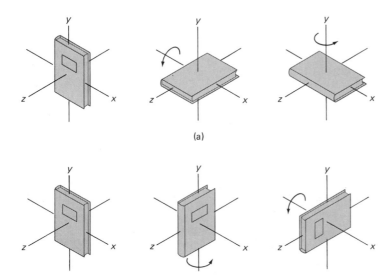

(a)

(b)

FIGURE 9-2

A book is rotated counterclockwise by 90°: (a) about the x axis and then about the y axis; (b) about the y axis and then about the x axis.

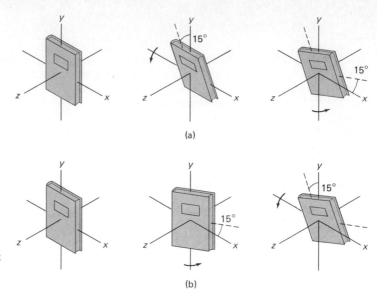

FIGURE 9–3

A book is rotated counterclockwise by 15°: (a) about the x axis and then about the y axis; (b) about the y axis and then about the x axis.

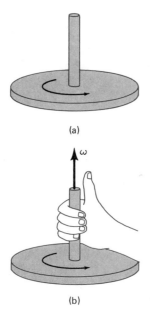

FIGURE 9–4

(a) Rotating wheel.
(b) Right-hand rule for obtaining direction of ω.

We have shown that ω and α are vectors. But what are the directions of ω and α? Consider the rotating wheel shown in Fig. 9–4a. The linear velocity of different particles of the wheel point in all different directions. The only unique direction in space associated with the rotation is along the axis of rotation, perpendicular to the actual motion. We therefore choose the axis of rotation to be the direction of the angular velocity vector, ω. Actually, there is still an ambiguity since ω could point in either direction along the axis of rotation (up or down in Fig. 9–4a). The convention we use, called the *right-hand rule*, is the following: when the fingers of the right hand are curled around the rotation axis and point in the direction of the rotation, then the thumb points in the direction of ω. This is shown in Fig. 9–4b. Note that ω points in the direction a right-handed screw would move when turned in the direction of rotation. Thus, if the rotation of the wheel in Fig. 9–4a is counterclockwise, the direction of ω is upward as shown in Fig. 9–4b; if the wheel rotates clockwise, then ω points in the opposite direction, downward.[†] Note that no part of the rotating body moves in the direction of ω.

If the axis of rotation is fixed, then ω can change only in magnitude. Thus $\alpha = d\omega/dt$ must also point along the axis of rotation. If the rotation is clockwise as in Fig. 9–4a and $|\omega|$ is increasing, α points upward, but if $|\omega|$ is decreasing (the wheel is slowing down), α points downward. If the rotation is clockwise, α will point downward if $|\omega|$ is increasing, and points upward if $|\omega|$ is decreasing.

Since ω always points along the axis of rotation, if the axis of rotation changes direction, ω changes direction. In this case α will not point along the axis of rotation. We will see examples of this in Chapter 10, but in this chapter we consider only motion about a fixed axis, so ω and α are both along the rotation axis.

[†] Strictly speaking, ω and α are not quite vectors. The problem is that they do not behave like vectors under reflection. Suppose, as we are looking directly into a mirror, a particle moving with velocity **v** to the right passes in front of and parallel to the mirror. In the reflection of the mirror, **v** still points to the right. Thus a true vector, like velocity, when pointing parallel to the face of mirror has the same direction in the reflection as in actuality. Now consider a wheel rotating in front of the mirror, so ω points to the right. (We will be looking at the edge of the wheel.) As viewed in the mirror, the wheel will be rotating in the opposite direction; so ω will point in the opposite direction (to the left) in the mirror. Because of this difference under reflection between ω and true vectors, ω is called a *pseudovector* or *axial vector*. The angular acceleration α is also a pseudovector, as are all cross products of true vectors (Section 10–1). The difference between true vectors and pseudovectors is of little importance in classical physics, and we will not distinguish between them in this book. The distinction is important in certain aspects of subatomic particle physics.

9-4 Torque

Up to now in this chapter, we have discussed rotational kinematics—the description of rotational motion in terms of angle, angular velocity, and angular acceleration. Now we discuss the dynamics, or causes, of rotational motion. Just as we found analogies between linear rotational motion for the description of motion, so rotational equivalents for dynamics exist as well. For example the rotational equivalent of Newton's first law states that a freely rotating body will continue to rotate with constant angular velocity as long as no forces (or, as we shall see shortly, no torques) act to change that motion. More difficult is the question of a rotational equivalent for Newton's second law; that is, what gives rise to angular acceleration? To make an object start rotating about an axis clearly requires a force; but the direction of this force, and where it is applied, are also important. Take, for example, an ordinary situation such as the door in Fig. 9-5. If you apply a force $\mathbf{F}_1$ to the door as shown, you will find that the greater F_1 is the more quickly the door opens. But now if you apply the same magnitude force at a point closer to the hinge, say $\mathbf{F}_2$ in Fig. 9-5, you will find that the door will not open so quickly. The effect of the force is less.

So the angular acceleration of the door is proportional not only to the magnitude of the force; but, assuming for the moment that only this one force acts (we ignore friction in the hinges, and so on), it is also proportional to the perpendicular distance from the axis of rotation to the line along which the force acts. Thus, if the distance R_1 in Fig. 9-5b is three times larger than R_2, then the angular acceleration of the door will be three times as great, assuming of course that the magnitudes of the forces are the same. To say it another way, if $R_1 = 3R_2$, then F_2 must be three times as large as F_1 to give the same angular acceleration. The distances R_1 and R_2 are called the *lever arms* or *moment arms* of the respective forces. The angular acceleration, then, will be proportional to the product of the force times the lever arm. This product is called the *moment* of the force about the axis; it is also called the **torque** and is abbreviated τ (Greek lowercase letter tau). The angular acceleration α of an object is found to be directly proportional to the applied torque, τ:

$$\alpha \propto \tau.$$

Thus we see that it is torque that gives rise to angular acceleration. This is the rotational analog of Newton's second law for linear motion, $a \propto F$. (In Section 9-5 we will see what factor is needed to make this proportionality an equation.)

We define the *lever arm* as the *perpendicular* distance of the axis of rotation from the line of action of the force (we mean the distance which is perpendicular to both the axis of rotation and to an imaginary line drawn along the direction of the force); we do this to take into account the effect of forces acting at an angle. It is clear that a force applied at an angle, such as $\mathbf{F}_3$ in Fig. 9-6, will be less effective than the same magnitude force applied straight on such as $\mathbf{F}_1$ (Fig. 9-6a).

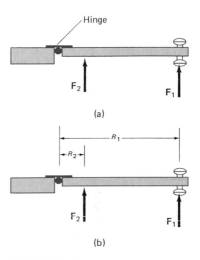

FIGURE 9-5

Applying the same force with different lever arms, R.

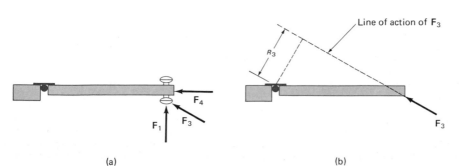

FIGURE 9-6

(a) Forces acting at different angles. (b) The lever arm is defined as the perpendicular distance from the axis of rotation to the line of action of the force.

And if you push on the end of the door so that the force is directed at the hinge (the axis of rotation), as indicated by $\mathbf{F}_4$, the door will not move at all.

The lever arm for a force such as $\mathbf{F}_3$ is found by drawing a line along the direction of $\mathbf{F}_3$ (this is the "line of action" of $\mathbf{F}_3$), and then drawing another line from the axis of rotation perpendicular to the first line (and also perpendicular to the rotation axis). The length of this second line is the lever arm for $\mathbf{F}_3$ and is labeled R_3 in Fig. 9–6b.

The torque associated with $\mathbf{F}_3$ is then $R_3 F_3$. This short lever arm and the corresponding smaller torque associated with $\mathbf{F}_3$ is consistent with the observed fact that $\mathbf{F}_3$ is less effective in opening the door than is $\mathbf{F}_1$. When the lever arm is defined in this way, experiment shows that the relation $\alpha \propto \tau$ is valid in general. Notice in Fig. 9–6a that the line action of the force $\mathbf{F}_4$ passes through the hinge and hence its lever arm is zero. Consequently zero torque is associated with $\mathbf{F}_4$ and it gives rise to no angular acceleration, in accord with everyday experience.

In general, then, we must write the torque about a given axis as

$$\tau = R_\perp F \qquad (9\text{–}8a)$$

where $R_\perp$ is the lever arm and the perpendicular symbol ($\perp$) reminds us that we must use the distance from the axis of rotation that is perpendicular to the line of action of the force (Fig. 9–7a). An alternate but equivalent way of determining the torque associated with a force is to resolve the force into components parallel and perpendicular to a line joining the point of application of the force to the axis as shown in Fig. 9–7b. Then the torque will be equal to $F_\perp$ times the distance R from the axis to the point of application of the force:

$$\tau = R F_\perp. \qquad (9\text{–}8b)$$

That this gives the same result as Eq. 9–8a can be seen from the fact that $F_\perp = F \sin \theta$ and $R_\perp = R \sin \theta$; so

$$\tau = R F \sin \theta \qquad (9\text{–}8c)$$

in either case. We can use any of Eqs. 9–8 to calculate the torque, whichever is easiest.

Since torque is a force times a distance, it is measured in units of $N \cdot m$ in SI units[†], dyne $\cdot$ cm in the cgs system, and lb $\cdot$ ft in the English system.

EXAMPLE 9–7 What is the torque applied by the biceps on the lower arm in Fig. 9–8a and b. The axis of rotation is through the elbow joint and the muscle is inserted 5.0 cm from the elbow joint.

SOLUTION (a) $F = 700$ N, $R_\perp = 0.050$ m, so $\tau = (0.050 \text{ m})(700 \text{ N}) = 35$ N$\cdot$m.

(b) $F = 700$ N, $R_\perp = (0.050 \text{ m})(\sin 45°)$, so $\tau = (0.050 \text{ m})(0.71)(700 \text{ N}) = 25$ N$\cdot$m.

Since we are interested only in rotation about a fixed axis in this chapter, we consider only forces that act in a plane perpendicular to the axis of rotation. If there is a force (or component of a force) acting parallel to the axis of rotation, it will tend to turn the axis of rotation—the component $\mathbf{F}_{\|}$ in Fig. 9–9 is an example. Since we are assuming the axis remains fixed in direction, either there can be no such forces or else the axis must be an axle (or the like) which is mounted in bearings (or hinges) that exert a compensating torque to keep the axis fixed. Thus, only a force, or component of a force ($\mathbf{F}_\perp$ in Fig. 9–9), in a plane perpendicular to the axis will give rise to rotation about the axis, and it is only these we consider in this chapter.

[†] Note that the unit for torque (N$\cdot$m in SI) is the same as that for energy. But the two quantities are very different. An obvious difference is that energy is a scalar whereas torque, as we shall see, is a vector. The special name *joule* (1 J = 1 N$\cdot$m) is used only for energy (and for work), never for torque.

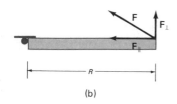

(a)

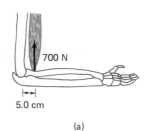

(b)

FIGURE 9–7

Torque $= R_\perp F = R F_\perp$.

FIGURE 9–8

Example 9–7.

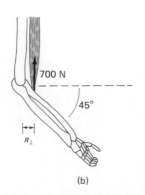

700 N

5.0 cm

(a)

700 N

45°

$R_\perp$

(b)

We have seen that the angular acceleration α of a rotating body is proportional to the torque τ applied to it:

$$\alpha \propto \tau.$$

This corresponds to Newton's second law for translational motion, $a \propto F$, where torque has taken the place of force and correspondingly the angular acceleration α takes the place of the linear acceleration a. In the linear case, the acceleration is not only proportional to the net force but is also inversely proportional to the inertia of the body, which we call its mass m; thus we could write $a = F/m$. But what plays the role of mass for the rotational case? That is what we now set out to determine. At the same time we will see that the relation $\alpha \propto \tau$ follows directly from Newton's second law, $F = ma$.

We first consider a very simple case: a particle of mass m rotating in a circle of radius R at the end of a string or rod whose mass we can ignore (Fig. 9–10). The torque which gives rise to its angular acceleration is $\tau = RF$. If we make use of Newton's second law for linear quantities, $F = ma$, and Eq. 9–5 relating the angular acceleration to the tangential linear acceleration, $a_T = R\alpha$, we have

$$F = ma$$
$$= mR\alpha.$$

When we multiply both sides by R we find that the torque $\tau = RF$ is given by

$$\tau = mR^2\alpha. \qquad \text{[single particle]} \qquad (9\text{--}9)$$

Here at last we have a direct relation between the angular acceleration and the applied torque τ. The quantity mR^2 represents the *rotational inertia* of the particle and is called its *moment of inertia*.

Now let us consider a rotating rigid body, such as a wheel, which we can think of as consisting of many particles located at various distances from the axis of rotation. We can apply Eq. 9–9 to each particle of the body and then sum over all the particles. The sum of the various torques is just the total torque, which we now call τ, so we obtain:

$$\tau = \left(\sum m_i R_i^2\right)\alpha \qquad \text{[axis fixed]} \qquad (9\text{--}10)$$

where we factored out the α since it is the same for all the particles of the body. The resultant torque τ actually represents the sum of all internal torques that each particle exerts on another plus all external torques; but, because of Newton's third law, the sum of the internal torques is zero.[†] Hence τ represents the resultant *external* torque.

The sum $\sum m_i R_i^2 = m_1 R_1^2 + m_2 R_2^2 + \cdots + m_n R_n^2$ in Eq. 9–10 represents the sum of the masses of each particle in the body multiplied by the perpendicular distance of that particle from the axis of rotation. This quantity is called the *rotational inertia* or **moment of inertia** of the body, I:

$$I = \sum m_i R_i^2. \qquad (9\text{--}11)$$

Combining Eqs. 9–10 and 9–11, we can write

$$\tau = I\alpha. \qquad \text{[axis fixed]} \qquad (9\text{--}12)$$

[†] This depends on the so-called "strong" form of Newton's third law in which not only is the force one particle exerts on a second equal and opposite to the force the second exerts on the first, but these two forces act along the same line. This strong form of Newton's third law holds for most forces, but does not hold for certain electromagnetic forces. Nonetheless, even in this case it is possible to show that the sum of the internal torques is still zero.

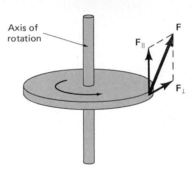

FIGURE 9–9

Only the component of **F** that acts in the plane perpendicular to the rotation axis, $\mathbf{F_\perp}$, acts to turn the wheel about the axis. The component parallel to the axis, $\mathbf{F_\parallel}$, would tend to move the axis itself, which we assume is fixed.

FIGURE 9–10

A mass m rotating in a circle of radius R about a fixed point.

This is the rotational equivalent of Newton's second law. It is valid for the rotation of a rigid body about a fixed axis.[†] It can be shown (see Chapter 10) that Eq. 9–12 is valid also when the body is translating with acceleration, but only if I and α are calculated about the center of mass of the body, and the rotation axis through the cm doesn't change direction. Then

$$\tau_{cm} = I_{cm}\alpha_{cm}. \qquad \text{[direction of axis fixed]} \qquad (9\text{--}13)$$

We see that the moment of inertia, I, which is a measure of the rotational inertia of a body, plays the same role for rotational motion that mass does for translational motion. As can be seen from Eq. 9–11, the rotational inertia of an object depends not only on its mass, but also on how that mass is distributed. For example, a large diameter cylinder will have greater rotational inertia than one of equal mass but smaller diameter (and therefore longer), Fig. 9–11. The former will be harder to start rotating, and harder to stop. When the mass is concentrated further from the axis of rotation, the rotational inertia is greater. For rotational motion, the mass of a body *cannot* be considered as concentrated at its center of mass.

Whenever Eq. 9–12 is used, it must be remembered to give α in rad/s^2 and other quantities in a consistent set of units. The moment of inertia, I, has units of kg·m^2 in SI units.

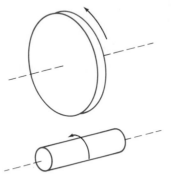

FIGURE 9–11

A large-diameter wheel has greater rotational inertia than one of smaller diameter but equal mass.

■ **EXAMPLE 9–8**   Two particles, of mass 5.0 kg and 7.0 kg, are mounted 4.0 m apart on a light rod (whose mass can be ignored) as shown in Fig. 9–12. Calculate the moment of inertia of the system (*a*) when rotated about an axis passing halfway between the masses, Fig. 9–12a, and (*b*) when the system rotates about an axis located 0.50 m to the left of the 5.0 kg mass (Fig. 9–12b).

SOLUTION   (*a*) Both particles are the same distance, 2.0 m, from the axis of rotation. Thus $I = \sum m_i R_i^2 = (5.0 \text{ kg})(2.0 \text{ m})^2 + (7.0 \text{ kg})(2.0 \text{ m})^2 = 48 \text{ kg} \cdot \text{m}^2$.

(*b*) The 5.0 kg mass is now 0.50 m from the axis and the 7.0 kg mass is 4.50 m from the axis. Then

$$
\begin{aligned}
I = \sum m_i R_i^2 &= (5.0 \text{ kg})(0.50 \text{ m})^2 + (7.0 \text{ kg})(4.5 \text{ m})^2 \\
&= 1.3 \text{ kg} \cdot \text{m}^2 + 142 \text{ kg} \cdot \text{m}^2 \\
&= 143 \text{ kg} \cdot \text{m}^2.
\end{aligned}
$$

FIGURE 9–12

Calculating moment of inertia (Example 9–8).

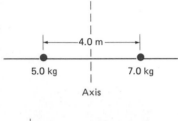

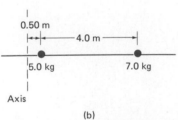

(b)

The above example illustrates two important points. First, the moment of inertia of a given system is different for different axes of rotation. Second, part (*b*) illustrates the fact that mass close to the axis of rotation contributes little to the total moment of inertia; in this example, the 5.0 kg object contributed less than 1 percent to the total.

For most ordinary bodies, the mass is distributed continuously and the sum $\sum m_i R_i^2$ can sometimes be calculated with the use of calculus (see Section 9–6). In this way expressions can be worked out for the moments of inertia of regularly shaped bodies in terms of their dimensions. Figure 9–13 gives these expressions for a number of solids rotated about the axes specified. The only one for which the result is obvious is the thin hoop of radius R_0 rotated about its center. For this object, all the mass is concentrated at the same fixed distance, R_0, from the axis. Thus $\sum m_i R_i^2 = (\sum m_i)R_0^2 = M R_0^2$ where M is the total mass of the ring.

When discussing moments of inertia, it is often convenient to work with the *radius of gyration*, k, which is a sort of average radius. In particular, the radius of gyration of an object is defined so that if all the mass of the object were concentrated at this distance from the axis, then it would have the same moment of inertia as the original object. For example (see Fig. 9–13), the radius of gyration of a solid cylinder is $(1/\sqrt{2})R_0 \approx 0.71 \, R_0$; this means that a solid cylinder of radius 10.0 cm has the same moment of inertia as an equal-mass thin hoop of radius 7.1 cm. The moment

_____

[†] That is, the axis is fixed relative to the body and is fixed in an inertial reference frame; this includes an axis moving at uniform velocity in an inertial frame, since the axis can be considered fixed in a second inertial frame that moves with respect to the first.

Object	Location of axis		Moment of inertia	Radius of gyration
(a) Thin hoop of radius R_0	Through center	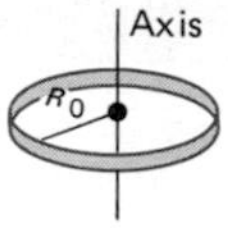 Axis	MR_0^2	R_0
(b) Thin hoop of radius R_0 and width w	Through central diameter	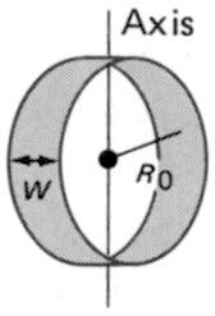 Axis	$\frac{1}{2}MR_0^2 + \frac{1}{12}Mw^2$	$\sqrt{\dfrac{R_0^2}{2} + \dfrac{w^2}{12}}$
(c) Solid cylinder of radius R_0	Through center	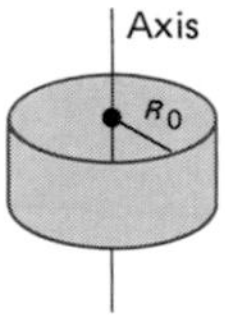 Axis	$\frac{1}{2}MR_0^2$	$\dfrac{R_0}{\sqrt{2}}$
(d) Hollow cylinder of inner radius R_1 and outer radius R_2	Through center	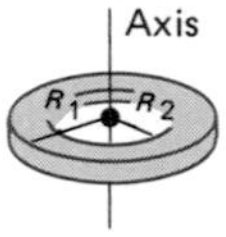 Axis	$\frac{1}{2}M(R_1^2 + R_2^2)$	$\sqrt{\dfrac{R_1^2 + R_2^2}{2}}$
(e) Solid sphere of radius r_0	Through center	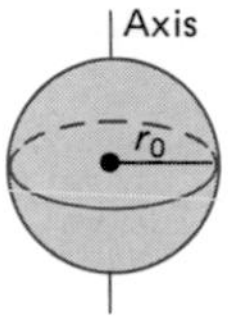 Axis	$\frac{2}{5}Mr_0^2$	$\sqrt{\dfrac{2}{5}}\, r_0$
(f) Thin rod of length L	Through center	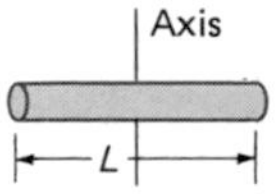 Axis	$\frac{1}{12}ML^2$	$\dfrac{L}{\sqrt{12}}$
(g) Thin rod of length L	Through end	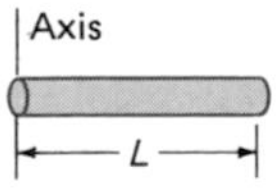 Axis	$\frac{1}{3}ML^2$	$\dfrac{L}{\sqrt{3}}$
(h) Rectangular thin plate, of length l and width w	Through center	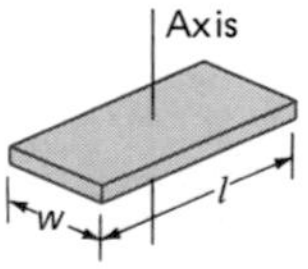 Axis	$\frac{1}{12}M(l^2 + w^2)$	$\sqrt{\dfrac{l^2 + w^2}{12}}$

FIGURE 9–13

Moments of inertia for various objects of uniform composition.

of inertia of any object can be written in terms of its radius of gyration as

$$I = Mk^2.$$

For unusual or irregularly shaped objects it is common to specify the radius of gyration.

FIGURE 9–14

Example 9–9.

EXAMPLE 9–9 A 15.0-N force (represented by **T**) is applied to a cord wrapped around a wheel of mass $M = 4.00$ kg, radius $R_0 = 33.0$ cm, and radius of gyration $k = 30.0$ cm, Fig. 9–14. If there is a frictional torque $\tau_{fr} = 1.10$ N·m at the hub, what is the acceleration of the wheel?

SOLUTION First we calculate I:

$$I = Mk^2 = (4.00\ \text{kg})(0.300\ \text{m})^2 = 0.360\ \text{kg} \cdot \text{m}^2.$$

There are two torques involved, that due to the 15.0-N force of magnitude $(0.330\ \text{m})(15.0\ \text{N}) = 4.95$ N·m, and the opposing friction torque of 1.10 N·m.

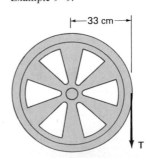

Thus, we use Eq. 9–12 and obtain

$$\alpha = \frac{\tau}{I} = \frac{TR_0 - \tau_{\text{fr}}}{I} = \frac{(4.95 \text{ N·m}) - (1.10 \text{ N·m})}{0.360 \text{ kg·m}^2} = 10.7 \text{ rad/s}^2.$$

EXAMPLE 9–10 Suppose in Fig. 9–14 that instead of a constant 15.0-N force being exerted on the cord hanging from the edge of the wheel, a block of weight 15.0 N (and mass $m = 1.53$ kg) hangs from the cord which we assume not to stretch. (a) Calculate the angular acceleration α of the wheel and the linear acceleration a of the mass m. (b) Determine the angular velocity ω of the wheel and the linear velocity v of the mass m at $t = 3.00$ s if the wheel starts from rest at $t = 0$.

SOLUTION (a) Let T be the tension in the cord. Then a force T acts at the edge of the wheel and we have (see Example 9–9) for the wheel:

$$\alpha = \frac{\tau}{I} = \frac{TR_0 - \tau_{\text{fr}}}{I}.$$

Two forces act on the block of mass m: the force of gravity, mg, acts downward, and the tension of the cord T pulls upward. So by $F = ma$ we have

$$ma = mg - T.$$

Note that the tension T, which is the force exerted on the edge of the wheel, is *not* in general equal to the weight of the block ($= mg = 15.0$ N); this is because the block is accelerating, so (by the last equation above) $T = mg - ma$. To obtain α, we eliminate T between the above two equations and use Eq. 9–5:

$$a = R_0\alpha$$

which is valid since the tangential acceleration of a point on the edge of the wheel is the same as the acceleration of the block if the cord doesn't stretch or slip. Thus

$$\alpha = \frac{(mg - mR_0\alpha)R_0 - \tau_{\text{fr}}}{I}$$

or (since α appears on both sides above)

$$\alpha = \frac{mgR_0 - \tau_{\text{fr}}}{I + mR_0^2}$$

$$= \frac{(15.0 \text{ N})(0.330 \text{ m}) - 1.10 \text{ N·m}}{0.360 \text{ kg·m}^2 + (1.53 \text{ kg})(0.330 \text{ m})^2} = 7.31 \text{ rad/s}^2.$$

The angular acceleration is somewhat less in this case than in Example 9–9; this reflects the fact that T is somewhat less than the weight of the block, mg. The linear acceleration of the block is

$$a = R_0\alpha = (0.330 \text{ m})(7.31 \text{ rad/s}^2) = 2.41 \text{ m/s}^2.$$

(b) Since the angular acceleration is constant,

$$\omega = \omega_0 + \alpha t$$
$$= 0 + (7.31 \text{ rad/s}^2)(3.00 \text{ s}) = 21.9 \text{ rad/s};$$

and the velocity of the block is the same as that for a point on the wheel's edge:

$$v = R_0\omega = (0.330 \text{ m})(21.9 \text{ rad/s}) = 7.23 \text{ m/s}.$$

This can also be obtained using the linear equation $v = v_0 + at = 0 + (2.41 \text{ m/s}^2)(3.00 \text{ s}) = 7.23 \text{ m/s}.$

⬛ *9-6 Calculation of Moments of Inertia

The moments of inertia of any body about any axis can be determined experimentally by measuring the net torque τ required to give the body an angular acceleration α. Then, from Eq. 9-12, $I = \tau/\alpha$.

For many bodies, or systems of particles, the moment of inertia can be calculated. A simple illustration was done in Example 9-8. Many bodies can be considered as a continuous distribution of mass. In this case, Eq. 9-11 defining moment of inertia becomes

$$I = \int R^2 \, dm \qquad (9-14)$$

where dm represents the mass of any infinitesimal particle of the body and R is the perpendicular distance of this particle from the axis of rotation; the integral is taken over the whole body. This is easily done only for bodies of simple geometric shape.

EXAMPLE 9-11 Show that the moment of inertia of a uniform hollow cylinder of inner radius R_1, outer radius R_2, and mass M, is $I = \frac{1}{2}M(R_1^2 + R_2^2)$, as stated in Fig. 9-13, if the rotation axis is the axis of symmetry.

SOLUTION We divide the cylinder into concentric cylindrical rings or hoops of thickness dR; one of these infinitesimal hoops is indicated in Fig. 9-15. If the density (mass per unit volume) is ρ, then

$$dm = \rho \, dV$$

where dV is the volume of the loop of radius R, thickness dR, and height h; since $dV = (2\pi R)(dR)(h)$, we have

$$dm = 2\pi\rho h R \, dR.$$

Then the moment of inertia is obtained by integrating (summing) over all these hoops:

$$I = \int R^2 \, dm$$
$$= \int_{R_1}^{R_2} 2\pi\rho h R^3 \, dR = 2\pi\rho h \left[\frac{R_2^4 - R_1^4}{4}\right]$$

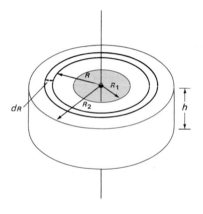

FIGURE 9-15
Determining the moment of inertia of a hollow cylinder (Example 9-11).

where we have assumed the cylinder has uniform density, ρ = constant. (If this were not so, we would have to know ρ as a function of R before the integration could be carried out.) The volume V of this hollow cylinder is $V = (\pi R_2^2 - \pi R_1^2)h$ so its mass M is

$$M = \rho V = \rho\pi(R_2^2 - R_1^2)h.$$

Since $(R_2^4 - R_1^4) = (R_2^2 - R_1^2)(R_2^2 + R_1^2)$ we have

$$I = \frac{\pi\rho h}{2}(R_2^2 - R_1^2)(R_2^2 + R_1^2)$$
$$= \frac{1}{2}M(R_1^2 + R_2^2),$$

as stated in Fig. 9-13. Note that for a solid cylinder, $R_1 = 0$ and we obtain, with $R_2 = R_0$:

$$I = \frac{1}{2}MR_0^2$$

which is that given in Fig. 9-13 for a solid cylinder of mass M and radius R_0.

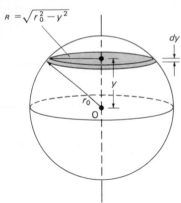

$R = \sqrt{r_0^2 - y^2}$

FIGURE 9–16

Determining the moment of inertia of a sphere of radius r_0 (Example 9–12).

EXAMPLE 9–12 Determine the moment of inertia of a uniform solid sphere, of radius r_0, about an axis through its center.

SOLUTION We divide the sphere into infinitesimal cylinders of thickness dy as shown in Fig. 9–16. Each cylinder has a radius

$$R = \sqrt{r_0^2 - y^2}$$

and a mass

$$dm = \rho \, dV = \rho \pi R^2 \, dy = \rho \pi (r_0^2 - y^2) \, dy.$$

Hence, the moment of inertia of each infinitesimal cylinder is

$$dI = \tfrac{1}{2} \, dm \, R^2 = \frac{\rho \pi}{2} (r_0^2 - y^2)^2 \, dy = \frac{\rho \pi}{2} (r_0^4 - 2r_0^2 y^2 + y^4) \, dy.$$

We sum, or rather integrate, over all these infinitesimal cylinders:

$$I = \int dI = \frac{\rho \pi}{2} \int_{-r_0}^{r_0} (r_0^4 - 2r_0^2 y^2 + y^4) \, dy$$

$$= \frac{\rho \pi}{2} \left(r_0^4 y - \tfrac{2}{3} r_0^2 y^3 + \frac{y^5}{5} \right) \Bigg|_{-r_0}^{r_0} = \tfrac{8}{15} \rho \pi r_0^5.$$

Since the volume of a sphere is $V = \tfrac{4}{3} \pi r_0^3$, then $M = \rho V = \tfrac{4}{3} \pi \rho r_0^3$; so

$$I = \tfrac{2}{5} M r_0^2,$$

as listed in Fig. 9–13.

There are two simple theorems that are helpful in obtaining moments of inertia. The first is called the *parallel-axis theorem*. It states that if I is the moment of inertia of a body of total mass M about any axis, and I_{cm} is the moment of inertia about an axis passing through the center of mass and parallel to the first axis a distance h away, then

$$I = I_{cm} + Mh^2. \tag{9–15}$$

Thus, for example, if the moment of inertia about an axis through the cm is known, the moment of inertia about any axis parallel to this is easily obtained. The proof of the parallel-axis theorem is as follows. We choose our coordinate system so the origin is at the cm, and I_{cm} is the moment of inertia about the z axis. Figure 9–17 shows a cross section of a body of arbitrary shape in the xy plane. We let I represent the moment of inertia of the body about an axis parallel to the z axis that passes through the point A in Fig. 9–17, where the point A has coordinates x_A and y_A. Let x_i, y_i, and m_i represent the coordinates and mass of an arbitrary point in the body. The square of the distance from this point to A is $[(x_i - x_A)^2 + (y_i - y_A)^2]$; so the moment of inertia, I, about the axis through A is

$$I = \sum m_i [(x_i - x_A)^2 + (y_i - y_A)^2]$$

$$= \sum m_i (x_i^2 + y_i^2) - 2x_A \sum m_i x_i - 2y_A \sum m_i y_i + (\sum m_i)(x_A^2 + y_A^2).$$

The first term on the right is just $I_{cm} = \sum m_i (x_i^2 + y_i^2)$ since the cm is at the origin. The second and third terms are zero since, by definition of the cm, $\sum m_i x_i = \sum m_i y_i = 0$ because $x_{cm} = y_{cm} = 0$. The last term is Mh^2 since $\sum m_i = M$ and $(x_A^2 + y_A^2) = h^2$ where h is the distance of A from the cm. Thus we have proved $I = I_{cm} + Mh^2$, which is Eq. 9–15.

The parallel-axis theorem can be applied to any body. The second theorem, the *perpendicular-axis theorem*, can be applied only to plane figures—that is, to two-dimensional bodies, or bodies of uniform thickness whose thickness can be neglected compared to the other dimensions. This theorem states that the sum of the moments of inertia of a plane body about any two perpendicular axes in the plane of the body

FIGURE 9–17

Derivation of parallel-axis theorem.

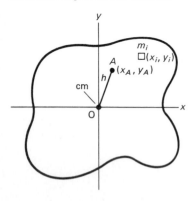

is equal to the moment of inertia about an axis through their point of intersection perpendicular to the plane of the object. That is, if the figure is in the xy plane,

$$I_z = I_x + I_y. \quad \text{[object in } xy \text{ plane]} \quad (9-16)$$

Here I_z, I_x, I_y are moments of inertia about the z, x, and y axes. The proof is simple: since $I_x = \sum m_i y_i^2$, $I_y = \sum m_i x_i^2$, and $I_z = \sum m_i(x_i^2 + y_i^2)$ Eq. 9–16 follows directly.

◆ **EXAMPLE 9–13** Determine the moment of inertia of a solid cylinder of radius R_0 and mass M about an axis tangent to its edge and parallel to its symmetry axis, Fig. 9–18a.

SOLUTION We use the parallel axis theorem with $I_{cm} = \frac{1}{2}MR_0^2$ (Fig. 9–13c). Since $h = R_0$, we have

$$I = I_{cm} + Mh^2 = \frac{3}{2}MR_0^2.$$

◆ **EXAMPLE 9–14** Determine the moment of inertia of a thin circular coin (a cylinder) about an axis through its center in the plane of the coin (Fig. 9–18b).

SOLUTION We use the perpendicular axis theorem and we want to calculate the moment of inertia about the x axis in Fig. 9–18b. From Fig. 9–13c we have $I_z = \frac{1}{2}MR_0^2$, and, from symmetry, $I_x = I_y$. Hence $2I_x = I_z$ or $I_x = \frac{1}{2}I_z = \frac{1}{4}MR_0^2$.

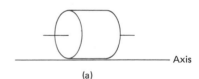

(a)

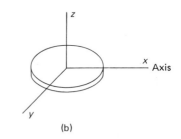

(b)

FIGURE 9–18

Determining moments of inertia in Examples 9–13 and 9–14.

◰ ***9–7 Why Does a Rolling Sphere Slow Down?**

A sphere of mass M and radius r_0 rolling on a horizontal flat table eventually comes to rest. What force causes it to come to rest? You might think it is friction but when you examine the problem from a simple, straightforward point of view, a serious contradiction seems to arise.

Suppose a sphere is rolling to the left as shown in Fig. 9–19, and is slowing down. By Newton's second law, $F = ma$, there must be a force **F** (presumably frictional) acting to the right as shown, so that the acceleration **a** will also point to the right and v will decrease. Curiously enough, though, if we now look at the torque equation (calculated about the center of mass), $\tau = I\alpha$, we see that the force **F** acts to increase the angular acceleration α, and thus to *increase* the velocity of the wheel. Thus the contradiction! The force **F** acts to decelerate the sphere if we look at the translational motion, but speeds it up if we look at the rotational motion.

The resolution of this apparent paradox is that some other force must be acting. The only other forces acting are gravity, $m\mathbf{g}$, and the normal force $\mathbf{F_N}$ ($= -m\mathbf{g}$). These act vertically and hence do not affect the horizontal translational motion. If we assume the sphere and plane are rigid so the sphere is in contact at only one point, they give rise to no torques about the cm either, since they act through the cm.

The only recourse we have to resolve the contradiction is to give up our idealization that the bodies are rigid. In fact, all bodies are deformable to some extent, and so our sphere flattens slightly and the level surface also acquires a slight depression where the two are in contact. Hence, there is an *area* of contact, not a point. Because this is true, there can be a torque at this area of contact which acts in the opposite direction to the torque associated with **F**, and hence acts to decelerate the sphere. This torque can be associated with the normal force $\mathbf{F_N}$ that the table exerts on the sphere over the whole area of contact. The net effect is that we can consider $\mathbf{F_N}$ acting vertically a distance l in front of the cm as shown in Fig. 9–20 (where the deformation is greatly exaggerated).

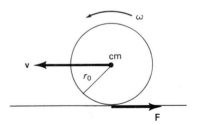

FIGURE 9–19

Sphere rolling to the left.

FIGURE 9–20

The normal force, $\mathbf{F_N}$, exerts a torque that slows down the rolling sphere.

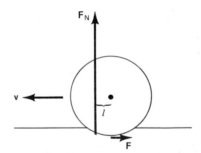

Is it reasonable that the normal force $\mathbf{F}_N$ should effectively act in *front* of the cm as shown in Fig. 9–20? Yes. The ball is rolling, and the leading edge strikes the surface with a slight impulse. The table therefore pushes upward a bit more strongly on the front part of the sphere than it would if the sphere were at rest. At the back part of the area of contact, the sphere is starting to move upward and so the table pushes upward on it less strongly than when the sphere is at rest. The table pushing up more strongly on the front part of the area of contact and less at the back, gives rise to the necessary torque and justifies the effective acting point of $\mathbf{F}_N$ being in front of the cm.

It is clear that the torque

$$\tau_N = l F_N$$

acts to reduce the angular velocity of the wheel, and thus slow it down. We can, in fact, determine the magnitude of τ_N in terms of either F or a. From $\tau = I\alpha$ we have

$$\tau_N - r_0 F = I\alpha = I\frac{a}{r_0}$$

since the angular acceleration α is related to the linear acceleration a by $\alpha = a/r_0$, or:

$$\tau_N = I\frac{a}{r_0} + r_0 F.$$

For the translational (horizontal) motion, only F acts so $F = Ma$ and thus

$$\tau_N = I\frac{F}{Mr_0} + r_0 F = F\left(\frac{I}{Mr_0} + r_0\right)$$

or

$$\tau_N = I\frac{a}{r_0} + r_0 Ma = a\left(\frac{I}{r_0} + r_0 M\right).$$

For a sphere, $I = \frac{2}{5}Mr_0^2$ and these relations become

$$\tau_N = \frac{7}{5}Fr_0$$
$$= \frac{7}{5}Mr_0 a.$$

Thus, for a sphere, the torque associated with $\mathbf{F}_N$ is $\frac{7}{5}$ times the torque due to $\mathbf{F}$, the friction force, no matter at what rate the deceleration takes place. For very hard surfaces, the deceleration rate is almost zero, so $\tau_N \approx 0$, which tells us that $l \approx 0$ so $\mathbf{F}_N$ acts nearly through the cm.

EXAMPLE 9–15 A 2.6-kg sphere of radius $r_0 = 35$ cm is initially rolling on a flat surface with a speed $v = 0.40$ m/s. It comes to rest in 80 m. Determine the force F and the distance l.

SOLUTION From Eq. 2–9c, $a = (v^2 - v_0^2)/2x = 1.0 \times 10^{-3}$ m/s². Hence $F = Ma = 2.6 \times 10^{-3}$ N. Since $\tau_N = \frac{7}{5}Fr_0 = 1.3 \times 10^{-3}$ N·m, then $l = \tau_N/F_N = \tau_N/Mg = 5.0 \times 10^{-5}$ m, or 0.05 mm, a very small distance. Note that we cannot determine τ_N or l unless we measure a (or F).

When other forces are present, the tiny torque due to $\mathbf{F}_N$ can usually be ignored. For example, when a sphere or cylinder rolls down an incline, the force of gravity has far more influence than τ_N, so the latter can be ignored. Thus, for most purposes, we can assume a hard sphere is in contact with a hard surface at essentially one point.

9–8   Angular Momentum and
Its Conservation

Equations 9–12 and 9–13,

$$\tau = I\alpha,$$

describe the rotation of a rigid body about a fixed axis, or about an axis that moves without changing direction as long as the axis passes through the cm of the body. This relation, $\tau = I\alpha$, is the rotational equivalent of Newton's second law for translational motion, $F = ma$. This latter relation can also be written $F = dp/dt = d(mv)/dt$ where p is the linear momentum. An analogous relation can be written for the rotational motion of a rigid body; since the angular acceleration $\alpha = d\omega/dt$ (Eq. 9–3), we have

$$\tau = I\alpha = I\frac{d\omega}{dt} = \frac{d(I\omega)}{dt}.$$

(This simple derivation assumed that the moment of inertia, I, remains constant. However, it is valid even if the moment of inertia changes, and this will be proved in Chapter 10.) The quantity $I\omega$ is called the **angular momentum**, L, of the body about its axis of rotation:[†]

$$L = I\omega. \tag{9–17}$$

So Newton's second law for rotational motion of a rigid body about a fixed axis (or a moving one that is fixed in direction but passes through the cm) becomes:

$$\tau = \frac{dL}{dt} = \frac{d(I\omega)}{dt}. \tag{9–18}$$

Equation 9–18 is simply another way of writing Eqs. 9–12 or 9–13, and the same restrictions apply as already mentioned.

Note that the angular momentum of a rigid body, $L = I\omega$, is completely analogous to ordinary linear momentum, $p = mv$. The rotational inertia (or moment of inertia) I, is the analog of ordinary (linear) inertia, namely the mass m; and angular velocity, ω, is the analog of the ordinary velocity, v.

Angular momentum is a very important concept in physics. We will deal with it for more general cases and in more detail, including its vector nature, in Chapter 10. Now we will concern ourselves with one of its most important aspects: that under certain conditions, angular momentum, like energy and linear momentum, is a *conserved* quantity. What are the conditions under which angular momentum is conserved? From Eq. 9–18 it is clear that if the net torque τ acting on the rotating body is zero, then:

$$\frac{dL}{dt} = 0 \quad \text{and } L = I\omega = \text{constant}.$$

This is the **law of conservation of angular momentum** for a rotating body:

> **The total angular momentum of a rotating body remains constant if the net torque acting on it is zero.**

This law, particularly in its more general form (see Chapter 10), is one of the great conservation laws of physics.

When there is zero net torque acting on a body, and the body is rotating about

---

[†] We will see in Chapter 10 that angular momentum is a vector and that the angular momentum discussed here, $L = I\omega$, is the component along the rotation axis. There may or may not be other components.

a fixed axis or about an axis through its cm such that its direction doesn't change, we can write

$$I\omega = I_0\omega_0 = \text{constant};$$

I_0 and ω_0 are the moment of inertia and angular velocity about that axis at some initial time ($t = 0$), and I and ω are their values at some other time. The parts of the body may alter their positions relative to one another, so that I changes; ω changes as well and the product $I\omega$ remains constant.

Many interesting everyday phenomena can be understood on the basis of conservation of angular momentum. For example, a skater doing a spin on ice rotates at a relatively low speed when her arms are outstretched; but when she brings her arms in close to her body she suddenly spins much faster. Remembering the definition of moment of inertia as $I = \sum_i m_i R_i^2$, it is clear that when she pulls her arms in closer to the axis of rotation her moment of inertia is reduced. Since the angular momentum $I\omega$ remains constant (we ignore the small torque due to friction), if I decreases then the angular velocity ω must increase. If she reduces her moment of inertia by a factor of two, she will then rotate with twice the angular velocity.

A similar example is the diver shown in Fig. 9–21. The push as he leaves the board gives him an initial angular momentum about his cm. When he curls himself into the tuck position he rotates quickly one or more times; he then stretches out again, increasing his moment of inertia, which reduces the angular velocity to a small value, and then he enters the water. The change in moment of inertia from the straight position to the tuck position can be a factor of as much as $3\frac{1}{2}$.

FIGURE 9–21

Diver rotates faster when arms and legs are tucked in than when outstretched. Angular momentum is conserved.

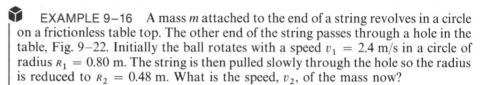

EXAMPLE 9–16 A mass m attached to the end of a string revolves in a circle on a frictionless table top. The other end of the string passes through a hole in the table, Fig. 9–22. Initially the ball rotates with a speed $v_1 = 2.4$ m/s in a circle of radius $R_1 = 0.80$ m. The string is then pulled slowly through the hole so the radius is reduced to $R_2 = 0.48$ m. What is the speed, v_2, of the mass now?

SOLUTION The force exerted by the string on the mass m does not alter its angular momentum about the axis of rotation, since the force is exerted toward the axis; so $\tau = 0$. Hence, from conservation of angular momentum:

$$I_1\omega_1 = I_2\omega_2$$
$$mR_1^2\omega_1 = mR_2^2\omega_2$$

since the moment of inertia of a single particle is $I = mR^2$. Since $v = R\omega$, we have

$$v_2 = R_2\omega_2 = R_2\omega_1\left(\frac{R_1^2}{R_2^2}\right) = R_2\frac{v_1}{R_1}\left(\frac{R_1^2}{R_2^2}\right) = v_1\frac{R_1}{R_2}$$

$$= (2.4 \text{ m/s})\left(\frac{0.80 \text{ m}}{0.48 \text{ m}}\right) = 4.0 \text{ m/s}.$$

Although we will not go into the vector nature of angular momentum in any detail until the next chapter, we can deal with some simple cases involving

FIGURE 9–22
Example 9–16.

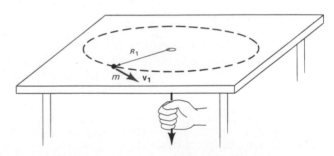

conservation of momentum here. For a body rotating about a fixed axis, the direction of the angular momentum can be taken as the direction of the angular velocity, ω. That is

$$\mathbf{L} = I\omega.$$

This is strictly true[†] only if the rotation axis is an axis of symmetry of the body or the body is thin and flat and rotates about an axis perpendicular to the plane of the body (such as a wheel rotating on an axle).

   As a simple example, consider a person standing at rest on a circular platform capable of rotating friction-free about an axis through its center (that is, a simplified merry-go-round). If the person now starts to walk along the edge of the platform, the platform starts rotating in the opposite direction. Why? Well, this is an example of the conservation of angular momentum. If the person starts walking counterclockwise, the person's angular momentum will be pointed upward along the axis of rotation (remember how we defined the direction of ω using the right-hand rule in Section 9–3). The magnitude of the person's angular momentum will be $L = I\omega = (mR^2)(v/R)$ where v is the person's speed (relative to earth, not the platform), R is his distance from the rotation axis, m is his mass and his moment of inertia is mR^2 if we consider him a particle (mass concentrated at one point). The platform rotates in the opposite direction, so its angular momentum points downward. If the initial total angular momentum was zero (person and platform at rest), it will remain zero after the person starts walking—that is, the upward angular momentum of the person just balances the oppositely directed angular momentum of the platform, so the total vector angular momentum is still zero. Even though the person exerts a force (and torque) on the platform, and vice versa, these are internal torques (internal to the system consisting of platform plus person); there are no external torques (assuming friction-free bearings of the platform) so from Eq. 9–18, the angular momentum remains constant.

 ## 9–9   Rotational Kinetic Energy

The quantity $\frac{1}{2}mv^2$ is the kinetic energy of a body undergoing translational motion. A body rotating about an axis is said to have **rotational kinetic energy**. By analogy with translational KE, we might expect the rotational KE of a body rotating about a fixed axis to be given by the formula $\frac{1}{2}I\omega^2$ where I is the moment of inertia of the body about that axis, and ω is its angular velocity. We can indeed show that this is true. Consider the rotating body as made up of tiny particles, each of mass m_i. If R_i represents the perpendicular distance of any one particle from the axis of rotation, then its linear velocity is $v_i = R_i\omega$. The total kinetic energy of the whole body will be the sum of the KE's of all its particles:

$$\text{KE} = \sum \left(\tfrac{1}{2}m_i v_i^2\right)$$
$$= \tfrac{1}{2}\left(\sum m_i R_i^2\right)\omega^2$$

where we have factored out the $\frac{1}{2}$ and the ω^2 since they are the same for every particle. Since $\sum m_i R_i^2 = I$, the moment of inertia, we see that the kinetic energy of an object rotating about a fixed axis is, as expected:

$$\text{Rotational KE} = \tfrac{1}{2}I\omega^2. \qquad \text{[fixed axis]} \qquad (9\text{–}19)$$

If the axis is not fixed in space, the rotational KE can take on a more complicated form.

[†] For more complicated situations of bodies rotating about a fixed axis, there will be a *component* of **L** along the direction of ω (and perhaps other components as well) and its magnitude will be equal to $I\omega$; if the total angular momentum is conserved, so will any component be conserved, and so what we have to say here can be applied to any rotation about a fixed axis.

The work done on a body rotating about a fixed axis can be written in terms of angular quantities. Suppose a force **F** is exerted at a point which is a perpendicular distance R from the axis of rotation. The work done by this force is

$$W = \int \mathbf{F} \cdot d\mathbf{l} = \int F_l R \, d\theta$$

where $d\mathbf{l}$ is an infinitesimal distance perpendicular to R (and hence in the direction of motion) whose magnitude is $dl = R \, d\theta$, and F_l is the component of **F** parallel to $d\mathbf{l}$; see Fig. 9–23. But $F_l R$ is the torque about the axis, so

$$W = \int \tau \, d\theta. \qquad (9\text{–}20)$$

The rate of work done, or power P, is

$$P = \frac{dW}{dt} = \tau \frac{d\theta}{dt} = \tau\omega. \qquad (9\text{–}21)$$

The work-energy theorem holds for rotation of a rigid body about a fixed axis. From Eq. 9–12 we have

$$\tau = I\alpha = I\frac{d\omega}{dt} = I\frac{d\omega}{d\theta}\frac{d\theta}{dt} = I\omega\frac{d\omega}{d\theta},$$

where we used the chain rule and $\omega = d\theta/dt$. Then $\tau \, d\theta = I\omega \, d\omega$ and

$$W = \int_{\theta_1}^{\theta_2} \tau \, d\theta = \int_{\omega_1}^{\omega_2} I\omega \, d\omega = \tfrac{1}{2}I\omega_2^2 - \tfrac{1}{2}I\omega_1^2.$$

This is the work-energy theorem for a body rotating about a fixed axis; it states that the work done in rotating the body through an angle $\theta_2 - \theta_1$ is equal to the change in rotational kinetic energy of the body.

An object that rotates while its cm undergoes translational motion will have both translational and rotational KE. Its *total kinetic energy will be equal to the translational KE of its cm, plus the kinetic energy of motion relative to the center of mass.* This is a general theorem and to prove it we proceed as follows. Let $\mathbf{r}_{cm} = x_{cm}\mathbf{i} + y_{cm}\mathbf{j} + z_{cm}\mathbf{k}$ represent the position of the cm at any moment in some inertial reference frame. Let $\mathbf{r}_i = x_i\mathbf{i} + y_i\mathbf{j} + z_i\mathbf{k}$ be the position vector of the ith particle of mass m_i in this inertial reference frame and $\mathbf{r}_i^* = x_i^*\mathbf{i} + y_i^*\mathbf{j} + z_i^*\mathbf{k}$ be the position vector of this particle with reference to the cm (not necessarily an inertial reference frame). Then (see Fig. 9–24)

$$x_i = x_{cm} + x_i^*, \quad y_i = y_{cm} + y_i^*, \quad \text{and} \quad z_i = z_{cm} + z_i^*.$$

The velocity of the ith particle in the inertial reference frame is

$$\boldsymbol{v}_i = \mathbf{v}_{cm} + \mathbf{v}_i^*$$

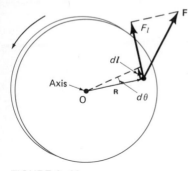

FIGURE 9–23

Calculating the work done by a torque acting on a rigid body rotating about a fixed axis.

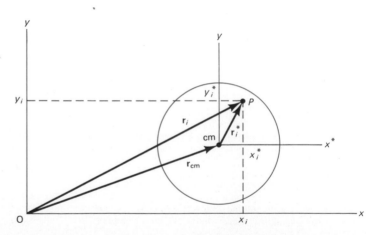

FIGURE 9–24

Point P has coordinates $x_i, y_i, z_i, (\mathbf{r}_i)$ relative to origin O of an inertial frame, and coordinates $x_i^*, y_i^*, z_i^*, (\mathbf{r}_i^*)$ relative to the center of mass (cm).

where $\mathbf{v}_{cm}$ is the velocity of the cm in this reference frame and $\mathbf{v}_i^*$ is the velocity of the ith particle relative to the cm. The total kinetic energy is, since we can write $v^2 = \mathbf{v} \cdot \mathbf{v}$ using the vector dot product,

$$\begin{aligned} \text{KE} &= \tfrac{1}{2}\sum m_i v_i^2 = \tfrac{1}{2}\sum m_i(\mathbf{v}_i \cdot \mathbf{v}_i) \\ &= \tfrac{1}{2}\sum m_i(\mathbf{v}_{cm} + \mathbf{v}_i^*)^2 \\ &= \tfrac{1}{2}\sum m_i v_{cm}^2 + \mathbf{v}_{cm}\cdot\left(\sum m_i\mathbf{v}_i^*\right) + \tfrac{1}{2}\sum m_i v_i^{*^2}. \end{aligned}$$

Since $\sum m_i = M$, the total mass of the body, and $\sum m_i v_i^* = 0$ by definition of the center of mass ($r_{cm} = (1/m)\sum m_i r_i^* = 0$, and taking derivatives of both sides gives us this result). Thus

$$\text{KE} = \tfrac{1}{2}Mv_{cm}^2 + \tfrac{1}{2}\sum m_i v_i^{*^2} \qquad (9\text{--}22)$$

and we have proved that the total KE is the sum of the translational KE of the center of mass, $\tfrac{1}{2}Mv_{cm}^2$, plus the kinetic energy of motion relative to the cm.

Our proof of Eq. 9–22 is general. We can apply this equation to the motion of a rigid body which moves in a plane, such as a wheel rolling down a hill. In this case the axis of rotation is fixed in direction (perpendicular to the plane in which the body moves) although it moves along with the center of mass and so is not fixed in position. For each particle $v_i^* = \omega R_i^*$, where R_i^* is the perpendicular distance of the ith particle from a line passing through the cm, perpendicular to the plane of motion. Then

and

$$\tfrac{1}{2}\sum m_i v_i^{*^2} = \tfrac{1}{2}\left(\sum m_i R_i^{*^2}\right)\omega^2 = \tfrac{1}{2}I_{cm}\omega^2$$

$$\text{KE} = \tfrac{1}{2}Mv_{cm}^2 + \tfrac{1}{2}I_{cm}\omega^2. \qquad \begin{bmatrix}\text{rigid body, axis} \\ \text{fixed in direction}\end{bmatrix} \qquad (9\text{--}23)$$

In this equation, I_{cm} is the moment of inertia of the body about an axis through its center of mass and perpendicular to the plane of motion. Thus the total KE of a body moving in a plane with both translational and rotational motion such that the rotation axis doesn't change direction is the sum of the translational KE of the cm plus the rotational KE about the cm, $\tfrac{1}{2}I_{cm}\omega^2$.

EXAMPLE 9–17   What will be the speed of a solid sphere of mass M and radius r_0 when it reaches the bottom of an incline if it starts from rest at a vertical height H and rolls without slipping? See Fig. 9–25. Ignore losses due to retarding forces.

FIGURE 9–25
Example 9–17.

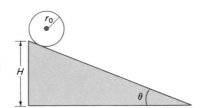

SOLUTION   We use the law of conservation of energy where we must now include rotational kinetic energy. The total energy at any point a vertical distance y above the base of the incline is

$$\tfrac{1}{2}Mv^2 + \tfrac{1}{2}I_{cm}\omega^2 + Mgy.$$

We equate the total energy at the top ($y = H$ and $v = \omega = 0$) to the total energy at the bottom ($y = 0$):

$$0 + 0 + MgH = \tfrac{1}{2}Mv^2 + \tfrac{1}{2}I_{cm}\omega^2 + 0.$$

From Fig. 9–13, the moment of inertia of a sphere about an axis through its cm is $I_{cm} = \tfrac{2}{5}Mr_0^2$. Since the sphere rolls without slipping, the speed, v, of the center of mass with respect to the point of contact (which is momentarily at rest at any instant) is equal to the speed of a point on the edge relative to the center; we therefore have $\omega = v/r_0$. Hence

$$\tfrac{1}{2}Mv^2 + \tfrac{1}{2}(\tfrac{2}{5}Mr_0^2)\left(\frac{v^2}{r_0^2}\right) = MgH.$$

Canceling the M's and r_0's, we obtain

$$(\tfrac{1}{2} + \tfrac{1}{5})v^2 = gH$$

or

$$v = \sqrt{\tfrac{10}{7}gH}.$$

This can be compared to the velocity of an object that slides down the same plane without rotating and without friction, in which case $v = \sqrt{2gH}$, which is greater.

If there were no friction between the sphere and plane in this example, the sphere would have slid instead of rolled. Friction must be present to make the sphere roll. We did not need to take it into account in the energy equation because it is static friction and does no work. For if we assume the sphere is in contact with the ground at a point, then the force of friction acts parallel to the plane as shown in Fig. 9–26; and the point of contact of the sphere at each instant does not slide but moves perpendicular to the plane (upward) as the sphere rolls. Thus, no work is done since the force and the motion are perpendicular. (The situation is more complicated if we take into account retarding forces due to elastic deformation, as in Section 9–7.) If we wish to examine friction and other forces involved, then we must use dynamical methods as in the next example.

EXAMPLE 9–18 Analyze the rolling sphere of Example 9–17 and Fig. 9–25 in terms of forces and torques; in particular, find the velocity v and the magnitude of the friction force, F_{fr}, Fig. 9–26.

SOLUTION We analyze the motion as translation of the cm plus rotation about the cm. For translation in the x direction we have from $F = ma$:

$$Mg \sin \theta - F_{fr} = Ma$$

and in the y direction

$$F_N - Mg \cos \theta = 0$$

since there is no acceleration perpendicular to the plane. This last equation merely tells us

$$F_N = Mg \cos \theta.$$

For the rotational motion about the cm, we use $\tau = I\alpha$;

$$F_{fr}r_0 = (\tfrac{2}{5}Mr_0^2)\alpha;$$

since the other forces, $\mathbf{F}_N$ and $\mathbf{Mg}$ have lever arms equal to zero they do not appear here. As we saw in Example 9–17, $\omega = v/r_0$ where v is the speed of the cm. Taking derivatives we have $\alpha = a/r_0$ and substituting this into the last equation we find

$$F_{fr} = \tfrac{2}{5}Ma.$$

When we substitute this into the top equation, we get:

$$Mg \sin \theta - \tfrac{2}{5}Ma = Ma,$$

or

$$a = \tfrac{5}{7}g \sin \theta.$$

We thus see that the acceleration of the cm of a rolling sphere is less than that for a (frictionless) sliding object ($a = g \sin \theta$). To find the speed v at the bottom we use Eq. 2–9c where the total distance traveled along the plane is $x = H/\sin \theta$

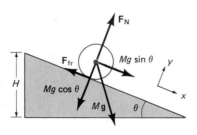

FIGURE 9–26
Example 9–18.

where H is the height of the plane (see Fig. 9–26); thus

$$v = \sqrt{2ax} = \sqrt{2\left(\frac{5}{7} g \sin \theta\right)\left(\frac{H}{\sin \theta}\right)} = \sqrt{\frac{10}{7} gH}.$$

This is the same result obtained in Example 9–17, although less effort was needed there. To get the magnitude of the force of friction, we use equations obtained above:

$$F_{\text{fr}} = \tfrac{2}{5}Ma = \tfrac{2}{5}M(\tfrac{5}{7}g \sin \theta) = \tfrac{2}{7}Mg \sin \theta.$$

If the coefficient of static friction is sufficiently small, or θ sufficiently large so that $F_{\text{fr}} > \mu_s F_N$ (that is, $\tan \theta > \tfrac{7}{2}\mu_s$), the sphere will not simply roll but will slip as it moves down the plane.

Summary

When a rigid body rotates about a fixed axis, each point of the body moves in a circular path. Lines drawn perpendicularly from the rotation axis to various points in the body all sweep out the same angle θ in any given time interval. Angles are conveniently measured in *radians* where one radian is the angle subtended by an arc whose length is equal to the radius.

Angular velocity, ω, is defined as the rate of change of angular position:

$$\omega = \frac{d\theta}{dt}.$$

All parts of a rigid body rotating about a fixed axis have the same angular velocity at any instant. *Angular acceleration*, α, is defined as the rate of change of angular velocity:

$$\alpha = \frac{d\omega}{dt}.$$

The linear velocity and acceleration of any point in a body rotating about a fixed axis are related to the angular velocity and acceleration by

$$v = R\omega, \quad a_T = R\alpha, \quad a_c = \frac{v^2}{R} = \omega^2 R$$

when R is the perpendicular distance of the point from the rotation axis and a_T and a_c are the tangential and centripetal accelerations. If a rigid body undergoes uniformly accelerated rotational motion ($\alpha = $ constant), equations analogous to those for linear motion are valid.

Although the angle θ is not a vector, both angular velocity and angular acceleration are vectors; for a rigid body rotating about a fixed axis, both $\boldsymbol{\omega}$ and $\boldsymbol{\alpha}$ point along the rotation axis, in the direction given by the right-hand rule.

The *torque* due to a force $\mathbf{F}$ exerted on a rigid body is equal to

$$\tau = R_\perp F = RF_\perp = RF \sin \theta$$

where $R_\perp$, called the *lever arm*, is the perpendicular distance from the axis of rotation to the line along which the force acts, and θ is the angle between $\mathbf{F}$ and $\mathbf{R}$.

The rotational equivalent of Newton's second law is

$$\tau = I\alpha$$

where $I = \sum m_i R_i^2$ is the *moment of inertia* of the body about the axis of rotation; this relation is valid for a rigid body rotating about an axis fixed in an inertial reference frame, or when τ, I, and α are calculated about the center of mass of a body even if the cm is moving.

The *angular momentum*, L, of a body about a fixed rotation axis is given by

$$L = I\omega.$$

Newton's second law, in terms of angular momentum, becomes

$$\tau = \frac{dL}{dt}.$$

If the net torque on the body is zero, $dL/dt = 0$, so $L = $ constant. This is the *law of conservation of angular momentum* for a rotating body.

The *rotational kinetic energy* of a body rotating about a fixed axis with angular velocity ω is

$$\text{KE} = \tfrac{1}{2}I\omega^2.$$

For a body both translating and rotating, the total kinetic energy is the sum of the translational KE of the body's cm plus the rotational KE of the body about its cm:

$$\text{KE} = \tfrac{1}{2}Mv_{\text{cm}}^2 + \tfrac{1}{2}I_{\text{cm}}\omega^2$$

as long as the rotation axis is fixed in direction.

1. You are standing a known distance from the Statue of Liberty. Describe how you could determine its height using only a meter stick, and without moving from your place.

2. A bicycle odometer (which measures distance traveled) is attached near the wheel hub and is designed for 27-in wheels. What happens if you use it on a bicycle with 24-in wheels?

3. Suppose a record turntable rotates at constant angular velocity. Does a point on the rim have radial and/or tangential acceleration? If the turntable accelerates uniformly, does the point have radial and/or tangential acceleration? For which cases would the magnitude of either (linear) acceleration change?

4. If the angular quantities θ, ω, and α were specified in terms of degrees rather than radians, how would Eqs. 9–7 for uniformly accelerated rotational motion have to be altered?

5. In what direction is the earth's angular velocity for its daily rotation on its axis?

6. The angular velocity of a wheel rotating on a horizontal axle points west. In what direction is the linear velocity of a point on the top of the wheel? If the angular acceleration points east, describe the tangential linear acceleration of this point. Is the angular speed increasing or decreasing?

7. Can a small force exert a greater torque than a larger force?

8. If a force **F** acts on a body such that its lever arm is zero, does it have any effect on the body's motion?

9. The moment of inertia of this textbook would be the least about what axis?

10. Two uniform wheels of the same mass and thickness are rotated about axes through their centers. If they are made of materials with different densities, which will have the greater moment of inertia?

11. What purpose does a walking stick serve when hiking in rough country? Be specific.

12. Why is it more difficult to do a situp with your hands behind your head than when they are outstretched in front of you? A diagram may help you to answer this.

13. Expert bicyclists use very lightweight "sew-up" tires. They claim that reducing the mass of the tires is far more significant than an equal reduction in mass elsewhere on the bicycle. Explain why this is true.

14. Mammals that depend on being able to run fast have slender lower legs with flesh and muscle concentrated high, near the shoulder. On the basis of rotational dynamics, explain why this distribution of mass is advantageous.

15. Why do tightrope walkers carry a long, narrow beam?

16. When a motorcyclist leaves the ground on a jump, if the throttle is left on (so the rear wheel spins) why does the front of the cycle rise up?

17. A quarterback leaps into the air to throw a forward pass. As he throws the ball, the upper part of his body rotates; if you look quickly you will notice that his hips and legs rotate in the opposite direction. Explain.

18. If there were a great migration of people toward the equator, how would this affect the length of the day?

19. A sphere, a cylinder, and a hoop, all of the same diameter, roll down an inclined plane. Which arrives at the bottom first? Which last?

20. Two inclines have the same height but make different angles with the horizontal. If the same steel ball is allowed to roll down each incline, on which one will it take the least time to reach the bottom? For which will the speed at the bottom be greatest? Explain. If the ball was made of a soft material and significant friction was present, how would this affect your responses?

21. Two spheres stand at rest at the top of an incline. One has twice the radius and twice the mass of the other. Which reaches the bottom of the incline first? Which has the greater speed there? Which has the greater total kinetic energy at the bottom?

22. A sphere and a cylinder have the same radius and same mass. They start from rest at the top of an incline. Which reaches the bottom first? Which has the greater speed at the bottom? Which has the greater kinetic energy at the bottom?

23. The total kinetic energy of a system of particles is equal to the translational kinetic energy of the center of mass of the system plus the KE of motion relative to the cm (Eq. 9–22). Is it possible or useful to make a similar statement about the total linear momentum?

 Problems _____

SECTION 9–1

1. (I) The Eiffel tower is 300 m tall. When you are standing at a certain place in Paris, it subtends an angle of 5°. How far are you then from the Eiffel Tower?

2. (I) A bicycle with 68-cm-diameter tires travels 2.0 km. How many revolutions do the wheels make?

3. (I) A 20-cm-diameter grinding wheel rotates at 2000 rpm (revolutions per minute). Calculate (a) its angular velocity in rad/s, and (b) the linear speed of a point on the edge of the grinding wheel.

4. (I) Estimate the angle subtended by the moon using a ruler and your finger or other object to just blot out the moon. Describe your measurement and the result obtained and then use it to estimate the diameter of the moon. The moon is about 380,000 km from the earth.

5. (I) A large spool of rope lies on the ground with the end of the rope lying on the top edge of the spool. A person grabs the end of the rope and walks a distance L holding onto it. The spool rolls behind him without slipping. What length of rope unwinds from the spool? How far does the spool's cm move?

6. (I) Calculate the angular velocity of the earth (*a*) in its orbit around the sun, and (*b*) about its axis.

7. (I) What is the linear speed of a point (*a*) on the equator, (*b*) at a latitude of 50°N, due to the earth's rotation? (See previous problem.)

8. (I) A record player turntable of radius R_1 is turned by a circular rubber roller of radius R_2 in contact with it at their outer edges. What is the ratio of their angular velocities, ω_1/ω_2?

9. (II) The angle through which a rotating wheel has turned in time *t* is given by $\theta = 5.0t + 3.0t^2 - 4.5t^4$, where θ is in radians and *t* in seconds. (*a*) What is the average angular velocity, and (*b*) the average angular acceleration, between $t = 2.0$ s and $t = 3.0$ s? (*c*) Determine an expression for the instantaneous angular velocity ω, and (*d*) for the instantaneous angular acceleration α. (*e*) Evaluate ω and α at $t = 3.0$ s.

10. (II) The angular acceleration of a wheel, as a function of time, is $\alpha = 8.0t - 2.5t^2$, where α is in radians per second and *t* in seconds. If the wheel starts from rest ($\theta = \omega = 0$ at $t = 0$) determine a formula for (*a*) the angular velocity ω, and (*b*) the angular position θ, as a function of time. (*c*) Evaluate ω and θ at $t = 2.0$ s.

11. (II) A 40-cm-diameter wheel accelerates uniformly from 80 rpm to 300 rpm in 3.6 s. Determine (*a*) its angular acceleration and (*b*) the radial and tangential components of the linear acceleration of a point on the edge of the wheel 2.0 s after it started accelerating.

SECTION 9–2

12. (I) An automobile engine slows down from 4500 rpm to 1000 rpm in 6.5 s. Calculate (*a*) its angular acceleration, assumed uniform, and (*b*) the total number of revolutions the engine makes in this time.

13. (I) Derive Eq. 9–7c.

14. (I) A phonograph turntable reaches its rated speed of 33 rpm after making 1.5 revolutions. What was its angular acceleration?

15. (II) Two rubber wheels are mounted next to one another so their circular edges touch. Wheel 1, of radius $R_1 = 3.0$ cm, accelerates at a rate 0.88 rad/s² and drives the second wheel, of radius $R_2 = 5.0$ cm, by contact (without slipping). (*a*) Starting from rest, how long does it take the second wheel to reach an angular speed of 33 rpm? (*b*) What was the angular acceleration of wheel 2?

16. (II) A wheel, starting from rest, undergoes uniform angular acceleration α about its fixed axle. (*a*) Write the components of the linear acceleration, a_T and a_c, for a point *P* which is a distance *R* from the axle in terms of α, *R* and time *t*. (*b*) Let ϕ be the angle between the linear acceleration vector, **a**, and the line drawn between *P* and the axis. Express ϕ in terms of the total number of revolutions of the wheel, *N*.

17. (II) The tires of a car make 55 revolutions as the car reduces its speed uniformly from 80 km/h to 55 km/h. The tires have a diameter of 1.0 m. (*a*) What was the angular acceleration? (*b*) If the car continues to decelerate at this rate, how much more time is required for it to stop?

18. (II) A cyclist accelerates uniformly from rest to a speed of 12 m/s in 10.0 s. How fast will a point on the rim of the tire (diameter = 68 cm) at the top be moving after 4.0 seconds? [Hint: at any moment, the lowest point on the tire is in contact with the ground, and hence is at rest.]

FIGURE 9–27

SECTION 9–3

19. (II) The axle of a wheel is mounted on supports that rest on a rotating turntable as shown in Fig. 9–27. The wheel has angular velocity $\omega_1 = 60$ rad/s about its axle, and the turntable has angular velocity $\omega_2 = 45$ rad/s about a vertical axis. (Note arrows showing these motions in the figure.) (*a*) What are the directions of ω_1 and ω_2? (*b*) What is the resultant angular velocity of the wheel, as seen by an outside observer, at the instant shown. Give magnitude and direction. (*c*) What is the magnitude and direction of the angular acceleration at the instant shown?

SECTION 9–4

20. (I) What is the maximum torque exerted by a 60-kg person riding a bicycle if all the person's weight is put on each pedal when climbing a hill? The pedals rotate in a circle of radius 18 cm.

21. (II) A wheel of diameter 24.0 units is constrained to rotate in the *xy* plane, about the *z* axis, which passes through its center. A force $\mathbf{F} = -41.2\mathbf{i} + 30.6\mathbf{j}$ acts at a point on the edge of the wheel which lies exactly on the *x* axis at a particular instant. What is the torque about the rotation axis at this instant?

SECTION 9–5

22. (I) A 4-kg grindstone in the shape of a uniform cylinder of radius 0.10 m acquires a rotational rate of 1800 rev/s from rest over a 5.0-s interval at constant angular acceleration. Calculate the torque delivered by the motor.

23. (I) Four equal masses (*M*) are located in a plane at the four corners of a square of side *s*. What is the moment of inertia of the system about an axis through one of the masses perpendicular to the plane?

24. (I) A person exerts a force of 18 N on the end of a door 84 cm wide. What is the magnitude of the torque if the force is exerted (*a*) perpendicular to the door, and (*b*) at a 60° angle to the face of the door?

25. (I) Calculate the moment of inertia of a 10-kg solid sphere of radius 0.20 m when the axis of rotation is through its center.

26. (I) A 200-kg solid cylindrical wheel 12.0 m in diameter is to start from rest and reach a speed of 200 rpm after 3.0 minutes. Calculate (*a*) the angular acceleration, (*b*) the force needed if applied by a rope wrapped around the wheel's rim.

27. (II) A 2.4-kg ball on the end of a light rod is rotated in a horizontal circle of radius 1.2 m. Calculate: (*a*) the moment of inertia of the ball, and (*b*) the torque needed to keep the ball rotating at constant angular velocity if air resistance exerts a force of 0.020 N on the ball.

28. (II) A grinding wheel is a uniform cylinder of radius 8.2 cm and mass 0.88 kg. Calculate: (a) its moment of inertia, and (b) the torque needed to accelerate it from rest to 1200 rpm in 4.0 s if a frictional torque of 0.014 N · m is also acting.

29. (II) What is the radius of gyration of a 13.6-kg wheel that accelerates from rest to 600 rpm in 10 s when a net torque of 3.2 N·m acts on it.

30. (II) A 0.84-m-diameter solid sphere can be rotated about an axis through its center by a torque of 12.3 N·m which accelerates it uniformly from rest through a total of 180 revolutions in 15.0 s. What is the mass of the sphere?

31. (II) A merry-go-round accelerates from rest to 3.0 rad/s in 34 s. Assuming the merry-go-round is a uniform disc of radius 8.0 m and mass 31,000 kg, calculate the net torque required to accelerate it.

32. (II) Four equal masses (M) are spaced at equal intervals (L) along a vertical straight rod whose mass can be ignored. The system is to be rotated about an axis passing through the mass at the bottom end of the rod and perpendicular to it. (a) What is the moment of inertia of the system about this axis? (b) What minimum force, applied to the farthest mass, will impart on angular acceleration α? (c) What is the direction of this force?

33. (II) A wheel of mass M has radius R_0. It is standing vertically on the floor and we want to exert a horizontal force F at its axle so that it will climb a step against which it rests. The step has height h where $h < R_0$. What minimum force F is needed?

34. (II) Suppose the force T in the cord hanging from the wheel of Example 9–9, Fig. 9–14, is given by the relation $T = 3.00t - 0.20t^2$ (newtons) where t is in seconds. If the wheel starts from rest, what is the linear speed of a point on its rim 8.0 s later?

35. (II) A centrifuge rotor rotating at 10,000 rpm is shut off and is eventually brought to rest by a frictional torque of 0.20 N·m. If the mass of the rotor is 4.3 kg and its radius of gyration is 0.070 m, through how many revolutions will the rotor turn before coming to rest and how long will it take?

36. (II) The forearm in Fig. 9–28 accelerates a 10-kg ball at 8.0 m/s² by means of the triceps muscle as shown. Calculate: (a) the torque needed, and (b) the force that must be exerted by the triceps muscle. Ignore the mass of the arm.

37. (III) A string passing over a pulley has a 3.20-kg mass attached to one end and a 3.40-kg mass attached to the other end. The pulley is a uniform solid cylinder of radius 3.0 cm and mass 0.80 kg. (a) If the pulley were frictionless, what would be the acceleration of the two masses? (b) In fact, it is found that if the heavier mass is given a downward speed of 0.20 m/s, it comes to rest in 6.2 s. What is the average frictional torque acting on the pulley?

38. (III) A thin rod of length L stands vertically on a table. The rod begins to fall, but its lower end does not slide. (a) Determine the angular velocity of the rod as a function of the angle ϕ it makes with the table top. (b) What is the speed of the tip of the rod just before it strikes the table?

39. (III) Assume that a 1.0-kg ball is thrown solely by the action of the forearm which rotates about the elbow joint under the action of the triceps muscle, Fig. 9–28. The ball is accelerated from rest to 10.0 m/s in 0.22 s at which point it is released. Calculate: (a) the angular acceleration of the arm, and (b) the force required of the triceps muscle. Assume the forearm has a mass of 3.2 kg and rotates like a uniform rod about an axis at its end.

40. (III) A hammer thrower accelerates the hammer (mass = 7.3 kg) from rest within four full turns (revolutions) and releases it at a speed of 27.2 m/s. Assuming a uniform rate of increase in angular velocity and a radius of 2.0 m, calculate: (a) the angular acceleration, (b) the (linear) tangential acceleration, (c) the centripetal acceleration just before release, (d) the net force being exerted on the hammer by the athlete just before release, and (e) the angle of this force with respect to the radius of the circular motion.

*SECTION 9–6

41. (I) Calculate the moment of inertia of a 67-cm-diameter bicycle wheel. The rim plus tire have a mass of 1.3 kg. Why can the mass of the hub be ignored?

*42. (I) An oxygen molecule consists of two oxygen atoms whose total mass is 5.3×10^{-26} kg and whose moment of inertia about an axis perpendicular to the line joining them at its center is 1.9×10^{-46} kg·m². Estimate, from these data, the effective distance between the two atoms.

*43. (I) Use the parallel-axis theorem to show that the moment of inertia of a thin rod about an axis perpendicular to the rod at one end is $I = \frac{1}{3}ML^2$, given that if the axis passes through the center, $I = \frac{1}{12}ML^2$ (Fig. 9–13f and g).

*44. (II) Use the perpendicular-axis theorem, and Fig. 9–13h, to determine a formula for the moment of inertia of a thin square plate of side s about an axis through its center (a) along a diagonal of the plate, (b) parallel to a side.

*45. (II) Determine a formula for the moment of inertia of a thin hopp of radius R_0 and mass M about an axis (a) passing through its center in the plane of the circle, (b) tangent to its circular outline.

*46. (II) Two uniform solid spheres of mass M and radius r_0 are connected by a thin (massless) rod of length r_0 so the centers are $3r_0$ apart. (a) Determine the moment of inertia of this system about an axis perpendicular to the rod at its center. (b) What would be the percentage error if the masses of each sphere were assumed to be concentrated at their centers, so a very simple calculation was made?

*47. (II) The flat sides of two uniform solid cylindrical wheels of radii R_1 and R_2 are placed next to each other with their centers superposed. They are of equal thickness and are made of the same density material. Find the moment of

FIGURE 9–28

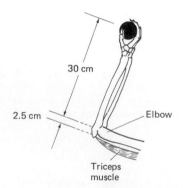

30 cm

2.5 cm

Elbow

Triceps muscle

inertia of the system in terms of R_1, R_2, and M, the total mass of the system, as calculated about an axis passing through their centers perpendicular to the faces of the wheels.

*48. (II) A thin 3.0-kg wheel of radius 26 cm is weighted to one side by a 0.50 kg weight, small in size, placed 22 cm from the center of the wheel. Calculate (a) the position of the center of mass of the weighted wheel, and (b) the moment of inertia about an axis through its center, perpendicular to its face.

*49. (III) Derive the formula for the moment of inertia of a uniform thin rod of length L about an axis through its center, perpendicular to the rod, Fig. 9–13f.

*50. (III) (a) Derive the formula for the moment of inertia of a uniform flat rectangular plate of dimensions $l \times w$, Fig. 9–13h, about an axis through the center, perpendicular to the plate. (b) What is the moment of inertia about each of the axes through the center that are parallel to the edges of the plate?

*51. (III) Determine the moment of inertia of a uniform cube of side s about an axis through its center of mass perpendicular to a face of the cube.

*52. (III) The density of a thin rod of length L increases uniformly from ρ_0 at one end to $2\rho_0$ at the other end. Determine the moment of inertia about an axis perpendicular to the rod through its geometric center.

SECTION 9–8
53. (I) (a) What is the angular momentum of a 2.3-kg uniform cylindrical grinding wheel of radius 12 cm when rotating at 1500 rpm? (b) How much torque is required to stop it in 7.0 s?

54. (I) Determine the angular momentum of the earth (a) about its rotation axis (assume earth is uniform sphere), (b) in its orbit around the sun.

55. (I) A person stands, hands at the side, on a platform that is rotating at a rate of 1.20 rev/s. If the person now raises her arms, the speed of rotation decreases to 0.80 rev/s. (a) Why does this occur? (b) By what factor has the moment of inertia of the person changed?

56. (I) A diver (such as that shown in Fig. 9–21) can reduce his moment of inertia by a factor of about 3.5 when changing from the straight position to the tuck position. If he makes two rotations in 1.5 s when in the tuck position, what is his angular speed (rev/s) when in the straight position?

57. (II) A 4.5-m-diameter merry-go-round is rotating freely with an angular velocity of 0.70 rad/s; its total moment of inertia is 1750 kg·m². Four people standing on the ground, each of 65-kg mass, suddenly step onto the edge of the merry-go-round. What will be the angular velocity of the merry-go-round now? What if the people were on it initially, and then jumped off?

58. (II) Suppose a 60-kg person stands at the edge of a 6.0-m-diameter merry-go-round turntable which is mounted on frictionless bearings and has a moment of inertia of 1800 kg·m². The turntable is at rest initially, but when the person begins running at a speed of 4.2 m/s (with respect to the turntable) around its edge, the turntable begins to rotate in the opposite direction. Calculate the angular velocity of the turntable.

59. (II) A toy train can run on a horizontal circular track of radius R mounted near the edge of a solid circular board of uniform thickness with mass M that is free to rotate, without friction, about a vertical axis through its center. If the system starts from rest, what mass should the train have so that it remains in the same position relative to the ground no matter how fast it travels relative to the tracks?

SECTION 9–9
60. (I) Calculate the translational speed of a cylinder when it reaches the foot of an incline 18 m high. Assume it starts from rest and rolls without slipping.

61. (I) A centrifuge rotor has a moment of inertia of 4.0×10^{-2} kg·m². How much energy is required to bring it from rest to 10,000 rpm?

62. (I) A merry-go-round has a mass of 1500 kg and a radius of gyration of 18 m. How much work is required to accelerate it from rest to a rotation rate of one revolution in 7.0 s?

63. (I) An automobile engine develops a torque of 280 N·m at 4,000 rpm. What is the horsepower of the engine?

64. (I) Estimate the kinetic energy of the earth with respect to the sun as the sum of two terms, (a) that due to its daily rotation about its axis and (b) that due to its yearly revolution about the sun (assume the earth is a uniform sphere, mass $= 6.0 \times 10^{24}$ kg, radius $= 6.4 \times 10^6$ m, and is 1.5×10^8 km from the sun).

65. (II) A rotating uniform cylindrical platform of mass 180 kg and radius 4.5 m slows down from 3.2 rev/s to rest in 18 s when the driving motor is disconnected. Estimate the power output of the motor (hp) to maintain a steady speed of 3.2 rev/s?

66. (II) Two masses, $m_1 = 25$ kg and $m_2 = 32$ kg, are connected by a rope that hangs over a pulley (as in Fig. 9–29). The pulley is a uniform cylinder of radius 0.30 m and mass 8.8 kg. Initially m_1 is on the ground and m_2 rests 2.5 m above the ground. If the system is now released, use conservation of energy to determine the speed of m_2 just before it strikes the ground. Assume the pulley is frictionless.

67. (II) A 5.0-m-long pole is balanced vertically on its tip. What will be the speed of the tip of the pole just before it hits the ground? Assume the lower end of the pole does not slip.

68. (II) A uniform thin rod of length L and mass M is suspended freely from one end. It is pulled to the side by an angle θ and released. If friction can be ignored, what will be its angular velocity, and the speed of its free end, at the lowest point?

FIGURE 9–29

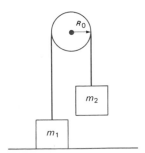

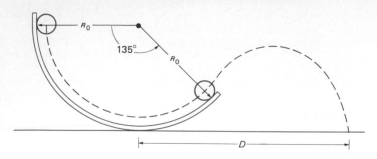

FIGURE 9–30

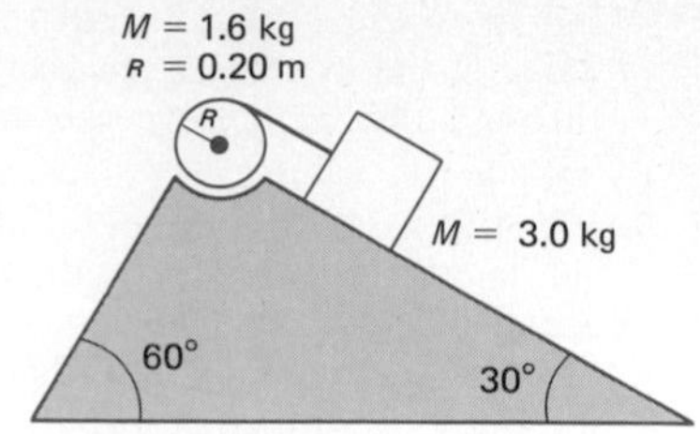

$M = 1.6$ kg
$R = 0.20$ m

$M = 3.0$ kg

FIGURE 9–31

69. (II) One possibility for a low-pollution automobile is for it to use energy stored in a heavy rotating flywheel. Suppose such a car has a total mass of 1400 kg, uses a 1.50-m diameter uniform cylindrical flywheel of mass 240 kg, and should be able to travel 300 km without needing a flywheel "spinup." (a) Make reasonable assumptions (average frictional retarding force = 500 N, twenty acceleration periods from rest to 90 km/h, equal uphill and downhill—assuming during downhill, energy can be put back into the flywheel), and show that the total energy needed to be stored in the flywheel is about 1.6×10^8 J. (b) What is the angular velocity of the flywheel when it has a full "energy charge"? (c) About how long would it take a 150 hp motor to give the flywheel a full energy charge before a trip?

70. (II) A narrow but solid spool of thread has radius R and mass M. If you pull up on the thread so that the cm of the spool remains suspended in the air at the same place, (a) what force must you exert on the thread? (b) How much work have you done by the time the spool turns with angular velocity ω?

71. (II) A hollow cylinder (hoop) is rolling on a horizontal surface at speed $v = 3.4$ m/s when it reaches a 20° incline. (a) How far up the incline will it go? (b) How long will it be on the incline before it arrives back at the bottom?

72. (II) Show that the sphere in Example 9–18 will slip if the angle of the incline exceeds $\theta = \tan^{-1}(7\mu_s/2)$ where μ_s is the coefficient of static friction.

73. (II) (a) In Example 9–16, determine the work a person must do to reduce the radius of the circular path from R_1 to R_2. (This requires an integral.) (b) Determine the change in kinetic energy of the mass m during this process and see if it is equal to the work done as calculated in part (a).

74. (II) A 70-kg person stands on a tiny rotating platform with arms outstretched. (a) Estimate the moment of inertia of the person using the following approximation: the body (including head and legs) is a 60-kg cylinder 12 cm in radius and 1.70 m high, and each arm is a 5.0-kg thin rod, 60 cm long, attached to the cyclinder. (b) Using the same approximation, estimate the moment of inertia when the arms are at the person's sides. (c) If one rotation takes 1.5 s when the person's arms are outstretched, how fast is the rotation with arms at the sides? Ignore the moment of inertia of the platform. (d) Determine the change in kinetic energy when the arms are lifted from the sides to the horizontal position. (e) From your answer to part (d), would you expect it to be harder or easier to lift your arms when rotating or when at rest?

75. (II) Suppose the person in problem 74 holds a 15-kg weight in each hand. What will be the ratio of the angular velocity with arms outstretched to that when arms and hands are at the sides, next to the body?

76. (II) A small sphere of radius $r = 5.0$ cm rolls without slipping on the track shown in Fig. 9–30 whose radius $R_0 = 30.0$ cm. It starts rolling at a height R_0 above the bottom of the track. When it leaves the track after passing through an angle of 135° as shown, at what distance D from the base of the track will the sphere hit the ground?

77. (III) In Examples 9–17 and 9–18 suppose the sphere does not simply roll, but slips as well. To be concrete, suppose $\theta = 30°$ and $\mu = \mu_s = \mu_k = 0.10$ with the incline having a height $H = 2.0$ m and the sphere a radius $r_0 = 10$ cm and mass $M = 850$ g. The sphere starts from rest at the top. Determine (a) the speed of the cm of the sphere when it reaches the foot of the incline; (b) the total kinetic energy of the sphere at the foot; and (c) the loss in mechanical energy. (d) Determine these same quantities for $\mu = 0.30$, for which there is no slipping, and compare.

78. (III) A cord connected at one end to a block which can slide on an inclined plane has its other end wrapped around a cylinder resting in a depression at the top of the plane as shown in Fig. 9–31. Determine the speed of the block after it has traveled 1.20 m along the plane, starting from rest. Assume (a) there is no friction, (b) the coefficient of friction between all surfaces is $\mu = 0.15$. (Hint: in part (b) first determine the normal force on the cylinder, and make any reasonable assumptions needed.)

General
Rotation

10

In Chapter 9 we dealt with the kinematics and dynamics of the rotation of a rigid body about an axis fixed in an inertial reference frame. We analyzed the motion in terms of the rotational equivalent of Newton's laws (where torque plays the role that force does for translational motion) as well as in terms of angular momentum and rotational kinetic energy.

To keep the axis of a rotating body fixed, the body must usually be constrained by external supports (such as bearings at the end of an axle). The motion of bodies that are not constrained to move about a fixed axis is more difficult to describe and analyze. Indeed, the complete analysis of the general rotational motion of a body (or system of bodies) in space is too long and complicated to be discussed in this book. Nonetheless, we will look at some aspects of general rotational motion in this chapter. We will prove some general theorems and apply them to some interesting types of motion that are not too complicated. In particular, we will deal with the vector nature of torque and angular momentum.

Much of this chapter may seem like a repeat of material in Chapter 9. But, as a careful reading will reveal, it is a more general and useful treatment of rotational motion, making use of the vector nature of rotational quantities, and deriving many assertions of Chapter 9. The material of this chapter, while important, is perhaps a little more complicated than most of the rest of this book. Except for the very important first section on the vector cross product, the rest of the book does not depend on the material in this chapter, and so it could be considered optional.

 ## 10–1 Vector Cross Product

To deal with the vector nature of angular momentum and torque in general, we will need the concept of the *vector cross product* (often called simply the *vector product* or *cross product*).

In general, the **vector** or **cross product** of two vectors, **A** and **B**, is defined as another vector $\mathbf{C} = \mathbf{A} \times \mathbf{B}$ *whose magnitude is*

$$C = |\mathbf{A} \times \mathbf{B}| = AB \sin \theta \qquad (10\text{–}1)$$

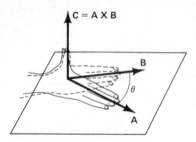

FIGURE 10–1

The vector $\mathbf{C} = \mathbf{A} \times \mathbf{B}$ is perpendicular to the plane containing $\mathbf{A}$ and $\mathbf{B}$; its direction is given by the right-hand rule.

where θ is the angle ($<180°$) between $\mathbf{A}$ and $\mathbf{B}$, and whose direction is perpendicular to both $\mathbf{A}$ and $\mathbf{B}$ in the sense of the right-hand rule, Fig. 10–1. According to the right-hand rule, as shown in the figure, you orient your hand so your fingers point along $\mathbf{A}$, and when you bend your fingers they point along $\mathbf{B}$. When your hand is correctly oriented in this way, your thumb will point along the direction of $\mathbf{C} = \mathbf{A} \times \mathbf{B}$.

The cross product of two vectors, as defined above (Eq. 10–1), is the second useful way to define the multiplication of two vectors. The first way, discussed in Section 6–2, is the scalar or dot product: $\mathbf{A} \cdot \mathbf{B} = AB \cos \theta$. Although the product of two vectors can be defined in other ways, these two ways have great practical value in physics since many quantities can be expressed as either a scalar or vector product of two other vector quantities.

Some properties of the cross product are the following:

$$\mathbf{A} \times \mathbf{A} = 0 \qquad (10-2a)$$

$$\mathbf{A} \times \mathbf{B} = -\mathbf{B} \times \mathbf{A} \qquad (10-2b)$$

$$\mathbf{A} \times (\mathbf{B} + \mathbf{C}) = \mathbf{A} \times \mathbf{B} + (\mathbf{A} \times \mathbf{C}) \qquad [\text{distributive law}] \qquad (10-2c)$$

$$\frac{d}{dt}(\mathbf{A} \times \mathbf{B}) = \frac{d\mathbf{A}}{dt} \times \mathbf{B} + \mathbf{A} \times \frac{d\mathbf{B}}{dt}. \qquad (10-2d)$$

Equation 10–2a follows from Eq. 10–1 (since $\theta = 0$); so does Eq. 10–2b, since the magnitude of $\mathbf{B} \times \mathbf{A}$ is the same as that for $\mathbf{A} \times \mathbf{B}$, but by the right-hand rule the direction is opposite. Thus the order of the two vectors is crucial; if you change the order, you change the result. That is, the commutative law does *not* hold for the cross product, although it does for the dot product of two vectors and for the product of scalars. The proofs for Eqs. 10–2c and d are left as problems. Note in Eq. 10–2d that the order of quantities in the two products on the right must not be changed (because of Eq. 10–2b).

10–2 The Torque Vector

Torque is an example of a cross product. To see this, let us take a simple example: the thin wheel shown in Fig. 10–2 which is free to rotate about an axis through its center at point O. A force $\mathbf{F}$ acts at the edge of the wheel, at a point whose position relative to the center O is given by the position vector $\mathbf{r}$ as shown. The force $\mathbf{F}$ tends to rotate the wheel (assumed initially at rest) counterclockwise, so the angular velocity ω will point out of the page toward the viewer (remember the right-hand rule from Section 9–3). The torque due to $\mathbf{F}$ will tend to increase ω so $\boldsymbol{\alpha}$ also points outward along the rotation axis. The relation between angular acceleration and torque we developed in Chapter 9 for a body rotating about a fixed axis is

$$\tau = \frac{dL}{dt} = \frac{d(I\omega)}{dt} = I\alpha,$$

FIGURE 10–2

The torque due to $\mathbf{F}$ starts the wheel rotating counterclockwise so ω and $\boldsymbol{\alpha}$ point out of the page.

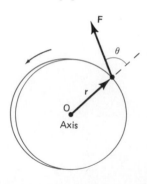

where L is the angular momentum and I is the moment of inertia (Eqs. 9–18 and 9–12). This scalar equation is the rotational equivalent of $F = dp/dt$, and we would like to make it a vector equation just as $\mathbf{F} = d\mathbf{p}/dt$ is a vector equation. To do so in the case of Fig. 10–2 we must have the direction of $\boldsymbol{\tau}$ point outward along the rotation axis, since $\boldsymbol{\alpha}$ ($= d\boldsymbol{\omega}/dt$) does, and the magnitude of the torque must be (see Eq. 9–8 and Fig. 10–2) $\tau = rF_\perp = rF \sin \theta$. We can achieve this by defining the torque vector to be the cross product of $\mathbf{r}$ and $\mathbf{F}$:

$$\boldsymbol{\tau} = \mathbf{r} \times \mathbf{F}. \qquad (10-3)$$

From the definition of the cross product above (Eq. 10–1) the magnitude of $\boldsymbol{\tau}$ will be $rF \sin \theta$ and the direction will be along the axis, as required for this special case.

We will see in Sections 10–3 through 10–6 that if we take Eq. 10–3 as the general definition of torque, then the vector relation $\boldsymbol{\tau} = d\mathbf{L}/dt$ will hold in general. Thus we state now that Eq. 10–3 is the general definition of torque. It contains both magnitude and direction information. It is important to note that this definition involves the position vector **r** (relative to the origin of a coordinate system); it does *not* involve R (the perpendicular distance from the rotation axis) which is used for determining moment of inertia. In other words, torque is calculated about a point.

For a particle of mass m on which a force **F** is applied, the torque about a point O is

$$\tau = \mathbf{r} \times \mathbf{F}$$

where **r** is the position vector of the particle relative to O (Fig. 10–3). If we have a system of particles (which could be the particles making up a rigid body) the total torque τ on the system will be the sum of the torques on the individual particles:

$$\tau = \sum \mathbf{r}_i \times \mathbf{F}_i$$

where $\mathbf{r}_i$ is the position vector of the ith particle and $\mathbf{F}_i$ is the net force on the ith particle.

 ## 10–3 Angular Momentum of a Particle

The most general way of writing Newton's second law for the translational motion of a particle (or systems of particles) is in terms of the linear momentum $\mathbf{p} = m\mathbf{v}$ as given by Eq. 8–9 (or 8–12)

$$\mathbf{F} = \frac{d\mathbf{p}}{dt}.$$

The rotational analog of linear momentum is *angular momentum*. Just as **p** is related to the force **F**, so we might expect the angular momentum to be related to the torque. Indeed, we saw this was true in Chapter 9 for the special case of a rigid body rotating about a fixed axis. Now we will see it is true in general. We first treat a single particle.

Suppose a particle of mass m has momentum **p** and position vector **r** with respect to the origin O in some chosen inertial reference frame. Then the angular momentum l of the particle about point O is defined as the vector cross product of **r** and **p**:

$$l = \mathbf{r} \times \mathbf{p}. \qquad \text{[particle]} \qquad (10\text{–}4)$$

Angular momentum is a vector.[†] Its direction is perpendicular to both **r** and **p** as given by the right-hand rule (Fig. 10–4); its magnitude is given by

$$l = rp \sin \theta$$

or

$$l = rp_\perp = r_\perp p$$

where θ is the angle between **r** and **p** and $p_\perp (= p \sin \theta)$ and $r_\perp = (r \sin \theta)$ are the components of **p** and **r** perpendicular to **r** and **p** respectively.

Now let us find the relation between angular momentum and torque for a particle. If we take the derivative of l with respect to time we have

$$\frac{dl}{dt} = \frac{d}{dt}(\mathbf{r} \times \mathbf{p}) = \frac{d\mathbf{r}}{dt} \times \mathbf{p} + \mathbf{r} \times \frac{d\mathbf{p}}{dt}.$$

But

$$\frac{d\mathbf{r}}{dt} \times \mathbf{p} = \mathbf{v} \times m\mathbf{v} = m(\mathbf{v} \times \mathbf{v}) = 0$$

[†] Actually a pseudovector; see footnote in Section 9–3. Also note that we are using small l for the angular momentum of a particle, and capital **L** for a collection of particles or an extended body (as in previous chapter and later in this chapter).

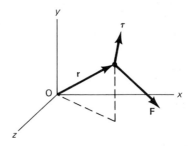

FIGURE 10–3

$\tau = \mathbf{r} \times \mathbf{F}$, where **r** is the position vector.

FIGURE 10–4

The angular momentum of a particle of mass m is given by $l = \mathbf{r} \times \mathbf{p} = \mathbf{r} \times m\mathbf{v}$.

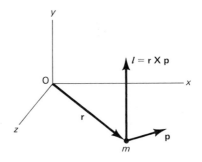

since $\sin\theta = 0$ for this case. Thus:

$$\frac{d\mathbf{l}}{dt} = \mathbf{r} \times \frac{d\mathbf{p}}{dt}.$$

If we let $\mathbf{F}$ represent the resultant force on the particle, then, in an inertial reference frame, $\mathbf{F} = d\mathbf{p}/dt$, and

$$\mathbf{r} \times \mathbf{F} = \mathbf{r} \times \frac{d\mathbf{p}}{dt} = \frac{d\mathbf{l}}{dt}.$$

But $\mathbf{r} \times \mathbf{F} = \tau$ is the net torque on our particle. Hence

$$\tau = \frac{d\mathbf{l}}{dt}. \qquad \text{[particle; inertial frame]} \qquad (10\text{-}5)$$

The time rate of change of angular momentum of a particle is equal to the net torque applied to it! Equation 10–5 is the rotational equivalent of Newton's second law for a particle, written in its most general form. It is valid only in an inertial frame since only then is it true that $\mathbf{F} = d\mathbf{p}/dt$ which was used in the proof.

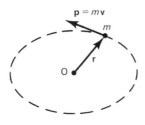

p = mv

FIGURE 10–5

The angular momentum of a particle of mass m rotating in a circle of radius **r** with velocity **v** is $\mathbf{l} = \mathbf{r} \times m\mathbf{v}$ (Example 10–1).

 EXAMPLE 10–1 Determine the angular momentum of a particle of mass m moving with speed v in a circle of radius r in a counterclockwise direction.

SOLUTION The value of the angular momentum depends on the choice of the point O. We calculate l with respect to the center of the circle (Fig. 10–5). Then $\mathbf{r}$ is perpendicular to $\mathbf{p}$ so $l = |\mathbf{r} \times \mathbf{p}| = rmv$. The direction of $\mathbf{l}$, by the right-hand rule, is perpendicular to the plane of the circle toward the viewer. Since $v = \omega r$, we can write

$$l = mvr = mr^2\omega = I\omega$$

since $I = mr^2$ for a single particle rotating about an axis a distance r away.

10–4 Angular Momentum and Torque for a System of Particles; General Motion

Consider a system of n particles which have angular momenta $l_1, l_2, \ldots l_n$. The system could be anything from a rigid body to a loose assembly of particles whose positions are not fixed relative to each other. The total angular momentum $\mathbf{L}$ of the system is defined as the vector sum of the angular momenta of all the particles in the system:

$$\mathbf{L} = \sum_{i=1}^{n} l_i. \qquad (10\text{-}6)$$

The resultant torque acting on the system is the sum of the net torques acting on all the particles:

$$\tau = \sum \tau_i.$$

This sum includes (1) internal torques due to internal forces that particles of the system exert on other particles of the system, and (2) external torques due to forces exerted by objects outside our system. By Newton's third law in its strong form (see footnote in Section 9–5), the force each particle exerts on another is equal and opposite and acts along the same line as the force the second particle exerts back on the first; hence the sum of all internal torques adds to zero, and

$$\tau = \sum_i \tau_i = \tau_{\text{ext}}.$$

Now we take the time derivative of Eq. 10–6 and use Eq. 10–5 for each particle to obtain

$$\frac{d\mathbf{L}}{dt} = \sum_i \frac{d\mathbf{l}_i}{dt} = \sum_i \boldsymbol{\tau}_i = \boldsymbol{\tau}$$

or

$$\frac{d\mathbf{L}}{dt} = \boldsymbol{\tau}. \qquad \text{[inertial reference frame]} \qquad (10\text{–}7a)$$

(We have dropped the subscript on τ_{ext} for convenience.) This fundamental result states that the time rate of change of the total angular momentum of a system of particles (or a rigid body) equals the resultant external torque on the system. It is the rotational equivalent of Eq. 8–12, $d\mathbf{P}/dt = \mathbf{F}_{\text{ext}}$ for translational motion. Equation 10–7a is valid when $\mathbf{L}$ and τ are calculated with reference to a point fixed in an inertial reference frame. (In the derivation, we used Eq. 10–5 which is valid only in this case.) It is also valid when τ and $\mathbf{L}$ are calculated about a point which is moving uniformly in an inertial reference frame since such a point can be considered the origin of a second inertial reference frame. It is *not* valid in general when τ and $\mathbf{L}$ are calculated about a point that is accelerating, except for one special (and very important) case—when that point is the center of mass of the system:

$$\frac{d\mathbf{L}_{\text{cm}}}{dt} = \boldsymbol{\tau}_{\text{cm}}. \qquad (10\text{–}7b)$$

Equation 10–7b is valid no matter how the cm moves, and τ_{cm} is the net external torque calculated about the cm.

It is because of the validity of Eq. 10–7b that we are justified in describing the general motion of a system of particles as the *translational* motion of the cm plus rotation about the cm.[†] We have used this principle previously (such as in Example 9–18). Now we prove Eq. 10–7b using Newton's laws in the next (optional) section.

 #### ∗ 10–5 Proof of General Relation Between τ and L

The proof of Eq. 10–7b is as follows. Let $\mathbf{r}_i$ be the position vector of the ith particle in an inertial reference frame, and $\mathbf{r}_{\text{cm}}$ be the position vector of the center of mass of the system in this reference frame. The position of the ith particle with respect to the cm is $\mathbf{r}_i^*$ where (see Fig. 10–6)

$$\mathbf{r}_i = \mathbf{r}_{\text{cm}} + \mathbf{r}_i^*.$$

When we take the derivative of this equation, we can write

$$\mathbf{p}_i = m_i \frac{d\mathbf{r}_i}{dt} = m_i \frac{d}{dt}(\mathbf{r}_i^* + \mathbf{r}_{\text{cm}}) = \mathbf{p}_i^* + m_i \mathbf{v}_{\text{cm}}.$$

The angular momentum with respect to the cm is then

$$\mathbf{L}_{\text{cm}} = \sum_i (\mathbf{r}_i^* \times \mathbf{p}_i^*) = \sum_i \mathbf{r}_i^* \times (\mathbf{p}_i - m_i \mathbf{v}_{\text{cm}}).$$

Then, taking the time derivative, we have

$$\frac{d\mathbf{L}_{\text{cm}}}{dt} = \sum_i \left(\frac{d\mathbf{r}_i^*}{dt} \times \mathbf{p}_i^* \right) + \sum_i \left(\mathbf{r}_i^* \times \frac{d\mathbf{p}_i^*}{dt} \right).$$

FIGURE 10–6

The position of m_i in the inertial frame is $\mathbf{r}_i$; with regard to the cm (which could be accelerating) it is $\mathbf{r}_i^*$, where $\mathbf{r}_i = \mathbf{r}_i^* + \mathbf{r}_{\text{cm}}$ and $\mathbf{r}_{\text{cm}}$ is the position of the cm in the inertial frame.

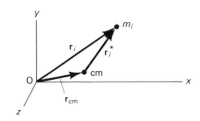

[†] Equation 9–22 or 9–23, which tells us the total $\text{KE} = \text{KE}_{\text{cm}} + \text{KE}_{\text{rot}}$ applies to the energy but not to the other aspects of the motion. Equations 10–7b plus 8–12 ($d\mathbf{P}_{\text{cm}}/dt = \mathbf{F}_{\text{ext}}$) provide the more general statement of this principle.

The first term is $\mathbf{v}_i^* \times m\mathbf{v}_i^* = 0$ since $\mathbf{v}_i^*$ is parallel to itself. Thus

$$\frac{d\mathbf{L}_{cm}}{dt} = \sum_i \mathbf{r}_i^* \times \frac{d}{dt}(\mathbf{p}_i - m_i\mathbf{v}_{cm})$$

$$= \sum_i \mathbf{r}_i^* \times \frac{d\mathbf{p}_i}{dt} - \left(\sum_i m_i\mathbf{r}_i^*\right) \times \frac{d\mathbf{v}_{cm}}{dt}.$$

The second term on the right is zero since, by Eq. 8–2, $\sum m_i\mathbf{r}_i^* = M\mathbf{r}_{cm}^*$, and $\mathbf{r}_{cm}^* = 0$ by definition because the position of the cm is at the origin of the cm reference frame. Furthermore, by Newton's second law we have

$$\frac{d\mathbf{p}_i}{dt} = \mathbf{F}_i$$

where $\mathbf{F}_i$ is the net force on m_i. (Note that $d\mathbf{p}_i^*/dt \neq \mathbf{F}_i$ since the cm may be accelerating and Newton's second law does not hold in a noninertial reference frame.) Consequently

$$\frac{d\mathbf{L}_{cm}}{dt} = \sum_i \mathbf{r}_i^* \times \mathbf{F}_i = \sum_i (\tau_i)_{cm} = \tau_{cm}$$

where τ_{cm} is the resultant external torque on the entire system calculated about the cm. The sum over all the τ_i eliminated the net torque due to internal forces by Newton's third law as discussed previously. This last equation is Eq. 10–7b, and this concludes its proof.

To summarize, the relation

$$\tau = \frac{d\mathbf{L}}{dt}$$

is *not* valid in general. It is true only when τ and $\mathbf{L}$ are calculated with respect to either (1) the origin of an inertial reference frame, or (2) the center of mass of a system of particles (or a rigid body).

10–6   Angular Momentum and Torque for a Rigid Body

Let us now consider the rotation of a rigid body about an axis that has a fixed direction in space. We will use the general principles developed earlier in this chapter and will arrive at the special conclusions already used in Chapter 9.

Let us calculate the component of angular momentum along the rotation axis. We will call this component L_ω since the angular momentum $\boldsymbol{\omega}$ points along the rotation axis. For each particle of the body,

$$\mathbf{l}_i = \mathbf{r}_i \times \mathbf{p}_i.$$

If θ is the angle between $\mathbf{l}_i$ and the rotation axis (see Fig. 10–7), then the component of $\mathbf{l}_i$ along the rotation axis is

$$l_{i\omega} = r_i p_i \cos\theta = m_i v_i r_i \cos\theta,$$

where m_i and v_i are the mass and velocity of the ith particle. Now $v_i = R_i\omega$ where ω is the angular velocity of the body and R_i is the perpendicular distance of m_i from the axis of rotation. Furthermore, $R_i = r_i \cos\theta$ as can be seen in Fig. 10–7, so

$$l_{i\omega} = m_i v_i(r_i \cos\theta) = m_i R_i^2 \omega.$$

We sum over all the particles to obtain

$$L_\omega = \sum_i l_{i\omega} = \left(\sum_i m_i R_i^2\right)\omega.$$

But $\sum_i m_i R_i^2$ is the moment of inertia I of the body about the axis of rotation.

Therefore the component of the total angular momentum along the rotation axis is given by

$$L_\omega = I\omega. \qquad (10\text{-}8)$$

Note that we would obtain Eq. 10–8 no matter where we choose the point O (for measuring r_i) as long as it is on the axis of rotation. Equation 10–8 is just Eq. 9–17 of Chapter 9, which we have now proved from the general definition of angular momentum.

The general relation between angular momentum and torque is Eq. 10–7:

$$\tau = \frac{d\mathbf{L}}{dt}$$

where τ and $\mathbf{L}$ are calculated either about (1) the origin of an inertial reference frame, or (2) the cm of the system. This is a vector relation, and must therefore be valid for each component. Hence, for a rigid body, the component along the rotation axis is

$$\tau_\omega = \frac{dL_\omega}{dt} = \frac{d}{dt}(I\omega) = I\frac{d\omega}{dt} = I\alpha,$$

which is valid for a rigid body rotating about an axis fixed relative to the body and also either (1) fixed in an inertial system or (2) passing through the cm of the body. This is equivalent to Eqs. 9–12 and 9–13, which we now see are special cases of Eq. 10–7, $\tau = d\mathbf{L}/dt$.

For a body which rotates about an axis of symmetry passing through its center of mass, we can write

$$\mathbf{L} = I\boldsymbol{\omega} \qquad [\text{rotation axis} = \text{symmetry axis, through cm}] \qquad (10\text{-}9)$$

where $\mathbf{L}$ is measured relative to the cm. To see why this is true, refer again to Fig. 10–7. Although $\boldsymbol{l}_i$ for the ith particle does not point along the axis of rotation, when $\boldsymbol{l}_i$ is added to the angular momentum of a similar particle exactly opposite it on the other side of the axis, the sum of the two will point along the rotation axis. Since a symmetrical body is made up of such pairs of particles, the sum over all of them will point along the rotation axis. Hence, for a body rotating about a symmetry axis the angular momentum points along the rotation axis and therefore $L = L_\omega$, so Eq. 10–9 is proved.

EXAMPLE 10–2 An *Atwood Machine* consists of two masses, m_1 and m_2, which are connected by a massless inelastic cord that passes over a pulley, Fig. 10–8. If the pulley has radius R_0 and moment of inertia I about its axle, determine the acceleration of the masses m_1 and m_2, and compare to the situation where the moment of inertia of the pulley is ignored.

SOLUTION The angular momentum is calculated about an axis along the axle through the center O of the pulley. The pulley has angular momentum $I\omega$, where $\omega = v/R_0$ and v is velocity of m_1 and m_2 at any instant. The angular momentum of m_1 is $R_0 m_1 v$ and that of m_2 is $R_0 m_2 v$. The total angular momentum is

$$L = (m_1 + m_2)vR_0 + I\frac{v}{R_0}.$$

The external torque on the system is:

$$\tau = m_2 g R_0 - m_1 g R_0$$

(since the net force on the pulley, exerted by the support on its axle, gives rise to no torque because the lever arm is zero). We apply Eq. 10–7a:

$$\tau = \frac{dL}{dt}$$

$$(m_2 - m_1)gR_0 = (m_1 + m_2)R_0\frac{dv}{dt} + \frac{I}{R_0}\frac{dv}{dt}.$$

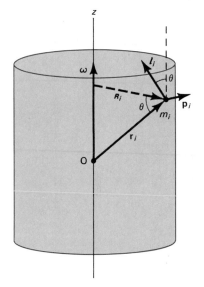

FIGURE 10–7
Calculating $\mathbf{L}_z = \Sigma(\boldsymbol{l}_z)_i$.

FIGURE 10–8
Atwood's machine.

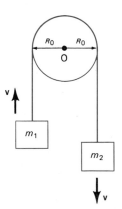

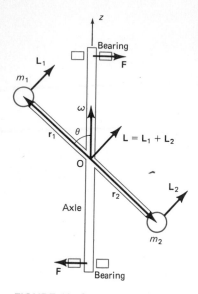

FIGURE 10–9

In this system $\mathbf{L}$ and ω are not parallel. This is an example of rotational imbalance.

Solving for $a = dv/dt$, we get:

$$a = \frac{dv}{dt} = \frac{(m_2 - m_1)g}{(m_1 + m_2) + I/R_0^2}.$$

If we were to ignore I, $a = (m_2 - m_1)g/(m_2 + m_1)$ and we see that the effect of the moment of inertia of the pulley is to slow down the system. This is just what we would expect.

Although Eq. 10–9 is often very useful, it is not valid in general. Nonetheless, it can be shown that every rigid body, no matter what its shape, has three "principal axes" about which Eq. 10–9 is valid (we won't go into the details here). As an example of a case where Eq. 10–9 is not valid, consider the nonsymmetrical body shown in Fig. 10–9. It consists of two equal masses, m_1 and m_2, attached to the ends of a rigid (massless) rod which makes an angle θ with the axis of rotation. We calculate the angular momentum about the cm at point O. At the moment shown, m_1 is coming toward the viewer, and m_2 is moving away, so $\mathbf{L}_1 = \mathbf{r}_1 \times \mathbf{p}_1$ and $\mathbf{L}_2 = \mathbf{r}_2 \times \mathbf{p}_2$ are as shown. The total angular momentum is $\mathbf{L} = \mathbf{L}_1 + \mathbf{L}_2$, which is clearly *not* along ω; since our system is *not* symmetrical, it is no surprise that Eq. 10–9 is not valid here.

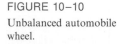 ***10–7   Rotational Imbalance**

Let us go one step further with the system shown in Fig. 10–9, since it is a fine illustration of $\tau = d\mathbf{L}/dt$. If the system rotates with constant angular velocity, ω, the magnitude of $\mathbf{L}$ will not change, but its direction will. As the rod and two masses rotate about the z axis, $\mathbf{L}$ also rotates about the axis. At the moment shown in Fig. 10–9, $\mathbf{L}$ is in the plane of the paper. A time dt later, when the rod has rotated through an angle $d\theta = \omega\, dt$, $\mathbf{L}$ will also have rotated through an angle $d\theta$ (it remains perpendicular to the rod). $\mathbf{L}$ will then have a component pointing into the page; thus $d\mathbf{L}$ points into the page and so must $d\mathbf{L}/dt$. Since

$$\tau = \frac{d\mathbf{L}}{dt}$$

we see that a torque, directed into the page at the moment shown, must be applied to the axle on which the rod is mounted. The torque is supplied by bearings (or other constraint) at the end of the axle. The forces $\mathbf{F}$ exerted by the bearings on the axle are shown in Fig. 10–9. The direction of each force $\mathbf{F}$ rotates as the system does, always being in the plane of $\mathbf{L}$ and ω for this system. If the torque due to these forces were not present, the system would not rotate about the fixed axis as desired.

The axle itself, by Newton's third law, must exert forces of $-\mathbf{F}$ on the bearings. Hence the axle tends to move in the direction of $-\mathbf{F}$ and thus tends to wobble as it rotates. This has many practical applications, such as the vibrations felt in a car whose wheels are not balanced. Consider an automobile wheel that is symmetrical except for an extra mass m_1 on one rim and an equal mass m_2 opposite it on the other rim, as shown in Fig. 10–10. Because of the nonsymmetry of m_1 and m_2, the wheel bearings would have to exert a force parallel to the wheel face at all times simply to keep the wheel rotating, just as in Fig. 10–9. The bearings would quickly wear down and the wobble of the wheel would be felt by occupants of the car. When the wheels are balanced, they rotate smoothly without wobble. This is why "dynamic balancing" of automobile wheels and tires is important. The wheel of Fig. 10–10 would balance *statically* just fine. If equal masses m_3 and m_4 are added symmetrically, below m_1 and above m_2, the wheel will also be balanced dynamically ($\mathbf{L}$ will be parallel to ω and $\tau_{\text{ext}} = 0$).

FIGURE 10–10

Unbalanced automobile wheel.

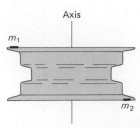

EXAMPLE 10–3   Determine the magnitude of the torque τ needed to keep the system turning in Fig. 10–9.

SOLUTION   Figure 10–11 is a view of the angular momentum vector, looking down on Fig. 10–9, as it rotates. $L \cos \theta$ is the component of **L** perpendicular to the axle. In a time dt, **L** changes in magnitude by an amount

$$dL = (L \cos \theta)\, d\theta = L \cos \theta\, \omega\, dt.$$

Hence

$$\tau = \frac{dL}{dt} = \omega L \cos \theta.$$

Now $L = L_1 + L_2 = r_1 m_1 v_1 + r_2 m_2 v_2 = r_1 m_1 (\omega r_1 \sin \theta) + r_2 m_2 (\omega r_2 \sin \theta) = (m_1 r_1^2 + m_2 r_2^2)\omega \sin \theta = I\omega/\sin \theta$; so

$$\tau = (m_1 r_1^2 + m_2 r_2^2)\, \omega^2 \sin \theta \cos \theta = I\omega^2/\tan \theta$$

where $I = (m_1 r_1^2 + m_2 r_2^2) \sin^2 \theta$ is the moment of inertia about the axis of rotation.

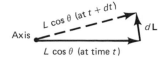

FIGURE 10–11

Angular momentum vector looking down along the rotation axis of the system of Fig. 10–9 as it rotates during a time dt.

The situation of Fig. 10–9 illustrates the usefulness of the vector nature of torque and angular momentum. If we had considered only the components of angular momentum and torque along the rotation axis, we could not have calculated the torque due to the bearings (since the forces **F** act at the axle and hence produce no torque along that axis). By using the concept of vector angular momentum we have a far more powerful technique for understanding and for attacking problems.

10–8   Conservation of Angular Momentum

In Chapter 8 we saw that the most general form of Newton's second law for the translational motion of a particle or system of particles is

$$\mathbf{F} = \frac{d\mathbf{P}}{dt}$$

where **P** is the (linear) momentum, defined as $m\mathbf{v}$ for a particle, or $M\mathbf{v}_{cm}$ for a system of particles of total mass M whose cm moves with velocity $\mathbf{v}_{cm}$, and **F** is the net force acting on the particle or system. This relation is valid only in an inertial reference frame.

In this chapter, we have found a similar relation to describe the general rotation of a system of particles (including rigid bodies):

$$\boldsymbol{\tau} = \frac{d\mathbf{L}}{dt}$$

where $\boldsymbol{\tau}$ is the net torque acting on the system, and **L** is the total angular momentum. This relation is valid when $\boldsymbol{\tau}$ and **L** are calculated about a point fixed in an inertial reference frame, or about the cm of the system. For a rigid body rotating with angular velocity ω about an axis fixed in direction, the angular momentum about this axis is (Eq. 10–8):

$$L = I\omega$$

where I is the moment of inertia of the body about this axis.

For translational motion, if the net force on the system is zero, $d\mathbf{P}/dt = 0$, so the total momentum of the system remains constant. This is the law of conservation of momentum. For rotational motion, if the net torque on the system is zero, then

$$\frac{d\mathbf{L}}{dt} = 0 \quad \text{and} \quad \mathbf{L} = \text{constant.} \qquad [\tau_{\text{net}} = 0] \qquad (10\text{–}10)$$

The total angular momentum of a system remains constant if the net torque acting on the system is zero. This is the *law of conservation of angular momentum.* This law ranks with the laws of conservation of energy and linear momentum (and others to be discussed later) as one of the great conservation laws of physics. In Chapter 9 we already saw some examples of this important law applied to the special case of a rigid body rotating about a fixed axis. Here we have it in general form. We use it now in an interesting example.

EXAMPLE 10–4   *Kepler's second law* states that each planet moves so that a line from the sun to the planet sweeps out equal areas in equal periods of time (see Section 5–7). Use the law of conservation of angular momentum to show this.

SOLUTION   The planet moves in an ellipse as shown in Fig. 10–12. In a time dt, the planet moves a distance $v\,dt$ and sweeps out an area equal to the area of a triangle of base r and height $v\,dt\,\sin\theta$ (shown exaggerated in Fig. 10–12); hence

$$dA = \tfrac{1}{2}(r)(v\,dt\,\sin\theta)$$

and

$$\frac{dA}{dt} = \tfrac{1}{2}rv\,\sin\theta.$$

The magnitude of the angular momentum **L** about the sun is

$$L = |\mathbf{r} \times m\mathbf{v}| = mrv\,\sin\theta,$$

so

$$\frac{dA}{dt} = \frac{1}{2m}\,L.$$

But L = constant, since the gravitational force **F** is directed toward the sun (we ignore the pull of the other planets) so $\boldsymbol{\tau} = \mathbf{r} \times \mathbf{F} = 0$. Hence dA/dt = constant, which is what we set out to prove.

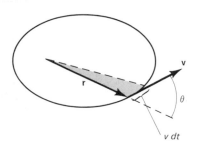

FIGURE 10–12

Kepler's second law of planetary motion (Example 10–4).

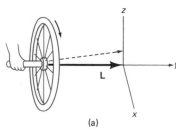

FIGURE 10–13

When you try to tilt a rotating bicycle wheel vertically, it swerves to the side.

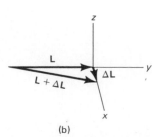

10–9   A Rotating Wheel

Let us consider now an experiment that is easily done, and gives a surprising result which illustrates the vector nature of the relation

$$\boldsymbol{\tau} = \frac{d\mathbf{L}}{dt}.$$

Suppose you are holding a bicycle wheel by a handle connected to its axle as in Fig. 10–13a. The wheel is spinning rapidly so its angular momentum **L** points horizontally as shown. Now you suddenly try to tilt the axle upward as shown by the dashed line in Fig. 10–13a (so the cm moves vertically). You expect the wheel to go up, but it unexpectedly swerves to the right! To explain this bizarre effect—you may need to do it to believe it—we only need to use the relation $\boldsymbol{\tau} = d\mathbf{L}/dt$. In the short time Δt, you exert a torque (about an axis through your wrist) that points along the x axis perpendicular to **L**. Thus the change in **L** is

$$\Delta\mathbf{L} \approx \boldsymbol{\tau}\,\Delta t$$

so $\Delta\mathbf{L}$ must also point (approximately) along the x axis, since $\boldsymbol{\tau}$ does (Fig. 10–13b). Thus the new angular momentum, $\mathbf{L} + \Delta\mathbf{L}$, points to the right of **L**. Since the angular momentum is directed along the axle of the wheel, we see that the axle, which now is along $\mathbf{L} + \Delta\mathbf{L}$, must move sideways, to the right, which is what we observe.

Note that if the wheel were not spinning, $\mathbf{L} = 0$ initially; $\Delta\mathbf{L}$ will still point along the x axis, and thus $\Delta\mathbf{L}$ corresponds to the rotational motion of the wheel about the x axis (passing through your wrist). That is, the axle tilts upward as expected with no sideways motion.

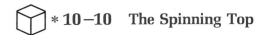

* 10–10 The Spinning Top

The motion of a rapidly spinning toy top, or a gyroscope, is an interesting example of rotational motion and of the use of the vector equation

$$\tau = \frac{d\mathbf{L}}{dt}.$$

Consider a symmetrical top of mass M spinning rapidly about its symmetry axis, as in Fig. 10–14. The top is balanced on its tip at point O in an inertial reference frame. If the axis of the top makes an angle θ to the vertical (z axis), when the top is released carefully its axis will move, sweeping out a cone about the vertical as shown by the dashed line in Fig. 10–14. This type of motion, in which a torque produces a change in the direction of a rotation axis, is called *precession*; and the rate at which the rotation axis moves about the vertical (z) axis is called the angular velocity of precession, Ω (capital Greek omega). Let us now try to understand the reasons for this motion, and also calculate Ω.

If the top were not spinning, it would immediately fall to the ground when released due to the pull of gravity. The apparent mystery of a top is that when spinning, it does not immediately fall to the ground but instead precesses—it moves sideways. But this is not really so mysterious if we examine it from the point of view of angular momentum and torque, which we calculate about the point O. When the top is spinning with angular velocity ω about its symmetry axis, it has an angular momentum $\mathbf{L}$ which is directed along its axis as shown in Fig. 10–14. (There is also angular momentum due to the precessional motion, so that the total $\mathbf{L}$ is not exactly along the axis of the top; but if $\Omega \ll \omega$, which is usually the case, we can ignore this.)

FIGURE 10–14

Spinning top.

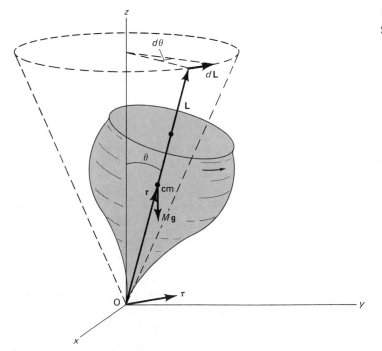

Now to change the angular momentum, a torque is required. If no torque were applied to the top, **L** would remain constant in magnitude and direction; the top would neither fall nor precess. But there is a torque about O, equal to $\tau = \mathbf{r} \times M\mathbf{g}$ where **r** is the position vector of the top's cm with respect to O. The direction of τ is perpendicular to both **r** and $M\mathbf{g}$ and by the right-hand rule is, as shown in Fig. 10–14, in the horizontal (xy) plane. The change in **L** in a time dt is

$$d\mathbf{L} = \tau \, dt$$

which is perpendicular to **L** and horizontal, as shown in Fig. 10–14. Since $d\mathbf{L}$ is perpendicular to **L**, the magnitude of **L** does not change. Only the direction of **L** changes. Since **L** points along the axis of the top, we see that the axis of the top moves to the right in the figure; that is, the upper end of the top's axis moves in a horizontal direction perpendicular to **L**. This explains why the top precesses rather than falls. The vector **L** and the top's axis move together in a horizontal circle. As they do so, τ and $d\mathbf{L}$ rotate as well so as to be horizontal and perpendicular to **L**.

To determine Ω, we see from Fig. 10–14 that the angle $d\phi$ (which is in a horizontal plane) is related to dL by

$$dL = L \sin \theta \, d\phi$$

since **L** makes an angle θ to the z axis. The angular velocity of precession is

$$\Omega = \frac{d\phi}{dt}$$

which becomes

$$\Omega = \frac{1}{L \sin \theta} \frac{dL}{dt} = \frac{\tau}{L \sin \theta}. \qquad \text{[spinning top]} \qquad (10\text{–}11a)$$

But $\tau = |\mathbf{r} \times M\mathbf{g}| = rMg \sin \theta$ (because $\sin(\pi - \theta) = \sin \theta$) so we can also write:

$$\Omega = \frac{Mgr}{L}. \qquad \text{[spinning top]} \qquad (10\text{–}11b)$$

Thus the rate of precession does not depend on the angle θ; but it is inversely proportional to the top's angular momentum. The faster the top spins, the greater L is and the slower the top precesses.

 Summary ———

The *vector* or *cross product* of two vectors **A** and **B** is another vector $\mathbf{C} = \mathbf{A} \times \mathbf{B}$ whose magnitude is $AB \sin \theta$ and whose direction is perpendicular to both **A** and **B** in the sense of the right-hand rule.

The *torque* τ due to a force **F** is a vector quantity and is always calculated about some point O (the origin of a coordinate system) as follows:

$$\tau = \mathbf{r} \times \mathbf{F}$$

where **r** is the position vector of the point at which the force **F** acts.

Angular momentum is also a vector. For a particle having momentum $\mathbf{p} = m\mathbf{v}$, the angular momentum l about some point O is

$$l = \mathbf{r} \times \mathbf{p}$$

where **r** is the position vector of the particle relative to the point O at any instant. The net torque τ on a particle is related to its angular momentum by

$$\tau = \frac{dl}{dt}.$$

For a system of particles, the total angular momentum $\mathbf{L} = \sum l_i$. The total angular momentum of the system is related to the total net torque τ on the system by

$$\tau = \frac{d\mathbf{L}}{dt}.$$

This last relation is the vector rotational equivalent of Newton's second law. It is valid when **L** and τ are calculated about an origin (1) fixed in an inertial reference system, or (2) situated at the cm of the system. For a rigid body rotating about a fixed axis, the component of angular momentum about the rotation axis is given by $L_\omega = I\omega$. If a body rotates about an axis of symmetry, then the vector relation $\mathbf{L} = I\boldsymbol{\omega}$ holds, but this is not true in general.

If the total net torque on a system is zero, then the total vector angular momentum $\mathbf{L}$ remains constant; this is the important *law of conservation of angular momentum*. It applies to the vector $\mathbf{L}$ and therefore also to each of its components.

Questions

1. If all the components of the vectors $\mathbf{V}_1$ and $\mathbf{V}_2$ were reversed, how would this alter $\mathbf{V}_1 \times \mathbf{V}_2$?

2. Name the three different conditions that could make $\mathbf{V}_1 \times \mathbf{V}_2 = 0$.

3. A force $\mathbf{F} = \mathbf{j}F$ is applied to a body at a position $\mathbf{r} = x\mathbf{i} + y\mathbf{j} + z\mathbf{k}$ where the origin is at the cm. Does the torque about the cm depend on x? On y? On z?

4. A particle moves with constant speed along a straight line. How does its angular momentum, calculated about any point not on its path, change in time?

5. Two particles have the same momentum $\mathbf{p}$ but are located at different positions. Can you choose an origin O so they have the same angular momentum? What if one particle has momentum $\mathbf{p}$ and the other $-\mathbf{p}$?

6. If the net force on a system is zero, is the net torque also zero? If the net torque on a system is zero, is the net force zero?

7. A stick stands vertically on its end on a frictionless surface. Describe its motion when it is tipped slightly to one side and falls.

8. Look at the face of a clock with a second hand. In what direction is the angular momentum of the second hand?

9. Try spinning a raw egg and a hard boiled egg. Why does one spin readily and the other not?

10. Explain how a child "pumps" on a swing to make it go higher.

11. Explain why six variables are needed to locate a rigid body in space. How many variables are needed to locate a rigid body confined to move in a plane?

12. A cyclist rides over the top of a hill. Is the bicycle's motion rotational, translational, or a combination of both?

*13. For the nonsymmetrical system of Fig. 10–9, is there a point about which the angular momentum ($\mathbf{L}$) could be calculated so that it would be in the same direction as the angular velocity $\boldsymbol{\omega}$? What if there were only one mass (say, $m_2 = 0$)? If your answer to either question is yes, how do you explain the need for forces at the bearings for both cases and the requirement $\boldsymbol{\tau} = d\mathbf{L}/dt$?

14. We claim that momentum and angular momentum are conserved. Yet most moving or rotating bodies eventually slow down and stop. Explain.

15. On the basis of the law of conservation of angular momentum, discuss why a helicopter must have more than one rotor (or propeller). Discuss one or more ways the second propeller can operate in order to keep the body stable.

16. A football is kicked end-over-end. Is its angular momentum about its cm conserved as it flies through the air? What if its angular momentum is calculated about the original position of the punter's toe?

17. Will angular momentum be conserved for any central force, $\mathbf{F} = F(r)\,\hat{\mathbf{r}}$? (See Section 7–5.) Will Kepler's second law apply for any central force?

18. A wheel is rotating freely about a vertical axis with constant angular velocity. Small parts of the wheel come loose and fly off. How does this affect the rotational speed of the wheel? Is angular momentum conserved? Is kinetic energy conserved? Explain.

19. Consider the following vector quantities: displacement, velocity, acceleration, momentum, angular momentum, torque. (a) Which of these are independent of the choice of origin of coordinates? (Consider different points as origin which are at rest with respect to each other.) (b) Which are independent of the velocity of the coordinate system?

20. Will a particle traveling with constant momentum $\mathbf{p}$ sweep out equal areas in equal times? Explain.

21. Describe the torque needed if the person in Fig. 10–13 is to tilt the axle of the rotating wheel directly upward with no swerving to the side.

22. How does a car make a right turn? Where does the torque come from?

23. Suppose you are standing on a turntable that can freely rotate. When you hold a rotating bicycle wheel over your head, with its axis vertical, you are at rest. If you now move the wheel so its axis is horizontal, what happens to you? What happens if you then point the axis of the wheel downward?

*24. The axis of the earth precesses with a period of about 25,000 years. This is much like the precession of a top. Explain how the earth's equatorial bulge gives rise to a torque exerted by the sun on the earth; see Fig. 10–15, which is drawn for the winter solstice (December 21). About what axis would you expect the earth's rotation axis to precess as a result of this torque? Does the torque exist 3 months later? Explain.

FIGURE 10–15

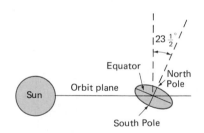

Problems

1. (I) Show that (a) $\mathbf{i} \times \mathbf{i} = \mathbf{j} \times \mathbf{j} = \mathbf{k} \times \mathbf{k} = 0$, (b) $\mathbf{i} \times \mathbf{j} = \mathbf{k}$, $\mathbf{i} \times \mathbf{k} = -\mathbf{j}$, $\mathbf{j} \times \mathbf{k} = \mathbf{i}$.

2. (I) Consider a particle of a rigid body rotating about a fixed axis. Show that the tangential and radial vector components of the linear acceleration are:

$$\mathbf{a}_T = \boldsymbol{\alpha} \times \mathbf{r} \quad \text{and} \quad \mathbf{a}_c = \boldsymbol{\omega} \times \mathbf{v}.$$

3. (II) Use the limiting process (see Chapter 2, particularly Example 2–3) to obtain Eq. 10–2d.

4. (II) Prove the distributive law for the cross product, Eq. 10–2c.

5. (II) (a) Show that the cross product of two vectors, $\mathbf{A} = A_x\mathbf{i} + A_y\mathbf{j} + A_z\mathbf{k}$, and $\mathbf{B} = B_x\mathbf{i} + B_y\mathbf{j} + B_z\mathbf{k}$ is

$$\mathbf{A} \times \mathbf{B} = (A_yB_z - A_zB_y)\mathbf{i} + (A_zB_x - A_xB_z)\mathbf{j} + (A_xB_y - A_yB_x)\mathbf{k}.$$

(b) Then show that the cross product can be written

$$\mathbf{A} \times \mathbf{B} = \begin{vmatrix} \mathbf{i} & \mathbf{j} & \mathbf{k} \\ A_x & A_y & A_z \\ B_x & B_y & B_z \end{vmatrix},$$

where we use the rules for evaluating a determinant. (Note, however, that this is not actually a determinant, but a memory aid.)

6. (III) Show that the velocity $\mathbf{v}$ of any point in a body rotating with angular velocity ω about a fixed axis can be written

$$\mathbf{v} = \boldsymbol{\omega} \times \mathbf{r}$$

where $\mathbf{r}$ is the position vector of the point relative to an origin O located on the axis of rotation. Can O be anywhere on the rotation axis? Will $\mathbf{v} = \boldsymbol{\omega} \times \mathbf{r}$ if O is located at a point not on the axis of rotation?

7. (III) Let $\mathbf{A}$, $\mathbf{B}$, $\mathbf{C}$ be three non-coplanar vectors. Show that $\mathbf{A} \cdot (\mathbf{B} \times \mathbf{C}) = \mathbf{B} \cdot (\mathbf{C} \times \mathbf{A}) = \mathbf{C} \cdot (\mathbf{A} \times \mathbf{B})$. [Hint: see previous problem.]

SECTION 10–2

8. (II) The origin of a coordinate system is at the center of a wheel. The wheel can rotate in the xy plane about an axis through the origin. A force $F = 83$ N, which acts at a $30°$ angle to the x axis, is applied to the wheel at the point $x = 10$ cm, $y = 35$ cm. What is the magnitude and direction of the torque produced by this force about the axis?

9. (II) A 30-cm-diameter thin wheel is constrained to rotate about an axis through its center, which we choose to be the z axis (the wheel rotates in the xy plane). A force $\mathbf{F} = 26\mathbf{j} - 15\mathbf{k}$ (newtons) is exerted at a point on the edge of the wheel which is exactly on the x axis. (a) Determine the torque (magnitude and direction) calculated about the wheel's center produced by this force. (b) Is the direction of this torque parallel to the direction of $\boldsymbol{\alpha}$? If not, explain how τ can be proportional to $\boldsymbol{\alpha}$.

SECTION 10–3

10. (I) What are the x, y, and z components of the angular momentum of a particle located at $\mathbf{r} = x\mathbf{i} + y\mathbf{j} + z\mathbf{k}$ which has momentum $\mathbf{p} = p_x\mathbf{i} + p_y\mathbf{j} + p_z\mathbf{k}$?

11. (I) Show that the kinetic energy a particle of mass m, moving in a circular path, is $\text{KE} = l^2/2I$ where l is its angular momentum, and I is its moment of inertia about the center of the circle.

12. (II) Two identical particles have equal but opposite momenta, $\mathbf{p}$ and $-\mathbf{p}$, but they are not traveling along the same line. Show that the total angular momentum of this system does not depend on the choice of origin.

13. (II) Two particles rest a distance d apart on the edge of a table. One of the particles, of mass m, slides off the edge and falls vertically. Using the other particle (still at rest) as the origin, calculate (a) the torque τ on m as a function of time (before hitting the ground below), (b) the angular momentum l of m as a function of time, and (c) show that $\tau = dl/dt$.

14. (III) In the Bohr model of the hydrogen atom, the electron (mass m) is held in a circular orbit about the nucleus (a proton) by the electric force on it, $F = ke^2/r^2$ (where e is the electric charge on the electron and on the proton, and k is a constant), and only certain orbits are allowed: those for which the angular momentum l of the electron about the nucleus is an integer multiple n of $h/2\pi$ where h is a constant called Planck's constant. That is, $l = nh/2\pi$ where n is an integer. Show that the possible radii for electron orbits are given by

$$r = \frac{n^2h^2}{4\pi^2kme^2} \qquad n = 1, 2, 3, \ldots.$$

To do this, first show that the radius of an orbit is $r = ke^2/mv^2$ and then apply $l = nh/2\pi$.

SECTIONS 10–4 TO 10–6

15. (II) Four identical particles of mass m are mounted at equal intervals on a thin rod of length L and mass M, with one mass at each end of the rod. If the system is rotated with angular velocity ω about an axis perpendicular to the rod through one of the end masses, determine (a) the kinetic energy and (b) the angular momentum of the system.

16. (II) Use $\tau = dL/dt$ to determine the acceleration of the block in Example 9–10.

17. (II) The moon orbits the earth so that the same side always faces the earth. Determine the ratio of its spin angular momentum (about its own axis) and its orbital angular momentum.

18. (III) A thin string is wrapped around a solid cylindrical spool of radius R and mass M. One end of the string is fixed, and the spool is allowed to fall vertically, starting from rest, as the string unwinds. (a) Determine the angular momentum of the spool about its cm as a function of time. (b) What is the tension in the string as function of time?

19. (III) A uniform thin rod, 80 cm long with a mass of 400 g, lies on a frictionless horizontal table. It is struck, at right angles to its length, at a point 20 cm from one end. If the

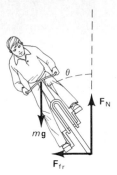

FIGURE 10–16

FIGURE 10–17
A wheel, rotating about a horizontal axle
supported at one end, precesses.

impulse is 8.5 N·s, describe the resulting motion of the stick.

20. (III) A thin rectangular plate, with sides of length a and b, is rotated about an axis along a diagonal. Determine the magnitude and direction of $\mathbf{L}$.

21. (III) A bicyclist traveling with speed $v = 5.2$ m/s on a flat road is making a turn with a radius $r = 3.8$ m. The forces acting on the cyclist and cycle are the normal force ($\mathbf{F}_N$) and friction force ($\mathbf{F}_{fr}$) exerted by the road on the tires, and $m\mathbf{g}$, the total weight of the cyclist and cycle. (a) Explain carefully why the angle θ the bicycle makes with the vertical (Fig. 10–16) must be given by $\tan \theta = F_{fr}/F_N$ if the cyclist is to maintain balance. (b) Calculate θ for the values given. (Hint: consider the "circular" translational motion of the bicycle and rider.) (c) If the coefficient of static friction between tires and road is $\mu_s = 0.65$, what is the minimum turning radius?

22. (III) Show that the total angular momentum $\mathbf{L} = \sum \mathbf{r}_i \times \mathbf{p}_i$ of a system of particles about the origin of an inertial reference frame can be written as the sum of the angular momentum about the cm (spin angular momentum) $\mathbf{L}^*$, plus the angular momentum of the cm about the origin (orbital angular momentum):

$$\mathbf{L} = \mathbf{L}^* + \mathbf{r}_{cm} \times M\mathbf{v}_{cm}.$$

(Hint: see the derivation of Eq. 9–22.)

*SECTION 10–7

*23. (I) What is the magnitude of the force $\mathbf{F}$ exerted by each bearing in Fig. 10–9 (Example 10–3)? The bearings are a distance d apart.

*24. (II) Suppose in Fig. 10–9 that $m_2 = 0$; that is, only one mass, m_1, is actually present. If the bearings are a distance d

apart, determine the forces F_1 and F_2 at the upper and lower bearings respectively. (Hint: choose an origin—different than 0 in Fig. 10–9—such that $\mathbf{L}$ is parallel to $\boldsymbol{\omega}$.)

*25. (II) A uniform cylindrical wheel 60 cm in diameter and of mass 18 kg rotates about an axle displaced 1.0 cm from its true center. (a) If the bearings are mounted 9.5 cm apart, what is the force exerted on them when the wheel rotates at 12 rev/s? (b) To balance this wheel, where should a 1.00 kg mass be placed?

SECTION 10–8

26. (II) A person whose moment of inertia is I_p stands on a platform that can freely rotate without friction (but isn't) and holds a spinning bicycle wheel with axis horizontal as in Fig. 10–13. The wheel has moment of inertia I_w and angular velocity, ω_w. What will be the angular velocity ω of the platform if the person moves the axis of the wheel so it points (a) vertically upward, (b) at a 60° angle to the vertical, (c) vertically downward. (d) What will ω be if the person reaches up and stops the wheel in part (a)?

*SECTION 10–10

*27. (II) A 120-g top spinning at 25 rev/s makes an angle of 30° to the vertical and precesses at a rate of 1 rev per 8.0 s. If its cm is 3.5 cm from its tip along its symmetry axis, what is the moment of inertia of the top?

*28. (II) The solid wheel of Fig. 10–17 rotates at 25 rad/s; it has radius 6.0 cm and is mounted at the center of a horizontal thin axle 20 cm long. At what rate does the axle precess?

*29. (II) If a mass, equal to half the mass of the wheel, is placed at the free end of the axle, in problem 28, what will be the precession rate now?

Equilibrium, Elasticity, and Fracture

11

 ## 11–1 Statics—The Study of Forces in Equilibrium

In this chapter we will study a special case of motion—when the net force and torque on an object or system of objects are zero. In this case either the object or system will be at rest, or the center of mass of the object or system will be moving at constant velocity. We will be concerned only with the first case, when the object or objects are all at rest. Now you may think that the study of objects at rest is not very interesting since the objects will have neither velocity nor acceleration; and the net force and the net torque will be zero. But it is just this last aspect that makes this subject of *statics* interesting: true, the net force on an object at rest is zero, but this does not imply that no forces act on it. In fact it is virtually impossible to find a body on which no forces act at all. Objects within our experience have at least one force acting on them (gravity) and if they are at rest then there must be other forces acting on them as well so that the total net force is zero. An object at rest on a table for example, has two forces acting on it, the downward force of gravity and the force the table exerts upward on it, Fig. 11–1. Since the net force is zero, the upward force exerted by the table must be equal to the force of gravity acting downward. (Do not confuse this with the equal and opposite forces of Newton's third law which act on different bodies; here both forces act on the same body.) Such a body is said to be in **equilibrium** (Latin for "equal forces") under the action of these two forces.[†]

The subject of statics is concerned with the calculation of the forces acting on bodies which are in equilibrium. The techniques for doing this can be applied in a wide range of fields. Architects and engineers must be able to calculate the forces on the structural members of buildings, bridges, machines, vehicles, and other structures, since any material will break or buckle if too much force is applied. In the human body, a knowledge of the forces in muscles and joints is of great value for medicine and physical therapy, and is also valuable for the study of athletic activity.

FIGURE 11–1

The book is in equilibrium; the net force on it is zero.

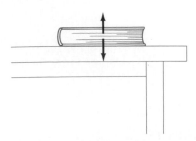

[†] A body in equilibrium at rest is said to be in *static equilibrium* whereas a body moving uniformly at constant velocity is in *dynamic equilibrium*; we consider mainly the former case.

11–2 Center of Gravity

One of the forces that acts on structures is the force of gravity, and our analysis in this chapter is greatly simplified by using the concept of center of gravity. Any extended body can be thought of as made up of many particles, each of mass m_i. Although gravity acts on each of these particles, we can show that the sum of all these individual forces has the equivalent effect of a single force which acts at a single point called the **center of gravity** (cg); this force is equal to $M\mathbf{g}$ where $M = \sum m_i$ is the total mass of the body and $\mathbf{g}$ is acceleration due to gravity. If $\mathbf{g}$ has the same value on all parts of the body, which is the usual case, the cg is the same as the center of mass (cm). We now set out to prove the above statements.

The total force of gravity on a body made up of n particles of mass m_1, $m_2, \ldots m_n$, is

$$\mathbf{F} = m_1\mathbf{g} + m_2\mathbf{g} + \ldots + m_n\mathbf{g} = \sum_i m_i\mathbf{g} = M\mathbf{g}$$

where M is the total mass of the body and we assume $\mathbf{g}$ is the same at all parts of the body. By Newton's second law, a single force $\mathbf{F} = M\mathbf{g}$ has the same effect on the translational motion of the body as does the sum of all the gravitational forces acting on the particles of the body. But where must this force act so that the rotational motion of the body also is the same? To determine this, let us calculate the sum of all the torques on the body about some arbitrary point 0, as shown in Fig. 11–2. If $\mathbf{r}_i$ is the position vector of the ith particle relative to 0, then the sum of all the torques due to gravity acting on the particles of the body is

$$\tau = \mathbf{r}_1 \times m_1\mathbf{g} + \mathbf{r}_2 \times m_2\mathbf{g} + \ldots + \mathbf{r}_n \times m_n\mathbf{g}$$
$$= \left(\sum m_i\mathbf{r}_i\right) \times \mathbf{g}.$$

The same torque could be applied by a single force, $M\mathbf{g}$, if it acts at a point whose position vector $\mathbf{r}_{\text{cg}}$ is given by

$$\tau = \mathbf{r}_{\text{cg}} \times (M\mathbf{g})$$

where, by comparison to the previous equation, $\mathbf{r}_{\text{cg}}$ must be given by

$$\mathbf{r}_{\text{cg}} = \frac{\sum m_i\mathbf{r}_i}{M}.$$

We recognize this as the definition of center of mass (Eq. 8–2): hence $\mathbf{r}_{\text{cg}} = \mathbf{r}_{\text{cm}}$. Thus, a single force $M\mathbf{g}$ acting at the cm will produce the same torque, and have the same rotational effects, as all the torques due to the force of gravity on all the particles.

We have thus shown that the force of gravity acting on all the particles of a body has the same translational and rotational effects on the body as a single force, $\mathbf{F} = M\mathbf{g}$, which acts at the center of gravity, and we have shown that the cg is the same as the cm for a uniform gravitational field $\mathbf{g}$.

If the gravitational field $\mathbf{g}$ were not the same over the whole body—and this can happen since $\mathbf{g}$ depends on the distance from the attracting body according to Newton's law of universal gravitation (Section 5–2)—then our equations above would be slightly different. For each particle of the body, we would have to replace $\mathbf{g}$ by $\mathbf{g}_i$, the value at the position of the ith particle. In this case, the cg would not be the same as the cm. For this difference to be significant in the earth's gravitational field, the body would have to be very large.

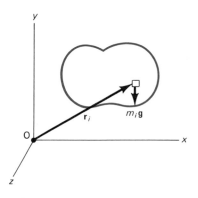

FIGURE 11–2

Determining the torque about O due to gravitational force.

 11–3 The Conditions for Equilibrium

For a body to be in equilibrium, the *vector sum of all external forces acting on the body must be zero*:

$$\sum \mathbf{F} = 0. \qquad (11\text{–}1a)$$

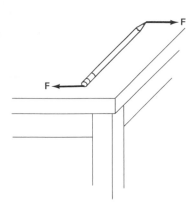

FIGURE 11–4

Example 11–1.

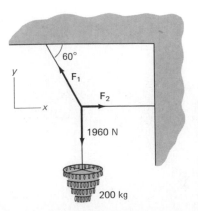

This is called the *first condition for equilibrium* and is equivalent to three component equations:

$$\sum F_x = 0, \quad \sum F_y = 0, \quad \sum F_z = 0. \qquad (11\text{–}1b)$$

Equations 11–1 assure that the center of mass of our object does not accelerate: $\mathbf{a}_{cm} = 0$. If the object is at rest initially ($v_{cm} = 0$), it will remain at rest.

For an object to be in equilibrium we need more than just this first condition. The net force on the pencil shown in Fig. 11–3, for example, is zero, but the pencil is not in equilibrium since it can rotate. The *second condition for equilibrium* is that the *vector sum of all external torques acting on the body must add up to zero*:

$$\sum \tau = 0. \qquad (11\text{–}2)$$

This will assure that the angular acceleration, $\boldsymbol{\alpha}$, about any point will be zero; if the body is not rotating initially ($\omega = 0$) it will not start rotating. Equations 11–1 and 11–2 are the only requirements for a body to be in equilibrium.

Equation 11–2 can also be written in component form

$$\sum \tau_x = 0, \quad \sum \tau_y = 0, \quad \sum \tau_z = 0.$$

The torques are calculated about some point 0, and τ_x, τ_y, and τ_z are the components along any three chosen axes. However, we will restrict ourselves in the rest of this chapter to a common situation that is simpler than the general case: the situation when all the external forces are acting in a plane. If we call this plane the xy plane, then we have only two force equations,

$$\sum F_x = 0, \qquad \sum F_y = 0,$$

and one torque equation,

$$\sum \tau_z = 0.$$

The torque is calculated about an axis that is perpendicular to the xy plane. The choice of this axis is arbitrary. Since $\tau = I\alpha$ for a rigid body about any fixed axis (Eq. 9–12), $\tau = 0$ about any fixed axis if the body is in equilibrium ($\alpha = 0$): therefore we can choose any axis that makes our calculation easier.

Since we have three equations, we can solve for at most three unknowns. If there are more than three unknowns, sometimes additional relations can be found (such as relations involving the elasticity of the materials), but we will not consider such cases here since they are often very complex. In certain cases, some of the forces can be measured experimentally; this is often done on the structure itself, or on a model of the structure, using strain gauges.

This subject of statics is important because it allows us to calculate certain forces on (or within) a structure when some of the forces on it are already known. There is no single technique for attacking such statics problems, but the following procedure is helpful. (1) Choose one body at a time for consideration, and on a diagram show all the forces acting on the body and the points at which these forces act; this is called a "free-body diagram." (2) Choose a convenient coordinate system and resolve the forces into their components. (3) Using letters to represent unknowns, write down equations for $\sum F_x = 0$, $\sum F_y = 0$, and $\sum \tau_z = 0$, and then solve these equations. Undoubtedly the hardest step is (1): all the forces on the body must be included, but the forces exerted by this body on other objects must not be included.

We now take some examples of calculating forces in structures that are in equilibrium.

EXAMPLE 11–1 Calculate the tension in the two cords used to support the 200-kg chandelier shown in Fig. 11–4.

SOLUTION The three forces, $\mathbf{F}_1$, $\mathbf{F}_2$, and the 200-kg weight of the chandelier, act at the point where the three cords join. We choose this junction point (it could be a knot) as the object for which we write $\sum F_x = 0$, $\sum F_y = 0$. (Note that we don't bother considering the chandelier itself since only two forces act on it, gravity downward and the equal but opposite force exerted upward by the cord, both of which equal $mg = 1960$ N.) Since our object (the knot) is essentially a point, there are no torques. Luckily there are only two unknowns and we can solve for them using $\sum \mathbf{F} = 0$. We first resolve $\mathbf{F}_1$ into its horizontal (x) and vertical (y) components. Although we don't know the value of F we can write $F_{1x} = F_1 \cos 60°$ and $F_{1y} = F \sin 60°$. $\mathbf{F}_2$ has no y component. In the vertical direction we have only the weight of the chandelier $= (200$ kg$)(g)$ acting downward and the vertical component of $\mathbf{F}_1$ upward. Since $\sum F_y = 0$, we have:

$$\sum F_y = F_1 \sin 60° - (200 \text{ kg})(g) = 0$$

so

$$F_1 = \frac{(200 \text{ kg})(g)}{\sin 60°} = (230 \text{ kg})(g) = 2260 \text{ N}.$$

In the horizontal direction,

$$\sum F_x = F_2 - F_1 \cos 60° = 0.$$

Thus

$$F_2 = F_1 \cos 60° = (230 \text{ kg})(g)(0.500) = (115 \text{ kg})(g) = 1130 \text{ N}.$$

The magnitudes of $\mathbf{F}_1$ and $\mathbf{F}_2$ determine the strength of cord or wire that must be used. In this case, the wire must be able to hold at least 230 kg. Note in this example that we didn't insert the value of g, the acceleration due to gravity, until the end; in this way we found the magnitude of the force in terms of the mass of an equivalent number of kilograms (which may be a more familiar quantity than newtons).

EXAMPLE 11–2 A uniform 1200-kg beam supports 15,000 kg of machinery, Fig. 11–5. Calculate the force on each of the supports.

SOLUTION We analyze the forces on the beam, since the force it exerts on the supports is equal and opposite to the forces exerted by the supports on the beam. We call the latter $\mathbf{F}_1$ and $\mathbf{F}_2$ in Fig. 11–5. The weight of the beam itself acts at its center of gravity, 10 m from either end. Since it doesn't matter which point we choose as the axis for writing the torque equation, we can choose one that is convenient. If we calculate the torques about the point of application of $\mathbf{F}_1$, then $\mathbf{F}_1$ will not enter (its lever arm will be zero) and we will have an equation in only one unknown, F_2. Thus $\sum \tau = 0$ gives

$$-(10 \text{ m})(1200 \text{ kg})(g) - (15 \text{ m})(15,000 \text{ kg})(g) + (20 \text{ m})F_2 = 0$$

Solving for F_2 we find $F_2 = (12,000 \text{ kg})(g) = 120,000$ N. To find F_1 we use $\sum F_y = 0$:

$$\sum F_y = F_1 - (1200 \text{ kg})(g) - (15,000 \text{ kg})(g) + F_2 = 0$$

Putting in $F_2 = (12,000 \text{ kg})(g)$ we find $F_1 = (4200 \text{ kg})(g) = 41,000$ N.

EXAMPLE 11–3 A uniform beam, 2.20 m long with mass $m = 25.0$ kg, is mounted on a hinge at a wall as shown in Fig. 11–6. It is held in a horizontal position by a wire that makes an angle $\theta = 30.0°$ as shown. The beam supports a mass $M = 280$ kg suspended from its end. Determine the components of the force $\mathbf{F}$ that the wall exerts on the beam at the hinge, and the components of the tension $\mathbf{T}$ in the supporting wire.

SOLUTION The sum of the forces in the vertical (y) direction is

$$F_y + T_y - mg - Mg = 0, \tag{i}$$

and in the horizontal (x) direction is

$$F_x - T_x = 0. \tag{ii}$$

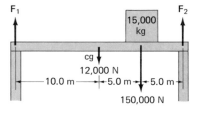

FIGURE 11–5
A 1200-kg beam supports a 15,000-kg machine (Example 11–2).

FIGURE 11–6
Example 11–3.

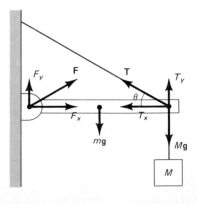

For the torque equation, we choose the origin at the point where $\mathbf{T}$ and $M\mathbf{g}$ act (so our equation then contains only one unknown and we can solve it more quickly):

$$-(F_y)(2.20 \text{ m}) + (mg)(1.10 \text{ m}) = 0 \qquad \text{(iii)}$$

since the cg is at the center of a uniform beam. We know that

$$mg = (25.0 \text{ kg})(9.80 \text{ m/s}^2) = 245 \text{ N}$$

$$Mg = (280 \text{ kg})(9.80 \text{ m/s}^2) = 2740 \text{ N}$$

but we still have only three equations and there are four unknowns (F_x, F_y, T_x, T_y). You might think we could get a fourth equation by calculating torques about another point, say the hinge: $T_y(2.20 \text{ m}) - (Mg)(2.20 \text{ m}) - (mg)(1.10 \text{ m}) = 0$; but this equation would not be independent of the other three—it is in fact only a combination of the other three. There can be only three independent equations involving $\sum \tau = \sum \mathbf{F} = 0$ when the forces lie in a plane. But we do know one other fact: the wire, since it is flexible, would bend if $\mathbf{T}$ had a component perpendicular to the wire. Thus $\mathbf{T}$ must act along the wire (a fact we used in Example 11–3 without mentioning it), so we have a fourth equation

$$T_y = T_x \tan \theta = 0.577 T_x, \qquad \text{(iv)}$$

since $\theta = 30.0°$. From Eq. iii above we have

$$F_y = \frac{mg}{2} = 122 \text{ N}.$$

From Eqs. i, ii, and iv we get

$$T_y = (m + M)g - F_y = 2870 \text{ N}$$

$$T_x = T_y/0.577 = 4970 \text{ N}$$

$$F_x = T_x = 4970 \text{ N}.$$

The tension in the wire is $T = \sqrt{T_x^2 + T_y^2} = 5730 \text{ N}$.

We next consider an example that relates to the muscles, bones, and joints of the human body.

EXAMPLE 11–4 How much force must the biceps muscle exert when a 5.0-kg mass is held in the hand (*a*) with the arm outstretched as in Fig. 11–7a and (*b*) when the arm is at a 45° angle as in Fig. 11–7b. Assume the mass of forearm and hand together is 2.0 kg and their cg is as shown.

SOLUTION (*a*) The forces acting on the forearm are shown in part (*a*) of the figure and include the upward force $\mathbf{F}_M$ exerted by the muscle and a force $\mathbf{F}_J$ exerted at the joint by the bone in the upper arm. We wish to find F_M, which is done most easily by using the torque equation, calculated about the joint so that F_J does not enter:

$$(0.050 \text{ m})(F_M) - (0.15 \text{ m})(2.0 \text{ kg})(g) - (0.35 \text{ m})(5.0 \text{ kg})(g) = 0.$$

We solve this for F_M and find $F_M = (41 \text{ kg})(g) = 400 \text{ N}$.

(*b*) The lever arm for the muscle is reduced to $(0.050 \text{ m}) \sin 45° = 0.035 \text{ m}$. But the lever arms for the downward forces are reduced by the same ratio; so the same result will be obtained, $F_M = 400 \text{ N}$. Note in each case that the force required of the muscle is quite large compared to the weight of the object lifted. Indeed, the muscles and joints of the body are generally subjected to quite large forces. The point of insertion of a muscle (where it attaches to the bone) varies from person to person. A slight increase in the point of insertion of the biceps muscle from 5.0 cm to 5.5 cm can be a considerable advantage for lifting and throwing. Indeed, top athletes are often found to have muscle insertions farther from the joint than the average person.

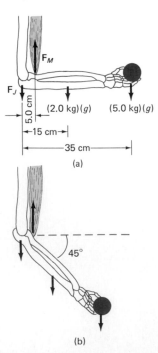

FIGURE 11–7

Example 11–4.

11–4 Elasticity and Elastic Moduli; Stress and Strain

In the first part of this chapter we studied how to calculate the forces on objects in equilibrium. In this section we study the effects of these forces, for any object changes shape under the action of applied forces. If the forces are great enough, the object will break or *fracture*, and we discuss this in Section 11–5.

If a force is exerted on an object, such as the vertically suspended metal bar shown in Fig. 11–8, the length of the object changes. If the amount of elongation, ΔL, is small compared to the length of the object, experiment shows that ΔL is proportional to the weight or force exerted on the object; this was first noted by Robert Hooke (1635–1703). This proportionality can be written as an equation and is sometimes called Hooke's law (although it is not truly a "law"—see footnote in Section 6–3:

$$F = k\,\Delta L. \qquad (11-3)$$

Here F represents the force (or weight) pulling on the object, ΔL is the increase in length, and k is a proportionality constant. Equation 11–3 is found to be valid for almost any solid material from iron to bone, but it is valid only up to a point. For if the force is too great, the object stretches excessively and eventually breaks. Figure 11–9 shows a typical graph of elongation versus applied force. Up to a point called the *elastic limit*, the object will return to its original length if the applied force is removed. This is called the *elastic region*. If the object is stretched beyond the elastic limit, it will be permanently deformed. For most common materials Eq. 11–3 is a good approximation almost up to the elastic limit and the graph is a straight line. Beyond this point, the graph deviates from a straight line and no simple relationship exists between F and ΔL. If the object is stretched much beyond the elastic limit, it will break. The maximum force that can be applied without breaking is called the *ultimate strength* of the material.

The amount of elongation of an object, such as the bar shown in Fig. 11–8, depends not only on the force applied to it, but also on the material from which it is made and on its dimensions. That is, the constant k in Eq. 11–3 can be written in terms of these factors. If we compare bars made of the same material but of different lengths and cross-sectional areas, it is found that for the same applied force the amount of stretch (again assumed small compared to the total length) is proportional to the original length and inversely proportional to the cross-sectional area. That is, the longer the object, the more it elongates for a given force; and the fatter it is, the less it elongates. These experimental findings can be combined with Eq. 11–3 to yield the relation

$$\Delta L = \frac{1}{E}\frac{F}{A}L_0, \qquad (11-4)$$

where L_0 is the original length of the object, A is the cross-sectional area, and ΔL is the change in length due to the applied force F; E is a constant of proportionality[†] known as the *elastic modulus*, or *Young's modulus*, and its value depends only on the material. The value of Young's modulus for various materials is given in Table 11–1. Because E is a property only of the material and is independent of the object's size or shape, Eq. 11–4 is far more useful for practical calculation than Eq. 11–3. From Eq. 11–4 we see that the change in length of an object is directly proportional to the product of the object's length L_0 and the force per unit area F/A applied to it. It is general practice to define the force per unit area, as the *stress*:

$$\text{stress} = \frac{\text{force}}{\text{area}} = \frac{F}{A}.$$

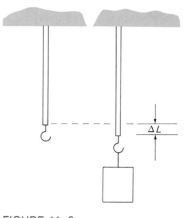

FIGURE 11–8
Hooke's law: $\Delta L \propto$ applied force.

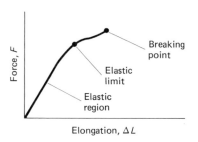

FIGURE 11–9
Applied force vs. elongation for a typical solid.

[†] The fact that E is in the denominator so that $1/E$ is the actual proportionality constant is merely a convention.

Also, the *strain* is defined to be the ratio of the change in length to the original length:

$$\text{strain} = \frac{\text{change in length}}{\text{original length}} = \frac{\Delta L}{L_0}.$$

Strain is thus the fractional change in length of the object and is a measure of how much the bar has been deformed. Equation 11–4 can be rewritten as

$$\frac{F}{A} = E\frac{\Delta L}{L_0} \qquad\qquad (11\text{–}5)$$

or

$$E = \frac{F/A}{\Delta L/L_0} = \frac{\text{stress}}{\text{strain}}.$$

Thus we see that the strain is directly proportional to the stress.

EXAMPLE 11–5   A 1.60-m-long steel piano wire has a diameter of 0.20 cm. How great is the tension in the wire if it stretches 0.30 cm when tightened?

SOLUTION   We solve for F in Eq. 11–4 and note that the area $A = \pi r^2 = (3.14)(0.0010\ \text{m})^2 = 3.1 \times 10^{-6}\ \text{m}^2$. Then

$$F = E\frac{\Delta L}{L_0}A$$

$$= (2.0 \times 10^{11}\ \text{N/m}^2)\left(\frac{0.0030\ \text{m}}{1.60\ \text{m}}\right)(3.1 \times 10^{-6}\ \text{m}^2) = 1200\ \text{N},$$

where we obtained the value for E from Table 11–1.

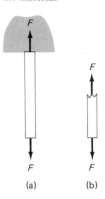

FIGURE 11–10

Stress exists *within* the material.

(a)   (b)

The bar shown in Fig. 11–8 is said to be under *tension* or *tensile stress*. For not only is there a force pulling down on the bar at its lower end, but since the bar is in equilibrium we know that the support at the top is exerting an equal upward force[†] on the bar at its upper end, Fig. 11–10a. In fact, this tensile stress exists throughout

[†] If we ignore the weight of the bar.

TABLE 11–1

Elastic moduli

Material	Young's Modulus, $E(\text{N/m}^2)$	Shear Modulus, $G\ (\text{N/m}^2)$	Bulk Modulus, $B\ (\text{N/m}^2)$
Solids			
Iron, cast	100×10^9	40×10^9	90×10^9
Steel	200×10^9	80×10^9	140×10^9
Brass	100×10^9	35×10^9	80×10^9
Aluminum	70×10^9	25×10^9	70×10^9
Concrete	20×10^9		
Brick	14×10^9		
Marble	50×10^9		70×10^9
Granite	45×10^9		45×10^9
Wood (pine)			
(parallel to grain)	10×10^9		
(perpendicular to grain)	1×10^9		
Nylon	5×10^9		
Bone (limb)	15×10^9	80×10^9	
Liquids			
Water			2.0×10^9
Alcohol (ethyl)			1.0×10^9
Mercury			2.5×10^9

the material. Consider for example the lower half of a suspended bar as shown in Fig. 11–10b. This lower half is in equilibrium so there must be an upward force on it to balance the downward force at its lower end. What exerts this upward force? It must be the upper part of the bar. Thus we see that external forces applied to an object give rise to internal forces, or stress, within the material itself.

Strain or deformation due to tensile stress is but one type of stress to which materials can be subjected. There are two other common types of stress: compressive and shear. *Compressive* stress is the exact opposite of tensile stress. Instead of being stretched, the material is compressed: the forces act inwardly on the body. Any sort of column that supports a weight, such as the columns of a Greek temple (Fig. 11–11) is subjected to a compressive stress. Equations 11–4 and 11–5 apply equally well to compression and tension, and the values for E are usually the same.

Figure 11–12 compares tensile and compressive stresses as well as the third type, shear stress. An object under *shear* stress has equal and opposite forces applied *across* its opposite faces. An example is a book or brick firmly attached to a table top with a force exerted parallel to the top surface; the table exerts an equal and opposite force along the bottom surface. Although the dimensions of the object do not change significantly, the shape of the object does change as shown in the figure. An equation similar to 11–4 can be applied to calculate shear strain:

$$\Delta L = \frac{1}{G} \frac{F}{A} L_0 \qquad (11–6)$$

but ΔL, L, and A must be reinterpreted as indicated in Fig. 11–12c. Note that A is the area of the surface *parallel* to the applied force (and not perpendicular as for tension

FIGURE 11–11
A Greek temple (in Paestum, Italy).

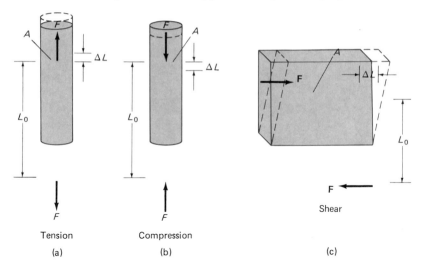

Tension
(a)

Compression
(b)

Shear
(c)

FIGURE 11–12
The three types of stress.

and compression), and ΔL is *perpendicular* to L_0. The constant of proportionality, G, is called the *shear modulus* and is generally one-half to one-third the value of the elastic modulus, E (see Table 11–1).

The rectangular object undergoing shear in Fig. 11–12c would not actually be in equilibrium under the forces shown, for a net torque would exist. If the object is in fact in equilibrium, there must be two more forces acting on it which balance out this torque. One acts vertically upward on the right, and the other acts vertically downward on the left as shown in Fig. 11–13. This is generally true of shear forces. If the object is a brick or book lying on a table, these two additional forces can be exerted by the table and whatever exerts the other horizontal force.

If an object is subjected to a pressure on all sides, its volume will be compressed. A common situation is a body submerged in a fluid; for in this case, the fluid exerts a pressure on the object in all directions, as we shall see in Chapter 12. Pressure is defined as force per unit area and thus is the equivalent of "stress." For

FIGURE 11–13
Balance of forces and torques for shear stress.

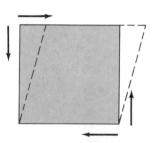

this situation the change in volume, ΔV, is found to be proportional to the original volume V_0 and to the increase in the pressure ΔP. We thus obtain a relation of the same form as Eq. 11−4 but with a proportionality constant called the *bulk modulus*, B:

$$\frac{\Delta V}{V_0} = -\frac{1}{B}\Delta P$$

or

$$B = -\frac{\Delta P}{\Delta V/V_0}. \qquad (11-7a)$$

The minus sign is included to indicate that the volume *decreases* with an increase in pressure. The bulk modulus is sometimes given a more general definition in terms of differentials:

$$B = -V\frac{dP}{dV}. \qquad (11-7b)$$

Values for the bulk modulus are given in Table 11−1. Since liquids and gases do not have a fixed shape, only the bulk modulus applies to them.

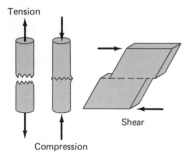

Tension

Shear

Compression

FIGURE 11−14

Fracture as a result of the three types of stress.

 11−5   Fracture

If the stress on a solid object is too great, the object fractures or breaks, Fig. 11−14. In Table 11−2 are listed the ultimate tensile strength, compressive strength, and shear strength for a variety of materials. These give the maximum force per unit area that an object can withstand under each of these three types of stress. They are, however, representative values only and the actual value for a given specimen can differ considerably. It is therefore necessary to maintain a safety factor of from 3 to perhaps 10 or more—that is, the actual stresses on a structure should not exceed one-tenth to one-third the values given in the table. One sometimes encounters tables of the "allowable stresses" in which an appropriate safety factor has already been included.

TABLE 11−2
Ultimate strengths of materials (force/area)

Material	Tensile Strength (N/m^2)	Compressive Strength (N/m^2)	Shear Strength (N/m^2)
Iron, cast	170×10^6	550×10^6	170×10^6
Steel	500×10^6	500×10^6	250×10^6
Brass	250×10^6	250×10^6	200×10^6
Aluminum	200×10^6	200×10^6	200×10^6
Concrete	2×10^6	20×10^6	2×10^6
Brick		35×10^6	
Marble		80×10^6	
Granite		170×10^6	
Wood (pine) (parallel to grain)	40×10^6	35×10^6	5×10^6
(perpendicular to grain)		10×10^6	
Nylon	500×10^6		
Bone (limb)	130×10^6	170×10^6	

EXAMPLE 11–6 (*a*) What minimum cross-sectional area should the supports have to support the beam of Example 11–2 (Fig. 11–5) assuming the supports are made of concrete and a safety factor of 6 is required? (*b*) How much will the chosen supports compress under the given load?

SOLUTION (*a*) The right-hand support receives the larger force, 1.2×10^5 N. It is clearly under compression and from Table 11–2 we see that the ultimate compressive strength of concrete is 2.0×10^7 N/m². Using a safety factor of 6, the maximum allowable stress is 3.3×10^6 N/m². Since $F/A = 3.3 \times 10^6$ N/m² and $F = 1.2 \times 10^5$ N we can solve for A and find

$$A = \frac{1.2 \times 10^5 \text{ N}}{3.3 \times 10^6 \text{ N/m}^2} = 3.6 \times 10^{-2} \text{ m}^2 \text{ or } 360 \text{ cm}^2.$$

A support 18 cm × 20 cm will be adequate.
   (*b*) We solve for

$$\frac{\Delta L}{L_0} = \frac{1}{E}\frac{F}{A} = \left(\frac{1}{2.0 \times 10^{10} \text{ N/m}^2}\right)(3.3 \times 10^6 \text{ N/m}^2) = 1.7 \times 10^{-4}.$$

Thus, if the support has a length $L_0 = 5.0$ m, $\Delta L = 0.85 \times 10^{-3}$ m, or about 1 mm. This calculation was for the right-hand support. If the left-hand support is made of the same cross-sectional area, it will compress less and this should be taken into account.

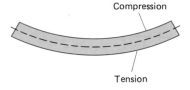

FIGURE 11–15

A beam sags, at least a little (but is exaggerated here), even under its own weight. The beam thus changes shape so that the upper portion is compressed, and the lower portion is elongated (under tension). Shearing stress also occurs within the beam.

As can be seen in Table 11–2, concrete (like stone and brick) is reasonably strong under compression but extremely weak under tension. Thus concrete can be used as vertical columns placed under compression but is of little value as a beam since it cannot withstand the tensile forces that arise (see Fig. 11–15). Reinforced concrete, in which iron rods are embedded in the concrete, is much stronger in terms of stability; but the concrete on the lower edge of a loaded beam still tends to crack since concrete is weak under tension. This problem is solved with prestressed concrete which also contains iron rods or a wire mesh, but during the pouring of the concrete, the rods or wire are held under tension. After the concrete dries, the tension on the iron is released and this puts the concrete under compression. The amount of compressive stress is carefully predetermined so that when the design loads are applied to the beam, they reduce the compression on the lower edge but never put the concrete into tension.

Summary

A body, or system of bodies, on which the total net force is zero is said to be in *equilibrium*. The subject of determining the forces on or in a body at rest in equilibrium is called *statics*.

As far as translational and rotational motion of a body is concerned, the effect of gravity acting on all parts of the body (assuming uniform **g**) is equivalent to a single force acting at the *center of gravity* (cg) of the body. The strength of this force is simply the product of the body's total mass and the acceleration due to gravity, $M\mathbf{g}$. If the acceleration of gravity is the same at all parts of the body, the cg and the cm of the body are the same.

For a body to be in equilibrium, the sum of all forces and all torques on the body must be zero:

$$\sum \mathbf{F} = 0, \quad \sum \boldsymbol{\tau} = 0.$$

These two vector equations are equivalent to the six component scalar equations (x, y, and z for each). If all the forces are in a plane (say the xy plane) these reduce to only three equations:

$$\sum F_x = 0, \quad \sum F_y = 0, \quad \sum \tau_z = 0.$$

All materials are elastic to some extent. Materials may undergo *tension* (stretching), *compression*, or *shear*. For each type, the *stress* is defined as the force per unit area and the *strain* as the change in length divided by the original length. The strain is directly proportional to the stress as long as the *elastic limit* is not exceeded. If the stress exceeds the *ultimate strength* of the material, the object fractures.

 Questions

1. Discuss what is meant by "the center of gravity of the moon."

2. If the cg and cm of a body were in different locations, describe what the motion of the body would be if it were thrown vertically into the air with no initial rotation? Explain and compare to the case of cm and cg having the same location.

3. Describe several situations where a body is not in equilibrium, even though the net force on it is zero.

4. A bear sling, Fig. 11–16, is used in some national parks for placing backpackers' food out of the reach of bears. Explain why the force needed to pull the backpack up increases as the backpack gets higher and higher.

5. A ladder, leaning against a wall, makes a 60° angle with the ground. When is it more likely to slip: when a person stands near the top or near the bottom?

6. If the net torque on a body is zero when calculated about a point 0, will it necessarily be zero when calculated about a different point 0'?

7. Explain why touching your toes while seated on the floor with outstretched legs produces less stress on the lower spinal column than when touching your toes from a standing position. Use a diagram.

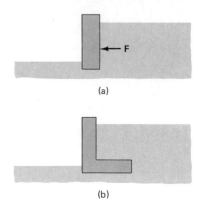

(a)

(b)

FIGURE 11–17

8. An earth retaining wall is shown in Fig. 11–17a. The earth, particularly when wet, can exert a significant force **F** on the wall. (*a*) What force produces the torque to keep the wall upright? (*b*) Explain why the retaining wall in Fig. 11–17b would be much less likely to overturn.

9. Examine how a pair of scissors or shears cuts through a piece of cardboard. Is the name "shears" justified?

10. Materials such as ordinary concrete and stone are very weak under tension or shear. Would it be wise to use such a material for either of the supports of the cantilever shown in Fig. 11–18? If so, which one(s)?

FIGURE 11–16

FIGURE 11–18

A cantilever is a beam that extends beyond its supports.

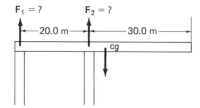

 Problems

SECTION 11–3

1. (I) What should be the tension in the wire of the orthodontic treatment of Fig. 11–19 if the net force exerted on the tooth indicated is to be 0.60 N?

2. (I) A 150-kg horizontal beam is supported at each end. A 200-kg piano rests a quarter of the way from one end. What is the vertical force on each of the supports?

3. (I) Two cords support a chandelier in the manner shown in Fig. 11–4 except that the upper wire makes an angle of 45° with the ceiling. If the cords can sustain a force of 1000 N without breaking, what is the maximum chandelier mass that can be supported?

4. (I) (*a*) Calculate the force required of the "deltoid" muscle, **F**_M, to hold up the outstretched arm shown in Fig. 11–20.

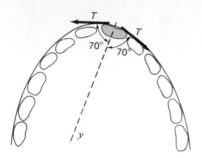

FIGURE 11–19

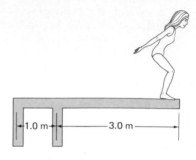

FIGURE 11–21

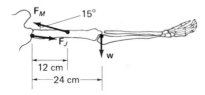

FIGURE 11–20

The total mass of the arm is 2.8 kg. (b) What would the force, $\mathbf{F_M}$, be if the hand in the above problem holds a 10-kg mass, assuming the mass is 50 cm from the shoulder joint?

5. (II) Calculate the magnitude and direction of the force $\mathbf{F_J}$ exerted by the shoulder on the upper arm at the joint for both parts of problem 4.

6. (II) A 20-lb ball is supported from the ceiling by rope A. Rope B pulls downward and to the side on the ball. If the angle of A to the vertical is 20° and if B makes an angle of 50° to the vertical, find the tensions in ropes A and B.

7. (II) A door 2.35 m high and 1.10 m wide has a mass of 13.0 kg. A hinge 0.35 m from the top and another 0.35 m from the bottom each support half the door's weight. Assume that the center of gravity is at the geometrical center of the door and determine the horizontal and vertical force components exerted by each hinge on the door.

8. (II) Calculate $\mathbf{F_1}$ and $\mathbf{F_2}$ for the 11,000-kg uniform cantilever shown in Fig. 11–18. Give magnitude and direction for each.

9. (II) Calculate the forces $\mathbf{F_1}$ and $\mathbf{F_2}$ that the supports exert on the diving board of Fig. 11–21 when a 60-kg person stands at its tip. (a) Ignore the weight of the board. (b) Take

into account the board's mass of 35 kg. Assume the board's cg is at its center.

10. (II) A tightly stretched high wire is 50 m long and sags 3.8 m when a 60 kg tightrope walker stands at its center. What is the tension in the wire? Is it possible to increase the tension in the wire so that there is no sag?

11. (II) A large 80-kg board leans at a 45° angle against the edge of a barn door, 2.6 m from the hinge. How great a horizontal force must a person behind the door exert (at the edge) in order to open it? Assume there is negligible friction between the door and the board, but that the board is firmly set against the ground.

12. (II) Repeat the above problem assuming the coefficient of friction between the board and the door is 0.45.

13. (II) A 160-cm tall person lies on a light (massless) board which is supported by two scales, one under the feet and one beneath the top of the head. The two scales read, respectively, 32.8 kg and 29.8 kg. Where is the center of gravity of this person?

14. (II) The two trees in Fig. 11–16 are 12.5 m apart. Calculate the magnitude of the force $\mathbf{F}$ a backpacker must exert to hold the 15-kg backpack so the rope sags at its midpoint by (a) 2.0 m, (b) 0.20 m.

15. (II) In Fig. 11–22, consider the northernmost section of the Golden Gate Bridge which has a length $d_1 = 343$ m. Assume the cg of the span shown is halfway between the tower and anchor. Determine T_1 and T_2 (which act on the northernmost cable) in terms of mg, the weight of the northernmost span, and calculate the height h needed for equilibrium. Assume the roadway is supported only by the suspension cable. (Hint: T_3 does not act on this section.)

16. (II) Assume that a single-span suspension bridge such as the Golden Gate Bridge has a configuration as indicated

FIGURE 11–22

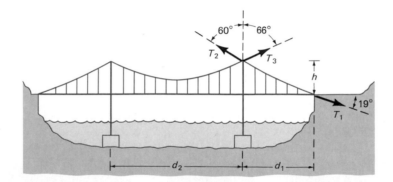

in Fig. 11–22. Assume that the roadway is uniform over the length of the bridge and that each segment of the suspension cable provides the sole support for the roadway directly below it. The ends of the cable are anchored to the ground only, not the roadway. What must the ratio of d_2 to d_1 be so that the suspension cable exerts no net horizontal force on the towers. Neglect the mass of the cables and the fact that the roadway isn't exactly horizontal.

17. (II) A 9.5-m-long ladder of mass 16.0 kg leans against a smooth wall (so the force exerted by the wall, $\mathbf{F_w}$, is perpendicular to the wall). The ladder makes a 20° angle with the vertical wall; and the ground is rough. (a) Calculate the components of the force exerted by the ground on the ladder at its base, and (b) determine what the coefficient of friction at the base of the ladder must be if the ladder is not to slip when a 75-kg person stands three-fourths of the way up the ladder.

18. (II) If the coefficient of friction between the ladder and the ground in the situation described in the preceding problem is 0.40, how far up the ladder can the person climb before the ladder starts to slip?

19. (II) Figure 11–23 shows a round arch and a pointed (Gothic) arch. They have the same 8.0-m span and support the wall (and roof) of a building. Consider each arch as made up of two "half arches," connected (or pivoted) at the uppermost point; in the diagram, the load of the wall above is indicated by a single force on each half arch. (a) Show that a horizontal buttressing force, F_H, is needed to keep either arch from falling. (b) Show that the buttressing force need be only half as great for the pointed arch as for the round one, assuming identical loads above. (c) What can you say about the technological advantages of the Gothic arch?

FIGURE 11–23

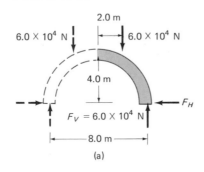

2.0 m

6.0×10^4 N 6.0×10^4 N

4.0 m

$F_V = 6.0 \times 10^4$ N F_H

8.0 m

(a)

2.0 m

6.0×10^4 N 6.0×10^4 N

8.0 m

$F_V = 6.0 \times 10^4$ N F_H

8.0 m

(b)

FIGURE 11–24

20. (II) A uniform flexible steel cable of weight mg is suspended between two equal elevation points as shown in Fig. 11–24. Determine the tension in the cable (a) at its lowest point, (b) at the points of attachment. What is the direction of the tension force in each case?

21. (II) A 50-story building is to be 210 m high with a base 40 m by 70 m. Its total mass will be about 1.60×10^7 kg. Assume the building's cg is at its center. (a) Will the building tip over in a 200 km/h wind striking its 70-m wide face? [Hint: assume the building rests on the ground without anchoring; assume the wind is brought to rest upon striking the building; the density of air is 1.29 kg/m³.] (b) What minimum wind velocity could cause the building to tip?

22. (II) A home mechanic wants to raise the 220-kg engine out of a car. The plan is to stretch a rope vertically from the engine to a branch of a tree 25 m above. The rope passes over a firmly placed pulley just above the engine so that when the mechanic climbs halfway up the tree and pulls horizontally on the rope at its midpoint, the engine rises out of the car. How much force must the mechanic exert to raise the engine 0.50 m?

23. (II) Two guy wires run from the top of a pole 2.4 m tall that supports a volleyball net. The two wires are anchored to the ground 2.0 m apart and each is 2.0 m from the pole. The tension in each wire is 65 N. What is the tension in the net, assumed horizontal and attached at the top of the pole?

24. (II) A 30-kg round table is supported by three legs placed equal distances apart on the edge. What minimum mass, placed on the table's edge, will cause the table to overturn?

25. (II) A person wants to push a lamp (mass 9.6 kg) across the floor. Assuming the person pushes at a height of 60 cm

FIGURE 11–25

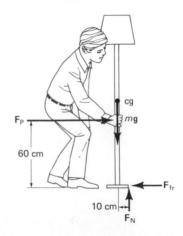

F_P

cg

mg

60 cm

F_{fr}

10 cm

F_N

above the ground and the coefficient of friction is 0.20, determine (a) whether the lamp will slide or tip over (Fig. 11–25). (b) Calculate the maximum height above the floor at which the person can push the lamp so it slides rather than tips.

26. (III) When walking, a person momentarily puts all the weight on one foot. The cg of the body lies over the supporting foot. Figure 11–26 shows the supporting leg and the forces on it. Calculate the force exerted by the "hip abductor" muscles, $\mathbf{F_M}$, and the x and y components of the force $\mathbf{F_J}$ acting on the joint. Take the whole leg as the object under discussion.

27. (III) (a) Using the information in problem 26 calculate $\mathbf{F_M}$ and $\mathbf{F_J}$, assuming the person carrries a 20-kg suitcase in each hand. Assume the cg of each suitcase lies directly below the edge of the hipbone. (b) Calculate $\mathbf{F_M}$ and $\mathbf{F_J}$ assuming the person carries a suitcase only in the hand on the side opposite the supporting leg. [Hint: first calculate the common cg of the person plus suitcase; this point will be above the foot, thus shifting the horizontal measurements from those shown in Fig. 11–26. Note that these forces are larger when a person is carrying one 20-kg bag than when carrying two 20-kg bags, one in each hand!]

28. (III) Four bricks are to be stacked at the edge of a table, each brick overhanging the one below it, so that the top brick extends as far as possible beyond the edge of the table. (a) To achieve this, show that the bricks must extend no more than (starting at the top) 1/2, 1/4, 1/6, and 1/8 of their length beyond the one below. (b) Is the top brick completely beyond the base? (c) Determine a general formula for the maximum total distance spanned by n bricks if they are to remain stable.

FIGURE 11–26

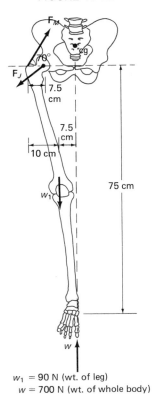

$w_1 = 90$ N (wt. of leg)
$w = 700$ N (wt. of whole body)

SECTION 11–4

29. (I) A marble column of cross-sectional area 2.0 m² supports a mass of 30,000 kg. (a) What is the stress within the column? (b) What is the strain? (c) By how much is the column shortened if it is 10.5 m high?

30. (II) How much force is required to stretch a steel piano wire of 0.052 cm diameter by 0.015 percent?

31. (II) How much pressure is needed to compress the volume of an iron block by 0.11 percent? Express answer in N/m², and compare it to atmospheric pressure (1.0×10^5 N/m²).

32. (II) One liter of alcohol (1000 cm³) in a flexible container is carried to the bottom of the sea where the pressure is 2.8×10^6 N/m². What will its volume be now?

33. (II) A 16-cm-long animal tendon was found to stretch 3.3 mm by a force of 12.4 N. The tendon was approximately round with an average diameter of 8.6 mm. Calculate the elastic modulus of this tendon.

34. (II) A scallop forces open its shell with an elastic material called abductin whose elastic modulus is about 2.0×10^6 N/m². If this piece of abductin is 3.0 mm thick and has a cross-sectional area of 0.50 cm², how much potential energy does it store when compressed 1.0 mm?

35. (III) A pole projects horizontally from the front wall of a shop. A 4.5-kg sign hangs from the pole at a point 2.1 m from the wall. (a) What is the torque due to this sign calculated about the point where the pole meets the wall? (b) If the pole is not to fall off, there must be another torque exerted to balance it. What exerts this torque? Use a diagram to show how this torque must act. (c) Discuss whether compression, tension, and/or shear play a role in part (b).

SECTION 11–5

36. (I) The femur bone in the leg has an average effective cross-sectional area of about 3.0 cm². How much compressive force can it withstand before breaking?

37. (II) What is the minimum cross-sectional area required of a vertical steel wire from which is suspended a 280-kg chandelier? Assume a safety factor of 5.0.

38. (II) An iron bolt is used to connect two iron plates together. The bolt must withstand forces up to about 2500 N. Calculate the minimum diameter for the bolt, based on a safety factor of 4.5.

39. (II) In Example 8–8 in Chapter 8 we calculated the impulse and average force on the leg of a person who jumps 5.0 m down to the ground. If the legs are not bent upon landing, so the body moves a distance d of only about $d = 1.0$ cm during collision, determine (a) the stress in the tibia bone (area $= 3.0 \times 10^{-4}$ m²) and (b) whether or not the bone will break. (c) Repeat for a bent-knees landing ($d = 50.0$ cm).

40. (II) Assume the supports of the cantilever shown in Fig. 11–18 are made of wood. Calculate the minimum cross-sectional area required of each, assuming a safety factor of 8.5 and a cantilever mass of 11,000 kg.

41. (II) The roof of a 7.5 m × 10.0 m room in a school is to support a uniformly distributed load of 10,000 kg (133 kg/m²) in addition to its own mass of 4100 kg. The roof is to be supported by "2 × 4s" (actually about 4.0 cm × 9.0 cm) along the 10.0-m sides. How many supports are

required on each side and how far apart must they be? Consider only compression and assume a safety factor of 15.

42. (II) There is a maximum height of a uniform vertical column made of any material that can support itself without buckling, and it is independent of the cross-sectional area (why?). Calculate this height for (a) steel (density 7.8×10^3 kg/m^3), (b) granite (density 2.7×10^3 kg/m^3).

43. (II) A steel cable is to support an elevator whose total (loaded) mass is not to exceed 2500 kg. If the maximum acceleration of the elevator is 1.5 m/s^2, calculate the diameter of cable required. Assume a safety factor of 5.0.

44. (III) From what height must a 1.2-kg rectangular brick 15.0 cm $\times$ 6.0 cm $\times$ 4.0 cm be dropped above a rigid steel floor in order to break the brick? Assume the brick strikes the floor directly on its largest face, and that the compression of the brick is much greater than that of the steel (that is, ignore compression of the steel).

Fluids at Rest

12

The three common states, or *phases*, of matter are solid, liquid, and gas. We normally distinguish these three phases as follows. A *solid* maintains a fixed shape and a fixed size; even if a large force is applied to a solid, it does not readily change its shape or volume. A *liquid* cannot sustain a shear stress and cannot maintain a fixed shape—it takes on the shape of its container—but like a solid it is not readily compressible and its volume can be changed significantly only by a very large force. A *gas* has neither a fixed shape nor a fixed volume—it will expand to fill its container. For example when air is pumped into an automobile tire, the air does not all run to the bottom of the tire as a liquid would; it fills the whole volume of the tire. Since liquids and gases do not maintain a fixed shape, they each have the ability to flow; they are thus often referred to collectively as **fluids**.

The division of matter into three states is not always simple. How, for example, should butter be classified? Furthermore a fourth state of matter can be distinguished, the *plasma* state, which occurs only at very high temperatures, where electrons exist free from their atoms which are therefore ions. Some scientists believe that so-called colloids (suspensions of tiny particles in a liquid) should also be considered a separate state of matter. However, for our present purposes we will mainly be interested in the three ordinary states of matter. In earlier chapters we discussed the motion of solid (rigid) bodies and some of the properties of solid materials (particularly in Chapter 11). In this chapter we will discuss the behavior of fluids which are at rest, and, in Chapter 13, the behavior of fluids that are in motion.

 ## 12–1 Density and Specific Gravity

It is sometimes said that iron is "heavier" than wood. This cannot really be true since a large log clearly weighs more than an iron nail. What we should say is that iron is more *dense* than wood.

The **density**, ρ, of a substance is defined as its mass per unit volume:

$$\rho = \frac{m}{V} \tag{12–1}$$

where m is the mass of an amount of the substance whose volume is V. Density is a characteristic property of a substance; objects made of a given substance, say pure iron, can have any size or mass, but the density will be the same for each.

The SI unit for density is kg/m³. Sometimes densities are given in g/cm³. Note that since 1 kg/m³ = 1000 g/(100 cm)³ = 10^{-3} g/cm³, a density given in g/cm³ must be multiplied by 1000 to give the result in kg/m³. Thus the density of aluminum is ρ = 2.70 g/cm³ which is equal to 2700 kg/m³. The densities of a variety of substances are given in Table 12–1. Temperature and pressure affect the density of substances (although the effect is slight for liquids and solids) and hence the temperature is specified in Table 12–1.

 EXAMPLE 12–1   What is the mass of a lead sphere of radius 0.500 m.

SOLUTION   The volume of the sphere is

$$V = \tfrac{4}{3}\pi r^3 = \tfrac{4}{3}(3.14)(0.500 \text{ m})^3 = 0.523 \text{ m}^3.$$

From Table 12–1, the density of lead is ρ = 11,300 kg/m³, so we have from Eq. 12–1:

$$m = \rho V = (11{,}300 \text{ kg/m}^3)(0.523 \text{ m}^3) = 5910 \text{ kg}.$$

TABLE 12–1

Densities of substances[†]

Substance	Density, ρ (kg/m³)
Solids	
Aluminum	2.70×10^3
Iron and Steel	7.8×10^3
Copper	8.9×10^3
Lead	11.3×10^3
Gold	19.3×10^3
Concrete	2.3×10^3
Granite	2.7×10^3
Wood (typical)	$0.3\text{–}0.9 \times 10^3$
Glass, common	$2.4\text{–}2.8 \times 10^3$
Ice	0.917×10^3
Bone	$1.7\text{–}2.0 \times 10^3$
Liquids	
Water (4°C)	1.00×10^3
Blood plasma	1.03×10^3
Blood whole	1.05×10^3
Sea water	1.025×10^3
Mercury	13.6×10^3
Alcohol, ethyl	0.79×10^3
Gasoline	0.68×10^3
Gases	
Air	1.29
Helium	0.179
Carbon dioxide	1.98
Water (steam) (100°C)	0.598

[†] Densities are given at 0°C and 1 atm pressure unless otherwise specified.

The **specific gravity** of a substance is defined as the ratio of the density of that substance to the density of water at 40°C. Specific gravity (abbreviated SG) is a pure number, without dimensions or units. Since the density of water is $1.00 \text{ g/cm}^3 = 1.00 \times 10^3 \text{ kg/m}^3$, the specific gravity of any substance will be precisely equal numerically to its density specified in g/cm³, or 10^{-3} times its density specified in kg/m³. For example (see Table 12–1) the specific gravity of lead is 11.3 and of alcohol 0.79.

12–2 Pressure in Fluids

Pressure is defined as force per unit area, where the force F is understood to be acting perpendicular to the surface area A:

$$\text{pressure} = P = \frac{F}{A}. \qquad (12\text{–}2)$$

The SI unit of pressure is N/m². This unit has the official name pascal (Pa): 1 Pa = 1 N/m²; however, for simplicity, we will often use N/m². Other units sometimes used are dynes/cm², 1b/in² (sometimes abbreviated "psi"), and kg/cm² (as if kilograms were a force: that is, 1 kg/cm² = 9.8 N/cm² = 9.8 × 10⁴ N/m²). The last two are often used on tire gauges. We will meet several other units shortly.

As an example of calculating pressure, a 60-kg person whose two feet cover an area of 500 cm² exerts a pressure of $F/A = mg/A = (60 \text{ kg})(9.8 \text{ m/s}^2)/(0.050 \text{ m}^2) = 12 \times 10^3 \text{ N/m}^2$ on the ground. If the person stands on one foot, the force is the same but the area will be half, so the pressure will be twice as much: 24 × 10³ N/m². The concept of pressure is particularly useful in dealing with fluids, which is why we introduced it in this chapter.

It is an experimental fact that *a fluid exerts a pressure in all directions*. This is well known to swimmers and divers who feel the water pressure on all parts of their bodies. At a particular point in a fluid at rest, the pressure is the same in all directions. This is illustrated in Fig. 12–1; consider a tiny cube of the fluid which is so small (infinitesimal) that we can ignore the force of gravity on it. Then the pressure on one side of it must equal the pressure on the opposite side. If this weren't true, the net force on this cube would not be zero, and it would move until the pressure did become equal; if the fluid is not flowing, then the pressures must be equal.

Another important property of a fluid at rest is that the force due to fluid pressure always acts *perpendicularly* to any surface it is in contact with. If there were a component of the force parallel to the surface as shown in Fig. 12–2 then according to Newton's third law, the surface would exert a force back on the fluid that also would have a component parallel to the surface; this component would cause the fluid to flow, in contradiction to our assumption that the fluid is at rest.

Let us now calculate quantitatively how the pressure in a liquid of uniform density varies with depth. Consider a point which is at a depth h below the surface of the liquid (that is, the surface is a height h above this point) as shown in Fig. 12–3. The pressure due to the liquid at this depth h is due to the weight of the column of liquid above it. Thus the force acting on the area is $F = mg = \rho Ahg$ where Ah is the volume of the column, ρ is the density of the liquid (assumed to be constant), and g is the acceleration of gravity. The pressure, P, is then

$$P = \frac{F}{A} = \frac{\rho Ahg}{A}$$

$$P = \rho gh. \qquad \text{[liquid]} \qquad (12\text{–}3)$$

Thus the pressure is directly proportional to the density of the liquid, and to the depth within the liquid. In general, the pressure at equal depths within a uniform liquid is the same.

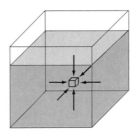

FIGURE 12–1

Pressure is the same in every direction in a fluid at a given depth; if it weren't, the fluid would be in motion.

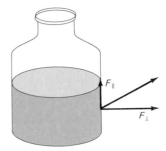

FIGURE 12–2

The force a non-moving fluid exerts on a solid surface is perpendicular to the surface; that is, $F_{\parallel} = 0$.

FIGURE 12–3

Calculating the pressure at a depth h in a liquid.

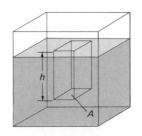

Equation 12–3 tells us what the pressure is at a depth h in the liquid, due to the liquid itself. But what if there is additional pressure exerted at the surface of the liquid, such as the pressure of the atmosphere? And what if the density of the fluid is not constant? Gases are quite compressible and hence their density can vary significantly with depth; and liquids, too, can be compressed, although we can often ignore the variation in density. (One exception is in the depths of the ocean where the great weight of water above significantly compresses the water and increases its density.) To cover these, and other cases, we now treat the general case of determining how the pressure in a fluid varies with depth.

Consider any fluid, and let us determine the pressure at any height y above some reference point, as shown in Fig. 12–4. Within this fluid, at the height y, we consider a tiny, flat, slablike volume of the fluid whose area is A and whose (infinitesimal) thickness is dy, as shown. Let the pressure acting upward on its lower surface (at height y) be P; the pressure acting downward on the top surface of our tiny slab (at height $y + dy$) is then $P + dP$. The fluid pressure acting on our slab thus exerts a force equal to PA upward on our slab, and $(P + dP)A$ downward on it; the only other force acting vertically on the slab is the (infinitesimal) force of gravity dw which is given by

$$dw = (dm)g = \rho g\, dV = \rho g A\, dy$$

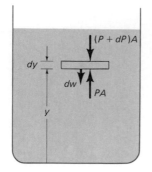

FIGURE 12–4

Forces on a flat, slablike volume of fluid for determining the pressure P at height y in the fluid.

where ρ is the density of the fluid at the height y. Since the fluid is assumed to be at rest, our slab is in equilibrium so the net force on it must be zero. Therefore we have

$$PA - (P + dP)A - \rho g A\, dy = 0,$$

which when simplified becomes

$$\frac{dP}{dy} = -\rho g. \tag{12–4}$$

This relation tells us how the pressure varies with height within the fluid. The minus sign indicates that the pressure decreases with an increase in height; or that the pressure increases with depth (reduced height).

If the pressure at a height y_1 in the fluid is P_1, and at height y_2 it is P_2, then we can integrate Eq. 12–4 to obtain

$$\int_{P_1}^{P_2} dP = -\int_{y_1}^{y_2} \rho g\, dy$$

$$P_2 - P_1 = -\int_{y_1}^{y_2} \rho g\, dy. \tag{12–5}$$

FIGURE 12–5

Pressure at a depth $h = (y_2 - y_1)$ in a liquid of density ρ is $P = P_0 + \rho g h$ where P_0 is the external pressure at the liquid's top surface.

This is a general relation and we apply it now to two special cases: (1) pressure in liquids of constant density, and (2) pressure variations in the earth's atmosphere.

For liquids in which any variation in density can be ignored, $\rho = $ constant and Eq. 12–5 is readily integrated:

$$P_2 - P_1 = -\rho g(y_2 - y_1). \tag{12–6a}$$

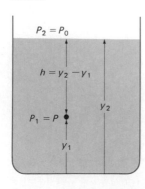

For the everyday situation of a liquid in an open container—such as water in a swimming pool, a lake, or the ocean—there is a free surface at the top. And it is convenient to measure distances from this top surface; that is, we let h be the *depth* in the liquid where $h = y_2 - y_1$ as shown in Fig. 12–5. If we let y_2 be the position of the top surface, then P_2 represents the atmospheric pressure, P_0, at the top surface. Then the pressure $P (= P_1)$ at a depth h in the fluid is, from Eq.12–6a:

$$P = P_0 + \rho g h. \tag{12–6b}$$

Note the Eq. 12–6b is the same as Eq. 12–3 for the liquid pressure, plus the pressure P_0 due to the atmosphere above.

EXAMPLE 12–2   The surface of the water in a storage tank is 30 m above a water faucet in the kitchen of a house. Calculate the water pressure at the faucet.

SOLUTION   Atmospheric pressure acts both at the surface of the water in the storage tank, and on the water leaving the faucet. The pressure difference between the inside and outside of the faucet will be $\Delta P = \rho g h = (1.0 \times 10^3 \text{ kg/m}^3)$ $(9.8 \text{ m/s}^2)(30 \text{ m}) = 2.9 \times 10^5 \text{ N/m}^2$. The height h is sometimes called the *pressure head*. In this example, the head of the water is 30 m.

Now let us apply Eq. 12–4 or 12–5 to gases. The density of gases is normally quite small, so the difference in pressure at different heights can usually be ignored if $y_2 - y_1$ is not large (which is why, in Example 12–2, we could ignore the difference in air pressure between the faucet and the top of the storage tank). Indeed, for most ordinary containers of gas, we can assume that the pressure is the same throughout. However, if $y_2 - y_1$ is very large, we cannot make this assumption. An interesting example is the earth's atmosphere whose pressure at sea level is about $1.013 \times 10^5 \text{ N/m}^2$ and decreases slowly with altitude.

EXAMPLE 12–3   (a) Determine the variation in pressure in the earth's atmosphere as a function of height y above sea level, assuming g is constant and that the density of the air is proportional to the pressure. (This last assumption is not terribly accurate, in part because temperature effects are important.) (b) At what elevation is the air pressure equal to half the pressure at sea level?

SOLUTION   (a) Since ρ is proportional to P, we have

$$\frac{\rho}{\rho_0} = \frac{P}{P_0}$$

where $P_0 = 1.013 \times 10^5 \text{ N/m}^2$ is atmospheric pressure at sea level and $\rho_0 = 1.29 \text{ kg/m}^3$ is the density of air at sea level at 0°C (Table 12–1). From Eq. 12–4 we have:

$$\frac{dP}{dy} = -\rho g$$

$$= -P\left(\frac{\rho_0}{P_0}\right)g$$

so

$$\frac{dP}{P} = -\frac{\rho_0}{P_0} g \, dy.$$

We integrate this from $y = 0$, $P = P_0$ to the height y where the pressure is P:

$$\int_{P_0}^{P} \frac{dP}{P} = -\frac{\rho_0}{P_0} g \int_0^y dy$$

$$\ln \frac{P}{P_0} = -\frac{\rho_0}{P_0} gy$$

or

$$P = P_0 e^{-(\rho_0 g/P_0)y}.$$

So, based on our assumptions, we find that the air pressure in our atmosphere decreases approximately exponentially with height. (Note that the atmosphere does not have a distinct upper limit, so there is no natural point from which to measure depth in the atmosphere, as we do for a liquid.)

(*b*) The constant $(\rho_0 g/P_0)$ has the value $(1.29 \text{ kg/m}^3)(9.80 \text{ m/s}^2)/(1.013 \times 10^5 \text{ N/m}^2) = 1.25 \times 10^{-4} \text{ m}^{-1}$. Then, when we set $P = \frac{1}{2}P_0$, we have

$$\tfrac{1}{2} = e^{-(1.25 \times 10^{-4} \text{ m}^{-1})y}$$

or

$$y = (ln \, 2.00)/(1.25 \times 10^{-4} \text{ m}^{-1}) = 5550 \text{ m}$$

since ln 2 = 0.693. Thus, atmospheric pressure drops to half what it is at sea level at an elevation of about 5500 m (about 18,000 ft); it is not surprising that mountain climbers often use oxygen tanks at very high altitudes,

12–3   Atmospheric Pressure and Gauge Pressure

The pressure of the earth's atmosphere varies with altitude, as we have seen. But even at a given place, it varies slightly according to the weather. As we've already mentioned, at sea level the pressure of the atmosphere on the average is 1.013×10^5 N/m^2 (or 14.7 lb/in^2). This value is used to define a commonly used unit of pressure, the *atmosphere* (abbreviated atm):

$$1 \text{ atm} = 1.013 \times 10^5 \text{ N/m}^2 = 1.013 \times 10^5 \text{ Pa}.$$

Another unit of pressure sometimes used (in meteorology and on weather maps) is the *bar* which is defined as 1 bar = 1.00×10^5 N/m^2; thus standard atmospheric pressure is slightly more than one bar.

The pressure due to the weight of the atmosphere is exerted on all objects immersed in this great sea of air, including our bodies. How does a human body withstand the enormous pressure on its surface? The answer lies in the fact that living cells maintain an internal pressure that just balances the external pressure. Likewise, the pressure inside a balloon balances the outside pressure of the atmosphere. An automobile tire, because of its rigidity, can maintain pressures much greater than the external pressure.

One must be careful, however, when determining the pressure in a tire or other container of gas since tire gauges and most other pressure gauges register the pressure over and above atmospheric pressure. This is called *gauge pressure*. Thus, to get the absolute pressure P, one must add the atmospheric pressure, P_a, to the gauge pressure, P_G:

$$P = P_a + P_G.$$

For example, if a tire gauge registers 220 kPa, the actual pressure within the tire is 220 kPa + 100 kPa = 320 kPa. This is equivalent to about 3.2 atm (2.2 atm gauge pressure).

12–4   Measurement of Pressure

Many devices have been invented to measure pressure, some of which are shown in Fig. 12–6. The simplest is the open-tube manometer (Fig. 12–6a) which is a U-shaped tube partially filled with a liquid, usually mercury or water. The pressure P being measured is related to the difference in height of the two levels of the liquid by the relation

$$P = P_0 + \rho g h$$

where P_0 is atmospheric pressure and ρ is the density of the liquid. Note that the quantity $\rho g h$ is the "gauge pressure"—the amount by which P exceeds atmospheric pressure (Section 12–3). If the liquid in the left-hand column were lower than that in the right-hand column, this would indicate that P was less than atmospheric pressure (and h would be negative).

FIGURE 12–6

Pressure gauges: (a) open-tube manometer; (b) Bourdon gauge; (c) aneroid gauge.

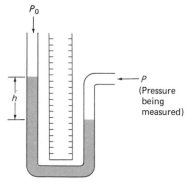

(a) Open–tube manometer

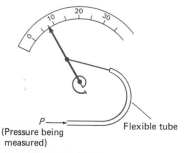

(b) Bourdon gauge

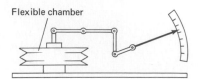

(c) Aneroid gauge (mainly used for air pressure and then is called an aneroid barometer)

Conversion factors between different units of pressure

In Terms of $1 \text{ Pa} = 1 \text{ N/m}^2$	Related to 1 atm
$1 \text{ atm} = 1.013 \times 10^5 \text{ N/m}^2$	$1 \text{ atm} = 1.013 \times 10^5 \text{ N/m}^2$
$= 1.013 \times 10^5 \text{ Pa} = 101.3 \text{ kPa}$	
$1 \text{ bar} = 1.000 \times 10^5 \text{ N/m}^2$	$= 1.013 \text{ bar}$
$1 \text{ dyne/cm}^2 = 0.1 \text{ N/m}^2$	$= 1.013 \times 10^6 \text{ dyne/cm}^2$
$1 \text{ kg/cm}^2 = 9.85 \times 10^4 \text{ N/m}^2$	$= 1.03 \text{ kg/cm}^2$
$1 \text{ lb/in}^2 = 6.90 \times 10^3 \text{ N/m}^2$	$= 14.7 \text{ lb/in}^2$
$1 \text{ lb/ft}^2 = 47.9 \text{ N/m}^2$	$= 2.12 \times 10^3 \text{ lb/ft}^2$
$1 \text{ cm Hg} = 1.33 \times 10^3 \text{ N/m}^2$	$= 76 \text{ cm Hg}$
$1 \text{ mm Hg} = 133 \text{ N/m}^2$	$= 760 \text{ mm Hg}$
$1 \text{ torr} = 133 \text{ N/m}^2$	$= 760 \text{ torr}$
$1 \text{ mm H}_2\text{O} (4°C) = 9.81 \text{ N/m}^2$	$= 1.03 \times 10^4 \text{ mm H}_2\text{O} (4°C)$

Instead of calculating the product $\rho g h$, it is common merely to specify the height h. In fact pressures are sometimes specified as so many "millimeters of mercury" (mm Hg), and sometimes as so many "mm of water." The unit mm Hg is equivalent to a pressure of 133 N/m², since $1.00 \text{ mm} = 1.00 \times 10^{-3} \text{ m}$ and

$$\rho g h = (13.6 \times 10^3 \text{ kg/m}^3)(9.8 \text{ m/s}^2)(1.00 \times 10^{-3} \text{ m})$$
$$= 1.33 \times 10^2 \text{ N/m}^2.$$

The unit mm Hg is also called the "torr" in honor of Evangelista Torricelli (1608–1647) who invented the barometer (see below). Conversion factors among the various units of pressure are given in Table 12–2. It is important that only N/m² = Pa, the proper SI unit, be used in calculations involving other quantities specified in SI units.

Other types of pressure gauge are the Bourdon gauge (Fig. 12–6b) in which increased pressure tends to straighten out the delicate tube which is attached to a pointer, and the aneroid gauge (Fig. 12–6c) in which the pointer is connected to the flexible ends of an evacuated thin metal chamber. In a more sophisticated type of gauge, the pressure to be measured is applied to a thin metal diaphragm and the resulting distortion of the diaphragm is detected electrically with the use of a so-called pressure transducer (see Chapter 27).

Atmospheric pressure is often measured by a modified kind of mercury manometer with the end closed. This is called a mercury *barometer*, Fig. 12–7. The glass tube is filled with mercury and then inverted into the bowl of mercury. If the tube is long enough, the level of the mercury will drop, leaving a vacuum at the top of the tube, since atmospheric pressure can support a column of mercury only about 76 cm high (exactly 76.0 cm at normal atmospheric pressure). That is, a column of mercury 76 cm high exerts the same pressure as the atmosphere: from the formula $P = \rho g h$, with $\rho = 13.6 \times 10^3 \text{ kg/m}^3$ for mercury and $h = 76.0 \text{ cm}$ we have

$$P = (13.6 \times 10^3 \text{ kg/m}^3)(9.80 \text{ m/s}^2)(0.760 \text{ m})$$
$$= 1.013 \times 10^5 \text{ N/m}^2$$
$$= 1.00 \text{ atm}.$$

Household barometers are usually of the aneroid type, Fig. 12–6c.

A calculation similar to that above will show that atmospheric pressure can maintain a column of water 10.3 m high in a tube whose top is under vacuum. It was a source of wonder and frustration a few centuries ago that no matter how good a vacuum pump was, it could not lift water more than about 10 m. It was, for example, a great practical difficulty in trying to pump water out of deep mine shafts which required multiple stages for depths greater than 10 m. This problem

FIGURE 12-7

A mercury barometer, when the air pressure is 76 cm Hg.

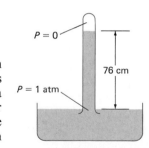

concerned Galileo, and the reason for it was first understood by Torricelli. The point is that a pump does not really suck water up a tube. It is air pressure that pushes the water up the tube if the other end is under a vacuum, just as it is air pressure that pushes (or keeps) the mercury 76 cm high in a barometer.

12–5 Pascal's Principle

The earth's atmosphere exerts a pressure on all objects with which it is in contact, including other fluids. Atmospheric pressure acting on a fluid is transmitted throughout that fluid. For instance, according to Eq. 12–6b the water pressure at a depth of 100 m below the surface of a lake is $P = \rho g h = (1000 \text{ kg/m}^3) \cdot (9.8 \text{ m/s}^2)(100 \text{ m}) = 9.8 \times 10^5 \text{ N/m}^2$ or 9.7 atm; and the total pressure at this point is due to the pressure of water plus the pressure of the air above it: the total pressure (if the lake is near sea level) is 9.7 atm + 1.0 atm = 10.7 atm. This is just one example of a general principle attributed to the French philosopher and scientist Blaise Pascal (1623–1662). **Pascal's principle** states that *pressure applied to a confined fluid increases the pressure throughout by the same amount.*

A number of practical devices make use of Pascal's principle. Two examples, hydraulic brakes in an automobile and the hydraulic lift, are illustrated in Fig. 12–8. In the case of a hydraulic lift, a small force can be used to exert a large force by making the area of one piston (the output) larger than the area of the other (the input). If the input quantities are represented by the subscript "i" and the output by "o," we have

$$P_o = P_i$$

$$\frac{F_o}{A_o} = \frac{F_i}{A_i}$$

or, finally

$$\frac{F_o}{F_i} = \frac{A_o}{A_i}.$$

FIGURE 12–8

Applications of Pascal's principle: (a) hydraulic brakes in a car; (b) hydraulic lift.

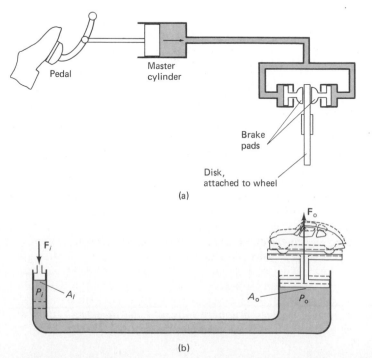

F_o/F_i is the "mechanical advantage" of the hydraulic press, and is equal to the ratio of the areas. For example if the area of the output piston is 20 times that of the input cylinder, the force is multiplied by a factor of 20; thus a force of 200 lb could lift a 4000-lb car.

 ## 12–6   Buoyancy and Archimedes' Principle

Objects submerged in a fluid appear to weigh less than they do when outside the fluid. For example, a large rock that you would have difficulty lifting off the ground can often be easily lifted from the bottom of a stream; when the rock breaks through the surface of the water it suddenly seems to be much heavier. Many objects, such as wood, float on the surface of the water. These are two examples of *buoyancy*. In each example, the force of gravity is acting downward; but in addition, an upward buoyant force is exerted by the liquid.

The buoyant force arises from the fact that the pressure in a fluid increases with depth. Thus the upward pressure on the bottom surface of a submerged object is greater than the downward pressure on its top surface. To see the effect of this, consider a cylinder of height h whose top and bottom ends have an area A and which is completely submerged in a fluid of density ρ_f as shown in Fig. 12–9. The fluid exerts a pressure $P_1 = \rho_f g h_1$ against the top surface of the cylinder; the force due to this pressure on top of the cylinder is $F_1 = P_1 A = \rho_f g h_1 A$, and it is directed downward. Similarly, the fluid exerts an upward force on the bottom of the cylinder equal to $F_2 = P_2 A = \rho g h_2 A$. The net force due to the fluid pressure, which is the *buoyant force* $\mathbf{F_B}$, acts upward and has the magnitude

$$
\begin{aligned}
F_B &= F_2 - F_1 \\
&= \rho_f g A (h_2 - h_1) \\
&= \rho_f g A h \\
&= \rho_f g V
\end{aligned}
$$

where $V = Ah$ is the volume of the cylinder. Since ρ_f is the density of the fluid, the product $\rho_f g V = m_f g$ is the weight of fluid which takes up a volume equal to the volume of the cylinder. Thus the buoyant force on the cylinder is equal to the weight of fluid displaced[†] by the cylinder. This result is valid no matter what the shape of the object. It was first discovered by Archimedes (287?–212 B.C.) and hence is called **Archimedes' principle**: *The buoyant force on a body immersed in a fluid is equal to the weight of the fluid displaced by that object.*

We can derive Archimedes' principle in general by the following simple but elegant argument. The irregularly shaped object D shown in Fig. 12–10a is acted on by the force of gravity (its weight, w) and the buoyant force, F_B. (If there is no other force acting on the object, such as a hand pulling up, the object shown will move downward since $w > F_B$.) We wish to determine F_B and to do so we next consider a body of (the same) fluid (D' in Fig. 12–10b) with the same shape and size as the original object, and located at the same depth; you might think of this body of fluid as being separated from the rest of the fluid by an imaginary transparent membrane. The buoyant force F_B on this body of fluid will be exactly the same as that on the original object since the surrounding fluid, which exerts F_B, is in exactly the same configuration. Now the body of fluid D' is in equilibrium (the fluid as a whole is at rest); therefore, $F_B = w'$, where w' is the weight of the body of fluid. Hence the buoyant force F_B is equal to the weight of the body of fluid whose volume is equal to the volume of the original submerged object, which is Archimedes' principle.

[†] By "fluid displaced" we mean a volume of fluid equal to the volume of the object, or that part of the object submerged if it floats (the fluid that used to be where the object was). If the object is placed in a glass initially filled to the brim with water, the water that flows over the top represents the water displaced by the object.

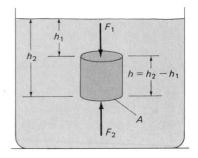

FIGURE 12–9
Determination of buoyant force.

FIGURE 12–10
Archimedes' principle.

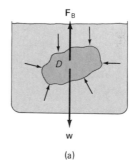

(a)

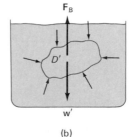

(b)

EXAMPLE 12-4   A 70-kg rock lies at the bottom of a lake. Its volume is 3.0×10^4 cm³. How much force is needed to lift it?

SOLUTION   The buoyant force on the rock due to the water is equal to the weight of 3.0×10^{-2} m³ of water:

$$F_B = \rho_{H_2O}gV$$
$$= (1.0 \times 10^3 \text{ kg/m}^3)(9.8 \text{ m/s}^2)(3.0 \times 10^{-2} \text{ m}^3)$$
$$= 2.9 \times 10^2 \text{ N}.$$

The weight of the rock is $mg = (70 \text{ kg})(9.8 \text{ m/s}^2) = 6.9 \times 10^2$ N. Hence the force needed to lift it is 690 N − 290 N = 400 N. It is as if the rock had a mass of only (400 N)/(9.8 m/s²) = 41 kg.

Archimedes is said to have discovered his principle in his bath while thinking how he might determine whether the king's new crown was pure gold or a fake. Gold has a specific gravity of 19.3, somewhat higher than most metals, but a determination of specific gravity or density is not readily done directly since the volume of an irregularly shaped object is not easily determined. However, if the object is weighed in air ($= w$) and also "weighed" while it is under water ($= w'$), the density can be determined using Archimedes' principle as the following example shows.

EXAMPLE 12-5   A 14.7-kg crown has an apparent weight when submerged in water corresponding to 13.4 kg. Is it gold?

SOLUTION   The apparent weight of the submerged object, w', equals its actual weight w minus the buoyant force F_B:

$$w' = w - F_B = \rho_o gV - \rho_f gV$$

where V is the volume of the object, ρ_o its density, and ρ_f the density of the fluid (water in this case). Then we can write

$$\frac{w}{w - w'} = \frac{\rho_o gV}{\rho_f gV} = \frac{\rho_o}{\rho_f}.$$

[Thus $w/(w - w')$ is equal to the specific gravity of the object if the fluid in which it is submerged is water.] For the crown we have

$$\frac{\rho_o}{\rho_{H_2O}} = \frac{w}{w - w'} = \frac{14.7 \text{ kg}}{1.3 \text{ kg}} = 11.3.$$

This corresponds to a density of 11,300 kg/m³. The crown seems to be made of lead!

FIGURE 12-11

A hydrometer.

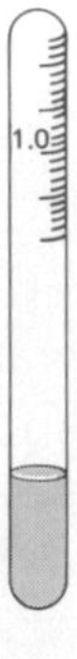

Archimedes' principle applies equally well to objects that float such as wood. In general, *an object floats on a fluid if its density is less than that of the fluid.* For example, a log whose specific gravity is 0.60 and whose volume is 2.0 m³ will have a mass of 1200 kg. If the log is fully submerged, it will displace a mass of water $m = \rho V = (1000 \text{ kg/m}^3)(2.0 \text{ m}^3) = 2000$ kg. Hence the buoyant force on it will be greater than its weight and it will float to the top. It will come to equilibrium when it displaces 1200 kg of water, which means that 1.2 m³ or 0.60 of its volume will be submerged. In general, the fraction of the object submerged is given by the ratio of the object's density to that of the fluid.

EXAMPLE 12-6   A hydrometer is a simple instrument used to measure specific gravity. A particular hydrometer (Fig. 12-11) consists of a glass tube, weighted at the bottom, which is 25.0 cm long, 2.00 cm² in cross-sectional area, and has a mass of 45.0 g. How far from the end should the 1.000 mark be placed?

SOLUTION   The hydrometer has a density

$$\rho = \frac{m}{V} = \frac{45.0 \text{ g}}{(2.00 \text{ cm}^2)(25.0 \text{ cm})} = 0.900 \text{ g/cm}^3.$$

Thus, when placed in water, it will come to equilibrium when 0.900 of its volume is submerged. Since it is of uniform cross section, $(0.900)(25.0 \text{ cm}) = 22.5 \text{ cm}$ of its length will be submerged. Since the specific gravity of water is defined to be 1.000, the mark should be placed 22.5 cm from the end.

Air is a fluid and it too exerts a buoyant force. Ordinary objects weigh less in air than they do if weighed in a vacuum. Because the density of air is so small, the effect for ordinary solids is slight (but see the problems at the end of the chapter). There are objects, however, that float in air—helium baloons, for example.

■ EXAMPLE 12–7   What volume of helium is needed if a balloon is to lift a load of 800 kg (including the weight of the empty balloon)?

SOLUTION   The buoyant force on the helium, F_B, which is equal to the weight of displaced air, must at least be equal to the weight of the helium plus the load:

$$F_B = (m_{He} + 800 \text{ kg})g$$

where g is the acceleration due to gravity. This equation can be written in terms of density:

$$\rho_{air} V g = (\rho_{He} V + 800 \text{ kg})g$$

Solving now for V we find

$$V = \frac{800 \text{ kg}}{\rho_{air} - \rho_{He}} = \frac{800 \text{ kg}}{(1.29 \text{ kg/m}^3 - 0.18 \text{ kg/m}^3)}$$

$$= 720 \text{ m}^3.$$

*12–7 Surface Tension

Up to now in this chapter we have mainly been studying what happens beneath the surface of a liquid (or gas). But the actual surface of a liquid also behaves in an interesting way. A number of common observations suggest that the surface of a liquid acts like a stretched membrane under tension. For example, a drop of water on the end of a dripping faucet, or hanging from a thin branch in the early morning dew, forms into a nearly spherical shape as if it were a tiny balloon filled with water; and a steel needle can be made to float on the surface of water even though it is denser than the water. The surface of a liquid acts like it is under tension, and this tension, acting parallel to the surface, arises from the attractive forces between the molecules. This effect is called *surface tension*. More specifically a quantity called the *surface tension, γ* (the Greek letter gamma), is defined as the force F per unit length L that acts across any line in a surface, tending to pull the surface open:

$$\gamma = \frac{F}{L}. \tag{12–7}$$

To understand this, consider the U-shaped apparatus shown in Fig. 12–12 which encloses a thin film of liquid; because of surface tension, a force F is required to pull the movable wire and thus increase the surface area of the liquid. The liquid contained by the wire apparatus is a thin film having both a top and a bottom surface; hence the length of the surface being increased is $2l$. Thus the surface tension

FIGURE 12–12

U-shaped wire apparatus holding film of liquid to measure surface tension $(\gamma = F/2l)$.

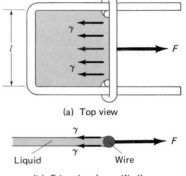

(a) Top view

(b) Edge view (magnified)

TABLE 12–3
Surface tension of some substances

Substance	Surface Tension (N/m)
Mercury (20°C)	0.44
Blood, whole (37°C)	0.058
Blood, plasma (37°C)	0.073
Alcohol, ethyl (20°C)	0.023
Water (0°C)	0.076
(20°C)	0.072
(100°C)	0.059
Benzene (20°C)	0.029
Soap solution (20°C)	≈ 0.025
Oxygen (−193°C)	0.016

$\gamma = F/2l$. A delicate apparatus of this type can be used to measure the surface tension of various liquids. The surface tension of water is 0.072 N/m at 20°C. Table 12–3 gives the values for other liquids; note that temperature has a considerable effect on the surface tension.

We can see how surface tension arises by examining the process from the molecular point of view. The molecules of a liquid exert attractive forces on each other; these attractive forces are shown acting, in Fig. 12–13, on a molecule deep within the liquid and on a second molecule at the surface. The molecule inside the liquid is in equilibrium due to the forces of other molecules acting in all directions. The molecule at the surface is also normally in equilibrium (the liquid is at rest); this is true even though the forces on a molecule at the surface can only be exerted by molecules below it (or at an equal height). Hence there is a net attractive force downward, which tends to compress the surface layer slightly—but only to the point where this downward force is balanced by an upward (repulsive) force due to close contact (or collision with) the molecules below.[†] This compression of the surface means that, in essence, the liquid tries to minimize its surface area. This is why water tends to form spherical droplets, for a sphere represents the minimum surface area for a given volume.

In order to increase the surface area of a liquid, a force is required and work must be done to bring molecules from the interior to the surface (Fig. 12–14). This work increases the potential energy of the molecules and is sometimes called *surface energy*. The greater the surface area, the greater the surface energy.

The amount of work needed to increase the surface area can be calculated from Fig. 12–12, and Eq. 12–7:

$$W = F\,\Delta x$$
$$= \gamma L\,\Delta x$$
$$= \gamma\,\Delta A$$

where Δx is the change in distance and ΔA is the total increase in area (at both surfaces in Fig. 12–12) and so we can write

$$\gamma = \frac{W}{\Delta A}.$$

Thus, the surface tension γ is not only equal to the force per unit length; it is also equal to the work done per unit increase in surface area. Hence, γ can be specified in N/m or J/m^2.

FIGURE 12–13

Molecular theory of surface tension, showing attractive forces (only) on a molecule at the surface, and on one deep inside the liquid.

FIGURE 12–14

Forces on a molecule being brought to the surface because of an increase in the surface area. The molecules at the surface must exert a strong force on the molecule being pulled up, and they in turn feel a strong force back (Newton's third law). This is the surface-tension force.

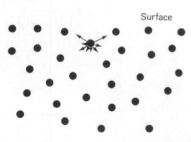

Surface

---

† The air molecules above exert a force, but this is a small effect since the air molecules are so far apart. The surface tension therefore does depend on the material above the surface but the effect is generally small if the latter is a dilute gas. But because of this small effect, surface tensions are specified for the boundary between two materials. If the second material is not mentioned, it is assumed to be air at atmospheric pressure.

Because of surface tension, insects can walk on water; and objects more dense than water, such as a steel needle, can actually float on the surface. Figure 12–15a shows how the surface tension can support the weight w of an object. Actually, w is the "effective weight" of the object—its true weight less the buoyant force—since the object sinks slightly into the fluid. If the object is spherical in shape, which is close for the base of an insect's leg (see Fig. 12–15b), the surface tension acts at all points along a circle of radius r. Only the vertical component, $\gamma \cos \theta$, acts to balance w. Hence the net upward force due to surface tension is $2\pi r \gamma \cos \theta$.

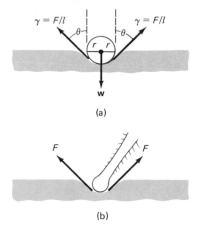

EXAMPLE 12–8 The base of an insect's leg is approximately spherical in shape with a radius of about 2.0×10^{-5} m. The 0.0030 g mass of the insect is supported equally by the six legs. Calculate the angle θ (see Fig. 12–15). Assume the water temperature is 20°C.

SOLUTION Since

$$2\pi r \gamma \cos \theta = w$$

where w is one-sixth the weight of the insect (since it has six legs) we have

$$(6.28)(2.0 \times 10^{-5} \text{ m})(0.072 \text{ N/m}) \cos \theta = \tfrac{1}{6}(3.0 \times 10^{-6} \text{ kg})(9.8 \text{ m/s}^2)$$

$$\cos \theta = \frac{0.49}{0.90} = 0.54$$

$$\theta = 57°.$$

Notice that if $\cos \theta$ were greater than 1, this would indicate that the surface tension would not be great enough to support the weight.

FIGURE 12–15
Surface tension acting on (a) a sphere, (b) an insect leg.

Calculations such as that above are not always accurate since the radius r of the surface depression is not precisely equal to the radius of the object. However, approximate estimates can be made as to whether an object will or will not remain on the surface.

Soaps and detergents have the effect of lowering the surface tension of water. This is desirable for washing and cleaning since the high surface tension of pure water prevents it from penetrating easily between the fibers of material and into tiny crevices. Substances which reduce the surface tension of a liquid are called *surfactants*.

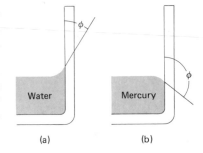

FIGURE 12–16
Water (a) "wets" the surface of glass, whereas mercury (b) does not.

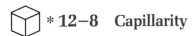 * 12–8 Capillarity

It is a common observation that water in a glass container rises up slightly where it touches the glass, Fig. 12–16a. The water is said to "wet" the glass. Mercury, on the other hand, is depressed when it touches the glass, Fig. 12–16b; the mercury does not wet the glass. Whether or not a liquid wets a solid surface is determined by the relative strengths of the cohesive forces between the molecules of the liquid compared to the adhesive forces between the molecules of the liquid and those of the solid. (*Cohesion* refers to the force between like molecules and *adhesion* to the force between unlike molecules.) Water wets glass because the water molecules are more strongly attracted to the glass molecules than they are to other water molecules. The opposite is true for mercury: the cohesive forces are stronger than the adhesive forces.

The angle that the tangent to the liquid surface makes with the solid surface is called the *angle of contact* ϕ. Its value depends on the relative strengths of the cohesive and adhesive forces (see Fig. 12–16 and Table 12–4). When ϕ is less than 90°, the liquid wets the solid; if ϕ is greater than 90°, the liquid does not wet the solid.

TABLE 12–4
Contact angle of some substances

Substance	Contact Angle
Water–glass	0°
Organic liquids (most)–glass	0°
Mercury–glass	140°
Water–parafin	107°
Kerosene–glass	26°

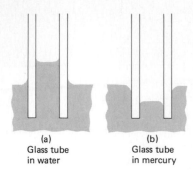

(a)
Glass tube
in water

(b)
Glass tube
in mercury

FIGURE 12–17

Capillarity.

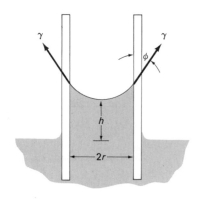

FIGURE 12–18

A liquid rises a height h in a capillary tube given by $h = 2\gamma \cos \phi / \rho gr$ (see text).

FIGURE 12–19

Producing a negative pressure.

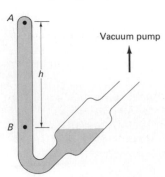

In tubes having very small diameters, liquids are observed to rise or fall relative to the level of the surrounding liquid. This phenomenon is called *capillarity*, and such thin tubes are called *capillaries*. Whether the liquid rises or falls (Fig. 12–17) depends on the relative strengths of the adhesive and cohesive forces. Thus water rises in a glass tube whereas mercury falls. The actual amount of rise (or fall) depends on the surface tension—since it is this which keeps the liquid surface from breaking apart—as well as on the contact angle ϕ and the radius r of the tube. To calculate h, the height of rise, we refer to Fig. 12–18. The surface tension γ acts at an angle ϕ all around a circle of radius r. The magnitude of the vertical force F due to surface tension is, by Eq. 12–7, $F = (\gamma \cos \phi)(L)$. Since $L = 2\pi r$, we have $F = 2\pi r\gamma \cos \phi$. This force is just balanced by the weight of the liquid below which is approximately a cylinder of height h and volume $V = \pi r^2 h$. Hence

$$2\pi r\gamma \cos \phi = mg$$
$$= \rho V g$$
$$= \rho \pi r^2 h g$$

where ρ is the density of the liquid. Solving for h we find

$$h = \frac{2\gamma \cos \phi}{\rho gr}. \qquad (12\text{–}8)$$

For many liquids, such as water in glass, ϕ is close to zero, and since $\cos 0° = 1$, Eq. 12–8 reduces to the simpler form $h = 2\gamma/\rho gr$.

Equation 12–8 is also valid when the liquid drops, as in the case of mercury in a glass tube. In this situation, the contact angle ϕ is greater than $90°$, and $\cos \phi$ will be negative; this makes h negative which corresponds to a decrease in level. Notice from Eq. 12–8 that the thinner the tube the greater will be the rise (or fall) of the liquid.

EXAMPLE 12–9 If the xylem (tiny tubes which carry nutrients upward) of a plant has a radius of 0.0010 cm, calculate how high the surface tension can be expected to pull a column of water. Assume $\phi = 0°$.

SOLUTION Using $\gamma = .072$ N/m in Eq. 12–8 we find

$$h = \frac{(2)(0.072 \text{ N/m})(1)}{(1.0 \times 10^3 \text{ kg/m}^3)(9.8 \text{ m/s}^2)(1.0 \times 10^{-5} \text{ m})}$$
$$= 1.5 \text{ m}.$$

*12–9   Negative Pressure and the Cohesion of Water

Normally, a fluid exerts an outward pressure on its container and the container exerts an inward pressure on the fluid. Pressures can range from zero up to very large positive values. But can fluid pressure be negative? The answer, strange as it may seem, is yes; under certain circumstances a negative pressure can exist in a liquid, but not in a gas.

Negative pressures have been obtained, although it is difficult, using the apparatus shown in Fig. 12–19. The closed tube is initially filled with liquid and then the reservoir on the right is pumped down with a vacuum pump. According to Eq. 12–6a the pressure difference between points A and B is ρgh where ρ is the density of the liquid:

$$P_B - P_A = \rho gh.$$

When the pump has reduced the pressure to near zero, $P_B = 0$ since point B is at the same level as the surface of the liquid in the reservoir. Hence the pressure at A must be negative:

$$P_A = -\rho g h.$$

Normally we would expect the liquid to fall out of the tube when the pump reduces the pressure. But if the tube is scrupulously clean and the liquid is free from impurities the liquid may remain in the tube. Negative pressures reaching -270 atmospheres have been achieved using this technique. The liquid under negative pressure is unstable, however, and the smallest disturbance will cause the liquid to shatter into droplets, and it falls.

A liquid under negative pressure actually pulls inward on its container; the liquid acts as if it were all under tension (not just its surface) much like a rope pulled at the two ends. How can this occur? It is due to the cohesive forces between the liquid molecules as well as adhesive forces between the liquid and the container. The cohesive forces between water molecules are quite strong—the tensile strength of pure water is about 30×10^6 N/m^2. These are the same forces that hold the water molecules together as ice at low temperatures. The difference is that in the liquid state the molecules have considerable kinetic energy and are free to move over one another. Liquid water does not normally sustain a tension force as a solid does because the slightest intrusion of an impurity (such as air) allows the water to flow and take on another shape; it can then fall under gravity.

It has long been wondered how water can rise to the top of tall trees such as redwoods which reach heights of over 100 m. Water, containing mineral nutrients from the soil, is carried upward by tiny capillaries (the xylem). The radius of these capillaries varies from about 0.01 mm to 0.3 mm. As we saw in Example 12–9, even the smallest sized capillaries can raise water by only about 1.5 m. On the other hand, atmospheric pressure can raise water only about 10 m even when the top end is under vacuum (Section 12–4). Neither of these effects account for the rise of water in tall trees. Today it is generally accepted that although the above effects may help, it is the cohesive forces between water molecules acting under tension at negative pressure that are responsible, particularly for very tall trees. This theory was first proposed in the last century, but until recently there was much opposition to it. Within the last few years actual measurements of the pressure within the xylem of trees have been made; they were indeed found to be negative, often as low as -25 atm at the top. When water evaporates from leaves, other water molecules move in to take their place. Through cohesive forces, the water below is pulled upward.

Summary

The three common states of matter are solid, liquid, and gas. Liquids and gases are both called *fluids*, meaning they have the ability to flow. The *density* of a substance is defined as its mass per unit volume. The *specific gravity* of a substance is the ratio of its density to that of water (at 4°C).

Pressure is defined as force per unit area. A fluid at rest exerts an equal pressure in all directions at any point within it. At a depth h below the surface of a liquid of uniform density, ρ, the pressure due to the liquid is equal to $\rho g h$ where g is the acceleration due to gravity. If the density of a fluid is not uniform, the pressure P varies with height y as $dP/dy = -\rho g$. If there is an external pressure applied to the surface of a confined fluid (such as the pressure of the atmosphere on an open container or body of water), this pressure is transmitted throughout the fluid; this is known as *Pascal's principle*.

Pressure is measured using a manometer or other type of gauge. A barometer is used to measure atmospheric pressure. Standard atmospheric pressure at sea level is 1.013×10^5 N/m^2. *Gauge pressure* is total pressure less atmospheric pressure.

Archimedes' principle states that an object submerged wholly or partially in a fluid is buoyed up by a force equal to the weight of fluid it displaces. This principle is used in methods to determine specific gravity, and explains why objects whose density is less than that of a fluid will float in that fluid.

Questions

1. What is the specific gravity of (a) air, (b) ice, (c) gold?

2. If one material has a higher density than another, does this mean the molecules of the first must be heavier than those of the second? Explain.

3. Devise a simple method for estimating the density of your body using a swimming pool.

4. Airplane travelers often note that their cosmetics bottles and other containers have leaked after a trip. What might cause this?

5. The three containers in Fig. 12–20 are filled with water to the same height and have the same surface area at the base; hence the water pressure, and the total force on the base of each, is the same. Yet the total weight of water is different for each. Explain why this "hydrostatic paradox" is not contradictory.

6. Consider what happens when you push both a pin and a stick against your skin with the same force. Decide what determines whether your skin suffers a cut—the net force applied to it or the pressure.

7. A cork floats in a bottle of water open to the atmosphere. If you used a pump to compress the air in the bottle, could you make the cork sink? Explain.

8. Devise a method for determining the mass of one of your legs, using a swimming pool.

9. Describe how panning for gold is done, using principles from this chapter.

10. Why do you float more easily in salt water than in fresh?

11. Is the buoyant force on a diving bell deep beneath the ocean precisely the same as when it is just beneath the surface? Explain.

12. Will a balloon rise indefinitely in the air? Explain.

13. A barge loaded with rocks approaches a low bridge over the river and cannot quite pass under it. Should rocks be added to or removed from the barge?

14. It is harder to pull the plug out of the drain of a bathtub when the tub is full of water than when it is empty. Is this a contradiction of Archimedes' principle? Explain.

15. An ice cube floats in a glass of water filled to the brim. As the ice melts, will the glass overflow?

16. Explain why a uniform wood pencil always floats horizontally rather than vertically. Explain why it can float vertically if weighted sufficiently at one end.

17. A beaker of water is accelerated uniformly to the right along a horizontal surface. Explain why the surface of the water makes an angle to the horizontal. In which direction does the surface slope?

18. (a) Show that the buoyant force on a partially submerged object acts at the center of gravity of the fluid before it is displaced. This point is called the *center of buoyancy*. (b) For a ship to be stable, should its center of buoyancy be above, below, or at the same point as its center of gravity? Explain.

19. Will an empty balloon have precisely the same weight on a scale as one that is filled with air? Explain.

20. A floating helium balloon is tied by a short string to the seat in the interior of a car. If the car makes a left turn, how does the balloon move relative to the car?

21. Why does a sinking ship sometimes roll on its side before becoming submerged? (Hint: see question 18.)

22. Is Archimedes' principle valid in an elevator accelerating (a) at $\frac{1}{2}g$, (b) in free fall?

23. A piece of wood is 60 percent submerged in a tub of water on earth. Will the wood float or sink if the tub is in an elevator accelerating at $\frac{1}{2}g$ (a) upward, (b) downward? (c) What if the elevator is in free fall?

24. A small amount of water is boiled in a 1-gallon gasoline can. The can is removed from the heat and the lid put on. Shortly thereafter, the can collapses. Explain.

25. It is often said that "water seeks its own level." Explain.

26. Is there a limit to the depth at which a skin diver can use a snorkel beneath the sea? Explain.

27. Explain how the tube in Fig. 12–21, known as a *siphon*, can transfer liquid from one container to a lower one even though the liquid must flow uphill for part of its journey. (Note that the tube must be filled with liquid to start with.) Why doesn't the liquid in each side of the tube flow back into its container?

FIGURE 12–20

FIGURE 12–21
A siphon.

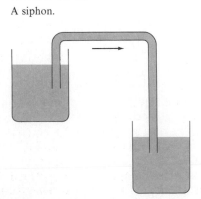

FIGURE 12–22

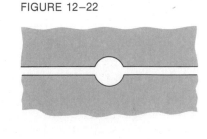

28. Approximately how high would you expect the column of mercury to be in a barometer used in a satellite orbiting the earth at a height of 6400 km?

29. When measuring blood pressure, why is the sleeve attached to the manometer placed on your arm at the level of the heart?

*30. Why does a window frame and/or the sill have a groove as in Fig. 12–22?

*31. A duck can float in water because it preens its feathers to apply a layer of grease. Explain how the increased surface tension allows the duck to float.

 Problems _____

SECTION 12–1

1. (I) The approximate volume of the granite monolith known as El Capitan in Yosemite National Park is about 10^8 m³. What is its approximate mass?

2. (I) What is the approximate mass of air in a living room 6.8 m × 3.4 m × 2.8 m?

3. (I) If 5.5 L of antifreeze solution (SG = 0.80) is added to 4.5 L of water to make a 10.0-L mixture, what is the specific gravity of the mixture?

4. (I) A bottle has a mass of 31.20 g when empty and 98.44 g when filled with water. When filled with another fluid the mass is 88.78 g. What is the specific gravity of the fluid?

SECTION 12–2

5. (I) The arm of a record player exerts a force of (1.0 g) × g on a record. If the diameter of the stylus is 0.0013 cm (= 0.5 mil = 0.5 × 10^{-3} in), estimate the pressure on the record groove in N/m² and in atmospheres.

6. (I) What is the difference in blood pressure between the top of the head and bottom of the feet of a 1.60-m-tall person standing vertically?

7. (I) What is the approximate difference in air pressure between the top and the bottom of the World Trade Center Building in New York City which is 410 m tall and is located at sea level? Express as a fraction of atmospheric pressure at sea level.

8. (I) When you drive up into the mountains, or descend rapidly from the mountains, your ears may "pop," which means that the pressure behind the eardrum is being equalized to that outside. If this did not happen, what would be the approximate force on an eardrum of area 0.50 cm² if a change in altitude of 1000 m takes place?

9. (II) Show that the work done when a constant pressure P acts to move a volume of fluid ΔV is $W = P \Delta V$.

10. (II) Estimate the air pressure on the summit of Mt. Everest (8,850 m above sea level).

11. (II) What is the total force on a rectangular shaped dam 75 m high and 120 m wide if the water is filled to the top?

12. (II) One arm of a U-shaped tube (open at both ends) contains water and the other alcohol. If the two fluids meet exactly at the bottom of the U, and the alcohol is at a height of 18.0 cm, at what height will the water be?

13. (II) Calculate the total mass of the earth's atmosphere using the known value of atmospheric pressure at sea level.

14. (II) Estimate the density of the water 10.0 km deep in the sea. (See Section 11–4 and Table 11–1.) By what fraction does it differ from the density at the surface?

15. (II) Determine the general pressure at a depth h in a fluid of density ρ if the fluid and its container have an acceler-

ation a that is (a) upward, (b) downward, (c) equal to g downward (free fall).

16. (II) A cylindrical bucket of liquid (density ρ) is rotated about its symmetry axis which is vertical. If the angular velocity is ω, show that the pressure at a distance r from the rotation axis is

$$P = P_a + \tfrac{1}{2}\rho\omega^2 r^2$$

where P_a is the pressure at the axis at the same depth.

17. (II) Pure water rests at equal levels in the bottom of a U-shaped tube whose two ends are open to the atmosphere. A second liquid, which does not mix with water, is poured into one side of the U-shaped tube. The water in the other side of the tube rises 8.3 cm and its surface is 2.1 cm above that of the second liquid. What is the density of the second liquid?

18. (III) A beaker of liquid accelerates from rest, on a horizontal surface, with acceleration a to the right. (a) Show that the surface of the liquid makes an angle $\theta = \tan^{-1} a/g$ with the horizontal (b) Which edge of the water surface is higher? (c) How does the pressure vary with depth below the surface?

19. (III) Water stands at a height h behind a vertical dam of uniform width b. (a) Use integration to show that the total force of the water on the dam is $F_T = \tfrac{1}{2}\rho g h^2 b$. (b) Show that the torque about the base of the dam due to this force can be considered to act with a lever arm equal to $h/3$. (c) For a freestanding concrete dam of uniform thickness t and height h, what minimum thickness is needed to prevent overturning? Do you need to add in atmospheric pressure for this last part? Explain.

SECTIONS 12–3 AND 12–4

20. (I) A typical value for systolic blood pressure is 120 mm Hg. Convert this to (a) torr, (b) N/m², (c) atm, (d) lb/in².

21. (I) (a) Calculate the total force of the atmosphere acting on the top of a table which measures 3.2 m × 1.2 m. (b) What is the total force acting upward on the underside of the table?

22. (I) Determine the minimum gauge pressure needed in the water pipe leading into a building if water is to come out of a faucet on the twelfth story 40 m above.

23. (I) The contraction of the left ventricle (chamber) of the heart pumps blood to the body. Assuming that the inner surface of the left ventricle has an area of 85 cm² and the maximum pressure in the blood is 120 mm Hg, calculate the net force exerted by the ventricle at the point of maximum pressure.

24. (II) Suppose a person can reduce the pressure in the lungs to −80 mm Hg gauge pressure. How high can water then be sucked up a straw?

25. (II) An open-tube mercury manometer is used to measure the pressure in an oxygen tank. On a day when the atmospheric pressure is 1040 mbars, what is the absolute pressure (in Pa) in the tank if the height of the mercury in the open tube is (a) 28.0 cm higher, (b) 4.2 cm lower, than the mercury in the tube connected to the tank?

26. (II) Intravenous infusions are often made by connecting the needle (inserted in the arm) by a tube to a raised bottle full of fluid. Assuming the fluid has a density of 1.00 g/cm^3, at what height h should the bottle be placed so the liquid pressure is (a) 60 mm Hg, (b) 600 mm H_2O? (c) If the blood pressure is 18 mm Hg above atmospheric pressure, how high should the bottle be placed so that the fluid just barely enters the vein? Let h refer to the vertical height of the fluid surface in the bottle above the needle.

27. (II) The gauge pressure in each of the four tires of an 1800-kg automobile is 210 kN/m^2. How much area of each tire is in contact with the ground?

28. (II) During each heartbeat, approximately 70 cm^3 of blood is pushed from the heart at an average pressure of 105 mm Hg. Calculate the power output of the heart, in watts, assuming 60 beats per minute.

29. (II) How high would the level be in an alcohol barometer at normal atmospheric pressure?

SECTION 12–5

30. (I) The gauge pressure in a hydraulic lift is 16 atm. What is the largest-size vehicle (kg) it can lift if the diameter of the output line is 17 cm?

31. (II) A 3.0-N force is applied to the plunger of a hypodermic needle. If the diameter of the plunger is 1.0 cm and that of the needle 0.20 mm, (a) with what force does the fluid leave the needle? (b) What force on the plunger would be needed to push fluid into a vein where the gauge pressure is 18 mm Hg?

32. (II) In working out his principle, Pascal showed dramatically how force can be multiplied with fluid pressure. He placed a long tube of 0.30-cm radius vertically into a 20-cm-radius wine barrel. He found that when the barrel was filled with water, and the tube filled to a height of 12 m, the barrel burst. Calculate (a) the mass of fluid in the tube, (b) the net force on the lid of the barrel.

SECTION 12–6

33. (I) The hydrometer of Example 12–6 sinks to a depth of 22.3 cm when placed in a fermenting vat. What is the density of the brewing liquid?

34. (I) A geologist finds that a moon rock whose mass is 7.20 kg has an apparent mass of 5.88 kg when submerged in water. What is the density of the rock?

35. (I) How large is the buoyant force, due to the atmosphere, on a 4700 m^3 tank of water?

36. (I) What fraction of a piece of iron will be submerged when it floats in mercury?

37. (II) A 0.40-kg piece of wood floats in water but is found to sink in alcohol (SG = 0.79), in which it has an apparent mass of 0.020 kg. What is the specific gravity of the wood?

38. (II) Use Table 12–1 to determine what fraction of a (pure) ice cube is submerged in a glass of (a) pure water, (b) sea water. (c) What are your answers if the experiment is done

on the moon where g is one-sixth its value on earth? (d) Would the answer to either (a) or (b) apply to an iceberg in the ocean? If so, which?

39. (II) A bucket of water is accelerated upward at $3.5 g$. What is the buoyant force on a 1.0-kg granite rock (SG = 2.7) submerged in the water? Will the rock float? Why or why not?

40. (II) A 4.0-kg cubical block of wood is 50 percent submerged in a lake. How much work is required to submerge it?

41. (II) A freighter has a horizontal cross-sectional area of 3100 m^2 at the water line. When loaded, the ship drops 6.1 m. What is the mass of its load?

42. (II) What is the SG of a 25-kg animal if, when floating in water, only 2.0 cm^3 remains above the surface?

43. (II) A small object is found to remain suspended in a mixture of 18 percent alcohol (by weight) and 82 percent water. What is its density?

44. (II) A rowboat has a volume of 1.5 m^3 and a mass of 35 kg. How many people, of mass 70 kg each, can the boat hold without sinking?

45. (II) Archimedes' principle can be used not only to determine the specific gravity of a solid using a known liquid (Example 12–5); the reverse can be done as well. (a) As an example, a 12.00-kg aluminum ball has an apparent mass of 9.40 kg when submerged in a particular liquid; calculate the density of the liquid. (b) Derive a simple formula for determining the density of a liquid using this procedure.

46. (II) Calculate the true mass (in vacuum) of a piece of aluminum whose apparent mass is 2.0000 kg when weighed in air.

47. (II) A 520-g piece of wood (SG = 0.50) floats on water. What minimum mass of lead, hung from it by a string, will cause it to sink?

48. (II) If an object floats in water, its density can be determined by tying a "sinker" on it so that the object and the weight sink. Show that the specific gravity is given by $w/(w_1 - w_2)$ where w is the weight of the object alone in air, w_1 is the apparent weight of the object when a "sinker" is tied to it and the sinker only is submerged, and w_2 is the apparent weight of the object when both the object and the sinker are submerged.

49. (III) Two solid cubes of wood, each of volume 1.00 m^3, have specific gravities of 0.35 and 0.60. Faces of equal area are placed together, flush, and glued. If this combination is now held horizontally (the two blocks side by side) in water so it floats, what net torque acts on it, attempting to turn it vertical?

*SECTION 12–7

*50. (I) If the force F needed to move the wire in Fig. 12–12 is $6.4 \times 10^{-3} \text{ N}$, calculate the surface tension γ of the enclosed fluid. Assume $l = 0.075 \text{ m}$.

*51. (II) The surface tension of a liquid can be determined by measuring the force F needed to just lift a circular platinum ring of radius r from the surface of the liquid. (a) Find a formula for γ in terms of F and r. (b) At 30°C, if $F = 9.40 \times 10^{-3} \text{ N}$ and $r = 3.5 \text{ cm}$, calculate γ for the tested liquid.

*52. (II) How much work is required to increase a soap bubble (see Table 12–3) in size from a 3.0-cm diameter to a 5.0-cm diameter?

*53. (II) A small pool of water on a table is broken into 50 droplets. By what factor does the surface energy change? Assume the original pool is flat and of depth h, and the droplets are hemispheres of radius h.

*54. (III) Show that inside a soap bubble there must be a pressure ΔP in excess of that outside equal to $\Delta P = 4\gamma/R$ where R is the radius of the bubble and γ is the surface tension. (Hint: think of the bubble as two hemispheres in contact with each other; and remember that there are two surfaces to the bubble. Note that this result applies to any kind of membrane where $T = 2\gamma$ is the tension per unit length in that membrane.)

*SECTION 12–8

*55. (I) How high will water rise in a glass tube 0.12 mm in radius?

*56. (I) A glass tube 0.85 mm in diameter is placed in mercury. Where will the level of the mercury be in the tube relative to the rest of the liquid?

*57. (I) When a glass tube is placed in a container of ethyl alcohol, the liquid rises 3.4 mm up the tube. What is the diameter of the tube?

*58. (II) A pencil 1.0 cm in diameter is held vertically in a glass of water. The water wets the pencil so that the contact angle is 0°. Calculate the magnitude and direction of the net force on the pencil due to surface tension.

*59. (II) How thin would the xylem tubes have to be if water were to reach the top of a 100-m tall tree by capillary action alone?

*60. (II) How high will water rise due to capillarity between two flat plates of glass 0.11 mm apart when placed vertically in water?

Hydrodynamics: Fluids in Motion

13

We turn now from the study of fluids at rest to the more complex subject of fluids in motion, which is called **hydrodynamics**. Many aspects of fluid motion are still not completely understood; nonetheless, with certain simplifying assumptions, a good comprehension of this subject can be obtained.

One approach to the study of fluid flow is to follow individual particles (or tiny elements of volume) of the fluid. The motion of each particle, as governed by Newton's laws, could in principle be calculated, but this turns out to be extremely complex and difficult. Instead, the more usual approach, and the one we take here, is to describe the properties of the fluid at each point in space. That is, instead of following the motion of each fluid particle as it moves through space as a function of time, we instead look at each point in space and describe the motion of the fluid (by giving the fluid velocity and density) at each point as a function of time.

FIGURE 13–1

(a) Streamline or laminar flow;
(b) turbulent flow.

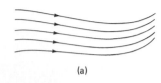

(a)

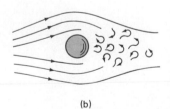

(b)

13–1 Characteristics of Flow

We can distinguish two main types of fluid flow. If the flow is smooth such that neighboring layers of the fluid slide by each other smoothly, the flow is said to be **streamline** or **laminar flow**. This kind of flow is characterized by the fact that each particle of the fluid follows a smooth path, and these paths do not cross over one another, Fig. 13–1a. Above a certain speed, which depends on a number of factors as we shall see later, the flow becomes turbulent. **Turbulent flow** is characterized by erratic, small whirlpool-like circles called eddy currents or eddies, Fig. 13–1b. Eddies absorb a great deal of energy and although a certain amount of internal friction, called *viscosity*, is present during streamline flow, it is much greater when the flow is turbulent. A few tiny drops of ink or food coloring dropped into a moving liquid can quickly reveal whether the flow is streamline or turbulent.

For both streamline and turbulent flow, we can consider four important characteristics of fluid flow. (1) The fluid can be considered *compressible* or *incompressible*; although no material is truly incompressible, the flow of many fluids is such that variations in density are so small that they can be ignored, and this greatly simplifies the analysis. (2) *Viscosity*, or internal friction, is always present in fluid motion, but it too is often small enough to be ignored; in the first part of this chapter we will assume *nonviscous* flow, and then in later sections we will investigate the effects of viscosity. (3) The flow can be *steady*, which means the velocity of the fluid at each point in space remains constant in time (which does not necessarily imply that the velocity is the same at all points in space). If the velocity at a point changes in time—such as when water starts to move in a pipe just as you turn a faucet on—the flow is *unsteady*; we will mainly be interested in steady flow. (4) The flow can be *rotational* or *irrotational*. It is irrotational if there is no net angular momentum of fluid about each point; that is, a tiny paddlewheel placed anywhere in the fluid would not rotate; if the wheel were to rotate, as it would in a whirlpool or eddy, the flow is rotational. We will not be directly concerned here with this rather complex characteristic.

13-2   Flow Rate and the Equation of Continuity

In the steady laminar flow of a fluid, the path taken by a given particle is called a **streamline** (see Fig. 13–1a). The fluid velocity at any point is tangent to the streamline at that point. A streamline can, in principle, be drawn through every point of the fluid, although normally we draw only a few of them. Two streamlines cannot cross over one another, since this would imply that at such a crossover point the velocity would not be definite.

   A bundle of streamlines, such as those shown in Fig. 13–2, is called a *tube of flow*. Since the streamlines represent the paths of particles, we see that no fluid can flow into or out of the sides of a tube of flow.

   Let us now examine the steady streamline flow of a tube of flow and determine how the speed of the fluid varies with the size of the tube. Let us choose the tube small enough so that the velocity across any cross section is essentially constant.[†] Thus, in Fig.13–2, v_1 represents the velocity as it passes through the cross-sectional area A_1, and v_2 is the velocity as it passes through the cross-sectional area A_2. The mass *flow rate* is defined as the mass Δm of fluid that passes a given point per unit time Δt: mass flow rate $= \Delta m/\Delta t$. In Fig. 13–2, the volume of fluid passing point 1 (that is, through area A_1) in a time Δt is just $A_1 \Delta l_1$ where Δl_1 is the distance the fluid moves in time Δt. Since the velocity of fluid passing point 1 is $v_1 = \Delta l_1/\Delta t$, the mass flow rate $\Delta m/\Delta t$ through area A_1 is (where $\Delta V_1 = A_1 \Delta l_1$ is the volume of mass Δm)

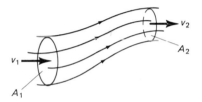

FIGURE 13-2
A tube of flow.

$$\frac{\Delta m}{\Delta t} = \frac{\rho_1 \Delta V_1}{\Delta t} = \frac{\rho_1 A_1 \Delta l_1}{\Delta t} = \rho_1 A_1 v_1.$$

Similarly, at point 2 (through area A_2), the flow rate is $\rho_2 A_2 v_2$. Since no fluid flows in or out the sides of a tube of flow, the flow rate through A_1 and A_2 must be equal. Thus

$$\rho_1 A_1 v_1 = \rho_2 A_2 v_2.$$

This is called the **equation of continuity**. If the fluid is incompressible, which is an excellent approximation for liquids under most circumstances (and often for gases as well), then $\rho_1 = \rho_2$ and the equation of continuity becomes

$$A_1 v_1 = A_2 v_2. \qquad \text{[incompressible fluid]} \qquad (13\text{--}1)$$

Notice that the product Av represents the *volume* rate of flow (volume of fluid passing a given point per second), since $\Delta V/\Delta t = A\Delta l/\Delta t = Av$. Equation 13–1 tells us that where the cross-sectional area of a flow tube (or of a pipe) is large the velocity

---

[†] Real fluids have viscosity and this internal friction causes different layers of the fluid to flow at different speeds. In this case we can let v_1 and v_2 represent the average speeds at each cross section.

is low; and where the area is small, the velocity is high. That this makes sense can be seen by looking at a river; a river flows slowly and languidly through a meadow where it is broad, but speeds up to torrential speed when passing through a narrow gorge. We note also from Eq. 13–1, and Fig. 13–2, that closely spaced streamlines (point 2) indicate higher fluid speed, and widely spaced lines (point 1) indicate lower fluid speed.

Equation 13–1 can be applied to the flow of blood in the body. Blood flows from the heart into the aorta from which it passes into the major arteries; these branch into the small arteries (arterioles), which in turn branch into myriads of tiny capillaries; the blood returns to the heart via the veins.

EXAMPLE 13–1 The radius of the aorta is about 1.0 cm and the blood flowing through it has a speed of about 30 cm/s. Calculate the average speed of the blood in the capillaries using the fact that although each capillary has a diameter of about 8×10^{-4} cm, there are literally billions of them so that their total cross section is about 2000 cm^2.

SOLUTION The speed of blood in the capillaries is

$$v_2 = \frac{v_1 A_1}{A_2} = \frac{(0.30 \text{ m/s})(3.14)(0.010 \text{ m})^2}{(2 \times 10^{-1} \text{ m}^2)}$$

$$= 5 \times 10^{-4} \text{ m/s}$$

or 0.5 mm/s.

Another example that makes use of the equation of continuity and the argument leading up to it is the following.

EXAMPLE 13–2 How large must a heating duct be if air moving 3.0 m/s along it can replenish the air in a room of 300 m^3 volume every 15 min? Assume the air's density remains constant.

SOLUTION We consider the room as a large section of duct for purposes of applying Eq. 13–1 (call it point 2). Reasoning in the same way we did to obtain Eq. 13–1 (changing Δt to t), we see that $A_2 v_2 = A_2 l_2/t = V_2/t$ where V_2 is the volume of the room. Then $A_1 v_1 = A_2 v_2 = V_2/t$ and

$$A_1 = \frac{V_2}{v_1 t} = \frac{300 \text{ m}^3}{(3.0 \text{ m/s})(900 \text{ s})} = 0.11 \text{ m}^2.$$

Since $A = \pi r^2$ we find that the radius must be 0.19 m or 19 cm.

 ## 13–3 Bernoulli's Equation

Have you ever wondered how air can circulate in a prairie dog's burrow, why smoke goes up a chimney, or why a car's convertible top bulges upward at high speeds? These are examples of a principle worked out by Daniel Bernoulli (1700–1782) in the early eighteenth century. In essence, *Bernoulli's principle* states that *where the velocity of a fluid is high, the pressure is low, and where the velocity is low, the pressure is high.* For example, if the pressures at points 1 and 2 in Fig. 13–2 are measured, it will be found that the pressure is lower at point 2, where the velocity is higher, than it is at point 1, where the velocity is lower. At first glance, this might seem strange; you might expect that the higher speed at point 2 would imply a greater pressure. But this cannot be the case; for if the pressure at point 2 were higher than at 1, this higher pressure would slow the fluid down, whereas in fact it has speeded up. Thus the pressure at 2 must be less than at 1 which will allow the fluid to accelerate.

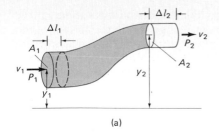

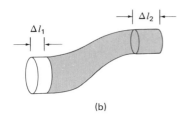

(a)

(b)

FIGURE 13–3

Fluid flow: for derivation of Bernoulli's equation.

Bernoulli developed an equation that expresses this principle quantitatively. To derive Bernoulli's equation, we assume the flow is steady and laminar, the fluid is incompressible, and the viscosity is small enough to be ignored. To be general, we consider a tube of flow which varies (along the tube length) in cross section and also in height above some reference level, Fig. 13–3. We will consider the amount of fluid shown in color and calculate the work done to move it from the position shown in (a) to that shown in (b). In this process fluid at point 1 flows a distance Δl_1 and forces the fluid at the point 2 to move a distance Δl_2. The fluid to the left of point 1 exerts a pressure P_1 on the fluid and does an amount of work $W_1 = F_1 \Delta l_1 = P_1 A_1 \Delta l_1$. At point 2, the work done is $W_2 = -P_2 A_2 \Delta l_2$; the negative sign is present because the force exerted on the fluid is opposite to the motion (thus the fluid shown in color does work on the fluid to the right of point 2). Work is also done on the fluid by the force of gravity; since the net effect of the process shown in Fig. 13–3 is to move a mass m of volume $A_1 \Delta l_1$ $(= A_2 \Delta l_2)$ from point 1 to point 2, the work done by gravity is

$$W_3 = -mg(y_2 - y_1).$$

Notice that in the case shown in Fig. 13–3 this term is negative since the motion is uphill against the force of gravity. The net work W done on the fluid is thus:

$$W = W_1 + W_2 + W_3$$

$$W = P_1 A_1 \Delta l_1 - P_2 A_2 \Delta l_2 - mgy_2 + mgy_1.$$

According to the work-energy theorem (Section 6–4), the net work done on a system is equal to its change in kinetic energy. Thus

$$\tfrac{1}{2}mv_2^2 - \tfrac{1}{2}mv_1^2 = P_1 A_1 \Delta l_1 - P_2 A_2 \Delta l_2 - mgy_2 + mgy_1.$$

The mass m has volume $A_1 \Delta l_1 = A_2 \Delta l_2$ and thus we can substitute $m = \rho A_1 \Delta l_1 = \rho A_2 \Delta l_2$ and obtain (after dividing through by $A_1 \Delta l_1 = A_2 \Delta l_2$ and rearranging):

$$P_1 + \tfrac{1}{2}\rho v_1^2 + \rho g y_1 = P_2 + \tfrac{1}{2}\rho v_2^2 + \rho g y_2. \tag{13–2}$$

This is **Bernoulli's equation**. Since points 1 and 2 can be any two points along a tube of flow, Bernoulli's equation can be written:

$$P + \tfrac{1}{2}\rho v^2 + \rho g y = \text{constant}$$

at every point in the fluid.

Bernoulli's equation can be applied to a great many situations. One example is to calculate the velocity, v_1, of a liquid flowing out of a spigot at the bottom of a

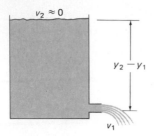

FIGURE 13–4

Torricelli's theorem.

reservoir, Fig. 13–4. We choose point 2 in Eq. 13–2 to be the top surface of the liquid, and assuming the diameter of the reservoir is large compared to that of the spigot, v_2 will be almost zero. The pressure at points 1 (the spigot) and 2 are each equal to atmospheric pressure so $P_1 = P_2$. Then Bernoulli's equation becomes

$$\tfrac{1}{2}\rho v_1^2 + \rho g y_1 = \rho g y_2$$

or

$$v_1 = \sqrt{2g(y_2 - y_1)}. \tag{13–3}$$

This result is called *Torricelli's theorem*. Although it is seen to be a special case of Bernoulli's equation, it was discovered a century before Bernoulli by Evangelista Torricelli (1608–1647); hence its name. Notice that the liquid leaves the spigot with the same speed that a freely falling object would attain falling the same height. This should not be too surprising since the derivation of Bernoulli's equation relies on the conservation of energy.

Another special case of Bernoulli's equation arises when the fluid is flowing but there is no appreciable change in height; that is, $y_1 = y_2$. Then Eq. 13–2 becomes

$$P_1 + \tfrac{1}{2}\rho v_1^2 = P_2 + \tfrac{1}{2}\rho v_2^2. \tag{13–4}$$

This tells us quantitatively that where the speed is high the pressure is low, and vice versa. It explains many everyday phenomena, some of which are illustrated in Fig. 13–5. The pressure in the air blown at high speed across the top of the vertical tube of a perfume atomizer (Fig. 13–5a) is less than the normal air pressure acting on the surface of the liquid in the bowl; thus perfume is pushed up the tube because of the reduced pressure at the top. A Ping-Pong ball can be made to float above a blowing jet of air (some vacuum cleaners can blow air), Fig. 13–5b; if the ball begins to leave the jet of air, the higher pressure outside the jet pushes the ball back in.

Airplane wings and other airfoils are designed to deflect the air so that although streamline flow is largely maintained, the streamlines are crowded together above the wing, Fig. 13–5c. Just as the flow lines are crowded together in a pipe constriction where the velocity is high, so the crowded streamlines above the

FIGURE 13–5

Examples of Bernoulli's principle.

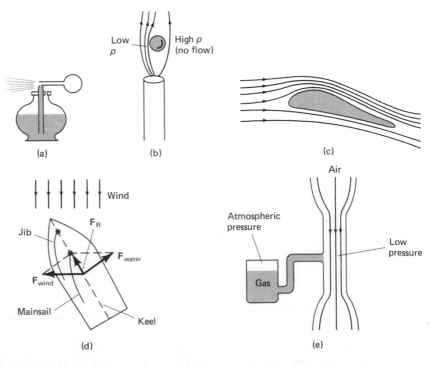

wing indicate the air speed is greater than below the wing. Hence the air pressure above the wing is less than that below and there is thus a net upward force; this is called *dynamic lift*. Actually, Bernoulli's principle is only one aspect of the lift on a wing. Wings are usually tilted slightly upward so that air striking the bottom surface is deflected downward; the change in momentum of the rebounding air molecules results in an additional upward force on the wing. Turbulence also plays an important role.

A sailboat can move against the wind, Fig. 13–5d, and the Bernoulli effect aids in this considerably if the sails are arranged so the air velocity increases in the narrow constriction between the two sails. (The normal pressure behind the mainsail is larger than the reduced pressure in front of it and this pushes the boat forward.) When going against the wind, the mainsail is set at an angle approximately midway between the wind direction and the boat's axis (keel line) as shown. The force of the wind on the sail (momentum change of wind bouncing off the sail), plus the Bernoulli effect, acts nearly perpendicular to the sail ($\mathbf{F}_{wind}$). This would tend to make the boat move sideways but the keel beneath prevents this—for the water exerts a force ($\mathbf{F}_{water}$) on the keel nearly perpendicular to it. The resultant of these two forces ($\mathbf{F}_{R}$) is almost directly forward as shown.

A *Venturi tube* is essentially a pipe with a narrow constriction (the throat). One example of a Venturi tube is the barrel of a carburetor in a car, Fig. 13–5e. The flowing air speeds up as it passes this constriction (Eq. 13–1) and so the pressure is lower. Because of the reduced pressure, gasoline under atmospheric pressure in the carburetor reservoir is forced into the air stream and mixes with the air before entering the cylinders.

The Venturi tube is also the basis of the *Venturi meter* which is used to measure the flow speed of fluids, Fig. 13–6. The flow speed can be shown (see the problems) to be given by the relation

$$v_1 = A_2\sqrt{2(P_1 - P_2)/\rho(A_1^2 - A_2^2)}$$

where ρ is the density of the fluid and P_1 and P_2 are the pressure readings at points 1 and 2 where the area of the tube is A_1 and A_2. If the manometer type is used (Fig. 13–6b) $P_1 - P_2$ is replaced by $(\rho_m - \rho)gh$ where ρ_m is the density of fluid in the manometer. Venturi meters can be used to measure the flow velocities of gases and liquids and have even been designed to measure blood velocity in arteries. The flow rate can also be measured since it equals $v_1 A_1$.

Why does smoke go up a chimney? It's partly because hot air rises (i.e., density). But Bernoulli's principle also plays a role. Because wind blows across the top of a chimney, the pressure is less there than inside the house. Hence, air and smoke are pushed up the chimney. Even on an apparently still night there is usually enough ambient air flow at the top of a chimney to allow upward flow of smoke.

If gophers, prairie dogs, rabbits, and other animals that live underground are to avoid suffocation, the air must circulate in their burrows. The burrows are always made to have at least two entrances. The speed of air flow across different holes will usually be slightly different. This results in a slight pressure difference which forces a flow of air through the burrow a la Bernoulli. The flow of air is enhanced if one hole

FIGURE 13–6

Venturi meters: (a) standard; (b) manometer type.

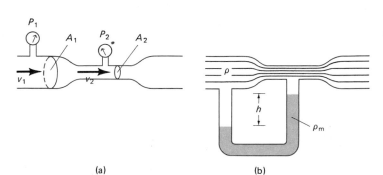

(a)   (b)

is higher than the other (and this is often done by animals) since wind speed tends to increase with height.

    EXAMPLE 13–3   Water circulates throughout a house in a hot water heating system. If the water is pumped at a speed of 0.50 m/s through a 4.0-cm-diameter pipe in the basement under a pressure of 3.0 atm, what will be the flow speed and pressure in a 2.6-cm-diameter pipe on the second floor 5.0 m above?

SOLUTION   We first calculate the velocity v_2 using the equation of continuity, Eq. 13–1:

$$v_2 = \frac{v_1 A_1}{A_2} = (0.50 \text{ m/s}) \frac{(\pi)(0.020 \text{ m})^2}{(\pi)(0.013 \text{ m})^2} = 1.2 \text{ m/s}.$$

To find the pressure we use Bernoulli's equation:

$$
\begin{aligned}
P_2 &= P_1 + \rho g(y_1 - y_2) + \tfrac{1}{2}\rho(v_1^2 - v_2^2) \\
&= (3.0 \times 10^5 \text{ N/m}^2) + (1.0 \times 10^3 \text{ kg/m}^3)(9.8 \text{ m/s}^2)(-5.0 \text{ m}) \\
&\quad + \tfrac{1}{2}(1.0 \times 10^3 \text{ kg/m}^3)[(0.50 \text{ m/s})^2 - (1.2 \text{ m/s})^2] \\
&= 3.0 \times 10^5 \text{ N/m}^2 - 4.9 \times 10^4 \text{ N/m}^2 - 6.0 \times 10^2 \text{ N/m}^2 \\
&= 2.5 \times 10^5 \text{ N/m}^2.
\end{aligned}
$$

Notice that the velocity term contributes very little in this case.

Bernoulli's equation ignores the effects of friction (viscosity) and the compressibility of the fluid. The energy that is transformed to internal (or potential) energy due to compression and to thermal energy by friction can be taken into account by adding terms to the right side of Eq. 13–2. These terms are difficult to calculate theoretically and are normally determined empirically. We will not pursue it here, but merely note that it does not significantly alter the explanations for the phenomena described above.

13–4   Viscosity

As already mentioned, real fluids have a certain amount of internal friction which is called **viscosity**. It exists in both liquids and gases and is essentially a frictional force between different layers of fluid as they move past one another. In liquids it is due to the cohesive forces between the molecules; in gases it arises from collisions between the molecules.

Different fluids possess different amounts of viscosity: syrup is more viscous than water; grease is more viscous than engine oil; liquids in general are much more viscous than gases. The viscosity of different fluids can be expressed quantitatively by a *coefficient of viscosity*, η (the Greek lowercase letter eta), which is defined in the following way. A thin layer of fluid is placed between two flat plates. One plate is stationary and the other is made to move, Fig. 13–7. The fluid directly in contact with either plate is held to the surface by the adhesive force between the molecules of the liquid and those of the plate. Thus the upper surface of the fluid moves with the same speed v as the upper plate, whereas the fluid in contact with the stationary plate remains stationary. The stationary layer of fluid retards the flow of the layer just

FIGURE 13–7

Determination of viscosity.

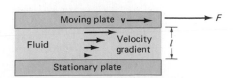

TABLE 13–1

13–5 Laminar Flow in 249
Tubes—Poiseuille's Equation

Coefficient of viscosity for various fluids

Fluid	Temperature (°C)	Coefficient of Viscosity, η (Pa·s)[†]
Water	0	1.8×10^{-3}
	20	1.0×10^{-3}
	100	0.3×10^{-3}
Whole blood	37	$\approx 4 \times 10^{-3}$
Blood plasma	37	$\approx 1.5 \times 10^{-3}$
Ethyl alcohol	20	1.2×10^{-3}
Engine oil (SAE 10)	30	200×10^{-3}
Glycerine	20	1500×10^{-3}
Air	20	0.018×10^{-3}
Hydrogen	0	0.009×10^{-3}
Water vapor	100	0.013×10^{-3}

[†] $1 \text{ Pa·s} = 10 \text{ P} = 1 \times 10^3 \text{ cP}$

above it; this layer retards the flow of the next layer and so on. Thus the velocity varies linearly from 0 to v as shown. The increase in velocity divided by the distance over which this change is made—equal to v/l—is called the *velocity gradient*. To move the upper plate requires a force, a fact you can verify by moving a flat plate across a puddle of syrup on a table. For a given fluid, it is found that the force required, F, is proportional to the area of either plate, A, and to the speed, v, and is inversely proportional to the separation of the plates, $l: F \propto vA/l$. For different fluids, the more viscous the fluid, the greater is the required force. Hence the proportionality constant for this equation is defined as the *coefficient of viscosity, η*:

$$F = \eta A \frac{v}{l}. \qquad (13\text{–}5a)$$

Solving for η, we find $\eta = Fl/vA$. Thus the SI units of η are $\text{N·s/m}^2 = \text{Pa·s}$ (pascal·second); in the cgs system the unit is dyne·s/cm^2 and this unit is called a *poise* (P). Viscosities are often stated in *centipoise* (cP), which is one-hundredth of a poise. Table 13–1 lists the coefficient of viscosity for various fluids; the temperature is also specified, since it has a strong effect—the viscosity of liquids such as motor oil, for example, decreases rapidly as temperature increases.

Equation 13–5a applies when the velocity gradient is uniform. In the general case of a nonuniform velocity gradient, Eq. 13–5a becomes

$$F = \eta A \frac{dv}{dy} \qquad (13\text{–}5b)$$

where the velocity gradient dv/dy is the rate the velocity changes per unit distance measured perpendicular to the direction of the velocity.

The direct proportionality between force and velocity given by Eq. 13–5a does not hold for all fluids (fluids for which it does hold are called *newtonian fluids*). Alternately we can say that η is a function of velocity for nonnewtonian fluids; examples are blood (which contains corpuscles) and other suspensions.

*13–5 Laminar Flow in Tubes— Poiseuille's Equation

If a fluid had no viscosity, it could flow through a level pipe without a force being applied. Because of viscosity, a pressure difference between the ends of a tube is necessary for the steady flow of any real fluid, be it water or oil in a pipe, or blood in the circulatory system of a human.

The rate of flow of a fluid in a round tube depends on the viscosity of the fluid, the pressure difference, and the dimensions of the tube. The French scientist J. L. Poiseuille (1799–1869), who was interested in the physics of blood circulation (and after whom the "poise" is named), determined how these variables affect the flow rate of an incompressible fluid undergoing laminar flow in a cylindrical tube. His result, known as *Poiseuille's equation*, is as follows:

$$Q = \frac{\pi R^4 (P_1 - P_2)}{8 \eta L} \qquad (13\text{-}6)$$

where R is the inside radius of the tube, L is its length, $P_1 - P_2$ is the pressure difference between the ends, η is the viscosity, and Q is the volume rate of flow (volume of fluid flowing past a given point per unit time).

We now derive Poiseuille's equation. Consider a fluid undergoing steady laminar flow through a cylindrical tube of inner radius R as shown in Fig. 13–8. Since a fluid tends to adhere to the walls of the tube, we expect that the fluid velocity will be zero (or near zero) at the walls; we therefore assume the cylindrical layer of fluid next to the tube wall has zero velocity; each successive layer[†] has only a slightly larger velocity because of the viscous friction with the previous layer. The velocity thus increases with distance from the wall and reaches a maximum at the center of the tube. Such a velocity gradient is indicated in Fig. 13–8. We first determine v as a function of r by considering a solid cylinder of fluid of radius r ($r < R$) whose center line is along the center of the tube, as shown in Fig. 13–9a. The force on this cylinder due to the difference in pressure at the ends of the tube is

$$F = (P_1 - P_2)\pi r^2$$

where πr^2 is the area of the ends of our cylinder. The motion of this cylinder of fluid is retarded by the viscous force exerted by the next layer of fluid just outside our cylinder; the magnitude of this viscous force is given by Eq. 13–5b where, for the area A, we must use the area of the sides of the cylinder, $A = (2\pi r)(L)$. The viscous force is thus

$$F = -\eta(2\pi r L)\frac{dv}{dr}$$

[†] The word laminar means "in layers." Thus "laminar flow" refers to our model of fluid flow as being in layers.

FIGURE 13–8

Fluid of density ρ, viscosity η, flowing to right in tube of radius R. The variation in fluid velocity at different positions in the fluid is shown.

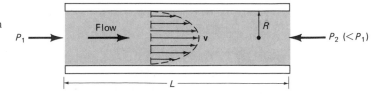

FIGURE 13–9

Deriving Poiseuille's equation: (a) consider solid cylinder of flow of radius r; (b) consider cylindrical shell of radius r and thickness dr (see text).

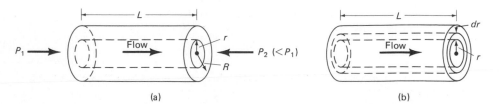

(a)   (b)

where we insert the minus sign since this force opposes the motion. Since the fluid is undergoing steady flow, there is no acceleration. Hence these two forces balance:

$$(P_1 - P_2)\pi r^2 = -2\pi r \eta L \frac{dv}{dr}.$$

We solve for the velocity gradient and obtain

$$\frac{dv}{dr} = -\frac{(P_1 - P_2)r}{2\eta L}.$$

We integrate this to obtain v as a function of r, the distance from the center of the tube, and note that $v = 0$ at $r = R$:

$$\int_0^v dv = -\frac{P_1 - P_2}{2\eta L} \int_R^r r \, dr$$

$$v = -\frac{P_1 - P_2}{2\eta L} \left[\frac{r^2}{2} \right]_R^r$$

$$v = \frac{P_1 - P_2}{4\eta L}(R^2 - r^2). \tag{13–7}$$

As expected, the maximum velocity occurs at the center of the tube ($r = 0$); its magnitude is proportional to the square of the tube radius and is also proportional to the *pressure gradient*, $\Delta P/\Delta L = (P_1 - P_2)/L$.

Now that we know v as a function of r, we can determine the total flow rate Q through the tube, where $Q = dV/dt$ is the total volume of fluid passing a cross section of the tube per unit time. Since the velocity v is not constant across the tube, we cannot simply write $Q = Av$ (as in Eq. 13–1). Instead, we divide up the cross section of the tube into thin rings, of thickness dr, as shown in Fig. 13–9b; we then calculate the flow through each of these thin rings and sum over them all to get the total flow rate. The area of the thin ring shown in Fig. 13–9b is the product of its circumference, $2\pi r$, and its width, dr:

$$dA = 2\pi r \, dr.$$

Since the fluid velocity v depends only on r (Eq. 13–7), v is the same through all parts of the thin ring. Thus the flow rate through the thin ring shown is

$$dQ = v \, dA$$

$$= \frac{P_1 - P_2}{4\eta L}(R^2 - r^2)2\pi r \, dr.$$

We sum over all the rings to get the total flow rate through the tube:

$$Q = \int_{r=0}^{r=R} dQ = \frac{\pi(P_1 - P_2)}{2\eta L} \int_0^R (R^2 r - r^3) \, dr$$

$$= \frac{\pi(P_1 - P_2)}{2\eta L} \left[\frac{R^2 r^2}{2} - \frac{r^4}{4} \right]_0^R$$

$$= \frac{\pi(P_1 - P_2)R^4}{8\eta L}$$

which is Eq. 13–6; so we have derived Poiseuille's equation from our assumption of simple laminar flow.

EXAMPLE 13–4   Engine oil (assume SAE 10—Table 13–1) passes through a fine 1.80-mm-diameter tube in a prototype engine. The tube is 5.5 cm long. What pressure difference is needed to maintain a flow rate of 5.6 mL/min?

SOLUTION   The flow rate in SI units is $Q = 5.6 \times 10^{-6}$ m³/60 s = 9.3×10^{-8} m³/s. We solve for $P_1 - P_2$ in Eq. 13–6 and put all terms in SI units:

$$P_1 - P_2 = \frac{8\eta L Q}{\pi R^4}$$

$$= \frac{8(2.0 \times 10^{-1} \text{ N·s/m}^2)(5.5 \times 10^{-2} \text{ m})(9.3 \times 10^{-8} \text{ m}^3/\text{s})}{3.14(0.90 \times 10^{-3} \text{ m})^4}$$

$$= 4.0 \times 10^3 \text{ N/m}^2$$

or about 0.04 atm.

Poiseuille's equation tells us that the flow rate Q is directly proportional to the pressure gradient, $(P_1 - P_2)/L$, and it is inversely proportional to the viscosity of the fluid. This is just what we might expect. It may be surprising, however, that Q also depends on the *fourth* power of the tube's radius. This means that for the same pressure gradient, if the tube radius is doubled the flow rate is increased by a factor of sixteen! Thus the rate of flow, or alternately the pressure required to maintain a given flow rate, is greatly affected by only a small change in tube radius.

An interesting example of this R^4 dependence is the flow of blood in the human body. However since Poiseuille's equation is valid only for the streamline flow of an incompressible fluid with constant viscosity η, it cannot be precisely accurate for blood whose flow is not without turbulence; blood also contains corpuscles (whose diameter is almost equal to that of a capillary), and η depends to a certain extent on the flow speed v. Nonetheless Poiseuille's equation does give a good first approximation. The body controls the flow of blood by means of tiny bands of muscle surrounding the arteries. Contraction of these muscles reduces the diameter of an artery and, due to the R^4 in Eq. 13–6, the flow rate is greatly reduced for only a small change in radius. Very small actions by these muscles can thus control precisely the flow of blood to different parts of the body. Another aspect is that the radius of arteries is reduced as a result of arteriosclerosis (hardening of the arteries) and by cholesterol buildup; when this happens, the pressure gradient must be increased to maintain the same flow rate. If the radius is reduced by half, the heart would have to increase the pressure by a factor of about 16 in order to maintain the same blood-flow rate. The heart must work much harder under these conditions but usually cannot maintain the original flow rate. Thus, high blood pressure is an indication both that the heart is working harder and that blood-flow rate is reduced.

*13–6   Turbulent Flow in Tubes—Reynolds Number

If the flow velocity is large, the flow through a tube will become turbulent and Poiseuille's equation will no longer hold. When the flow is turbulent, the flow rate Q for a given pressure difference will be less than for laminar flow as given in Eq. 13–6. This is a result of the fact that friction forces are much greater when turbulence is present.

The onset of turbulence is often abrupt and can be characterized approximately by the so-called *Reynolds number*, *Re*:

$$Re = \frac{2\bar{v}r\rho}{\eta}, \tag{13–8}$$

where $\bar{v}$ is the average speed[†] of the fluid, ρ is its density, η is its viscosity, and r is the radius of the tube in which the fluid is flowing. Experiments show that the flow is laminar if Re has a value less than about 2000 but is turbulent if Re exceeds this value.

■ **EXAMPLE 13–5** The average speed of blood in the aorta ($r = 1.0$ cm) during the resting part of the heart's cycle is about 30 cm/s. Is the flow laminar or turbulent?

SOLUTION To answer this we calculate the Reynolds number using the values of ρ and η from Tables 12–1 and 13–1:

$$Re = \frac{(2)(0.30 \text{ m/s})(0.010 \text{ m})(1.05 \times 10^3 \text{ kg/m}^3)}{(4.0 \times 10^{-3} \text{ N} \cdot \text{s/m}^2)} = 1600.$$

The flow will probably be laminar, but it is close to turbulence.

Notice in this example that since 1 N = 1 kg·m/s², Re has no units. Hence we see that Reynolds number is always a *dimensionless* quantity; its value is the same in any consistent set of units.

*13–7 Object Moving in a Fluid; Sedimentation and Drag

In the previous section we saw how viscosity (and other factors) affect the flow of a fluid through a tube. In this section we will examine a slightly different situation, that of an object moving relative to a fluid. It could be an obstacle that obstructs the flow of a fluid, such as a large rock in a river, or it could be an object moving in a fluid, such as a glider or car moving through air, a submarine in water, or a molecule sedimenting in a centrifuge.

When an object moves relative to a fluid, the fluid exerts a force on the object. This force, which is referred to as a *drag force*, is due to the viscosity of the fluid and also, at high speeds, to turbulence behind the object.

To characterize the motion of an object relative to a fluid, it is useful to define another Reynolds number

$$Re' = \frac{vL\rho}{\eta} \qquad (13-9)$$

where ρ and η are the density and viscosity of the fluid, v is the object's velocity relative to the fluid, and L is a characteristic length of the object. This Reynolds number must be clearly distinguished from the one used for fluid flow in a tube (although the form is similar) since the phenomena are quite different.

When the Reynolds number for our present case is less than about 1,[‡] the flow around an object is essentially laminar and it is found experimentally that the viscous force F_v is directly proportional to the speed of the object:

$$F_v = kv. \qquad (13-10)$$

[†] The average speed, $\bar{v}$, is defined as that velocity, uniform over the whole cross section of the tube, that would give the same volume flow rate, Q.

[‡] An object 1 mm long moving at a speed of 1 mm/s through water has a Reynolds number equal to 1. So does an object 2 mm long traveling 7 mm/s in air. This situation thus applies mainly to fairly small objects such as raindrops, pollen grains, and molecules in a centrifuge.

The magnitude of k depends on the size and shape of the object and on the viscosity of the fluid. For a sphere of radius r it has been calculated to be

$$k = 6\pi r\eta. \qquad \text{[sphere]}$$

Thus the viscous force on a small sphere, when the flow is laminar, is given by an equation known as Stokes's equation:

$$F_v = 6\pi r\eta v. \qquad \text{[for a sphere]}$$

For larger Reynolds numbers (usually above a value between about 1 and 10), there will be turbulence behind the body known as the wake (see Fig. 13–1b), and the drag force will be larger than that given by Stokes's equation for a sphere. For more streamlined objects, however, there will be less turbulence and hence less drag. When turbulence is present, experiments show that the drag force increases as the square of the speed, $F_v \propto v^2$. The increase with speed is thus much more rapid than in the case of strictly laminar flow. When the Reynolds number approaches a value around 10^6, the drag force increases abruptly. For above this value, turbulence exists not only behind the object but also in the layer of fluid lying next to the body (called the *boundary layer*) all along its sides.

Sedimentation refers to small objects falling in a fluid—examples are tiny particles of rock or minerals sedimenting under the sea, and red blood cells sedimenting in the fluid plasma in a laboratory.

An object of mass m falling through a fluid under the action of gravity has several forces on it as shown in Fig. 13–10: the force of gravity, mg; F_B, the buoyant force of the fluid; and F_v the viscous force. By Newton's second law, the net force is equal to the mass times the acceleration of the object:

$$mg - F_B - F_v = ma.$$

The buoyant force F_B is equal to the weight of fluid displaced; that is, $F_B = \rho_f Vg$, where ρ_f is the density of the fluid, V is the volume of the object (and hence the volume displaced), and g is the acceleration due to gravity. We can also write $mg = \rho_o Vg$ where ρ_o is the density of the object. Using Eq. 13–10, we can write the above equation as:

$$(\rho_o - \rho_f)Vg - kv = ma. \qquad (13\text{–}11)$$

The first term is the effective weight of the object in the fluid. As the object increases in speed, the viscous force increases until it just balances the effective weight of the object. At this point the acceleration is zero and the speed increases no further. This maximum speed, v_T, is called the *terminal velocity* or *sedimentation velocity* and is obtained from Eq. 13–11 by setting $a = 0$:

$$v_T = \frac{(\rho_o - \rho_f)Vg}{k}. \qquad (13\text{–}12)$$

FIGURE 13–10

Forces on a small object falling through a fluid.

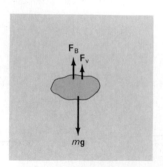

The sedimentation velocity of tiny objects such as macromolecules and other constituents of cells is extremely small. It can be increased using a centrifuge (Chapter 5) since the effect of a centrifuge is to exert a force on a particle as if the acceleration of gravity were increased to a value of $\omega^2 r$ (see Eq. 9–6); here, ω is the angular velocity of the rotor and r is the distance of the object from the axis of rotation. Thus, we can still use Eq. 13–12 for a centrifuge if we substitute the effective acceleration of gravity, equal to $\omega^2 r$, for g:

$$v_T = \frac{(\rho_o - \rho_f)V\omega^2 r}{k}. \qquad (13\text{–}13)$$

Centrifugation is often used to separate similar but slightly different particles or macromolecules (such as two types of nucleic acid) as well as to obtain valuable information on size and mass.

Summary

Fluid flow can be characterized as *laminar*, in which the layers of fluid move smoothly and regularly along paths called *streamlines*, or as *turbulent*, in which the flow is irregular and contains whirlpools. *Viscosity* refers to friction within a fluid that prevents the fluid from flowing freely, and can be visualized as a frictional force between different layers of fluid as they pass over one another.

Fluid *flow rate* is the mass or volume of fluid that passes a given point per unit time. For steady laminar flow of a fluid in an enclosed (not necessarily uniform) tube, the *equation of continuity* states that the mass flow rate, which is the product of fluid density, flow velocity, and tube cross-sectional area, is the same at all points in the tube: $\rho Av = $ constant. If the fluid is incompressible, then $Av = $ constant.

Bernoulli's principle states that where the velocity of a fluid is high, the pressure is low, and where the velocity is low, the pressure is high. Quantitatively, *Bernoulli's equation* for the steady laminar flow of an incompressible and nonviscous fluid in a tube of flow (of nonuniform cross section and varying height) is

$$P_1 + \tfrac{1}{2}\rho v_1^2 + \rho g y_1 = P_2 + \tfrac{1}{2}\rho v_2^2 + \rho g y_2,$$

for any two points along the tube.

Questions

1. Can streamline flow be either compressible or incompressible? Viscous or nonviscous? Steady or nonsteady? Rotational or irrotational? For each "yes" answer, give an example. Explain any "no" answers.
2. Can turbulent flow be either compressible or incompressible? Viscous or nonviscous? Steady or nonsteady? Rotational or irrotational? For each "yes" answer, give an example. Explain any "no" answers.
3. What forces act to accelerate the fluid in Fig. 13–2 when it moves from region 1 to region 2?
4. Why does the stream of water from a faucet become narrower as it falls?
5. If you dangle two pieces of paper vertically, a few inches apart (Fig. 13–11), and blow between them, how do you think the paper will move? Try it and see. Explain.

FIGURE 13–11

6. A baseball pitcher puts spin on the ball when throwing a curve. Use Bernoulli's principle to explain carefully why the ball curves. Explain why a spinning ball with a very smooth surface curves in the opposite direction to one with a rough surface (such as a baseball or tennis ball).
7. Why do airplanes normally take off into the wind?
8. Children are told to avoid standing too close to a rapidly moving train because they might get sucked under it. Is this possible? Explain.
9. Roofs of houses are sometimes pushed off during a tornado or hurricane. Explain, using Bernoulli's principle.
10. Why does the canvas top of a convertible bulge out when the car is traveling at high speed?
11. Two ships moving in parallel paths close to one another risk colliding. Why?
12. Explain why the speed of wind increases with increased height above the earth's surface. (Hint: see Fig. 13–7.) Explain how this affects a mole that builds its burrow with one of its two entrances higher than the other.
13. Identical steel ball bearings are dropped into tubs of water at 10°C and then at 40°C. In which tub will the ball bearing reach the bottom more quickly?
14. Why does oil (or any other liquid) flow more smoothly and quickly from a can if two holes (on opposite sides of the top) are used rather than only one? Under what conditions will smooth flow be obtained if only a single hole is made?
15. Hummingbirds expend 20 times as much energy to hover in front of a flower as they do in normal flight. Explain.
*16. Blood corpuscles tend to flow in the center of blood vessels. Explain.
*17. Show that the viscous flow of a fluid in a tube, Fig. 13–8, is rotational.

Problems

SECTION 13–2

1. (I) Using the data of Example 13–1 calculate the average speed of blood flow in the major arteries of the body which have a total cross-sectional area of about 2.0 cm².
2. (I) A 15-cm-radius air duct is used to replenish the air of a room 10 m × 5.1 m × 3.2 m every 10 min. What is the average speed of air flow in the duct?

3. (I) How long does it take to fill a swimming pool 9.5 m × 21.0 m whose average depth is 3.1 m if water flows from a 1.9-cm-diameter hose at a speed of 1.5 m/s?

4. (II) The stream of water from a faucet decreases in diameter as it falls. Derive an equation for the diameter of the stream as a function of the distance y below the faucet, given that the water has speed v_0 when it leaves the faucet whose diameter is D.

SECTION 13–3

5. (I) Show that Bernoulli's equation reduces to the hydrostatic variation of pressure with depth (Eq. 12–6) when there is no flow ($v_1 = v_2 = 0$).

6. (I) How fast does water flow from a hole at the bottom of a 4.6-m-deep storage tank filled with water? Ignore viscosity.

7. (I) What gauge pressure in the water mains is necessary if a firehose is to spray water to a height of 25 m?

8. (I) How high should the pressure head be if water is to come from a faucet at a speed of 8.0 m/s? Ignore viscosity.

9. (II) If wind blows at 25 m/s over your house, what is the net force on the roof if its area is 250 m²?

10. (II) If the absolute water pressure at street level is 2.75 × 10⁵ N/m², will this be sufficient to lift water to the top floor of a proposed 50-m-high apartment building? What maximum height can the building be?

11. (II) What is the volume rate of flow of water from a 1.8-cm-diameter faucet if the pressure head is 10 m?

12. (II) What is the lift (in newtons) due to Bernoulli's principle on a wing of area 50 m² if the air passes over the top and bottom surfaces at speeds of 320 m/s and 290 m/s respectively?

13. (II) Show that the power needed to drive a fluid through a pipe is equal to the volume rate of flow, Q, times the pressure difference, $P_1 - P_2$.

14. (II) Water at a pressure of 3.3 atm at street level flows into an office building at a speed of 0.50 m/s through a pipe 5.0 cm in diameter. The pipes taper down to 2.5-cm diameter by the top floor, 25 m above. Calculate the flow velocity and the pressure in such a pipe on the top floor. Ignore viscosity.

15. (II) An airplane has a mass of 2.0 × 10⁶ kg and the air flows past the lower surface of the wings at 100 m/s. If the wings have a surface area of 1200 m², how fast must the air flow over the upper surface of the wing if the plane is to stay in the air? Consider only the Bernoulli effect.

16. (II) A particular four-cylinder automobile engine has a displacement of 2000 cm³; this is approximately how much air is brought into the four cylinders per revolution. If the engine runs at a speed of 1500 rpm, and the Venturi of the carburetor has a radius of 2.5 cm, (a) what is the speed of the air flow through the Venturi; (b) what is the pressure (in atmospheres) in the Venturi?

17. (II) In Fig. 13–4, take into account the speed of the top surface of the tank and show that the speed of fluid leaving the opening at the bottom is

$$v_1 = \sqrt{2gh/(1 - A_1^2/A_2^2)}$$

where $h = y_2 - y_1$ and A_1 and A_2 are the areas of the opening and the top surface, respectively.

18. (II) Suppose the top surface of the vessel in Fig. 13–4 is subjected to an external pressure P_0. (a) Derive a formula for the speed, v_1, at which the liquid flows from the opening at the bottom into atmospheric pressure, P_a. Assume the velocity of the liquid surface, v_2, is approximately zero. (b) If $P_0 = 0.85$ atm and $y_2 - y_1 = 2.1$ m, determine v_1 for water.

19. (II) The *Pitot tube* is a device used to measure flow velocity, Fig. 13–12. The fluid passes the opening at B with speed v whereas at A the fluid is at rest. The manometer registers the difference of pressure at these two points, $P_A - P_B$. (a) Derive an equation for the velocity of flow v in terms of the height h and the density ρ_f of the moving fluid and the density ρ_M of the fluid in the manometer. (b) A tiny Pitot tube used to measure the flow of blood indicates a height $h = 20$ mm Hg in the manometer; calculate the flow velocity v of the blood.

20. (II) A 6.0-cm-diameter pipe gradually narrows to 4.0 cm. When water flows through this pipe at a certain rate, the gauge pressure in these two sections is 32 kPa and 24 kPa. What is the volume rate of flow?

21. (II) *Thrust of a rocket.* (a) Use Bernoulli's equation and the equation of continuity to show that the emission speed of the propelling gases of a rocket is

$$v = \sqrt{2(P - P_0)/\rho}$$

where ρ is the density of the gas, P is the pressure of the gas inside the rocket, and P_0 is atmospheric pressure just outside the exit orifice. Assume that the gas density remains constant, that the area of the exit orifice, A_0, is much smaller than the cross-sectional area, A, of the inside of the rocket (take it to be a large cylinder), and that the gas speed is not so high that significant turbulence or nonsteady flow sets in. (b) Show that the thrust force on the rocket due to the emitted gases is

$$F = 2A_0(P - P_0).$$

22. (II) (a) Show that the flow velocity measured by a Venturi meter is given by the relation

$$v_1 = A_2 \sqrt{\frac{2(P_1 - P_2)}{\rho(A_1^2 - A_2^2)}}.$$

See Fig. 13–6a. (b) For the manometer type, Fig. 13–6b, find v_1 in terms of h and other relevant quantities. (c) A Venturi tube is measuring the flow of water. It has a main diameter of 3.0 cm tapering down to a throat diameter of 1.0 cm. The pressure difference $P_1 - P_2$ is measured to be 18 mm Hg. Calculate the velocity of the fluid.

23. (III) Suppose the opening in the tank of Fig. 13–4 is a height h_1 above the base and the liquid surface is a height

FIGURE 13–12

Pitot tube.

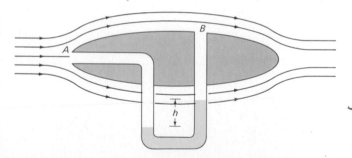

h_2 above the base. The tank rests on level ground. (a) At what horizontal distance from the base of the tank will the fluid strike the ground? (b) At what other height, h_1', can a hole be placed and the emerging liquid have the same "range"?

24. (III) (a) In Fig. 13–4, show that the level of the liquid, $h = y_2 - y_1$, drops at a rate

$$\frac{dh}{dt} = \sqrt{\frac{2ghA_1^2}{A_2^2 - A_1^2}},$$

where A_1 and A_2 are the areas of the opening and the top surface, and viscosity is ignored. (b) Determine h as a function of time by integrating. Let $h = h_0$ at $t = 0$. (c) How long will it take to empty a 9.4-cm-tall cylinder filled with 1.0 L of water if the opening is at the bottom and has a 0.50-cm diameter?

SECTION 13–4

25. (II) A viscometer consists of two concentric cylinders, 10.20 cm and 10.60 cm in diameter. A particular liquid fills the space between them to a depth of 12.0 cm. The outer cylinder is fixed and a torque of 0.024 N·m keeps the inner cylinder turning at a steady rotational speed of 62 rev/min. What is the viscosity of the liquid?

26. (II) (a) Show that the viscous liquid of Fig. 13–7 is undergoing shearing strain much like that for a solid (Section 11–4). (b) Show that the time rate of change of shearing strain (change in shearing strain per unit time—see Section 11–4) is given by v/l. (c) Show that the coefficient of viscosity is defined as

$$\eta = \frac{\text{shearing stress}}{\text{rate of change of shearing strain}}.$$

*SECTION 13–5

*27. (I) Estimate the volume rate of flow of water through a 15-m length of 1.0-cm-diameter tubing if the pressure difference between its ends is 0.35 atm and the temperature is 20°C.

*28. (I) A gardener feels it is taking him too long to water a garden with a $\frac{3}{8}$-inch-diameter hose. By what factor will his time be cut if he uses a $\frac{5}{8}$-inch-diameter hose? Assume nothing else is changed.

*29. (II) Calculate the pressure drop per centimeter along the aorta using the data of Example 13–1 and Table 13–1.

*30. (II) Assuming a constant pressure gradient, by what factor must a blood vessel decrease in radius if the blood flow is reduced by 80 percent?

*31. (II) A patient is to be given a blood transfusion. The blood is to flow through a tube from a raised bottle to a needle inserted in the vein. The inside diameter of the needle is 0.50 mm, its length is 4.0 cm, and the required flow rate is 4.0 cm³ of blood per minute. How high should the bottle be placed above the needle? Obtain ρ and η from Tables 12–1 and 13–1. Assume the blood pressure is 20 torr above atmospheric pressure.

*32. (II) (a) What must be the pressure difference between the two ends of a 2.0-km section of pipe 40 cm in diameter if it is to transport oil ($\rho = 950$ kg/m³, $\eta = 2.0$ P) at a rate of 400 cm³/s? (b) At what rate is thermal energy produced under these conditions?

*33. (II) What diameter must a 30-m-long air duct have if the pressure of a ventilation and heating system is to replenish the air in a room 10 m × 18 m × 4.0 m every 10 min? Assume the pump can exert a gauge pressure of 4.0×10^{-4} atm.

*34. (II) Water is to be pumped through a 10.0-cm-diameter pipe over a distance of 300 m. The far end of the pipe is 20 m above the pump and is at normal atmospheric pressure. What gauge pressure must the pump develop for there to be any flow at all?

*35. (II) Show that the curve of v versus r, the laminar flow velocity of a viscous fluid in a tube versus the distance from the center of the tube, is a parabola (Fig. 13–8 and Eq. 13–7).

*36. (II) The velocity of water at the center of a 5.2-cm-diameter pipe 20 m long is 18 cm/s. Determine (a) the pressure difference between the ends, (b) the volume rate of flow.

*37. (II) When a viscous liquid flows through a circular tube, as in Fig. 13–8, show that the average velocity of flow—that velocity which if uniform across the tube would give the same flow rate—is equal to half the fluid velocity at the center (the maximum velocity).

*38. (II) Water shoots 14.6 m high from a 1.0-cm-diameter pipe in a fountain. What must be the pressure at the pump, which is 4.2 m below the nozzle (in the ground)? Take into account viscosity, but ignore air resistance. Note any simplifying assumptions you make.

*39. (II) Consider a siphon which transfers water (20°C) from one vessel to a second (lower) one as in Fig. 12–21. (a) Determine the rate of flow if the hose has a 1.2-cm diameter and the difference in water levels of the two containers is 64 cm. (b) What is the maximum height that the hose can reach and still act as a siphon?

*SECTION 13–6

*40. (I) During heavy exercise, the flow speed of blood increases by perhaps a factor of two. Referring to Example 13–5, calculate the Reynolds number and determine what type of flow you would expect in the aorta.

*41. (I) Calculate the Reynolds number for the flow of blood through a capillary if its speed is 4.7×10^{-2} cm/s.

*42. (II) What is the (approximate) maximum flow rate, Q, of water in a 10-cm-diameter pipe if turbulence is to be avoided?

*SECTION 13–7

*43. (I) Calculate the magnitude of the terminal velocity of a 1.0-mm-radius (assume constant) air bubble rising in oil of viscosity 0.20 Pa·s and specific gravity 0.90.

*44. (I) What is the approximate terminal speed of a steel ball of radius 2.0 cm moving in the air if the flow is laminar?

*45. (II) If an object requires 30 min to sediment in an ultracentrifuge rotating at 30,000 rpm at an average distance from the axis of 8.0 cm, how long would it take to sediment under gravity in the same tube standing vertically in the lab?

46. (II) (a) Show that the terminal velocity of a small sphere of density ρ_0 falling through a fluid of density ρ_f and viscosity η is

$$v_T = \frac{2}{9}\frac{(\rho_0 - \rho_f)r^2 g}{\eta}.$$

(b) What is the terminal velocity of a spherical raindrop of radius $r = 0.020$ cm falling in air?

*47. (II) The viscosity, η, of a fluid can be determined by measuring the terminal velocity, v_T, of a sphere falling through the liquid. Determine a formula for η in terms of the radius, r, and density, ρ, of the sphere, and the density, ρ_f, of the fluid. Assume no turbulence.

48. (II) (a) Show that the buoyant force on a small object in the liquid of a centrifuge rotating with angular velocity ω is given by

$$F_B = \rho_f V \omega^2 r$$

where V is the volume of the object, r is its distance from the axis of rotation, ρ_f is the density of the liquid. (b) Compare the dependence of F_B on position with that for the buoyant force on an object sedimenting under gravity.

49. (II) Show that the fluid pressure in a rotating centrifuge, a distance r from the axis of rotation, is given by

$$P = \tfrac{1}{2}\rho_f\omega^2(r^2 - r_0^2)$$

where ρ_f is the fluid density, r_0 is the distance from the axis of rotation of the surface of fluid in the tube, and ω is the angular velocity.

Oscillations

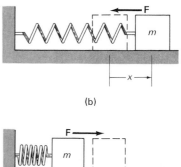

14

Many objects vibrate or oscillate—an object on the end of a spring, a tuning fork, the balance wheel of a watch, a pendulum, a plastic ruler held firmly over the edge of a table and gently struck, the strings of a guitar or piano. Spiders detect prey by the vibrations of their webs, cars oscillate up and down when they hit a bump, buildings and bridges vibrate when heavy trucks pass or the wind is fierce. Indeed, because most solids are elastic (see Chapter 11), most material objects vibrate (at least briefly) when given an impulse. Electrical oscillations occur in radio and TV sets. And at the atomic level, atoms vibrate within a molecule, and the atoms of a solid vibrate about their relatively fixed positions. Because it is so common and occurs in so many areas of physics, oscillatory (or vibrational) motion is of great importance. Vibrational motion is not really a "new" phenomenon since vibrations of mechanical systems are fully described on the basis of Newtonian mechanics.

FIGURE 14–1

Mass vibrating at the end of a spring.

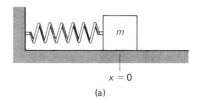

$x = 0$

(a)

14–1 Oscillations of a Spring

When we speak of a *vibration* or an *oscillation* we mean the motion of an object that repeats itself, back and forth, over the same path. That is, the motion is **periodic**. The simplest form of periodic motion is represented by an object oscillating on the end of a spring. Because many other types of vibrational motion closely resemble this system, we will look at it in detail. We assume that the mass of the spring can be ignored, and that the spring is mounted horizontally as shown in Fig. 14–1a, so that the object of mass *m* slides without friction on the horizontal surface. Any spring has a natural length at which it exerts no force on the mass *m*, and this is called the **equilibrium position**. If the mass is moved either to the left, which compresses the spring, or to the right, which stretches it, the spring exerts a force on the mass which acts in the direction of returning it to the equilibrium position; hence it is called a "restoring force." The magnitude of the restoring force *F* is found to be directly proportional to the distance *x* the spring has been stretched or compressed, Fig. 14–1b and c:

$$F = -kx. \qquad (14\text{–}1)$$

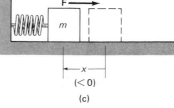

(b)

(c)

Equation 14–1 is accurate as long as the spring is not compressed to the point where the coils touch, or stretched beyond the elastic region (see Fig. 11–9). The minus sign indicates that the restoring force is always in the direction opposite to the displacement x. For example, if we choose the positive direction to the right in Fig. 14–1, x is positive when the spring is stretched, but the direction of the restoring force is to the left. Notice that we have taken $x = 0$ at the equilibrium position. If the spring is compressed, x is negative but the force F acts toward the right (Fig. 14–1c). The proportionality constant k in Eq. 14–1 is called the "spring constant." Notice that in order to stretch the spring a distance x one has to exert an (external) force on the spring at least equal to $F = +kx$. The greater the value of k, the greater the force needed to stretch a spring a given distance. That is, the stiffer the spring, the greater the spring constant k.

Let us examine what happens when the spring is initially stretched a distance $x = A$ as shown in Fig. 14–2a and then released. The spring exerts a force on the mass that pulls it toward the equilibrium position; but because the mass has been accelerated by the force, it passes the equilibrium position with considerable speed. Notice that as the mass reaches the equilibrium position, the force on it decreases to zero, but its speed at this point is a maximum, Fig. 14–2b. As it moves farther to the left, the force on it acts to slow it down, and it stops momentarily at $x = -A$, Fig. 14–2c. It then begins moving back in the opposite direction, Fig. 14–2d, until it reaches the original starting point, $x = A$, Fig. 14–2e. It then repeats the motion.

To discuss vibrational motion, we need to define a few terms. The distance x of the mass from the equilibrium point at any moment is called the **displacement**. The maximum displacement—the greatest distance from the equilibrium point—is called the **amplitude**, A. One *cycle* refers to the complete to-and-fro motion from some initial point back to that same point, say from $x = A$ to $x = -A$ back to $x = A$. The **period**, T, is defined as the time required for one complete cycle. Finally, the **frequency**, f, is the number of complete cycles per second. Frequency is usually specified in hertz (Hz) where 1 Hz = 1 cycle per second. It is evident that

$$f = \frac{1}{T} \quad \text{and} \quad T = \frac{1}{f}. \tag{14–2}$$

For example, if the frequency is 5 cycles per second then each cycle takes $\frac{1}{5}$ s.

The oscillation of a spring hung vertically is essentially the same as that of a horizontal spring. Because of the force of gravity, the length of the vertical spring at equilibrium will be longer than when it is horizontal. If x is measured from the new equilibrium position, Eq. 14–1 can be used directly with the same value of k. (The proof is left as problem 9 at the end of the chapter.)

14–2 Simple Harmonic Motion

Any vibrating system for which the restoring force is directly proportional to the negative of the displacement (as in Eq. 14–1, $F = -kx$) is said to exhibit **simple harmonic motion** (SHM). Such a system is often called a **simple harmonic oscillator** (SHO). We saw in Chapter 11 (Section 11–4) that most solid materials stretch or compress according to Eq. 14–1 as long as the displacement is not too great. Because of this, many natural vibrations are simple harmonic or close to it.

Let us now determine the position x as a function of time. To do so we make use of Newton's second law, $F = ma$. Since the acceleration $a = d^2x/dt^2$, we have

$$m \frac{d^2x}{dt^2} = -kx \tag{14–3a}$$

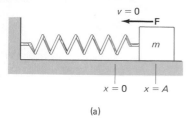

FIGURE 14–2

Force on, and velocity of, mass at different positions of its oscillation.

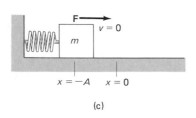

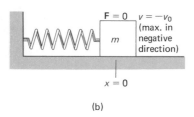

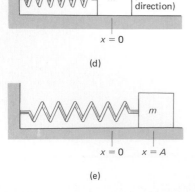

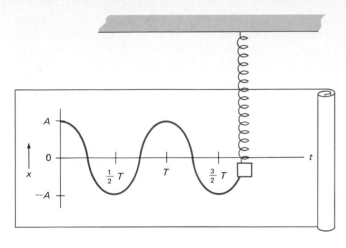

FIGURE 14-3

Sinusoidal nature of SHM as a function of time; in this case, $x = A \cos(2\pi t/T)$.

where m is the mass[†] which is oscillating. We arrange this to obtain

$$\frac{d^2x}{dt^2} + \frac{k}{m}x = 0 \qquad (14\text{-}3b)$$

which is known as the *equation of motion* for the simple harmonic oscillator. Mathematically it is called a *differential equation*, since it involves derivatives. We want to determine what function of time, $x(t)$, satisfies this equation. We might guess the form of the solution by noting that if a pen were attached to a vibrating mass (Fig. 14-3) and a sheet of paper moved at a steady rate beneath it, the pen would trace the curve shown. The shape of this curve is *sinusoidal* (cosine or sine, depending on the position at $t = 0$) as a function of time, and is multiplied by the amplitude A. Indeed, from Eq. 14-3a, we see that the second derivative of x is equal to the negative of x itself, times the constant k/m; both the sine and cosine have this property:

$$\frac{d^2}{dt^2}(\sin \omega t) = -\omega^2 \sin \omega t$$

$$\frac{d^2}{dt^2}(\cos \omega t) = -\omega^2 \cos \omega t$$

where ω is a constant. Thus $x = \cos \omega t$ or $x = \sin \omega t$ would satisfy the equation of motion, Eq. 14-3a or b, if the constant ω is chosen correctly. But to be general, let us try a solution, $x(t)$, that has the general form

$$x = a \cos \omega t + b \sin \omega t$$

where a and b are arbitrary constants. We differentiate this $x = x(t)$ twice:

$$\frac{dx}{dt} = -a\omega \sin \omega t + b\omega \cos \omega t$$

$$\frac{d^2x}{dt^2} = -a\omega^2 \cos \omega t - b\omega^2 \sin \omega t = -\omega^2(a \cos \omega t + b \sin \omega t).$$

When we substitute these relations for x and d^2x/dt^2 into Eq. 14-3b we get:

$$-\omega^2(a \cos \omega t + b \sin \omega t) + \frac{k}{m}(a \cos \omega t + b \sin \omega t) = 0$$

[†] In the case of a mass, m', on the end of a spring, the spring itself also oscillates and at least a part of its mass must be included; in fact, it can be shown—see the problems—that approximately one-third the mass of the spring, m_s, must be included; so $m = m' + \frac{1}{3}m_s$ in our equation. Often m_s is small enough to be ignored.

or

$$\left(\frac{k}{m} - \omega^2\right)(a \cos \omega t + b \sin \omega t) = 0.$$

Thus the equation of motion is satisfied for all times by our trial solution if

$$\frac{k}{m} - \omega^2 = 0$$

or

$$\omega^2 = \frac{k}{m}. \tag{14-4}$$

Thus

$$x = a \cos \omega t + b \sin \omega t \tag{14-5}$$

is a solution of the equation of motion if and only if Eq. 14–4 holds.

Equation 14–5 is in fact the general solution since it contains two arbitrary constants, a and b. This is necessary since the second derivative, d^2x/dt^2, in Eq. 14–3 implies that two integrations must be made to obtain $x(t)$, and each integration yields one arbitrary constant. In real physical situations, the constants a and b are determined by the **initial conditions**. If, for example, the mass is started at its maximum displacement $x = A$ and released without a push, the motion is a pure cosine[†] curve (as in Fig. 14–3) with $a = A$ and $b = O$:

$$x = A \cos \omega t.$$

On the other hand, suppose at $t = 0$ the mass m was at $x = 0$ and it was struck, giving it an initial velocity toward increasing values of x; since $x = 0$ at $t = 0$ in this case, we must have $a = 0$ in Eq. 14–5, and b must then be the amplitude A, so

$$x = A \sin \omega t.$$

There are many other situations in which neither a nor b is zero—such as when the spring is stretched a certain distance at $t = 0$ and is given a push so, at $t = 0$, x is less than A. In all cases, specifying two properties such as displacement and velocity at a given instant will uniquely determine a and b.

Equation 14–5 for x can be written in the following more convenient form:

$$x = A \cos (\omega t + \phi). \tag{14-6}$$

That Eqs. 14–6 and 14–5 are equivalent can be seen from the trigonometric identity

$$\cos (\omega t + \phi) = \cos \omega t \cos \phi - \sin \omega t \sin \phi;$$

the constants A and ϕ in Eq. 14–6 are related to the constants a and b in Eq. 14–5 by $A \cos \phi = a$ and $-A \sin \phi = b$. The physical interpretation of Eq. 14–6 is simpler than for Eq. 14–5. As shown in Fig. 14–4, A is simply the amplitude (which occurs when the cosine in Eq. 14–6 has its maximum value of 1); and ϕ, called the **phase angle**, tells how long after (or before) $t = 0$ the peak ($x = A$) is reached. For $\phi = 0$, we have $x = A \cos \omega t$, as in Fig. 14–3. For $\phi = -\pi/2$ we have

$$x = A \cos \left(\omega t - \frac{\pi}{2}\right) = A \sin \omega t$$

or a pure sine wave.

Notice that the value of ϕ does not affect the shape of the $x(t)$ curve, but only the placement at some arbitrary time, $t = 0$. Simple harmonic motion is thus always *sinusoidal*. Indeed, simple harmonic motion is often *defined* as motion that is purely sinusoidal.

[†] From Eq. 14–5, $v = -a \sin \omega t + b \cos \omega t = b$ at $t = 0$, so b must be zero if $v(t = 0) = 0$.

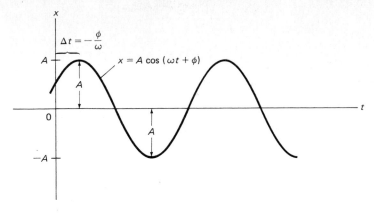

FIGURE 14–4

A plot of $x = A \cos(\omega t + \phi)$ when $\phi < 0$.

Since our oscillating mass repeats its motion after a time equal to its period T, it must be at the same position and moving in the same direction at $t = T$ as it was at $t = 0$. And since a sine or cosine function repeats itself after every 2π radians, then we must have, from Eq. 14–6, that

$$\omega T = 2\pi.$$

Hence

$$\omega = \frac{2\pi}{T} = 2\pi f$$

where f is the frequency of the motion. (ω is usually called the *angular frequency* to distinguish it from the frequency f.) Thus we can also write Eq. 14–6 as

$$x = A \cos\left(\frac{2\pi t}{T} + \phi\right) \qquad (14\text{–}7\text{a})$$

or

$$x = A \cos(2\pi f t + \phi) \qquad (14\text{–}7\text{b})$$

where, because of Eq. 14–4,

$$f = \frac{1}{2\pi}\sqrt{\frac{k}{m}}, \qquad (14\text{–}8\text{a})$$

$$T = 2\pi\sqrt{\frac{m}{k}}. \qquad (14\text{–}8\text{b})$$

Note that *the frequency and period do not depend on the amplitude*. Changing the amplitude of an oscillating spring does not affect its frequency! Equation 14-8a tells us that the greater the mass, the lower the frequency; and the stiffer the spring, the higher the frequency. This makes sense since a greater mass means more inertia and therefore slower response (or acceleration); and larger k means greater force and therefore quicker response.

The velocity and acceleration of the oscillating mass can be obtained by differentiation of Eq. 14–6:

$$v = \frac{dx}{dt} = -\omega A \sin(\omega t + \phi) \qquad (14\text{–}9)$$

$$a = \frac{d^2 x}{dt^2} = \frac{dv}{dt} = -\omega^2 A \cos(\omega t + \phi). \qquad (14\text{–}10)$$

The velocity and acceleration of a SHO also vary sinusoidally. In Fig. 14–5 we plot the displacement, velocity, and acceleration of a SHO as a function of time for

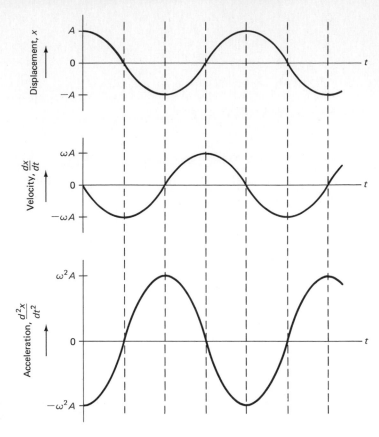

FIGURE 14–5

Displacement, x, velocity, dx/dt, and acceleration, d^2x/dt^2, of a simple harmonic oscillator when $\phi = 0$.

the case when $\phi = 0$. As can be seen, the speed reaches its maximum

$$v_{max} = \omega A = \sqrt{\frac{k}{m}}\, A$$

when the oscillating object is passing through its equilibrium point, $x = 0$; and it is zero at points of maximum displacement, $x = \pm A$. This is in accord with our discussion of Fig. 14–2. Similarly, the acceleration has its maximum value

$$a_{max} = \omega^2 A = \frac{k}{m}\, A$$

where $x = \pm A$, and is zero at $x = 0$, as we expect, since $ma = F = -kx$.

For the general case when $\phi \neq 0$, we can relate the constants A and ϕ to the initial values of x, v, and a by setting $t = 0$ in Eqs. 14–6, 14–9, and 14–10:

$$x_0 = x(0) = A \cos \phi$$

$$v_0 = v(0) = -\omega A \sin \phi = -v_{max} \sin \phi$$

$$a_0 = a(0) = -\omega^2 A \cos \phi = -a_{max} \cos \phi.$$

EXAMPLE 14–1 A spring stretches 0.150 m when a 0.300-kg mass is hung from it. The spring is then stretched an additional 0.100 m from this equilibrium point and released. Determine (a) the spring constant k, (b) the amplitude of the oscillation, (c) the maximum velocity, v_{max}, (d) the maximum acceleration of the mass, (e) the period T and frequency f, (f) the displacement x as a function of time, (g) the velocity at $t = 0.150$ s.

SOLUTION (*a*) Since the spring stretches 0.150 m when 0.300 kg is hung from it, we find *k* from Eq. 14–1 to be

$$k = \frac{F}{x} = \frac{mg}{x} = \frac{(0.300 \text{ kg})(9.80 \text{ m/s}^2)}{(0.150 \text{ m})} = 19.6 \text{ N/m}.$$

(*b*) Since the spring is stretched 0.100 m from equilibrium and is given no initial speed, $A = 0.100$ m.

(*c*) From Eq. 14–9, the maximum velocity is

$$v_{max} = \omega A = \sqrt{\frac{k}{m}} A = \sqrt{\frac{19.6 \text{ N/m}}{0.300 \text{ kg}}} (0.100 \text{ m}) = 0.808 \text{ m/s}.$$

(*d*) Since $F = ma$, the maximum acceleration occurs where the force is greatest—that is, when $x = A = 0.100$ m. Thus

$$a_{max} = \frac{kA}{m} = \frac{(19.6 \text{ N/m})(0.100 \text{ m})}{(0.300 \text{ kg})} = 6.53 \text{ m/s}^2.$$

(*e*) Equations 14–8a and b give

$$T = 2\pi \sqrt{\frac{m}{k}} = 6.28 \sqrt{\frac{0.300 \text{ kg}}{19.6 \text{ N/m}}} = 0.777 \text{ s}$$

$$f = \frac{1}{T} = 1.29 \text{ Hz}.$$

(*f*) The motion begins at a point of maximum displacement downward. If we take *x* positive upward, then $x = -A$ at $t = 0$. So we need a sinusoidal curve that has its maximum negative value at $t = 0$; this is just a negative cosine:

$$x = -A \cos 2\pi ft$$

which gives us $x = -A \cos 0 = -A$ at $t = 0$. Putting in numbers

$$x = -0.100 \cos 8.10t$$

where *t* is in seconds and *x* is in meters. Note that the phase angle (Eq. 14–6) is $\phi = \pi$ or 180°.

(*g*) The velocity at any time *t* is

$$v = \frac{dx}{dt} = -A(-2\pi f) \sin 2\pi ft = 0.810 \sin 8.10t.$$

At $t = 0.150$ s, $v = (0.810) \sin (1.22) = 0.761$ m/s.

We have found, in this section, the general analytic solution to the differential equation, Eq. 14–3b, that describes simple harmonic motion:

$$\frac{d^2x}{dt^2} + \frac{k}{m} x = 0.$$

Other differential equations are not so easily solved. But solutions can generally be found for given initial conditions using the techniques of numerical integration (discussed in Section 2–10). Even for a simple equation like the one above, solving it numerically can provide added insight (see the problems).

14–3 Energy in the Simple Harmonic Oscillator

For a simple harmonic oscillator, such as a mass *m* oscillating on the end of a massless spring, the restoring force is given by

$$F = -kx.$$

The potential energy function, as we have already seen in Chapters 6 and 7, is given by

$$U = -\int F\,dx = \tfrac{1}{2}kx^2$$

where we set the constant of integration equal to zero so $U = 0$ at $x = 0$.

The total mechanical energy is the sum of the kinetic and potential energies:

$$E = \tfrac{1}{2}mv^2 + \tfrac{1}{2}kx^2,$$

where v is the velocity of the mass m when it is a distance x from the equilibrium position. For SHM there is no friction, so the total mechanical energy E remains constant. As the mass oscillates back and forth, the energy continuously changes from potential energy to kinetic energy, and back again. At the extreme points, $x = A$ and $x = -A$, the velocity $v = 0$, so all the energy is potential energy and we have:

$$E = \tfrac{1}{2}m(0)^2 + \tfrac{1}{2}kA^2 = \tfrac{1}{2}kA^2.$$

Thus, the *total mechanical energy of a SHO is proportional to the square of the amplitude*. At the equilibrium point, $x = 0$, all the energy is kinetic:

$$E = \tfrac{1}{2}mv_0^2 + \tfrac{1}{2}k(0)^2 = \tfrac{1}{2}mv_{max}^2,$$

where v_{max} represents the maximum velocity during the motion. At intermediate points the energy is part kinetic and part potential and because energy is conserved we have:

$$E = \tfrac{1}{2}mv^2 + \tfrac{1}{2}kx^2 = \tfrac{1}{2}kA^2 = \tfrac{1}{2}mv_{max}^2. \qquad (14\text{--}11)$$

We can now obtain a useful equation for the velocity v as a function of x by solving for v^2 in Eq. 14–11:

$$v = \pm\sqrt{\frac{k}{m}(A^2 - x^2)} \qquad (14\text{--}12a)$$

or, since $v_{max} = A\sqrt{k/m}$,

$$v = \pm v_{max}\sqrt{1 - \frac{x^2}{A^2}}. \qquad (14\text{--}12b)$$

Again we see that v is a maximum at $x = 0$, and is zero at $x = \pm A$.

The potential energy, $U = \tfrac{1}{2}kx^2$, is plotted in Fig. 14–6. The horizontal line represents a particular value of the total energy $E = \tfrac{1}{2}kA^2$. As we discussed in Section 7–6 (see Fig. 7–8) the distance between this line and the U curve represents the kinetic energy, and the motion is restricted to x values between $-A$ and $+A$. These results are, of course, consistent with our full solution of the previous section.

EXAMPLE 14–2   For the SHO of Example 14–1, determine (*a*) the total energy, (*b*) the kinetic and potential energies as a function of time, (*c*) the velocity when the mass is 0.050 m from equilibrium, (*d*) the kinetic and potential energies at half amplitude ($x = \pm A/2$).

SOLUTION   (*a*) Since $k = 19.6$ N/m and $A = 0.100$ m, the total energy E from Eq. 14–11 is

$$E = \tfrac{1}{2}kA^2 = \tfrac{1}{2}(19.6 \text{ N/m})(0.100 \text{ m})^2 = 9.80 \times 10^{-2} \text{ J}.$$

(*b*) We have, from parts (*f*) and (*g*) of Example 14–1, $x = -0.100 \cos 8.10t$ and $v = 0.810 \sin 8.10t$, so

$$\text{PE} = \tfrac{1}{2}kx^2 = (9.80 \times 10^{-2} \text{ J}) \cos^2 8.10t$$

$$\text{KE} = \tfrac{1}{2}mv^2 = (9.80 \times 10^{-2} \text{ J}) \sin^2 8.10t.$$

FIGURE 14–6

Graph of PE $= U = \tfrac{1}{2}kx^2$. KE + PE $= E =$ constant for any point x where $-A \le x \le A$.

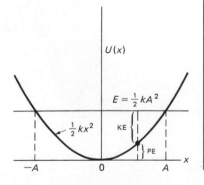

(c) We use Eq. 14–12 and find

$$v = v_{max} \sqrt{1 - x^2/A^2} = 0.70 \text{ m/s}.$$

(d) At $x = A/2 = 0.050$ m, we have

$$\text{PE} = \tfrac{1}{2}kx^2 = 2.5 \times 10^{-2} \text{ J}$$

$$\text{KE} = E - \text{PE} = 7.3 \times 10^{-2} \text{ J}.$$

14–4 Simple Harmonic Motion Related to Uniform Circular Motion

Simple harmonic motion has an interesting simple relationship to a particle rotating in a circle with uniform speed. Consider a mass m rotating in a circle of radius A with speed v_M on top of a table as shown in Fig. 14–7. As viewed from above, the motion is a circle. But a person who looks at the motion from the edge of the table, sees an oscillatory motion back and forth, and this corresponds precisely to SHM as we shall now see. What the person sees, and what we are interested in, is the projection of the circulatory motion onto the x axis, Fig. 14–7b. To see that this motion is analogous to SHM, let us calculate the x component of the velocity v_M which is labeled v in Fig. 14–7. The two triangles shown are similar, so

$$\frac{v}{v_M} = \sqrt{\frac{A^2 - x^2}{A}}$$

or

$$v = v_M \sqrt{1 - \frac{x^2}{A^2}}.$$

This is exactly the equation for the speed of a mass oscillating with SHM, Eq. 14–12b where $v_M = v_{max}$. Furthermore, we can see from the diagram that if the angular displacement at $t = 0$ is ϕ, then after a time t the particle will have rotated through an angle $\theta = \omega t$ where $\omega = v_M/A$. Thus

$$x = A \cos(\theta + \phi) = A \cos(\omega t + \phi);$$

and, since $v_M = 2\pi A/T$, then $\omega = v_M/A = 2\pi/T = 2\pi f$ where T is the time required for one rotation and f is the frequency. This corresponds precisely to the back-and-forth motion of a simple harmonic oscillator. Thus, the projection on the x axis of a particle rotating in a circle has the same motion as a mass undergoing SHM. Indeed, we can say that the projection of circular motion onto a straight line is SHM.

The projection of uniform circular motion onto the y axis is also simple harmonic. Thus uniform circular motion can be thought of as two simple harmonic motions operating at right angles. (More on this in Section 14–9.)

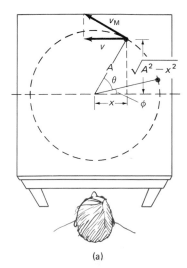

(a)

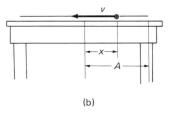

(b)

FIGURE 14–7

Analysis of simple harmonic motion as a side view (b) of circular motion (a).

FIGURE 14–8

Simple pendulum.

14–5 The Simple Pendulum

A simple pendulum consists of a small object (the pendulum "bob") suspended from the end of a light cord, Fig. 14–8. We assume that the cord doesn't stretch and that its mass can be ignored relative to that of the bob. A simple pendulum moving back and forth resembles simple harmonic motion: it oscillates along the arc of a circle with equal amplitude on either side of its equilibrium point (where it hangs vertically) and as it passes through the equilibrium point it has its maximum speed. But is it really undergoing SHM? That is, is the restoring force proportional to its displacement? Let us find out.

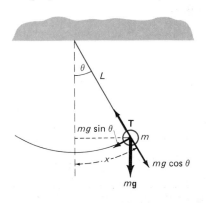

The displacement of the pendulum along the arc, x, is given by $x = L\theta$ where θ is the angle the string makes with the vertical and L is the distance from the string's point of attachment to the cm of the bob, Fig. 14–8. Thus, if the restoring force is proportional to x or to θ, then the motion will be simple harmonic. The restoring force is the component of the weight tangent to the arc:

$$F = -mg \sin \theta.$$

Since F is proportional to the sine of θ and not to θ itself, the motion is *not* SHM. However, if θ is small, then $\sin \theta$ is very nearly equal to θ if the latter is specified in radians. This can be seen from the series expansion[†] of $\sin \theta$, by looking at the trigonometric tables (inside back cover), or by noting that in Fig. 14–8 the arc length $x (= L\theta)$ is nearly the same length as the chord $(= L \sin \theta)$ indicated by the dashed line. For angles less than $15°$, the difference between θ and $\sin \theta$ is less than 1 percent. Thus, to a very good approximation for small angles,

$$F \approx -mg\theta.$$

Using the fact that $x = L\theta$ we have

$$F \approx -\frac{mg}{L} x.$$

Thus, for small displacements, the motion is essentially simple harmonic, with an effective force constant of $k = mg/L$ (see Eq. 14–1). The period of a simple pendulum can be found using Eq. 14–8b where for k we substitute mg/L:

$$T = 2\pi \sqrt{\frac{m}{mg/L}}$$

or

$$T = 2\pi \sqrt{\frac{L}{g}}. \qquad \text{[simple pendulum, small displacement]} \qquad (14\text{–}13)$$

A surprising result is that the period does not depend on the mass of the pendulum bob! You may have noticed this if you pushed a small child and a large one on the same swing.

We saw in Section 14–2 that the period of any SHM does not depend on the amplitude. Galileo is said to have first noted this fact with regard to the pendulum while watching a swinging lamp in the cathedral at Pisa; this discovery led to the pendulum clock, the first really precise timepiece which became the standard for centuries.

Because pendulum motion is not precisely SHM, the period does in fact depend slightly on the amplitude. The general formula is given by an infinite series:

$$T = 2\pi \sqrt{\frac{L}{g}} \left(1 + \frac{1}{2^2} \sin^2 \frac{\theta_M}{2} + \frac{1}{2^2} \frac{3^2}{4^2} \sin^4 \frac{\theta_M}{2} + \dots \right) \qquad (14\text{–}14)$$

where θ_M is the maximum angular displacement. Each succeeding term is smaller than preceding ones, so we need to keep only those terms that contribute to the desired degree of accuracy. For $\theta_M = 15°$, Eq. 14–13 is in error by less than 0.5 percent; at larger angles, the discrepancy increases rapidly.

The accuracy of a pendulum clock would be affected, after many swings, by the decrease in amplitude due to friction; but the mainspring in a pendulum clock (or the falling weight in a grandfather clock) supplies energy to compensate for the friction and to maintain the amplitude constant, so that the timing remains accurate.

The pendulum finds use in geology, for geologists are interested in surface irregularities of the earth and frequently need to measure the acceleration of gravity at a given location very accurately. They often use a carefully designed pendulum to do this as illustrated in the next example.

[†] $\sin \theta = \theta - \dfrac{\theta^3}{3!} + \dfrac{\theta^5}{5!} - \dfrac{\theta^7}{7!} + \dots.$

EXAMPLE 14–3 A geologist's simple pendulum, whose length is 37.10 cm and amplitude is 6.0°, has a frequency of 0.8152 Hz at a particular location on the earth. What is the acceleration of gravity at this location?

SOLUTION The series expansion (in parentheses) of Eq. 14–14 gives

$$1 + \frac{1}{4}(.0523)^2 + \frac{9}{64}(.0523)^4 + \ldots = 1 + 7 \times 10^{-4} + 1 \times 10^{-6} + \ldots \approx 1.0007$$

to the four-place accuracy we need. The frequency is

$$f = \frac{1}{T} = \frac{1}{2\pi} \sqrt{\frac{g}{L}} \left(\frac{1}{1.0007} \right)$$

and solving for g we obtain

$$g = (2\pi f)^2 L (1.0007)^2$$
$$= (6.283 \times 0.8152 \text{ s}^{-1})^2 (0.3710 \text{ m})(1.0014)$$
$$= 9.795 \text{ m/s}^2.$$

*14–6 The Physical Pendulum

The term physical pendulum refers to any real extended body which oscillates back and forth, in contrast to the rather idealized simple pendulum where all the mass is assumed concentrated in the tiny pendulum bob. An example of a physical pendulum is shown in Fig. 14–9 suspended from the point O. The force of gravity acts at the center of gravity (cg) of the body located a distance h from the pivot point. The physical pendulum is best analyzed using the equations of rotational motion. The torque on a physical pendulum, calculated about point O, is

$$\tau = -mgh \sin \theta.$$

Newton's second law for rotational motion, Eq. 9–12, states that

$$\tau = I\alpha = I \frac{d^2\theta}{dt^2}$$

where I is the moment of inertia of the body and $\alpha = d^2\theta/dt^2$ is the angular acceleration. Thus we have

$$I \frac{d^2\theta}{dt^2} = -mgh \sin \theta$$

or

$$\frac{d^2\theta}{dt^2} + \frac{mgh}{I} \sin \theta = 0$$

where I is calculated about an axis through point O. For small angular amplitude, $\sin \theta \approx \theta$, so we have

$$\frac{d^2\theta}{dt^2} + \left(\frac{mgh}{I} \right) \theta = 0.$$

This is just the equation for SHM, Eq. 14–3b, except that θ replaces x and mgh/I replaces k/m. Thus, for small angular displacements, a physical pendulum undergoes SHM, given by

$$\theta = \theta_M \cos \left(\frac{2\pi t}{T} + \phi \right)$$

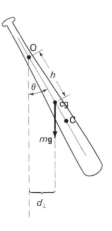

FIGURE 14–9

A physical pendulum suspended from point O.

where θ_M is the maximum angular displacement. The period, T, is (see Eq. 14–8b):

$$T = 2\pi \sqrt{\frac{I}{mgh}}. \qquad \text{[small angular displacement]} \qquad (14\text{–}15)$$

For larger angular displacements, Eq. 14–15 must be modified by the same factor that is used for the simple pendulum (the term in parentheses in Eq. 14–14).

Notice that for rotation, the motion will be SHM if the restoring torque is proportional to the negative of the angular displacement,

$$\tau = -k\theta,$$

where k is a constant that depends on the properties of the system.

EXAMPLE 14–4 *Moment of inertia measurement.* An easy way to measure the moment of inertia of an object about any axis is to measure the period of oscillation about that axis. For example, suppose a nonuniform 1.6-kg stick can be balanced at a point 42 cm from one end. If it is pivoted about that end, it oscillates at a frequency of 2.5 Hz. What is its moment of inertia about this end?

SOLUTION Given $T = 1/f = 0.40$ s, and $h = 0.42$ m, we solve Eq. 14–15 for I:

$$I = mghT^2/4\pi^2 = 0.27 \text{ kg·m}^2.$$

Since $I = \frac{1}{3}Ml^2$ for a uniform stick of length l pivoted about one end (Fig. 9–13), do you think our stick is longer or shorter than 84 cm?

EXAMPLE 14–5 A thin, straight, uniform rod of length $l = 1.00$ m and mass $m = 160$ g is pivoted at one end. (*a*) What is its period for small amplitude oscillations? (*b*) What is the length of a simple pendulum that will have the same period?

SOLUTION (*a*) The moment of inertia of a thin rod about an axis through one end (see Fig. 9–13) is $I = \frac{1}{3}ml^2$. Since the center of mass is at the center, $h = l/2$, then the period is

$$T = 2\pi \sqrt{\frac{I}{mgh}} = 2\pi \sqrt{\frac{2l}{3g}}$$

$$= 6.28 \sqrt{\frac{2(1.00 \text{ m})}{3(9.8 \text{ m/s}^2)}} = 1.64 \text{ s}.$$

(*b*) To have the same period, a simple pendulum must have a length

$$L = \frac{gT^2}{4\pi^2} = \frac{gI}{mgh} = \frac{I}{mh}$$

which, for a uniform rod pivoted at one end, is $L = \frac{2}{3}l$, or 0.67 m in our case.

That point in a physical pendulum which is a distance

$$L = I/mh$$

from the pivot point O along a line through the cg is called the *center of oscillation* and is labeled C in Fig. 14–9. (For a uniform rod of length l, pivoted about one end, C is $\frac{2}{3}l$ from that end.) As we saw in the last example, a simple pendulum of length $L = I/mh$ has the same period as a physical pendulum. Thus, a physical pendulum has the same period as if all its mass were concentrated at its center of oscillation (hence the name). The center of oscillation has two other interesting properties: (1) If C represents the center of oscillation when O is the pivot point, then when C is the pivot point, O is the center of oscillation and the period is the same. (2) If a pivoted object is struck a horizontal blow in the plane of oscillation at its center of oscillation (Fig. 14–10), no "reaction" force is felt at the pivot point. An interesting example of this second property occurs in baseball. Batters can feel a "sting" in their hands when they hit the ball unless they hit it at the center of oscillation. The center of oscillation is thus often called the *center of percussion*. Both of these properties are derived in the problems. Note that the position of the center of oscillation is different for a different pivot point.

It is interesting to note that the pace of ordinary walking is determined by the functioning of the leg as a physical pendulum. Each step is basically a half period of a swinging pendulum. In an effortless walk, the leg is allowed to swing at its natural frequency. To walk at a faster or slower rate, or to run, requires extra muscle exertion. However, one can cover more ground, often with less effort, by using the natural gait (frequency) with a lengthened stride since the period is nearly independent of amplitude.

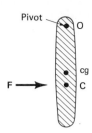

FIGURE 14–10
A horizontal impulsive force, **F**, strikes a body (pivoted at O) at its center of oscillation.

 14–7 Damped Harmonic Motion

The amplitude of any real oscillating spring or swinging pendulum slowly decreases in time until the oscillations stop altogether. Figure 14–11 shows a typical graph of the displacement as a function of time. This is called **damped harmonic motion**. The damping† is generally due to the resistance of air and to internal friction within the oscillating system. The energy that is thus dissipated to thermal energy is reflected in a decreased amplitude of oscillation.

Since natural oscillating systems are damped in general, why do we even talk about (undamped) simple harmonic motion? The answer is that SHM is much easier to deal with mathematically. And if the damping is not large, the oscillations can be thought of as simple harmonic motion on which is superposed the damping—that is, the decrease in amplitude represented by the dashed curves in Fig. 14–11. Although frictional damping does alter the frequency of vibration, the effect is usually small if the damping is small. Let us look at this in more detail. The damping force depends on the speed of the oscillating object; it opposes the motion and in many cases can be considered to be directly proportional to the speed:

$$F_{\text{damping}} = -bv,$$

where b is a constant. For a mass oscillating on the end of a spring, the restoring force of the spring is $F_x = -kx$; so Newton's second law ($F = ma$) becomes

$$ma = -kx - bv.$$

We bring all terms to the left side of the equation and substitute $v = dx/dt$ and $a = d^2x/dt^2$ and obtain

$$m\frac{d^2x}{dt^2} + b\frac{dx}{dt} + kx = 0 \qquad (14\text{--}16)$$

---

† To "damp" means to diminish, restrain, or extinguish, as to "dampen one's spirits."

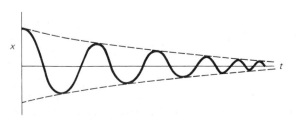

FIGURE 14–11
Damped harmonic motion.

which is the equation of motion. To solve this equation, we guess at a solution. If the damping constant b is small, x as a function of t is as plotted in Fig. 14–11 which looks like a cosine function times a factor (represented by the dashed lines) that decreases in time. A simple function that does this is the exponential, $e^{-\alpha t}$, so let us try as our solution

$$x = Ae^{-\alpha t} \cos \omega't \qquad (14\text{–}17)$$

where A, α, and ω' are assumed to be constants, and $A = x$ at $t = 0$. (We choose the phase constant $\phi = 0$ for convenience.) We take the first and second derivatives of Eq. 14–17:

$$\frac{dx}{dt} = -\alpha Ae^{-\alpha t} \cos \omega't - \omega'Ae^{-\alpha t} \sin \omega't$$

$$\frac{d^2x}{dt^2} = \alpha^2 Ae^{-\alpha t} \cos \omega't + \alpha A\omega'e^{-\alpha t} \sin \omega't + \omega'\alpha Ae^{-\alpha t} \sin \omega't - \omega'^2 Ae^{-\alpha t} \cos \omega't.$$

We next substitute these relations back into Eq. 14–16 and reorganize to obtain:

$$Ae^{-\alpha t}[(m\alpha^2 - m\omega'^2 - b\alpha + k) \cos \omega't + (2\omega'\alpha m - b\omega') \sin \omega't] = 0. \qquad (14\text{–}18)$$

The left side of this equation must equal zero for all times t, but this can only be so for certain values of α and ω'. To determine α and ω', we choose two values of t that will make their evaluation easy. At $t = 0$, $\sin \omega't = 0$, so the above relation reduces to $A(m\alpha^2 - m\omega'^2 - b\alpha + k) = 0$ which means[†] that

$$m\alpha^2 - m\omega'^2 - b\alpha + k = 0. \qquad (14\text{–}19a)$$

And at $t = \pi/2\omega'$, $\cos \omega't = 0$ so Eq. 14–18 can be valid only if

$$2\alpha m - b = 0. \qquad (14\text{–}19b)$$

From Eq. 14–19b we have

$$\alpha = \frac{b}{2m}$$

and from 14–19a

$$\omega' = \sqrt{\alpha^2 - \frac{b\alpha}{m} + \frac{k}{m}} = \sqrt{\frac{k}{m} - \frac{b^2}{4m}}.$$

Thus we see that Eq. 14–17 is a solution to the equation of motion for the damped harmonic oscillator as long as α and ω' have these specific values.

In summary, the displacement x of a damped harmonic oscillator, when the damping constant, b, is not too large, is given by

$$x = Ae^{-(b/2m)t} \cos \omega't \qquad (14\text{–}20)$$

where

$$\omega' = \sqrt{\frac{k}{m} - \frac{b^2}{4m^2}}. \qquad (14\text{–}21)$$

Of course a phase constant, ϕ, can be added to the argument of the cosine in Eq. 14–20. As it stands (with $\phi = 0$), it is clear that the constant A in Eq. 14–20 is simply the initial displacement, $x = A$ at $t = 0$. The frequency f is

$$f = \frac{\omega'}{2\pi} = \frac{1}{2\pi} \sqrt{\frac{k}{m} - \frac{b^2}{4m^2}}. \qquad (14\text{–}22)$$

[†] It would also be satisfied by $A = 0$, but this gives the trivial and uninteresting solution $x = 0$ for all t—that is, no oscillation.

So the frequency is less, and the period longer, than for undamped SHM. (In most practical cases of light damping, however, ω' differs only slightly from $\omega = \sqrt{k/m}$.) This makes sense since we expect friction to slow down the motion. Equation 14–22 reduces to Eq. 14–8a, as it should, when there is no friction ($b = 0$). The constant $\alpha = b/2m$ is a measure of how quickly the oscillations decrease toward zero (Fig. 14–11). The time $t_l = 2m/b$ is the time taken for the oscillations to drop to $1/e$ of the original amplitude; t_l is called the "mean lifetime" of the oscillations. Note that the larger b is, the more quickly the oscillations die away.

The solution, Eq. 14–20, is not valid if b is so large that

$$b^2 > 4mk$$

since then ω' (Eq. 14–21) would become imaginary. What happens in this case is that the system does not oscillate at all but returns directly to its equilibrium position, as we now discuss. Three common cases of heavily damped systems are shown in Fig. 14–12. Curve C represents the situation when the damping is so large ($b^2 \gg 4mk$) it takes a long time to reach equilibrium; the system is *overdamped*. Curve A represents an *underdamped* situation in which the system makes several swings before coming to rest ($b^2 < 4mk$). Curve B represents *critical damping*: $b^2 = 4mk$; in this case equilibrium is reached in the shortest time. The terms used above derive from the use of practical damped systems such as door closing mechanisms and shock absorbers in a car. These are usually designed to give critical damping; but as they wear out, underdamping occurs: a door slams and a car bounces up and down several times whenever it hits a bump. Needles on electronic instruments (voltmeters, ammeters, level indicators on tape recorders) are usually critically damped or slightly underdamped; if they were very under-damped, they would swing back and forth excessively before arriving at the correct value; and if overdamped they would take too long to reach equilibrium and rapid changes in the signal (say recording level) would not be detected.

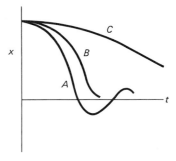

FIGURE 14–12
Underdamped (A), critically damped (B), and overdamped (C) motion.

 ## 14–8 Forced Vibrations; Resonance

When a vibrating system is set into motion it vibrates at its natural frequency. Earlier in this chapter we developed formulas that relate the natural frequency (or period) to the properties of the system for elastic objects (like springs) and pendulums.

However, a system is often not left to merely oscillate on its own, but may have an external force applied to it which itself oscillates at a particular frequency. For example, we might pull the mass on the spring of Fig. 14–1 back and forth at an angular frequency $\omega = 2\pi f$. The mass thus vibrates at the frequency ω of the external force, even if this frequency is different from the **natural frequency** of the undamped spring, which we now denote by ω_0 where

$$\omega_0 = \sqrt{\frac{k}{m}}.$$

This is an example of **forced vibration**. Another example is a child's swing, which is basically a pendulum with its own natural frequency; when we push a child on the swing, we have a "driven" or "forced" harmonic oscillator. Many types of forced harmonic motion occur, and we will discuss some of them in this and later chapters.

In a forced vibration, the amplitude of vibration and hence the energy trans-ferred to the vibrating system is found to depend on the difference between ω and ω_0 as well as on the amount of damping. Suppose the external force is sinusoidal and can be represented by

$$F_{\text{ext}} = F_0 \cos \omega t$$

where $\omega = 2\pi f$ is the angular frequency applied externally to the oscillator. Then the equation of motion (with damping) is

$$ma = -kx - bv + F_0 \cos \omega t.$$

This can be written as

$$m\frac{d^2x}{dt^2} + b\frac{dx}{dt} + kx = F_0 \cos \omega t. \qquad (14\text{--}23)$$

The external force, on the right of the equation, is the only term that does not involve x or one of its derivatives. It is left as an exercise (problem 54) to show that the solution to Eq. 14–23 is

$$x = A_0 \sin (\omega t + \phi_0), \qquad (14\text{--}24)$$

where

$$A_0 = \frac{F_0}{m\sqrt{(\omega^2 - \omega_0^2)^2 + b^2\omega^2/m^2}} \qquad (14\text{--}25a)$$

and

$$\phi_0 = \tan^{-1}\frac{\omega_0^2 - \omega^2}{\omega(b/m)}. \qquad (14\text{--}25b)$$

Actually, the general solution to Eq. 14–23 is Eq. 14–24 plus another term of the form of Eq. 14–20 for the natural damped motion of the oscillator; this second term approaches zero in time, so in most cases we need to be concerned only with Eq. 14–24.

The amplitude of forced harmonic motion, A_0, depends strongly on the difference between the applied and the natural frequency. A plot of A_0 (Eq. 14–25a) as a function of the applied frequency, ω, is shown in Fig. 14–13 for three specific values of the damping constant b. Curve A ($b = \frac{1}{6}m\omega_0$) represents light damping, curve B ($b = \frac{1}{2}m\omega_0$) fairly heavy damping, and curve C ($b = \sqrt{2}\,m\omega_0$) overdamped

FIGURE 14–13

Amplitude of a forced harmonic oscillator as a function of ω. Curves A, B, and C correspond to light, heavy, and overdamped systems respectively ($Q = m\omega_0/b = 6$, 2, 0.71).

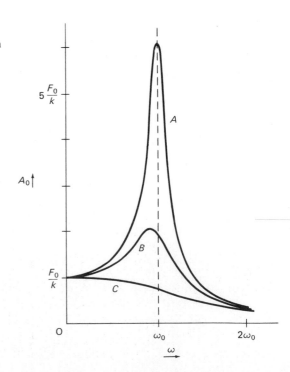

motion. The amplitude can become large when the driving frequency ω is near the natural frequency, $\omega \approx \omega_0$, as long as the damping is not too large. When the damping is small, the increase in amplitude near $\omega = \omega_0$ is very large (and often dramatic). This is known as **resonance**. The natural vibrating frequency ω_0 of a system is called its **resonant frequency**.[†] If $b = 0$, resonance occurs at $\omega = \omega_0$ and the resonant peak (of A_0) becomes infinite; in such a case, energy is being continuously transferred into the system and none is dissipated. For real systems, b is never precisely zero, and the resonant peak is finite; the peak does not occur precisely at $\omega = \omega_0$ (because of the term $b^2\omega^2/m^2$ in the denominator of Eq. 14-25a), although it is quite close to ω_0 unless the damping is very large. If the damping is large, there is little or no peak (curve C in Fig. 14-13).

The height and narrowness of a resonant peak is often specified by its *quality factor* or Q *value*, defined as

$$Q = \frac{m\omega_0}{b}. \qquad (14-26)$$

In Fig. 14-13, curve A has $Q = 6$, curve B has $Q = 2$, and curve C has $Q = 1/\sqrt{2}$. The smaller the damping constant b, the larger the Q value becomes, and the higher the resonance peak. The Q value is also a measure of the width of the peak: let ω_1 and ω_2 be the frequencies where the square of the amplitude A_0 has half its maximum value (we use the square because the power transferred to the system is proportional to A_0^2—see the problems); then $\Delta\omega = \omega_1 - \omega_2$, which is called the *width* of the resonance peak, is related to Q by

$$\frac{\Delta\omega}{\omega_0} = \frac{1}{Q}. \qquad (14-27)$$

(The proof of this relation, which is accurate only for weak damping, is the subject of problem 58.) The larger the Q value, the narrower will be the resonance peak relative to its height. Thus a large Q value, representing a system of high quality, has a high, narrow resonance peak.

A simple illustration of resonance is pushing a child on a swing. A swing, like any pendulum, has a natural frequency of oscillation. If you were to close your eyes

[†] Sometimes the resonant frequency is defined as the actual value of ω at which the amplitude has its maximum value, and this depends somewhat on the damping constant. Except for very heavy damping, this value is quite close to ω_0.

FIGURE 14-14

Large amplitude oscillation of the Tacoma Narrows Bridge due to heavy gusty winds just prior to collapse (November 7, 1940). (Wide World Photo.)

FIGURE 14–15

Combinations of two
simple harmonic motions
at right angles.

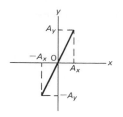

(a) $\omega_x = \omega_y$, $\phi_x = \phi_y$, $A_y = 2A_x$

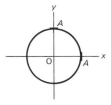

(b) $\omega_x = \omega_y$, $\phi_y - \phi_x = -\frac{\pi}{2}$,
$A_x = A_y = A$

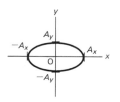

(c) $\omega_x = \omega_y$, $\phi_y - \phi_x = -\frac{\pi}{2}$,
$A_x = 2A_y$

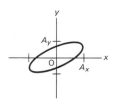

(d) $\omega_x = \omega_y$, $\phi_y - \phi_x = \frac{\pi}{4}$,
$A_x = 2A_y$

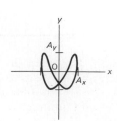

(e) $\omega_y = 2\omega_x$, $\phi_y - \phi_x = \frac{\pi}{4}$,
$A_x = A_y$

and push on the swing at some random frequency, the swing would bounce around and reach no great amplitude. But if you push with a frequency equal to the natural frequency of the swing, the amplitude increases greatly. This illustrates the fact that at resonance, relatively little effort is required to obtain a large amplitude.

The great tenor Enrico Caruso was said to be able to break a goblet by singing a note of just the right frequency at full voice. This is an example of resonance, for the sound waves emitted by the voice act as a forced vibration on the glass. At resonance, the resulting vibration of the goblet may be large enough in amplitude that the glass exceeds its elastic limit and breaks.

Since material objects are in general elastic, resonance is an important phenomenon in a variety of situations. It is particularly important in building, although the effects are not always to be foreseen. For example, it has been reported that a railway bridge collapsed because a nick in one of the wheels of a passing train set up a resonant vibration in the bridge. Indeed, marching soldiers break step when crossing a bridge to avoid the possibility of a similar catastrophe. And the famous collapse of the Tacoma Narrows Bridge (Fig. 14–14) in 1940 was due in part to resonance of the bridge.

Important examples of resonance occur in other areas of physics such as electricity and magnetism, and in atoms and molecules. We will discuss many of these in later chapters. We will also see that vibrating objects often have not one, but many resonant frequencies.

*14–9 Combinations of Two Harmonic Motions

Consider a particle which is undergoing simple harmonic motion along two perpendicular directions, say the x and y axes. For such a combination of two linear simple harmonic motions we can write

$$x = A_x \cos(\omega_x t + \phi_x)$$
$$y = A_y \cos(\omega_y t + \phi_y).$$

The resulting motion in the xy plane depends on the relative values of the frequencies, amplitudes, and phases. So let us deal with some particular cases.

1. *Equal frequency, $\omega_x = \omega_y = \omega$.*
   a. *Equal phases, $\phi_x = \phi_y = \phi$.* The motion will be a straight line in the xy plane, of slope A_y/A_x. This is easily seen since

   $$x = A_x \cos(\omega t + \phi)$$

   and

   $$y = A_y \cos(\omega t + \phi) = \frac{A_y}{A_x} x$$

   which is the equation of a straight line of slope (A_y/A_x). Figure 14–15a shows the case when $A_y = 2A_x$.
   b. $\phi_y - \phi_x = \pm\pi/2$, $A_x = A_y = A$. Here the phases differ by $\pi/2$. The motion will be a circle (either clockwise or counterclockwise) which we can prove as follows. Let $\phi_x = \phi$, then $\phi_y = \phi - \pi/2$, where we chose the minus sign to be concrete. Then

   $$x = A \cos(\omega t + \phi)$$

   $$y = A \cos\left(\omega t + \phi - \frac{\pi}{2}\right).$$

   Since $\cos(\omega t + \phi - \pi/2) = \sin(\omega t + \phi)$ we can write

   $$x^2 + y^2 = A^2 \cos^2(\omega t + \phi) + A^2 \sin^2(\omega t + \phi) = A^2$$

which is the equation of a circle in the xy plane of radius A. This is shown in Fig. 14–15b. This agrees with our conclusion in Section 14–4 that uniform circular motion can be considered as a combination of two simple harmonic motions at right angles.

   c. $\phi_y - \phi_x = \pm\pi/2$, $A_x \neq A_y$. When the phase differs by $\pi/2$ but the amplitudes are *not* equal, the motion is an ellipse, as shown in Fig. 14–15c, with major and minor axes equal to $2A_x$ and $2A_y$. The example shown has $A_x = 2A_y$.

   d. $\phi_y - \phi_x \neq 0$, $\neq \pi/2$, $\neq \pi$. If the phase difference is neither zero, nor $\pi/2$, nor π, the resulting motion is an ellipse whether A_x and A_y are equal or not. An example is shown in Fig. 14–15d, in which $\phi_y - \phi_x = \pi/4$ and $A_x = 2A_y$.

  2. *Nonequal frequency*, $\omega_x \neq \omega_y$. When the frequencies are not equal, the motion can be very complex. Generally the curve does not even close on itself and thus is not periodic. However, if the ratio ω_x/ω_y is equal to the ratio of two integers, the curve does close on itself and the result is periodic motion (though often complex). An example is shown in Fig. 14–15e, which has $\omega_y = 2\omega_x$ and a phase difference $\phi_y - \phi_x = \pi/4$. These types of curves are referred to as *Lissajous figures*. They are nicely displayed on the face of an oscilloscope by putting in sinusoidal signals to the horizontal and vertical inputs. The amplitudes, frequencies, and phases can be altered to produce a wide variety of patterns.

Combinations of simple harmonic motions along the same direction are also interesting, and we will deal with them in later chapters.

Summary

A vibrating object undergoes *simple harmonic motion* (SHM) if the restoring force is proportional to the displacement, $F = -kx$. The maximum displacement during oscillation is called the *amplitude*. The *period, T*, is the time required for one complete cycle (back and forth), and the *frequency, f*, is the number of cycles per second; they are related by $T = 1/f$. The period of the SHM of a mass m at the end of a spring is given by $T = 2\pi\sqrt{m/k}$; and the displacement as a function of time is $x = A\cos(2\pi ft + \phi)$ where A is the amplitude and ϕ is the *phase angle*; the value of A and ϕ depend on the *initial conditions* (x and v at $t = 0$). During SHM, the total energy $E = \frac{1}{2}mv^2 + \frac{1}{2}kx^2$ is continually changing from potential to kinetic and back again.

A *simple pendulum* of length L approximates SHM if its amplitude is small and friction can be ignored; its period is then given by $T = 2\pi\sqrt{L/g}$ where g is the acceleration of gravity.

When friction is present (for all real springs and pendulums), the motion is said to be *damped*. The maximum displacement decreases in time and the energy is eventually all transformed to heat. If the friction is very large, so no oscillations occur, the system is said to be *overdamped*; if the friction is small enough that one or more oscillations occur, the system is *underdamped*, and the displacement is given by $x = Ae^{-\alpha t}\cos\omega' t$ where α and ω' are constants; for a *critically damped* system, no oscillations occur and equilibrium is reached in the shortest time.

If an oscillating force is applied to a system capable of vibrating, the amplitude of vibration can be very large if the frequency of the applied force is near the *natural* (or *resonant*) frequency of the oscillator; this is called *resonance*.

Questions

  1. Give some everyday examples of vibrating objects. Which follow SHM, at least approximately?

  2. Contrast the equations for x, v, and a for uniformly accelerated linear motion with those for simple harmonic motion. Discuss their similarities and differences.

  3. If a particle undergoes SHM with amplitude A, what is the total distance it travels in one period?

  4. Real springs have mass. How will the true period and frequency differ from those given by the equations for a mass oscillating on the end of an idealized massless spring?

  5. For a simple harmonic oscillator, when (if ever) are the displacement and velocity vectors in the same direction? The displacement and acceleration vectors?

  6. How could you double the maximum speed of a SHO?

7. What is the value of the phase constant ϕ in Eq. 14–6 if, at $t = 0$, the oscillating particle is at (a) $x = A$, (b) $x = 0$, (c) $x = -A$, (d) $x = \frac{1}{2}A$?

8. If Eq. 14–1, $F = -kx$, applies to the elasticity of a solid material, what does this say about the forces between molecules of the material?

9. A mass m hangs from a spring with stiffness constant k. The spring is cut in half and the same mass hung from it. By what factor does the frequency of oscillation change?

10. Two equal masses are attached to separate identical springs next to one another. One mass is pulled so its spring stretches 20 cm and the other pulled so its spring stretches only 10 cm. The masses are released simultaneously. Which mass reaches the equilibrium point first?

11. A 10-kg fish is attached to the hook of a vertical spring scale, and is then released. Describe the scale reading as a function of time.

12. Is the motion of a piston in an automobile engine simple harmonic? Explain.

13. If a pendulum clock is accurate at sea level, will it gain or lose time when taken to the mountains?

14. According to Eq. 14–14, the period of a pendulum is longer for larger amplitudes. Give physical reasons for why this should be so.

15. Do you expect that Eq. 14–14 is accurate for a simple pendulum when $\theta > 90°$? Explain.

16. If you double the amplitude of a SHO, how does this change the frequency, maximum velocity, maximum acceleration, total mechanical energy?

17. A child on a swing starts out very high but doesn't "pump." How does the frequency of the swing change in time?

*18. Describe the possible motion of a solid object that is suspended so it is free to rotate about its center of gravity. Is it a physical pendulum?

*19. A thin uniform rod of mass m is suspended from one end and oscillates with a frequency f. If a small sphere of mass $2m$ is attached to the other end, does the frequency increase or decrease? Explain.

20. Is the acceleration of a simple harmonic oscillator ever zero? If so, when? What about a damped harmonic oscillator?

21. A tuning fork of natural frequency 264 Hz sits on a table at the front of a room. At the back of the room, two tuning forks, one of natural frequency 260 Hz and one of 420 Hz are initially silent, but when the tuning fork at the front of the room is set into vibration, the 260-Hz fork spontaneously begins to vibrate but the 420-Hz fork does not. Explain.

22. Give several everyday examples of resonance.

23. Is a rattle in a car ever a resonance phenomenon? Explain.

24. Over the years, buildings have been able to be built out of lighter and lighter materials. How has this affected the natural vibration frequencies of buildings and the problems of resonance due to passing trucks, airplanes, or natural sources of vibration?

*25. For the example discussed in Section 14–9 as case 1b, in what direction is the motion: clockwise or counterclockwise? What change would you make so the motion is along the same path but in the reverse direction?

*26. For two harmonic motions at right angles, suppose $\omega_x = \omega_y$ and $\phi_x = \phi_y \pm \pi$. Describe the combined motion.

 Problems

SECTION 14–2

1. (I) A piece of rubber is 45 cm long when a weight of 18.0 N hangs from it and is 68 cm long when a weight of 22.5 N hangs from it. What is the "spring" constant of this piece of rubber?

2. (I) When an 80-kg person climbs into a 1200-kg car, the car's springs compress vertically by 1.40 cm. What will be the frequency of vibration when the car hits a bump? (Ignore damping.)

3. (I) (a) What is the equation describing the motion of a spring that is stretched 20 cm from equilibrium and then released, and whose period is 1.5 s? (b) What will be its displacement after 1.8 s?

4. (I) A small cockroach of mass 0.30 g is caught in a spider's web. The web vibrates predominately with a frequency of 15 Hz. (a) What is the value of the spring constant k for the web? (b) At what frequency would you expect the web to vibrate if an insect of mass 0.10 g were trapped?

5. (I) A spring vibrates with a frequency of 2.4 Hz when 0.80 kg hangs from it. What will its frequency be if only 0.50 kg hangs from it?

6. (II) A mass on the end of a spring is stretched 8.0 cm from equilibrium and released. At what distance from equilib-

rium will it have (a) velocity equal to half its maximum velocity and (b) acceleration equal to half its maximum acceleration?

7. (II) In some diatomic molecules, the force each atom exerts on the other can be approximated by $F = -C/r^2 + D/r^3$ where C and D are positive constants. (a) Graph F versus r from $r = 0$ to $r = 2D/C$. (b) Show that equilibrium occurs at $r = r_0 = D/C$. (c) Let $\Delta r = r - r_0$ be a small displacement from equilibrium, where $\Delta r \ll r_0$. Show that for such small displacements the motion is approximately simple harmonic, and (d) determine the force constant. (e) What is the period of such motion? (Hint: assume one atom is kept at rest.)

8. (II) The water in a U-shaped tube is displaced an amount Δx from equilibrium. (The level in one side is $2\Delta x$ above the level in the other side.) If friction is neglected, will the water oscillate harmonically? Determine a formula for the equivalent of the spring constant k. Does k depend on the density of the liquid, the cross section of the tube, or the length of the water column?

9. (II) If a mass m hangs from a vertical spring, show that the equilibrium length of the spring is a distance $l = mg/k$ longer than when the spring is horizontal as in Fig. 14-1. Show also that Eq. 14–1, $F = -kx$, holds for a vertical

spring where x is the displacement from the (vertical position) equilibrium point, and that k has the same value when the spring oscillates vertically as when it is horizontal.

10. (II) A mass m at the end of a spring vibrates with a frequency of 0.62 Hz; when an additional 700-g mass is added to m, the frequency is 0.48 Hz. What is the value of m?

11. (II) A block of mass m is supported by two identical parallel vertical springs, each with spring constant k. What will be the frequency of vibration?

12. (II) A mass m is at rest on the end of a spring of spring constant k. At $t = 0$ it is given an impulse J by a hammer. Write the formula for the subsequent motion in terms of m, k, J, and t.

13. (II) The position of a SHO as a function of time is given by $x = 2.4 \cos (5\pi t/4 + \pi/6)$ where t is in seconds and x in meters. Find (a) the period and frequency, (b) the position and velocity at $t = 0$, (c) the velocity and acceleration at $t = 1.0$ s.

14. (II) A tuning fork vibrates at a frequency of 264 Hz and the tip of each prong moves 1.5 mm to either side of center. Calculate (a) the maximum speed and (b) the maximum acceleration of the tip of a prong.

15. (II) A mass m is gently placed on the end of a freely hanging spring. The mass then falls 30 cm before it stops and begins to rise. What is the frequency of the motion?

16. (II) A rectangular block of wood floats in a calm lake. Show that, if friction is ignored, when the block is pushed gently down into the water, it will then vibrate with SHM. Also, determine an equation for the force constant.

17. (III) A mass m is connected to two springs, with spring constants k_1 and k_2, in two different ways as shown in Fig. 14–16a and b. Show that the period for the configuration shown in part (a) is given by

$$T = 2\pi \sqrt{m \left(\frac{1}{k_1} + \frac{1}{k_2} \right)}$$

and for that in part (b) is given by

$$T = 2\pi \sqrt{\frac{m}{k_1 + k_2}}.$$

Ignore friction.

FIGURE 14–16

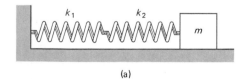

(a)

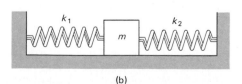

(b)

18. (III) Two equal masses, m_1 and m_2, are connected by three identical springs of spring constant k as shown in Fig. 14–17. (a) Apply $F = ma$ to each mass and obtain two differential equations for the displacements x_1 and x_2. (b) Determine the possible frequencies of vibration by assuming a solution of the form $x_1 = A_1 \cos \omega t$, $x_2 = A_2 \cos \omega t$.

19. (III) A spring with spring constant 250 N/m vibrates with an amplitude of 8.00 cm when 0.300 kg hangs from it. (a) What is the equation describing this motion as a function of time? Assume the mass passes through the equilibrium point, toward positive x, at $t = 0.060$ s. (b) At what times will the spring have its maximum and minimum lengths? (c) What is the force exerted by the spring at $t = 0$? (d) What is the displacement at $t = 0$? (e) What is the maximum speed and when is it first reached after $t = 0$?

SECTION 14–3

20. (I) A 1.0-kg mass vibrates according to the equation $x = 0.42 \cos 7.40t$ where x is in meters and t in seconds. Determine (a) the amplitude, (b) the frequency, (c) the total energy, (d) the kinetic energy and potential energy when $x = 0.16$ m.

21. (I) (a) At what displacement of a SHO is the energy half kinetic and half potential? (b) What fraction of the total energy of a SHO is kinetic and what fraction potential when the displacement is half the amplitude?

22. (II) A 0.350-kg mass at the end of a spring vibrates 2.0 times per second with an amplitude of 0.18 m. Determine (a) the velocity when it passes the equilibrium point, (b) the velocity when it is 0.10 m from equilibrium, (c) the total energy of the system, and (d) the equation describing the motion of the mass.

23. (II) It takes a force of 60 N to compress the spring of a popgun 0.10 m to "load" a 0.200-kg ball. With what speed will the ball leave the gun?

24. (II) A 300-kg wooden raft floats on a lake. When a 75-kg man stands on the raft, it sinks deeper into the water by 5.0 cm. When the man jumps off, the raft vibrates briefly. (a) What is the frequency of vibration? (b) What is the total energy of vibration (ignoring damping)?

25. (II) A 0.012-kg bullet strikes a 0.300-kg block attached to a fixed horizontal spring whose spring constant is 5.2×10^3 N/m and sets it into vibration with an amplitude of 12.4 cm. What was the speed of the bullet if the two objects move together after impact?

26. (II) If one vibration has 10 times the energy of a second one of equal frequency, but the first's spring constant k is twice as large as the second's, how do their amplitudes compare?

27. (II) A 70-kg person jumps from a window to a fire net 15 m below, which stretches the net 1.2 m. Assume the net behaves like a simple spring and calculate how much it

FIGURE 14–17

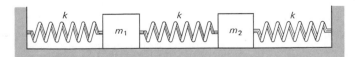

would stretch if the same person were lying in it. How much would it stretch if the person jumped from 30 m?

28. (II) At $t = 0$, a 650-g mass at rest on the end of a horizontal spring ($k = 64$ N/m) is struck by a hammer which gives it an initial speed of 1.16 m/s. Determine (a) the period and frequency of the motion, (b) the amplitude, (c) the maximum acceleration, (d) the position as a function of time, (e) the total energy, (f) the kinetic energy when $x = 0.65\ A$.

29. (II) Obtain the displacement x as a function of time for the simple harmonic oscillator using the conservation of energy, Eq. 14–11. (Hint: integrate Eq. 14–12a with $v = dx/dt$.)

30. (III) Consider a mass m oscillating on the end of a spring whose mass, m_s, is small compared to m but is not negligible. Show that the "equivalent mass" of the vibrating system is $(m + \frac{1}{3}m_s)$ so that the period of vibration is

$$T = 2\pi\sqrt{\frac{m + \frac{1}{3}m_s}{k}}$$

where k is the spring constant. (Hint: assume the spring stretches and compresses uniformly along its length and all portions oscillate in phase.)

SECTION 14–5

31. (I) How long must a simple pendulum be if it is to make exactly one complete vibration per second?

32. (I) What is the period of a simple pendulum on Mars, where the acceleration of gravity is about 0.37 that on Earth, if the pendulum has a period of 0.60 s on Earth?

33. (II) The length of a simple pendulum is 0.36 m and it is released at an angle of 10° to the vertical. (a) With what frequency does it vibrate? (b) What is the pendulum bob's speed when it passes through the lowest point of the swing?

34. (II) What is the period of a simple pendulum 80 cm long (a) on the earth, and (b) when it is in a freely falling elevator?

35. (II) A simple pendulum is 0.24 m long. At $t = 0$ it is released starting at an angle of 14°. Ignoring friction, what will be the angular position of the pendulum at (a) $t = 0.25$ s, (b) $t = 1.60$ s, and (c) $t = 5.00$ s?

36. (II) Derive a formula for the maximum speed v_0 of a simple pendulum bob in terms of g, the length L, and the angle of swing θ.

37. (II) A simple pendulum vibrates with an amplitude of 10.0°. What fraction of the time does it spend between $+5.0°$ and $-5.0°$? Assume SHM.

38. (II) A simple pendulum oscillates with frequency f. What is its frequency if it accelerates at $\frac{1}{2}g$ (a) upward, (b) downward, (c) horizontally?

39. (II) What is the maximum amplitude of a pendulum if Eq. 14–13 is to be accurate to (a) 1.0 percent, (b) 0.1 percent?

40. (II) The pendulum of an accurate clock oscillates with an amplitude of $\pm 12.0°$. If, due to a faulty mechanism, the amplitude is instead maintained at $\pm 1.0°$, what will be the clock error per day? Does it gain or lose?

*SECTION 14–6

*41. (II) A uniform circular wheel of radius R is pivoted at its edge. What is the frequency for small oscillations about the pivot?

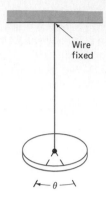

Wire fixed

FIGURE 14–18

*42. (II) A pendulum consists of a tiny bob of mass M and a uniform cord of mass m and length L. (a) Determine a formula for the period. (b) What would be the fractional error if one used the formula for a simple pendulum, Eq. 14–13?

*43. (II) *Torsion pendulum.* A cylindrical disc whose moment of inertia is I is firmly attached to a wire from which it is suspended as shown in Fig. 14–18; the disc can oscillate when twisted an angle θ from equilibrium as shown. The twisted wire exerts a restoring torque

$$\tau = -K\theta$$

where K is a constant and θ is not too large. (a) Determine the equation of motion (for θ as a function of time) and show that the motion is simple harmonic. (b) Determine a formula for the period. [The balance wheel of a watch is another example of a torsion pendulum in which the restoring torque is applied by a coil spring.]

*44. (II) The balance wheel of a watch is a thin ring of radius 0.95 cm and oscillates with a frequency of 3.10 Hz. If a torque of 1.1×10^{-5} N·m causes the wheel to rotate 60°, calculate the mass of the balance wheel.

*45. (II) Treating the leg as a physical pendulum, determine the period (a) by measuring it for your own leg, and (b) by using Eq. 14–15 assuming the leg is a long rod pivoted about one end. For part (b), assume the leg weighs 12.0 kg, is 0.80 m long, and its cg is 0.50 m from the floor.

*46. (III) Suppose that C represents the center of oscillation for a body oscillating about a pivot at O. If the body is allowed to oscillate about a parallel axis through C, show that (a) the period is the same, and (b) that point O is now the center of oscillation. (Hint: use the parallel-axis theorem for moments of inertia.)

*47. (III) Show that if a body receives an impulsive blow at its center of oscillation C (see Fig. 14–10), no reaction force is exerted at the pivot point O during the impulse. (Hint: consider translation of the cm, and rotation about the cm.)

SECTION 14–7

48. (II) A 750-g block oscillates on the end of a spring whose force constant is $k = 56.0$ N/m. The mass moves in a fluid which offers a resistive force $F = -bv$ where $b =$

0.162 N·s/m. (a) What is the period of the motion? (b) What is the fractional decrease in amplitude per cycle? (c) Write the displacement as a function of time if at $t = 0$, $x = 0$, and at $t = 1.00$ s, $x = 0.120$ m.

49. (II) (a) Show that the total mechanical energy, $E = \frac{1}{2}mv^2 + \frac{1}{2}kx^2$, as a function of time for a lightly damped harmonic oscillator is

$$E = \frac{1}{2}kA^2 e^{-(b/m)t} = E_0 e^{-(b/m)t}$$

where E_0 is the total mechanical energy at $t = 0$. (Assume $\omega' \gg b/2m$.) (b) Show that the fractional energy lost per period is

$$\frac{\Delta E}{E} = \frac{2\pi b}{m\omega_0} = \frac{2\pi}{Q}$$

where $\omega_0 = \sqrt{k/m}$ and $Q = m\omega_0/b$ is called the *quality factor* or *Q value* of the system. A large Q value means the system can undergo oscillations for a longer time. The Q value is discussed further in Section 14–8.

50. (II) A damped harmonic oscillator loses 5.0 percent of its mechanical energy per cycle. (a) By what percentage does its frequency differ from the natural frequency $\omega_0 = \sqrt{k/m}$? (b) After how many periods will the amplitude have decreased to $1/e$ of its original value?

SECTION 14–8

51. (II) (a) At resonance ($\omega = \omega_0$), what is the value of the phase angle ϕ, which is the phase difference between the driving force and the displacement? (b) What, then, is the displacement at a time when the driving force F is a maximum? At a time when $F = 0$?

52. (II) Construct an accurate resonance curve, from $\omega = 0$ to $\omega = 2\omega_0$, for $Q = 4.0$.

53. (II) The amplitude of a driven harmonic oscillator reaches a value of $28.6 F_0/m$ at a resonant frequency of 382 Hz. What is the Q value of this system?

54. (II) By direct substitution, show that Eq. 14–24 (with Eq. 14–25) is a solution of the equation of motion (14–23) for the forced oscillator.

55. (II) Differentiate Eq. 14–25a to show that the resonant amplitude peaks at

$$\omega = \sqrt{\omega_0^2 - b^2/2m^2}.$$

56. (III) Consider a simple pendulum (point mass bob) 0.50 m long with a Q of 400. (a) How long does it take for the amplitude (assumed small) to decrease by two-thirds? (b) If the amplitude is 2.0 cm and the bob has mass 0.20 kg, what is the initial energy loss rate of the pendulum in watts? (c) If we are to stimulate resonance with a sinusoidal driving force, how close must the driving frequency be to the natural frequency of the pendulum?

57. (III) *Power transferred to driven oscillator.* (a) Show that the power input to a forced oscillator due to the external force F_{ext} is

$$P = F_{ext}v$$

$$= \frac{F_0^2\omega \cos\phi_0 \cos^2\omega t - \frac{1}{2}F_0^2\omega \sin\phi_0 \sin 2\omega t}{m\sqrt{(\omega^2 - \omega_0^2)^2 + \omega^2 b^2/m^2}}.$$

(b) Show that average power input, averaged over one (or many) full cycles, is

$$\bar{P} = \frac{\omega F_0^2 \cos\phi_0}{2m\sqrt{(\omega^2 - \omega_0^2)^2 + \omega^2 b^2/m^2}} = \frac{1}{2}\omega A_0 F_0 \cos\phi_0$$

or

$$\bar{P} = \frac{1}{2}F_0 v_{max} \cos\phi_0$$

where v_{max} is the maximum value of dx/dt. (c) Plot $\bar{P}$ versus ω from $\omega = 0$ to $\omega = 2\omega_0$ for $Q = 6.0$. Note that although the amplitude is not zero at $\omega = 0$, $\bar{P}$ is zero at $\omega = 0$.

58. (III) Derive Eq. 14–27.

*SECTION 14–9

*59. (II) Write down the x and y equations for the motions shown (a) in Fig. 14–15d, and (b) in Fig. 14–15e. Take $\phi_y = 0$.

*60. (II) Show analytically that the combination of two simple harmonic motions at right angles is in the form of an ellipse $(x^2/a^2 + y^2/b^2 = 1)$ when $\omega_x = \omega_y$, $\phi_y - \phi_x = \pm\pi/2$, $A_x \neq A_y$.

*61. (II) A particle moves with a constant speed of 24 m/s in a circle centered at the origin. At $t = 0$, the particle is at $x = 3.0$ m, $y = 4.2$ m. (a) Find equations for its x and y motion. (b) What is the frequency of rotation? (c) What is the initial phase angle?

*62. (II) Signals fed to the horizontal and vertical inputs of an oscilloscope are given by

$$x = A \cos(\omega t + \phi)$$

$$y = A \cos\omega t.$$

Describe the motion (shape of path) for the following cases: (a) $\phi = 0°$, (b) $\phi = 60°$, (c) $\phi = -60°$, (d) $\phi = 90°$.

*63. (II) Draw the Lissajous figure for the case $\omega_x = 3\omega_y$, $A_x = A_y$, $\phi_y - \phi_x = \pi/2$.

*NUMERICAL/PROGRAMMABLE CALCULATOR

*64. (III) *Simple Harmonic Motion.* Eq. 14–3b

$$\frac{d^2x}{dt^2} + \frac{k}{m}x = 0$$

is a differential equation. In Section 14–2 we found the general analytic solution to this equation. The solution to a differential equation is not always so easy to find as in this case, and numerical techniques may have to be used. As an example of how to do this, take the spring and initial conditions of Example 14–1 and use numerical integration[†] to determine x as a function of time from $t = 0$ to $t = 1.20$ s. (Hint: two integrations are necessary: first numerically integrate

$$dv = -\frac{k}{m}x\,dt,$$

and then $dx = v\,dt$.) Plot your solution and be sure it is within 1 percent of the analytic solution at each point. (If it's not, choose smaller time intervals and repeat.)

† See Section 2–10.

*65. (III) *Damped Harmonic Motion.* Equation 14–16 is the differential equation for damped harmonic motion when the damping force is directly proportional to the speed v. Suppose the spring of Example 14–1 has a damping constant b which may take on the values (a) 1.90 kg/s (b) 7.25 kg/s; use numerical integration (see problem 64) to plot the displacement x as a function of time from $t = 0$ to $t = 2.50$ s for each case. (c) Determine the value of b needed for critical damping; then plot x versus t for this case as well, using numerical integration. In all three cases, compare your plot to the analytic solution, and be sure they correspond within 2 percent at all points.

*66. (III) *Damping proportional to v^2.* Suppose the oscillator of Example 14–1 is damped by a force proportional to the velocity squared, $F_{damping} = -cv^2$ where $c = 0.275$ kg·s/m is a constant. Numerically integrate the differential equation from $t = 0$ to $t = 2.00$ s to an accuracy of 2 percent, and plot your results.

*67. (III) *Simple pendulum.* (a) For a simple pendulum (Section 14–5) show that the differential equation of motion is

$$\frac{d^2\theta}{dt^2} + \frac{g}{L}\sin\theta = 0.$$

(b) Suppose that the bob of a simple pendulum is 1.20 m below its suspension point, and at $t = 0$ the bob is released from rest at an angle $\theta_M = 45°$. Numerically integrate the equation of motion to find θ as a function of time from $t = 0$ to $t = 3.50$ s, with an accuracy of at least 1 percent. (c) Plot your solution, and on the same graph plot the analytic solution obtained with the assumption of θ_M being very small. (d) What is the period of the motion as determined by your numerical calculations? What value do you get using Eq. 14–14? What would be the period for small oscillations (θ_M small)? What can you say about how the period T depends on the maximum angle θ_M?

Wave Motion

15

When you throw a stone into a lake or pool of water, circular waves form and move outward, Fig. 15–1. Waves will also travel along a cord that is stretched out straight on a table if you vibrate one end back and forth as shown in Fig. 15–2. Water waves, and waves on a cord, are two common examples of wave motion. Sound also travels as a wave, and light can be considered as a wave of electromagnetic fields. The ultimate constituents of matter, such as electrons, also behave in certain ways as waves. Thus the study of wave phenomena is very important since it occurs in so many areas of physics. In this chapter we will concentrate on *mechanical waves*— waves that travel in a material medium, such as water waves, or waves on a string. We will discuss other types of wave motion in later chapters.

Waves—whether ocean waves, waves on a string, earthquake waves, or sound waves in air—have as their source a vibration. In the case of sound, not only is the source a vibrating object, but so is the detector—the ear drum or the membrane of a microphone. Indeed, the medium through which a wave travels itself vibrates.

If you have ever watched ocean waves moving toward shore, you may have wondered if the waves were carrying water into the beach. This is, in fact, not the case.[†] Water waves move with a recognizable velocity. But each particle of the water itself merely oscillates about an equilibrium point. This is clearly demonstrated by observing leaves on a pond as waves move by. The leaves (or a cork) are not carried forward by the waves, but simply oscillate about an equilibrium point because this is the motion of the water itself. Similarly, the wave on the rope of Fig. 15–2 moves to the right, but each piece of the rope only vibrates to and fro. This is a general feature of waves: waves can move over large distances but the medium (the water or the rope) itself has only a limited movement. Thus, although a wave is not matter, the wave pattern can travel in matter. A wave consists of oscillations that move without carrying matter with them.

Waves carry energy from one place to another. Energy is given to a water wave, for example, by a rock thrown into the water, or by wind far out at sea. The

---

[†] Do not be confused by the "breaking" of ocean waves which occurs when the wave interacts with the ground and hence is no longer a simple wave. Also, objects can be blown by the wind to or from the shore.

FIGURE 15–1
Water waves spreading outward from
a source. (D.C. Giancoli.)

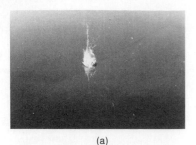

(a)

(b)

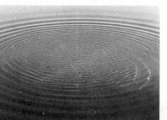

(c)

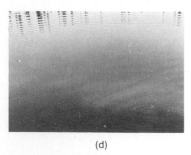

(d)

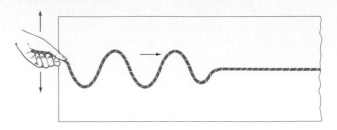

FIGURE 15–2
Wave traveling on a rope.

energy is transported by waves to the shore. If you have been under an ocean wave
when it breaks, you know the energy it carries. The oscillating hand in Fig. 15–2
transfers energy to the rope which is then transported down the rope and can be
transferred to an object at the other end. All forms of wave motion transport energy.

15–1   Characteristics of Wave Motion

Let us look at how a wave is formed and how it comes to "travel." We first look at a
single wave bump or *pulse*. A single pulse can be formed on a rope by a quick up and
down motion of the hand, Fig. 15–3. The hand pulls up on one end of the rope and
because the end piece is attached to adjacent pieces, these also feel an upward force
and they too begin to move upward. As each succeeding piece of rope moves
upward, the wave crest moves outward along the rope. Meanwhile, the end piece of
rope has been returned to its original position by the hand, and as each succeeding
piece of rope reaches its peak position, it too is pulled back down again. Thus, the
source of a traveling wave pulse is a disturbance, and cohesive forces between
adjacent pieces of rope cause the pulse to travel outward. Waves in other media are
created and propagate outward in a similar fashion.

A *continuous* or *periodic wave*, such as that shown starting in Fig. 15–2 has as
its source a disturbance that is continuous and oscillating; that is, the source is a
vibration or *oscillation*. In Fig. 15–2, a hand oscillates one end of the rope. Water
waves may be produced by any vibrating object placed at the surface such as your
hand; or the water itself is made to vibrate when wind blows across it or a rock or
tennis ball is thrown into it. A vibrating tuning fork or drum membrane gives rise to
sound waves in air; and we will see later that oscillating electric charges give rise to
light waves. Indeed, almost any vibrating object sends out waves.

FIGURE 15–3
Motion of a wave pulse.
Arrows indicate velocity
of rope particles.

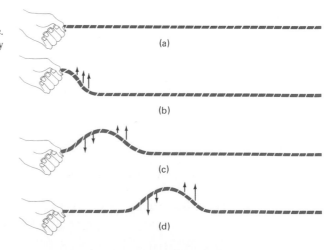

The source of any wave, then, is a vibration. And it is the *vibration* that propagates outward and thus constitutes the wave. If the source vibrates sinusoidally in SHM, then the wave itself—if the medium is perfectly elastic—will have a sinusoidal shape both in space and in time. That is, if you take a picture of the wave spread throughout space at a given instant of time, the wave will have the shape of a sine or cosine function. On the other hand, if you look at the motion of the medium at one place over a long period of time—for example, if you look between two closely spaced posts of a pier or out of a ship's porthole as water waves pass by—the up and down motion of that small segment of water will be simple harmonic motion—the water moves up and down sinusoidally in time.

Some of the important quantities used to describe a periodic sinusoidal wave are shown in Fig. 15–4. The high points on a wave are called crests, the low points troughs. The **amplitude** is the maximum height of a crest, or depth of a trough, relative to the normal (or equilibrium) level; the total swing from a crest to a trough is twice the amplitude. The distance between two successive crests is called the **wavelength**, λ (the Greek letter lambda). The wavelength is also equal to the distance between *any* two successive identical points on the wave. The **frequency**, f (sometimes called v, the Greek letter nu) is the number of crests—or complete cycles—that pass a given point per unit time. The period, T, of course, is just $1/f$.

The **wave velocity**, v, is the velocity at which wave crests appear to move. (The wave velocity must be distinguished from the velocity of a particle of the medium itself. For example, for a wave traveling along a string as in Fig. 15–2, the wave velocity is parallel to the string, whereas the velocity of a particle of the string is perpendicular to it.) Since a wave crest travels a distance of one wavelength, λ, in one period, T, the wave velocity $v = (\lambda/T)$ or

$$v = \lambda f. \tag{15-1}$$

For example, suppose a wave has a wavelength of 5 m and a frequency of 3 Hz. Since 3 crests pass a given point per second, and the crests are 5 m apart, the first crest (or any other part of the wave) must travel a distance of 15 m during the 1 s; so its speed is 15 m/s.

The velocity of a wave depends on the properties of the medium in which it travels. The velocity of a wave on a stretched string, for example, depends on the tension in the string, F_T, and on the mass per unit length, μ (the Greek letter mu); for waves of small amplitude, the relationship is

$$v = \sqrt{\frac{F_T}{\mu}}. \tag{15-2}$$

Before giving a derivation of this formula, it is worth noting that at least qualitatively it makes sense on the basis of Newtonian mechanics: that is, we do expect the tension to be in the numerator and the mass per unit length in the denominator because when the tension is greater we expect the velocity to be greater since each segment of string is in better contact with its neighbor; and, the greater the mass per unit length, the more inertia the string has and the more slowly the wave would be expected to propagate.

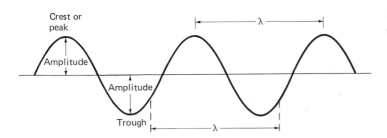

FIGURE 15–4

Characteristics of a continuous wave.

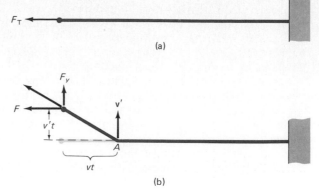

FIGURE 15–5

Diagram of simple wave pulse for derivation of Equation 15–2.

We can make a simple derivation of Eq. 15–2 using a simple model of a string under a tension F_T as shown in Fig. 15–5a. The string is pulled upward at a speed v' by the force F_y; as shown in Fig. 15–5b, all points of the string to the left of point A move upward at the speed v', and those to the right are still at rest. The speed of propagation, v, of this wave pulse is the speed of point A, the leading edge of the pulse. Point A moves to the right a distance vt in a time t, whereas the end of the rope moves upward a distance $v't$. By similar triangles we have the approximate relation

$$\frac{F_T}{F_y} = \frac{v}{v'}$$

which is accurate for small displacements ($v't \ll vt$) so that F_T does not change appreciably. As we saw in Chapter 8, the impulse given to an object is equal to its change in momentum. During the time t the total upward impulse is $F_y t = (v'/v)F_T t$; the change in momentum of the string, Δp, is the mass of rope moving upward times its velocity. Since the upward moving segment of rope has mass equal to the mass per unit length μ times its length vt we have

$$F_y t = \Delta p$$

$$\frac{v'}{v} F_T t = (\mu vt)v'.$$

Solving for v we find $v = \sqrt{F_T/\mu}$ which is Eq. 15–2. Although it was derived for a special case, it is valid for any wave shape since other shapes can be considered to be made up of many tiny such lengths. But it is valid only for small displacements (as was our derivation). Experiment is in accord with this result derived from Newtonian mechanics.

EXAMPLE 15–1   A wave whose wavelength is 0.50 m is traveling down a 300-m-long wire whose total mass is 30 kg. If the wire is under a tension of 4000 N, what is the velocity and frequency of this wave?

SOLUTION   From Eq. 15–2 the velocity is

$$v = \sqrt{\frac{4000 \text{ N}}{(30 \text{ kg})/(300 \text{ m})}} = 200 \text{ m/s}.$$

The frequency then is

$$f = \frac{v}{\lambda} = \frac{200 \text{ m/s}}{0.50 \text{ m}} = 400 \text{ Hz}.$$

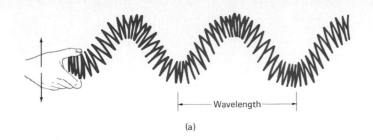

Wavelength

(a)

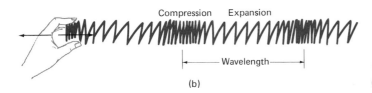

Compression   Expansion

Wavelength

(b)

FIGURE 15–6

(a) Transverse wave; (b) longitudinal wave.

15–2   Types of Waves

We saw earlier that although waves may travel over long distances, the particles of the medium vibrate only over a limited region of space. When a wave travels down a rope, say from left to right, the particles of the rope vibrate up and down in a direction transverse (or perpendicular) to the motion of the wave itself. Such a wave is called a **transverse wave**. There exists another type of wave known as a **longitudinal wave**. In a longitudinal wave, the vibration of the particles of the medium is along the *same* direction as the motion of the wave. Longitudinal waves are readily formed on a stretched spring or "slinky" by alternately compressing and expanding one end. This is shown in Fig. 15–6b, and can be compared to the transverse wave in Fig. 15–6a. A series of compressions and expansions propagate along the spring. The *compressions* are those areas where the coils are momentarily close together. *Expansions* (sometimes called *rarefactions*) are regions where the coils are momentarily far apart. Compressions and expansions correspond to the crests and troughs of a transverse wave.

An important example of a longitudinal wave is a sound wave in air. A vibrating drum head, for example, alternately compresses and rarefies the air and produces a longitudinal wave that travels outward in the air as shown in Fig. 15–7.

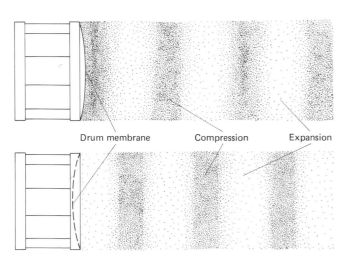

Drum membrane   Compression   Expansion

FIGURE 15–7

Production of a sound wave, which is longitudinal.

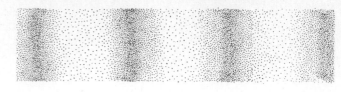

(a)

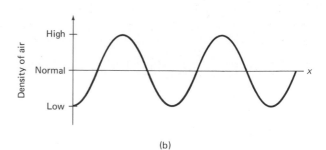

(b)

FIGURE 15–8

(a) A longitudinal wave with (b) its graphical representation.

As in the case of transverse waves, each section of the medium in which a longitudinal wave passes oscillates over a very small distance whereas the wave itself can travel large distances. Wavelength, frequency, and wave velocity all have meaning for a longitudinal wave. The wavelength is the distance between successive compressions (or between successive expansions), and frequency is the number of compressions that pass a given point per second. The wave velocity is the velocity with which each compression appears to move and is equal to the product of wavelength and frequency.

A longitudinal wave can be represented graphically by plotting the density of air molecules (or coils of a slinky) versus position as shown in Fig. 15–8. We will often use such a graphical representation because it is much easier to illustrate what is happening. Note that the graph looks much like a transverse wave.

The velocity of a longitudinal wave has a form similar to that for a transverse wave on a string (Eq. 15–2); that is

$$v = \sqrt{\frac{\text{elastic force factor}}{\text{inertia factor}}}.$$

FIGURE 15–9

Determining the speed of a one-dimensional longitudinal wave in a fluid contained in a long narrow tube.

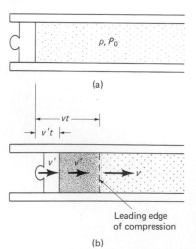

In particular, for a longitudinal wave traveling down a long solid rod,

$$v = \sqrt{\frac{E}{\rho}} \qquad (15\text{–}3)$$

where E is the elastic modulus (Section 11–4) of the material and ρ is its density. For a longitudinal wave traveling in a liquid or gas,

$$v = \sqrt{\frac{B}{\rho}} \qquad (15\text{–}4)$$

where B is the bulk modulus (Section 11–4) and ρ again the density.

We now derive Eq. 15–4. We consider a wave pulse traveling in a fluid in a long tube, so that the wave motion is one dimensional. The tube is filled with a fluid which, at $t = 0$, is of uniform density ρ and at uniform pressure P_0, Fig. 15–9a. At this moment the piston in the end of the tube is suddenly started moving to the right with speed v', compressing the fluid in front of it. In the (short) time t the piston moves a distance $v't$. The compressed fluid itself also moves with speed v', but the leading edge of the compressed region moves to the right at the characteristic speed v of compression waves in that fluid; we assume the wave speed v is much larger than the speed v'. The leading edge of the compression in the fluid (which at $t = 0$ was at the piston face) thus moves a distance vt in time t as shown in Fig. 15–9b. Let the pressure in the compression be $P_0 + \Delta P$, which is ΔP higher than in the uncompressed fluid. To move the piston to the right requires an external force $(P_0 + \Delta P)A$ acting to the right, where A is the area of the tube. The *net* force on the compressed region of the fluid is

$$(P_0 + \Delta P)A - P_0 A = A \, \Delta P$$

since the uncompressed fluid exerts a force $P_0 A$ to the left at the leading edge. Hence the impulse given to the compressed fluid, which equals its change in momentum, is

$$A \, \Delta P t = (\rho A v t)v'$$

where $(\rho A v t)$ represents the mass of fluid which is given the speed v'. Hence we have

$$\Delta P = \rho v v'.$$

From the definition of the bulk modulus, B (Eq. 11–7a)

$$B = -\frac{\Delta P}{\Delta V / V_0} = -\frac{\rho v v'}{\Delta V / V_0}$$

where $\Delta V / V_0$ is the fractional change in volume due to compression. The original volume of the compressed fluid is $V_0 = Avt$, and it has been compressed by an amount $\Delta V = -Av't$ (Fig. 15–9b). Thus

$$B = -\frac{\rho v v'}{\Delta V / V_0} = -\rho v v' \left(\frac{Avt}{-Av't} \right) = \rho v^2,$$

and

$$v = \sqrt{\frac{B}{\rho}},$$

which is Eq. 15–4. The derivation of Eq. 15–3 follows similar lines, but takes into account the fact that when the end of a rod is compressed, the sides expand slightly.

EXAMPLE 15–2 You can often hear an approaching train by putting your ear to the track. How long does it take for the wave to travel down the steel track if the train is 1.0 km away?

SOLUTION Referring to Tables 11–1 and 12–1 for the elastic modulus and density of steel, respectively, we have

$$v = \sqrt{\frac{2.0 \times 10^{11} \text{ N/m}^2}{7.8 \times 10^3 \text{ kg/m}^3}} = 5.1 \times 10^3 \text{ m/s}.$$

Then the time $t = $ distance/velocity $= (1.0 \times 10^3 \text{ m})/(5.1 \times 10^3 \text{ m/s}) = 0.20$ seconds.

Both transverse and longitudinal waves are produced when a disturbance known as an earthquake occurs. The transverse waves that travel through the body of the earth are called S waves and the longitudinal waves are called P waves. Both longitudinal and transverse waves can travel through a solid since the atoms or molecules can vibrate about their relatively fixed positions in any direction. But in a fluid, only longitudinal waves can propagate; this is because any transverse motion would experience no restoring force since a fluid can flow. This fact was used by geophysicists to infer that the earth's core is molten: longitudinal waves are detected diametrically across the earth, but never transverse waves; the only explanation for this was that the core of the earth must be liquid.

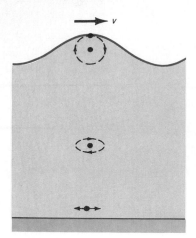

FIGURE 15–10

A water wave is an example of a *surface wave*, which is a combination of transverse and longitudinal wave motion.

There is a third kind of wave called a *surface wave* that occurs at the boundary between two materials. A wave on water is actually a surface wave that moves on the boundary between water and air. The motion of each particle of water at the surface is elliptical (Fig. 15–10) if the wavelength is less than the depth, so it is a combination of transverse and longitudinal motions. Below the surface, there is also transverse plus longitudinal wave motion, in an elliptical shape as shown; at the bottom, the motion is only longitudinal. Surface waves are also set up on the earth when an earthquake occurs. The waves that travel along the surface are mainly responsible for the damage caused by earthquakes.

Waves which travel along a line in one dimension, such as transverse waves on a stretched string or longitudinal waves in a rod or fluid-filled tube, are *linear* or *one-dimensional waves*. Surface waves, such as the water waves of Fig. 15–1, are *two-dimensional waves*. Finally, waves that move out from a source in all directions, such as sound from a speaker or earthquake waves through the earth, are *three-dimensional waves*. We will be concerned with all three types, but especially linear waves, since they are simpler and clearer.

15–3   Energy Transmitted by Waves

Waves transmit energy from one place to another. As waves travel through a medium, the energy is transmitted as vibrational energy from particle to particle of the medium. For a sinusoidal wave of frequency f, the particles move in SHM as a wave passes, so each particle has an energy $E = \frac{1}{2}kD_M^2$ where D_M is the maximum displacement (amplitude) of its motion, either transversely or longitudinally (see Eq. 14–11, in which we have replaced A by D_M). Using Eq. 14–8 we can write k in terms of the frequency, $k = 4\pi^2 mf^2$, so that

$$E = 2\pi^2 mf^2 D_M^2.$$

The mass $m = \rho V$, where ρ is the density of the medium and V is its volume. Also, the volume $V = Al$ where A is the cross-sectional area through which the wave travels, and we can write l as the distance the wave travels in a time t as $l = vt$; here v is the speed of the wave. Thus $m = \rho V = \rho Al = \rho Avt$ and

$$E = 2\pi^2 \rho Avtf^2 D_M. \tag{15–5}$$

If we consider the leading edge of a sinusoidal wave about to enter a region where there is no wave motion (as in Fig. 15–2), then we see that E in Eq. 15–5 represents the average energy that crosses the boundary into this region in a time t. From this equation we have the important result that the *energy transported by a wave is proportional to the square of the amplitude*. The average *rate* of energy transferred is the average power $\bar{P}$:

$$\bar{P} = \frac{E}{t} = 2\pi^2 \rho Avf^2 D_M^2. \tag{15–6}$$

Finally, the **intensity**, I, of a wave is defined as the average power transferred across unit area perpendicular to the direction of energy flow:

$$I = \frac{\bar{P}}{A} = 2\pi^2 v\rho f^2 D_M^2 \tag{15–7}$$

and we see that the *intensity of a wave is proportional to the square of its amplitude*.

If the waves flow out from the source in all directions it is a three-dimensional wave. Examples are sound traveling in the open air, earthquake waves, and light waves. If the medium is isotropic (the same in all directions) then the wave is spherical in shape and is said to be a *spherical wave*, Fig. 15–11. As the wave moves outward, it is spread over a larger and larger area since the surface area of a sphere of radius r is $4\pi r^2$. Because energy is conserved, we can see from

FIGURE 15–11

Wave traveling outward from source has spherical shape; two different crests (or compressions) are shown, of radius r_1 and r_2.

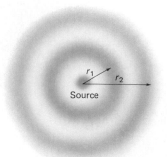

Eq. 15-5 or 15-6 that as the area A increases, the amplitude D_M must decrease. That is, at two different distances from the source, r_1 and r_2 (see Fig. 15-11), $A_1 D_{M1}^2 = A_2 D_{M2}^2$ where D_{M1} and D_{M2} are the amplitudes of the wave at r_1 and r_2 respectively. Since $A_1 = 4\pi r_1^2$ and $A_2 = 4\pi r_2^2$, we have $(D_{M1}^2 r_1^2) = (D_{M2}^2 r_2^2)$, or

$$\frac{D_{M2}}{D_{M1}} = \frac{r_1}{r_2}.$$

Thus the amplitude decreases inversely as the distance from the source. When the wave is twice as far from the source, the amplitude is half as large, and so on (ignoring damping due to friction).

The intensity I also decreases with distance. Since I is proportional to D_M^2 (Eq. 15-7), then I must decrease as the *square* of the distance from the source. This inverse square law applies to sound and light and other types of waves. Another way to view this is to consider two points r_1 and r_2 at the same time. If the power output remains constant, then the intensity at r_1 is $I_1 = \bar{P}/4\pi r_1^2$ and $I_2 = \bar{P}/4\pi r_2^2$. Thus

$$\frac{I_2}{I_1} = \frac{r_1^2}{r_2^2}. \tag{15-8}$$

 EXAMPLE 15-3   If the intensity of an earthquake P wave is $1.0 \times 10^6 \, \text{W/m}^2$ at 100 km from the source, what is the intensity 400 km from the source?

SOLUTION   The intensity decreases as the square of the distance from the source. Therefore, at 400 km, the intensity should be $(1/4)^2 = 1/16$ what it is at 100 km, or $6.2 \times 10^4 \, \text{W/m}^2$.

The situation is different for a one-dimensional wave, such as a transverse wave on a string or a longitudinal wave pulse traveling down a uniform metal rod. The area A remains constant so the amplitude D_M also remains constant; thus the intensity as well as the amplitude do not decrease with distance.

In practice, frictional damping is generally present and some of the energy is transformed into thermal energy. Thus the amplitude and intensity of a one-dimensional wave decreases with distance from the source, and for a three-dimensional wave the decrease will be greater than that discussed above, although the effect is often small.

15-4   Mathematical Representation of a Traveling Wave

Let us now consider a one-dimensional wave traveling along the x axis. It could be, for example, a transverse wave on a string or a longitudinal wave traveling in a rod or fluid-filled tube. Let us assume the wave shape is sinusoidal of a particular wavelength λ and frequency f. At $t = 0$, suppose the wave shape is given by

$$D = D_M \sin \frac{2\pi}{\lambda} x \tag{15-9}$$

as shown by the solid curve in Fig. 15-12; D is the displacement of the wave at position x, and D_M is the amplitude (maximum displacement) of the wave. This relation gives a shape that repeats itself every wavelength (which is what we want) since the displacement is the same at, for example, $x = 0$, $x = \lambda$, $x = 2\lambda$, and so on (since $\sin 4\pi = \sin 2\pi = \sin 0$).

Now suppose the wave is moving to the right with velocity v. Then, after a time t, each part of the wave (indeed, the whole wave "shape") has moved to the right a distance vt; see the dashed curve in Fig. 15-12. Consider any point on the wave at

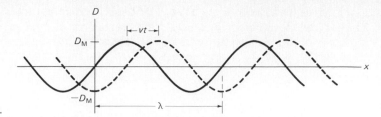

FIGURE 15–12

A traveling wave. In time t, the wave moves a distance vt.

$t = 0$—say, a crest which is at some position x. After a time t, that crest will have traveled a distance vt so its new position is vt greater than its old position. To describe this same point on the wave shape, the argument of the sine function must be the same, so we replace x in Eq. 15–9 by $(x - vt)$:

$$D = D_M \sin\left[\frac{2\pi}{\lambda}(x - vt)\right]. \tag{15–10a}$$

Said another way, if you are riding on a crest, the argument of the sine function remains the same ($= \pi/2, 5\pi/2$, and so on); as t increases, x must increase at the same rate so that $(x - vt)$ remains constant.

Equation 15–10a is the mathematical representation of a sinusoidal wave traveling along the x axis to the right (increasing x). It gives the displacement D of the wave at any point x at any time t. Since $v = \lambda f$ (Eq. 15–1), we can write Eq. 15–10a in other ways that are often convenient:

$$D = D_M \sin\left(\frac{2\pi x}{\lambda} - \frac{2\pi t}{T}\right) \tag{15–10b}$$

where $T = 1/f = \lambda/v$ is the period; and

$$D = D_M \sin(kx - \omega t) \tag{15–10c}$$

where $\omega = 2\pi f = 2\pi/T$ is the angular frequency and

$$k = \frac{2\pi}{\lambda} \tag{15–11}$$

is called the **wave number**. (Do not confuse the wave number k with the spring constant k; they are very different quantities.) All three forms, Eq. 15–10a, b, and c, are equivalent; Eq. 15–10c is the simplest and is perhaps the most common. The quantity $(kx - \omega t)$, and its equivalent in the other two equations, is called the **phase** of the wave. The velocity v of the wave, which is often called the **phase velocity** since it describes the velocity of the phase (or shape) of the wave, can be written in terms of ω and k:

$$v = \lambda f = \left(\frac{2\pi}{k}\right)\left(\frac{\omega}{2\pi}\right) = \frac{\omega}{k}. \tag{15–12}$$

For a wave traveling along the x axis to the left (decreasing values of x), we start again with Eq. 15–9 and note that a particular point on the wave changes position by $-vt$ in a time t; so x in Eq. 15–9 must be replaced by $(x + vt)$. Thus, for a wave traveling to the left with velocity v,

$$D = D_M \sin\left[\frac{2\pi}{\lambda}(x + vt)\right] \tag{15–13a}$$

$$= D_M \sin\left(\frac{2\pi x}{\lambda} + \frac{2\pi t}{T}\right) \tag{15–13b}$$

$$= D_M \sin(kx + \omega t). \tag{15–13c}$$

In other words, we simply replace v in Eq. 15–10 by $-v$.

Let us look for a moment at Eq. 15–13 (or, just as well, at Eq. 15–10). At $t = 0$ we have

$$D = D_M \sin kx,$$

which is what we started with, a sinusoidal wave shape. If we look at the wave shape in space at a particular later time t_0, then we have

$$D = D_M \sin (kx + \omega t_0).$$

That is, if we took a picture of the wave at $t = t_0$, we would see a sine wave with a phase constant ωt_0. Thus, for fixed $t = t_0$, the wave has a sinusoidal shape in space. On the other hand, if we consider a fixed point in space, say $x = 0$, we can see how the wave varies in time:

$$D = D_M \sin \omega t$$

where we used Eq. 15–13c. This is just the equation for simple harmonic motion (see Section 14–2, Eq. 14–6). For any other fixed value of x, say $x = x_0$, $D = D_M \sin (\omega t + kx_0)$ which differs only by a phase constant, kx_0. Thus, at any fixed point in space, the displacement undergoes the oscillations of simple harmonic motion in time. Equations 15–10 and 15–13 combine both these aspects to give us the representation for a *traveling sinusoidal wave* (also called a *harmonic wave*).

Now let us consider a general wave (or wave pulse) of any shape. If frictional losses are small, experiment shows that the wave maintains its shape as it travels. Thus we can make the same arguments as we did right after Eq. 15–9. Suppose our wave has some shape given by

$$D = f(x)$$

at $t = 0$, where D is the displacement of the wave at x and $f(x)$ is some function of x. Then at some later time, if the wave is traveling to the right along the x axis, the wave will have the same shape but all parts will have moved a distance vt where v is the phase velocity of the wave. Hence we must replace x by $x - vt$ to obtain the amplitude at time t:

$$D = f(x - vt). \tag{15–14}$$

Similarly, if the wave moves to the left, we must replace x by $x + vt$ so

$$D = f(x + vt). \tag{15–15}$$

Thus, any wave traveling along the x axis must have the form of Eq. 15–14 or 15–15. The function f gives the actual shape of the wave.

EXAMPLE 15–4   It can be shown that any one-dimensional wave satisfies the following equation

$$\frac{\partial^2 D}{\partial t^2} = v^2 \frac{\partial^2 D}{\partial x^2} \tag{15–16}$$

called the **wave equation**, where v is the velocity of the wave. The quantity $\partial^2 D/\partial t^2$ means to take the second derivative of D with respect to t while holding x constant; and $\partial^2 D/\partial x^2$ means take the second derivative of D with respect to x while holding t constant. These are called "partial derivatives" and are used when you have a function of two or more variables. (*a*) Show that the sinusoidal wave, Eq. 15–10c, satisfies the wave equation. (*b*) Show the same for the general wave relation, Eq. 15–14.

SOLUTION   (*a*) We take the derivative of Eq. 15–10c twice with respect to t:

$$\frac{\partial D}{\partial t} = -\omega D_M \cos (kx - \omega t)$$

$$\frac{\partial^2 D}{\partial t^2} = -\omega^2 D_M \sin (kx - \omega t).$$

With respect to x the derivatives are

$$\frac{\partial D}{\partial x} = kD_M \cos (kx - \omega t)$$

$$\frac{\partial^2 D}{\partial x^2} = -k^2 D_M \sin (kx - \omega t).$$

If we now divide the second derivatives we get

$$\frac{\partial^2 D/\partial t^2}{\partial^2 D/\partial x^2} = \frac{-\omega^2 D_M \sin (kx - \omega t)}{-k^2 D_M \sin (kx - \omega t)} = \frac{\omega^2}{k^2}.$$

From Eq. 15–12 we have $\omega^2/k^2 = v^2$, so we see that Eq. 15–10 does satisfy the wave equation (15–16). (b) Let the quantity $(x - vt)$ be represented by z (so $z = x - vt$). Then if $D = f(x - vt)$, we use the chain rule for derivatives and find:

$$\frac{\partial D}{\partial t} = \frac{\partial f}{\partial z}\frac{\partial z}{\partial t} = \frac{\partial f}{\partial z}(-v)$$

since $\partial z/\partial t = -v$; and

$$\frac{\partial^2 D}{\partial t^2} = \frac{\partial}{\partial t}\left(-v\frac{\partial f}{\partial z}\right) = -v\frac{\partial^2 f}{\partial z^2}\left(\frac{\partial z}{\partial t}\right) = v^2\frac{\partial^2 f}{\partial z^2}.$$

Similarly,

$$\frac{\partial D}{\partial x} = \frac{\partial f}{\partial z}\frac{\partial z}{\partial x} = \frac{\partial f}{\partial z}$$

since $\partial z/\partial x = 1$; and

$$\frac{\partial^2 D}{\partial x^2} = \frac{\partial^2 f}{\partial z^2}.$$

Since $\partial^2 D/\partial t^2 = v^2(\partial^2 f/\partial z^2)$ and $\partial^2 D/\partial x^2 = \partial^2 f/\partial z^2$, then $\partial^2 D/\partial t^2 = v^2(\partial^2 D/\partial x^2)$, which is the wave equation.

EXAMPLE 15–5 The left-hand end of a long horizontal stretched string is oscillated transversely in SHM with frequency $f = 250$ Hz and amplitude 2.6 cm. The string is under a tension of 140 N and has a linear density $\mu = 0.12$ kg/m. At $t = 0$, the end of the string is displaced upward 1.6 cm and is rising. Determine (a) the wavelength of waves produced, and (b) the equation for the traveling wave.

SOLUTION (a) The wave velocity is $v = \sqrt{F_T/\mu} = \sqrt{140\text{ N}/0.12\text{ kg/m}} = 34$ m/s. Then $\lambda = v/f = (34\text{ m/s})/(250\text{ Hz}) = 0.14$ m or 14 cm. (b) Let $x = 0$ at the left-hand end of the string. The phase of the wave at $t = 0$ is not zero in general as was assumed in Eqs. 15–9, 10, 11, and 13. The general form for a wave traveling to the right is then

$$D = D_M \sin (kx - \omega t + \phi)$$

where ϕ is the phase angle. In our case, the amplitude $D_M = 2.6$ cm; and at $t = 0$, $x = 0$, we have $D = 1.6$ cm. Thus

$$1.6 = 2.6 \sin \phi,$$

so $\phi = 38° = 0.66$ rad. We also have $\omega = 2\pi f = 1570$ s^{-1} and $k = 2\pi/\lambda = 45$ m^{-1}. Hence

$$D = 0.026 \sin (45x - 1570t + 0.66)$$

where D and x are in meters and t in seconds.

15–5 The Principle of Superposition

When two or more waves pass through the same region of space at the same time, it is found that for many waves *the actual displacement is the vector (or algebraic) sum of the separate displacements*. This is called the **principle of superposition**. It is valid for mechanical waves as long as the displacements are not too large and there is a linear relationship between the displacement and the restoring force of the oscillating medium[†]. If the amplitude of a mechanical wave, for example, is so large that it goes beyond the elastic region of the medium, and Hooke's law is no longer operative, the superposition principle is no longer accurate.[‡] For the most part, we will consider systems for which the superposition principle can be assumed to hold.

One result of the superposition principle is that if two waves pass through the same region of space, they continue to move independently of one another. You may have noticed, for example, that the ripples on the surface of water (two-dimensional waves) that form from two rocks striking the water at different places will pass through each other.

Figure 15–13 shows an example of the superposition principle. In this case there are three waves present, on a stretched string, each of different amplitude and

---

[†] For electromagnetic waves in vacuum, Chapter 33, the superposition principle always holds.

[‡] Intermodulation distortion in high-fidelity equipment is an example of the superposition principle not holding when two frequencies do not combine linearly in the electronics.

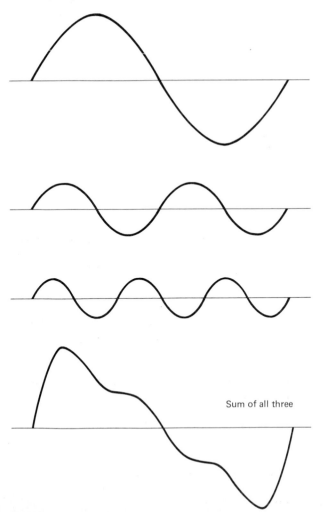

FIGURE 15–13

The superposition principle (one-dimensional wave). Composite wave formed from three sine waves of different amplitudes and frequencies (f_0, $2f_0$, $3f_0$). The amplitude of the composite wave at each point in space, at any time, is the algebraic sum of the amplitudes of the component waves.

Sum of all three

frequency. At any time, such as at the instant shown, the actual amplitude of any position is the algebraic sum of the amplitude of the three waves at that position. The actual wave is not a simple sinusoidal wave and is called a *composite* (or *complex*) *wave*. (Amplitudes are exaggerated in Fig. 15–13.)

It can be shown that any complex wave whatever can be considered as composed of many simple sinusoidal waves of different amplitudes, wavelengths, and frequencies. This is known as *Fourier's theorem*. A complex periodic wave of period T can be represented as a sum of pure sinusoidal terms whose frequencies are integral multiples of $f = 1/T$. If the wave is not periodic, the sum becomes an integral (called a Fourier integral). Although we will not go into the details here, we see the importance of considering sinusoidal waves (and simple harmonic motion): because any other wave shape can be considered a sum of such pure sinusoidal waves.

When the restoring force is not precisely proportional to the displacement for mechanical waves in some medium, the speed of sinusoidal waves depends on the frequency. This is called *dispersion*. The different sinusoidal waves that compose a complex wave will travel with slightly different speeds in such a case. Consequently, a complex wave will change shape as it travels if the medium is "dispersive." A pure sine wave will not change shape under these conditions, however, except by the influence of friction or dissipative forces. If there is no dispersion (or friction), even a complex linear wave doesn't change shape.

 ## 15–6   Reflection of Waves

When a wave strikes an obstacle, or comes to the end of the medium it is traveling in, at least a part of the wave is reflected. You have probably seen water waves reflect off of a rock or the side of a swimming pool. And you may have heard a shout reflected from a distant cliff—which we call an "echo."

A wave pulse traveling down a rope is reflected as shown in Fig. 15–14. You can observe this for yourself and see that the reflected pulse is inverted as in part a if the end of the rope is fixed, and returns right side up if the end is free as in part b.

FIGURE 15–14

Reflection of a wave pulse on a rope when the end of the rope is (a) fixed and (b) free.

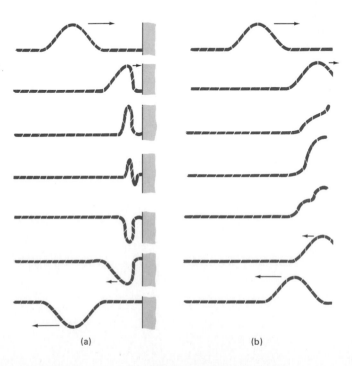

(a)   (b)

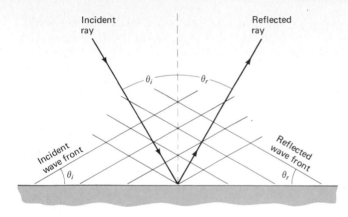

FIGURE 15-15
Law of reflection.

When the end is fixed to a support, as in part a, the pulse reaching that fixed end exerts a force (upward) on the support. The support exerts an equal but opposite force (Newton's third law) down on the cord. This downward force on the cord is what "generates" the inverted pulse which is the (inverted) reflected pulse. The inversion of the reflected pulse in part a is said to be a phase change of $180°$. (It is as if the phase moved up by $\frac{1}{2}\lambda$, or $180°$, from a crest to a trough.) If a wave is traveling down the cord with fixed end, the reflected wave will thus be $180°$ out of phase with the incident wave. In Fig. 15-14b, the free end is constrained by neither a support nor by additional cord. It therefore tends to overshoot—its displacement being momentarily greater than that of the traveling pulse. The overshooting end exerts an upward pull on the cord, and this is what generates the reflected pulse, which is not inverted (no phase change).

For a two- or three-dimensional wave, such as a water wave, we are concerned with **wave fronts**, by which we mean the whole width of a wave crest—that is, the line or surface of all points having the same total phase. A line drawn in the direction of motion, perpendicular to the wave front, is called a *ray*. As shown in Fig. 15-15, the angle that the incoming, or *incident*, wave makes with the reflecting surface is equal to the angle made by the reflected wave. That is, *the angle of reflection equals the angle of incidence*; the "angle of incidence" is defined as the angle the incident ray makes with the perpendicular to the reflecting surface (or the wave front makes with a tangent to the surface) and the "angle of reflection" is the corresponding angle for the reflected wave.

When the wave pulse in Fig. 15-14a reaches the wall, not all of the energy is reflected. Some of it is absorbed by the wall. Part of the absorbed energy is transformed into thermal energy, and part continues to propagate through the material of the wall. This is perhaps more clearly illustrated by considering a pulse that travels down a rope that consists of a light section and a heavy section as shown in Fig. 15-16. When the wave reaches the boundary between the two sections, part of the pulse is reflected and part is transmitted as shown. The heavier the second section, the less is transmitted; and when the second section is a wall or rigid support, very little is transmitted. If the second section has a greater linear density than the first, the reflected pulse undergoes a phase change of $180°$; if the second section has a lower density, the reflected pulse does not change phase.

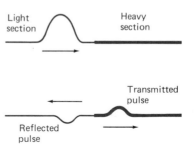

FIGURE 15-16
When a wave pulse reaches a discontinuity, part is reflected and part transmitted.

 ## 15-7 Refraction

When any wave strikes a boundary, some of the energy is reflected and some is transmitted or absorbed. When a two- or three-dimensional wave traveling in one medium crosses a boundary into a medium where its velocity is different, the transmitted wave may move in a different direction than the incident wave, as shown

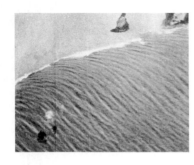

FIGURE 15–17

Refraction of waves passing a boundary.

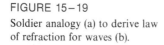

FIGURE 15–18

Water waves refracting. (Courtesy of U.S. Air Force, Cambridge Research Laboratories.)

in Fig. 15–17. This phenomenon is known as **refraction**. One example is a water wave; the velocity decreases in shallow water and the waves refract, Fig. 15–18. In Fig. 15–17, the velocity of the wave in medium 2 is less than in medium 1. In this case, the direction of the wave bends so it moves more nearly perpendicular to the boundary. That is, the angle of refraction, θ_r, is less than the angle of incidence, θ_i. To see why this is so, and to help us get a quantitative relation between θ_r and θ_i, let us think of each wave front as a row of soldiers; the soldiers are marching from firm ground (medium 1) into mud (medium 2) and hence are slowed down. The soldiers that reach the mud first are slowed down first and the row bends as shown in Fig. 15–19a. Let us consider the wave front (or row of soldiers) labeled A in Fig. 15–19b. In the same time t that A_1 moves a distance $l_1 = v_1 t$, we see that A_2 moves a distance $l_2 = v_2 t$. The two triangles shown have the side labeled a in common. Thus

$$\sin \theta_1 = \frac{l_1}{a} = \frac{v_1 t}{a}$$

and

$$\sin \theta_2 = \frac{l_2}{a} = \frac{v_2 t}{a}.$$

FIGURE 15–19

Soldier analogy (a) to derive law of refraction for waves (b).

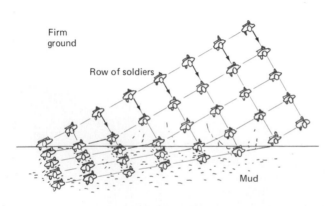

(a)

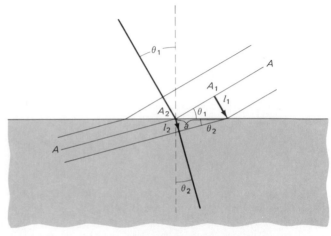

(b)

Dividing these two equations we find

$$\frac{\sin \theta_2}{\sin \theta_1} = \frac{v_2}{v_1}. \qquad (15-17)$$

Since θ_1 is the angle of incidence (θ_i), and θ_2 is the angle of refraction (θ_r), Eq. 15–17 gives the quantitative relation between the two. Of course, if the wave were going in the opposite direction, the argument would not be changed; only θ_1 and θ_2 would change roles: θ_1 would be the angle of refraction and θ_2 the angle of incidence. Clearly then, if the wave travels into a medium where it can move faster, it will bend in the opposite way, $\theta_r > \theta_i$. We see from Eq. 15–17 that if the velocity increases, the angle increases and vice versa.

Earthquake waves refract within the earth as they travel through rock of different densities (and therefore the velocity is different) just as water waves do. Light waves refract as well, and when we discuss light we shall find Eq. 15–17 very useful.

EXAMPLE 15–6 An earthquake P wave passes across a boundary in rock where its velocity increases from 6.5 km/s to 8.0 km/s. If it strikes this boundary at 30°, what is the angle of refraction?

SOLUTION Since $\sin 30° = 0.50$, Eq. 15–17 yields

$$\sin \theta_2 = \frac{(8.0 \text{ m/s})}{(6.5 \text{ m/s})} (0.50) = 0.62,$$

so $\theta_2 = 38°$.

15–8 Interference

Interference refers to what happens when two waves pass through the same region of space at the same time; it is an example of the superposition principle. Consider for example the two wave pulses on a string traveling toward each other as shown in Fig. 15–20. In part a the two pulses have the same amplitude but one is a crest and the other a trough, and in part b they are both crests. In both cases, the waves meet and pass right on by each other. However, when they overlap, the resultant

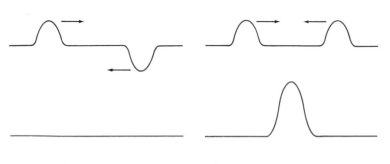

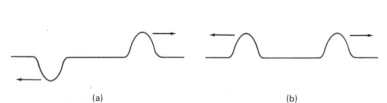

(a)   (b)

FIGURE 15–20

Two waves pulses pass each other. Where they overlap, interference occurs: (a) destructive; (b) constructive.

(a)

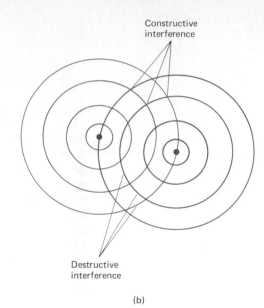

Constructive
interference

Destructive
interference

(b)

FIGURE 15–21
Interference of water waves.

displacement is the algebraic sum of their separate displacements as we know from the principle of superposition. In part a, the two wave amplitudes are opposite one another as they pass by and the result is called **destructive interference**. In part b, the resultant displacement is greater than that of either pulse and the result is called **constructive interference**.

When two rocks are thrown into a pond simultaneously, the two sets of circular waves interfere with one another as shown in Fig. 15–21. In some areas of overlap, crests of one wave meet crests of the other (and troughs meet troughs); this is constructive interference and the water oscillates up and down with greater amplitude than either wave separately. In other areas, destructive interference occurs where the water actually does not move at all—this is where crests of one wave meet troughs of the other, and vice versa. In the first case, of constructive interference, the two waves are *in phase*, whereas in destructive interference the two waves are *out of phase* by one-half wavelength or 180°. Of course the relative phases of the two waves in most areas is intermediate between these two extremes, and this results in *partially destructive* interference. All three of these situations are shown in Fig. 15–22 where the amplitudes are plotted versus time at a given point in space. We will deal with interference in more detail when we discuss sound and light.

FIGURE 15–22
Two waves interfere: (a) constructively, (b) destructively, (c) partially destructively.

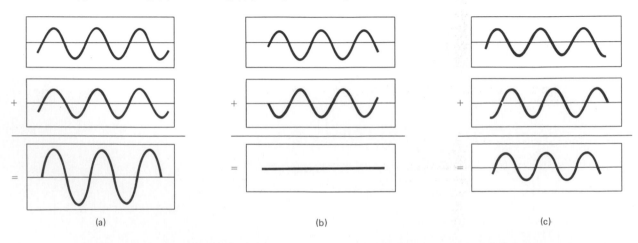

(a)   (b)   (c)

15–9   Diffraction

Waves exhibit another important characteristic known as **diffraction**, which we deal with only qualitatively now (a mathematical treatment will be given when we discuss light). Diffraction refers to the fact that waves spread as they travel, and when they encounter an obstacle they bend around it somewhat and pass into the region behind as shown in Fig. 15–23 for water waves.

The amount of diffraction depends on the wavelength of the wave and on the size of the obstacle. This is shown in Fig. 15–24. If the wavelength is much larger than the object, such as the grass blades of Fig. 15–24a, the wave bends around them almost as if they weren't even there. For larger size objects, parts b and c, there is more of a "shadow" region behind the obstacle. But notice in part d, where the obstacle is the same as in part c but the wavelength is longer, that there is more diffraction into the shadow region. As a rule of thumb, only if the wavelength is less than the size of the object, will there be a significant shadow region.

It is worth noting that this rule applies to *reflection* from the obstacle as well. Very little of the wave is reflected in Fig. 15–24a and b. Only if the wavelength is less than the size of the obstacle, as in part c, will there be significant reflection (but it isn't shown on the diagram).

The fact that waves can bend around obstacles, and thus can carry energy to areas behind obstacles, is in clear distinction to energy carried by material particles. A clear example is the following: if you are standing behind a wall, you can't be hit by a baseball thrown from the other side, but you can hear a shout or other sound because the sound waves diffract around the edges.

Both interference and diffraction occur only for energy carried by waves and not for energy carried by material particles. This distinction was important for understanding the nature of light, and of matter itself, as we shall see in later chapters.

FIGURE 15–23

Wave diffraction. (From R. L. Wiegel, *Oceanographical Engineering*, Prentice-Hall, Inc., Englewood Cliffs, N.J., 1964.)

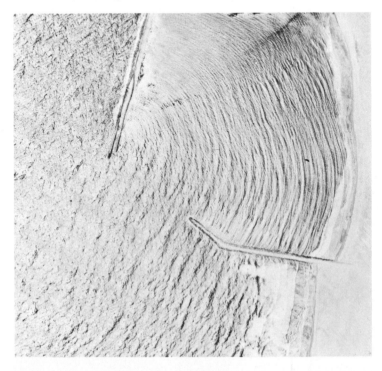

(a) Water waves passing blades of grass

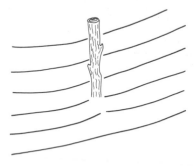

(b) Stick in water

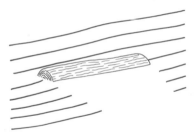

(c) Short-wavelength waves passing log

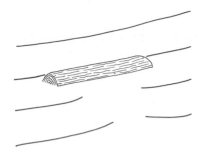

(d) Long-wavelength waves passing log

FIGURE 15–24

Water waves passing objects of various sizes. Note that the larger the wavelength compared to the size of the object, the more diffraction there is into the "shadow region."

15–10 Standing Waves; Resonance

If you shake one end of a rope (or slinky) and the other end is kept fixed, a continuous wave will travel down to the fixed end and be reflected back. As you continue to vibrate the rope, there will be waves traveling in both directions, and the wave traveling down the rope will interfere with the reflected wave coming back. Usually there will be quite a jumble. But if you vibrate the rope at just the right frequency, these two waves will interfere in such a way that a large amplitude **standing wave** will be produced, Fig. 15–25. It is called a "standing wave" because it doesn't appear to be moving. The points of destructive interference, called **nodes**, and of constructive interference, called **antinodes**, remain in fixed positions. Standing waves occur at more than one frequency. The lowest frequency of vibration that produces a standing wave gives rise to the pattern shown in Fig. 15–25a. The standing waves shown in parts b and c are produced at precisely twice and three times the lowest frequency, assuming the tension in the rope is the same. The rope can also vibrate with four loops at four times the lowest frequency, and so on.

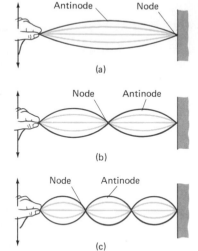

The frequencies at which standing waves are produced are the **natural frequencies** or **resonant frequencies** of the rope and the different standing wave patterns shown in Fig. 15–25 are different "resonant modes of vibration." For although a standing wave is the result of the interference of two waves traveling in opposite directions, it is also an example of a vibrating object at resonance. When a standing wave exists on a rope, the rope is vibrating in place; and at the frequencies at which resonance occurs, little effort is required to achieve a large amplitude. Standing waves then represent the same phenomenon as the resonance of a vibrating spring or pendulum which we discussed in the previous chapter. The only difference is that a spring or pendulum has only one resonant frequency, whereas the rope has an infinite number of resonant frequencies, each of which is a whole number multiple of the lowest resonant frequency.

FIGURE 15–25

Standing waves corresponding to three resonant frequencies.

Now let us consider a string stretched between two supports that is plucked like a guitar or violin string, Fig. 15–26a. Waves of a great variety of frequencies[†] will travel in both directions along the string, will be reflected at the ends and travel back in the opposite direction. Most of these waves interfere in a random way with each other and quickly die away. However, those waves that correspond to the resonant frequencies of the string will persist. The ends of the string, since they are fixed, will be nodes. There may be other nodes as well. Some of the possible resonant modes of vibration (standing waves) are shown in Fig. 15–26b. Generally, the motion will be a combination of these different resonant modes; but only those frequencies that correspond to a resonant frequency will be present.

To determine the resonant frequencies, we first note that the wavelengths of the standing waves bear a simple relationship to the length L of the string. The lowest frequency, called the **fundamental frequency**, or **first harmonic**,[‡] corresponds to one antinode (or loop); and as can be seen in Fig. 15–26b, the whole length corresponds to one-half wavelength. Thus $L = \frac{1}{2}\lambda_1$, where λ_1 stands for the wavelength of the fundamental. The next mode has two loops and is called the **second harmonic**; the length of the string L corresponds to one complete wavelength: $L = \lambda_2$. For the third and fourth harmonics, $L = \frac{3}{2}\lambda_3$ and $L = 2\lambda_4$, respectively, and so on. In general, we can write

$$L = \frac{n\lambda_n}{2}, \qquad \text{where} \quad n = 1, 2, 3, \ldots.$$

The integer n labels the number of the harmonic; $n = 1$ for the fundamental, $n = 2$ corresponds to the second harmonic, and so on. The second harmonic is also called

---

† Fourier analysis (see Section 15–5) shows that the triangular-shaped pulse of Fig. 15–26a can be considered to be the sum of sinusoidal waves of different frequencies.

‡ The term "harmonic" comes from music, because such integral multiples of frequencies "harmonize."

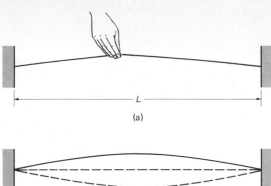

(a)

$L = \frac{1}{2}\lambda_1$

(Fundamental or first harmonic)

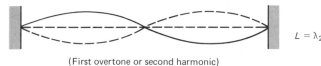

$L = \lambda_2$

(First overtone or second harmonic)

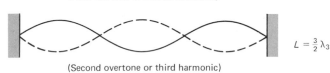

$L = \frac{3}{2}\lambda_3$

(Second overtone or third harmonic)

(b)

FIGURE 15–26

(a) A string is plucked. (b) Only standing waves
corresponding to resonant frequencies persist for long.

the first *overtone*, the third harmonic, the second overtone, and so on. We solve for λ_n
and find:

$$\lambda_n = \frac{2L}{n}, \qquad n = 1, 2, 3, \ldots . \qquad (15\text{--}18)$$

In order to find the frequency f of each vibration we use Eq. 15–1, $f = v/\lambda$.

Because a standing wave is equivalent to two traveling waves moving in
opposite directions, the concept of velocity still makes sense and is given by Eq.
15–2 in terms of the tension F_T in the string and its mass per unit length μ:
$v = \sqrt{F_T/\mu}$.

EXAMPLE 15–7 A piano string is 1.10 m long and has a mass of 9.0 g. (*a*)
How much tension must the string be under if it is to vibrate at a fundamental
frequency of 131 Hz? (*b*) What are the frequencies of the first four harmonics?

SOLUTION (*a*) The wavelength of the fundamental is $\lambda = 2L = 2.20$ m (Eq.
15–18). The velocity is then $v = \lambda f = (2.20 \text{ m})(131 \text{ s}^{-1}) = 288$ m/s. Then, from
Eq. 15–2 we have

$$F_T = \mu v^2$$

$$= \left(\frac{0.0090 \text{ kg}}{1.10 \text{ m}}\right)(288 \text{ m/s})^2 = 679 \text{ N}.$$

(*b*) The frequencies of the second, third, and fourth harmonics are 2, 3, and
4 times the fundamental frequency: 262, 393, and 524 Hz.

In Section 15–4, we saw how to write an equation for the displacement D of
a linear traveling wave as a function of position x and time t. We can do the same
for a standing wave on a string. We already discussed the fact that a standing
wave can be considered to consist of two traveling waves that move in opposite
directions; these can be written (see Eqs. 15–10c and 15–13c).

$$D_1 = D_M \sin (kx - \omega t)$$

$$D_2 = D_M \sin (kx + \omega t)$$

since, assuming no damping, the amplitudes are equal as are the frequencies and wavelengths. The sum of these two traveling waves produces a standing wave which can be written mathematically as

$$D = D_1 + D_2 = D_M[\sin (kx - \omega t) + \sin (kx + \omega t)].$$

From the trigonometric identity $\sin \theta_1 + \sin \theta_2 = 2 \sin \frac{1}{2}(\theta_1 + \theta_2) \cos \frac{1}{2}(\theta_2 - \theta_1)$, we can rewrite this as

$$D = 2D_M \sin kx \cos \omega t. \tag{15-19}$$

If we let $x = 0$ at the left-hand end of the string, then the right-hand end is $x = L$ where L is the length of the string. Since the string is fixed at its two ends (Fig. 15–26), D must be zero at $x = 0$ and at $x = L$. Equation 15–19 already satisfies the first condition ($D = 0$ at $x = 0$) and satisfies the second condition if

$$kL = \pi, 2\pi, 3\pi, \ldots, n\pi, \ldots$$

where $n =$ integer, or, since $k = 2\pi/\lambda$,

$$\lambda = \frac{2L}{n}. \qquad (n = \text{integer})$$

This is just Eq. 15–18.

Equation 15–19, with the condition $\lambda = 2L/n$, is the mathematical representation of a standing wave. We see that a particle at any position x vibrates in simple harmonic motion (because of the factor $\cos \omega t$). All particles of the string vibrate with the same frequency $f = \omega/2\pi$, but the amplitude depends on x and equals $2D_M \sin kx$. (Compare this to a traveling wave for which all particles vibrate with the same amplitude.) The amplitude has a maximum, equal to $2D_M$, when $kx = \pi/2, 3\pi/2, 5\pi/2$, and so on—that is, at

$$x = \frac{\lambda}{4}, \frac{3\lambda}{4}, \frac{5\lambda}{4}, \cdots$$

These are, of course, the positions of the antinodes (see Fig. 15–26).

A standing wave does appear to be standing still (and a traveling wave appears to move). The term "standing" wave is also meaningful from the point of view of energy. Since the string is at rest at the nodes, no energy flows past these points. Hence the energy "stands" in place in the string and is not transmitted down the string.

Standing waves are produced not only on strings, but on any object that is set into vibration. Even when a rock or a piece of wood is struck with hammer, standing waves are set up that correspond to the natural resonant frequencies of that object. In general, the resonant frequencies depend on the dimensions of the object, just as for a string they depend on its length. For example a small object does not have as low resonant frequencies as a large object. All musical instruments depend on standing waves to produce their musical sounds, from string instruments to wind instruments (in which a column of air vibrates as a standing wave) to drums and other percussion instruments. We will look at this in more detail in the next chapter.

 Summary _____

Vibrating objects act as sources of waves that travel outward from the source. Waves on water and on a string are examples. The wave may be a *pulse* (a single crest) or it may be continuous (many crests and troughs). The *wavelength* of a continuous wave is the distance between two successive crests (or any two identical points on the wave shape). The *frequency* is the number of full wavelengths (or crests) that pass a given point per unit time. The *wave velocity* (how fast a crest, or any other part of the wave, moves) is equal to the product of wavelength and frequency, $v = \lambda f$. The *amplitude* of a wave is the maximum height of a crest, or depth of a trough, relative to the normal (or equilibrium) level.

In a *transverse wave*, the oscillations are perpendicular to the direction in which the wave travels; an example is a wave on a string. In a *longitudinal wave*, the oscillations are along (parallel to) the line of travel; sound is an example of a longitudinal wave. The velocity of both longitudinal and transverse waves in matter is proportional to the square root of an elastic force factor divided by an inertia factor (or density).

Waves carry energy from place to place without matter being carried. The energy transported by a wave, the power (energy transported per unit time), and the *intensity* of a wave (energy transported across unit area per unit time) are all proportional to the square of the amplitude of the wave. For a wave traveling outward in three dimensions from a point source, the intensity (ignoring damping) decreases with the square of the distance from the source, and the amplitude decreases linearly with distance from the source.

A one-dimensional transverse wave traveling to the right along the x axis can be represented by a formula for its amplitude as a function of position and time as $D = D_\mathrm{M} \sin [(2\pi/\lambda)(x - vt)] = D_\mathrm{M} \sin (kx - \omega t)$ where $k = 2\pi/\lambda$ and $\omega = 2\pi f$. If it is traveling in the opposite direction, $D = D_\mathrm{M} \sin (kx + \omega t)$.

When two or more waves pass through the same region of space at the same time, the displacement at any given point will be the vector sum of the displacements of the separate waves. This is the *principle of superposition*. It is valid for mechanical waves only if the amplitudes are not so great that the restoring force of the medium is not proportional to displacement.

Waves reflect off objects in their path. When the *wave front* (of a two- or three-dimensional wave) strikes an object, the angle of reflection is equal to the angle of incidence. When a wave strikes a boundary between two materials in which it can travel, part of the wave is reflected and part is transmitted. The transmitted wave front undergoes *refraction* or bending. The angles that the incident and refracted wave fronts make with the boundary between the two materials are related to the wave velocities in the two media by $\sin \theta_1/\sin \theta_2 = v_1/v_2$. When two waves pass through the same region of space at the same time, they *interfere*. From the superposition principle, the resultant displacement at any point and time is the sum of their separate displacements; this can result in *constructive interference*, *destructive interference*, or something in between depending on the amplitudes and relative phases of the waves. Waves also undergo *diffraction*, meaning they can bend around an obstacle, so there is some wave motion behind the obstacle. The smaller the wavelength relative to the size of the object, the less the diffraction.

Waves traveling on a string (or other medium) of fixed length interfere with waves that have reflected off the end and are traveling back in the opposite direction. At certain frequencies, *standing waves* can be produced in which the waves seem to be standing still rather than traveling. The string (or other medium) is vibrating as a whole. This is a resonance phenomenon and the frequencies at which standing waves occur are called *resonant frequencies*. The points of destructive interference (no vibration) are called *nodes*; points of constructive interference (maximum amplitude of vibration) are called *antinodes*.

 Questions _____

1. Is the frequency of a simple periodic wave equal to the frequency of its source? Why or why not?

2. Explain the difference between the speed of a transverse wave traveling down a rope and the speed of a tiny piece of the rope.

3. Why do the strings used for the lowest-frequency notes on a piano normally have wire wrapped around them?

4. What kind of waves do you think will travel down a horizontal metal rod if you strike its end (a) vertically from above and (b) horizontally parallel to its length?

5. Since the density of air decreases with an increase in temperature, but the bulk modulus B is nearly independent of temperature, how would you expect the speed of sound waves in air to vary with temperature?

6. Two solid rods have the same bulk modulus but one is twice as dense as the other. In which rod will the speed of longitudinal waves be greater, and by what factor?

7. Give examples, other than those already mentioned, of one-, two-, and three-dimensional waves.

8. The speed of sound in most solids is somewhat greater than in air, yet the density of solids is much greater (10^3–10^4 times). Explain.

9. Give two reasons why circular water waves decrease in amplitude as they travel away from the source.

10. Two linear waves have the same amplitude and otherwise are identical, except one has half the wavelength of the other. Which transmits more energy? How much more energy does it transfer?

11. The intensity of a sound in real-life situations does not decrease precisely with the square of the distance from the source as we might expect from Eq. 15–8. Why not?

12. Will any function of $(x - vt)$—see Eq. 15–14—represent a wave motion? Why or why not? If not, give an example.

13. When a sinusoidal wave crosses the boundary between two sections of rope as in Fig. 15–16, the frequency does not

change (although the wavelength and velocity do change). Explain why.

14. If a sinusoidal wave on a two-section string (Fig. 15–16) is inverted upon reflection, does the transmitted wave have a longer or shorter wavelength?

15. Is energy always conserved when two waves interfere? Explain.

16. If we knew that energy was being transmitted from one place to another, how might we determine whether the energy was being carried by particles (material bodies) or by waves?

17. AM radio signals can usually be heard behind a hill, but FM often cannot. That is, AM signals bend more than FM. Explain. (Radio signals, as we shall see, are carried by electromagnetic waves whose wavelength for AM is typically 200 to 600 m and for FM about 3 m.)

18. If a string is vibrating in three segments, are there any places one can touch it with a knife blade without disturbing the motion?

19. When a standing wave exists on a string, the vibrations of incident and reflected waves cancel at the nodes. Does this mean that energy was destroyed?

20. Why can you make water slosh back and forth in a pan only if you shake the pan at a certain frequency?

21. Can the amplitude of the standing waves in Fig. 15–25 be greater than the amplitude of the vibrations that cause it (up and down motion of the hand or mechanical vibrator)?

22. When a cord is vibrated as in Fig. 15–25 by hand or by a mechanical vibrator, the "nodes" are not quite true nodes (at rest). Explain. (Hint: consider damping and energy flow from hand or vibrator.)

 Problems

SECTION 15–1

1. (I) A fisherman notices that wave crests pass the bow of his anchored boat every 5.0 s. He measures the distance between two crests to be 15 m. How fast are the waves traveling?

2. (I) AM radio signals have frequencies between 550 kHz and 1600 kHz (kilohertz) and travel with a speed of 3.0×10^8 m/s. What are the wavelengths of these signals? On FM the frequencies range from 88 MHz to 108 MHz (megahertz) and travel at the same speed; what are their wavelengths?

3. (I) A rope of mass 0.85 kg is stretched between two supports 30 m apart. If the tension in the rope is 1950 N, how long will it take a pulse to travel from one support to the other?

4. (II) A 0.40-kg rope is stretched between two supports, 4.8 m apart. When one support is struck by a hammer, a transverse wave travels down the rope and reaches the other support in 0.85 s. What is the tension in the rope?

5. (II) The ripples in a certain groove 12.5 cm from the center of a 33-rpm phonograph record have a wavelength of 2.45 mm. What will be the frequency of the sound emitted?

6. (II) The wave on a string shown in Fig. 15–27 is moving to the right with a speed of 1.20 m/s. (a) Draw the shape of the string 1.20 s later and indicate which parts of the string are moving up and which down at that instant. (b) What is the vertical speed of point A on the string at the instant shown in the figure?

SECTION 15–2

7. (I) Calculate the speed of longitudinal waves in (a) water, (b) granite.

8. (I) Determine the wavelength of a 7000-Hz-sound wave traveling along an iron rod.

9. (II) A sailor strikes the side of his ship just below the surface of the sea. He hears the echo of the wave reflected from the ocean floor directly below 2.1 s later. How deep is the ocean at this point?

10. (II) S and P waves from an earthquake travel at different speeds and this fact helps in the determination of the earthquake "focus" (where the disturbance took place). (a) Assuming typical speeds of 9.0 km/s and 5.0 km/s for P and S waves, respectively, how far away did the earthquake occur if a particular seismic station detects the arrival of these two types of waves exactly 2.0 min apart? (b) Is one seismic station sufficient to determine the position of the focus? Explain.

11. (II) A uniform cord of length L and mass m is hung vertically from a support. (a) Show that the speed of transverse waves in this cord is $\sqrt{gh}$ where h is the height above the lower end. (b) How long does it take for a pulse to travel upward from one end to the other?

SECTION 15–3

12. (I) Two earthquake waves have the same frequency as they travel through the same portion of the earth, but one is

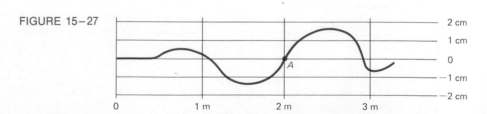

FIGURE 15–27

carrying twice the energy. What is the ratio of the amplitudes of the two waves?

13. (I) Compare (a) the intensities and (b) the amplitudes of an earthquake P wave as it passes two points 10 km and 20 km from the source.

14. (I) The intensity of a particular earthquake wave is measured to be 1.4×10^6 J/m²·s at a distance of 100 km from the source. (a) What was the intensity when it passed a point only 2.0 km from the source? (b) What was the total power passing through an area of 5.0 m² at a distance of 2.0 km?

15. (II) Show that if damping is ignored the amplitude D_M of circular water waves decreases as the square root of the distance r from the source: $D_M \propto 1/\sqrt{r}$.

16. (II) (a) Show that the intensity of a wave is equal to the energy density (energy per unit volume) in the wave times the wave speed. (b) What is the energy density 10 m from a 100-W light bulb? Light travels 3.0×10^8 m/s.

17. (II) For a spherical wave traveling uniformly away from a point source, show that the displacement can be represented by

$$D = \left(\frac{A}{r}\right) \sin{(kr - \omega t)}$$

where r is the radial distance from the source and A is a constant. (See also Section 15–4.)

SECTION 15–4

18. (I) Suppose at $t = 0$, a wave shape is represented by $D = D_M \sin{(2\pi x/\lambda + \phi)}$; that is, it differs from Eq. 15–9 by a constant phase factor ϕ. What then will be the equation for a wave traveling to the left along the x axis as a function of x and t?

19. (I) Show that the wave forms of Eqs. 15–13 and 15–15 satisfy the wave equation, Eq. 15–16.

20. (II) A transverse traveling wave on a cord is represented by the relation $D = 0.42 \sin{(7.6x + 94t)}$ where D and x are in meters and t in seconds. For this wave determine (a) the wavelength, (b) frequency, (c) velocity (magnitude and direction), (d) amplitude, and (e) maximum and minimum speeds of particles of the cord.

21. (II) Consider a point on the string of Example 15–5 that is 1.00 m from the left-hand end. Determine (a) the maximum velocity of this point, and (b) its maximum acceleration. (c) What is its velocity and acceleration at $t = 2.0$ s?

22. (II) A 262-Hz longitudinal wave in air has a speed of 345 m/s. (a) What is the wavelength? (b) How much time is required for the phase to change by 90° at a given point in space? (c) At a particular instant, what is the phase difference (in degrees) between two points 6.4 cm apart?

23. (II) Write down the equation for the wave in problem 22 if its amplitude is 0.020 cm and at $t = 0$, $x = 0$, and $D = -0.020$ cm.

24. (II) Show, for a sinusoidal transverse wave traveling on a string, that the slope of the string at any point x is equal to the ratio of the transverse speed of the particle to the speed of the wave at that point.

SECTION 15–5

25. (II) Let two linear waves be represented by $D_1 = f_1(x, t)$ and $D_2 = f_2(x, t)$. If both these waves satisfy the wave equation (Eq. 15–16), show that any combination $D = C_1 D_1 + C_2 D_2$ does as well. (C_1 and C_2 are constants.)

SECTION 15–6

26. (II) Consider a sine wave traveling down the stretched two-part cord of Fig. 15–16. Determine a formula (a) for the ratio of the speeds of the wave in the two sections, v_2/v_1, and (b) for the ratio of the wavelengths in the two sections. (The frequency is the same in both sections. Why?) (c) Is the wavelength greater in the heavier cord or the lighter?

27. (III) A cord stretched to a tension F_T, consists of two sections (as in Fig. 15–16) whose linear densities are μ_1 and μ_2. Take $x = 0$ to be the point (a knot) where they are joined, with μ_1 referring to that section of cord to the left and μ_2 that to the right. A sinusoidal wave, $D = A \sin[k_1(x - v_1 t)]$, starts at the left end of the cord. When it reaches the knot, part of it is reflected and part is transmitted. Let the equation of the reflected wave be $D = A_R \sin[k_1(x + v_1 t)]$ and that for the transmitted wave be $D = A_T \sin[k_2(x - v_2 t)]$. Since the frequency must be the same in both sections, we have $\omega_1 = \omega_2$ or $k_1 v_1 = k_2 v_2$. (a) Using the fact that the rope is continuous—so that a point an infinitesimal distance to the left of the knot has the same displacement at any moment (due to incident plus reflected waves) as a point just to the right of the knot (due to the transmitted wave)—show that $A = A_T + A_R$. (b) Assuming that the slope (dD/dx) of the string just to the left of the knot is the same as the slope just to the right of the knot, show that the amplitude of the reflected wave is given by

$$A_R = \left(\frac{v_1 - v_2}{v_1 + v_2}\right) A = \left(\frac{k_2 - k_1}{k_2 + k_1}\right) A.$$

(c) What is A_T in terms of A?

SECTION 15–7

28. (I) An earthquake P wave traveling 14.5 km/s strikes a boundary within the earth between two kinds of material. If it approaches the boundary at an incident angle of 42° and the angle of refraction is 26°, what is the speed in the second medium?

29. (II) A longitudinal earthquake wave strikes a boundary between two types of rock at a 10° angle. As it crosses the boundary the specific gravity of the rock changes from 3.6 to 4.9. Assuming the elastic modulus is the same for both types of rock, determine the angle of refraction.

30. (II) It is found for any type of wave, say an earthquake wave, that if it reaches a boundary beyond which its speed is increased, there is a maximum incident angle if there is to be a transmitted refracted wave. This maximum incident angle θ_{iM} corresponds to an angle of refraction equal to 90°. If $\theta_i > \theta_{iM}$, all the wave is reflected at the boundary and none is refracted (because this would correspond to $\sin{\theta_r} > 1$, where θ_r is the angle of refraction, which is impossible); this is referred to as total internal reflection. (a) Find a formula for θ_{iM} using Eq. 15–17. (b) At what angles of incidence will there be only reflection and no transmission for an earthquake P wave traveling 6.5 km/s when it reaches a different kind of rock where its speed is 8.2 km/s?

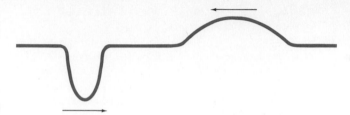

FIGURE 15–28

31. (I) The two pulses shown in Fig. 15–28 are moving toward each other. (a) Sketch the shape of the string at the moment they directly overlap (b) Sketch the shape of the string a few moments later. (c) In Fig. 15–20a, at the moment the pulses pass each other, the string is straight. What has happened to the energy at this moment?

32. (II) Suppose two linear waves of equal amplitude and frequency have a phase difference ϕ as they travel in the same medium. They can be represented by

$$D_1 = D_M \sin (kx - \omega t)$$

$$D_2 = D_M \sin (kx - \omega t + \phi).$$

(a) Use the trigonometric identity $\sin \theta_1 + \sin \theta_2 = 2 \sin \frac{1}{2}(\theta_1 + \theta_2)\cos \frac{1}{2}(\theta_1 - \theta_2)$ to show that the resultant wave is given by

$$D = \left[2 D_M \cos \frac{\phi}{2} \right] \sin \left(kx - \omega t + \frac{\phi}{2} \right).$$

(b) What is the amplitude of this resultant wave? Is the wave purely sinusoidal, or not? (c) Show that constructive interference occurs if $\phi = 0$, 2π, 4π, and so on, and destructive interference occurs if $\phi = \pi$, 3π, 5π, etc. (d) Describe the resultant wave, by equation and in words, if $\phi = \pi/2$.

33. (I) A violin string vibrates at 196 Hz when unfingered. At what frequency will it vibrate if it is fingered one-fourth of the way down from the end?

34. (I) A particular string resonates in four loops at a frequency of 220 Hz. Name at least three other frequencies at which it will resonate.

35. (I) The velocity of waves on a string is 480 m/s. If the frequency of standing waves is 86.0 Hz, how far apart are the nodes?

36. (II) If two successive overtones of a vibrating string are 320 Hz and 360 Hz, what is the frequency of the fundamental?

37. (II) Show that the frequency of standing waves on a string of length L and linear density μ, which is stretched to a tension F_T, is given by

$$f = \frac{n}{2L} \sqrt{\frac{F_T}{\mu}}$$

where n is an integer.

38. (II) One end of a horizontal string of linear density 4.2×10^{-4} kg/m is attached to a small amplitude mechanical 60 Hz-vibrator. The string passes over a pulley, a distance $L = 2.40$ m away, and weights are hung from this end. What mass must be hung from this end of the string to produce (a) one loop (b) two loops, (c) five loops of a standing wave? Assume the string at the vibrator is a node, which is nearly true. Why can the amplitude of the standing wave be much greater than the vibrator amplitude?

39. (II) The displacement of a standing wave is given by $D = 5.6 \sin (0.66x) \cos (53t)$, where x and D are in centimeters and t is in seconds. (a) What is the distance (cm) between nodes? (b) Give the amplitude, frequency, and speed of each of the component waves. (c) Find the speed of a particle of the string at $x = 2.10$ cm when $t = 1.25$ s.

40. (II) The displacement of a transverse wave traveling on a string is represented by $D = 4.2 \sin (0.71x - 47t + 2.1)$. Find the equation that represents a wave which, when traveling in the opposite direction, will produce a standing wave when added to this one.

41. (II) When you slosh the water back and forth in a tub at just the right frequency, the water alternately rises and falls at each end. Suppose the frequency to produce such a standing wave in a 50-cm wide tub is 0.85 Hz. What is the speed of the water wave?

42. (II) A guitar string is 90 cm long and has a mass of 3.6 g. From the bridge to the support post ($= L$) is 60 cm and the string is under a tension of 520 N. What are the frequencies of the fundamental and first two overtones?

43. (II) A particular violin string plays at a frequency of 294 Hz. If the tension is increased 10 percent, what will the frequency be now?

Sound

16

Sound is associated with our sense of hearing and, therefore, with the physiology of our ears and the psychology of our brain which interprets the sensations that reach our ears. The term *sound* also refers to the physical sensation that stimulates our ears: namely, longitudinal waves.

We can distinguish three aspects of any sound. First, there must be a *source* for a sound; and as with any wave, the source of a sound wave is a vibrating object. Second, the energy is transferred from the source in the form of longitudinal sound *waves*. And third, the sound is *detected* by an ear or an instrument. We will discuss sources and detectors of sound later in this chapter, but now we look at some aspects of sound waves themselves.

 ## 16-1 Characteristics of Sound

We already saw in Chapter 15, Fig. 15–7, how a vibrating drumhead produces a sound wave in air. Indeed, we usually think of sound waves traveling in the air, for normally it is air that is in contact with our eardrums and the vibrations of the air force our eardrums to vibrate. But sound waves can also travel in other materials. Two stones struck together under water can be heard by a swimmer beneath the surface, for the vibrations are carried to the ear by the water. When you put your ear flat against the ground, you can hear an approaching train or truck. In this case the ground does not actually touch your eardrum but we call the longitudinal wave transmitted by the ground a sound wave just the same, for its vibrations cause the outer ear and the air within it to vibrate. Indeed, longitudinal waves traveling in any material medium are often referred to as sound waves. Clearly sound cannot travel in the absence of matter. For example, a bell ringing inside an evacuated jar cannot be heard.

The *speed of sound* is different in different materials. In air at 0°C and 1 atm, sound travels at a speed of 331.3 m/s. As we saw in Chapter 15, Eqs. 15–3 and 15–4, the speed depends on the elastic (or bulk) modulus and the density of the material. In air and other fluids,

$$v = \sqrt{\frac{B}{\rho}}$$

where B is the bulk modulus and ρ the density of the fluid. In helium, whose density is much less than that of air but whose bulk modulus is not greatly different, the speed is about three times as large. In liquids and solids, which are much less compressible and, therefore, have much greater elastic moduli, the speed is larger still. The speed of sound in various materials is given in Table 16–1. The values depend somewhat on temperature, but this is significant mainly for gases. For example, in air, the speed increases approximately 0.60 m/s for each Celsius degree increase in temperature:

$$v \approx (331 + 0.60T) \text{ m/s}$$

where T is the temperature in °C. At 20°C for instance, $v \approx [331 + (0.60) \cdot (20)]$ m/s = 343 m/s.

Two aspects of any sound are immediately evident to a human listener. These are "loudness" and "pitch," and each refers to a sensation in the consciousness of the listener. But to each of these subjective sensations there corresponds a physically measurable quantity. **Loudness** is related to the intensity of the sound wave and we'll discuss it in Section 16–3.

The **pitch** of a sound refers to whether it is high like the sound of a piccolo or violin, or low like the sound of a bass drum or string bass. The physical quantity that determines pitch is the frequency, a fact that was first noted by Galileo. The lower the frequency, the lower the pitch, and the higher the frequency, the higher the pitch.[†] The human ear responds to frequencies in the range from about 20 Hz to about 20,000 Hz. This is called the **audible range**. These limits vary somewhat from one individual to another. One general trend is that as people age, they are less able to hear the high frequencies, so that the high-frequency limit may be 10,000 Hz or less.

Sound waves whose frequency is outside the audible range may reach the ear, but we are not generally aware of them. Frequencies above 20,000 Hz are called *ultrasonic* (do not confuse with *supersonic* which is used to refer to an object moving with a speed faster than the speed of sound). Many animals can hear ultrasonic frequencies; dogs, for example, can hear sounds as high as 50,000 Hz and bats can detect frequencies as high as 100,000 Hz. Ultrasonic waves have a number of applications in medicine and other fields.

Sound waves whose frequencies are below the audible range (that is, less than 20 Hz) are called *infrasonic*. Sources of infrasonic waves are earthquakes, thunder, volcanoes, and waves produced by vibrating heavy machinery. This last source can be particularly troublesome to workers, for infrasonic waves—even though inaudible—can cause damage to the human body. These low-frequency waves act in a resonant fashion, causing motion and irritation of internal organs of the body.

TABLE 16–1

Speed of sound in various materials at 20°C

Material	Speed (m/s)
Air	343
Air (0°C)	331
Helium	1005
Hydrogen	1300
Water	1440
Sea water	1560
Iron and steel	≈ 5000
Glass	≈ 4500
Aluminum	≈ 5100
Hard wood	≈ 4000

16–2 Mathematical Representation of Longitudinal Waves

In Section 15–4, we saw that a one-dimensional sinusoidal wave traveling along the x axis can be represented by the relation (Eq. 15–10c)

$$D = D_M \sin (kx - \omega t); \tag{16–1}$$

here the wave number k is related to the wavelength λ by $k = 2\pi/\lambda$, and $\omega = 2\pi f$

[†] Although pitch is determined mainly by frequency, it also depends to a slight extent on loudness; for example, a very loud sound may seem slightly lower in pitch than a quiet sound of the same frequency.

where f is the frequency; D is the displacement at position x and time t, and D_M is its maximum value (the *amplitude*). For a transverse wave—such as a wave on a string—the displacement D is perpendicular to the direction of wave propagation along the x axis. But for a longitudinal wave the displacement D is *along the direction of wave propagation*. That is, D is parallel to x and represents the displacement of a tiny volume element from its equilibrium position.

Longitudinal (sound) waves can also be considered from the point of view of variations in pressure rather than displacement; indeed, longitudinal waves are often called *pressure waves*. The pressure variation is usually easier to measure than the displacement (see Example 16–2). As can be seen in Fig. 15–7, in a wave "compression" (where molecules are closest together), the pressure is higher than normal, whereas in an expansion (or rarefaction) the pressure is less than normal. Fig. 16–1 shows a graphical representation of a sound wave in air in terms of (a) displacement and (b) pressure. Note that the displacement wave is a quarter wavelength, or 90°, out of phase with the pressure wave: where the pressure is a maximum or minimum, the displacement from equilibrium is zero; and where the pressure variation is zero, the displacement is a maximum or minimum. Why this is so will become clear in a moment.

Let us now derive the mathematical representation of the pressure variation in a traveling longitudinal wave. From the definition of the bulk modulus, B (Eq. 11–7a),

$$\Delta P = -B(\Delta V/V)$$

where $\Delta V/V$ is the fractional change in volume of the medium due to a pressure change ΔP. For convenience, we let p represent the pressure difference from the normal pressure P in the absence of a wave; that is, we let $p = \Delta P$ so

$$p = -B(\Delta V/V).$$

The negative sign reflects the fact that the volume decreases ($\Delta V < 0$) if the pressure is increased. Consider now a layer of fluid through which the longitudinal wave is passing. If this layer has thickness Δx and area A, then its volume is $V = A\,\Delta x$. As a result of pressure variation in the wave, the volume will change by an amount $\Delta V = A\,\Delta D$ where ΔD is the change in thickness of this layer as it compresses or expands. (Remember that D represents the displacement of the medium.) Thus we have

$$p = -B\frac{A\,\Delta D}{A\,\Delta x}.$$

To be precise, we take the limit of $\Delta x \to 0$, so we obtain

$$p = -B\frac{\partial D}{\partial x} \tag{16–2}$$

where we use the partial derivative notation since D is a function of both x and t. If the displacement D is sinusoidal as given by Eq. 16–1, then we have from Eq. 16–2 that

$$p = -(BD_M k)\cos(kx - \omega t). \tag{16–3}$$

Thus the pressure varies sinusoidally as well, but is out of phase from the displacement by 90° or a quarter wavelength; see Fig. 16–1. The quantity $BD_M k$ is called the **pressure amplitude**, p_M; it represents the maximum and minimum amounts by which the pressure varies from the normal ambient pressure. Since the wave velocity is given by $v = \sqrt{B/\rho}$ we can write the pressure amplitude as

$$p_M = BD_M k$$
$$= \rho v^2 D_M k$$
$$= 2\pi\rho v D_M f, \tag{16–4}$$

and

$$p = -p_M \cos(kx - \omega t). \tag{16–5}$$

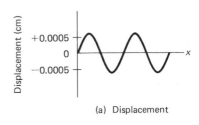

(a) Displacement

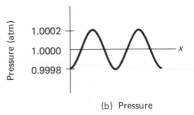

(b) Pressure

FIGURE 16–1

Representation of a sound wave in terms of (a) displacement; (b) pressure.

16–3   Intensity of Sound

Like pitch, *loudness* is a sensation in the consciousness of a human being. It too is related to a physically measurable quantity, the **intensity** of the wave. Intensity is defined as the energy transported by a wave per unit time across unit area and, as we saw in the previous chapter, is proportional to the square of the wave amplitude.

The human ear can detect sounds with an intensity as low as 10^{-12} W/m^2 and as high as 1 W/m^2 (and even higher, although above this it is painful). This is an incredibly wide range of intensity, spanning a factor of 10^{12} from lowest to highest. Presumably because of this wide range, what we perceive as loudness is not directly proportional to the intensity. True, the greater the intensity, the louder the sound. But to produce a sound that sounds twice as loud requires a sound wave that has about ten times the intensity. This is true to a first approximation at any sound level. For example, a sound wave of intensity 10^{-9} W/m^2 sounds to an average human like it is about twice as loud as one whose intensity is 10^{-10} W/m^2; and an intensity of 10^{-2} W/m^2 sounds about twice as loud as 10^{-3} W/m^2, and four times as loud as 10^{-4} W/m^2.

Because of this relationship between the subjective sensation of loudness and the physically measurable quantity intensity, it is usual to specify sound intensity levels using a logarithmic scale. The unit on this scale is a *bel*, or much more commonly, the *decibel* (dB) which is $\frac{1}{10}$ bel (1 dB = 0.1 bel). The **intensity level**, β, of any sound is defined in terms of its intensity, I, as follows

$$\beta \text{ (in dB)} = 10 \log \frac{I}{I_0} \qquad (16-6)$$

where I_0 is the intensity of some reference level and the logarithm is to the base 10. I_0 is usually taken as the minimum intensity audible to an average person, the "threshold of hearing," which is $I_0 = 1.0 \times 10^{-12}$ W/m^2. Thus, for example, the intensity level of a sound whose intensity $I = 1.0 \times 10^{-10}$ W/m^2 will be

$$\beta = 10 \log \frac{10^{-10}}{10^{-12}} = 10 \log 100 = 20 \text{ dB},$$

since log 100 is equal to 2.0. Notice that the intensity level at the threshold of hearing is 0 dB; that is $\beta = 10 \log 10^{-12}/10^{-12} = 10 \log 1 = 0$ since log 1 = 0. Notice too that an increase in intensity by a factor of 10 corresponds to a level increase of 10 dB. An increase in intensity by a factor of 100 corresponds to a level increase of 20 dB. Thus a 50-dB sound is 100 times more intense than a 30-dB sound, and so on.

TABLE 16–2
Intensity of various sounds

Source of the Sound	Intensity Level (in decibels)	Intensity (W/m^2)
Jet plane, at 30 m	140	100
Threshold of pain	120	1
Loud indoor rock concert	120	1
Siren, at 30 m	100	1×10^{-2}
Auto interior, moving at 60 mph	75	3.2×10^{-5}
Busy street traffic	70	1×10^{-5}
Ordinary conversation, at 50 cm	65	3.2×10^{-6}
Quiet radio	40	1×10^{-8}
Whisper	20	1×10^{-10}
Rustle of leaves	10	1×10^{-11}
Threshold of hearing	0	1×10^{-12}

The intensities and intensity levels for a number of common sounds are listed in Table 16–2.

EXAMPLE 16–1   A high-quality loudspeaker is advertised to reproduce, at full volume, frequencies from 30 Hz to 18,000 Hz with uniform intensity ± 3 dB. That is, over this frequency range, the intensity level does not vary by more than 3 dB from the average. By what factor does the intensity change for the maximum intensity-level change of 3 dB?

SOLUTION   Let us call the average intensity I_1 and the average level β_1. Then the maximum intensity, I_2, corresponds to a level $\beta_2 = \beta_1 + 3$ dB. Thus

$$\beta_2 - \beta_1 = 10 \log \frac{I_2}{I_0} - 10 \log \frac{I_1}{I_0}$$

$$3 \text{ dB} = 10 \left(\log \frac{I_2}{I_0} - \log \frac{I_1}{I_0} \right)$$

$$= 10 \log \frac{I_2}{I_1}$$

since $(\log a - \log b) = \log a/b$. Then

$$\log \frac{I_2}{I_1} = 0.30.$$

Using a calculator we calculate 10^x with $x = 0.30$, or from a log table we find the logarithm of what number equals 0.30. The result is 2.0, so

$$\frac{I_2}{I_1} = 2.0$$

or I_2 is twice as intense as I_1.

It is worth noting that a sound level difference of 3 dB (which corresponds to a doubled intensity as we just saw) corresponds to only a very small change in the subjective sensation of apparent loudness. Indeed, the average human can distinguish a difference in level of only about 1 dB.

The intensity I is proportional to the square of the amplitude as we saw in Chapter 15. Indeed, using Eq. 15–7 we can relate amplitude quantitatively to the intensity I or level β as the following example shows.

EXAMPLE 16–2   (a) Calculate the maximum displacement of air molecules for a sound at the threshold of hearing, having a frequency of 1000 Hz. (b) Determine the maximum pressure variation in such a sound wave.

SOLUTION   (a) We use Eq. 15–7 of Chapter 15 and solve for D_M:

$$D_M = \frac{1}{\pi f} \sqrt{\frac{I}{2\rho v}}$$

$$= \frac{1}{(3.14)(1.0 \times 10^3 \text{ s}^{-1})} \sqrt{\frac{1.0 \times 10^{-12} \text{ W/m}^2}{(2)(1.29 \text{ kg/m}^3)(331 \text{ m/s})}}$$

where we have taken the density of air to be 1.29 kg/m^3 and the speed of sound in air (assumed $0°$C) as 331 m/s. Doing the arithmetic we find $D_M = 1.0 \times 10^{-11}$ m.

(b) From Eq. 16–4,

$$p_M = 2\pi \rho v D_M f = 2.7 \times 10^{-5} \text{ Pa}$$

or 2.7×10^{-10} atm.

The result of this example illustrates just how sensitive the human ear is. For it can detect displacements of air molecules which are actually less than the diameter of atoms (about 10^{-10} m)!

By combining Eqs. 15–7 and 16–4, we can write the intensity in terms of the pressure amplitude, p_M:

$$I = 2\pi^2 v\rho f^2 D_M^2 = 2\pi^2 v\rho f^2 (p_M/2\pi\rho vf)^2$$

$$= \frac{p_M^2}{2v\rho}. \tag{16–7}$$

The intensity, when given in terms of pressure amplitude, thus does not depend on frequency. Intensities of sounds of different frequencies can be directly compared using instruments that measure pressure variations. (This would not be the case for an instrument that detected displacement—see Eq. 15–7.)

Normally, the loudness or intensity of a sound decreases as you get farther from the source of the sound. In interior rooms, this effect is reduced because of reflections from the walls. However, if a source is in the open so that sound can radiate freely in all directions, the intensity decreases as the inverse square of the distance,

$$I \propto \frac{1}{r^2}$$

as we saw in Section 15–3, Eq. 15–8. Of course, if there is significant reflection from structures or the ground, the situation will be more complicated.

 EXAMPLE 16–3   The intensity level of a jet plane at a distance of 30 m is 140 dB. What is the intensity level at 300 m? (Ignore reflections from the ground.)

SOLUTION   The intensity I at 30 m is found from Eq. 16–6: 140 dB = $10 \log I/(10^{-12} \text{ W/m}^2)$. Reversing the log equation to solve for I we have: $I = 10^2 \text{ W/m}^2$. At 300 m, ten times as far away, the intensity will be $(\frac{1}{10})^2 = 1/100$ as much, or 1 W/m^2. Hence, the intensity level $\beta = 10 \log(1/10^{-12}) = 120$ dB. Even at 300 m, the sound is at the threshold of pain. This is why workers at airports wear ear covers to protect their ears from damage.

16–4   Sources of Sound: Vibrating Strings and Air Columns

The source of any sound is a vibrating object. Almost any object can vibrate and hence be a source of sound. We now discuss some simple sources of sound, particularly musical instruments. In musical instruments, the source is set into vibration by striking, plucking, bowing, or blowing. Standing waves are produced and the object vibrates at its natural resonant frequencies. A drum has a stretched membrane that vibrates. Xylophones and marimbas have metal or wood bars that can be set into vibration. Bells, cymbals, and gongs also make use of a vibrating metal. The most widely used instruments make use of vibrating strings, such as the violin, guitar, and piano, or make use of vibrating columns of air, such as the flute, trumpet, and pipe organ.

We saw in Chapter 15, Fig. 15–26, how standing waves are established on a string. This is the basis for all stringed instruments. The pitch is normally determined by the lowest resonant frequency, the *fundamental*, which corresponds to nodes occurring only at the ends. The wavelength of the fundamental is equal to twice the length of the string; therefore, the fundamental frequency $f = v/\lambda = v/2L$ where v is the velocity of the wave on the string. When a finger is placed on the string of, say, a guitar or violin, the effective length of the string is shortened; so its pitch is higher since the wavelength of the fundamental is shorter. The strings on a guitar or violin are all the same length. They sound at a different pitch because the strings have different mass per unit length, μ, which affects the velocity as seen in Eq. 15–2, $v = \sqrt{F_T/\mu}$. (The tension may also be different; adjusting the tension is the means for tuning the instrument.) Thus the velocity on a heavier string is less and the frequency will be less for the same wavelength. In pianos and harps, the strings are each of different length; for the lower notes they are not only longer but heavier as well, and the reason why is illustrated in the following example.

EXAMPLE 16–4 The highest key on a piano corresponds to a frequency about 150 times that of the lowest key. If the string for the highest note is 5.0 cm long, how long would the string for the lowest note have to be if it had the same mass per unit length and was under the same tension?

SOLUTION The velocity would be the same on each string so the frequency is inversely proportional to the length L of the string ($f = v/\lambda = v/2L$). Thus

$$\frac{L_l}{L_h} = \frac{f_h}{f_l}$$

where the subscripts l and h refer to the lowest and highest notes, respectively. Thus $L_l = L_h(f_h/f_l) = (5.0 \text{ cm})(150) = 750$ cm, or 7.5 m. This would be ridiculously long for a piano. The longer lower strings are made heavier partly to avoid this, so that even on grand pianos the strings are no longer than about 3 m.

Stringed instruments would not be very loud if they relied on their vibrating strings to produce the sound waves since the strings are simply too thin to compress and expand much air. Stringed instruments therefore make use of a kind of a mechanical amplifier known as a *sounding board* (piano) or *sounding box* (guitar, violin) which acts to amplify the sound by putting a greater surface area in contact with the air. When the strings are set into vibration, the sounding board or box is set into vibration as well. Since it has much greater area in contact with the air, it can produce a much stronger sound wave; it thus acts to amplify the sound. On an electric guitar the sounding box is not so important since the vibrations of the strings are amplified electrically.

Instruments such as woodwinds, the brasses, and the pipe organ produce sound from the vibrations of standing waves in a column of air within a tube or pipes. Standing waves can occur in the air of any cavity, but the frequencies are difficult to calculate for any but very simple shapes such as a long, narrow tube. Fortunately, and for good reason, this is the situation for most wind instruments. In some instruments, a vibrating reed or the vibrating lip of the player helps to set up vibrations of the air column. In others a stream of air is directed against one edge of the opening or mouthpiece, leading to turbulence which sets up the vibrations. Because of the disturbance, whatever its source, the air within the tube vibrates with a variety of frequencies; but only certain frequencies persist, which correspond to standing waves. For a string fixed at both ends, we saw that the standing waves have nodes (no movement) at the two ends, and one or more antinodes (large amplitude of vibration) in between; a node also separates each antinode. The lowest-frequency standing wave corresponds to a single antinode and is called the fundamental. The higher-frequency standing waves are called overtones or harmonics; specifically, the first harmonic is the fundamental, the second harmonic has twice the frequency of the fundamental, and so on; see Fig. 15–26.

The situation is similar for a column of air, but we must remember that it is now air itself that is vibrating. Thus the air at the closed end of a tube must be a (displacement) node since the air is not free to move there, whereas at the open end of a tube there will be an antinode since the air can move freely. The air within the tube vibrates in the form of longitudinal standing waves. The possible modes of vibration for a tube open at both ends (called an *open tube*) and for one that is open at one end but closed at the other (called a *closed tube*) are shown graphically in Fig. 16–2. The graphs represent the displacement amplitude of the vibrating air within the tube. The antinodes do not occur precisely at the open ends of tubes; their position depends on the diameter of the tube, but if this is small compared to the length, which is the usual case, the antinode occurs very close to the end as shown, and we assume this is the case in what follows. (Wavelength and other factors can also affect the position of the antinode.)

Let us look first at the open pipe, Fig. 16–2a. An open pipe has displacement antinodes at both ends. Notice that there must be at least one node within an

FIGURE 16–2

Modes of vibration (standing waves) for (a) an open tube; (b) a closed tube.

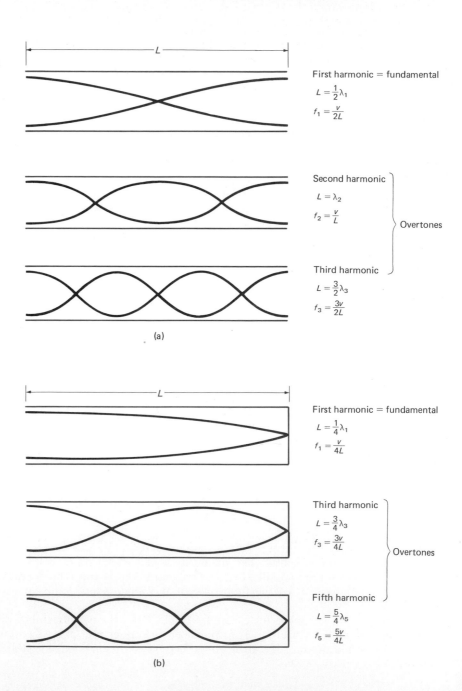

First harmonic = fundamental

$L = \frac{1}{2}\lambda_1$

$f_1 = \frac{v}{2L}$

Second harmonic

$L = \lambda_2$

$f_2 = \frac{v}{L}$

$\Big\}$ Overtones

Third harmonic

$L = \frac{3}{2}\lambda_3$

$f_3 = \frac{3v}{2L}$

(a)

First harmonic = fundamental

$L = \frac{1}{4}\lambda_1$

$f_1 = \frac{v}{4L}$

Third harmonic

$L = \frac{3}{4}\lambda_3$

$f_3 = \frac{3v}{4L}$

$\Big\}$ Overtones

Fifth harmonic

$L = \frac{5}{4}\lambda_5$

$f_5 = \frac{5v}{4L}$

(b)

open pipe if there is to be a standing wave at all. This corresponds to the fundamental frequency of the tube; since the distance between two successive nodes, or between two successive antinodes, is $\frac{1}{2}\lambda$, there is one-half a wavelength within the length of the tube in this case: $L = \frac{1}{2}\lambda$, so the fundamental frequency is $f_1 = v/\lambda = v/2L$, where v is the velocity of sound in air. The standing wave with two nodes is the first overtone or second harmonic and has half the wavelength ($L = \lambda$) and twice the frequency. Indeed, the frequency of each overtone is an integral multiple of the fundamental frequency. This is just what is found for a string.

For a closed tube, Fig. 16–2b, there is always a displacement node at the closed end and an antinode at the open end. Since the distance between a node and the nearest antinode is $\frac{1}{4}\lambda$, we see that the fundamental in this case corresponds to only one-fourth a wavelength within the length of the tube: $L = \lambda/4$. The fundamental frequency is thus $f_1 = v/4L$, or half what it is for an open pipe of the same length. There is another difference, for as we can see from Fig. 16–2b only the odd harmonics are present in a closed pipe: the overtones have frequencies equal to 3, 5, 7, . . . times the fundamental frequency. There is no way for a wave with 2, 4, . . . times the fundamental to have a node at one end and an antinode at the other, and thus they cannot exist as standing waves in a closed tube.

Organs make use of both open and closed pipes. Notes of different pitch are sounded using different pipes which vary in length from a few centimeters to 5 m or more. Other musical instruments act either like a closed tube or an open tube. A flute for example is an open tube for it is open not only where one blows into it, but also at the opposite end as well. The different notes on a flute and many other instruments are obtained by shortening the length of the tube—that is, by uncovering holes along its length. In a trumpet, on the other hand, the pushing down of the valves opens additional lengths of tube. In all these instruments, the longer the length of the vibrating air column, the lower the pitch.

The diagrams of Fig. 16–2 represent the *displacement* of the standing waves. The amplitude of the *pressure*, on the other hand, will be 90° out of phase with the displacement, just as for a traveling wave. Thus there is a pressure node at an open end of a tube (which makes sense since it is open to the atmosphere) and a pressure antinode at a closed end of a tube.

EXAMPLE 16–5   What will be the fundamental frequencies and first three overtones for a 26-cm long organ pipe at 20°C if it is (a) open and (b) closed?

SOLUTION   At 20°C, the speed of sound in air is 343 m/s (Section 16–1). (a) For the open pipe, the fundamental frequency is:

$$f_1 = \frac{v}{2L} = \frac{343 \text{ m/s}}{2(0.26 \text{ m})} = 660 \text{ Hz}.$$

The overtones, which include all harmonics, are 1320 Hz, 1980 Hz, 2640 Hz, and so on.

(b) Referring to Fig. 16–2 we see that

$$f_1 = \frac{v}{4L} = \frac{343 \text{ m/s}}{4(0.26 \text{ m})} = 330 \text{ Hz}.$$

But only the odd harmonics will be present, so the first three overtones will be 990 Hz, 1650 Hz, and 2310 Hz.

EXAMPLE 16–6   A flute is designed to play middle C (264 Hz) as the fundamental frequency when all the holes are covered. Approximately how long should the distance be from the mouthpiece to end of the flute. (Note: this is only approximate since the antinode does not occur precisely at the mouthpiece). Assume the temperature is 20°C.

SOLUTION   The speed of sound in air at 20°C is 343 m/s (Section 16–1). Then, from Fig. 16–2 the fundamental frequency f_1 is related to the length of the vibrating air column by $f = v/2L$. Solving for L we find

$$L = \frac{v}{2f} = \frac{343 \text{ m/s}}{(2)(264 \text{ s}^{-1})} = 0.650 \text{ m}.$$

EXAMPLE 16–7   If the temperature is only 10°C, what will be the frequency of the note played when all the openings are covered in the flute of Example 16–6?

SOLUTION   The length L is still 65.0 cm. But now the velocity of sound is less since it changes by 0.60 m/s per each C°. For a drop of 10 C°, the velocity decreases by 6 m/s to 337 m/s. The frequency will be

$$f = \frac{v}{2L} = \frac{337 \text{ m/s}}{2(0.650 \text{ m})} = 259 \text{ Hz}.$$

This example illustrates why players of wind instruments take time to "warm up" their instruments so they will be in tune. The effect of temperature on stringed instruments is much smaller.

* 16–5   Quality of Sound

Whenever we hear a sound, particularly a musical sound, we are aware of its loudness, its pitch, and also of a third aspect called "quality." For example, when a piano and then an oboe play a note of the same loudness and the same pitch (say middle C), there is a clear difference in the overall sound. We would never mistake a piano for an oboe. This is what is meant by the **quality**[†] of a sound; for musical instruments the terms *timbre* or *tone color* are also used.

Just as loudness and pitch can be related to physically measurable quantities, so too can quality. The quality of a sound depends on the presence of over-tones—their number and their relative amplitudes. Generally, when a note is played on a musical instrument, the fundamental as well as the overtones are present simultaneously. We saw in Fig. 15–13 how the superposition of three wave forms, in that case the fundamental and first two overtones (of particular amplitudes), would combine to give a composite *wave form*. Of course more than two overtones are usually present.

The relative amplitudes of the various overtones are different for different musical instruments, and this is what gives each instrument its characteristic quality or timbre. A graph showing the relative amplitudes of the harmonics produced by an instrument is called a "sound spectrum"; several typical examples for different instruments are shown in Fig. 16–3. Normally, the fundamental has the greatest amplitude and its frequency is what is heard as the pitch.

The manner in which an instrument is played strongly influences the sound quality. Plucking a violin string, for example, makes a very different sound than pulling a bow across it. The sound spectrum at the very start (or end) of a note (as when a hammer strikes a piano string) can be very different from the subsequent sustained tone. This too affects the subjective tone quality of an instrument.

An ordinary sound, like that made by striking two stones together, is a noise that has a certain quality, but a clear pitch is not discernible. A noise such as this is a mixture of many frequencies which bear little relation to one another. If a sound spectrum were made of this noise, it would not show discrete lines like

FIGURE 16–3

Sound spectra for several instruments.

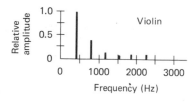

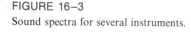

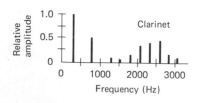

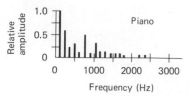

[†] Note that quality in this sense does not refer to the goodness or badness of a sound or to the craftsmanship that went into building the instrument.

those of Fig. 16–3; instead it would show a continuous, or nearly continuous, spectrum of frequencies.

16–6 Interference of Sound Waves; Beats

We saw in Section 15–8 that when two waves simultaneously pass through the same region of space, they interfere with one another. Since this is true of any kind of wave, we should expect that interference will occur with sound waves, and indeed it does.

As a simple example, consider two large speakers, A and B, a distance d apart on the stage of an auditorium as shown in Fig. 16–4. Let us assume the two speakers are emitting sound waves of the same single frequency and that they are in phase: that is, when one speaker is forming a compression, so is the other. The curved lines in the diagram represent the crests of sound waves from each speaker. Of course we must remember that for a sound wave, a crest is a compression in the air whereas a trough—which falls between two crests—is a rarefaction. A person or detector at a point such as C, which is the same distance from each speaker, will experience a loud sound because the interference will be constructive. On the other hand, at a point such as D in the diagram, little if any sound will be heard because destructive interference occurs—compressions of one wave meet rarefactions of the other and vice versa (see Fig. 15–21 and related discussion on water waves in Section 15–8). An analysis of this situation is clearer if we graphically represent the wave forms as in Fig. 16–5. There it can be seen that at point C constructive interference occurs since both waves simultaneously have crests or simultaneously have troughs. In Fig. 16–5b we see the situation for point D. The wave from speaker B must travel a greater distance than the wave from A. Thus the wave from B lags behind that from A. In this diagram, point E is chosen so that the distance ED is equal to AD. Thus we see that if the distance BE is equal to precisely one-half the wavelength of the sound, the two waves will be exactly out of phase when they reach D, and destructive interference occurs. This then is the criterion for determining at what point(s) destructive interference occurs: destructive interference occurs at any point whose distance from one speaker is greater than its distance from the other speaker by exactly one-half wavelength. Notice that if this extra distance (BE in the diagram) is equal to a whole wavelength (or 2, 3, ... wavelengths) then the two waves will be in phase and *constructive interference* occurs. If the distance BE equals $\frac{1}{2}$, $1\frac{1}{2}$, $2\frac{1}{2}$, ... wavelengths, *destructive interference* occurs.

If a speaker emits a whole range of frequencies, not all wavelengths will destructively interfere at any one point such as D; rather only specific wavelengths will destructively interfere completely according to the criteria above.

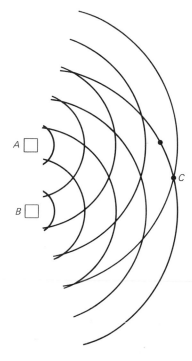

FIGURE 16–4

Sound waves from two loudspeakers interfere.

FIGURE 16–5

Sound waves from loudspeakers A and B (see Fig. 16–4) constructively interfere at C and destructively interfere at D.

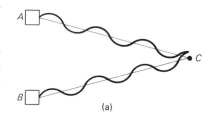

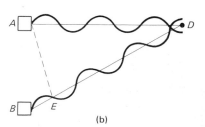

EXAMPLE 16–8 The two speakers in Fig. 16–4 are 1.00 m apart. A person stands 4.00 m from one speaker. How far must this person be from the second speaker in order to detect destructive interference when the speakers emit an 1150-Hz sound? Assume the temperature is 20°C.

SOLUTION The wavelength of this sound is

$$\lambda = \frac{v}{f} = \frac{343 \text{ m/s}}{1150 \text{ Hz}} = 0.30 \text{ m}.$$

For destructive interference to occur, the person must be one-half wavelength further from one speaker than the other, or 0.15 m. Thus the person must be 4.15 m (or 3.85 m) from the second speaker. Notice in this example that if the speakers were less than 0.15 m apart, there would be no point that was 0.15 m farther from one speaker than the other and there would be no point where destructive interference would occur.

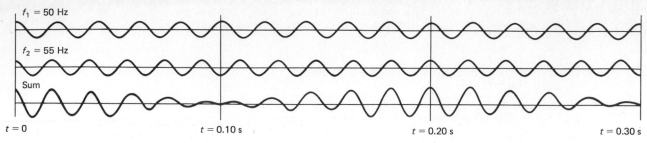

FIGURE 16–6

Beats occur as a result of the superposition of two sound waves of slightly different frequency.

An interesting and important example of interference occurs in the phenomenon known as *beats*. This is the phenomenon that occurs if two sources of sound—say, two tuning forks—are close in frequency but not exactly the same. Sound waves from the two sources interfere with each other and the sound level alternately rises and falls; the regularly spaced intensity changes are called beats and often sound eerie.

To see how beats arise, consider two equal-amplitude[†] sound waves of frequency $f_1 = 50$ Hz and $f_2 = 55$ Hz, respectively. In 1.00 s, the first source makes 50 vibrations whereas the second makes 55. We now examine the waves at one point in space equidistant from the two sources. The wave forms for each wave as a function of time are shown on the first two lines of Fig. 16–6; the third line shows the sum of the two waves. At the time $t = 0$ the two waves are shown to be in phase and interfere constructively. Because the two waves vibrate at different rates, at the time $t = 0.10$ s they are completely out of phase and destructive interference occurs. At $t = 0.20$ s they are again in phase and the resultant amplitude again large. Thus the resultant amplitude is large every 0.20 s and in between it drops drastically. This rising and falling of the intensity is what is heard as beats. In this case the beats are 0.20 s apart. That is, the *beat frequency* is five per second or 5 Hz. This result, that the beat frequency equals the difference in frequency of the two waves, can be shown in general as follows.

Let the two waves, of frequencies f_1 and f_2, be represented at a fixed point in space by:

$$D_1 = D_M \sin 2\pi f_1 t$$

and

$$D_2 = D_M \sin 2\pi f_2 t.$$

The resultant displacement, by the principle of superposition, is

$$D = D_1 + D_2 = D_M (\sin 2\pi f_1 t + \sin 2\pi f_2 t)$$

or, using the trigonometric identity $\sin A + \sin B = 2 \sin \frac{1}{2}(A + B) \cos \frac{1}{2}(A - B)$, we have

$$D = \left[2D_M \cos 2\pi \left(\frac{f_1 - f_2}{2} \right) t \right] \sin 2\pi \left(\frac{f_1 + f_2}{2} \right) t. \qquad (16\text{–}8)$$

We can interpret Eq. 16–8 as follows. The superposition of the two waves results in a wave that vibrates at the average frequency of the two components, $(f_1 + f_2)/2$; this vibration has an amplitude given by the expression in brackets, and this amplitude varies in time, from zero to a maximum of $2D_M$ (the sum of the separate amplitudes), with a frequency of $(f_1 - f_2)/2$. A beat occurs whenever $\cos 2\pi[(f_1 - f_2)/2]t$ equals $+1$ or -1 (see Fig. 16–6); that is, two beats occur per cycle, so the beat frequency is twice $(f_1 - f_2)/2$ which is just $f_1 - f_2$, the difference in frequency of the component waves.

[†] Beats will be heard even if the amplitudes are not equal, as long as the difference is not great.

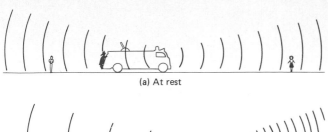

(a) At rest

(b) Firetruck moving

FIGURE 16–7

(a) Both observers on the sidewalk hear the same frequency from the firetruck at rest. (b) Doppler effect: observer toward whom the firetruck moves hears a higher-frequency sound, and observer behind the firetruck hears a lower frequency.

The phenomenon of beats can occur with any kind of wave and is a very sensitive method for comparing frequencies.

 16–7 Doppler Effect

You may have noticed that the pitch of the siren on a speeding firetruck drops abruptly as it passes you. Or you may have noticed the change in pitch of a blaring horn on a fast moving car as it passes by. The pitch of the sound from the engine of a race car changes too as it passes an observer. When a source of sound is moving toward an observer, the pitch is higher than when the source is at rest; and when the source is traveling away from the observer, the pitch is lower. This phenomenon is known as the **Doppler effect** and occurs for all types of waves. Let us now see why it occurs, and calculate the change in frequency for sound waves.

To be concrete, consider the siren of a firetruck at rest which is emitting sound of a particular frequency in all directions as shown in Fig. 16–7a. If the firetruck is moving, the siren emits sound at the same frequency. But the sound waves it emits forward are closer together than normal as shown in Fig. 16–7b. This is because the fire engine, as it moves, is "catching up" with the previously emitted waves. Thus an observer on the sidewalk will detect more wave crests passing per second so the frequency is higher. The waves emitted behind the truck are, on the other hand, farther apart than normal because the truck is speeding away from them. Hence, fewer wave crests per second pass by an observer behind the truck and the pitch is lower.

To calculate the change in frequency, we make use of Fig. 16–8, and we assume the air (or other medium) is at rest in our reference frame (that is, the air is still). In Fig. 16–8a, the source of the sound (say a siren) is at rest; two successive wave crests are shown, the second of which is just in the process of being emitted. The distance between these crests is λ, the wavelength. If the frequency of the source is f, then the time between emissions of wave crests is

$$T = \frac{1}{f}.$$

In Fig. 16–8b the source is moving with a velocity v_s. In a time T (as just defined) the first wave crest has moved a distance $d = vT$ where v is the velocity of the sound wave in air (which is of course the same whether the source is moving or not). In this same time, the source has moved a distance $d_s = v_s T$. Then the distance between successive wave crests, which is the new wavelength λ', is

$$\lambda' = d - d_s$$
$$= (v - v_s)T$$
$$= (v - v_s)\frac{1}{f}$$

FIGURE 16–8

Determination of frequency change in the Doppler effect (see text).

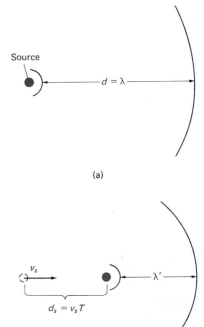

Source

$d = \lambda$

(a)

v_s

λ'

$d_s = v_s T$

(b)

since $T = 1/f$. The frequency f' of the wave is

$$f' = \frac{v}{\lambda'}$$

$$= \left(\frac{v}{v - v_s}\right)f$$

or

$$f' = \frac{1}{\left(1 - \dfrac{v_s}{v}\right)}f. \qquad \left[\begin{array}{l} \text{source moving toward} \\ \text{stationary observer} \end{array}\right] \quad (16\text{--}9a)$$

Because the denominator is less than one, $f' > f$. For example, if a source emits a sound of frequency 400 Hz when at rest, then when the source moves toward a fixed observer with a speed of 30 m/s, the observer hears a frequency (at 0°C) of

$$f' = \frac{400 \text{ Hz}}{\left(1 - \dfrac{30 \text{ m/s}}{331 \text{ m/s}}\right)} = 440 \text{ Hz}.$$

For a source which is moving *away* from the observer at a speed v_s, the new wavelength will be

$$\lambda' = d + d_s,$$

and the frequency f' will then be

$$f' = \frac{1}{(1 + v_s/v)}f. \qquad \left[\begin{array}{l} \text{source moving away from} \\ \text{stationary observer} \end{array}\right] \quad (16\text{--}9b)$$

The Doppler effect also occurs when the source is at rest (with respect to the medium for the sound waves) and the observer is in motion. If the observer is traveling toward the source, the pitch is higher; and if the observer is traveling away from the source, the pitch is lower. Quantitatively, the change in frequency is slightly different than for the case of a moving source. In this case the distance between wave crests, the wavelength λ, is not changed. But the velocity of the crests with respect to the observer is changed. If the observer is moving toward the source, the speed of the waves relative to the observer is $v' = v + v_o$ where v is the velocity of the sound in the air (we assume the air is still) and v_o is the velocity of the observer. Hence, the new frequency is

$$f' = \frac{v'}{\lambda} = \frac{v + v_o}{\lambda}$$

or, since $\lambda = v/f$

$$f' = \left(1 + \frac{v_o}{v}\right)f. \qquad \left[\begin{array}{l} \text{observer moving toward} \\ \text{stationary source} \end{array}\right] \quad (16\text{--}10a)$$

If the observer is moving away from the source, the relative velocity is $v' = v - v_o$ and

$$f' = \left(1 - \frac{v_o}{v}\right)f. \qquad \left[\begin{array}{l} \text{observer moving away} \\ \text{from stationary source} \end{array}\right] \quad (16\text{--}10b)$$

When a sound wave is reflected from a moving obstacle, the frequency of the reflected wave will, because of the Doppler effect, be different from that of the incident wave. This is illustrated in the following example.

EXAMPLE 16–9 A 5000-Hz sound wave is directed toward an object moving 3.30 m/s toward the source. What is the frequency of the reflected wave?

SOLUTION There are actually two Doppler shifts in this situation. First, the object acts like a moving observer and "detects" a sound wave of frequency

$$f' = \left(1 + \frac{v_o}{v}\right)f = \left(1 + \frac{3.30 \text{ m/s}}{331 \text{ m/s}}\right)(5000 \text{ Hz}) = 5050 \text{ Hz}.$$

Secondly, the object acts like a moving source in reemitting (reflecting) the sound, so the reflected frequency is

$$f'' = \frac{1}{(1 - v_s/v)}f' = \frac{5050 \text{ Hz}}{\left(\dfrac{1 - 3.30 \text{ m/s}}{331 \text{ m/s}}\right)} = 5100 \text{ Hz}.$$

Thus the frequency shifts by 100 Hz.

The incident wave and the reflected wave, when mixed together (say, electronically), interfere with one another and beats are produced. The beat frequency is equal to the difference in the two frequencies, and in the above example would be 100 Hz. This Doppler technique is used in a variety of medical applications, usually with ultrasonic waves in the megahertz frequency range. For example, ultrasonic waves reflected from red blood cells can be used to determine the velocity of blood flow. Similarly, the technique can be used to detect the movement of the chest of a young fetus and also to monitor its heartbeat. The Doppler effect is also the basis for detecting speeding automobiles by radar, but here the waves used are electromagnetic (radio) waves.

The accuracy of Eqs. 16–9 and 16–10 declines if v_s or v_o approaches the speed of sound. This is due mainly to the fact that the displacement of the medium is no longer proportional to the restoring force (Hooke's law), so much of our theoretical development will not be valid.

*16–8 Shock Waves and the Sonic Boom

An object such as an airplane traveling faster than the speed of sound is said to have a *supersonic speed*. Such a speed is often given as a *Mach number* which is defined as the ratio of the object's speed to that of sound in the medium at that location. For example, a plane traveling 900 m/s high in the atmosphere where the speed of sound is only 300 m/s has a speed of Mach 3.

When a source of sound moves at subsonic speeds, the pitch of the sound is altered as we have seen (the Doppler effect); see also Fig. 16–9a and b. But if a source of sound moves faster than the speed of sound, a more dramatic effect known as a *shock wave* occurs. In this case the source is actually "outrunning" the waves it produces. As shown in Fig. 16–9c, when the source is traveling at the speed of sound, the waves it emits in the forward direction "pile up" directly in front of it. When the object moves at a supersonic speed, the waves pile up on one another along the sides as shown in Fig. 16–9d. The different wave crests overlap one another and form a single very large crest which is the shock wave. Behind this very large crest there is usually a very large trough. A shock wave is essentially due to the constructive interference of a large number of waves. A shock wave in air is analogous to the bow wave of a boat traveling faster than the speed of the water waves it produces, Fig. 16–10.

FIGURE 16–9

Sound waves emitted by an object at rest (a) or moving (b, c, d). If the object's velocity is less than the velocity of sound, the Doppler effect occurs (b); if its velocity is greater than the velocity of sound, a shock wave is produced (d).

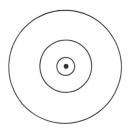

(a) $v_{obj} = 0$

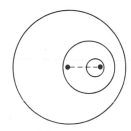

(b) $v_{obj} < v_{snd}$

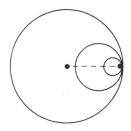

(c) $v_{obj} = v_{snd}$

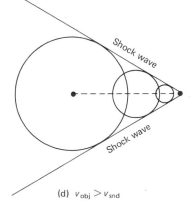

Shock wave

Shock wave

(d) $v_{obj} > v_{snd}$

FIGURE 16–10

Bow wave produced by a boat. (Courtesy of Naval Photographic Center.)

When an airplane travels at supersonic speeds, the noise it makes and its disturbance of the air form into a shock wave containing a tremendous amount of sound energy. When the shock wave passes a listener, it is heard as a loud "sonic boom." A sonic boom lasts only a fraction of a second, but the energy it contains is often sufficient to break windows and cause other damage. It can be psychologically unnerving as well. Actually the sonic boom from a supersonic aircraft is a double boom since a shock wave forms at both the front and the rear of the aircraft (Fig. 16–11).

When an aircraft approaches the speed of sound, it encounters a barrier of sound waves in front of it (see Fig. 16–9c). In order to exceed the speed of sound, extra thrust is needed to pass through this "sound barrier." This is called "breaking the sound barrier." Once a supersonic speed is attained, this barrier no longer impedes the motion. It is sometimes erroneously thought that a sonic boom is produced only at the moment an aircraft is breaking through the sound barrier. Actually, a shock wave follows the aircraft at all times it is traveling at supersonic speeds. A series of observers on the ground will each hear a loud "boom" as the shock wave passes, Fig. 16–11. The shock wave consists of a cone whose apex is at the aircraft. The angle of this cone, θ, (see Fig. 16–9d) is given by

$$\sin \theta = \frac{v_{\text{snd}}}{v_{\text{obj}}}, \qquad (16\text{–}11)$$

where v_{obj} is the velocity of the object (the aircraft) and v_{snd} is the velocity of sound in the medium (the proof is left as problem 57.)

FIGURE 16–11

The (double) sonic boom has already been heard by the person on the right; it is just being heard by the person in the center; and it will shortly be heard by the person on the left.

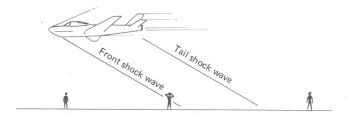

Summary

Sound travels as a longitudinal wave in air and other materials. In air, the speed of sound increases with temperature; at 0°C it is about 331 m/s. A sound wave can be represented both in terms of the displacement of molecules or particles of the medium, or in terms of the pressure variation in the medium; the pressure wave is 90° out of phase with the displacement wave, meaning the crests of the one follow the crests of the other by one-fourth a wavelength.

The *pitch* of a sound is determined by the frequency; the higher the frequency, the higher the pitch. The *audible range* of frequencies is roughly 20 to 20,000 Hz (1 Hz = 1 cycle per second). The *loudness* or *intensity* of a sound is proportional to the square of the amplitude of the wave. Because the human ear can detect sound intensities from 10^{-12} W/m^2 to over 1 W/m^2, intensity levels are specified on a logarithmic scale. The *intensity level*, β, specified in decibels, is defined in terms of the intensity I as $\beta = 10 \log (I/I_0)$, where the reference intensity I_0 is usually taken to be 1.0×10^{-12} W/m^2. An increase in intensity by a factor of 100, for example, corresponds to a level increase of 20 dB.

Musical instruments are simple sources of sound in which standing waves are produced. The strings of a stringed instrument may vibrate as a whole with nodes only at the ends; the frequency at which this occurs is called the *fundamental*. The string can also vibrate at higher frequencies, called overtones or *harmonics*, in which there are one or more additional nodes. The frequency of each harmonic is a whole-number multiple of the fundamental. In wind instruments, standing waves are set up in the column of air within the tube. The vibrating air in an open tube (open at both ends) has antinodes at both ends; the fundamental frequency corresponds to a wavelength equal to twice the tube length. The harmonics have frequencies that are 2, 3, 4, . . . times the fundamental frequency. For a closed tube (closed at one end) the fundamental corresponds to a wavelength four times the length of the tube; only the odd harmonics are present, equal to 1, 3, 5, 7, . . . times the fundamental frequency.

Sound waves from different sources can interfere with each other. If two sounds are at slightly different frequencies, *beats* can be heard at a frequency equal to the difference in frequency of the two sources.

The *Doppler effect* refers to the change in pitch of a sound due to the motion either of the source or of the listener. If they are approaching each other, the pitch is higher; if they are moving apart, the pitch is lower.

 Questions ─────────────────────────────────────

1. What is the evidence that sound travels as a wave?

2. What is the evidence that sound is a form of energy?

3. Country folk have a rule of thumb that the time delay between seeing lightning and hearing the thunder is an indication of how far away the lightning hit, and in particular, that each 5 s corresponds to 1 mi. Explain. What would be the rule for kilometers?

4. When boating on a lake or river at night, one can often hear clearly the voices or radios of people quite a distance away on the shore. Yet this rarely happens in the daytime. This phenomenon can be explained by considering the refraction of sound due to different layers of air having different densities (because of differences in temperature). Draw a diagram and explain this phenomenon and determine if the layer of air nearest the water surface is denser than the one above at nighttime or in the daytime.

5. When a sound wave passes from air into water, do you expect the frequency or wavelength to change?

6. What evidence can you give that the speed of sound in air does not depend significantly on frequency?

7. What is the main reason the speed of sound in hydrogen is less than in air?

8. The molecules of a gas, such as air, move around randomly at fairly high speeds (Chapter 18). The distance between molecules, on the average, is many times their diameter. When a wave passes through a gas the impulse given to one molecule is given to another only when this distance is traveled and the two collide. Would you therefore expect the speed of sound in a gas to be limited by the average molecular speed?

9. Two tuning forks oscillate with the same amplitude, but one has twice the frequency. Which (if either) produces the more intense sound?

10. How does a rise in air temperature affect the loudness of sound coming from a source of fixed frequency and amplitude? (Assume that atmospheric pressure doesn't change.)

11. The voice of a person who has inhaled helium sounds very high pitched. Why?

12. What is the reason that catgut strings on some musical instruments are wrapped with fine wire?

13. Whistle through your lips and describe how the pitch of the whistle is controlled.

14. Draw a diagram, corresponding to each of those in Fig. 16–2, for the standing *pressure* waves in (*a*) an open tube, and (*b*) a closed tube.

15. Explain how a tube might be used as a filter to reduce the amplitude of sounds in various frequency ranges. (An example is a car muffler.) Is there a low-frequency "cut-off" for transmission of sound? If so, how is it determined?

16. How will the air temperature in a room affect the pitch of organ pipes?

17. Noise control is an important goal today. One mode of attack is to reduce the area of vibration of noisy machinery, for example by keeping it as small as possible or isolating it (acoustically) from the floor and walls. A second method is to make the surface out of a thicker material. Explain how each of these can reduce the noise level.

18. Why are the frets on a guitar spaced closer together as you move down the fingerboard toward the bridge?

19. Standing waves can be said to be due to "interference in space," whereas beats can be said to be due to "interference in time." Explain.

20. Under what conditions might you hear beats between the sounds from two tuning forks that have precisely the same natural frequency?

21. Suppose a source of sound moves at right angles to the line of sight of a listener at rest in still air. Will there be a Doppler effect? Explain.

22. If a wind is blowing, will this alter the frequency of the sound heard by a person at rest with respect to the source? Is the wavelength or velocity changed?

*23. A sonic boom sounds much like an explosion. Explain the similarity between the two.

 Problems ─────────────────────────────────────

SECTION 16–1

1. (I) Ultrasonic waves with frequencies as high as 250,000 Hz are emitted by dolphins. What would be the wavelength of such a wave (a) in water and (b) in air? Take $T = 20°C$.

2. (I) A hiker determines the length of a lake by listening for the echo of her shout reflected by a cliff at the far end of the lake. She hears the echo 1.20 s after shouting. How long is the lake?

3. (II) A stone is dropped from the top of a cliff. The splash it makes when striking the water below is heard 4.0 s later. How high is the cliff? (Let $T = 20°C$.)

4. (II) A person sees a heavy stone strike the concrete pavement. A moment later two sounds are heard from the impact: one travels in the air and the other in the concrete, and they are 1.2 s apart. How far away did the impact occur?

SECTION 16–2

5. (II) The pressure variation in a sound wave is given by

$$p = 2.2 \sin\left(\frac{\pi}{3}x - 1700\pi t\right)$$

where p is in pascals, x in meters, and t in seconds. Determine (a) the wavelength, (b) the frequency, (c) the speed, and (d) the displacement amplitude of the wave. Assume the density of the medium to be $\rho = 2.7 \times 10^3$ kg/m^3.

6. (II) Ultrasonic waves in the megahertz frequency range are often used for medical diagnostic purposes. Suppose a 5.0-MHz ultrasonic wave passes from muscle tissue where its speed is 1200 m/s to bone where the speed is 2800 m/s. Write down the two mathematical expressions for $p(x, t)$, the pressure variation of the waves, in muscle and bone which contain λ and v, and the two which contain λ and f. Designate the respective pressure amplitudes of the waves in the two media by $p_{M-muscle}$ and p_{M-bone}.

7. (II) Show that the pressure amplitude of a one-dimensional longitudinal wave can be written as the product of the bulk modulus of the medium and the ratio of the maximum particle speed $(\partial D/\partial t)$ to the wave speed.

SECTION 16–3

8. (I) Two sound waves have equal displacement amplitudes, but one has twice the frequency of the other. (a) Which has the greater pressure amplitude and by what factor is it greater? (b) What is the ratio of their intensities?

9. (I) (a) What is the intensity level of a sound whose intensity is 7.5×10^{-8} W/m^2? (b) What is the intensity of a sound whose intensity level is 35 dB?

10. (I) If the amplitude of a sound wave is tripled, (a) by what factor will the intensity increase? (b) By how many dB will the intensity level increase?

11. (II) What would be the intensity level (in dB) of a sound wave in air that corresponds to a displacement amplitude of vibrating air molecules of 1.2 mm at 80 Hz?

12. (II) A stereo tape recorder is said to have a signal-to-noise ratio of 58 dB. What is the ratio of intensities of the signal and the background noise?

13. (II) Human beings can detect a difference in sound intensity level of 1.0 dB. What is the ratio of the amplitudes of two sounds whose levels differ by this amount?

14. (II) (a) Calculate the maximum displacement of air molecules (at 20°C) when a 120-Hz sound wave passes whose intensity is at the threshold of pain (120 dB). (b) What is the pressure amplitude in this wave?

15. (II) If two firecrackers produce an intensity level of 95 dB at a certain place, what will be the intensity level if only one is exploded?

16. (II) A single mosquito 10 m from a person makes a sound close to the threshold of human hearing (0 dB). What will be the intensity level of 1000 such mosquitoes?

17. (II) A 75-dB sound wave strikes an eardrum whose area is 5.0×10^{-5} m^2. How much energy is absorbed by the eardrum per second?

18. (II) (a) Estimate the power output of sound from a person speaking in normal conversation. Use Table 16–2. Assume the sound spreads roughly uniformly over a hemisphere in front of the mouth. (b) How many people would produce a total sound output of 100 W of ordinary conversation?

19. (II) A stereo amplifier is rated at 25 W output at 1000 Hz. The output drops by 2 dB at 20 Hz. What is the power output at 20 Hz?

20. (II) What is the resultant intensity level when an 80-dB sound and an 85-dB sound are heard simultaneously?

21. (II) In audio and communications systems, the *gain*, β, in decibels is defined as:

$$\beta = 10 \log \frac{P_{out}}{P_{in}},$$

where P_{in} is the power input to the system and P_{out} is the power output. A particular stereo amplifier puts out 35 W of power for an input of 1 mW. What is its gain in dB?

22. (II) (a) Show that the intensity level, β, can be written in terms of the pressure amplitude, p_M, as

$$\beta(dB) = 20 \log \frac{p_M}{p_{MO}},$$

where p_{MO} is the pressure amplitude at some reference level. (b) The reference pressure p_{MO} is often taken to be 3.0×10^{-5} N/m^2 corresponding to an intensity of 1.0×10^{-12} W/m^2. What would the intensity level be if p_M were 1 atm?

23. (II) The intensity level 12.0 m from a loudspeaker, placed in the open, is 100 dB. What is the acoustic power output (W) of the speaker?

24. (II) A jet plane emits 2.0×10^5 J of sound energy per second. (a) What is the intensity level 40 m away? Air absorbs sound at a rate of about 7 dB/km; calculate what the intensity level will be (b) 1 km and (c) 5 km away from this jet plane, taking into account air absorption.

SECTION 16-4

25. (I) The G string on a violin has a fundamental frequency of 196 Hz. The length of the vibrating portion is 32 cm and has a mass of 0.50 g. Under what tension must the string be placed?

26. (I) An unfingered guitar string is 0.70 m long and is tuned to play E above middle C (330 Hz). How far from the end of this string must the finger be placed to play A above middle C (440 Hz)?

27. (I) How far from the end of the flute in Example 16-6 should the hole be that must be uncovered to play D above middle C at 294 Hz?

28. (I) Determine the length of a closed organ pipe that emits middle C (264 Hz) when the temperature is $15°C$.

29. (II) An organ is in tune at $20°C$. By what fraction will the frequency be off at $10°C$?

30. (II) Calculate the resonant frequency of the column of air in the outer ear of a human being, which is about 2.5 cm long.

31. (II) Each string on a violin is tuned to a frequency $1\frac{1}{2}$ times that of its neighbor. If all the strings are to be placed under the same tension, what must be the mass per unit length of each string relative to that of the lowest string?

32. (II) The A string of a violin is 32 cm long between fixed points with a fundamental frequency of 440 Hz and a linear density of 5.0×10^{-4} kg/m. (a) What are the wave speed and tension in the string? (b) What is the length of the tube of a simple wind instrument (say an organ pipe) closed at one end whose fundamental is also 440 Hz if the speed of sound is 331 m/s in air? (c) What is the frequency of the first overtone of each instrument?

33. (II) (a) At $T = 20°C$, how long must an open organ pipe be if it is to have a fundamental frequency of 264 Hz? (b) If this pipe was filled with helium, what would its fundamental frequency be?

34. (II) A tuning fork is set into vibration above a vertical open tube filled with water. The water level is allowed to drop slowly; as it does so the air in the tube above the water level is heard to resonate with the tuning fork when the distance from the tube opening to the water level is 0.125 m and again at 0.395 m. What is the frequency of the tuning fork? Assume $T = 20°C$.

35. (II) How many overtones are present within the audible range for a 100-cm-long organ pipe at $20°C$ (a) if it is open and (b) if it is closed?

36. (II) An 80-cm-long guitar string of mass 1.50 g is placed near a tube open at one end, and also 80 cm long. How much tension should be in the string so that its third harmonic has the same frequency as the fourth harmonic of the tube? Take $T = 20°C$.

*SECTION 16-5

37. (II) Approximately what are the intensities of the first two overtones of a violin compared to the fundamental? How many decibels softer than the fundamental are the first and second overtones? (See Fig. 16-3.)

SECTION 16-6

38. (I) What will be the "beat frequency" if middle C (262 Hz) and C# (277 Hz) are played together? Will this be audible? What if each is played two octaves lower (each frequency reduced by a factor of 4)?

39. (II) A person hears a pure tone coming from two sources that seems to be in the 500–1000 Hz range. The sound is loudest at points equidistant from the two sources. In order to determine exactly what the frequency is, the person moves about and finds that the sound level is minimal at a point 0.22 m farther from one source than the other. What is the frequency of the sound assuming $T = 20°C$?

40. (II) Two loudspeakers are 2.5 m apart. A person stands 3.0 m from one speaker and 3.5 m from the other. (a) What is the lowest frequency at which destructive interference will occur at this point? (b) Calculate two other frequencies that also result in destructive interference at this point (give the next two highest). Let $T = 20°C$.

41. (II) Two piano strings are supposed to be vibrating at 132 Hz but a piano tuner hears one beat every 2 s when they are played together. (a) If one is vibrating at 132 Hz, what must be the frequency of the other (is there only one answer)? (b) By how much (in percent) must the tension be increased or decreased to bring them in tune?

42. (II) How many beats will be heard if two identical flutes each try to play middle C (262 Hz), but one is at $0.0°C$ and the other at $20.0°C$?

43. (II) The two sources of sound in Fig. 16-4 face each other and emit sounds of equal amplitude and equal frequency (250 Hz) but 180° out of phase. For what minimum separation of the two speakers will there be complete (a) constructive interference and (b) destructive interference, at some point ($T = 20°C$)?

44. (II) Show that the two speakers in Fig. 16-4 must be separated by at least a distance d equal to one-half the wavelength λ of sound if there is to be any place where complete destructive interference occurs. The speakers are in phase.

45. (II) A source emits sound of wavelength 2.80 m and 3.10 m. (a) How many beats per second will be heard (assume $T = 20°C$)? (b) How far apart in space are the regions of maximum intensity?

46. (II) The two sources shown in Fig. 16-4 emit sound waves, in phase, each of wavelength λ and amplitude D_M. Consider a point such as C or D in the diagram, and let r_A and r_B be the distances of this point from source A and source B, respectively. Show that if r_A and r_B are nearly equal ($r_A - r_B \ll r_A$) then the amplitude varies approximately with position as

$$\left(\frac{2D_M}{r_A}\right) \cos \frac{\pi}{\lambda} (r_A - r_B).$$

47. (III) A source of sound waves (wavelength λ) is a distance l from a detector. Sound reaches the detector directly, and also by reflecting off an obstacle as shown in Fig. 16-12. The obstacle is equidistant from source and detector. When the obstacle is a distance d to the right of the line of sight between source and detector, as shown, the two waves

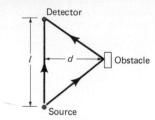

Detector

l d Obstacle

Source

FIGURE 16–12

arrive in phase. How much farther to the right must the obstacle be moved if the two waves are to be out of phase by 180°, so destructive interference occurs? (Assume $\lambda \ll l, d$.)

SECTION 16–7

48. (I) The predominant frequency of a certain police car's siren is 1800 Hz when at rest. What is the detected frequency if the car is (a) moving toward a stationary observer at 90 km/h and (b) is moving away at the same speed? (c) What are the frequencies if the police car is at rest and the observer's car moves at 90 km/h toward or away from the source?

49. (II) Derive a general formula for the changed frequency f' due to the Doppler effect when both the source and the observer are moving.

50. (II) Two trains emit whistles of the same frequency, 260 Hz. If one train is traveling 80 km/h away from an observer at rest and the other train is at rest, what will the observer detect as the beat frequency?

51. (II) If the velocity of blood flow in the aorta is normally about 0.28 m/s, what beat frequency would you expect if 4.20 MHz ultrasound waves were directed along the flow and reflected from the red blood cells? Assume the waves travel with a speed of 1.5×10^3 m/s, which is close to that of sound in water.

52. (II) (a) Use the binomial expansion to show that Eqs. 16–9a and 16–10a become essentially the same for small relative velocity between source and observer. (b) What percent error would result if Eq. 16–10a were used instead of Eq. 16–9a for a relative velocity of 25 m/s?

53. (II) The Doppler effect using ultrasonic waves of frequency 1.8 MHz is used to monitor the heartbeat of a fetus. A (maximum) beat frequency of 600 Hz is observed. Assuming that the speed of sound in tissue is 1.5×10^3 m/s calculate the maximum velocity of the surface of the beating heart.

54. (II) In the above problem, the beat frequency is found to appear and then disappear 180 times per minute, which reflects the fact that the heart is beating and its surface changes speed. What is the heartbeat rate?

55. (III) A factory whistle emits sound of frequency 650 Hz. On a day when the wind velocity is 12.0 m/s from the north, what frequency will observers hear who are located, at rest, (a) due north, (b) due south, (c) due east, (d) due west, of the whistle? What frequency is heard by a cyclist heading (e) north, or (f) west, toward the whistle at 15.0 m/s? ($T = 20°C$.)

*SECTION 16–8

*56. (I) A boat traveling 9.2 m/s makes a bow wave at an angle of 18° to its direction of motion. What is the speed of the water waves?

*57. (II) Show that the angle θ a sonic boom makes with the path of a supersonic object is given by Eq. 16–11.

*58. (II) An airplane travels at Mach 2.8 where the speed of sound is 310 m/s. (a) What is the angle the shock wave makes with the direction of the airplane's motion? (b) If the plane is flying at a height of 6500 m, how long after it is directly overhead will a person on the ground hear the shock wave?

*59. (II) The excess pressure p in a sonic boom decreases with the altitude h of an aircraft approximately as

$$p \propto \frac{1}{h^{3/4}}$$

By what factor does the pressure change if a plane flies twice as high? Compare to an inverse-square law.

Temperature, Thermal Expansion, and the Ideal Gas Law

17

In this and the next few chapters (Chapters 17 through 21), we will study the intimately related topics of temperature, heat, thermodynamics, and kinetic (atomic) theory. We will often consider a particular **system**, by which we mean a particular object or set of objects; everything else in the universe is called the "environment." In order to describe the *state* (or condition) of a particular system—such as a gas in a container—we can do so from either a microscopic or macroscopic point of view. A **microscopic** description would involve details of the motion of all the atoms or molecules making up the system and this could be very complicated; this is the subject matter of kinetic theory (and statistical mechanics), which we will discuss mainly in the next chapter. A **macroscopic** description is given in terms of quantities that are more or less detectable directly by our senses, such as volume, mass, pressure, and temperature. The description of processes in terms of macroscopic quantities is the field of **thermodynamics**. The number of macroscopic variables required to describe the state of a system at any time depends on the type of system. To describe the state of a pure gas in a container, for example, we need only three variables, which could be the volume, the pressure, and the temperature; quantities such as these that can be used to describe the state of a system are called *state variables*.

The emphasis in this chapter is on the concept of temperature. We will begin, however, with a brief discussion of the theory that matter is made up of atoms and that these atoms are in continuous random motion. This theory is called the *kinetic theory* ("kinetic," you may remember, comes from the Greek word for "moving"), and we will discuss it in more detail in Chapter 18.

 ## 17–1 Atoms

The idea that matter is made of atoms dates back to the ancient Greeks. According to the Greek philosopher Democritus, if a given substance—say, a piece of iron—were cut into smaller and smaller bits, eventually a smallest piece of that substance would be obtained which could not be divided further. This smallest piece was called an **atom**, which comes from the Greek *atomos*, "indivisible."[†] An alternative to the atomic theory of matter was the idea that matter is continuous and can be subdivided indefinitely.

Today the atomic theory is generally accepted by scientists. The experimental evidence in its favor, however, came mainly in the eighteenth and nineteenth centuries, and much of it was obtained from the analysis of chemical reactions. A crucial piece of evidence was the *law of definite proportions*, which is a summation of experimental results collected during the half century prior to 1800; it states that when two or more elements combine to form a compound, they always do so in the same proportions by weight. For example, table salt is always formed from 23 parts sodium and 35 parts chlorine; and water is formed from one part hydrogen and eight parts oxygen. A continuous theory of matter could hardly account for the law of definite proportions but, as John Dalton (1766–1844) pointed out, the atomic theory could: the weight proportions of each element required to form a compound corresponded to the relative weights of the combining atoms; one atom of sodium (Na), for example, could combine with one atom of chlorine (Cl) to form one molecule of salt (NaCl), and one atom of sodium would have a mass 23/35 times as large as one of chlorine. By measuring the relative amounts of each element needed to form a large variety of compounds, experimenters established the relative weights of atoms. Hydrogen, the lightest atom, was arbitrarily assigned the relative weight of 1; on this scale, carbon was about 12, oxygen 16, sodium 23, and so on.[‡]

Today, we speak of the relative masses of atoms and molecules—what we call the **atomic mass** or **molecular mass**, respectively[§]—and these are based on assigning an ordinary carbon atom, ^{12}C, the value of exactly 12.0000 atomic mass units (u). The atomic mass of hydrogen is then 1.0078 u, and the values for other atoms are as listed in Appendix D.

Another important piece of evidence for the atomic theory is the so-called *Brownian movement*, named after the biologist Robert Brown, who is credited with its discovery in 1827. While he was observing tiny pollen grains suspended in water under his microscope, Brown noticed that the tiny grains moved about in tortuous paths even though the water appeared to be perfectly still. The atomic theory easily explains Brownian movement if the further reasonable assumption is made that the atoms of any substance are continually in motion. Then Brown's tiny pollen grains are jostled about by the vigorous barrage of rapidly moving molecules of water.

In 1905, Albert Einstein examined[‖] Brownian movement from a theoretical point of view and was able to calculate from the experimental data the approximate size and mass of atoms and molecules. His calculations showed that the diameter of a typical atom is about 10^{-10} m.

At the start of Chapter 12 we distinguished the three common states of matter—solid, liquid, gas—based on *macroscopic*, or "large-scale," properties. Now let us see how these three phases of matter differ from the atomic, or *microscopic*, point of view. Clearly, atoms and molecules must exert attractive forces on each other. For how else could a brick or a piece of aluminum stay together in one piece? These forces are of an electrical nature (more on this in

---

[†] Today, of course, we don't consider the atom as indivisible, but rather as consisting of a nucleus (containing protons and neutrons) and electrons. See Chapters 40–43.

[‡] It was not quite so simple, however. For example, from the various compounds oxygen formed, its relative weight was judged to be 16; but this was inconsistent with the weight ratio in water of oxygen to hydrogen (only 8 to 1). This difficulty was explained by assuming two H atoms combine with one O atom to form a water molecule. To make a self-consistent scheme, many other molecules also had to be judged as containing more than one atom of a given type.

[§] The terms *atomic weight* and *molecular weight* are popularly used for these quantities, but properly speaking we are comparing masses.

[‖] It is possible that Einstein was unaware of Brown's work, and so independently predicted, from theoretical ideas, the existence of Brownian movement.

later chapters). On the other hand, if the molecules come too close together, the force becomes repulsive (electric repulsion between their outer electrons); they thus maintain a minimum distance from each other. In a solid material, the attractive forces are strong enough that the atoms or molecules are held in more or less fixed positions, usually in an array known as a crystal lattice, as shown in Fig. 17–1a. The atoms or molecules in a solid are in motion—they vibrate about their nearly fixed positions. In a liquid, the atoms or molecules are moving more rapidly, or the forces between them are weaker, so that they are sufficiently free to roll over one another as in Fig. 17–1b. In a gas, the forces are so weak, or the speeds so high, that the molecules do not even stay close together. They move rapidly every which way, Fig. 17–1c, filling any container and occasionally colliding with one another. On the average, the speeds are sufficiently high in a gas that when two molecules collide, the force of attraction is not strong enough to keep them close together and they fly off in new directions.

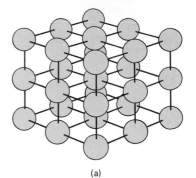

(a)

 ## 17–2 Temperature; Thermometers and Temperature Scales

In everyday life, **temperature** refers to how hot or cold an object is. A hot oven is said to have a high temperature whereas a cold tray of ice is said to have a low temperature.

Many properties of matter change with temperature. For example, most materials expand when heated. An iron beam is longer when hot than when cold; concrete roads and sidewalks expand and contract slightly according to temperature, which is why compressible spacers are placed at regular intervals. The electrical resistance of matter changes with temperature (see Chapter 26). And so too does the color radiated by objects, at least at high temperatures: you may have noticed that the heating element of an electric stove glows with a red color when hot; at higher temperatures, solids such as iron glow orange or even white; the white light from an ordinary incandescent light bulb comes from an extremely hot tungsten wire.

An instrument designed to measure temperature is called a **thermometer**. There are many kinds of thermometers, but they all have in common the fact that they depend on some property of matter that changes with temperature. Most common thermometers rely on the expansion of a material with an increase in temperature. The first thermometer, invented by Galileo, made use of the expansion of a gas. Common thermometers today consist of a hollow glass tube filled with mercury or with alcohol colored with a red dye. The liquid expands more than the glass when the temperature is increased, so the liquid level rises in the tube. Although metals also expand with temperature, the change in length of, say, a metal rod is generally too small to measure accurately for ordinary changes in temperature. However, a useful thermometer can be made by bonding together two dissimilar metals whose rate of expansion is different. When the temperature is increased, the different amounts of expansion cause the bimetallic strip to bend. Often the bimetallic strip is in the form of a coil, one end of which is fixed and the other is attached to a pointer. This kind of thermometer is used as ordinary air thermometers, oven thermometers, and in automobiles as the automatic choke.

In order to measure temperature quantitatively, some sort of numerical scale must be defined. The most common scale today is the *Celsius* scale, sometimes called the *centigrade* scale. In the United States, the *Fahrenheit* scale is also common. The most important scale in scientific work is the absolute, or Kelvin, scale, and it will be discussed in Section 17–7.

One way to define a temperature scale is to assign arbitrary values to two readily reproducible temperatures. For both the Celsius and Fahrenheit scales

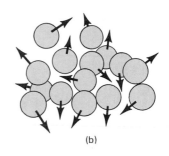

(b)

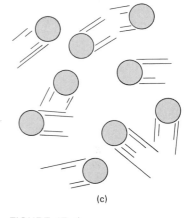

(c)

FIGURE 17–1

Atomic arrangements in (a) a crystalline solid, (b) a liquid, (c) a gas.

these two fixed points are chosen to be the freezing point and the boiling point[†] of water, both taken at atmospheric pressure. On the Celsius scale, the freezing point of water is chosen to be 0°C ("zero degrees Celsius") and the boiling point 100°C. On the Fahrenheit scale, the freezing point is defined as 32°F and the boiling point 212°F. A practical thermometer is calibrated by placing it in carefully prepared environments at each of the two temperatures and marking the position of the mercury or pointer. For a Celsius scale, the distance between the two marks is then divided into one hundred equal intervals separated by small marks representing each degree between 0°C and 100°C (hence the name "centigrade scale" meaning "hundred steps"). For a Fahrenheit scale, the two points are labeled 32°F and 212°F and the distance between them is divided into 180 equal intervals. For temperatures below the freezing point of water and above the boiling point of water the scales can be extended using the same equally spaced intervals. However, ordinary thermometers can be used only over a limited temperature range because of their own limitations—for example, the mercury in a mercury-in-glass thermometer solidifies at some point (39°C), below which the thermometer will be useless; it is also rendered useless above temperatures where the fluid vaporizes (357°C for mercury). For very low or very high temperatures, specialized thermometers are required, some of which we will mention later.

Every temperature on the Celsius scale corresponds to a particular temperature on the Fahrenheit scale, Fig. 17–2. It is easy to convert from one to the other if you remember that 0°C corresponds to 32°F and that a range of 100°C on the Celsius scale corresponds to a range of 180° on the Fahrenheit scale. Thus, one Fahrenheit degree (1 F°) corresponds to $100/180 = \frac{5}{9}$ of a Celsius degree (1 C°). That is, $1\ \text{F}° = \frac{5}{9}\text{C}°$. (Notice that when we refer to a specific temperature, we say "degrees Celsius," as in 20°C; but when we merely refer to a change in temperature or a temperature interval, we say "Celsius degrees," as in "1 C°.")

 EXAMPLE 17–1 Normal body temperature is 98.6°F. What is this on the Celsius scale?

SOLUTION First we note that 98.6°F is $98.6 - 32.0 = 66.6\ \text{F}°$ above the freezing point of water. Since each F° is equal to $\frac{5}{9}$ C°, this corresponds to $66.6 \times \frac{5}{9} = 37.0$ Celsius degrees above the freezing point; as the freezing point is 0°C, the temperature is 37.0°C.

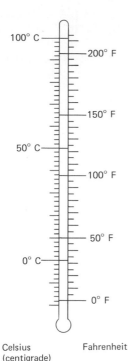

Celsius   Fahrenheit
(centigrade)

FIGURE 17–2

Celsius and Fahrenheit scales compared.

17–3   The Constant-Volume Gas Thermometer

Different materials do not expand in quite the same way over a wide temperature range. Consequently, if we calibrate different kinds of thermometers exactly as described above, they will not usually agree precisely. Because of how we calibrated them, they will agree at 0°C and at 100°C. But because of the different expansion properties, they may not agree at intermediate temperatures (remember we arbitrarily divided the thermometer scale into 100 equal divisions between 0°C and 100°C). Thus a carefully calibrated mercury-in-glass thermometer might register a temperature of 52.0°C whereas a carefully calibrated thermometer of another type might read 52.6°C.

Because of this discrepancy, some standard kind of thermometer must be chosen so that these intermediate temperatures can be precisely defined. The chosen standard for this purpose is the so-called *constant-volume gas thermometer*. As shown schematically in Fig. 17–3, this thermometer consists of a bulb filled

FIGURE 17–3

A constant-volume gas thermometer.

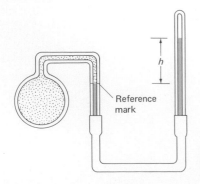

Reference
mark

[†] To be specific, the freezing point of a substance is defined as that temperature at which the solid and liquid phases coexist in equilibrium—that is, without the liquid changing into the solid or vice versa. Experimentally, this is found to occur at only one definite temperature, for a given pressure. Similarly, the boiling point is defined as that temperature at which the liquid and gas coexist in equilibrium. Since these points vary with pressure, the pressure must be specified (usually it is 1 atm).

with a dilute gas connected by a thin tube to a mercury manometer.[†] The volume of the gas is kept constant by raising or lowering the right-hand tube of the manometer so that the mercury in the left tube coincides with the reference mark. An increase in temperature causes a proportional increase in pressure in the bulb; so the tube must be lifted higher to keep the gas volume constant. The height of the mercury in the right-hand column is then a measure of the temperature. This thermometer can be calibrated and the resulting scale is defined as the standard temperature scale. One great advantage of this thermometer is that at low densities and pressures, all gases behave in the same way; and in the limit as the pressure in the gas is allowed to approach zero, all gases yield the same precise temperature when used in a constant-volume gas thermometer. Thus the constant-volume gas thermometer gives the same reading independent of the gas used. It is for this reason that it has been chosen to define the temperature scale precisely. We will discuss thermometers and a standard temperature scale in Section 17–11.

⬚ * 17–4   Thermal Equilibrium; Zeroth Law of Thermodynamics

Although the concept of temperature is obviously useful in everyday life, its value as a precise and useful quantity in physics is not totally obvious. We now investigate this.

A system is said to be in **thermal equilibrium** when the variables that describe it are the same throughout the system[‡] and are not changing in time. Indeed, if a system is not in thermal equilibrium at a given moment, we cannot even assign a pressure or temperature to the system. For example, if a pot of water is being heated on a stove, different parts of the water will register different temperatures (which may not be clearly defined and will be changing), and we cannot assign a temperature to the pot as a whole until the heating stops and the water eventually settles down to a uniform temperature; only then will it be in thermal equilibrium. Or consider a long, air-filled tube, closed at one end and fitted with a movable piston at the other end. If you quickly push the cylinder in a short distance, the pressure at that moment will be higher right next to the cylinder than elsewhere in the tube; this compression will then propagate as a pressure wave pulse down the tube. As it travels back and forth along the tube, it eventually dampens out and the whole tube again reaches a uniform pressure, and only then is it in thermal equilibrium.

Now consider two different systems whose pressures and temperatures are different. If they are kept apart so that they do not interact and thus cannot influence each other, they can remain at their different pressures and temperatures. If, on the other hand, they are put into contact so that they do influence one another, they are said to be in *thermal contact*.[§] Two metal blocks, one at 300°C and the other at 50°C, will influence each other if placed in thermal contact; the temperature of the hotter block will decrease whereas the temperature of the cooler block will rise until the two blocks reach the same temperature. At this point, the two objects will be in *thermal equilibrium with each other*. Suppose now that we have two separate systems that are not in contact, but if they were to be put into thermal contact with each other, no observable changes in their state variables

---

[†] The actual apparatus is considerably more complicated than is shown here. Many corrections are also required, such as for expansion or contraction of the bulb holding the gas.

[‡] Sometimes in this case we say the system is in "thermal equilibrium with itself."

[§] Two systems that are kept apart so they cannot influence each other are said to be separated, ideally, by an *adiabatic wall*; in essence, an adiabatic wall would be a perfect thermal insulator that allowed no flow of heat. On the other hand, if two systems are put in thermal contact so that they do influence each other, they are sometimes said to be connected by a *diathermic wall*; a diathermic wall is a good conductor of heat—a thin metal plate would be a good diathermic wall.

would occur. We say that these two separate systems are in thermal equilibrium. That is, two systems are said to be in thermal equilibrium even when not in contact, if, when put into contact, they would be in equilibrium.

One way to determine if two systems, A and B, are in thermal equilibrium would be to make use of a third system, C (which could be considered a thermometer). Suppose that C and A are in thermal equilibrium and that C and B are in thermal equilibrium. Does this imply that A and B are necessarily in thermal equilibrium with each other? You may say, obviously yes. Actually, it isn't completely obvious; after all, if Jane likes George and Joan likes George, it is not necessarily true that Jane likes Joan. Nonetheless, our experience, and a great deal of experiment, indicates that it is true for thermal equilibrium: *if two systems are in thermal equilibrium with a third system, then they are in thermal equilibrium with each other.* This postulate is sometimes called the **zeroth law of thermodynamics**. (It has this rather odd name since it was not until after the great first and second laws of thermodynamics were worked out that scientists realized that this apparently obvious postulate needed to be stated first.)

The quantity we have called temperature is a property of a system that determines whether the system will be in thermal equilibrium with other systems. When two systems are in thermal equilibrium, their temperatures are, by definition, equal. This is consistent with our everyday notion of temperature, since when a hot body and a cold one are put into contact, they eventually come to the same temperature. Thus the importance of the zeroth law is that it allows a useful definition of temperature. Suppose that the zeroth law were not valid. That is, suppose A and C are in thermal equilibrium and so are B and C, but A and B are not in thermal equilibrium; this would imply that $T_A = T_C$ and $T_B = T_C$, but that $T_A \neq T_B$, which clearly would not be consistent, so temperature would be a useless variable. But no experiment has contradicted the zeroth law, so we assume it is valid, and therefore temperature is a useful quantity.

 17–5   Thermal Expansion

Most substances expand when heated and contract when cooled. However, the amount of expansion or contraction varies, depending on the material.

Experiments indicate that the change in length ΔL of almost all solids is, to a very good approximation, directly proportional to the change in temperature ΔT. As might be expected, the change in length is also proportional to the original length of the object, L_0. That is, for the same temperature change, a 4-m-long iron rod will increase twice as much in length as a 2-m-long iron rod. We can write this proportionality as an equation:

$$\Delta L = \alpha L_0 \Delta T, \tag{17–1}$$

where α, the proportionality constant, is called the *coefficient of linear expansion* for the particular material and has units of $(C°)^{-1}$, meaning "per degree." The values of α for various materials[†] at 20°C are listed in Table 17–1. It should be noted that α does vary slightly with temperature (which is why thermometers made of different materials do not agree precisely). However, if the temperature range is not too great, the variation can usually be ignored.

EXAMPLE 17–2   An iron ring is to fit snugly on a cylindrical iron rod. At 20°C the diameter of the rod is 6.453 cm and the inside diameter of the ring is 6.420 cm. To what temperature must the ring be brought if its hole is to be large enough so it will slip over the rod?

[†] For certain crystalline materials, α may be different for the three spatial directions. We won't be concerned with this.

TABLE 17–1

17–5 Thermal Expansion **335**

Coefficients of expansion at 20°C

Material	Coefficient of Linear Expansion, α $(C°)^{-1}$	Coefficient of Volume Expansion, β $(C°)^{-1}$
Solids		
Aluminum	25×10^{-6}	75×10^{-6}
Brass	19×10^{-6}	56×10^{-6}
Iron or steel	12×10^{-6}	35×10^{-6}
Lead	29×10^{-6}	87×10^{-6}
Glass (Pyrex)	3×10^{-6}	9×10^{-6}
Glass (ordinary)	9×10^{-6}	27×10^{-6}
Quartz	0.4×10^{-6}	1×10^{-6}
Concrete and brick	$\approx 12 \times 10^{-6}$	$\approx 36 \times 10^{-6}$
Marble	$1.4–3.5 \times 10^{-6}$	$4–10 \times 10^{-6}$
Liquids		
Gasoline		950×10^{-6}
Mercury		180×10^{-6}
Ethyl alcohol		1100×10^{-6}
Glycerin		500×10^{-6}
Water		210×10^{-6}
Gases		
Air (and most other gases at atmospheric pressure)		3400×10^{-6}

SOLUTION   The hole in the ring must be increased from a diameter of 6.420 cm to 6.453 cm. The ring must be heated since the hole diameter will increase linearly with temperature. (Note that the material does *not* expand to fill the hole; in a solid object, without a hole, all sections of the object—the ring—increase with temperature; the presence of the hole doesn't affect this, so the diameter increases with increased temperature.) We solve for ΔT in Eq. 17–1 and find

$$\Delta T = \frac{\Delta L}{\alpha L_0} = \frac{(6.453 \text{ cm} - 6.420 \text{ cm})}{(12 \times 10^{-6} \text{ C}°^{-1})(6.420 \text{ cm})} = 430 \text{ C}°.$$

So it must be raised to at least 450°C.

The change in volume of a material which undergoes a temperature change is given by a relation which is similar to Eq. 17–1, namely

$$\Delta V = \beta V_0 \Delta T, \tag{17–2}$$

where ΔT is the change in temperature, V_0 is the original volume, ΔV is the change in volume, and β is the *coefficient of volume expansion*. The units of β are $(C°)^{-1}$. The value of β for a number of materials is given in Table 17–1. Notice that for solids β is normally equal to approximately 3α (why this should be so is the subject of problem 18); but note that this is not true for solids that are not isotropic (isotropic means having the same properties in all directions). Notice also that linear expansion has no meaning for liquids and gases since they do not have fixed shapes.

Eqs. 17–1 and 17–2 are accurate only if ΔL or ΔV is small compared to L_0 or V_0. This is of particular concern for liquids and even more so for gases because of the large values of β. Furthermore, the variation of β with temperature is also quite large for gases. Therefore a more convenient way of dealing with gases is needed and this is discussed starting in Section 17–7.

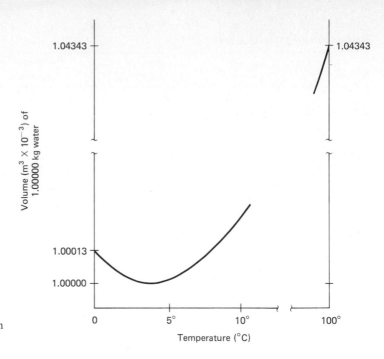

FIGURE 17–4

Volume (m³ × 10⁻³) of 1.00000 kg water as a function
of temperature (note the change of scale).

EXAMPLE 17–3 The 70-L steel gas tank of a car is filled to the top with gasoline at 20°C. The car is then left to sit in the sun and the tank reaches a temperature of 50°C. How much gasoline do you expect to overflow from the tank?

SOLUTION The gasoline expands by

$$\Delta V = \beta V \Delta T = (950 \times 10^{-6} \, \text{C}^{\circ -1})(70 \, \text{L})(30 \, \text{C}^{\circ}) = 2.0 \, \text{L}.$$

The tank too expands. We can think of it as a shell that undergoes volume expansion ($\beta \approx 3\alpha = 36 \times 10^{-6} \, \text{C}^{\circ -1}$); but if it were solid, the exterior surface layer (the shell) would expand just the same. Thus the tank increases in volume by

$$\Delta V = (36 \times 10^{-6} \, \text{C}^{\circ -1})(70 \, \text{L})(30 \, \text{C}^{\circ}) = 0.075 \, \text{L},$$

so the tank expansion has little effect. If this full tank were left in the sun, a couple of liters of gas could spill out into the road.

FIGURE 17–5

Typical curve of potential energy vs. separation of atoms, r, for atoms in a crystal solid (simplified).

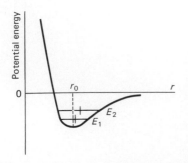

Most substances expand more or less uniformly with an increase in temperature. Water, however, does not follow the usual pattern. If water at 0°C is heated, it actually *decreases* in volume until it reaches 4°C; above 4°C water behaves normally and expands as the temperature is increased, Fig. 17–4. Water thus has its greatest density at 4°C. It is for this reason that water in a lake freezes first at its top surface; when the water cools below 4°C, the less dense cooler water rises (or remains) at the top whereas the denser 4°C water remains below; the cooler water at the top freezes first since it drops below 0°C first.

How can we understand thermal expansion from a microscopic point of view? Let us assume that the atoms in a solid are always in motion, vibrating about their equilibrium positions; and let us assume that their average kinetic energy increases with temperature, as we shall discuss in the next chapter. But as temperature increases, does this mean that the average distance between atoms increases? Experimentally a solid rod gets longer when you raise its temperature, so we conclude that the average distance between atoms must increase. To understand this, let us look at a typical simplified potential-energy diagram as shown in Fig. 17–5, which represents the PE of two atoms versus their separation r. At large

r, we assume the PE ≈ 0, and as r decreases, the PE decreases, indicating an attractive force as discussed in Section 7–6. For r less than r_0 (the equilibrium position) the PE curve rises, indicating a repulsive force. The horizontal lines in Fig. 17–5 labeled E_2 and E_1 represent the total energy for two different temperatures, T_2 and T_1, where $T_2 > T_1$. The short vertical lines for E_1 and E_2 on the diagram represent the midpoints of the motion at these two temperatures. Because the PE curve is not symmetrical, the average separation of atoms is greater for the higher temperature, as shown. Thus thermal expansion is due to the nonsymmetry of the potential-energy function. If the PE curve were symmetrical, there would be no thermal expansion at all. What might the PE diagram look like for water between 0°C and 4°C?

*17–6 Thermal Stresses

In some situations, the ends of a rod or slab of material are rigidly fixed, which prevents expansion or contraction. If the temperature should change, large compressive or tensile stresses will be set up; these are sometimes referred to as *thermal stresses*. The magnitude of these stresses can be calculated using the concept of elastic modulus developed in Chapter 11. To calculate the internal stress, we can think of this process as occurring in two steps. The rod expands (or contracts) by an amount ΔL given by Eq. 17–1, and then a force is applied to compress (or expand) the material back to its original length. The force F required is given by Eq. 11–5:

$$\Delta L = \frac{1}{E}\frac{F}{A}L_0$$

where E is Young's modulus for the material. To calculate the internal stress, F/A, we then set ΔL in Eq. 17–1 equal to ΔL in the equation above and find

$$\alpha L_0 \Delta T = \frac{1}{E}\frac{F}{A}L_0.$$

EXAMPLE 17–4 Blocks of concrete 10 m long are placed end to end with no space in between them to allow for expansion. If the blocks were placed at a temperature of 10°C, what will be the force of compression when the temperature reaches 40°C? The contact area between each block is $0.20\,\text{m}^2$. Will fracture occur?

SOLUTION We solve for F in the equation above and use the value of E found in Table 11–1:

$$F = \alpha \Delta T\, E A$$
$$= (12 \times 10^{-6}/\text{C}°)(30\ \text{C}°)(20 \times 10^9\ \text{N/m}^2)(0.20\ \text{m}^2)$$
$$= 1.4 \times 10^6\ \text{N}.$$

The stress, F/A, is $(1.4 \times 10^6\ \text{N})/(0.20\ \text{m}^2) = 7.0 \times 10^6\ \text{N/m}^2$. This is not far from the ultimate strength of concrete (Table 11–2) under compression and exceeds it for tension and shear. Hence, assuming the concrete is not perfectly aligned, part of the force will act in shear and fracture is likely.

17–7 The Gas Laws and Absolute Temperature

The use of Eq. 17–2 to describe the expansion of a gas is not very useful, partly because the expansion can be so great, and partly because gases generally expand to fill whatever container they are in. Indeed, Eq. 17–2 is meaningful only if the pressure is kept constant. The volume of a gas depends very much on the pressure as well as on the temperature. It is therefore valuable to determine a relation between the volume, the pressure, the temperature, and the mass of a gas. Such a relation is called an **equation of state**.[†]

If the state of a system is changed we will always wait until the pressure and temperature have reached the same values throughout. We thus consider only equilibrium states of a gas. We also note that the results of this section are accurate only for gases that are not too dense (the pressure is not too high, on the order of 1 atm or less) and not close to their condensation (boiling) point.

For a given quantity of gas it is found experimentally that to a good approximation *the volume of a gas is inversely proportional to the pressure applied to it when the temperature is kept constant*. That is,

$$V \propto \frac{1}{P}. \qquad \text{[constant } T\text{]}$$

For example, if the pressure on a gas is doubled, the volume is reduced to half its original volume. This relation is known as *Boyle's law*, after Robert Boyle (1627–1691), who first stated it on the basis of his own experiments. Boyle's law can also be written:

$$PV = \text{constant.} \qquad \text{[constant } T\text{]}$$

That is, at constant temperature, if either the pressure or volume of the gas is allowed to vary, the other variable also changes so that the product PV remains constant.

Temperature also affects the volume of a gas, but a quantitative relationship between V and T was not found until more than a century after Boyle's work. The Frenchman Jacques Charles (1746–1823) found that when the pressure is not too high and is kept constant, the volume of a gas increases with temperature at a nearly constant rate as shown in Fig. 17–6a. However, all gases liquefy at low temperatures

---

[†] An equation of state can also be sought for solids and liquids since for them too the volume depends on the mass, temperature, and external pressure, although the temperature and pressure have much less effect than on gases. However, the situation with solids and liquids, because of their complexity, is much more complicated.

FIGURE 17–6

Volume of a gas as a function of (a) Celsius temperature, (b) Kelvin temperature, when the pressure is kept constant.

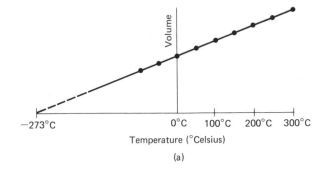

(a)

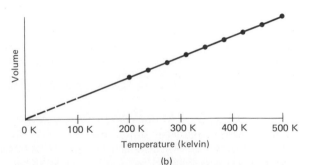

(b)

(for example oxygen liquefies at $-183°$C) and so the graph cannot be extended below the liquefaction point. Nonetheless, the graph is essentially a straight line and if projected to lower temperatures as shown by the dashed line, it crosses the axis at about $-273°$C.

This kind of graph can be drawn for any gas, and the straight line always projects back to $-273°$C at zero volume. This seems to imply that if a gas could be cooled to $-273°$C it would have zero volume, and at lower temperatures a negative volume—which makes no sense, of course. This led to the suggestion that perhaps $-273°$C was the lowest temperature possible, and many other more recent experiments indicate that this is so. This temperature is referred to as the *absolute zero* of temperature. Its value has been measured to be about $-273.15°$C.

Absolute zero forms the basis of a temperature scale known as the **absolute**, or **Kelvin, scale**, and it is used extensively in scientific work. On this scale the temperature is specified as degrees Kelvin or, preferably, simply as kelvins (K) without the degree sign. The intervals are the same as for the Celsius scale, but the zero on this scale (0 K) is chosen as absolute zero itself. Thus the freezing point of water ($0°$C) is 273.15 K and the boiling point 373.15 K. Indeed, any temperature on the Celsius scale can be changed to kelvins by adding 273.15 to it:

$$T(K) = T(°C) + 273.15.$$

We will discuss this scale more precisely later.

Now let us look at Fig. 17–6b where we see that the graph of the volume of a gas versus absolute temperature is essentially a straight line. Thus, to a good approximation, *the volume of a given amount of gas is directly proportional to the absolute temperature when the pressure is kept constant.* This is known as *Charles's law,* and can be written

$$V \propto T. \qquad \qquad \text{[constant } P \text{]}$$

A third gas law, known as *Gay-Lussac's law,* after Joseph Gay-Lussac (1778–1850), states that *at constant volume, the pressure of a gas is directly proportional to the absolute temperature:*

$$P \propto T. \qquad \qquad \text{[constant } V \text{]}$$

A familiar example of this is the fact that a closed jar or aerosol can thrown into a fire will explode due to the increase in gas pressure inside. This relation is also the basis for the constant-volume gas thermometer (Section 17–3).

The laws of Boyle, Charles, and Gay-Lussac are not really laws in the sense that we use this term today (precise, deep, wide-ranging validity—see Section 4–5). They are really only approximations that are accurate for real gases only as long as the pressure and density of the gas are not too high, and the gas is not too close to condensation. Unfortunately, the term law applied to these three relationships has become traditional, so we have stuck with this usage.

 ## 17–8   The Ideal Gas Law

The gas laws of Boyle, Charles, and Gay-Lussac were obtained by using a technique that is very useful in science: namely, to hold one or more variables constant in order to see clearly the effects of changing only one of the variables. These laws can now be combined into a single more general relation between the pressure, volume, and temperature of a fixed quantity of gas:

$$PV \propto T.$$

This relation indicates how any of the quantities P, V, or T will vary when the other two quantities change. This relation reduces to Boyle's, Charles's, or Gay-Lussac's law when either the temperature, the pressure, or the volume, respectively, is held constant.

Finally, we must incorporate the effect of the amount (or mass) of gas that is present. Anyone who has blown up a balloon knows that the more air forced into the balloon, the bigger it gets. Indeed, careful experiment shows that at constant temperature and pressure, the volume V of an enclosed gas increases in direct proportion to the mass m of gas present. Hence we can write

$$PV \propto mT.$$

This proportion relates the significant variables for gases. It can be made into an equation by inserting a constant of proportionality. Experiment shows that this constant has a different value for different gases. However, the constant of proportionality turns out to be the same for all gases if instead of the mass m we use the number of moles; one **mole** (abbreviated mol) is defined as the amount of substance which contains as many atoms or molecules as there are in 0.012 kg of carbon 12 (whose atomic mass is exactly 12 u). A simpler but equivalent definition is: 1 mol is that number of grams numerically equal to the molecular mass of the substance. For example, the molecular mass of hydrogen gas (H_2) is 2.0 u (since each molecule contains two atoms of hydrogen and each atom has an atomic mass of 1.0 u); thus 1 mol of H_2 is equal to 0.0020 kg. Similarly, 1 mol of neon gas has a mass of 0.010 kg, and 1 mol of CO_2 has a mass of $[12 + (2 \times 16)] \times 10^{-3}$ kg = 0.044 kg. The mole is the official unit in the SI system. Sometimes the kilogram-mole (kg·mol) is used, which is that number of kilograms numerically equal to the molecular mass of the substance; 1 kg·mol = 10^3 mol.

Now we write the above proportion as an equation:

$$PV = nRT, \tag{17-3}$$

where n represents the number of moles and R is the constant of proportionality. R is called the *universal gas constant* because its value is found experimentally to be the same for all gases; to the best accuracy today its value is

$$R = (8.31441 \pm 0.00026) \text{ J/mol·K}.$$

Rounded off, and given in other (sometimes useful) sets of units (only the first is the proper SI unit), its value is:

$$R = 8.314 \text{ J/(mol·K)} \qquad [\text{SI units}]$$
$$= 0.0821 \text{ (L·atm)/(mol·K)}$$
$$= 1.986 \text{ calories/(mol·K)}.$$

Equation 17–3 is called the **ideal gas law**, or the **equation of state for an ideal gas**. We use the term "ideal" because real gases do not follow Eq. 17–3 precisely, particularly at high pressures or when the gas is near the liquefaction point. However, at pressures on the order of 1 atm or less, and when T is not close to the boiling point of the gas, Eq. 17–3 is quite accurate. (A more accurate equation of state for real gases will be discussed in Section 18–6.)

The ideal gas law is an extremely useful tool, and we now consider some examples. We will often refer to "standard conditions" or "standard temperature and pressure" (STP) which means $T = 273 \text{ K} (0°C)$ and $P = 1.00 \text{ atm} = 1.01 \times 10^5 \text{ N/m}^2$.

EXAMPLE 17–5   Determine the volume of 1 mol of any gas at STP, assuming it behaves like an ideal gas.

SOLUTION   We solve for V in Eq. 17–3:

$$V = \frac{nRT}{P} = \frac{(1.00 \text{ mol})(8.31 \text{ J/mol·K})(273 \text{ K})}{(1.01 \times 10^5 \text{ N/m}^2)}$$
$$= 22.4 \times 10^{-3} \text{ m}^3.$$

Since 1 L is 1000 cm^3 = 1×10^{-3} m^3, 1 mol of any gas has a volume of 22.4 L at STP.

The value for the volume of 1 mol of gas at STP (22.4 L) is worth remembering for it sometimes makes calculation simpler as the following example shows.

EXAMPLE 17–6 A flexible container of oxygen (O_2) at STP has a volume of 10.0 m³. What is the mass of gas enclosed?

SOLUTION Since 1 mol occupies a volume of 22.4×10^{-3} m³ the 10.0 m³ of oxygen corresponds to $(10.0)/(22.4 \times 10^{-3}) = 446$ mol. Since 1 mol has a mass of 0.0320 kg, the mass of oxygen is $(446 \text{ mol})(0.0320 \text{ kg/mol}) = 14.3$ kg.

Frequently, volume is specified in liters and pressure in atmospheres. Rather than convert these to SI units, we can instead use the value of R given above as 0.0821 L·atm/mol·K. In many situations it is not necessary to use the values of R at all. For example, many problems involve a change in the pressure, temperature, and volume of a fixed amount of gas. In this case, $PV/T = nR = $ constant, since n and R remain constant. If we now let P_1, V_1, and T_1 represent the appropriate variables initially, and P_2, V_2, T_2 represent the variables after the change is made, then we can write

$$\frac{P_1 V_1}{T_1} = \frac{P_2 V_2}{T_2}.$$

If we know any five of the quantities in this equation, we can solve for the sixth.

EXAMPLE 17–7 An automobile tire is filled to a gauge pressure of 200 kPa at 10°C. After driving 100 km the temperature within the tire rises to 40°C. What is the pressure within the tire now?

SOLUTION Since the volume remains essentially constant, $V_1 = V_2$ and therefore

$$\frac{P_1}{T_1} = \frac{P_2}{T_2}.$$

This is, incidentally, a statement of Gay-Lussac's law. Since the pressure given is the gauge pressure, we must add atmospheric pressure = 101 kPa to get the absolute pressure $P_1 = 301$ kPa. Then

$$P_2 = \frac{P_1}{T_1} T_2 = \frac{(3.01 \times 10^5 \text{ Pa})(313 \text{ K})}{(283 \text{ K})} = 3.33 \times 10^5 \text{ Pa}.$$

Subtracting out atmospheric pressure we find the resulting gauge pressure to be 232 kPa, which is a 15 percent higher increase. This example illustrates why car manuals suggest checking the pressure when the tires are cold.

17–9 Ideal Gas Law in Terms of Molecules— Avogadro's Number

The fact that the constant R has the same value for all gases is a remarkable reflection of simplicity in nature. This fact was first recognized, although in a slightly different form, by the Italian scientist Amedeo Avogadro (1776–1856). Avogadro stated that *equal volumes of gas at the same pressure and temperature contain equal numbers of molecules.* This is sometimes called *Avogadro's hypothesis.* That this is consistent with R being the same for all gases can be seen as follows. First of all from Eq. 17–3 we see that for the same number of moles, n, and the same pressure and temperature, the volume will be the same for all gases as long as R is the same. And secondly the number of molecules in 1 mol is the same for all gases; this follows directly from the definition of a mole.[†] Thus Avogadro's hypothesis is equivalent to R being the same for all gases.

The number of molecules in a mole is known as **Avogadro's number**, N_A. Although Avogadro conceived the notion, he was not able actually to determine the value of N_A. Indeed, precise measurements had to await the twentieth century. A number of methods have been devised to measure N_A and the accepted value today is

$$N_A = (6.022045 \pm 0.000031) \times 10^{23} \quad [\text{molecules/mole}]$$

or, rounded off,

$$N_A = 6.02 \times 10^{23} \quad [\text{molecules/mole}].$$

Since the total number of molecules, N, in a gas is equal to the number per mole times the number of moles ($N = nN_A$), the ideal gas law, Eq. 17–3, can be written in terms of the number of molecules present:

$$PV = nRT = \frac{N}{N_A} RT,$$

or

$$PV = NkT, \tag{17–4}$$

where $k = R/N_A$ is called *Boltzmann's constant* and has the value

$$k = \frac{R}{N_A} = (1.380662 \pm 0.000044) \times 10^{-23} \text{ J/K}$$

or rounded off,

$$k = 1.38 \times 10^{-23} \text{ J/K}.$$

 EXAMPLE 17–8 Use Avogadro's number to determine the mass of a hydrogen atom.

SOLUTION One mole of hydrogen (atomic mass = 1.008) has a mass of 1.008×10^{-3} kg and contains 6.02×10^{23} atoms. Thus one atom has a mass

$$m = \frac{1.008 \times 10^{-3} \text{ kg}}{6.02 \times 10^{23}} = 1.67 \times 10^{-27} \text{ kg}.$$

Historically, the reverse process was one method used to obtain N_A—that is, from the measured mass of the hydrogen atom.

EXAMPLE 17–9 How many molecules do you breathe in a 1.0-L breath of air?

SOLUTION One mole corresponds to 22.4 L (Example 17–5), so 1.0 L of air is $1/22.4 = 0.045$ mol. Then 1.0 L of air contains $(0.045 \text{ mol})(6.02 \times 10^{23}$ molecules/mole$) = 2.7 \times 10^{22}$ molecules.

17–10 Partial Pressure

When two or more gases occupy the same volume, the total pressure can be written as the sum of the partial pressures of the individual gases. The *partial pressure* of a gas is defined as the pressure each gas would exert if it alone occupied the whole

---

[†] For example, the molecular mass of H_2 gas is 2.0 atomic mass units (u) while that of O_2 gas is 32.0 u. Thus 1 mol of H_2 has a mass of 0.0020 kg and 1 mol of O_2 gas, 0.032 kg. Now the number of molecules in a mole is equal to the total mass M of a mole divided by the mass m of one molecule; since this ratio (M/m) is the same for all gases by definition of the mole, a mole of any gas must contain the same number of molecules.

volume. An experimental law due to Dalton, and known as *Dalton's law of partial pressures*, states that *each gas in a mixture exerts a partial pressure proportional to its molecular concentration*. These ideas are consistent with the ideal gas law, Eq. 17–4. For suppose we have a mixture of three gases with N_1, N_2, and N_3 molecules respectively. The total pressure is

$$P = \frac{NkT}{V}$$

where N is the total number of molecules and $N = N_1 + N_2 + N_3$. Then we can write

$$P = \frac{N_1 kT}{V} + \frac{N_2 kT}{V} + \frac{N_3 kT}{V}$$

$$= P_1 + P_2 + P_3$$

where P_1, P_2, and P_3 are the partial pressures of each of the three gases, and N_1/V, N_2/V, and N_3/V represent their concentrations. Thus the total pressure is equal to the sum of the partial pressures and each of the latter is proportional to its molecular concentration.

For example, 78 percent (by volume) of dry air is nitrogen and 21 percent is oxygen, with much smaller amounts of argon and other gases. At an air pressure of 1 atm, oxygen exerts a partial pressure of 0.21 atm and nitrogen 0.78 atm.

*17–11 Ideal Gas Temperature Scale— a Standard

It is important to have a very precisely defined temperature scale so that measurements of temperature made at different laboratories around the world can be accurately compared. We now discuss such a scale that has been accepted by the general scientific community.

The standard thermometer for this scale is the constant-volume gas thermometer already discussed in Section 17–3. The scale itself is called the *ideal gas temperature scale*, since it is based on the property of an ideal gas that the pressure is directly proportional to the absolute temperature (Gay-Lussac's law). A real gas, which would have to be used in any real constant-volume gas thermometer, approaches this ideal at low density. In other words, the temperature at any point in space is *defined* as being proportional to the pressure in the (nearly) ideal gas used in the thermometer. To set up a scale we need two fixed points. One fixed point will be $P = 0$ at $T = 0$ K. The second fixed point is chosen to be the *triple point* of water, which can be reproduced at different laboratories with great precision. The triple point of water is that point where water in the solid, liquid, and gas states can coexist in equilibrium. This occurs only at a unique temperature and pressure.[†] The pressure at the triple point of water is 4.58 torr and the temperature is approximately 0.01°C. This temperature corresponds to about 273.16 K (since absolute zero is about −273.15°C). In fact, the triple point is now *defined* to be exactly 273.16 K. (Note again the lack of degree sign for temperatures in kelvins.)

The absolute or Kelvin temperature T at any point is then defined, using a constant-volume gas thermometer for an ideal gas, as

$$T = (273.16 \text{ K})\left(\frac{P}{P_{tp}}\right); \qquad \text{[ideal gas; constant volume]} \qquad (17-5a)$$

[†] Liquid water and steam can coexist (the boiling point) at a range of temperatures depending on the pressure. Water boils at a lower temperature when the pressure is less, such as high in the mountains. The triple point represents a more precisely reproducible fixed point than does either the freezing point or boiling point of water at, say, 1 atm.

In this relation, P_{tp} is the pressure in the thermometer at the triple point temperature of water, and P is the pressure in the thermometer when it is at the point where T is being determined. Note that if we let $P = P_{tp}$ in this relation, then $T = 273.16$ K, as it must.

If we use this definition of temperature (Eq. 17–5a) with a constant-volume gas thermometer filled with a real gas (which is all we can do since a perfect ideal gas does not exist), we find that we get different results for the temperature depending on the type of gas that is used in the thermometer. Temperatures determined in this way also vary depending on the amount of gas in the bulb of the thermometer: for example, the boiling point of water at 1.00 atm is found from Eq. 17–5a to be 373.87 K when the gas is O_2 and $P_{tp} = 1000$ torr; if the amount of O_2 in the bulb is reduced so that at the triple point $P_{tp} = 500$ torr, the boiling point of water from Eq. 17–5a is then found to be 373.51 K. If H_2 gas is used instead, the corresponding values are 373.07 K and 373.11 K (see Fig. 17–7). But now suppose we use a particular real gas and make a series of measurements in which the amount of gas in the thermometer bulb is reduced to smaller and smaller amounts, so that P_{tp} becomes smaller and smaller. It is found experimentally that an extrapolation of such data to $P_{tp} = 0$ always gives the same value for the temperature of a given system (such as $T = 373.15$ K for the boiling point of water at 1.00 atm) as shown in Fig. 17–7. Thus the temperature T at any point in space, determined using a constant-volume gas thermometer containing a real gas, is defined using this limiting process:

$$T = (273.16 \text{ K}) \lim_{P_{tp} \to 0} \left(\frac{P}{P_{tp}} \right). \quad \text{[constant volume]} \quad (17\text{–}5b)$$

This defines the *ideal gas temperature scale*. One of the great advantages of this scale is that the value for T does not depend on the kind of gas used. But the scale does depend on the properties of gases in general. Helium has the lowest condensation point of all gases; at very low pressures it liquefies at about 1 K, so temperatures below this cannot be defined on this scale.

It would be advantageous to have a scale that could be used below 1 K, which would also be independent of the properties of any particular material used. Such a scale has been defined which is based on certain thermodynamic properties as discussed in Chapter 21. It is called the "absolute thermodynamic scale" and, strictly speaking, only this scale is called the Kelvin scale with temperatures given in kelvins. However, it is identical to the ideal gas temperature scale over the whole range where the latter can be used (>1 K).

FIGURE 17–7

Temperature readings of a constant-volume gas thermometer for the boiling point of water at 1.00 atm are plotted, for different gases, as a function of the gas pressure in the thermometer at the triple point (P_{tp}). Note that as the amount of gas in the thermometer is reduced, so that $P_{tp} \to 0$, all gases give the same reading, 373.15 K. For pressure less than 0.10 atm (76 torr), the variation shown is less than 0.07 K.

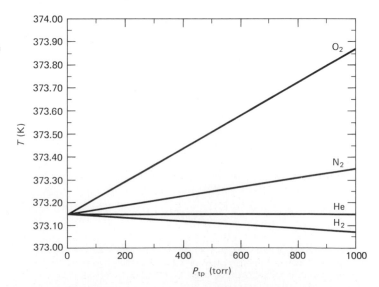

Finally we note that determining ideal gas temperature is time consuming and difficult. Therefore an "international practical temperature scale" has been devised which is easier to use in practice and gives results that are extremely close to the ideal gas scale. It consists of a large number of fixed-point temperatures (such as boiling and freezing points of different materials) along with procedures for determining intermediate temperatures.

Summary

The atomic theory of matter postulates that all matter is made up of tiny entities called *atoms* which are typically 10^{-10} m in diameter. *Atomic* and *molecular masses* are specified on a scale where ordinary carbon (^{12}C) is arbitrarily given the value 12.0000 u (atomic mass units). The distinction between solids, liquids, and gases can be attributed to the strength of the attractive forces between the atoms or molecules and to their average speed.

Temperature is a measure of how hot or cold a body is. *Thermometers* are used to measure temperature on the Celsius (°C), Fahrenheit (°F), and Kelvin (K) scales. Two standard points on each scale are the freezing point of water (0°C, 32°F, 273.15 K) and the boiling point of water (100°C, 212°F, 373.15 K). A change in temperature of one kelvin equals a change of one Celsius degree or 9/5 Fahrenheit degrees.

The change in length, ΔL, of a solid when its temperature changes by an amount ΔT is directly proportional to the temperature change and to its original length L_0. That is, $\Delta L = \alpha L_0 \Delta T$ where α is the coefficient of linear expansion. The change in volume of most solids, liquids, and gases is proportional to the temperature change and to the original volume V_0: $\Delta V = \beta V_0 \Delta T$; the coefficient of volume expansion, β, is approximately equal to 3α for solids. Water is unusual because, unlike most materials whose volume increases with temperature, its volume actually decreases as the temperature increases in the range from 0°C to 4°C.

The *ideal gas law*, or *equation of state for an ideal gas*, relates the pressure, P, volume, V, and temperature, T (in kelvins), of n moles of gas by the equation

$$PV = nRT$$

where $R = 8.314$ J/mol·K for all gases. Real gases obey the ideal gas law quite accurately if they are not at too high a pressure or near their liquefaction point. One *mole* of a substance is defined as the number of grams which is numerically equal to its atomic or molecular mass. *Avogadro's number*, $N_A = 6.02 \times 10^{23}$, is the number of atoms or molecules in 1 mol of any pure substance. The ideal gas law can be written in terms of the number of molecules, N, in the gas as

$$PV = NkT$$

where $k = R/N_A = 1.38 \times 10^{-23}$ J/K is Boltzmann's constant.

Dalton's law of partial pressures states that each gas in a mixture exerts a partial pressure proportional to its molecular concentration; the total pressure is the sum of the partial pressures.

Questions

1. Which has more atoms: one kg of aluminum or one kg of iron (see Appendix D).

2. Suppose a temperature scale was defined in terms of the square of a property, rather than linearly, $T = ax^2$. Let x be the length L of the mercury column in a mercury-in-glass thermometer. Discuss the problems we would have with such a scale.

*3. Suppose system C is not in equilibrium with system A nor in equilibrium with system B. Does this imply that A and B are not in equilibrium? What can you infer regarding the temperatures of A, B, and C?

*4. If system A is in equilibrium with system B, but B is not in equilibrium with system C, what can you say about the temperatures of A, B, and C?

5. Figure 17–8 shows a diagram of a typical *thermostat* used to control a furnace (or other heating or cooling system). The bimetallic strip consists of two strips of different metals bonded together. Explain why this strip bends when the temperature changes, and how this controls the furnace.

6. A circular ring is heated from 20°C to 80°C. Will the hole in the ring become larger or smaller? Explain from the microscopic (molecular) point of view.

7. In the relation $\Delta L = \alpha L_0 \Delta T$, should L_0 be the initial length, the final length, or does it matter?

8. Explain why it is sometimes easier to remove the lid from a tightly closed jar after warming it under hot running water.

FIGURE 17–8

A typical thermostat.

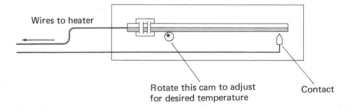

Wires to heater

Rotate this cam to adjust for desired temperature

Contact

9. Long steam pipes often have a section in the shape of a U. Why?

10. Why is it incorrect to say that air expands at a rate of 3.4 mL/L per Celsius degree? When is this true?

11. In view of the fact that fireplaces become hot, why should chimneys not be used for support of a house or building?

12. Explain why it is advisable to add water to an overheated automobile engine only slowly, and only with the engine running.

13. A glass container may break if one part of it is heated or cooled more rapidly than adjacent parts. Explain.

14. The principal virtue of Pyrex glass is that its coefficient of linear expansion is much smaller than that for ordinary glass (Table 17–1). Explain why this gives rise to the increased heat resistance of Pyrex.

15. Why might you expect an alcohol-in-glass thermometer to be more precise than a mercury-in-glass thermometer?

16. When a cold mercury-in-glass thermometer is first placed in a hot tub of water, the mercury initially descends a bit and then rises. Explain.

17. A flat bimetallic strip consists of a strip of aluminum riveted to a strip of iron. When heated, which metal will be on the outside of the curve?

18. Give an explanation for why rivers freeze first at their surface.

19. What temperature do you expect the water to be at the bottom of a deep, cold lake? (Hint: see Fig. 17–4.)

20. Will the buoyant force on an aluminum sphere submerged in water increase or decrease if the temperature is increased from $20°C$ to $40°C$?

21. A flat, uniform cylinder of lead floats in mercury at $0°C$. Will the lead float higher or lower when the temperature is raised?

22. Which scale, Fahrenheit, Celsius, or Kelvin, might be considered most "natural" from a scientific point of view? Discuss.

23. If an atom is measured to have a mass of 6.7×10^{-27} kg, what atom do you think it is?

*24. From a practical point of view, does it really matter what gas is used in a constant-volume gas thermometer? If so, explain. (Hint: see Fig. 17–7.)

 Problems

SECTION 17–1

1. (I) What mass of gold has the same number of atoms as 1 kg of iron?

SECTION 17–2

2. (I) (a) "Room temperature" is often taken to be $68°F$; what is this on the Celsius scale? (b) The temperature of the filament in a light bulb is about $1800°C$; what is this on the Fahrenheit scale?

3. (I) The original Celsius temperature scale [due to Anders Celsius (1701–1744)] defined the freezing point of water as $100°$ and the boiling point as $0°$. What temperature on this scale corresponds to $25°C$?

4. (I) In an alcohol-in-glass thermometer, the alcohol column has length 12.45 cm at $0.0°C$ and length 21.30 cm at $100.0°C$. What is the temperature if the column has length (a) 15.10 cm, and (b) 22.95 cm.

5. (II) At what temperature will the Fahrenheit and Centigrade scales yield the same numerical value?

SECTION 17–5

6. (I) A concrete highway is built of slabs 26 m long. How wide should the expansion cracks between the slabs be to prevent buckling if the range of temperature is $-20°C$ to $+50°C$?

7. (I) A precise steel tape measure has been calibrated at $20°C$. At $40°C$, (a) will it read high or low, and (b) what will be the percentage error?

8. (I) By what factor would the coefficients of linear expansions in Table 17–1 change if British units (ft, lb, °F) were used?

9. (II) To make a secure fit, rivets that are larger than the rivet hole are often used and the rivet is cooled (usually in dry ice) before it is placed in the hole. A steel rivet 2.385 cm in diameter is to be placed in a hole 2.382 cm in diameter. To what temperature must the rivet be cooled if it is to fit in the hole (at $20°C$)?

10. (II) If the density of mercury is 13.59×10^3 kg/m^3 at $20°C$, what will its density be at $65°C$?

11. (II) A uniform rectangular plate of length l and width w has coefficient of linear expansion α. Show that, if we neglect very small quantities, the change in area of the plate due to a temperature change ΔT is $\Delta A = 2\alpha lw \Delta T$.

12. (II) An iron sphere is 28.0 cm in diameter. What will be its change in volume if it is heated from $20°C$ to $200°C$?

13. (II) The pendulum of a pendulum clock is made of solid brass. At $25°C$ its period is 1.12 s and it keeps perfect time. When its owners are on a winter vacation for 2 weeks, the average temperature in the house is $-5°C$. How far off will the clock be when they return? Will it be fast or slow?

14. (II) (a) Show that the change in the density ρ of a substance when the temperature changes by ΔT is given by $\Delta \rho = -\beta \rho \Delta T$. (b) What is the fractional change in density of a lead sphere whose temperature decreases from $30°C$ to $-30°C$?

15. (II) If a rod of original length L_1 has its temperature changed from T_1 to T_2, determine a formula for its new length L_2 in terms of T_1, T_2, and α. Assume (a) $\alpha =$ constant, (b) $\alpha = \alpha(T)$ is some function of temperature, (c) $\alpha = \alpha_0 + bT$ where α_0 and b are constants.

16. (II) A brass plug is to be placed in a ring made of iron. At room temperature the diameter of the plug is 9.12 cm and that of the inside of the ring is 9.09 cm. They must both be brought to what common temperature in order to fit?

17. (II) If a fluid is contained in a long narrow vessel so it can expand in essentially one direction only, show that the effective coefficient of linear expansion α is approximately equal to the coefficient of volume expansion β.

18. (II) Show that for an isotropic solid, $\beta = 3\alpha$ as long as the amount of expansion is small. β and α are the coefficients of volume and linear expansion, respectively.

19. (II) An ordinary glass is filled to the brim with 288.3 mL of water at 10°C. If the temperature is now increased by 30°C, about how much water (if any) will spill from the glass?

20. (II) (a) Determine a formula for the change in surface area of a uniform solid sphere of radius r if its coefficient of linear expansion is α (assumed constant) and its temperature is changed by ΔT. (b) What is the increase in area of a solid iron sphere of radius 88.0 cm if its temperature is raised from 20° to 200°C?

21. (III) An iron cube floats in a bowl of liquid mercury at 0°C. (a) If the temperature is raised to 30°C, will the cube float higher or lower in the mercury? (b) By what percent will the fraction of volume submerged change?

22. (III) A 12.4-kg solid iron cylindrical wheel of radius 0.45 m is rotating about its axle in frictionless bearings with angular velocity $\omega = 32.8$ rad/s. If its temperature is now raised from 20°C to 80°C, what is the fractional change in ω?

23. (III) (a) The tube of a mercury thermometer has an inside diameter of 0.120 mm. The bulb has a volume of 0.250 cm³. How far will the thread of mercury move when the temperature changes from 10.0°C to 20.0°C? Take into account expansion of the glass (Pyrex). (b) Determine a formula for the length of the mercury column in terms of relevant variables.

*SECTION 17-6

24. (I) At what temperature will the ultimate compressive strength of concrete be exceeded for the blocks discussed in Example 17-4?

25. (II) (a) A horizontal steel I-beam of cross-sectional area 0.016 m² is rigidly connected to two vertical steel girders. If the beam was installed when the temperature was 25°C, what stress is developed in the beam when the temperature drops to −14°C? (b) Is the ultimate strength of the steel exceeded? (c) What stress is developed if the beam is concrete and has a cross-sectional area of 0.13 m²? Will it fracture?

26. (II) A wine barrel of diameter 122.860 cm at 20°C is to be enclosed by an iron band. The circulator band has an inside diameter of 122.848 cm at 20°C. It is 8.7 cm wide and 0.55 cm thick. (a) To what temperature must the rim be heated so that it will fit over the barrel? (b) What will be the tension in the rim when it cools to 20°C?

SECTION 17-7

27. (I) What are the following temperatures on the Kelvin scale: (a) 37°C, (b) 80°F, (c) −196°C?

28. (I) The *Rankine temperature scale* is the Fahrenheit equivalent of the Kelvin scale. That is, the size of the degree is the same size as on the Fahrenheit scale but the zero is shifted so that zero (0°R) is at absolute zero. What is (a) the freezing point and (b) the boiling point of water in °R? (c) What is absolute zero in °F?

29. (II) Typical temperatures in the interior of the earth and sun are about 4×10^3°C and 1.5×10^7°C, respectively. (a) What are these temperatures in kelvins? (b) What percent error is made in each case if a person forgets to change °C to K?

SECTION 17-8

30. (I) If 5.00 m³ of a gas initially at STP is placed under a pressure of 4.0 atm, the temperature of the gas rises to 25°C. What is the volume?

31. (I) The pressure in a helium gas cylinder is initially 30 atmospheres. After many balloons have been blown up, the pressure has decreased to 6 atm. What fraction of the original gas remains in the cylinder?

32. (II) Write the ideal gas law in terms of the density of the gas.

33. (II) Calculate the density of oxygen at STP using the ideal gas law.

34. (II) A storage tank contains 25.7 kg of nitrogen (N_2) at an absolute pressure of 2.60 atm. What will the pressure be if the nitrogen is replaced by an equal mass of CO_2?

35. (II) A tank contains 28.0 kg of O_2 gas at a gauge pressure of 6.70 atm. If the oxygen is replaced by helium, how many kg of the latter will be needed to produce a gauge pressure of 8.25 atm?

36. (II) If 8.10 mol of helium gas is at 20°C and a gauge pressure of 0.190 atm, calculate (a) the volume of the helium gas under these conditions and (b) the temperature if the gas is compressed to half the volume at a gauge pressure of 2.10 atm.

37. (II) A house has a volume of 600 m³. (a) What is the total mass of air inside the house at 0°C? (b) If the temperature rises to 25°C, what mass of air enters or leaves the house?

38. (II) What is the pressure inside a 20-L container holding 24 kg of argon gas at 20°C?

39. (II) A tire is filled with air at 15°C to a gauge pressure of 190 kPa. If the tire reaches a temperature of 40°C, what fraction of the original air must be removed if the original pressure of 190 kPa is to be maintained?

40. (II) If 50.0 L of oxygen at 10°C and an absolute pressure of 1.88 atm are compressed to 36.6 L and at the same time the temperature is raised to 80°C, what will the new pressure be?

41. (II) If a skin diver fills his lungs to full capacity of 5.5 L when 12 m below the surface, to what volume would his lungs expand if he quickly rose to the surface? Is this advisable?

42. (II) An ocean decompression chamber for divers is essentially an inverted cylinder, open at the bottom, as shown in Fig. 17-9. When fully submerged near the surface, the volume of trapped air is 8.2 m³ and the temperature is

FIGURE 17-9

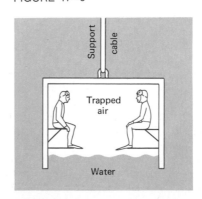

25°C. (*a*) What is the minimum mass of the cylinder so that it does not rise? (*b*) Assuming this is the actual mass, suppose the cylinder now descends to a depth of 50 m where the temperature is 12°C; what will be the tension in the support cable? (No air is added or lost.) Assume the air (actually a special gas mixture for divers) behaves like an ideal gas, that its weight can be ignored, and that the density of the seawater is uniform at 1.025×10^3 kg/m^3.

43. (II) Determine a formula for the coefficient of volume expansion, β, for an ideal gas at constant pressure in terms of P, V, T and/or n.

44. (II) A hot-air balloon achieves its buoyant lift by heating the air inside the balloon which makes it less dense than the air outside. Suppose the volume of a balloon is 1500 m^3 and the required lift is 2500 N (rough estimate of the weight of the equipment and passenger). Calculate the temperature of the air inside the balloon which will produce the required lift. Assume that the outside air temperature is 0° and that air is an ideal gas under these conditions. What factors limit the maximum altitude attainable by this method for a given load? (Neglect variables like wind.)

45. (II) An air bubble at the bottom of a lake 16 m deep has a volume of 1.10 cm^3. If the temperature at the bottom is 5°C and at the top it is 16°C, what is the volume of the bubble just before it reaches the surface?

46. (II) Compare the value for the density of water vapor at 100°C and 1 atm with the value predicted from the ideal gas law. Why would you expect a difference?

SECTION 17–9

47. (I) Calculate the number of molecules/m^3 in an ideal gas at STP.

48. (I) How many moles of water are there in 1.000 L? How many molecules?

49. (II) The lowest pressure attainable using the best available vacuum techniques is about 10^{-12} N/m^2. At such a pressure, how many molecules are there per cm^3 at 0°C?

50. (II) From the known value of atmospheric pressure at the surface of the earth, estimate the total number of air molecules in the earth's atmosphere.

51. (II) A cubic box of volume 8.0×10^{-3} m^3 is filled with air at atmospheric pressure at 20°C. The box is closed and heated to 150°C. What is the net force on each side of the box?

SECTION 17–10

52. (II) The partial pressure of CO_2 in the lungs is about 35 torr, somewhat higher than outside in the air. What is the percentage of CO_2 in air in the lungs?

*SECTION 17–11

*53. (I) At the boiling point of sulfur (444.6°C) the pressure in a constant-volume gas the thermometer is 187 torr. Estimate (*a*) the pressure at the triple point of water, (*b*) the temperature when the pressure in the thermometer is 112 torr.

*54. (I) In a constant-volume gas thermometer, what is the limiting ratio of the pressure at the boiling point of water at 1 atm to that at the triple point? (Keep five significant figures.)

*55. (II) Use Fig. 17–7 to determine the inaccuracy of a constant-volume gas thermometer using oxygen if it reads a pressure $P = 268$ torr at the boiling point of water at 1 atm. Express answer (*a*) in K, (*b*) as a percentage.

56. (II) A constant-volume gas thermometer is being used to determine the temperature of the melting point of a substance. The pressure in the thermometer at this temperature is 218 torr; at the triple point of water, the pressure is 286 torr. Some gas is now released from the thermometer bulb so that the pressure at the triple point of water becomes 163 torr. At the temperature of the melting substance, the pressure is 128 torr. Estimate, as accurately as possible, the melting-point temperature of the substance.

Kinetic Theory

 (chapter number 18 in the right margin)

The concept that matter is made up of atoms which are in continuous random motion is called the **kinetic theory**. We now investigate the properties of a gas from the point of view of kinetic theory, which is based on the laws of classical mechanics. But to apply Newton's laws to each one of the vast number of molecules in a gas ($> 10^{25}/m^3$ at STP) is far beyond the capability of any present computer. Instead we take a statistical approach and determine averages of certain quantities; and these averages correspond to macroscopic variables. We will, of course, demand that our microscopic description correspond to the macroscopic properties of gases; otherwise our theory would be of little value.

First we will calculate the pressure in a gas in terms of its molecular properties. We will also arrive at an important relation between the average kinetic energy of molecules in a gas and the absolute temperature. We then will go on to investigate more properties of gases in terms of the kinetic theory.

18–1 The Ideal Gas Law and the Molecular Interpretation of Temperature

We make the following assumptions about the molecules in a gas. These assumptions reflect a simple view of a gas but nonetheless correspond well to the essential features of real gases that are at low pressures and far from the liquefaction point. Under these circumstances gases follow the ideal gas law closely, and indeed we shall refer to such a gas as an *ideal gas*. The assumptions, which represent the basic postulates of the kinetic theory, are: (1) There are a large number of molecules, N, each of mass m, moving in random directions with a variety of speeds. This assumption is in accord with our observation that a gas fills its container and, in the case of air on the earth, is kept from escaping only by the force of gravity. (2) The molecules are, on the average, far apart from one another; that is, their average separation is much greater than the diameter of each molecule. (3) The molecules are assumed to obey the laws of classical mechanics, and are presumed to interact with one another only when they collide. Although molecules exert weak attractive forces on each other between collisions, the potential energy associated with these forces is small compared to the kinetic energy, and we ignore it for now. (4) Collisions with another molecule or the wall of the vessel are assumed to be perfectly elastic, like the collisions of perfectly elastic billiard balls (Chapter 8). We assume the collisions are of very short duration compared to the time between collisions; then we can ignore the potential energy associated with collisions in comparison to the kinetic energy between collisions.

We can see immediately how this kinetic view of a gas can explain Boyle's law: the pressure exerted on a wall of a container of gas is due to the constant bombardment of molecules; if the volume is reduced by (say) half, the molecules are closer together and twice as many will be striking a given area of the wall per second. Hence we expect the pressure to be twice as great, which is Boyle's law.

Now let us calculate quantitatively the pressure in a gas based on kinetic theory. For purposes of argument, we imagine that the molecules are contained in a rectangular vessel whose ends have area A and whose length is l as shown in Fig. 18–1. The pressure exerted by the gas on the walls of its container is, according to our model, due to the collisions of the molecules with the walls. Let us focus our attention on the wall, of area A, at the left end of the container and examine what happens when one molecule strikes this wall, as shown in Fig. 18–2. This molecule exerts a force on the wall and the wall exerts an equal and opposite force back on the molecule. The magnitude of this force, according to Newton's second law, is equal to the molecule's rate of change of momentum, $F = dp/dt$. Assuming the collision is elastic, only the x component of the molecule's momentum changes, and it changes from $-mv_x$ (it is moving in the negative x direction) to $+mv_x$. Thus the change in momentum, $\Delta(mv)$, which is the final momentum minus the initial momentum, is

$$\Delta(mv) = mv_x - (-mv_x) = 2mv_x$$

for one collision. This molecule will make many collisions with the wall, each separated by a time Δt, which is the time it takes the molecule to travel across the box and back again, a distance equal to $2l$. Thus $2l = v_x \Delta t$ or $\Delta t = 2l/v_x$. The time Δt between collisions is very small so the number of collisions per second is very large. Thus the average force—averaged over many collisions—will be equal to the force exerted during one collision divided by the time between collisions (Newton's second law):

$$F = \frac{\Delta(mv)}{\Delta t}$$

$$= \frac{2mv_x}{2l/v_x} = \frac{mv_x^2}{l}. \qquad \text{[due to one molecule]}$$

During its passage back and forth across the container, the molecule may collide with the tops and sides of the container, but this does not alter its x component of momentum and thus does not alter our result. It may also collide with other molecules, which may change its v_x; however, any loss (or gain) of momentum is acquired by the other molecule and because we will sum over all the molecules in the end, this effect will be included; so our result above is not altered.

Of course the actual force due to one molecule is intermittent, but because a huge number of molecules are striking the wall per second, the force is pretty steady. To calculate the force due to *all* the molecules in the box, we have to add the contributions of each. Thus the average net force on the wall is:

$$F = \frac{m}{l}(v_{x1}^2 + v_{x2}^2 + \ldots + v_{xN}^2),$$

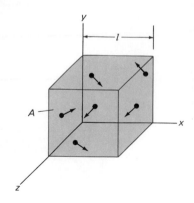

FIGURE 18–1

Molecules of a gas moving about in a cubical container.

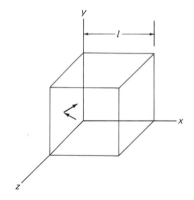

FIGURE 18–2

Arrows indicate the momentum of one molecule as it rebounds from the end wall.

where v_{x1} means v_x for particle number 1 (and so on) and the sum extends over the total of N molecules. Now the average value of the square of the x component of velocity is

$$\overline{v_x^2} = \frac{v_{x1}^2 + v_{x2}^2 + \ldots + v_{xN}^2}{N}. \qquad (18\text{--}1)$$

Thus we can write the average force as

$$F = \frac{m}{l} N \overline{v_x^2}.$$

We know that the square of any vector is equal to the sum of the squares of its components (theorem of Pythagoras). Thus $v^2 = v_x^2 + v_y^2 + v_z^2$ for any velocity v. Taking averages we obtain

$$\overline{v^2} = \overline{v_x^2} + \overline{v_y^2} + \overline{v_z^2}.$$

Since the velocities of the molecules in our gas are assumed to be random, there is no preference for one direction or another; hence

$$\overline{v_x^2} = \overline{v_y^2} = \overline{v_z^2},$$

and therefore

$$\overline{v^2} = 3\overline{v_x^2}.$$

We substitute this into the equation for the average force F:

$$F = \frac{m}{l} N \frac{\overline{v^2}}{3}.$$

The pressure on the wall is then

$$P = \frac{F}{A} = \frac{1}{3} \frac{Nm\overline{v^2}}{Al}$$

or

$$P = \frac{1}{3} \frac{Nm\overline{v^2}}{V} \qquad (18\text{--}2)$$

where $V = lA$ is the volume of the container. This is the result we were seeking, the pressure in a gas expressed in terms of molecular properties.

Equation 18–2 can be rewritten in a clearer form by multiplying both sides by V and slightly rearranging the right side:

$$PV = \tfrac{2}{3} N (\tfrac{1}{2}m\overline{v^2}). \qquad (18\text{--}3)$$

The quantity $\tfrac{1}{2}m\overline{v^2}$ is the average kinetic energy ($\overline{\text{KE}}$) of the molecules in the gas. If we compare Eq. 18–3 with Eq. 17–4, the ideal gas law, we see that the two agree if

$$\tfrac{2}{3}(\tfrac{1}{2}m\overline{v^2}) = kT,$$

or,

$$\tfrac{1}{2}m\overline{v^2} = \tfrac{3}{2}kT, \qquad (18\text{--}4)$$

where k, the Boltzmann constant (see Chapter 17), has the value $k \approx 1.38 \times 10^{-23}$ J/K. This equation tells us that the *average translational kinetic energy of molecules in a gas is directly proportional to the absolute temperature*. The higher the temperature, according to kinetic theory, the faster the molecules are moving on the average. This relation is one of the triumphs of the kinetic theory.

EXAMPLE 18–1 What is the average translational kinetic energy of molecules in a gas at 37°C?

SOLUTION We use Eq. 18–4 and change 37°C to 310 K:

$$\overline{KE} = \tfrac{3}{2}kT$$
$$= \tfrac{3}{2}(1.38 \times 10^{-23} \text{ J/K})(310 \text{ K}) = 6.42 \times 10^{-21} \text{ J}.$$

Equation 18–4 implies that as the temperature approaches absolute zero the kinetic energy of molecules approaches zero; modern quantum theory, however, tells us this is not quite so; instead, as absolute zero is approached, the kinetic energy approaches a very small nonzero minimum value. Even though all real gases become liquid or solid above 0 K, molecular motion does not cease, even at absolute zero.

We can use Eq. 18–4 to calculate how fast molecules are moving on the average. Notice that the average in Eqs. 18–1 through 18–4 is over the *square* of the velocity. The square root of $\overline{v^2}$ is called the "root mean square" velocity, v_{rms} (since we are taking the square *root* of the *mean* of the *square* of the velocity):

$$v_{rms} = \sqrt{\overline{v^2}} = \sqrt{\frac{3kT}{m}}. \qquad (18\text{–}5)$$

The *mean speed*, $\bar{v}$, is the average of the magnitudes of the speeds themselves; $\bar{v}$ is generally not equal to v_{rms}. The difference between the mean speed and the rms speed can be seen in the following example.

EXAMPLE 18–2 Eight particles have the following speeds, given in m/s: 1.0, 6.0, 4.0, 2.0, 6.0, 3.0, 2.0, 5.0. Calculate (*a*) the mean speed, and (*b*) the rms speed.

SOLUTION (*a*) The mean speed is

$$\bar{v} = \frac{1.0 + 6.0 + 4.0 + 2.0 + 6.0 + 3.0 + 2.0 + 5.0}{8} = 3.6 \text{ m/s}.$$

(*b*) The rms speed is

$$v_{rms} = \sqrt{\frac{(1.0)^2 + (6.0)^2 + (4.0)^2 + (2.0)^2 + (6.0)^2 + (3.0)^2 + (2.0)^2 + (5.0)^2}{8}} \text{ m/s}$$
$$= 4.1 \text{ m/s}.$$

We see in this example that $\bar{v}$ and v_{rms} are not necessarily equal. In fact, for an ideal gas they differ by about 8 percent. We will see in the next section how to calculate $\bar{v}$ for an ideal gas; we already have the tool to calculate v_{rms} (Eq. 18–5).

EXAMPLE 18–3 What is the rms speed of air molecules at room temperature (20°C)?

SOLUTION We must apply Eq. 18–5 to oxygen and nitrogen separately since they have different masses. Using the result of Example 17–8, the masses of one molecule of O_2 (molecular mass = 32 u) and N_2 (molecular mass = 28 u) are:

$$m(O_2) = (32)(1.67 \times 10^{-27} \text{ kg}) = 5.3 \times 10^{-26} \text{ kg}$$
$$m(N_2) = (28)(1.67 \times 10^{-27} \text{ kg}) = 4.7 \times 10^{-26} \text{ kg}.$$

Thus, for oxygen

$$v_{rms} = \sqrt{\frac{3kT}{m}} = \sqrt{\frac{(3)(1.38 \times 10^{-23} \text{ J/K})(293 \text{ K})}{(5.3 \times 10^{-26} \text{ kg})}}$$

$$= 480 \text{ m/s}.$$

For nitrogen the result is $v_{rms} = 510$ m/s. (This is over 1500 km/h or 1000 mi/h.)

18–2 Distribution of Molecular Speeds

The molecules in a gas are assumed to be in random motion, which means that many molecules have speeds less than the average speed and others have speeds greater than the average. In 1859, James Clerk Maxwell (1831–1879) worked out a formula for the most probable distribution of speeds in a gas containing N molecules. We will not give a derivation here but merely quote his result:

$$f(v) = 4\pi N \left(\frac{m}{2\pi kT}\right)^{3/2} v^2 e^{-\frac{1}{2}\frac{mv^2}{kT}}. \qquad (18-6)$$

$f(v)$ is called the **Maxwell distribution of speeds**, and is plotted in Fig. 18–3. The quantity $f(v)\,dv$ represents the number of molecules that have speed between v and $v + dv$. In this formula, m is the mass of a single molecule, T is the absolute temperature, and k is Boltzmann's constant. Since N is the total number of molecules in the gas, when we sum over all the molecules in the gas we must get N; thus we must have

$$\int_0^\infty f(v)\,dv = N.$$

(Problem 17 is an exercise to show this is true.)

EXAMPLE 18–4 Determine formulas for (a) the average speed, $\bar{v}$, and (b) the most probable speed, v_p, of molecules in an ideal gas at temperature T.

SOLUTION (a) The average value of any quantity is found by multiplying each possible value of that quantity (say, speed) by the number of molecules that have that value, and then summing all these numbers and dividing by N (the total number). We are given a continuous distribution of speeds (Eq. 18–6), so the sum

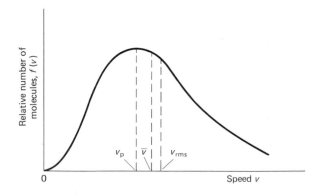

FIGURE 18–3

Distribution of speeds of molecules in an ideal gas. Note that $\bar{v}$ and v_{rms} are not at the peak of the curve (that speed is called the "most probable speed," v_p). This is because the curve is skewed to the right: it is not symmetrical.

becomes an integral:

$$\bar{v} = \frac{\int_0^\infty v f(v) \, dv}{N} = 4\pi \left(\frac{m}{2\pi kT}\right)^{3/2} \int_0^\infty v^3 e^{-\frac{1}{2}\frac{mv^2}{kT}} \, dv.$$

We can look up the definite integral in the tables, or integrate by parts,[†] and obtain

$$\bar{v} = 4\pi \left(\frac{m}{2\pi kT}\right)^{3/2} \left(\frac{2k^2 T^2}{m^2}\right) = \sqrt{\frac{8}{\pi}\frac{kT}{m}} \approx 1.60 \sqrt{\frac{kT}{m}}.$$

(b) The most probable speed is that speed which occurs more than any others, and thus is that speed where $f(v)$ has its maximum value. Since $df(v)/dv = 0$ at this point, we have

$$\frac{df(v)}{dv} = 4\pi \left(\frac{m}{2\pi kT}\right)^{3/2} \left(2v e^{-\frac{mv^2}{2kT}} - \frac{2mv^3}{2kT} e^{-\frac{mv^2}{2kT}}\right) = 0.$$

Solving for v, we get

$$v_p = \sqrt{\frac{2kT}{m}} \approx 1.41 \sqrt{\frac{kT}{m}}.$$

(Another solution is $v = 0$, but this corresponds to a minimum, not a maximum.) In summary,

$$v_p = \sqrt{2\frac{kT}{m}} \approx 1.41 \sqrt{\frac{kT}{m}}$$

$$\bar{v} = \sqrt{\frac{8}{\pi}\frac{kT}{m}} \approx 1.60 \sqrt{\frac{kT}{m}}$$

$$v_{rms} = \sqrt{3\frac{kT}{m}} \approx 1.73 \sqrt{\frac{kT}{m}}.$$

These are all indicated in Fig. 18–3.

From Eq. 18–6 and Fig. 18–3, it is clear that the speeds of molecules in a gas vary from zero up to many times the average speed, but as can be seen from the graph, most molecules have speeds that are not far from the average. Less than 1 percent of the molecules exceed four times v_{rms}.

Experiments to determine the distribution of speeds in real gases were first performed in the 1920s. They confirmed with considerable accuracy the Maxwell distribution (for gases at not too high a pressure) and the direct proportion between average kinetic energy and absolute temperature, Eq. 18–4.

The Maxwell distribution for a given gas depends only on the absolute temperature. Figure 18–4 shows the distribution for two different temperatures; just as v_{rms} increases with temperature, so the whole distribution curve shifts to the right at higher temperatures.

[†] For integration by parts,

$$\int_0^\infty f \, dg = fg \Big|_0^\infty - \int_0^\infty g \, df.$$

We let $f = v^2$ and $dg = v e^{-av^2}$ where $a = m/2kT$. Then

$$\int_0^\infty v^3 e^{-av^2} \, dv = (2v)\left(-\frac{1}{2a} e^{-av^2}\right)\Big|_0^\infty - \int_0^\infty (2v)\left(-\frac{1}{2a} e^{-av^2}\right) dv$$

$$= 0 - \frac{1}{2a^2} e^{-av^2}\Big|_0^\infty = \frac{1}{2a^2}.$$

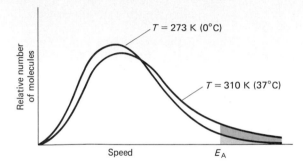

FIGURE 18–4

Distribution of molecular speeds for two different temperatures.

Figure 18–4 illustrates how kinetic theory can be used to explain the fact that many chemical reactions, including those in biological cells, take place more rapidly as the temperature increases. Most chemical reactions take place in a liquid solution, and the molecules in a liquid have a distribution of speeds close to the Maxwell distribution. Two molecules may chemically react only if their kinetic energy is great enough so that when they collide they penetrate into each other somewhat. The minimum energy required is called the *activation energy*, E_A, and it has a specific value for each chemical reaction. The molecular speed corresponding to a kinetic energy of E_A for a particular reaction is indicated in Fig. 18–4. The number of molecules with energy greater than this value is given by the area under the curve above E_A. In Fig. 18–4 the respective areas for two different temperatures are indicated by the two different shadings in the figure. It is clear that the number of molecules that have kinetic energies in excess of E_A increases greatly for only a small increase in temperature. The rate at which a chemical reaction occurs is proportional to the number of molecules with energy greater than E_A, and thus we see why reaction rates increase rapidly with increased temperature.

 ## 18–3 Evaporation, Vapor Pressure, and Boiling

If a glass of water is left out overnight, the water level will have dropped by morning. We say the water has evaporated, meaning that some of the water has changed to the vapor or gas phase.

This process of *evaporation* can be explained on the basis of kinetic theory. The molecules in a liquid move past one another with a variety of speeds that follow, approximately, the Maxwell distribution. There are strong attractive forces between these molecules, which is what keeps them close together in the liquid phase. A molecule in the upper regions of the liquid may, because of its speed, leave the liquid momentarily. But just as a rock thrown into the air returns to the earth, so the attractive forces of the other molecules can pull the vagabond molecule back to the liquid surface—that is, if its velocity is not too large. If the molecule has a high enough velocity, it will escape from the liquid entirely, like a rocket escaping the earth, and become part of the gas phase. Only those molecules that have kinetic energy above a particular value can escape to the gas phase. We have already seen that kinetic theory predicts that the number of molecules with kinetic energy above a particular value (such as E_A in Fig. 18–4) increases with temperature, and this is in accord with the well-known observation that the evaporation rate is greater at higher temperatures.

Since it is the fastest molecules that escape from the surface, the average speed of those remaining is less. When the average speed is less, the absolute temperature is less. Thus, evaporation is a cooling process. You have no doubt noticed this when you stepped out of a warm shower and you felt cold as the water on your body began to evaporate; and after working up a sweat on a hot day even a slight breeze makes you feel cool.

Air normally contains water vapor (water in the gas phase) and it comes mainly from evaporation. To look at this process in a little more detail, consider a closed container that is partially filled with water (it could just as well be any other liquid), and from which the air has been removed (Fig. 18–5). The fastest moving molecules quickly evaporate into the space above. As they move about, some of these molecules will strike the liquid surface and return to the liquid phase; this is called *condensation*. The number of molecules in the vapor increases for a time, until a point is reached where the number returning to the liquid equals the number leaving in the same time interval. Equilibrium then exists and the space is said to be *saturated*. The pressure of the vapor when it is saturated is called the *saturated vapor pressure* (or sometimes simply the vapor pressure).

The saturated vapor pressure does not depend on the volume of the container. If the volume above the liquid were reduced, the density of molecules in the vapor phase would be increased. More molecules would then be striking the liquid surface per second. There would be a net flow of molecules back to the liquid phase until equilibrium was again reached and this would occur at the same value of the saturated vapor pressure.

The saturated vapor pressure of any substance depends on the temperature. At higher temperatures, a greater number of molecules have sufficient kinetic energy to break from the liquid surface into the vapor phase. Hence equilibrium will be reached at a higher pressure. The saturated vapor pressure of water at various temperatures is given in Table 18–1. Notice that even solids—for example, ice—have a measurable saturated vapor pressure.

In everyday situations, evaporation from a liquid takes place into the air above it rather than into a vacuum. This does not materially alter the discussion above relating to Fig. 18–5. Equilibrium will still be reached when there are sufficient molecules in the gas phase that the number reentering the liquid equals the number leaving. This number is not affected by the presence of air although collisions with air molecules may lengthen the time needed to reach equilibrium. Thus equilibrium occurs at the same value of the saturated vapor pressure as if air weren't there.

Of course, if the container is large or is not closed, all the liquid may evaporate before saturation is reached. And if the container is not sealed—as, for example, a room in your house—it is not likely that the air will become saturated with water vapor; unless of course, it is raining outside.

The saturated vapor pressure of a liquid increases with temperature. When the temperature is raised to the point where the saturated vapor pressure at that temperature equals the external pressure, boiling occurs. Let us examine this process. As the boiling point is approached, tiny bubbles tend to form in the liquid, which indicates a change from the liquid to the gas phase. However, if the vapor pressure inside the bubbles is less than the external pressure, the bubbles immediately are crushed. As the temperature is increased, the saturated vapor pressure inside a bubble eventually becomes equal to or exceeds the external air pressure. The bubble will then not collapse but will increase in size and rise to the surface. Boiling has then begun. A liquid boils when its saturated vapor pressure equals the external pressure. This occurs for water under 1 atm (760 torr) of pressure at 100°C as can be seen from Table 18–1.

The boiling point of a liquid clearly depends on the external pressure. At high elevations, the boiling point of water is somewhat less than at sea level since the air pressure is less. For example, on the summit of Mt. Everest (8850 m) the air pressure is about one third of what it is at sea level, and from Table 18–1 we can see that water will boil at about 70°C. Cooking food by boiling takes longer at high elevations, since the temperature is less. Pressure cookers however, reduce cooking time, since they build up a pressure as high as 2 atm.

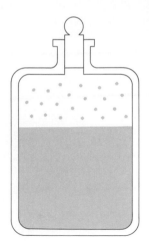

FIGURE 18–5

Vapor appears above a liquid in a closed container.

TABLE 18–1

Saturated vapor pressure of water

Temperature (°C)	Saturated Vapor Pressure	
	torr (= mm Hg)	Pa (= N/m²)
−50	0.030	4.0
−10	1.95	2.60×10^2
0	4.58	6.11×10^2
5	6.54	8.72×10^2
10	9.21	1.23×10^3
15	12.8	1.71×10^3
20	17.5	2.33×10^3
25	23.8	3.17×10^3
30	31.8	4.24×10^3
40	55.3	7.37×10^3
50	92.5	1.23×10^4
60	149	1.99×10^4
70	234	3.12×10^4
80	355	4.73×10^4
90	526	7.01×10^4
100	760	1.01×10^5
120	1489	1.99×10^5
150	3570	4.76×10^5

18–4   Humidity

When we refer to the weather as being dry or being humid, we are referring to the water vapor content of the air. We use the concept of partial pressure in order to specify this quantitatively. The partial pressure of water in the air can be as low as zero, and can vary up to a maximum equal to the saturated vapor pressure of water at the given temperature. Thus at 20°C the partial pressure cannot exceed 17.5 torr (see Table 18–1). The *relative humidity* is defined as the ratio of the partial pressure to the saturated vapor pressure at a given temperature. It is usually expressed as a percentage:

$$\text{Relative humidity} = \frac{\text{partial pressure of } H_2O}{\text{saturated vapor pressure of } H_2O} \times 100.$$

Thus, when the humidity is close to 100%, the air holds nearly all the water vapor it can.

 EXAMPLE 18–5   On a particular hot day, the temperature is 30°C and the partial pressure of water vapor in the air is 21.0 torr. What is the relative humidity?

SOLUTION   From Table 18–1, the saturated vapor pressure of water at 30°C is 31.9 torr. Hence the relative humidity is

$$\frac{21.0 \text{ torr}}{31.9 \text{ torr}} \times 100 = 66\%.$$

Humans are sensitive to humidity. A relative humidity of 40–50 percent is generally optimum for both health and comfort. High humidity, particularly on a hot day, reduces the evaporation of moisture from the skin, which is one of the body's vital mechanisms for regulating body temperature. Very low humidity, on the other hand, can have a drying effect on the skin and mucous membranes.

Proper humidity must be maintained to prevent damage to paintings, tape recordings, and a wide variety of sensitive objects. Thus the design of heating and air-conditioning systems for buildings must take into account not only heating and cooling but the control of relative humidity as well.

Air is saturated with water vapor when the partial pressure of water in the air is equal to the saturated vapor pressure at that temperature. If the partial pressure of water exceeds the saturated vapor pressure, the air is said to be *supersaturated*. This situation can occur when a temperature decrease occurs. For example, suppose the temperature is 30°C and the partial pressure of water is 21 torr, which is a humidity of 66 percent as we saw above. Suppose now, the temperature falls to, say, 20°C, such as might happen at nightfall. From Table 18–1, we see that the saturated vapor pressure of water at 20°C is 17.5 torr. Hence the relative humidity would be greater than 100 percent and the supersaturated air cannot hold this much water. The excess water condenses and appears as dew; this process is also responsible for the formation of fog, clouds, and rain.

When air containing a given amount of water is cooled, a temperature is reached where the partial pressure of water equals the saturated vapor pressure. This is called the *dew point*. Measurement of the dew point is the most accurate means of determining the relative humidity. One method uses a polished metal surface in contact with air which is gradually cooled down. The temperature at which moisture begins to appear on the surface is the dew point, and the partial pressure of water can then be obtained from saturated vapor pressure tables. If, for example, on a given day the temperature is 20°C and the dew point is 5°C, then the partial pressure of water (Table 18–1) in the original air was 6.51 torr whereas its saturated vapor pressure was 17.5 torr; hence the relative humidity was 6.51/17.5 = 37 percent.

A more convenient but less accurate method for measuring relative humidity is the so-called wet bulb–dry bulb technique which makes use of two thermometers. One thermometer bulb is fitted with a snug cloth jacket that is soaking wet. The apparatus is usually swung in the air, and the lower the humidity, the more evaporation takes place from the wet bulb; this causes its temperature reading to be less. A comparison of the temperature readings on the wet bulb thermometer and the dry (ordinary) thermometer can then be compared to special tables that have been compiled to obtain the relative humidity.

 ## 18–5   Real Gases and Changes of Phase; the Critical Point

The ideal gas law, as mentioned before, is an accurate description of the behavior of a gas as long as the pressure is not too high, and as long as the temperature is far from the liquefaction point. But what happens to real gases when these two criteria are not satisfied?

To answer this question, let us look at a graph of pressure plotted against volume for a given amount of gas. On such a "*PV* diagram," Fig. 18–6, each point represents an equilibrium state of the given substance. The various curves show how the pressure varies as the volume is changed at constant temperature for several different values of the temperature. The dashed curve *A'* represents the behavior of a gas as predicted by the ideal gas law; that is, *PV* = constant. The solid curve *A* represents the behavior of a real gas at the same temperature. Notice that at high pressure, the volume in a real gas is less than that predicted by the ideal gas law. At lower temperatures, curves *B* and *C* in Fig. 18–6, the

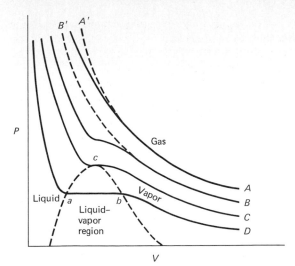

FIGURE 18—6

A *PV* diagram for a real substance.

behavior deviates even more from the curves predicted by the ideal gas law (for example, *B'*), and the deviation is greater the closer the gas is to liquefying.

To explain this, we note that at higher pressures we expect the molecules to be closer together; and, particularly at lower temperatures, the potential energy associated with the attractive forces between the molecules that we ignored before is no longer negligible compared to the now reduced kinetic energy of the molecules. These forces tend to pull the molecules closer together so that at a given pressure, the volume is less than expected from the ideal gas law. (More on this in Section 18–6.) At still lower temperatures, these forces will cause liquefaction and the molecules become *very* close together.

Indeed, curve *D* represents the situation when liquefaction does occur. At low pressure on curve *D* (on the right in Fig. 18–6), the substance is a gas and occupies a large volume. As the pressure is increased, the volume decreases until point *b* is reached. Beyond *b*, the volume decreases with no change in pressure; the substance is gradually changing from the gas to the liquid phase. At point *a*, all of the substance has changed to liquid. Further increase in pressure reduces the volume only slightly—liquids are nearly incompressible—so the curve is very steep as shown. The area within the tongue-shaped dashed line represents the region where the gas and liquid phases exist together in equilibrium.

Curve *C* in Fig. 18–6 represents the behavior of the substance at its *critical temperature*; and the point *c* (the one point where this curve is horizontal) is called the **critical point**. At temperatures less than the critical temperature (and this is the definition of the term), a gas will change to the liquid phase if sufficient pressure is applied. Above the critical temperature, no amount of pressure can cause a gas to change phase and become a liquid—what happens instead is that the gas becomes denser and denser as the pressure is increased and gradually it acquires properties resembling a liquid, but no liquid surface forms. The critical temperatures for various gases are given in Table 18–2. Scientists tried for many years to liquefy oxygen without success. It was not until the discovery of the behavior of substances associated with the critical point that it was realized that oxygen can be liquefied only if first cooled below its critical temperature of −118°C.

Often a distinction is made between the terms "gas" and "vapor": a substance below its critical temperature in the gaseous state is called a *vapor*; when above the critical temperature, it is called a *gas*; this is indicated in Fig. 18–6.

The behavior of a substance can be diagrammed not only on a *PV* diagram but also on a *PT* diagram. A *PT* diagram, often called a *phase diagram*, is particularly convenient for comparing the different phases of a substance. Figure 18–7 is

TABLE 18–2
Critical temperatures and pressures

Substance	Critical Temperature		Critical Pressure (atm)
	°C	K	
Water	374	647	218
Carbon dioxide	31	304	72.8
Oxygen	−118	155	50
Nitrogen	−147	126	33.5
Hydrogen	−239.9	33.3	12.8
Helium	−267.9	5.3	2.3

the phase diagram for water. The curve labeled *l-v* represents those points where the liquid and vapor phases are in equilibrium—it is thus a graph of the boiling point versus pressure. Note that the curve correctly shows that at a pressure of 1 atm the boiling point is 100°C and that the boiling point is lowered for a decreased pressure. The curve *s-l* represents points where solid and liquid exist in equilibrium and thus is a graph of the freezing point versus pressure. At 1 atm, the freezing point of water is of course 0°C as shown. Notice also in Fig. 18–7 that at a pressure of 1 atm, the substance is in the liquid phase if the temperature is between 0°C and 100°C but is in the solid or vapor phase if the temperature is below 0°C or above 100°C. The curve labeled *s-v* is the *sublimation point* versus pressure curve. *Sublimation* refers to the process whereby at low pressures (in the case of water, at pressures less than 0.0060 atm) a solid changes directly into the vapor phase without passing through the liquid phase. Carbon dioxide, for example, which in the solid phase is called dry ice, sublimates even at atmospheric pressure.

The intersection of the three curves is the **triple point**. The triple point represents a unique temperature and pressure (see Table 18–3) and it is only at this point that the three phases can exist together in equilibrium. Because the triple point corresponds to a unique value of temperature and pressure, it is precisely reproducible and is often used as a point of reference (as was discussed, for example, in Section 17–11).

Because the *l-v* curve in Fig. 18–7 is the boiling point curve, it also represents the vapor pressure of the substance at a given temperature (remember boiling

FIGURE 18–7

Phase diagram for water (note that scales are not linear).

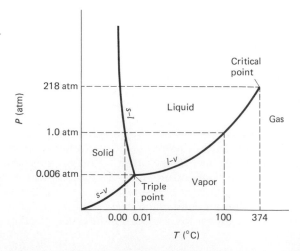

TABLE 18–3
Triple-point data

Substance	Temperature (K)	Pressure	
		N/m²	atm
Water	273.16 (0.01°C)	6.10×10^2	6.03×10^{-3}
Carbon dioxide	216.6	5.16×10^5	5.10
Ammonia	195.40	6.06×10^3	6.00×10^{-2}
Nitrogen	63.2	1.25×10^4	1.24×10^{-1}
Oxygen	54.4	1.52×10^2	1.50×10^{-3}
Hydrogen	13.8	7.03×10^3	6.95×10^{-2}

takes place when the external pressure equals the vapor pressure). Below the triple point, the *s-v* curve represents the vapor pressure as a function of temperature. Thus the vapor pressure at any temperature can be read off an accurate phase diagram.

Notice that the *s-l* curve for water slopes upward to the left. This is true only of substances that *expand* upon freezing; for at a higher pressure, a lower temperature is needed to cause the liquid to freeze. More commonly, substances contract upon freezing and the *s-l* curve slopes upward to the right, as shown for CO_2 in Fig. 18–8.

The phase transitions we have been discussing are the common ones. Some substances, however, can exist in several forms in the solid phase. A transition from one of these phases to another occurs at a particular temperature and pressure, just like ordinary phase changes. For example, ice has been observed in at least eight different modifications at very high pressure. Ordinary helium is a unique substance in that it has two distinct liquid phases, called helium I and II. They exist only at temperatures within a few degrees of absolute zero. Helium II exhibits very unusual properties referred to as *superfluidity*. It has extremely low viscosity and exhibits strange properties such as actually climbing up the sides of an open container and over the top. Some materials, like glass, certain resins, and sulfur, are called *amorphous solids*. They do not have a crystal structure like most solids and they do not have a definite melting point. When heated, they gradually soften and there is no phase transition. Amorphous solids are thus often considered to be extremely viscous liquids rather than true solids.

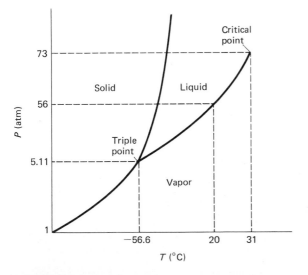

FIGURE 18–8
Phase diagram for carbon dioxide.

⬡ *18–6   Van der Waals Equation of State

We saw in the previous section how real gases deviate from ideal gas behavior particularly at high densities (or when near condensing to a liquid). We would like to understand these deviations from a microscopic (molecular) point of view. J. D. van der Waals (1837–1923) analyzed this problem and in 1873 arrived at a more complicated equation of state which fits real gases more accurately. His analysis is based on kinetic theory but takes into account: (1) the finite size of molecules (we previously neglected the actual volume of the molecules themselves, compared to the total volume of the container, and this assumption becomes poorer as the density increases and molecules become closer together); (2) the range of forces between molecules may be greater than the size of the molecules (we previously assumed that intermolecular forces act only during collision, when the molecules are "in contact"). Let us now look at this analysis and derive the van der Waals equation of state.

Suppose the molecules in a gas are spherical with radius r. If we assume these molecules behave like hard spheres, then two molecules collide and bounce off one another if the distance between their centers gets as small as $2r$. Thus the actual volume in which the molecules can move about is somewhat less than the volume V of the container holding the gas. The amount of "unavailable volume" depends on the number of molecules and on their size (see the problems). Let b represent the "unavailable volume per mole" of gas. Then in the ideal gas law we replace V by $(V - nb)$, where n is the number of moles, and we obtain

$$P(V - nb) = nRT.$$

If we divide through by n and let $v = V/n$ be the volume occupied by one mole of the gas (v is called the *specific volume*—do not confuse it with velocity), then we get

$$P(v - b) = RT. \tag{18–7}$$

This relation (sometimes called the *Clausius equation of state*) predicts that at a given temperature, the pressure $P = RT/(v - b)$ will be greater than for an ideal gas; this makes sense since the reduced volume means the number of collisions with the walls is increased.

Van der Waals then went on to include the effects of attractive forces between molecules. We expect such forces to exist since they must be responsible for holding molecules in the liquid and solid states at lower temperatures. These forces are electrical in nature and although they act even when molecules are not touching, we assume their range is small—that is, they act mainly between nearest neighbors. In the interior of a gas, the attractive forces on a given molecule act in all directions. But for a molecule at the edge of the gas, there is a net inward force. (Figs. 12–13 and 12–14 also apply here.) Molecules that are headed toward a wall of the container are slowed down by this net inward force, and thus will exert less force and less pressure on the wall than if there were no attractive forces. The reduced pressure will be proportional to the number per unit volume (N/V) of molecules in the layer of gas at the surface, and also to the number per unit volume in the next layer, which exerts the inward force.[†] Therefore we expect the pressure to be reduced by a factor proportional to $(N/V)^2$. Since $N = nN_A$ where N_A is Avogadro's number, we can write $(N/V)^2 = (nN_A/V)^2 = N_A^2/v^2$ and so the pressure is reduced by an amount proportional to $(1/v^2)$. If the pressure is given by Eq. 18–7, then we should reduce this by an amount a/v^2 where a is a

---

[†] This is similar to the gravitational force in which the force on mass m_1 due to mass m_2 is proportional to the product of their masses (Eq. 5–1).

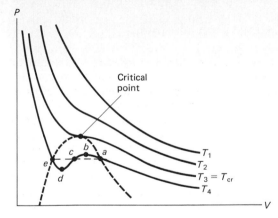

P

Critical
point

T_1
T_2
$T_3 = T_{cr}$
T_4

V

FIGURE 18–9

PV diagram for a van der Waals gas.

proportionality constant. Thus we have

$$P = \frac{RT}{v - b} - \frac{a}{v^2}$$

or

$$\left(P + \frac{a}{v^2}\right)(v - b) = RT \qquad (18\text{–}8)$$

which is the *van der Waals equation of state.*

The constants a and b in the van der Waals equation are different for different gases and are determined by fitting to experimental data for each gas. For CO_2 gas, the best fit is obtained for $a = 3.6 \times 10^{-3} \text{ N·m}^4/\text{mol}^2$ and $b = 4.2 \times 10^{-5} \text{ m}^3/\text{mol}$. Figure 18–9 shows a typical PV diagram for Eq. 18–8 (a "van der Waals gas") for four different temperatures. For T_1, T_2, and T_3 (T_3 is chosen equal to the critical temperature) the curves fit experimental data very well for most gases. The curve labeled T_4, a temperature below the critical point, passes through the liquid-gas region. The maximum (point b) and minimum (point d) are not normally observed; rather, we usually see constant pressure as indicated by the horizontal dashed line. However, for very pure substances that are supersaturated vapors or supercooled liquids, the sections ab and ed, respectively, have been observed. (The section bd would be unstable and has not been observed.)

Neither the van der Waals equation of state nor the many other equations of state that have been proposed are accurate for all gases under all conditions. Yet Eq. 18–8 is a very useful relation, and because it is quite accurate for many situations, its derivation gives us further insight into the nature of gases at the microscopic level. Note that at low densities $(a/v^2) \ll P$ and $b \ll v$, so that the van der Waals equation reduces to the equation of state for an ideal gas, $Pv = RT$ or $PV = nRT$.

*18–7   Mean Free Path

If gas molecules were truly point particles, they would never collide with one another. Thus, when a person opened a perfume bottle, you would be able to smell it almost instantaneously across the room (since molecules travel hundreds of meters per second). In fact it takes some time before you detect an odor, and according to kinetic theory, this must be due to collisions of molecules of nonzero size.

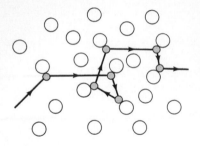

FIGURE 18–10

Zigzag path of a molecule colliding with other molecules.

If we were to follow the path of a particular molecule, we would expect to see it follow a zigzag path as shown in Fig. 18–10. Between each collision the molecule would move in a straight-line path. (Not quite true if we take account of the small intermolecular forces that act between collisions.) An important parameter for a given situation is the *mean free path*, which is defined as the average distance a molecule travels between collisions. We would expect that the greater the gas density, and the larger the molecules, the shorter the mean free path would be. We now determine the nature of this relationship for an ideal gas.

Suppose our gas is made up of molecules which are hard spheres of radius r. A collision will occur whenever the centers of two molecules come within a distance $2r$ of one another. Let us follow a molecule as it traces a straight-line path. In Fig. 18–11, the dashed line represents the path of our particle if it made no collisions. Also shown is a cylinder of radius $2r$; if the center of another molecule lies within this cylinder, a collision will occur. (Of course, when a collision occurs the particle's path would change direction, as would our imagined cylinder, but our calculation won't be altered by unbending a zigzag cylinder into a straight one for purposes of calculation.) Assume our molecule is an average one, moving at the mean speed in the gas, $\bar{v}$. For the moment, let us assume that the other molecules are not moving, and that the concentration of molecules (number per unit volume) is n_V.[†] Then the number of molecules whose center lies within the cylinder of Fig. 18–11 is n_V times the volume of this cylinder, and this also represents the number of collisions that will occur. In a time Δt, our molecule travels a distance $\bar{v}\,\Delta t$; so the length of the cylinder is $\bar{v}\,\Delta t$ and its volume is $\pi(2r)^2\bar{v}\,\Delta t$. Hence the number of collisions that occur in a time Δt is $n_V\pi(2r)^2\bar{v}\,\Delta t$. We define the *mean free path*, l_{m}, as the average distance between collisions; this distance is equal to the distance traveled ($\bar{v}\,\Delta t$) in a time Δt divided by the number of collisions made in time Δt:

$$l_{\mathrm{m}} = \frac{\bar{v}\,\Delta t}{n_V\pi(2r)^2\bar{v}\,\Delta t} = \frac{1}{4\pi r^2 n_V}. \qquad (18\text{–}9\mathrm{a})$$

Thus we see that l_{m} is inversely proportional to the cross-sectional area ($= \pi r^2$) of the molecules and to their concentration (number/volume), n_V. However, Eq. 18–9a is not fully correct since we assumed the other molecules are all at rest. In fact, they are moving, and the number of collisions in a time Δt must depend on the *relative* speed of the colliding molecules, rather than on $\bar{v}$; hence the number of collisions per second is $n_V\pi(2r)^2 v_{\mathrm{rel}}\Delta t$ (rather than $n_V\pi(2r)^2\bar{v}\,\Delta t$) where v_{rel} is the average relative speed of colliding molecules. A careful calculation shows that for a Maxwellian distribution of speeds $v_{\mathrm{rel}} = \sqrt{2}\bar{v}$. Hence the mean free path is

$$l_{\mathrm{m}} = \frac{1}{4\pi\sqrt{2}\,r^2 n_V}. \qquad (18\text{–}9\mathrm{b})$$

[†] Do not confuse n_V with n, the number of moles.

FIGURE 18–11

Molecule at left moves to the right with speed $\bar{v}$. It collides with any molecule whose center is within the cylinder of radius $2r$.

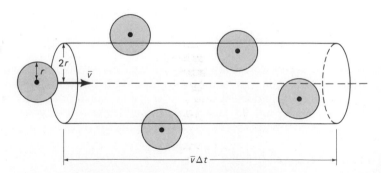

EXAMPLE 18–6 Estimate the mean free path of air molecules at STP. The diameter of O_2 and N_2 molecules is about 3×10^{-10} m.

SOLUTION We saw in Example 17–5 that 1 mol of an ideal gas occupies a volume of 22.4×10^{-3} m^3. Hence

$$n_V = \frac{6.02 \times 10^{23} \text{ molecules}}{22.4 \times 10^{-3} \text{ m}^3} = 2.69 \times 10^{25} \text{ molecules/m}^3.$$

Then

$$l_m = \frac{1}{4\pi\sqrt{2}(1.5 \times 10^{-10} \text{ m})^2(2.7 \times 10^{25} \text{ m}^{-3})} \approx 9 \times 10^{-8} \text{ m}$$

which is about 100 times the diameter of a molecule.

At very low densities, such as in an evacuated vessel, the concept of mean free path loses meaning since collisions with walls of the container may occur more frequently than collisions with other molecules. For example, in a cubical box 20 cm on a side containing air at 10^{-7} torr, the mean free path is about 700 m, which means many more collisions are made with the walls than with other molecules. (Note, nonetheless, that the box contains over 10^{12} molecules.) If the concept of mean free path were to include any type of collision, it would be closer to 0.2 m than to the 700 m calculated from Eq. 18–9.

*18–8 Diffusion

If you carefully place a drop of food coloring in a glass of water, you will find that the color spreads throughout the water. The process may take several hours (assuming you don't shake the glass), but eventually the color will become uniform. This mixing occurs because of the random movement of the molecules, and is called *diffusion*. Diffusion occurs in gases too. Common examples include perfume or smoke (or the odor of something burned on the stove) diffusing in air, although convection often plays a greater role in spreading the odor than does diffusion. In each case the diffusing substance moves from a region where its concentration is high to one where its concentration is low.

Diffusion can be readily understood on the basis of kinetic theory and the random motion of molecules. Consider a tube of cross-sectional area A containing one type of molecule which has higher concentration on the left than on the right, Fig. 18–12. We also assume that there are a large number of a second type of molecule which we refer to as background molecules. (The situation could be ink molecules in a background of water molecules.) The background molecules are not shown in Fig. 18–12. So that we can focus on diffusion, let us assume that the total pressure is uniform (so there will be no hydrodynamic flow due to a pressure gradient) and the temperature is uniform (so there is no convection). The concentration of our "type one" molecules is assumed to change only along one direction, the x direction, and not along y or z. The molecules are in random motion. Yet there will be a net flow of type one molecules to the right in Fig. 18–12 because of the difference in concentration. To see why, consider the small section of tube of length Δx as shown. Molecules from both regions 1 and 2 cross into this central section as a result of their random motion. The more molecules there are in a region, the more will strike a given area or cross a boundary. Since there is a greater concentration of molecules in region 1 than in region 2, more molecules cross into the central section from region 1 than from region 2. There is, then, a net flow of molecules from left to right, from high concentration toward low concentration. The flow stops only when the concentrations become equal.

You might expect that the greater the difference in concentration, the greater the flow rate. This is indeed the case. In 1855, the physiologist Adolf Fick (1829–

FIGURE 18–12

Diffusion occurs from a region of high concentration to one of lower concentration (only one type of molecule is shown).

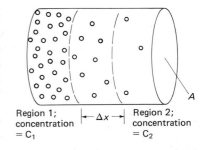

Region 1; concentration $= C_1$ |← Δx →| Region 2; concentration $= C_2$

1901) determined experimentally that the rate of flow across unit area, J, is directly proportional to the change in concentration per unit distance $(n_{V2} - n_{V1})/\Delta x$ (which is called the *concentration gradient*):

$$J = -D \frac{n_{V2} - n_{V1}}{\Delta x}; \tag{18-10}$$

this is called *Fick's law* and is written more precisely in terms of derivatives:

$$J = -D \frac{dn_V}{dx}. \tag{18-11}$$

D is a constant of proportionality called the *diffusion constant*. If x is given in meters, and the concentration n_V is given as the number of molecules per cubic meter, then J is the number of molecules crossing unit area per second. (If n_V is given in mol/m^3, then J is the number of moles crossing unit area per second (mol/m^2·s); if the concentration is given in kg/m^3, then J has units kg/m^2·s.) The minus signs in Eqs. 18–10 and 18–11 merely remind us that the flow is in the direction opposite to the concentration gradient; that is, if the concentration increases to the left ($dn_V/dx < 0$), the flow is to the right.

If the pressure is uniform throughout, as we assumed, there will also be a gradient in the concentration of the background molecules. Thus this second type of molecule will also diffuse, in the opposite direction, according to Eq. 18–11; the diffusion constant D may be different, however, for this second type of molecule. The diffusion process is simplest to analyze if both types of molecule are of the same type. That is, we follow molecules that are "tagged" (say, by being radioactive) as they diffuse through otherwise identical background molecules. This is called *self-diffusion*, and we now assume this is the case, or at least that different types of molecules are quite similar.

EXAMPLE 18–7   To get an idea of the time required for diffusion, estimate how long it might take for ammonia (NH_3) to be detected 10 cm from a bottle after it is opened, assuming only diffusion.

SOLUTION   This will be an order of magnitude calculation. The rate of diffusion J can be set equal to the number of molecules N diffusing across area A in a time t: $J = N/At$. We solve for t:

$$t = \frac{N}{AJ}$$

and then use Eq. 18–10:

$$t = \frac{N}{AD} \frac{\Delta x}{\Delta n_V}.$$

The average concentration (midway between bottle and nose) can be approximated by $\bar{n}_V \approx N/V$ where V is the volume over which the molecules move; V is roughly of the order of $V \approx A \Delta x$ where Δx is 10 cm, so we substitute $N = \bar{n}_V A \Delta x$ into the above equation:

$$t \approx \frac{(\bar{n}_V A \Delta x) \Delta x}{AD \Delta n_V} = \frac{\bar{n}_V}{\Delta n_V} \frac{(\Delta x)^2}{D}.$$

The concentration of ammonia is high near the bottle and low near the detecting nose, so $\bar{n}_V \approx \Delta n_V / 2$ or $(\bar{n}_V/\Delta n_V) \approx \frac{1}{2}$. Since NH_3 has a size somewhere between H_2 and O_2, from Table 18–4 we can choose $D \approx 4 \times 10^{-5}$ m^2/s. Then

$$t \approx \frac{1}{2} \frac{(0.10 \text{ m})^2}{(4 \times 10^{-5} \text{ m}^2/\text{s})} \approx 100 \text{ s},$$

or about a minute or two. This seems a bit long from experience, probably because air currents (convection) usually help it along.

It is possible to relate the diffusion constant D to the mean free path l_m of the diffusing gas molecules. We derive the relationship now using simple arguments[†] and at the same time justify the proportionality between J and dn_V/dx on the basis of kinetic theory. Let us assume that the mean free path is much smaller than the size of the containing vessel so we can ignore collision with the walls. Consider a plane surface at position $x = x_0$ in the gas as shown in Fig. 18–13. The concentration of molecules (per cubic meter) at $x = x_0$ is $n_V = n_{V0}$. The molecules have velocities in random directions in three-dimensional space; but for simplicity we can consider that $\frac{1}{3}$ of them move along the x axis, $\frac{1}{3}$ along the y axis, and $\frac{1}{3}$ along the z axis. We are concerned only with those that move along the x axis. Of the $\frac{1}{3}n_V$ that move along the x axis, half of these, or $\frac{1}{6}n_V$, move to the right $(+\bar{v})$ and half, or $\frac{1}{6}n_V$, move to the left $(-\bar{v})$. Consider those moving to the right; in a time Δt, each molecule moves a distance $\bar{v}\,\Delta t$; the number that cross an area A in the plane x_0 in a time Δt had to occupy a volume $A\bar{v}\,\Delta t$, so the number that cross x_0 per unit per unit time is

$$\frac{\frac{1}{6}n_V A(\bar{v}\,\Delta t)}{A\,\Delta t} = \frac{1}{6}n_V \bar{v}.$$

Similarly, $\frac{1}{6}n_V \bar{v}$ move to the left. However, the concentrations to the left and right of $x = x_0$ are not the same. The molecules moving to the right had their last collision approximately one mean free path to the left of x_0, at $x = x_0 - l_m$; and their concentration there was $n_V = n_0 - (dn_V/dx)\, l_m$. Hence the flow rate to the right is

$$J_\rightarrow = \frac{1}{6}\left(n_{V0} - \frac{dn_V}{dx}\, l_m\right)\bar{v}.$$

The molecules moving to the left had their last collision at $x = x_0 + l_m$ where $n_V = n_{V0} + (dn_V/dx)l_m$, so the flow rate to the left is

$$J_\leftarrow = \frac{1}{6}\left(n_{V0} + \frac{dn_V}{dx}\, l_m\right)\bar{v}.$$

The net flow rate is then

$$J = J_\rightarrow - J_\leftarrow = \frac{1}{6}\left(n_{V0} - \frac{dn_V}{dx}\, l_m\right)\bar{v} - \frac{1}{6}\left(n_{V0} + \frac{dn_V}{dx}\, l_m\right)\bar{v}$$

or

$$J = \frac{1}{3}\,\bar{v}l_m \frac{dn_V}{dx}.$$

This is just Fick's law, Eq. 18–11. Thus kinetic theory predicts that the flow rate J is proportional to the concentration gradient, dn_V/dx. And it gives the diffusion coefficient as

$$D = \tfrac{1}{3}\bar{v}l_m. \qquad (18\text{–}12)$$

EXAMPLE 18–8 Estimate the diffusion constant for O_2 molecules in air at STP. (This is close to self-diffusion since N_2 and O_2 are close in mass and size).

SOLUTION In Example 18–6 we found $l_m \approx 9 \times 10^{-8}$ m for air molecules, and from Example 18–4 (see also Example 18–3).

$$\bar{v} = \sqrt{\frac{8}{\pi}\frac{kT}{m}} = 430 \text{ m/s}.$$

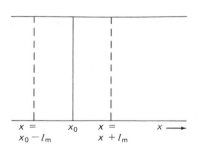

FIGURE 18–13

Derivation of Fick's law and form of diffusion constant for self-diffusion; consider molecules crossing imaginary surface at $x = x_0$.

[†] A more detailed treatment can be found in more advanced books on thermodynamics and kinetic theory, such as F. W. Sears and G. L. Salinger, *Thermodynamics, Kinetic Theory and Statistical Mechanics* (Reading MA: Addison-Wesley, 1975).

Hence

$$D \approx \tfrac{1}{3}(430 \text{ m/s})(9 \times 10^{-8} \text{ m}) \approx 1.3 \times 10^{-5} \text{ m}^2/\text{s}.$$

The measured value is about[†] $1.6 \times 10^{-5} \text{ m}^2/\text{s}.$

This analysis of diffusion in terms of the kinetic theory can also be applied to a number of other *transport processes*. In diffusion, it is molecules or matter that is transported. The processes of thermal conductivity and electrical conductivity involve the transport of energy and electric charge, respectively, and can be treated in a similar way. So too can the viscosity of gases (see the problems) for which it is momentum that is transported across a surface.

Diffusion also takes place in liquids and Fick's law applies as well; however, the diffusion constant will be quite different than for gases, although it still depends on molecular properties. The values for D for a variety of substances are given in Table 18–4.

For some purposes it is convenient to write Fick's law, Eq. 18–11, in terms of partial pressures rather than concentrations. We saw earlier (Section 17–10) that the partial pressure of each component in a mixture of gases is directly related to its concentration. This followed from the ideal gas law. Letting P_i be the partial pressure of a particular component, and n_{V_i} its concentration in molecules per cubic meter ($n_{V_i} = N_i/V$), we can write

$$P_i = \frac{N_i kT}{V} = n_{V_i} kT.$$

Thus Fick's law can be written as

$$J_i = -\frac{D}{kT}\left(\frac{\Delta P_i}{\Delta x}\right) \quad \text{or} \quad J_i = -\frac{D}{kT}\frac{dP_i}{dx} \tag{18–13}$$

where $\Delta P_i/\Delta x$ (and dP_i/dx) is the pressure gradient for substance i. Thus a gas diffuses from a region where its partial pressure is high to a region where its partial pressure is low, irrespective of the pressure due to the other components.

[†] See Table 18–4, and problem 43 for the temperature dependence of D.

TABLE 18–4
Diffusion constants, D (20°C, 1 atm)

Diffusing Molecules	Medium	$D(\text{m}^2/\text{s})$
H_2	Air	6.3×10^{-5}
O_2	Air	1.8×10^{-5}
O_2	Water	100×10^{-11}
Blood hemoglobin	Water	6.9×10^{-11}
Glycine (an amino acid)	Water	95×10^{-11}
DNA (molecular mass 6×10^6 u)	Water	0.13×10^{-11}

 Summary _____

According to the kinetic theory of gases, which is based on the idea that a gas is made up of molecules that are moving rapidly and at random, the average kinetic energy of the molecules is proportional to the Kelvin temperature T:

$$\overline{\text{KE}} = \tfrac{3}{2}kT$$

where k is Boltzmann's constant. At any moment, there exists a wide distribution of molecular speeds within a gas. The Maxwell distribution of speeds is derived from simple kinetic theory assumptions, and is in good accord with experiment for gases at not too high a pressure.

Evaporation of a liquid is the result of the fastest moving molecules escaping from the surface. Because the average molecular velocity is less after the fastest molecules escape, the temperature decreases when evaporation takes place. *Saturated vapor pressure* refers to the pressure of the vapor above a liquid when the two phases are in equilibrium. The vapor pressure of a substance (such as water) depends strongly on temperature and is equal to atmospheric pressure at the boiling point. *Relative humidity* of air at a given place is the ratio of the partial pressure of water vapor in the air to the vapor pressure at that temperature; it is usually expressed as a percentage.

The behavior of real gases at high pressure, and when near their liquefaction point, deviates from the ideal gas law. These deviations are due to the finite size of molecules and to the attractive forces between molecules which become more important in these situations. Below the *critical temperature*, a gas can change to a liquid if sufficient pressure is applied; but if the temperature is higher than the critical temperature, no amount of pressure will cause a liquid surface to form. The *triple point* of a substance is that unique temperature and pressure at which all three phases—solid, liquid, and gas—can coexist in equilibrium. Because of its precise reproducibility, the triple point of water is often taken as a standard reference point.

 ## Questions

1. Why doesn't the size of different molecules enter into the gas laws?
2. When a gas is rapidly compressed—say, by pushing down a piston—its temperature increases. When a gas expands against a piston, it cools. Explain these changes in temperature using the kinetic theory, in particular noting what happens to the momentum of molecules when they strike the moving piston.
3. In Section 18–1 we assumed the gas molecules made perfectly elastic collisions with the walls of the container. This assumption is not necessary as long as the walls are at the same temperature as the gas. Why?
4. Explain in words how Charles's law follows from kinetic theory and the relation between average kinetic energy and the absolute temperature.
5. Explain in words how Gay-Lussac's law follows from kinetic theory.
6. As you go higher in the earth's atmosphere, the ratio of N_2 molecules to O_2 molecules increases. Why?
7. Can you determine the temperature of a vacuum?
8. Is temperature a macroscopic or microscopic variable?
9. Discuss why the Maxwell distribution of speeds, Fig. 18–3, is not a symmetrical curve.
10. Explain why the peak of the curve for 310 K in Fig. 18–4 is not as high as for 273 K. (Assume the total number of molecules is the same for both.)
11. Explain, using the Maxwell distribution of speeds, (a) why the moon has very little atmosphere, (b) why hydrogen, if at one time in the earth's atmosphere, would probably have escaped.
12. Explain why putting food in the freezer retards spoilage.
13. A thermal gradient precipitator is a device for removing particulate matter from polluted air. It consists, basically, of two solid surfaces, one hot and one cold, close to each other. Particles in air that is allowed to pass between the two surfaces tend to collect on the cold surface and thus are removed. Explain. (This effect can be noticed on the wall behind a heating radiator, particularly if the wall is an outside wall and therefore quite cool.)
14. Would the average kinetic energy of molecules in an ideal gas correspond to $\bar{v}$, v_{rms}, v_p, or to some other value?

15. If the pressure in a gas is doubled while its volume is held constant, by what factor do (a) v_{rms}, and (b) $\bar{v}$, change?
16. If a container of gas is at rest, the average velocity of molecules must be zero. Yet the average speed is not zero. Explain.
17. A baseball, initially at rest, is thrown by a pitcher. Compare the average kinetic energy of the molecules of the baseball, in the reference frame of the pitcher, before and after being thrown. Explain any difference.
18. Draw, roughly, the Maxwellian distribution of *velocities*. Discuss what the values would be for mean, rms, and most probable velocities. Would the curve be symmetrical?
19. What everyday observation would tell you that not all molecules in a material have the same speed?
20. We saw that the saturated vapor pressure of a liquid (say, water) does not depend on the external pressure. Yet the temperature of boiling does depend on the external pressure. Is there a contradiction? Explain.
21. Alcohol evaporates more quickly than water at room temperature. What can you infer about the molecular properties of one relative to the other?
22. Explain why a hot humid day is far more uncomfortable than a hot dry day at the same temperature.
23. Is it possible to boil water at room temperature (20°C) without heating it? Explain.
24. What exactly does it mean when we say that oxygen boils at −183°C?
25. A length of thin wire is placed over a block of ice (or an ice cube) at 0°C and weights are hung from the ends of the wire. It is found that the wire cuts its way through the ice cube, but leaves a solid block of ice behind it. This process is called *regelation*. Explain how this happens by inferring how the freezing point of water depends on pressure.
26. How do a gas and a vapor differ?
27. Under what conditions can liquid CO_2 exist? Be specific. Can it exist as a liquid at normal room temperature?
28. Why does dry ice not last long at room temperature?
29. (a) At atmospheric pressure, in what phases can CO_2 exist? (b) At what temperature(s) does it change phase? (c) For what range of pressures and temperatures can CO_2 be a liquid? Refer to Fig. 18–8.

30. In what phase does CO_2 exist when the pressure is 30 atm and the temperature is 30°C?

31. In which phase is water when the temperature is 50°C and the pressure is 0.01 atm?

*32. Name several ways to reduce the mean free path in a gas.

*33. Discuss why sound waves can travel in a gas only if their wavelength is somewhat larger than the mean free path.

*34. How might we modify Eq. 18–9 for the mean free path if the molecules are not spherical?

*35. Suppose the force acting between two molecules was of long range, say, inversely proportional to the square of the distance between them rather than being large only when they are in "contact." What meaning could you then attach to the mean free path?

*36. Are we justified, in our derivation of the mean free path (Eq. 18–9a), in assuming that our molecule maintains the speed $\bar{v}$ even after a collision? Discuss.

*37. Explain qualitatively why you would expect the relative velocity of two molecules, v_{rel}, to be greater than the mean speed of a molecule, $\bar{v}$.

*38. What might be some reasons why the result of Example 18–8 is nearly 30% different from the measured value?

*39. It is possible to define a *drift velocity* v_{dr} for diffusion by the relation:

$$v_{dr} = \frac{J}{n_V}.$$

Discuss the meaning of this term.

*40. Why must we breathe? That is, why can't we simply rely on diffusion of oxygen into our lungs?

 Problems _____

SECTION 18–1

1. (I) (a) What is the average kinetic energy of an oxygen molecule at STP? (b) What is the total translational kinetic energy of 1 mol of O_2 molecules at 20°C?

2. (I) Calculate the rms speed of helium atoms near the surface of the sun at a temperature of about 6000°K.

3. (I) By what factor will the rms speed of gas molecules increase if the temperature is increased from 0°C to 100°C?

4. (I) A gas is at 20°C. To what temperature must it be raised to double the rms speed of its molecules?

5. (II) Show that the rms speed of molecules in a gas is given by

$$v_{rms} = \sqrt{3P/\rho}$$

where P is the pressure in the gas, and ρ is the gas density.

6. (II) Show that for a mixture of two gases at the same temperature, the ratio of their rms speeds is equal to the inverse ratio of the square roots of their molecular masses.

7. (II) What is the rms speed of nitrogen molecules contained in an 8.0-m^3 volume at 2.1 atm if the total amount of nitrogen is 1300 mol?

8. (II) (a) Calculate the approximate rms speed of an amino acid whose molecular mass is 89 u in a living cell at 37°C. (b) What would be the average speed of a protein of molecular mass 50,000 u at 37°C?

9. (II) A space vehicle returning from the moon enters the atmosphere at a speed of about 40,000 km/h. Molecules (assume nitrogen) striking the nose of the vehicle with this speed correspond to what temperature? (Because of this high temperature, the nose of a space vehicle requires a very special design; indeed, part of it does vaporize, and this is seen as a bright blaze upon reentry.)

10. (II) In outer space the density of matter is about one atom per cm^3, mainly hydrogen atoms, and the temperature is about 3.4 K. Calculate the average speed of these hydrogen atoms, and the pressure (in atmospheres).

11. (II) Calculate (a) the rms speed of an oxygen molecule at 0°C, and (b) determine how many times per second it would move back and forth across a 5.0-m-long room on the average, assuming it made very few collisions with other molecules.

12. (II) What is the average distance between oxygen molecules at STP?

13. (II) The two isotopes of uranium, ^{235}U and ^{238}U (the superscripts refer to their atomic mass), can be separated by a gas diffusion process by combining them with fluorine to make the gaseous compound UF_6. Calculate the ratio of the rms speeds of these molecules for the two isotopes.

14. (II) Calculate approximately the total translational kinetic energy of all the molecules in an *E. coli* bacterium of mass 2.0×10^{-15} kg at 37°C. Assume 70% of the cell, by weight, is water, and the other molecules have an average molecular weight on the order of 10^5.

SECTION 18–2

15. (II) A group of twenty-two particles have the following speeds: two have speed 10 m/s, seven have 15 m/s, four have 20 m/s, one has 25 m/s, five have 30 m/s, one has 35 m/s, and two have 40 m/s. Determine (a) the average speed, (b) the rms speed, and (c) the most probable speed.

16. (III) At room temperature, it takes approximately 2.45×10^3 J to evaporate 1.00 g of water. Estimate the average speed of evaporating molecules. What multiple of v_{rms} (at 20°C) for all water molecules is this? (Assume Eq. 18–4 holds.)

17. (III) Starting from the Maxwell distribution of speeds, Eq. 18–6, show (a) $\int_0^\infty f(v)\,dv = N$, and (b) $\int_0^\infty v^2 f(v)\,dv/N = 3kT/m$.

SECTION 18–3

18. (I) If the air pressure at a particular place in the mountains is 0.80 atm, at what temperature does water boil?

19. (I) What is the air pressure at a place where water boils at 70°C?

20. (II) What is the approximate pressure inside a pressure cooker if the water is boiling at a temperature of 120°C? Assume no air escaped during the heating process which started at 20°C.

21. (II) An *autoclave* is a device used to sterilize laboratory instruments. It is essentially a high-pressure steam boiler and operates on the same principle as a pressure cooker. However, because hot steam under pressure is more effective in killing microorganisms than moist air at the same temperature and pressure, the air is removed and replaced by steam. Typically, the gauge pressure inside the autoclave is 1.0 atm; what is the temperature of the steam? Assume the steam is in equilibrium with boiling water.

SECTION 18–4

22. (I) What is the dew point (approximately) if the relative humidity is 25 percent on a day when the temperature is 27°C.

23. (I) What is the partial pressure of water on a day when the temperature is 25°C and the relative humidity is 55%?

24. (II) If the relative humidity in a room 3.5 m × 6.1 m × 8.0 m is 80 percent, what mass of water can still evaporate from an open pan if the temperature is 20°C?

25. (II) If the humidity in a room of volume 520 m³ at 25°C is 90 percent, what mass of water must be removed to reduce the humidity to 50 percent?

26. (II) Air that is at its dew point of 5°C is drawn into a building where it is heated to 25°C. What will be the relative humidity at this temperature? Take into account the expansion of the air.

27. (II) On a day when the temperature is 30°C, the wet bulb thermometer of a humidity tester falls to 10°C. What is the relative humidity?

*SECTION 18–6

*28. (I) Write the van der Waals equation of state in terms of the volume V rather than specific volume v.

*29. (II) For oxygen gas, the van der Waals equation of state achieves its best fit for $a = 0.14$ N·m⁴/mol² and $b = 3.2 \times 10^{-5}$ m³/mol. Determine the pressure in the gas at 0°C for a specific volume of 0.40 L/mol calculated using (a) the van der Waals equation, (b) the ideal gas law.

*30. (III) (a) From the van der Waals equation of state, show that the critical temperature and pressure are given by

$$T_{cr} = \frac{8a}{27bR}, \qquad P_{cr} = \frac{a}{27b^2}.$$

(Hint: use the fact that the P versus V curve has an inflection point at the critical point so that the first and second derivatives are zero). (b) Determine a and b for CO_2 from the measured values of $T_{cr} = 304$ K and $P_{cr} = 72.8$ atm.

*31. (III) (a) Show that a collision between two spherical molecules, each of radius r, is equivalent to a collision between a point particle and a sphere of radius $2r$, and thus the center of one molecule cannot penetrate a volume equal to the volume of a sphere of radius $2r$ when two molecules collide. (b) Show that the total unavailable volume per mole is $b = 16\pi r^3 N_A/3$ where N_A is Avogadro's number. (Hint: in summing the total excluded volume, multiply by $\frac{1}{2}$ to avoid counting each pair of molecules twice.) Note that the unavailable volume is four times the actual molecular volume. (c) Estimate the diameter of a CO_2 molecule (see problem 30).

*SECTION 18–7

*32. (II) At about what pressure would the mean free path of air molecules be (a) 1.0 m, (b) equal to the diameter of air molecules, $\approx 3 \times 10^{-10}$ m?

*33. (II) (a) The mean free path of CO_2 molecules at STP is measured to be about 5.6×10^{-8} m. Estimate the diameter of a CO_2 molecule. (b) Do the same for He gas for which $I_m \approx 25 \times 10^{-8}$ m at STP.

*34. (II) A cubical box of volume 2.0×10^{-3} m³ contains 55 marbles, each 1.2 cm in diameter. What is the mean free path for a marble (a) when the box is shaken vigorously, and (b) when it is slightly shaken.

*35. (II) A cubical box 20 cm on a side is evacuated so the pressure of air inside is 10^{-6} torr. Estimate how many molecular collisions there are per each collision with a wall (0°C).

*36. (II) A very small amount of hydrogen gas is released into the air. If the air is at 1.0 atm and 25°C, estimate the mean free path for a H_2 molecule. What assumptions did you make?

*37. (II) Estimate the maximum allowable pressure in a 45-cm-long cathode ray tube if 98 percent of all electrons must hit the screen without first striking an air molecule.

*38. (II) (a) Show that the number of collisions a molecule makes per second, called the *collision frequency*, v, is given by $v = 4\sqrt{2}\pi r^2 n_v \bar{v}$, and thus $v = \bar{v}/l_m$. (b) What is the collision frequency for N_2 molecules in air at 20°C and $P = 10^{-2}$ atm?

*39. (II) Suppose that a gas contains two types of molecules in concentrations n_1 and n_2. Their radii are r_1 and r_2. Show that the mean free path for type 1 molecules is given by

$$l_{m1} = \frac{1}{4\pi r_1^2 n_1 + \pi(r_1 + r_2)^2 n_2}$$

if we assume the molecules are hard spheres.

*40. (III) At some instant, suppose we have N_0 identical molecules; show that the number N of molecules that travel a distance x or more before the next collision is given by $N = N_0 e^{-x/l_m}$ where l_m is the mean free path. This is called the *survival equation*.

*SECTION 18–8

*41. (II) Estimate the radius of an oxygen molecule from measurements of (a) the mean free path $= 9.05 \times 10^{-8}$ m and (b) the diffusion coeffcient $= 1.8 \times 10^{-5}$ m²/s, at 0°C and 1 atm.

*42. (II) Estimate the diffusion coefficient for low-concentration H_2 gas diffusing in air. (Hint: most collisions are with air molecules.) The observed value is about 6.3×10^{-5} m²/s at 20°C and 1 atm.

*43. (II) (a) Show that the diffusion constant for an ideal gas varies with temperature as $T^{3/2}$. (b) Estimate the diffusion constant for O_2 in air at 0°C and 1 atm starting from the value given in Table 18–4.

*44. (II) Oxygen diffuses from the surface of insects to the interior through tiny tubes called tracheae. An average trachea is about 2 mm long and has cross-sectional area of 2×10^{-9} m². Assuming the concentration of oxygen inside is half what it is outside in the atmosphere, calculate (a) the diffusion rate J and (b) estimate the average time for a molecule to diffuse in. Assume the diffusion constant is 1×10^{-5} m²/s.

45. (II) (a) Derive *Graham's law*, which states that, "the rate of diffusion of gas molecules is inversely proportional to the square root of the molecular mass." (b) Show that it is also inversely proportional to the square of the density. (c) Which would diffuse faster, N_2 gas or O_2 gas, and by how much (percentage)?

*46. (III) Make a derivation, similar to that for Eq. 18–12, to show that the coefficient of viscosity, η, of a dilute gas is given by

$$\eta = \tfrac{1}{3} n_V \bar{v} m l_m$$

where m is the mass of a molecule. In this case, it is the speed of the fluid (assumed much less than $\bar{v}$, the average molecular speed), rather than molecules, that is transferred across the surface of Fig. 18–13; see Eq. 13–5b and Fig. 13–7.

Heat

19

When a pot of cold water is placed on a hot burner of a stove, the temperature of the water increases. We say that heat flows from the hot burner to the cold water. Whenever two objects at different temperatures are put in contact, heat flows from the hotter one to the colder one. The flow of heat is in the direction tending to equalize the temperature. If the two objects are kept in contact long enough for their temperatures to become equal, the two bodies are said to be in equilibrium, and there is no further heat flow between them. For example, when the mercury in a fever thermometer is still rising, heat is flowing from the patient's mouth to the thermometer; when the mercury stops, the thermometer is then in equilibrium with the person's mouth, and they are at the same temperature.

 ## 19–1 Early Theory of Heat; the Calorie

It is common to speak of the flow of heat—heat flows from a stove burner to a pot of coffee, from the sun to the earth, from a person's mouth into a fever thermometer. It flows from an object at higher temperature to one at lower temperature. Indeed, an eighteenth-century theory of heat pictured heat flow as movement of a fluid substance called *caloric*. According to the caloric theory, any object contained a certain amount of caloric; if more caloric flowed into the object, its temperature increased; and if caloric flowed out, the object's temperature decreased. When a material was broken apart, such as during burning, a great deal of caloric was believed to be released.

However, no change in mass was ever detected as a result of the flow of heat, nor could the caloric be detected by any other means. So caloric was assumed to be massless, odorless, tasteless, and transparent. In spite of the mysterious nature of this fluid, the caloric theory did explain many observations such as the "flow" of heat from a hot object to a cold one. But other phenomena came to light that could not be satisfactorily explained as we shall see shortly.

Although the caloric theory has long since been discarded, remnants of this theory still remain, such as the expression "flow of heat," as if heat were a fluid. A common unit for heat that is still used today is named after caloric. It is called the *calorie* and is defined as *the amount of heat necessary to raise the temperature of 1 g of water by 1 Celsius degree, from 14.5°C to 15.5°C.* This particular temperature range is specified since the heat required is very slightly different at different temperatures. (The difference is less than 1 percent over the whole range from 0 to 100°C and for most purposes can be ignored.) More often used than the calorie is the *kilocalorie* (kcal) which is 1000 calories. Thus, 1 kcal is the heat needed to raise 1 kg of water by 1 C° from 14.5°C to 15.5°C. Sometimes a kilocalorie is called a *Calorie* (with a capital C), and it is with this unit that the energy value of food is specified.

19–2   Heat as Energy Transfer; the Mechanical Equivalent of Heat

One of the main problems with the caloric theory was its inability to account for all the heat[†] generated by friction. You can, for example, rub your hands or two pieces of metal together for a long time and generate heat indefinitely. The American, Benjamin Thompson (1753–1814), who later became Count Rumford of Bavaria, became acutely aware of this problem when he was supervising the boring out of cannon barrels. Water was placed in the bore of the cannon to keep it cool during the cutting process, and as the water boiled away, it was replenished. The "caloric" which caused the boiling of the water was assumed to arise from the breaking apart of the metal. But Rumford noticed that even when the cutting tools were so dull that they didn't cut the metal, heat was still generated and the water boiled away. Thus, caloric was being released even though subdivision of matter was not occurring. Furthermore, this process could go on indefinitely and produce a limitless amount of heat. This was not consistent with the idea that heat is a substance and therefore only a finite amount of it could be contained within an object. Rumford therefore rejected the caloric theory and proposed instead that heat is a kind of motion. He claimed that in some circumstances, at least, heat is produced by doing mechanical work (for example, rubbing two objects together). This idea was pursued by others in the early 1800s, particularly by an English brewer, James Prescott Joule (1818–1889).

Joule performed a number of experiments that were crucial in establishing our present-day view that heat, like work, represents a transfer of energy. One of Joule's experiments is shown (simplified) in Fig. 19–1. The falling weight causes the paddle wheel to turn; the friction between the water and the paddle wheel causes the temperature of the water to rise slightly (barely measurable, in fact, by Joule). Of course the same temperature rise could also be obtained by heating the water on a hot stove. In this and a great many other experiments (some involving electrical energy) Joule found that a given amount of work was always equivalent to a particular amount of heat. Quantitatively, 4.186 joules (J) of work was found to be equivalent to 1 calorie (cal) of heat. This is known as the *mechanical equivalent of heat*:

$$4.186 \text{ J} = 1 \text{ cal}$$

$$4.186 \times 10^3 \text{ J} = 1 \text{ kcal}.$$

As a result of these and other experiments, scientists came to interpret heat not as a substance, and not even as a form of energy, but rather as a transfer of

FIGURE 19–1

Joule's experiment on the mechanical equivalent of heat.

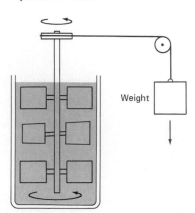

Weight

[†] We use the word "heat" here rather colloquially. We refine the concept later in this section and give our present-day definition of it. See also Section 19–3.

energy: when heat flows from a hot object to a cooler one, it is energy that is being transferred from the hot to the cold object. Thus, **heat** *is energy that is transferred from one body to another because of a difference in temperature.* In SI units the unit for heat, as for any form of energy, is the joule. Nonetheless, calories and kcal are still sometimes used. Today the calorie is *defined* in terms of the joule (rather than in terms of the properties of water, as given in Section 19–1). The definition above (1 cal = 4.186 J) is consistent with the older definition. However, a second definition is often used:

$$1 \text{ cal} = 4.184 \text{ J};$$

this is called the *thermochemical calorie*. We will not be concerned with the slight difference in these two definitions (0.05 percent).

The development of kinetic theory fully supported, and indeed nicely explains, the idea of heat as a transfer of energy. Let us examine the process of heating a pot of water on a hot stove. According to kinetic theory, the average kinetic energy of molecules increases with temperature; thus the molecules of the stove burner have much more kinetic energy on the average than those of the cold water or the pot. When the high-KE stove molecules collide with the lower-KE molecules of the pot, some of their kinetic energy is transferred to the pot molecules, just as a fast-moving billiard ball transfers some of its kinetic energy to a ball it collides with. The molecules of the pot gain in kinetic energy (those of the stove loose energy). The now-higher-KE pot molecules, in turn, transfer some of their kinetic energy, by collision, to the lower-KE water molecules. The temperature of the water and the pot consequently increases. Thus we see how a transfer of heat is a transfer of energy.

19–3 Distinction Between Temperature, Heat, and Internal Energy

We introduce the concept of internal energy now since it will help clarify ideas about heat. The sum total of all the energy of all the molecules in an object is called its **thermal energy** or **internal energy**. (We will use the two terms interchangeably.) Occasionally the term "heat content" of a body is used for this purpose; however, this is not a good term to use for it can be confused with heat itself. Heat, as we have seen, is not the energy a body contains, but rather refers to the amount of energy transferred from a hot to a cold body.

Using the kinetic theory, we can make a clear distinction between temperature, heat, and internal energy. Temperature is a measure of the *average* kinetic energy of individual molecules. Thermal or internal energy refers to the *total* energy of all the molecules in the object. (Thus two equal-mass hot ingots of iron may have the same temperature, but two of them have twice as much thermal energy as one does.) Heat, finally, refers to a *transfer* of energy (usually thermal energy) from one object to another because of a difference in temperature.

Notice that the direction of heat flow between two objects depends on their temperatures, not on how much internal energy they each have. Thus, if 50 g of water at 30°C is placed in contact (or mixed) with 200 g of water at 25°C, heat flows *from* the water at 30°C *to* the water at 25°C even though the internal energy of the 25°C water is much greater because there is so much more of it.

19–4 Internal Energy of an Ideal Gas

Let us calculate the internal energy of *n* moles of an ideal monatomic (one atom per molecule) gas. The internal energy, *U*, is the sum of the translational kinetic energies of all the atoms. This sum is just equal to the average kinetic energy per molecule times the total number of molecules, N:

$$U = N(\tfrac{1}{2}m\overline{v^2}).$$

From Eq. 18–4 we then have

$$U = \tfrac{3}{2}NkT$$

or

$$U = \tfrac{3}{2}nRT \qquad\qquad \text{[monatomic ideal gas]}$$

where n is the number of moles. Thus, the internal energy of an ideal gas depends only on temperature and the number of moles of gas.

If the gas molecules contain more than one atom, then the rotational and vibrational energy of the molecules must also be taken into account. The internal energy will be greater at a given temperature than for a monatomic gas, but it will still be a function only of temperature.

The internal energy of real gases also depends mainly on temperature, but where they deviate from ideal gas behavior, it also depends somewhat on pressure and volume.

The internal energy of liquids and solids is quite complicated, for it includes electrical potential energy associated with the forces (or "chemical" bonds) between atoms and molecules.

 ## 19–5   Specific Heat

The amount of heat Q required to change the temperature of a system is found to be proportional to the mass m of the system and to the temperature change ΔT, a fact known as early as the eighteenth century. This is expressed in the equation

$$Q = mc\,\Delta T \qquad\qquad (19\text{–}1)$$

where c is a quantity characteristic of the material called *specific heat*. For water at 15°C and a constant pressure of 1 atm, $c = 1.00$ kcal/kg·C° or 4.18×10^3 J/kg·C°; this corresponds to the fact that it takes 1 kcal of heat to raise the temperature of 1 kg of water by 1 C°. Table 19–1 gives the values of specific heat for other substances at 20°C. The values of c depend to some extent on temperature (as well as slightly on pressure), but for temperature changes that are not too great, c can often be considered constant.[†]

The specific heat as defined by Eq. 19–1 also depends somewhat on how the process of heating is carried out. We assumed the process was carried out at constant (atmospheric) pressure; for such a process, we usually call c the "specific heat at constant pressure" and use the symbol c_p. This is what is given in Table 19–1, since it is easiest to measure for solids and liquids. There are many other possible conditions at which the heat could be added; for example, the volume of the material might be kept constant (allowing the pressure to vary), and then c is written c_v, the specific heat at constant volume. The difference between c_p and c_v for solids and liquids is typically a few percent; for gases the difference is much greater and we discuss it in Section 20–4. For now we assume we are dealing with c_p (unless otherwise mentioned) and drop the subscript.

[†] To take into account the dependence of c on T, we can write Eq. 19–1 in differential form: $dQ = mc\,dT$; then the heat Q required to change the temperature from T_1 to T_2 is

$$Q = \int_{T_1}^{T_2} mc\,dT$$

where c is a function of temperature.

TABLE 19–1

19–5 Specific Heat   **377**

Specific Heats† **(20°C and constant, 1 atm, pressure)**

Substance	Specific Heat, c_p	
	kcal/kg · C°	J/kg·C°
Aluminum	0.22	900
Copper	0.093	390
Glass	0.20	840
Ice (-5°C)	0.50	2100
Iron or steel	0.11	450
Lead	0.031	130
Marble	0.21	860
Silver	0.056	230
Wood	0.4	1700
Alcohol (ethyl)	0.58	2400
Mercury	0.033	140
Water (15°C)	1.00	4186
Steam (110°C)	0.48	2010
Human body (average)	0.83	3470
Protein	0.4	1700

† For gases, see Table 20–1

EXAMPLE 19–1 How much heat is required to raise the temperature of 20 kg of iron from 10°C to 90°C?

SOLUTION From Table 19–1, the specific heat of iron is 0.11 kcal/kg·C°. The change in temperature is (90°C − 10°C) = 80 C°. Thus

$$Q = mc\,\Delta T = (20 \text{ kg})(0.11 \text{ kcal/kg·C°})(80 \text{ C°}) = 180 \text{ kcal.}$$

If the iron had been cooled from 90°C to 10°C, 180 kcal of heat would have flowed out of the iron. In other words, Eq. 19–1 is valid for heat flow either in or out, with a corresponding increase or decrease in temperature. If 20 kg of water, rather than iron, had been heated from 10°C to 90°C in the above example, the heat required would have been 1600 kcal. Water has one of the highest specific heats of all substances, which makes it an ideal substance for hot-water space-heating systems and other uses that require a minimal drop in temperature.

When different parts of an isolated system are at different temperatures, heat will flow from the part at higher temperature to the part at lower temperature. If the system is completely isolated, no energy can flow into or out of it; so, from the conservation of energy, the heat lost by one part of the system is equal to the heat gained by the other part:

$$\text{heat lost} = \text{heat gained.}$$

Let us take an example.

EXAMPLE 19–2 If 200 cm^3 of tea at 95°C is poured into a 300-g glass cup initially at 25°C, what will be the final temperature of the mixture when equilibrium is reached, assuming that no heat flows to the surroundings?

SOLUTION Since tea is mainly water, its specific heat is 1.00 kcal/kg · C° and its mass m is its density times its volume: $m = \rho V = (1.0 \times 10^3 \text{ kg/m}^3) \cdot$

$(200 \times 10^{-6} \text{ m}^3) = 0.20$ kg. We set

$$\text{heat lost by tea} = \text{heat gained by cup}$$

$$m_{\text{tea}}c_{\text{tea}} (95°C - T) = m_{\text{cup}}c_{\text{cup}} (T - 25°C)$$

where T is the as yet unknown final temperature. Putting in numbers and using Table 19–1, we solve for T, and find

$$(0.20 \text{ kg})(1.00 \text{ kcal/kg·C°})(95°C - T) = (0.30 \text{ kg})(0.20 \text{ kcal/kg·C°})(T - 25°C)$$

$$19 - 0.20T = 0.060T - 1.5$$

$$T = 79°C.$$

The exchange of energy as discussed in this example is the basis for a technique known as *calorimetry*, which is the quantitative measurement of heat exchange. To make such measurements, a *calorimeter* is used; a simple water calorimeter is shown in Fig. 19–2. It is very important that the calorimeter be well insulated so that only a minimal amount of heat is exchanged with the outside. One important use of the calorimeter is in the determination of specific heats of substances. In the technique known as the "method of mixtures," a sample of the substance is heated to a high temperature, which is accurately measured, and then quickly placed in the cool water of the calorimeter. The heat lost by the sample will be gained by the water and the calorimeter; by measuring the final temperature of the mixture, the specific heat can be calculated as illustrated in the following example.

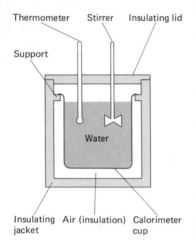

Thermometer   Stirrer   Insulating lid

Support

Water

Insulating   Air (insulation)   Calorimeter
jacket   cup

FIGURE 19–2

Simple water calorimeter.

EXAMPLE 19–3   We wish to determine the specific heat of a new alloy. A 0.150-kg sample of the alloy is heated to 540°C. It is then quickly placed in 400 g of water at 10.0°C which is contained in a 200-g aluminum calorimeter cup. (We do not need to know the mass of the insulating jacket since we assume it does not change significantly in temperature.) The final temperature of the mixture is 30.5°C. Calculate the specific heat of the alloy.

SOLUTION   We use the fact that the heat lost equals the heat gained; that is

$$\begin{pmatrix} \text{heat lost} \\ \text{by sample} \end{pmatrix} = \begin{pmatrix} \text{heat gained} \\ \text{by water} \end{pmatrix} + \begin{pmatrix} \text{heat gained by} \\ \text{calorimeter cup} \end{pmatrix}$$

$$m_s c_s \Delta T_s = m_w c_w \Delta T_w + m_{\text{cal}} c_{\text{cal}} \Delta T_{\text{cal}}.$$

Putting in values and using Table 19–1 this equation becomes

$$(0.150 \text{ kg})(c_s)(540°C - 30.5°C) = (0.40 \text{ kg})(1.0 \text{ kcal/kg·C°})(30.5°C - 10.0°C)$$
$$+ (0.20 \text{ kg})(0.22 \text{ kcal/kg·C°})(30.5°C - 10.0°C)$$

$$76.5c_s = (8.2 + 0.9) \text{ kcal/kg·C°}$$

$$c_s = 0.12 \text{ kcal/kg·C°}.$$

In making this calculation, we have ignored any heat transferred to the thermometer and the stirrer (which is needed to quicken the heat transfer process and thus reduce heat loss to the outside). These can be taken into account by adding additional terms to the right side of the above equation and will result in a slight correction to the value of c_s (see the problems). It should be noted that the quantity $m_{\text{cal}} c_{\text{cal}}$ is often called the *water equivalent* of the calorimeter—that is, $m_{\text{cal}} c_{\text{cal}}$ is numerically equal to the mass of water (in kilograms) that would absorb the same amount of heat.

19–6 Latent Heat

When a material changes phase from solid to liquid, or liquid to gas (see also Section 18–5), a certain amount of energy is involved in this *change of phase*. For example, let us trace what happens when 1.0 kg of water is heated at a steady rate from, say, $-20°C$, where it is ice, until it has all changed to water and then to steam above $100°C$, all at 1 atm pressure. As shown in the graph of Fig. 19–3, as heat is added to the ice, its temperature rises at a rate of about 2 C°/kcal of heat added (since for ice $c \approx 0.50$ kcal/kg·C°). However, when $0°C$ is reached, the temperature stops increasing even though heat is still being added. Instead, as heat is added, the ice is gradually observed to change to water in the liquid state without any change in temperature. After about 40 kcal have been added at $0°C$, half the ice remains and half has changed to water. After about 80 kcal has been added, all the ice has changed to water, still at $0°C$. Further addition of heat causes the temperature to again increase (now at a rate of 1 C°/kcal). When $100°C$ is reached, the temperature again remains constant as the heat added changes the water to vapor (steam). It requires about 540 kcal to change the 1.0 kg of water completely to steam, after which the graph rises again indicating that the temperature of the steam now rises as heat is added.

The heat required to change 1.0 kg of a substance from the solid to the liquid state is called the *heat of fusion*; it is denoted by l_F. The heat of fusion of water is 79.7 kcal/kg. The heat required to change a substance from the liquid to the vapor phase is called the *heat of vaporization*, l_V, and for water it is 539 kcal/kg. Other substances follow graphs similar to Fig. 19–3, although the melting-point and boiling-point temperatures are different, as are the specific heats and heats of fusion and vaporization. Values for the heats of fusion and vaporization, which are also called the *latent heats*, are given in Table 19–2 for a number of substances.

The heats of vaporization and fusion also refer to the amount of heat released by a substance when it changes from a gas to a liquid, or from a liquid to a solid. Thus, steam releases 539 kcal/kg when it changes to water, and water releases 79.7 kcal/kg when it becomes ice.

Of course, the heat involved in a change of phase depends not only on the latent heat but also on the total mass of the substance. That is

$$Q = ml$$

where l is the latent heat of the particular process, m is the mass of substance, and Q is the heat required or given off. For example, when 5.00 kg of water freezes at $0°C$, (5.00 kg) × (79.7 kcal/kg) = 398 kcal of energy is released.

Calorimetry sometimes involves a change of state as the following examples show. Indeed, latent heats are often measured using calorimetry.

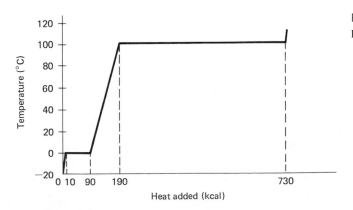

FIGURE 19–3
Heat added to bring 1.0 kg of ice at $-20°C$ to steam.

TABLE 19–2
Latent heats (at 1 atm)

Substance	Melting Point (°C)	Heat of Fusion kcal/kg[†]	Heat of Fusion J/kg	Boiling Point (°C)	Heat of Vaporization kcal/kg[†]	Heat of Vaporization J/kg
Oxygen	−218.8	3.3	0.14×10^5	−183	51	2.1×10^5
Ethyl alcohol	−114	25	1.04×10^5	78	204	8.5×10^5
Water	0	79.7	3.33×10^5	100	539	22.6×10^5
Lead	327	5.9	0.25×10^5	1750	208	8.7×10^5
Silver	961	21	0.88×10^5	2193	558	23×10^5
Tungsten	3410	44	1.84×10^5	5900	1150	48×10^5

[†] Numerical values in kcal/kg are the same in cal/g

EXAMPLE 19–4 How much energy does a refrigerator have to remove from 1.5 kg of water at 20°C to make ice at −12°C?

SOLUTION Heat must flow out to reduce the water from 20°C to 0°C, to change it to ice, and then to lower the ice from 0°C to −12°C:

$$Q = mc_{water}(20°C - 0°C) + ml_F + mc_{ice}[0° - (-12°C)]$$
$$= (1.5 \text{ kg})(1.0 \text{ kcal/kg·C°})(20 \text{ C°}) + (1.5 \text{ kg})(80 \text{ kcal/kg})$$
$$+ (1.5 \text{ kg})(0.5 \text{ kcal/kg·C°})(12 \text{ C°})$$
$$= 160 \text{ kcal,}$$

or 6.7×10^5 J.

EXAMPLE 19–5 A 0.50-kg chunk of ice at −10°C is placed in 3.0 kg of water at 20°C. At what temperature and in what phase will the final mixture be?

SOLUTION In this situation, before we can write down an equation, we must first check to see if the final mixture will be ice, a mixture of ice and water at 0°C, or all water. To bring the 3.0 kg of water at 20°C down to 0°C would require an energy release of $mc \Delta T = (3.0 \text{ kg})(1.0 \text{ kcal/kg·C°})(20 \text{ C°}) = 60$ kcal. To raise the ice from −10°C to 0°C would require $(0.50 \text{ kg})(0.50 \text{ kcal/kg·C°})·(10 \text{ C°}) = 2.5$ kcal, and to change the ice to water at 0°C will require $(0.50 \text{ kg})(80 \text{ kcal/kg}) = 40$ kcal, for a total of 42.5 kcal. This is not enough energy to bring the 3.0 kg of water at 20°C down to 0°C, so we know that the mixture must end up all water, somewhere between 0°C and 20°C. Now we can determine the final temperature by writing

$$\begin{pmatrix} \text{heat to raise} \\ \text{ice from} \\ -10°C \text{ to } 0°C \end{pmatrix} + \begin{pmatrix} \text{heat to} \\ \text{change ice} \\ \text{to water} \end{pmatrix} + \begin{pmatrix} \text{heat to raise} \\ \text{the 0.50 kg water} \\ \text{from } 0°C \text{ to } T \end{pmatrix} = \begin{pmatrix} \text{heat lost by 3.0 kg} \\ \text{of water cooling} \\ \text{from } 20°C \text{ to } T \end{pmatrix};$$

then

$$2.5 \text{ kcal} + 40 \text{ kcal} + (0.50 \text{ kg})(1.0 \text{ kcal/kg·C°})(T)$$
$$= (3.0 \text{ kg})(1.0 \text{ kcal/kg·C°})(20°C - T)$$

or

$$42.5 + 0.50T = 60 - 3.0T$$
$$T = 5.0°C.$$

The latent heat to change a liquid to a gas is needed not only at the boiling point. Water can change from the liquid to the gas phase even at room temperature. This process is called *evaporation* (and we already discussed it in Section 18–3). The value of the heat of vaporization increases slightly with a decrease in temperature: at 20°C, for example, it is 585 kcal/kg compared to 539 kcal/kg at 100°C.

We can make use of kinetic theory to see why energy is needed to melt or vaporize a substance. At the melting point, the latent heat of fusion does not increase the kinetic energy (and the temperature) of the molecules in the solid but instead is used to overcome the potential energy associated with the forces between the molecules. That is, work must be done against these attractive forces to break the molecules loose from their relatively fixed positions in the solid so they can freely roll over one another in the liquid phase. Similarly, energy is required for molecules held close together in the liquid phase to escape into the gaseous phase. This process is a more violent reorganization of the molecules than is melting (the average distance between the molecules is greatly increased) and hence the heat of vaporization is generally much greater than the heat of fusion for a given substance.

19–7 Heat Transfer: Conduction

Heat is transferred from one place or body to another in three different ways: by *conduction*, *convection*, and *radiation*. We will now discuss each of these in turn; but in practical situations, any two or all three may be operating at the same time. We start with conduction.

When a metal poker is put in a hot fire, or a silver spoon is placed in a hot bowl of soup, the exposed end of the poker or spoon soon becomes hot as well, even though it is not directly in contact with the source of heat. We say that heat has been conducted from the hot end to the cold end.

Heat **conduction** can be visualized as the result of molecular collisions. As one end of the object is heated, the molecules there move faster and faster. As they collide with their slower moving neighbors, they transfer some of their energy to these molecules whose speeds thus increase. These in turn transfer some of their energy by collision with molecules still farther down the object. Thus the energy of thermal motion is transferred by molecular collision down the object.

Heat conduction takes place only if there is a difference in temperature. Indeed, it is found experimentally that the rate of heat flow through a substance is proportional to the difference in temperature between its ends. It also depends on the size and shape of the object, and to investigate this quantitatively, let us consider the heat flow through a uniform object as illustrated in Fig. 19–4. It is found experimentally that the heat flow ΔQ per time interval Δt is given by the relation

$$\frac{\Delta Q}{\Delta t} = kA \frac{T_2 - T_1}{L} \qquad (19\text{–}2a)$$

where A is the cross-sectional area of the object, L is the distance between the two ends, which are at temperatures T_1 and T_2, and k is a proportionality constant called the *thermal conductivity*, which is characteristic of the material. In some cases (such as when k or A cannot be considered constant) we need to consider the limit of an infinitesimally thin slab of thickness dx. Then Eq. 19–2a becomes

$$\frac{dQ}{dt} = -kA \frac{dT}{dx} \qquad (19\text{–}2b)$$

FIGURE 19–4

Heat conduction.

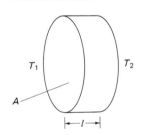

TABLE 19–3
Thermal conductivities

Substance	Thermal Conductivity, k	
	kcal/s·m·C°	J/s·m·C°
Silver	10×10^{-2}	420
Copper	9.2×10^{-2}	380
Aluminum	5.0×10^{-2}	200
Steel	1.1×10^{-2}	40
Glass (typical)	2.0×10^{-4}	0.84
Brick and concrete	2.0×10^{-4}	0.84
Water	1.4×10^{-4}	0.56
Human tissue (without blood)	0.5×10^{-4}	0.2
Asbestos	0.4×10^{-4}	0.16
Wood	$0.2–0.4 \times 10^{-4}$	0.08–0.16
Cork and glass wool	0.1×10^{-4}	0.042
Down	$.06 \times 10^{-4}$	0.025
Air	$.055 \times 10^{-4}$	0.023

where dT/dx is the temperature gradient and the negative sign is included since the heat flow is in the direction opposite to the temperature gradient.[†]

The thermal conductivities, k, for a variety of substances are given in Table 19–3. Substances for which k is large conduct heat rapidly and are said to be *good conductors*. Most metals fall in this category, although there is a wide range even among them as you may observe by holding the ends of a silver spoon and a stainless steel spoon immersed in the same hot cup of soup. Substances for which k is small, such as asbestos and down, are poor conductors of heat and are therefore good *insulators*. The relative magnitude of k can explain simple phenomena such as why a tile floor is much colder on the feet than a rug-covered floor at the same temperature: tile is a better conductor of heat than the rug; heat transferred from your foot to the rug is not conducted away rapidly, so the rug quickly heats up on its surface to the temperature of your foot; but the tile conducts the heat away rapidly and thus can take more heat from your foot, so its surface temperature drops.

EXAMPLE 19–6 A major source of heat loss from a house is through the windows. Calculate the rate of heat flow through a glass window 2.0 m × 1.5 m in area and 3.2 mm thick if the temperatures at the inner and outer surfaces are 15.0°C and 14.0°C, respectively.

SOLUTION Since $A = (2.0 \text{ m})(1.5 \text{ m}) = 3.0 \text{ m}^2$, $l = 3.2 \times 10^{-3} \text{ m}$, and using Table 19-3 to get k, we have from Eq. 19–2a:

$$\frac{\Delta Q}{\Delta t} = \frac{(0.84 \text{ J/s·m·C°})(3.0 \text{ m}^2)(15.0°C - 14.0°C)}{(3.2 \times 10^{-3} \text{ m})}$$

$$= 790 \text{ W}.$$

This is equivalent to $(790 \text{ J/s})/(4.18 \times 10^3 \text{ J/kcal}) = 0.19 \text{ kcal/s}$, or $(0.19 \text{ kcal/s}) \cdot (3600 \text{ s/h}) = 680 \text{ kcal/h}$.

You might notice that 15°C is not very warm for the living room of a house. The room itself may indeed be much warmer, and the outside might be colder than

[†] That the rate of heat flow is directly proportional to the cross-sectional area and to the temperature gradient is quite similar to the relations describing diffusion (Chapter 18) and the flow of fluids through a pipe (Chapter 13). In those cases, the flow of matter per unit area was found to be proportional to the concentration gradient or to the pressure gradient. This close similarity is one reason we speak of the "flow" of heat; yet we must keep in mind that no substance is flowing in this case—it is energy that is being transferred.

14°C. But the temperatures of 15°C and 14°C were specified as those at the window surfaces and there is usually a considerable drop in temperature of the air in the vicinity of the window both on the inside and the outside.† That is, the layer of air on either side of the window acts as an insulator and normally the major part of the temperature drop between the inside and outside of the house takes place across the air layer. Thus increasing the width of the air layer (such as using two panes of glass separated by an air gap) is far more helpful than simply increasing the glass thickness.

The insulating properties of clothing come from the insulating properties of air. Without clothes, our bodies would heat the air in contact with the skin and soon become reasonably comfortable since air is a very good insulator; but since air moves—there are breezes and drafts, and people themselves move about—the warm air would be replaced by cold air, thus increasing the heat loss from the body. Clothes keep us warm by holding air so it cannot move readily. It is not the cloth that insulates us, but the air that the cloth traps. Down is a very good insulator because even a small amount of it fluffs up and traps a great amount of air. On this basis, can you see one reason why drapes in front of a window reduce heat loss from a house?

19–8 Heat Transfer: Convection

Although liquids and gases are generally not very good conductors of heat, they can transfer heat quite rapidly by convection. **Convection** is the process whereby heat is transferred by the mass movement of molecules from one place to another. Whereas conduction involves molecules moving only over very small distances and colliding, convection involves the movement of molecules over large distances.

A forced-air furnace, in which air is heated and then blown by a fan into a room, is an example of *forced convection*. *Natural convection* occurs as well, and one familiar example is that hot air rises. For instance the air above a radiator (or other type of heater) expands as it is heated and hence its density decreases; because its density is less, it rises. Warm or cold ocean currents, such as the Gulf Stream, represent natural convection on a large scale. Wind is another example of convection, and weather in general is a result of convective air currents.

When a pot of water is heated, Fig. 19–5, convection currents are set up as the heated water at the bottom of the pot rises because of its reduced density and is replaced by cooler water from above. This principle is used in many heating systems in houses and other buildings. We will not deal with convection quantitatively here since it is complicated and is more in the realm of engineering than physics.

FIGURE 19–5
Convection currents in a pot of water being heated on a stove.

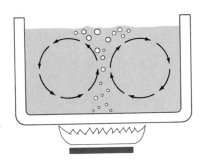

19–9 Heat Transfer: Radiation

Convection and conduction require the presence of matter. Yet all life on earth depends on the transfer of energy from the sun, and this energy is transferred to the earth over empty (or nearly empty) space. This form of energy transfer is heat—since the sun's temperature is much higher (6000 K) than earth's—and is referred to as **radiation**. The warmth we receive from a fire is mainly radiant

† If there is a heavy wind, the air outside a window will constantly be replaced with cold air; the temperature gradient across the glass will be greater and there will be a much greater rate of heat loss.

energy (most of the air heated by a fire rises by convection up the chimney and does not reach us).

As we shall see in later chapters, radiation consists essentially of electromagnetic waves. Suffice it to say for now that radiation from the sun consists of visible light plus many other wavelengths that the eye is not sensitive to; this includes infrared (IR) radiation, which is mainly responsible for heating the earth.

The rate at which an object radiates energy has been found to be proportional to the fourth power of the Kelvin temperature, T. That is, a body at 2000 K as compared to one at 1000 K radiates energy at a rate $2^4 = 16$ times as much. The rate of radiation is also proportional to the area A of the emitting object, so the rate at which energy leaves the object, $\Delta Q/\Delta t$, is

$$\frac{\Delta Q}{\Delta t} = e\sigma AT^4; \tag{19-3a}$$

here σ is a universal constant called the *Stefan-Boltzmann constant* which has the value

$$\sigma = (5.67032 \pm 0.00071) \times 10^{-8} \text{ W/m}^2 \cdot \text{K}^4$$

or, rounded off,

$$\sigma = 5.67 \times 10^{-8} \text{ W/m}^2 \cdot \text{K}^4;$$

and e, called the *emissivity*, is a number between 0 and 1 that is characteristic of the material. Very black surfaces, such as charcoal, have emissivity close to 1 whereas shiny surfaces have e close to zero and thus emit correspondingly less radiation. The value of e depends somewhat on the temperature of the body.

Any object not only emits energy by radiation, but it also absorbs energy radiated by other bodies. If an object of emissivity e and area A is at a temperature T_1, it radiates energy at a rate $e\sigma AT_1^4$. If the object is surrounded by an environment at temperature T_2, then the rate the surroundings radiate energy is proportional to T_2^4 and thus the rate that energy is absorbed by the object is proportional to T_2^4. The *net* rate of radiant heat flow from the object is then given by the equation

$$\frac{\Delta Q}{\Delta t} = e\sigma A(T_1^4 - T_2^4), \tag{19-4}$$

where A is the surface area of the object, T_1 its temperature and e its emissivity (at temperature T_1), and T_2 is the temperature of the surroundings. Notice in this equation that the rate of heat absorption by an object was taken to be $e\sigma AT_2^4$; that is, the proportionality constant is the same for both emission and absorption. This must be true to correspond with the experimental fact that equilibrium between object and surroundings is reached when they come to the same temperature; that is, $\Delta Q/\Delta t$ must equal zero when $T_1 = T_2$, so the coefficient of each term must be the same. Thus a good emitter is a good absorber. Black and very dark objects, which are good emitters, absorb nearly all the radiation that falls on them (which is why light-colored clothing is preferable to dark clothing on a hot day). On the other hand, not only do shiny surfaces emit little radiation, but they absorb little of the radiation that falls on them (most is reflected).

Because both an object and its surroundings radiate energy, there is a net transfer of energy from one to the other unless everything is at the same temperature. From Eq. 19-4 it is clear that if $T_1 > T_2$, the net flow of heat is from the body to the surroundings, so the body cools; but if $T_1 < T_2$, the net heat flow is from the surroundings into the body and its temperature rises. If different parts of the surroundings are at different temperatures, then Eq. 19-4 becomes more complicated.

◆ EXAMPLE 19-7   A ceramic teapot ($e = 0.70$), and a shiny one ($e = 0.10$), each hold 0.75 L of tea at 95°C. (*a*) Estimate the rate of heat loss from each and (*b*) estimate the temperature drop after 30 min for each. Consider only radiation and assume the surroundings are at 20°C.

SOLUTION (a) A teapot that holds 0.75 L can be approximated by a cube 10 cm on a side, so its surface area would be about 5×10^{-2} m². The rate of heat loss would be about

$$\frac{\Delta Q}{\Delta t} = e\sigma A(T_1^4 - T_2^4)$$

$$= e(5.67 \times 10^{-8} \text{ W/m}^2 \cdot \text{K}^4)(5 \times 10^{-2} \text{ m}^2)([368 \text{ K}]^4 - [293 \text{ K}]^4)$$

$$= e(30) \text{ J/s},$$

or about 20 W for the ceramic pot ($e = 0.70$) and 3 W for the shiny one ($e = 0.10$).

(b) To estimate the temperature drop, we use the concept of specific heat and ignore the contribution of the pots compared to that of the 0.75 L of water. Then, using Eq. 19–1,

$$\frac{\Delta T}{\Delta t} = \frac{(\Delta Q/\Delta t)}{mc} = \frac{e(30) \text{ J/s}}{(0.75 \text{ kg})(4.18 \times 10^3 \text{ J/kg} \cdot \text{C}°)} = e(0.01) \text{ C}°/\text{s}$$

which, for 30 min (1800 s), represents about 12 C° for the ceramic pot and about 2 C° for the shiny one. The shiny one clearly has an edge, at least as far as radiation is concerned. However convection and conduction could play a greater role than radiation.

Heating of an object by radiation from the sun cannot be calculated using Eq. 19–4 since this equation assumes a uniform temperature, T_2, of the environment surrounding the object, whereas the sun is essentially a point source; hence it must be treated as an additional source of energy. Heating by the sun is calculated using the fact that about 1350 J of energy strikes the atmosphere of the earth from the sun per second per meter squared of area at right angles to the sun's rays. This number, 1350 W/m², is called the *solar constant*. The atmosphere may absorb as much as 70 percent of this energy before it reaches the ground, depending on the cloud cover. On a clear day, about 1000 W/m² reaches the earth's surface; an object of emissivity e with area A facing the sun absorbs heat at a rate, in watts, of about $1000eA\cos\theta$ where θ is the angle between the sun's rays and a line perpendicular to the area A (Fig. 19–6). That is, $A\cos\theta$ is the "effective" area, at right angles to the sun's rays. The explanations for the seasons, the polar ice caps, and why the sun heats the earth more at midday than at sunrise or sunset are also related to this $\cos\theta$ factor. (The length of the day is also important.)

FIGURE 19–6

Radiant energy striking a body at an angle θ.

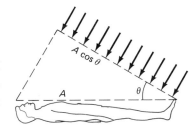

Summary

Thermal energy, or *internal energy*, U, refers to the total energy of all the molecules in a body. *Heat* refers to the transfer of energy from one body to another because of a difference of temperature. Heat is measured in energy units, such as joules; heat and thermal energy are also sometimes specified in calories or kilocalories, where 1 cal = 4.186 J is the amount of heat needed to raise the temperature of 1 g of water by 1 C°.

The *specific heat*, c, of a substance is defined as the energy (or heat) required to change the temperature of unit mass of substance by 1 degree; as an equation, $Q = mc\,\Delta T$, where Q is the heat absorbed or given off, ΔT the temperature rise or decline, and m the mass of the substance. When heat flows within an isolated system, the heat gained by one part of the system is equal to the heat lost by the other part of the system; this is the basis of *calorimetry*, which is the quantitative measurement of heat exchange.

An exchange of energy occurs, without a change in temperature, whenever a substance changes phase; this happens because the potential energy of the molecules changes as a result of the changes in the relative positions of the molecules. The *heat of fusion* is the heat required to melt 1 kg of a substance from the solid into the liquid phase; it is also equal to the heat given off when the substance changes from liquid to solid. The *heat of vaporization* is the energy required to change 1 kg of a substance from the liquid to the vapor phase; it is also the energy given off when the substance changes from vapor to liquid.

Heat is transferred from one place (or body) to another in three different ways. In *conduction*, energy is transferred from higher-KE molecules to lower-KE neighboring molecules when they collide. *Convection* is the transfer of energy by the mass movement of molecules over considerable distances. *Radiation*, which does not require the presence of matter, is energy transfer by electromagnetic waves, such as from the sun. All bodies radiate energy in an amount that is proportional to their surface area and to the fourth power of their Kelvin temperature; the energy radiated (or absorbed) also depends on the nature of the surface (dark and absorbing versus brightly reflecting), which is characterized by the emissivity.

 Questions _____

1. What happens to the work done when a jar of orange juice is vigorously shaken?

2. Is heat really involved in the Joule experiment of Fig. 19–1?

3. When a hot object warms a cooler object, does temperature flow between them? Are the temperature changes of the two objects equal?

4. Explain why cities situated on the ocean tend to have less extreme temperatures than inland cities at the same latitude.

5. In warmer areas where tropical plants grow but the temperature may drop below freezing a few times in the winter, the destruction of sensitive plants due to freezing can be reduced by watering them in the evening. Explain.

6. The specific heat of water is quite large. Explain why this fact makes water particularly good for heating systems (that is, hot-water radiators).

7. Why does water in a canteen stay cooler if the cloth jacket surrounding the canteen is kept moist?

8. Explain why burns caused by steam on the skin are often so severe.

9. Explain, using the concepts of latent heat and internal energy, why water cools (its temperature drops) when it evaporates.

10. Why is it important, when hot air furnaces are used to heat a house, that there be a vent for air to return to the furnace? What happens if this vent is blocked by a bookcase?

11. Early in the day, after the sun has reached the slope of a mountain, there tends to be a gentle upward movement of air. Later, after a slope goes into shadow, there is a gentle downdraft. Explain.

12. Will potatoes cook faster if the water is boiling faster?

13. Does an ordinary electric fan cool the air? Why or why not? If not, why use it?

14. The temperature very high in the earth's atmosphere can be as high as 700°C. Yet an animal there would freeze to death rather than boil. Explain.

15. Why is the liner of a thermos bottle silvered, and why does it have a vacuum between its two walls?

16. Explain why air-temperature readings are always taken with the thermometer in the shade.

17. The floor of a house on a foundation under which the air can flow is often cooler than a floor which rests directly on the ground (such as a concrete slab foundation). Explain.

18. The earth cools off at night much more quickly when the weather is clear than when cloudy. Why?

19. Sea breezes are often encountered on sunny days at the shore of a large body of water. Explain in light of the fact that the temperature of the land rises more rapidly than that of the nearby water.

20. Down sleeping bags and parkas are often specified as so many inches or centimeters of *loft*, the actual thickness of the garment when it is fluffed up. Explain.

21. Why are light-colored clothes more comfortable in hot climates than are dark clothes?

22. A premature baby in an incubator can be dangerously cooled even when the air temperature in the incubator is warm. Explain.

23. In the Northern Hemisphere the amount of heat required to heat a room where the windows face north is much higher than that required where the windows face south. Explain.

24. Suppose you are designing one of the following (choose one): house, concert hall, medical office building. List as many sources of heat as you can think of. Estimate the heat produced by each.

25. Heat loss occurs through windows by the following processes: (1) ventilation around edges, (2) through the frame, particularly if it is metal, (3) through the glass panes, (4) radiation. (a) For the first three, what is (are) the mechanism(s): conduction, convection, or radiation? (b) Heavy curtains reduce which of these heat losses? Explain in detail.

26. A piece of wood lying in the sun absorbs more heat than a piece of shiny metal. Yet the wood feels less hot than the metal when you pick it up. Explain.

 Problems _____

SECTION 19–2

1. (I) How much work must a person do to offset eating a 400 Cal piece of cake?

2. (I) A British thermal unit (Btu) is a unit of heat in the British system of units. One Btu is defined as the heat needed to raise 1 lb of water by 1 F°. Show that

$$1 \text{ Btu} = 252 \text{ cal} = 1055 \text{ J}.$$

3. (I) A water heater can generate 7500 kcal/h. How much water can it heat from 20°C to 60°C per hour?

4. (II) How many kilocalories of heat are generated when the brakes of a 1200-kg car are used to bring it to rest from a speed of 90 km/h?

SECTION 19–5

5. (I) What is the specific heat capacity of a metal substance if 36 kcal of heat is needed to raise 4.5 kg of the metal from 20°C to 42°C?

6. (I) What is the water equivalent of 0.228 kg of glass?

7. (I) What is the specific heat capacity of water in Btu/lb·F°? (See problem 2.)

8. (I) The *heat capacity*, C, of an object is defined as the amount of heat needed to raise its temperature by 1 C°. Thus, to raise the temperature by ΔT requires heat Q given by

$$Q = C \, \Delta T.$$

(a) Write the heat capacity C in terms of the specific heat, c, of the material. (b) What is the heat capacity (at constant pressure) of 1 kg of water? (c) Of 35 kg of water?

9. (II) Determine the Calorie content of 100 grams of a certain kind of cake from the following measurements. A 10-g sample of the cake is allowed to dry before putting it into a sealed container called a *bomb calorimeter* containing O_2 gas. The 0.615-kg aluminum "bomb" is placed in 2.00 kg of water, itself contained in a 0.524 kg aluminum calorimeter cup. The cake is then burned in the O_2 of the bomb, and the temperature of the whole apparatus is observed to rise from 12°C to 31°C. (Ignore the heat capacity of the gases.)

10. (II) (a) Show that if the specific heat varies as a function of temperature, $c(T)$, the heat needed to raise the temperature of a substance from T_1 to T_2 is given by

$$Q = \int_{T_1}^{T_2} mc(T) \, dT.$$

(b) Suppose $c(T) = c_0(1 + aT)$ for some substance, where $a = 2.0 \times 10^{-3} \, c_0$ and T is the Celsius temperature. Determine the heat required to raise the temperature from T_1 to T_2. (c) What is the mean value of c over the range T_1 to T_2 for part (b), expressed in terms of c_0, the heat capacity at 0°C?

11. (II) A 45-g glass thermometer reads 19.0°C before it is placed in 220 mL of water. When the water and thermometer come to equilibrium, the thermometer reads 38.5°C. What was the original temperature of the water?

12. (II) The 0.50-kg head of a hammer has a speed of 5.0 m/s just before it strikes a nail and is brought to rest. Estimate the temperature rise of a 15-g iron nail generated by ten such hammer blows done in quick succession. Assume the nail absorbs all the "heat."

13. (II) What will be the equilibrium temperature when a 200-g block of copper at 210°C is placed in a 180-g aluminum calorimeter cup containing 800 g of water at 11.0°C?

14. (II) When a 290-g piece of iron at 190°C is placed in a 100-g aluminum calorimeter cup containing 250-g of glycerin at 10°C, the final temperature is observed to be 38°C. What is the specific heat of glycerin?

15. (II) During light activity, a 70-kg person may generate 200 kcal/h. Asssuming that 20 percent of this goes into useful work and the other 80 percent is converted to heat, cal-

culate the temperature rise of the body after 1.00 h if none of this heat were transferred to the environment.

16. (II) How long does it take a 500-W coffee pot to bring to a boil 0.45 L of water initially at 10°C? Assume that the part of the pot which is heated with the water is made of 400 g of aluminum.

17. (II) When 220 g of a substance is heated to 330°C and then plunged into a 90-g aluminum calorimeter cup containing 150 g of water at 11.5°C, the final temperature, as registered by a 17-g glass thermometer, is 33.8°C. What is the specific heat of the substance?

18. (II) A 1.50-cm-diameter lead rod absorbs 82 kcal of heat. By how much does its length change? What would happen if the rod were only 2.0 cm long?

SECTION 19–6

19. (I) During exercise, a person may give off 180 kcal of heat in 30 min by evaporation of water from the skin. How much water has been lost?

20. (I) How much heat is needed to melt 13.00 kg of silver that is initially at 20°C?

21. (I) If 1.70×10^5 J of energy is supplied to a flask of oxygen at -183°C, how much oxygen will evaporate?

22. (II) An iron boiler of mass 128 kg contains 780 kg of water at 20°C. A heater supplies energy at the rate of 16,000 kcal/h. How long does it take for the water (a) to reach the boiling point and (b) to all have changed to steam?

23. (II) The specific heat of mercury is 0.033 kcal/kg·°C. Determine the latent heat of fusion of mercury using the following calorimeter data: 1.00 kg of solid Hg at its melting point of -39.0°C is placed in a 0.620-kg aluminum calorimeter with 0.400 kg of water at 12.80°C; the resulting equilibrium temperature is 5.06°C.

24. (II) What will be the final result when equal amounts of ice at 0°C and steam at 100°C are mixed together?

25. (II) A 55.0-kg ice skater moving at 8.5 m/s glides to a stop. Assuming the ice is at 0°C and that 50 percent of the heat generated by friction is absorbed by the ice, how much ice melts?

26. (II) A 25-g lead bullet traveling at 400 m/s passes through a thin iron wall and emerges at a speed of 250 m/s. If the bullet absorbs 50 percent of the heat generated, (a) what will be the temperature rise of the bullet? (b) If the ambient temperature is 20°C, will any of the bullet melt, and if so, how much?

SECTIONS 19–7 TO 19–9

27. (I) (a) How much power is radiated by a tungsten sphere (emissivity $e = 0.35$) of radius 10 cm at a temperature of 20°C? (b) If the sphere is enclosed in a room whose walls are kept at -5°C, what is the *net* flow of energy out of the sphere?

28. (I) Over what distance must there be heat flow by conduction from the blood capillaries beneath the skin to the surface if the temperature difference is 0.50°C? Assume 200 kcal/h must be transferred through the whole body's surface area of 1.5 m².

29. (I) Approximately how much radiation does a person, of total area 1.5 m², absorb per hour from the sun when it makes a 40° angle to the vertical on a clear day? Assume $e = 0.80$.

30. (II) Estimate the rate that heat can be conducted from the interior of the body to the surface. Assume that the thickness of tissue is 4.0 cm, that the skin is at 34°C and the interior at 37°C, and that the surface area is 1.5 m². Compare this to the measured value of 200 kcal/h that must be dissipated by a person working lightly. This clearly shows the necessity of convective cooling by the blood.

31. (II) A 100-W light bulb generates 95 W of heat which is dissipated through a glass bulb that has a radius of 3.0 cm and is 1.0 mm thick. What is the difference in temperature between the inner and outer surfaces of the glass?

32. (II) A mountain climber wears down clothing 2.8 cm thick whose total surface area is 1.8 m². The temperature at the surface of the clothing is 0°C and at the skin is 34°C. Determine the rate of heat flow by conduction through the clothing (a) assuming it is dry and k, the thermal conductivity, is that of down, and (b) assuming the clothing is wet, so that k is that of water and the jacket has matted down to 0.5-cm thickness.

33. (II) Write an equation for the total rate of heat flow through the wall of a house if the wall consists of material with thermal conductivity k_1, total area A_1, and thickness l_1, and of windows with thermal conductivity k_2, area A_2, and thickness l_2. The temperature difference is ΔT.

34. (II) *Newton's law of cooling* states that for small temperature differences, if a body at a temperature T_1 is in surroundings at a temperature T_2, the body cools at a rate given by:

$$\frac{\Delta Q}{\Delta t} = k(T_1 - T_2)$$

where k is a constant. It includes the effects of conduction, convection, and radiation. That this linear relationship should hold is obvious if only conduction is considered. Show that it is also approximately true for radiation by showing that Eq. 19–4 reduces to

$$\Delta Q/\Delta t = 4\sigma e A T_2^3 (T_1 - T_2) = \text{constant} \times (T_1 - T_2)$$

if $(T_1 - T_2)$ is small.

35. (II) A *thermal transmission coefficient*, U, defined by the equation:

$$\frac{\Delta Q}{\Delta t} = AU(T_2 - T_1),$$

is often used in practical work. Note that U is the rate of heat flow per unit area per degree of temperature. (a) How is U related to the thermal conductivity coefficient k? (b) Is U a characteristic property of a material? If not, what else must be known about the material before U can be determined? (c) Write an equation describing the total heat loss by conduction from a room in terms of the values of U for the different materials; assume the windows are of total area A_1, and have thermal transmission coefficient U_1. The walls have a total area A_2 and consist of a layer of brick (U_2) separated by an air gap (U_3) from a layer of wood (U_4). (d) What is the thermal transmission coefficient U for a brick wall 12 cm thick and 8.0 m wide by 2.5 m high?

36. (II) A house has well-insulated walls 22.0 cm thick (as-sume conductivity of air) and area 350 m², a roof of wood 5.5 cm thick and area 280 m², and uncovered windows 0.65 cm thick and total area 28 m². (a) Assuming that the heat loss is only by condution, calculate the rate at which heat must be supplied to this house to maintain its temperature at 20°C if the outside temperature is −5°C. (b) If the house is initially at 10°C, estimate how much heat must be supplied to raise the temperature to 20°C within 30 min. Assume that only the air needs to be heated and that its volume is 700 m³. (c) If natural gas costs $0.055 per kilogram and its heat of combustion is 5.4 × 10⁷ J/kg, how much is the monthly cost to maintain the house as in part (a) above for 12 h each day assuming 90 percent of the heat produced is used to heat the house? Take the specific heat of air to be 0.17 kcal/kg · C°.

37. (II) A double-glazed window is one with two panes of glass separated by an air space. (a) Show that the rate of heat flow by conduction is given by:

$$\frac{\Delta Q}{\Delta t} = \frac{A(T_2 - T_1)}{l_1/k_1 + l_2/k_2 + l_3/k_3},$$

where k_1, k_2, and k_3 are the thermal conductivities for glass, air, and glass, respectively. (b) Generalize this expresssion for any number of materials placed next to one another.

38. (II) Approximately how long should it take 20 kg of ice at 0°C to melt when it is placed in a carefully sealed styrofoam "icebox" of dimensions 30 cm × 20 cm × 50 cm whose walls are 1.5 cm thick. Assume that the conductivity of styrofoam is equal to that of air and that the outside temperature is 25°C.

39. (II) A leaf of area 40 cm² and mass $4.5 × 10^{-4}$ kg directly faces the sun on a clear day. The leaf has an emissivity of 0.85, and a specific heat of 0.80 kcal/kg · K. (a) Estimate the rate of rise of the leaf's temperature. (b) Calculate the temperature the leaf would reach if it lost all its heat by radiation (the surroundings are at 20°C). (c) In what other ways can the heat be dissipated by the leaf?

40. (II) A house thermostat is normally set to 22°C but at night it is turned down to 12°C for 8 h. Estimate how much more heat would be needed (state as a percentage of daily usage) if the thermostat were not turned down at night. Assume that the outside temperature averages 0°C for the 8 h at night and 8°C for the remainder of the day, and that the heat loss from the house is proportional to the difference in temperature inside and out. To obtain an estimate from the data, you will have to make other simplifying assumptions; state what these are.

FIGURE 19–7

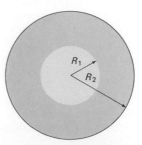

41. (III) A sphere of radius R_2 has a concentric central cavity of radius R_1 (Fig. 19–7). The cavity is filled with material at temperature T_1 whereas the outer surface of the sphere is at T_2. Show that the rate of heat flow is

$$\frac{dQ}{dt} = \frac{4\pi k(T_1 - T_2)R_1 R_2}{R_2 - R_1}.$$

42. (III) A cylindrical pipe has inner radius R_1 and outer radius R_2. The interior of the pipe carries hot water at temperature T_1. The temperature outside is T_2 ($< T_1$). (a)

Show that the rate of heat loss per length L of pipe is

$$\frac{dQ}{dt} = \frac{2\pi k(T_1 - T_2)L}{\ln(R_2/R_1)}.$$

(b) Suppose the pipe is iron with $R_1 = 1.50$ cm, $R_2 = 2.00$ cm, and $T_2 = 30°$C. If the pipe holds water at $T_1 = 98.00°$C, what will be its temperature decrease per second? (c) Suppose water at $98.00°$C enters the pipe and moves at a speed of 4.0 cm/s. What will be its temperature drop per centimeter of travel?

The First Law of Thermodynamics

20

We have seen that *heat represents a transfer of energy from one body to another due to a temperature difference between them.* Heat is thus much like work; for as we saw in Chapter 6, work represents a transfer of energy from one body to another by mechanical means. We generalize this definition of work to include all transfers of energy except those we call heat. That is, *work refers to a transfer of energy that is not caused by a temperature difference.* **Thermodynamics** is concerned with processes in which energy is transferred as heat or work. We normally focus on some kind of system, and we consider the transfer of energy into or out of that system.

We can define several kinds of systems in thermodynamics. A *closed system* is one whose mass is constant. In an *open system*, mass may enter or leave. Many (idealized) systems we study in physics are closed systems. But many systems, including plants and animals, are open systems since they exchange materials with the environment (food, oxygen, waste products). A closed system is said to be *isolated* if no energy in any form passes across its boundaries; otherwise it is *nonisolated*.

The equilibrium state of a system is defined by variables such as (for a gas) P, V, T, and n (number of moles). Work and heat are not used to describe a state. A system in a given state does not "have" a certain amount of heat or work. Rather, when work is done on a system (such as compressing a gas), or when heat is added or removed from a system, the state of the system *changes*. Thus work and heat are involved in *thermodynamic processes* which can change the system from one state to another; they are not characteristic of the state itself, as are pressure, volume, temperature, and mass.

 ## 20–1 Work Done in Volume Changes; Isothermal and Isobaric Processes

Let us now see how to calculate the work done in a very simple and common type of thermodynamic process: when there is a change in volume, such as when a gas expands or is compressed. Suppose we have a gas confined to a cylindrical container fitted with a movable piston (Fig. 20–1). We must always be careful to define exactly what our system is. In this case we choose our system to be the gas; so the container's walls and the piston are parts of the environment. Now let us calculate the work done by the gas when it expands **quasistatically** ("almost statically"), by which we mean that the process is carried out extremely slowly—ideally, infinitely slowly—so the system passes through a succession of infinitesimally close equilibrium states; in this way P and T are defined for the system at all instants.[†] The gas expands against the piston, whose area is A. The gas exerts a force $F = PA$ on the piston, where P is the pressure in the gas. (Since the piston is assumed to move at a very slow and constant speed, some external pressure, or friction, exerts an equal force on the piston in the opposite direction.) The work done by the gas to move the piston an infinitesimal displacement dl is

$$dW = \mathbf{F} \cdot d\mathbf{l} = PA\,dl = P\,dV \tag{20–1}$$

since the infinitesimal increase in volume is $dV = A\,dl$. If the gas were *compressed*, so that dl pointed into the gas, the volume would decrease and $dV < 0$. The work done by the gas in this case would then be negative, which is equivalent to saying that positive work was done *on* the gas, not by it. For a finite change in volume from V_1 to V_2, the work W done by the gas will be

$$W = \int dW = \int_{V_1}^{V_2} P\,dV. \tag{20–2}$$

Equations 20–1 and 20–2 are valid for the work done in any volume change—by a gas, a liquid, or a solid—as long as it is done quasistatically.

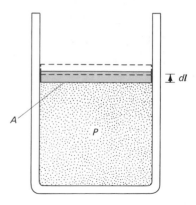

FIGURE 20–1

The work done by a gas when its volume increases by $dV = A\,dl$ is $dW = P\,dV$.

In order to integrate Eq. 20–2 we need to know how the pressure varies during the process, and this depends on the type of process. Suppose we have an ideal gas and we want to expand its volume from V_1 to V_2, but we want the temperature at the beginning to be the same as at the end, $T_1 = T_2$. One way of doing this would be to allow the gas to expand quasistatically at constant temperature. Such a process is called an **isothermal process**. To be sure the temperature remains constant, we assume our gas is in thermal contact with a **heat reservoir** (a body whose mass is so large, ideally, that its temperature does not change significantly when heat is exchanged with our system); and the expansion process is done very slowly so that the whole gas stays at the same constant temperature. This process is represented by the curve between points 1 and 2 on the PV diagram of Fig. 20–2. The work done in this process, according to Eq. 20–2, is just the area between the PV curve and the V axis; this is shown shaded in Fig. 20–2. We can do the integral in Eq. 20–2 for an ideal gas by using the ideal gas law:

$$P = \frac{nRT}{V};$$

the work done is

$$W = \int P\,dV = nRT \int_{V_1}^{V_2} \frac{dV}{V} = nRT \ln \frac{V_2}{V_1} \cdot \quad \begin{bmatrix} \text{isothermal process;} \\ \text{ideal gas} \end{bmatrix} \tag{20–3}$$

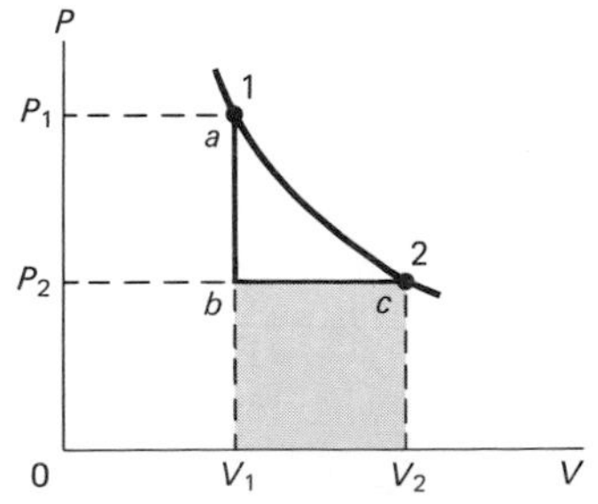

Let us next consider a different way of taking the gas between the same states 1 and 2; this time, let us lower the pressure in the gas from P_1 to P_2 as indicated by the line ab in Fig. 20–3; then let the gas expand from V_1 to V_2 at constant pressure ($= P_2$), which is indicated by the line bc in Fig. 20–3. (A process done at constant volume, such as ab, is called an *isochoric process*; a process carried out at constant pressure, such as bc, is called an *isobaric process*.) No work is done in the first part, ab, since $dV = 0$. In the second part, bc, the pressure remains constant so

$$W = \int_{V_1}^{V_2} P\,dV = P_2(V_2 - V_1)$$

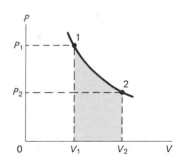

FIGURE 20–2

PV diagram for an ideal gas undergoing an isothermal process.

FIGURE 20–3

Process abc consists of an isochoric (ab) and an isobaric (bc) process.

[†] If the gas expanded or were compressed quickly, there would be turbulence, and different parts would be at different pressure (and temperature).

and the work done is again represented by the area between the curve (*abc*) on the *PV* diagram and the *V* axis which is indicated by the shading in Fig. 20–3. Using the ideal gas law, we can also write

$$W = P_2(V_2 - V_1) = nRT_2\left(1 - \frac{V_1}{V_2}\right). \qquad \begin{bmatrix} \text{isobaric process;} \\ \text{ideal gas} \end{bmatrix} \quad (20\text{–}4)$$

(Note that the temperature does *not* remain constant during the isobaric process, although it is the same at the end points of the isochoric plus isobaric process, [*abc* in Fig. 20–3: $T_1 = T_2$].)

As can be seen from the shaded areas in Figs. 20–2 and 20–3, or by putting in numbers in Eqs. 20–3 and 20–4 (try it for $V_2 = 2V_1$), the work done in these two processes is different. This is a general result: *the work done in taking a system from one state to another depends not only on the initial and final states, but also on the type of process (or "path")*.

This result reemphasizes the fact that work cannot be considered a property of a system.

The same result is true of heat. The heat input required to change the gas from state 1 to state 2 depends on the process; for the isothermal process of Fig. 20–2 it turns out to be greater than for the process *abc* of Fig. 20–3. In general, *the amount of heat added or removed in taking a system from one state to another depends not only on the initial and final states but also on the path or process*.

 ## 20–2   The First Law of Thermodynamics

In Section 19–3, we defined the internal energy of a system as the sum total of all the energy of the molecules of the system. We would expect that the internal energy of a system would be increased either by doing work on the system, or by adding heat to it; and the internal energy would be decreased if heat flows out of the system or if work is done by the system on something else.

As a result of Joule's experiments (and a great many others), it is reasonable to propose an important law: the change in internal energy of a closed system, ΔU, will be given by

$$\Delta U = Q - W \qquad (20\text{–}5)$$

where Q is the heat *added* to the system and W is the work done *by* the system. Note that if work is done *on* the system, W will be negative and ΔU will increase. (Of course we could have defined W as the work done on the system, in which case there would be a plus sign in Eq. 20–5; but it is conventional to define W and Q as we have done.) Similarly, if heat leaves the system, Q is negative. Equation 20–5 is known as the **first law of thermodynamics**. It is one of the great laws of physics, and its validity rests on experiment in which no exceptions have been seen. Since Q and W represent energy transferred into or out of the system, the internal energy changes accordingly; thus the first law of thermodynamics is simply a statement of the *law of conservation of energy*. It is worth noting that the conservation of energy law was not formulated until the nineteenth century, for it depended on the interpretation of heat as a transfer of energy.

Equation 20–5 applies to a closed system. It also applies to an open system if we take into account the change in internal energy due to the increase or decrease in the amount of matter. For an isolated system, we have (by definition) $W = Q = 0$, and so $\Delta U = 0$.

We have arrived at the first law of thermodynamics, Eq. 20–5, in an intuitive way, appealing to a microscopic (molecular) point of view. But the validity of the first law of thermodynamics rests on experiment. And since experiments are done in the macroscopic world, it is important to examine the first law of thermo-dynamics from the macroscopic (thermodynamic) point of view. We saw in the previous section that when a system is moved from a given state (1) to a second state (2), the amount of heat added to the system, Q, and the amount of work done by the system, W, depends on the particular process (or path) used; Q and W are different for different processes even though the initial and final states of the system are the same for each process. However, experiments (and a great many have been done to confirm this) show that the quantity $Q - W$ is the same for *all* processes that take the system between the same two states. That is, Q and W may differ for different processes, but $Q - W$ is the same for all. (The value of $Q - W$ of course depends on what the system is and what the initial and final states are.) This means we can define a state variable, which we will call the **internal energy** U, by the relation

$$\Delta U = U_2 - U_1 = Q - W$$

where U_1 and U_2 represent the internal energy of the system in states 1 and 2, and Q and W are the heat added to the system and work done by the system in going from state 1 to state 2. Since $Q - W$ depends only on the initial and final states of a system, so then does $\Delta U = U_2 - U_1$. That is, U_1 and U_2 are functions of the variables that describe the system in states 1 and 2, respectively. Therefore U is a function only of the state of the system and does not depend on how the system got to that state (U doesn't depend on the system's "past history"). From a thermodynamic point of view, then, the first law is a statement that we can define a function, called the internal energy, which is a property of the state of the system.

This is all rather abstract. But it shows the validity of the first law of thermo-dynamics without recourse to microscopic models. Of course when we interpret the internal energy as the total energy of all the molecules of the system, we get an intuitive grasp of the concept of internal energy.

It is sometimes useful to write the first law of thermodynamics in differential form:

$$dU = dQ - dW.$$

Here, dU represents an infinitesimal change in internal energy when an infinitesimal[†] amount of heat dQ is added to the system, and the system does an infinitesimal amount of work dW.

20–3 Applications of the First Law of Thermodynamics to Some Simple Thermodynamic Processes

Let us now apply the first law of thermodynamics to some simple processes. In Section 20–1 we saw that an isothermal process is one carried out at constant temperature, an isobaric process is one carried out at constant pressure, and an isochoric process is one carried out at constant volume. We also saw how to calculate the work done in each of these processes.

[†] The differential form of the first law is often written

$$dU = dQ - dW$$

where the bars on the differential sign (d) are used to remind us that W and Q are not functions of the state variables (such as P, V, T, n). Internal energy, U, is a function of the state variables, and so dU represents the differential (called an *exact differential*) of some function U. The differentials dW and dQ are not exact differentials (they are not the differential of some mathematical function); they thus only represent infinitesimal amounts. We won't worry about this point in this book.

EXAMPLE 20–1   Suppose 2.00 mol of an ideal gas of volume $V_1 = 3.50 \text{ m}^3$ at $T_1 = 300 \text{ K}$ is allowed to expand to $V_2 = 7.00 \text{ m}^3$ at $T_2 = 300 \text{ K}$. The process is done (a) isothermally; (b) along the path abc in Fig. 20–3 so that the pressure is allowed to drop at constant volume, then the volume increases at constant pressure. For each process, (a) and (b), determine the work done by the gas, the heat added to the gas, and the change in internal energy of the gas.

SOLUTION   (a) The work done by the gas as it expands at constant temperature is, by Eq. 20–3,

$$W = nRT \ln \frac{V_2}{V_1} = (2.00 \text{ mol})(8.314 \text{ J/mol} \cdot \text{K})(300 \text{ K})(\ln 2.00)$$
$$= 3460 \text{ J}.$$

There is no obvious way to calculate the heat flow for this situation (for example, we might try to use Eq. 19–1, but specific heat at constant temperature is not defined). However, we can calculate the change in internal energy, and then from the first law obtain Q. The internal energy of an ideal gas depends only on temperature (Section 19–4); since the temperature doesn't change, neither does the internal energy; hence

$$\Delta U = 0$$

for an isothermal process (not quite true for real gases). From the first law of thermodynamics, the heat added to the gas is

$$Q = \Delta U + W = W = 3460 \text{ J}.$$

(b) This process has two parts (see Fig. 20–3); path ab at constant volume, and path bc at constant pressure. The work done along path ab at constant volume is $W_{ab} = 0$, since $\int P\,dV = 0$ when $dV = 0$; for path bc at constant pressure the work done is (Eq. 20–4):

$$W_{bc} = nRT_2\left(1 - \frac{V_1}{V_2}\right) = (2.00 \text{ mol})(8.314 \text{ J/mol} \cdot \text{K})(300 \text{ K})(1 - 0.50) = 2490 \text{ J}.$$

The total work done in process abc is then $W = 0 + 2490 \text{ J} = 2490 \text{ J}$. The total change in internal energy is $\Delta U = 0$ since the initial and final states have the same temperature, and thus $Q = \Delta U + W = 2490 \text{ J}$. We can check this value for Q for the process abc since we can make use of the specific heat equation, 19–1. We need to anticipate a result from the next section, namely that the molar heat capacity (that is, specific heats per mole rather than per kilogram) for an ideal gas are, at constant pressure $C_p = 4.97 \text{ cal/mol} \cdot \text{K}$ and at constant volume, $C_v = 2.98 \text{ cal/mol} \cdot \text{K}$. From the ideal gas law, the volume and temperature at point b (Fig. 20–3) are $V_b = 3.50 \text{ m}^3$ and $T_b = T_2(V_b/V_2) = 150 \text{ K}$. Thus, for the path ab,

$$Q_{ab} = nC_v\Delta T = (2.00 \text{ mol})(2.98 \text{ cal/mol} \cdot \text{K})(150 \text{ K} - 300 \text{ K}) = -894 \text{ cal}.$$

(The minus sign means heat was lost by the gas along path ab.) Along path bc we have

$$Q_{bc} = nC_p\,\Delta T = (2.00 \text{ mol})(4.97 \text{ cal/mol} \cdot \text{K})(300 \text{ K} - 150 \text{ K}) = 1490 \text{ cal}$$

Thus

$$Q = Q_{ab} + Q_{bc} = -894 \text{ cal} + 1490 \text{ cal} = (596 \text{ cal})(4.18 \text{ J/cal}) = 2490 \text{ J}.$$

This is the result we expected, but note that because we calculated both W and Q, we could predict the change in internal energy to be $\Delta U = Q - W = 2490 \text{ J} - 2490 \text{ J} = 0$, which is consistent with our result.

EXAMPLE 20–2   Determine (a) the work done, and (b) the change in internal energy of 1.00 kg of water when it is all boiled to steam at 100°C. Assume a constant pressure of 1.00 atm.

SOLUTION   (a) The volume of 1.00 kg of water at 100°C is 1000 cm³ or 1.00×10^{-3} m³; 1.00 kg of steam at 100°C has a volume of 1.67 m³ (see Table 12–1). Hence the work done is

$$W = P(V_2 - V_1) = (1.01 \times 10^5 \ \text{N/m}^2)(1.67 \ \text{m}^3 - 1.00 \times 10^{-3} \ \text{m}^3) = 1.69 \times 10^5 \ \text{J}.$$

(b) The heat required to boil 1.00 kg of water is the heat of vaporization, $Q = 539$ kcal $= 22.6 \times 10^5$ J. From the first law of thermodynamics

$$\Delta U = Q - W = 22.6 \times 10^5 \ \text{J} - 1.7 \times 10^5 \ \text{J} = 20.9 \times 10^5 \ \text{J}.$$

Thus, only about 8 percent of the heat added is used to do work. The other 92 percent increases the internal energy of the water.

An **adiabatic process** is one in which no heat is allowed to flow into or out of the system: $Q = 0$. This situation can occur if the system is extremely well insulated, or the process happens so quickly[†] that heat—which flows slowly—has no time to flow in or out. The expansion of gases in an internal combustion engine is one example of a process that is very nearly adiabatic. A slow adiabatic expansion of an ideal gas follows a curve like that labeled AC in Fig. 20–4. Since $Q = 0$, we have from Eq. 20–5 that $\Delta U = -W$. That is, the internal energy decreases; hence the temperature decreases as well. This should be evident in Fig. 20–4 where the product $PV \ (= nRT)$ is less at point C than at point B (curve AB is for an isothermal process, for which $\Delta U = 0$ and $\Delta T = 0$). In an adiabatic compression, work is done on the gas and hence the internal energy increases and the temperature rises. In a diesel engine, the air is rapidly compressed adiabatically by a factor of 15 or more; the temperature rise is so great that when the fuel is injected, the mixture ignites spontaneously.

One type of adiabatic process is a so-called *free expansion* in which a gas is allowed to expand in volume adiabatically without doing any work. The apparatus to accomplish a free expansion is shown in Fig. 20–5; it consists of two well-insulated compartments (to assure no heat flow in or out) connected by a valve or stopcock; one compartment is filled with gas, the other is empty. When the valve is opened, the gas expands to fill both containers. No heat flows in or out ($Q = 0$), and no work is done because the gas does not move any other object. Thus $Q = W = 0$ and by the first law of thermodynamics, $\Delta U = 0$. The *internal energy of a gas does not change in a free expansion*. For an ideal gas, $\Delta T = 0$ also since U depends only on T (Section 19–4). Experimentally, the free expansion has been used to determine if the internal energy of *real gases* depends only on T. The experiment is very difficult to do accurately, but it has been found that the temperature of real gases drops very slightly in a free expansion. Thus the internal energy of real gases must depend, at least a little, on pressure or volume as well as on temperature.

Note, incidentally, that a free expansion could not be plotted on a PV diagram, because the process is rapid, not quasistatic. The intermediate states are not equilibrium states, and hence the pressure (and even the volume at some instants) is not clearly defined.

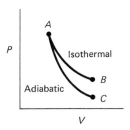

FIGURE 20–4

PV diagram for adiabatic (AC) and isothermal (AB) processes on an ideal gas.

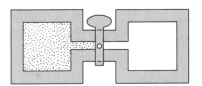

FIGURE 20–5

Free expansion.

20–4   Heat Capacities of Gases and the Equipartition of Energy

In Section 19–5 we discussed the concept of specific heat and applied it to solids and liquids. Much more than for solids and liquids, the values of the specific heat for gases depends on how the process is carried out. Two important processes are

[†] It would not be quasistatic and would therefore not be plottable on a PV diagram.

those in which either the volume or the pressure is kept constant. Although for solids and liquids it matters little, Table 20–1 shows that the specific heats of gases at constant volume (c_v) and at constant pressure (c_p) are quite different.

This is easily explained in terms of the first law of thermodynamics and kinetic theory. Indeed, the values of the specific heats can be calculated using the kinetic theory and the results are in close agreement with experiment. Before showing this, we introduce the *molar heat capacities*, C_v and C_p, which are defined as the heat required to raise 1 mol of the gas by 1 C° at constant volume and at constant pressure, respectively. That is, in analogy to Eq. 19–1, the heat Q needed to raise n moles of gas by ΔT degrees is:

$$Q = nC_v\Delta T \qquad \text{[volume constant]} \qquad (20\text{–}6a)$$

$$Q = nC_p\Delta T. \qquad \text{[pressure constant]} \qquad (20\text{–}6b)$$

It is clear from the definition of molar heat capacity (or by comparing Eqs. 19–1 and 20–6 that

$$C_v = Mc_v$$

$$C_p = Mc_p$$

where M is the molecular mass of the gas. The values for molar heat capacities are included in Table 20–1, and we see that the values are nearly the same for different gases that have the same number of atoms per molecule.

Let us now make use of the kinetic theory, and see first why the heat capacities of gases are higher for constant-pressure processes than for constant-volume processes. Let us imagine that an ideal gas is slowly heated via these two different processes—first at constant volume, and then at constant pressure. In both of these processes, we let the temperature increase by the same amount, ΔT. In the process done at constant volume, no work is done since $\Delta U = 0$. Thus, according to the first law of thermodynamics the heat added (which we now denote by Q_v) all goes into increasing the internal energy of the gas:

$$Q_v = \Delta U.$$

In the process carried out at constant pressure, work is done and hence the heat added, Q_p, must not only increase the internal energy but also is used to do the work $W = P\Delta V$. Thus, more heat must be added in this process than in the first process at constant volume, and from the first law of thermodynamics

$$Q_p = \Delta U + P\Delta V.$$

TABLE 20–1

Heat capacities of gases at 15°C

Gas	Specific Heats (kcal/kg·K)		Molar Heat Capacities (cal/mol·K)		$C_p - C_v$ (cal/mol·K)	$\gamma = \dfrac{C_p}{C_v}$
	c_v	c_p	C_v	C_p		
Monatomic						
He	0.75	1.15	2.98	4.97	1.99	1.67
Ne	0.148	0.246	2.98	4.97	1.99	1.67
Diatomic						
N_2	0.177	0.248	4.96	6.95	1.99	1.40
O_2	0.155	0.218	5.03	7.03	2.00	1.40
Triatomic						
CO_2	0.153	0.199	6.80	8.83	2.03	1.30
H_2O (100°C)	0.350	0.482	6.20	8.20	2.00	1.32
Polyatomic						
C_2H_6	0.343	0.412	10.30	12.35	2.05	1.20

Since ΔU is the same in the two processes (ΔT was chosen to be the same) we have:

$$Q_p - Q_v = P \Delta V.$$

From the ideal gas law, $V = nRT/P$, so for a process at constant pressure we have $\Delta V = nR \Delta T/P$. Putting this into the above equation and using Eq. 20–6, we find

$$nC_p \Delta T - nC_v \Delta T = P\left(\frac{nR \Delta T}{P}\right)$$

or

$$C_p - C_v = R. \tag{20–7}$$

Since the gas constant $R = 8.314$ J/mol·K $= 1.986$ cal/mol·K our prediction is that the C_p will be larger than C_v by about 1.99 cal/mol·K. Indeed, this is very close to what is obtained experimentally as can be seen from the next to last column in Table 20–1.

Let us now calculate the molar heat capacity of a monatomic gas using the kinetic theory model of gases. First we consider a process carried out at constant volume. Since no work is done in this process the first law of thermodynamics tells us that if heat Q is added to the gas, the internal energy of the gas changes by

$$\Delta U = Q.$$

For an ideal monatomic gas, the internal energy U is the total kinetic energy of all the molecules,

$$U = N(\tfrac{1}{2}m\overline{v^2}) = \tfrac{3}{2}nRT$$

as we saw in Section 19–4. Then, using Eq. 20–6a, we can write $\Delta U = Q$ in the form

$$\Delta U = \tfrac{3}{2}nR \Delta T = nC_v \Delta T \tag{20–8}$$

or

$$C_v = \tfrac{3}{2}R. \tag{20–9}$$

Since $R = 8.314$ J/mol·K $= 1.986$ cal/mol·K, kinetic theory predicts that $C_v = 2.98$ cal/mol·K for an ideal monatomic gas. This is very close to the experimental values for monatomic gases such as helium and neon (Table 20–1). From Eq. 20–7, C_p is predicted to be about 4.97 cal/mol·K, also in agreement with experiment.

The measured molar heat capacities for more complex gases (Table 20–1) such as diatomic (two atoms) and triatomic (three atoms) gases increase with increased number of atoms per molecule. This is apparently due to the fact that the internal energy includes not only translational kinetic energy but other forms of energy as well. Take, for example, a diatomic gas. As shown in Fig. 20–6, the two atoms can rotate about two different axes (but rotation about a third axis passing through the two atoms would give rise to very little energy since the moment of inertia is so small). Thus the molecules can have rotational as well as translational kinetic energy. It is useful to introduce the idea of **degrees of freedom**, by which we mean the number of independent ways molecules can possess energy. For example, a monatomic gas is said to have three degrees of freedom, since an atom can have velocity along the x axis, the y axis, and the z axis; these are considered to be three independent motions since change in any one of the components would not affect one of the others. A diatomic molecule has the same three degrees of freedom associated with translational kinetic energy plus two more degrees of freedom associated with rotational kinetic energy, for a total of five degrees of freedom. A quick look at Table 20–1 indicates that the C_v for diatomic gases is about $\tfrac{5}{3}$ times as great as for a monatomic gas—that is, in the same ratio as

FIGURE 20–6

A diatomic molecule can rotate about two different axes.

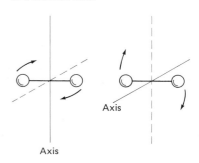

their degrees of freedom. This led nineteenth-century physicists to an important idea, the **principle of equipartition of energy**. This principle[†] states that energy is shared equally among the active degrees of freedom, and in particular each active degree of freedom of a molecule has on the average an energy equal to $\frac{1}{2}kT$. Thus, the average energy for a molecule of a monatomic gas would be $\frac{3}{2}kT$ (which we already knew) and of a diatomic gas $\frac{5}{2}kT$. Hence the internal energy of a diatomic gas would be $U = N(\frac{5}{2}kT) = \frac{5}{2}nRT$, where n is the number of moles; using the same argument we did for monatomic gases, we find that for diatomic gases the molar heat capacity at constant volume would be $\frac{5}{2}R = 4.97$ cal/mol·K in accordance with measured values. More complex molecules have even more degrees of freedom and thus greater molar heat capacities.

The situation was complicated, however, by measurements that showed that for diatomic gases at very low temperatures, C_v has a value of only $\frac{3}{2}R$ as if it had only three degrees of freedom. And at very high temperatures, C_v was about $\frac{7}{2}R$ as if there were seven degrees of freedom. The explanation for this is that at low temperatures, the molecules have mainly translational kinetic energy; that is, no energy goes into rotational energy, so only three degrees of freedom are "active." At very high temperatures, on the other hand, all five degrees of freedom are active plus two additional ones. We can interpret the two new degrees of freedom as being associated with the two atoms vibrating as if they were connected by a spring, as shown in Fig. 20–7: one degree of freedom comes from the kinetic energy of the vibrational motion, and the second from the potential energy of vibrational motion ($\frac{1}{2}kx^2$). At room temperature, these two degrees of freedom are apparently not active. (Just why fewer degrees of freedom are "active" at lower temperatures was eventually explained by Einstein using the quantum theory.) Thus, calculations based on kinetic theory and the principle of equipartition of energy (as modified by the quantum theory) give numerical results in accord with experiment.

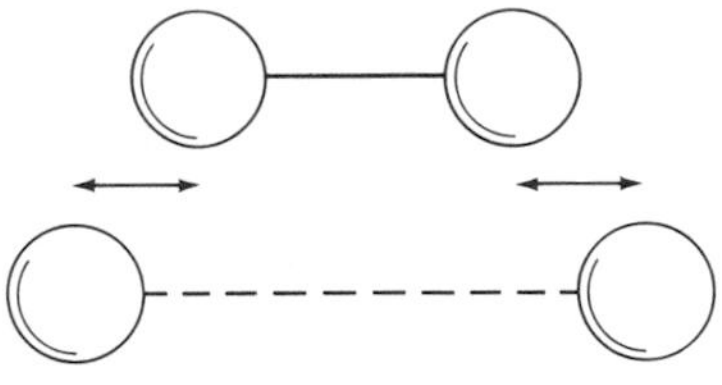

The principle of equipartition of energy can be applied to solids as well. The molar heat capacities of all solids, at high temperatures, is close to $3R$ (6.0 cal/mol·K), Fig. 20–8. This is called the Dulong and Petit value after the scientists who first measured it in 1819. (Note that Table 19–1 gave the specific heats per kilogram, not per mole.) At high temperatures, each atom apparently has six degrees of freedom, although some are not active at low temperatures. Each atom in a crystalline solid can vibrate about its equilibrium position as if it were connected by springs to each of its neighbors (Fig. 20–9). Thus it can have three degrees of freedom for kinetic energy and three more associated with potential energy of vibration in each of the x, y, and z directions, which is in accord with the measured values.

[†] Maxwell theoretically derived the principle using statistical mechanics methods.

FIGURE 20–7

A diatomic molecule can vibrate.

FIGURE 20–8

Molar heat capacities (of solids) as a function of temperature.

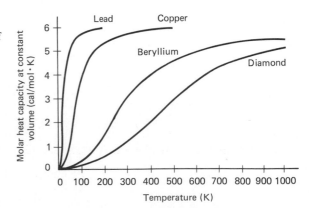

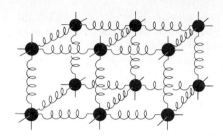

FIGURE 20-9
The atoms in a crystalline solid can vibrate about their equilibrium positions as if they were connected to their neighbors by springs. (The forces between atoms are actually electrical in nature.)

 ## 20–5 Adiabatic Expansion of a Gas

The PV curve for the quasistatic (slow) adiabatic expansion of an ideal gas was shown in Fig. 20–4 (curve AC). It is somewhat steeper than for an isothermal process, which indicates that for the same change in volume the change in pressure will be greater. Hence the temperature of the gas must drop during an adiabatic expansion. Conversely, the temperature rises during an adiabatic compression.

We can derive the relation between the pressure P and the volume V of an ideal gas that is allowed to slowly expand adiabatically. We begin with the first law of thermodynamics, written in differential form:

$$dU = dQ - dW = -dW = -P\,dV$$

since $dQ = 0$ for an adiabatic process. Equation 20–8 gives us a relation between ΔU and C_v, which is valid for any ideal gas process since U is a function only of T for an ideal gas; we write this in differential form:

$$dU = nC_v\,dT.$$

When we combine these last two equations we obtain

$$nC_v\,dT + P\,dV = 0.$$

We next take the differential of the ideal gas law, $PV = nRT$, allowing P, V, and T to vary:

$$P\,dV + V\,dP = nR\,dT.$$

We solve for dT in this relation and substitute it into the previous relation and get

$$(C_v + R)P\,dV + C_v V\,dP = 0.$$

We note from Eq. 20–7 that $C_v + R = C_p$ so we have

$$C_p P\,dV + C_v V\,dP = 0.$$

We define

$$\gamma = \frac{C_p}{C_v} \qquad (20-10)$$

so that our last equation becomes

$$\frac{dP}{P} + \gamma\frac{dV}{V} = 0.$$

This is integrated to become

$$\ln P + \gamma \ln V = \text{constant}.$$

This simplifies (using the rules for addition and multiplication of logarithms) to

$$PV^\gamma = \text{constant}. \qquad \left[\begin{array}{l}\text{Quasistatic adiabatic process;}\\ \text{ideal gas}\end{array}\right] \qquad (20-11)$$

This is the relation between P and V for a quasistatic adiabatic expansion or contraction. We will find it very useful when we discuss heat engines in the next chapter. Table 20–1 gives values of γ for some real gases.

> **EXAMPLE 20–3** An ideal monatomic gas is allowed to expand slowly until its pressure is reduced to exactly half its original value. By what factor does the volume change if the process is (a) adiabatic, (b) isothermal?
>
> **SOLUTION**   (a) From Eq. 20–11, $P_1 V_1^\gamma = P_2 V_2^\gamma$; hence
>
> $$\frac{V_2}{V_1} = \left(\frac{P_1}{P_2}\right)^{1/\gamma} = (2)^{3/5} = 1.52$$
>
> since $\gamma = C_p/C_v = (5/2)/(3/2) = 5/3$.
>
> (b) For constant temperature, the ideal gas law tells us simply that $P_1 V_1 = P_2 V_2$ since $T_1 = T_2$. Hence
>
> $$\frac{V_2}{V_1} = \frac{P_1}{P_2} = 2.0.$$

*20–6   Adiabatic Character of Sound Waves

The compressions and expansions of air in a sound wave are very nearly adiabatic for audible frequencies. To see why, note that when a gas is compressed, its temperature rises unless heat can flow out; when a gas expands, its temperature drops unless heat flows in. In an audible sound wave in air, the thermal conductivity is rather low and the distance between adjacent compressions and expansions is rather large ($\frac{1}{2}\lambda$); this, coupled with the fact that alternate compressions and expansions at each location occur very quickly, means that heat has little chance to flow. Hence the process is adiabatic. We can now use this fact to determine the speed of sound in terms of fundamental quantities. In Section 15–2, we saw that the speed of a longitudinal sound wave is given by

$$v_{\text{snd}} = \sqrt{\frac{B}{\rho}}$$

where ρ is the density of the medium and B is its bulk modulus; from Eq. 11–7b, $B = -V(dP/dV)$. The change in pressure when the volume is changed, dP/dV, depends on how the process is carried out, and thus B depends on the process. For an adiabatic process, we can determine B (we now call it B_{ad}) by differentiating Eq. 20–11 with respect to V:

$$V^\gamma \left(\frac{dP}{dV}\right)_{\text{ad}} + \gamma P V^{\gamma-1} = 0.$$

Hence

$$B_{\text{ad}} = -V\left(\frac{dP}{dV}\right)_{\text{ad}} = -V\left(-\frac{\gamma P}{V}\right) = \gamma P.$$

Thus the velocity of sound in a gas is given by

$$v_{\text{snd}} = \sqrt{\frac{B_{\text{ad}}}{\rho}} = \sqrt{\frac{\gamma P}{\rho}}. \tag{20–12}$$

For air, $\gamma = 1.40$, so at 0°C and 1 atm we have

$$v_{\text{snd}} = \sqrt{(1.40)(1.01 \times 10^5 \text{ Pa})/(1.29 \text{ kg/m}^3)} = 331 \text{ m/s}$$

which is an excellent agreement with the measured value. This accurate agreement substantiates our argument that the compressing and expanding of air is essentially adiabatic. As a comparison, if the process were isothermal, the predicted value for v_{snd} is about 280 m/s, rather far from the experimental value (see problem 35).

Summary

The work done by (or on) a gas to change its volume by dV is $dW = P\,dV$ where P is the pressure. Work and heat are not functions of the state of a system (as are P, V, T, n, and U) but depend on the type of process that takes a system from one state to another.

The *first law of thermodynamics* states that the change in internal energy, ΔU, of a system is equal to the heat added to the system, Q, minus the work, W, done by the system:

$$\Delta U = Q - W.$$

This important law is simply a restatement of the conservation of energy, and is found to hold for all processes. Two simple thermodynamic processes are *isothermal*, which is a process carried out at constant temperature, and *adiabatic*, a process in which no heat is exchanged.

The *molar heat capacity* of an ideal gas at constant volume, C_v, and at constant pressure, C_p, are related by $C_p - C_v = R$ where R is the gas constant. For a monatomic ideal gas, $C_v = \frac{3}{2}R$. For ideal gases made up of diatomic or more complex molecules, C_v is equal to $\frac{1}{2}R$ times the number of *degrees of freedom* of the molecule; unless the temperature is very high, some of the degrees of freedom may not be active and so do not contribute. According to the *principle* of *equipartition of energy*, energy is shared equally among the active degrees of freedom in an amount $\frac{1}{2}kT$ on the average.

When an ideal gas expands (or contracts) adiabatically ($Q = 0$), the relation PV^γ = constant holds, where $\gamma = C_p/C_v$.

Questions

1. What happens to the internal energy of water vapor in the air that condenses on the outside of a cold glass of water? Is work done or heat exchanged? Explain.

2. Use the conservation of energy to explain why the temperature of a gas increases when it is compressed—say, by pushing down on a cylinder—whereas the temperature decreases when the gas expands.

3. In Fig. 20–4, will more work be done in the isothermal process AB or in the adiabatic process AC? In which process will there be a greater change in internal energy? In which will there be a greater flow of heat?

4. In an isothermal process, 3700 J of work is done by an ideal gas. Is this enough information to tell how much heat has been added to the system? If so, how much?

5. An ideal gas was slowly compressed at constant temperature to one-half its original volume. In the process, 80 kcal of heat was given off. (*a*) How much work was done? (*b*) What was the change in internal energy of the gas?

6. An ideal gas is allowed to expand adiabatically to twice its volume. In doing so, the gas does 850 J of work. (*a*) How much heat flowed into the gas? (*b*) What is the change in internal energy of the gas? (*c*) Did its temperature rise or fall?

7. One liter of air is cooled at constant pressure until its volume is halved and then it is allowed to expand isothermally back to its original volume. Draw the process on a PV diagram.

8. Give examples of systems that do work even though their volume doesn't change.

9. Is it possible for the temperature of a system to remain constant even though heat flows into or out of it? If so, give examples.

10. Is it possible to determine whether the change in internal energy of a system was done by heat transfer or by work?

11. Does the temperature of an isolated system always remain constant?

12. We saw that in the free expansion of a real gas (Fig. 20–5) the temperature drops very slightly. This can be interpreted as due to attractive forces between molecules. Explain, using a potential energy diagram and Eq. 18–4.

13. Discuss how the first law of thermodynamics can apply to metabolism in humans. In particular, note that a person does work W, but very little heat Q is added to the body (rather, it tends to flow out); why then doesn't the internal energy drop drastically in time?

14. Explain in words why C_p is greater than C_v.

15. Hot air rises, yet the air is usually cooler at high elevations than at sea level. Explain.

16. Explain why the temperature of a gas increases when it is adiabatically compressed.

17. An ideal monatomic gas is allowed to slowly expand to twice its volume (1) isothermally, (2) adiabatically, (3) isobarically. Plot each on a PV diagram. In which process is ΔU the greatest, and in which is ΔU the least? In which is W the greatest and the least? In which is Q the greatest and the least?

Problems

SECTION 20–1

1. (I) How much work is done by 8.0 mol of O_2 gas initially at 0°C and 1 atm when its volume doubles (a) isothermally, and (b) at constant pressure.

2. (II) Write a formula for the density of a gas when it is allowed to expand (a) as a function of temperature when the pressure is kept constant, (b) as a function of pressure when temperature is kept constant.

3. (II) Show that Eqs. 20–1 and 20–2 are valid for any shape volume that changes. To do so, draw an arbitrary closed curve to represent the boundary of the volume; then draw a slightly larger curve to represent an increase in volume; choose a small section of the original boundary, of area ΔA, and show that $dW = P \Delta A\, dl = P\, dV$ for this section; then integrate over the whole boundary, and finally integrate over a finite volume.

4. (II) How much work is done to slowly compress, isothermally, 2.50 L of nitrogen at 0°C and 1 atm to 1.50 L at 0°C?

*5. (III) Determine the work done by 1 mol of a van der Waals gas (Section 18–6) when it expands from volume V_1 to V_2 isothermally.

SECTION 20–3

6. (I) When 1400 kcal of heat is added to a gas enclosed in a cylinder fitted with a light frictionless piston maintained at atmospheric pressure, the volume is observed to slowly increase from 12.0 m³ to 18.2 m³. Calculate (a) the work done by the gas, and (b) the change in internal energy of the gas.

7. (II) An ideal gas is compressed at a constant pressure of 2.0 atm from 10.0 L to 2.0 L. (In this process some heat flows out and the temperature drops.) Heat is then added to the gas, holding the volume constant, and the pressure and temperature are allowed to rise until the temperature reaches its original value. Calculate (a) the total work done on the gas in the process, and (b) the total heat flow into the gas.

8. (II) When a gas is taken from a to c along the curved path in Fig. 20–10, the work done by the gas is $W = -35$ J and the heat added to the gas is $Q = -63$ J. Along path abc, the work done is $W = -48$ J. (a) What is Q for path abc? (b) If $P_c = \frac{1}{2}P_b$, what is W for path cda? (c) What is

FIGURE 20–10

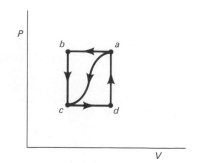

Q for path cda? (d) What is $U_a - U_c$? (e) If $U_d - U_c = 5$ J, what is Q for path da?

9. (II) A vertical, steel I-beam at the base of a building is 6.0 m tall, has a mass of 300 kg, and supports a load of 3.0×10^5 N. If the beam's temperature decreases by 4.0°C, calculate the change in its internal energy using the facts that for steel c_p is 0.11 kcal/kg·C° and the coefficient of linear expansion is 11×10^{-6} C°⁻¹.

[More problems dealing with the first law are found in the next two sections.]

SECTION 20–4

10. (I) Show that the work done by n moles of an ideal gas when it expands adiabatically is $W = nC_v(T_1 - T_2)$ where T_1 and T_2 are the initial and final temperatures, and C_v is the molar heat capacity at constant volume.

11. (I) What is the internal energy of 3.0 mol of an ideal diatomic gas at 600 K, assuming all degrees of freedom are active?

12. (I) A certain gas has specific heat $c_v = 0.0356$ kcal/kg·C°, which changes little over a wide temperature range. What is the atomic mass of this gas? What gas is it?

13. (I) By how much will the temperature rise if 80 kcal of heat is added to 300 mol of CO_2 held at constant pressure?

14. (I) If a heater supplies 1.5×10^6 J/h to a room 6.5 m × 5.0 m × 3.0 m containing air at 20°C and 1 atm, by how much will the temperature rise in one hour, assuming no heat losses to the outside?

15. (I) Show that if the molecules of a gas have n degrees of freedom, then theory predicts $C_v = (n/2)R$ and $C_p = [(n + 2)/2]R$.

16. (I) Estimate the molar heat capacity and specific heat at both constant pressure and constant volume for hydrogen gas (H_2) at room temperature.

17. (II) In problem 9 of Chapter 19 we ignored the heat absorbed by the gases evolved in the burning of the cake (CO_2 and H_2O). Calculate approximately how big a correction (in percent) this would make. Assume 2.5 g of each gas is produced.

18. (II) The specific heat at constant volume of a particular gas is 0.182 kcal/kg·K at room temperature, and its molecular mass is 34. (a) What is its specific heat at constant pressure? (b) What do you think is the molecular structure of this gas?

19. (II) How much heat must be added to 12.0 m³ of nitrogen gas initially at 20°C to double its volume at a pressure of 1.00 atm?

20. (II) An audience of 2500 fills a concert hall of volume 30,000 m³. If there were no ventilation, by how much would the temperature rise over a period of 2.0 h due to the metabolism of the people (70 W/person)?

21. (II) A sample of 800 mol of nitrogen gas is maintained at a constant pressure of 1.00 atm in a flexible container. The gas is heated from 40°C to 180°C. Calculate (a) the heat added to the gas, (b) the work done by the gas, and (c) the change in internal energy.

22. (II) One mole of N_2 gas at 0°C is heated to 100°C at constant pressure (1.00 atm). Determine (a) the change in internal energy, (b) the work the gas does, and (c) the heat added to it.

23. (II) At very low temperatures, the molar heat capacity of many substances varies as the cube of the absolute temperature:

$$C = k \frac{T^3}{T_0^3}$$

which is sometimes called Debye's law. For rock salt, $T_0 = 281$ K and $k = 1940$ J/mol·K. Determine the heat needed to raise 3.5 mol of salt from 12.0 K to 38.0 K.

24. (II) A hot-air balloon achieves its buoyant lift by heating the air inside it thus making it less dense than the air outside. Suppose a hot-air balloon is filled with 1800 m³ of air at 345 K and that this was accomplished by heating it from the outside temperature of 290 K to 345 K at the constant pressure of 1.0 atm. Take the density of air at 290 K and 1 atm to be 1.22 kg/m³ and its average molecular weight to be 29. Assuming that air is a diatomic ideal gas with five degrees of freedom, calculate the following: (a) The total number of moles of air in the balloon; (b) The internal energy of the air in the balloon at 290 K and 345 K, and the change in its internal energy upon heating; (c) The work done by the air in the balloon on the outside air during its expansion upon heating. (d) Use the first law of thermodynamics to calculate the amount of heat delivered to the air in the balloon during heating.

25. (III) A 1.00-mol sample of an ideal diatomic gas at a pressure of 1.00 atm and temperature of 580 K undergoes a process in which its pressure increases linearly with temperature. The final temperature and pressure are 720 K and 1.60 atm. Determine (a) the change in internal energy, (b) the work done by the gas, (c) the heat added to the gas. [Assume 5 active degrees of freedom.]

SECTION 20–5

26. (I) A 1.00-mol sample of an ideal diatomic gas, originally at 1.00 atm and room temperature, expands adiabatically to twice its volume; what is the final pressure in the gas? (Assume no molecular vibration.)

27. (II) Show, using Eqs. 20–1 and 20–11, that the work done by a gas that slowly expands adiabatically from pressure P_1 and volume V_1 to P_2, V_2, is given by $W = (P_1V_1 - P_2V_2)/(\gamma - 1)$.

28. (II) An ideal gas at 400 K is expanded adiabatically to 5.0 times its original volume. Determine its resulting temperature if the gas is (a) monatomic, (b) diatomic (no vibrations), (c) diatomic (molecules do vibrate).

29. (II) Prove that the slope of the PV diagram for an adiabatic process is steeper (more negative) than that for an isothermal process at any point.

30. (II) A diesel engine accomplishes ignition without a spark plug by an adiabatic compression of air to a temperature above the ignition temperature of the diesel fuel which is injected into the cylinder at the peak of the compression. Suppose air is taken into the cylinder at 300 K and volume V_1, and is compressed adiabatically to 560°C ($\approx 1000°$F) and volume V_2. Assuming that the air behaves as an ideal gas whose ratio of C_p to C_v is 1.4, calculate the compression ratio V_1/V_2 of the engine.

*SECTION 20–6

***31. (II)** A 5.00-mol sample of an ideal diatomic gas expands adiabatically from a volume of 0.1210 m³ to 0.750 m³. Initially the pressure was 1.00 atm. Determine (a) the initial and final temperatures, (b) the change in internal energy, (c) the work done on the gas, (d) the heat lost by the gas. (Assume no molecular vibration.)

***32. (II)** (a) Show that the speed of sound in an ideal gas is given by

$$v_{snd} = \sqrt{\gamma RT/m}$$

where M is the molecular mass of the gas. (b) What is the ratio of the speed of sound in two different gases at the same temperature?

***33. (II)** Use the result of problem 32 and the binomial expansion, to show that the speed of sound in air near 0°C increases about 0.61 m/s per each C° increase in temperature.

***34. (II)** An ideal monatomic gas, consisting of 2.4 mol of volume 0.084 m³, expands adiabatically. The initial and final temperatures are 25°C and -58°C. What is the final volume of the gas?

***35. (II)** (a) Show that the bulk modulus $B = -V(dP/dV)$ is equivalent to P for an isothermal process. (b) Predict the speed of sound in air at 0°C and 1 atm if sound waves were to travel isothermally (rather than adiabatically).

***36. (III)** A 1.00-mol sample of an ideal monatomic gas, originally at a pressure of 1.00 atm, undergoes a three-step process: (1) it is expanded adiabatically from $T_1 = 550$ K to $T_2 = 389$ K; (2) it is compressed at constant pressure until its temperature reaches T_3; (3) it then returns to its original pressure and temperature by a constant-volume process. (a) Plot these processes on a PV diagram. (b) Determine T_3. (c) Calculate the change in internal energy, the work done by the gas, and the heat added to the gas for each process, and (d) for the complete cycle.

The Second Law
of Thermodynamics

21

 21–1 Need for a New Law of Thermodynamics

The first law of thermodynamics states that energy is conserved. There are, however, many processes we can imagine that conserve energy but are not observed to occur in nature. For example, when a hot object is placed in contact with a cold object, heat flows from the hotter one to the colder one, never the reverse. If heat were to leave the colder and pass to the hotter one, energy would still be conserved. Yet it doesn't happen. As a second example, consider what happens when you drop a rock and it hits the ground. The initial potential energy of the rock changes to kinetic as the rock falls and when the rock hits the ground this energy in turn is transformed into internal energy of the rock and the ground in the vicinity of the impact; the molecules move faster and the temperature rises slightly. But have you seen the reverse happen—a rock at rest on the ground suddenly rise up in the air because the thermal energy of molecules is transformed into kinetic energy of the rock as a whole? Energy would be conserved in this process, yet we never see it happen.

There are many other examples of processes that occur in nature but whose reverse does not. Here are two more. If you put a layer of salt in a jar and cover it with a layer of pepper, when you shake it you get a thorough mixture; no matter how long you shake it, the mixture is unlikely to separate into two layers again. Coffee cups and glasses break spontaneously if you drop them; but they don't go back together spontaneously.

The first law of thermodynamics would not be violated if any of these processes occurred in reverse. To explain this lack of reversibility, scientists in the latter half of the nineteenth century came to formulate a new principle known as the **second law of thermodynamics**. This law is a statement about which processes occur in nature and which do not. It can be stated in a variety of ways, all of which are equivalent. One statement, due to R. J. E. Clausius (1822–1888), is that *heat flows naturally from a hot object to a cold object; heat will not flow spontaneously from a cold object to a hot object.* Since this statement applies to one particular process, it is not obvious how it applies to other processes. A more general statement is needed that will include other possible processes in a more obvious way.

The development of a general statement of the second law was based partly on the study of heat engines. A **heat engine** is any device that changes thermal energy into mechanical work. We now examine heat engines, both from a practical point of view and to show their importance in developing the second law of thermodynamics.

21–2 Heat Engines and Refrigerators

It is easy to produce thermal energy by doing work—for example, by simply rubbing your hands together briskly, or indeed in any frictional process. But to get work from thermal energy is more difficult, and the invention of a practical device to do this came only about 1700 with the development of the steam engine.

The basic idea behind any heat engine is that mechanical energy can be obtained from heat only when heat is allowed to flow from a high temperature to a low temperature; in the process some of the heat can then be transformed to mechanical work, as diagrammed in Fig. 21–1. The high and low temperatures, T_H and T_L, are called the *operating temperatures* of the engine, and for simplicity we often assume these temperatures are maintained by two heat reservoirs at uniform temperature T_H and T_L. We will be interested only in engines that run in a repeating *cycle* (that is, the system returns repeatedly to its starting point) and thus can run continuously.

The operation of two practical engines, the steam engine and the internal combustion engine (used in most automobiles), is illustrated in Figs. 21–2 and 21–3. Steam engines today are of two main types. In the so-called reciprocating type, Fig. 21–2a, the heated steam passes through the intake valve and expands against a piston, forcing it to move; as the piston returns to its original position, it forces the gases out the exhaust valve. In a steam turbine, Fig. 21–2b, everything is essentially the same, except that the reciprocating piston is replaced by a rotating turbine that resembles a paddlewheel with many sets of blades. Most of our electricity today is generated using steam turbines.[†] The material which is

---

[†] Even nuclear power plants utilize steam turbines; the nuclear fuel—uranium—merely serves as fuel to heat the steam.

FIGURE 21–1

Schematic diagram of a heat engine.

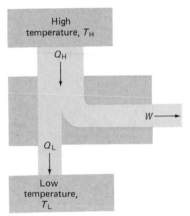

FIGURE 21–2

Steam engines.

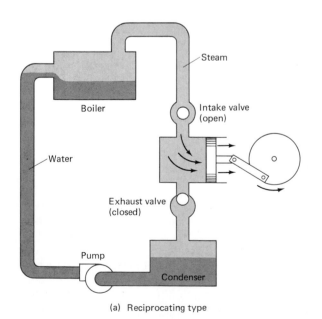

(a) Reciprocating type

(b) Turbine (boiler, condenser not shown)

FIGURE 21–3

Four-cycle internal combustion engine:
(a) the gasoline–air mixture flows into the
cylinder as the cylinder moves down;
(b) the piston moves upward and
compresses the gas; (c) firing of the spark
plug ignites the gasoline–air mixture,
raising it to a high temperature; (d) the
gases, now at high temperature and
pressure, expand against the piston in this,
the power stroke; (e) the burned gases are
pushed out to the exhaust pipe; the intake
valve then opens, and the whole cycle repeats.

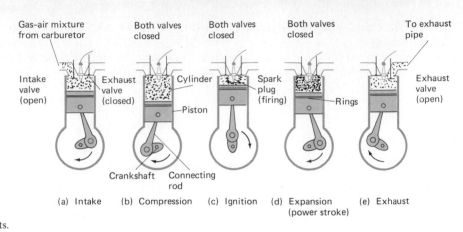

(a) Intake　(b) Compression　(c) Ignition　(d) Expansion　(e) Exhaust
(power stroke)

FIGURE 21–4

Schematic diagram of the operation
of a refrigerator or air conditioner.

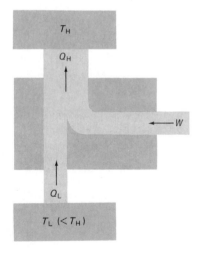

heated and cooled, steam in this case, is called the *working fluid*. In a steam
engine, the high temperature is obtained by burning coal, oil, or other fuel to heat
the steam. In an internal combustion engine, the high temperature is achieved by
burning the gasoline–air mixture in the cylinder itself (ignited by the spark plug).

To see why, from a practical point of view, a *temperature difference* is required to run an engine, let us examine the steam engine. In the reciprocating
engine, for example (Fig. 21–2a), suppose there were no condenser or pump, and
that the steam was at the same temperature throughout the system. This would
mean that the pressure of the gas being exhausted would be the same as that on
intake. Thus, although work would be done by the gas *on* the piston when it
expands, an equal amount of work would have to be done *by* the piston to force
the steam out the exhaust; hence no net work would be done. In a real engine,
the exhausted gas is cooled to a lower temperature and condensed so that the
exhaust pressure is less than the intake pressure. Thus, although the piston must
do work on the gas to expel it on the exhaust stroke, it is less than the work done
by the gas on the piston during the intake. So a net amount of work can be
obtained—but only if there is a difference of temperature. Similarly, in the gas
turbine if the gas were not cooled, the pressure on each side of the blades would
be the same; by cooling the gas on the exhaust side, the pressure on the front side
of the blade is greater and hence the turbine turns.

The operating principle of a refrigerator, or other *heat pump* (such as one
to produce a flow of heat into or out of a house—the latter is called an air
conditioner) is just the reverse of a heat engine. As diagrammed in Fig. 21–4,
by doing work W heat is taken from a low-temperature region, T_L (inside of a

FIGURE 21–5

Typical refrigeration system.

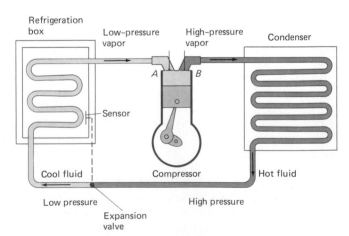

refrigerator, say), and a greater amount of heat is exhausted at a high temperature, T_H (the room). You can often feel this heat blowing out beneath a refrigerator. The work W is usually done by a compressor motor that compresses the working fluid as illustrated in Fig. 21–5.

21–3 Efficiency of Heat Engines and the Second Law of Thermodynamics

In the analysis of heat engines that follows, we will be mainly interested in the *magnitudes* of heat flow; so to avoid worrying about the sign convention of Section 20–2 for heat flow into or out of a system, we will use absolute value signs ($|Q|$) and the proper $+$ or $-$ sign as needed. In Section 21–6 we will again use the original convention. The efficiency, e, of any heat engine can be defined as the ratio of the work it does, W, to the heat input at the high temperature, $|Q_H|$ (Fig. 21–1);

$$e = \frac{W}{|Q_H|}.$$

This is a very practical definition since W is the output (what you get from the engine) whereas $|Q_H|$ is what you put in and pay for in burned fuel. Since energy is conserved, the heat input $|Q_H|$ must equal the work done plus the heat that flows out at the low temperature ($|Q_L|$):

$$|Q_H| = W + |Q_L|.$$

Thus $W = |Q_H| - |Q_L|$ and the efficiency of an engine is

$$e = \frac{W}{|Q_H|} = \frac{|Q_H| - |Q_L|}{|Q_H|} = 1 - \frac{|Q_L|}{|Q_H|}. \tag{21–1}$$

It is clear from Eq. 21–1 that the efficiency of an engine will be greater if $|Q_L|$ can be made small. However, from experience with a wide variety of systems, it has not been found possible to reduce $|Q_L|$ to zero. If $|Q_L|$ could be reduced to zero we would have a 100 percent efficient engine, as diagrammed in Fig. 21–6. That such a perfect engine (running continuously in a cycle) is not possible is another way of expressing the second law of thermodynamics. This can be stated formally as follows:

No cyclic process is possible whose only result is to transfer an amount of heat Q from a source at a single temperature totally into work W (so $W = Q$).

This is known as the *Kelvin-Planck statement of the second law of thermodynamics.* If the second law were not true, so that a perfect engine could be built, some rather remarkable things could happen; for example, if the engine of a ship did not need a low-temperature reservoir to exhaust heat into, the ship could sail across the ocean using the vast resources of the internal energy of the ocean water; indeed, we would have no fuel problems at all!

It has not been found possible to construct a perfect engine. It has also not been found possible to construct its reverse, a perfect refrigerator—which would be a device to take heat from a low-temperature region to a high-temperature region without requiring any work ($W = 0$, $|Q_L| = |Q_H|$ in Fig. 21–4). This can be stated formally as follows:

No cyclic process is possible whose only result is a flow of heat out of a system at one temperature and an equal magnitude flow of heat into a second system at a higher temperature.

FIGURE 21–6

Schematic diagram of a hypothetical perfect heat engine in which all the heat input is used to do work, $W = |Q|$. Such a perfect heat engine is not possible to construct.

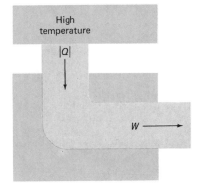

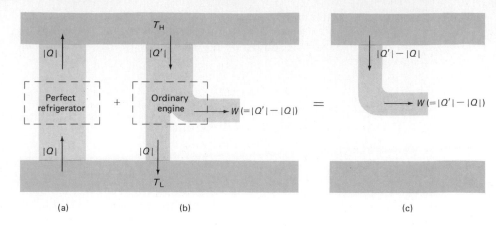

FIGURE 21-7

Equivalence of Clausius and Kelvin-Planck statements of the second law of thermodynamics. A perfect refrigerator (a) working in conjunction with an ordinary engine (b) would be equivalent to a perfect engine (c).

This is the *Clausius statement of the second law of thermodynamics.* It is a more formal statement of the one given in Section 21–1, that heat does not flow spontaneously from a cold object to a hot object. To accomplish such a task, work must be done. The Clausius statement can also be stated as: *there can be no perfect refrigerator.*

We now show that the two different statements of the second law, the Clausius and the Kelvin-Planck, are equivalent. To do so, we show that if one statement is false then the other must be as well. Thus both would have to be false or both true (not one true and one false), and thus equivalent.

Suppose the Clausius statement is not true—so that a perfect refrigerator would be possible; as shown in Fig. 21–7a it could take heat $|Q|$ from a low-temperature reservoir and deliver it to a high temperature reservoir with no work being done. Now consider an ordinary engine that takes heat $|Q'|$ from the high-temperature reservoir, does work W, and exhausts heat $|Q|$ to the low-temperature reservoir, Fig. 21–7b. The net effect of these two devices is to take an amount of heat $|Q'| - |Q|$ from the high temperature reservoir and turn it all into work $W = |Q'| - |Q|$ as shown in Fig. 21–7c. The net effect is a perfect engine that violates the Kelvin-Planck statement.

Now let us assume the Kelvin-Planck statement is false and show that this implies that the Clausius statement is false. Suppose a perfect engine (Fig. 21–8a) takes heat $|Q|$ from a high-temperature reservoir and turns it all into useful work $W(W = |Q|)$. An ordinary refrigerator (Fig. 21–8b) then uses this work to extract heat $|Q'|$ from a low-temperature reservoir and deposits heat Q'' into the high-temperature reservoir; note that $|Q''| = |Q'| + W = |Q'| + |Q|$. This combination removes heat $|Q|$ from the high-temperature reservoir and adds heat $|Q''|$ to it; the

FIGURE 21-8

Equivalence of Kelvin-Planck and Clausius statements of the second law of thermodynamics. A perfect engine (a) operated in conjunction with an ordinary refrigerator (b) would produce a perfect refrigerator (c) since $|Q''| - |Q| = W + |Q'| - |Q| = |Q'|$ because $W = |Q|$.

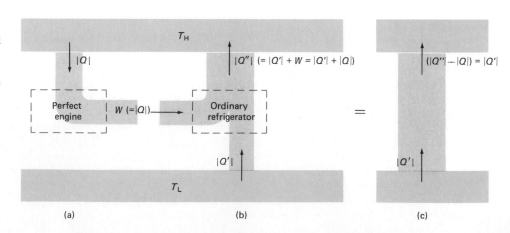

net flow of heat into this reservoir is then $|Q''| - |Q| = (|Q'| + |Q|) - |Q| = |Q'|$. Thus the net effect of this combination (Fig. 21–8c) is to remove heat $|Q'|$ from the low-temperature reservoir and deliver an equal amount of heat $|Q'|$ to the high-temperature reservoir. And this would violate the Clausius statement of the second law.

We thus see that if either the Clausius or the Kelvin-Planck statement is false, the other must be false also. Hence, if either is true, the other must be as well, and so they are equivalent statements of the second law of thermodynamics.

21–4 The Carnot Engine; Reversible and Irreversible Processes

In the early nineteenth century, the French scientist N. L. Sadi Carnot (1796–1832) studied in detail the process of transforming heat into mechanical energy. His aim had been to determine how to increase the efficiency of heat engines, but his studies soon led him to investigate the foundations of thermodynamics itself.

As an aid in this pursuit, Carnot, in 1824, invented (on paper) an idealized type of engine which we now call the "Carnot engine." The importance of the Carnot engine is not really as a practical engine, but rather as an aid to the understanding of heat engines in general, and also because Carnot and his engine contributed to the establishment and understanding of the second law of thermodynamics.

The Carnot engine involves *reversible processes*, so before we discuss it we must discuss what is meant by reversible and irreversible processes. A **reversible process** is one that is carried out infinitely slowly, so that the process can be considered as a series of equilibrium states, and the whole process could be done in reverse with no change in magnitude of the work done or heat exchanged. For example, a gas contained in a cylinder fitted with a tight, movable, but frictionless piston could be compressed isothermally in a reversible way if done infinitely slowly. Not all very slow (quasistatic) processes are reversible, however. If there is friction present, for example (as between the movable piston and cylinder just mentioned), the work done in one direction (going from some state A to state B) will not be the negative of the work done in the reverse direction (state B to state A). Such a process would not be considered reversible. Of course a perfectly reversible process is not possible in reality since it would require an infinite time; reversible processes can be approached arbitrarily closely, however and they are very important theoretically. All real processes are **irreversible**: they are not done infinitely slowly. There could be turbulence in the gas, friction would be present, and so on; any process could not be done precisely in reverse since the heat lost to friction would not reverse itself, the turbulence would be different, and so on. For any given volume there would not be a well-defined pressure P and temperature T since the system would not always be in an equilibrium state. Thus a real, irreversible, process cannot be plotted on a PV diagram (except insofar as it may approach an ideal reversible process). But a reversible process (since it is a quasistatic series of equilibrium states) always can be plotted on a PV diagram; and a reversible process, when done in reverse, retraces the same path on a PV diagram. Although all real processes are irreversible, reversible processes are conceptually important, just as the concept of an ideal gas is.

Now let us look at Carnot's idealized engine. The Carnot engine makes use of a *reversible cycle*, by which we mean a series of reversible processes that take a given substance (the *working substance*) from an initial equilibrium state through many other equilibrium states and returns it again to the same initial state. In particular, the Carnot engine utilizes the **Carnot cycle**, which is illustrated in Fig. 21–9, with the working substance assumed to be an ideal gas. (For a real gas, the PV diagram would be slightly different.) Let us take point a as the initial state. The gas is first expanded isothermally, and reversibly,

FIGURE 21–9

The Carnot cycle.

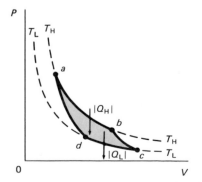

path *ab*, at temperature T_H; to do so, we can imagine the gas to be in contact with a heat reservoir at temperature T_H which delivers heat $|Q_H|$ to our working substance. Next the gas is expanded adiabatically and reversibly, path *bc*; no heat is exchanged and the temperature of the gas is reduced to T_L. The third step is a reversible isothermal compression, path *cd*, in contact with a heat reservoir at the low temperature, T_L, during which heat $|Q_L|$ flows out of the working substance. Finally, the gas is compressed adiabatically, path *da*, back to its original state. Thus a Carnot cycle consists of two isothermal and two adiabatic processes.

It is easy to show that the net work done in one cycle by a Carnot engine (or any other type of engine using a reversible cycle) is equal to the area enclosed by the curve representing the cycle on the *PV* diagram, the curve *abcd* in Fig. 21–9. The proof is left as problem 7 (see also Section 20–1).

21–5 Carnot Efficiency and the Second Law of Thermodynamics

The efficiency of a Carnot engine, like any heat engine, is given by Eq. 21–1:

$$e = 1 - \frac{|Q_L|}{|Q_H|}.$$

For a Carnot engine, however, we can show that the efficiency depends only on the temperatures of the heat reservoirs, T_H and T_L. In the first isothermal process, *ab* in Fig. 21–9, the work done by the gas is (see Eq. 20–3):

$$W_{ab} = nRT_H \ln \frac{V_b}{V_a}$$

where *n* is the number of moles of the ideal gas used as working substance. Since the internal energy of an ideal gas does not change when the temperature remains constant, the heat added to the gas equals the work done by the gas (the first law of thermodynamics):

$$|Q_H| = nRT_H \ln \frac{V_b}{V_a}.$$

Similarly, the heat lost by the gas in the isothermal process *cd* is

$$|Q_L| = nRT_L \ln \frac{V_c}{V_d}.$$

Since paths *bc* and *da* are adiabatic, we have (see Eq. 20–11)

$$P_b V_b^\gamma = P_c V_c^\gamma \quad \text{and} \quad P_d V_d^\gamma = P_a V_a^\gamma.$$

Also, from the ideal gas law,

$$\frac{P_b V_b}{T_H} = \frac{P_c V_c}{T_L} \quad \text{and} \quad \frac{P_d V_d}{T_L} = \frac{P_a V_a}{T_H}.$$

When we divide these last equations, term by term, into the corresponding set of equations on the line above, we obtain

$$T_H V_b^{\gamma-1} = T_L V_c^{\gamma-1} \quad \text{and} \quad T_L V_d^{\gamma-1} = T_H V_a^{\gamma-1}.$$

Next we divide the equation on the left by the one on the right and get:

$$\left(\frac{V_b}{V_a}\right)^{\gamma-1} = \left(\frac{V_c}{V_d}\right)^{\gamma-1}.$$

Hence

$$\frac{V_b}{V_a} = \frac{V_c}{V_d}$$

or

$$\ln \frac{V_b}{V_a} = \ln \frac{V_c}{V_d}.$$

Using this result in our equations for $|Q_H|$ and $|Q_L|$ above we have

$$\frac{|Q_L|}{|Q_H|} = \frac{T_L}{T_H}.$$ [Carnot cycle]   (21–2)

Hence the efficiency of a reversible Carnot engine is

$$e = 1 - \frac{|Q_L|}{|Q_H|} = 1 - \frac{T_L}{T_H}.$$ ["Carnot efficiency"]   (21–3)

The temperatures T_L and T_H are the absolute or Kelvin temperatures as measured on the ideal gas temperature scale.[†] Thus the efficiency of a Carnot engine depends only on the temperatures T_L and T_H.

We could imagine other possible reversible cycles that could be used for an ideal reversible engine. And, according to a theorem stated by Carnot:

All reversible engines operating between the same two temperatures have the same efficiency; no irreversible engine operating between the same two temperatures can have an efficiency greater than this.

This is known as *Carnot's theorem*. It tells us that Eq. 21–3, $e = 1 - (T_L/T_H)$, applies to *any* reversible engine, and that this equation represents a maximum possible efficiency for a real (irreversible) engine.

Carnot's theorem can be shown to follow from either the Clausius or Kelvin-Planck statements of the second law of thermodynamics. We use the former. Suppose we have two reversible heat engines that run between two heat reservoirs at temperatures T_L and T_H; let's assume they have efficiencies e and e', and we run them so they have the same work output, W. We now reverse one of the engines (e') so it runs backward, as a refrigerator. The work output W of the engine is made to become the work input to the refrigerator as shown in Fig. 21–10. Since $W = |Q_H| - |Q_L|$ for the engine, and $W = |Q_H'| - |Q_L'|$ for the refrigerator, we have $|Q_H| - |Q_L| = |Q_H'| - |Q_L'|$, or

$$|Q_L'| - |Q_L| = |Q_H'| - |Q_H|.$$

[†] Section 17–11; see also Section 21–11.

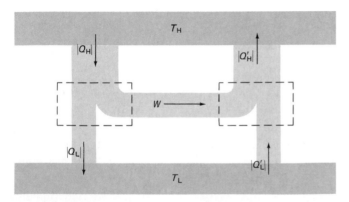

FIGURE 21–10

Work output of engine on left is used to drive refrigerator (a reversible engine) on the right. Used to show that Carnot's theorem follows from the second law of thermodynamics.

Now suppose the efficiencies are not equal, and $e > e'$. This means

$$\frac{W}{|Q_H|} > \frac{W}{|Q'_H|}$$

so

$$|Q'_H| > |Q_H|.$$

There is thus a net flow of heat, $|Q'_H| - |Q_H|$ into the high-temperature reservoir; there is a net flow of heat, $|Q'_L| - |Q_L|$, out of the low-temperature reservoir and, as we saw above, $|Q'_L| - |Q_L| = |Q'_H| - |Q_H|$. Thus the net result of our combination is a flow of heat from a low-temperature to a high-temperature reservoir with no net work being done. This contradicts the Clausius statement of the second law. Thus $e > e'$ is not consistent with the second law. We can next reverse the roles of the two engines and use the same arguments to show that $e' > e$ is also not consistent with the Clausius statement. To be consistent with the Clausius statement we must therefore have:

$$e = e'. \qquad \text{[reversible engines]}$$

Now suppose one of the engines (e) is irreversible, and the other (e') is reversible. Thus the same argument applies as in Fig. 21–10, that e cannot be greater than e'. However, we cannot reverse the irreversible engine and make a refrigerator with the reverse characteristics; hence we cannot show that e' cannot be greater than e. We are thus left with

$$e_{irr} \leq e'_{rev};$$

the efficiency of an irreversible engine is less than or equal to the efficiency of a reversible engine operating between the same two temperatures. We have thus shown that Carnot's theorem follows from the second law of thermodynamics. In fact, it can be considered as a third way to state this law.

In practice, the efficiency of real engines is always less than the Carnot efficiency; well-designed engines reach perhaps 60 to 80 percent of Carnot efficiency.

EXAMPLE 21–1   A steam engine operates between 500°C and 270°C. What is the maximum possible efficiency of this engine?

SOLUTION   We must first change the temperature to degrees Kelvin: $T_H = $ 773 K and $T_L = 543$ K; then

$$e = 1 - \frac{543}{773} = 0.30.$$

Thus the maximum (or Carnot) efficiency is 30 percent. Realistically, an engine might attain 70 percent of this value or about 21 percent. Note in this example that the exhaust temperature is still rather high, 270°C. This is typical, and steam engines are often arranged in series so that the exhaust of one engine is used as intake by a second or third engine.

 21–6   Entropy

We have seen several aspects of the second law of thermodynamics; but we have not yet arrived at a general statement of it. Both the Clausius and Kelvin-Planck statements deal with rather specific situations. Yet, as mentioned at the beginning of this chapter, there are a great many processes that simply are not observed in nature, even though they would not violate the first law of thermodynamics if they did occur. To cover all these processes, a more general statement of the second law of thermodynamics is needed. This general statement can be made in terms of a quantity, introduced by Clausius in the 1860s, called *entropy*, which we now discuss.

In our study of the Carnot cycle we found (Eq. 21–2) that $|Q_L|/|Q_H| = T_L/T_H$. We rewrite this as

$$\frac{|Q_H|}{T_H} = \frac{|Q_L|}{T_L}.$$

In this relation, both $|Q_H|$ and $|Q_L|$ are positive since they are absolute values. Let us now remove the absolute value signs and recall our original convention as used in the first law (see Section 20–2), that Q is positive when it represents a heat flow into the system (as Q_H) and negative for a heat flow out of the system (as Q_L). Then this relation becomes

$$\frac{Q_H}{T_H} + \frac{Q_L}{T_L} = 0. \qquad \text{[Carnot cycle]} \quad (21–4)$$

Now consider *any* reversible cycle, as represented by the smooth (oval-shaped) curve in Fig. 21–11. Any reversible cycle can be approximated as a series of Carnot cycles. Figure 21–11 shows only six—the isotherms (dashed lines) are connected by adiabatic paths for each—and the approximation becomes better and better if we increase the number of Carnot cycles. Equation 21–4 is valid for each of these cycles, so we can write

$$\sum \frac{Q}{T} = 0 \qquad \text{[Carnot cycles]} \quad (21–5)$$

for the sum of all these cycles. But note that the heat output Q_L of one cycle is approximately equal to the negative of the heat input, Q_H, of the cycle below it (actual equality in the limit of an infinite number of infinitely thin Carnot cycles); hence the heat flows on the inner paths of all these Carnot cycles cancel out, so the net heat transferred, and the work done, is the same for the series of Carnot cycles as for the original cycle. Hence, in the limit of infinitely many Carnot cycles, Eq. 21–5 applies to any reversible cycle; in this case Eq. 21–5 becomes

$$\oint \frac{dQ}{T} = 0, \qquad \text{[reversible cycle]} \quad (21–6)$$

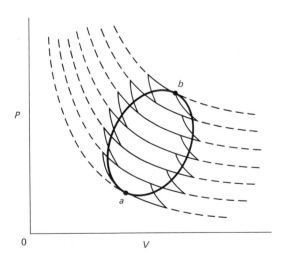

FIGURE 21–11

Any reversible cycle can be approximated as a series of Carnot cycles. (The dashed lines represent isotherms.)

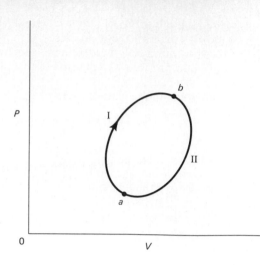

FIGURE 21–12
The integral, $\oint dS$, of the entropy for a reversible cycle is zero. Hence the difference in entropy between states a and b, $S_b - S_a = \int_a^b dS$, is the same for path I as for path II.

where dQ represents an infinitesimal heat flow.[†] The symbol $\oint$ means take the integral around a closed path; the integral can be started at any point on the path such as at a or b in Fig. 21–11, and proceed in either direction. Let us divide the cycle of Fig. 21–11 into two parts as indicated in Fig. 21–12, and then rewrite Eq. 21–6 as

$$\int_{a \atop I}^b \frac{dQ}{T} + \int_{b \atop II}^a \frac{dQ}{T} = 0;$$

the first term is the integral from point a to point b along path I in Fig. 21–12, and the second term is the integral from b back to a along path II. (Path I plus path II is the whole cycle.) If one path is taken in reverse, say path II, dQ at each point becomes $-dQ$, since the path is reversible. We can therefore write:

$$\int_{a \atop I}^b \frac{dQ}{T} = \int_{a \atop II}^b \frac{dQ}{T}. \qquad \text{[reversible paths]} \quad (21\text{--}7)$$

Since our cycle is arbitrary, Eq. 21–7 tells us that the integral of dQ/T between any two equilibrium states, a and b, does not depend on the path of the process. We therefore define a new quantity, called the **entropy**, S, by the relation:

$$dS = \frac{dQ}{T}. \qquad\qquad (21\text{--}8)$$

Then we have, from Eq. 21–6;

$$\oint dS = 0; \qquad \text{[reversible cycle]} \quad (21\text{--}9)$$

and from Eq. 21–7, the quantity

$$\Delta S = S_b - S_a = \int_a^b dS = \int_a^b \frac{dQ}{T} \qquad \text{[reversible processes]} \quad (21\text{--}10)$$

is *independent of the path between the two points*. This is an important result. It tells us that the difference in entropy, $S_b - S_a$, between two equilibrium states of a system does not depend on how you get from one state to the other. Thus *entropy is a state variable*—its value depends only on the state of the system,

[†] dQ is often written $đQ$: see footnote at end of Section 20–2.

and not on the process or the past history of how it got there.[†] This is in clear distinction to Q and W which are *not* state variables; their values do depend on the processes undertaken.

 ## 21–7 Entropy and the Second Law of Thermodynamics

So we have defined a new quantity, S, the entropy, which can be used to describe the state of the system, along with P, T, V, U, and n. But what does this rather abstract quantity have to do with the second law of thermodynamics? To answer this, let us take some examples in which we calculate the entropy changes during particular processes. But note first that Eq. 21–10 can be applied only to reversible processes.[‡] How then do we calculate $\Delta S = S_b - S_a$ for a real process that is irreversible? What we can do is this: we figure out some other *reversible* process that takes the system between the same two states, and calculate ΔS for this reversible process. This will equal ΔS for the irreversible process since ΔS depends only on the initial and final states of the system.

EXAMPLE 21–2   A 1.00-kg piece of ice at $0°C$ melts very slowly to water at $0°C$. Assume the ice is in contact with a heat reservoir whose temperature is only infinitesimally greater than $0°C$. Determine the entropy change of (a) the ice cube, (b) the heat reservoir.

SOLUTION   (a) The process is carried out at a constant temperature $T = 273$ K and is done reversibly, so we can use Eq. 21–10.

$$\Delta S_{\text{ice}} = \int \frac{dQ}{T} = \frac{1}{T}\int dQ = \frac{Q}{T}.$$

Since the heat needed to melt the ice is $Q = ml$, where the heat of fusion $l = 79.7$ kcal/kg $= 3.33 \times 10^5$ J/kg, we have

$$\Delta S_{\text{ice}} = \frac{(1.00 \text{ kg})(79.7 \text{ kcal/kg})}{273 \text{ K}} = 0.292 \text{ kcal/K},$$

or 1220 J/K.
(b) Heat $Q = ml$ is *removed* from the heat reservoir, so (since $T = 273$ K and is constant):

$$\Delta S_{\text{res}} = -\frac{Q}{T} = -0.292 \text{ kcal/K} = -1220 \text{ J/K}.$$

Note that the *total* entropy change, $\Delta S_{\text{ice}} + \Delta S_{\text{res}}$, is zero.

EXAMPLE 21–3   Consider the *adiabatic free expansion* of n moles of an ideal gas from volume V_1 to volume V_2 ($V_2 > V_1$) as was discussed in Section 20–3, Fig. 20–5. Calculate the change in entropy (a) of the gas, (b) of the surrounding environment.

SOLUTION   As we saw in Section 20–3, the gas is initially in a closed container of volume V_1, and, with the opening of a valve, adiabatically it expands into a previously empty container; the total volume of the two containers is V_2. The

---

[†] Equation 21–10 says nothing about the absolute value of S; it only gives the change in S. This is much like potential energy—Section 6–5. However, one form of the so-called *third law of thermodynamics* (see also Section 21–11) states that as $T \to 0$, $S \to 0$.

[‡] The integral $\int_a^b dQ/T$ can sometimes be evaluated—perhaps numerically—but it does not give the change in entropy if the process is irreversible since Eq. 21–10 does not work for irreversible processes.

whole apparatus is thermally insulated from the surroundings, so no heat flows into the gas, $Q = 0$. The gas does no work, $W = 0$, so there is no change in internal energy, $\Delta U = 0$, and the temperature of the initial and final states is the same, $T_2 = T_1 = T$.

(a) The process takes place very quickly, and so is irreversible. So we cannot apply Eq. 21–10 to this process. Instead we must think of a reversible process that will take the gas from volume V_1 to V_2 at the same temperature, and use Eq. 21–10 on this process to get ΔS. A reversible isothermal process will do the trick; in such a process, the internal energy does not change, so (from the first law):

$$dQ = dW = P\,dV.$$

Then

$$\Delta S_{\text{gas}} = \int \frac{dQ}{T} = \frac{1}{T} \int_{V_1}^{V_2} P\,dV;$$

from the ideal gas law $P = nRT/V$, so

$$\Delta S_{\text{gas}} = \frac{nRT}{T} \int_{V_1}^{V_2} \frac{dV}{V} = nR \ln \frac{V_2}{V_1}.$$

Since $V_2 > V_1$, $\Delta S_{\text{gas}} > 0$.

(b) Since no heat is transferred to the surrounding environment, there is no change of the state of the environment due to this process. Hence $\Delta S_{\text{env}} = 0$. Note that the total change in entropy, $\Delta S_{\text{gas}} + \Delta S_{\text{env}}$ is greater than zero.

◼ EXAMPLE 21–4 *Heat Conduction.* A red-hot 2.0-kg piece of iron at temperature $T_1 = 880$ K is thrown into a huge lake whose temperature is $T_2 = 280$ K. Assume the lake is so large that its temperature rise is insignificant. Determine the change in entropy (a) of the iron, (b) of the surrounding environment (the lake).

SOLUTION (a) The process is irreversible, but the same entropy change will occur for a reversible process, and we use the concept of specific heat, Eq. 19–1. We assume the specific heat of the iron is constant at $c = 0.11$ kcal/kg·K. Then $dQ = mc\,dT$ and

$$\Delta S_{\text{iron}} = \int \frac{dQ}{T} = mc \int_{T_1}^{T_2} \frac{dT}{T} = mc \ln \frac{T_2}{T_1} = -mc \ln \frac{T_1}{T_2}.$$

Putting in numbers, we find

$$\Delta S_{\text{iron}} = -(2.0 \text{ kg})(0.11 \text{ kcal/kg·K}) \ln \frac{880 \text{ K}}{280 \text{ K}} = -0.25 \text{ kcal/K}$$

or $\Delta S = -1100$ J/K.

(b) The initial and final temperatures of the lake are the same, $T = 280$ K. The lake receives from the iron an amount of heat $Q = mc(T_2 - T_1) = (2.0 \text{ kg})(0.11 \text{ kcal/kg·K})(880 \text{ K} - 280 \text{ K}) = 130$ kcal. Strictly speaking, this is an irreversible process (the lake heats up locally before equilibrium is reached), but is equivalent to a reversible isothermal transfer of heat $Q = 130$ kcal at $T = 280$ K. Hence

$$\Delta S_{\text{env}} = \frac{130 \text{ kcal}}{280 \text{ K}} = 0.46 \text{ kcal/K}$$

or 1900 J/K. Thus, although the entropy of the iron actually decreases, the *total* change in entropy of iron plus environment is positive: 0.46 kcal/K − 0.25 kcal/K = +0.21 kcal/K.

We can now easily show in general that for an isolated system of two bodies, the flow of heat from the higher-temperature (T_H) body to the lower-temperature (T_L) body always results in an increase in the total entropy. The two bodies eventually come to some intermediate temperature, T_M; the heat lost by the hotter body ($Q_H = -Q$ where Q is positive) is equal to the heat gained by the colder one ($Q_L = Q$), so the total change in entropy is

$$\Delta S = \Delta S_H + \Delta S_L = -\frac{Q}{T_{HM}} + \frac{Q}{T_{LM}}$$

where T_{HM} is some intermediate temperature between T_H and T_M for the hot body as it cools from T_H to T_M, and T_{LM} is the counterpart for the cold body; since the temperature of the hot body, at all times during the process, is greater than that of the cold body, $T_{HM} > T_{LM}$. Hence

$$\Delta S = Q\left(\frac{1}{T_{LM}} - \frac{1}{T_{HM}}\right) > 0.$$

One body decreases in entropy, while the other gains in entropy, but the *total* change is positive.

EXAMPLE 21–5 A falling rock has kinetic energy E_K just before striking the ground and coming to rest. What is the total change in entropy of the rock plus environment as a result of this collision?

SOLUTION The energy E_K is transformed into internal energy of the rock plus ground. Their temperature T (assumed equal) won't change significantly. This process is equivalent to a reversible transfer of heat $Q = E_K$ into the rock and environment at constant temperature T. Hence $\Delta S = E_K/T$, and again $\Delta S > 0$.

In each of these examples, the entropy of our system plus that of the environment (or surroundings) either stayed constant or increased. For any *reversible* process, such as that in Example 21–2, the total entropy change is zero. This can be seen as follows: any reversible process can be considered as a series of quasistatic isothermal transfers of heat ΔQ between a system and the environment, which differ in temperature only by an infinitesimal amount; hence the change in entropy of either the system or environment is $\Delta Q/T$ and that of the other is $-\Delta Q/T$, so the total is

$$\Delta S = \Delta S_{syst} + \Delta S_{env} = 0. \qquad \text{[any reversible process]}$$

In Examples 21–3, 21–4, and 21–5, we found that the total entropy of system plus environment increases. Indeed, it has been found that for all real (irreversible) processes, the entropy increases. No exceptions have been found. This, then, is the *general statement of the second law of thermodynamics*:

The total entropy of any system plus that of its environment increases as a result of any natural process:

$$\Delta S > 0.$$

Although the entropy of one part of the universe may decrease in any process (see the examples above), the entropy of some other part of the universe always increases by a greater amount, so the total entropy always increases.

Now that we finally have a quantitative general statement of the second law of thermodynamics, we can see that it is an unusual law. It differs considerably from other laws of physics, which are typically equalities (such as $F = ma$) or conservation laws (such as for energy and momentum). The second law introduces a new quantity, the entropy, S, but does not tell us it is conserved. Quite the opposite. Entropy is not conserved; it always increases in time.

The second law summarizes very succinctly which processes are observed in nature, and which are not. Or, said another way, it tells us about the *direction* processes go. For the reverse of any of the processes in the last three examples, the entropy would decrease; and we never observe them. For example, we never observe heat flowing spontaneously from a cold body to a hot body, the reverse of Example 21–4; nor do we ever observe a gas spontaneously compressing itself into a smaller volume, the reverse of Example 21–3 (gases always expand to fill their containers); nor do we see thermal energy transform into kinetic energy of a rock so the rock rises spontaneously from the ground. Any of these processes would be consistent with the first law of thermodynamics (conservation of energy); it is the second law they are not consistent with, and this is why we need the second law. If you were to see a movie run backward, you would probably realize it immediately because you would see odd occurrences—such as rocks rising spontaneously from the ground, or air rushing in from the atmosphere to fill an empty balloon (the reverse of free expansion). We thus can tell whether time is passing normally or is being simulated to run backward. This is why entropy is sometimes called "time's arrow": it tells in which direction time is going.

Is the general statement of the second law of thermodynamics—"the principle of entropy increase"—consistent with the Clausius and Kelvin-Planck statements? Yes, it is. This is easy to see since (see discussion right after Example 21–4), if a process occurred in which heat flowed spontaneously out of a low-temperature (T_L) reservoir (so its entropy decreased), and all of it flowed into a high-temperature (T_H) reservoir (which increased in entropy), in violation of the Clausius statement, then the total change in entropy would be $\Delta S = Q/T_H - Q/T_L$ which is less than zero, since $T_L < T_H$. Thus the principle of entropy implies the Clausius statement. Can you show the equivalence of the entropy principle to the Kelvin-Planck statement?

21–8   Order to Disorder

The concept of entropy, as we have discussed it so far, may seem rather abstract. But we can relate it to the more ordinary concepts of *order* and *disorder*. In fact, the entropy of a system can be considered a *measure of the disorder of the system*. Then the second law of thermodynamics can be stated simply as:

Natural processes tend to move toward a state of greater disorder.

To make this clear, let us now consider a few examples: some of these will show us how this very general statement of the second law actually applies beyond what we usually consider as thermodynamics.

Let us first look at the simple processes mentioned at the beginning of this chapter in Section 21–1. A jar containing separate layers of salt and pepper is more orderly than when the salt and pepper are all mixed up. Shaking a jar containing separate layers results in a mixture, and no amount of shaking brings the layers back again. The natural process is from a state of relative order (layers) to one of relative disorder (a mixture), not the reverse. That is, disorder increases. Similarly, a solid coffee cup is a more "orderly" object than the pieces of a broken cup. Cups break when they fall, but they do not spontaneously mend themselves. Again, the normal course of events is an increase of disorder.

Now let us consider some processes for which we have actually calculated the entropy change, and see that an increase in entropy results in an increase in disorder (or vice versa). When ice melts to water at 0°C, the entropy of the water increases, as we saw in Example 21–2. Intuitively, we can think of solid water, ice, as being more ordered than the less orderly fluid state which can flow all over the place. This change from order to disorder can be seen more clearly from the molecular point of view: the orderly arrangement of water molecules in an ice crystal has changed to the disorderly and somewhat random motion of the molecules in the fluid state.

When a hot object is put in contact with a cold object, heat flows from the high temperature to the low until the two objects reach the same intermediate temperature. The total entropy increases in this process (see discussion after Example 21–4), and this too is a process in which order goes to disorder. At the beginning of the process we can distinguish two classes of molecules—those with a high average kinetic energy and those with a low average kinetic energy. After the process, all the molecules are in one class with the same average kinetic energy; we no longer have the more orderly arrangement of molecules in two classes. Order has gone to disorder; entropy has increased.

These examples illustrate the general concept that an increase in entropy corresponds to an increase in disorder. (We discuss this connection quantitatively in Section 21–10). In general, we associate disorder with uniformity, with randomness: salt and pepper in layers is more orderly than a random mixture;[†] a neat stack of numbered pages is more orderly than pages strewn randomly about on the floor.

Next consider a falling rock. The velocity of each molecule of the rock can be written as a sum of two terms, $v_i = v_{cm} + v_i^*$ where v_i^* is its random motion relative to the rock's center of mass (cm) and v_{cm} is that part of its velocity due to the motion of the rock as a whole (v_{cm} is the velocity of the rock's cm). The total kinetic energy of this rock at a given moment can be written as a sum of two terms (Chapter 9), the kinetic energy of the center of mass of the rock, and the kinetic energy of all the molecules relative to the center of mass:

$$KE_{cm} + \sum KE_i^*$$

The first term, KE_{cm}, is the *mechanical* kinetic energy of the rock as a whole; it can be considered *ordered* energy since all of it can be used to do work (such as moving something else, turning a generator, and so on). The second term, $\sum KE_i^*$, is the sum of the kinetic energies of all the moving molecules within the rock—it is energy they would have even if the rock were at rest. It represents the random, disordered motion of the molecules and hence corresponds to the *internal energy* of the molecules; only a part of this energy can be used to do work, but only if there is a lower-temperature region and if a heat engine is available. Thus ordinary *mechanical energy is ordered energy*, whereas *internal energy is disordered*. When the rock strikes the ground, though, the mechanical energy is transformed to internal energy. Order goes to disorder, and entropy increases, as in all natural processes.

An interesting example of the increase in entropy relates to biological evolution and to growth of organisms. Clearly, a human being is a highly ordered organism. The process of evolution from the early macromolecules and simple forms of life to *Homo sapiens* is a process of increasing order. So, too, the development of an individual from a single cell to a grown person is a process of increasing order. Do these processes violate the second law of thermodynamics? No, they do not. In the processes of evolution and growth, and even during the mature life of an individual, waste products are eliminated. These small molecules that remain as a result of metabolism are simple molecules without much order. Thus they represent relatively great disorder or entropy. Indeed, the total entropy of the molecules cast aside by organisms during the processes of evolution and growth is greater than the decrease in entropy associated with the order of the growing individual or evolving species.

[†] A more orderly arrangement is thus one that requires more *information* to specify or classify it. When we have one hot and one cold body, we have two classes of molecules and two pieces of information; when they come to the same temperature there is only one class and one piece of information. When salt and pepper are mixed, there is only one (uniform) class; when they are in layers, there are two classes. In this sense, information is connected to order, or low entropy. This is the foundation upon which the modern field of *information theory* is built.

21–9 Unavailability of Energy

Let us consider again the natural process of heat conduction from a hot body to a cold one. We have seen that entropy increases and that order goes to disorder. The interpretation that disorder has increased as a result of such a process is strengthened by the following consideration: separate hot and cold objects could serve as the high- and low-temperature regions for a heat engine and thus could be used to obtain useful work. But after the two objects are put in contact and reach the same uniform temperature, no work can be obtained from them. With regard to being able to do useful work, order has gone to disorder in this process.

The same can be said about a falling rock that comes to rest upon striking the ground. Just before hitting the ground, all the kinetic energy of the rock could have been used to do useful work. But once the rock's mechanical kinetic energy becomes internal energy, this is no longer possible.

Both these examples illustrate another important aspect of the second law of thermodynamics—that *in any natural (irreversible) process, some energy becomes unavailable to do useful work.*[†] In any process no energy is ever lost (it is always conserved). Rather, it becomes less useful—it can do less useful work. As time goes on, energy is, in a sense, *degraded*; it goes from more orderly forms (such as mechanical) eventually to the least orderly form, internal or thermal energy.

A natural outcome of this is the prediction that as time goes on, the universe will approach a state of maximum disorder. Matter will become a uniform mixture, heat will have flowed from high temperature regions to low temperature regions until the whole universe is at one temperature. No work can then be done. All the energy of the universe will have become degraded to thermal energy. All change will cease. This, the so-called *heat death* of the universe, has been much discussed by philosophers. This final state seems an inevitable consequence of the second law of thermodynamics, although it lies very far in the future. Yet it is based on the assumption that the universe is finite, which cosmologists are not really sure of. Furthermore, there is some question as to whether the second law of thermodynamics, as we know it, actually applies in the vast reaches of the universe. The answers are not yet in.

*21–10 Statistical Interpretation of Entropy and the Second Law

The ideas of entropy and disorder are made clearer with the use of a statistical or probabilistic analysis of the molecular state of a system. This statistical approach, which was first applied toward the end of the nineteenth century by Ludwig Boltzmann (1844–1906), makes a distinction between the "macrostate" and the "microstate" of a system. The *microstate* of a system would be specified when the position and velocity of every particle (or molecule) is given. The *macrostate* of a system is specified by giving the macroscopic properties of the system—the temperature, pressure, number of moles, and so on. In reality, we can know only the macrostate of a system. There are generally far too many molecules in a system to be able to know the velocity and position of every one at a given moment. Nonetheless, it is important to recognize that a great many different microstates can correspond to the *same* macrostate.

Let us take a simple example. Suppose you repeatedly shake four coins in your hand and drop them on the table. The number of heads and the number of tails that appear on a given throw is the macrostate of this system. Specifying each coin as being a head or a tail is the microstate of the system. In the following table we see the number of microstates that correspond to each macrostate:

[†] It can be shown that the amount of energy that becomes unavailable to do useful work is equal to $T_L \Delta S$ where T_L is the lowest available temperature and ΔS is the total increase in entropy during the process.

Macrostate	Possible Microstates (H = heads, T = tails)	Number of Microstates
4 heads	H H H H	1
3 heads, 1 tail	H H H T, H H T H, H T H H, T H H H	4
2 heads, 2 tails	H H T T, H T H T, T H H T, H T T H, T H T H, T T H H	6
1 head, 3 tails	T T T H, T T H T, T H T T, H T T T	4
4 tails	T T T T	1

A basic principle behind the statistical approach is that *each microstate is equally probable.* Thus the number of microstates that give the same macrostate corresponds to the relative probability of that macrostate occurring. The macrostate of two heads and two tails is the most probable one in our case of tossing four coins; out of the total of 16 possible microstates, six correspond to two heads and two tails, so the probability of throwing two heads and two tails is 6 out of 16, or 38 percent. The probability of throwing one head and three tails is 4 out of 16, or 25 percent; the probability of four heads is only 1 in 16, or 6 percent. Of course if you threw the coins 16 times, you might not find that two heads and two tails appears exactly 6 times, or four tails exactly once. These are only probabilities or averages. But if you made 1600 throws, very nearly 38 percent of them would be two heads and two tails. The greater the number of tries, the closer are the percentages to the calculated probabilities.

If we consider tossing more coins, say 100, the relative probability of throwing all heads (or all tails) is greatly reduced. There is only one microstate corresponding to all heads. For 99 heads and 1 tail, there are 100 microstates since each of the coins could be the one tail. The relative probabilities for other macrostates are given in Table 21–1. There are a total of about 10^{30} microstates possible.[†] Thus the relative probability of finding all heads is 1 in 10^{30}, an incredibly unlikely event! The probability of obtaining 50 heads and 50 tails (see Table 21–1) is $1.0 \times 10^{29}/10^{30} = 0.10$ or 10 percent. The probability of obtaining between 45 and 55 heads is 90 percent.

Thus we see that as the number of coins increases, the probability of obtaining an orderly arrangement (all heads or all tails) becomes extremely unlikely. The least orderly arrangement (half heads, half tails) is the most probable and the probability of being within a certain percentage (say 5 percent) of the most probable greatly increases as the number of coins increases. These same ideas can be applied to the molecules of a system. For example the most probable state of a gas (say the air in a room) is one in which the molecules take up the whole space and move about randomly; this corresponds to the Maxwellian distribution, Fig. 21–13a (and see Chapter 18). On the other hand, the very orderly arrangement

TABLE 21–1

Probabilities of various macrostates for 100 coin tosses

Macrostate		Number of Microstates,
Heads	Tails	W
100	0	1
99	1	1.0×10^2
90	10	1.7×10^{13}
80	20	5.4×10^{20}
60	40	1.4×10^{28}
55	45	6.1×10^{28}
50	50	1.0×10^{29}
45	55	6.1×10^{28}
40	60	1.4×10^{28}
20	80	5.4×10^{20}
10	90	1.7×10^{13}
1	99	1.0×10^2
0	100	1

[†] Each coin has two possibilities, heads or tails. Then the possible number of microstates is $2 \times 2 \times 2 \times \ldots = 2^{100}$. Using logarithms, we see that $2 = 10^{\log 2} = 10^{0.301}$ so that $2^{100} = (10^{0.301})^{100} = 10^{30.1} = 1.3 \times 10^{30}$.

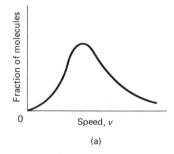

(a)

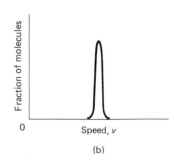

(b)

FIGURE 21–13

(a) Most probable distribution of molecular speeds in a gas (Maxwellian, or random); (b) orderly, but highly unlikely, distribution of speeds in which all molecules have nearly the same speed.

of all the molecules located in one corner of the room and all moving with the same velocity (Fig. 21–13b) is extremely unlikely; and indeed it is never observed.

From these examples, it is clear that probability is directly related to disorder and hence to entropy. That is, the most probable state is the one with greatest entropy or greatest disorder and randomness. Boltzmann showed that, consistent with Clausius's definition ($dS = dQ/T$) that the entropy of a system in a given (macro) state can be written

$$S = k \ln W \qquad (21–11)$$

where k is Boltzmann's constant ($k = 1.38 \times 10^{-23}$ J/K) and W is the number of microstates corresponding to the given macrostate; that is, W is proportional to the probability of occurrence of that state (W is called the *thermodynamic probability*, or, sometimes, the *disorder parameter*).

In terms of probability, the second law of thermodynamics—which tells us that entropy increases in any process—reduces to the statement that those processes occur which are most probable. The second law thus becomes a trivial statement. However there is an additional element now. The second law in terms of probability does not *forbid* a decrease in entropy. Rather, it says the probability is extremely low. It is not impossible that salt and pepper should separate spontaneously into layers, or that a broken tea cup should mend itself. It is even possible that a lake should freeze over on a hot summer day (that is, heat flow out of the cold lake into the warmer surroundings). But the probability for such events occurring is extremely small. In our coin examples above, we saw that increasing the number of coins from 4 to 100 reduced drastically the probability of large deviations from the average, or most probable, arrangement. In ordinary systems we are dealing with incredibly large numbers of molecules: in 1 mol alone there are 6×10^{23} molecules. Hence the probability of deviation far from the average is incredibly tiny. For example, it has been calculated that the probability that a stone resting on the ground should transform 1 cal of thermal energy into mechanical energy and rise up into the air is much less likely than the probability that a group of monkeys typing randomly would by chance produce the complete works of Shakespeare.

*21–11   Thermodynamic Temperature Scale; Absolute Zero

In Sections 21–4 and 21–5 we discussed the Carnot engine and other (ideal) reversible engines. We saw that the efficiency of any reversible engine operating between two heat reservoirs depends only on the temperatures of these two reservoirs; and the efficiency does not depend on the working substance—it could be helium, water, or something else, and the efficiency would be the same. The efficiency is given by

$$e = 1 - \frac{|Q_L|}{|Q_H|}$$

where $|Q_H|$ is the heat absorbed from the high-temperature reservoir and $|Q_L|$ is the heat exhausted to the low-temperature reservoir. Since $|Q_L/Q_H|$ is the same for any reversible engine operating between the same two temperatures, Kelvin suggested using this fact to define an absolute temperature scale; that is, the ratio of the temperatures of the two reservoirs, T_H and T_L, is defined as the ratio of the heats, $|Q_H|$ and $|Q_L|$, exchanged with them by a Carnot or other reversible engine:

$$\frac{T_H}{T_L} = \frac{|Q_H|}{|Q_L|}. \qquad (21–12)$$

This is the basis for the *Kelvin* or *thermodynamic temperature scale*.

In Section 21–5, we saw that this same relation, $T_H/T_L = |Q_H/Q_L|$, holds for a Carnot engine (Eq. 21–2) when the temperatures are based on the ideal gas temperature scale (Section 17–11) which we have been using up to now. Indeed, to complete the definition of the thermodynamic scale, we assign the value $T_{tp} = 273.16$ K to the triple point of water so that

$$T = (273.16 \text{ K})\left(\frac{|Q|}{|Q_{tp}|}\right) \qquad (21\text{–}13)$$

where $|Q|$ and $|Q_{tp}|$ are the heats exchanged by a Carnot engine with reservoirs at temperatures T and T_{tp}. This corresponds precisely to the definition of the ideal gas scale. Thus, the thermodynamic scale is identical to the ideal gas scale over the range of validity of the latter (below about 1 K, no substance is a gas, so the ideal gas scale cannot be used). The thermodynamic scale is now considered the standard scale since it can be used over the entire range of possible temperatures, and it is also independent of the substance used.

You might think the thermodynamic scale is not very practical, and indeed its importance lies to a great extent on the fact that it gives us a fundamental definition of temperature. But it is actually used in scientific work, particularly at very low temperatures where temperatures are determined by measuring the heats Q_H and Q_L of a Carnot cycle.

Very low temperatures are difficult to obtain experimentally. In fact, it is found experimentally that the closer the temperature is to absolute zero, the more difficult it is to reduce the temperature further. And it is generally accepted that it is not possible to actually reach absolute zero in any finite number of processes. This last statement is one way to state[†] the **third law of thermodynamics**. Since the maximum efficiency that any heat engine can have is the Carnot efficiency

$$e = 1 - \frac{T_L}{T_H},$$

and since T_L can never be zero, we see that a 100-percent efficient heat engine is not possible.

[†] See also the statement in the footnote in Section 21–6. We won't go into the equivalence of these two statements here.

Summary

A *heat engine* is a device for changing thermal energy, by means of heat flow, into useful work. The efficiency of a heat engine is defined as the ratio of the work W done by the engine to the heat input $|Q_H|$. Because of conservation of energy, the work input equals $|Q_H| - |Q_L|$ where $|Q_L|$ is the heat exhausted to the environment; hence the efficiency $e = W/|Q_H| = 1 - |Q_L/Q_H|$. Carnot's (idealized) engine consists of two isothermal and two adiabatic processes in a reversible cycle. For a Carnot engine, or any reversible engine operating between two temperatures, T_H and T_L, the efficiency is $e = 1 - T_L/T_H$. Irreversible (real) engines always have an efficiency less than this. All heat engines give rise to thermal pollution because they exhaust heat to the environment. The operation of refrigerators and air conditioners is the reverse of that of a heat engine: work is done to extract heat from a cold region and exhaust it to a region at a higher temperature.

The *second law of thermodynamics* can be stated in several equivalent ways: (1) heat flows spontaneously from a hot object to a cold one, but not the reverse; (2) there can be no 100-percent efficient heat engine—that is, one that can change a given amount of heat completely into work; and (3) natural processes tend to move toward a state of greater disorder or greater entropy. This last is the most general statement and can be restated most precisely as: the total entropy, S, of any system plus that of its environment increases as a result of any natural process: $\Delta S > 0$. Entropy, which is a state variable, is a quantitative measure of the disorder of a system. The change in entropy of a system during a reversible process is given by $\Delta S = \int dQ/T$. From a statistical point of view, the most probable state of a system is that with the most entropy or disorder. The second law of thermodynamics tells us in which direction processes tend to proceed; hence entropy is called "time's arrow." As time goes on, energy is degraded to less useful forms—that is, it is less available to do useful work.

Questions

1. Is it possible to cool down a room on a hot summer day by leaving the refrigerator door open?

2. The expansion valve in a refrigeration system, Fig. 21–5, is crucial for cooling the fluid. Explain how the cooling occurs.

3. Would a definition of heat engine efficiency as $e = W/Q_L$ be a useful one? Explain.

4. Can mechanical energy ever be transformed completely into heat or internal energy? Can the reverse happen? In each case, if your answer is no, explain why not; if yes, give examples.

5. Describe a process in nature that is nearly reversible.

6. (a) Describe how heat could be added to a system reversibly. (b) Could you use a stove burner to add heat to a system reversibly? Explain.

7. Powdered milk is very slowly (quasistatically) added to water while being stirred. Is this a reversible process?

8. Must a Carnot engine be started at point *a* in Fig. 21–9? Explain.

9. What plays the role of high-temperature and low-temperature reservoirs in (a) an internal combustion engine, (b) a steam engine? Are they, strictly speaking, heat reservoirs?

10. Discuss the factors that keep real engines from reaching Carnot efficiency.

11. Which will give the greater improvement in the efficiency of a Carnot engine, a 10°C increase in the high-temperature reservoir, or a 10°C decrease in the low-temperature reservoir?

12. The oceans contain a tremendous amount of thermal energy. Why, in general, is it not possible to put this energy to useful work?

13. Two identical systems are taken from state *a* to state *b* by two different *irreversible* processes. Will the change in entropy be the same for each process? Answer carefully and completely.

14. It can be said that the *total change in entropy during a process is a measure of the irreversibility of the process.* Discuss why this is valid, starting with the fact that $\Delta S = 0$ for a reversible process.

15. Use arguments, other than the principle of entropy increase, to show that for an adiabatic process, $\Delta S = 0$ if it is done reversibly and $\Delta S > 0$ if done irreversibly.

16. A gas is allowed to expand (a) adiabatically and (b) isothermally. In each process, does the entropy increase, decrease, or stay the same?

17. Entropy is often called "time's arrow" because it tells us in which direction natural processes occur. If a movie film were run backward, name some processes that you might see that would tell you that time was "running backward."

18. Give three examples, other than those mentioned in this chapter, of naturally occurring processes in which order goes to disorder. Discuss the observability of the reverse process.

19. Which do you think has the greater entropy, 1 kg of solid iron or 1 kg of liquid iron? Why?

20. What happens if you remove the lid of a bottle containing chlorine gas? Does the reverse process ever happen? Why or why not?

21. Think up several processes (other than those already mentioned) that would obey the first law of thermodynamics, but, if they actually occurred, would violate the second law.

22. Describe how a free expansion, for which the entropy increases, can be considered as a process in which order goes to disorder. (Hint: consider a stack of papers dropped into a large box versus dropped all over the floor.)

23. Suppose you collect a lot of papers strewn all over the floor and put them in a neat stack; does this violate the second law of thermodynamics? Explain.

24. The first law of thermodynamics is sometimes whimsically stated as, "You can't get something for nothing," and the second law as, "You can't even break even." Explain how these statements could be equivalent to the formal statements.

25. Give three examples of naturally occurring processes that illustrate the degradation of usable energy into internal energy.

26. List the following five-card hands in order of increasing probability: (a) four aces and a king; (b) six of hearts, eight of diamonds, queen of clubs, three of hearts, jack of spades; (c) two jacks, two queens, and an ace; (d) any hand having no two equal-value cards. Explain ranking, using microstates and macrostates.

Problems

SECTION 21–3

1. (I) A heat engine produces 7250 J of heat while performing 2250 J of useful work. What is the efficiency of this engine?

2. (II) The burning of gasoline in a car releases about 3.0×10^4 kcal/gal. If a car averages 33 km/gal when driving 90 km/h, which requires 20 hp, what is the efficiency of the engine under those conditions?

3. (II) A 40-percent efficient power plant puts out 850 MW (megawatts) of work (electrical energy). Cooling towers are used to take away the exhaust heat. If the air temperature is allowed to rise 7.5 C°, what volume of air (km³) is heated per day? Will the local climate be heated significantly? If the heated air were to form a layer 200 m thick, how large an area would it cover for 24 h of operation? (The heat capacity of air is about 7.0 cal/mol·C° at constant pressure.)

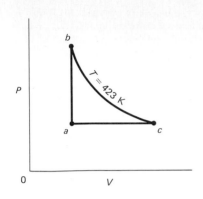

FIGURE 21–14

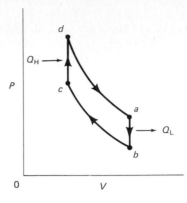

FIGURE 21–15

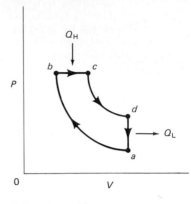

FIGURE 21–16

4. (II) Figure 21–14 is a PV diagram for a reversible heat engine in which 1.0 mol of argon, a nearly ideal monotomic gas, is initially at STP (point a). Points b and c are on an isotherm at $T = 423$ K. Process ab is at constant volume, process ac at constant pressure. (a) Is the path of the cycle carried out clockwise or counterclockwise? (b) What is the efficiency of this engine?

5. (III) The operation of an internal combustion engine (see Fig. 21–3) can be approximated by a reversible cycle known as the *Otto cycle*, which is shown in Fig. 21–15. Paths ab and cd are at constant volume, paths bc and da are adiabatic. (a) Which of these paths correspond to the compression stroke, burning of the gases, power stroke, and exhaust stroke of an internal combustion engine? (Intake is not represented on the diagram.) (b) For an ideal gas as working substance, show that the (idealized) efficiency of the engine is given by

$$e = 1 - \left(\frac{V_b}{V_c}\right)^{1-\gamma}$$

where γ is the ratio of specific heats (Sections 20–4 and 20–5). (c) Determine the efficiency for a compression ratio $V_b/V_c = 6.0$ assuming the gas is diatomic (like N_2 and O_2). (d) Is the heat exchange done at constant temperature in the Otto cycle as it is in the Carnot cycle (Section 21–5)? Could we use Eq. 21–3 for the reversible Otto cycle? (e) Real efficiencies for an internal combustion engine are less than that predicted above. What factors cause this?

6. (III) The operation of a *Diesel engine* can be idealized by the cycle shown in Fig. 21–16. Air is drawn into the cylinder during the intake stroke (not part of the idealized cycle). The air is compressed adiabatically, path ab. At point b diesel oil is injected into the cylinder which immediately burns since the temperature is very high. Combustion is slow, and during the first part of the power stroke, the gas expands at (nearly) constant pressure, path bc; after burning, the rest of the power stroke is adiabatic, path cd. Path da corresponds to the exhaust stroke. (a) Show that, for a quasistatic reversible engine undergoing this cycle using an ideal gas, the ideal efficiency is

$$e = 1 - \frac{(V_a/V_c)^{-\gamma} - (V_a/V_b)^{-\gamma}}{\gamma[(V_a/V_c)^{-1} - (V_a/V_b)^{-1}]}$$

where V_a/V_b is the "compression ratio" and V_a/V_c is the "expansion ratio." (b) If $V_a/V_b = 15$ and $V_a/V_c = 5.0$, calculate the efficiency assuming the gas is diatomic (like N_2 and O_2) and ideal.

SECTION 21–4

7. (I) (a) Show that the work done by a Carnot engine is equal to the area enclosed by the Carnot cycle on a PV diagram, Fig. 21–9. (See Section 20–1.) (b) Generalize this to any reversible cycle.

SECTION 21–5

8. (I) What is the maximum efficiency of a heat engine whose operating temperatures are 480°C and 305°C?

9. (I) The exhaust temperature of a heat engine is 280°C. What must be the high temperature if the Carnot efficiency is to be 32 percent?

10. (I) It has been suggested that a heat engine could be developed that made use of the fact that the temperature several hundred meters beneath the surface of the ocean is several degrees cooler than the temperature at the surface. In the tropics, the temperatures may be 5°C and 25°C, respectively. What is the maximum efficiency such an engine could have? Why might such an engine be feasible in spite of the low efficiency? Can you imagine any adverse environmental effects that might occur?

11. (II) An engine that operates at half its theoretical (Carnot) efficiency operates between 525°C and 290°C while producing work at the rate of 850 kW. How much heat is wasted per hour?

12. (II) A Carnot engine performs work at the rate of 650 kW while using 1250 kcal of heat per second. If the temperature of the heat source is 590°C, at what temperature is the waste heat exhausted?

13. (II) A heat engine utilizes a heat source at 610°C and has a Carnot efficiency of 27 percent. To increase the efficiency to 35 percent, what must be the temperature of the heat source?

14. (II) At a steam power plant, steam engines work in pairs, the heat output of one being the approximate heat input of the second. The operating temperatures of the first are 670°C and 430°C, and of the second 420°C and 280°C. If the heat of combustion of coal is 2.8×10^7 J/kg, at what rate must coal be burned if the plant is to put out 450 MW of power? Assume the efficiency of the engines is 65 percent of the ideal (Carnot) efficiency.

15. (II) Water is used to cool the power plant in problem 14. If the water temperature is allowed to increase by no more than 7.5 C°, estimate how much water must pass through the plant per hour.

16. (II) A "Carnot" refrigerator (reverse of a Carnot engine) absorbs heat from the freezer compartment at a temperature of $-17°C$ and exhausts it into the room at $25°C$. (a) How much work must be done by the refrigerator to change 0.50 kg of water at $25°C$ into ice at $-17°C$? (b) If the compressor output is 200 W, what minimum time is needed to take 0.50 kg of $25°C$ water and freeze it at $0°C$?

17. (II) The coefficient of performance, κ, of a refrigerator is defined as the ratio of the heat $|Q_L|$ taken from the low-temperature region to the work W needed to perform this task:

$$\kappa = \frac{|Q_L|}{W}.$$

(a) Show that for an ideal (Carnot) refrigerator,

$$\kappa = \frac{T_L}{T_H - T_L}.$$

(b) Write κ in terms of the efficiency e of the reversible heat engine obtained by running the refrigerator backward. (c) What is the coefficient of performance for an ideal refrigerator that maintains a freezer compartment at $-16°C$ when the compressor's temperature is $22°C$?

18. (II) Show that if two different adiabatic paths intersected at a single point on a PV diagram, they could be connected by an isotherm to form a cycle, and that an engine run on this cycle would violate the second law of thermodynamics. What do you conclude about the crossing of adiabatic lines?

SECTIONS 21–6 AND 21–7

19. (I) A 3.0-kg box having an initial speed of 2.2 m/s slides along a rough table and comes to rest. Estimate the total change in entropy of the universe. Assume all objects are at room temperature (293 K).

20. (I) If 1.00 kg of water at $100°C$ is changed by a reversible process to steam at $100°C$, determine the change in entropy of (a) the water, (b) the surroundings, (c) the universe as a whole. (d) How would your answers differ if the process were irreversible?

21. (I) An aluminum rod conducts 160 cal/s from a heat source maintained at $280°C$ to a large body of water at $22°C$. Calculate the rate entropy increases in this process.

22. (II) What is the total change in entropy when 2.5 kg of water at $0°C$ is frozen to ice at $0°C$ by being in contact with 45 kg of ice at $-10°C$?

23. (II) When 2.0 kg of water at $20°C$ is mixed with 1.0 kg of water at $80°C$ in a well-insulated container, what is the change in entropy of the system?

24. (II) Show that the principle of entropy increase is equivalent to the Kelvin-Planck statement of the second law of thermodynamics.

25. (II) Two 1400-kg cars are traveling 30 km/h in opposite directions when they collide and are brought to rest. What is the change in entropy of the universe as a result of this collision?

26. (II) (a) Calculate the change in entropy of 1.00 kg of water when it is heated from $0°C$ to $100°C$. (b) Does the entropy of the surroundings change? If so, by how much?

27. (II) A 120-g insulated aluminum cup at $20°C$ is filled with 210 g of water at $100°C$. (a) What is the final temperature of the mixture? (b) What is the total change in entropy as a result of the mixing process?

28. (II) Two samples of an ideal gas are initially at the same temperature and pressure; they are each compressed reversibly from a volume V to volume $V/2$, one isothermally, the other adiabatically. (a) In which sample is the final pressure greater? (b) Determine the change in entropy of the gas for each process. (c) What is the entropy change of the environment for each process?

29. (II) One mole of nitrogen (N_2) gas and one mole of argon (Ar) gas are in separate, equal-sized, insulated containers at the same temperature. The containers are then connected and the gases (assumed ideal) allowed to mix. (a) What is the change in entropy of the system, and (b) of the environment? (c) Repeat part (a) but assume one container is twice as large as the other.

30. (II) (a) Why would you expect the total entropy change in a Carnot cycle to be zero? (b) Do a calculation to show that it is zero.

31. (II) Thermodynamic processes can be represented not only on PV and PT diagrams, but another useful one is a TS (temperature-entropy) diagram. (a) Draw a TS diagram for a Carnot cycle. (b) What does the area within the curve represent?

32. (II) A real heat engine working between heat reservoirs at 400 K and 650 K produces 500 J of work per cycle for a heat input of 1400 J. (a) Compare the efficiency of this real engine to that of a Carnot engine. (b) Calculate the total entropy change of the universe for each cycle of the real engine. (c) Calculate the total entropy change of the universe for a Carnot engine operating between the same two temperatures. (d) Show that the difference in work done by these two engines per cycle is $T_L \Delta S$ where T_L is the temperature of the low-temperature reservoir (400 K) and ΔS is the entropy increase per cycle of the real engine. (See also problem 34, and Section 21–9.)

33. (II) Suppose a power plant delivers energy at 1000 MW using steam turbines. The steam goes into the turbines superheated at 520 K and deposits its unused heat in river water at 290 K. Assume that the turbine operates as a reversible Carnot engine. (a) If the river flow rate is 40 m³/s, calculate the temperature increase of the river water downstream from the power plant. (b) What is the entropy increase per kilogram of the downstream river water in J/kg·K?

SECTION 21–9

34. (III) A general theorem states that the amount of energy that becomes unavailable to do useful work in any process is equal to $T_L \Delta S$ where T_L is the lowest temperature available and ΔS is the total change in entropy during the process. Show that this is valid in the specific cases of (a) a falling rock that comes to rest when it hits the ground,

as in Example 21–5; (b) the free adiabatic expansion of an ideal gas, and (c) the conduction of heat, Q, from a high-temperature (T_H) reservoir to a low-temperature (T_L) reservoir. (Hint: in part (c) compare to a Carnot engine.)

*SECTION 21–10

*35. (II) Suppose you repeatedly shake six coins in your hand and drop them on the table. Construct a table showing the number of microstates that correspond to each macrostate. What is the probability of obtaining (a) three heads and three tails, and (b) six heads?

*36. (II) Calculate the relative probabilities, when you throw two dice, of obtaining (a) a 7, and (b) an 11.

*37. (II) (a) Suppose you have four coins, all with tails up. You now rearrange them so two heads and two tails are up. What was the change in entropy of the coins? (b) Suppose your system is the 100 coins of Table 21–1; what is the change in entropy of the coins if they are mixed randomly initially, 50 heads and 50 tails, and you arrange them so all 100 are heads? (c) Compare these entropy changes to ordinary thermodynamic entropy changes.

Electric Charge
and Electric Field

22

The word "electricity" may evoke an image of complex modern technology: computers, lights, motors, electric power. But the electric force plays an even deeper role in our lives, since according to atomic theory, the forces that act between atoms and molecules to hold them together to form liquids and solids are electrical forces. Similarly the electric force is responsible for the metabolic processes that occur within our bodies. Even ordinary pushes and pulls are the result of the electric force between the molecules of your hand and those of the object being pushed or pulled. Indeed, most of the forces we have dealt with beginning in Chapter 4, such as elastic forces and the normal force acting on a body, are now considered to be electric forces acting at the atomic level. This does not include gravity, however, which is a separate force.[†]

The earliest studies on electricity date back to the ancients, but it has been only in the past two centuries that electricity was studied in detail. We will discuss the development of ideas about electricity, including practical devices, as well as the relation to magnetism, in the next twelve chapters.

 ## 22–1 Static Electricity; Electric Charge and Its Conservation

The word *electricity* comes from the Greek word *elektron*, which means "amber." Amber is petrified tree resin, and the ancients knew that if you rub an amber rod with a piece of cloth, the amber attracts small pieces of leaves or dust. A piece of hard rubber, a glass rod, or a plastic ruler rubbed with a cloth will also display this "amber effect," or *static electricity* as we call it today. You

---

[†] In this century physicists came to recognize only four different forces in nature: (1) gravitational force, (2) electromagnetic force (we will see later that electric and magnetic forces are intimately related), (3) strong nuclear force, and (4) weak nuclear force. The last two forces operate at the level of the nucleus of an atom and although they manifest themselves in such phenomena as radioactivity and nuclear energy, they are much less obvious in our daily lives. We will discuss them in Chapters 42 and 43.

can readily pick up small pieces of paper with a plastic ruler that you've just vigorously rubbed with even a paper towel, Fig. 22–1. You have probably experienced static electricity when combing your hair or upon taking a synthetic blouse or shirt from a clothes dryer. And you may have felt a shock when you touched a metal door knob after sliding across a car seat or walking across a nylon carpet. In each case, an object becomes "charged" due to a rubbing process and is said to possess an **electric charge**.

Is all electric charge the same, or is it possible that there is more than one type? In fact, there are two types of electric charge as the following simple experiments show. A plastic ruler is suspended by a thread and rubbed vigorously to charge it; when a second ruler, which has also been charged by rubbing, is brought close to the first, it is found that the one ruler *repels* the other. This is shown in Fig. 22–2a. Similarly, if a rubbed glass rod is brought close to a second charged glass rod, again a repulsive force is seen to act, Fig. 22–2b. However, if the charged glass rod is brought close to the charged plastic ruler, it is found that they *attract* each other, Fig. 22–2c. The charge on the glass must therefore be different than that on the plastic. Indeed, it is found experimentally that all charged objects fall into one of two categories. Either they are attracted to the plastic and repelled by the glass, just as glass is; or they are repelled by the plastic and attracted to the glass, just as the plastic ruler is. Thus there seems to be two, and only two, types of electric charge. Each type of charge repels the same type but attracts the opposite type. That is: *unlike charges attract; like charges repel.*

These two types of charge were referred to as *positive* and *negative* by the American statesman, philosopher, and scientist Benjamin Franklin (1706–1790). The choice of which name went with what type of charge was of course arbitrary. Franklin's choice sets the charge on the rubbed glass rod to be positive charge, so the charge on a rubbed plastic ruler (or amber) is called negative charge; we still follow this convention today.

Franklin's theory of electric charge was actually a "single-fluid" theory that viewed a positive charge as an excess of the electric fluid beyond an object's normal content of electricity and a negative charge as a deficiency. Franklin argued that whenever a certain amount of charge is produced on one body in a process, an equal amount of the opposite type of charge is produced on another body. The names positive and negative are to be taken *algebraically*, so that during any process the net change in the amount of charge produced is zero. For example, when a plastic ruler is rubbed with a paper towel, the plastic acquires a negative charge and the towel an equal amount of positive charge; the charge is separated, but the sum of the two is zero. This is an example of a law that is now well established: the **law of conservation of electric charge**, which states that

the net amount of electric charge produced in any process is zero.

No violations have ever been found, and this conservation law is as firmly established as those for energy and momentum.

22–2   Electric Charge in the Atom

Only within the past century has it become clear that electric charge has its origin within the atom itself. In later chapters we will discuss atomic structure and the ideas that led to our present view of the atom in more detail; but it will help our understanding of electricity if we discuss it briefly now.

Today's view, slightly simplified, shows the atom as having a heavy, positively charged nucleus surrounded by one or more negatively charged electrons. In its normal state, the positive and negative charges within the atom are equal,

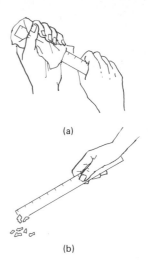

FIGURE 22–1

Rub a plastic ruler (a) and bring it close to some tiny pieces of paper (b).

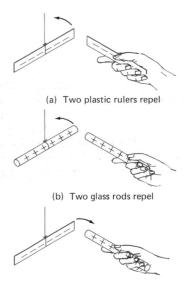

(a) Two plastic rulers repel

(b) Two glass rods repel

(c) Glass rod attracts plastic ruler

FIGURE 22–2

Unlike charges attract, whereas like charges repel one another.

and the atom is electrically neutral. Sometimes, however, an atom may lose one or more of its electrons, or may gain extra electrons. In this case the atom will have a net positive or negative charge, and is called an *ion*.

The nuclei in a solid material can vibrate, but they remain close to fixed positions, whereas some of the electrons move quite freely. The charging of an object by rubbing is explained by considering how strongly the electrons are attracted to the nuclei of the material. When a plastic ruler becomes negatively charged by rubbing with a paper towel, the electrons in the towel are evidently held less tightly than those in the plastic; thus some of the electrons are pulled off the towel onto the plastic. This leaves the towel with a positive charge equal in magnitude to the negative charge given the plastic.

Normally when objects are charged by rubbing, they hold their charge only for a limited time and eventually return to the neutral state. Where does the charge go? It "leaks off" onto water molecules in the air. This is because water molecules are *polar*—that is, even though they are neutral, their charge is not distributed uniformly, Fig. 22–3. Thus the extra electrons on, say, a charged plastic ruler "leak off" into the air because they are attracted to the positive end of water molecules. A positively charged object, on the other hand, is neutralized by transfer of loosely held electrons from water molecules in the air. On dry days, static electricity is much more noticeable since the air contains fewer water molecules to allow leakage. On humid or rainy days, it is difficult to make any object hold its charge for long.

FIGURE 22–3

Diagram of a water molecule. Because it has opposite charges on different ends, it is called a "polar" molecule.

22–3   Insulators and Conductors

Suppose we have two metal spheres, one highly charged and the other electrically neutral. If we now place an iron nail so that it touches both the spheres, it is found that the previously uncharged sphere quickly becomes charged. If, however, we had connected the two spheres together with a wooden rod or a piece of rubber, the uncharged ball would not have become noticeably charged. Materials like the iron nail are said to be **conductors** of electricity, whereas wood and rubber are called **nonconductors** or **insulators**.

Metals are generally good conductors whereas most other materials are insulators (although even insulators conduct electricity very slightly). It is interesting that nearly all natural materials fall into one or the other of these two quite distinct categories. There are a few materials, however, (notably silicon, germanium, and carbon) that fall into an intermediate (but distinct) category known as *semiconductors*.

From the atomic point of view the electrons in an insulating material are bound very tightly to the nuclei, whereas in a conductor many of them are bound very loosely and can move about freely[†] within the material. When a positively charged object is brought close to or touches a conductor, the free electrons move quickly toward this positive charge. On the other hand, the free electrons move swiftly away from a negative charge that is brought close. In a semiconductor, there are very few free electrons, and in an insulator, almost none.

FIGURE 22–4

Charging by induction.

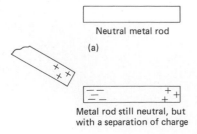

Neutral metal rod

(a)

Metal rod still neutral, but with a separation of charge

(b)

22–4   Induced Charge; the Electroscope

Suppose a positively charged metal object is brought close to a second (neutral) metal object. If the two touch, the free electrons in the neutral one are attracted to the positively charged object and some will pass over to it. Since the second

---

[†] The free electrons move about within the metal rather easily. But they are not usually able to *leave* the metal easily. Indeed, rubbing a metal object rarely gives it a static charge, as can be done with nonconductors such as glass and plastic.

object is now missing some of its negative electrons, it will have a net positive charge. This process is called "charging by conduction."

Now suppose a positively charged object is brought close to a neutral metal rod, but does not touch it. Although the electrons of the metal rod do not leave the rod, they still move within the metal toward the charged object; this leaves a positive charge at the opposite end, Fig. 22–4. A charge is said to have been *induced* at the two ends of the metal rod. Of course no charge has been created; it has merely been *separated*; the net charge on the metal rod is still zero. However, if the metal were now cut in half, we would have two charged objects, one charged positively and one charged negatively.

Another way to induce a net charge on a metal object is to connect it with a conducting wire to the ground (or a pipe leading into the ground) as shown in Fig. 22–5a (⏚ means "ground"). The object is then said to be "grounded" or "earthed." Now the earth, since it is so large, can easily accept or give up electrons; hence it acts like a reservoir for charge. If a charged object—let's say negative this time—is brought up close to the metal, free electrons in the metal are repelled and many of them move down the wire into the earth, Fig. 22–5b. This leaves the metal positively charged. If the wire is now cut, the metal will have a positive induced charge on it. If the wire were cut after the negative object is moved away, the electrons would all have moved back into the metal and it would be neutral.

An *electroscope* (or simple *electrometer*) is a device that can be used for detecting charge. As shown in Fig. 22–6, it consists of a case inside of which are two movable leaves, often made of gold. (Sometimes only one leaf is movable.) The leaves are connected by a conductor to a metal ball on the outside of the case, but is insulated from the case itself. If a charged object is brought close to the knob, a separation of charge is induced in it, Fig. 22–7a; the two leaves become charged and repel each other as shown. If, instead, the knob is charged by conduction the whole apparatus becomes charged as shown in Fig. 22–7b. In either case, the greater the amount of charge, the greater the separation of the leaves.

Note however, that you cannot tell the sign of the charge in this way since a negative charge will cause the leaves to separate just as much as an equal magnitude positive charge—in either case the two leaves repel each other. An electroscope can, however, be used to determine the sign of the charge if it is first charged by conduction, say negatively, as in Fig. 22–8a. Now if a negative object is brought close, as in Fig. 22–8b, electrons are induced to move further down into the leaves and they separate further. On the other hand, if a positive object is brought close, the electrons are induced to flow upward, leaving the leaves less negative and their separation is reduced, Fig. 22–8c.

The electroscope was much used in the early days of electricity. The same principle, aided by some electronics, is used in much more sensitive modern *electrometers*.

 22–5   Coulomb's Law

We have seen that an electric charge exerts a force on other electric charges. But how does the magnitude of the charges and other factors affect the magnitude of this force? To answer this, the French physicist Charles Coulomb (1736–1806) investigated electric forces in the 1780s using a torsion balance (Fig. 22–9) much like that used by Cavendish for his studies of the gravitational force (Section 5–2).

Although precise instruments for the measurement of electric charge were not available in Coulomb's time, he was able to prepare small spheres with different amounts of charge in which the *ratio* of the charges was known. He reasoned that if a charged conducting sphere is placed in contact with an identical

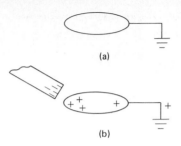

FIGURE 22–5
Inducing a charge on an object connected to ground.

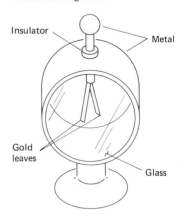

FIGURE 22–6
Electroscope.

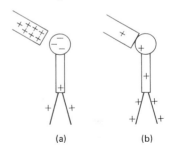

FIGURE 22–7
Electroscope charged (a) by induction, (b) by conduction.

FIGURE 22–8
A previously charged electroscope can be used to determine the sign of a given charge.

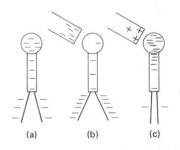

431

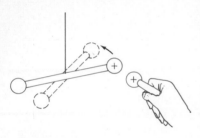

FIGURE 22–9

Schematic diagram of Coulomb's apparatus. It is similar to Cavendish's which was used to measure the gravitational force. When a charged sphere is placed close to the one on the suspended bar, the bar rotates slightly. The suspending fiber resists the twisting motion and the angle of twist is proportional to the force applied. By use of this apparatus, Coulomb was able to investigate how the electric force varied as a function of the magnitude of the charges and of the distance between them.

uncharged sphere, the charge on the first would then be shared equally by the two of them because of symmetry. He thus had a way to produce charges equal to $\frac{1}{2}$, $\frac{1}{4}$, and so on, of the original charge. Although he had some difficulty with induced charges, Coulomb was able to argue that the force one charged object exerted on a second tiny charged object is directly proportional to the charge on each of them. That is, if the charge on either of the objects was doubled, the force was doubled; and if the charge on both of the objects was doubled the force increased to four times the original value. This was the case when the distance between the two spheres remained the same. If the distance between them was allowed to change, he found that the force decreased with the square of the distance between them. That is, if the distance was doubled, the force fell to only one-fourth of its original value. Thus, Coulomb concluded, the force one tiny charged object (ideally a *point charge*—one which, like an idealized point particle, has no spatial extent) exerts on a second one is proportional to the product of the amount of charge on one, Q_1, times the amount of charge on the other, Q_2, and inversely proportional to the square of the distance r between them; that is,

$$F = k\frac{Q_1 Q_2}{r^2} \qquad (22-1)$$

where k is a proportionality constant. This is known as **Coulomb's law**. Its validity rests on careful experiments that are much more sophisticated than Coulomb's difficult-to-perform experiment. The exponent 2 has been shown to be accurate to 1 part in 10^{16} (that is, $2 \pm 2 \times 10^{-16}$).

Since we are dealing here with a new quantity (electric charge) we could choose its unit so that the proportionality constant k in Eq. 22–1 would be one. Indeed, such a system of units was once common.[†] However, the most widely used unit now is the *coulomb* (C) which is the SI unit. The precise definition of the coulomb today is in terms of electric current and magnetic field and will be discussed later (Section 29–4). In SI units, k has the value

$$k = 8.988 \times 10^9 \text{ N·m}^2/\text{C}^2 \approx 9.0 \times 10^9 \text{ N·m}^2/\text{C}^2.$$

Thus, 1 C is that amount of charge which, if it exists on each of two tiny objects placed 1 m apart, will result in each object exerting a force of about $(9.0 \times 10^9 \text{ N·m}^2/\text{C}^2)(1.0 \text{ C})(1.0 \text{ C})/(1.0 \text{ m})^2 = 9.0 \times 10^9 \text{ N}$ on the other.

Charges produced by rubbing ordinary objects (such as a comb or plastic ruler) are typically a microcoulomb (1 μC $= 10^{-6}$ C) or less. The magnitude of the charge on one electron, on the other hand, is measured to be about 1.602×10^{-19} C (but is negative). This is the smallest known charge.[‡] and because of its fundamental nature it is given the symbol e and is often referred to as the *elementary charge*:

$$e = (1.6021892 \pm 0.0000046) \times 10^{-19} \text{ C}$$

or

$$e \approx 1.602 \times 10^{-19} \text{ C}.$$

Since an object cannot gain or lose a fraction of an electron, the net charge on any object must be an integral multiple of this charge. Electric charge is thus said to be *quantized* (existing only in discrete amounts); because e is so small, however, we normally don't notice this discreteness in macroscopic charges (1 μC requires about 10^{13} electrons) which thus seem continuous.

[†] This is a cgs system of units, and the unit of electric charge is called the *electrostatic unit* (esu) or the *statcoulomb*. One esu is defined as that charge, on each of two tiny objects 1 cm apart, that gives rise to a force of 1 dyne.

[‡] Elementary-particle physicists theorize the existence of smaller particles, called quarks, that would have a smaller charge equal to $\frac{1}{3}$ or $\frac{2}{3}e$. They have not been detected for sure experimentally, and theory indicates that free quarks may not be detectable (see Chapter 43).

Equation 22–1 gives the magnitude of the force that either object exerts on the other. The direction of this force is along the line joining the two objects. If the two charges have the same sign, the force on each object is directed away from the other; if the two objects have opposite charges, the force on one is directed toward the other, Fig. 22–10. Notice that the force one charge exerts on the second is equal but opposite to that exerted by the second on the first; this is in accord with Newton's third law. Coulomb's law can be written in vector form (as we did for Newton's law of universal gravitation in Chapter 5) as

$$\mathbf{F}_{12} = k\frac{Q_1 Q_2}{r_{21}^2}\hat{\mathbf{r}}_{21}$$

where $\mathbf{F}_{12}$ is the vector force[†] on charge Q_1 due to Q_2 and $\hat{\mathbf{r}}_{21}$ is the unit vector pointing from Q_2 toward Q_1.

It should be recognized that Eq. 22–1 applies to objects whose size is much smaller than the distance between them. Ideally, it applies to point charges. For finite-sized objects, it is not always clear what value to use for r, particularly since the charge may not be distributed uniformly on the objects. If the two objects are spheres and the charge is known to be distributed uniformly on each, then r is the distance between their centers. (This will be discussed later in this chapter and in the next.) It is important to keep in mind that Eq. 22–1 gives the force on a charge due to only *one* other charge. If several (or many) charges are present, the net force on any one of them will be the vector sum of the forces due to each of the others.

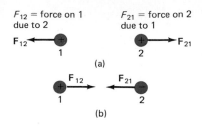

FIGURE 22–10
Direction of force depends on whether the charges have (a) same sign or (b) opposite sign.

EXAMPLE 22–1   What is the magnitude of the electric force on the electron of a hydrogen atom exerted by the single proton ($Q_2 = e$) that is its nucleus, when the electron "orbits" the proton at its average distance of 0.53×10^{-10} m.

SOLUTION   We use Eq. 22–1 with $Q_2 = +1.6 \times 10^{-19}$ C, $Q_1 = -Q_2$, and $r = 0.53 \times 10^{-10}$ m:

$$F = \frac{(9.0 \times 10^9\,\mathrm{N\cdot m^2/C^2})(+1.6 \times 10^{-19}\,\mathrm{C})(-1.6 \times 10^{-19}\,\mathrm{C})}{(0.53 \times 10^{-10}\,\mathrm{m})^2}$$

$$= -8.2 \times 10^{-8}\,\mathrm{N}.$$

The minus sign indicates that the force is attractive.

EXAMPLE 22–2   Calculate the net electric force on particle 3 in Fig. 22–11a (the $-4.0\ \mu$C on the right) due to the other two charges.

SOLUTION   The net force on particle 3 will be the sum of the force $\mathbf{F}_{31}$ exerted by particle 1 and the force $\mathbf{F}_{32}$ exerted by particle 2. The magnitudes of these two forces are

$$F_{31} = \frac{(9.0 \times 10^9\,\mathrm{N\cdot m^2/C^2})(-4.0 \times 10^{-6}\,\mathrm{C})(-3.0 \times 10^{-6}\,\mathrm{C})}{(0.30\,\mathrm{m})^2} = 1.2\,\mathrm{N},$$

$$F_{32} = \frac{(9.0 \times 10^9\,\mathrm{N\cdot m^2/C^2})(5.0 \times 10^{-6}\,\mathrm{C})(-4.0 \times 10^{-6}\,\mathrm{C})}{(0.20\,\mathrm{m})^2} = -4.5\,\mathrm{N}.$$

$\mathbf{F}_{31}$ is repulsive and $\mathbf{F}_{32}$ is attractive, so the direction of the forces is as shown in Fig. 22–11b. The net force on particle 3 is then

$$F = F_{32} + F_{31} = -4.5\,\mathrm{N} + 1.2\,\mathrm{N} = -3.3\,\mathrm{N};$$

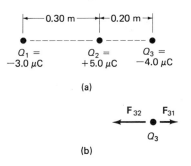

FIGURE 22–11
Diagram for Example 22–2.

[†] Note that if Q_1 and Q_2 have the same sign, the product $Q_1 Q_2 > 0$ and the force on Q_1 points away from Q_2—that is, it is repulsive. If Q_1 and Q_2 have opposite signs, $Q_1 Q_2 < 0$ and $\mathbf{F}_{12}$ points toward Q_2—that is, it is attractive.

The magnitude is 3.3 N, and it points to the left. (Notice that the charge in the middle (Q_2) in no way blocks the effect of the other one on the end (Q_1); it does exert its own force, of course.)

EXAMPLE 22–3   Calculate the force on charge Q_3 shown in Fig. 22–12 due to the charges Q_1 and Q_2.

SOLUTION   The forces $\mathbf{F}_1$ and $\mathbf{F}_2$ have the directions shown in the diagram since Q_1 exerts an attractive force and Q_2 a repulsive force. The magnitudes of $\mathbf{F}_1$ and $\mathbf{F}_2$ are (ignoring signs since we know the directions)

$$F_1 = \frac{(9.0 \times 10^9 \text{ N·m}^2/\text{C}^2)(6.5 \times 10^{-5} \text{ C})(8.6 \times 10^{-5} \text{ C})}{(0.60 \text{ m})^2} = 140 \text{ N},$$

$$F_2 = \frac{(9.0 \times 10^9 \text{ N·m}^2/\text{C}^2)(6.5 \times 10^{-5} \text{ C})(5.0 \times 10^{-5} \text{ C})}{(0.30 \text{ m})^2} = 330 \text{ N}.$$

We resolve $\mathbf{F}_1$ into its components along the x and y axes as shown:

$$F_{1x} = F_1 \cos 30° = 120 \text{ N},$$

$$F_{1y} = -F_1 \sin 30° = -70 \text{ N}.$$

The force $\mathbf{F}_2$ has only a y component. So the net force $\mathbf{F}$ on Q_3 has components $F_x = F_{1x} = 120$ N, and $F_y = F_2 + F_{1y} = 330 \text{ N} - 70 \text{ N} = 260$ N. Thus the magnitude of the net force $F = \sqrt{F_x^2 + F_y^2} = \sqrt{(120 \text{ N})^2 + (260 \text{ N})^2} = 290$ N; and it acts at an angle θ (see Fig. 22–12b) given by $\tan \theta = F_y/F_x = 260 \text{ N}/120 \text{ N} = 2.2$; so $\theta = 65°$.

The constant k in Eq. 22–1 is usually written in terms of another constant, ε_0, called the **permittivity** *of free space*; it is related to k by: $k = 1/4\pi\varepsilon_0$. Coulomb's law can then be written

$$F = \frac{1}{4\pi\varepsilon_0} \frac{Q_1 Q_2}{r^2}, \tag{22–2}$$

where, given to its maximum known accuracy,

$$\varepsilon_0 = \frac{1}{4\pi k} = (8.85418782 \pm 0.00000007) \times 10^{-12} \text{ C}^2/\text{N·m}^2,$$

FIGURE 22–12   Determining the forces for Example 22–3.

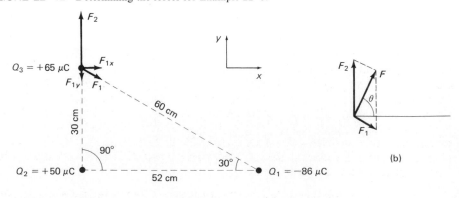

(a)

(b)

or, rounded off,

$$\varepsilon_0 \approx 8.85 \times 10^{-12} \ \text{C}^2/\text{N}\cdot\text{m}^2.$$

Although Eq. 22–2 looks more complicated than Eq. 22–1, most other equations in electromagnetic theory are simpler when written in terms of ε_0 rather than k (since the 4π often cancels in the final result). Hence we will usually use Eq. 22–2 remembering that

$$\frac{1}{4\pi\varepsilon_0} \approx 9.0 \times 10^9 \ \text{N}\cdot\text{m}^2/\text{C}^2.$$

Coulomb's law describes the force between two charges when they are at rest. Additional forces come into play when charges are in motion, and these will be discussed in later chapters. In this chapter we discuss only charges at rest, and this is called **electrostatics**.

22–6 The Electric Field

In Chapter 5 (Section 5–9), we saw that the concept of "action at a distance" was a difficult one for early thinkers. That is, how can one electric charge exert a force on a second one, even though they are not touching? Newton himself felt uneasy with this idea which he applied to his law of universal gravitation. The conceptual difficulties can be overcome, as we saw, with the idea of the *field*, developed by the British scientist Michael Faraday (1791–1867). In the electrical case, according to Faraday, an *electric field* extends outward from every charge and permeates all of space. When a second charge is placed near the first charge, it feels a force because of the electric field that is there. The electric field at the location of the second charge is considered to interact directly with this charge to produce the force. It must be emphasized, however, that a field is *not* a kind of matter. It is, rather, a concept[†]—and a very useful one.

We can investigate the electric field surrounding a charge or group of charges by measuring the force on a small positive **test charge**. By a test charge we mean a charge so small that the force it exerts does not significantly alter the distribution of the other charges, the ones that cause the field being measured. The force on a tiny positive test charge q placed at various locations in the vicinity of a single positive charge Q would be as shown in Fig. 22–13. The force at b is less than at a because the distance is greater (Coulomb's law); and the force at c is smaller still. In each case, the force is directed radially outward from Q. The electric field is defined in terms of the force on such a positive test charge. In particular, the **electric field**, **E**, at any point in space is defined as the force **F** exerted on a tiny positive test charge at that point divided by the magnitude of the test charge q:

$$\mathbf{E} = \frac{\mathbf{F}}{q}. \tag{22–3}$$

(Ideally, **E** is defined as the limit of $\mathbf{F}/q$ as q is taken smaller and smaller, approaching zero.) From this definition (Eq. 22–3) we see that the direction of the electric field at any point in space is defined as the direction of the force on a positive test charge at that point. And the magnitude of the electric field is the *force per unit charge*. Thus **E** is measured in units of newtons per coulomb (N/C).

The reason for defining **E** as $\mathbf{F}/q$ is so that **E** does not depend on the magnitude of the test charge q. This means that **E** describes only the effect of the charges creating the electric field at that point. Since **E** is a vector, the electric field is referred to as a *vector field*.

[†] Whether the electric field is "real," and really exists, is a philosophical, even metaphysical, question. In physics it is a very useful idea, in fact a great invention of the human mind.

FIGURE 22–13

Force exerted by charge $+Q$ on a small test charge, q, placed at points a, b, and c.

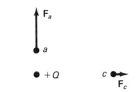

22–7   Calculation of Electric Field, E

For many simple situations we can calculate what the electric field will be at a given point in space using Eq. 22–3. For example, the magnitude of the electric field due to a single point charge Q at a distance r from that charge would be

$$E = \frac{1}{4\pi\varepsilon_0} \frac{qQ}{r^2} \cdot \frac{1}{q}$$

$$= \frac{1}{4\pi\varepsilon_0} \frac{Q}{r^2}. \qquad \text{[single point charge]} \qquad (22\text{–}4)$$

This relation for the electric field due to a single point charge is also (in addition to Eq. 22–2) referred to as Coulomb's law. Notice that E is independent of q—that is, it depends only on the charge Q which produces the field, and not on the value of the test charge q.

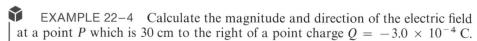 **EXAMPLE 22–4**   Calculate the magnitude and direction of the electric field at a point P which is 30 cm to the right of a point charge $Q = -3.0 \times 10^{-4}$ C.

SOLUTION   The magnitude of the electric field is

$$E = \frac{1}{4\pi\varepsilon_0} \frac{Q}{r^2} = \frac{(9.0 \times 10^9 \text{ N·m}^2/\text{C}^2)(3.0 \times 10^{-4} \text{ C})}{(0.30 \text{ m})^2} = 3.0 \times 10^7 \text{ N/C}.$$

The direction of the electric field is *toward* the charge Q as shown in Fig. 22–14a since we defined the direction as that of the force on a positive test charge. If Q had been positive, the electric field would have pointed away as in Fig. 22–14b.

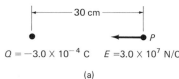

$Q = -3.0 \times 10^{-4}$ C $E = 3.0 \times 10^7$ N/C

(a)

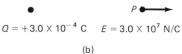

$Q = +3.0 \times 10^{-4}$ C $E = 3.0 \times 10^7$ N/C

(b)

FIGURE 22–14

Electric field at point P due to (a) a negative charge Q and (b) a positive charge Q (Example 22–4).

If the field is due to more than one charge, the contributions from each are calculated as a vector sum. The electric field $\mathbf{E}$ at any point due to n point charges is

$$\mathbf{E} = \mathbf{E}_1 + \mathbf{E}_2 + \ldots + \mathbf{E}_n = \sum_{i=1}^{n} \mathbf{E}_i \qquad (22\text{–}5)$$

where $\mathbf{E}_i$ is the electric field at that point due to the ith charge. The validity of this *superposition principle* for the electric field derives from experiment, and no exceptions have been observed.

 EXAMPLE 22–5   Calculate the electric fields at points A and B in Fig. 22–15.

SOLUTION   (a) The calculation is much like that of Example 22–3, but now we are dealing with electric fields. The electric field at A is the vector sum of the fields $\mathbf{E}_{A1}$ due to Q_1, and $\mathbf{E}_{A2}$ due to Q_2: $E_{A1} = (9.0 \times 10^9 \text{ N·m}^2/\text{C}^2) \times (50 \times 10^{-6} \text{ C})/(0.60 \text{ m})^2 = 1.25 \times 10^6 \text{ N/C}$, and similarly $E_{A2} = 5.0 \times 10^6 \text{ N/C}$. The directions are as shown, so the total electric field at A, $\mathbf{E}_A$, has components $E_{Ax} = E_{A1} \cos 30° = 1.1 \times 10^6 \text{ N/C}$ and $E_{Ay} = E_{A2} - E_{A1} \sin 30° = 4.4 \times 10^6 \text{ N/C}$. Thus the magnitude of $\mathbf{E}_A$ is $E_A = \sqrt{(1.1)^2 + (4.4)^2} \times 10^6 \text{ N/C} = 4.5 \times 10^6 \text{ N/C}$. It will act at an angle ϕ (see diagram) where $\tan \phi = 4.4/1.1 = 4.0$; so $\phi = 76°$.

(b) Since B is equidistant (40 cm by Pythagorean theorem) from the two equal charges, the magnitudes of E_{B1} and E_{B2} are the same; that is, $E_{B1} = E_{B2} = (9.0 \times 10^9 \text{ N·m}^2/\text{C}^2)(50 \times 10^{-6} \text{ N})/(0.40 \text{ m})^2 = 2.8 \times 10^6 \text{ N/C}$. Also, because of the symmetry, the y components are equal and opposite. Hence the total field E_B is horizontal and equals $E_{B1} \cos \theta + E_{B2} \cos \theta = 2E_{B1} \cos \theta$; from the diagram, $\cos \theta = 26 \text{ cm}/40 \text{ cm} = 0.65$. Then $E_B = 2(2.8 \times 10^6 \text{ N/C})(0.65) = 3.6 \times 10^6 \text{ N/C}$, and is along the x direction.

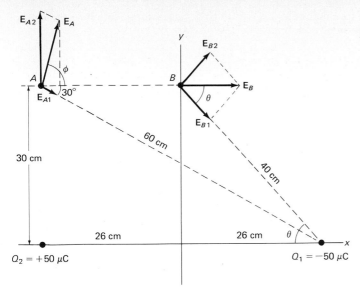

FIGURE 22–15

Calculation of the electric field at points A and B for Example 22–5.

In many cases, we can treat[†] charge as being distributed continuously. What we can do then is divide up the charge distribution into infinitesimal charges dQ. The electric field at a distance r from each dQ is

$$dE = \frac{1}{4\pi\varepsilon_0} \frac{dQ}{r^2}. \tag{22–6a}$$

Then the electric field, **E**, at any point is obtained by summing over all the infinitesimal contributions, which is the integral

$$\mathbf{E} = \int d\mathbf{E}. \tag{22–6b}$$

Note that $d\mathbf{E}$ is a vector (Eq. 22–6a gives its magnitude), so the summation process can be rather complicated. Because of this, other techniques (discussed in the next two chapters) are often used to calculate **E**. Nonetheless, Eqs. 22–6 are often useful in relatively simple situations to obtain **E** analytically. Numerical integration can also be used in many cases.

EXAMPLE 22–6   A thin, ring-shaped object of radius a holds a total charge Q distributed uniformly around it. Determine the electric field at a point P on its axis, a distance x from the center. See Fig. 22–16.

[†] Because we believe there is a minimum charge (e), the treatment here is only for convenience; it is nonetheless useful and accurate since e is usually very much smaller than macroscopic charges.

FIGURE 22–16

Example 22–6.

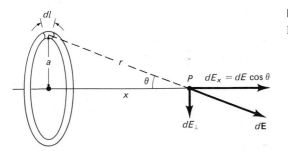

SOLUTION   The electric field, dE, due to a particular segment of the ring of length dl has magnitude

$$dE = \frac{1}{4\pi\varepsilon_0}\frac{dQ}{r^2}$$

where, since the whole ring has length (circumference) of $2\pi a$, $dQ = Q(dl/2\pi a)$; hence

$$dE = \frac{1}{4\pi\varepsilon_0}\frac{Q\,dl}{r^2 2\pi a}.$$

The vector $d\mathbf{E}$ has components dE_x along the x axis and $dE_\perp$ perpendicular to the x axis (Fig. 22–16). We are going to sum (integrate) around the entire ring; and we note that an equal-length segment diametrically opposite the dl shown will produce a $d\mathbf{E}$ whose component perpendicular to the x axis will just cancel the $dE_\perp$ shown. This is true for all segments of the ring, so $\mathbf{E}$ will be directed along the x axis and we need only sum the x components, dE_x. The total field is then

$$E = \int dE_x = \int dE\cos\theta = \frac{Q}{4\pi\varepsilon_0}\frac{1}{2\pi a}\int\frac{dl}{r^2}\cos\theta.$$

Since $\cos\theta = x/r$, where $r = (x^2 + a^2)^{1/2}$, we have

$$E = \frac{Q}{(4\pi\varepsilon_0)(2\pi a)}\frac{x}{(x^2 + a^2)^{3/2}}\int_0^{2\pi a} dl = \frac{1}{4\pi\varepsilon_0}\frac{Qx}{(x^2 + a^2)^{3/2}}.$$

At great distances, $x \gg a$, this reduces to $E = Q/4\pi\varepsilon_0 x^2$. We would expect this result since at great distances the ring would appear to be a point charge ($1/r^2$ dependence). In fact this "checking" at the limit of large r serves as an indication (but not proof) of the correctness of the answer (if the result did not check at large r, the result no doubt would be wrong entirely).

EXAMPLE 22–7   *Long line of charge.*   Determine the magnitude of the electric field at any point P a distance x from a very long line (a wire, say) of uniformly distributed charge, Fig. 22–17. Assume x is much smaller than the length of the wire, and let λ be the charge per unit length (C/m).

SOLUTION   We set up a coordinate system so the wire is on the y axis with origin O as shown. A length of wire dy has charge $dQ = \lambda\,dy$; the field $d\mathbf{E}$ at P due to such a length of wire at y has magnitude

$$dE = \frac{1}{4\pi\varepsilon_0}\frac{\lambda\,dy}{(x^2 + y^2)}.$$

The vector $d\mathbf{E}$ has components dE_x and dE_y as shown where $dE_x = dE\cos\theta$ and $dE_y = dE\sin\theta$. If the wire is extremely long in both directions from O (so distant contributions have little effect compared to nearby ones), or if O is at the midpoint of the wire (even if the wire is short), the y component of $\mathbf{E}$ will be zero since there will be equal contributions to $E_y = \int dE_y$ from above and below point O, so

$$E_y = \int dE\sin\theta = 0.$$

Then we have

$$E = E_x = \int dE\cos\theta = \frac{\lambda}{4\pi\varepsilon_0}\int\frac{\cos\theta\,dy}{(x^2 + y^2)}.$$

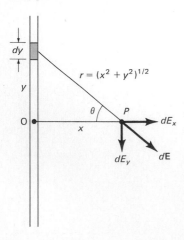

FIGURE 22–17

Example 22–7.

We must now write θ as a function of y, or y as a function of θ. We do the latter: since $y = x \tan \theta$, $dy = x\, d\theta/\cos^2 \theta$ and $(x^2 + y^2) = x^2/\cos^2 \theta$. Then

$$E = \frac{\lambda}{4\pi\varepsilon_0} \frac{1}{x} \int_{-\pi/2}^{\pi/2} \cos \theta \, d\theta = \frac{\lambda}{4\pi\varepsilon_0 x} (\sin \theta)\bigg|_{-\pi/2}^{\pi/2} = \frac{1}{2\pi\varepsilon_0} \frac{\lambda}{x},$$

where we have assumed the wire is extremely long in both directions ($y \to \pm\infty$) which corresponds to the limits $\theta = \pm\pi/2$. Thus the field of a long straight line of charge decreases inversely as the first power of the distance from the wire. This result, obtained for an infinite wire, is a good approximation for a wire of finite length as long as x is small compared to the distance of P from the ends of the wire.

EXAMPLE 22–8 *Uniform plane of charge.* Charge is distributed uniformly over a large square plane of side L, as shown in Fig. 22–18. The charge per unit area is σ. Calculate the electric field at a point P a distance z above the center of the plane, where z is much less than L.

SOLUTION Since we already know the field due to a long line of charge (Example 22–7), let us divide the plane into long narrow strips of width dy and length L, and then sum the fields due to each strip to get the total field. Since σ is the charge per unit area, for each strip $\sigma = dQ/L\,dy$ where dQ is the charge on each strip and $L\,dy$ is the area of the strip. Thus $dQ = \sigma L\,dy$, and each strip can be considered a line of charge with charge per unit length of

$$\lambda = \frac{dQ}{L} = \frac{\sigma L\,dy}{L} = \sigma\,dy.$$

The field due to the strip shown is, using the result of Example 22–7 (since $z \ll L$),

$$dE = \frac{1}{2\pi\varepsilon_0} \frac{\lambda}{(z^2 + y^2)^{1/2}} = \frac{1}{2\pi\varepsilon_0} \frac{\sigma\,dy}{(z^2 + y^2)^{1/2}}$$

where the distance of P to the center of the strip is $(z^2 + y^2)^{1/2}$. The plane is symmetric about a line through the center, so when we sum over all the strips that make up the plane, the y components will vanish. Hence we need sum over only the z components, where

$$dE_z = dE \cos \theta = dE \frac{z}{(z^2 + y^2)^{1/2}}.$$

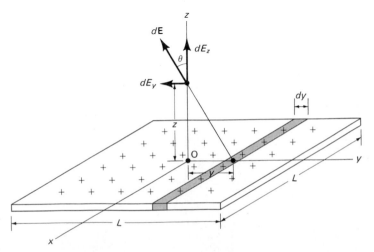

FIGURE 22–18

Calculation of electric field above a uniformly charged plane.

The total field at z is thus

$$E = \int dE_z = \frac{\sigma z}{2\pi\varepsilon_0} \int_{-L/2}^{L/2} \frac{dy}{(z^2 + y^2)} = \frac{\sigma z}{2\pi\varepsilon_0} \left(\frac{1}{z} \tan^{-1} \frac{y}{z} \right)\Bigg|_{y=-L/2}^{y=L/2}.$$

If we consider only points for which $z \ll L$, distant contributions will be small (Coulomb's law) and we can effectively let $y = \pm L/2 \to \pm\infty$ compared to z; so[†]

$$E = \frac{\sigma}{2\pi\varepsilon_0} \left(\frac{\pi}{2} + \frac{\pi}{2} \right) = \frac{\sigma}{2\varepsilon_0}.$$

This result is valid for any point above (or below) an infinite plane. It is also valid for points close to a finite plane, as long as the point is close to the plane compared to the distance to the plane's edge. Thus the field near a large, uniformly charged plane is uniform and directed outward.

EXAMPLE 22–9   Determine the electric field between two large parallel plates, separated by a distance d, which is small compared to the dimensions of the plates, when one plate carries a uniform surface charge density σ and the other carries a uniform surface charge density $-\sigma$, Fig. 22–19.

SOLUTION   From Example 22–8, each plate sets up an electric field of magnitude $E = \pm\sigma/2\varepsilon_0$. The field due to the positive plate points outward whereas that due to the negative plate points inward. Hence, in the region between the plates, the fields add together as shown:

$$E = E_+ + E_- = \frac{\sigma}{2\varepsilon_0} + \frac{\sigma}{2\varepsilon_0} = \frac{\sigma}{\varepsilon_0}. \qquad \text{[between parallel plates]}$$

Since the field is uniform, this result is valid for any point, whether near one or the other of the plates, or midway between them. Outside the plates, the fields cancel,

$$E = E_+ + E_- = \frac{\sigma}{2\varepsilon_0} - \frac{\sigma}{2\varepsilon_0} = 0,$$

as shown in the diagram. These results are valid ideally for infinitely large plates; they are a good approximation for finite plates if the separation is much less than the dimensions of the plate and for points not too close to the edge. (These useful and extraordinary results illustrate the principle of superposition and its great power.)

[†] The accuracy of this result for large L depends on the value of $\tan^{-1}(L/2z)$. For example, if $(z/L) \lesssim 0.008$ the result above (for $L = \infty$ and $z/L = 0$) will be accurate to about 1 percent since in this case $\tan^{-1}(L/2z)$ is within 1 percent of $\pi/2$.

FIGURE 22–19

Example 22–9.

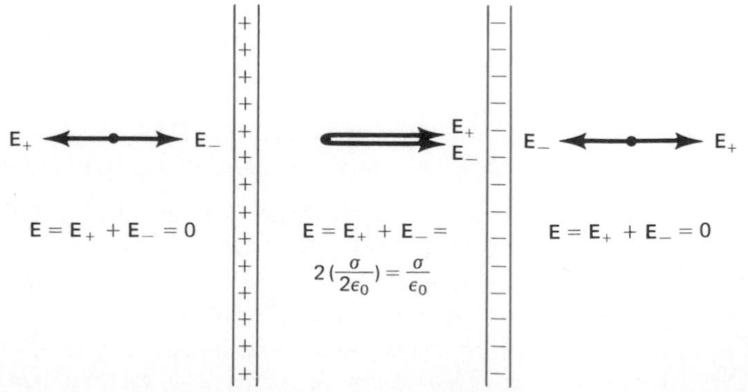

22–8 Lines of Force

Since the electric field is a *vector field*, we could indicate the electric field with arrows at various points in a given situation, such as at *a*, *b*, and *c* in Fig. 22–13; the directions of $\mathbf{E}_a$, $\mathbf{E}_b$, and $\mathbf{E}_c$ would be the same as that of the forces already shown, but the lengths (magnitudes) would be different (since we divide by *q*); however, the relative lengths of $\mathbf{E}_a$, $\mathbf{E}_b$, and $\mathbf{E}_c$ would be the same as for the forces since we divide by the same *q* each time. However, to indicate the electric field in such a way at many points would result in many arrows, which would appear confusing. To avoid this, we use another technique, that of lines of force.

In order to visualize the electric field, we draw a series of lines to indicate the direction of the electric field at various points in space. These **electric field lines**, or **lines of force**, are drawn so that they indicate the direction of the force due to the given field on a positive test charge. The lines of force due to a single positive charge are shown in Fig. 22–20a and for a single negative charge in Fig. 22–20b. In part *a*, the lines point radially outward from the charge, and in part *b* they point radially inward toward the charge; these are the directions of the force that would be exerted on a positive test charge in each case. Only a few representative lines have been shown. One could just as well draw lines in between those shown since the electric field exists there as well. However, we can (and will) always draw the lines so the number of lines starting on a positive charge, or ending on a negative charge, is proportional to the magnitude of the charge. Notice that near the charge, where the force is greatest, the lines are closer together. This is a general property of electric field lines; *the closer the lines are together, the stronger the electric field in that region.* In fact the lines can always be drawn so that the number crossing unit area perpendicular to $\mathbf{E}$ is proportional to the magnitude of the electric field. For a single point charge (Fig. 22–20), for example, the magnitude of the field decreases as $1/r^2$; so too does the number of uniformly spaced field lines per unit area since the total number of lines doesn't change with *r* but the total area increases as $4\pi r^2$ (surface area of sphere of radius *r*). Thus the number of lines per unit area is proportional to $1/r^2$ as well.

Figure 22–21a shows the electric field lines surrounding two charges of opposite sign. The electric field lines are curved in this case and they are directed from the positive charge to the negative charge. The direction of the field at any point is directed tangentially as shown by the faint arrow at point *P*. To satisfy yourself that this is the correct pattern for the electric field lines, you can make a few calculations such as those done in Example 22–5 for just this case (see Fig. 22–15). Figures 22–21b and c show the electric field surrounding two positive charges and between two oppositely charged parallel plates. Notice that the electric field lines between the two plates are parallel and equally spaced, except near the edges. Thus, in the central region, the electric field has the same magnitude at all points and we can write

$$E = \text{constant}. \qquad \text{[between two closely spaced parallel plates]} \quad (22\text{–}7)$$

Although the field fringes near the edges (the lines curve), we can often ignore this, particularly if the separation of the plates is small compared to their size. This should be compared to the field of a single point charge where the field decreases as the square of the distance, Eq. 22–4.

We summarize the properties of lines of force as follows:

1. The lines of force indicate the direction of the electric field; the field points in the direction tangent to the field lines at any point.

2. The lines are drawn so that the electric field, *E*, is proportional to the number of lines crossing unit area perpendicular to the lines.

3. Electric field lines start only on positive charges and end only on negative charges; and the number starting or ending is proportional to the magnitude of the charge.

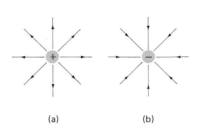

FIGURE 22–20

Electric field lines near (a) a positive point charge and (b) a negative point charge.

FIGURE 22–21

Electric field lines for three arrangements of charge.

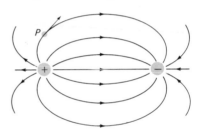

(a)

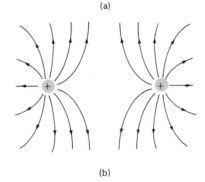

(b)

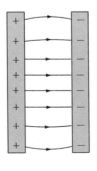

(c)

Another way of visualizing the electric field lines is that each represents the path that would be followed by a tiny test charge placed on it. (This is strictly true only if the test charge has no inertia, or moves slowly because of, say, a frictional force.) Note that no two field lines can cross. (If they did, it would mean the electric field would point in two different directions at the same point, which doesn't make sense.)

22–9 Electric Fields and Conductors

The electric field inside a good conductor is zero in the static situation—that is, when the charges are at rest. If there were an electric field within a conductor, there would be a force on its free electrons; so they would move until they reached positions where the electric field, and therefore the force on them, did become zero.

This reasoning has some interesting consequences. For one, any net charge on a good conductor distributes itself on the outer surface. We can see why this is so from another point of view: For a negatively charged conductor, for example, we can imagine that the negative charges repel one another and race to the surface to get as far away from one another as possible. Another consequence is the following. Suppose that a positive charge Q is placed at the center of a cavity within an uncharged isolated metal conductor whose shape is a spherical shell, Fig. 22–22. Because there can be no field within the metal, the lines leaving the positive charge must end on negative charges on the inner surface of the metal. Thus an equal amount of negative charge, $-Q$, is induced on the inner surface of the spherical shell; and a positive charge, $+Q$, of the same magnitude must then exist on the outer surface (since the shell is neutral). Thus, although no field exists in the metal itself, an electric field exists outside of it as shown in Fig. 22–22 as if the metal were not even there.

A related property of electric fields and conductors is that the electric field is always perpendicular to the surface of a conductor. If there were a component of **E** parallel to the surface, electrons at the surface would move along the surface in response to this force, until they reached positions where no force was exerted on them; that is, until the electric field was perpendicular to the surface.

These properties pertain only to conductors. In a nonconductor, which does not have free electrons, an electric field can exist; and the electric field does not necessarily make an angle of 90° to its surface.

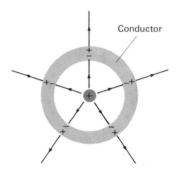

Conductor

FIGURE 22–22

A charge placed inside a spherical shell. Charges are induced on the conductor surfaces. The electric field exists even beyond the shell but not within the conductor itself.

22–10 Motion of a Charged Particle in an Electric Field

The field concept, as we have seen, is useful for describing how one or more electrically charged objects exert a force on another charged object: the force on an object of charge q at any point in space is given by

$$\mathbf{F} = q\mathbf{E}$$

(see Eq. 22–3) where **E** is the electric field at that point due to all other charged objects.

In the past few sections we have seen how to determine **E** for some particular situations. Now let us suppose we know **E** and we want to find the force on a charged object and its subsequent motion.

EXAMPLE 22–10 An electron (mass $m = 9.1 \times 10^{-31}$ kg) is accelerated in the uniform field **E** ($E = 2.0 \times 10^4$ N/C) between two parallel charged plates. The separation of the plates is 1.5 cm. The electron is accelerated from rest near the negative plate and passes through a tiny hole in the positive plate, Fig. 22–23. (a) With what speed does it leave the hole? (b) Show that the gravitational force can be ignored.

SOLUTION   (a) The magnitude of the force on the electron is

$$F = qE$$

and is directed to the right. The magnitude of the electron's acceleration is

$$a = \frac{F}{m} = \frac{qE}{m}.$$

Between the plates **E** is uniform so the electron undergoes uniformly accelerated motion with acceleration

$$a = \frac{(1.6 \times 10^{-19}\ \text{C})(2.0 \times 10^4\ \text{N/C})}{(9.1 \times 10^{-31}\ \text{kg})} = 3.5 \times 10^{15}\ \text{m/s}^2.$$

It travels a distance $x = 1.5 \times 10^{-2}$ m before reaching the hole, and since its initial speed was zero, we can use Eq. 2–9c with $v_0 = 0$:

$$v = \sqrt{2ax} = \sqrt{2(3.5 \times 10^{15}\ \text{m/s}^2)(1.5 \times 10^{-2}\ \text{m})} = 1.0 \times 10^7\ \text{m/s}.$$

There is no electric field outside the plates, so after passing through the hole, the electron moves with this speed, which is now constant.

(b) The magnitude of the electric force on the electron is $qE = (1.6 \times 10^{-19}\ \text{C})(2.0 \times 10^4\ \text{N/C}) = 3.2 \times 10^{-15}$ N. The gravitational force is $mg = (9.1 \times 10^{-31}\ \text{kg})(9.8\ \text{m/s}^2) = 8.9 \times 10^{-30}$ N, which is 10^{14} times smaller! Note that the electric field due to the electron does not enter the problem. (It would if we wanted to know the force on the parallel plates.)

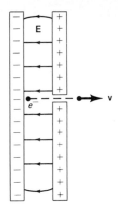

FIGURE 22–23
Example 22–10.

EXAMPLE 22–11   Suppose an electron (say from the above example) traveling with speed $v_0 = 1.0 \times 10^7$ m/s enters a uniform electric field **E** at right angles to $\mathbf{v}_0$ as shown in Fig. 22–24. Describe its motion by giving the equation of its path while in the electric field.

SOLUTION   When the electron enters the electric field (at $x = y = 0$) it has velocity $\mathbf{v}_0 = v_0\mathbf{i}$ in the x direction. The electric field **E**, pointing vertically downward, imparts a uniform vertical acceleration to the electron of

$$a_y = \frac{F}{m} = -\frac{qE}{m}.$$

Its vertical position is given by

$$y = \frac{1}{2} a_y t^2 = -\frac{qE}{2m} t^2$$

and its horizontal position by

$$x = v_0 t$$

since $a_x = 0$. We eliminate t between these two equations and obtain

$$y = -\frac{qE}{2mv_0^2} x^2$$

which is the equation of a parabola.

FIGURE 22–24
Examples 22–11 and 22–12.

EXAMPLE 22–12   At what angle will the electrons in Example 22–11 leave the uniform electric field at the end of the parallel plates (point P in Fig. 22–24)? Assume the plates are 6.0 cm long and $E = 5.0 \times 10^3$ N/C.

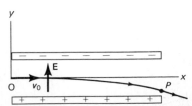

SOLUTION At point P, the y component of velocity has magnitude (see Eq. 2–9c)

$$v_y = \sqrt{2a_y y} = \sqrt{\left(\frac{2qE}{m}\right)\left(\frac{qEx^2}{2mv_0^2}\right)} = \frac{qEx}{mv_0}$$

whereas

$$v_x = v_0 = \text{constant.}$$

Then

$$\tan\theta = \frac{v_y}{v_x} = \frac{qEx}{mv_0^2}$$

$$= \frac{(1.6 \times 10^{-19}\,\text{C})(5.0 \times 10^3\,\text{N/C})(6.0 \times 10^{-2}\,\text{m})}{(9.1 \times 10^{-31}\,\text{kg})(1.0 \times 10^7\,\text{m/s})^2} = 0.53,$$

so $\theta = 28°$ below the horizontal.

22–11 Electric Dipoles

The combination of two equal charges of opposite sign, $+Q$ and $-Q$, separated by a distance l, is referred to as an **electric dipole**. The quantity Ql is called the **dipole moment** and is represented by the symbol p. Many molecules, such as the diatomic molecule CO, have a dipole moment (C has a small positive charge and O a small negative charge); even though the molecule as a whole is neutral, there is a separation of charge that results from an uneven sharing of electrons by the two atoms. (Symmetric diatomic molecules, like O_2, have no dipole moment.)

First let us consider a dipole, of dipole moment $p = Ql$, that is placed in a uniform electric field $\mathbf{E}$, as shown in Fig. 22–25. The dipole moment can be considered to be a vector $\mathbf{p}$, of magnitude Ql, that points from the negative to the positive charge as shown. If the field is uniform, the force $Q\mathbf{E}$ on the positive charge, and the force $-Q\mathbf{E}$ on the negative charge results in no net force on the dipole. There will, however, be a *torque* on the dipole which has magnitude (calculated about the center, O, of the dipole)

$$\tau = QE\frac{l}{2}\sin\theta + QE\frac{l}{2}\sin\theta = pE\sin\theta; \qquad (22\text{–}8a)$$

this can be written in vector notation as

$$\boldsymbol{\tau} = \mathbf{p} \times \mathbf{E}. \qquad (22\text{–}8b)$$

The effect of the torque is to try to turn the dipole so $\mathbf{p}$ is parallel to $\mathbf{E}$. The work, W, done on the dipole by the electric field to change the angle θ from θ_1 to θ_2, is given by (see Eq. 9–20):

$$W = \int_{\theta_1}^{\theta_2} \tau\,d\theta = pE\int_{\theta_1}^{\theta_2} \sin\theta\,d\theta = pE(\cos\theta_1 - \cos\theta_2).$$

Work done by the field decreases the potential energy, U, of the dipole in this field; if we choose $U = 0$ when $\mathbf{p}$ is perpendicular to $\mathbf{E}$ ($\theta = 90°$), then

$$U = -W = -pE\cos\theta = -\mathbf{p}\cdot\mathbf{E}. \qquad (22\text{–}9)$$

If the electric field is *not* uniform, the force on the $+Q$ of the dipole may not have the same magnitude as the force on the $-Q$, so there may be a net force, as well as a torque.

We have just seen how an external electric field affects an electric dipole. Now let us look at another aspect of dipoles. Let us suppose that there is no external field, and let us determine the nature of the electric field produced *by* the

FIGURE 22–25

An electric dipole in a uniform electric field.

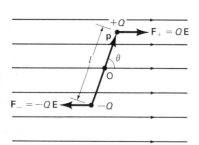

dipole (and which can affect something else). For brevity, we restrict ourselves to points that are on the perpendicular bisector of the dipole, such as point P in Fig. 22–26 which is a distance r above the midpoint of the dipole. [Note that r in Fig. 22–26 is not the distance from either charge to P; the latter distance is $(r^2 + l^2/4)^{1/2}$ and this is what must be used in Eq. 22–4.] The total field at P is

$$\mathbf{E} = \mathbf{E}_+ + \mathbf{E}_-,$$

where $\mathbf{E}_+$ and $\mathbf{E}_-$ are the fields due to the $+$ and $-$ charge respectively; E_+ and E_- have equal magnitude:

$$E_+ = E_- = \frac{1}{4\pi\varepsilon_0} \frac{Q}{(r^2 + l^2/4)}.$$

Their y components cancel at point P, so the magnitude of $\mathbf{E}$ is

$$E = 2E_+ \cos\phi = \frac{1}{2\pi\varepsilon_0} \frac{Q}{(r^2 + l^2/4)} \frac{l}{2(r^2 + l^2/4)^{1/2}}$$

or

$$E = \frac{1}{4\pi\varepsilon_0} \frac{p}{(r^2 + l^2/4)^{3/2}}. \qquad \begin{bmatrix} \text{on perpendicular bisector} \\ \text{of dipole} \end{bmatrix} \quad (22\text{–}10)$$

Far from the dipole, $r \gg l$, this reduces to

$$E = \frac{1}{4\pi\varepsilon_0} \frac{p}{r^3}. \qquad \begin{bmatrix} \text{on perpendicular bisector} \\ \text{of dipole; } r \gg l \end{bmatrix} \quad (22\text{–}11)$$

So the field decreases more rapidly for a dipole than for a single point charge ($1/r^3$ versus $1/r^2$), which we expect since at large distances the two opposite charges appear so close together as to neutralize each other. This $1/r^3$ dependence also applies for points not on the perpendicular bisector (see the problems).

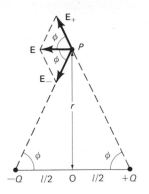

FIGURE 22–26
Electric field due to an electric dipole.

 Summary ⎯⎯⎯⎯⎯⎯⎯⎯⎯⎯⎯⎯⎯⎯⎯⎯⎯⎯⎯⎯⎯⎯⎯⎯⎯⎯⎯⎯⎯⎯⎯⎯⎯⎯⎯⎯

There are two kinds of *electric charge*, positive and negative. These designations are to be taken algebraically— that is, any charge is plus or minus so many coulombs (C), in SI units. Electric charge is *conserved*: if a certain amount of one type of charge is produced in a process, an equal amount of the opposite type is also produced on the same body or on a different body; thus the *net charge* produced is zero. According to the atomic theory, electric charge originates in the atom, which consists of a positively charged nucleus surrounded by negatively charged electrons. Each electron has a charge $-e = -1.6 \times 10^{-19}$ C. Conductors are those materials in which many electrons are relatively free to move, whereas electric insulators are those in which very few electrons are free to move. An object is negatively charged when it has an excess of electrons, and positively charged when it has less than its normal amount of electrons. An object can become charged in three ways: by rubbing, in which electrons are transferred from one material to another; by conduction, which is transfer of charge from one charged object to another by touching; or by induction, the separation of charge within an object because of the close approach of another charged object but without touching.

Electric charges exert a force on each other. If two charges are of opposite types, one positive and one negative, they each exert an attractive force on the other. If the two charges are the same type, each repels the other. The magnitude of the force one point charge exerts on another is proportional to the product of their charges and inversely proportional to the square of the distance between them:

$$F = \frac{1}{4\pi\varepsilon_0} \frac{Q_1 Q_2}{r^2};$$

this is *Coulomb's law*.

An *electric field* is imagined to exist in space due to any charge or group of charges. The force on another charged object is then conceived to be due to the electric field present at its location. The *electric field*, $\mathbf{E}$, at any point in space due to one or more charges is defined as the force per unit charge that would act on a test charge q placed at that point: $\mathbf{E} = \mathbf{F}/q$. Electric fields are represented by *lines of force* which start on positive charges and end on negative charges. Their direction indicates the direction the force would be on a tiny positive test charge placed at a point; the lines can be drawn so that the number per unit area is proportional to E. The static electric field (that is, no charges moving) inside a good conductor is zero, and the electric field lines just outside a charged conductor are perpendicular to its surface.

An *electric dipole* is a combination of two equal but opposite charges, $+Q$ and $-Q$, separated by a distance l. The dipole moment is $p = Ql$. A dipole placed in a uniform electric field feels no net force but does feel a net torque (unless **p** is parallel to **E**). The electric field produced by a dipole decreases as the third power of the distance r from the dipole ($E \propto 1/r^3$) for r large compared to l.

 Questions ───

1. If you charge a pocket comb by rubbing with a silk scarf, how can you determine if the comb is positively or negatively charged?

2. Why does a phonograph record attract dust just after it has been wiped clean?

3. Explain why fog or rain droplets tend to form around ions or electrons in the air.

4. Can you guess why trucks carrying flammable fluids drag a chain along the ground?

5. Contrast the *net charge* on a conductor to the "free charges" in the conductor.

6. Can a charge be induced in an insulator as it can be for a conductor (Figs. 22–4 and 22–5)?

7. Figures 22–4 and 22–5 show how a charged rod placed near an uncharged metal object can attract (or repel) electrons. There are a great many electrons in the metal, yet only some of them move as shown. Why not all of them?

8. When an electroscope is charged, the two leaves repel each other and remain at an angle. What balances the electric force of repulsion so that the leaves don't separate further?

9. Why does a shirt or blouse taken from a clothes dryer sometimes cling to your body?

10. A plastic ruler, rubbed with a cloth, is brought close to a neutral piece of paper, which it attracts. Draw a diagram showing the separation of charge and explain why attraction occurs.

11. The form of Coulomb's law is very similar to that for Newton's law of universal gravitation. What are the differences between these two laws? Compare also gravitational mass and electric charge.

12. We are not normally aware of the gravitational or electrical force between two ordinary objects. What is the reason in each case? Give an example where we are aware of each one and why.

13. Is the electric force a conservative force? Why or why not? (See Chapter 7.)

14. Is there an attraction between the two objects in Fig. 22–5b? Why or why not?

15. What experimental observations mentioned in the text rule out the possibility that the numerator in Coulomb's law contains the quantity $(Q_1 + Q_2)$ rather than the product $Q_1 Q_2$?

16. A negatively charged ruler attracts an object suspended as in Fig. 22–2. Does the object necessarily have a positive charge? If the negative ruler repels a suspended object, must the object be negatively charged?

17. When a charged ruler attracts small pieces of paper, sometimes a piece jumps quickly away after touching the ruler. Explain.

18. When determining an electric field, must we use a *positive* test charge, or would a negative one do as well? Explain.

19. We wish to determine the electric field at a point near a positively charged metal sphere (a good conductor). We do so by bringing a small test charge, q_0, to this point and measure the force F_0 on it. Will F_0/q_0 be greater than, less than, or equal to, the electric field **E** as it was at that point before the test charge was present? What about **E** at that point when q_0 *is* there?

20. Explain why we use *small* test charges when measuring electric fields.

21. Given two point charges, Q and $2Q$, a distance l apart, is there a point along the straight line that passes through them where $E = 0$ when their signs are (a) opposite, (b) the same? If yes, state roughly where this point will be.

22. Suppose the ring of Fig. 22–16, has a uniformly distributed negative charge Q. What is the magnitude and direction of **E** at point P?

23. Explain why the superposition principle holds for the electric field, Eq. 22–5.

24. Consider a small positive test charge located on an electric field line at some point, such as point P in Fig. 22–21a. Is the direction of the velocity and/or acceleration of the test charge along this line? Discuss.

25. Show, using the three rules for field lines given in Section 22–8, that the electric field lines starting or ending on a single point charge must be symmetrically spaced around the charge.

26. A negative point charge is placed exactly at the midpoint of the line joining two equal positive point charges. What will be the motion of the negative charge? Is it in equilibrium? If so, what type? What if the negative charge was replaced by a positive one?

27. Draw the electric field lines surrounding two negative electric charges a distance l apart.

28. Assume that the two opposite charges in Fig. 22–21a are 12.0 cm apart. Consider the magnitude of the electric field 2.5 cm from the positive charge. On which side of this charge—top, bottom, left, or right—is the electric field the strongest? The weakest?

29. Why can lines of force never cross?

30. In what ways does the electron motion in Example 22–11 resemble projectile motion (Section 3–8)? In which ways not?

31. Describe the motion of the dipole shown in Fig. 22–25 if it is released from rest at the position shown.

32. Explain why there can be a net force on an electric dipole placed in a nonuniform electric field.

SECTION 22–5

1. (I) How many electrons make up a charge of 100 μC?

2. (I) Two charged bodies exert a force of 480 mN on each other. What will be the force if they are moved so they are only one-eighth as far apart?

3. (I) What is the magnitude of the electric force of attraction between an iron nucleus ($q = +26e$) and its innermost electron if the distance between them is 1.0×10^{-12} m?

4. (I) What is the total charge on all the electrons in 1.0 kg of H_2O?

5. (II) How close must two electrons be if the electric force between them is equal to the weight of either at the earth's surface?

6. (II) Three charged particles are placed at the corners of an equilateral triangle of side 1.40 m. The charges are $+4.0$ μC, -3.0 μC, and -5.0 μC. Calulate the magnitude and direction of the net force on each due to the other two.

7. (II) Particles of charge $+88$ μC, -55 μC, and $+70$ μC are placed in a line. The center one is 0.75 m from each of the others. Calculate the net force on each due to the other two.

8. (II) A charge of $+0.0050$ C is placed at each corner of a square 1.15 m on a side. Determine the magnitude and direction of the force on each charge.

9. (II) Repeat problem 8 for the case when two of the charges, on opposite corners, are replaced by negative charges of the same magnitude.

10. (II) In a simple model of the hydrogen atom, the electron revolves in a circular orbit around the proton with a speed of 1.1×10^6 m/s. What is the radius of the electron's orbit?

11. (II) A -8.0-μC and a $+1.8$-μC charge are placed 11.8 cm apart. Where can a third charge be placed so that it experiences no net force?

12. (II) Two point charges have a total charge of 880 μC. When placed 1.10 m apart, the force each exerts on the other is 22.8 N and is repulsive. What is the charge on each? What if the force were attractive?

13. (II) Two point charges are a fixed distance apart. The sum of their charges is Q_T. What charge must each have to maximize the electric force between them? To minimize it?

14. (II) A 3.0-g copper penny has a positive charge of 0.55 mC. What fraction of its electrons has it lost?

15. (II) A large electroscope is made with "leaves" that are 60 cm long wires with 25-g balls at the ends. When charged, nearly all the charge resides on the balls. If the wires each make a 30° angle with the vertical, what total charge must have been applied to the electroscope?

16. (II) Two charges, $-Q_0$ and $-3Q_0$, are a distance l apart. These two charges are free to move but do not because there is a third charge nearby. What must be the charge and placement of the third charge for the first two to be in equilibrium?

17. (III) On each corner of a cube of side l there is a point charge Q. What is the force on each charge due to the others?

18. (III) The two strands of the helix-shaped DNA molecule (the genetic material in all cells) are held together by elec-

trostatic forces as shown in Fig. 22–27. Assume that the charge indicated on H and N atoms is 0.2e, and on the indicated C and O atoms is 0.4e, that atoms on each molecule are separated by 1.0×10^{-10} m, and that all relevant angles are 120°. Estimate the net force between (a) a thymine and an adenine; and (b) between a cytosine and a guanine. (c) Estimate the total force for a DNA molecule containing 10^5 pairs of such molecules.

SECTION 22–7

19. (I) What is the magnitude and direction of the electric field 35.0 cm directly above a 35.0×10^{-4} C charge?

FIGURE 22–27

(a)

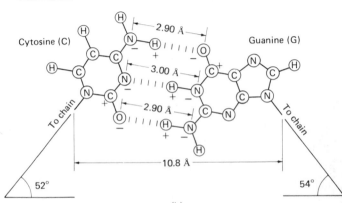

(b)

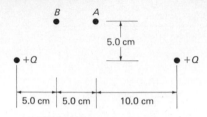

FIGURE 22–28

20. (I) What is the magnitude and direction of the electric field at a point midway between a $-20\text{-}\mu C$ and a $+60\text{-}\mu C$ charge 40 cm apart?

21. (I) A proton ($m = 1.67 \times 10^{-27}$ kg) is suspended at rest in a uniform electric field $\mathbf{E}$. Take into account gravity and determine $\mathbf{E}$.

22. (II) You are given two unknown point charges, Q_1 and Q_2. At a point on the line joining them, one-third of the way from Q_1 to Q_2, the electric field is zero. What can you say about these two charges?

23. (II) Use Coulomb's law to determine the electric field at A and B in Fig. 22–28 due to the two positive charges ($Q = 4.0 \ \mu C$) shown. Calculate magnitude and direction. Is your result consistent with Fig. 22–21b?

24. (II) Calculate the electric field at the center of a square 25 cm on a side if one corner is occupied by a $+33.0\text{-}\mu C$ charge and the other three are occupied by $-21.0\text{-}\mu C$ charges.

25. (II) Calculate the electric field at one corner of a square 80 cm on a side if the other three corners are occupied by 18.2×10^{-4} C charges.

26. (II) Calculate the magnitude of the electric field at the center of a square with 10 cm sides if the corners, taken in rotation, have charges of 1.0 μC, 2.0 μC, 3.0 μC, and 4.0 μC.

27. (II) At what position, $x = x_M$, is the field along the axis of the ring of Example 22–6 a maximum?

28. (II) Show that if the uniformly charged straight wire in Fig. 22–17 has the length L, and point O is at the mid-point, the field at point P, a perpendicular distance x from O, is given by

$$E = \frac{\lambda}{2\pi\varepsilon_0} \frac{L}{x(L^2 + 4x^2)^{1/2}},$$

where λ is the charge per unit length.

29. (II) A circular disc of radius a carries a uniformly distributed charge Q. (a) Determine $\mathbf{E}$ as a function of r along the axis of the disc. (Hint: use the result of Example 22–6.) (b) Show, for $a \to \infty$, that the electric field has magnitude $E = \sigma/2\varepsilon_0$, where σ is the charge per unit area, independent of distance (that is, $\mathbf{E}$ is uniform).

30. (II) Suppose the charge on the ring of Fig. 22–16 was all distributed uniformly on only the upper half of the ring. Determine the electric field $\mathbf{E}$ at P. (Take y vertically upward.)

31. (III) Derive a formula for the electric field at a distance z above the center of the plane in Fig. 22–18 for the general case when z is not much less than L.

32. (III) Suppose the uniformly charged wire of Fig. 22–17 starts at point O and rises vertically to a length L. (a) De-

termine the magnitude and direction of the electric field at point P, a distance x from O. (That is, calculate $\mathbf{E}$ near one end of a long wire.) (b) For $L = \infty$, show that $\mathbf{E}$ makes a $45°$ angle to the horizontal for any x.

33. (III) Suppose in Example 22–7 that $x = 0.250$ m, $Q = 2.0$ μC, and that the uniformly charged wire rises 2.0 m above point O and reaches 4.0 m below point O. (a) Calculate E_x and E_y. (b) Determine what the error would be if you simply used the result of Example 22–7, $E = \lambda/2\pi\varepsilon_0 x$; express this error as $(E_x - E)/E$ and E_y/E.

SECTION 22–8

34. (I) Electric field lines can always be drawn so that their number per unit area perpendicular to the lines is proportional to the electric field, intensity, E. However, if Coulomb's law were not valid—that is, if the field due to a single point did not fall off precisely as $1/r^2$ (so the exponent on r was not precisely 2)—this property of field lines would not be true. Show why. (Hint: take the example of a single point charge.)

35. (II) Draw, approximately, the electric field lines about two point charges, $+q$ and $-2q$, which are a distance l apart.

36. (II) Draw, approximately, the electric field lines emanating from a uniformly charged straight wire whose length l is not great. The spacing of lines near the wire should be somewhat less than l. (Hint: also consider points very far from the wire.)

SECTION 22–9

37. (II) Consider two thin, concentric, cylindrical metal shells (like a coaxial cable). The outer one has charge Q_1, the inner one $-Q_2$. On a diagram looking down the axis, show the electric field lines if (a) $Q_1 = Q_2$, (b) $Q_1 < Q_2$, (c) $Q_1 > Q_2$.

SECTION 22–10

38. (I) What is the electric field strength at a point in space where a proton ($m = 1.67 \times 10^{-27}$ kg) experiences an acceleration of 7.6×10^4 m/s^2?

39. (II) An electron with speed $v_0 = 2.4 \times 10^6$ m/s is traveling parallel to an electric field ($\mathbf{v}_0 \| \mathbf{E}$) of magnitude $E = 8.4 \times 10^3$ N/C. (a) How far will it travel before turning around? (b) How much time will elapse before it returns to its starting point?

40. (II) Suppose an electron enters the uniform electric field of Fig. 22–24 midway between the plates but moving at an upward $45°$ angle. What maximum speed can it have if it is to avoid striking the upper plate?

41. (II) Suppose the electron of Example 22–12 enters the electric field midway between the plates at an angle θ_0 to the horizontal. Its path is symmetrical, so it leaves at the same angle θ_0, and it just barely misses the top plate. What is θ_0?

42. (II) A water droplet of radius 0.020 mm remains stationary in the air. If the electric field of the earth is 100 N/C, how many excess electron charges must the water droplet have?

SECTION 22–11

43. (II) The H Cl molecule has a dipole moment of about 3.4×10^{-30} C·m. The two atoms are separated by about 1.0×10^{-10} m. (a) What is the net charge on each atom? (b) Is this equal to an integral multiple of e? If not, explain.

(c) What maximum torque would this dipole experience in a 2.5×10^4 N/C electric field? (d) How much energy would be needed to rotate one molecule $45°$ from its equilibrium position of lowest PE?

44. (II) Suppose both charges in Fig. 22–26 were positive. (a) Show that the field on the perpendicular bisector, for $r \gg l$, is given by $(1/4\pi\varepsilon_0)(2Q/r^2)$. (b) Explain why the field decreases at $1/r^2$ here whereas for a dipole it decreases as $1/r^3$.

45. (II) An electric dipole, of dipole moment p and moment of inertia I, is placed in a uniform electric field **E**. (a) If displaced by an angle θ as shown in Fig. 22–25 and released, under what conditions will it oscillate in simple harmonic motion? (b) What will be its frequency?

46. (II) Suppose a dipole **p** is placed in a nonuniform electric field $\mathbf{E} = E\mathbf{i}$ that points along the x axis. If E depends only on x, show that the net force on the dipole is

$$\mathbf{F} = \left(\mathbf{p} \cdot \frac{d\mathbf{E}}{dx}\right)\mathbf{i}$$

where $d\mathbf{E}/dx$ is the gradient of the field in the x direction.

47. (II) (a) Show that at points along the axis of a dipole (along the same line that contains $+Q$ and $-Q$), the electric field has magnitude

$$E = \frac{1}{4\pi\varepsilon_0}\frac{2p}{r^3}$$

for $r \gg l$ (Fig. 22–26) where r is the distance from the point to the center of the dipole. (b) In what direction does **E** point?

48. (II) One type of *electric quadrupole* consists of two dipoles placed end to end with their negative charges (say) overlapping; that is, in the center is $-2Q$ flanked (on a line) by a $+Q$ to either side. Determine the electric field **E** at points along the perpendicular bisector and show that E decreases as $1/r^4$.

49. (III) *Field of a dipole at any point.* Let point 0 in Fig. 22–26 be the origin of a rectangular xy coordinate system with x horizontal to the right and y vertically upward. (a) Show that at any distant point $P(x, y)$,

$$E_x = \frac{1}{4\pi\varepsilon_0}\frac{p(2x^2 - y^2)}{(x^2 + y^2)^{5/2}}$$

$$E_y = \frac{1}{4\pi\varepsilon_0}\frac{3pxy}{(x^2 + y^2)^{5/2}}.$$

(b) Show, in polar coordinates (r, θ, see Appendix C), that for $r \gg l$, the electric field has components

$$E_r = \frac{1}{4\pi\varepsilon_0}\frac{2p\cos\theta}{r^3},$$

$$E_\theta = \frac{1}{4\pi\varepsilon_0}\frac{p\sin\theta}{r^3}.$$

NUMERICAL/PROGRAMMABLE CALCULATOR

*50. (III) *Plotting electric field lines.* Write a program, along the following lines, that will enable you to plot the field lines for a given charge distribution in a plane. We will restrict ourselves to at most three point charges in the xy plane. The values of the charges (in coulombs) and their positions in the xy plane must be input variables: $(Q_a, x_a, y_a; Q_b, x_b, y_b; Q_c, x_c, y_c)$. Since we will want the position of points along a given field line, another input variable will be Δl, the interval of distance at which we will calculate the points along the line; Δl should be no more than $\frac{1}{50}$ of the distance between the nearest charges, and if the plot doesn't make sense, Δl will have to be made smaller. Also as input must be a point (x_1, y_1) near one of the charges. At this first point, the program should calculate the values of E_{1x} and E_{1y} from Coulomb's law, and then the angle $\theta_1 = \tan^{-1}(E_{1y}/E_{1x})$. This angle θ_1 will give the direction of the electric field at this point. The position of the next point along the line (call it (x_2, y_2)) will be a distance Δl from (x_1, y_1) at the angle θ (that is, $x_2 = x_1 + \Delta l \cos\theta_1$, $y_2 = y_1 + \Delta l \sin\theta_1$). Once (x_2, y_2) is determined the x and y components of **E** are determined. Then the next point (x_3, y_3) can be obtained in the same way, and so on up to some last point (x_N, y_N). (Should the value of N be an input of your choosing?) The calculated points (x_1, y_1), (x_2, y_2), (x_3, y_3), and so on, are then plotted (or more practically every fifth or tenth) and connected by a line. This, then, is one field line. To obtain other field lines, a new initial point (x_1, y_1) must be chosen as input. It is suggested that several different points, equally spaced around a given charge, be used as starting points. This can be done for each of the charges present (but remember that lines starting on one charge may end on another). In the table below are a number of possible configurations for which to run the program. (Consider having a check in the program to stop if a given line comes very close to a charge, or is beyond the region of interest.)

*51. (III) *Line of charge.* Design a program (along the lines of problem 50) that will give data points to plot the electric field lines emanating from a line of charge (a thin wire) like that of Fig. 22–17, but of finite length L. Plot at least six lines starting at equidistant points between one end and the center of the wire. This will give one-fourth of the field, but the other three-fourths follows from symmetry. The lines need to be calculated up to a distance of about L from the wire.

	Q_a	x_a	y_a	Q_b	x_b	y_b	Q_c	x_c	y_c	Comments
a.	+1	0	0	−1	10.0	0	0	0	0	This is a dipole. (Compare to Fig. 22–21a)
b.	+1	0	0	+1	10.0	0	0	0	0	Compare to Fig. 22–21b
c.	+2	0	0	−1	10.0	0	0	0	0	
d.	+1	0	0	+1	10.0	0	+1	5.0	8.66	An equilateral triangle.
e.	−3	0	0	−1	0	5.0	+2	5.0	0.0	

Gauss's Law

23

Gauss's law, which we develop and discuss in this chapter, is a statement of the relation between electric charge and electric field. It is a more general and elegant form of Coulomb's law.

We can, in principle, always determine the electric field due to a given distribution of electric charge using Coulomb's law. The total electric field at any point will be the vector sum (or integral) of contributions from all charges present (see Eqs. 22–5 and 22–6). Except for some simple cases, the sum or integral can be quite complicated to evaluate; for situations in which an analytic solution (such as we carried out in the examples of Section 22–7) is not possible, a computer can of course be used.

In some cases, however, the electric field due to a given charge distribution can be calculated more easily or more elegantly using Gauss's law, as we shall see (Section 23–3). But the major importance of Gauss's law is that it gives us additional insight into the nature of electrostatic fields, and a more general relationship between charge and field.[†]

Before discussing Gauss's law itself, we first discuss the concept of *flux*.

 ## 23–1 Electric Flux

Consider first a surface of area A through which a uniform electric field $\mathbf{E}$ passes, Fig. 23–1. The area could be a rectangle (as shown), a circle, or any other shape. If the electric field direction is perpendicular to the surface, as in Fig. 23–1a, the **electric flux**, Φ_E, through this surface is defined as the product

$$\Phi_E = EA.$$

If the (same) area A is not perpendicular to $\mathbf{E}$, but rather makes an angle θ as shown in Fig. 23–1b, fewer lines of force will pass through the area. In this case

<hr/>

[†] For this reason, this chapter may seem somewhat formal.

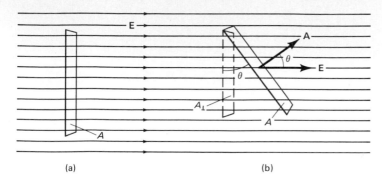

(a)   (b)

FIGURE 23–1

A uniform field **E** (indicated by the parallel field lines) passing through a surface of area A: (a) perpendicular to **E**; (b) not perpendicular to **E**. The dashed surface of area $A_\perp$, in (b), is the projection of A perpendicular to the field **E**.

we define the electric flux through the surface as

$$\Phi_E = EA_\perp = EA \cos\theta, \qquad [\textbf{E uniform}] \quad (23\text{–}1a)$$

where $A_\perp$ is the projection of the area A on a surface perpendicular to **E** as shown. The area A of a surface can be represented by a vector **A** whose magnitude is A and whose direction is perpendicular to the surface, as shown in Fig. 23–1b. The angle θ is the angle between **E** and **A**, so the electric flux can also be written

$$\Phi_E = \mathbf{E} \cdot \mathbf{A}. \qquad [\textbf{E uniform}] \quad (23\text{–}1b)$$

Because of how we defined it, the electric flux has a simple intuitive interpretation in terms of lines of force. We saw in Section 22–8 that field lines can always be drawn so the number (N) passing through unit area perpendicular to the field ($A_\perp$) is proportional to the magnitude of the field (E): $E \propto N/A_\perp$. Hence,

$$N \propto EA_\perp = \Phi_E,$$

so the flux through an area is proportional to the number of lines passing through the area.

Now let us consider the more general case, when the electric field **E** is not uniform and the surface is not flat, Fig. 23–2. We divide up the surface into n small elements of surface whose areas are $\Delta A_1, \Delta A_2, \dots \Delta A_n$. We choose the division so that (1) each ΔA_i is small enough that it can be considered flat, and (2) the electric field varies so little over this small area that it can be considered uniform. Then the electric flux through the entire surface is approximately

$$\Phi_E \approx \sum_{i=1}^{n} \mathbf{E}_i \cdot \Delta \mathbf{A}_i$$

where $\mathbf{E}_i$ is the field passing through $\Delta \mathbf{A}_i$. In the limit as we let $\Delta \mathbf{A}_i \to 0$, the sum becomes an integral over the entire surface and the relation becomes mathematically exact:

$$\Phi_E = \int \mathbf{E} \cdot d\mathbf{A}. \qquad (23\text{–}2)$$

In many cases (in particular, for Gauss's law) we deal with the flux through a closed surface—that is, a surface that completely encloses a volume (like a sphere or the surface of a football), Fig. 23–3. In this case, the net flux through the surface is given by

$$\Phi_E = \oint \mathbf{E} \cdot d\mathbf{A} \qquad [\text{closed surface}] \quad (23\text{–}3)$$

where the integral sign is written $\oint$ to indicate that the integral is over an enclosing surface.

FIGURE 23–2

Electric flux through a curved surface. One small area of the surface, $\Delta\mathbf{A}_i$, is indicated.

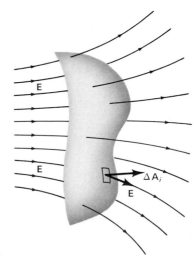

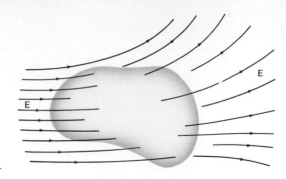

FIGURE 23–3

Electric flux through a closed surface.

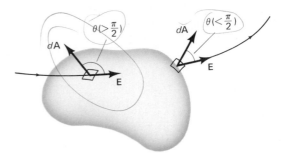

FIGURE 23–4

The direction of an element of area $d\mathbf{A}$ is taken to point outward from an enclosed surface.

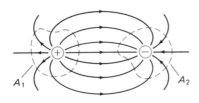

FIGURE 23–5

An electric dipole. Flux through surface A_1 is positive, through A_2 is negative.

FIGURE 23–6

Net flux through surface A_1 is negative.

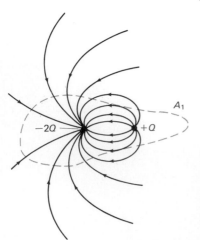

Up to now we have not been concerned with the fact that there is an ambiguity in the direction of the vector $\mathbf{A}$ that represents a surface; for example, in Fig. 23–1, the vector $\mathbf{A}$ could point upward and to the right (as shown) or downward to the left and still be perpendicular to the surface. For a closed surface, we define (arbitrarily) the direction of $\mathbf{A}$, or of $d\mathbf{A}$, to point *outward* from the enclosed volume, Fig. 23–4. For a line leaving the enclosed volume (on the right in Fig. 23–4), the angle θ between $\mathbf{E}$ and $d\mathbf{A}$ must be less than $\pi/2$ ($=90°$), so $\cos\theta > 0$; for a line entering the volume (on the left in Fig. 23–4) $\theta > \pi/2$, so $\cos\theta < 0$. Hence, *flux entering the enclosed volume is negative* ($\int E\cos\theta\,dA < 0$), whereas *flux leaving the volume is positive*. Consequently, Eq. 23–3 gives the net flux *out* of the volume. If Φ_E is negative, there is a net flux into the volume.

In Fig. 23–3, the number of lines that enter the volume is equal to the number of lines that leave. Hence $\Phi_E = 0$; there is no net flux out of this surface. The flux, $\oint \mathbf{E}\cdot d\mathbf{A}$ will be nonzero only if some lines start or end within the surface; since electric field lines start and stop only on electric charges, the flux will be nonzero only if the surface encloses a net charge. For example, the surface labeled A_1 in Fig. 23–5 encloses a positive charge and there is a net outward flux through this surface ($\Phi_E > 0$); the surface A_2 encloses an equal magnitude negative charge and there is a net inward flux ($\Phi_E < 0$). For the configuration shown in Fig. 23–6, the flux through the surface shown is negative (count the lines). The value of Φ_E depends on the charge, and this is what Gauss's law is all about.[†]

[†] The concept of flux applies equally well to fluid flow, and thus makes an interesting analogy. (Indeed, the word "flux" comes from the Latin word for "flow".) The electric field $\mathbf{E}$ at each point corresponds to the fluid flow velocity $\mathbf{v}$, so the electric field lines correspond to streamlines of a fluid flow. The flux, Φ, through a surface for a fluid is the volume rate of flow and is given by $\Phi = \int \mathbf{v}\cdot d\mathbf{A}$. In Figs. 23–1, 23–2, and 23–3, the lines can correspond to the steady streamline flow of a fluid with no sources (such as a faucet) or sinks (such as a leak or drain). In this case, the net flux through a closed surface, as in Fig. 23–3, is zero: what flows in also flows out. In Figs. 23–5 and 23–6, there are sources (corresponding to a positive charge) where flow lines start, and also sinks (corresponding to negative charge) where flow lines end. Although this comparison between electric flux and fluid flux is interesting, and perhaps offers some insight, do not get them confused—an electric field is not a flow of any substance. Flux can be defined for any vector field, and we will later use it also for the magnetic field.

23–2 Gauss's Law

The precise relation between the electric flux through a closed surface and the net charge Q within that surface is given by **Gauss's Law**:

$$\oint \mathbf{E} \cdot d\mathbf{A} = \frac{Q}{\varepsilon_0} \tag{23–4}$$

where ε_0 is the same constant (permittivity of free space) that appears in Coulomb's law. Note that Q is the net charge enclosed by the surface over which the integral on the left side is taken. It doesn't matter where or how the charge is distributed within the surface. Any charge outside this surface must not be included. (Such an outside charge may affect the position of the electric field lines, but will not affect the net number of lines entering or leaving the surface.)[†]

Before we discuss the validity of Gauss's law, we note that the integral is often rather difficult to carry out in practice; we rarely need to do it except for some fairly simple situations that will be discussed later (mainly in Section 23–3).

Now let us see how Gauss's law is related to Coulomb's law. First, we show that Coulomb's law follows from Gauss's law. Consider a single point charge, Q; Gauss's law is supposed to hold for any enclosing surface, so let us choose one that is easy to work with—namely, the very symmetrical surface of a sphere of radius r with the charge Q at the center, Fig. 23–7. Because of the symmetry of this (imaginary) sphere about the charge at its center, we know that $\mathbf{E}$ must have the same magnitude at any point on the surface, and that $\mathbf{E}$ points outward (or inward) parallel to $d\mathbf{A}$, an element of the surface area. Hence

$$\oint \mathbf{E} \cdot d\mathbf{A} = \frac{Q}{\varepsilon_0}$$

becomes

$$\frac{Q}{\varepsilon_0} = \oint \mathbf{E} \cdot d\mathbf{A} = \mathbf{E} \cdot \oint d\mathbf{A} = E(4\pi r^2)$$

since the surface area of a sphere of radius r is $4\pi r^2$. Solving for E we obtain

$$E = \frac{Q}{4\pi\varepsilon_0 r^2}$$

which is the electric field form of Coulomb's law, Eq. 22–4.

Now for the reverse: We cannot actually derive Gauss's law from Coulomb's law in general; in fact, Gauss's law is a more general (and more elegant) law than Coulomb's. But we can show that it follows from Coulomb's law in some special cases; we use an intuitive argument based on lines of force. First we consider a single point charge surrounded by a spherical surface as in Fig. 23–7.

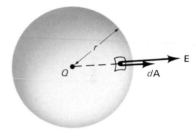

FIGURE 23–7
Spherical surface around a single point charge Q.

[†] Note that Gauss's law would look more complicated in terms of the constant $k = 1/4\pi\varepsilon_0$ that we originally used in Coulomb's law (Eq. 22–1):

Coulomb's law	Gauss's law
$E = k\dfrac{Q}{r^2}$	$\oint \mathbf{E} \cdot d\mathbf{A} = 4\pi k Q$
$E = \dfrac{1}{4\pi\varepsilon_0}\dfrac{Q}{r^2}$	$\oint \mathbf{E} \cdot d\mathbf{A} = Q/\varepsilon_0.$

Gauss's law has a simpler form using ε_0; Coulomb's law is simpler using k. The normal convention is to use ε_0 rather than k because Gauss's law is considered more general and therefore it is preferable to have it in simpler form.

Coulomb's law tells us that the electric field at the spherical surface is $E = (1/4\pi\varepsilon_0)(Q/r^2)$. Reversing the argument we just used, we have

$$\oint \mathbf{E} \cdot d\mathbf{A} = \oint \frac{1}{4\pi\varepsilon_0} \frac{Q}{r^2} \, d\mathbf{A} = \frac{Q}{4\pi\varepsilon_0 r^2} (4\pi r^2) = \frac{Q}{\varepsilon_0}.$$

This is Gauss's law, and we derived it for the special case of a spherical surface enclosing a point charge at its center. But what about some other surface, such as the irregular surface labeled A_2 in Fig. 23–8. The same number of field lines (due to our charge Q) pass through this surface as pass through the spherical surface, A_1. Since, as we saw in Section 23–1, the flux through a surface is proportional to the number of lines through it, the flux through A_2 is the same as through A_1:

$$\oint_{A_2} \mathbf{E} \cdot d\mathbf{A} = \oint_{A_1} \mathbf{E} \cdot d\mathbf{A} = \frac{Q}{\varepsilon_0}.$$

Hence we can expect that

$$\oint \mathbf{E} \cdot d\mathbf{A} = \frac{Q}{\varepsilon_0}$$

would be valid for *any* surface around a single point charge.

Finally, let us look at the case of more than one charge. For each charge individually,

$$\oint \mathbf{E}_i \cdot d\mathbf{A} = \frac{Q_i}{\varepsilon_0}.$$

Because the total field $\mathbf{E}$ is equal to the sum of the fields due to each separate charge, $\mathbf{E} = \sum \mathbf{E}_i$, we have

$$\oint \mathbf{E} \cdot d\mathbf{A} = \oint \left(\sum \mathbf{E}_i \right) \cdot d\mathbf{A} = \sum \frac{Q_i}{\varepsilon_0} = \frac{Q}{\varepsilon_0}$$

where $Q = \sum Q_i$ is the total net charge within the surface. Thus we see, based on this simple argument, that Gauss's law is valid for any distribution of electric charge enclosed within a closed surface of any shape. However it is important to recognize that $\mathbf{E}$ on the left side of Gauss's law is not necessarily due only to the charge Q that appears on the right. For example in Fig. 23–3 there is an electric field $\mathbf{E}$ at all points on the surface, but it is not due to the charge enclosed by

FIGURE 23–8

A single point charge surrounded by a spherical surface A_1, and an irregular surface, A_2.

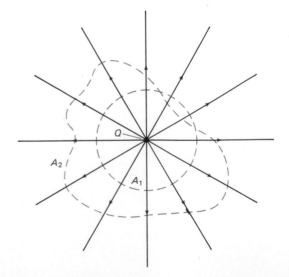

the surface (which is $Q = 0$ in this case). Gauss's law has been found to be valid for the total field at any surface. It tells us that any difference between the input and output flux of the electric field over any surface is due to charge within that surface.

Gauss's law holds for any inverse square vector field, such as the gravitational field. It does not hold for other types of field. For example, suppose the field due to a point charge Q decreased as kQ/r; then the flux through a sphere of radius r due to the point charge Q would be

$$\oint \mathbf{E} \cdot d\mathbf{A} = \left(\frac{kQ}{r}\right)(4\pi r^2) = (4\pi kQ)r.$$

The bigger the sphere the greater the flux even though the charge inside didn't change.

23–3 Applications of Gauss's Law

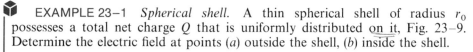

Gauss's law is a very compact and elegant way to write the relation between electric charge and electric field. It also offers a simple way to determine the electric field when the charge distribution is simple and symmetrical. In order to do this, however, we must choose the "gaussian" surface (for the integral on the left side of Gauss's law) very carefully so we can determine **E**. We normally try to think of a surface that has just the symmetry needed so **E** will be constant on all or on parts of its surface.

EXAMPLE 23–1 *Spherical shell.* A thin spherical shell of radius r_0 possesses a total net charge Q that is uniformly distributed on it, Fig. 23–9. Determine the electric field at points (*a*) outside the shell, (*b*) inside the shell.

SOLUTION (*a*) Because the charge is distributed symmetrically, the electric field must also be symmetric. Thus the field must be directed radially outward (inward if $Q < 0$) and must depend only on r, not on θ (in polar coordinates). The electric field will thus have the same magnitude at all points on an imaginary gaussian surface which we draw as a sphere of radius r, concentric with the shell,

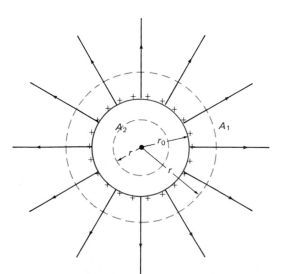

FIGURE 23–9

Uniform distribution of charge on a thin spherical shell of radius r_0.

and labeled A_1 in Fig. 23–9. Since **E** is perpendicular to this surface, Gauss's law gives

$$\oint \mathbf{E} \cdot d\mathbf{A} = E(4\pi r^2) = \frac{Q}{\varepsilon_0}$$

or

$$E = \frac{1}{4\pi\varepsilon_0} \frac{Q}{r^2}. \qquad\qquad [r > r_0]$$

Thus the field outside a uniform spherical shell of charge is the same as if all the charge were concentrated at the center as a point charge.

(b) Inside the shell, the field must also be symmetric; so E must again have the same value at all points on a spherical gaussian surface (A_2 in Fig. 23–9) concentric with the shell. Thus E can be factored out of the integral and we have

$$\oint \mathbf{E} \cdot d\mathbf{A} = E(4\pi r^2) = 0$$

since the charge Q enclosed by the shell is zero. Hence

$$E = 0 \qquad\qquad [r < r_0]$$

inside a uniform spherical shell of charge.

The above result, which is very useful, also applies to a uniformly charged spherical conductor since all the charge would lie in a thin shell at the surface.

EXAMPLE 23–2 *Solid sphere of charge.* An electric charge Q is distributed uniformly throughout a nonconducting sphere of radius r_0, Fig. 23–10. Determine the electric field (a) outside the sphere ($r > r_0$) and (b) inside the sphere ($r < r_0$).

SOLUTION Since the charge is distributed symmetrically in the sphere, the electric field at all points must again be symmetric. **E** depends only on r and is directed radially outward (or inward if $Q < 0$). (a) For our gaussian surface we choose a sphere of radius r ($r > r_0$), labeled A_1 in the diagram. Since E depends only on r, Gauss's law gives

$$\oint \mathbf{E} \cdot d\mathbf{A} = E(4\pi r^2) = \frac{Q}{\varepsilon_0}$$

or

$$E = \frac{1}{4\pi\varepsilon_0} \frac{Q}{r^2}.$$

Again, the field outside a spherically symmetric distribution of charge is the same as that for a point charge of the same magnitude located at the center of the sphere.

(b) Inside the sphere, we choose for our gaussian surface a concentric sphere of radius r ($r < r_0$), labeled A_2 in Fig. 23–10. From symmetry, **E** is the same at all points on A_2 and is perpendicular to the surface, so

$$\oint \mathbf{E} \cdot d\mathbf{A} = E(4\pi r^2).$$

We must equate this to the charge enclosed by A_2 (and divide by ε_0), but this is not the total charge Q but only a portion of it. Since the charge per unit volume (the *charge density* ρ) is uniform, the charge enclosed by A_2 is

$$\left(\frac{\frac{4}{3}\pi r^3 \rho}{\frac{4}{3}\pi r_0^3 \rho} \right) Q = \frac{r^3}{r_0^3} Q.$$

FIGURE 23–10

A solid sphere of uniform charge density.

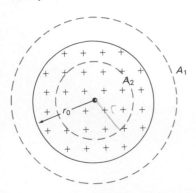

Hence, from Gauss's law

$$E(4\pi r^2) = \frac{r^3}{r_0^3}\frac{Q}{\varepsilon_0}$$

or

$$E = \frac{1}{4\pi\varepsilon_0}\frac{Q}{r_0^3}r. \qquad\qquad [r < r_0]$$

Thus the field increases linearly with r, until $r = r_0$. It then decreases as $1/r^2$, as plotted in Fig. 23–11.

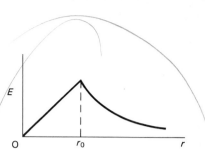

FIGURE 23–11
Magnitude of the electric field as a function of the distance r from the center of a uniformly charged solid sphere.

The results above would have been difficult to obtain from Coulomb's law by integrating over the sphere. Using Gauss's law and the symmetry of the situation, this result is obtained almost trivially. This shows the great power of Gauss's law. However, its use in this way is limited mainly to cases where the charge distribution has a high degree of symmetry. In such cases, we *choose* a simple surface on which E = constant, so the integration is simple. Gauss's law holds, of course, for any surface; we choose simple surfaces to make calculation easier. The next two examples are symmetric cases that we did treat before, using Coulomb's law, but we get the result more easily using Gauss's law.

EXAMPLE 23–3 *Long uniform line of charge.* A very long straight wire possesses a uniform charge per unit length, λ. Calculate the electric field at points near (but outside) the wire, far from the ends.

SOLUTION Because of the symmetry, we expect the field to be directed radially outward (if $Q > 0$) and to depend only on the perpendicular distance, r, from the wire. Because of this cylindrical symmetry, the field will be the same at all points on a gaussian surface that is a cylinder with the wire along its axis, Fig. 23–12. **E** is perpendicular to this surface at all points. But for Gauss's law, we need a closed surface, so we must include the flat ends of the cylinder; since **E** is parallel to the ends, there is no flux through the ends, so

$$\oint \mathbf{E} \cdot d\mathbf{A} = E(2\pi rl) = \frac{Q}{\varepsilon_0} = \frac{\lambda l}{\varepsilon_0}$$

where l is the length of our chosen gaussian surface ($l \ll$ length of wire). Hence

$$E = \frac{1}{2\pi\varepsilon_0}\frac{\lambda}{r}.$$

This is the same result as we got in Example 22–7 using Coulomb's law (we used x there instead of r), but here it took much less effort. Again we see the great power of Gauss's law.

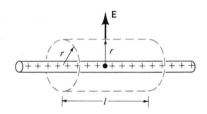

FIGURE 23–12
Calculation of **E** due to a very long line of charge.

FIGURE 23–13
Calculation of the electric field outside a large uniformly charged plane surface.

EXAMPLE 23–4 *Infinite plane of charge.* Charge is distributed uniformly, with a surface charge density (charge per unit area) σ, over a very large nonconducting flat plane surface. Determine the electric field at points near the plane.

SOLUTION We choose as our gaussian surface a small, closed cylinder whose axis is perpendicular to the plane and which extends through the plane as shown in Fig. 23–13. Because of the symmetry, we expect **E** to be directed perpendicular to the plane on both sides as shown, and to be uniform over the end caps of the cylinder, each of whose area is A. Since no flux passes through the curved sides of the cylinder, all the flux is through the two end caps. So Gauss's law gives

$$\oint \mathbf{E} \cdot d\mathbf{A} = 2EA = \frac{Q}{\varepsilon_0} = \frac{\sigma A}{\varepsilon_0}$$

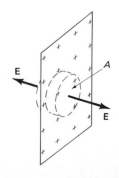

where $Q = \sigma A$ is the charge enclosed by our gaussian cylinder. The electric field is then:

$$E = \frac{\sigma}{2\varepsilon_0}.$$

This is the same result we obtained much more laboriously in Example 22–8. The field is uniform for points far from the ends of the plane, and close to its surface.

 EXAMPLE 23–5 Show that the electric field just outside the surface of any good conductor of arbitrary shape is given by

$$E = \frac{\sigma}{\varepsilon_0}$$

where σ is the surface charge density on the conductor at that point.

SOLUTION We choose as our gaussian surface a small cylindrical surface, as we did in the previous example. We choose the cylinder to be very small in height, so that one of its circular ends is just above the conductor (Fig. 23–14); the other end is just below the conductor's surface, and the sides are perpendicular to it. The electric field is zero inside a conductor and is perpendicular to the surface just outside it (Section 22–9), so electric flux passes only through the outside end of our cylinder. We choose the area A (of the flat cylinder end) small enough so that E is essentially uniform over it; then Gauss's law gives

$$\oint \mathbf{E} \cdot d\mathbf{A} = EA = \frac{Q}{\varepsilon_0} = \frac{\sigma A}{\varepsilon_0}$$

so that

$$E = \frac{\sigma}{\varepsilon_0}. \qquad \text{[at surface of conductor]} \quad (23\text{–}5)$$

This is a useful result which applies for any shape conductor.

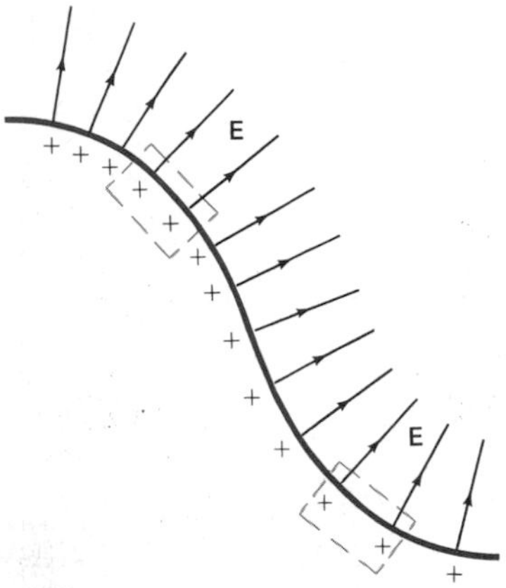

FIGURE 23–14

Electric field near surface of a conductor.

You may wonder why the field outside a large plane nonconductor is $E = \sigma/2\varepsilon_0$ (Example 23–4) whereas outside a conductor it is $E = \sigma/\varepsilon_0$. The reason for the factor of 2 comes not so much from conductor versus nonconductor as from what we mean by the surface charge density σ. For a conductor, the charge lies at the surface and all the electric field lines leave on one side of the surface. For a plane nonconductor, the lines leave both sides, Fig. 23–13. Now, if we had a large, thin, flat conducting plane, the charge would accumulate on both surfaces and the field would emanate from both sides. If we called σ the surface charge for the plane as a *whole*, each face of the plane would have surface charge $\sigma/2$, and so the result of Example 23–5 would give $E = (\sigma/2)/\varepsilon_0 = \sigma/2\varepsilon_0$, the same as for a nonconducting plane. Normally, however, we use σ to apply to each face of the plane and then we would get $E = \sigma/\varepsilon_0$. Thus the factor of 2 between Examples 23–4 and 23–5 comes from two different ways of defining σ.

Summary

The *electric flux* passing through a flat area A for a uniform electric field $\mathbf{E}$ is $\Phi_E = \mathbf{E} \cdot \mathbf{A}$. If the field is not uniform, the flux is determined from the integral $\Phi_E = \int \mathbf{E} \cdot d\mathbf{A}$. The direction of the vector $\mathbf{A}$ or $d\mathbf{A}$ is chosen to be perpendicular to the surface whose area is A or dA; it points outward from an enclosed surface. The flux through a surface is proportional to the number of lines of force passing through it.

Gauss's law states that the net flux passing out of any closed surface is equal to the net charge enclosed by the surface divided by ε_0:

$$\oint \mathbf{E} \cdot d\mathbf{A} = \frac{Q}{\varepsilon_0}.$$

Gauss's law can in principle be used to determine the electric field due to a given charge distribution, but its usefulness is mainly limited to a small number of cases, usually where the charge distribution displays much symmetry. The real importance of Gauss's law is that it is a more general and elegant statement (than Coulomb's law) for the relation between electric charge and electric field. It is one of the basic equations of electromagnetism.

 Questions _____

1. If the electric flux through a closed surface is zero, is the electric field necessarily zero at all points on the surface? What about the converse: if $\mathbf{E} = 0$ at all points on the surface is the flux through the surface zero?

2. Is the electric field $\mathbf{E}$ in Gauss's law, $\oint \mathbf{E} \cdot d\mathbf{A} = Q/\varepsilon_0$, that due only to the charge Q?

3. A point charge is surrounded by a spherical gaussian surface of radius r. If the sphere is replaced by a cube of side $r/2$, will Φ_E be larger, smaller, or the same?

4. What can you say about the flux through a closed surface that encloses an electric dipole?

5. The electric field $\mathbf{E}$ is zero at all points on a closed surface; is there necessarily no net charge within the surface? If a surface encloses zero net charge, is the electric field necessarily zero at all points on the surface?

6. Define gravitational flux in analogy to electric flux. Are there "sources" and "sinks" for the gravitational field as there are for the electric field? Discuss.

7. Would Gauss's law be helpful in determining the electric field due to an electric dipole?

8. A spherical basketball (a nonconductor) is given a charge Q distributed uniformly over its surface. What can you say about the electric field inside the ball? A person now steps on the ball and forces most of the air out without altering the charge. What can you say about the field inside now?

9. In Example 23–3, it may seem that the electric field calculated is due only to the charge enclosed by the cylinder chosen as our gaussian surface. In fact, the entire charge along the whole length of the wire contributes to the field. Explain how the charge outside the cylindrical surface of Fig. 23–12 does contribute to E inside that cylinder. (Hint: compare to the field due to a wire of short length.)

10. Refer to the discussion after Example 23–5. Could you define a surface charge density for each face of a non-conducting plane? If so, under what circumstances?

11. A point charge Q is surrounded by a spherical surface of radius r_0, whose center is at Q. Later, the charge is moved to the right a distance $r_0/2$, but the sphere remains where it was. How is the electric flux Φ_E through the sphere changed? Is the electric field at the surface of the sphere changed? For each "yes" answer, describe the change.

12. Suppose the line of charge in Example 23–3 was not very long but extended only a short way beyond the ends of the cylinder shown in Fig. 23–12. How would the result of Example 23–3 be altered?

13. A conductor carries a net charge Q. There is a hollow cavity within the conductor, at whose center is a point charge q. What is the charge on (a) the outer surface of the conductor, (b) the inner surface of the conductor?

14. A point charge q is placed at the center of the cavity of a thin metal shell which is neutral. Will a charge Q placed outside the shell feel an electric force? Explain.

 Problems _____

SECTION 23–1

1. (I) A circular ring of radius 15 cm is placed in a uniform electric field of magnitude 3.6×10^2 N/C. What is the electric flux through the ring when its face is (a) perpendicular to the field lines, (b) at 45° to the field lines, (c) parallel to the field lines.

2. (I) A cube of side l is placed in a uniform field $E = 8.1 \times 10^3$ N/C with edges parallel to the field lines. What is the net flux through the cube?

3. (II) A point charge Q is placed at the center of a cube of side l. What is the flux through one face of the cube?

4. (II) A uniform field $\mathbf{E}$ is parallel to the axis of a hemisphere of radius R. What is the electric flux through the hemispherical surface? What is the result if $\mathbf{E}$ is perpendicular to the axis?

SECTION 23–2

5. (I) The total electric flux from a cubical box 18.0 cm on a side is 1.45×10^3 N·m²/C. What charge is enclosed by the box?

6. (II) Write Gauss's law for the gravitational field $\mathscr{G}$ (see Section 5–9).

7. (II) Use Gauss's law to prove that any net charge on any insulated conductor must be at its outer surface. (Hint: choose a gaussian surface just inside the conductor's surface.)

8. (II) The concept of flux applies equally well to fluid flow as to the electric field (indeed, the word flux comes from the Latin word for "flow"). The fluid flow velocity $\mathbf{v}$ corresponds to $\mathbf{E}$, so flow streamlines correspond to lines of force. (a) What does the flux Φ represent for fluid flow? (b) Write Gauss's law for the streamline flow of an incompressible fluid.

9. (I) The field just outside a 3.0-cm-radius metal ball is 2.1×10^2 N/C and points toward the ball. What charge resides on the ball?

10. (I) Starting from the result of Example 23–1, show that the electric field just outside a uniformly charged spherical conductor is $E = \sigma/\varepsilon_0$, consistent with Example 23–5.

11. (II) The charge density ρ (charge/volume) within a sphere of radius r_0 is uniform. Determine the electric field in terms of ρ for (a) $r < r_0$ and (b) $r > r_0$. Do they agree at $r = r_0$?

12. (II) A point charge Q is placed a distance $r_0/2$ above the center of a spherical surface of radius r_0. (a) What is the electric flux through the sphere? (b) What range of values does E have at the surface of the sphere? (c) Is **E** perpendicular to the sphere at all points? (d) Is Gauss's law useful for obtaining E at the surface of the sphere?

13. (II) The earth is surrounded by an electric field, pointing inward, of magnitude $E \approx 150$ N/C near the surface. (a) What is the net charge on the earth? (b) How many excess electrons per square meter on the earth's surface does this correspond to?

14. (II) Two large, flat, round metal plates are separated by a distance that is very small compared to their diameters. The conductors are given equal but opposite uniform surface charge densities σ. Ignore edge effects and use Gauss's law to show that (a) for points far from the edges, the electric field between the plates is $E = \sigma/\varepsilon_0$, and (b) that outside the plates on either side the field is zero. (c) How would your results be altered if the two plates were nonconductors?

15. (II) Suppose the two conducting plates in problem 14 have the *same* sign of charge. What then will be the electric field (a) between them, (b) outside them on either side? (c) What if the planes are nonconducting?

16. (II) A charge Q is located at one corner of a cube of side a. Determine the electric flux through the cube. (Hint: integration or complex calculation of angles is not necessary.)

17. (II) A cube of side l has one corner at the origin of coordinates, and extends along the positive x, y, and z axes. Suppose the electric field in this region is given by $E = (a + by)\mathbf{j}$. Determine the charge inside the cube.

18. (II) A thin cylindrical shell of radius R_1 is surrounded by a second concentric cylindrical shell of radius R_2. The inner shell has a total charge $+Q$ and the outer shell $-Q$. Assuming the length L of the shells is much greater than R_1 or R_2, determine the electric field as a function of r (the perpendicular distance from the common axis of the cylinders) for (a) $r < R_1$, (b) $R_1 < r < R_2$, (c) $r > R_2$. (d) What is the kinetic energy of an electron if it moves between (and concentric with) the shells in a circle of radius $(R_1 + R_2)/2$?

19. (II) In problem 18, (a) under what conditions will $E = 0$ for $r > R_2$? (b) Under what conditions will $E = 0$ for $R_1 < r < R_2$?

20. (II) Two concentric spherical shells of radii R_1 and R_2 ($R_1 < R_2$) contain uniform surface charge densities σ_1 and σ_2, respectively. Determine the electric field for (a) $r < R_1$, (b) $R_1 < r < R_2$, (c) $r > R_2$. (d) Under what conditions will $E = 0$ for $r > R_2$? (e) Under what conditions will $E = 0$ for $R_1 < r < R_2$?

21. (II) A spherical rubber balloon carries a total charge Q uniformly distributed on its surface. At $t = 0$ the balloon has radius r_0 and the balloon is then slowly blown up so that r increases linearly to $2r_0$ in 30 s. Determine the electric field as a function of time (a) just outside the balloon surface, (b) at $r = 4r_0$.

22. (II) Suppose the sphere of Example 23–2 has a spherical cavity of radius r_1 centered at the sphere's center. Assuming the charge Q is distributed uniformly in the "shell" (between $r = r_1$ and $r = r_0$), determine the electric field as a function of r for (a) $r < r_1$, (b) $r_1 < r < r_0$, (c) $r > r_0$.

23. (II) Suppose the density of charge between r_1 and r_0 of the hollow sphere of problem 22 varies as $\rho = \rho_0 r_1/r$. Determine the electric field as a function of r for (a) $r < r_1$, (b) $r_1 < r < r_0$, (c) $r > r_0$. (d) Plot E versus r from $r = 0$ to $r = 2r_0$.

24. (II) Suppose at the center of the cavity inside the shell of Fig. 23–9 (and Example 23–1), there is a point charge q ($\neq Q$). Determine the electric field for (a) $r < r_0$, (b) $r > r_0$. What are your answers if (c) $q = Q$, (d) $q = -Q$?

25. (II) A long cylindrical shell of radius R_0 and length L ($R_0 \ll L$) possesses a uniform surface charge density (charge per unit area) σ. Determine the electric field at points (a) outside the cylinder ($r > R_0$) and (b) inside the cylinder ($r < R_0$); assume the points are far from the ends and not too far from the shell ($r \ll L$). (c) Compare to the result for a long line charge, Example 23–3.

26. (II) A very long solid nonconducting cylinder of radius R_0 and length L ($R_0 \ll L$) possesses a uniform volume charge density ρ (C/m^3). Determine the electric field at points (a) outside the cylinder ($r > R_0$) and (b) inside the cylinder ($r < R_0$). Do only for points far from the ends and for which $r \ll L$.

27. (II) A flat square sheet of aluminum foil, 20 cm on a side, carries a uniformly distributed 35 mC charge. What, approximately, is the electric field (a) 1.0 cm above the sheet, (b) 15 m above the sheet?

28. (II) A point charge Q is on the axis of a cylinder at its center. The diameter of the cylinder is equal to its length l. What is the total flux through the curved sides of the cylinder? (Hint: first calculate the flux through the ends.)

29. (III) A solid nonconducting sphere of radius r_0 has a total charge Q which is distributed according to

$$\rho = br$$

where ρ is the charge per unit volume, or charge density (C/m^3), and b is a constant. Determine (a) b in terms of Q, (b) the electric field at points inside the sphere, (c) the electric field at points outside the sphere.

Electric Potential

24

We saw in Chapters 6 and 7 that the concept of energy was extremely valuable in dealing with mechanical problems. For one thing energy is a conserved quantity and thus an important aspect of nature. Furthermore we saw that many problems could be solved using the energy concept even though a detailed knowledge of the forces involved was not possible, or when a calculation involving Newton's laws would have been too difficult.

The energy point of view can be used in electricity, and it is especially useful. It not only extends the law of conservation of energy, but it gives us another way to view electrical phenomena and is a tool in solving problems more easily, in many cases, than by using forces and electric fields.

24–1 Electric Potential and Potential Difference

It is possible to define potential energy only for a conservative force as we saw in Section 7–1; the work done by such a force in moving a particle between any two positions is independent of the path taken. It is easy to see that the electrostatic force is a conservative force: the force one point charge exerts on a second point charge is given by Coulomb's law, $F = kQ_1Q_2/r^2$; this has the same $1/r^2$ form as the law of gravitation, $F = Gm_1m_2/r^2$; and, as we saw in Section 7–5, such a $1/r^2$ force is conservative. The force on a particular charge due to any charge distribution can always be written as a sum of Coulomb-like terms (Section 22–7); hence the force due to any charge distribution is conservative. Since the electrostatic force is conservative, we can define electric potential energy.

The difference in electric potential energy of a point charge q between two different points in an electric field can be defined as the work done by an external force to move the charge ("against" the electric force) from one point to the other. This is equivalent to defining the change in electric potential energy as the negative of the work done by the electric field itself to move the charge from one point to the other (See Section 6–5).

For example, consider the electric field between two equally but oppositely charged parallel plates which are large compared to their separation so the field will be uniform over most of the region, Fig. 24–1. Now consider a positive point charge q placed at point a very near the positive plate as shown. The electric force on the charge will tend to move the charge toward the negative plate and will do work on the charge to move it from point a to point b (near the negative plate). In the process the charged particle will be accelerated and its kinetic energy increased; and the potential energy will be decreased by an equal amount, which is equal to the work done by the electric force in moving the particle from a to b. In accord with the conservation of energy, electric potential energy is transformed into kinetic energy and the total energy is conserved. Note that the positive charge q has its greatest potential energy U when near the positive plate.[†] The reverse is true for a negative charge: its potential energy is greatest near the negative plate.

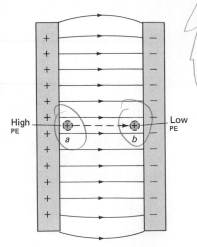

High
PE

Low
PE

a   b

FIGURE 24–1

Work is done by the electric field in moving the positive charge from position a to position b.

We defined the electric field (Chapter 22) as the force per unit charge; similarly, it is useful to define the **electric potential** (or simply the **potential** when "electric" is understood) as the *potential energy per unit charge*. Electric potential is given the symbol V; so if a point charge q has electric potential energy U_a at some point a, the electric potential at this point is:

$$V_a = \frac{U_a}{q}.$$

As we saw in Chapter 6, only differences in potential energy are physically measurable. Hence only the **difference in potential**, or the **potential difference**, between two points a and b (such as between a and b in Fig. 24–1) is measurable. Since the difference in potential energy between points a and b is equal to the negative of the work, W_{ba}, done by the electric field in moving the charge from a to b, we have that the potential difference V_{ba} is

$$V_{ba} = V_b - V_a = -\frac{W_{ba}}{q}. \tag{24–1}$$

The unit of electric potential, and of potential difference, is joules/coulomb and is given a special name, the *volt*, in honor of Alessandro Volta (1745–1827; he is best known for having invented the electric battery); the volt is abbreviated V, so 1 V = 1 J/C. Note from our definition that the positive plate in Fig. 24–1 is at a higher potential than the negative plate. Thus a positively charged object moves naturally from a high potential to a low potential; a negative charge does the reverse. Potential difference, since it is measured in volts, is often referred to as **voltage**.

If we wish to speak of the potential, V_a, at some point a, we must be aware that V_a depends on where the potential is chosen to be zero; and the zero point for electric potential in a given situation, just as for potential energy, can be chosen arbitrarily since only differences in potential energy can be measured. Often the ground, or a conductor connected directly to the ground, is taken as zero potential, and other potentials are given with respect to ground. (That is, a point where the voltage is 50 V is one where the difference of potential between it and ground is 50 V.) In other cases, as we shall see, we may choose the potential to be zero at infinity.

Since the electric potential is defined as the potential energy per unit charge, then the change in potential energy of a charge q when moved between two points a and b is

$$\Delta U = U_b - U_a = qV_{ba}. \tag{24–2}$$

That is, if an object with charge q moves through a potential difference V_{ba}, its potential energy changes by an amount qV_{ba}. For example, if the potential

[†] At this point it has its greatest ability to do work (on some other object or system).

difference between the two plates in Fig. 24–1 is 6 V, then if a 1-C charge is moved (say by an external force) from b to a, it will gain (1 C)(6 V) = 6 J of electric potential energy. (And it will lose 6 J of electric PE as it moves from a to b.) Similarly, a 2-C charge will gain 12 J, and so on. Thus, electric potential is a measure of how much energy an electric charge can acquire in a given situation. And, since energy is the ability to do work, the electric potential is also a measure of how much work a given charge can do. The exact amount depends both on the potential and on the charge.

$$V = \frac{KQ}{r}$$

To better understand electric potential, let's make a comparison to the gravitational case when a rock falls from the top of a cliff. The greater the height of a cliff, the more potential energy the rock has and the more kinetic energy it will have when it reaches the bottom. The actual amount of kinetic energy it will acquire, and the amount of work it can do, depends both on the height of the cliff and the mass of the rock. Similarly in the electrical case: the potential energy change, or the work that can be done, depends both on the potential difference (corresponding to the height of the cliff) and on the charge (corresponding to mass), Eq. 24–1.

Practical sources of electrical energy such as batteries and electric generators are intended to maintain a particular potential difference; the actual amount of energy used or transformed depends on how much charge flows. For example, consider an automobile headlight connected to a 12.0-V battery. The amount of energy transformed (into light, and of course heat) is proportional to how much charge flows, which in turn depends on how long the light is on. If over a given period 5.0 C of charge flows through the light, the total energy transformed is (5.0 C)(12.0 V) = 60 J. If the headlight is left on twice as long, 10.0 C of charge will flow and the energy transformed is (10.0 C)(12.0 V) = 120 J.

■ EXAMPLE 24–1 An electron in the picture tube of a TV set is accelerated from rest through a potential difference V_{ba} = 5000 V. (a) What is the change in potential energy of the electron? (b) What is the speed of the electron as a result of this acceleration?

SOLUTION (a) The charge on an electron is $e = -1.6 \times 10^{-19}$ C. Therefore its change in potential energy is equal to

$$\Delta U = qV_{ba} = (-1.6 \times 10^{-19} \text{ C})(+5000 \text{ V})$$

$$= -8.0 \times 10^{-16} \text{ J}.$$

(The potential difference, V_{ba}, has a positive sign since the final potential is higher than the initial potential; that is, negative electrons are attracted from a negative electrode to a positive one.)

(b) The potential energy lost by the electron becomes kinetic energy. Thus, since $\Delta KE + \Delta U = 0$ from conservation of energy (see Eq. 7–3),

$$\Delta KE = -\Delta U$$

$$\tfrac{1}{2}mv^2 - 0 = -qV_{ba}$$

where the initial KE = 0 since we assume the electron started from rest. We solve for v and put in the mass of the electron $m = 9.1 \times 10^{-31}$ kg:

$$v = \sqrt{-\frac{2qV_{ba}}{m}}$$

$$= \sqrt{\frac{-2(-1.6 \times 10^{-19} \text{ C})(5000 \text{ V})}{9.1 \times 10^{-31} \text{ kg}}}$$

$$= 4.2 \times 10^7 \text{ m/s}.$$

24-2 Relation Between Electric Potential and Electric Field

The effects of any charge distribution can be described either in terms of electric field or in terms of electric potential. Furthermore, there is an intimate connection between the potential and the field. Let us first examine this relation for the case of a uniform electric field, such as that between the plates of Fig. 24-1 whose difference of potential is V_{ba}. The work done by the electric field to move a positive charge q from a to b is, from Eq. 24-1,

$$W = -qV_{ba}.$$

Notice that $V_{ba} = V_b - V_a$ is negative (< 0) since point a is at the higher potential (it is positive relative to point b). Hence, the work done is positive. We can also write the work done as the force times distance and use the fact that the force on q is $\mathbf{F} = q\mathbf{E}$ where $\mathbf{E}$ is the uniform electric field between the plates. Thus

$$W = Fd = qEd$$

where d is the distance (parallel to the field lines) between a and b. We now set these two expressions for W equal and find $-qV_{ba} = qEd$, or

$$V_b - V_a = V_{ba} = -Ed. \qquad [E \text{ uniform}] \quad (24\text{-}3)$$

The minus sign on the right merely indicates that $V_a > V_b$, the potential of the positive plate being higher than that of the negative plate as we already discussed; and positive charges move spontaneously from high potential to low potential. If we solve for E, we find $E = -V_{ba}/d$.

From this equation we can see that the units for electric field intensity can be written as volts per meter (V/m) as well as newtons per coulomb (N/C). These are equivalent in general, since $1 \text{ N/C} = 1 \text{ N·m/C·m} = 1 \text{ J/C·m} = 1 \text{ V/m}$.

EXAMPLE 24-2 Two parallel plates are charged to a voltage of 50 V. If the separation between the plates is 5.0 cm, calculate the electric field between them.

SOLUTION We have from Eq. 24-3, dealing only with magnitudes for convenience,

$$E = V/d = 50 \text{ V}/0.050 \text{ m} = 1000 \text{ V/m}.$$

To treat the general case of a nonuniform electric field, let us recall the relation between force $\mathbf{F}$ and the potential energy U due to that force. As discussed in Section 6-5, the difference in potential energy between any two points in space, a and b, is given by Eq. 6-13:

$$U_b - U_a = -\int_a^b \mathbf{F} \cdot d\mathbf{l}$$

where $d\mathbf{l}$ is an infinitesimal increment of displacement and the integral is taken along any path in space from point a to point b. For the electrical case, we are more interested in the potential difference, given by Eq. 24-2, $V_{ba} = V_b - V_a = (U_b - U_a)/q$, than in the potential energy itself. Also, the electric field $\mathbf{E}$ at any point in space is defined as the force per unit charge (Eq. 22-3): $\mathbf{E} = \mathbf{F}/q$. Putting these two relations in the above equation gives us

$$V_{ba} = V_b - V_a = -\int_a^b \mathbf{E} \cdot d\mathbf{l}. \qquad (24\text{-}4)$$

This is the general relation between electric field and potential difference.

Equation 24-4 reduces to Eq. 24-3 when the field is uniform; in Fig. 24-1, for example, a path parallel to the electric field lines from point a at the positive plate to point b at the negative plate gives (since $\mathbf{E}$ and $d\mathbf{l}$ are in the same direction at each point),

$$V_b - V_a = -\int_a^b \mathbf{E} \cdot d\mathbf{l} = -E \int_a^b dl = -Ed$$

where d is the distance, parallel to the field lines, between points a and b. Again, the negative sign on the right simply tells us that $V_a > V_b$ in Fig. 24–1.

24–3   Equipotential Surfaces

The electric potential can be represented diagrammatically by drawing **equipotential lines** or, in three dimension, **equipotential surfaces**. An equipotential surface is one on which all points are at the same potential. That is, the potential differences between any two points on the surface is zero, and no work is required to move a charge from one point to the other. An *equipotential surface must be perpendicular to the electric field* at any point. If this were not so—that is, if there were a component of $\mathbf{E}$ parallel to the surface—it would require work to move the charge along the surface against this component of $\mathbf{E}$; and this would contradict the idea that it is an *equi*potential surface.

The fact that the electric field lines and equipotential surfaces are mutually perpendicular helps us locate the equipotentials when the electric field lines are known. In Fig. 24–2, a few of the equipotential lines are drawn (dashed lines) for the field between two parallel plates at a potential difference of 20 V. The lines are only a part of the equipotential surface that extends into and out of the paper. The negative plate is arbitrarily chosen to be zero volts and the potential of each equipotential line is indicated. The equipotential lines for the case of two equal but oppositely charged particles are shown in Fig. 24–3 as dashed lines.

We saw in Section 22–9 that there can be no electric field within a conductor in the static case, for otherwise the free electrons would feel a force and would move. Indeed *a conductor must be entirely at the same potential in the static case*, and the surface of a conductor is then an equipotential surface. (If it weren't, the free electrons at the surface would move since whenever there is a potential difference between two points, work can be done on charged particles to move them.) This is fully consistent with the fact, discussed earlier, that the electric field at the surface of a conductor must be perpendicular to the surface.

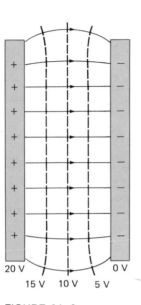

FIGURE 24–2

Equipotential lines (the dashed lines) between two charged parallel plates; note that they are perpendicular to the electric field lines (shown solid).

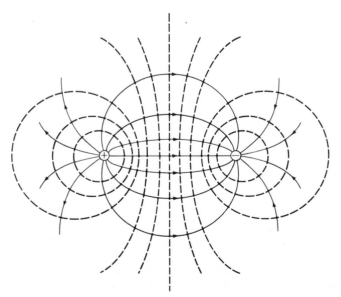

FIGURE 24–3

Equipotential lines (the dashed lines) and electric field lines (the solid lines) for two oppositely charged particles.

24-4   The Electron Volt, a Unit of Energy

As we shall see, the joule is a very large unit for dealing with energies of electrons, atoms, or molecules, whether in atomic and nuclear physics or in chemistry and molecular biology. For this purpose, the *electron volt* (eV) is used. One electron volt is defined as the energy acquired by an electron as a result of moving through a potential difference of 1 V. Since the charge on an electron is 1.6×10^{-19} C, and since the change in potential energy equals qV, 1 eV is equal to $(1.6 \times 10^{-19}$ C) $\cdot$ $(1.0 \text{ V}) = 1.6 \times 10^{-19}$ J:

$$1 \text{ eV} = 1.6 \times 10^{-19} \text{ J}.$$

An electron that accelerates through a potential difference of 1000 V will lose 1000 eV of potential energy and will thus gain 1000 eV or 1 keV (kilo-electron volt) of kinetic energy. On the other hand, if a particle has a charge equal to twice the charge on the electron ($= 2e = 3.2 \times 10^{-19}$ C), when it moves through a potential difference of 1000 V its energy will change by 2000 eV.

Although the electron volt is handy for *stating* the energies of molecules and elementary particles, it is not a proper SI unit; for calculations it should be converted to joules using the conversion factor given above. In Example 24-1, the electron acquired a kinetic energy of 8.0×10^{-16} J. We normally would quote this energy as 5000 eV ($= 8.0 \times 10^{-16}$ J/1.6×10^{-19} J/eV); but in determining its speed in SI units we had to use the KE in J.

24-5   Electric Potential Due to Single Point Charges

The electric potential at a distance r from a single point charge Q can be derived directly from Eq. 24-4. The electric field due to a single point charge has magnitude (Eq. 22-4)

$$E = \frac{1}{4\pi\varepsilon_0} \frac{Q}{r^2}$$

and is directed radially outward from the charge (inward if $Q < 0$). We take the integral in Eq. 24-4 along a (straight) field line (Fig. 24-4) from point a, a distance r_a from Q, to point b, a distance r_b from Q. Then $d\mathbf{l}$ will be parallel to $\mathbf{E}$ and

FIGURE 24-4

We integrate Eq. 24-4 along the straight line (shown dashed) from point a to point b; the line ab is parallel to a field line.

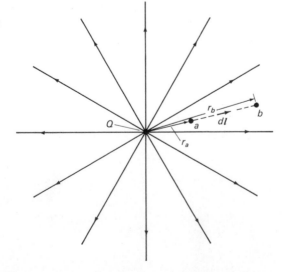

dl = dr. Thus

$$V_b - V_a = -\int_{r_a}^{r_b} \mathbf{E} \cdot d\mathbf{l} = -\frac{Q}{4\pi\varepsilon_0} \int_{r_a}^{r_b} \frac{1}{r^2}\, dr$$

$$= \frac{1}{4\pi\varepsilon_0}\left(\frac{Q}{r_b} - \frac{Q}{r_a}\right).$$

As mentioned earlier, only differences in potential have physical meaning. We are free, therefore, to choose the value of the potential at some one point to be whatever we please. It is common to choose the potential to be zero at infinity (let, say, $V_b = 0$ at $r_b = \infty$); then the electric potential V at a distance r from a single point charge is

$$V = \frac{1}{4\pi\varepsilon_0}\frac{Q}{r}. \qquad \text{[single point charge]} \quad (24\text{–}5)$$

(handwritten: $V = \frac{kQ}{r}$)

This is the electric potential at r relative to infinity, and is sometimes called the "absolute potential" due to a single point charge. Notice that the potential V decreases with the first power of the distance whereas the electric field (Eq. 22–4) decreases as the *square* of the distance. The potential near a positive charge is large and it decreases toward zero at very large distances. For a negative charge, the potential is *less* than zero (negative) and increases toward zero at large distances.

(handwritten margin notes: $V \approx \frac{1}{r}$ $E = \frac{1}{r^2}$)

EXAMPLE 24–3 How much work is required to bring a charge $q = 3.0\ \mu C$ from a great distance away (take $r = \infty$) to a point 0.50 m from a charge $Q = 20.0\ \mu C$?

SOLUTION The work required is

$$W = qV_{ba} = \frac{q}{4\pi\varepsilon_0}\left(\frac{Q}{r_b} - \frac{Q}{r_a}\right)$$

where $r_b = 0.50$ m and $r_a = \infty$. Therefore

$$W = (3.0 \times 10^{-6}\ C)\frac{(9.0 \times 10^9\ N\cdot m^2/C^2)(2.00 \times 10^{-5}\ C)}{(0.50\ m)} = 1.08\ J.$$

(Remember that $1/4\pi\varepsilon_0 = 9.0 \times 10^9\ N\cdot m^2/C^2$.)

To determine the electric field surrounding a collection of two or more point charges requires adding up the electric fields due to each charge. Since the electric field is a vector, this can often be a chore. To find the electric potential due to a collection of point charges is far easier since the electric potential is a scalar and hence you only need to add together scalars without concern for direction. This is a major advantage in using electric potential.

EXAMPLE 24–4 Calculate the electric potential at points A and B in Fig. 22–15 due to the two charges shown. (See also Example 22–5 where we calculated the electric field at these points.)

SOLUTION The potential at point A is the sum of the potentials due to the + and − charges, and we use Eq. 24–5 for each:

$$V_A = V_{A2} + V_{A1}$$

$$= \frac{(9.0 \times 10^9\ N\cdot m^2/C^2)(5.0 \times 10^{-5}\ C)}{(0.30\ m)}$$

$$+ \frac{(9.0 \times 10^9\ N\cdot m^2/C^2)(-5.0 \times 10^{-5}\ C)}{(0.60\ m)}$$

$$= 7.5 \times 10^5\ V.$$

At point *B*:

$$V_B = V_{B2} + V_{B1}$$

$$= \frac{(9.0 \times 10^9 \ \text{N·m}^2/\text{C}^2)(5.0 \times 10^{-5} \ \text{C})}{(0.40 \ \text{m})}$$

$$+ \frac{(9.0 \times 10^9 \ \text{N·m}^2/\text{C}^2)(-5.0 \times 10^{-5} \ \text{C})}{(0.40 \ \text{m})}$$

$$= 0 \ \text{volts}.$$

The potential will be zero not only for this point but for any point on the plane between the two charges that is equidistant from each.

A simple summation like this can easily be performed for any number of point charges.

24–6 Potential of an Electric Dipole

Two equal point charges *Q*, of opposite sign, separated by a distance *l*, are called an *electric dipole* (Section 22–11). The electric field lines and equipotential surfaces for a dipole are shown in Fig. 24–3.

Let us calculate the electric potential due to a dipole at an arbitrary point *P*, as shown in Fig. 24–5. Since *V* is the sum of the potentials due to each of the two charges, we have

$$V = \frac{1}{4\pi\varepsilon_0} \frac{Q}{r} + \frac{1}{4\pi\varepsilon_0} \frac{(-Q)}{r + \Delta r} = \frac{Q}{4\pi\varepsilon_0} \left(\frac{1}{r} - \frac{1}{r + \Delta r} \right) = \frac{Q}{4\pi\varepsilon_0} \frac{\Delta r}{r(r + \Delta r)},$$

where *r* is the distance from *P* to the positive charge and *r* + Δ*r* is the distance to the negative charge. This equation becomes simpler if we consider points *P* whose distance from the dipole is much larger than the separation of the two charges; that is, for *r* ≫ *l*. From the diagram we can see that in this case Δ*r* ≈ *l* cos *θ*; then *r* ≫ Δ*r* = *l* cos *θ* so we can neglect Δ*r* in the denominator as compared to *r* in this case. This type of approximation is often useful, and in this case we obtain a simple form for the potential:

$$V = \frac{1}{4\pi\varepsilon_0} \frac{p \cos \theta}{r^2} \qquad \text{[dipole; } r \gg l \text{]} \quad (24\text{–}6)$$

where *p* = *Ql* is the *dipole moment*. When *θ* is between 0° and 90°, *V* is positive; if *θ* is between 90° and 180°, *V* is negative (since cos *θ* is then negative). This makes sense since in the first case *P* is closer to the positive charge and in the second case it is closer to the negative charge. At *θ* = 90°, the potential is zero (cos 90° = 0) in accordance with the result of Example 24–4. From Eq. 24–6 we see that the potential decreases as the *square* of the distance from the dipole whereas for a single point charge the potential decreases with the first power of the distance (Eq. 24–5). It is not surprising that the potential should fall off faster for a dipole; for when you are far from a dipole, the two equal but opposite charges appear so close together as to tend to neutralize each other.

In many molecules, even though they are electrically neutral, the electrons spend more time in the vicinity of one atom rather than another, which results in a separation of charge. Such molecules have a dipole moment and are called *polar molecules*.

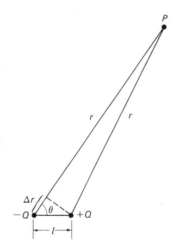

FIGURE 24–5

Electric dipole. Calculation of potential *V* at point *P*.

EXAMPLE 24–5 The distance between the carbon (C) and oxygen (O) atoms in the group C=O, which occurs in many organic molecules is about 1.2×10^{-10} m and the dipole moment of this group is about 8.0×10^{-30} C·m. Calculate (a) the net charge Q on the C(+) and O(−) atoms, (b) the potential 9.0×10^{-10} m from the dipole along its axis with the oxygen being the nearest atom (that is, to the left in Fig. 24–5, so $\theta = 180°$); (c) what would the potential be at this point if only the oxygen (O⁻) were charged?

SOLUTION (a) The dipole moment is $p = Ql$. Therefore

$$Q = \frac{p}{l} = \frac{(8.0 \times 10^{-30} \text{ C·m})}{(1.2 \times 10^{-10} \text{ m})} = 6.6 \times 10^{-20} \text{ C}.$$

(b) Since $\theta = 180°$, we have, using Eq. 24–6,

$$V = \frac{1}{4\pi\varepsilon_0} \frac{p \cos \theta}{r^2} = \frac{(9.0 \times 10^9 \text{ N·m}^2/\text{C}^2)(8.0 \times 10^{-30} \text{ C·m})(-1.0)}{(9.0 \times 10^{-10} \text{ m})^2}$$

$$= -0.088 \text{ V}.$$

(c) If we assume that O has charge $Q = -6.6 \times 10^{-20}$ C, as in part (a) above, and that C is not charged, we use the formula for a single charge, Eq. 24–5:

$$V = \frac{1}{4\pi\varepsilon_0} \frac{Q}{r} = \frac{(9.0 \times 10^9 \text{ N·m}^2/\text{C}^2)(-6.6 \times 10^{-20} \text{ C})}{(9.0 \times 10^{-10} \text{ m})}$$

$$= -0.66 \text{ V}.$$

Of course we expect the potential of a single charge to have greater magnitude than that of a dipole of equal charge at the same distance.

24–7 Potential Due to Any Charge Distribution

If we know the electric field in a region of space due to any distribution of electric charge, we can determine the difference in potential between two points in the region using Eq. 24–4. In many cases, of course, we don't know **E** and it may be difficult to calculate. We can calculate the potential V due to a given charge distribution in another, often easier, way using the potential due to a single point charge (Eq. 24–5)

$$V = \frac{1}{4\pi\varepsilon_0} \frac{Q}{r}$$

and summing over all the charges. If we have n individual point charges, the potential at some point c is then

$$V_c = \sum_{i=1}^{n} V_i = \frac{1}{4\pi\varepsilon_0} \sum_{i=1}^{n} \frac{Q_i}{r_{ic}} \tag{24–7a}$$

where r_{ic} is the distance from the ith charge (q_i) to the point c. (We already used this approach in Example 24–4 and in Section 24–6 for a dipole.) If the charge distribution can be considered continuous, then

$$V = \frac{1}{4\pi\varepsilon_0} \int \frac{dq}{r} \tag{24–7b}$$

where r is the distance from a tiny element of charge, dq, to the point where V is being determined.

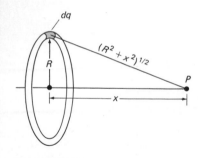

FIGURE 24–6

Calculating the potential at point P, a distance x from the center of a uniform ring of charge (Example 24–6).

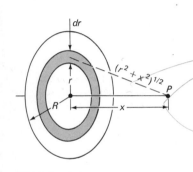

FIGURE 24–7

Calculating the electric potential at point P on the axis of a uniformly charged thin disk.

EXAMPLE 24–6 A thin circular ring of radius R carries a uniformly distributed charge Q. Determine the electric potential at a point P on the axis of the ring a distance x from its center, Fig. 24–6.

SOLUTION Each point on the ring is equidistant from point P, and this distance is $(x^2 + R^2)^{1/2}$. So the potential at P is, using Eq. 24–7b,

$$V = \frac{1}{4\pi\varepsilon_0} \int \frac{dq}{r} = \frac{1}{4\pi\varepsilon_0} \frac{1}{(x^2 + R^2)^{1/2}} \int dq = \frac{1}{4\pi\varepsilon_0} \frac{Q}{(x^2 + R^2)^{1/2}}.$$

EXAMPLE 24–7 A thin flat disk, of radius R, carries a uniformly distributed charge Q, Fig. 24–7. Determine the potential at a point P on the axis of the disk, a distance x from its center.

SOLUTION Divide the disk into thin rings of radius r and thickness dr. The charge Q is distributed uniformly, so the charge contained in each ring is proportional to its area:

$$dq = Q \frac{(2\pi r)(dr)}{\pi R^2} = \frac{2Qr\,dr}{R^2}.$$

Then the potential at P is, using Eq. 24–7b

$$V = \frac{2Q}{4\pi\varepsilon_0 R^2} \int_0^R \frac{r\,dr}{(x^2 + r^2)^{1/2}} = \frac{Q}{2\pi\varepsilon_0 R^2} (x^2 + r^2)^{1/2} \Big|_{r=0}^{r=R}$$

$$= \frac{Q}{2\pi\varepsilon_0 R^2} \left[(x^2 + R^2)^{1/2} - x \right].$$

EXAMPLE 24–8 *Uniformly charged conducting sphere.* Determine the potential at a distance r from the center of a uniformly charged conducting sphere of radius R for (a) $r > R$, (b) $r = R$, (c) $r < R$. The total charge on the sphere is Q.

SOLUTION (a) The charge Q is distributed over the surface of the sphere since it is a conductor. We saw in Example 23–1 that the electric field outside the sphere in this case is

$$E = \frac{1}{4\pi\varepsilon_0} \frac{Q}{r^2} \qquad\qquad [r > R]$$

and points radially outward (inward if $Q < 0$). Since we know **E**, we use Eq. 24–4; we integrate along a radial line, so dl is parallel to **E** (as in Fig. 24–4) between two points which are distances r_a and r_b from the sphere's center:

$$V_b - V_a = -\int_{r_a}^{r_b} \mathbf{E} \cdot dl = -\frac{Q}{4\pi\varepsilon_0} \int_{r_a}^{r_b} \frac{dr}{r^2} = \frac{Q}{4\pi\varepsilon_0} \left(\frac{1}{r_b} - \frac{1}{r_a} \right).$$

If we let $V = 0$ for $r = \infty$ (say $V_b = 0$ at $r_b = \infty$), then at any other point r $(r > R)$ we have

$$V = \frac{1}{4\pi\varepsilon_0} \frac{Q}{r}. \qquad\qquad [r > R]$$

(b) As r approaches R, we see that

$$V = \frac{1}{4\pi\varepsilon_0} \frac{Q}{R} \qquad\qquad [r = R]$$

at the surface of the conductor. (c) For points within the conductor, $E = 0$. Thus the integral, $\int \mathbf{E} \cdot dl$, between $r = R$ and any point within the conductor gives zero change in V. Hence V is constant within the conductor:

$$V = \frac{1}{4\pi\varepsilon_0} \frac{Q}{R}. \qquad [r \leq R]$$

The whole conductor, not just its surface, is at this same potential. Plots of both E and V as a function of r are shown in Fig. 24–8 for a conducting sphere.

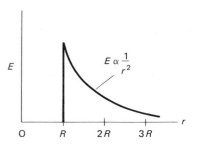

EXAMPLE 24–9 *Breakdown voltage.* In many kinds of equipment, very high voltages are used. A problem with high voltage is that the air can become ionized due to the high electric fields—that is, the electric field pulls electrons off the N_2 or O_2 atoms; the air then becomes conducting and the high voltage cannot be maintained as charge flows. The breakdown of air occurs for electric fields of about 3×10^6 V/m. (*a*) Show that the breakdown voltage for a spherical conductor in air is proportional to the radius of the sphere, and (*b*) calculate the breakdown voltage in air for a sphere of diameter 1.0 cm.

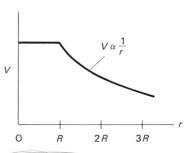

SOLUTION (*a*) The electric potential at the surface of a spherical conductor (Example 24–8) of radius R is

$$V = \frac{1}{4\pi\varepsilon_0} \frac{Q}{R}$$

and the electric field just outside the surface (Example 23–1) is

$$E = \frac{1}{4\pi\varepsilon_0} \frac{Q}{R^2}.$$

Hence

$$V = RE. \qquad [\text{at surface of spherical conductor}]$$

(*b*) For $R = 5 \times 10^{-3}$ m, the breakdown voltage in air is

$$V = (5 \times 10^{-3} \text{ m})(3 \times 10^6 \text{ V/m}) \approx 15{,}000 \text{ V}.$$

FIGURE 24–8
E vs. r and V vs. r for a uniformly charged solid conducting sphere of radius R (the charge distributes itself on the surface); r is the distance from the center of the sphere.

This example makes clear why large terminals are used for high-voltage equipment. It also explains why breakdown, or sparks, occur at rough edges or points (regions of small radius of curvature) on a conductor and why conductors are usually made very smooth.

 ## 24–8 **E** Determined from V

Equation 24–4 can be used to determine the difference in potential between two points if the electric field is known in the region between those two points. By inverting Eq. 24–4, we can write the electric field in terms of the potential; then the electric field can be determined from a knowledge of V. Let us see how to do this.

We can write Equation 24–4 in infinitesimal form as

$$dV = -\mathbf{E} \cdot d\mathbf{l} = -E_l \, dl$$

where dV is the infinitesimal difference in potential between two points a distance dl apart, and E_l is the component of the electric field in the direction of the infinitesimal displacement $d\mathbf{l}$. We can then write

$$E_l = -\frac{dV}{dl}. \qquad (24\text{–}8)$$

Thus *the component of the electric field in any direction is equal to the negative of the rate of change of the electric potential with distance in that direction.* The quantity dV/dl is called the gradient of V in a particular direction. If the direction is not specified, the term *gradient* refers to that direction in which V changes most rapidly; this would be the direction of **E** at that point, for in this direction the component of **E** would be the full magnitude E:

$$E = -\frac{dV}{dl}. \qquad [dl \parallel \mathbf{E}]$$

If **E** is written as a function of x, y, and z, and we let l refer to the x, y, and z axes, then Eq. 24–8 becomes

$$E_x = -\frac{\partial V}{\partial x}, \quad E_y = -\frac{\partial V}{\partial y}, \quad E_z = -\frac{\partial V}{\partial z}. \qquad (24\text{–}9)$$

Here, $\partial V/\partial x$ is the "partial derivative" of V with respect to x, with y and z held constant.

EXAMPLE 24–10 Determine the electric field at point P on the axis of (a) a circular ring of charge (Fig. 24–6) and (b) a uniformly charged disk (Fig. 24–7).

SOLUTION (a) From Example 24–6,

$$V = \frac{1}{4\pi\varepsilon_0} \frac{Q}{(x^2 + R^2)^{1/2}}.$$

Then

$$E_x = -\frac{\partial V}{\partial x} = \frac{1}{4\pi\varepsilon_0} \frac{Qx}{(x^2 + R^2)^{3/2}}$$

$$E_y = E_z = 0.$$

This is the same result we obtained in Example 22–6.
(b) From Example 24–7,

$$V = \frac{Q}{2\pi\varepsilon_0 R^2} \left[(x^2 + R^2)^{1/2} - x \right]$$

so

$$E_x = -\frac{\partial V}{\partial x} = \frac{Q}{2\pi\varepsilon_0 R^2} \left[1 - \frac{x}{(x^2 + R^2)^{1/2}} \right]$$

$$E_y = E_z = 0.$$

For points very close to the disk, $x \ll R$, this can be approximated by

$$E_x \approx \frac{Q}{2\pi\varepsilon_0 R^2} = \frac{\sigma}{2\varepsilon_0}$$

where $\sigma = Q/\pi R^2$ is the surface charge density. This is the result we obtained in Example 22–8 for a plane of charge; why would you expect this correspondence?

EXAMPLE 24–11 Determine the components of the electric field, E_x and E_y, at any point P in the xy plane due to a dipole, Fig. 24–9. Assume $r = (x^2 + y^2)^{1/2} \gg l$.

SOLUTION From Eq. 24–5,

$$V = \frac{1}{4\pi\varepsilon_0} \frac{p \cos\theta}{r^2}$$

$$= \frac{1}{4\pi\varepsilon_0} \frac{px}{(x^2 + y^2)^{3/2}}$$

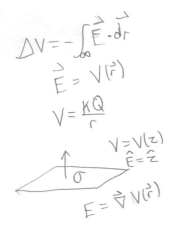

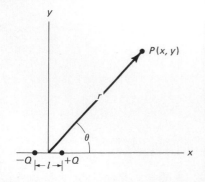

FIGURE 24–9

Determining the electric field at any point due to a dipole (Example 24–11).

since $r^2 = x^2 + y^2$ and $\cos \theta = x/r = x/(x^2 + y^2)^{1/2}$. Thus

$$E_x = -\frac{\partial V}{\partial x} = -\frac{p}{4\pi\varepsilon_0}\left(\frac{1}{(x^2 + y^2)^{3/2}} - \frac{3x^2}{(x^2 + y^2)^{5/2}}\right)$$

$$= -\frac{p}{4\pi\varepsilon_0}\frac{(y^2 - 2x^2)}{(x^2 + y^2)^{5/2}}$$

and

$$E_y = -\frac{\partial V}{\partial y} = \frac{p}{4\pi\varepsilon_0}\frac{3xy}{(x^2 + y^2)^{5/2}}.$$

In Section 22–11 we saw that $E = E_x = -(P/4\pi\varepsilon_0)(1/r^3)$ for points on the y axis ($x = 0$), which is consistent with our more general results here.

In this last example we obtained **E** for a dipole at any point in space. This would have been much more difficult if we had added the electric field vectors due to the two charges as in Section 22–11. (There we only calculated **E** on the y axis.) Indeed, for many charge distributions, it is much easier to calculate V first, and then **E** from Eq. 24–9, than to calculate **E** due to each charge from Coulomb's law. This is because V due to many charges is a scalar sum, whereas **E** is a vector sum.

 ## 24–9   Electrostatic Potential Energy

Suppose a point charge q is moved between two points in space, a and b, where the electric potential due to other charges is V_a and V_b, respectively. The change in electrostatic potential energy of q in the field of these other charges is, according to Eq. 24–2,

$$\Delta U = U_b - U_a = q(V_b - V_a) = qV_{ba}.$$

Now suppose we have a system of several point charges. What is the electrostatic potential energy of the system? It is most convenient to choose the potential energy to be zero when the charges are very far (ideally infinitely far) apart. A single point charge, Q_1, in isolation, has no potential energy, because if there are no other charges around, no force can be exerted on it. If a second point charge Q_2 is brought close to the first charge, the potential at the position of this second charge is

$$V = \frac{1}{4\pi\varepsilon_0}\frac{Q_1}{r_{12}}$$

where r_{12} is the distance between the two; so the potential energy of the two charges is

$$U = \frac{1}{4\pi\varepsilon_0}\frac{Q_1Q_2}{r_{12}}. \tag{24–10}$$

This represents the work needed to bring Q_2 from infinity ($V = 0$) to a distance r_{12} from Q_1. Or, it is the negative of the work needed to separate them to infinity.

If the system consists of three charges, the total potential energy will be the work needed to bring all three together. Equation 24–10 represents the work needed to bring Q_2 close to Q_1; to bring a third charge Q_3 so that it is a distance r_{13} from Q_1 and r_{23} from Q_2 requires work equal to

$$\frac{1}{4\pi\varepsilon_0}\frac{Q_1Q_3}{r_{13}} + \frac{1}{4\pi\varepsilon_0}\frac{Q_2Q_3}{r_{23}}.$$

So the potential energy of a system of three point charges is

$$U = \frac{1}{4\pi\varepsilon_0}\left(\frac{Q_1 Q_2}{r_{12}} + \frac{Q_1 Q_3}{r_{13}} + \frac{Q_2 Q_3}{r_{23}}\right).$$

For a system of four charges, the potential energy would contain six such terms, and so on. (Caution must be used in making such sums to avoid double counting of the different pairs.)

Often we are not interested in the total electrostatic potential energy, but only a portion of it. For example, we may want to find the potential energy of one dipole in the presence of a second dipole. There are four charges involved, Q_1 and $-Q_1$ of one dipole and Q_2 and $-Q_2$ of the second dipole. The potential energy of one dipole in the presence of the other (sometimes called the *interaction energy*) is the work needed to bring the dipoles together from infinity. In this case we are not interested in the potential energy of the Q_1 and $-Q_1$ pair or of the Q_2 and $-Q_2$ pair; the interaction energy of the two dipoles would involve only the four terms between Q_1 and Q_2, Q_1 and $-Q_2$, $-Q_1$ and Q_2, and $-Q_1$ and $-Q_2$.

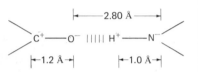

FIGURE 24–10

Diagram for calculation of interaction energy between C=O dipole of thymine and H—N dipole of adenine.

EXAMPLE 24–12 The two strands of a DNA molecule (the genetic material in human and other biological cells) are held together by electrostatic forces. (See Fig. 22–27 and its caption.) These forces include dipole-dipole interactions, such as that shown in Fig. 24–10. Calculate the interaction energy between the C=O dipole (on a thymine molecule) and the H—N dipole (on an adenine molecule) when they are aligned as shown in Fig. 24–10 (all distances are from experimental measurements, with 1 Å $= 10^{-10}$ m). The dipole moments of C=O and H—N are about 8.0×10^{-30} C·m and 3.0×10^{-30} C·m, respectively.

SOLUTION The interaction energy, U, will be equal to the potential energy of one dipole in the presence of the other since this will be equal to the negative of the work needed to pull them infinitely far apart. However, we cannot use the dipole approximation (Eq. 24–6) for the potential since the distances from the dipole are not much larger than the dipoles themselves. Hence we must use Eq. 24–10 for the potential between two point charges and we will have four terms:

$$U = U_{CH} + U_{CN} + U_{OH} + U_{ON}.$$

Here, U_{CH} means the potential energy of C in the presence of H, and similarly for the the other terms. We do not have terms corresponding to C and O or N and H since the two dipoles are assumed to be stable entities. This relation becomes:

$$U = \frac{1}{4\pi\varepsilon_0}\left(\frac{Q_C Q_H}{r_{CH}} + \frac{Q_C Q_N}{r_{CN}} + \frac{Q_O Q_H}{r_{OH}} + \frac{Q_O Q_N}{r_{ON}}\right).$$

Now $Q_C = -Q_O = p/l = (8.0 \times 10^{-30}$ C·m$)/(1.2 \times 10^{-10}$ m$) = 6.6 \times 10^{-20}$ C, and $Q_H = -Q_N = p/l = (3.0 \times 10^{-30}$ C·m$)/(1.0 \times 10^{-10}$ m$) = 3.0 \times 10^{-20}$ C. The distances are obtained from Fig. 24–10: $r_{CH} = 3.0 \times 10^{-10}$ m, $r_{CN} = 4.0 \times 10^{-10}$ m, $r_{OH} = 1.8 \times 10^{-10}$ m and $r_{ON} = 2.8 \times 10^{-10}$ m. Putting in these numbers and factoring out the powers of ten, we find

$$U = (9 \times 10^9 \text{ N·m}^2/\text{C}^2)\left(\frac{(6.6)(3.0)}{(3.0)} + \frac{(6.6)(-3.0)}{(4.0)}\right.$$
$$\left. + \frac{(-6.6)(3.0)}{(1.8)} + \frac{(-6.6)(-3.0)}{(2.8)}\right)\frac{10^{-40} \text{ C}^2}{10^{-10} \text{ m}}$$
$$= (5.9 - 4.5 - 9.9 + 6.4) \times 10^{-20} \text{ J} = -2.1 \times 10^{-20} \text{ J}$$
$$= -0.13 \text{ eV}.$$

The potential energy is negative, which means that work is required to separate the molecules.

The *electric potential* at any point in space is defined as the electric potential energy per unit charge. The *electric potential difference* between any two points is defined as the negative of the work done by the field in moving a unit electric charge between the two points. Potential difference is measured in volts (1 V = 1 J/C) and sometimes referred to as *voltage*. The change in potential energy of a charge q when it moves through a potential difference V_{ba} is $\Delta U = qV_{ba}$. The potential difference V_{ba} between two points (a and b) where a uniform electric field $\mathbf{E}$ exists is given by $V = -Ed$, where d is the distance (parallel to the field lines) between the two points. For a nonuniform field $\mathbf{E}$ the relation is $V_{ba} = -\int_a^b \mathbf{E} \cdot d\mathbf{l}$. Thus V_{ba} can be found in any region where $\mathbf{E}$ is known.

If V is known, the components of $\mathbf{E}$ can be found from the inverse of the above relation, namely $E_x = -\partial V/\partial x$, $E_y = -\partial V/\partial y$, $E_z = -\partial V/\partial z$.

An *equipotential line* or surface is all at the same potential, and is perpendicular to the electric field at all points.

The electric potential due to a single point charge Q, relative to zero potential at infinity, is given by

$$V = \frac{1}{4\pi\varepsilon_0} \frac{Q}{r}.$$

The potential due to any charge distribution can be obtained by summing (or integrating) over the potentials for all the charges.

Questions

1. An electron is accelerated by a potential difference of, say, 100 V. How much greater would its final speed be if it were accelerated with four times as much voltage?

2. Is it possible for a plastic comb to acquire a charge of 0.5 V?

3. If two points are at the same potential, does this mean that no work is done in moving a test charge from one point to the other? Does this imply that no force must be exerted?

4. If a negative charge is initially at rest in an electric field, will it move toward a region of higher potential or lower potential? What about a positive charge? How does the potential energy of the charge change in each case?

5. Can a particle ever move from a region of low electric potential to one of high potential and yet have its electric potential energy decrease? Explain.

6. State clearly the difference between (a) electric potential and electric field, (b) electric potential and electric potential energy.

7. If $V = 0$ at a point in space, must $\mathbf{E} = 0$? If $\mathbf{E} = 0$ at some point, must $V = 0$ at that point? Explain. Give examples for each.

8. When dealing with practical devices, we often take the ground (the earth) to be 0 V. If instead we said the ground was -10 V, how would this affect (a) V, (b) E, at other points? (c) Does the fact that the earth carries a net charge affect the choice of V at its surface?

9. Draw in a few equipotential lines in Fig. 22–21b.

10. Can two equipotential lines cross? Explain.

11. What can you say about the electric field in a region of space that has the same potential throughout?

12. Is there a point between two equal positive charges where the electric field is zero? Where the electric potential is zero? Explain.

13. If the potential at a point is zero, must the electric field also be zero? Give an example.

14. A satellite orbits the earth along a gravitational equipotential line. What shape must the orbit be?

15. Suppose the charged ring of Example 24–6 was not uniformly charged, so that the density of charge was twice as high near the top as near the bottom. Would this affect the potential at point P on the axis (Fig. 24–6)? Would it affect the value of $\mathbf{E}$ at that point? Is there a discrepancy here? Explain.

16. Consider an oblong conductor, such as a chunk of metal, in the shape of a football. If it carries a total charge Q, where would you expect the charge density σ to be greatest, at the ends or along the flatter sides? Explain. (Hint: near the surface of a conductor, $E = \sigma/\varepsilon_0$.)

17. A conducting sphere carries a charge Q and a second identical conducting sphere is neutral. The two are initially insulated, but then they are placed in contact. (a) What can you say about the potential of each when they are in contact? (b) Will charge flow from one to the other? If so, how much? (c) If the spheres do not have the same radius, how are your answers to parts (a) and (b) altered?

18. Why is the electric field near the center of a disk of charge, as calculated at the end of Example 24–10, the same as that calculated for a plane of charge (Example 22–8)?

19. At a particular point, the electric field points due north. In what direction(s) will the rate of change of potential be (a) greatest, (b) least, (c) zero?

20. If you know V at a point in space, can you calculate $\mathbf{E}$ at that point? If you know $\mathbf{E}$ at a point can you calculate V at that point? If not, what else must be known in each case?

21. Equipotential lines are spaced 1.00 V apart. Does the distance between the lines in different regions of space tell you anything about the relative strengths of $\mathbf{E}$ in those regions? If so, what?

22. If the electric field $\mathbf{E}$ is uniform in a region, what can you infer about the electric potential V? If V is uniform in a region of space, what can you infer about $\mathbf{E}$?

23. Is the electric potential energy of two unlike charges positive or negative? What about two like charges? What is the significance of the sign of the PE in each case?

SECTION 24–1

1. (I) How much work is needed to move a -8.0-μC charge from ground to a point whose potential is $+600$ V higher?

2. (I) An electron acquires 6.4×10^{-16} J of kinetic energy when it is accelerated by an electric field from plate A to plate B. What is the potential difference between the plates, and which plate is at the higher potential?

3. (II) A lightning flash transfers 30 C of charge to earth through a potential difference of 3.5×10^7 V. (a) How much energy is dissipated? (b) How much water at $0°$C could be brought to boiling?

4. (II) The work done by an external force to move a -2.0-μC charge from point a to point b is 8.0×10^{-4} J. If the charge was started from rest and had 1.0×10^{-4} J of kinetic energy when it reached point b, what must be the potential difference between a and b?

5. (II) Show analytically that the electrostatic force, as given by Coulomb's law, is a conservative force.

SECTION 24–2

6. (I) The electric field between two parallel plates connected to a 45-V battery is 1500 V/m. How far apart are the plates?

7. (I) How strong is the electric field between two parallel plates 5.0 mm apart if the potential difference between them is 110 V?

8. (II) In a TV picture tube, electrons are accelerated by thousands of volts through a vacuum. If a TV set were laid on its back, would electrons be able to move upward against the force of gravity? What potential difference, acting over a distance of 20 cm, would be needed to balance the downward force of gravity so that an electron would remain stationary? Assume that the electric field is uniform.

9. (II) An electron is accelerated horizontally from rest in a TV picture tube by a potential difference of 20,000 V. It then passes between two horizontal plates 6.0 cm long and 1.0 cm apart which have a potential difference of 200 V Fig. 24–11. At what angle θ will the electron be traveling after it passes between the plates?

SECTION 24–3

10. (II) Equipotential surfaces are to be drawn 1.00 V apart near a very large uniformly charged plate carrying a charge density $= 0.55$ mC/m^2. How far apart (in space) are the equipotential surfaces?

SECTION 24–4

11. (I) What potential difference is needed to give a helium nucleus ($Q = 3.2 \times 10^{-19}$ C) 48 keV of kinetic energy?

12. (I) What is the average kinetic energy (in electron volts) of an oxygen molecule in air at STP ($0°$C, 1 atm)?

13. (I) What is the speed of a proton whose kinetic energy is 20 MeV?

SECTION 24–5

14. (I) (a) What is the electric potential 0.50×10^{-10} m from a proton (charge $+e$)? (b) What is the potential energy of an electron at this point?

15. (II) Consider point a which is 85 cm north of a -45 mC point charge and point b which is 60 cm west of the charge. Determine (a) $V_{ba} = V_b - V_a$, (b) $\mathbf{E}_b - \mathbf{E}_a$ (magnitude and direction).

16. (II) A $+25$-μC charge is placed 5.0 cm from an identical $+25$-μC charge. How much work would be required by an external force to move a $+0.12$-μC test charge from a point midway between them to a point 1.0 cm closer to either of the charges?

17. (II) A 3.0-μC and a -2.0-μC charge are placed 2.0 cm apart. At what points along the line joining them is (a) the electric field zero, and (b) the potential zero?

SECTION 24–6

18. (I) An electron and a proton are 0.53×10^{-10} cm apart. (a) What is their dipole moment if they are at rest? (b) What is the average dipole moment if the electron revolves about the proton in a circular orbit?

19. (I) Calculate the electric potential due to a dipole whose dipole moment is 4.8×10^{-30} C·m at a point 1.0×10^{-9} m away if this point is (a) along the axis of the dipole nearer the positive charge; (b) $45°$ above the axis but nearer the positive charge; (c) $45°$ above the axis but nearer the negative charge.

20. (II) (a) In Example 24–5, part (b), calculate the electric potential without using the dipole approximation, Eq. 24–6; that is, don't assume $r \gg l$. (b) What is the percent error in this case when the dipole approximation is used?

21. (II) One type of electric quadrupole consists of two dipoles placed end to end with their negative charges touching; that is, a $-2Q$ point charge is flanked on either side by a $+Q$ charge each a distance l away. (a) Determine a precise formula for the potential V at a distance r from the central charge along the line joining the three charges. (Let $r > l$ but not $\gg l$.) (b) Show that for $r \gg l$,

$$V = \frac{1}{4\pi\varepsilon_0} \frac{2Ql^2}{r^3}.$$

The quantity ($2Ql^2$) is called the quadrupole moment. Compare the dependence on r to that for a dipole and to that for a single point charge. (c) What dependence on r would you expect for the potential of an electric octopole? How might you form an electric octopole?

22. (II) The dipole moment, considered as a vector, points from the negative to the positive charge. The water molecule, Fig. 24–12, has a dipole moment $\mathbf{p}$ which can be considered as the vector sum of the two dipole moments $\mathbf{p}_1$ and $\mathbf{p}_2$ as shown. The distance between each H and the

FIGURE 24–11

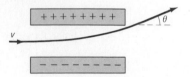

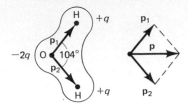

FIGURE 24–12

O is about 0.96×10^{-10} m; the lines joining the center of the O atom with each H atom make an angle of 104° as shown, and the net dipole moment has been measured to be $p = 6.1 \times 10^{-30}$ C·m. (a) Determine the charge q on each H atom. (b) Determine the electric potential, far from the molecule, due to each dipole, $\mathbf{p}_1$ and $\mathbf{p}_2$, and show that

$$V = \frac{1}{4\pi\varepsilon_0} \frac{p \cos\theta}{r^2}$$

where p is the magnitude of the net dipole moment, $\mathbf{p} = \mathbf{p}_1 + \mathbf{p}_2$, and V is the total potential due to both $\mathbf{p}_1$ and $\mathbf{p}_2$.

SECTION 24–7

23. (I) Draw an oblong conductor in the shape of a football. This conductor carries a net negative charge, $-Q$. Draw in a dozen or so electric field lines and equipotential lines.

24. (II) The electric potential of a very large flat metal plate is V_0; it carries a uniform distribution of charge of surface density $\sigma(C/m^2)$. Determine V at a distance x from the plate. Consider the point x to be far from the edges and assume x is much smaller than the plate dimensions.

25. (II) The earth produces an inwardly directed electric field of magnitude 150 V/m near its surface. (a) What is the potential of the earth's surface relative to $V = 0$ at $r = \infty$? (b) If the potential of the earth is chosen to be zero, what is the potential at infinity? (Ignore the fact that positive charge in the ionosphere approximately cancels the earth's net charge; how would this affect your answer?)

26. (II) What is the maximum amount of charge that a spherical conductor of radius 5.0 cm can hold?

27. (II) What minimum radius must a large conducting sphere of an electrostatic generating machine have if it is to carry 30,000 V without discharge into the air? How much charge will it carry?

28. (II) A 32-cm-diameter conducting sphere is charged to 500 V. (a) What is the surface charge density σ? (b) At what distance will the potential due to the sphere be only 10 V?

29. (II) How much voltage must be used to accelerate a proton so that it has just sufficient energy to touch the surface of an iron nucleus? An iron nucleus has a charge 26 times that of the proton ($= e$) and its radius is about 4.0×10^{-15} m. Assume the nucleus is spherical and uniformly charged.

30. (II) An insulated spherical conductor of radius R_1 carries a charge Q. A second conducting sphere of radius R_2 and initially uncharged is then connected to the first by a long conducting wire. (a) After the connection, what can you say about the electric potential of each sphere? (b) How much charge is transferred to the second sphere? Assume the connected spheres are far apart compared to their radii. (Why make this assumption?)

31. (II) Suppose a uniform layer of electrons was held near the surface of the earth by the earth's gravity. What is the maximum number of electrons that could be held in this manner? Ignore all other electric charges and fields except those of the electrons themselves.

32. (II) Determine the difference in potential between two points that are distances R_a and R_b from a very long ($\gg R_a$ or R_b) straight wire carrying a uniform charge per unit length λ.

33. (II) A very long conducting cylinder (length L) of radius R_0 ($R_0 \ll L$) carries a uniform surface charge density $\sigma(C/m^2)$. The cylinder is at an electric potential V_0. What is the potential, at points far from the end, at a distance r from the center of the cylinder. Determine for (a) $r > R_0$. (b) $r < R_0$. (c) Is $V = 0$ at $r = \infty$ (assume $L = \infty$)? Explain.

34. (III) A hollow spherical conductor, carrying a net charge $+Q$, has inner radius R_1 and outer radius $R_2 = 2R_1$. At the center of the sphere is a point charge $+Q/2$. Determine the potential as a function of r, the distance from the center for (a) $0 < r < R_1$, (b) $R_1 < r < R_2$, (c) $r > R_2$. (d) Determine the electric field strength E in all three regions as a function of r. (e) Plot both V and E as a function of r from $r = 0$ to $r = 2R_2$.

35. (III) A nonconducting sphere of radius R carries a total charge Q distributed uniformly throughout its volume. Determine the electric potential as a function of the distance r from the center of the sphere for (a) $r > R$, (b) $r < R$. Take $V = 0$ at $r = \infty$. (c) Plot V vs. r and E vs. r. Compare to Example 24–8.

36. (III) A nonconducting sphere of radius R_2 contains a concentric spherical cavity of radius R_1. The material between R_1 and R_2 carries a uniform charge density ρ (C/m^3). Determine the electric potential V, relative to $V = 0$ at $r = \infty$, as a function of the distance r from the center for (a) $r > R_2$, (b) $R_1 < r < R_2$, (c) $r < R_1$. Is V continuous at R_1 and R_2?

37. (III) Repeat problem 35 assuming the charge density ρ increases as the square of the distance from the center of the sphere, and $\rho = 0$ at the center.

SECTION 24–8

38. (I) What is the potential gradient near the surface of a uranium nucleus ($Q = +92e$) whose diameter is about 15×10^{-15} m.

39. (I) Show that the electric field of a single point charge (Eq. 22–4) follows from Eq. 24–5, $V = (1/4\pi\varepsilon_0)(Q/r)$.

40. (II) The electric potential in a region of space varies as $V = ay/(b^2 + y^2)$. Determine $\mathbf{E}$.

41. (II) In polar coordinates (see Appendix C) the relation between $\mathbf{E}$ and V is

$$E_r = -\frac{\partial V}{\partial r}, \qquad E_\theta = -\frac{1}{r}\frac{\partial V}{\partial \theta}.$$

(a) Show, starting with Eq. 24–8, that these relations are plausible. (b) Determine the components of the electric field, E_r and E_θ, at any point in space (r, θ), for an electric dipole (Fig. 24–5). Assume $r \gg l$.

42. (III) A thin short horizontal length l of wire carries a uniformly distributed charge Q, Fig. 24–13. Determine the electric potential (a) at a distance y above its midpoint, (b) at a distance x from the center of the wire on a line

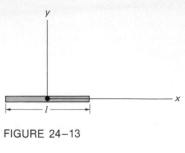

FIGURE 24–13

p_1 p_2

FIGURE 24–14

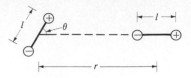

FIGURE 24–15

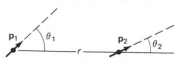

FIGURE 24–16

through the length of the wire ($x > l/2$). (Choose $V = 0$ at infinity.) (*c*) Determine the *x* and *y* components of **E** for each of the points $(0, y)$ and $(x, 0)$.

SECTION 24–9

43. (I) Determine the mutual electrostatic potential energy (in electron volts) of two protons in a uranium (^{235}U) nucleus (*a*) if they are at the surface, on opposite sides of the nucleus, (*b*) if one is at the center and the other is at the surface. The diameter of a ^{235}U nucleus is about 15×10^{-15} m.

44. (I) Write the total electrostatic energy, U, for (*a*) four point charges, (*b*) five point charges.

45. (I) How much work must be done to bring three electrons from a great distance apart to within 1.0×10^{-10} m from one another?

46. (II) Show that if two dipoles with dipole moments p_1 and p_2 are in line with one another (Fig. 24–14), the potential energy of one in the presence of the other (their "interaction energy") is given by

$$U = -\frac{1}{2\pi\varepsilon_0} \frac{p_1 p_2}{r^3}$$

where r is the distance between the two dipoles. Assume that r is much greater than the length of either dipole.

47. (II) Two dipoles interact most strongly when they are directly lined up. To see that this is so, calculate the interaction energy for the two dipoles shown in Fig. 24–15 (*a*) when they are aligned ($\theta = 0$); (*b*) when the angle θ between their axes is 45°; (*c*) when θ is 90°; (*d*) when θ is 180°. Assume that the distance from the center of one dipole to the other is 3.0 Å in each case, that the charge on each is $0.40e$, and that $l = 1.0$ Å.

48. (II) Four point charges are located at the corners of a square that is 8.0 cm on a side. The charges, going in rotation around the square, are Q, $2Q$, $-3Q$, and $2Q$, where $Q = 4.8$ μC. What is the total electric potential energy stored in the system?

49. (II) Four equal point charges, Q, are fixed at corners of a square of side b. (*a*) What is their total electrostatic potential energy? (*b*) How much potential energy will a fifth charge, Q, have at the center of the square (relative to $V = 0$ at ∞)? (*c*) Is the fifth charge in stable or unstable equilibrium? If unstable, what maximum kinetic energy could it acquire? (*d*) Repeat part (*c*) for a negative ($-Q$) charge.

50. (II) Repeat problem 49 assuming that two of the charges, on diagonally opposite corners, are replaced by $-Q$ charges.

51. (II) In the Bohr model of the hydrogen atom, the electron orbits the nucleus (a single proton) in a circular orbit of radius r. Determine r from the measured value of the ionization energy (energy needed to remove the electron) of 13.6 eV.

52. (II) Determine the total electrostatic potential energy of a conducting sphere of radius R that carries a total charge Q distributed uniformly on its surface.

53. (III) Determine the total electrostatic potential energy of a nonconducting sphere of radius R carrying a total charge Q distributed uniformly throughout its volume.

54. (III) Show that the electrostatic potential energy of two dipoles in a plane as shown in Fig. 24–16 is given by

$$U = \frac{1}{4\pi\varepsilon_0} \frac{p_1 p_2}{r^3} \left[\cos(\theta_1 - \theta_2) - 3\cos\theta_1 \cos\theta_2\right].$$

Assume r is much larger than the length of each dipole. The vector dipole moments, $\mathbf{p}_1$ and $\mathbf{p}_2$, point from the negative charge toward the positive charge of the dipole.

Capacitance, Dielectrics, Electric Energy Storage

25

This chapter will complete our study of electrostatics. It deals first of all with an important device, the capacitor, which is used in nearly all electronic circuits. We will also discuss the possibility of electric energy storage and the effects of a so-called dielectric (nonconductor) on electric fields and potential differences.

 ## 25–1   Capacitors

A **capacitor** is a device for storing electric charge, and consists of two conducting objects placed near one another but not touching. A typical capacitor consists of a pair of parallel plates of area A separated by a small distance d, Fig. 25–1a. Often the

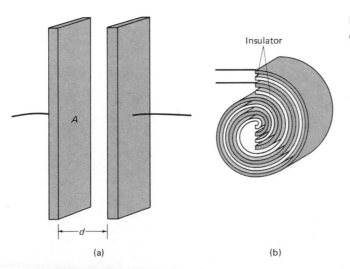

FIGURE 25–1
Capacitors.

Insulator

A

d

(a)   (b)

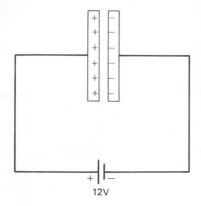

FIGURE 25–2

Parallel-plate capacitor connected
to a battery.

two plates are rolled into the form of a cylinder with paper or other insulator
between the plates (Fig. 25–1b).

    Suppose a voltage is applied to a capacitor, say by connecting the capacitor to
a battery, as in Fig. 25–2. (A battery is a device which maintains a relatively constant
voltage across its terminals.) When connected to a battery, the capacitor quickly
becomes charged. One plate acquires a negative charge, the other an equal amount
of positive charge. For a given capacitor, it is found that the amount of charge Q
acquired by each plate is proportional to the potential difference V_{ba}:

$$Q = CV_{ba}. \qquad (25\text{–}1)$$

The constant of proportionality, C, in the above relation is called the **capacitance**
of the capacitor. The units of capacitance are coulombs per volt and this unit
is called a *farad* (F). Most capacitors have capacitance in the range 1 pF
(picofarad $= 10^{-12}$ F) to 1 μF (microfarad $= 10^{-6}$ F). The relation, Eq. 25–1, was
first suggested by Volta (see Section 26–1) in the late eighteenth century.

25–2   Determination of Capacitance

The capacitance C is a constant for a given capacitor. The value of the capacitance C
depends on the size, shape, and relative position of the two conductors, and also on
the material that separates them. In this section, we assume the conductors are
separated by a vacuum, or air.

    The capacitance of a given capacitor can be determined experimentally
directly from Eq. 25–1, by measuring the charge Q on either conductor for a given
potential difference V_{ba}.

    For capacitors whose geometry is simple, we can determine C analytically. To
illustrate this, we now determine C for a parallel-plate capacitor, Fig. 25–3. Each
plate has area A and the two plates are separated by a distance d. We assume d is
small compared to the dimensions of each plate so that the electric field $\mathbf{E}$ is uniform
between them and we can ignore fringing (lines of $\mathbf{E}$ not straight) at the edges. We
saw earlier (Example 22–9) that the electric field between two closely spaced parallel
plates has magnitude $E = \sigma/\varepsilon_0$ and its direction is perpendicular to the plates. Since
σ is the charge per unit area, $\sigma = Q/A$, and

$$E = \frac{Q}{\varepsilon_0 A}.$$

The relation between electric field and electric potential, as given by Eq. 24–4, is

$$V_{ba} = -\int_a^b \mathbf{E} \cdot d\mathbf{l}.$$

We can take the line integral along a path antiparallel to the field lines, from one
plate to the other; so

$$V_{ba} = +\int_a^b E\, dl = \frac{Q}{\varepsilon_0 A} \int_a^b dl = \frac{Qd}{\varepsilon_0 A}.$$

This relates Q to V_{ba}, and from it we can get the capacitance C in terms of the
geometry of the plates:

$$C = \frac{Q}{V_{ba}} = \varepsilon_0 \frac{A}{d}. \qquad \begin{bmatrix} \text{parallel-plate} \\ \text{capacitor} \end{bmatrix} \quad (25\text{–}2)$$

This relation makes sense intuitively: a larger area A means that for a given number
of charges, there will be less repulsion between them (they're further apart), so we
expect that more charge can be held on each plate; and a greater separation d means
the charge on each plate exerts less attractive force on the other plate, so less charge
is drawn from the battery, and the capacitance is less.

FIGURE 25–3

Parallel-plate
capacitor whose
plates have
area A.

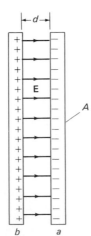

Note also from Eq. 25–2 that the value of C does not depend on Q or V, so Q is predicted to be proportional to V as is found experimentally.

EXAMPLE 25–1 (a) Calculate the capacitance of a capacitor whose plates are 20 cm $\times$ 3.0 cm and are separated by a 1.0-mm air gap. (b) What is the charge on each plate if the capacitor is connected to a 12-V battery?

SOLUTION (a) The area $A = (20 \times 10^{-2} \text{ m})(3.0 \times 10^{-2} \text{ m}) = 6.0 \times 10^{-3} \text{ m}^2$. The capacitance C is then

$$C = (8.85 \times 10^{-12} \text{ C}^2/\text{N·m}^2)\frac{(6.0 \times 10^{-3} \text{ m}^2)}{(1.0 \times 10^{-3} \text{ m})} = 53 \text{ pF}.$$

(b) The charge on each plate is

$$Q = CV = (53 \times 10^{-12} \text{ F})(12 \text{ V}) = 6.4 \times 10^{-10} \text{ C}.$$

The proportionality $C \propto A/d$ in Eq. 25–2 is valid also for a parallel-plate capacitor that is rolled up into a spiral cylinder, as in Fig. 25–1b; however, the constant factor, ε_0, must be replaced if an insulator separates the plates, as is usual, and this is discussed in Section 25–5. For a true cylindrical capacitor—consisting of two long coaxial cylinders—the result is somewhat different as the next example shows.

EXAMPLE 25–2 A cylindrical capacitor consists of a cylinder (or wire) of radius R_1 surrounded by a coaxial cylindrical shell of inner radius R_2, Fig. 25–4a. Both cylinders have length L which we assume is much greater than the separation of the cylinders, $R_2 - R_1$. The capacitor is charged (say, by connecting it to a battery) so that one cylinder has a charge $+Q$ (say, the inner one) and the other one a charge $-Q$. Determine a formula for the capacitance.

SOLUTION We need to determine the potential difference between the cylinders, V_{ba}, in terms of Q. We might feel safe in using our earlier result (Example 22–7) that the electric field outside a long wire is directed radially outward and has magnitude $E = (1/2\pi\varepsilon_0)(\lambda/r)$ where r is the distance from the axis and λ is the charge per unit length, Q/L; then we would guess that $E = (1/2\pi\varepsilon_0)(Q/Lr)$ for points between the cylinders. But let us start from scratch and carefully determine the electric field using Gauss's law. We assume the cylinders are charged uniformly, and so the electric field depends only on the radial distance r from the center of the

FIGURE 25–4

(a) Cylindrical capacitor consists of two coaxial cylindrical conductors. (b) The electric field lines (the solid lines) and a cylindrical gaussian surface (the dashed line) are shown in the cross-sectional view.

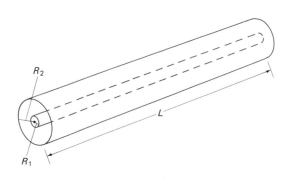

(a)

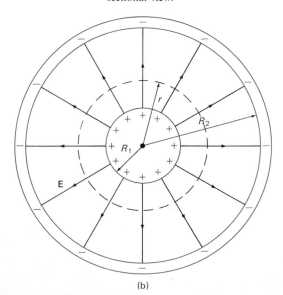

(b)

cylinder, and due to the symmetry it points radially outward as shown in Fig. 25–4b. (Since $R_2 - R_1 \ll L$, we can ignore fringing of the field near the ends.) We imagine a closed cylindrical surface, of radius r, where $R_1 < r < R_2$, coaxial with the two conductors as shown dashed in Fig. 25–4b. We use this as our gaussian surface. No flux passes through the end caps, so it all passes through the curved surface. Since E is uniform over this surface, whose area is $(2\pi r)(L)$, Gauss's law gives

$$\oint \mathbf{E} \cdot d\mathbf{A} = Q/\varepsilon_0$$

$$E(2\pi r L) = Q/\varepsilon_0$$

since the surface encloses the total charge Q of the inner cylinder. Then

$$E = \frac{1}{2\pi\varepsilon_0 L} \frac{Q}{r}. \qquad [R_1 < r < R_2]$$

This confirms our earlier result. To obtain V_{ba} in terms of Q, we use the above result in Eq. 24–4 and take the line integral from the outer cylinder to the inner one along a radial line:

$$V_{ba} = -\int_a^b \mathbf{E} \cdot d\boldsymbol{l}$$

$$= -\frac{Q}{2\pi\varepsilon_0 L} \int_{R_2}^{R_1} \frac{dr}{r} = \frac{Q}{2\pi\varepsilon_0 L} \ln \frac{R_2}{R_1},$$

where, since our path is "inward", $d\boldsymbol{l} = -d\mathbf{r}$ and $\mathbf{E}$ is in the opposite direction to $d\boldsymbol{l}$. Again, Q and V_{ba} are proportional, and the capacitance C is

$$C = \frac{Q}{V_{ba}} = \frac{2\pi\varepsilon_0 L}{\ln(R_2/R_1)}. \qquad \begin{bmatrix} \text{cylindrical} \\ \text{capacitor} \end{bmatrix}$$

Does the dependence on L, R_1, and R_2 make sense intuitively? (See discussion immediately after Eq. 25–2.)

A single isolated conductor can also be said to have a capacitance, C. In this case, C is defined as the ratio of the charge to absolute potential V on the conductor (relative to $V = 0$ at $r = \infty$), so that the relation

$$Q = CV$$

remains valid. For example, the potential of a conducting sphere of radius r_0 (Example 24–8) carrying a net charge Q is

$$V = \frac{1}{4\pi\varepsilon_0} \frac{Q}{r_0};$$

so its capacitance is

$$C = \frac{Q}{V} = 4\pi\varepsilon_0 r_0.$$

But note that a single conductor alone is not considered a capacitor. In practical cases, a single conductor may be near other conductors or the earth (which can be thought of as the other "plate" of a capacitor), and these will affect the value of the capacitance.

EXAMPLE 25–3 *Electrometer.* Show that a simple leaf electroscope or electrometer (Section 22–4) when used for measurements on other objects is a potential-measuring device and not a charge-measuring device. Let all potentials and capacitances be relative to earth.

SOLUTION   As we saw in Section 22–4, the deflection of the leaves on an electroscope is a measure of the charge on the electroscope. Since $Q = CV$, the leaf deflection could also be calibrated for the potential V on the electroscope. (Neither scale would be linear, nor would they be identical since the capacitance C_e of the electroscope itself would depend at least slightly on the leaf separation.) If we wish to use an electroscope as a practical device to measure an electrical state of some other object, we note first of all that when the object is connected to the electroscope, the two come to the same potential V (relative to earth, say). When the two are connected, some of the charge on the object will flow to the previously uncharged electroscope. If more than a small fraction of the object's charge Q flows thereby to the electroscope, the object's own potential (and charge, of course) will change—that is, the act of measuring will seriously affect the system and the quantity being measured. We can minimize this by requiring the capacitance C_e of the electroscope to be much less than the capacitance C of the charged object; in this way, the charge Q_e transferred will be minimized: that is, since V is the same for both and $Q = CV$ and $Q_e = C_eV$, then $Q_e \ll Q$ if $C_e \ll C$. From this it follows that the potential V of the object is not seriously affected when connected to the electroscope. The charge Q_e transferred to the electroscope, and therefore the separation of its leaves, will depend only on the potential V (since $Q_e = C_eV$) as long as our object is kept far enough away (and the connecting wire is thin enough so that it carries negligible charge) so as not to exert an external influence on the electroscope (such as inducing a charge on it). Thus an electroscope can be calibrated to measure the potential V. But it *cannot* be calibrated to measure charge on different objects. This can easily be seen by considering a second object at the same potential but whose capacitance, call it C', is twice that of our original object (C): $C' = 2C$. The electroscope will give the same leaf deflection, since this is governed by $Q_e = C_eV$ and V is the same. But note that the second object has charge $Q' = C'V = 2CV = 2Q$ or twice that of our first object. The electroscope gives the same reading for both these charges (Q and $2Q$), and so clearly it is not useful for measuring charge on various external objects. There is one exception to this last statement: If the object on which charge is to be measured is always the same—or at least always has the same capacitance—then the electroscope could be calibrated to measure charge for this special system. But such a severe restriction prevents us from saying that an electroscope can measure charge in general.

This example of what appears to be a simple device should serve as a caution about the use of any measuring device. It is always important to investigate (1) just what the device can be expected to measure, and how accurately, and (2) how much the device itself affects the system on which measurements are being made. (We will see other examples when we discuss voltmeters and ammeters in Sections 27–6 and 27–7.)

25–3   Capacitors in Series and Parallel

Capacitors can be connected together in a variety of ways. This is often done in practical circuits and the effective capacitance of the combination depends on how they are connected. The two basic ways are to connect them in parallel or in series, as illustrated in Fig. 25–5. (When drawing electric circuits, a capacitor is represented by the symbol ─││─ and a battery by ─│├─.) We first consider a parallel connection as shown in Fig. 25–5a. If a battery of voltage V is connected to points a and b, this voltage exists across each of the capacitors; that is, since the left-hand plates of all the capacitors are connected by conductors, they all reach the same potential when connected to the battery; the same can be said for the right-hand plates. Each capacitor plate acquires a charge given by $Q_1 = C_1V$, $Q_2 = C_2V$, and $Q_3 = C_3V$. The total charge Q that must leave the battery is

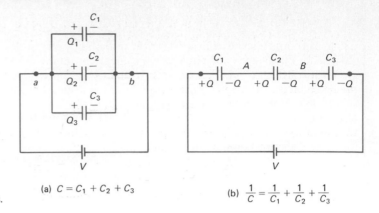

FIGURE 25–5

Capacitors (a) in parallel, (b) in series.

(a) $C = C_1 + C_2 + C_3$

(b) $\dfrac{1}{C} = \dfrac{1}{C_1} + \dfrac{1}{C_2} + \dfrac{1}{C_3}$

then

$$Q = Q_1 + Q_2 + Q_3 = C_1 V + C_2 V + C_3 V.$$

A single equivalent capacitor that will hold the same charge Q at the same voltage V will have a capacitance C given by

$$Q = CV.$$

Thus we have

$$CV = C_1 V + C_2 V + C_3 V,$$

or

$$C = C_1 + C_2 + C_3. \qquad \text{[capacitors in parallel]} \quad (25\text{--}3)$$

The net effect of connecting capacitors in parallel is thus to increase the capacitance. This is just what we should expect since we are essentially increasing the area of the plates for charge to accumulate on (see, for example, Eq. 25–2).

If the capacitors are connected in series, as in Fig. 25–5b, a charge $+Q$ flows from the battery to one plate of C_1, and $-Q$ flows to one plate of C_3. The regions A and B between the capacitors were originally neutral; so the net charge there must still be zero. The $+Q$ on the left plate of C_1 attracts a charge of $-Q$ on the opposite plate; because region A must have a zero net charge, there is thus $+Q$ on the left plate of C_2. The same considerations apply to the other capacitors, so we see the charge on each capacitor is the same, namely Q. A single capacitor that could replace these three in series without affecting the circuit would have a capacitance C where

$$Q = CV.$$

Now the total voltage V across the three capacitors in series must equal the sum of the voltages across each capacitor:

$$V = V_1 + V_2 + V_3.$$

We also have $Q = C_1 V_1$, $Q = C_2 V_2$, and $Q = C_3 V_3$, so we substitute for V_1, V_2, and V_3 into the last equation and get

$$\frac{Q}{C} = \frac{Q}{C_1} + \frac{Q}{C_2} + \frac{Q}{C_3}$$

or

$$\frac{1}{C} = \frac{1}{C_1} + \frac{1}{C_2} + \frac{1}{C_3}. \qquad \begin{bmatrix} \text{capacitors} \\ \text{in series} \end{bmatrix} \quad (25\text{--}4)$$

Other connections of capacitors can be analyzed in terms of series and parallel connections.

EXAMPLE 25-4 Determine the capacitance of a single capacitor that will have the same effect as the combination shown in Fig. 25-6. Take $C_1 = C_2 = C_3 = C$.

SOLUTION C_2 and C_3 are connected in parallel, so they are equivalent to a single capacitor having capacitance

$$C_{23} = C_2 + C_3 = 2C.$$

C_{23} is in series with C_1, so the equivalent capacitance, C_e, is given by

$$\frac{1}{C_e} = \frac{1}{C_1} + \frac{1}{C_{23}} = \frac{1}{C} + \frac{1}{2C} = \frac{3}{2C}.$$

Hence $C_e = \frac{2}{3}C$, which is the equivalent capacitance of the entire combination.

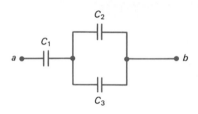

FIGURE 25-6
Example 25-4.

25-4 Electric Energy Storage

A charged capacitor stores electrical energy. The energy stored in a capacitor will be equal to the work done to charge it. The net effect of charging a capacitor is to remove charge from one plate and add it to the other plate. This is what a battery does when it is connected to a capacitor. Initially, when the capacitor is uncharged, it requires no work to move the first bit of charge over. When some charge is on each plate, it requires work to add more charge of the same sign because of the electric repulsion. The more charge already on a plate, the more work is required to add more. The work needed to add a small amount of charge dq, when a potential difference V is across the plates, is $dW = V\,dq$. Since $V = q/C$ at any moment (Eq. 25-1), where C is the capacitance, the work done is

$$W = \int_0^Q V\,dq = \frac{1}{C}\int_0^Q q\,dq = \frac{1}{2}\frac{Q^2}{C}.$$

Thus we can say that the energy "stored" in a capacitor is

$$U = \frac{1}{2}\frac{Q^2}{C}$$

energy stored in the capacitor **in terms of the capacitance**

when the capacitor C carries charges $+Q$ and $-Q$ on its two conductors. Since $Q = CV$, where V is the potential difference across the capacitor, we can also write

$$U = \frac{1}{2}\frac{Q^2}{C} = \frac{1}{2}CV^2 = \frac{1}{2}QV. \qquad (25-5)$$

in terms of the voltage

$$U = \frac{1}{2}\frac{Q^2}{C} = \frac{1}{2}CV^2 + \frac{1}{2}QV$$

EXAMPLE 25-5 A 12-V battery is connected to a 20-μF capacitor. How much electric energy can be stored in the capacitor?

SOLUTION From Eq. 25-5 we have

$$\text{energy} = \tfrac{1}{2}CV^2 = \tfrac{1}{2}(20 \times 10^{-6}\ \text{F})(12\ \text{V})^2 = 1.4 \times 10^{-3}\ \text{J}.$$

Energy is not a substance and does not have a definite location. Nonetheless, it is often useful to think of it as being stored in the electric field between the plates. As an example let us calculate the energy stored in a parallel-plate capacitor in terms of the electric field.

We have seen (Eq. 24-3) that the electric field **E** between two parallel plates is (approximately) uniform and its magnitude is related to the potential difference by $V = Ed$ where d is the separation. Also, Eq. 25–2 tells us $C = \varepsilon_0 A/d$ for a parallel-plate capacitor. Thus

$$U = \frac{1}{2}CV^2 = \frac{1}{2}\left(\frac{\varepsilon_0 A}{d}\right)(E^2 d^2)$$

$$= \frac{1}{2}\varepsilon_0 E^2 Ad.$$

The quantity Ad is the volume between the plates in which the electric field E exists. If we divide both sides by the volume, we obtain an expression for the energy per unit volume or **energy density**, u:

$$u = \text{energy density} = \frac{1}{2}\varepsilon_0 E^2. \tag{25–6}$$

The *electric energy stored per unit volume in any region of space is proportional to the square of the electric field* in that region. We derived Eq. 25–6 for the special case of a parallel-plate capacitor. But it can be shown to be true for any region of space where there is an electric field.

25–5   Dielectrics

In most capacitors there is an insulating sheet of material called a **dielectric** (such as paper or plastic) between the plates. This serves several purposes. First of all, dielectrics break down (allowing electric charge to flow) less readily than air, so higher voltages can be applied without charge passing across the gap. Furthermore, a dielectric allows the plates to be placed closer together to increase the capacitance without fear of them touching.[†] Finally, it is found experimentally that if the dielectric fills the space between the two conductors, it increases the capacitance by a factor K which is called the **dielectric constant**. Thus

$$C = KC_0 \tag{25–7}$$

where C_0 is the capacitance when the space between the two conductors of the capacitor is a vacuum and C is the capacitance when the space is filled with a material whose dielectric constant is K. The values of the dielectric constant for various materials are given in Table 25–1. Note that for air (at 1 atm pressure) $K = 1.0006$ which differs so little from 1.0000 that the capacitance for an air-filled capacitor hardly differs from that for a vacuum.

For a parallel plate capacitor (see Eq. 25–2),

$$C = K\varepsilon_0\frac{A}{d}, \qquad \left[\begin{array}{c}\text{parallel plate}\\\text{capacitor}\end{array}\right] \tag{25–8}$$

when the space between the plates is completely filled with a dielectric whose dielectric constant is K. (The situation when the dielectric only partially fills the space is dealt with later—see Example 25–6.) The quantity $K\varepsilon_0$ appears so often in formulas that we define a new quantity

$$\varepsilon = K\varepsilon_0 \tag{25–9}$$

called the *permittivity* of a material. Then the capacitance of a parallel-plate capacitor becomes

$$C = \varepsilon\frac{A}{d}.$$

Note that ε_0 represents the permittivity of free space (a vacuum). Also, the energy density stored in an electric field E (Section 25–4) in a dielectric is given by

TABLE 25–1

Dielectric constants (at 20°C)

Material	Dielectric Constant K
Vacuum	1.0000
Air (1 atm)	1.0006
Paraffin	2.2
Rubber, hard	2.8
Vinyl (plastic)	2.8–4.5
Paper	3–7
Quartz	4.3
Glass	4–7
Porcelain	6–8
Mica	7
Ethyl alcohol	24
Water	80

[†] As we saw in Example 25–1 above, a simple parallel-plate capacitor of ordinary dimensions has a capacitance of only a few picofarads. A capacitance of 1 μF or more would require huge plates if dielectrics weren't used.

$$u = \tfrac{1}{2}K\varepsilon_0 E^2 = \tfrac{1}{2}\varepsilon E^2.$$

The effect of a dielectric on capacitance was first explored thoroughly by Faraday. He found that when the space between the plates of a capacitor was filled with a dielectric, the charge that accumulated on each plate for a given applied voltage was somewhat greater than when air occupied the space between the plates. That is, if the charge on each plate of an air-filled capacitor is Q_0, then when a dielectric is inserted between the plates and the same voltage V_0 is applied, the charge that the battery supplies to each plate is increased to

$$Q = KQ_0. \qquad\qquad \text{[voltage constant]}$$

This is consistent with Eq. 25–7 since the capacitance has increased by

$$C = \frac{Q}{V_0} = \frac{KQ_0}{V_0} = KC_0$$

where $C_0 = Q_0/V_0$ is the capacitance in the absence of a dielectric.

Now let us consider a slightly different case from that just described (voltage kept constant while dielectric inserted). This time we let the plates of the capacitor acquire a charge

$$Q_0 = CV_0$$

when connected to a battery of voltage V_0, and then, before inserting the dielectric, the battery is disconnected; now when the dielectric is inserted, and we assume it fills the space between the plates, the charge Q_0 on each plate is unchanged. In this case it is found that the potential difference across the plates decreases by a factor K:

$$V = \frac{V_0}{K}.$$

Again, the capacitance is

$$C = \frac{Q_0}{V} = \frac{Q_0}{V_0/K} = K\frac{Q_0}{V_0} = KC_0.$$

Both of these results are consistent with Eq. 25–7.

The electric field within a dielectric is also altered. When no dielectric is present, the electric field between the plates of a parallel-plate capacitor is given by Eq. 24–3:

$$E_0 = \frac{V_0}{d}$$

where V_0 is the potential difference between the plates and d is their separation. If the capacitor is isolated so that the charge remains fixed on the plates when a dielectric is inserted, the potential difference drops to $V = V_0/K$. So the electric field in the dielectric is now

$$E = \frac{V}{d} = \frac{V_0}{Kd}$$

or

$$E = \frac{E_0}{K}. \qquad\qquad \text{[in a dielectric]} \quad (25\text{--}10)$$

The electric field within the dielectric is thus also reduced by a factor equal to the dielectric constant. Although the field is reduced in a dielectric (or insulator), it is not reduced all the way to zero as in a conductor.

FIGURE 25–7

Molecular view of the effects
of a dielectric.

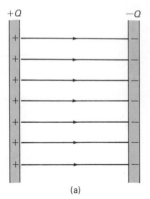

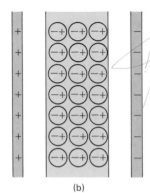

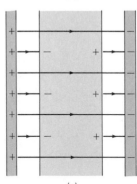

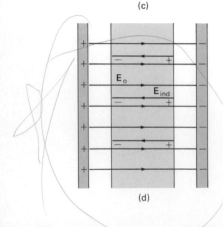

Dielectric behavior can be understood from the molecular point of view. Consider a capacitor whose plates are separated by an air gap. This capacitor has a charge $+Q$ on one plate and $-Q$ on the other, Fig. 25–7a. The capacitor is isolated (not connected to a battery). The potential difference between the plates, V_0, is given by Eq. 25–1: $Q = C_0 V_0$. (The zero subscripts refer to the situation when only air is between the plates.) Now we insert a dielectric between the plates, Fig. 25–7b. The molecules of the dielectric may be *polar*; that is, although the molecules are neutral they have a permanent dipole moment. The electric field will exert a torque on the dipoles, tending to rotate them so they are parallel to the field as shown in Fig. 25–7b; thermal motion of the molecules will prevent perfect alignment of all molecules, but the greater the field, the better the alignment. Even if the molecules are not polar, the electric field between the plates will induce a separation of charge in the molecules, so they will have an *induced dipole moment*; although the electrons do not leave the molecules, they will move slightly within the molecule toward the positive plate. So the situation is still as illustrated in Fig. 25–7b. The net effect in either case is as if there were a net negative charge on the outer edge of the dielectric facing the positive plate, and a net positive charge on the opposite side (Fig. 25–7c). Because of this induced charge on the dielectric, some of the electric field lines actually don't pass through the dielectric but end (and restart) on those induced charges, as shown in Fig. 25–7c. Hence the electric field within the dielectric is less than in the air.

Another way to visualize this is shown in Fig. 25–7d. The electric field within the dielectric can be considered as the vector sum of the electric field $\mathbf{E}_0$ due to the "free" charges on the conducting plates, and the field $\mathbf{E}_{ind}$ due to the induced charge on the dielectric; since they are in opposite directions, the net field within the dielectric, $E_0 - E_{ind}$, is less than E_0. The precise relationship is given by Eq. 25–10:

$$E_0 - E_{ind} = \frac{E_0}{K},$$

or

$$E_{ind} = E_0\left(1 - \frac{1}{K}\right).$$

(Here, from symmetry, it should be clear that for parallel plates that are large compared to their separation, the induced charge on the surface of a dielectric will be independent of whether the dielectric fills the space between the plates or not, as long as its faces are parallel to the plates; thus Eq. 25–10 remains valid in this case, although $V = V_0/K$ would not remain valid [why not?].)

The electric field between two parallel plates is related to the surface charge density, σ (see Section 23–3), by $E = \sigma/\varepsilon_0$. Thus

$$E_0 = \frac{\sigma}{\varepsilon_0}$$

where $\sigma = Q/A$ is the surface charge density on the conductor; Q is the net charge on the conductor and is often called the *free charge* (since charge is free to move in a conductor). Similarly, we define an equivalent induced surface charge density σ_{ind} and

$$E_{ind} = \frac{\sigma_{ind}}{\varepsilon_0}$$

where E_{ind} is the electric field due to the induced charge $Q_{ind} = \sigma_{ind}A$ on the surface of the dielectric, Fig. 25–7d; Q_{ind} is often called the *bound charge* (since it is on an insulator and is not free to move). Since $E_{ind} = E_0(1 - 1/K)$ as shown above, we now have

$$\sigma_{ind} = \sigma\left(1 - \frac{1}{K}\right) \tag{25–11a}$$

$$Q_{\text{ind}} = Q \left(1 - \frac{1}{K}\right). \qquad (25\text{-}11\text{b})$$

Since K is always greater than 1, we see that the charge induced on the dielectric is always less than the free charge on each of the capacitor plates.

EXAMPLE 25–6 A parallel-plate capacitor has plates of area $A = 250$ cm^2 and separation $d = 2.00$ mm. The capacitor is charged to a potential difference $V_0 = 150$ V. Then the battery is disconnected (the charge Q on the plates then won't change), and a dielectric sheet ($K = 3.50$) of the same area A but thickness $l = 1.00$ mm is placed between the plates as shown in Fig. 25–8. Determine (a) the initial capacitance of the air-filled capacitor, (b) the charge on each plate before the dielectric is inserted, (c) the charge induced on each face of the dielectric, (d) the electric field in the space between each plate and the dielectric, (e) the electric field in the dielectric, (f) the potential difference between the plates after the dielectric is added, (g) the capacitance after the dielectric is in place.

SOLUTION (a) Before the dielectric is in place,

$$C_0 = \varepsilon_0 \frac{A}{d} = (8.85 \times 10^{-12} \text{ C}^2/\text{N}\cdot\text{m}^2) \frac{(2.50 \times 10^{-2} \text{ m}^2)}{(2.00 \times 10^{-3} \text{ m})} = 111 \text{ pF}.$$

(b) The charge on each plate is

$$Q = C_0 V_0 = (1.11 \times 10^{-10} \text{ F})(150 \text{ V}) = 1.66 \times 10^{-8} \text{ C}.$$

(c) From Eq. 25–11b

$$Q_{\text{ind}} = Q\left(1 - \frac{1}{K}\right) = (1.66 \times 10^{-8} \text{ C})\left(1 - \frac{1}{3.5}\right) = 1.19 \times 10^{-8} \text{ C}.$$

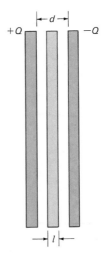

FIGURE 25–8
Example 25–6.

(d) The electric field in the gaps between the plates and the dielectric (see Fig. 25–7c) is the same as in the absence of the dielectric since the charge on the plates has not been altered. Gauss's law, as applied in Example 23–5, could be used here, which gives $E_0 = \sigma/\varepsilon_0$; or we can note that, in the absence of the dielectric, $E_0 = V_0/d = Q/C_0 d$ (since $V_0 = Q/C_0$) $= Q/\varepsilon_0 A$ (since $C_0 = \varepsilon_0 A/d$) which is the same result; thus

$$E_0 = \frac{Q}{\varepsilon_0 A} = \frac{1.66 \times 10^{-8} \text{ C}}{(8.85 \times 10^{-12} \text{ C}/\text{N}\cdot\text{m}^2)(2.50 \times 10^{-2} \text{ m}^2)} = 7.50 \times 10^4 \text{ V/m}.$$

(e) In the dielectric the electric field is (Eq. 25–10):

$$E_d = \frac{E_0}{K} = \frac{7.50 \times 10^4 \text{ V/m}}{3.50} = 2.14 \times 10^4 \text{ V/m}.$$

(f) To obtain the potential difference in the presence of the dielectric we use Eq. 24–4 and integrate along a straight line parallel to the field lines:

$$V = -\int \mathbf{E} \cdot d\mathbf{l} = E(d - l) + E_d l$$

which can be simplified to

$$V = E_0 \left(d - l + \frac{l}{K}\right)$$

$$= (7.50 \times 10^4 \text{ V/m})(1.00 \times 10^{-3} \text{ m} + 1.00 \times 10^{-3} \text{ m}/3.50)$$

$$= 96.4 \text{ V}.$$

(g) In the presence of the dielectric

$$C = \frac{Q}{V} = \frac{1.66 \times 10^{-8} \text{ C}}{96.4 \text{ V}} = 172 \text{ pF}.$$

Note that if the dielectric filled the space between the plates, the answers to (f) and (g) would be 42.9 V and 387 pF, respectively.

 EXAMPLE 25–7 A dielectric, whose dielectric constant is K, fills the space between the plates of a parallel-plate capacitor. The plates have area A and are separated by a distance d; they are charged to a potential difference V and then the battery is disconnected (so the charge on the plates won't change). The dielectric is now slowly pulled out of the capacitor. How much work is required? Explain why work is needed.

SOLUTION The energy stored in the capacitor initially is (see Eq. 25–5):

$$U_1 = \tfrac{1}{2}CV^2$$

where $C = K\varepsilon_0 A/d$. After the dielectric is removed, the capacitance drops by a factor K but the voltage is increased by a factor K. Hence

$$U_2 = \frac{1}{2}\left(\frac{C}{K}\right)(KV)^2 = K(\tfrac{1}{2}CV^2) = KU_1.$$

So the energy stored in the capacitor has increased by a factor K. Work is therefore required to remove the dielectric and the amount needed (neglecting friction) is

$$W = U_2 - U_1 = \tfrac{1}{2}CV^2(K - 1).$$

That work is needed to remove the dielectric can be seen intuitively from the fact that there will be a force of attraction between the induced charge on the dielectric and the charges on the plates (Fig. 25–7c) as it is pulled out. Hence an external force must be exerted to overcome this, and work must be done.

FIGURE 25–9
Gauss's law in a dielectric.

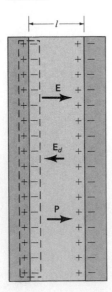

* 25–6 Gauss's Law in Dielectrics

We now discuss the use of Gauss's law in a situation when a dielectric is present. Consider a parallel-plate capacitor containing a dielectric that fills the space between the plates as shown in Fig. 25–9. We assume the plates are large (of area A) compared to the separation l so that **E** is uniform and perpendicular to the plates. For our gaussian surface, we choose the long rectangular box indicated by the dashed lines in Fig. 25–9, which just barely reaches into the dielectric. The surface encloses both the free charge Q on the conductor, and the induced (bound) charge Q_{ind} on the dielectric, so

$$\oint \mathbf{E} \cdot d\mathbf{A} = \frac{Q - Q_{\text{ind}}}{\varepsilon_0}. \tag{25–12}$$

Since, from Eq. 25–11b, $Q_{\text{ind}} = Q(1 - 1/K)$, we have $Q - Q_{\text{ind}} = Q(1 - 1 + 1/K) = Q/K$, so

$$\oint \mathbf{E} \cdot d\mathbf{A} = \frac{Q}{K\varepsilon_0} = \frac{Q}{\varepsilon}. \tag{25–13}$$

This relation (Eq. 25–13), although obtained for a special case, is valid in general when dielectrics are present. Note that Q in Eq. 25–13 is the *free* charge only. The induced bound charge is not included since it is accounted for by the factor K (or ε).

For the surface shown in Fig. 25–9, we have $\mathbf{E} = 0$ within the conductor, so there is no flux through that part of the surface inside the conductor; also there is essentially no flux through the short sides of the box, since $\mathbf{E}$ is nearly parallel to the sides, and besides the sides are very short so the contribution would be very small anyway. So the only flux is that through the surface within the dielectric. Thus, Gauss's law gives

$$E_d A = \frac{Q}{\varepsilon_0} - \frac{Q_{\text{ind}}}{\varepsilon_0} = \frac{Q}{\varepsilon}$$

where we have used both Eqs. 25–12 and 25–13. E_d represents the field inside the dielectric and is given by either

$$E_d = \frac{Q - Q_{\text{ind}}}{\varepsilon_0 A}$$

or

$$E_d = \frac{Q}{K\varepsilon_0 A} = \frac{Q}{\varepsilon A} = \frac{\sigma}{\varepsilon}.$$

In this result, as well as in the general form of Gauss's law, Eq. 25–13, we see that our earlier relations for E are altered only by replacing ε_0 by $K\varepsilon_0 = \varepsilon$.

*25–7 The Polarization and Electric Displacement Vectors (**P** and **D**)

The rectangular dielectric between the plates of the charged parallel plates in Fig. 25–9 has a dipole moment whose magnitude is

$$Q_{\text{ind}} l$$

where l is the thickness of the dielectric and Q_{ind} is the charge induced on the surface of the dielectric. For any dielectric, we can define a new quantity, the *polarization vector*, **P**, which is the *dipole moment per unit volume*. For a rectangular dielectric of thickness l and whose faces have area A,

$$P = \frac{Q_{\text{ind}}\, l}{Al} = \frac{Q_{\text{ind}}}{A} = \sigma_{\text{ind}}.$$

Thus the magnitude of the polarization vector, in this case, is equal to the surface charge density induced on the dielectric.[†]

The polarization vector points from the negative charge sheet on one side of the dielectric to the positive charge sheet on the other (just as a dipole moment does), as shown in Fig. 25–9. For the surface shown, we can write

$$\oint \mathbf{P} \cdot d\mathbf{A} = PA = Q_{\text{ind}} \tag{25–14}$$

since **P** is zero in the conductor and is parallel to the short sides of the rectangular box chosen as our gaussian surface. Equation 25–14 is valid in general and we can combine it with Eq. 25–12 to obtain

$$\oint \mathbf{E} \cdot d\mathbf{A} = \frac{Q}{\varepsilon_0} - \frac{1}{\varepsilon_0} \oint \mathbf{P} \cdot d\mathbf{A}$$

or

$$\oint (\varepsilon_0 \mathbf{E} + \mathbf{P}) \cdot d\mathbf{A} = \mathbf{Q}. \tag{25–15}$$

This is another way to write Gauss's law when dielectrics are present (see also Eqs. 25–12 and 25–13), and it too is a general result. It can be written in terms

[†] For more complicated cases, σ is equal to the component of **P** perpendicular to the surface.

of a new vector called the *electric displacement*, **D**, defined as

$$\mathbf{D} = \varepsilon_0 \mathbf{E} + \mathbf{P}. \qquad (25\text{–}16)$$

Then Gauss's law becomes

$$\oint \mathbf{D} \cdot d\mathbf{A} = Q. \qquad (25\text{–}17)$$

In the dielectric between the plates of the parallel-plate capacitor of Fig. 25–9, this relation gives

$$D = \frac{Q}{A} \qquad \left[\begin{array}{c}\text{parallel-plate}\\\text{capacitor}\end{array}\right] \quad (25\text{–}18)$$

where Q is the free charge.

There is a simple interpretation of the three vectors **E**, **D**, and **P**. The electric field **E** is due to *all* charges, whether free or bound, as implied by Eq. 25–12. The polarization vector **P**, as can be seen from Eq. 25–14, is connected only with the induced bound charge. Finally, the electric displacement vector **D** is associated with the free charge only, as implied by Eqs. 25–17 and 25–18.

The vector **E** is, still, the basic electric field vector. The vectors **P** and **D** are useful auxiliaries for more advanced work, although we will not use them much here.

 ## Summary

A *capacitor* is a device used to store charge and consists of two separated conductors. The two conductors generally carry equal and opposite charges, Q, and the ratio of this charge to the potential difference V between the conductors is called the *capacitance*, C; so $Q = CV$. The capacitance of a parallel-plate capacitor is proportional to the area of each plate and inversely proportional to their separation. The space between the conductors contains a nonconducting material such as air, paper, or plastic; the latter materials are referred as *dielectrics*, and the capacitance is proportional to a property of dielectrics called the dielectric constant, K (nearly equal to 1 for air).

If two or more capacitors are connected in *parallel*, the equivalent capacitance C of the combination is the sum of the individual capacitances. If several capacitors are connected in *series*, the reciprocal of the equivalent capacitance C is equal to the sum of the reciprocals of the individual capacitances.

A charged capacitor stores an amount of electric energy given by $\frac{1}{2}QV = \frac{1}{2}CV^2 = \frac{1}{2}Q^2/C$. This energy can be thought of as stored in the electric field between the plates. In any electric field **E** in free space the *energy density* (energy per unit volume) is $\frac{1}{2}\varepsilon_0 E^2$. If a dielectric is present, the energy density is $\frac{1}{2}K\varepsilon_0 E^2 = \frac{1}{2}\varepsilon E^2$, where $\varepsilon = K\varepsilon_0$ is the permittivity of the dielectric material.

 ## Questions

1. Suppose two nearby conductors carry the same negative charge. Can there be a potential difference between them? If so, can the definition of capacitance, $C = Q/V$, be used here?

2. Suppose the separation of plates d in a parallel-plate capacitor is not very small compared to the dimensions of the plates. Would you expect Eq. 25–2 to give an overestimate or underestimate of the true capacitance? Explain.

3. Suppose one of the plates of a parallel-plate capacitor was slid so that the area of overlap was reduced by half, but they are still parallel. How would this affect the capacitance?

4. Suppose one plate of a parallel-plate capacitor is tilted at one end away from the other plate so the separation at that end is $2d$. How would this affect the capacitance?

5. Explain how the relation for the capacitance of a cylindrical capacitor, Example 25–2, makes sense intuitively. Use arguments such as those just after Eq. 25–2.

6. Describe a simple method of measuring ε_0 using a capacitor.

7. When a battery is connected to a capacitor, why do the two plates acquire charges of the same magnitude? Will this be true if the two conductors are different sizes or shapes?

8. A large copper sheet of thickness l is placed between the parallel plates of a capacitor, but does not touch the plates. How will this affect the capacitance?

9. Suppose three identical capacitors are connected to a battery. Will they store more energy if connected in series or in parallel?

10. The parallel plates of an isolated capacitor carry opposite charges, Q. If the separation of the plates is increased, is a force required? Is the potential difference changed? What happens to the work done in the pulling process?

11. How does the energy in a capacitor change if (a) the potential difference is doubled, (b) the charge on each plate is doubled, (c) the separation of the plates is doubled, as the capacitor remains connected to a battery?

12. For dielectrics consisting of polar molecules, how would you expect the dielectric constant to change with temperature?

13. An isolated charged capacitor has horizontal plates. If a thin dielectric is inserted a short way between the plates, how will it move when it is then released?

14. Suppose a battery remains connected to the capacitor in the above question. What then will happen when the dielectric is released?

15. A dielectric is pulled from between the plates of a capacitor which remains connected to a battery. What changes occur to the capacitance, charge on the plates, potential difference, energy stored, and electric field?

16. How does the energy stored in a capacitor change when a a dielectric is inserted if (a) the capacitor is isolated so Q doesn't change, (b) the capacitor remains connected to a battery so V doesn't change?

17. We have seen that the capacitance C depends on the size, shape, and position of the two conductors, as well as on the dielectric constant K. What then did we mean when we said that C is a constant in Eq. 25–1?

18. What value might we assign to the dielectric constant for a good conductor? Explain.

19. *Dissolving Power of Water.* The very high dielectric constant of water, $K = 80$ (see Table 25–1), has a profound effect on materials in that it allows many of them to be dissolved in water. For example, ordinary table salt, NaCl (sodium chloride), whose crystal structure (Fig. 25–10a) is held together by the attractive forces between the ions Na$^+$ and Cl$^-$, is easily dissolved when placed in water. Explain why we would expect that the electric field produced by each ion would be reduced by a factor equal to the dielectric constant; that is, discuss the extension of Eq. 25–10 to the field of a point charge in a dielectric, and thus explain (using this simple model) how salt is dissolved (see Fig. 25–10b).

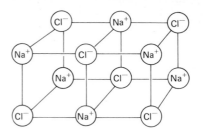

(a)

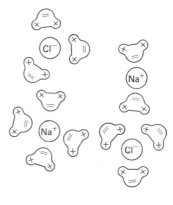

(b)

FIGURE 25–10

(a) Sodium chloride crystal; (b) sodium chloride dissolving in water.

 Problems _____

SECTION 25–1

1. (I) How much charge flows from a 12-V battery when it is connected to a 25-μF capacitor?

2. (I) A 12,000-pF capacitor holds 8.0×10^{-8} C of charge. What is the voltage across the capacitor?

3. (II) A capacitor C_1 carries a charge Q_0. It is then connected directly to a second, uncharged, capacitor C_2. What charge will each carry now? What will be the potential difference across each?

4. (II) A 2.5-μF capacitor is charged by a 35-V battery and then is disconnected from the battery. When it is then connected to a second capacitor, C_2, the voltage on the first drops to 16 V. What is the value of C_2?

5. (II) It takes 16.0 J of energy to move a 0.20-mC charge from one plate of a 60-μF capacitor to the other. How much charge is on each plate?

6. (II) A 2.4-μF capacitor is charged to 200 V and a 1.10-μF capacitor is charged to 60 V. (a) The positive plates are now connected to each other and the negative plates are connected to each other; what will be the potential difference across each and the charge on each? (b) What is the voltage and charge for each if plates of opposite sign are connected?

SECTION 25–2

7. (I) A 2.0-F capacitor is desired. What should the area of the plates be if they are to be separated by a 4.5-mm air gap?

8. (I) Use Gauss's law to show that $\mathbf{E} = 0$ inside the inner conductor of a cylindrical capacitor (see Fig. 25–4 and Example 25–2) as well as outside the outer cylinder.

9. (I) Determine the capacitance of the earth, assuming it to be a spherical conductor.

10. (II) An electric field of 8.0×10^6 V/m is desired between two parallel plates each of area 210 cm² and separated by 2.50 cm of air. What charge must be on each plate?

11. (II) In the limit of a very small separation of the two cylinders of a cylindrical capacitor ($R_2 - R_1 \ll R_1$ in Fig. 25–4) show that the relation derived in Example 25–2 reduces to that of a parallel-plate capacitor (Eq. 25–2).

12. (II) A large metal sheet of thickness l is placed between, and parallel to, the plates of the parallel-plate capacitor of Fig. 25–3. It does not touch the plates, and extends beyond their edges. (a) What is now the net capacitance in terms of A, d, and l? (b) If $l = \frac{2}{3}d$, by what factor does the capacitance change when the sheet is inserted?

13. (II) A *spherical capacitor* consists of two thin concentric spherical shells of radii R_1 and R_2 ($> R_1$). (a) Determine a formula for its capacitance. (b) Show that for $R_2 - R_1 \ll R_1$, the formula reduces to that for a parallel-plate capacitor.

14. (II) Dry air will break down if the electric field exceeds about 3.0×10^6 V/m. What amount of charge can be placed on a capacitor if the area of each plate is 8.5 cm²?

SECTION 25–3

15. (I) Six 1.5-μF capacitors are connected in parallel. What is the equivalent capacitance? What is their equivalent capacitance if connected in series?

16. (I) A circuit contains a 3.0-μF capacitor. However, a technician decides 4.8 μF would be better. What size capacitor should be added to the circuit and how should it be connected?

17. (I) The capacitance of a portion of a circuit is to be reduced from 3600 pF to 1000 pF. What capacitance can be added to the circuit to produce this effect without removing anything from the circuit? How should the extra capacitor be connected?

18. (I) Suppose three parallel-plate capacitors, whose plates have areas A_1, A_2, and A_3 and separations d_1, d_2, and d_3 are connected in parallel. Show, using only Eq. 25–2, that Eq. 25–3 is valid.

19. (II) (a) Determine the equivalent capacitance of the circuit shown in Fig. 25–11. (b) If $C_1 = C_2 = 2C_3 = 4.0$ μF, how much charge is stored on each capacitor when $V = 50$ V?

20. (II) Three conducting plates, each of area A, are connected as shown in Fig. 25–12. (a) Are the two capacitors thus formed connected in series or in parallel? (b) The middle plate can be moved (changing the values of d_1 and d_2), so as to vary the capacitance. What are the minimum and maximum values of the net capacitance? (c) Determine C

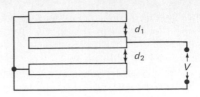

FIGURE 25–12

as a function of d_1, d_2, and A. Assume $d_1 + d_2$ is much less than the dimensions of the plates.

21. (II) You have three capacitors, of capacitance 2000 pF, 5000 pF, and 0.010 μF. What is the maximum and minimum capacitance that you can form from these? How do you make the connection in each case?

22. (II) A 0.20-μF and a 0.10-μF capacitor are connected in series to a 9.0-V battery. Calculate (a) the potential difference across each capacitor, and (b) the charge on each. (c) Repeat parts (a) and (b) assuming the two capacitors are in parallel.

23. (II) A 3.0-μF and a 4.0-μF capacitor are connected in series and this combination is connected in parallel with a 2.0-μF capacitor. (a) What is the net capacitance? (b) If 50 V is applied across the whole network, calculate the voltage across each capacitor.

24. (II) The variable capacitance of a radio tuner consists of six plates connected together placed alternately between six other plates, also connected together. Each plate is separated from its neighbor by 1.0 mm of air. One set of plates can move so the area of overlap varies from 1.0 cm² to 4.0 cm². (a) Are these eleven capacitors connected in series or in parallel? (b) Determine the range of capacitance values.

25. (III) Suppose the left-hand plate of the capacitor in Fig. 25–3 were tilted to the left at the top so it made a small angle θ with the vertical (and with the other plate); the separation at the bottom remains d. Determine a formula for C in terms of A, d, and θ. Assume the plates are square. (Hint: imagine the capacitor as many infinitesimal capacitors in parallel.)

26. (III) A voltage V is applied to the capacitor network shown in Fig. 25–13. (a) What is the equivalent capaci-

FIGURE 25–13

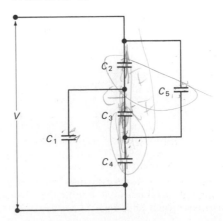

FIGURE 25–11

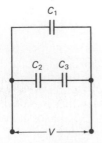

tance? (b) Determine the equivalent capacitance if $C_2 = C_4 = 3.0\ \mu F$ and $C_1 = C_3 = C_5 = 6.0\ \mu F$.

SECTION 25–4

27. (I) If 200 V is applied to a 200 pF capacitor, how much electric energy is stored?

28. (I) There is an electric field near the earth's surface whose intensity is about 150 V/m. How much energy is stored per cubic meter in this field?

29. (I) How much energy is stored by the electric field between two square plates, 11.0 cm on a side, separated by a 2.0 mm air gap? The charges on the plates are equal and opposite and of magnitude 300 μC.

30. (II) A parallel-plate capacitor has a fixed charge Q. The separation of the plates is then doubled. (a) By what factor does the energy stored in the electric field change? (b) How much work must be done to double the separation of the plates each of whose area is A?

31. (II) In Fig. 25–11, let $V = 100$ V and $C_1 = C_2 = C_3 = 1200$ pF. How much energy is stored in the capacitor network?

32. (II) How much energy must a 12-V battery expend to fully charge a 0.10-μF and a 0.20-μF capacitor when they are placed (a) in parallel, (b) in series? (c) How much charge flowed from the battery in each case?

33. (II) (a) Suppose the outer radius (R_2) of a cylindrical capacitor was doubled, but the charge was kept constant. By what factor would the stored energy change? Where would the energy come from? (b) Repeat, assuming the voltage remains constant.

34. (II) A 2.0-μF capacitor is charged by a 12-V battery. It is disconnected from the battery and then connected to an uncharged 5.0-μF capacitor. Determine the total stored energy (a) before the two capacitors are connected and (b) after they are connected. (c) What is the change in energy? (d) Is energy conserved? Explain why.

35. (II) How much work would be required to remove the metal sheet from between the plates of the capacitor in problem 12, assuming (a) the battery remains connected so the voltage remains constant, (b) the battery is disconnected so the charge remains constant?

36. (II) (a) Show that each plate of a parallel-plate capacitor exerts a force

$$F = \frac{1}{2}\frac{Q^2}{\varepsilon_0 A}$$

on the other, by calculating dW/dx where dW is the work needed to increase the separation by dx. (b) Why does using $F = QE$ give the wrong answer?

37. (II) Show that the electrostatic energy stored in the electric field outside an isolated spherical conductor of radius R carrying a net charge Q is

$$U = \frac{1}{8\pi\varepsilon_0}\frac{Q^2}{R}.$$

Do this in three ways: (a) Using Eq. 25–6 for the energy density in an electric field (Hint: consider spherical shells of thickness dr); (b) using Eq. 25–5 together with the capacitance of an isolated sphere (Section 25–2); (c) by calculating the work needed to bring all the charge Q up from infinity in infinitesimal bits dq.

SECTION 25–5

38. (I) The *dielectric strength* of a material is the maximum electric field an insulating material can withstand without breaking down and conducting electric charge through it. Typical dielectrics used in capacitors have dielectric strengths of about 10^7 V/m. (a) By what factor is this an improvement over air? (See Example 24–9.) (b) What minimum thickness of dielectric is needed in a capacitor if it is to withstand 1000 V, or (c) 50,000 V?

39. (I) What is the capacitance of two square parallel plates 5.5 cm on a side that are separated by 1.8 mm of paraffin?

40. (II) A 4500-pF air-gap capacitor is connected to a 12-V battery. If a piece of mica is placed between the plates, how much charge will flow from the battery?

41. (II) Suppose the capacitor in Example 25–7 remains connected to the battery as the dielectric is removed. What will be the work required to remove the dielectric in this case?

42. (II) How much energy would be stored in the capacitor of problem 29 if a mica dielectric is placed between the plates? Assume (a) the mica is 2.0 mm thick (and therefore fills the space between the plates), (b) the mica is 1.0 mm thick.

43. (II) Repeat Example 25–6 assuming the battery remains connected when the dielectric is inserted. Also, what is the free charge on the plates after the dielectric is added (let this be part (h) of the problem)?

44. (II) Two different dielectrics each fill half the space between the plates of a parallel-plate capacitor as shown in Fig. 25–14. Determine a formula for the capacitance in terms of K_1, K_2, the area A of the plates, and the separation d. (Hint: can you consider this capacitor as two capacitors in series or in parallel?)

45. (II) Two different dielectrics fill the space between the plates of a parallel-plate capacitor as shown in Fig. 25–15. Determine a formula for the capacitance in terms of K_1, K_2, the area A of the plates, and the separation $d_1 = d_2 = d/2$.

FIGURE 25–14

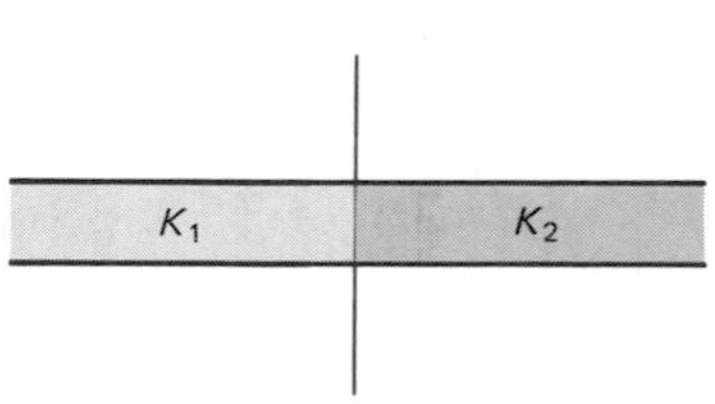

FIGURE 25–15

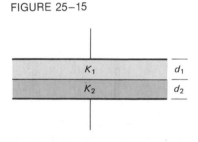

46. (II) Repeat problem 45 (Fig. 25–15) but assume the separation $d_1 \neq d_2$.

47. (II) In Example 25–6, what percent of the stored energy is stored in the dielectric?

48. (II) Using Example 25–6 as a model, derive a formula for the capacitance of a parallel-plate capacitor whose plates have area A, separation d, with a dielectric of thickness l ($l < d$) with dielectric constant K placed between the plates.

*SECTION 25–6

*49. (II) In a cylindrical capacitor (Fig. 25–4), a dielectric ($K = 3.7$) completely fills the space between the conductors whose radii are 3.0 mm and 6.0 mm, and whose length $L = 12.0$ cm. The capacitor plates each carry 1.60 μC of free charge. Determine (a) the surface charge density on each plate, (b) the induced charge on each surface of the dielectric, (c) the maximum and minimum values of the electric field in the dielectric, (d) the potential difference across the capacitor.

50. (III) Repeat problem 49, assuming the dielectric does not fully fill the space but is a cylindrical shell of inner and outer radii equal to 3.0 mm and 4.0 mm. Calculate also (e) the maximum and minimum values of the electric field in the air gap.

*SECTION 25–7

51. (II) For the capacitor of Example 25–6, determine **E**, **D**, and **P** (a) in the dielectric, (b) in the air space between the dielectric and the plates.

Electric
Current

26

Until the year 1800, the technical development of electricity consisted mainly of producing a static charge by friction. In the preceding century a number of machines had been built that could produce rather large potential by frictional means; one type of such apparatus is shown in Fig. 26–1. Large sparks could be produced by these machines, but they had little practical value.

In nature itself there were grander displays of electricity such as lightning and "St. Elmo's fire," which is a glow that appeared around the yardarms of ships during storms. That these phenomena were electrical in origin was not recognized until the eighteenth century. For example, it was only in 1752 that Franklin, in his famous kite experiment, showed that lightning was an electric discharge—a giant electric spark.

Finally, in 1800, an event of great practical importance occurred: Alessandro Volta (1745–1827) invented the electric battery, and with it produced the first steady flow of electric charge—that is, a steady electric current. This discovery opened a new era, which transformed our civilization, for today's electrical technology is based on electric current.

 ## 26–1 The Electric Battery

The events that led to the discovery of the battery are interesting; for not only was this an important discovery, but it also gave rise to a famous scientific debate between Volta and Luigi Galvani (1737–1798), eventually involving many others in the scientific world.

In the 1780s Galvani, a professor at the University of Bologna (thought to be the world's oldest university still in existence), carried out a long series of experiments on the contraction of frog's leg muscle through electricity produced by a static-electricity machine. In the course of these investigations, Galvani found, much to his surprise, that contraction of the muscle could be produced by other means as well: when a brass hook was pressed into the frog's spinal cord and then hung from an iron railing that also touched the frog, the leg muscles again would contract. With further investigation, Galvani found that this strange but important phenomenon occurred for other pairs of metals as well.

FIGURE 26–1
Early electrostatic generator.

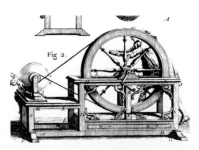

497

(a)

(b)

FIGURE 26–2

Two types of voltaic battery: (a) a pile; (b) "crown of cups." Z stands for zinc and A for silver (*argentum* in Latin). Taken from Volta's original publication. (Burndy Library, Norwalk, CT)

What was the source of this unusual phenomenon? Galvani believed that the source of the electric charge was in the frog muscle or nerve itself, and that the wire merely transmitted the charge to the proper points; and when he published his work in 1791, he termed it "animal electricity." Many wondered, including Galvani himself, if he had discovered the long-sought "life-force."

Volta, at the University of Pavia 200 km away, was at first skeptical of Galvani's results; but at the urging of his colleagues he soon confirmed and extended those experiments. But Volta doubted Galvani's idea of animal electricity. Instead he came to believe that the source of the electricity was not in the animal itself but rather in the *contact between the two metals*. Volta made public his views and soon had many followers, although others still sided with Galvani.

Volta was both a strong theoretician and a careful and skillful experimenter. He soon realized that a moist conductor, such as a frog muscle or moisture at the contact point of the two dissimilar metals, was necessary in the circuit if it was to be effective. He also saw that the contracting frog muscle was a sensitive instrument for detecting electric "tension" or "electromotive force" (his words for what we now call potential), in fact more sensitive than the best available electrometers that he and others had developed. Most important, he recognized that a decisive answer to Galvani could be given only if the sensitive frog leg was replaced by an inorganic detector; that is, to cement his view that it was the contact of two dissimilar metals that caused the frog muscle to contract, he would have to connect the two dissimilar metals directly to an electrometer and observe a separation of the leaves representing a potential difference. This proved difficult since his most sensitive[†] electrometers were much less sensitive than the frog muscle. But the eventual success of this experiment vindicated Volta's theory.

Volta's research found that certain combinations of metals produced a greater effect than others, and, using his measurements, he listed them in order of effectiveness. (This "electrochemical series" is still used by chemists today.) And he found that carbon could be used in place of one of the metals.

Volta then conceived what is his greatest contribution to science. Between a disc of zinc and one of silver, he placed a piece of cloth or paper soaked in salt solution or dilute acid and piled a "battery" of such couplings, one on top of another, as shown in Fig. 26–2; this "pile" or "battery" produced a much increased potential difference. Indeed, when strips of metal connected to the two ends of the pile were brought close, a spark was produced. Volta had designed and built the first electric battery. A second design, known as the "crown of cups" is also shown in Fig. 26–2. Volta made public this great discovery in 1800.

The potential produced by Volta's battery was still weak compared to that produced by the best friction machines of the time, although it could produce considerable charge. (The electrostatic machines were high-potential, low-charge devices.) But it had a great advantage: it was "self-renewing"—it could produce a

[†]Volta's most sensitive electrometer measured about 40 V per degree (of leaf separation). Nonetheless, he was finally able to estimate the potential differences produced by dissimilar metals in contact: for a silver-zinc contact he got about 0.7 V, remarkably close to today's value of 0.78 V.

flow of electric charge continuously for a relatively long period of time. It was not long before even more powerful batteries were constructed.

After Volta's discovery of the electric battery, it was eventually recognized that a battery produces electricity by transforming chemical energy into electrical energy. Today a great variety of electric cells and batteries are available, from flashlight batteries (sometimes called "dry cells") to the storage battery of a car. The simplest batteries contain two plates or rods made of dissimilar metals (one can be carbon) called *electrodes*. The electrodes are immersed in a solution, such as a dilute acid, called the *electrolyte*. (In a dry cell, the electrolyte is absorbed in a powdery paste.) Such a device is properly called an *electric cell*, and several cells connected together is a battery. The chemical reactions involved in most electric cells are quite complicated. A detailed description of different electric cells can be found in chemistry textbooks. Here we describe how one very simple cell works, emphasizing the physical aspects.

The simple cell shown in Fig. 26–3 uses dilute sulfuric acid as the electrolyte. One of the electrodes is made of carbon, the other of zinc. That part of each electrode remaining outside the solution is called the *terminal*, and connections to wires and circuits are made here. The acid attacks the zinc electrode and tends to dissolve it. But each zinc atom leaves two electrons behind, so it enters the solution as a positive ion. The zinc electrode thus acquires a negative charge. As more zinc ions enter solution, the electrolyte becomes increasingly positively charged. Because of this, and through other chemical reactions, electrons are pulled off the carbon electrode. Thus the carbon electrode becomes positively charged. The positive electrode is called the *anode*, the negative electrode, the *cathode*. Because there is an opposite charge on the two electrodes, there is a potential difference between the two terminals. In a cell whose terminals are not connected, only a small amount of the zinc is dissolved; for as the zinc electrode becomes increasingly negative, any new positive zinc ions produced are attracted back to the electrode. Thus a particular potential difference or voltage is maintained between the two terminals. If charge is allowed to flow between the terminals, say through a wire (or a light bulb), then more zinc can be dissolved. The carbon too suffers disintegration. After a time, one or the other electrode is used up and the cell becomes "dead."

The voltage that exists between the terminals of a battery depends on what the electrodes are made of and their relative ability to be dissolved or give up electrons. The potential difference when no charge flows to an external circuit is called the **electromotive force**, or **emf**, of the battery. The symbol $\mathscr{E}$ is often used for emf (don't confuse it with E for electric field). The emf of typical cells is 1.0 to 2.0 V. When charge is allowed to flow from a battery to an external circuit, the voltage at the terminals drops below the emf due to "internal resistance" (this is discussed in Section 27–2). When two or more cells are connected so that the positive terminal of one is connected to the negative terminal of the next, they are said to be connected in *series* and their voltages add up. Thus the voltage between the ends of two flashlight batteries so connected is 3.0 V, while the six 2-V cells of an automobile storage battery give 12 V.

Any device, such as a battery, that can maintain a potential difference and supply charge to an external circuit is called a *source*, or *seat*, of emf. Other sources of emf are electric generators, photocells, thermocouples, and so on (more on them later).

 ## 26–2 Electric Current

When a continuous conducting path, such as a wire, is connected to the terminals of a battery, we have an electric *circuit*, Fig. 26–4a. On any diagram of a circuit as in Fig. 26–4b, we will represent a battery by the symbol "—|⊢—"; the longer line on this symbol represents the positive terminal and the shorter line the negative terminal.

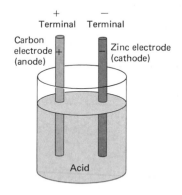

FIGURE 26–3

Simple electric cell.

FIGURE 26–4

(a) Very simple electric circuit.
(b) Schematic drawing of the circuit in part (a).

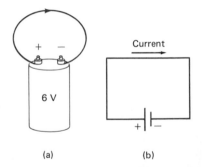

When such a circuit is formed, charge can flow through the circuit from one terminal of the battery to the other. A flow of charge such as this is called an **electric current**.

More precisely, the electric current in a wire is defined as the net amount of charge that passes through it at a given point per unit time. Thus, the average current I is defined as

$$I = \frac{\Delta Q}{\Delta t} \qquad (26-1)$$

where ΔQ is the amount of charge that passes through a cross-section of conductor at a given point during the time interval Δt. Electric current is measured in coulombs per second; this is given a special name, the *ampere* (abbreviated amp or A), after the French physicist André Ampère (1775–1836). Thus 1 A = 1 C/s. If the current is not constant in time, then we can define the instantaneous current at any moment as the infinitesimal limit as $t \to 0$:

$$I = \frac{dQ}{dt}. \qquad (26-2)$$

However, we will mainly be concerned with constant currents in this chapter.

EXAMPLE 26-1   A steady current of 2.5 A flows in a wire connected to a battery. But after 4.0 min, the current suddenly ceases because the wire is disconnected. How much charge passed through the circuit?

SOLUTION   Since the current was 2.5 A or 2.5 C/s, then in 4.0 minutes (240 s) the total charge that flowed was, from Eq. 26-1,

$$\Delta Q = I \, \Delta t$$
$$= (2.5 \text{ C/s})(240 \text{ s}) = 600 \text{ C}.$$

We saw in Chapter 22 that conductors contain many free electrons; thus, when a conducting wire is connected to the terminals of a battery as in Fig. 26–4, it is actually the negatively charged electrons that flow in the wire. When the wire is first connected, free electrons at one end of the wire are attracted into the positive terminal. At the same time, electrons leave the negative terminal of the battery and enter the wire at the other end. Thus there is a continuous flow of electrons through the wire that begins as soon as the wire is connected to *both* terminals. However, when the conventions of positive and negative charge were invented two centuries ago, it was assumed positive charge flowed in a wire. Actually, for nearly all purposes, positive charge flowing in one direction is exactly equivalent to negative charge flowing in the opposite direction, Fig. 26–5 (but see Section 28–9.) Today, we still use the historical convention of positive current flow when discussing the direction of a current. (It also fits in well with our definition of electric potential in terms of a positive test charge.) Thus when we speak of the current flowing in a circuit we mean the direction positive charge would flow. This is sometimes referred to as *conventional current*. When we want to speak of the direction of electron flow, we will specifically state it is the electron current. In liquids and gases, both positive and negative charges (ions) can move. (The microscopic view of current flow is discussed in Section 26–5.)

26–3   Ohm's Law; Resistance and Resistors

In order to produce an electric current in a circuit, a difference in potential is required. One way of producing a potential difference is by a battery. It was Georg Simon Ohm (1787–1854) who established experimentally that the current in a metal wire is proportional to the potential difference V applied to its ends:

$$I \propto V.$$

FIGURE 26-5

Conventional current from + to − is equivalent to a negative (electron) current flowing from − to +.

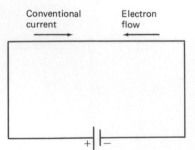

Conventional
current

Electron
flow

+ −

This is known as **Ohm's law**. If, for example, we connect a wire to a 6-V battery, the current flow will be twice what it would be if the wire were connected to a 3-V battery.

It is helpful to compare an electric current to the flow of water in a river or a pipe. If the pipe (or river) is nearly level, the flow rate is small. But if one end is somewhat higher than the other, the flow rate—or current—is much greater. The greater the difference in height, the greater the current. We saw in Chapter 24 that electric potential is analogous, in the gravitational case, to the height of a cliff; and this applies in the present case to the height through which the fluid flows. Just as an increase in height causes a greater flow of water, so a greater electric potential difference, or voltage, causes a greater current flow.

Exactly how much current flows in a wire depends not only on the voltage but also on the resistance the wire offers to the flow of electrons. The walls of a pipe, or the banks of a river and rocks in the middle, offer resistance to the flow of current. Similarly, electrons are slowed down because of interactions with the atoms of the wire. The higher this resistance, the less the current for a given voltage V. We then define resistance so that the current is inversely proportional to the resistance. When we combine this with the above proportion we have

$$I = \frac{V}{R};$$ (26-3)

R is the **resistance** of some object or device, V is the potential difference across the device, and I is the current that flows through it. This relation (Eq. 26-3) is often written

$$V = IR,$$

and is often referred to as Ohm's law. Actually, it is not a law, but rather is the *definition of resistance*. If we want to call something Ohm's law, it would be the statement that the current through a *metal conductor* is proportional to the applied voltage, $I \propto V$. But this relation does *not* apply generally for other substances such as semiconductors, vacuum tubes, transistors, and so on. Thus "Ohm's law" is not a fundamental law like Newton's laws, the laws of thermodynamics, or Gauss's law. Rather, it is a description of a certain class of materials (metal conductors). Materials or devices that do not follow Ohm's law are said to be *nonohmic*. The definition of resistance

$$R = V/I$$

(Eq. 26-3) can be applied also to nonohmic cases; in these cases, R would not be constant and would depend on the applied voltage. What Ohm's law says is that R is a constant, independent of V, for metal conductors. Since it is not really a fundamental law, we probably ought to call it "Ohm's equation" or some such thing; but the habit of calling it Ohm's law is so ingrained that we won't quibble with continuing this usage, as long as we keep in mind its limitations.

The unit for resistance is called the *ohm* and is abbreviated Ω (Greek capital omega). Because $R = V/I$, we see that 1.0 Ω is equivalent to 1.0 V/A.

EXAMPLE 26-2   A plate on the bottom of a small tape recorder specifies that it should be connected to 6.0 V and will draw 300 mA. (*a*) What is the net resistance of the recorder? (*b*) If the voltage dropped to 5.0 V, how would the current change?

SOLUTION   (*a*) From Eq. 26-3 we have

$$R = \frac{V}{I} = \frac{6.0 \text{ V}}{0.30 \text{ A}} = 20 \text{ }\Omega.$$

(*b*) If the resistance stayed constant, the current would be approximately $I = V/R = 5.0 \text{ V}/20 \text{ }\Omega = 0.25 \text{ A}$, or a drop of 50 mA. Actually, resistance depends on temperature (Section 26-4), and so this is only an approximation.

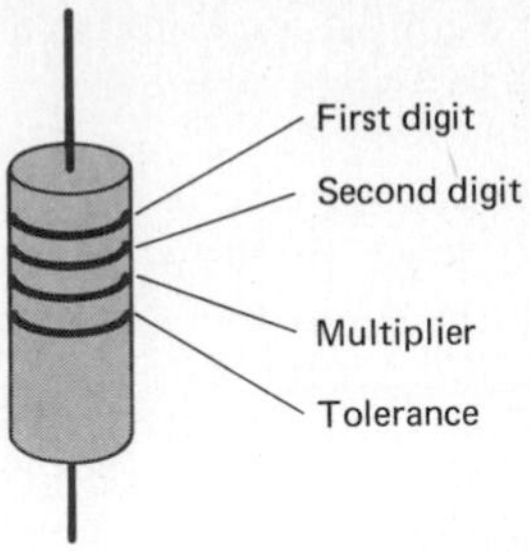

Resistor color code

First digit
Second digit
Multiplier
Tolerance

Color	Number	Multi-plier	Toler-ance (%)
Black	0	1	
Brown	1	10^1	
Red	2	10^2	
Orange	3	10^3	
Yellow	4	10^4	
Green	5	10^5	
Blue	6	10^6	
Violet	7	10^7	
Gray	8	10^8	
White	9	10^9	
Gold		10^{-1}	5%
Silver		10^{-2}	10%
No color			20%

FIGURE 26-6

Some resistors. The resistance value of a given resistor is written on the exterior or may be given a color code as shown in the figure. The first two colors represent the first two digits in the value of the resistance, the third represents the power of ten that it must be multiplied by, and the fourth is the manufactured tolerance. For example, a resistor whose four colors are red, green, orange, and silver has a resistance of 25,000Ω (25kΩ), give or take 10 percent.

All electric devices, from heaters to light bulbs to stereo amplifiers, offer resistance to the flow of current. Generally, the connecting wires have very low resistance. In many circuits, particularly in electric devices, *resistors* are used to control the amount of current. Resistors have resistances from less than an ohm to millions of ohms. Several kinds are shown in Fig. 26–6. The two main types are "wire wound" resistors, which consist of a coil of fine wire, and "composition" resistors, which are usually made of the semiconductor carbon.

When we draw a diagram of a circuit, we indicate a resistance with the symbol ⟋⟍⟋⟍. Wires whose resistance is negligible, however, are shown simply as straight lines.

26–4 Resistivity and Superconductivity

It is found experimentally that the resistance R of a metal wire is directly proportional to its length L and inversely proportional to the cross-sectional area A. That is

$$R = \rho \frac{L}{A}, \qquad (26-4)$$

where ρ, the constant of proportionality, is called the *resistivity* and depends on the material used. This relation makes sense since we expect that the resistance of a thick wire would be less than that of a thin one because a thicker wire has more area for the electrons to pass through. And you might expect the resistance to be greater if the length is greater since there would be more obstacles to electron flow.

Typical values of ρ for various materials are given in the first column of Table 26–1. (The values depend somewhat on purity, heat treatment, temperature, and other factors.) Notice that silver has the lowest resistivity and is thus the best conductor; but it is expensive. Copper is not far behind, so it is clear why most wires are made of copper. Aluminum, although it has a higher resistivity, is much less dense than copper; it is thus preferable to copper in some situations, such as transmission lines, because its resistance for the same weight is less than that for copper.

TABLE 26–1
Resistivity and temperature coefficients (at 20°C)

Material	Resistivity, ρ ($\Omega \cdot$ m)	Temperature Coefficient, α (C°)$^{-1}$
Conductors		
Silver	1.59×10^{-8}	0.0061
Copper	1.68×10^{-8}	0.0068
Aluminum	2.65×10^{-8}	0.00429
Tungsten	5.6×10^{-8}	0.0045
Iron	9.71×10^{-8}	0.00651
Platinum	10.6×10^{-8}	0.003927
Mercury	98×10^{-8}	0.0009
Nichrome (alloy of Ni, Fe, Cr)	100×10^{-8}	0.0004
Semiconductors[†]		
Carbon (graphite)	$(3-60) \times 10^{-5}$	−0.0005
Germanium	$(1-500) \times 10^{-3}$	−0.05
Silicon	0.1–60	−0.07
Insulators		
Glass	$10^9 - 10^{12}$	
Hard rubber	$10^{13} - 10^{15}$	

[†] Values depend strongly on presence of even slight amounts of impurities.

The reciprocal of the resistivity, called the *conductivity*, σ

$$\sigma = \frac{1}{\rho}, \qquad (26-5)$$

is often used. The conductivity has units of $(\Omega \cdot m)^{-1}$.

EXAMPLE 26–3 Suppose you want to connect your stereo set to a remote speaker. If each wire must be 20 m long, what diameter copper wire should you use to keep the resistance less than 0.10 Ω per wire?

SOLUTION We solve Eq. 26–4 for A and use Table 26–1:

$$A = \rho \frac{L}{R} = \frac{(1.7 \times 10^{-8} \, \Omega \cdot m)(20 \, m)}{(0.10 \, \Omega)} = 3.4 \times 10^{-6} \, m^2.$$

The cross-sectional area A of a circular wire is related to its diameter d by $A = \pi d^2/4$. The diameter must then be at least $d = \sqrt{4A/\pi} = 2.1 \times 10^{-3} \, m = 2.1$ mm.

The resistivity of a material depends somewhat on temperature. In general, the resistance of metals increases with temperature. This is not surprising, for at higher temperatures the atoms are moving more rapidly and are arranged in a less orderly fashion. So they might be expected to interfere more with the flow of electrons. If the temperature change is not too great, the resistivity of metals increases nearly linearly with temperature. That is,

$$\rho_T = \rho_0(1 + \alpha[T - T_0]) \qquad (26-6)$$

where ρ_T is the resistivity at temperature T, ρ_0 is the known resistivity at a standard temperature T_0, and α is the *temperature coefficient of resistance*. Values for α are given in Table 26–1. Note that the temperature coefficient for semiconductors can be negative; this is apparently because at higher temperatures, some of the electrons that are not normally free become free and can contribute to the current. Thus the resistance of a semiconductor can decrease with an increase in temperature, although this is not always the case.

EXAMPLE 26–4 *Resistance thermometer.* The variation in electrical resistance with temperature can be used to make precise temperature measurements. Platinum is usually used since it is relatively free from corrosive effects and has a high melting point. To be specific, suppose at 0°C the resistance of a platinum resistance thermometer is 164.2 Ω. When placed in a particular solution, the resistance is 187.4 Ω. What is the temperature of this solution?

SOLUTION Since the resistance R is directly proportional to the resistivity α, we can combine Eq. 26–4 with 26–6 and write (setting $T_0 = 0°$C)

$$R = R_0(1 + \alpha T).$$

Here $R_0 = \rho_0 L/A$ is the resistance of the wire at 0°C. We solve this equation for T and find

$$T = \frac{R - R_0}{\alpha R_0} = \frac{187.4\ \Omega - 164.2\ \Omega}{(3.927 \times 10^{-3}(\text{C}°)^{-1})(164.2\ \Omega)} = 35.9°\text{C}.$$

More convenient for some applications is a *thermistor*, which consists of a metal oxide or semiconductor whose resistance also varies in a repeatable way with temperature. They can be made quite small and respond very quickly to temperature changes.

The value of α depends on temperature, so it is important to check the temperature range of validity of any value (say in a handbook of physical data). If the temperature range is wide, Eq. 26–6 is not adequate and terms proportional to the square and cube of the temperature are needed: $\rho_T = \rho_0(1 + \alpha T + \beta T^2 + \gamma T^3)$ where the coefficients β and γ are generally very small and we set $T_0 = 0°$C; but when T is large their terms become significant.

At very low temperatures, the resistivity of certain metals and their compounds or alloys becomes essentially zero to within experimental measuring techniques. Materials in such a state are said to be **superconducting**. This phenomenon was first observed by H. K. Onnes (1853–1926) in 1911 when he cooled mercury below 4.2 K. He found that at this temperature the resistance of mercury suddenly dropped to zero. In general, superconductors become superconducting only below a certain *transition temperature* which is usually within a few degrees of absolute zero. Current in a ring-shaped superconducting material has been observed to flow for years in the absence of a potential difference, with no measurable decrease.

Much research has been done on superconductivity in recent years to try to understand why it occurs and to find materials that superconduct at more reasonable temperatures to reduce the cost and inconvenience of refrigeration at the required very low temperature. The first successful superconductivity theory was published by Bardeen, Cooper, and Schrieffer (the BCS theory) in 1957.

Superconductors are already being used in large magnets (we shall see in Chapter 28 that electric currents produce magnetic fields) which sharply reduce the electric power needs. Energy is needed, of course, to refrigerate the superconducting material at a low temperature.[†]

[†] Superconductors are being planned for use in large generating plants and for large motors where the negligible resistance will allow very large currents to be used. Also being developed are superconducting transmission cables that will reduce energy loss tremendously and will allow electric generating plants to be placed close to fuel sources (coal, gas, and so on) rather than close to population centers, thus reducing fuel transportation costs. Superconductors may also be used in high-speed ground transportation: the magnetic fields produced by superconducting magnets would be used to "levitate" vehicles over the tracks so there is essentially no friction; the levitation arises from the repulsive force between the magnet (on the train) and the eddy currents (see Section 30–6) produced in the track below.

26–5   Microscopic View of Electric Current: Current Density and Drift Velocity

Up to now in this chapter we have dealt mainly with a macroscopic, everyday-world view of electric current. We did see, however, that in metal wires, according to atomic theory, the electric current is carried by negatively charged electrons, and that in solutions current can also be carried by positive and/or negative ions. Let us now look at this microscopic picture in more detail.

When a potential difference is applied to the two ends of a wire of uniform cross section, the direction of **E** is parallel to the walls of the wire (Fig. 26–7). The existence of **E** within the conducting wire does not contradict our earlier result that **E** = 0 inside a conductor in the electrostatic case. We are no longer dealing with the static case. If **E** is not zero, charges are free to move in a conductor, and hence can move under the action of the electric field; if all the charges are at rest, then **E** must be zero (electrostatics).

We now define a new microscopic quantity, the **current density**, **j**. It is defined as the *electric current per unit cross-sectional area* at any point in space. If the current density **j** in a wire of cross-sectional area A is uniform over the cross section, then j is related to the electric current by

$$j = \frac{I}{A}, \quad \text{or} \quad I = jA. \tag{26–7}$$

If the current density is not uniform, then the general relation is

$$I = \int \mathbf{j} \cdot d\mathbf{A} \tag{26–8}$$

where $d\mathbf{A}$ is an element of surface and I is the current through the surface over which the integration is taken. The direction of the current density at any point is the direction that a positive charge would move when placed at that point—that is, the direction of **j** at any point is generally the same as the direction of **E**, Fig. 26–7. (Inertial effects can usually be ignored.) The current density exists for any *point* in space. The current I, on the other hand, refers to a conductor as a whole, and hence is a macroscopic quantity.

The direction of **j** is chosen to represent the direction of flow of positive charge (as in our earlier convention). In a conductor, it is negatively charged electrons that move, so they move in the direction of −**j**, or −**E** (to the left in Fig. 26–7). The force on an electron in a conductor due to the electric field **E** is **F** = −e**E**, where $q = -e$ is the charge on one electron ($e = +1.6 \times 10^{-19}$ C). However, the electrons have no average acceleration (except for a very short time just after the battery is connected) because of collisions with atoms of the conductor. The "retarding" forces due to collisions balance the electric force, so the electrons quickly reach a steady velocity known as their *average drift velocity*, $\mathbf{v}_d$. This is much like the terminal velocity reached by a light object falling in air. We use the term "drift" velocity because this speed is much less than the average thermal speed of the electrons: according to the "classical model" of conduction,[†] the free electrons in a metal are treated like the atoms in an ideal gas; these conduction electrons move with very high speeds in every direction, and superposed

---

[†] The classical model, while useful, gives poor quantitative predictions and has been significantly modified by the quantum theory; see discussion directly after Example 26–5.

FIGURE 26–7

Electric field **E** in a uniform wire of cross-sectional area A carrying a current I; the current density $j = I/A$.

on this random thermal motion is a slow "drift" in the direction of the electric field **E** (opposite to **E** for electrons).

We can relate the current in a wire to the drift velocity $\mathbf{v}_d$. In a time t, electrons will travel a distance $l = v_d t$ along the wire. Suppose there are n conduction electrons per unit volume. The number of electrons that pass through a cross section of the wire in time t is nAl where A is the cross-sectional area of the wire. Hence the amount of charge that passes through area A in a time t is

$$Q = -enAl = -enAv_d t$$

where $-e$ is the charge on each electron. The current I is $I = Q/t$, so

$$I = -nev_d A. \tag{26–9}$$

The current density, $j = I/A$, is

$$j = -nev_d. \tag{26–10}$$

In vector form, this is written

$$\mathbf{j} = -ne\mathbf{v}_d \tag{26–11}$$

where the minus sign indicates that the direction of (positive) current flow is opposite to the drift velocity of electrons.

We can generalize Eq. 26–11 to any type of charge flow, such as flow of ions in an electrolyte. If there are several types of ions (which can include free electrons), each of density n_i (number per unit volume), charge q_i ($q_i = -e$ for electrons) and drift velocity $\mathbf{v}_{di}$, then the net current density at any point is

$$\mathbf{j} = \sum_i n_i q_i \mathbf{v}_{di}. \tag{26–12}$$

The total current I passing through an area A perpendicular to $\mathbf{j}$ is then

$$I = \sum_i n_i q_i v_{di} A.$$

EXAMPLE 26–5 A copper wire, 3.20 mm in diameter, carries a 5.00-A current. Determine (a) the current density in the wire, (b) the drift speed of the free electrons, and (c) the rms speed of electrons assuming they behave like an ideal gas at 20°C. Assume that one electron per Cu atom is free to move (the others remain bound to the atom).

SOLUTION (a) The cross-sectional area of the wire is $A = \pi r^2 = (3.14)\cdot(1.60 \times 10^{-3}\text{ m})^2 = 8.04 \times 10^{-6}\text{ m}^2$. The current density is then

$$j = I/A = (5.00\text{ A})/(8.04 \times 10^{-6}\text{ m}^2) = 6.22 \times 10^5\text{ A/m}^2.$$

(b) Since there is one free electron per atom, the density of free electrons, n, is the same as the density of Cu atoms. The atomic mass of Cu is 63.5u, so 63.5 g of Cu contains 6.02×10^{23} free electrons. The density of copper is $\rho = 8.89 \times 10^3\text{ kg/m}^3$, so the number of free electrons per unit volume is

$$n = \left(\frac{6.02 \times 10^{23}\text{ electrons}}{63.5 \times 10^{-3}\text{ kg}}\right)(8.89 \times 10^3\text{ kg/m}^3) = 8.43 \times 10^{28}\text{ m}^{-3}.$$

Then, by Eq. 26–10, the drift speed is

$$v_d = \frac{j}{ne} = \frac{6.22 \times 10^5\text{ A/m}^2}{(8.43 \times 10^{28}\text{ m}^{-3})(1.60 \times 10^{-19}\text{ J})} = 4.61 \times 10^{-5}\text{ m/s},$$

which is only about 0.05 mm/s.

(c) If we treat the free electrons as an ideal gas (a rough approximation), we use Eq. 18–5:

$$v_{rms} = \sqrt{\frac{3kT}{m}} = \sqrt{\frac{3(1.38 \times 10^{-23} \text{ J/K})(293 \text{ K})}{(9.11 \times 10^{-31} \text{ kg})}} = 1.15 \times 10^5 \text{ m/s}.$$

Thus we see that the drift speed is very much less than the rms thermal speed of the electrons (by a factor of about 10^9).

The numerical result of part (c) in Example 26–5 is not reliable. The classical model of electric conduction, in which the electrons are treated as an ideal gas, predicts values for macroscopic quantities that are often far from measured values. According to the quantum theory, whose predictions are much closer to experiment, the free electrons do *not* obey the equipartition theorem (Section 20–4), so the calculation above, $v_{rms} = \sqrt{3kT/m}$, is not accurate. Instead, the speed of free electrons is about 15 times larger, about 1.6×10^6 m/s in copper at 20°C. We will not go into this complex theory here.

Equation 26–3, $V = IR$, can be written in terms of microscopic quantities as follows. We write the resistance R in terms of the resistivity ρ:

$$R = \rho \frac{L}{A};$$

and we write V and I as

$$I = jA$$

and

$$V = EL$$

where the last relation follows from Eq. 24–4 where we assume the electric field is uniform within the wire and L is the length of the wire (or a portion of the wire) between whose ends the potential difference is V. Thus, from $V = IR$, we have

$$EL = (jA)\left(\rho \frac{L}{A}\right) = j\rho L$$

or

$$j = \frac{1}{\rho} E = \sigma E, \tag{26-13}$$

where $\sigma = 1/\rho$ is the *conductivity*. For a metal conductor, ρ and σ do not depend on V (and hence not on E). Therefore the current density **j** is proportional to the electrical field **E** in the conductor. This is the "microscopic" statement of Ohm's law. Equation 26–13, which can be written in vector form as

$$\mathbf{j} = \sigma \mathbf{E} = \frac{1}{\rho} \mathbf{E},$$

is sometimes taken as the definition of conductivity σ and resistivity ρ.

EXAMPLE 26–6 What is the electric field inside the wire of Example 26–5?

SOLUTION Table 26–1 gives $\rho = 1.68 \times 10^{-8}$) Ω·m for copper. Since $j = 6.22 \times 10^5$ A/m², $E = \rho j = (1.68 \times 10^{-8} \text{ Ω·m})(6.22 \times 10^5 \text{ A/m}^2) = 1.04 \times 10^{-2}$ V/m. For comparison, the electric field between the plates of a capacitor is often much larger; in Example 25–6, for example, E is on the order of 10^5 V/m. Thus we see that only a modest electric field is needed for current flow in practical cases.

⬡ 26–6   Electric Power

Electric energy is useful to us because it can easily be transformed into other forms of energy. Motors, whose operation we will examine in Chapter 28, transform electric energy into mechanical work.

In other devices such as electric heaters, stoves, toasters, and hair dryers, electric energy is transformed into thermal energy in a wire resistance known as a "heating element." And in an ordinary light bulb, the tiny wire filament becomes so hot it glows; only a few percent of the energy is transformed into light and the rest, over 90 percent, into thermal energy. Light-bulb filaments and heating elements in household appliances have a resistance typically of a few ohms to a few hundred ohms.

Electric energy is transformed into thermal energy or light in such devices because the current is usually rather large, and there are many collisions between the moving electrons and the atoms of the wire. In each collision, part of the electron's kinetic energy is transferred to the atom with which it collides. As a result, the kinetic energy of the atoms increases and hence the temperature of the wire element increases. The increased thermal energy (internal energy) can be transferred as heat by conduction and convection to the air in a heater or to food in a pan, by radiation to toast in a toaster, or radiated as visible light. To find the power transformed by an electric device we use the fact that the energy transformed when an infinitesimal charge q moves through a potential difference V is $dU = dqV$ (Eq. 24–2). If dt is the time required for an amount of charge dq to move through V, the power P, which is the rate energy is transformed, is:

$$P = \frac{dU}{dt} = \frac{dq}{dt} V.$$

The charge that flows per second, dq/dt, is simply the electric current I. Thus we have

$$P = IV. \tag{26–14}$$

This general relation gives us the instantaneous power transformed by any device, where I is the current passing through it and V is the potential difference across it. It also gives the power delivered by a source such as a battery. The SI unit of electric power is the same as for any kind of power, namely the *watt* (1 W = 1 J/s).

The rate of energy transformation in a resistance R can be written, by combining Ohm's law ($V = IR$) with Eq. 26–14 ($P = IV$), in two other ways:

$$P = I(IR) = I^2R \tag{26–15a}$$

$$P = \left(\frac{V}{R}\right) V = \frac{V^2}{R}. \tag{26–15b}$$

Equations 26–15a and b apply only to resistors, whereas Eq. 26–14, $P = IV$, applies to any device.

⬡ **EXAMPLE 26–7**   Calculate the resistance of a 40-W automobile headlight designed for 12 V.

SOLUTION   Since $P = 40$ W and $V = 12$ V, we use Eq. 26–15b and solve for R:

$$R = \frac{V^2}{P} = \frac{(12 \text{ V})^2}{(40 \text{ W})} = 3.6 \ \Omega.$$

This is the resistance when the bulb is burning brightly at 40 W; when the bulb is cold, the resistance is much lower.

It is energy, not power, you must pay for on your electric bill. Since power is the *rate* energy is transformed, the total energy used by any device is simply its power consumption multiplied by the time it is on. If the power is in watts and the time is in seconds, the energy will be in joules since 1 W = 1 J/s. Electric companies usually specify the energy with a much larger unit, the kilowatt-hour (kWh). One kWh = (1000 W)(3600 s) = 3.60×10^6 J.

EXAMPLE 26–8 An electric heater draws 15 A on a 120-V line. How much power does it use and how much does it cost per month (30 days) if it operates 3.0 h per day and the electric company charges $0.060 per kWh? (For simplicity, assume the current flows steadily in one direction.)

SOLUTION The power is $P = IV = $ (15 A)(120 V) = 1800 W or 1.8 kW. To operate it for 90 h would cost (1.8 kW)(90 h)($0.060) = $9.72.

The electric wires that carry electricity to lights and other electric appliances have some resistance, although usually it is quite small. Nonetheless, if the current is large enough, the wires will heat up and produce thermal energy at a rate equal to I^2R where R is the wire's resistance. One possible hazard is that the current carrying wires in the wall of a building may become so hot as to start a fire. Thicker wires of course have less resistance (see Eq. 26–4) and thus can carry more current without becoming too hot. When a wire carries more current than is safe, it is said to be "overloaded." A building should, of course, be designed with wiring heavy enough for any expected load. To prevent overloading, fuses or circuit breakers are installed in circuits. They are basically switches (Fig. 26–8) that open the circuit when the current exceeds some particular value. A 20-A fuse or circuit breaker, for example, opens when the current passing through it exceeds 20 A. If a circuit repeatedly burns out a fuse or opens a circuit breaker, there are two possibilities: there may be too many devices drawing current in that circuit; or there is a fault somewhere, such as a "short." A short, or "short circuit," means that two wires have crossed, perhaps because the insulation has worn down; so the path of the current is shortened. The resistance of the circuit is then very small so the current will be very large. Short circuits should of course be remedied immediately.

Household circuits are designed with the various devices connected so that each receives the 120 V or so from the electric company. When you have trouble with a blowing fuse or circuit breaker, first check the total current you are drawing. For example suppose a household circuit draws the following currents: a light bulb draws $I = P/V = $ 100 W/120 V = 0.8 A, a heater 1800 W/120 V = 15.0 A, and a frying pan 1300 W/120 V = 10.8 A for a total of 26.6 A. If the circuit has a 20-A fuse, you might expect it to blow. If it has a 30-A fuse, it shouldn't blow, so a short may be the problem. (The most likely place is in the cord of one of the appliances.)

In electric circuits, heat dissipation by resistors must be considered. The physical size of a resistor is a rough indicator of the maximum permissible

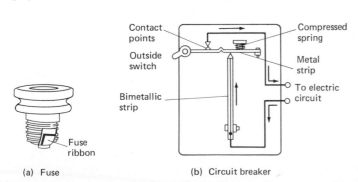

(a) Fuse   (b) Circuit breaker

FIGURE 26–8

(a) A fuse. When the current exceeds a certain value, the ribbon melts and the circuit opens. Then the fuse must be replaced. (b) A circuit breaker. Electric current passes through a bimetallic strip. When the current is great enough, the bimetallic strip heats sufficiently to bend so far to the left that the notch in the spring-loaded metal strip drops down over the end of the bimetallic strip. The circuit then opens at the contact points (one is attached to the metal strip) and the outside switch is also flipped. As the device cools down, it can be reset using the outside switch.

power it can dissipate $(= I^2R)$ without appreciable rise in temperature. Common values are $\frac{1}{4}$W, $\frac{1}{2}$W, and 1 W; the higher the wattage, the larger the physical size.

26–7 Alternating Current

When a battery is connected to a circuit, the current flows steadily in one direction. This is called a *direct current* or dc. Electric generators at electric power plants, however, produce *alternating current* or ac.[†] An alternating current reverses direction many times per second and is usually sinusoidal as shown in Fig. 26–9b. The current supplied to homes and businesses by electric companies is ac throughout virtually the entire world.

The voltage produced by an ac electric generator is sinusoidal, as we shall see in Chapter 30. The current it produces is thus sinusoidal (Fig. 26–9b). We can write the voltage as a function of time as

$$V = V_0 \sin 2\pi ft.$$

The potential V oscillates between $+V_0$ and $-V_0$; V_0 is referred to as the *peak voltage*. The frequency f is the number of complete oscillations made per second. In most areas of the United States and Canada, f is 60 Hz (cycles per second). In some countries 50 Hz is used.

From Ohm's law, if a voltage V exists across a resistance R, then the current I is

$$I = \frac{V}{R} = \frac{V_0}{R} \sin 2\pi ft = I_0 \sin 2\pi ft. \qquad (26\text{–}16)$$

The quantity $I_0 = V_0/R$ is the *peak* current. The current is considered positive when the electrons flow in one direction and negative when they flow in the opposite direction. It is clear from Fig. 26–9b that an alternating current is as often positive as it is negative. Thus the average current is zero. This does not mean, however, that no power is needed or that no heat is produced in a resistor. Electrons do move back and forth in a resistor and do produce heat. Indeed, the power delivered to a resistance R at any instant is

$$P = I^2R = I_0^2 R \sin^2 2\pi ft.$$

Because the current is squared, we see that the power is always positive; see Fig. 26–10. The quantity $\sin^2 2\pi ft$ varies between 0 and 1; and it is not too difficult to show (see problem 49) that its average value is $\frac{1}{2}$. Thus the *average power* developed, $\bar{P}$, is

$$\bar{P} = \frac{1}{2} I_0^2 R.$$

Since power can also be written $P = V^2/R = (V_0^2/R) \sin^2 2\pi ft$, we also have that the average power is

$$\bar{P} = \frac{1}{2} \frac{V_0^2}{R}.$$

The average value of the *square* of the current or voltage is thus what is important for calculating average power: $\overline{I^2} = \frac{1}{2} I_0^2$ and $\overline{V^2} = \frac{1}{2} V_0^2$. The square root of each of these is the rms (root mean square) value of the current or voltage:

$$I_{\text{rms}} = \sqrt{\overline{I^2}} = \frac{1}{\sqrt{2}} I_0 = 0.707 I_0, \qquad (26\text{–}17a)$$

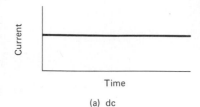

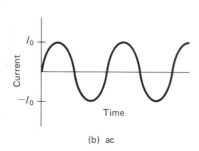

FIGURE 26–9

(a) Direct current. (b) Alternating current.

FIGURE 26–10

Power delivered to a resistor in an ac circuit.

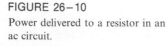

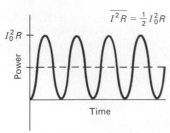

[†] Although it is redundant, we sometimes say "ac current" or "ac voltage" which really means "alternating current" or "alternating voltage".

$$V_{rms} = \sqrt{\overline{V^2}} = \frac{1}{\sqrt{2}} V_0 = 0.707 V_0. \qquad\qquad (26\text{--}17b)$$

The rms values of V and I are sometimes called the "effective values." They are useful because they can be substituted directly into the power formulas, Eqs. 26–14 and 26–15, to get the average power. For example, $\overline{P} = \frac{1}{2}I_0^2 R = I_{rms}^2 R$. Thus a direct current whose values of I and V equal the rms values of I and V for an alternating current will produce the same power. Hence it is usually the rms value of current which is specified or measured. For example, in the United States and Canada, standard line voltage[†] is 120 V ac. The 120 V is V_{rms}; the peak voltage V_0 is

$$V_0 = \sqrt{2} V_{rms} = 170 \text{ V}.$$

In most of Europe the rms voltage is 240 V, so the peak voltage is 340 V.

EXAMPLE 26–9 Calculate the resistance and the peak current in a 1000-W hair dryer connected to a 120-V line. What happens if it is connected to a 240-V line in Britain?

SOLUTION The rms current is

$$I_{rms} = \frac{\overline{P}}{V_{rms}} = \frac{1000 \text{ W}}{120 \text{ V}} = 8.33 \text{ A}.$$

Thus $I_0 = \sqrt{2}I_{rms} = 11.8\text{A}$. The resistance is

$$R = \frac{V_{rms}}{I_{rms}} = \frac{120 \text{ V}}{8.33 \text{ A}} = 14.4 \ \Omega.$$

The resistance could equally well be calculated using peak values: $R = V_0/I_0 = 170 \text{ V}/11.8 \text{ A} = 14.4 \ \Omega$. When connected to a 240-V line, the average power delivered would be

$$\overline{P} = \frac{V^2}{R} = \frac{(240 \text{ V})^2}{(14.4 \ \Omega)} = 4000 \text{ watts}.$$

This would undoubtedly melt the heating element or the wire coils of the motor.

This section has given a brief introduction to the simpler aspects of alternating currents. We will discuss ac circuits in more detail in Chapter 32. In Chapter 27 we will deal with the details of dc circuits only.

Summary

By transforming chemical energy into electric energy, an electric battery serves as a source of potential difference or *electromotive force* (emf). A simple battery consists of two electrodes made of different metals and immersed in a solution or paste known as an electrolyte.

Electric current, I, refers to the rate of a flow of electric charge and is measured in amperes (A): 1 A equals a flow of 1 C/s past a given point. The direction of current flow is generally taken as being that of positive charge; in a wire, it is actually negatively charged electrons that move, so they flow in a direction opposite to the direction of the "conventional current." Positive conventional current always flows from a high potential to a low potential.

Ohm's law states that the current in a good conductor is proportional to the potential difference applied to its two ends; the proportionality constant is called the *resistance R* of the material, so $V = IR$. The unit of resistance is the ohm (Ω), where $1 \ \Omega = 1 \text{ V/A}$.

[†] The line voltage can vary, depending on the total load; the frequency of 60 Hz, however, remains extremely steady.

The resistance R of a wire is inversely proportional to its cross-sectional area A and directly proportional to its length L and to a property of the material called its resistivity, $R = \rho L/A$. The resistivity, ρ, increases with temperature for metals, but for semiconductors it may decrease. A superconductor is a material for which the electrical resistance is essentially zero.

Current density $\mathbf{j}$ is the current per cross-sectional area. From a microscopic point of view, the current density is related to the number of charge carriers per unit volume, n, their charge, q, and their drift velocity, $\mathbf{v_d}$, by $\mathbf{j} = ne\mathbf{v_d}$. The electric field within a wire is related to $\mathbf{j}$ by $\mathbf{j} = \sigma\mathbf{E}$ where $\sigma = 1/\rho$.

The rate at which energy is transformed in a resistance R from electric to other forms of energy (such as heat and light) is equal to the product of current and voltage; that is, the power transformed, measured in watts, is given by $P = IV$ and can be written with the help of Ohm's law as $P = I^2R = V^2/R$. The total electric energy transformed in any device equals the product of power and the time during which the device is operated. In SI units, energy is given in joules (1 J = 1 W·s), but electric companies use a larger unit, the kilowatt-hour (1 kWh = 3.6×10^6 J).

Electric current can be direct (dc), in which the current is steady in one direction; or it can be alternating (ac), in which the currents reverse direction at a particular frequency, typically 60 Hz. Alternating currents are often sinusoidal in time, $I = I_0 \sin 2\pi ft$, and are produced by an alternating voltage. The rms values of sinusoidally alternating currents and voltages are given by $I_{rms} = I_0/\sqrt{2}$ and $V_{rms} = V_0/\sqrt{2}$, where I_0 and V_0 are the peak values. The power relationship, $P = IV = I^2R = V^2/R$, is valid for the average power in alternating currents when the rms values are used.

 Questions

1. When you turn on a water faucet, the water usually flows immediately. You don't have to wait for water to flow from the faucet valve to the spout. Why not? Is the same thing true when you connect a wire to the terminals of a battery?

2. Car batteries are often rated in ampere-hours (A·h). What does this rating mean?

3. When an electric cell is connected to a circuit, electrons flow away from the negative terminal in the circuit. But within the cell, electrons flow *to* the negative terminal. Explain.

4. Is current used up in a resistor?

5. Develop an analogy between blood circulation and an electrical circuit. Discuss what plays the role of the heart for the electric case, and so on.

6. Design a circuit in which two different switches of the type shown in Fig. 26–11 can be used to operate the same light bulb from opposite sides of a room.

7. Can a copper wire and an aluminum wire of the same length have the same resistance? Explain.

8. If a large rectangular block of carbon has sides a, $2a$, $3a$, how would you connect the wires from a battery so as to obtain (a) the least resistance, (b) the greatest resistance?

9. Describe carefully how you would set up an experiment to measure the resistivity of a large rectangular block of material.

10. We saw in Example 26–5 that the drift speed of electrons is typically a fraction of 1 mm/s. That is, it may take an electron several hours to travel a meter. Why then does a light go on immediately after you flip the switch?

11. A wire of length L and radius R is connected to a voltage V. How is the electron drift velocity affected if (a) L is doubled, (b) R is doubled, (c) V is doubled.

12. Compare the drift speeds and electric currents in two wires that are geometrically identical, the density of atoms is similar, but the number of free electrons per atom in the material of one wire is twice that in the other.

13. In a car, one terminal of the battery is said to be connected to "ground." Since it is not really connected to the ground, what is meant by this expression?

14. The equation $P = V^2/R$ indicates that the power dissipated in a resistor decreases if the resistance is increased whereas the equation $P = I^2R$ implies the opposite. Is there a contradiction here? Explain.

15. What happens when a light bulb burns out?

16. Which draws more current, a 100-W light bulb or a 75-W bulb?

17. Electric power is transferred over large distances at very high voltages. Explain how the high voltage reduces power losses in the transmission lines.

18. Why is it dangerous to replace a 15-A fuse that blows repeatedly with a 25-A fuse?

19. Electric lights operated on low frequency ac (say 10 Hz) flicker noticeably. Why?

20. Suppose Franklin's original convention had been the reverse, so that electrons were considered positive. How would this affect the various results and analyses discussed in this chapter?

FIGURE 26–11

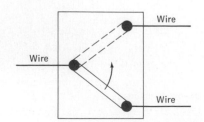

Problems

1. (I) A service station charges a battery using a current of 6.5 A for 5.0 h. How much charge passes through the battery?

2. (I) A current of 1.00 A flows in a wire. How many electrons are flowing past any point in the wire per second?

3. (I) What is the current in amperes if 1000 Na$^+$ ions ($Q = +e$) were to flow across a cell membrane in 4.0 μs?

SECTION 26–3

4. (I) What voltage will produce 1.2 A of current through a 25-Ω resistor?

5. (I) Two identical 1.5-V batteries are connected in series to a light bulb whose resistance is 10 Ω. How many electrons leave each battery per minute?

6. (II) A bird stands on an electric transmission line carrying 1800 A. The line has 2.0×10^{-5} Ω resistance per meter and the bird's feet are 2.5 cm apart. What voltage does the bird feel?

7. (II) The *conductance* G of an object is defined as the reciprocal of the resistance R: $G = 1/R$. The unit of conductance is the mho ($= \text{ohm}^{-1}$) which is also called the siemens (S). What is the conductance (in siemens) of an object that draws 800 mA of current at 12.0 V?

SECTION 26–4

8. (I) What is the resistance of a 2.2-m length of copper wire 1.8 mm in diameter?

9. (I) A 20.0-m length of wire 1.50 mm in diameter has a resistance of 2.50 Ω. What is the resistance of a 35.0-m length of wire 3.00 mm in diameter made of the same material?

10. (I) Can a 2.0-mm-diameter copper wire have the same resistance as a tungsten wire of the same length? Give numerical details.

11. (II) A length of wire is cut in half and the two lengths are wrapped together side by side to make a thicker wire. How does the resistance of this new combination compare to the resistance of the original wire?

12. (II) A 100-W light bulb has a resistance of about 12 Ω when cold and 140 Ω when on (hot). Estimate the temperature of the filament when "on" assuming an average temperature coefficient of resistivity $\alpha = 0.0060\,(\text{C}°)^{-1}$.

13. (II) A rectangular solid made of carbon has sides lying along the x, y, and z axes, whose lengths are 1.0 cm, 2.0 cm, and 4.0 cm, respectively. Determine the resistance for current that flows through the solid in (a) the x direction, (b) the y direction, (c) the z direction. Assume the resistivity is $\rho = 3.0 \times 10^{-5}\,\Omega \cdot \text{m}$.

14. (II) A 12.0-m length of wire consists of two equal diameter (3.0 mm) and equal length portions, one of copper and one of aluminum, welded together. What voltage is required to produce a current of 6.0 A?

15. (II) (a) Show that if a straight wire of cross-sectional area A lies along the x axis, the rate at which charge flows is given by

$$\frac{dq}{dt} = -\sigma A \frac{dV}{dx}$$

where dV/dx is the potential gradient, and σ is the conductivity. (b) Make an analogy to heat conduction (Section 19–7 and Eq. 19–2). Would you expect σ and K (thermal conductivity) to be related?

16. (II) A hollow cylindrical resistor with inner radius R_1 and outer radius R_2, and length L, is made of a material whose resistivity is ρ. (a) Show that the resistance is given by

$$R = \frac{\rho}{2\pi L} \ln \frac{R_2}{R_1}$$

for current that flows radially outward. (Hint: divide the resistor into concentric cylindrical shells and integrate.) (b) Evaluate R for such a resistor made of carbon whose inner and outer radii are 1.0 mm and 1.8 mm and whose length is 1.0 cm. (c) What is the resistance in part (b) for current flowing *parallel* to the axis?

17. (II) A 1.00-Ω wire is drawn out to 3.0 times its original length. What is its resistance now?

18. (II) For some applications it is important that the value of a resistance not change with temperature. For example, suppose you made a 2.2-kΩ resistor from a carbon resistor and a nichrome wire-wound resistor connected together so the total resistance will be the sum of their separate resistances. What value should each of these resistors have (at 0°C) so that the combination is temperature independent?

19. (II) Determine a formula for the total resistance of a spherical shell made of material whose conductivity is σ and whose inner and outer radii are r_1 and r_2. Assume the current flows radially outward.

20. (III) The filament of a light bulb has a resistance of 12 Ω at 20°C and 140 Ω when hot (as in problem 12). (a) Calculate the temperature of the filament when it is hot, and take into account the change in length and area of the filament due to thermal expansion (assume tungsten for which $\alpha \approx 5 \times 10^{-6}\,\text{C}°^{-1}$). (b) In this temperature range, what percentage of the change in resistance is due to thermal expansion, and what percentage is due solely to the change in ρ? Use Eq. 26–6.

21. (III) A 38.0-Ω resistor is made from a coil of copper wire whose total mass is 11.2 g. What is the diameter of the wire and how long is it?

SECTION 26–5

22. (II) A 0.40-mm-diameter copper wire carries a tiny current of 3.0 μA. What is (a) the electron drift speed, (b) the current density, and (c) the electric field in the wire?

23. (II) In a solution, Na$^+$ ions in a concentration of 5.00 mol/m^3 move with a drift speed of 5.00×10^{-4} m/s; SO$_4^{2-}$ ions move with a speed of 2.00×10^{-4} m/s; the current density is zero. What is the concentration of SO$_4^{2-}$ ions?

24. (II) At a point high in the earth's atmosphere, He^{2+} ions in a concentration of 3.5×10^{12}/m^3 are moving due north at a speed of 2.0×10^6 m/s; and an 8.0×10^{11}/m^3 concentration of O$_2^-$ ions is moving due south at a speed of 8.5×10^6 m/s. Determine the magnitude and direction of the current density **j** at this point.

25. (III) Treat the free electrons in a conductor as an ideal

gas and assume they have an acceleration $a = eE/m$ between collisions with atoms of the conductor. Show that the resistivity ρ is given by

$$\rho = \frac{2mv}{ne^2 l_m}$$

where v is the mean speed of the electrons (not the drift speed) and l_m is the mean free path between collisions.

SECTION 26–6

26. (I) What is the current through a 6.0-W light bulb if it is connected to its proper source voltage of 12 V?

27. (I) What is the maximum power consumption of a 9.0-V transistor radio that draws a maximum of 400 mA of current?

28. (I) What is the maximum voltage that can be applied to a 200-Ω resistor rated at $\frac{1}{4}$ watt?

29. (I) At \$0.065 per kilowatt-hour, what does it cost if you forget to turn off a 200-W (input) dining room lamp before going to sleep for 8 h?

30. (I) What is the total amount of energy stored in a 12-V, 50-A·h car battery when it is fully charged?

31. (I) At \$0.08 per kWh, what does it cost to leave a 25-W porch light on all day for a year?

32. (II) A person accidentally leaves a car with the lights on. If each of the two front lights uses 40 W and each of the two rear lights 6 W, for a total of 92 W, how long will a fresh 12-V battery last if it is rated at 45 A·h? Assume the full 12 V appears across each bulb.

33. (II) The heating element of a 120-V, 1800-W heater is 6.8 m long. If it is made of iron, what must its diameter be?

34. (II) (a) A particular household uses a 3.5-kW heater 2.5 h/day ("on" time), six 100-W light bulbs 5.0 h/day, a 3.3-kW electric stove element for a total of 1.2 h, and miscellaneous power amounting to 1.6 kWh/day. If electricity costs \$0.070 per kWh, what will be their monthly bill (30 d)? (b) How much coal (which produces 7000 kcal/kg) must be burned by a 35-percent-efficient power plant to provide the yearly needs of this household?

35. (II) What is the efficiency of a 0.50 hp electric motor that draws 4.4 A from a 120-V line?

36. (II) A power station delivers 360 kW of power to a factory through 3.2-Ω lines. How much less power is wasted if the electricity is delivered at 40,000 V rather than 12,000 V?

37. (II) The wiring in a house must be thick enough so it doesn't become so hot as to start a fire. What diameter must a copper wire be if it is to carry a maximum current of 40 A and produce no more than 1.8 W of heat per meter of length?

38. (II) A 1200-W hair dryer is designed for 117 V. What will be the percentage change in output if the voltage drops to 105 V? Assume no change in resistance. How would the actual change in resistivity with temperature affect your answer?

39. (II) The current in an electromagnet connected to a 240-V line is 6.60 A. At what rate must cooling water pass over the coils if the water temperature is to rise by no more than 8.00°C?

40. (II) A small immersion heater can be used in a car to heat a cup of water for coffee. If the heater can heat 200 ml of water from 5°C to 95°C in 5 min, approximately how much current does it draw from the 12-V battery?

41. (II) An electric heater is used to heat a room of volume 36 m³. Air is brought into the room at 5°C and is changed completely twice per hour. Heat loss through the walls amounts to approximately 500 kcal/h. If the air is to be maintained at 20°C, what minimum wattage must the heater have? (The specific heat capacity of air is about 0.17 kcal/kg·C°.)

42. (II) An electric car makes use of a storage battery as its source of energy. Suppose that a small 650-kg postman's delivery car is to be powered by ten 70-A·h, 12-V batteries. Assume that the car is driven on the level at an average speed of 30 km/h, and the average friction force is 200 N. Assume 100 percent efficiency and neglect energy used for acceleration; note that no energy is consumed when the vehicle is stopped since the engine doesn't need to idle. (a) Determine the horsepower required; (b) after approximately how many kilometers must the batteries be recharged?

SECTION 26–7

43. (I) An ac voltage, whose peak value is 80 V, is across a 60-Ω resistor. What is the value of the rms and peak currents in the resistor?

44. (I) What is the resistance of an ordinary 60-W, 120-V_{rms} light bulb when it is on?

45. (I) Calculate the peak current in a 2.2-kΩ resistor connected to a 240-V ac source.

46. (II) The peak value of an alternating current passing through a 600-W device is 3.0 A. What is the rms voltage across it?

47. (II) A 10-Ω heater coil is connected to a 240-V ac line. What is the average power used? What are the maximum and minimum values of the instantaneous power?

48. (II) Suppose a current is given by the equation $I = 2.5 \sin 120t$, where I is in amperes and t in seconds. (a) What is the frequency? (b) What is the rms value of the current? (c) If this is the current through a 50-Ω resistor, what is the equation that describes the voltage as a function of time?

49. (II) Show that the value of $\sin^2 2\pi ft$, averaged over many cycles, is $\frac{1}{2}$. (Hint: use integration.)

DC Circuits and Instruments

27

In Chapter 26 we discussed the basic principles of current electricity. Now we will apply these principles to analyze dc circuits and to understand the operation of a number of useful instruments.[†]

When we draw a diagram of a circuit, we represent a battery by the symbol ⊣⊢, a capacitor by the symbol ⊣⊢, and a resistor by the symbol ⌁. Wires whose resistance is negligible compared to other resistance in the circuit are drawn simply as straight lines. For the most part in this chapter, except in Section 27-5, we will be interested in circuits operating in their steady state—that is, we won't be looking at a circuit at the moment any change is made in it, such as a battery or resistor connected or disconnected, but rather a short time later when the currents have reached their steady values.

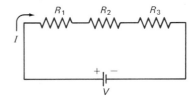

FIGURE 27–1
Resistors connected in series: $R = R_1 + R_2 + R_3$.

27–1   Resistors in Series and in Parallel

FIGURE 27–2
Resistors connected in parallel: $1/R = 1/R_1 + 1/R_2 + 1/R_3$.

When two or more resistors are connected end to end so the same current passes through each in turn, as in Fig. 27–1, they are said to be connected in *series*. On the other hand, if the resistors are connected so that the current from the source splits into separate branches as shown in Fig. 27–2, the resistors are said to be in *parallel*; in this case, the same potential difference exists across each. In either case, we will want to calculate the net resistance of the set of resistors. In other words, we want to find what single resistance R could replace the combination of given resistors without altering the rest of the circuit. The resistors could be like those shown in Fig. 26–6, or they could be light bulbs or other resistive electrical devices.

First we consider the series case, Fig. 27–1. The same current I passes through each resistor. If it did not, this would imply that charge was accumulating at

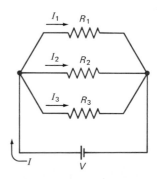

[†] Ac circuits that contain only a source of emf and resistors can be analyzed like the dc circuits in this chapter. However, ac circuits in general are more complex—for example, if they contain capacitors—and we will discuss them in Chapter 32.

some point in the circuit, which does not happen in the steady state. We let V represent the voltage across all three resistors; we assume all other resistance in the circuit can be ignored, and so V equals the emf of the battery. We let V_1, V_2, and V_3 be the potential differences across each of the resistors R_1, R_2, and R_3. By Ohm's law, $V_1 = IR_1$, $V_2 = IR_2$, and $V_3 = IR_3$. Because of conservation of energy, the total voltage V is equal to the sum of the voltages across each resistor, so we have

$$V = V_1 + V_2 + V_3 = IR_1 + IR_2 + IR_3.$$

For the equivalent single resistor R that would draw the same current we would have

$$V = IR.$$

We equate these two expressions for V and find

$$R = R_1 + R_2 + R_3. \qquad \text{[series]} \quad (27\text{--}1)$$

This is, in fact, what we expect. When we put several resistances in series, the total resistance is the sum of the separate resistances. This applies to any number of resistances, not simply for three. Clearly, when you add more resistance to the circuit, the current will decrease. If a 12-V battery is connected to a 4-Ω resistor, the current will be 3 A. But if the 12-V battery is connected to three 4-Ω resistors in series, the total resistance is 12 Ω and the current only 1 A.

The situation is quite different for the parallel case, Fig. 27–2. Again we want to find what single resistance R is equivalent to the three in parallel. In this situation, the total current I that leaves the battery breaks into three branches. We let I_1, I_2, and I_3 be the currents through each of the resistors R_1, R_2, and R_3. Because charge is conserved, the current flowing into a junction must equal the current flowing out, so

$$I = I_1 + I_2 + I_3$$

The full voltage of the battery is applied to each resistor, so

$$I_1 = \frac{V}{R_1}, \qquad I_2 = \frac{V}{R_2}, \quad \text{and} \quad I_3 = \frac{V}{R_3}.$$

Furthermore, for the single resistor R that will draw the same current I as these three in parallel, we must have

$$I = \frac{V}{R}.$$

We now combine these equations:

$$I = I_1 + I_2 + I_3$$

$$\frac{V}{R} = \frac{V}{R_1} + \frac{V}{R_2} + \frac{V}{R_3}.$$

When we divide out the V from each term we have

$$\frac{1}{R} = \frac{1}{R_1} + \frac{1}{R_2} + \frac{1}{R_3}. \qquad \text{[parallel]} \quad (27\text{--}2)$$

For example, if three 30-Ω resistors are put in parallel, the net resistance R offered by this network is

$$\frac{1}{R} = \frac{1}{30\,\Omega} + \frac{1}{30\,\Omega} + \frac{1}{30\,\Omega} = \frac{3}{30\,\Omega} = \frac{1}{10\,\Omega},$$

and so $R = 10\,\Omega$. Thus the net resistance is *less* than that of each single resistance. This may at first seem surprising. But remember that when you put resistors in parallel, you are giving the current additional paths to follow; hence the net resistance will be less.

EXAMPLE 27–1   How much current flows from the battery shown in Fig. 27–3a?

SOLUTION   First we find the equivalent resistance, R_p, of the 500-Ω and 700-Ω resistors that are in parallel:

$$\frac{1}{R_p} = \frac{1}{500\ \Omega} + \frac{1}{700\ \Omega} = 0.0020\ \Omega^{-1} + 0.0014\ \Omega^{-1}$$

$$= 0.0034\ \Omega^{-1}.$$

This is $1/R$ so we must take the reciprocal to find R. (It is a common mistake to forget to do this. Notice that the units of reciprocal ohms, Ω^{-1}, help to remind us of this.) Thus

$$R_p = \frac{1}{0.0034\ \Omega^{-1}} = 290\ \Omega.$$

This 290 Ω is the equivalent resistance of the two parallel resistors and is in series with the 400-Ω resistor; the equivalent circuit is shown in Fig. 27–3b. To find the total resistance R_T we add the 400-Ω and 290-Ω resistances together, since they are in series, and find

$$R_T = 400\ \Omega + 290\ \Omega = 690\ \Omega.$$

The total current flowing from the battery is then

$$I = \frac{V}{R} = \frac{12.0\ V}{690\ \Omega} = 17\ mA.$$

EXAMPLE 27–2   What is the current flowing through the 500-Ω resistor in Fig. 27–3a?

SOLUTION   To solve this problem, we must find the voltage across the 500-Ω resistor; this is the voltage between points b and c in the diagram and we call it V_{bc}; once V_{bc} is known, we can apply Ohm's law to get the current. First we find the voltage across the 400-Ω resistor, V_{ab}. Since 17 mA passes through this resistor, the voltage across it can be found using Ohm's law, $V = IR$:

$$V_{ab} = (0.017\ A)\ (400\ \Omega) = 6.8\ V.$$

Since the total voltage across the network of resistors is $V_{ac} = 12\ V$, then V_{bc} must be 12 V − 6.8 V = 5.2 V. Then Ohm's law tells us that the current I_1 through the 500-Ω resistor is

$$I_1 = \frac{5.2\ V}{500\ \Omega} = 10\ mA.$$

This is the answer we wanted. However, we can also calculate the current I_2 through the 700-Ω resistor since the voltage across it is also 5.2 V:

$$I_2 = \frac{5.2\ V}{700\ \Omega} = 7\ mA.$$

Notice that when I_1 combines with I_2 to form the total current I (at point c in Fig. 27–3a), their sum is 10 mA + 7 mA = 17 mA. This is, of course, the total current as calculated in Example 27–1.

FIGURE 27–3

(a) Circuit for Examples 27–1 and 27–2. (b) Equivalent circuit, showing the equivalent resistance of 290Ω for the two parallel resistors in (a).

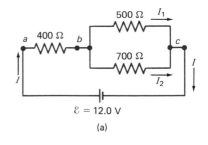

(a)

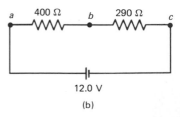

(b)

27–2   EMF and Terminal Voltage

As mentioned in Chapter 26, batteries, electric generators, and any other type of device that transforms one type of energy (chemical, mechanical, light, and so on) into electric energy is called a *source of emf* or source of electromotive force. (The term "electromotive force" is a misnomer since it does not refer to a "force" that is measured in newtons; hence to avoid confusion we will stick to the abbreviation, emf.) You may have noticed in your own experience that when a current is drawn from a battery, the voltage across its terminals drops below its rated emf. For example, when you have started a car with the headlights on, you may have noticed the headlights dim. This happens because the starter draws a great deal of current and the battery voltage drops as a result. The voltage drop occurs because the chemical reactions in a battery cannot supply charge fast enough to maintain the full emf. For one thing, charge must flow within the electrolyte between the electrodes of the battery, and there is always some hindrance to completely free flow. Thus, a battery itself has some resistance, which is called its *internal resistance*; it is usually designated r. The internal resistance can most simply be represented as if it were in series with the emf as shown in Fig. 27–4. Since this resistance r is inside the battery, we can never separate it from the battery. The two points a and b in the diagram represent the two terminals of the battery. What we measure is the **terminal voltage** V_{ab}. When no current is drawn from the battery the terminal voltage equals the emf, which is determined by the chemical reactions in the battery: $V_{ab} = \mathcal{E}$. However, when a current I flows from the battery, there is a drop in voltage equal to Ir. Thus the terminal voltage is:

$$V_{ab} = \mathcal{E} - Ir.$$

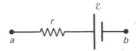

FIGURE 27–4

Diagram for an electric cell or battery.

For example, if a 12-V battery has an internal resistance of 0.1 Ω, then when 10 A flows from the battery, the terminal voltage is 12 V − (10 A)(0.1 Ω) = 11 V. The internal resistance of a battery is usually small.[†] For example an ordinary flashlight battery when fresh may have an internal resistance of perhaps 0.05 Ω. However, as it ages and the electrolyte dries out, the internal resistance increases to many ohms. Car batteries have even lower internal resistance.

EXAMPLE 27–3   A 9.0-V battery whose internal resistance r is 0.50 Ω is connected in the circuit shown in Fig. 27–5a. (*a*) How much current is drawn from the battery? (*b*) What is the terminal voltage of the battery?

SOLUTION   (*a*) The 4.0-Ω and 8.0-Ω resistors in parallel have an equivalent resistance R_1 given by

$$\frac{1}{R_1} = \frac{1}{8.0\ \Omega} + \frac{1}{4.0\ \Omega} = \frac{3}{8.0\ \Omega},$$

so $R_1 = 2.7\ \Omega$. This is in series with the 6.0 Ω resistor so the net resistance of the lower arm of the circuit is 6.0 Ω + 2.7 Ω = 8.7 Ω as shown in Fig. 27–5b. The equivalent resistance R_{II} of the 10.0-Ω and the 8.7-Ω resistors in parallel is given by

$$\frac{1}{R_{\text{II}}} = \frac{1}{10.0\ \Omega} + \frac{1}{8.7\ \Omega} = 0.21\ \Omega^{-1}.$$

So $R_{\text{II}} = 4.8\ \Omega$. This 4.8 Ω is in series with the 5.0-Ω resistor and the 0.50-Ω internal resistance of the battery so the total resistance R of the circuit is

---

[†] When a battery is being charged, a current is forced to pass through it (this happens in Fig. 27–8b), and we have to write

$$V_{ab} = \mathcal{E} + Ir.$$

See, for example, problem 23 and Fig. 27–21.

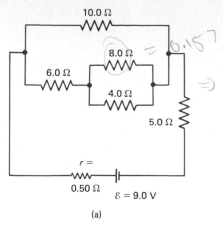

(a)

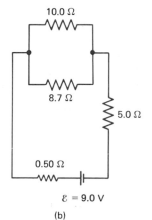

$\mathscr{E}$ = 9.0 V

(b)

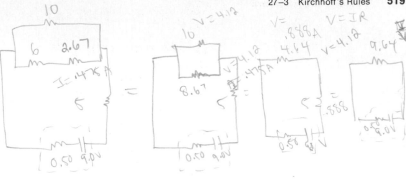

$$I_T = \frac{\mathscr{E}}{R+r} = 0.888\ A$$

$$V_T = 8.56V$$

FIGURE 27-5

Circuit for Example 27-3; r is the internal resistance of the battery.

$R = 4.8\ \Omega + 5.0\ \Omega + 0.50\ \Omega = 10.3\ \Omega$. Hence the current drawn is

$$I = \frac{\mathscr{E}}{R} = \frac{9.0\ V}{10.3\ \Omega} = 0.87\ A.$$

(b) The terminal voltage of the battery is $V_{ab} = \mathscr{E} - Ir = 9.0\ V - (0.87\ A)(0.50\ \Omega) = 8.6\ V.$

27-3 Kirchhoff's Rules

In the last few examples we have been able to find the currents flowing in circuits by combining resistances in series and parallel, and using Ohm's law. This technique can be used for many circuits. However, we sometimes encounter a circuit that is too complicated for this analysis. For example, we cannot find the currents flowing in each part of the circuit shown in Fig. 27-6 simply by combining resistances as we did before.

To deal with such complicated circuits, we use Kirchhoff's rules, invented by G. R. Kirchhoff (1824-1887) in the mid-nineteenth century). There are two of them, and they are simply convenient applications of the laws of conservation of charge and energy. **Kirchhoff's first** or **junction rule** is based on the conservation of charge, and we already used it in deriving the rule for parallel resistors; it states that *at any junction point, the sum of all currents entering the junction must equal the sum of all currents leaving the junction.* For example, at point *a* in Fig. 27-6,

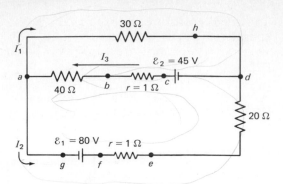

FIGURE 27–6

Currents in this circuit can be calculated using Kirchhoff's rules.

I_3 is entering whereas I_1 and I_2 are leaving. Thus Kirchhoff's junction rule states that $I_3 = I_1 + I_2$. We already saw an instance of this at the end of Example 27–2 where the currents passing through the 400-Ω and 700-Ω resistors were 10 mA and 7 mA, respectively, and they added at point c in Fig. 27–3a to give the outgoing current of 17 mA.

Kirchhoff's junction rule is based on the conservation of charge. Charges that enter a junction must also leave—none is lost or gained. **Kirchhoff's second** or **loop rule** is based on the conservation of energy. It states that *the algebraic sum of the changes in potential around any closed path of the circuit must be zero.* To see why this should hold, consider the analogy of a roller coaster on its track. When it starts from the station, it has a particular potential energy. As it climbs the first hill, its potential energy increases and reaches a peak at the top. As it descends the other side, its potential energy decreases and reaches a local minimum at the bottom of the hill. As the roller coaster continues on its path, its potential energy goes through more changes. But when it arrives back at the starting point, it has exactly as much potential energy as it had when it started at this point. Another way of saying this is that there was as much uphill as there was downhill.

The same reasoning can be applied to an electric circuit. As an example consider the simple circuit in Fig. 27–7. We have chosen it to be the same as the equivalent circuit of Fig. 27–3b already dealt with. The current flowing in this circuit is $I = (12.0 \text{ V})/(690 \ \Omega) = 0.017$ A as we calculated in Example 27–1. (Note that we are ignoring the internal resistance of the battery.) The positive side of the battery, point e in the figure, is at a high potential compared to point d at the negative side of the battery. That is, point e is like the top of a hill for a roller coaster. We can now follow the current around the circuit starting at any point we choose. Let us start at point e and follow a positive test charge completely around this circuit. As we go, we will note all changes in potential; when the charge returns to point e the potential there will be the same as when we started so the total change in potential will be zero. It is helpful to plot the changes in voltage around the circuit as in Fig. 27–7b. (Point d is arbitrarily taken as zero.) As our positive test charge goes from point e to point a, there is no change in potential since there is no source of potential nor any resistance. However, as the charge passes through the 400-Ω resistor to get to point b there is a decrease in potential of $V = IR = (0.017 \text{ A})(400 \ \Omega) = 6.8$ V. In effect, the charge is flowing "downhill" since it is heading toward the negative terminal of the battery.

FIGURE 27–7

Changes in potential around the circuit in (a) are plotted in (b).

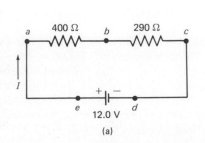

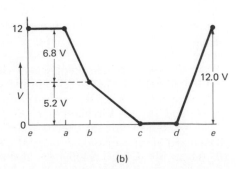

This is indicated in the graph of Fig. 27–7b. The decrease in voltage between the two ends of a resistor ($= IR$) is called a *voltage drop*. Because this is a *decrease* in voltage, we use a negative sign when applying Kirchhoff's loop rule; that is

$$V_{ba} = -6.8 \text{ V}.$$

As the charge proceeds from b to c there is a further voltage drop of (0.017 A) × (290 Ω) = 5.2 V and since this is a decrease we write

$$V_{cb} = -5.2 \text{ V}.$$

There is no change in potential as our test charge moves from c to d. But when it moves from d, which is the negative or low potential side of the battery, to point e which is the positive terminal, the voltage *increases* by 12.0 V. That is,

$$V_{ed} = +12.0 \text{ V}.$$

The sum of all the changes in potential in going around the circuit of Fig. 27–7 is then

$$-6.8 \text{ V} - 5.2 \text{ V} + 12.0 \text{ V} = 0.$$

And this is exactly what Kirchhoff's loop rule said it would be.

We already knew the details of this simple circuit, and we merely gave this example to show how the loop rule is applied. Before we use Kirchhoff's rules to determine the currents in a more difficult circuit, we must point out that when using Kirchhoff's rules, the current in each separate branch of the circuit is designated by a different subscript, such as I_1, I_2, and I_3 in Fig. 27–6. You do not have to know in advance in which direction these currents actually are moving. You make a guess and calculate the potentials around the circuit as if you were right; if you made the wrong guess for the direction of a current, your answer will merely have a negative sign. This and other details of using Kirchhoff's rules will become clearer in the following example.

EXAMPLE 27–4 Calculate the currents I_1, I_2, and I_3 in each of the branches of the circuit in Fig. 27–6.

SOLUTION We choose the directions of the currents as shown in the figure. Since (positive) current moves away from the positive terminal of a battery, we expect I_2 and I_3 to have the directions shown. It is hard to tell the direction of I_1 in advance so we chose the direction shown arbitrarily. We have three unknowns and therefore we need three equations. We first apply Kirchhoff's junction rule to the currents at point a: I_3 enters, whereas I_2 and I_1 leave. Therefore

$$I_3 = I_1 + I_2. \tag{a}$$

This same equation holds at point d, so we get no new information there. We now apply Kirchhoff's loop rule to two different closed loops. First we apply it to the loop $ahdcba$. From a to h we have a voltage drop $V_{ha} = -(I_1)(30 \text{ Ω})$. From h to d there is no change but from d to c the potential increases by 45 V: $V_{cd} = +45$ V. From c to a the voltage drops through the two resistors by an amount $V_{ac} = -(I_3)(40 \text{ Ω} + 1 \text{ Ω})$. Thus we have $V_{ha} + V_{cd} + V_{ac} = 0$, or

$$-30I_1 - 41I_3 + 45 = 0, \tag{b}$$

where we have omitted the units. For our second loop we take the complete circuit $ahdefga$. (We could have just as well taken $abcdefg$ instead.) Again we have $V_{ha} = -(I_1)(30 \text{ Ω})$, and $V_{dh} = 0$. But when we take our positive test charge from d to e, it actually is going uphill, against the natural flow of the current—or at least against the *assumed* direction of the current, which is what counts in this calculation. Thus $V_{ed} = I_2(20 \text{ Ω})$ has a *positive* sign. Similarly, $V_{fe} = I_2(1 \text{ Ω})$. From f to g there is a decrease in potential of 80 V since we go from the high potential terminal to the low. Thus $V_{fg} = -80$ V. Finally, $V_{ag} = 0$ and the sum of the potentials around this loop is then

$$-30I_1 + 21I_2 - 80 = 0. \tag{c}$$

We now have three equations—labeled (a), (b), and (c)—in three unknowns. From Eq. (c) we have

$$I_2 = \frac{80 + 30I_1}{21} = 3.8 + 1.4I_1. \tag{d}$$

From Eq. (b) we have

$$I_3 = \frac{45 - 30I_1}{41} = 1.1 - 0.73I_1. \tag{e}$$

We substitute these into Eq. (a) and solve for I_1:

$$I_1 = I_3 - I_2$$
$$= -2.7 - 2.1I_1,$$

so

$$3.1I_1 = -2.7$$
$$I_1 = -0.87A.$$

I_1 has magnitude 0.87A; the negative sign indicates that its direction is actually opposite to that initially assumed and shown in the figure. Note that the answer automatically comes out in amperes since all values were in volts and ohms. From Eq. (d) we have

$$I_2 = 3.8 + 1.4I_1 = 2.6A,$$

and from (e)

$$I_3 = 1.1 - 0.73I_1 = 1.7A.$$

This completes the solution.

27–4 EMFs in Series and in Parallel

When two or more sources of emf such as batteries are arranged in series, the total voltage is the algebraic sum of their respective voltages. For example, if two 1.5-V flashlight batteries are connected as shown in Fig. 27–8a, the voltage V_{ca} across the light bulb, represented by the resistor R is 3.0 V. (To be absolutely correct, we should also take into account the internal resistance of the batteries, but we assume it to be small.) On the other hand, when a 20-V and a 12-V battery are connected oppositely as shown in Fig. 27–8b, the net voltage V_{ca} is 8 V; that is, a positive test charge moved from a to b gains in potential by 20 V but when it passes from b to c it drops by 12 V. So the net change is 20 V − 12 V = 8 V. You might think that connecting batteries in reverse like this would be wasteful. And for most purposes that would be true. But such a reverse arrangement is precisely how a battery charger works: in Fig. 27–8b, the 20-V source is charging up the 12-V battery. Because of its greater voltage, the 20-V source is forcing charge back into the 12-V battery: electrons are being forced into its negative terminal and removed from its positive terminal. Not all batteries can be recharged, however, since the chemical reaction in many cannot be reversed; in this case, the arrangement of Fig. 27–8b would simply waste energy.

Sources of emf can also be arranged in parallel, Fig. 27–8c; this is not done to increase voltage but rather to provide more energy when large currents are needed. Each of the cells in parallel only has to produce a fraction of the total current so the loss due to internal resistance is less than for a single cell.

FIGURE 27–8

Batteries in series, (a) and (b), and in parallel, (c).

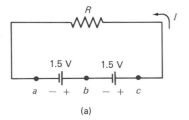

(a)

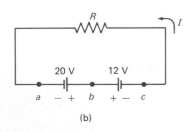

(b)

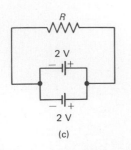

(c)

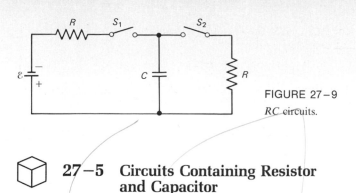

FIGURE 27-9

RC circuits.

![cube icon] **27-5 Circuits Containing Resistor and Capacitor**

Capacitors and resistors are often found together in a circuit. A simple example is shown in Fig. 27–9. We now analyze this *RC circuit*. First we leave the switch S_2 open so we can ignore the right-hand part of the circuit. We will then have the simple circuit of Fig. 27–10a when the switch S_1 is closed. Current immediately begins to flow through the circuit. Electrons will flow out the negative terminal of the battery, through the resistor R, and accumulate on the upper plate of the capacitor. And electrons will flow into the positive terminal of the battery leaving a positive charge on the other plate of the capacitor. As the charge accumulates on the capacitor, the current is reduced until eventually the voltage across the capacitor equals the emf of the battery and no further current flows. The charge Q on the capacitor thus increases gradually as shown in Fig. 27–10b and reaches a maximum value equal to $C\mathscr{E}$ (Eq. 25–1, $Q = CV_{ba} = C\mathscr{E}$). The mathematical form of this curve—that is, Q as a function of time—can be derived using conservation of energy (or Kirchhoff's loop rule). The emf $\mathscr{E}$ of the battery will equal the sum of the voltage drops across the resistor (IR) and the capacitor (Q/C):

$$\mathscr{E} = IR + \frac{Q}{C}. \tag{27-3}$$

The resistance R includes all resistance in the circuit, including the internal resistance of the battery, I is the current in the circuit at any instant, and Q is the charge on the capacitor at that same instant. Although $\mathscr{E}$, R, and C are constants, both Q and I are functions of time. The rate at which charge flows through the resistor ($I = dQ/dt$) is equal to the rate at which charge accumulates on the capacitor. Thus we can write,

$$\mathscr{E} = R\frac{dQ}{dt} + \frac{1}{C}Q.$$

FIGURE 27-10

For the *RC* circuit shown in (a), the charge on the capacitor varies as shown in (b) and the current through the resistor varies with time as shown in (c).

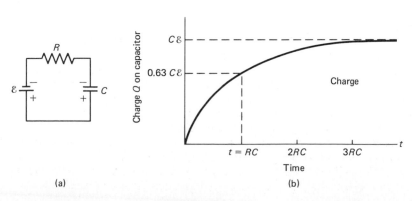

(a)

(b)

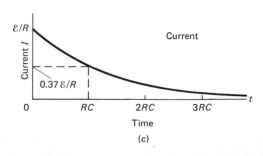

(c)

This equation can be solved by rearranging it:

$$\frac{dQ}{Q - C\mathscr{E}} = -\frac{dt}{RC}$$

and then integrating

$$\int \frac{dQ}{Q - C\mathscr{E}} = -\frac{1}{RC} \int dt$$

$$\ln(Q - C\mathscr{E}) = -\frac{t}{RC} + K$$

where K is a constant of integration. We can evaluate K by noting that at $t = 0$, $Q = 0$, so

$$\ln(-C\mathscr{E}) = K.$$

We substitute this value for K into the above relation and obtain

$$\ln(Q - C\mathscr{E}) - \ln(-C\mathscr{E}) = -\frac{t}{RC}$$

or

$$\ln\left(1 - \frac{Q}{C\mathscr{E}}\right) = -\frac{t}{RC}.$$

We take the exponential of both sides

$$1 - \frac{Q}{C\mathscr{E}} = e^{-t/RC}$$

or

$$Q = C\mathscr{E}(1 - e^{-t/RC}). \tag{27-4}$$

From this equation we see that the charge Q on the capacitor increases from $Q = 0$ at $t = 0$ to a maximum value $Q = C\mathscr{E}$ after a very long time. The quantity RC is called the *time constant* of the circuit. (The units of RC are $\Omega \cdot F = (V/A)(C/V) = C/(C/s) = s$.) It represents the time required for the capacitor to reach $(1 - e^{-1})$ or 63 percent of its full charge $[Q = C\mathscr{E}(1 - e^{-1}) \approx 0.63C\mathscr{E}]$. Thus the product RC is a measure of how quickly the capacitor gets charged. All circuits contain some resistance (if only in the connecting wires), so a capacitor never can be charged instantaneously when connected to a battery.

From Eq. 27–4, it appears that Q never quite reaches its maximum value $Q = C\mathscr{E}$ (except after infinite time). However, it reaches 86 percent of this value in $2RC$, 95 percent in $3RC$, 98 percent in $4RC$, and so on; so Q quickly approaches arbitrarily closely to its maximum value. For example, if $R = 20$ kΩ and $C = 0.30$ μF, the time constant is $(2.0 \times 10^4 \, \Omega)(3.0 \times 10^{-7} \, F) = 6.0 \times 10^{-3}$ s. So the capacitor is more than 98 percent charged in less than $\frac{1}{40}$ of a second.

The current I through the circuit of Fig. 27–10a at any time t can be obtained by differentiating Eq. 27–4:

$$I = \frac{dQ}{dt} = \frac{\mathscr{E}}{R} e^{-t/RC}. \tag{27-5}$$

Thus, at $t = 0$, the current is $I = \mathscr{E}/R$, as expected for a circuit containing only a resistor (there is not yet a potential difference across the capacitor). The current then drops exponentially in time with a time constant equal to RC. This is shown in Fig. 27–10c. The time constant RC represents the time required for the current to drop to $1/e \approx 0.37$ of its initial value.

EXAMPLE 27–6 The capacitance in the circuit of Fig. 27–10a is $C = 0.30\ \mu F$, the total resistance is 20 kΩ, and the battery emf is 12 V. Determine (a) the time constant, (b) the maximum charge the capacitor could acquire, (c) the time it takes for the charge to reach 99 percent of this value.

SOLUTION (a) The time constant is $RC = (2.0 \times 10^4\ \Omega)(3.0 \times 10^{-7}\ F) = 6.0 \times 10^{-3}\ s$.

(b) The maximum charge would be $Q = C\mathcal{E} = (3.0 \times 10^{-7}\ F)(12\ V) = 3.6\ \mu C$.

(c) In Eq. 27–4, we set $Q = 0.99 C\mathcal{E}$:

$$0.99 C\mathcal{E} = C\mathcal{E}(1 - e^{-t/RC}),$$

or

$$e^{-t/RC} = 1 - 0.99 = 0.01.$$

Then

$$\frac{t}{RC} = -\ln(0.01) = 4.6$$

so

$$t = 4.6 RC = 28 \times 10^{-3}\ s$$

or 28 ms (less than $\frac{1}{30}$ s).

Now let us return to the circuit of Fig. 27–9. This time, suppose the capacitor is charged to some value Q_0, and we open the switch S_1 and close S_2. We then have the circuit shown in Fig. 27–11a. As soon as S_2 is closed, charge begins to flow through the resistance R from one plate of the capacitor toward the other, until the capacitor is fully discharged. Conservation of energy (or Kirchhoff's loop rule) applies again, so Eq. 27–3 is valid with $\mathcal{E} = 0$:

$$IR + \frac{Q}{C} = 0.$$

Since $I = dQ/dt$, we have

$$\frac{dQ}{dt} R + \frac{Q}{C} = 0.$$

This can be rearranged to

$$\frac{dQ}{Q} = -\frac{dt}{RC}$$

and integrated to obtain

$$\ln Q = -\frac{t}{RC} + K.$$

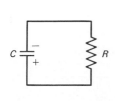

(a)

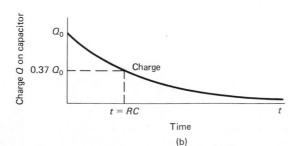

Time

(b)

FIGURE 27–11

For the RC circuit shown in (a), the charge on the capacitor varies as shown in (b).

The integration constant K can be evaluated from the fact that $Q = Q_0$ at $t = 0$, so $K = \ln Q_0$. Thus we have

$$\ln Q - \ln Q_0 = \ln \frac{Q}{Q_0} = -\frac{t}{RC}$$

or

$$Q = Q_0 e^{-t/RC}. \tag{27-6}$$

Thus the charge on the capacitor decreases exponentially in time with a time constant RC. This is shown in Fig. 27–11b. The current is

$$I = \frac{dQ}{dt} = -\frac{Q_0}{RC} e^{-t/RC} = I_0 e^{-t/RC}, \tag{27-7}$$

and it too is seen to decrease exponentially in time with the same time constant RC.

* 27–6   Ammeters and Voltmeters

An *ammeter* is used to measure current, and a *voltmeter* measures potential difference or voltage. The crucial part of each of these meters is a *galvanometer*, which works on the principle of the force between a magnetic field and a current-carrying coil of wire; we will discuss this in Chapter 28. For now we merely need to know that the deflection of the needle of a galvanometer is proportional to the current flowing through it. The *full-scale current sensitivity* I_m of a galvanometer is the current needed to make the needle deflect full scale, Fig. 27–12. For example, if the sensitivity I_m is 50 μA, a current of 50 μA will cause the needle to go to the end of the scale; a current of 25 μA will make it go only half way; if there is no current, the needle should be on zero and usually there is an adjustment screw to make it so.

Many meters we meet with in everyday life are galvanometers connected as ammeters or voltmeters. These include the VU meter on a tape recorder, and some of the meters on the dashboards of some cars.

A galvanometer can be used directly to measure small currents. For example, a galvanometer whose sensitivity I_m is 50 μA can measure currents from about 1 μA (currents smaller than this would be hard to read on the scale) up to 50 μA. To measure larger currents, a resistor is placed in parallel with the galvanometer. Thus an ammeter (represented by the symbol —Ⓐ—) consists of a galvanometer (—Ⓖ—) in parallel with a resistor called the *shunt resistor*. ("Shunt" is a synonym

FIGURE 27–12

Multimeter used as an ammeter. (Lester V. Bergman & Assoc., Inc.)

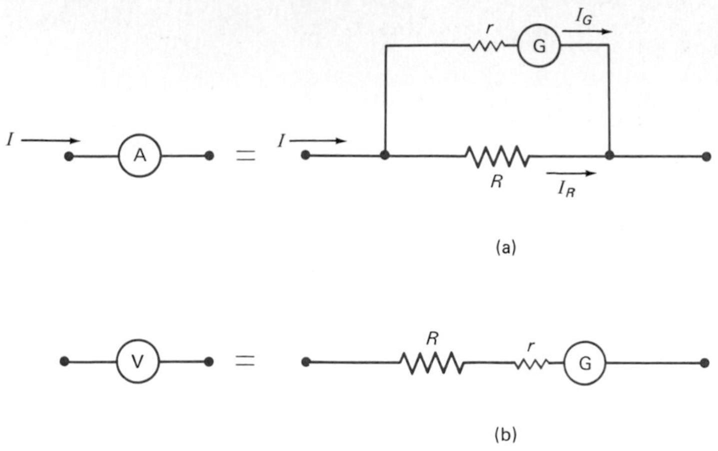

(a)

(b)

FIGURE 27–13

(a) An ammeter is a galvanometer in parallel with a small (shunt) resistor, R. (b) A voltmeter is a galvanometer in series with a large resistor, R.

for "in parallel.") This is shown in Fig. 27–13a. The shunt resistance is R and the resistance of the galvanometer coil (which carries the current) is r. The value of R is chosen according to what full-scale deflection is wanted.

Suppose we want to design a meter to read 1.0 A at full scale using a galvanometer with a full-scale sensitivity of 50 μA and resistance $r = 30\,\Omega$. This means that when the total current I entering the ammeter is 1.0 A, we want the current I_G through the galvanometer to be precisely 50 μA (to give full-scale deflection). Thus, when 1.0 A flows into the meter, we want 0.999950 A ($= I_R$) to pass through the shunt resistor R. Since the potential difference across the shunt is the same as across the galvanometer,

$$I_R R = I_G r,$$

then

$$R = \frac{I_G r}{I_R} = \frac{(5.0 \times 10^{-5}\,\text{A})(30\,\Omega)}{(0.999950\,\text{A})} = 1.5 \times 10^{-3}\,\Omega$$

or $0.0015\,\Omega$. The shunt resistor must thus have a very low resistance so that most of the current passes through it. If the current I into the meter is 0.50 A, this will produce a 25 μA current in the galvanometer and give a deflection half of full scale, as required.

A *voltmeter* (•—Ⓥ—•) also consists of a galvanometer and a resistor. But the resistor R is connected in series, Fig. 27–13b; and it is usually large. Suppose, using the same galvanometer with internal resistance $r = 30\,\Omega$ and full-scale current sensitivity of 50 μA, we want to make a voltmeter that reads from 0 to 15 V. When a potential difference of 15 V exists across the terminals of our voltmeter, we want 50 μA to be passing through it so as to give a full-scale deflection. From Ohm's law we have

$$15\,\text{V} = (50\,\mu\text{A})(r + R),$$

so

$$R = \frac{15\,\text{V}}{5.0 \times 10^{-5}\,\text{A}} - r = 300\,\text{k}\Omega.$$

Notice that $r = 30\,\Omega$ is so small compared to the value of R that it doesn't influence the calculation significantly. The scale will again be linear: if the voltage to be measured is 6.0 V, the current flowing through the voltmeter will be $6.0\,\text{V}/3.0 \times 10^5\,\Omega = 2.0 \times 10^{-5}\,\text{A}$ or 20 μA. This will produce two-fifths of full-scale deflection, as required.

＊27–7　Use of Voltmeters and Ammeters; Correcting for Meter Resistance

Suppose you wish to determine the current I in the circuit shown in Fig. 27–14a, and the voltage V across the resistor R_1. How exactly are ammeters and voltmeters hooked up?

An ammeter, because it is to measure the current flowing in the circuit, must be inserted directly into the circuit, in series with the other elements, Fig. 27–14b.

A voltmeter, on the other hand, is connected in parallel with the circuit element across which the voltage is to be measured. It is used to measure the potential difference between two points and its two wire leads (connecting wires) are connected to the two points, as shown in Fig. 27–14c where the voltage across R_1 is being measured.

Voltmeters and ammeters can have several series or shunt resistors to offer a choice of range. *Multimeters* can measure voltage, current, and resistance. Sometimes they are called VOMs (Volt-Ohm-Meter). Meters with digital readout are called digital voltmeters (DVM) or digital multimeters (DMM). To measure resistance, the meter must contain a battery of known voltage connected to an ammeter (galvanometer with a shunt). This makes an *ohmmeter* (see problem 51). The resistor whose resistance is to be measured completes the circuit. The deflection in this case is inversely proportional to the resistance. Thus, if the resistance is small the deflection is large, and vice versa. The calibration of the scale depends on the value of the series resistor. Since an ohmmeter sends a current through the device whose resistance is to be measured, it should not be used on very delicate devices that could be damaged by the current.

The *sensitivity* of a meter is generally specified on the face. It may be given as so many ohms per volt, which indicates how many ohms of resistance there are in the meter per volt of full-scale reading. For example, if the sensitivity is 30,000 Ω/V, this means that on the 10-V scale the meter has a resistance of 300,000 Ω. The full-scale current sensitivity, I_m, discussed earlier, is just the inverse of the sensitivity in Ω/V (note that the units $\Omega/\mathrm{V} = \mathrm{A}^{-1}$). For example a meter with sensitivity 30,000 Ω/V produces a full-scale deflection at 1.0 V when 30,000 Ω is in series with the galvanometer. Thus the current sensitivity is 1.0 V/3.0 $\times$ 10^4 Ω = 33 μA in this case.

It is important to know the sensitivity, for in many cases the resistance of the meter can seriously affect your results. Take the following example.

FIGURE 27–14

Measuring the current and voltage.

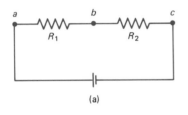

(a)

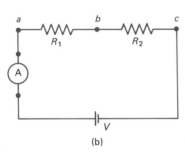

(b)

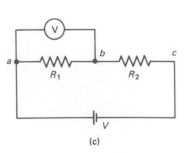

(c)

EXAMPLE 27–5 Suppose R_1 and R_2 are each 15 kΩ and are connected in series as shown in Fig. 27–14a. The battery maintains 8.0 V across them and has negligible internal resistance. A voltmeter whose sensitivity is 10,000 Ω/V is put on the 5.0-V scale. What does the meter read when connected across R_1, and what error is caused by the finite resistance of the meter?

SOLUTION On the 5.0-V scale, the voltmeter has an internal resistance of (5.0 V)(10,000 Ω/V) = 50,000 Ω. When connected across R_1 as in Fig. 27–14c we have this 50 kΩ in parallel with $R_1 = 15$ kΩ; the net resistance R of these two is given by $1/R = 1/50$ kΩ + $1/15$ kΩ = 13/150 kΩ; so $R = 11.5$ kΩ. This $R = 11.5$ kΩ is in series with $R_2 = 15$ kΩ so the total resistance of the circuit is now 26.5 kΩ. Hence the current from the battery is $I = 8.0$ V/26.5 kΩ = 0.30 mA, and the voltage drop across R_1 and the voltmeter is (3.0 $\times$ 10^{-4} A)(11.5 $\times$ 10^3 Ω) = 3.5 V. [That across R_2 is (3.0 $\times$ 10^{-4} A)(15 $\times$ 10^3 Ω) = 4.5 V, for a total of 8.0 V.] If we assume the meter is precise, it will thus read 3.5 V. In the normal circuit, without the meter, $R_1 = R_2$ so the voltage across R_1 is half that of the battery or 4.0 V. Thus the voltmeter, because of its internal resistance, gives a low reading. In this case it is off by 0.5 V, or more than 10 percent.

This example illustrates how seriously a meter can affect a circuit and give a misleading reading. If the resistance of a voltmeter is much higher than the resistance of the circuit, however, it will have little effect and its readings can be trusted, at least to the manufactured accuracy of the meter, which for ordinary meters is typically 3 to 4 percent of full-scale deflection. An ammeter also can interfere with a circuit, but the effect is minimal if its resistance is much less than that of the circuit as a whole. For both voltmeters and ammeters, the more sensitive the galvanometer the less effect it will have. A 50,000-Ω/V meter is far better than a 1,000-Ω/V meter.

Electronic meters using transistors have sensitivities in the range 10^6 to 10^7 Ω/V. Hence they have very little effect on most circuits and their readings are reliable for nearly any circuit. Very high resistance voltmeters, called *electrometers*, may have input resistances $\gtrsim 10^{14}$ Ω.

☐ *27–8 The Potentiometer

The *potentiometer* is a device to precisely measure potential differences. It uses a "null" method, as described below, and has the further advantage of being able to measure a voltage *without drawing current*. Hence it can be used to measure an emf directly (with no voltage drop due to internal resistance). And it won't alter the currents in an operating circuit.

A diagram of a simple potentiometer is shown in Fig. 27–15. A fresh battery, termed the "working battery," is connected in series to a variable resistor R_v(–⌁⌁–) which can be varied to give a convenient voltage between the points A and B. The potential difference between A and B, which we call V, produces a steady current through the precision resistor R' which has a slider (point C) to obtain an accurate variable resistance R. (This resistor R' could be a uniform wire, called a "slide wire," with a sliding contact C that moves along it.)[†] The emf to be measured, $\mathcal{E}_x$, can be placed in the circuit as shown; the slider is moved until the galvanometer indicates zero current ("null point") when the switch S is closed momentarily. In this position no current flows from the unknown source so the voltage between points A and C will be equal to $\mathcal{E}_x$, the emf of the unknown:

$$V_{AC} = \mathcal{E}_x.$$

This relationship holds only when no current flows from the unknown emf through the galvanometer, since otherwise there would be potential drops across the internal resistance of the galvanometer and the unknown source. There is a current I through R', of course, and if we let R_x represent the resistance between points A and C of R' when we are at the null point with $\mathcal{E}_x$ in the circuit, then

$$\mathcal{E}_x = IR_x.$$

A second measurement is necessary, this time with a standard cell—one whose emf $\mathcal{E}_s$ is accurately known in advance. (The standard cell should never be used to draw current, for this will affect its emf over time.) The slider is again adjusted to a new position C' so that again no current passes through the galvanometer; the resistance then between A and C' is R_s, given by

$$\mathcal{E}_s = IR_s$$

since the same current I is flowing through R'. Then

$$\frac{\mathcal{E}_x}{\mathcal{E}_s} = \frac{IR_x}{IR_s} = \frac{R_x}{R_s}$$

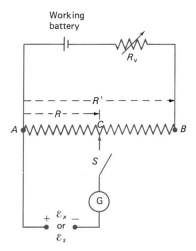

FIGURE 27–15
Potentiometer circuit.

[†] R' can also be a set of fixed precision resistors, with several taps leading out, plus a variable resistor—so that a wide range of precise values of R is available.

or

$$\mathcal{E}_x = \left(\frac{R_x}{R_s}\right)\mathcal{E}_s.$$

(If R' is a precisely uniform wire, then the ratio R_x/R_s is equal to the ratio of lengths AC/AC'.)

It should be noted that the word "potentiometer" is often used to refer simply to a variable resistance; and colloquially it is sometimes called a "pot."

*27-9 The Wheatstone Bridge

The Wheatstone bridge is probably the most common type of so-called *bridge circuits* which are used to make precise measurements of resistance. Fig. 27–16 is a diagram of a Wheatstone bridge circuit. The resistances of R_1, R_2, and R_3 are accurately known. At least one of these—R_3 in our diagram—is variable; its resistance is usually read from a dial. The resistance R_x which is to be measured (the unknown resistance) is connected so as to form the fourth arm of the bridge. When the battery is connected, current flows through each of the resistors. In operation, the variable resistor R_3 is adjusted until the points B and D are at the same potential. This is checked by momentarily closing the switch to the galvanometer which will not deflect when B and D are at the same potential. In this situation the bridge is said to be "balanced" and we have $V_{AB} = V_{AD}$, or (see Fig. 27–16)

$$I_3R_3 = I_1R_1.$$

I_1 is the current that passes through R_1 and also through R_2 when the bridge is balanced; I_3 is the current passing through R_3 and R_x. When the bridge is balanced the voltage across R_x equals that across R_2, so

$$I_3R_x = I_1R_2.$$

We now divide these two equations and find

$$\frac{I_3R_x}{I_3R_3} = \frac{I_1R_2}{I_1R_1}$$

or

$$R_x = \frac{R_2}{R_1}\,R_3.$$

This equation allows a precise calculation of R_x when the other three resistances are precisely known. In practice, the galvanometer is very sensitive, and so when R_3 is being adjusted the switch is closed only momentarily to check if the current is zero or not.

FIGURE 27–16

Wheatstone bridge.

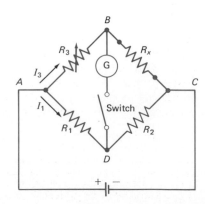

A Wheatstone bridge is used in many situations where resistance must be accurately measured, such as the resistance of a resistance thermometer, Example 26–4.

* 27–10   Transducers and the Thermocouple

A *transducer* is a device that converts one type of energy into another. A high fidelity loudspeaker is one kind of transducer—it transforms electric energy into sound energy (see Chapter 28); so is a microphone, which changes sound into an electrical signal.

Transducers are often used for measuring particular quantities. In this case they can be thought of as changing one kind of signal into another kind, usually electrical. The *resistance thermometer* and the *thermistor* mentioned in the previous chapter are examples; basically they convert temperature into an electrical property, resistance.

We now discuss some other transducers that are useful in a variety of fields. We will meet other examples later.

A *strain gauge*, Fig. 27–17, makes use of the fact that a wire will stretch an amount proportional to the stress (or force) applied to it. When stretched, its resistance must increase since it is longer and its cross-sectional area is reduced (see Eq. 26–4). The fine wire of a strain gauge is usually bonded to a flexible backing material. When fixed tightly to a structure, the resistance of the strain gauge changes in direct proportion to any change in stress on the structure. Since the wire must not be strained beyond its elastic limit, the change in length is generally quite small. Hence the change in resistance is quite small (less than one part per thousand) and a very sensitive Wheatstone bridge is used to measure it. Of course, strain gauges must be carefully calibrated before use. Strain gauges are used in many applications, such as by architects and engineers on models of proposed structures. In this way they can determine the stress at critical points.

A wire strain gauge can be attached to a membrane or diaphragm. A change in pressure against the membrane causes strain in the wire. Thus a strain gauge can be used to measure pressure. In this case it is called a *pressure transducer*. Strain gauges can be made very small (2 to 3 mm wide). When attached to a diaphragm, they can be swallowed to measure internal stomach pressure, or inserted into a vein or artery to measure blood pressure. In these and other situations there will also be variation in resistance due to temperature changes. To correct for this, an identical gauge must also be used which is subjected to the same temperature but no stress.

Another kind of pressure transducer makes use of the *piezoelectric effect*. This effect occurs in certain crystals, such as quartz, that become polarized (Sections 25–5 and 25–7) when a mechanical force is applied and produce an emf proportional to the force. Piezoelectric crystals are used in some phonograph cartridges where they transform the movement of the stylus in a record groove into an electric signal (an emf).

The *thermocouple* is a device that produces an electrical signal when subjected to differing temperatures. Unlike the resistance thermometer (in which it is the resistance that changes with temperature), a thermocouple produces an emf. Indeed, a thermocouple is one of the most useful of thermometers. The thermocouple is based on the "thermoelectric effect." When two dissimilar metals, say iron and copper, are joined at the ends, as shown in Fig. 27–18, it is found that an emf is produced if the two junctions are at different temperatures.[†] The magnitude of

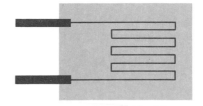

FIGURE 27–17
Wire strain gauge.

[†] According to theory this is due in part to electrons in one metal occupying lower energy states than in the other, so some of them flow across the junction. This leaves one metal slightly more positive than the other, and so a *contact potential* exists between them. If the two junctions are at the same temperature, the same contact potential exists at each; they balance each other and no current will flow. However, when one of the junctions is at a higher temperature, the energy states are altered and the contact potential will be different. In this case there will be a net emf and a current will flow.

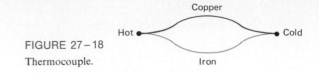

FIGURE 27–18

Thermocouple.

this emf depends on the temperature difference. In operation, one junction of a thermocouple is kept at a known temperature. This "reference temperature" is often 0°C. The other junction, called the "test junction," is placed where the desired temperature is to be measured. The emf is measured most accurately with a potentiometer. It is important that the potentiometer terminals be of the same metal, and kept at the same temperature, so that no additional emfs are produced. Often the thermocouple junctions are connected to lead wires, and these additional connections must also be kept at the same temperature.

A number of different combinations of dissimilar metals are used in thermocouples. The emf as a function of temperature can be found in tables for each set of metals and for a particular reference temperature. Over small temperature ranges the emf is approximately proportional to the temperature differences. However, this linear relationship cannot be assumed to be true in general and tables should be consulted. Incorrect measurements can also result from heat flow along the wires which can affect the temperature of the test junction.

A *microphone* is a transducer that changes a sound wave into an electrical signal. One type of microphone transducer is the capacitor (or condenser) microphone, Fig. 27–19. The changing air pressure in a sound wave causes one plate of the capacitor C to move back and forth. We saw in Chapter 25 that the capacitance is inversely proportional to the separation of the plates. Thus a sound wave causes the capacitance to change. This in turn causes the charge Q on the plates to change ($Q = CV$) so that an electric current is generated at the same frequencies as the incoming sound wave.

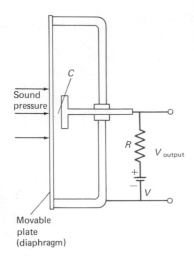

FIGURE 27–19

Diagram of a capacitor microphone.

 Summary _____

When resistances are connected in series, the net resistance is the sum of the individual resistances. When connected in parallel, the reciprocal of the total resistance equals the sum of the reciprocals of the individual resistances; in a parallel connection, the net resistance is less than any of the individual resistances.

A battery behaves like a source of emf in series with an internal resistance. The emf is the potential difference determined by the chemical reactions in the battery and equals the terminal voltage when no current is drawn. When a current is drawn, the voltage at the battery's terminals is less than its emf by an amount equal to the Ir drop across the internal resistance.

Kirchhoff's rules are helpful in determining the currents and voltages in a complex circuit. Kirchhoff's first, or "junction," rule is based on conservation of electric charges and states that the sum of all currents entering any junction equals the sum of all currents leaving that junction; the second, or "loop," rule is based on conservation of energy and states that the algebraic sum of the voltage changes around any closed path of the circuit must be zero.

When a circuit containing a resistor R in series with a capacitance C is connected to a dc source of emf, the voltage across the capacitor rises gradually in time according to an exponential of the form $(1 - e^{-t/RC})$ where the time constant RC is the time it takes for the voltage to reach 63 percent of its maximum value. A capacitor discharging through a resistor is characterized by the same time constant: in a time equal to RC, the voltage across the capacitor drops to 37 percent of its initial value. The charge on, or voltage across, the capacitor, as well as the current out of it, decreases as $e^{-t/RC}$.

 Questions _____

1. Discuss the advantages and disadvantages of Christmas tree lights connected in parallel versus those connected in series.

2. If all you have is a 120-V line, would it be possible to light up several 6-V lamps without burning them out? How?

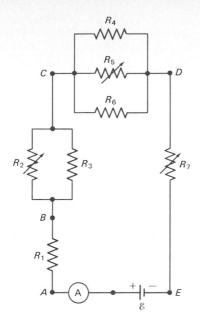

FIGURE 27–20

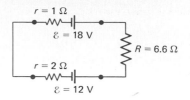

FIGURE 27–21

3. Compare and discuss the formulas for resistors and for capacitors (Section 25–3) when connected in series and in parallel.

4. Two light bulbs of resistance R_1 and R_2 $(>R_1)$ are connected in series. Which is brighter? What if they are connected in parallel?

5. Describe carefully the difference between emf and potential difference.

6. Can a resistor have an emf?

7. The internal resistance of an electric cell is not actually constant. Why not?

8. Explain why Kirchhoff's first (junction) rule is equivalent to conservation of electric charge.

9. Explain why Kirchhoff's second (loop) rule is a result of the conservation of energy.

10. How might Kirchhoff's junction rule be modified to include a point where charge could collect, such as one plate of a capacitor?

11. Given the circuit shown in Fig. 27–20, use the words "increases," "decreases," or "stays the same" to complete the following statements:

(a) If R_7 increases, the potential difference between A and E (assume no resistance in Ⓐ and $\mathscr{E}$)_____ .

(b) If R_7 increases, the potential difference between A and E (assume Ⓐ and $\mathscr{E}$ have resistance)_____.

(c) If R_7 increases, the voltage drop across R_4_____.

(d) If R_2 decreases, the current through R_1_____.

(e) If R_2 decreases, the current through R_6_____.

(f) If R_2 decreases, the current through R_3_____.

(g) If R_5 increases, the voltage drop across R_2_____.

(h) If R_5 increases, the voltage drop across R_4_____.

(i) If R_2, R_5, and R_7 increase, $\mathscr{E}$_____ .

12. Why are batteries connected in series? Why in parallel? Does it matter if the batteries are nearly identical or not in either case?

13. Explain the minus sign in Eq. 27–7; that is, why is the current of the discharging capacitor negative?

14. The 18-V source in Fig. 27–21 is "charging" the 12-V battery. Explain how it does this.

15. Explain in detail how you could go about measuring the internal resistance of a battery.

16. Can the terminal voltage of a battery ever exceed its emf? Explain.

17. When applying Kirchhoff's loop rule (such as in Fig. 27–21) does the sign (or direction) of a battery's emf depend on the direction of current through the battery?

*18. What is the main difference between a voltmeter and an ammeter?

*19. Explain why an ideal ammeter would have zero resistance and an ideal voltmeter infinite resistance.

*20. What are the advantages of a potentiometer over a voltmeter? What are the advantages of a voltmeter over a potentiometer?

*21. Explain why "null" measurements allow greater accuracy than meter-reading methods.

*22. If the battery of a Wheatstone bridge wears down so that its terminal voltage drops a little, will this affect the determination of the unknown resistance?

 Problems

In these problems neglect the internal resistance of a battery unless it is given.

SECTION 27–1

1. (I) Five 30-Ω light bulbs are connected in series. What is the total resistance of the circuit? What is their resistance if they are connected in parallel?

2. (I) Suppose you have a 600-Ω, an 800-Ω and a 1.2-kΩ resistor. What is (a) the maximum, and (b) the minimum, resistance you can obtain by combining these?

3. (I) Suppose you have a 6.0-V battery and you wish to apply a voltage of only 2.1 V. How could you connect resistors so that you could make a "voltage divider" that produced a 2.1-V output for a 6.0-V input?

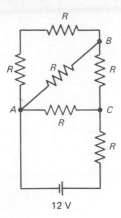

FIGURE 27–22

4. (II) Three 200-Ω resistors can be connected together in four different ways making combinations of series and/or parallel circuits. What are these four ways and what is the net resistance in each case?

5. (II) What is the net resistance of the circuit connected to the battery in Fig. 27–22? Each resistor has $R = 2.1 \text{ k}\Omega$.

6. (II) Eight lights are connected in series to a 120-V source by two leads of total resistance 1.7 Ω. If 120 mA flows through each bulb, what is the resistance of each, and what fraction of the total power is wasted in the leads?

7. (II) Seven 8.0-W Christmas tree lights are designed to be connected in series to a 120-V source. What is the resistance of each bulb?

8. (II) Suppose a person's body resistance is 900 Ω. (a) What current passes through the body when the person accidentally is connected to 100 V? (b) If there is an alternative path to ground whose resistance is 20 Ω, what current passes through the person? (c) If the voltage source can produce at most 1.2 A, how much current passes through the person in case (b)?

9. (II) Two resistors when connected in series to a 120-V line use one-fourth the power that is used when they are connected in parallel. If one resistor has $R = 1.8 \text{ k}\Omega$, what is the resistance of the other?

10. (II) A three-way light bulb can produce 50 W, 100 W, or 150 W at 120 V. Such a bulb contains two filaments that can be connected to the 120 V individually or in parallel. Describe how the connections to the two filaments are made to give each of the three wattages; what must be the resistance of each filament?

11. (II) A 75-W, 120-V bulb is connected in parallel with a 40-W, 120-V bulb. What is the net resistance?

12. (II) Suppose you want to run some apparatus that is 200 m from an electric outlet. Each of the wires connecting your apparatus to the 120-V source has a resistance per unit length of 0.0055 Ω/m. If your apparatus draws 3.2 A, what will be the voltage drop across the connecting wires and what voltage will be applied to your apparatus?

13. (II) A 1.2-kΩ and a 1.8-kΩ resistor are connected in parallel; this combination is connected in series with a 1.9-kΩ resistor. If each resistor is rated at $\frac{1}{2}$ W, what is the maximum voltage that can be applied across the whole network?

14. (II) How many $\frac{1}{2}$-W resistors, each of the same resistance, must be used to produce an equivalent 1.2-kΩ, 5-W resistor? What is the resistance of each, and how must they be connected?

15. (II) Calculate the current through each resistor in Fig. 27–22 if each resistance $R = 2.10 \text{ k}\Omega$. What is the potential difference between points A and B?

SECTION 27–2

16. (I) A battery whose emf is 6.0 V and whose internal resistance is 0.50 Ω is connected to a circuit whose net resistance is 11.6 Ω. What is the terminal voltage of the battery?

17. (I) Four 1.5-V cells are joined in series to a 9.2-Ω device. The internal resistance of each cell is 0.30 Ω. What current flows to the device?

18. (I) What is the internal resistance of a 12-V car battery whose terminal voltage drops to 7.8 V when the starter draws 70 A?

19. (I) A 1.5-V dry cell can be tested by connecting it to a low-resistance ammeter; it should be able to supply at least 30 A. What is the internal resistance of the cell in this case?

20. (II) The internal resistance of a 1.35-V mercury cell is 0.030 Ω whereas that of a 1.5-V dry cell is 0.35 Ω. Explain why three mercury cells can more effectively power a 2-W hearing aid that requires 4.0 V than can three dry cells.

21. (II) What is the current in the 8.0-Ω resistor in Fig. 27–5a.

22. (II) A battery produces 50.0 V when 5.5 A is drawn from it and 58.2 V when 1.8 A is drawn. What is the emf and internal resistance of the battery?

SECTION 27–3

23. (I) Determine the terminal voltage of each battery in Fig. 27–21.

24. (I) What is the potential difference between points a and d in Fig. 27–6.

25. (II) What is the terminal voltage of each battery in Fig. 27–6?

26. (II) Determine the current through each of the resistors in Fig. 27–23.

27. (II) If the 30-Ω resistor in Fig. 27–23 were shorted out (resistance = 0), what then would be the current through the 10-Ω resistor?

FIGURE 27–23

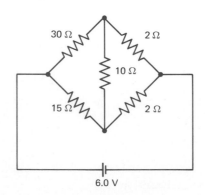

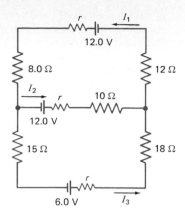

FIGURE 27–24

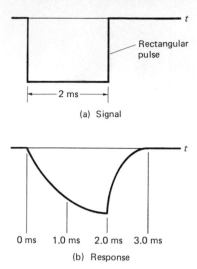

(a) Signal

(b) Response

0 ms   1.0 ms   2.0 ms   3.0 ms

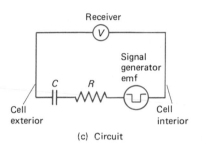

(c) Circuit

FIGURE 27–26

28. (II) Determine the currents I_1, I_2, and I_3 in Fig. 27–24. Assume the internal resistance of each battery is $r = 1.0\ \Omega$. What is the terminal voltage of the 6.0-V battery?

29. (II) What would the current I_1 be in Fig. 27–24 if the 12-Ω resistor were shorted out ($r = 1.0\ \Omega$)?

SECTION 27–5

30. (II) A pacemaker is designed to operate at 70 beats/min using a 9.0-μF capacitor. What value of resistance should be used if the pacemaker is to "fire" (capacitor discharge) when the voltage reaches 25 percent of maximum?

31. (II) Two resistors and two uncharged capacitors are arranged as shown in Fig. 27–25. With a potential difference of 24 V across the combination, (a) what is the potential at point a with S open? (Let $V = 0$ at negative terminal of source.) (b) What is the potential at point b with the switch open? (c) When the switch is closed, what is the final potential of point b? (d) How much charge flows through the switch S when it is closed? (e) What is the time constant for the circuit when the switch is open?

32. (II) How long does it take for the energy stored in a capacitor in a series RC circuit (Fig. 27–10a) to reach half its maximum value? Express answer in terms of the time constant $\tau = RC$.

33. (II) Two 6.0-μF capacitors, two 2.2-kΩ resistors, and a 12.0-V source are connected in series. Starting from the uncharged state, how long does it take for the current to drop from its initial value to 1.20 mA?

34. (II) In experiments on nerve axon and muscle fiber cells two electrodes are inserted into the cell (axon) close to each other. One electrode serves as a signal generator of rectangular pulses. The other serves as a receiver of the cell response to the signal pulses. The signal and response are shown in Fig. 27–26a and b. The electrical characteristics of the cell in such an experiment may be represented by the equivalent circuit illustrated in Fig. 27–26c where C is the capacitance of the cell membrane, R is the resistance of the cell membrane, the emf is the signal electrode, and the voltmeter is the receiver electrode. (a) Use the response graph to estimate the time constant of the cell. (b) If the area of the membrane is about $10^{-7}\ m^2$, its thickness is about $10^{-8}\ m$, and its dielectric constant about 3, what is the capacitance C? (c) Using the result of parts (a) and (b), calculate the value of R.

35. (III) (a) Determine the time constant for charging the capacitor in the circuit of Fig. 27–27. (Hint: use Kirchhoff's rules.) (b) What is the maximum charge on the capacitor?

FIGURE 27–25

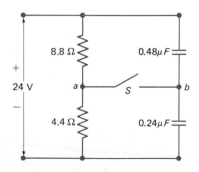

FIGURE 27–27

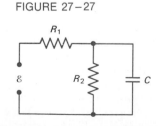

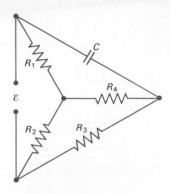

FIGURE 27–28

36. (III) (*a*) Determine the time constant for charging the capacitor in the circuit of Fig. 27–28. (Hint: use Kirchhoff's rules.) (*b*) What is the maximum charge on the capacitor?

37. (III) Suppose the switch *S* in Fig. 27–25 is closed. What is the time constant (or time constants) for charging the capacitors?

*SECTION 27–6

*38. (I) An ammeter has a sensitivity of 5000 Ω/V. What current passing through the galvanometer produces full-scale deflection?

*39. (I) What is the resistance of a voltmeter on the 50-V scale if the meter sensitivity is 30,000 Ω/V?

*40. (II) A galvanometer has a sensitivity of 50,000 Ω/V and internal resistance 24 Ω. How could you make this into (*a*) an ammeter reading 10 mA full scale; (*b*) a voltmeter reading 100 mV full scale?

*41. (II) How could you make an ammeter with a full-scale deflection of 10 A from a galvanometer that has a 200-μA full-scale deflection when 10 mV is across it?

*42. (II) A milliammeter reads 10 mA full scale. It consists of a 0.20-Ω resistor in parallel with a 30-Ω galvanometer. How can you change this ammeter to a voltmeter giving a full-scale reading of 10 V without taking the ammeter apart? What will be the sensitivity (Ω/V) of your voltmeter?

*43. (II) A galvanometer has an internal resistance of 30 Ω and deflects full scale for a 60-μA current. Describe how to use this galvanometer to make (*a*) an ammeter to read currents up to 15A; and (*b*) a voltmeter to give a full-scale deflection of 3000V.

*SECTION 27–7

*44. (II) A 45-V battery of negligible internal resistance is connected to a 37-kΩ and a 22-kΩ resistor in series. What reading will a voltmeter give when used to measure the voltage across each resistor if its internal resistance is 100 kΩ? What is the percent inaccuracy due to meter resistance for each case?

*45. (II) An ammeter whose internal resistance is 90 Ω reads 1.60 A when connected in a circuit containing a battery and two resistors in series whose values are 300 Ω and 500 Ω. What is the actual current when the ammeter is absent?

*46. (II) A battery with $\mathscr{E} = 9.0$ V and internal resistance $r = 1.2$ Ω is connected to two 5.0-kΩ resistors in series. An ammeter of internal resistance 0.50 Ω measures the current

and at the same time a voltmeter with internal resistance 15 kΩ measures the voltage across one of the 5.0-kΩ resistors in the circuit. What do the ammeter and voltmeter read?

*47. (II) Two 5.9-kΩ resistors are placed in series and connected to a battery. A voltmeter of sensitivity 1000 Ω/V is on the 3.0-V scale and reads 2.0 V when placed across either of the resistors. What is the emf of the battery? (Ignore the battery's internal resistance.)

*48. (II) What internal resistance should the voltmeter have to be in error by less than 3 percent for the situation of Example 27–5?

49. (II) (*a*) A voltmeter and an ammeter can be connected as shown in Fig. 27–29a to measure a resistance *R*. The value of *R* will not quite be *V/I* where *V* is the voltmeter reading and *I* is the ammeter reading since some of the current actually goes through the voltmeter. Show that the actual value of *R* is given by

$$\frac{1}{R} = \frac{I}{V} - \frac{1}{R_\text{v}}$$

where R_v is the voltmeter resistance. Note that $R \approx V/I$ if $R_\text{v} \gg R$. (*b*) A voltmeter and an ammeter can also be connected as shown in Fig. 27–29b to measure a resistance *R*. Show in this case that

$$R = \frac{V}{I} - R_\text{A}$$

where *V* and *I* are the voltmeter and ammeter readings and R_A is the resistance of the ammeter. Note that $R \approx V/I$ if $R_\text{A} \ll R$.

50. (II) The voltage across a 120-kΩ resistor in a circuit containing additional resistance (R_2) in series with a battery (*V*) is measured, by a 20,000-Ω/V meter on the 100-V scale, to be 25 V. On the 30-V scale, the reading is 23 V. What is the actual voltage in the absence of the voltmeter? What is R_2?

FIGURE 27–29

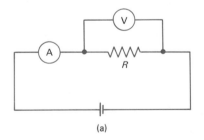

(a)

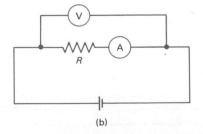

(b)

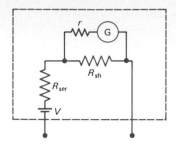

FIGURE 27–30

Ohmmeter.

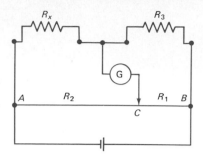

FIGURE 27–31

Slide-wire form of Wheatstone bridge.

*51. (II) One type of ohmmeter consists of an ammeter connected to a series resistor and a battery, Fig. 27–30. The scale is different from ammeters and voltmeters. Zero resistance corresponds to full-scale deflection (because maximum current flows from the battery) whereas infinite resistance corresponds to no deflection (no current flow). Suppose a 9.0-V battery is used, and the galvanometer has internal resistance of 30 Ω and deflects full scale for a current of 30 μA. What values of shunt resistance, R_{sh}, and series resistance, R_{ser}, are needed to make an ohmmeter that registers a midscale deflection (that is, half of maximum) for a resistance of 30 kΩ? (Note: an additional series resistor (variable) is also needed so the meter can be zeroed; this should be checked frequently by touching the leads together since the battery voltage can vary. Because of battery voltage variation, such ohmmeters are not precision instruments, but they are useful to obtain approximate values.)

*52. (II) A 9.0-V battery (internal resistance $r = 1.0$ Ω) is connected to two resistors in series. A voltmeter whose internal resistance is 10.0 kΩ measures 3.0 V and 4.0 V, respectively, when connected across each of the resistors. What is the resistance of each resistor?

*SECTION 27–8

*53. (I) A slide-wire potentiometer is balanced against a 1.0182-V standard cell when the slide wire is set at 40.2 cm out of a total length of 100.0 cm. For an unknown source, the setting is 11.9 cm. What is the emf of the unknown?

*54. (I) The galvanometer of a potentiometer has an internal resistance of 40 Ω and can detect a current as small as 0.015 mA. What is the minimum uncertainty possible in measuring an unknown voltage?

*55. (II) The uniform 30.00-cm-long slide wire of a potentiometer has a total resistance of 3.224 Ω. When a 1.0183-V standard cell is connected, a 22.40-cm length of wire is needed to achieve balance. When an unknown source is connected, balance is achieved only when a 9.050-Ω resistor previously connected between the working battery and point A in Fig. 27–15 is now included along with the slide-wire resistance, and the slide wire is set at 14.16 cm. What is the value of the unknown emf?

*56. (II) A potentiometer slide wire must be set at a 48.0-cm length to balance the emf of a cell. When a 5.0-Ω resistor is placed in parallel across the terminals of the cell, a 44.0-cm length is needed. What is the internal resistance of the cell?

*SECTION 27–9

*57. (I) Figure 27–31 shows a slide-wire form of Wheatstone bridge. If $R_3 = 34.5$ Ω and the lengths $AC = 28.2$ cm and $CB = 71.8$ cm when the galvanometer reads zero current, what is the value of R_x? Assume the slide wire is uniform.

*58. (II) An unknown length of platinum wire 1.2 mm in diameter is placed as the unknown resistance in a Wheatstone bridge. Arms 1 and 2 have resistance of 36.0 Ω and 84.0 Ω, respectively. Balance is achieved when R_3 is 4.16 Ω. How long is the platinum wire?

*SECTION 27–10

*59. (I) A copper-constantan thermocouple produces emfs of about 40 μV/C°. If the reference temperature is 25°C, what must be the temperature of the test junction if the emf produced is 1.72 mV?

*60. (I) For iron-copper junctions near room temperature, the emf produced by a thermocouple is about 14 μV/C°. If a potentiometer can detect emfs as low as 0.50 μV, to what accuracy can the temperature be read?

*61. (II) The emf of a thermocouple can be written $\mathscr{E} = aT + bT^2$, where T is the temperature (°C) to be measured and the reference temperature is 0°C; a and b are constants, and for a purely linear response b would be zero (usually it is small). (a) Determine the values for a and b if the emf is 3.62 mV at $T = 80$°C and is 6.80 mV at $T = 170$°C. (b) What is the temperature if the emf is 4.90 mV?

*62. (II) The *strain factor* K of a strain gauge is defined as the fractional change in resistance ($\Delta R/R$) divided by the fractional change in length:

$$K = \frac{\Delta R/R}{\Delta L/L},$$

and is relatively constant at a given temperature. In other words, the change in resistance ΔR is proportional to the change in length of the wire ΔL. (a) Show that this linear relationship is reasonable for small $\Delta L/L$. (b) A strain gauge with a strain factor of 1.8 is connected crosswise to a small muscle about 4.5 mm wide. The gauge is connected as the unknown arm in a Wheatstone bridge. When the muscle is relaxed, the Wheatstone bridge is balanced when $R_2/R_1 = 1.4800$ and $R_3 = 40.700$ Ω. When the muscle contracts, balance occurs for $R_3 = 40.736$ Ω. How much has the muscle widened?

Magnetism

28

Today it is clear that magnetism and electricity are closely related. This relationship was not discovered, however, until the nineteenth century. The history of magnetism begins much earlier with the ancient civilizations in Asia Minor. It was in a region of Asia Minor known as Magnesia that rocks were found that would attract each other. These rocks were called "magnets" after their place of discovery.

 ## 28–1 Magnets and Magnetic Fields

A magnet will attract paper clips, nails, and other objects made of iron. Any magnet, whether it is in the shape of a bar or a horseshoe, has two ends or faces, called poles; this is where the magnetic effect is strongest. If a magnet is suspended from a fine thread, it is found that one pole of the magnet will always point toward the north. It is not known for sure when this fact was discovered, but it is known that the Chinese were making use of it as an aid to navigation by the eleventh century and perhaps earlier. This is, of course, the principle of a compass. A compass needle is simply a magnet that is supported at its center of gravity so it can rotate freely. That pole of a freely suspended magnet which points toward the north is called the *north pole* of the magnet. The other pole points toward the south and is called the *south pole*.

It is a familiar fact that when two magnets are brought near one another, each exerts a force on the other. The force can be either attractive or repulsive and can be felt even when the magnets don't touch. If the north pole of one magnet is brought near the north pole of a second magnet, the force is repulsive. Similarly, if two south poles are brought close, the force is repulsive. But when a north pole is brought near a south pole, the force is attractive, Fig. 28–1. This is reminiscent of the force between electric charges; like poles repel and unlike poles attract. But do not confuse magnetic poles with electric charge. They are not the same thing.

Only iron and a few other materials such as cobalt, nickel and gadolinium show strong magnetic effects. They are said to be *ferromagnetic* (from the Latin word *ferrum* for iron). All other materials show some slight magnetic effect, but it is

FIGURE 28–1

Like poles of a magnet repel; unlike poles attract.

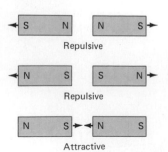

Repulsive

Repulsive

Attractive

extremely small and can be detected only with delicate instruments (more on this in Chapter 29.)

We found it useful to speak of an electric field surrounding an electric charge. In the same way we can imagine a *magnetic field* surrounding a magnet. The force one magnet exerts on another can then be described as the interaction between one magnet and the magnetic field of the other. Just as we drew electric field lines, we can also draw magnetic field lines. They can be drawn, as for electric field lines, so that (1) the direction of the magnetic field is tangent to a line at any point, and (2) the number of lines per unit area is proportional to the magnitude of the magnetic field.

The *direction* of the magnetic field at a given point can be defined as the direction that the north pole of a compass needle would point when placed at that point. Figure 28–2 shows how one magnetic field line around a bar magnet is found using compass needles. The magnetic fields determined in this way for a horseshoe magnet and a bar magnet are shown in Fig. 28–3. Notice that because of our definition, the lines always point from the north toward the south pole of a magnet.

We can define the **magnetic field** at any point as a vector, represented by the symbol **B**, whose direction is as defined above; the *magnitude* of **B** can be defined in terms of the torque exerted on a compass needle when it is *not* aligned with the magnetic field, as in Fig. 28–4. That is, the greater the torque, the greater the magnetic field strength. We can use this (rather rough) definition for now, but a more precise definition will be given in Section 28–3. The terms "magnetic flux density" and "magnetic induction" are often used for **B** rather than our simple term "magnetic field."

The earth's magnetic field is shown in Fig. 28–5. Since the north pole of a compass needle points north, the magnetic pole which is in the geographic north is magnetically a south pole (remember, that the north pole of one magnet is attracted to the south pole of a second). And the earth's magnetic pole near the geographic south pole is magnetically a north pole. The earth's magnetic poles do not, however, coincide with the geographic poles (which are on the earth's axis of rotation). The magnetic south pole, for example, is in northern Canada, about 1500 km from the geographic north pole. This must be taken into account when using a compass. The

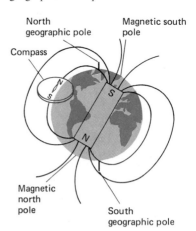

FIGURE 28–2

Plotting a magnetic field line of a bar magnet.

FIGURE 28–3

Magnetic field lines of
(a) a horseshoe magnet,
and (b) a bar magnet.

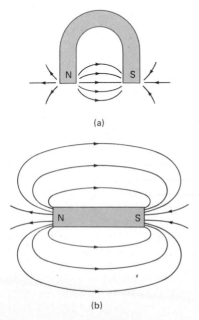

(a)

(b)

FIGURE 28–4

Forces on a compass needle that produce a torque to orient it parallel to the magnetic field lines. The torque will be zero when the needle is parallel to the magnetic field line at that point. (Only the attractive forces are shown; try drawing in the repulsive forces and show that they produce a similar torque.)

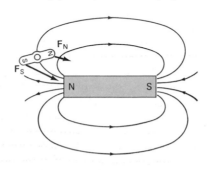

FIGURE 28–5

The earth acts like a huge magnet with its magnetic south pole near the geographic north pole.

angular difference between magnetic north and true (geographical) north is called the "magnetic declination." In the United States it varies from 0° to about 25°, depending on location.

 ## 28–2   Electric Currents Produce Magnetism

During the eighteenth century many natural philosophers sought to find a connection between electricity and magnetism. The first to uncover a significant connection was Hans Christian Oersted (1777–1851) in 1820. Oersted had believed for a long time in the unity of nature. Philosophically, he felt there ought to be a connection between magnetism and electricity. However, a stationary electric charge and a magnet had been shown not to have any influence on each other. But Oersted found that when a compass needle is placed near an electric wire, the needle deflects as soon as the wire is connected to a battery and a current flows.

As we have seen, a compass needle can be deflected by a magnetic field. What Oersted found was that *an electric current produces a magnetic field*. He had found a connection between electricity and magnetism.

A compass needle placed near a straight section of wire aligns itself so it is tangent to a circle drawn around the wire, Fig. 28–6. Thus, the magnetic field lines of the wire are in the form of circles with the wire at their center, Fig. 28–7a. The direction of these lines is indicated by the north pole of the compass in Fig. 28–6. There is a simple way to remember the direction of the magnetic field lines in this case. It is called a *right-hand rule*: You grasp the wire with your right hand so that your thumb points in the direction of the conventional (positive) current; then your fingers will encircle the wire in the direction of the magnetic field, Fig. 28–7b.

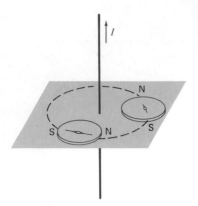

FIGURE 28–6

Deflection of a compass needle near a current-carrying wire, showing the presence and direction of the magnetic field.

 ## 28–3   Magnetic Force on a Current; Definition of **B**

In the previous section we saw that an electric current exerts a force on a magnet, such as a compass needle. By Newton's third law we might expect the reverse to be true as well: a magnet should exert a force on a current carrying wire. This is indeed the case and this effect was also discovered by Oersted.

Let us look at the force exerted on a wire in detail. Suppose a straight wire is placed between the pole pieces of a magnet as shown in Fig. 28–8. When a current flows in the wire, a force is exerted on the wire; but this force is *not* toward one or the other poles of the magnet. Instead, the force is directed *at right angles to the magnetic field direction*. If the current is reversed in direction, the force is in the opposite direction. It is found that the direction of the force is always perpendicular to the direction of the current-carrying wire and also perpendicular to the direction of the magnetic field, **B**. This description leaves an ambiguity, however (the force could be either up or down in Fig. 28–8a and still be perpendicular to both the wire and to **B**). Experimentally the direction of the force is given by the right-hand rule illustrated in Fig. 28–8b. First you orient your right hand so that the outstretched fingers point in the direction of the (conventional) current; from this position, when you bend your fingers they should then point in the direction of the magnetic field lines; if they do not, rotate your hand and arm about the wrist until they do, remembering that straightened fingers must point along the direction of the current. When your hand is oriented in this way, then the extended thumb points in the direction of the force on the wire.

This describes the direction of the force. Now what about its magnitude? It is found experimentally that the magnitude of the force is directly proportional to the

FIGURE 28–7

(a) Magnetic field lines around a straight wire. (b) Right-hand rule for remembering the direction of the magnetic field: when the thumb points in the direction of the conventional current, the fingers wrapped around the wire point in the direction of the magnetic field.

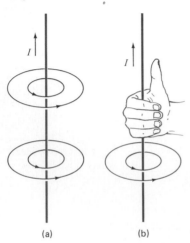

(a)   (b)

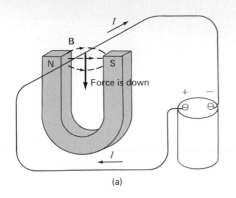

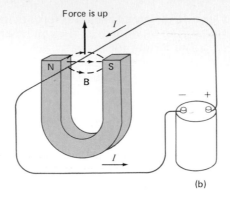

Force is up

(a)   (b)

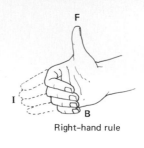

Right-hand rule

FIGURE 28–8

Force on a current-carrying wire placed in a magnetic field **B**.

current I in the wire, to the length l of wire in the magnetic field (assumed uniform), and to the magnetic field B. The force also depends on the angle θ between the wire and the magnetic field. When the wire is perpendicular to the field lines the force is strongest; when the wire is parallel to the magnetic field lines there is no force at all. And at other angles it is proportional to $\sin\theta$ (Fig. 28–9). Thus we have

$$F \propto IlB \sin\theta.$$

Up to now we have talked of magnetic field strengths in terms of the torque exerted by the field on a compass needle (Fig. 28–4). But we have not defined the magnetic field strength precisely. In fact, the magnetic field strength B is defined in terms of the above proportion so that the proportionality constant is precisely 1. Thus we have

$$F = IlB \sin\theta. \qquad (28\text{–}1)$$

If the wire is perpendicular to the field ($\theta = 90°$) then the force is

$$F_{\max} = IlB. \qquad [l \perp \mathbf{B}] \qquad (28\text{–}2)$$

And if the wire is parallel to the field ($\theta = 0°$), the force is zero.

In summary, the magnetic field vector **B** is defined as follows. The direction of **B** is the direction in which the force acting on a straight section of current-carrying wire is zero ($\theta = 0$ in Eq. 28–1). The magnitude of **B** is defined (from Eq. 28–2) as

$$B = \frac{F_{\max}}{Il}$$

where $F_{\max}$ is the maximum value of the force on a straight length l of wire carrying a current I; $F_{\max}$ is obtained when the wire is perpendicular to **B**.

The relation between the force **F** on a wire carrying current I, and the magnetic field **B** that causes the force can be written as a vector equation. To do so, we recall that the direction of **F** is given by the right-hand rule (Fig. 28–8b), and the magnitude by Eq. 28–1; this is consistent with the definition of the vector cross product (see Section 10–1), so we can write

$$\mathbf{F} = I\mathbf{l} \times \mathbf{B}; \qquad (28\text{–}3)$$

here, $\mathbf{l}$ is a vector whose magnitude is the length of the wire and its direction is along the wire (assumed straight) in the direction of the conventional (positive) current.

The above discussion applies if the magnetic field is uniform and the wire is straight. If **B** is not uniform, or if the wire does not everywhere make the same angle θ with **B**, then Eq. 28–3 can be written

$$d\mathbf{F} = I\,d\mathbf{l} \times \mathbf{B} \qquad (28\text{–}4)$$

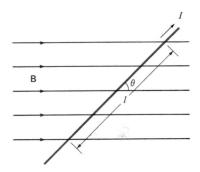

FIGURE 28–9

Current-carrying wire in a magnetic field.

where $d\mathbf{F}$ is the infinitesimal force acting on a differential length dl of the conductor. The total force on the conductor is then found by integrating.

The SI unit for magnetic field strength B is the *tesla* (T). From Eqs. 28–1, 2, 3, or 4, it is clear that $1\text{ T} = 1\text{ N/A·m}$. An older name for the tesla is the "weber per meter squared" ($1\text{ Wb/m}^2 = 1\text{ T}$). Another unit commonly used to specify magnetic field strength is a cgs unit, the gauss (G): $1\text{ G} = 10^{-4}\text{ T}$. A field given in gauss should always be changed to teslas before using with other SI units. To get a "feel" for these units, we note that the magnetic field of the earth at its surface is about $\frac{1}{2}$ G or 0.5×10^{-4} T. On the other hand, strong electromagnets can produce fields on the order of a couple teslas and superconducting magnets over 10 T.

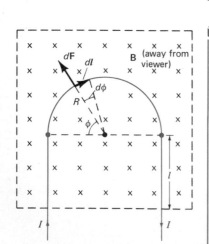

FIGURE 28–10

Measuring a magnetic field B (Example 28–1).

EXAMPLE 28–1   A rectangular loop of wire hangs vertically as shown in Fig. 28–10. A magnetic field **B** is directed horizontally, perpendicular to the wire, and points out of the paper at all points as represented by the symbol $\odot$. (The $\odot$ is meant to resemble the tip of an arrow heading at us; if the field were pointing into the paper, we would represent this using the symbol X to resemble the tail of an arrow going away.) The magnetic field **B** is very nearly uniform along the horizontal portion of wire ab (length $l = 10.0$ cm) which is near the center of the large magnet producing the field; the magnet is aligned symmetrically so that the nonuniformity of **B** along the section of vertical wire is the same for both vertical lengths; the top portion of the wire loop is entirely free of the field. The loop hangs from a balance which measures a downward force $F = 3.48 \times 10^{-2}$ N when the wire carries a current $I = 0.245$ A. What is the magnitude of the magnetic field B at the center of the magnet?

SOLUTION   The magnetic forces on the two vertical sections of the wire loop are in equal and opposite directions. Hence, the net magnetic force on the loop is that on the horizontal section ab whose length $l = 0.100$ m:

$$B = \frac{F}{Il} = \frac{3.48 \times 10^{-2}\text{ N}}{(0.245\text{ A})(0.100\text{ m})} = 1.42\text{ T}.$$

This is, incidentally, a highly precise means of determining magnetic fields.

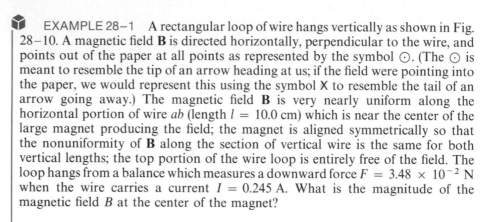

FIGURE 28–11

Example 28–2.

EXAMPLE 28–2   A rigid wire, carrying a current I, consists of a semicircle of radius R and two straight portions as shown in Fig. 28–11. The wire lies in a plane perpendicular to a uniform magnetic field **B**. The straight portions each have length l within the field. Determine the net force on the wire due to the magnetic field.

SOLUTION   The forces on the two straight sections are equal ($= IlB$) and opposite so they cancel. Hence the net force is that on the semicircular portion. We divide the semicircle into short lengths $dl = R\,d\phi$ as indicated and make use of Eq. 28–4:

$$dF = IBR\,d\phi.$$

where dF is the force on the length $dl = R\,d\phi$, and the angle between $d\mathbf{l}$ and **B** is 90° (so $\sin\theta = 1$ in the cross product). Because of the symmetry, equal-length portions on either side of the semicircle will experience equal horizontal components of $d\mathbf{F}$, which will cancel. Hence we need be concerned only with the vertical components, $dF\sin\phi$, and the total force will have magnitude

$$F = \int_0^\pi dF\sin\phi = IBR\int_0^\pi \sin\phi\,d\phi = -IBR\cos\phi\Big|_0^\pi = 2IBR,$$

with direction vertically upward.

28–4   Moving Electric Charge in a Magnetic Field

We have seen that a current-carrying wire experiences a force when placed in a magnetic field. Since a current in a wire consists of moving electric charges, we might expect that freely moving charged particles (not in a wire) would also experience a force when passing through a magnetic field. Indeed, this is the case.

From what we already know we can predict the force on a single moving electric charge. If n such particles of charge q pass by a given point in time t, they constitute a current $I = nq/t$. We let t be the time for a charge q to travel a distance l in a magnetic field $\mathbf{B}$; then $l = vt$ where v is the velocity of the particle. Thus, the force on these n particles is, by Eq. 28–3, $\mathbf{F} = I\mathbf{l} \times \mathbf{B} = (nq/t)(vt) \times \mathbf{B} = nq\mathbf{v} \times \mathbf{B}$. The force on *one* of the n particles is then

$$\mathbf{F} = q\mathbf{v} \times \mathbf{B}, \tag{28–5}$$

and this result is confirmed by experiment. The magnitude of the force is

$$F = qvB \sin \theta. \tag{28–6}$$

This gives the magnitude of the force on a particle of charge q moving with velocity $\mathbf{v}$ in a magnetic field of strength B; the angle between $\mathbf{v}$ and $\mathbf{B}$ is θ. The force is greatest when the particle moves perpendicular to $\mathbf{B}$ ($\theta = 90°$):

$$F = qvB. \qquad [\mathbf{v} \perp \mathbf{B}]$$

And the force is *zero* if the particle moves *parallel* to the field lines ($\theta = 0°$). The *direction* of the force is perpendicular to the magnetic field $\mathbf{B}$ and to the velocity $\mathbf{v}$ of the particle, and it is given by the right-hand rule: you orient your right hand so that your outstretched fingers point along the direction of motion of the particle ($\mathbf{v}$) and when you bend your fingers they must point along the direction of $\mathbf{B}$; then your thumb will point in the direction of the force. This is true only for *positively* charged particles, and will be "down" for the situation of Fig. 28–12. For negatively charged particles the force is in exactly the opposite direction ("up" in Fig. 28–12).

FIGURE 28–12

Force on charged particles due to a magnetic field.

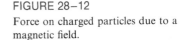
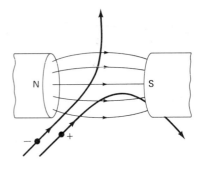

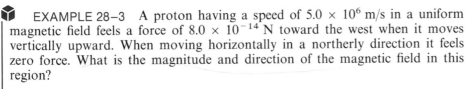

EXAMPLE 28–3   A proton having a speed of 5.0×10^6 m/s in a uniform magnetic field feels a force of 8.0×10^{-14} N toward the west when it moves vertically upward. When moving horizontally in a northerly direction it feels zero force. What is the magnitude and direction of the magnetic field in this region?

SOLUTION   Since the proton feels no force when moving north, the field must be in a north-south direction. The right-hand rule tells us that $\mathbf{B}$ must point toward the north in order to produce a force to the west when the proton moves upward. (Your thumb points west when your right hand is oriented so that your outstretched fingers point upward and your bent fingers point north.) The magnitude of $\mathbf{B}$, from Eq. 28–6, with $\theta = 90°$, is

$$B = \frac{F}{qv} = \frac{8.0 \times 10^{-14} \text{ N}}{(1.6 \times 10^{-19} \text{ C})(5.0 \times 10^6 \text{ m/s})} = 0.10 \text{ T}.$$

The path of a charged particle moving in a plane perpendicular to a uniform magnetic field is a circle (or the arc of a circle if the particle leaves the magnetic field region). This can be seen from Fig. 28–13. In this drawing, the magnetic field is directed *into* the paper as represented by x's. An electron at point P is moving to the right, and the force on it is therefore downward as shown. The electron is

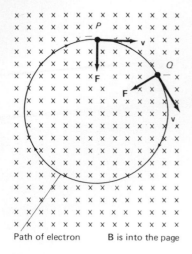

Path of electron   B is into the page

FIGURE 28–13

Force exerted by a uniform magnetic field on a moving charged particle (in this case an electron) produces a curved path.

thus deflected downward. A moment later, say when it reaches point Q, the force is still perpendicular to the velocity and is in the direction shown. Since the force is always perpendicular to $\mathbf{v}$, the magnitude of $\mathbf{v}$ does not change and the particle moves in a circular path. The force is directed toward the center of this circle at all points. Note that the electron moves clockwise in Fig. 28–13. A positive particle would feel a force in the opposite direction and would thus move in a counterclockwise circle.

EXAMPLE 28–4 An electron travels at 2.0×10^7 m/s in a plane perpendicular to a 0.10-T magnetic field. Describe its path.

SOLUTION The radius of the curved path is found using Newton's second law, $F = ma$, and the fact that the acceleration of a particle moving in a circle is $a = v^2/r$. Thus

$$F = ma$$

$$qvB = \frac{mv^2}{r}.$$

We solve for r and find:

$$r = \frac{mv}{qB}.$$

Since $\mathbf{F}$ is perpendicular to $\mathbf{v}$, the magnitude of $\mathbf{v}$ doesn't change. From this equation we see that if $\mathbf{B} =$ constant, then $r =$ constant and the curve must be a circle as claimed above. To get r we put in the numbers:

$$r = \frac{(9.1 \times 10^{-31} \text{ kg})(2.0 \times 10^7 \text{ m/s})}{(1.6 \times 10^{-19} \text{ C})(0.10 \text{ T})} = 1.1 \times 10^{-3} \text{ m},$$

or 1.1 mm.

The time T required for a particle of charge q moving with constant speed v to make one circular revolution in a uniform magnetic field $\mathbf{B}$ ($\perp \mathbf{v}$) is

$$T = \frac{2\pi r}{v}$$

where $2\pi r$ is the circumference of its circular path. From Example 28–4, $r = mv/qB$, so

$$T = \frac{2\pi m}{qB}.$$

Since T is the period of rotation, the frequency of rotation is

$$f = \frac{1}{T} = \frac{qB}{2\pi m}. \tag{28–7}$$

This is often called the *cyclotron frequency* of a particle in a field because this is the rotation frequency of particles in a cyclotron (Section 28–8). Note that f does not depend on the speed v; if v is large, r is large ($r = mv/qB$) for a given B, but the frequency is independent of v and r.

If a particle of charge q moves with velocity v in the presence of both a magnetic field $\mathbf{B}$ and an electric field $\mathbf{E}$, it will feel a force

$$\mathbf{F} = q(\mathbf{E} + \mathbf{v} \times \mathbf{B}) \tag{28–8}$$

where we have made use of Eqs. 22–3 and 28–5. Equation 28–8 is often called the *Lorentz equation*. For the case when $\mathbf{E}$ and $\mathbf{B}$ are uniform and parallel, convince yourself that the general motion of a charged particle will be helix-shaped (the shape of a coil spring).

28–5 Discovery and Properties of the Electron

The electron plays a basic role in our understanding of electricity and magnetism today. But its existence was not suggested until the 1890s. We discuss it here because magnetic fields were crucial for measuring its properties.

Toward the end of the nineteenth century, studies were being done on the discharge of electricity through rarefied gases. One apparatus, diagrammed in Fig. 28–14, was a glass tube fitted with electrodes and evacuated so only a small amount of gas remained inside. When a very high voltage was applied to the electrodes, a dark space seemed to extend outward from the cathode toward the opposite end of the tube; and that far end of the tube would glow. If one or more screens containing a small hole were inserted as shown, the glow was restricted to a tiny spot on the end of the tube. It seemed as though something being emitted by the cathode traveled to the opposite end of the tube; these were named *cathode rays*.

There was much discussion at the time about what these rays might be. Some scientists thought they might resemble light. But the observation that the bright spot at the end of the tube could be deflected to one side by an electric or magnetic field suggested that cathode rays could be charged particles; and the direction of the deflection was consistent with a negative charge. Furthermore, if the tube contained certain types of rarefied gas, the path of the cathode rays was made visible by a slight glow.

Estimates of the charge e of the (assumed) cathode-ray particles as well as of their charge-to-mass ratio, e/m, had been made by 1897; but in that year, J. J. Thomson (1856–1940) was able to measure e/m directly, using the apparatus shown in Fig. 28–15. Cathode rays are accelerated by a high voltage and then pass between a pair of parallel plates built into the tube; the voltage applied to the plates produces an electric field, and a pair of coils produces a magnetic field. When only the electric field is present, say with the upper plate positive, the cathode rays are deflected upward as in path a in the figure. If only a magnetic field exists, say inward in the figure, the rays are deflected downward along path c. These observations are just what is expected for a negatively charged particle.

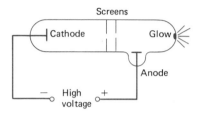

FIGURE 28–14

Discharge tube. In some models, one of the screens is the anode (postive plate).

FIGURE 28–15

Cathode rays deflected by electric and magnetic fields.

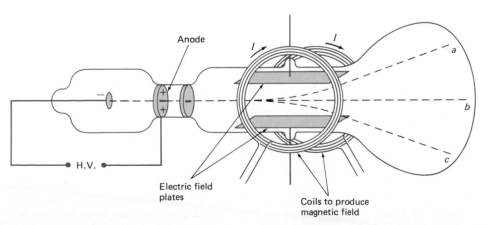

The force on the rays due to the magnetic field is

$$F = evB$$

where e is the charge and v is the velocity of the cathode rays. In the absence of an electric field, the rays are bent into a curved path, so we have, from $F = ma$,

$$evB = \frac{mv^2}{r},$$

and thus

$$\frac{e}{m} = \frac{v}{Br}.$$

The radius of curvature r can be measured and so can B. The velocity v is found by applying an electric field in addition to the magnetic field. The electric field E is adjusted so that the cathode rays are undeflected and follow path b in Fig. 28–15. In this situation, the force due to the electric field, $F = eE$, is just balanced by the force due to the magnetic field, $F = evB$. Thus we have $eE = evB$ and

$$v = \frac{E}{B}.$$

Combining this with the above equation we have

$$\frac{e}{m} = \frac{E}{B^2 r}. \tag{28–9}$$

The quantities on the right side can all be measured so that although e and m could not be determined separately, the ratio e/m could be determined. The accepted value today is $e/m = 1.76 \times 10^{11}$ C/kg. Cathode rays soon came to be called **electrons**.

It is worth noting that the "discovery" of the electron, like many others in science, is not quite so obvious as discovering gold or oil. Should the discovery of the electron be credited to the person who first saw a glow in the tube? Or to the person who first called them cathode rays? Perhaps neither one, for they had no conception of the electron as we know it today. In fact, the credit for the discovery is generally given to Thomson, but not because he was the first to see the glow in the tube. Rather it is because he believed not only that this phenomenon was due to tiny negatively charged particles and made careful measurements on them; but he also argued that these particles were constituents of atoms and not ions or atoms themselves as many thought, and he developed an electron theory of matter. His view is close to what we accept today, and this is why Thomson is credited with the "discovery." Note, however, that neither he nor anyone else ever actually saw an electron itself. We discuss this briefly, for it illustrates the fact that discovery in science is not always a clear-cut matter. In fact some philosophers of science think the word "discovery" is not always appropriate, such as in this case.

Thomson believed that an electron was not an atom, but rather a constituent, or part, of an atom. Convincing evidence for this came soon with the determination of the charge and the mass of the cathode rays. Thomson's student J. S. Townsend made the first direct (but rough) measurements of e in 1897. But it was the more refined *oil-drop experiment* of Robert A. Millikan (1868–1953) that yielded a precise value for the charge on the electron and showed that charge comes in discrete amounts. In this experiment, tiny droplets of mineral oil carrying an electric charge were allowed to fall under gravity between two parallel plates, Fig. 28–16. The electric field E between the plates was adjusted until the drop was suspended in midair; the downward pull of gravity, mg, was then just balanced by the upward force due to the electric field. Thus $qE = mg$ so the charge $q = mg/E$. The mass of the droplet was determined by measuring its terminal velocity in the absence of the electric field and using Stoke's equation. (See Section 13–7; the drop

FIGURE 28–16

Millikan's oil-drop experiment.

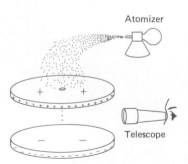

is too small to permit direct measurement of its radius). Sometimes the charged drop was negative and sometimes positive, suggesting that the drop had acquired or lost electrons (presumably through friction when ejected by the atomizer). Millikan's painstaking observations and analysis presented convincing evidence that any charge was an integral multiple of a smallest charge, e, that was ascribed to the electron, and that the value of e was 1.6×10^{-19} C. (Today's precise value of e, as mentioned in Chapter 22, is $e = (1.6021892 \pm 0.0000046) \times 10^{-19}$ C or, rounded off, 1.602×10^{-19} C.) This result, combined with the measurement of e/m, gives the mass of the electron to be $(1.6 \times 10^{-19} \text{ C})/(1.76 \times 10^{11} \text{ C/kg}) = 9.1 \times 10^{-31}$ kg. This is less than a thousandth the mass of the smallest atom, and thus confirmed the idea that the electron is only a part of an atom. The accepted value today for the mass of the electron is

$$m_e = (9.109534 \pm 0.000047) \times 10^{-31} \text{ kg}$$

or, rounded off,

$$m_e = 9.11 \times 10^{-31} \text{ kg.}$$

* 28-6 Thermionic Emission and the Cathode-Ray Tube

In the course of experiments on the electric light bulb, Thomas Edison (1847–1931) in 1883 made an interesting discovery. Into an evacuated glass bulb he inserted an electrode or plate in addition to the bulb filament. When a positive potential was applied by a battery to the plate, he found that a current would flow in the external circuit as long as the filament was hot and glowing (heated by a separate current passing through it). However, if a negative potential was applied to the plate, no current was observed to flow. When the filament was cold, no current flowed in either case. This phenomenon is known as *thermionic emission*.

Although Edison was unaware of the significance of his discovery, it eventually led to the invention of the electronic vacuum tube. Indeed, an explanation for this effect had to await the discovery of the electron over a decade later.

J. J. Thomson, in 1899, measured the value of e/m for the rays emitted in thermionic emission; he obtained the same value as that for the rays in the discharge tube produced by high voltage (Fig. 28–15), and concluded that it is electrons that are emitted in thermionic emission. Apparently what happens in thermionic emission is that electrons are being "boiled off" the filament when it is hot. When the plate is positive, the electrons are attracted to it and so a current flows. If the plate is negative, the electrons are repelled and no current flows.

We can understand how electrons might be "boiled off" a hot metal filament if we treat the electrons like molecules in a gas. This makes sense if the electrons are relatively free to move about, which is consistent with the fact that metals are good conductors. However, electrons don't readily escape from the metal. There are forces that keep them in. For example, if an electron were to escape outside the metal surface, a net positive charge would remain behind, and this would attract the electron back. To escape, an electron would have to have a certain minimum kinetic energy, just as molecules in a liquid must have a minimum KE to "evaporate" into the gaseous state. We saw in Chapter 18 that the average kinetic energy ($\overline{\text{KE}}$) of molecules in a gas is proportional to the absolute temperature T. We can apply this idea, but only very roughly, to free electrons in a metal as if they made up an "electron gas." Of course some electrons have more KE than average and others less. At room temperature, very few electrons would have sufficient energy to escape. At high temperatures, $\overline{\text{KE}}$ is larger and many escape; this is just like evaporation from liquids, which occurs more readily at high temperatures. Thus, significant thermionic emission occurs only at elevated temperatures. And Edison's results are readily understood.

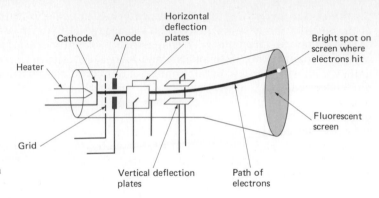

FIGURE 28–17

Cathode-ray tube. Magnetic deflection coils are often used in place of the electric deflection plates.

An important vacuum tube, whose operation depends on thermionic emission, is the *cathode-ray tube* (CRT). It is the picture tube of television sets, oscilloscopes, and computer display terminals. It derives its name from the fact that inside an evacuated glass tube a beam of cathode rays (electrons) is directed to various parts of a screen to produce a "picture." A simple CRT is diagrammed in Fig. 28–17. Electrons emitted by the heated cathode are accelerated by a high voltage applied to the anode (5000–50,000 V). The electrons pass out of this "electron gun" through a small hole in the anode. The inside of the tube face is coated with a fluorescent material that glows when struck by electrons. A tiny bright spot is thus visible where the electron beam strikes the screen. Two horizontal and two vertical plates deflect the beam of electrons when a voltage is applied to them. The electrons are deflected toward whichever plate is positive. By varying the voltage on the deflection plates, the bright spot can be placed at any point on the screen.

In the picture tube of a television set, magnetic deflection coils are usually used instead of electric plates. The electron beam is made to sweep over the screen in the manner shown in Fig. 28–18. The beam is swept horizontally by the horizontal deflection plates or coils. When the horizontal deflecting field is maximum in one direction, the beam is at one edge of the screen. As the field decreases to zero, the beam moves to the center; and as it increases to a maximum in the opposite direction the beam approaches the opposite edge. When it reaches this edge, the voltage or current abruptly changes to return the beam to the opposite side of the screen. Simultaneously, the beam is deflected downward slightly by the vertical deflection plates or coils, and then another horizontal sweep is made. In the United States 525 lines constitutes a complete sweep over the entire screen. (Some European systems use almost twice that number, giving greater sharpness.) The complete picture of 525 lines is swept out in $\frac{1}{30}$ s. Actually, a single vertical sweep takes $\frac{1}{60}$ s and involves every other line; the lines in between are then swept out over the next $\frac{1}{60}$ s. We see a picture because the image is retained by

FIGURE 28–18

Electrons sweep across a television screen in a succession of horizontal lines.

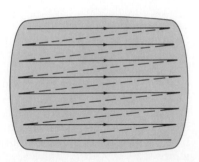

the fluorescent screen and by our eyes for about $\frac{1}{20}$ s. The picture we see consists of the varied brightness of the spots on the screen. The brightness at any point is controlled by the grid (a "porous" electrode, such as a wire grid, that allows passage of electrons) which can limit the flow of electrons by means of the voltage applied to it: the more negative this voltage, the fewer electrons pass through. The voltage on the grid is determined by the video signal sent out by the station and received by the set. Accompanying this signal are signals that synchronize the grid voltage to the horizontal and vertical sweeps.

An *oscilloscope* is a device for amplifying, measuring, and visually observing an electrical signal. It provides a principle means for measuring rapidly changing signals. The signal is displayed on the screen of a CRT. In normal operation, the electron beam is swept horizontally at a uniform rate in time by the horizontal deflection plates. The signal to be displayed is applied, after amplification, to the vertical deflection plates. The visible "trace" on the screen is thus a plot of the signal voltage (vertically) versus time (horizontally).

* 28–7   Mass Spectrometer

A number of methods were developed in the early part of this century to measure the masses of atoms. One of the most accurate was the *mass spectrometer*[†] of Fig. 28–19. Ions are produced by heating, or by an electric current, in the source S. Those that pass through slit S_1 enter a region where there are crossed electric and magnetic fields as in Thomson's device. Only those ions whose speed is $v = E/B$ will pass through undeflected and emerge through slit S_2. (This arrangement is called a velocity selector; see equation and discussion just before Eq. 28–9.) In this second region there is only a magnetic field B' and the ions follow a circular path. The radius of their path can be measured because the ions expose the photographic plate where they strike. Since $qvB' = mv^2/r$ and $v = E/B$, then we have

$$m = \frac{qB'r}{v} = \frac{qBB'r}{E}.$$

All the quantities on the right can be measured and thus m can be determined. Note that for ions of the same charge, the mass of each is proportional to the radius of its path.

The masses of many atoms were measured in this way. When a pure substance was used, it was sometimes found that two or more closely spaced marks would appear on the film. For example, neon produced two marks whose radii corresponded to atoms of mass 20 and 22 atomic mass units (u). Impurities were ruled out and it was concluded that there must be two types of neon with different mass. These different forms were called *isotopes*. It was soon found that most elements are mixtures of isotopes. We shall see in Chapter 42 that the difference in mass is due to different numbers of neutrons.

Mass spectrometers can be used not only to separate different elements and isotopes, but molecules as well. They are used in physics and chemistry, and in biological and biomedical laboratories.

EXAMPLE 28–5 Carbon atoms of atomic mass 12.0 u are found to be mixed with another, unknown, element. In a mass spectrometer, the carbon traverses a path of radius 22.4 cm and the unknown's path has a 26.2 cm radius. What is the unknown element? Assume they have the same charge.

[†] The term *mass spectrograph* is also used.

FIGURE 28–19

Bainbridge mass spectrometer. The magnetic fields B and B' point out of the paper (indicated by the dots).

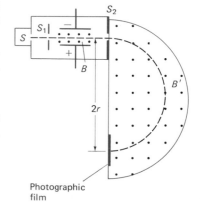

Photographic film

SOLUTION   Since mass is proportional to the radius, we have

$$\frac{m_x}{m_C} = \frac{26.2 \text{ cm}}{22.4 \text{ cm}} = 1.17$$

Thus $m_x = 1.17 \times 12.0 \text{ u} = 14.0 \text{ u}$. The other element is probably nitrogen (see appendix D). However, it could also be an isotope of carbon or oxygen. Further physical or chemical analysis would be needed.

 * 28–8   **The Cyclotron and Synchrotron**

The *cyclotron* was developed in 1932 by E. O. Lawrence (1901–1958) at the University of California, Berkeley, as a means to accelerate charged elementary particles such as protons or charged ions. The high-speed particles could then be used as projectiles in high-energy nuclear collisions, from which a great deal has been learned about nuclei and elementary particles (see Chapter 43). The cyclotron uses a magnetic field to maintain the charged particles in nearly circular paths. The charged particles move within two D-shaped cavities made of metal (the "dees") as shown in Fig. 28–20. Each time the particles pass into the gap between the dees, a voltage is applied that accelerates them. This increases their speed and also the radius of curvature of their path (see Example 28–4, $r = mv/qB$). After many revolutions, the protons acquire high energy and reach the outer edge of the cyclotron. They then either strike a target placed inside the cyclotron or leave the cyclotron with the help of a carefully placed "bending magnet" and are directed to an external target.

The voltage applied to the dees to produce the acceleration must be alternating. When protons are moving to the right across the gap in Fig. 28–20, the right dee must be negative and the left one positive. A half cycle later, the protons are moving to the left, so the left dee must be negative in order to accelerate them. The frequency of the applied voltage must be equal to that of the circulating protons, and this is just the cyclotron frequency determined in Section 28–4 (Equation 28–7).

    EXAMPLE 28–6   A small cyclotron of maximum radius $R = 0.50$ m accelerates protons in a 1.7-T magnetic field. Calculate (a) the frequency needed for the applied alternating voltage and (b) the kinetic energy of protons when they leave the cyclotron.

SOLUTION   (a) From Eq. 28–7

$$f = \frac{qB}{2\pi m} = \frac{(1.6 \times 10^{-19} \text{ C})(1.7 \text{ T})}{(6.28)(1.67 \times 10^{-27} \text{ kg})} = 2.6 \times 10^7 \text{ Hz}$$

(b) The protons leave the cyclotron at $r = R = 0.50$ m. Then, since $v = qBr/m$,

$$\text{KE} = \frac{1}{2} mv^2 = \frac{1}{2} m \frac{q^2 B^2 R^2}{m^2} = \frac{q^2 B^2 R^2}{2m}$$

$$= \frac{(1.6 \times 10^{-19} \text{ C})^2(1.7 \text{ T})^2(0.50 \text{ m})^2}{(2)(1.67 \times 10^{-27} \text{ kg})} = 5.5 \times 10^{-12} \text{ J},$$

or 34 MeV. Note that the magnitude of the voltage applied to the dees does not affect the final energy; but the higher this voltage, the fewer revolutions are required to bring the protons to full energy.

FIGURE 28–20

Diagram of a cyclotron. The magnetic field, applied by a large electromagnet, points into the paper. A is the ion source. The field lines shown are for the electric field in the gap.

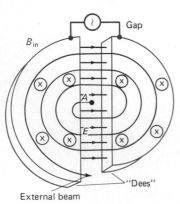

An important aspect of the cyclotron is that the frequency of the applied voltage, as given by Eq. 28–7, does not depend on the radius r. That is, the frequency does not have to be changed as the ions start from the source and are accelerated to paths of larger and larger radii. Unfortunately, this is only true for particle speeds not too close to the speed of light. For at higher speeds, the mass of the particle will increase (as we shall see in Chapter 39) according to Einstein's formula $m = m_0/\sqrt{1 - v^2/c^2}$ where m_0 is the rest mass. This is already a problem for the 34-MeV protons in the above example, for the KE is about 3 percent of the mass. As can be seen from Eq. 28–7, as the mass increases, the frequency of the applied voltage must be reduced. To achieve large energies, complex electronics is needed that decreases the frequency as a packet of protons increases in speed and reaches larger orbits. Such a modified cyclotron is called a *synchrocyclotron*.

Another way to deal with the increase in mass with speed is to increase the magnetic field B as the particles speed up. Such a device is called a *synchrotron*. The largest circular accelerators today fall into this category, and they are enormous. The Fermi National Accelerator Laboratory (Fermilab) at Batavia, Illinois, has a radius of 1.0 km and that at CERN (European Center for Nuclear Research) is 1.1 km in radius. They can accelerate protons to 500 GeV and that at Fermilab is expected soon to reach 1000 GeV using superconducting magnets. These large synchrotrons do not use enormous magnets a kilometer in radius. Instead, a narrow ring of magnets is used (see Fig. 28–21) which are all placed at the same radius from the center of the circle. The magnets are interrupted by gaps where high voltage accelerates the particles. After the particles are injected, they must then move in a circle of constant radius. This is done by giving them considerable energy initially in a much smaller accelerator, and then slowly increasing the magnetic field as they speed up in the large synchrotron. In Chapter 43 we will discuss many of the results of elementary particle physics investigated with the use of these machines.

(a)

(b)

FIGURE 28–21

(a) Aerial view of Fermilab at Batavia, Illinois; the accelerator is a circular ring 2.0 km in diameter. (Courtesy of Fermilab.) (b) Interior photograph, showing magnets that keep ions moving in a circle. (Courtesy of Fermilab.)

 * **28–9 The Hall Effect**

When a current-carrying conductor is held firmly in a magnetic field, the field exerts a sideways force on the charges moving in the conductor. For example, if electrons move to the right in the rectangular conductor shown in Fig. 28–22a, the inward magnetic field will exert a downward force on the electrons; so they will tend to move nearer face S than face R. There will thus be a potential difference between faces R and S of the conductor. This potential difference builds up until the electric field it produces exerts a force on the moving charges that is equal and opposite to the magnetic force. This effect is called the *Hall effect* after E. H. Hall, who discovered it in 1879. The difference of potential produced is called the *Hall emf*.

Now a current of negative charges moving to the right is equivalent to positive charges moving to the left, at least for most purposes. But the Hall effect can distinguish these two. As can be seen in Fig. 28–22b, positive particles moving to the left are deflected downward, so that the bottom surface is positive relative to the top surface. This is the reverse of part (a). Indeed, the direction of the emf in the Hall effect first revealed that it is negative particles that move in most conductors. In some semiconductors, however, the Hall effect reveals that the carriers of current are positive. (More on this in Chapter 41.)

The magnitude of the Hall emf is proportional to the current and the strength of the magnetic field. The Hall effect can thus be used to measure magnetic field strengths. First the conductor, called a *Hall probe*, is calibrated with known magnetic fields. Then, for the same current, its emf output will be

FIGURE 28–22
The Hall effect.

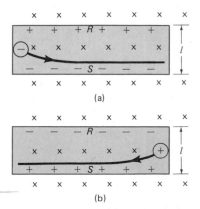

(a)

(b)

a measure of B. Hall probes can be made very small and are convenient and accurate to use.

* 28–10   Torque on a Current Loop; Magnetic Dipole Moment

When an electric current flows in a closed loop of wire placed in a magnetic field, as shown in Fig. 28–23, the magnetic force on the current can produce a torque. This is the basic principle behind a number of important practical devices including meters, and motors. (We discuss these applications in the next section.) The interaction between a current and a magnetic field is important in other areas as well, including atomic physics.

Let us consider now the rectangular loop of Fig. 28–23 whose length and width are a and b as shown. The loop carries a current I and is placed in a horizontal uniform magnetic field **B** which makes an angle θ with the normal to the plane surface of the loop. The loop is suspended so it is free to rotate about a vertical axis. The lead-in wires to the loop are very close together (and twisted) so that we can ignore any force on them due to **B** (the current flowing in, and that flowing out, effectively cancel since they are equal but in opposite directions, and very close together). The magnetic forces on the loop's upper and lower arms, of length b, are of equal magnitude and in opposite directions (vertically upward and downward); hence they cancel and produce no net force or torque on the loop. The force on each of the vertical sections, of length a, is

$$F = IaB$$

(from Eq. 28–2) since these sections are perpendicular to **B**. The directions of these two forces are inward and outward as shown (right-hand rule, Eq. 28–3), but not along the same line; hence they can produce a torque. The lever arm for each of these forces is

$$\frac{b}{2} \sin \theta;$$

so the net torque is

$$\tau = IaB\frac{b}{2}\sin\theta + IaB\frac{b}{2}\sin\theta = IAB\sin\theta,$$

where $A = ab$ is the area of the loop. If the loop contains N coils, the current in each section is then NI, so the torque will have magnitude

$$\tau = NIAB\sin\theta. \tag{28–10}$$

This formula, derived here for a rectangular coil, is valid for any shape of flat coil. The direction of the torque (right-hand rule: $\tau = \mathbf{r} \times \mathbf{F}$) is downward in this case.

The quantity NIA is called the *magnetic dipole moment* of the coil and is considered a vector:

$$\boldsymbol{\mu} = NIA \tag{28–11}$$

where the direction of **A** (and therefore of $\boldsymbol{\mu}$) is perpendicular to the plane of the coil consistent with the right-hand rule (cup your right hand so your fingers wrap around the loop in the direction of current flow, then your thumb points in the direction of $\boldsymbol{\mu}$ and **A**). With this definition of $\boldsymbol{\mu}$, we can rewrite Eq. 28–10 in vector form:

$$\tau = NIA \times \mathbf{B}$$

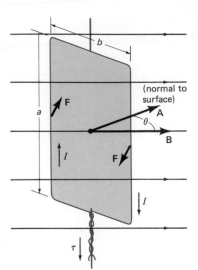

FIGURE 28–23

Calculating the torque on a current loop in a magnetic field **B**.

or

$$\tau = \mu \times B \qquad (28-12)$$

which gives the correct magnitude and direction for τ.

Equation 28–12 has the same form as Eq. 22–8 for an electric dipole (with electric dipole moment $\mathbf{p}$) in an electric field $\mathbf{E}$: $\tau = \mathbf{p} \times \mathbf{E}$. And just as an electric dipole has potential energy given by $U = -\mathbf{p} \cdot \mathbf{E}$ when in an electric field, we expect a similar form for a magnetic dipole in a magnetic field. In order to rotate a current loop (Fig. 28–23) so as to increase θ, we must do work against the magnetic force. Hence the potential energy depends on angle according to

$$U = \int \tau \, d\theta = \int NIAB \sin \theta \, d\theta = -\mu B \cos \theta + C.$$

If we choose $U = 0$ at $\theta = \pi/2$, then the arbitrary constant C is zero and the potential energy is

$$U = -\mu B \cos \theta = -\mu \cdot \mathbf{B}, \qquad (28-13)$$

as expected. Bar magnets and compass needles, as well as current loops, can be considered as magnetic dipoles.

EXAMPLE 28–7 Determine the magnetic dipole moment of the electrons orbiting the proton of a hydrogen atom, assuming (in the Bohr model) it is in its ground state with a circular orbit of radius 0.53×10^{-10} m. (Note: this is a very rough picture of atomic structure no longer accepted today—see Chapters 40 and 41—but this model nonetheless gives an accurate result.)

SOLUTION From Newton's second law, $F = ma$, we have, since the electron is held in its orbit by the coulomb force,

$$\frac{e^2}{4\pi\varepsilon_0 r^2} = \frac{mv^2}{r};$$

so

$$v = \sqrt{\frac{e^2}{4\pi\varepsilon_0 mr}} = \sqrt{\frac{(9.0 \times 10^9 \ \text{N}\cdot\text{m}^2/\text{C}^2)(1.6 \times 10^{-19} \ \text{C})^2}{(9.1 \times 10^{-31} \ \text{kg})(0.53 \times 10^{-10} \ \text{m})}} = 2.2 \times 10^6 \ \text{m/s}.$$

Since current is the electric charge that passes a given point per unit time, the rotating electron is equivalent to a current

$$I = \frac{e}{T} = \frac{ev}{2\pi r}$$

where $T = 2\pi r/v$ is the time required for one orbit. Since the area of the orbit is $A = \pi r^2$, the magnetic dipole moment is

$$\mu = IA = \left(\frac{ev}{2\pi r}\right)(\pi r^2) = \frac{evr}{2} = 9.3 \times 10^{-24} \ \text{A}\cdot\text{m}^2.$$

FIGURE 28–24
Galvanometer.

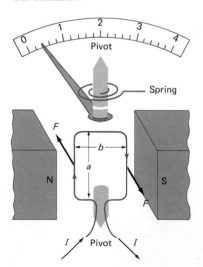

*28–11   Applications—Galvanometers, Motors, Loudspeakers

The basic component of most meters, including ammeters, voltmeters, and ohmmeters, is a *galvanometer*. We have already seen how these meters are designed (Section 27–6), and now we can examine how the crucial element, a galvanometer, itself works. As shown in Fig. 28–24, a galvanometer consists of a coil

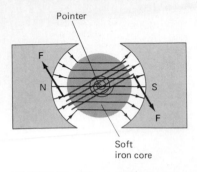

FIGURE 28-25

Galvanometer coil wrapped on an iron core.

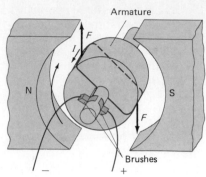

FIGURE 28-26

Diagram of a simple motor.

Voltage source

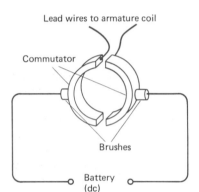

FIGURE 28-27

This commutator-brush arrangement in a dc motor assures alternation of the current in armature to keep rotation continuous. The commutators are attached to the motor shaft and turn with it while the brushes remain stationary.

FIGURE 28-28

Motor with many windings.

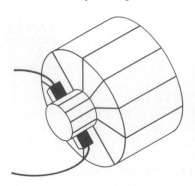

of wire suspended in the magnetic field of a permanent magnet. When current flows through the loop of wire, which is usually rectangular, the magnetic field exerts a torque on the loop, as given by Eq. 28–10, $\tau = NIAB \sin \theta$. This torque is opposed by a spring, which exerts a torque τ_s approximately proportional to the angle ϕ through which it is turned (Chapter 14). That is

$$\tau_s = k\phi$$

where k is the stiffness constant of the spring. Thus the coil and the attached pointer will rotate only to the point where the spring torque balances that due to the magnetic field. From Eq. 28–10 we then have $k\phi = NIAB \sin \theta$ or

$$\phi = \frac{NIAB \sin \theta}{k}.$$

Thus the deflection of the pointer, ϕ, is directly proportional to the current I flowing in the coil. But, according to this equation it also depends on the angle θ the coil makes with the field. For a useful meter we need ϕ to depend only on I, independent of θ, so a uniform field B is not sufficient. To solve this problem, curved pole pieces are used and the galvanometer coil is wrapped around a cylindrical iron core as shown in Fig. 28–25. The iron tends to concentrate the magnetic field lines so that B always points across the face of the coil at the wire. The force is then always perpendicular to the face of the coil and the torque will not vary with angle. Thus ϕ will be proportional to I as required.

An *electric motor* changes electric energy into (rotational) mechanical energy. A motor works on the same principle as a galvanometer, except that the coil is larger and is mounted on a large cylinder called the *rotor* or *armature*, Fig. 28–26. Actually, there are several coils, although only one is indicated in the figure. The armature is mounted on a shaft or axle. The permanent magnet is replaced in some motors by an electromagnet. Unlike a galvanometer, a motor must turn continuously in one direction. This presents a problem, for when the coil, which is rotating clockwise in Fig. 28–26, passes beyond the vertical position the forces would then act to return the coil back to vertical. Thus alternation of the current is necessary if a motor is to turn continuously in one direction. This can be achieved in a *DC motor* with the use of *commutators* and *brushes*: as shown in Fig. 28–27, the brushes are stationary contacts that rub against the conducting commutators mounted on the motor shaft. Every half revolution each commutator changes its connection to the other brush. Thus the current in the coil reverses every half revolution as required for continuous rotation. Most motors contain several coils, called "windings," each located in a different place on the armature, Fig. 28–28. Current flows through each coil only during a small part of a revolution, at the time when its orientation results in the maximum torque. In this way, a motor produces a much steadier torque than can be obtained from a single coil. The design of most practical motors is more complex than described here, but the general principles remain the same.

A *loudspeaker* also works on the principle that a magnet exerts a force on a current-carrying wire. The electrical output of a radio or TV set is connected to the wire leads of the speaker. The speaker leads are connected internally to a coil of wire, which is itself attached to the speaker cone, Fig. 28–29. The speaker cone is usually made of stiffened cardboard and is mounted so that it can move back and forth freely. A permanent magnet is mounted directly in line with the coil of wire. When the alternating current of an audio signal flows through the wire coil, the coil and the attached speaker cone feel a force due to the magnetic field of the magnet. As the current alternates at the frequency of the audio signal, the speaker cone moves back and forth at the same frequency, causing alternate compressions and rarefactions of the adjacent air, and sound waves are produced. A speaker thus changes electrical energy into sound energy, and the frequencies of the emitted sound waves are an accurate reproduction of the electrical input.

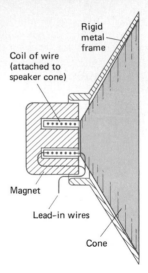

FIGURE 28–29
Loudspeaker.

Summary

A magnet has two poles, north and south. The north pole is that end which points toward the north when the magnet is freely suspended. Unlike poles of two magnets attract each other, whereas like poles repel. We can imagine that a *magnetic field* surrounds every magnet. The SI unit for *magnetic field strength* is the tesla (T). The force one magnet exerts on another is said to be an interaction between one magnet and the magnetic field produced by the other.

Electric currents produce magnetic fields. For example, the lines of magnetic field due to a current in a straight wire form circles around the wire and the field exerts a force on magnets placed near it.

A magnetic field exerts a force on an electric current. For a straight wire of length l carrying a current I the force has magnitude $F = IlB \sin \theta$; θ is the angle between the magnetic field **B** and the wire; the direction of the force is perpendicular to the wire and to the magnetic field, and is given by the right-hand rule. This relation serves as the definition of magnetic field **B**. Similarly, a magnetic field **B** exerts a force on a charge q moving with velocity **v** of magnitude $F = qvB \sin \theta$ where θ is the angle between **v** and **B**; in vector form $\mathbf{F} = q\mathbf{v} \times \mathbf{B}$. The path of a charged particle moving perpendicular to a uniform magnetic field is a circle. If both electric and magnetic fields are present, the force is

$$\mathbf{F} = q\mathbf{E} + q\mathbf{v} \times \mathbf{B}.$$

The measurement of the charge-to-mass ratio (e/m) of the electron was done using magnetic and electric fields. The charge e on the electron was first measured in the Millikan oil-drop experiment and then its mass was obtained from the measured value of the e/m ratio.

Questions

1. A compass needle is not always balanced parallel to the earth's surface, but one end may dip downward. Explain.
2. How do you suppose the first magnets found in Magnesia were formed?
3. Why will either pole of a magnet attract an unmagnetized piece of iron?
4. Suppose you have three iron rods, two of which are magnetized but the third is not. How would you determine which two are the magnets without using any additional objects?
5. Two iron bars attract each other no matter which ends are placed close together. Are both magnets?
6. Note that the pattern of magnetic field lines surrounding a bar magnet is similar to that of the electric field around an electric dipole. From this fact, predict how the magnetic field will change with distance (*a*) when near one pole of a very long bar magnet, and (*b*) when far from a magnet as a whole.
7. Draw the magnetic field lines around a straight section of wire carrying a current horizontally to the left.
8. In what direction are the magnetic field lines surrounding a straight wire carrying a current that is moving directly toward you?
9. A horseshoe magnet is held vertically with the north pole

on the left and south pole on the right. A wire passing between the poles carries a current away from you. In what direction is the force on the wire?

10. Charged cosmic ray particles from outside the earth tend to strike the earth more frequently near the poles than at lower latitudes. Explain. (See Fig. 28–5.)

11. In the relation, $\mathbf{F} = I\mathbf{l} \times \mathbf{B}$, which pairs of the vectors $(\mathbf{E}, \mathbf{l}, \mathbf{B})$ are always at 90°? Which can be at other angles?

12. The right-hand rules you learned in this chapter can be changed to "left-hand rules" if you are specifying the direction of movement of *negative* particles, such as electrons in a wire. Show, for each right-hand rule, that the same operations using the left hand give the same results if the direction of charge flow is for negative charges.

13. Can you set a resting electron into motion with a magnetic field? With an electric field?

14. A charged particle is moving in a circle under the influence of a uniform magnetic field. If an electric field that points in the same direction as the magnetic field is turned on, describe the path the charged particle will take.

15. The force on a particle in a magnetic field is the idea behind *electromagnetic pumping*. It is used to pump metallic fluids (such as sodium) and more recently to pump blood in artificial heart machines. The basic design is shown in Fig. 28–30. An electric field is applied perpendicular to a blood vessel and to a magnetic field. Explain how ions are caused to move. Do positive and negative ions feel a force in the same direction?

FIGURE 28–30

Electromagnetic pumping in a blood vessel.

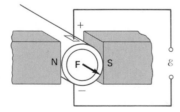

16. A beam of electrons is directed toward a horizontal wire carrying a current from left to right. In what direction is the beam deflected?

17. What kind of field or fields surround a moving electric charge?

18. Could we have defined the direction of the magnetic field $\mathbf{B}$ to be in the direction of the force on a moving charged particle? Explain.

19. A charged particle moves in a straight line through a particular region of space. Could there be a nonzero magnetic field in this region? If so, give two possible situations.

20. If a moving charged particle is deflected sideways in some region of space, can we conclude, for certain, that $\mathbf{B} \neq 0$ in that region?

21. In a particular region of space there is a uniform magnetic field $\mathbf{B}$; outside this region, $B = 0$. Can you inject an electron into the field so it will move in a closed circular path in the field?

22. How could you tell whether moving electrons in a certain region of space are being deflected by an electric field or by a magnetic field (or by both)?

23. Bringing a magnet close to a television screen will distort the picture. Why? (This can damage a color set.)

*24. Two ions have the same mass, but one is singly ionized and the other is doubly ionized. How will their positions on the film of the mass spectrometer of Fig. 28–19 differ?

*25. How can you make a compass without using iron or other ferromagnetic material?

*26. The dees of a cyclotron are good conductors. What then can you say about the electric field, due to the alternating voltage, felt by protons when inside the dees?

*27. What limits the maximum energy attainable for protons in an ordinary cyclotron? How is this overcome in a synchrotron?

*28. Describe how you could determine the dipole moment of a bar magnet or compass needle.

*29. In what positions (if any) will a current loop placed in a uniform magnetic field be in (a) stable equilibrium, (b) unstable equilibrium?

*30. What factors determine the sensitivity of a galvanometer?

 Problems

SECTION 28–3

1. (I) (a) What is the force per meter of length on a straight wire carrying at 10.5-A current when perpendicular to a 1.70-T magnetic field? (b) What if the angle between the wire and field is 45.0°?

2. (I) The force on a wire carrying 12.0 A is a maximum of 8.16 N when placed between the pole faces of a magnet. If the pole faces are 15 cm in diameter, what is the approximate strength of the magnetic field?

3. (I) Calculate the magnetic force on a 240-m length of wire stretched between two towers carrying a 150-A current. The earth's magnetic field of 5.0×10^{-5} T makes an angle of 60° with the wire.

4. (II) Suppose the straight wires connected to the semicircular conductor in Fig. 28–11 were bent so they were horizontal at the base of the semicircle. If a length l of each remained in the field $\mathbf{B}$, what would be the total force on the conductor as a whole?

5. (II) A circular loop of wire, of radius R, carries current I. It is placed in a magnetic field whose straight lines seem to diverge from a point a distance d below the ring on its axis. (That is, the field makes an angle θ with the loop at all points (see Fig. 28–31), where $\tan \theta = R/d$.) Determine the force on the loop.

6. (II) Two stiff parallel wires a distance l apart in a horizontal plane act as rails to support a light metal rod of mass m (perpendicular to each rail). A magnetic field B, directed vertically upward, acts throughout. At $t = 0$, wires connected to the rails are connected to a constant current source and a current I begins to flow through the system.

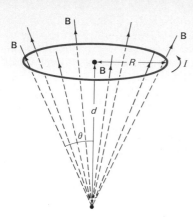

FIGURE 28–31

Determine the speed of the rod as a function of time (a) assuming no friction, (b) if the coefficient of friction is μ_k. (c) In which direction does the rod move, east or west, if the current through it heads north?

7. (II) A curved wire, connecting two points a and b, lies in a plane perpendicular to a uniform magnetic field **B** and carries a current I. Show that the resultant magnetic force on the wire, no matter what its shape, is the same as that on a straight wire connecting the two points carrying the same current I.

SECTION 28–4

8. (I) Determine the magnitude and direction of the force on an electron traveling 8.7×10^5 m/s horizontally to the east in a vertically upward magnetic field of strength 0.75 T.

9. (I) Describe the path of an electron that moves perpendicular to a 0.012-T magnetic field with a speed of 8.1×10^6 m/s. The field points directly toward the observer.

10. (I) Calculate the force on an airplane which has acquired a net charge of 180 C and moves with a speed of 280 m/s perpendicular to the earth's magnetic field of 5.0×10^{-5} T.

11. (I) A particle of charge q moves in a circular path of radius r in a uniform magnetic field **B**. Show that its momentum is $p = qBr$.

12. (II) A 50-MeV (kinetic energy) proton enters a 0.20-T field. What is the radius of its path?

13. (II) An electron experiences the greatest force as it travels 3.9×10^5 m/s in a magnetic field when it is moving westward. The force is upward and of magnitude 8.2×10^{-13} N. What is the magnitude and direction of the magnetic field?

14. (II) A doubly charged helium atom whose mass is 6.7×10^{-27} kg is accelerated by a voltage of 2800 V. (a) What will be its radius of curvature in a uniform 0.240-T field? (b) What is its period of revolution?

15. (II) A 7.0-gram bullet moves with a speed of 300 m/s perpendicular to the earth's magnetic field of 5.0×10^{-5} T. If the bullet possesses a net charge of 3.5×10^{-9} C, by what distance will it be deflected from its path due to the earth's magnetic field after it has traveled 600 m?

16. (II) Suppose the earth's magnetic field at the equator had magnitude 0.40×10^{-4} T and a northerly direction at all points. How fast must a uranium atom ($m = 238$ u, $q = e$) move so as to circle the earth 5.0 km above the equator?

17. (II) An electron experiences a force $\mathbf{F} = (3.2\mathbf{i} - 2.7\mathbf{j}) \times 10^{-13}$ N when passing through a magnetic field $\mathbf{B} = 0.72\mathbf{k}$ T. What is the electron's velocity?

18. (II) An electron enters a uniform magnetic field $B = 0.23$ T at a 45° angle to **B**. Determine the radius and pitch (distance between loops) of the electron's helical path assuming its speed is 3.0×10^7 m/s.

19. (II) Protons emerging from an accelerator must be bent at right angles by a "bending magnet" so as not to strike a barrier in their path a distance l from their exit hole in the accelerator. Show that the field **B** in the bending magnet must be at least $B \geq (2m\, \text{KE}/e^2 l^2)^{1/2}$ where m is the mass of the protons and KE is their kinetic energy.

20. (II) A straight 2.0-mm diameter copper wire can just "float" horizontally in air because of the force of the earth's magnetic field **B** which is horizontal and of magnitude 5.0×10^{-5} T. What current does the wire carry?

21. (III) Zeeman effect: In the Bohr model of the hydrogen atom, the electron is held in its circular orbit of radius r about its proton nucleus by electrostatic attraction. If the atoms are placed in a weak magnetic field **B**, the rotation frequency of electrons rotating in a plane perpendicular to **B** is changed by an amount

$$\Delta f = \pm \frac{eB}{4\pi m}$$

where e and m are the charge and mass of an electron. (a) Derive this result, assuming the force due to **B** is much less than that due to electrostatic attraction. (b) What does the ± sign indicate?

SECTION 28–5

22. (I) What is the velocity of a beam of electrons that go undeflected when passing through crossed electric and magnetic fields of magnitude 8.8×10^3 V/m and 3.5×10^{-3} T, respectively? What is the radius of the electron orbit if the electric field is turned off?

23. (I) Protons move in a circle of radius 8.10 cm in a 0.58-T magnetic field. What value of electric field could make their path straight? In what direction must it point?

24. (I) What is the value of e/m for a particle that moves in a circle of radius 8.0 mm in a 0.46-T magnetic field if a crossed 200-V/m electric field will make the path straight?

25. (II) An oil drop whose mass is determined to be 3.3×10^{-15} kg is held at rest between two large plates separated by 1.0 cm when the potential difference between them is 340 V. How many excess electrons does this drop have?

26. (III) In Millikan's oil-drop experiment, the mass of the oil drop is obtained by observing the terminal speed v_T of the freely falling drop in the absence of an electric field. Under these circumstances, the "effective" weight equals the viscous force given by Stokes's law (Section 13–7), $F = 6\pi\eta r v_T$, where η is the viscosity of air and r the radius of the drop. Also, the actual weight $mg = \frac{4}{3}\pi r^3 \rho g$ must be corrected for the buoyant force of the air; this is done by replacing ρ with $\rho - \rho_A$, where ρ is the density of the oil and ρ_A the density of air. With these preliminaries, show that the charge on the drop is given by

$$q = 18\pi \frac{d}{V} \sqrt{\frac{\eta^3 v_T^3}{2(\rho - \rho_A)g}},$$

where d is the separation of the plates (Fig. 28–16) and V is the voltage across them that just keeps the drop stationary. All of the quantities on the right side of this equation are known or can be measured. The terminal velocity v_T is determined by measuring the time it takes the drop to fall a measured distance, which is observed through a small telescope.

*SECTION 28–6

*27. (I) Use the ideal gas as a model to estimate the rms speed of a free electron in a metal at 300 K, and at 2500 K (the typical temperature of the cathode in a tube).

*28. (II) What would be the approximate maximum deflection of the electron beam near the center of a TV screen due to the earth's 5.0×10^{-5}-T field? Assume the screen is 20 cm from the electron gun where the electrons are accelerated (a) by 2.0 kV, (b) by 30 kV. Note that in color TV sets, the beam must be directed accurately to within less than 1 mm, in order to strike the correct phosphor. Because the earth's field is significant here, mu-metal shields are used to reduce the earth's field in the CRT.

*29. (II) Electrons are accelerated by 10.0 kV in a CRT. The screen is 24 cm wide and is 25 cm from the 2.8-cm-long deflection plates. Over what range must the horizontally deflecting electric field vary to sweep the beam fully across the screen?

*30. (II) In a given CRT, electrons are accelerated horizontally by 25 kV. They then pass through a uniform magnetic field B for a distance of 3.5 cm which deflects them upward so they reach the top of the screen 22 cm away, 11 cm above the center. Estimate the value of B.

*SECTION 28–7

*31. (I) In a mass spectrometer, germanium atoms have radii of curvature equal to 21.0, 21.6, 21.9, 22.2 and 22.8 cm. The largest radius corresponds to an atomic mass of 76 u. What are the atomic masses of the other isotopes?

*32. (II) Suppose the electric field between the electric plates in the mass spectrometer of Fig. 28–19 is 2.18×10^4 V/m and the magnetic fields $B = B' = 0.57$ T. The source contains boron isotopes of mass numbers 10 and 11 (to get their masses, multiply by 1.67×10^{-27} kg). How far apart are the lines formed by the singly charged ions of each type on the photographic film?

*33. (II) A mass spectrometer is being used to monitor air pollutants. It is difficult, however, to separate molecules with nearly equal mass such as CO (28.0106 u) and N_2 (28.0134 u). How large a radius of curvature must a spectrometer have if these two molecules are to be separated on the film by 0.24 mm?

*34. (II) One form of mass spectrometer accelerates ions by a voltage V before they enter a magnetic field B. The ions are assumed to start from rest. Show that the mass of an ion is $m = qB^2R^2/2V$ where R is the radius of the ions' path in the magnetic field and q is their charge.

*SECTION 28–8

*35. (I) What strength of magnetic field is used in a cyclotron in which protons make 2.1×10^7 revolutions per second?

*36. (I) If alpha particles are accelerated by the cyclotron of

Example 28–6, what must be the frequency of voltage applied to the dees?

*37. (I) If the cyclotron of Example 28–6 accelerated alpha particles, what maximum energy could they attain? What would their speed be?

*38. (I) The voltage across the dees of a cyclotron is 60 kV. How many revolutions do protons make to reach a kinetic energy of 20 MeV?

*39. (II) What would be the maximum kinetic energy of deuterons (^2_1H) accelerated by the cyclotron of Example 28–6? Of $^7_3Li^{3+}$ ions? What frequency of voltage is required?

*40. (II) Protons are injected into the 1.0-km-radius Fermilab synchrotron with an energy of 8.0 GeV. If they are accelerated by 2.5 MV each revolution, how far do they travel and how long does it take for them to reach 400 GeV? At these energies, the protons are traveling close to the speed of light: 3.0×10^8 m/s.

*SECTION 28–9

41. (II) (a) Show that the emf produced by the Hall effect is given by $\mathscr{E} = v_d Bl$, where v_d is the drift speed of the charged particles in a flat conductor of width l (Fig. 28–22). (b) Write $\mathscr{E}$ in terms of the electric current in the wire using the simple microscopic view of current given in Section 26–5. (c) Determine the density n, of electrons in a conductor in terms of measured values of $\mathscr{E}$, B, l, the current I, and the cross sectional area of the conductor.

*42. (II) The Hall effect can be used to measure blood flow rate because the blood contains ions that constitute an electric current; the apparatus is basically the same as for electromagnetic pumping, Fig. 28–30 (question 15), except that the blood flows in the artery of its own accord and the external emf is replaced by a potentiometer that measures the Hall emf. (a) Does the sign of the ions influence the emf? (b) Use the result of the previous problem to find the flow velocity in an artery 3.3 mm in diameter if the measured emf is 0.10 mV and B is 0.070 T. (In actual practice, an alternating magnetic field is used.)

*SECTION 28–10

*43. (I) A 10.0-cm diameter circular loop of wire is placed with its face parallel to the uniform magnetic field between the pole pieces of a large magnet. When 8.10 A flows in the coil, the torque on it is 0.116 N·m. What is the magnetic field strength?

*44. (I) How much work is required to rotate the current loop (Fig. 28–23) in a uniform magnetic field **B** from (a) $\theta = 0°$ ($\boldsymbol{\mu} \| \mathbf{B}$) to $\theta = 180°$, (b) $\theta = 90°$ to $\theta = -90°$?

*45. (II) Show that the magnetic dipole moment μ of an electron orbiting the proton nucleus of a hydrogen atom is related to the orbital momentum L of the electron by

$$\mu = \frac{e}{2m} L.$$

*46. (II) A circular coil 12.0 cm in diameter and containing 8 loops lies flat on the ground. The earth's magnetic field at this location is 5.50×10^{-5} T; it points into the earth at an angle of 66.0° below a line pointing due north. If a 6.10-A clockwise current passes through the coil, (a) determine the torque on the coil; (b) which edge of the coil rises up, north, east, south, or west?

*47. (III) A square loop of aluminum wire is 22.5 cm on a side. It is to carry 12.0 A and rotate in a 1.50-T magnetic field. (a) Determine the minimum diameter of the wire so it will not fracture from tension or shear. Assume a safety factor of 10. (See Table 11–2.) (b) What is the resistance of a single loop of this wire?

*48. (II) Show that the torque on a flat loop of any shape is given by Eq. 28–10. (Hint: approximate the arbitrarily shaped loop by a series of very thin rectangular loops and use a technique similar to that in Fig. 21–11 for calculating entropy changes.)

*49. (III) Suppose a nonconducting rod of length L carries a uniformly distributed charge Q. It is rotated with angular velocity ω about an axis perpendicular to the rod at one end. Show that the magnetic dipole moment of this rod is $\frac{1}{6}Q\omega L^2$. (Hint: consider the motion of each infinitesimal length of the rod.)

*SECTION 28–11

*50. (I) A galvanometer needle deflects full scale for a 34.0-μA current. What current will give full-scale deflection if the magnetic field weakens to 0.900 of its original value?

*51. (I) If the restoring spring of a galvanometer weakens by 12.0 percent over the years, what current will give full-scale deflection if originally it required 45.0 μA?

*52. (I) If the current to a motor drops by 20 percent, by what factor does the output torque change?

Sources of Magnetic Field

29

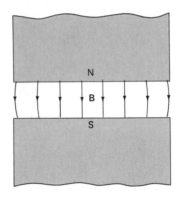

FIGURE 29–1

Magnetic field between two large poles of a magnet is nearly uniform except at the edges.

In the previous chapter, we discussed primarily the effects (forces and torques) that a magnetic field has on electric currents and on moving electric charges. We did see, however, that magnetic fields are produced not only by magnets but also by electric currents (Oersted's great discovery). It is this aspect of magnetism, the production of magnetic fields, that we discuss in this chapter.

The simplest magnetic field is one that is uniform—it doesn't change from one point to another. A perfectly uniform field over a large area is not easy to produce. But the field between two flat pole pieces of a magnet is nearly uniform if the area of the pole faces is large compared to their separation, Fig. 29–1. At the edges, the field "fringes" out somewhat and is not uniform. The parallel, evenly spaced field lines in the drawing indicate the field is uniform at points not too near the edge.

Nonuniform magnetic fields are more common, and it is often important to know how a magnetic field varies from point to point. We will now see how magnetic field strengths are determined for some simple situations, and discuss some general relations between magnetic fields and their sources.

 ## 29–1 Magnetic Field of a Straight Wire

We saw in the previous chapter (Fig. 28–7) that the magnetic field due to the electric current in a long straight wire is such that the field lines are circles with the wires at the center. You might expect that the field strength at a given point would be greater if the current flowing in the wire were greater; and that the field would be less at points further from the wire. This is indeed the case. Careful experiments show that the magnetic field strength B at a point near the wire is directly proportional to the current I in the wire and inversely proportional to the distance r from the wire:

$$B \propto \frac{I}{r}.$$

This relation is valid as long as r, the perpendicular distance to the wire, is much less than the distance to the ends of the wire.

The proportionality constant is written as $\mu_0/2\pi$;[†] thus

$$B = \frac{\mu_0}{2\pi}\frac{I}{r}. \qquad \text{[outside a long straight wire]} \quad (29\text{–}1)$$

The value of the constant μ_0, which is called the *permeability of free space*, is

$$\mu_0 = 4\pi \times 10^{-7}\ \text{T·m/A}.$$

(Do not confuse this μ_0 with $\boldsymbol{\mu}$, the magnetic dipole moment (Section 28–10); they are not connected.)

 EXAMPLE 29–1 A vertical electric wire in the wall of a building carries a dc current of 25 A upward. What is the magnetic field at a point 10 cm due north of this wire?

SOLUTION According to Eq. 29–1:

$$B = \frac{(4\pi \times 10^{-7}\ \text{T·m/A})(25\ \text{A})}{(2\pi)(0.10\ \text{m})} = 5.0 \times 10^{-5}\ \text{T}$$

or 0.50 G. By the right-hand rule (Fig. 28–7b) the field points to the west at this point. Since this field has about the same magnitude as the earth's, a compass would not point north but rather in a northwesterly direction.

29–2 Ampère's Law

Equation 29–1 gives the relation between the current in a long straight wire and the magnetic field it produces. This equation is valid only for a long straight wire. The following question arises: is there a general relation between a current in a wire of any shape and the magnetic field around it? The answer is yes: the French scientist André Marie Ampère (1775–1836), proposed such a relation shortly after Oersted's discovery. Consider any (arbitrary) closed path around a current as shown in Fig. 29–2, and imagine this path as being made up of short segments each of length Δl. First, we take the product of the length of each segment times the component of **B** parallel to that segment (call this component $B_\parallel$). If we now sum all these terms, according to Ampère, the result will be equal to μ_0 times the net current I that passes through the surface enclosed by the path:

$$\sum B_\parallel\, \Delta l = \mu_0 I.$$

The lengths Δl are chosen so that $B_\parallel$ is essentially constant along each length. The sum must be made over a *closed path*; and I is the net current passing through the surface bounded by this closed path. In the limit $\Delta l \to 0$, this relation becomes

$$\oint \mathbf{B} \cdot d\mathbf{l} = \mu_0 I \qquad (29\text{–}2)$$

where $d\mathbf{l}$ is an infinitesimal length vector and the vector dot product assures that the parallel component of **B** is taken. Equation 29–2 is known as **Ampère's law**. It is one of the great laws of electromagnetism. The integrand in Eq. 29–2 is taken around a closed path, and I is the current passing through the area enclosed by the chosen path.

To understand Ampère's law better, let us apply it to the simple case of a long straight wire carrying a current I which we've already examined, and which served as an inspiration for Ampère himself. Suppose we want to find the magnitude of **B** at some point A which is a distance r from the wire (Fig. 29–3). We know the

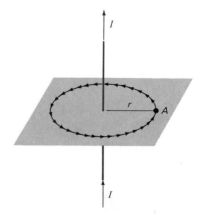

FIGURE 29–2

Arbitrary path enclosing a current, for Ampère's law. The path is broken down into segments of equal length Δl.

Closed path made up of segments of length Δl

Area enclosed by the path

FIGURE 29–3

Circular path of radius r.

[†] The constant is chosen like this so that Ampère's law—see next section—has a simple and elegant form.

magnetic field lines are circles with the wire at their center. So to apply Eq. 29–2 we choose as our path of integration a circle of radius r. (The choice of path is ours—so we choose one that will be convenient.) We choose this path because at any point on this path, $\mathbf{B}$ will be tangent to this circle. Thus for any short segment of the circle (Fig. 29–3), $\mathbf{B}$ will be parallel to that segment, and hence

$$\mu_0 I = \oint \mathbf{B} \cdot d\mathbf{l} = \oint B\, dl = B \oint dl = B(2\pi r).$$

We solve for B and obtain

$$B = \frac{\mu_0 I}{2\pi r}.$$

This is just Eq. 29–1 for the field near a long straight wire as discussed earlier.

Ampère's law thus works for this simple case. A great many experiments indicate that Ampère's law is valid in general. However, as with Gauss's law for the electric field, its practical value as a means to calculate the magnetic field is limited mainly to simple cases. Its real importance lies in the fact that it relates the magnetic field to the current in a direct and mathematically elegant way. Ampère's law is thus considered one of the basic laws of electricity and magnetism. It is valid for any situation where the currents and fields are steady and not changing in time, and no magnetic materials are present.

We now can see why the constant in Eq. 29–1 is written $\mu_0/2\pi$; this is done so that only μ_0 appears in Eq. 28–2 (rather than, say, $2\pi k$ if we had used k in Eq. 29–1). In this way, the more fundamental equation, Ampère's law, has the simpler form.

It should be noted, however, that the $\mathbf{B}$ in Ampère's law is not necessarily due only to the current I. Ampère's law, like Gauss's law for the electric field, is valid in general. $\mathbf{B}$ is the field at each point in space along the chosen path due to all sources—including the current I enclosed by the path, but also due to any other sources. For example, the field surrounding two parallel current-carrying wires is the vector sum of the fields produced by each, and the field lines are as shown in Fig. 29–4; if the path chosen for the integral (Eq. 29–2) is a circle centered on one of the wires with radius less than the distance between the wires (the dashed line in Fig. 29–4), only the current in the encircled wire (say, I_1) is included on the right side of Eq. 29–2; yet $\mathbf{B}$ on the left side of the equation must be the total $\mathbf{B}$ at each point due to both wires. This example brings up another point: that $\oint \mathbf{B} \cdot d\mathbf{l}$ for the path shown in Fig. 29–4 is the same whether the second wire is present or not (in both cases, it equals $\mu_0 I_1$). How can this be? It can be so because although the fields due to the two wires tend to cancel one another at points between them, such as point N in the diagram ($\mathbf{B} = 0$ at a point midway between the wires if $I_1 = I_2$), at a point such as M in the figure, the fields add together to produce a larger field. In the *sum*, $\oint \mathbf{B} \cdot d\mathbf{l}$, these effects just balance so that $\oint \mathbf{B} \cdot d\mathbf{l} = \mu_0 I$, whether the second wire is there or not. The integral $\oint \mathbf{B} \cdot d\mathbf{l}$ will be the same in each case, even though $\mathbf{B}$ will not be the same at every point for each of the two cases.

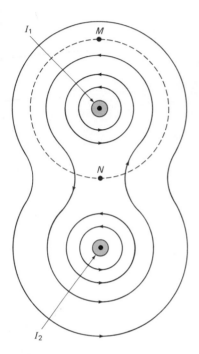

FIGURE 29–4

Magnetic field lines around two long parallel wires whose equal currents, I_1 and I_2, are coming out of the paper toward the viewer.

FIGURE 29–5

Magnetic field inside and outside a cyclindrical conductor (Example 29–2).

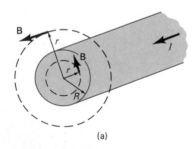

(a)

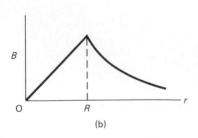

(b)

EXAMPLE 29–2 A long straight cylindrical wire conductor of radius R carries a current I of uniform current density in the conductor. Determine the magnetic field at (*a*) points outside the conductor ($r > R$), and (*b*) points inside the conductor ($r < R$). (See Fig. 29–5a.) Assume that r, the radial distance from the axis, is much less than the length of the wire.

SOLUTION (*a*) Because the wire is long, straight, and cylindrical, we expect from the symmetry of the situation that the magnetic field must be the same at all points that are the same distance from the center of the conductor.[†] We also expect $\mathbf{B}$ to be tangent to circles around the wire (Fig. 28–7), so let us choose a circular

---

† There is no reason why any such point should have preference over the others at the same distance from the wire (they are physically equivalent), and so B must have the same value at all these points.

path of integration outside the wire $(r > R)$, but concentric with it. Then

$$\oint \mathbf{B} \cdot d\mathbf{l} = B(2\pi r) = \mu_0 I$$

or

$$B = \frac{\mu_0 I}{2\pi r}. \qquad\qquad [r > R]$$

which is the same result as for a thin wire.

(b) Inside the wire $(r < R)$, we again choose a circular path concentric with the cylinder; we expect $\mathbf{B}$ to be tangential to this path, and again, because of the symmetry, it will have the same magnitude at all points on the circle. The current enclosed in this case is less than I by a factor of the ratio of the areas: $\pi r^2 / \pi R^2$. So Ampère's law gives

$$\oint \mathbf{B} \cdot d\mathbf{l} = B(2\pi r) = \mu_0 I \left(\frac{\pi r^2}{\pi R^2}\right)$$

or

$$B = \frac{\mu_0 I r}{2\pi R^2}. \qquad\qquad [r < R]$$

The field is zero at the center of the conductor and increases linearly with r until $r = R$; beyond $r = R$, B decreases as $1/r$. This is shown in Fig. 29–5b. Note that these results are valid only for points close to the conductor compared to its length. For a current to flow, there must be connecting wires (to a battery, say), and the field due to these conducting wires, if not very far away, will destroy the assumed symmetry.

 EXAMPLE 29–3 Use Ampère's law to show that a magnetic field in any region of space cannot be both unidirectional and nonuniform, as shown in Fig. 29–6a.

SOLUTION The wider spacing of lines near the top of Fig. 29–6a indicates the field has a smaller magnitude at the top than it does lower down. We now apply Ampère's law to the rectangular path $abcd$ shown dashed in the diagram. Since no current is enclosed by this path,

$$\oint \mathbf{B} \cdot d\mathbf{l} = 0.$$

The integral along sections ab and cd is zero, since $\mathbf{B} \perp d\mathbf{l}$. Thus

$$\oint \mathbf{B} \cdot d\mathbf{l} = B_{bc} l - B_{ad} l = (B_{bc} - B_{ad}) l$$

which is not zero since the field B_{bc} along the path bc is less than the field B_{ad} along path ad. Hence we have a contradiction: $\oint \mathbf{B} \cdot d\mathbf{l}$ cannot be both zero (since $I = 0$) and nonzero. Thus we have proved that a nonuniform unidirectional field is not consistent with Ampère's law. A nonuniform field whose direction also changes, as in Fig. 29–6b, is consistent (convince yourself this is so), and possible. The fringing of a permanent magnet's field (Fig. 29–1) has this shape.

FIGURE 29–6
Example 29–3.

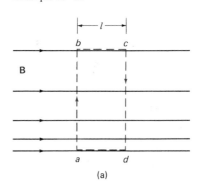

(a)

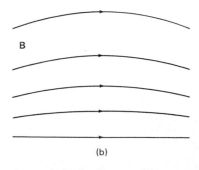

(b)

29–3 Magnetic Field of a Solenoid and a Torus

A long coil of wire consisting of many loops is called a *solenoid*. A solenoid has many uses. We shall see that its magnetic field resembles that of a bar magnet. The magnetic field inside a solenoid can be fairly large since it will be the sum of

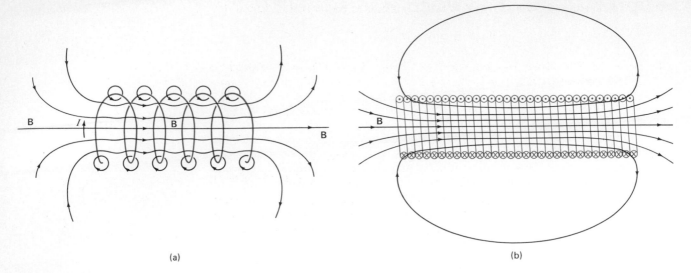

(a)

(b)

FIGURE 29–7

Magnetic field due to a solenoid:
(a) loosely spaced turns, (b) closely
spaced turns.

the fields due to the current in each loop. In Fig. 29–7a, we see the field due to a solenoid when the coils are far apart. In the immediate vicinity of each wire, the field lines are very nearly circles as for a straight wire (that is, at distances that are small compared to the curvature of the wire). Between any two wires, the fields due to each loop tend to cancel. Toward the center of the solenoid, the fields add up to give a nearly uniform field. For a long solenoid, with closely packed coils, the field is nearly uniform and parallel to the solenoid axes within the entire cross section, as shown in Fig. 29–7b. The field outside the solenoid is very small compared to the field inside (except near the ends); this should be clear from the fact that the same number of field lines that are concentrated inside the solenoid, spread out into the vast open space outside.

We now use Ampère's law to determine the magnetic field inside a very long (ideally, infinitely long) closely packed solenoid. We choose the path $abcd$ shown in Fig. 29–8, far from either end, for applying Ampère's law. We will consider this path as made up of four segments, the sides of the rectangle: ab, bc, cd, da. Then the left side of Eq. 29–2 becomes

$$\oint \mathbf{B} \cdot d\mathbf{l} = \int_a^b \mathbf{B} \cdot d\mathbf{l} + \int_b^c \mathbf{B} \cdot d\mathbf{l} + \int_c^d \mathbf{B} \cdot d\mathbf{l} + \int_d^a \mathbf{B} \cdot d\mathbf{l}.$$

The field outside the solenoid is so small as to be negligible compared to the field inside; thus the first term will be zero. Furthermore, $\mathbf{B}$ is perpendicular to the segments bc and da inside the solenoid, and is nearly zero between and outside the

FIGURE 29–8

Magnetic field inside a long solenoid is uniform.
Dashed lines indicate the path chosen for use in
Ampère's law.

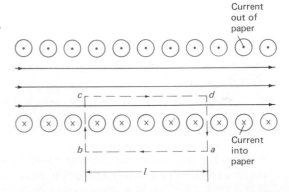

coils. Thus these terms are also zero. Therefore we have reduced the integral to

$$\oint \mathbf{B} \cdot d\mathbf{l} = \int_c^d \mathbf{B} \cdot d\mathbf{l} = Bl$$

where B is the field inside the solenoid and l is the length cd. Now we determine the current enclosed by this loop for the right side of Eq. 29–2. If a current I flows in the wires of the solenoid, the total current enclosed by our path $abcd$ is NI where N is the number of loops our path encircles. Thus Ampère's law gives us

$$Bl = \mu_0 NI.$$

If we let $n = N/l$ be the *number of loops per unit length*, then

$$B = \mu_0 nI. \qquad \text{[solenoid]} \quad (29{-}3)$$

This is the magnitude of the magnetic field within a solenoid. Note that B depends only on the number of loops per unit length, n, and the current I. The field does not depend on the position within the solenoid, so B is uniform. This is strictly true only for an infinite solenoid, but it is a good approximation for real ones for points not close to the ends.

EXAMPLE 29–4   A 10-cm-long solenoid has a total of 400 turns of wire and carries a current of 2.0 A. Calculate the field inside near the center.

SOLUTION   The number of turns per unit length is $n = 400/10\ \text{cm} = 4.0 \times 10^3\ \text{m}^{-1}$. Thus

$$B = \mu_0 nI = (12.57 \times 10^{-7}\ \text{T·m/A})(4.0 \times 10^3\ \text{m}^{-1})(2.0\ \text{A})$$
$$= 1.0 \times 10^{-2}\ \text{T}.$$

Notice that the field outside of a solenoid (Fig. 29–7) is much like that of a bar magnet (Fig. 28–4). Indeed, a solenoid acts like a magnet, with one end acting as a north pole and the other as south pole, depending on the direction of the current in the loops. Since magnetic field lines leave the north pole of a magnet, the north poles of the solenoids in Fig. 29–7 are on the right.

A useful device consists of a solenoid into which a rod of iron is partially inserted. This combination is also referred to as a solenoid. One simple use is as a doorbell, Fig. 29–9. When the circuit is closed by pushing the button, the coil effectively becomes a magnet and exerts a force on the iron rod as any magnet would. The rod is pulled into the coil and strikes the bell. A larger solenoid is used in the starters of cars; when you engage the starter, you are closing a circuit that not only turns the starter motor but activates a solenoid that first moves the starter into contact with the engine. Solenoids are used as switches in many other mechanical devices such as tape recorders. They have the advantage of moving mechanical parts quickly and accurately.

EXAMPLE 29–5   *Torus.*   Use Ampère's law to determine the magnetic field (*a*) inside and (*b*) outside a torus, which is a solenoid in the shape of a circle as shown in Fig. 29–10a.

SOLUTION   The magnetic field lines inside the torus will be circles concentric with the torus. (Thinking of the torus as a solenoid bent into a circle, the field lines bend along with the solenoid.) We choose as our path of integration one of these field lines of radius r inside the torus as shown by the dashed line labeled "path 1" in Fig. 29–10a. We make this choice to use the symmetry of the situation, so B must be the same at all points along the path (although not necessarily the same across

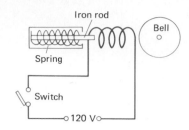

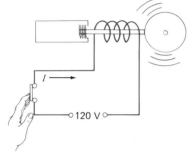

FIGURE 29–9

Solenoid used as a doorbell.

FIGURE 29–10

(a) A torus. (b) A section of the torus showing direction of the current for three loops: $\odot$ means current toward viewer, $\otimes$ means current away from viewer.

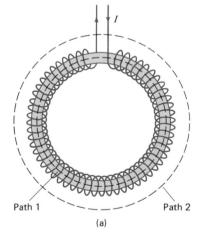

Path 1   Path 2

(a)

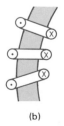

(b)

the whole cross-section of the torus); so Ampère's law

$$\oint \mathbf{B} \cdot d\mathbf{l} = \mu_0 I$$

becomes

$$B(2\pi r) = \mu_0 N I$$

where N is the total number of coils and I is the current in each of the coils. Thus

$$B = \frac{\mu_0 N I}{2\pi r}.$$

The magnetic field B is not uniform within the torus: it is largest along the inner edge (where r is smallest) and smallest at the outer edge. However, if the torus is large, but thin (so that the difference between the inner and outer radii is small compared to the average radius) the field will be essentially uniform within the torus. In this case, the formula for B reduces to that for a straight solenoid $B = \mu_0 n I$ where $n = N/(2\pi r)$ is the number of coils per unit length.

(b) Outside the torus, we choose as our path of integration a circle concentric with the torus, "path 2" in Fig. 29–10a. This path encloses N loops carrying current I in one direction and N loops carrying the same current in the opposite direction. (Figure 29–10b shows the directions of the current for the parts of the loop on the inside and outside of the torus.) Thus the net current enclosed by path 2 is zero. For a perfectly tightly packed torus, all points on path 2 are equidistant from the torus and equivalent, so we expect B to be the same at all points along the path. Hence, Ampère's law gives

$$\oint \mathbf{B} \cdot d\mathbf{l} = \mu_0 I$$

$$B 2\pi r = 0$$

or

$$B = 0.$$

So there is no field outside a perfectly tightly wound torus. It is all inside.

FIGURE 29–11

(a) Two parallel conductors carrying currents I_1 and I_2. (b) Magnetic field produced by I_1.

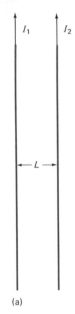

(a)

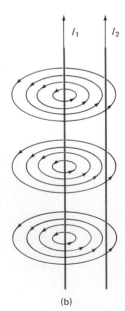

(b)

29–4   Force Between Two Parallel Wires; Definition of the Ampere and the Coulomb

You may have wondered how the constant μ_0 could be exactly $4\pi \times 10^{-7}$ T·m/A. With an older definition of the ampere, μ_0 was measured experimentally to be very close to this value. Today, however, μ_0 is *defined* to be exactly $4\pi \times 10^{-7}$ T·m/A. This of course could not be done if the ampere were defined independently. The ampere, the unit of current, is now defined in terms of the magnetic field B it produces using the defined value of μ_0.

Let us be more precise. Consider two long parallel conductors separated by a distance L, Fig. 29–11a. They carry currents I_1 and I_2, respectively. Each current produces a magnetic field that is "felt" by the other so that each should be expected to exert a force on the other, as Ampère first pointed out. For example, the magnetic field B_1 produced by I_1 is given by Eq. 29–1; at the location of the second conductor, the magnitude of this field is

$$B_1 = \frac{\mu_0}{2\pi} \frac{I_1}{L}.$$

See Fig. 29–11b where the field due *only* to I_1 is shown. (The total field due to both conductors was shown in Fig. 29–4.) According to Eq. 28–2, the force F per unit length l on the conductor carrying current I_2 is

$$\frac{F}{l} = I_2 B_1.$$

29–4 Force Between
Two Parallel Wires;
Definition of the Ampere
and the Coulomb   **567**

Note that the force on I_2 is due only to the field of I_1; I_2 also produces a field, but it does not exert a force on itself. We substitute in the above formula for B_1 and find

$$\frac{F}{l} = \frac{\mu_0}{2\pi} \frac{I_1 I_2}{L}. \tag{29–4}$$

If we use the right-hand rule of Fig. 28–7b, we see that the lines of B are as shown in Fig. 29–11b. Then using the right-hand rule of Fig. 28–8 (or Eq. 28–3) we see that the force exerted on I_2 will be to the left in the figure. That is, I_1 exerts an attractive force on I_2. This is true as long as the currents are in the same direction. If I_2 is in the opposite direction, the right-hand rule indicates that the force is in the opposite direction; that is, I_1 exerts a repulsive force on I_2.

Reasoning similar to that above shows that the magnetic field produced by I_2 exerts an equal but opposite force on I_1. We expect this to be true also, of course, from Newton's third law.

Equation 29–4 is used to define the ampere precisely. If $I_1 = I_2 = 1\,\text{A}$ exactly and the two wires are exactly 1 m apart, then

$$\frac{F}{l} = \frac{(4\pi \times 10^{-7}\ \text{T} \cdot \text{m/A})}{(2\pi)} \frac{(1\ \text{A})(1\ \text{A})}{(1\ \text{m})} = 2 \times 10^{-7}\ \text{N/m}.$$

Thus, *one ampere is defined as that current flowing in each of two long parallel conductors 1 m apart, which results in a force of exactly 2×10^{-7} N/m of length of each conductor.*

This is the precise definition of the ampere. The *coulomb* is then defined as being *exactly* one ampere-second: $1\ \text{C} = 1\ \text{A} \cdot \text{s}$. The value of k or ε_0 in Coulomb's law (Section 22–5) is obtained from experiment.

This may seem a rather roundabout way of defining quantities. The reason behind it is the desire for *operational definitions* of quantities—that is, definitions of quantities that can actually be measured given a definite set of operations to carry out. For example, the unit of charge, the coulomb, could be defined in terms of the force between two equal charges after defining a value for ε_0 or k in Eqs. 22–1 or 22–2. However, to carry out an actual experiment to measure the force between two charges is very difficult. For one thing, any desired amount of charge is not easily obtained precisely; and charge tends to leak from objects into the air. On the other hand, the amount of current in a wire can be varied accurately and continuously (by putting a variable resistor in a circuit). Thus the force between two current-carrying conductors is far easier to measure precisely. And this is why the ampere is defined first and then the coulomb in terms of the ampere. At the National Bureau of Standards, Washington, DC, precise measurement of current is made using circular coils of wire rather than straight lengths because it is more convenient and accurate.

Electric and magnetic field strengths are also defined operationally: the electric field in terms of the measurable force on a charge, via Eq. 22–3; and the magnetic field in terms of the force per unit length on a current-carrying wire, via Eq. 28–2.

EXAMPLE 29–6 The two wires of a 2.0-m long appliance cord are 3.0 mm apart and carry 8.0 A dc. Calculate the force between these wires.

SOLUTION Equation 29–4 gives us

$$F = \frac{(2.0 \times 10^{-7}\ \text{T} \cdot \text{m/A})(8.0\ \text{A})(8.0\ \text{A})(2.0\ \text{m})}{(3.0 \times 10^{-3}\ \text{m})} = 8.5 \times 10^{-3}\ \text{N}.$$

Since the currents are in opposite directions, the force would tend to spread them apart.

29–5 Biot-Savart Law

The usefulness of Ampère's law for determining the magnetic field **B** due to particular electric currents is restricted to situations where the symmetry of the given currents is great so that $\oint \mathbf{B} \cdot d\mathbf{l}$ can be readily evaluated. This difficulty of applicability does not, of course, invalidate Ampère's law nor does it reduce its fundamental standing. This is much like the electric case, where Gauss's law is also considered fundamental but is limited in its use for actually calculating **E**; so we must often determine the electric field **E** by summing over contributions due to infinitesimal charge elements dq via Coulomb's law: $dE = (1/4\pi\varepsilon_0)(dq/r^2)$. A magnetic equivalent to this infinitesimal form of Coulomb's law would be helpful for cases when the currents did not have great symmetry. Such a law was developed by Jean Baptiste Biot (1774–1862) and Felix Savart (1791–1841) shortly after Oersted's discovery in 1820 that a current produces a magnetic field.

According to Biot and Savart, a current I flowing in any path can be considered as many tiny (infinitesimal) current elements, such as in the wire of Fig. 29–12. If $d\mathbf{l}$ represents any infinitesimal length along which the current is flowing, then the magnetic field, $d\mathbf{B}$, at any point P in space, due to this element of current is given by

$$d\mathbf{B} = \frac{\mu_0 I}{4\pi} \frac{d\mathbf{l} \times \hat{\mathbf{r}}}{r^2} \tag{29–5}$$

where **r** is the displacement vector from the element $d\mathbf{l}$ to the point P, and $\hat{\mathbf{r}} = \mathbf{r}/r$ is the unit vector in the direction of **r** (see Fig. 29–12). Equation 29–5 is known as the **Biot-Savart law**. The magnitude of $d\mathbf{B}$ is

$$dB = \frac{\mu_0 I \, dl \sin \theta}{4\pi r^2} \tag{29–6}$$

where θ is the angle between $d\mathbf{l}$ and **r** (Fig. 29–12). The total magnetic field at point P is then found by summing (integrating) over all current elements:

$$\mathbf{B} = \int d\mathbf{B}.$$

The Biot-Savart law is thus the magnetic equivalent of Coulomb's law in its infinitesimal form; it is even an inverse square law, like Coulomb's law.

An important difference between the Biot-Savart law and Ampère's law (Eq. 29–2) is that in Ampère's law, $\oint \mathbf{B} \cdot d\mathbf{l} = \mu_0 I$, **B** is not necessarily due only to the current I enclosed by the path of integration. But in the Biot-Savart law the field $d\mathbf{B}$ in Eq. 29–5 is due only, and entirely, to the current element $I \, d\mathbf{l}$. (To find the total **B** at any point in space, it is necessary to include all currents.)

The Biot-Savart law (like Coulomb's law) is not considered a fundamental law, as is Ampère's law (and Gauss's law). Nonetheless it is very valuable for determining the magnetic field due to particular currents. The predictions it makes for **B** have always been confirmed by experiment, and its validity rests on this agreement.

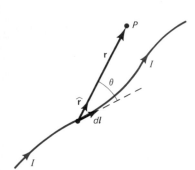

FIGURE 29–12

Biot-Savart law: the field at P due to current element $I d\mathbf{l}$ is $d\mathbf{B} = (\mu_0 I/4\pi)(d\mathbf{l} \times \mathbf{r}/r^2)$.

EXAMPLE 29–7   For the field near a long straight wire carrying a current I, show that the Biot-Savart law gives the same result as Eq. 29–1, $B = \mu_0 I/2\pi r$.

SOLUTION   We calculate the magnetic field at point P in Fig. 29–13 which is a perpendicular distance x from the infinitely long wire shown. The current is moving upwards, so the direction of the field $d\mathbf{B}$ due to each element must be directed into the plane of the page (see Eq. 29–5). Thus all the $d\mathbf{B}$ have the same direction and this direction is consistent with our previous result (see, for example,

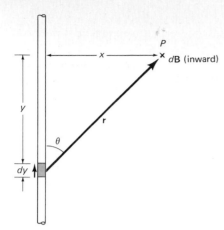

FIGURE 29–13

Determining **B** due to a long straight wire using the Biot-Savart law.

Fig. 28–7). The magnitude of **B** will be

$$B = \frac{\mu_0 I}{4\pi} \int_{y=-\infty}^{+\infty} \frac{dy \sin\theta}{r^2}$$

where $dy = dl$ and $r^2 = x^2 + y^2$. Note that we are integrating over y (the length of the wire) so x is considered constant. Both y and θ are variables, but they are not independent. In fact, $y = -x/\tan\theta$; so $dy = -x\csc^2\theta\,d\theta = -x\,d\theta/\sin^2\theta = -x\,d\theta/(x/r)^2 = r^2\,d\theta/x$. So our integral becomes

$$B = \frac{\mu_0 I}{4\pi}\frac{1}{x}\int_{\theta=0}^{\pi}\sin\theta\,d\theta = -\frac{\mu_0 I}{4\pi x}\cos\theta\Big|_0^{\pi} = \frac{\mu_0 I}{2\pi x}.$$

This is just Eq. 29–1 for the field near a long wire, where x has been used instead of r.

EXAMPLE 29–8 Determine **B** for points on the axis of a circular loop of wire of radius R carrying a current I, Fig. 29–14.

FIGURE 29–14

Determining **B** due to a current loop.

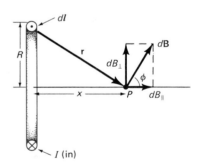

SOLUTION For an element of current at the top of the loop, the magnetic field $d\mathbf{B}$ at point P on the axis has the direction shown, and magnitude

$$dB = \frac{\mu_0 I\,dl}{4\pi r^2}$$

since $\theta = 90°$. We can break $d\mathbf{B}$ down into components $dB_\parallel$ and $dB_\perp$, which are parallel and perpendicular to the axis as shown. When we sum over all the elements of the loop, the perpendicular components will cancel. Hence, the total **B** will point along the axis, and will have magnitude

$$B = \int dB \cos\phi = \int dB\frac{R}{r} = \int dB\frac{R}{(R^2+x^2)^{1/2}}$$

where x is the distance of P from the center of the ring, and $r^2 = R^2 + x^2$. Now we put in dB from the equation above and integrate around the loop:

$$B = \frac{\mu_0 I}{4\pi}\frac{R}{(R^2+x^2)^{3/2}}\int dl = \frac{\mu_0 I}{2}\frac{R^2}{(R^2+x^2)^{3/2}}$$

since $\int dl = 2\pi R$, the circumference of the loop.

Recall from Section 28–10 that a current loop, such as that just discussed (Fig. 29–14) is considered a *magnetic dipole*. We saw there that a magnetic dipole has a dipole moment

$$\mu = NIA$$

where A is the area of the loop and N is the number of coils in the loop, each carrying current I. We also saw in Chapter 28 that a magnetic dipole placed in an external magnetic field experiences a torque and possesses potential energy, just like an electric dipole. In Example 29–8, we have looked at another aspect of a magnetic dipole: the magnetic field *produced by* a magnetic dipole has magnitude, along the dipole axis, of

$$B = \frac{\mu_0 I R^2}{2(R^2 + x^2)^{3/2}}.$$

We can write this in terms of the magnetic dipole moment $\mu = IA = I\pi R^2$ (for a single loop $N = 1$):

$$B = \frac{\mu_0}{2\pi} \frac{\mu}{(R^2 + x^2)^{3/2}}. \qquad \text{[magnetic dipole]} \quad (29\text{–}7a)$$

(Be careful to distinguish μ for dipole moment from μ_0, the magnetic permeability constant.) For distances far from the loop, $x \gg R$, this becomes

$$B \approx \frac{\mu_0}{2\pi} \frac{\mu}{x^3}. \qquad \left[\begin{array}{l} \text{magnetic dipole,} \\ x \gg R \end{array}\right] \quad (29\text{–}7b)$$

The magnetic field on the axis of a magnetic dipole decreases with the cube of the distance, just as for an electric dipole. The $1/x^3$ dependence occurs for points not on the axis as well, if $x \gg R$, although the multiplying factor is not the same. The magnetic field due to a current loop can be determined at various points using the Biot-Savart law and the results are in accord with experiment. The field lines around a current loop are shown in Fig. 29–15.

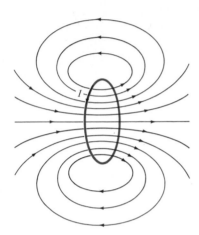

FIGURE 29–15

Magnetic field due to a circular loop of wire.

29–6 Ferromagnetism

As we saw in Section 28–1 iron (and a few other materials) can be made into strong magnets; they are said to be *ferromagnetic*. We now look more deeply into the sources of ferromagnetism.

A bar magnet, with its two opposite poles at either end, resembles an electric dipole with its two opposite charges. Indeed, a bar magnet is sometimes referred to as a "magnetic dipole." There are opposite "poles" separated by a distance; and the magnetic field lines of a bar magnet form a pattern much like that for the electric field of an electric dipole: compare Fig. 22–21a with Fig. 28–3b. One important difference, however, is that a positive or negative electric charge can easily be isolated. But the isolation of a single magnetic pole has proved much more difficult. If a bar magnet is cut in half, you do not obtain isolated north and south poles; instead, two new magnets are produced, Fig. 29–16. If the cutting operation is repeated, more magnets are produced, each with a north and a south pole. Physicists have tried various ways to isolate a single magnetic pole, and this is an active research field today since certain theories suggest they ought to exist. But so far there is no firm experimental evidence for their existence.

Microscopic examination reveals that a magnet is actually made up of tiny regions known as *domains*, which are at most about 1 mm in length or width. Each domain behaves like a tiny magnet with a north and a south pole. In an unmagnetized piece of iron, these domains are arranged randomly as shown in

FIGURE 29–16

If you break a magnet in half, you do not obtain isolated north and south poles; instead, two new magnets are produced, each with a north and a south pole.

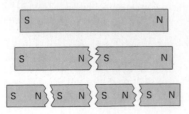

Fig. 29–17a. The magnetic effects of the domains cancel each other out, so this piece of iron is not a magnet. In a magnet, the domains are preferentially aligned in one direction as shown in Fig. 29–17b (downward in this case). A magnet can be made from an unmagnetized piece of iron by placing it in a strong magnetic field. (You can make a needle magnetic, for example by stroking it with one pole of a strong magnet.) Careful observations show in this case that domains may actually rotate slightly so they are more nearly parallel to the external field; or, more commonly, the borders of domains move so that those domains whose magnetic orientation is parallel to the external field grow in size at the expense of other domains. This can be seen by comparing Fig. 29–17a and b. This explains how a magnet can pick up unmagnetized pieces of iron like paper clips or bobby pins. The magnet's field causes a slight alignment of the domains in the unmagnetized object so that the object becomes a temporary magnet with its north pole facing the south pole of the permanent magnet, and vice versa; thus attraction results. In the same way, elongated iron filings will arrange themselves in a magnetic field just as a compass needle does, and will reveal the shape of the magnetic field, Fig. 29–18.

An iron magnet can remain magnetized for a long time, and thus it is referred to as a "permanent magnet." However, if you drop a magnet on the floor or strike it with a hammer, you may jar the domains into randomness; the magnet can thus lose some or all of its magnetism. Heating a magnet too can cause a loss of magnetism; for raising the temperature increases the random thermal motion of the atoms which tends to randomize the domains. Above a certain temperature known as the *Curie temperature* (1043 K for iron), a magnet cannot be made at all.[†]

A bar magnet is considered a magnetic dipole not only because its field resembles that of an electric dipole; but its field also closely resembles that of a current loop which we also called a magnetic dipole (see Section 28–10 and the end of Section 29–5). Bar magnets, current loops, and short solenoids (which, as we saw in Section 29–3, resemble a bar magnet) can all be considered magnetic dipoles. At great distances, the field due to any of these magnetic dipoles falls off as $1/r^3$.

The striking similarity between the fields produced by a bar magnet (Fig. 28–3b) and by a loop of electric current (Fig. 29–15) suggests that the magnetic field produced by a current may have something to do with ferromagnetism, an idea proposed by Ampère in the nineteenth century. According to modern atomic theory, the atoms that make up any material can be roughly visualized as containing electrons that orbit around a central nucleus. Since the electrons are charged, they constitute an electric current and therefore produce a magnetic field. But if there is no external field, the electron orbits in different atoms are arranged randomly so the magnetic effects due to the many orbits of all the atoms in a material will cancel out. However, electrons produce an additional intrinsic magnetic field—they have an intrinsic magnetic moment referred to as their "spin" magnetic moment.[‡] It is the magnetic field due to electron spin that is believed to produce ferromagnetism. In most materials, the magnetic fields due to electron spin cancel out because they are oriented at random. But in iron and other ferromagnetic materials, a complicated cooperative mechanism, known as "exchange coupling," operates; the result is that the electrons contributing to the ferromagnetism in a domain "spin" in the same direction. Thus the tiny magnetic fields due to each of the electrons add up to give the magnetic field of a domain. And when the domains are aligned, as we have seen, a strong magnet results.

It is believed possible today that *all* magnetic fields are caused by electric currents. This would explain why it has proved difficult to find a single magnetic

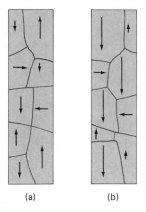

(a)   (b)

FIGURE 29–17
(a) Unmagnetized pieces of iron are made up of domains that are randomly arranged. (b) In a magnet, the domains are preferentially aligned in one direction. (The tip of each arrow represents the north pole of the domain, and the length of each arrow represents the magnitude of the magnetic moment of each domain.)

FIGURE 29–18
Iron filings line up along magnetic field lines.

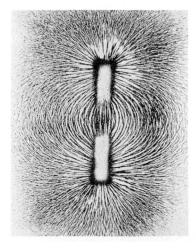

[†] Iron, nickel, cobalt, gadolinium, and certain alloys are ferromagnetic at room temperature; several other elements and alloys have low Curie temperature and thus are ferromagnetic only at low temperatures.

[‡] The name "spin" comes from an early suggestion that this intrinsic magnetic moment arises from the electron "spinning" on its axis (as well as "orbiting" the nucleus) to produce the extra field. However this view of a spinning electron is completely discredited today. See Chapter 41.

pole. There is no way to divide up a current and obtain a single magnetic pole. Of course if an isolated pole is found, we will have to alter the idea that all magnetic fields are produced by currents.

 ## 29–7   Hysteresis

The field inside a long solenoid or torus (Example 29–5) is directly proportional to the current (Eq. 29–3). If we put a piece of iron or other ferromagnetic material inside a solenoid, or a torus, the field may be greatly increased, often by hundreds or thousands of times. This occurs because the domains in the iron become preferentially aligned by the external field. The resulting magnetic field is the sum of that due to the current and that due to the iron. It is sometimes convenient to write the total field in this case as a sum of two terms:

$$\mathbf{B} = \mathbf{B_0} + \mathbf{B_M}. \tag{29–8}$$

Here, $\mathbf{B_0}$ refers to the field due only to the current in the wire (the "external field"); it is equal to the field that would be present in the absence of a ferromagnetic material. Then $\mathbf{B_M}$ represents the additional field due to the ferromagnetic material; often $\mathbf{B_M} \gg \mathbf{B_0}$. The quantitative study of the relation between $\mathbf{B_0}$, $\mathbf{B_M}$, and $\mathbf{B}$ is generally done using an iron-filled torus, Fig. 29–19, because in this device the lines of $\mathbf{B}$ are all essentially inside the iron core of the torus, and there are no air gaps through which $\mathbf{B}$ passes to complicate the analysis (as there would be for a solenoid, say). For convenience, we assume the torus is large, but thin, so the field inside in the absence of iron has magnitude

$$B = \mu_0 n I,$$

as discussed in Example 29–5, where I is the current in the coils and n is the number of coils per unit length. When iron is present, the ratio of the total field B, to the external field, B_0, is called the *relative permeability*, K_m:

$$K_m = \frac{B}{B_0}. \tag{29–9}$$

We can also define the so-called *magnetic permeability* μ (do not confuse this with the magnetic dipole moment) as

$$\mu = K_m \mu_0. \tag{29–10}$$

Then the total field within the torus is

$$B = \mu_0 K_m n I = \mu n I. \quad \text{[large thin torus]} \tag{29–11}$$

For nonferromagnetic materials, μ and K_m are constants as long as B_0 is not too large, and μ differs from μ_0 (and K_m differs from 1) by only a small amount (see Section 29–8). The value of μ for ferromagnetic materials can be much greater than μ_0 (and $K_m \gg 1$). Furthermore, μ and K_m are *not* constant for ferromagnetic materials; they depend on the value of the external field B_0 as the following experiment shows.

Suppose the iron core is initially unmagnetized and there is no current in the windings of the torus. Then the current I is slowly increased, and B_0 increases linearly with I. The total field B also increases, but follows the curved line shown in the graph of Fig. 29–20. (Note the different scales: $B \gg B_0$.) Initially (point a), no domains are aligned. As B_0 increases, the domains become more and more aligned until at point b, nearly all are aligned. The iron is said to be approaching *saturation*. (Point b is typically 70 percent of full saturation; the curve continues to rise very slowly, and reaches 98 percent saturation only when B_0 is increased by about a thousandfold above that at point b; the last few domains are very difficult to align.) Now suppose the external field B_0 is reduced by decreasing the current

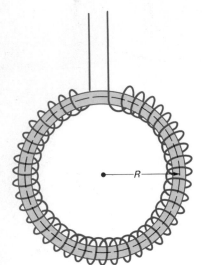

FIGURE 29–19
Iron-core torus.

FIGURE 29–20
Total magnetic field of an iron-core torus as a function of the external field B_0.

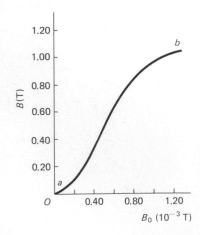

in the torus. As the current is reduced to zero, point c in Fig. 29–21, the domains do not become completely unaligned. Some permanent magnetism remains. If the current is then reversed in direction, enough domains can be turned around so $B = 0$ (point d). As the reverse current is increased further, the iron approaches saturation in the opposite direction (point e). Finally, if the current is again reduced to zero and then increased in the original direction, the total field follows the path $efgb$, again approaching saturation at point b.

Notice that the field did not pass through the origin (point a) in this cycle. The fact that the curves do not retrace themselves on the same path is called *hysteresis*. The curve $bcdefgb$ is called a *hysteresis loop*. In such a cycle, much energy is lost to heat (friction) due to realigning of the domains; it can be shown that the energy lost in this way is proportional to the area of the hysteresis loop.[†]

At points c and f, the iron core is magnetized even though there is no current in the torus. These points correspond to a permanent magnet. For a permanent magnet, it is desired that ac and af be as large as possible. Materials for which this is true are said to have high *retentivity*, and may be referred to as "hard." On the other hand, a hysteresis curve such as that in Fig. 29–22 occurs for so-called "soft iron" (it is soft only from a magnetic point of view). This is preferred for *electromagnets* (which consist of a piece of iron inside a solenoid or torus with an air gap, and can produce very high magnetic fields) since the field can be more readily switched off, and the field can be reversed with less loss of energy. Whether iron is "soft" or "hard" depends on how it is alloyed, heat treatment, and other factors.

You may wonder how a material can be demagnetized—that is, made unmagnetized. This can be done by reversing the magnetizing current repeatedly while decreasing its magnitude. This results in the curve of Fig. 29–23. If your watch becomes magnetized, it can be demagnetized by placing it in a coil of wire carrying an ac current, and then slowly pulling the watch away.

*29–8 Paramagnetism and Diamagnetism

All materials are magnetic to some extent. Nonferromagnetic materials fall into two principal classes: *paramagnetic*, in which the relative permeability K_m (Eq. 29–9) is slightly greater than 1; and *diamagnetic*, in which K_m is slightly less than 1.

The difference between paramagnetic and diamagnetic materials can be understood theoretically at the molecular level on the basis of whether or not the molecules have a permanent magnetic dipole moment. Paramagnetism occurs in materials whose molecules (or ions) have a permanent magnetic dipole moment.[‡] In the absence of an external field, the molecules are randomly oriented and no magnetic effects are observed. However, when an external magnetic field is applied (say, by putting the material in a solenoid), the applied field exerts a torque on the magnetic dipoles (Section 28–10), tending to align them parallel to the field; the total magnetic field (external plus that due to aligned magnetic dipoles) will be slightly greater than B_0. The thermal motion of the molecules reduces the alignment, however. A useful quantity is the *magnetization*, $\mathbf{M}$, defined as the magnetic dipole moment per unit volume,

$$\mathbf{M} = \mu/V,$$

[†] A similar hysteresis curve would be obtained for an iron core solenoid, although the ratio of B to B_0, along the two axes in the diagram would be somewhat less—perhaps between 10 and 100, rather than about 1000 for the torus as shown.

[‡] Other types of paramagnetism also occur whose origin is different from that described here, such as in metals where free electrons can contribute.

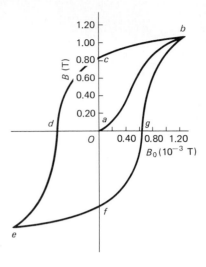

FIGURE 29–21
Hysteresis curve.

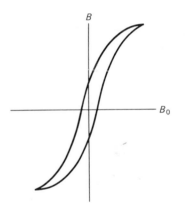

FIGURE 29–22
Hysteresis curve for soft iron.

FIGURE 29–23
Successive hysteresis loops during demagnetization.

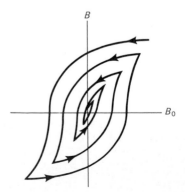

where μ is the magnetic dipole moment of the sample and V its volume; it is found experimentally that M is directly proportional to the external magnetic field (tending to align the dipoles) and inversely proportional to the Kelvin temperature T (tending to randomize dipole directions); this is called *Curie's law*, after Pierre Curie (1859–1906), who first noted it:

$$M = C\frac{B}{T}$$

where C is a constant. If the ratio B/T is very large (B very large or T very small) Curie's law is no longer accurate; as B is increased (or T decreased), the magnetization approaches some maximum value, M_{max}. This makes sense, of course, since M_{max} corresponds to complete alignment of all the permanent magnetic dipoles. However, even for very large magnetic fields, ≈ 2.0 T, deviations from Curie's law are normally noted only at very low temperatures, on the order of a few kelvins.

Ferromagnetic materials, as mentioned in Section 29–6, are no longer ferromagnetic above a characteristic temperature called the Curie temperature (1043 K for iron); above this Curie temperature, they generally are paramagnetic.

Diamagnetic materials (for which K_m is slightly less than 1) are made up of molecules that have no permanent magnetic dipole moment. When an external magnetic field is applied, magnetic dipoles are induced, but the induced magnetic dipole moment is in the direction opposite to that of the field. Hence the total field will be slightly less than the external field. (The effect of the external field—in the crude model of electrons orbiting nuclei—is to increase the "orbital" speed of electrons revolving in one direction, and to decrease the speed of electrons revolving in the other direction; the net result is a net dipole moment opposing the external field.) Diamagnetism is present in all materials, but is weaker even than paramagnetism and so is overwhelmed by paramagnetic and ferromagnetic effects in materials that display these other forms of magnetism.

* 29–9 Magnetization Vector and the Extension of Ampère's Law

The *magnetization vector* **M**, *defined as the magnetic dipole moment per unit volume* in Section 29–8, is a useful concept for any magnetic material (ferro-, para-, or diamagnetic). It is useful to write the magnetic field in terms of **M** when magnetic materials are present, just as we wrote the electric field in terms of the polarization vector **P** (Section 25–7).

Let us consider a torus, as in Section 29–7, which is filled with some material (not necessarily ferromagnetic). The total field **B** is given by Eq. 29–8:

$$\mathbf{B} = \mathbf{B_0} + \mathbf{B_M}$$

where $\mathbf{B_0}$ is the field due to the current in the coils and $\mathbf{B_M}$ is the field due to the magnetic material. (B_M is usually small compared to B_0 for para- and diamagnetic materials, but much larger for ferromagnetic materials.) We have seen that B_0 is given by

$$B_0 = \mu_0 nI = \mu_0\frac{N}{l}I$$

where now N is the total number of coils in length l. The field $\mathbf{B_M}$ due to the material can be imagined as arising from currents within the atoms of the material. The net effect of all these atomic currents can be thought of as a current I_M (the "magnetization current") around the outer surface of the material, Fig. 29–24, in analogy to the induced electric charge on the surface of a dielectric, Fig. 25–7. (In fact one would not be able to measure a current I_M at the

FIGURE 29–24

Effective current, I_M, in a magnetic material inside a coil of wire (small section of a torus) carrying current I.

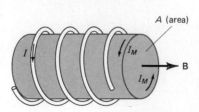

surface; the concept of a magnetization current I_M is useful, nonetheless, and to distinguish it from a real conduction current, I, we call the latter a "real current.") In analogy to our relation above for B_0 we write

$$B_M = \mu_0 \frac{N_M I_M}{l}$$

where N_M/l is the effective number of loops per unit length (or, $N_M I_M/l$ is the effective magnetization current per unit length). But the magnetic dipole moment is $N_M I_M A$ (Eq. 28–11) where A is the cross-sectional area of the material, Fig. 29–24. Hence the magnetic dipole moment per unit volume is

$$M = \frac{N_M I_M A}{V} = \frac{N_M I_M A}{Al} = \frac{N_M I_M}{l}$$

where $V = Al$ is the volume around which the total effective current $N_M I_M$ flows. We combine these last two relations and write $\mathbf{B_M}$ in terms of $\mathbf{M}$:

$$\mathbf{B_M} = \mu_0 \mathbf{M}.$$

Then the total field $\mathbf{B}$ is

$$\mathbf{B} = \mathbf{B_0} + \mu_0 \mathbf{M}. \qquad (29\text{–}12)$$

Ampère's law can be extended to include magnetic materials by including the magnetization current, I_M, on the right side:

$$\oint \mathbf{B} \cdot d\mathbf{l} = \mu_0(I + I_M) \qquad (29\text{–}13)$$

where I is the net real current and I_M is the equivalent magnetization current enclosed by the path of integration. If we now use Eq. 29–12, Ampère's law becomes

$$\oint (\mathbf{B_0} + \mu_0 \mathbf{M}) \cdot d\mathbf{l} = \mu_0(I + I_M).$$

In the absence of any material, $\oint \mathbf{B_0} \cdot d\mathbf{l} = \mu_0 I$ (remember $\mathbf{B_0}$ refers to $\mathbf{B}$ when no material is present), so from the last equation we obtain

$$\oint \mathbf{M} \cdot d\mathbf{l} = I_M, \qquad (29\text{–}14)$$

a useful relation between $\mathbf{M}$ and the magnetization current. (This could serve as the definition of I_M.) Again using $\oint \mathbf{B_0} \cdot d\mathbf{l} = \mu_0 I$, and combining it with Eq. 29–12 we obtain (after dividing through by μ_0):

$$\oint \left(\frac{\mathbf{B} - \mu_0 \mathbf{M}}{\mu_0} \right) \cdot d\mathbf{l} = I. \qquad (29\text{–}15)$$

This result, derived for a special case, is valid in general. It is usually written in terms of a new vector $\mathbf{H}$ defined as

$$\mathbf{H} = \frac{\mathbf{B} - \mu_0 \mathbf{M}}{\mu_0}, \qquad (29\text{–}16a)$$

which can be rewritten

$$\mathbf{B} = \mu_0 \mathbf{H} + \mu_0 \mathbf{M}. \qquad (29\text{–}16b)$$

From Eq. 29–12 we can also write

$$\mathbf{B_0} = \mu_0 \mathbf{H}. \qquad (29\text{–}17)$$

In terms of $\mathbf{H}$, Eq. 29–15 becomes

$$\oint \mathbf{H} \cdot d\mathbf{l} = I. \qquad (29\text{–}18)$$

The vector **H** is called the *magnetic field strength* and is to be distinguished from **B** which is generally referred to as the **magnetic induction**. Equation 29–18 tells us that the line integral of **H** around any closed path is equal to the total real current (only) enclosed, even when magnetic materials are present. Thus **H** is much like the vector **D** in electrostatics which is due only to free charges. The vectors **M**, **H**, and **B** are the counterparts, respectively, of **P**, **D**, and **E** for the electric case in dielectrics.

The vector **H** is often associated only with free currents, and **M** only with "magnetization currents," and **B** with all currents. This "association" really refers to the line integrals, Eqs. 29–13, 29–14, and 29–18. It does not mean, for example, that **H** is produced only by real currents. For example, at any point in free space (a vacuum, no magnetic materials), we have $B = \mu_0 H$. This is true just outside the pole piece of a permanent magnet, so $H \neq 0$ (since $B \neq 0$ there) even though no real currents are present. Using Eq. 29–18, it is easy to see that **H** must oppose **B** inside the magnet.

This analysis in terms of **H**, **M**, and **B** is necessary for more advanced treatments; we have included it here simply to give you a little familiarity with these ideas. It is not expected, however, that you will acquire a working knowledge of **M** and **H** from this brief treatment, and we will not need these ideas for most of what follows in this book. We do mention here, for completeness, the *magnetic susceptibility*, χ_m, defined as:

$$\mathbf{M} = \chi_m \mathbf{H}.$$

That is, χ_m is the ratio of magnetization to magnetic field strength. For paramagnetic and diamagnetic materials, χ_m is constant as long as B is not too great; but it is not constant for ferromagnetic materials. From Eqs. 29–16, 29–17, 29–9, and 29–10,

$$\chi_m = K_m - 1$$

and

$$\mu = (1 + \chi_m)\mu_0.$$

In a vacuum, $K_m = 1$, $\chi_m = 0$, and $\mu = \mu_0$.

Summary

Ampère's law states that the line integral of the magnetic field strength **B** around any closed loop is equal to μ_0 times the total net current I enclosed by the loop:

$$\oint \mathbf{B} \cdot d\mathbf{l} = \mu_0 I.$$

The magnetic field strength B at a distance r from a long straight wire is directly proportional to the current I in the wire and inversely proportional to r. The magnetic field lines are circles centered at the wire. The magnetic field inside a long tightly wound solenoid is $B = \mu_0 n I$ where n is the number of coils per unit length and I is the current in each coil.

The force that one long current-carrying wire exerts on a second parallel current-carrying wire a distance l away serves as the definition of the ampere unit, and ultimately of the coulomb as well.

The *Biot-Savart law* is useful for determining the mag-

netic field due to a known arrangement of currents. It states that

$$d\mathbf{B} = \frac{\mu_0 I}{4\pi} \frac{d\mathbf{l} \times \hat{\mathbf{r}}}{r^2}$$

where $d\mathbf{B}$ is the contribution to the total field at some point P due to a current I along an infinitesimal length $d\mathbf{l}$, and $\hat{\mathbf{r}}$ is the displacement vector from $d\mathbf{l}$ to P. The total field **B** will be the integral over all $d\mathbf{B}$.

Iron and a few other materials can be made into strong permanent magnets. They are said to be *ferromagnetic*. Ferromagnetic materials are made up of tiny *domains*—each a tiny magnet—which are preferentially aligned in a permanent magnet, but randomly aligned in a nonmagnetized sample. When a ferromagnetic material is placed in the magnetic field B_0 due to a current, say inside a solenoid or torus, the material becomes magnetized. When the current is turned off, however, the material remains magnetized, and when the current is increased in the opposite direction (and then again reversed), the total field B does not follow B_0 due to the current; instead, the plot of B vs. B_0 is a *hysteresis loop*, and the fact that the curves do not retrace themselves is called *hysteresis*.

 Questions _____

1. The magnetic field due to current in wires in your home can affect a compass. Discuss the problem in terms of currents, depending on whether they are ac or dc, and their distance away.

2. Compare and contrast the magnetic field due to a long straight current and the electric field due to a long straight line of electric charge at rest (Section 22–7).

3. Explain why a field such as that shown in Fig. 29–6b is consistent with Ampère's law. Could the lines curve upward instead of downward?

4. Compare Ampère's law to Gauss's law.

5. (a) Write Ampère's law for a path that surrounds both conductors in Fig. 29–4. (b) Repeat, assuming the lower current, I_2, is in the opposite direction ($I_2 = -I_1$).

6. Can the integral in Ampère's law be carried out over a surface?

7. Suppose the cylindrical conductor of Fig. 29–5 has a concentric cylindrical hollow cavity inside it (so it looks like a pipe). What can you say about B in the cavity?

8. What would be the effect on B inside a long solenoid if (a) the diameter of the loops was doubled, or (b) the spacing between loops was doubled, or (c) the solenoid's length was doubled along with a doubling in the total number of loops.

9. Another type of magnetic switch similar to a solenoid is a *relay*. A relay is an electromagnet (the iron rod inside the coil doesn't move) which, when activated, attracts a piece of soft iron on a pivot. Design a relay (a) to make a doorbell, (b) to close an electrical switch. A relay is used in the latter case when you need to switch on a circuit carrying a very large current but you don't want that large current flowing through the main switch. For example many electrical devices in a car are connected to relays so the large currents needed don't pass to the dashboard switches.

10. Two long wires carrying equal currents I are at right angles to each other, but don't quite touch. Describe the magnetic force one exerts on the other.

11. A horizontal wire carries a large current. A second wire carrying a current in the same direction is suspended below. Can the current in the upper wire hold the lower wire in suspension? Under what conditions will it be in equilibrium? Under what conditions will the equilibrium be stable?

12. A horizontal current-carrying wire, free to move, is suspended directly above a second, parallel, current-carrying wire. (a) In what direction is the current in the lower wire? (b) Can the upper wire be held in stable equilibrium due to the magnetic force of the lower wire? Explain.

13. Use the Biot-Savart law to convince yourself the field of the current loop in Fig. 29–15 at points off the axis is correct as shown.

14. Do you think B will be the same for all points in the plane of the current loop of Fig. 29–15?

15. Why does twisting the lead-in wires to electrical devices reduce the magnetic effects of the leads?

16. Compare the Biot-Savart law with Coulomb's law: what are the similarities and differences?

17. The neutron, though it has zero net charge, does have a magnetic moment. How might you explain this?

18. How might you measure the magnetic dipole moment of the earth?

19. (a) Apply the right-hand rule of Fig. 28–7 to find the direction of B at the center of the current loop of Fig. 29–15. (b) Think up another right-hand rule, in which your fingers point in the direction of I, to find the direction of B at the loop's center.

20. Comment on the following statement: "Since magnetic poles do not seem to exist, the simplest source of the magnetic field is a magnetic dipole."

21. How might you define or determine the magnetic pole strength (the magnetic equivalent of a single electric charge) for (a) a bar magnet, (b) a current loop?

22. A heavy magnet attracts, from rest, a heavy block of iron. Before striking the magnet the block has acquired considerable kinetic energy. (a) What is the source of this kinetic energy? (b) When the block strikes the magnet, some of the latter's domains are jarred into randomness; describe the energy transformations.

23. Will a magnet attract any metallic object or only those made of iron? (Try it and see.) Why is this so?

24. An unmagnetized nail will not attract an unmagnetized paper clip. However, if one end of the nail is in contact with a magnet, the other end *will* attract a paper clip. Explain.

25. The presence of iron inside a solenoid can greatly increase B. Yet without iron, greater fields could be obtained. Explain.

*26. Describe the magnetization curve for (a) a paramagnetic substance and (b) a diamagnetic substance, and compare to that for a ferromagnetic substance (Fig. 29–21).

*27. Can all materials be considered (a) diamagnetic, (b) paramagnetic, (c) ferromagnetic?

*28. Are B and H always in the same direction? If not, give an example.

*29. Is χ_m always positive? If not, for what condition is it negative?

*30. Can H ever be zero when B is not zero? If so, give an example.

*31. Consider (a) a point inside a permanent bar magnet, and (b) a point just outside the magnet near a pole. Is B, H, or M zero at either point? Explain.

Problems

1. (I) If a magnetic field of no more than 10^{-3} T is to be allowed 30 cm from an electrical wire, what is the maximum current the wire can carry?

2. (I) How strong is the magnetic field 11.0 cm from a long straight wire carrying 2.8 A?

3. (II) What is the acceleration (in g's) of a 280-g model airplane charged to 8.0 C and traveling at 1.5 m/s as it passes within 15 cm of a wire, nearly parallel to its path, carrying a 20-A current?

4. (II) Two long thin parallel wires 11.2 cm apart carry 25-A currents in the same direction. Determine the magnetic field strength at a point 12.0 cm from one wire and 7.0 cm from the other.

5. (II) A horizontal compass is placed 15 cm due south from a straight vertical wire carrying a 30-A current downward. In what direction does the compass needle point at this location? Assume the horizontal component of the earth's field at this point is 0.45×10^{-4} T and the magnetic declination is 0°.

6. (II) Lightning strikes a 7.0-m tall, 9.8-cm-diameter iron flagpole such that electrons flow up the pole. If the potential difference between the top and bottom of the pole is 20 kV, what will be the magnitude and direction of the magnetic field 50 cm north of the pole?

7. (II) Determine the magnetic field midway between two long straight wires 10 cm apart in terms of the current I in one when the other carries 10 A. Assume these currents are (a) in the same direction, and (b) in opposite directions.

8. (II) A compass needle in a particular location points 15°E of N outdoors. However, when it is placed 12 cm to the east of a vertical wire inside a building it points 60°E of N. What is the magnitude and direction of the current in the wire? The earth's field there is 0.50×10^{-4} T and is horizontal.

9. (II) Let two long parallel wires, a distance d apart, carry equal currents I in the same direction. Find **B** as a function of x along a perpendicular line connecting them far from the ends; one wire is at $x = 0$, the other at $x = d$.

10. (II) Repeat problem 9 if the wire at $x = 0$ carries twice the current ($2I$) as the other wire, and in the opposite direction.

11. (II) An electron traveling 4.50×10^6 m/s is moving at a 45° angle to a straight wire carrying a 15.0-A current. If the electron passes within 12.0 cm of the wire, what maximum force does it feel?

12. (II) A very long flat conducting strip of width l and negligible thickness lies in a horizontal plane and carries a uniform current I across its cross-section. (a) Show that at points a distance y directly above its center, the field is given by

$$B = \frac{u_0 I}{\pi l} \tan^{-1} \frac{l}{2y},$$

assuming the wire is infinitely long. [Hint: divide the strip into many thin "wires," and sum (integrate) over these.] (b) What value does B approach for $y \gg l$? Does this make sense? Explain.

13. (III) An electron is moving in a plane with a long straight current-carrying wire. The electron is heading at a 45° angle directly toward the wire when it is 50 cm away; the electron reaches only as close as 1.0 cm, before being repelled away, always moving in the same plane. If the electron has an initial speed of 3.4×10^6 m/s, determine (a) the current in the wire, and (b) the speed of the electron at its point of closest approach.

14. (I) Use Ampère's law to show that a uniform magnetic field, such as between the pole pieces of a magnet, Fig. 29–1, cannot drop abruptly to zero outside the magnet.

15. (II) (a) Use Eq. 29–1, and the vector nature of **B**, to show that the magnetic field lines around two long parallel wires carrying equal currents $I_1 = I_2$ are as shown in Fig. 29–4. (b) Draw the equipotential lines around two stationary positive electric charges. (c) Are these two diagrams similar? Identical? Why or why not?

16. (II) A coaxial cable consists of a solid inner conductor of radius r_1, surrounded by a concentric cylindrical tube of inner radius r_2 and outer radius r_3. The conductors carry equal and opposite currents I distributed uniformly across their cross-sections. Determine the magnetic field at a distance r from the axis for (a) $r < r_1$, (b) $r_1 < r < r_2$, (c) $r_2 < r < r_3$, (d) $r > r_3$.

17. (II) A very large flat conducting sheet of thickness t carries a uniform current density **j** throughout. Determine the magnetic field (magnitude and direction) at a distance y above the plane. (Assume the plane is infinitely long and wide.)

18. (III) For two long parallel wires carrying currents I_1 and I_2, as in Fig. 29–4, show that for the circular path of radius r ($r < l$, the separation of the wires) centered on I_1, that

$$\oint \mathbf{B} \cdot d\mathbf{l} = \mu_0 I_1$$

in accord with Ampère's law. (But do not use Ampère's law.)

19. (I) A 32-cm-long solenoid, 1.2 cm in diameter, is to produce a 0.20-T magnetic field at its center. If the maximum current is 3.7 A, how many turns must the solenoid have?

20. (II) You have 1.0 kg of copper and want to make a practical solenoid that produces the greatest possible magnetic field. Should you make your copper wire long and thin, short and fat, or what? Consider other variables, such as solenoid diameter, length, and so on.

21. (I) What is the magnitude and direction of the force between two parallel wires 80 m long and 30 cm apart, each carrying 65 A in the same direction?

22. (I) A vertical straight wire carrying a 5.0-A current exerts an attractive force per unit length of 6.0×10^{-4} N/m on a second parallel wire 8.0 cm away. What current (magnitude and direction) flows in the second wire?

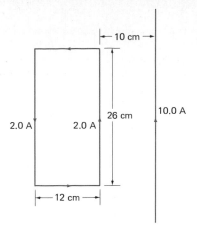

FIGURE 29–25

23. (II) Three long parallel wires are 12.0 cm from one another. (Looking along them, they are at three corners of an equilateral triangle.) The current in each wire is 3.0 A but that in wire A is opposite to that in wires B and C. Determine the magnetic force per unit length on each wire due to the other two.

24. (II) A rectangular loop of wire carries a 2.0-A current and lies in a plane which also contains a very long straight wire carrying a 10.0-A current as shown in Fig. 29–25. Determine (a) the net force and (b) the net torque on the loop due to the straight wire.

25. (II) A long horizontal wire carries a current of 78 A. A second wire, made of 3.5-mm-diameter copper wire and parallel to the first, but 18 cm below it, is held in suspension magnetically. (a) What is the magnitude and direction of the current in the lower wire? (b) Is it in stable equilibrium? (c) Repeat parts (a) and (b) if the second wire is suspended 20 cm above the first due to the latter's field.

SECTION 29–5

26. (I) The earth's magnetic field is essentially that of a magnetic dipole. If the field near the earth's surface is about 5.0×10^{-5} T, what will it be (approximately) 13,000 km above the surface at the North Pole?

27. (II) A wire, in a plane, has the shape shown in Fig. 29–26: two arcs of a circle connected by radial lengths of wire. Determine **B** at point C in terms of R_1, R_2, θ, and the current I.

28. (II) Use the Biot-Savart law to show that the magnetic field **B**, due to a single point charge q moving with velocity **v**, at a point P whose position vector relative to the charge is **r** is given by

$$\mathbf{B} = \frac{\mu_0}{4\pi} \frac{q\mathbf{v} \times \mathbf{r}}{r^3}.$$

29. (II) Start with the result of Example 29–8 for the field along the axis of a single loop to obtain the field inside a very long solenoid (Eq. 29–3).

30. (II) A circular conducting ring of radius R is connected to two exterior straight wires ending at two ends of a diameter, Fig. 29–27. The current I splits into equal portions while passing through the ring as shown. What is **B** at the center of the ring?

31. (II) A small loop of wire of radius 1.0 cm is placed at the center of a 15.0-cm wire loop. The planes of the loops are perpendicular to each other, and a 3.0-A current flows in each. Estimate the torque the large loop exerts on the smaller one. What simplifying assumption did you make?

32. (II) A square loop of wire, of side l, carries a current I. Show that the magnetic field at the center of the square is

$$B = 2\sqrt{2}\mu_0 I/\pi l.$$

(Hint: determine **B** for each segment of length l.)

33. (II) In problem 32, if you reshaped the square wire into a circle, would B increase or decrease at the center?

34. (II) A nonconducting circular plate, of radius R, carries a uniformly distributed electric charge Q. The plate is set spinning with angular velocity ω about an axis perpendicular to the plate through its center. Determine (a) its magnetic dipole moment, and (b) the magnetic field at points on its axis a distance x from its center; (c) does Eq. 29–7b apply in this case for $x \gg R$?

35. (II) (a) Show that the magnetic field strength B at the center of a circular coil of wire carrying a current I is

$$B = \frac{\mu_0 N I}{2R}$$

where N is the number of loops in the coil and R is its radius. (b) Suppose that an electromagnet uses a coil 2.0 m in diameter made from square copper wire 2.0 mm on a side; the power supply produces 50 V at a maximum power output of 1.0 kW. How many turns are needed to run the power supply at maximum power? (c) What is the magnetic field strength at the center of the coil? (d) If you

FIGURE 29–26

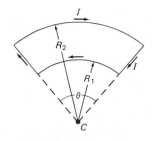

FIGURE 29–27

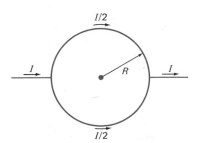

use a greater number of turns and this same power supply, will a greater magnetic field strength result? Explain.

36. (II) A solenoid has length L and radius R and contains N closely spaced coils. Determine the field on its axis at one of its ends. (Hint: use the result of Example 29–8 and perform a sum or integration.)

37. (II) Two large coils of wire, each carrying a current I in their N turns and separated by a distance equal to the radius R of each, are called *Helmholtz coils*. (a) Determine B at points x along the line joining their centers. Let $x = 0$ at the center of one coil, $x = R$ at the center of the other. (b) Plot B vs. x from $x = 0$ to $x = R$. (c) Determine B on the axis midway between the coils ($x = R/2$) if $R = 20.0$ cm, $I = 35$ A, and each coil contains $N = 350$ turns. (d) Show that the field midway between the coils is particularly uniform by showing that if the separation of the coils is s, then $dB/dx = 0$ and $d^2B/dx^2 = 0$ at the midpoint for $s = R$.

38. (III) A single rectangular loop of wire, with sides a and b, carries a current I. An xy coordinate system has its origin at the center of the rectangle with the x axis parallel to side b. Determine the magnetic field B at all points, x, y, within the loop.

39. (III) A square loop of wire, of side l, carries a current I. (a) Determine the magnetic field B at points on a line perpendicular to the plane of the square which passes through the center of the square. Express B as a function of x, the distance along the line from the center of the square. (b) For $x \gg l$, does the square appear to be a dipole? If so, what is its moment?

SECTION 29–6

40. (II) An iron atom has a magnetic dipole moment of about 1.8×10^{-23} A·m². (a) Determine the dipole moment of an iron bar 1.2 cm long, 1.2 cm wide, and 1.2 cm thick, if it is 100 percent saturated. (b) What torque would be exerted on this rod when placed in a 1.2-T field acting at right angles to the bar?

SECTION 29–7

41. (I) The following are some values of B and B_0 for a piece of annealed iron as it is being magnetized:

$B_0(10^{-4}$T)	0	0.13	0.25	0.50	0.63	0.78	1.0	1.3
B(T)	0	0.0042	0.010	0.028	0.043	0.095	0.45	0.67
$B_0(10^{-4}$T)		1.9	2.5	6.3	13.0	130	1,300	10,000
B(T)		1.01	1.18	1.44	1.58	1.72	2.26	3.15

Determine K_m and μ for each value and plot a graph of each versus B_0.

42. (II) An iron-core solenoid is 30 cm long, 1.8 cm in diameter, and has 500 turns of wire. A magnetic field of 2.2 T is produced when 48 A flows in the wire. What is the permeability μ at this high field strength?

*SECTION 29–9

*43. (II) What is the magnitude of the magnetization for the bar in problem 40?

*44. (II) For each value of B_0 given in problem 41, determine M and χ_m.

*45. (II) (a) Draw a few of the **B** field lines inside and outside a permanent bar magnet. (b) Draw a few of the **H** field lines inside and outside the magnet. (c) Outside the magnet, **H** is in the direction of **B**, but inside **H** is opposite to **B**. Explain.

*46. (II) In the previous problem, explain how it can be true (since $I = 0$) that

$$\oint \mathbf{H} \cdot d\mathbf{l} = 0$$

is valid for a path that (a) is only in empty space and goes around the magnet, (b) is entirely inside the magnet, (c) passes along the axis of the magnet and loops around outside to close on itself.

Electromagnetic Induction and Faraday's Law

In Chapter 28 we discussed two ways in which electricity and magnetism are related: (1) an electric current produces a magnetic field; and (2) a magnetic field exerts a force on an electric current or moving electric charge. These discoveries were made in 1820–1821. Scientists then began to wonder: if electric currents produce a magnetic field, is it possible that a magnetic field can produce an electric current? Ten years later the American Joseph Henry (1797–1878) and the Englishman Michael Faraday (1791–1867) independently found that it was possible. Henry actually made the discovery first; but Faraday published his results earlier and investigated the subject in more detail. We now discuss this phenomenon and some of its world-changing applications.

30–1 Induced EMF

In his attempt to produce an electric current from a magnetic field, Faraday used the apparatus shown in Fig. 30–1. A coil of wire, *X*, was connected to a battery. The current that flowed through *X* produced a magnetic field that was intensified by the iron core. Faraday hoped that by using a strong enough battery, a steady current in *X* would produce a great enough magnetic field to produce a current in a second coil *Y*. This second circuit, *Y*, contained a galvanometer to detect any current but contained no battery. He met no success with steady currents. But

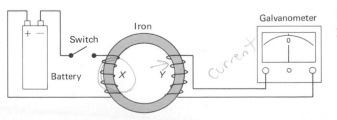

FIGURE 30–1

Faraday's experiment to induce an emf.

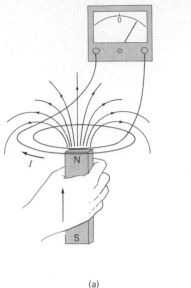

FIGURE 30–2

(a) A current is induced when a magnet is moved toward a coil (b) The induced current is opposite when the magnet is removed. Note that the galvanometer zero is in the center of the scale and deflects left or right, depending on direction of the current.

(a)   (b)

the long-sought effect was finally observed when Faraday saw the galvanometer in circuit Y deflect strongly at the moment he closed the switch in circuit X. And the galvanometer deflected strongly in the opposite direction when he opened the switch. A *steady* current in X had produced *no* current in Y. Only when the current in X was starting or stopping was a current produced in Y.

Faraday concluded that although a steady magnetic field produced no current, a changing magnetic field can produce an electric current! Such a current is called an *induced current*. When the magnetic field through coil Y changes, a current flows as if there were a source of emf in the circuit. We therefore say that an *induced emf is produced by a changing magnetic field*.

Faraday did further experiments on **electromagnetic induction**, as this phenomenon is called. For example, Fig. 30–2 shows that if a magnet is moved quickly into a coil of wire, a current is induced in the wire. If the magnet is quickly removed, a current is induced in the opposite direction. Furthermore, if the magnet is held steady and the coil of wire is moved toward or away from the magnet, again an emf is induced and a current flows. Motion or change is required to induce an emf. It doesn't matter whether the magnet or the coil moves.

30–2 Faraday's Law of Induction; Lenz's Law

Faraday investigated quantitatively what factors influence the magnitude of the emf induced. He found first of all that it depends on time: the more rapidly the magnetic field changes, the greater the induced emf. But the emf is not simply proportional to the rate of change of the magnetic field, **B**. Rather it is proportional to the rate of change of the **magnetic flux**, Φ_B, which is defined analogously to electric flux (Section 23–1) as

$$\Phi_B = \int \mathbf{B} \cdot d\mathbf{A}; \tag{30–1}$$

the integral of the magnetic field **B** is taken over any chosen surface. The unit of magnetic flux is the tesla $\cdot$ meter^2; this is called a weber: 1 Wb = 1 T $\cdot$ m^2. (Note also that 1 T = 1 Wb/m^2; this is where the old SI unit for magnetic field strength, the Wb/m², comes from.) The magnetic field B, since it equals flux per unit area (Φ_B/A), is often called the **flux density**.

Now that we have the concept of magnetic flux we can write down the results of Faraday's investigations: namely, that the emf induced in a circuit is equal to the rate of change of magnetic flux through the circuit:

$$\mathcal{E} = -\frac{d\Phi_B}{dt}. \qquad (30\text{--}2a)$$

The induced emf, $\mathcal{E}$, is in volts if the rate of change of magnetic flux, $d\Phi_B/dt$, is in Wb/s (T·m²/s). This result, Eq. 30–2a, is known as **Faraday's law of induction**, and is one of the fundamental relations of electromagnetism. If the circuit contains N closely wrapped loops, the emf's induced in each add together, and Eq. 30–2a becomes

$$\mathcal{E} = -N\frac{d\Phi_B}{dt}. \qquad (30\text{--}2b)$$

The minus signs in Eqs. 30–2a and b are placed there to remind us in which direction the induced emf acts. Experiment shows that *an induced emf always gives rise to a current whose magnetic field opposes the original change in magnetic flux*. This is known as **Lenz's law**. Let us apply it to the case of relative motion between a magnet and a coil, Fig. 30–2. The changing flux induces an emf, which produces a current in the coil; and this induced current produces its own magnetic field. In Fig. 30–2a the distance between the coil and the magnet decreases; so the magnetic field, and therefore the flux through the coil, increases. The magnetic field of the magnet points upward. To oppose this upward increase, the field produced by the induced current must point *downward*. Thus Lenz's law tells us that the current must move as shown (use the right-hand rule). In Fig. 30–2b, the flux *decreases*, so the induced current produces an *upward* magnetic field that is "trying" to maintain the status quo. Thus the current must be as shown.

Let us consider what would happen if Lenz's law were not true, but was just the reverse. The induced current in this pretend situation would produce a flux in the same direction as the original change; this greater change in flux would produce an even larger current followed by a still greater change in flux, and so on. The current would continue to grow indefinitely, producing power ($= I^2R$) even after the original stimulus ended. This would violate the conservation of energy. Such "perpetual motion" devices do not exist. Thus Lenz's law as stated above (and not its opposite) is consistent with the law of conservation of energy.

It is important to note that an emf is induced whenever there is a change in flux. Since magnetic flux $\Phi_B = \int \mathbf{B} \cdot d\mathbf{A} = \int B \cos \theta \, dA$, then Faraday's law can be written

$$\mathcal{E} = -\frac{d}{dt}\int B \cos \theta \, dA.$$

Thus, an emf can be induced in two ways: (1) by a changing magnetic field B; or (2) by changing the area of the loop or its orientation θ with respect to the field. Figures 30–1 and 30–2 illustrated case 1. Examples of case 2 are illustrated in Figs. 30–3 and 30–4 and in the following example.

[handwritten margin note: Induced EMF — changing B — changing the area of the loop or its orientation θ w/ respect to the field.]

◆ EXAMPLE 30–1 A square coil of side 5.0 cm contains 100 loops and is positioned perpendicular to a uniform 0.60-T magnetic field. It is quickly and uniformly pulled from the field (moving perpendicularly to **B**) to a region where B drops abruptly to zero. (We assume that initially one side of the square is at the boundary of the field.) It takes 0.10 s for the whole coil to reach the field-free region. How much energy is dissipated in the coil if its resistance is 100 Ω?

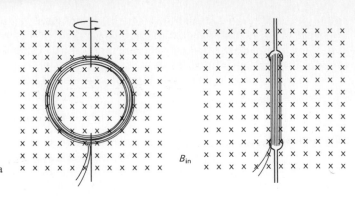

FIGURE 30–3

A current can be induced by rotating a coil in a magnetic field.

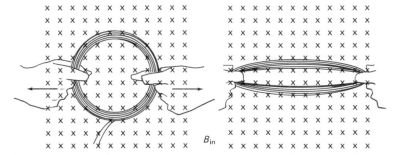

FIGURE 30–4

A current can be induced by changing the area of the coil. In both this case and that of Fig. 30–3, the flux through the coil is reduced.

SOLUTION The area of the coil is $A = (0.050 \text{ m})^2 = 2.5 \times 10^{-3} \text{ m}^2$. The flux is initially $\Phi_B = BA = (0.60 \text{ T})(2.5 \times 10^{-3} \text{ m}^2) = 1.5 \times 10^{-3} \text{ Wb}$. After 0.10 s the flux is zero. The rate of change of flux is constant during the 0.10 s, so the emf induced during this period is

$$\mathscr{E} = -(100) \frac{(0 - 1.5 \times 10^{-3} \text{ Wb})}{(0.10 \text{ s})} = 1.5 \text{ V}.$$

The current $I = \mathscr{E}/R = 1.5 \text{ V}/100 \, \Omega = 15 \text{ mA}$. The total energy dissipated is $I^2Rt = (1.5 \times 10^{-2} \text{ A})^2(100 \, \Omega)(0.10 \text{ s}) = 2.3 \times 10^{-3} \text{ J}$. From the conservation of energy principle, this is just equal to the work needed to pull the coil out of the field.

30–3 Emf Induced in a Moving Conductor

Another way to induce an emf is shown in Fig. 30–5 and this situation helps illuminate the nature of the induced emf. Assume that a uniform magnetic field **B** is perpendicular to the area bounded by the U-shaped conductor and the movable

FIGURE 30–5

A conducting rod is moved to the right on a U-shaped conductor in a uniform magnetic field **B** that points out of the paper.

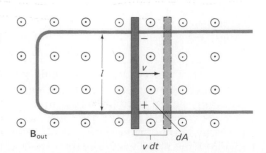

rod is resting on it. If the rod moves at a speed v, it travels a distance $dx = v\,dt$ in a time dt. Therefore the area of the loop increases by an amount $dA = lv\,dt$ in a time dt. By Faraday's law, there is an induced emf $\mathscr{E}$ whose magnitude is given by

$$\mathscr{E} = \frac{d\Phi}{dt} = \frac{B\,dA}{dt} = \frac{Blv\,dt}{dt} = Blv. \qquad (30\text{--}3)$$

This equation is valid as long as B, l, and v are mutually perpendicular. (If they are not, we use only the components of each that are mutually perpendicular.)

We can obtain this same relation in another way without the use of Faraday's law. We saw in Chapter 28 that a charged particle moving in a magnetic field $\mathbf{B}$ with velocity $\mathbf{v}$ experiences a force $\mathbf{F} = q\mathbf{v} \times \mathbf{B}$. When the rod of Fig. 30–5 moves to the right with speed v, the electrons in the rod move to the right with this same speed. Therefore, since $\mathbf{v} \perp \mathbf{B}$, each feels a force $F = qvB$ which acts upward in the figure.[†] If the rod were not in contact with the U-shaped conductor, electrons would collect at the upper end of the rod, leaving the lower end positive. If the rod does slide on the U-shaped conductor, the electrons will flow into it; there will thus be a clockwise (conventional) current flowing in the loop. To calculate the emf, we determine the work W needed to move a charge q from one end of the rod to the other: $W = \text{force} \times \text{distance} = (qvB)(l)$. The emf equals the work done per unit charge, so $\mathscr{E} = qvBl/q = Blv$, just as above.[‡]

 EXAMPLE 30–2   An airplane travels 1000 km/h in a region where the earth's field is 5.0×10^{-5} T and is nearly vertical. What is the potential difference induced between the wing tips which are 70 m apart?

SOLUTION   Since $v = 1000$ km/h $= 280$ m/s, we have $\mathscr{E} = Blv = (5.0 \times 10^{-5}\ \text{T})(70\ \text{m})(280\ \text{m/s}) = 1.0$ V. Not much to worry about.

30–4   A Changing Magnetic Flux Produces an Electric Field

In our microscopic view of electric current, we have seen (Chapter 26) that when an electric current flows in a conductor, there is an electric field in the conductor that causes the electrons to acquire a drift velocity, $\mathbf{v}_d$. In the case of the conductor moving in a magnetic field which we just discussed (Fig. 30–5), there is an induced current in the circuit, and this implies there is an electric field in the conductor. Within the moving rod of Fig. 30–5, the effective electric field E would be equal to the force per unit charge, $E = F/q$. Since $F = qvB$, then

$$E = \frac{qvB}{q} = vB. \qquad (30\text{--}4a)$$

This can be generalized to situations in which $\mathbf{v}$ and $\mathbf{B}$ are not perpendicular. From Eq. 28–5, the force on a charge q moving with velocity $\mathbf{v}$ in a magnetic field $\mathbf{B}$ is $\mathbf{F} = q\mathbf{v} \times \mathbf{B}$. Hence the effective electric field ($\mathbf{E} = \mathbf{F}/q$) would be

$$\mathbf{E} = \mathbf{v} \times \mathbf{B}. \qquad (30\text{--}4b)$$

Conductor moving in magnetic field

[†] Once a current is flowing, the electrons have a drift velocity $\mathbf{v}_d$, which is upward in Fig. 30–5. The force $\mathbf{F}_d = q\mathbf{v}_d \times \mathbf{B}$ on the electrons due to this component of their velocity acts to the left and thus contributes nothing to the induced emf.

[‡] This argument, which is basically the same as for the Hall effect, explains this one type of induced emf. It doesn't explain the general case, however.

In the situation in which a changing magnetic field (rather than a moving conductor) induces an emf (as, for example, in Fig. 30–2), there also is an induced current. And again this implies that there is an electric field in the wire. Thus we come to the important conclusion that *a changing magnetic flux produces an electric field*. This applies not only to wires and other conductors; it is actually a general result that applies to any region in space. Indeed, an electric field will be produced at any point in space where there is a changing magnetic field.

We can put these ideas into mathematical form by generalizing our relation between electric field and the potential difference between two points *a* and *b*: $V_{ab} = \int_a^b \mathbf{E} \cdot d\mathbf{l}$ (Eq. 24–4) where $d\mathbf{l}$ is an element of displacement along the path of integration. The emf $\mathscr{E}$ induced in a circuit is equal to the work done per unit charge by the electric field, which equals the integral of $\mathbf{E} \cdot d\mathbf{l}$ around the closed path. Thus

$$\mathscr{E} = \oint \mathbf{E} \cdot d\mathbf{l}. \qquad (30\text{–}5)$$

We combine this with Faraday's law, Eq. 30–2a, to obtain

$$\oint \mathbf{E} \cdot d\mathbf{l} = -\frac{d\Phi_B}{dt} \qquad (30\text{–}6)$$

which relates the changing magnetic flux to the electric field it produces. The integral on the left is taken around a path enclosing the area through which the magnetic flux Φ_B is changing. Equation 30–6 is a more elegant statement of Faraday's law. It is valid not only in conductors, but in any region of space. To illustrate this, let us take an example.

EXAMPLE 30–3 A magnetic field **B** between the pole faces of an electromagnet is nearly uniform at any instant over a circular area of radius *R* as shown in Fig. 30–6a. The current in the windings of the electromagnet is increasing in time so that **B** changes in time at a constant rate $d\mathbf{B}/dt$ at each point. Beyond the circular region ($r > R$), we assume **B** = 0 at all times. Determine the electric field **E** at any point *P* a distance *r* from the center of the circular area.

SOLUTION The changing magnetic flux through a circle of radius *r*, shown dashed in Fig. 30–6a, will produce an emf around this loop. Because all points on the dashed circle are equivalent physically, the electric field too must show this symmetry. Thus we can expect **E** to be perpendicular to **B** and to be tangent

FIGURE 30–6

Example 30–3. (a) Determining electric field **E** at point *P*. (b) Lines of **E** produced by increasing **B** (pointing outward).

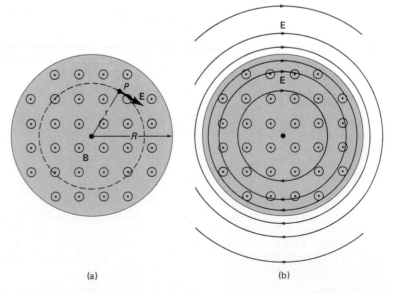

(a)   (b)

to the circle of radius r, as shown, and to have the same magnitude at all points on the circle of radius r. We therefore take this circle as our path of integration in Eq. 30–6 and obtain

$$E(2\pi r) = -(\pi r^2)\frac{dB}{dt} \qquad [r < R]$$

since $\Phi_B = BA = B(\pi r^2)$ at any instant. We solve for E and obtain

$$E = \frac{r}{2}\frac{dB}{dt}. \qquad [r < R]$$

This is valid up to the edge of the circle ($r \le R$) beyond which $\mathbf{B} = 0$. If we now consider a point P where $r > R$, the flux through a circle of radius r is $\Phi_B = \pi R^2$. Hence

$$E(2\pi r) = \pi R^2 \frac{dB}{dt} \qquad [r > R]$$

or

$$E = \frac{R^2}{2r}\frac{dB}{dt}. \qquad [r > R]$$

Thus the magnitude of the electric field increases linearly from zero at the center of the field to $B = (dB/dt)(R/2)$ at the edge, and then decreases inversely with distance in the region beyond the edge of the field. The field lines are circles as shown in Fig. 30–6b.

This example shows an important difference between electric fields produced by changing magnetic fields and electric fields produced by electric charges at rest (electrostatic fields). Electric field lines produced in the electrostatic case (Chapters 22–25) start and stop on electric charges. But the electric field lines produced by a changing magnetic field are continuous; they form closed loops. This distinction goes even further and is an important one. In the electrostatic case, the potential difference between two points is given by

$$V_{ab} = V_a - V_b = \int_a^b \mathbf{E} \cdot d\mathbf{l}.$$

If the integral is around a closed loop, so points a and b are the same, then $V_{ab} = 0$. Hence the integral of $\mathbf{E} \cdot d\mathbf{l}$ around a closed path is zero:

$$\oint \mathbf{E} \cdot d\mathbf{l} = 0. \qquad \text{[electrostatic field]}$$

This, of course, followed from the fact that the electrostatic force (Coulomb's law) is a conservative force, and so a potential energy function could be defined. Indeed, the relation above, $\oint \mathbf{E} \cdot d\mathbf{l} = 0$, tells us that the work done per unit charge around any closed path is zero (or the work done between any two points is independent of path—see Section 7–1), which is a property only of a conservative force. But in the nonelectrostatic case, when the electric field is produced by a changing magnetic field, the integral around a closed path is *not* zero, but is given by Eq. 30–6:

$$\oint \mathbf{E} \cdot d\mathbf{l} = -\frac{d\Phi_B}{dt}.$$

We thus come to the conclusion that the forces due to changing magnetic fields are *nonconservative*. We are not able therefore to define a potential energy, or potential function, in the nonelectrostatic case. Although static electric fields are *conservative fields*, the electric field produced by a changing magnetic field is a *nonconservative* field.

30-5   The Electric Generator

Probably the most important practical result of Faraday's great discovery was the development of the *electric generator* or *dynamo*. A generator transforms mechanical energy into electric energy. This is just the opposite of what a motor does. Indeed, a generator is basically the inverse of a motor.[†] A simplified diagram of an ac generator is shown in Fig. 30-7; a generator consists of many coils of wire (only one is shown) wound on an armature that can rotate in a magnetic field. The axle is turned by some mechanical means and an emf is induced in the rotating coil. An electric current is thus the *output* of a generator. In Fig. 30-7, the right-hand rule ($\mathbf{F} = q\mathbf{v} \times \mathbf{B}$) tells us that the (conventional) current in the wire labeled N on the armature is outward; therefore it is outward at brush N. (The brushes press against continuous slip-rings.) After one-half revolution, wire N will be where wire M is now in the drawing and the current then at brush N will be inward. Thus the current produced is alternating. Let us look at this in more detail.

In Fig. 30-8, the loop is being made to rotate clockwise in a uniform magnetic field $\mathbf{B}$ with constant angular velocity ω. From Faraday's law, the induced emf is

$$\mathscr{E} = -\frac{d\Phi_B}{dt} = -\frac{d}{dt}\int \mathbf{B} \cdot d\mathbf{A} = -\frac{d}{dt}\left[BA\cos\theta\right]$$

where A is the area of the loop and θ is the angle between $\mathbf{B}$ and $\mathbf{A}$. Since $\omega = d\theta/dt$, then $\theta = \theta_0 + \omega t$. We arbitrarily take $\theta_0 = 0$, so

$$\mathscr{E} = BA\frac{d}{dt}(\cos\omega t) = BA\omega\sin\omega t.$$

If the rotating coil contains N loops,

$$\mathscr{E} = NBA\,\omega\sin\omega t$$
$$= \mathscr{E}_0\sin\omega t. \tag{30-7}$$

Thus the output emf is sinusoidal with amplitude $\mathscr{E}_0 = NBA\omega$. Such a rotating coil in a magnetic field is the basic operating principle of an ac generator.

Over 99 percent of the electricity used in the United States is produced from generators. The frequency $f = \omega/2\pi$ is 60 Hz for general use in the United States and Canada, although 50 Hz is used in some countries. In plants that generate electric power, the armature is mounted on a heavy axle connected to a turbine, which is the modern equivalent of a waterwheel. Falling water from behind a dam can turn the turbine at a hydroelectric plant. Most of the power generation at present in the United States, however, is done at steam plants where the burning of fossil fuels (coal, oil, natural gas) boils water to produce high-pressure steam that turns the turbines. Thus, a heat engine (Chapter 21) connected to a generator is the principal means of producing electric power today.

EXAMPLE 30-4   The armature of a 60-Hz ac generator rotates in a 0.15-T magnetic field. If the area of the coil is 2.0×10^{-2} m², how many loops must the coil contain if the peak output is to be $\mathscr{E}_0 = 170$ V?

SOLUTION   From Eq. 30-7 we see that the maximum emf is $\mathscr{E}_0 = NAB\omega$. Since $\omega = 2\pi f = (6.28)(60\ \text{s}^{-1}) = 377\ \text{s}^{-1}$, we have

$$N = \frac{\mathscr{E}_0}{AB\omega} = \frac{170\ \text{V}}{(2.0 \times 10^{-2}\ \text{m}^2)(0.15\ \text{T})(377\ \text{s}^{-1})} = 150\ \text{turns}.$$

[†] You can, for example, actually run a car generator backward as a motor by connecting its output terminals to a battery.

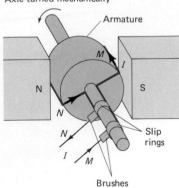

Axle turned mechanically

Armature

Slip rings

Brushes

FIGURE 30-7

Ac generator.

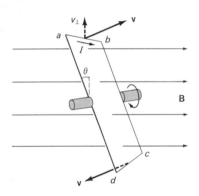

FIGURE 30-8

The emf is induced in the segments *ab* and *cd* whose velocity components perpendicular to field $\mathbf{B}$ equal $v\sin\theta$.

A dc generator is much like an ac generator, except the slip rings are replaced by split-ring commutators, Fig. 30–9a, just as in a dc motor. The output of such a generator is as shown and can be smoothed out by placing a capacitor in parallel with the output. More common is the use of many armature windings as in Fig. 30–9b which produces a smoother output.

In the past, automobiles used dc generators. More common now, however, are ac generators or *alternators* (Fig. 30–10). The ac output is changed to dc for charging the battery with the use of diodes (Chapter 41). Most alternators differ from the generator discussed above in that the magnetic field is made to rotate within a stationary armature. The brushes press against a continuous ring instead of a slotted commutator and are on the input instead of the output; so large output currents are carried in solid conductors rather than through sliding-ring commutators which are subject to wear and arcing. In a car, the rotation speed can be faster so the battery can be charged even at idling speed and there will still be no problem with electrical arcing across the rings when the car travels at high speeds.

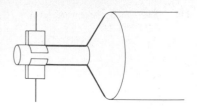
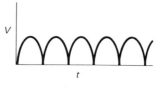
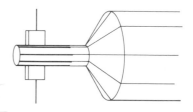

(a)

(b)

 **30–6   Counter Emf and Torque;
Eddy Currents**

A motor turns and produces mechanical energy when a current is made to flow in it. From our description in Sec. 28–11, you might expect that the armature would accelerate indefinitely due to the torque on it. However, as the armature of a motor turns the magnetic flux through the coil changes and an emf is generated. This induced emf acts to oppose the motion (Lenz's law) and is called the *back*, or *counter*, *emf*. The greater the speed of the motor, the greater the counter emf. A motor normally turns something, but if there were no load, its speed would increase until the counter emf equaled the input voltage. In the normal situation, when there is a mechanical load, the speed of the motor is limited also by the load. The counter emf will then be less than the external voltage. The greater the load, the slower the motor rotates and the lower is the counter emf.

EXAMPLE 30–5 The armature windings of a dc motor have a resistance of 5.0 Ω. The motor is connected to a 120-V line and when the motor reaches full speed against its normal load the counter emf is 108-V. Calculate (*a*) the current into the motor when it is just starting up; (*b*) the current when it reaches full speed.

FIGURE 30–9

Dc generator (a) with one set of commutators; (b) with many sets of commutators and windings.

FIGURE 30–10

Simplified diagram of an alternator. Electromagnet current connected through continuous slip rings is preferable to output of dc generator through many sets of split commutators (Fig. 30–9b), where poor connection or arcing can occur at high speeds. Sometimes the electromagnet is replaced by a permanent magnet.

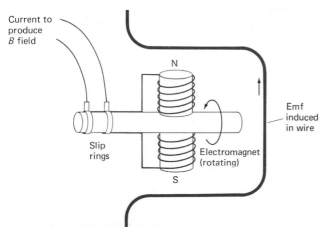

Current to produce B field

N

Slip rings

Electromagnet (rotating)

S

Emf induced in wire

+

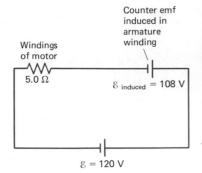

FIGURE 30–11

Circuit of a motor showing induced counter emf.

SOLUTION (a) Initially, the motor is not turning (or turning very slowly) so there is no induced counter emf. Hence from Ohm's law the current is

$$I = \frac{V}{R} = \frac{120 \text{ V}}{5.0 \ \Omega} = 24 \text{A}.$$

(b) At full speed, the counter emf is a source of emf that opposes the exterior source. We represent this counter emf as a battery in the equivalent circuit shown in Fig. 30–11. In this case, Ohm's law (or Kirchhoff's rule) gives

$$120 \text{ V} - 108 \text{ V} = I(5.0 \ \Omega).$$

Therefore

$$I = 12 \text{ V}/5.0 \ \Omega = 2.4 \text{ A}.$$

This example illustrates the fact that the current is very high when a motor first starts up. This is why the lights in your house may dim when the motor of the refrigerator (or other large motor) starts up. The large initial current causes the voltage at the outlets to drop (the house wiring has resistance so there is some voltage drop across it when large currents are drawn). If a motor is overloaded so it turns only at a slow speed, the counter emf will be reduced because of the reduced rotation speed. The current in the motor may then be sufficiently large to burn it out.

In a generator, the situation is the reverse of that for a motor. As we saw, the mechanical turning of the armature induces an emf in the loops which is the output. If the generator is not connected to an external circuit, the emf exists at the terminals but no current flows. In this case, it takes little effort to turn the armature. But if the generator *is* connected to a device that draws current, then a current flows in the coils of the armature. Because this current-carrying coil is in a magnetic field there will be a torque exerted on it (as in a motor), and this torque opposes the motion (use the right-hand rule for the force on a wire, in Fig. 30–7). This is called a *counter torque.* The greater the load—that is, the more current that is drawn—the greater will be the counter torque. Hence the external applied torque will have to be greater to keep the generator turning. This of course makes sense from the conservation of energy principle. More mechanical-energy input is needed to produce more electrical-energy output.

Induced currents are not always confined to well-defined paths such as in wires. Consider, for example, the rotating metal wheel in Fig. 30–12a. A magnetic field is applied to a limited area as shown and points into the paper. The section of wheel in the magnetic field has an emf induced in it since the conductor is moving (carrying electrons with it). The flow of (conventional) current is upward (Fig. 30–12b) and it follows a downward return path outside the region of the magnetic field. These currents are referred to as *eddy currents* and can be present in any conductor which is moving across a magnetic field or through which the magnetic flux is changing. In Fig. 30–12, the magnetic field exerts a force on the induced currents that opposes (use the right-hand rule) the rotational motion. Eddy currents can be used in this way as a smooth braking device on, say, a rapid-transit car. In order to stop the car, an electromagnet can be turned on that applies its field either to the wheels or to the moving steel rail below. Eddy currents can dampen (reduce) the oscillation of a vibrating system. A common example is in a galvanometer where induced eddy currents keep the needle from overshooting or oscillating violently. Eddy currents, however, can be a problem. For example, eddy currents induced in the armature of a motor or generator produce heat ($P = I\mathscr{E}$) and waste energy. To reduce the eddy currents the armatures are *laminated*; that is, they are made of very thin sheets of iron that are well insulated from one another. Thus the total path length of the eddy currents is confined to each slab which increases the total resistance; hence the current is less and there is less wasted energy.

FIGURE 30–12

Production of eddy currents in a rotating wheel.

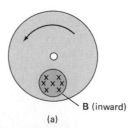

(a)

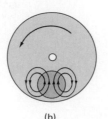

(b)

30–7 Transformers; Transmission of Power

A transformer is a device for increasing or decreasing an ac voltage. A transformer consists of two coils of wire known as the *primary* and *secondary* coils. The two coils can be interwoven; or they can be linked by a soft iron core (laminated to prevent eddy current losses), Fig. 30–13. The idea in either case is that the magnetic flux produced by a current in the primary should pass through the secondary coil. Transformers are designed so that (nearly) all the flux that passes through the primary also passes through the secondary coil, and we assume this is true in what follows. We also assume that energy losses in the resistance of the coils and hysteresis in the iron can be ignored—a good approximation for real transformers which are typically 99 percent efficient.

When an ac voltage is applied to the primary, the changing magnetic field it produces will induce an ac voltage of the same frequency in the secondary. However, the voltage will be different according to the number of loops in each coil. From Faraday's law, the voltage or emf induced in the secondary is

$$V_s = N_s \frac{d\Phi_B}{dt}$$

where N_s is the number of turns in the secondary coil, and $d\Phi_B/dt$ is the rate at which the magnetic flux changes. The input primary voltage, V_p, is also related to the rate at which the flux changes[†]

$$V_p = N_p \frac{d\Phi_B}{dt}$$

where N_p is the number of turns in the primary coil. We divide these two equations to find

$$\frac{V_s}{V_p} = \frac{N_s}{N_p}. \qquad (30\text{–}8)$$

This *transformer equation* tells how the secondary (output) voltage is related to the primary (input) voltage; V_s and V_p in Eq. 30–8 can be the rms values for both or peak values for both.

If N_s is greater than N_p, we have a *step-up transformer*. The secondary voltage is greater than the primary voltage. For example, if the secondary has twice as many turns as the primary, then the secondary voltage will be twice that of the primary. If N_s is less than N_p, we have a *step-down transformer*.

Although voltage can be increased (or decreased) with a transformer, we don't get something for nothing. Energy conservation tells us that the power output can be no greater than the power input. A well-designed transformer can be greater than 99 percent efficient so little energy is lost to heat. The power input thus essentially equals the power output; since power $P = VI$, we have

$$V_p I_p = V_s I_s$$

or

$$\frac{I_s}{I_p} = \frac{N_p}{N_s}. \qquad (30\text{–}9)$$

EXAMPLE 30–6 A transformer for a transistor radio reduces 120-V ac to 9.0-V ac. (Such a device also contains diodes to change the 9.0-V ac to dc. See Chapter 41.) The secondary contains 30 turns and the radio draws 400 mA. Calculate (*a*) the number of turns in the primary; (*b*) the current in the primary; (*c*) the power transformed.

[†] This follows from the fact that the changing flux produces a counter emf, $N_p d\Phi_B/dt$, in the primary that exactly balances the applied voltage V_p if the resistance of the primary can be ignored (Kirchhoff's rules).

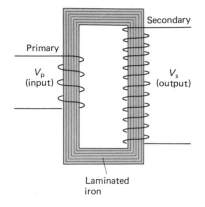

FIGURE 30–13

Step-up transformer ($N_p = 4$, $N_s = 12$).

SOLUTION   (*a*) This is a step-down transformer, and from Eq. 30–8 we have

$$N_p = N_s \frac{V_p}{V_s} = (30)(120 \text{ V})/(9.0 \text{ V}) = 400 \text{ turns.}$$

(*b*) From Eq. 30–9:

$$I_p = I_s \frac{N_s}{N_p} = (0.40 \text{ A})\left(\frac{30}{400}\right) = 0.030 \text{ A.}$$

(*c*) The power transformed is

$$P = I_s V_s = (9.0 \text{ V})(0.40 \text{ A}) = 3.6 \text{ W}$$

which is, assuming 100 percent efficiency, the same as the power in the primary, $P = (120 \text{ V})(0.030 \text{ A}) = 3.6 \text{ W.}$

It is important to recognize that a transformer operates only on ac. A dc current in the primary does not produce a changing flux and therefore induces no emf in the secondary. However, if a dc voltage is applied to the primary through a switch, at the instant the switch is opened or closed there will be an induced current in the secondary. For example, if the dc is turned on and off as shown in Fig. 30–14a, the voltage induced in the secondary is as shown in (*b*). Notice that the secondary voltage drops to zero when the dc voltage is steady.

Transformers play an important role in the transmission of electricity. Power plants are often situated some distance from metropolitan areas; hydroelectric plants are located at a dam site and nuclear plants need much cooling water; fossil-fuel plants too are often situated far from a city because of lack of availability of land or to avoid contributing to air pollution. In any case, electricity must often be transmitted over long distances, and there is always some power loss in the transmission lines. This loss can be minimized if the power is transmitted at high voltage, as the following example shows.

EXAMPLE 30–7   An average of 120 kW of electric power is supplied to a small town from a power plant 10 km away. The transmission lines have a total resistance of 0.40 Ω. Calculate the power loss if the power is transmitted at (*a*) 240 V and (*b*) 24,000 V.

SOLUTION   (*a*) If 120 kW is supplied at 240 V, the total current will be

$$I = \frac{1.2 \times 10^5 \text{ W}}{2.4 \times 10^2 \text{ V}} = 500 \text{ A}$$

The power loss in the lines, P_L, is then

$$P_L = I^2 R = (500 \text{ A})^2 (0.40 \text{ Ω}) = 100 \text{ kW.}$$

Thus over 80 percent of all the power would be wasted as heat in the power lines!
(*b*) When $V = 24,000$ V

$$I = \frac{P}{V} = \frac{1.2 \times 10^5 \text{ W}}{2.4 \times 10^4 \text{ V}} = 5.0 \text{ A.}$$

The power loss is then

$$P_L = I^2 R = (5.0 \text{A})^2 (0.40 \text{ Ω}) = 10 \text{ W,}$$

which is less than 1 percent.

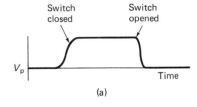

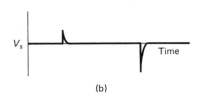

FIGURE 30–14
Dc voltage turned on and off as shown in (a) produces voltage pulses in the secondary (b). Voltage scales in (a) and (b) are not necessarily the same.

It should be clear that the greater the voltage, the less the current and thus the less power is wasted in the transmission lines. It is for this reason that power is usually transmitted at very high voltages, as high as 700 kV.

Power is generated at somewhat lower voltages than this, and the voltages in homes and factories is also much lower. The great advantage of ac, and a major reason it is in nearly universal use, is that the voltage can easily be stepped up and down by a transformer. The output voltage of an electric generating plant is stepped up prior to transmission. Upon arrival in a city, it is stepped down in stages at electric substations prior to distribution. The voltage in lines along city streets is typically 2400 V and is stepped down to 240 V or 120 V for home use by transformers.

Dc transmission has gained in popularity recently. Although changing voltage with dc is more difficult and expensive, it offers some advantages over ac. A few of these are as follows. Ac produces alternating magnetic fields which induce current in nearby wires and reduces transmitted power; this is absent in dc. Dc can be transmitted at a higher average voltage than ac since for dc the rms value equals the peak; and breakdown of insulation or of air must be avoided for the peak voltage.

* 30-8   Magnetic Microphone and Phonograph Cartridges

Many microphones work on the principle of induction. In one form, a microphone is just the inverse of a loudspeaker (Section 28–11). A small coil connected to a membrane is suspended close to a small permanent magnet. The coil moves in the magnetic field when sound waves strike the membrane; the frequency of the induced emf will be just that of the impinging sound waves. In a "ribbon" microphone, a thin metal ribbon is suspended between the poles of a permanent magnet. The ribbon vibrates in response to sound waves and the emf induced in the ribbon is proportional to its velocity.

In one type of phonograph cartridge, the needle which follows the grooves of a phonograph record is connected to a tiny magnet inside the cartridge in the arm of the record player. The magnet is suspended inside a tiny coil of wire and an emf is induced in the coil according to the motion of the magnet. This tiny emf, at frequencies corresponding to the wavy shape of the record grooves, is amplified electronically before sending it to a loudspeaker. Most good-quality cartridges are of the magnetic type. Others, called "crystal pickups," make use of the piezoelectric effect (see Section 27–10).

Microphones and cartridges are two more types of transducer. Another is the head of a tape recorder. Recorded tape has been magnetized at different intensities along its length corresponding to the program material. As tape passes the head, a tiny emf is induced in an induction coil due to the changing magnetic field passing the head. This signal too is amplified before going to the speaker.

Summary

If the magnetic flux through a loop of wire changes in time, an emf is induced in the loop; the magnitude of the induced emf in a loop equals the time rate of change of the magnetic flux $\Phi_B \ (= \int \mathbf{B} \cdot d\mathbf{A})$ through the loop:

$$\mathscr{E} = -\frac{d\Phi_B}{dt}.$$

This result is called *Faraday's law of induction*. The induced emf produces a current whose magnetic field opposes the original change in flux (*Lenz's law*). It also follows from Faraday's law that a straight wire of length l moving with speed v perpendicular to a magnetic field of strength B has an emf induced between its ends equal to Blv, and that a changing magnetic field produces an electric field. The general relation in the latter case is:

$$\oint \mathbf{E} \cdot d\mathbf{l} = -\frac{d\Phi_B}{dt}$$

where the integral on the left is around the loop through which the magnetic flux Φ_B is changing.

An electric generator changes mechanical energy into electrical energy. Its operation is based on Faraday's law:

a coil of wire is made to rotate by mechanical means in a uniform magnetic field, and the changing flux through the coil induces a sinusoidal current, which is the output of the generator.

A transformer, which is a device to change the magnitude of an ac voltage, consists of a primary and a secondary coil. The changing flux due to a changing voltage in the primary induces a voltage in the secondary. In a 100 percent efficient transformer, the ratio of output to input voltages equals the ratio of the number of turns in the secondary to the number in the primary: $V_s/V_p = N_s/N_p$. The ratio of secondary to primary current is in the inverse ratio of turns: $I_s/I_p = N_p/N_s$.

Questions

1. What would be the advantage, in Faraday's experiments (Fig. 30–1), of using coils with many turns?

2. What is the difference between magnetic flux and magnetic field strength? Discuss in detail.

3. Suppose you are holding a circular piece of wire and suddenly thrust a magnet, south pole first, toward the center of the circle; is a current induced in the wire? Is a current induced when the magnet is held steady within the loop? Is a current induced when you withdraw the magnet? In each case, if your answer is yes, specify the direction.

4. When you detect a current in a wire, can you tell if it is an induced current? That is, do induced currents differ in any way from ordinary currents produced, say, by a battery? Do induced emf's differ from the emf produced by a battery?

5. Suppose you are looking along a line through the centers of two circular (but separate) wire loops, one behind the other. A battery is suddenly connected to the front loop, establishing a clockwise current. Will a current be induced in the second loop? If so, when does this current start? When does it stop? In what direction is this current?

6. Is there a force between the two loops discussed in question 5? If so, in what direction?

7. The battery discussed in question 5 is disconnected. Will a current be induced in the second loop? If so, when does it start and stop? In what direction is this current?

8. If the solenoid in Fig. 30–15 is being pulled away from the loop shown, in what direction is the induced current in the part of the loop closest to the viewer?

9. A region where no magnetic field is desired is surrounded by a sheet of low-resistivity metal. (a) Will this sheet shield the interior from a rapidly changing magnetic field outside? Explain. (b) Will it act as a shield to a static magnetic field? (c) What if the sheet is superconducting (resistivity = 0)?

10. If the resistance of the resistor in Fig. 30–16 is slowly increased, what is the direction of the current induced in the small circular loop inside the larger loop?

11. Show, using Lenz's law, that the emf induced in the moving rod in Fig. 30–5 is positive at the bottom and negative at

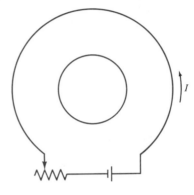

FIGURE 30–16

the top so that the current flows clockwise in the circuit loop on the left.

12. In what direction will the current flow in Fig. 30–5 if the rod moves to the left, which decreases the area of the loop to the left?

13. The rectangular loop shown in Fig. 30–17 is being pulled to the left, out of the magnetic field which points inward as shown. In what direction is the induced current?

14. Is a force needed to move the conducting rod in Fig. 30–5 at constant speed? Explain.

15. A car generator produces 12 V when the armature turns at 600 rev/min. What will be its output at 2000 rev/min assuming nothing else changes?

16. In some early automobiles, the starter motor doubled as a generator to keep the battery charged once the car was started. Explain how this might work.

17. Explain why, exactly, the lights may dim briefly when a refrigerator motor starts up. (Hint: consider "terminal voltage.")

18. When a refrigerator starts up, the lights in older houses may dim briefly. When an electric heater is turned on, the lights may stay dimmed as long as it is on. Explain the difference.

19. Explain what is meant by the statement "A motor acts as

FIGURE 30–15

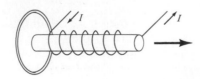

FIGURE 30–17

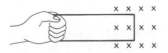

a motor and generator at the same time." Can the same be said for a generator?

20. Use Fig. 30–7 and a right-hand rule to show why the counter torque in a generator *opposes* the motion.

21. Will an eddy current brake (Fig. 30–12) work on a copper or aluminum wheel, or must it be ferromagnetic?

22. It has been proposed that eddy currents be used to help sort solid waste for recycling. The waste is first ground into tiny pieces and iron removed with a dc magnet. The waste then is allowed to slide down an incline over permanent magnets. How will this aid in the separation of nonferrous metals (aluminum, copper, lead, brass) from nonmetallic materials?

23. If an aluminum sheet is held between the poles of a large bar magnet, it requires some force to pull it out of the magnetic field even though the sheet is not ferromagnetic and does not touch the pole faces. Explain.

24. The pivoted metal bar with slots in Fig. 30–18 falls much more quickly through a magnetic field than does a solid bar. Why?

25. A bar magnet falling inside a vertical metal tube reaches a terminal velocity even if the tube is evacuated so that there is no air resistance. Explain.

26. A metal bar, pivoted at one end, oscillates freely in the absence of a magnetic field; but in a magnetic field its os-

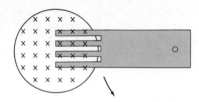

FIGURE 30–18

cillations are quickly damped out. Explain. (This *magnetic damping* is used in a number of practical devices.)

27. An enclosed transformer has four wire leads coming from it. How could you determine the ratio of turns on the two coils without taking the transformer apart? How would you know which wires paired with which?

28. The use of larger voltage lines in homes, say 600 V or 1200 V, would reduce energy waste. Why are they not used?

*29. Since a magnetic microphone is basically like a loudspeaker, could a loudspeaker actually serve as a microphone? That is, could you speak into a loudspeaker and obtain an output signal that could be amplified? Explain. Discuss, in light of your response, how a microphone and loudspeaker differ in construction.

 Problems

SECTION 30–2

1. (I) A 10-cm-diameter circular loop of wire is perpendicular to a uniform 0.35-T magnetic field. It is removed from the field in 0.12 s. What is the average induced emf?

2. (I) The magnetic field perpendicular to a circular loop of wire 22 cm in diameter is changed from -0.40 T to $+0.55$ T in 80 ms, where $+$ means the field points away from an observer and $-$ toward the observer. (a) Calculate the induced emf; (b) in what direction does the induced current flow?

3. (I) The magnetic flux through a coil of wire containing two loops changes from -8.6 Wb to $+4.7$ Wb in 0.74 s. What is the emf induced in the coil?

4. (II) A square loop of wire 21.0 cm on a side is rotated uniformly $360°$ in a magnetic field B in 45 ms. If the average induced emf is 180 mV, what is the average value of B?

5. (II) A 20-cm-diameter circular loop of wire has a resistance of 8.5 Ω. It is initially in a 0.40-T magnetic field, with its plane perpendicular to **B**, but is removed from the field in 100 ms. Calculate the electric energy dissipated in this process.

6. (II) The magnetic field perpendicular to a single 12.0-cm-diameter circular loop of copper wire decreases uniformly from 0.350 T to zero. If the wire is 1.25 mm in diameter, how much charge moves through the coil during this operation?

7. (II) The magnetic flux through each loop of a 35-loop coil is given by $(3.6 t - 0.71 t^3) \times 10^{-2}$ T·m², where the time

t is in seconds. (a) Determine the emf $\mathscr{E}$ as a function of time. (b) What is $\mathscr{E}$ at $t = 1.0$ s and $t = 5.0$s?

8. (II) Apply Faraday's law, in the form of Eq. 30–6, to show that the static electric field between the plates of a parallel-plate capacitor cannot drop abruptly to zero at the edges, but must, in fact, fringe. Use the path shown dashed in Fig. 30–19.

9. (II) A 25.0-cm-diameter coil consists of 20 turns of circular copper wire 2.0 mm in diameter. A uniform magnetic field, perpendicular to the plane of the coil, changes at a rate of 6.55×10^{-3} T/s. Determine (a) the current in the loop, and (b) the rate at which thermal energy is produced.

FIGURE 30–19

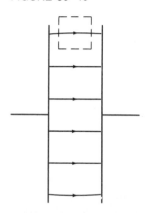

10. (II) A single circular loop of wire is placed inside a long solenoid with its plane perpendicular to the axis of the solenoid. The area of the loop is A_1 and that of the solenoid, which has n turns per unit length, is A_2. A current $I = I_0 \cos \omega t$ flows in the solenoid turns. What is the induced emf in the small loop?

11. (II) The area of an elastic circular loop decreases at a constant rate, $dA/dt = 6.50 \times 10^{-2} \text{ m}^2/\text{s}$. The loop is in a magnetic field $B = 0.42$ T whose direction is perpendicular to the plane of the loop. At $t = 0$, the loop has area $A = 0.285 \text{ m}^2$. Determine the induced emf at $t = 0$, and at $t = 2.00$ s.

12. (II) Suppose the radius of the elastic loop in problem 11 increases at a constant rate, $dr/dt = 46.0$ cm/s. Determine the emf induced in the loop at $t = 0$ and at $t = 2.00$ s.

13. (II) A *search coil for measuring B* (also called a *flip coil*) is a small coil with N turns, each of cross-sectional area A. It is connected to a so-called *ballistic galvanometer*, which is a device to measure the total charge Q that passes through it in a short time. The flip coil is placed in the magnetic field to be measured with its face perpendicular to the field. It is then quickly rotated $180°$. Show that the total charge Q that flows in the induced current during this short "flip" time is proportional to the magnetic field strength B; in particular, show that B is given by

$$B = \frac{QR}{2NA}$$

where R is the total resistance of the circuit including that of the coil and the ballistic galvanometer which measures the charge Q.

SECTION 30–3

14. (I) The moving rod in Fig. 30–5 is 22.0 cm long and moves with a speed of 35.0 cm/s. If the magnetic field is 1.15 T, calculate the emf developed.

15. (I) In Fig. 30–5 the rod moves with a speed of 2.3 m/s, is 34.0 cm long, and has negligible resistance. The magnetic field is 0.25 T and the resistance of the U-shaped conductor is 25.0 Ω at a given instant. Calculate (a) the emf induced; (b) the current flowing in the U-shaped conductor.

16. (II) If the U-shaped conductor in Fig. 30–5 has resistivity ρ, whereas that of the moving rod is negligible, derive a formula for the current I as a function of time. Assume the rod has length l, starts at the bottom of the U at $t = 0$, and moves with uniform speed v in the magnetic field B. The cross-sectional area of the rod and all parts of the U is A.

17. (II) (a) Show that the power $P = Fv$ needed to move the conducting rod to the right in Fig. 30–5 is equal to $B^2l^2v^2/R$ where R is the total resistance of the circuit. (b) Show that this equals the power dissipated in the resistance, I^2R.

18. (II) A thin metal rod of length L rotates with angular velocity ω about an axis through one end. The rotation axis is perpendicular to the rod and is parallel to a magnetic field **B**. Determine the emf developed between the ends of the rod.

19. (II) A circular metal disk of radius R rotates with angular velocity ω about an axis through its center perpendicular to its face. The disk rotates in a uniform magnetic field B

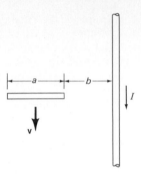

FIGURE 30–20

whose direction is parallel to the rotation axis. Determine the emf induced between the center and the edges.

20. (II) A short section of wire, of length a, is moving with velocity **v**, parallel to a very long wire carrying a current I as shown in Fig. 30–20. The near end of the wire section is a distance b from the long wire. Assuming the vertical wire is very long compared to $a + b$, determine the emf between the ends of the short section. Assume **v** is (a) in the same direction as I, (b) in the opposite direction to I.

21. (II) A conducting rod rests on two long frictionless parallel rails in a magnetic field B ($\perp$ to the rails and rod) as in Fig. 30–21. (a) If the rails are horizontal, will the rod travel at constant speed even though a magnetic field is present? (b) Suppose at $t = 0$, when the rod has speed $v = v_0$, the two rails are connected electrically by a wire from point a to point b. Assuming the rod has resistance R and the rails are resistanceless, determine the speed of the rod as a function of time. Discuss your answer.

22. (III) Suppose a conducting rod (mass m, resistance R) rests on two frictionless and resistanceless parallel rails a distance l apart in a uniform magnetic field B ($\perp$ to the rails and the rod) as in Fig. 30–21. At $t = 0$, the rod is at rest and a source of emf is connected to the points a and b. Determine the speed of the rod as a function of time (a) if the source puts out a constant current I, (b) the source puts out a constant emf ε_0. (c) Does the rod reach a terminal speed in either case? If so, what is it?

SECTION 30–4

23. (I) Determine the electric field in the moving rod of problem 14.

24. (II) What is the magnitude and direction of the electric field at each point in the rotating disk of problem 19?

25. (II) A square coil of wire is placed symmetrically in the magnetic field of Fig. 30–6, so its center is at the center of

FIGURE 30–21

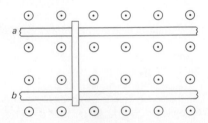

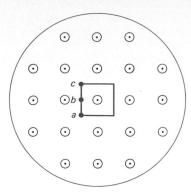

FIGURE 30–22

the circular region of the magnetic field, as shown in Fig. 30–22. If at any instant the field B is changing at a rate dB/dt, (a) show that the component of the electric field parallel to the wire, $E_\parallel$, is the same at all points; (b) find the current in the loop if its resistance is R; (c) what is the emf between points a and b if the length of each side of the square is l.

26. (II) In Fig. 30–6b, consider two identical small (imaginary) circles, one placed near the center of the magnetic field region and the other near its edge (but not beyond the edge). Show that the emf around each circle is the same, even though E is greater in the region of the outer circle.

27. (II) The *betatron*, a device used to accelerate electrons to high energy, consists of a circular vacuum tube placed in a magnetic field (Fig. 30–23), into which electrons are injected. The electromagnet produces a field that (1) keeps the electrons in their circular orbit inside the tube, and (2) increases the speed of the electrons when B changes. (a) Explain how the electrons are accelerated. (See Fig. 30–6). (b) In what direction are the electrons moving in Fig. 30–23 (give directions as if looking down from above). (c) Should B increase or decrease to accelerate the electrons? (d) The magnetic field is actually 60 Hz ac; show that the electrons can be accelerated only during $\frac{1}{4}$ of a cycle ($\frac{1}{240}$ s). (During this time they make hundreds of thousands of revolutions and acquire very high energy.)

28. (III) Show that the electrons in a betatron, problem 27 and Fig. 30–23, are accelerated at constant radius if the magnetic field B_0 at the position of the electron orbit in the tube is equal to half the average value of the magnetic field (B_{av}) over the area of the circular orbit at each moment: $B_0 = \frac{1}{2}B_{av}$. (This is the reason the pole faces have a rather odd shape, as indicated in Fig. 30–23.)

FIGURE 30–23

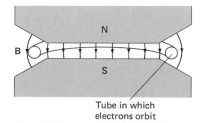

Tube in which
electrons orbit

29. (III) Find a formula for the net electric field in the moving rod of problem 22 as a function of time for each case, (a) and (b).

SECTION 30–5

30. (I) Calculate the peak output voltage of a simple generator whose square armature windings are 8.50 cm on a side if the armature contains 75 loops and rotates in a field of 0.123 T at a rate of 60.0 Hz.

31. (I) Show that the rms output of an ac generator is $V_{rms} = NAB\omega/\sqrt{2}$.

32. (I) A 100-loop square armature coil 12.0 cm on a side rotates at 60 rev/s in a uniform magnetic field. If the rms output is 120 V, what is the strength of the magnetic field?

33. (I) A simple generator has a 100-loop square coil 8.0 cm on a side. How fast must it turn in a 0.350-T field to produce a 12.0-V peak output?

SECTION 30–6

34. (I) A motor has an armature resistance of 2.50 Ω. If it draws 4.00 A when running at full speed and connected to a 120-V line, how large is the counter emf?

35. (I) The counter emf in a motor is 80 V when operating at 1200 rpm. What would the counter emf be at 1800 rpm if the magnetic field is unchanged?

36. (II) What will be the current in the motor of Example 30–5 if the load causes it to run at half speed?

37. (II) The magnetic field of a "shunt wound" dc motor is produced by field coils placed in parallel to the armature coils. Suppose the field coils have a resistance of 46.0 Ω and the armature coils 4.50 Ω. The back emf at full speed is 105 V when the motor is connected to a 115-V line. (a) Draw the equivalent circuits for the situations when the motor is just starting and when it is running full speed. (b) What is the total current drawn by the motor at start up? (c) What is the total current drawn when the motor runs at full speed?

38. (II) A small electric car overcomes a 200-N friction force when traveling 30 km/h. The electric motor is powered by ten 12-V batteries connected in series and is coupled directly to the wheels, whose diameters are 45 cm. The 200 armature coils are rectangular, 12.0 cm by 15.0 cm, and rotate in a 0.385-T magnetic field. (a) How much current does the motor draw to produce the required torque? (b) What is the back emf? (c) How much power is dissipated in the coils? (d) What percent of the input power is used to drive the car?

39. (III) Suppose the "magnetic brake" of Fig. 30–12 acts on a circular metal disk of radius R and thickness d, whose electrical resistivity is ρ. The magnetic field **B**, perpendicular to the disk, acts over a small area A whose center is a distance l from the center of the wheel. At a moment when the disk is rotating with angular speed ω about an axis through its center, determine an approximate formula for the torque acting to slow it down.

SECTION 30–7

40. (I) A transformer has 185 turns in the primary coil and 65 in the secondary. What kind of transformer is this and, assuming 100 percent efficiency, by what factor does it change the voltage?

41. (I) A step-up transformer increases 80 V to 180 V. What is the current in the secondary coil as compared to the primary? Assume 100 percent efficiency.

42. (I) Neon signs require 12 kV for their operation. To operate from a 120-V line, what must be the ratio of secondary to primary turns of the transformer? What would the voltage be if the transformer were connected backward?

43. (I) A transformer has 1800 primary turns and 120 secondary turns. The input voltage is 120 V and the output current is 8.0 A. What is the secondary voltage and primary current?

44. (II) The output voltage of a 280-W transformer is 6.0 V and the input current is 20 A. (a) Is this a step-up or step-down transformer? (b) By what factor is the voltage multiplied?

45. (II) A pair of power transmission lines each have a 0.46 Ω resistance and carry 580 A a distance of 18 km. If the input voltage is 18 kV, calculate (a) the voltage at the other end; (b) the power input; (c) power loss in the lines; (d) the power output.

46. (II) Show that the power loss in transmission lines, P_L, is given by $P_L = (P_T)^2 R_L / V^2$ where P_T is the power transmitted to the user, V is the delivered voltage, and R_L is the resistance of the power lines.

47. (II) If 80 kW is to be transmitted over two 0.055 Ω lines, estimate how much power is saved if the voltage is stepped up from 120 V to 1200 V and then down again, rather than simply transmitting at 120 V. Assume the transformers are each 99 percent efficient.

48. (II) Design a dc transmission line that can transmit 350 MW of electricity 600 km with only a 2 percent loss. The wires are to be made of aluminum and the voltage is 600 kV.

Inductance; Energy Storage in Magnetic Fields; and Electromagnetic Oscillations

31

We discussed in the last chapter how a changing magnetic flux through a circuit induces an emf in that circuit. We saw earlier that an electric current produces a magnetic field. Combining these two ideas, we expect that a changing current in one circuit ought to induce an emf (and a current) in a second nearby circuit and even induce an emf in itself. We already saw an example in the previous chapter (transformers), but now we will treat this effect in a more general way in terms of what we will call mutual inductance and self-inductance. The concept of inductance then gives us a springboard to treat energy storage in a magnetic field. This chapter then concludes with an analysis of circuits that contain inductance as well as resistance and/or capacitance.

 ## 31–1 Mutual Inductance

If two coils of wire are placed near one another, as in Fig. 31–1, a changing current in one will induce an emf in the other. According to Faraday's law, the emf $\mathscr{E}_2$ induced in coil 2 is proportional to the rate of change of flux passing through it. This flux is due to the current I_1 in coil 1, and it is often convenient to express the emf in coil 2 in terms of the current in coil 1.

We let Φ_{21} be the magnetic flux in each loop of coil 2 due to the current in coil 1. If coil 2 contains N_2 closely packed loops, then $N_2\Phi_{21}$ is called the "number of flux linkages" in coil 2. If the two coils are fixed in space, $N_2\Phi_{21}$ is proportional to the current I_1 in coil 1; the proportionality constant is called the **mutual inductance**, M_{21}, defined by

$$M_{21} = \frac{N_2\Phi_{21}}{I_1}. \qquad (31\text{–}1)$$

The emf $\mathscr{E}_2$ induced in coil 2 due to a changing current in coil 1 is, by Faraday's law,

$$\mathscr{E}_2 = -N_2\frac{d\Phi_{21}}{dt}.$$

FIGURE 31–1

A changing current in one coil will induce a current in the second coil.

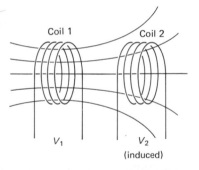

Coil 1   Coil 2

V_1 V_2
(induced)

We combine this with Eq. 31–1 to obtain

$$\mathscr{E}_2 = -M_{21} \frac{dI_1}{dt}. \tag{31–2}$$

This relates the current in coil 1 to the emf it induces in coil 2. The mutual inductance of coil 2 with respect to coil 1, M_{21}, is a constant that depends on the size, shape, number of turns, and relative positions of the two coils, and also on whether iron (or other ferromagnetic material) is present. For example, the closer the two coils in Fig. 31–1, the more lines of flux will pass through coil 2 so M_{21} will be greater. For some arrangements, the mutual inductance can be calculated (see Example 31–1); more usually it is determined experimentally.

Suppose, now, we consider the reverse situation: when a changing current in coil 2 induces an emf in coil 1. In this case,

$$\mathscr{E}_1 = -M_{12} \frac{dI_2}{dt} \tag{31–3}$$

where M_{12} is the mutual inductance of coil 2 with respect to coil 1. It is possible to show, although we will not prove it here, that $M_{12} = M_{21}$; hence, for a given arrangement we do not need the subscripts and we can let

$$M = M_{12} = M_{21} \tag{31–4}$$

so that

$$\mathscr{E}_1 = -M \frac{dI_2}{dt} \tag{31–5a}$$

and

$$\mathscr{E}_2 = -M \frac{dI_1}{dt}. \tag{31–5b}$$

The SI unit for mutual inductance is the henry (H), where $1 \text{ H} = 1 \text{ V·s/A} = 1 \text{ }\Omega\text{·s}$.

EXAMPLE 31–1   A long thin solenoid of length l and cross-sectional area A contains N_1 closely packed turns of wire. Wrapped tightly around it is an insulated coil of N_2 turns, Fig. 31–2. Assume all the flux from coil 1 (the solenoid) passes through coil 2, and calculate the mutual inductance.

SOLUTION   The magnetic field inside the solenoid is (Eq. 29–3)

$$B = \mu_0 \frac{N_1}{l} I_1.$$

Since all flux in the solenoid links the coil, the flux Φ_{21} through coil 2 is

$$\Phi_{21} = BA = \mu_0 \frac{N_1}{l} I_1 A.$$

Hence the mutual inductance is

$$M = \frac{N_2 \Phi_{21}}{I_1} = \frac{\mu_0 N_1 N_2 A}{l}.$$

Note that we calculated M_{21}; if we had tried to calculate M_{12}, it would have been very difficult. But $M_{12} = M_{21} = M$, so we can always do the simpler calculation to obtain M.

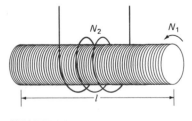

FIGURE 31–2

Example 31–1.

A transformer is an example of mutual inductance in which the coupling is maximized so nearly all flux lines pass through both coils. However, mutual inductance has other uses as well. For example some pacemakers, which are used to regulate the heartbeat so that blood flow is maintained in heart patients, are powered externally. Power in an external coil is transmitted via mutual inductance to a second coil in the pacemaker at the heart. This has the advantage over battery-powered pacemakers in that surgery is not needed to replace a battery when it wears out.

Mutual inductance can sometimes be a problem however. Any alternating current in a circuit can induce an emf in another part of the same circuit or in a different circuit even though the conductors are not in the shape of a coil. The mutual inductance M is usually small unless coils with many turns and/or iron cores are involved. However, in situations where small signals are present, problems due to mutual inductance often arise; shielded cable, in which an inner conductor is surrounded by a cylindrical grounded conductor, is often used to reduce the problem.

31–2   Self-Inductance

The concept of inductance applies also to an isolated single coil of N turns. When a changing current passes through a coil (or solenoid), a changing magnetic flux is produced inside the coil, and this in turn induces an emf. This induced emf opposes the change in flux (Lenz's law); it is much like the counter emf generated in a motor. For example, if the current through the coil is increasing, the increasing magnetic flux induces an emf that opposes the original current and tends to retard its increase. If the current is decreasing in the coil, the decreasing flux induces an emf in the same direction as the current, thus tending to maintain the original current.

Since the magnetic flux Φ passing through the N turns of a coil is proportional to the current I in the coil, we define the **self-inductance** L (in analogy to mutual inductance) as

$$L = \frac{N\Phi}{I}.$$ (31–6)

Then the emf $\mathscr{E}$ induced in a coil of inductance L is, from Faraday's law,

$$\mathscr{E} = -N\frac{d\Phi}{dt} = -L\frac{dI}{dt}.$$ (31–7)

Like mutual inductance, self-inductance is measured in henrys. The magnitude of L depends on the geometry and on the presence of a ferromagnetic material. Self-inductance can be defined, as above, for any circuit or part of a circuit.

An ac circuit always contains some inductance, but often it is quite small unless the circuit contains a coil of many turns. A coil that has significant self-inductance L is called an **inductor** or a *choke coil*; it is shown on circuit diagrams by the symbol ⁗⁗. It can serve a useful purpose in certain circuits. But often inductance is to be avoided in a circuit. Precision resistors are normally wire wound and thus would have inductance as well as resistance; the inductance can be minimized by winding the wire back on itself so little net magnetic flux is produced; this is called a "noninductive winding."

If an inductor has negligible resistance, it is the inductance (or its "back emf") that controls the current. If a source of alternating voltage is applied to the coil, this applied voltage will just be balanced by the induced emf of the coil given by Eq. 31–7. Thus we can see from Eq. 31–7 that if the inductance L is large, the change in the current—and therefore the current itself—will be small. The greater the inductance, the less the current. An inductance thus acts something like a resistance to impede the flow of alternating current; we use the term *impedance* for this quality of an inductor. We will discuss impedance more fully in Chapter 32, and we shall see that it depends not only on L but also on the frequency. Here we

602   Inductance; Energy Storage
in Magnetic Fields; and
Electromagnetic Oscillations

mention one example of its importance. The resistance of the primary in a transformer is usually quite small, perhaps less than an ohm. If resistance alone limited the current in a transformer, tremendous currents would flow when a high voltage was applied. Indeed, a dc voltage applied to a transformer can burn it out. It is the impedance of the coil to an alternating current (or its back emf) that limits the current to a reasonable value.

EXAMPLE 31–2 Determine a formula for the self-inductance L of a solenoid (a long coil) containing N turns of wire in its length l and whose cross-sectional area is A. (b) Calculate the value of L if $N = 100$, $l = 5.0$ cm, $A = 0.30$ cm^2 and the solenoid is air filled. (c) Calculate L if the solenoid has an iron core with $\mu = 4000\,\mu_0$.

SOLUTION (a) According to Eq. 29–3, the magnetic field inside a solenoid is $B = \mu_0 n I$ where $n = N/l$. The flux is $\Phi = BA = \mu_0 NIA/l$, so

$$L = \frac{N\Phi}{I} = \frac{\mu_0 N^2 A}{l}.$$

(b) Since $\mu_0 = 4\pi \times 10^{-7}$ T·m/A

$$L = \frac{(4\pi \times 10^{-7}\ \text{T·m/A})(100)^2(3.0 \times 10^{-5}\ \text{m}^2)}{(5.0 \times 10^{-2}\ \text{m})^2} = 7.5\ \mu\text{H}.$$

(c) Here we replace μ by $\mu_0 = 4000\mu_0$ so L will be 4000 times larger: $L = 0.030$ H $= 30$ mH.

EXAMPLE 31–3 Determine the inductance per unit length of a coaxial cable whose inner conductor has a radius R_1 and outer conductor a radius R_2, Fig. 31–3. Assume the conductors are thin, so that the magnetic field within them can be ignored. The conductors carry equal currents I in opposite directions.

SOLUTION From Ampère's law (see also Example 29–2),

$$\oint \mathbf{B} \cdot d\mathbf{l} = \mu_0 I.$$

The lines of $\mathbf{B}$ are circles surrounding the inner conductor (one is shown dashed); the magnitude of the field at a distance r from the center, when the inner conductor carries a current I, is

$$B = \frac{\mu_0 I}{2\pi r}.$$

FIGURE 31–3

Example 31–3. Coaxial cable: (a) end-on view, (b) side view (cross-section of).

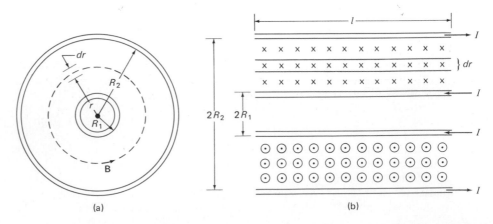

(a)   (b)

The magnetic flux through a rectangle of width dr and length l (along the cable Fig. 31–3b) is

$$d\Phi = Bl\,dr = \frac{\mu_0 I}{2\pi r}\,l\,dr.$$

The total flux in a length l of cable is

$$\Phi = \frac{\mu_0 Il}{2\pi}\int_{R_1}^{R_2}\frac{dr}{r} = \frac{\mu_0 Il}{2\pi}\ln\frac{R_2}{R_1}.$$

Since the current I all flows in one direction in the inner conductor, and the same current I all flows in the opposite direction in the outer conductor, we have only one turn, so $N = 1$. Hence the self-inductance for a length l is

$$L = \frac{\Phi}{I} = \frac{\mu_0 l}{2\pi}\ln\frac{R_2}{R_1}.$$

The inductance per unit length is

$$\frac{L}{l} = \frac{\mu_0}{2\pi}\ln\frac{R_2}{R_1}.$$

31–3 Energy Stored in a Magnetic Field

When an inductor of inductance L is carrying a current I which is changing at a rate dI/dt, energy is being supplied to the inductor at a rate

$$P = I\mathscr{E} = LI\frac{dI}{dt}$$

where P stands for power and we used Eq. 31–7. Let us calculate the work needed to increase the current in an inductor from zero to some value I. From the above relation, the work dW done in a time dt is

$$dW = P\,dt = LI\,dI.$$

Then the total work done to increase the current from zero to I is

$$W = \int dW = \int_0^I LI\,dI = \tfrac{1}{2}LI^2.$$

This work done is equal to the energy U stored in the inductor when it is carrying a current I (and we take $U = 0$ when $I = 0$):

$$U = \tfrac{1}{2}LI^2. \qquad (31\text{–}8)$$

This can be compared to the energy stored in a capacitor, C, when the potential difference across it is V (see Section 25–4):

$$U = \tfrac{1}{2}CV^2.$$

Just as the energy stored in a capacitor can be considered to reside in the electric field between its plates, so the energy in an inductor can be considered to be stored in its magnetic field. To write the energy in terms of the magnetic field, let us use the result of Example 31–2 that the inductance of a solenoid is $L = \mu_0 N^2 A/l$. Now the magnetic field B in a solenoid is related to the current I by $B = \mu_0 NI/l$. Thus

$$U = \tfrac{1}{2}LI^2 = \frac{1}{2}\left(\frac{\mu_0 N^2 A}{l}\right)\left(\frac{Bl}{\mu_0 N}\right)^2$$

$$= \frac{1}{2}\frac{B^2}{\mu_0}Al.$$

We can think of this energy as residing in the volume enclosed by the windings, which is Al. Then the energy per unit volume, or *energy density* is

$$u = \text{energy density} = \frac{1}{2}\frac{B^2}{\mu_0}. \qquad (31-9)$$

This formula, which was derived for the special case of a solenoid, can be shown to be valid for any region of space where a magnetic field exists. If a ferromagnetic material is present, μ_0 is replaced by μ. This equation is analogous to that for an electric field, $\frac{1}{2}\varepsilon_0 E^2$, Eq. 25–6.

 EXAMPLE 31–4 (a) How much energy is being stored per unit length in a coaxial cable whose conductors have radii R_1 and R_2 (Fig. 31–3) and which carry a current I? (b) Where is the energy density highest?

SOLUTION (a) From Example 31–3,

$$\frac{L}{l} = \frac{\mu_0}{2\pi}\ln\frac{R_2}{R_1}.$$

Hence

$$\frac{U}{l} = \frac{1}{2}LI^2 = \frac{\mu_0 I^2}{4\pi}\ln\frac{R_2}{R_1}.$$

(b) Since $B = \mu_0 I/2\pi r$, the field is greatest near the surface of the inner conductor, so the energy density, $u = B^2/2\mu_0$ will be greatest there.

31–4 *LR* Circuit

Any inductor will have some resistance. So we represent an inductor by drawing its inductance L and its resistance R separately, as in Fig. 31–4a. The resistance R could also include a separate resistor connected in series. Now we ask, what happens when a dc source of voltage V is connected in series to such an *LR* circuit. At the instant the switch connecting the battery is closed, the current starts to flow. It is of course opposed by the induced emf in the inductor. However, as soon as current starts to flow, there is a voltage drop across the resistance. Hence the voltage drop across the inductance is reduced and there is then less impedance to the current flow from the inductance. The current thus rises gradually as shown in Fig. 31–4b; it approaches the steady value $I_{max} = V/R$ when all the voltage drop is across the resistance.

We can show this analytically by applying Kirchhoff's loop rule to the circuit of Fig. 31–4a. The emf's in the circuit are the battery V, and the emf $\mathscr{E} = -L(dI/dt)$ in the inductor. These must equal the potential drop across the resistor:

$$V - L\frac{dI}{dt} = IR$$

FIGURE 31–4

(a) *LR* circuit; (b) growth of current when connected to battery; (c) decay of current when battery is removed from the circuit.

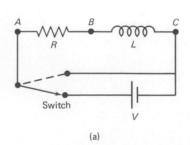

(a)

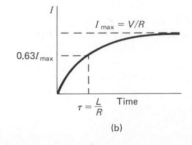

(b)

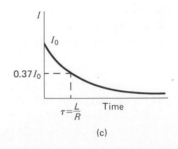

(c)

where I is the current in the circuit at any instant. We rearrange this to obtain

$$L\frac{dI}{dt} + RI = V. \tag{31-10}$$

This is a linear differential equation and can be integrated in the same way we did in Section 27–5 for an *RC* circuit:

$$\int_{I=0}^{I} \frac{dI}{V - IR} = \int_{0}^{t} \frac{dt}{L}.$$

Then

$$-\frac{1}{R} \ln\left(\frac{V - IR}{V}\right) = \frac{t}{L}$$

or,

$$I = \frac{V}{R}(1 - e^{-t/\tau}) \tag{31-11}$$

where

$$\tau = L/R \tag{31-12}$$

is the *inductive time constant* of the *LR* circuit. The symbol τ represents the time required for I to reach $(1 - 1/e) = 0.63$ or 63 percent of its maximum value (V/R). Equation 31–11 is plotted in Fig. 31–4b. (Compare to the *RC* circuit, Section 27–5.)

If the switch in Fig. 31–4a is flipped so that the battery is taken out of the circuit (and points A and C are connected together) at an instant when the current is I_0, then the differential equation (Eq. 31–10) becomes (since $V = 0$):

$$L\frac{dI}{dt} + RI = 0.$$

This is easily rearranged and integrated:

$$\int_{I_0}^{I} \frac{dI}{I} = -\int_{0}^{t} \frac{R}{L}dt$$

where $I = I_0$ at $t = 0$, and $I = I$ at time t. This is integrated to

$$\ln\frac{I}{I_0} = -\frac{R}{L}t$$

or

$$I = I_0 e^{-t/\tau}$$

where again the time constant is $\tau = L/R$. The current thus decays exponentially to zero as shown in Fig. 31–4c.

This analysis shows that there is always some "reaction time" when an electromagnet, for example, is turned on or off. We also see that an *LR* circuit has properties similar to an *RC* circuit (Section 27–5). Unlike the capacitor case, however, the time constant here is *inversely* proportional to R.

◆ **EXAMPLE 31–5** At $t = 0$, a 12.0-V battery is connected in series with a 30-Ω resistor and a 220-mH inductor. (*a*) What is the time constant? (*b*) How long will it take the current to reach half its maximum possible value? At this instant, at what rate is (*c*) energy being delivered by the battery, and (*d*) energy being stored in the inductor's magnetic field?

SOLUTION   (a) From Eq. 31–12, $\tau = L/R = (0.22\text{ H})/(30\ \Omega) = 7.3$ ms.

(b) The maximum current is $I_{max} = V/R$. So in Eq. 31–11 we set $I = \frac{1}{2}I_{max} = V/2R$, which gives us

$$1 - e^{-t/\tau} = \frac{1}{2}$$

or

$$t = \tau \ln 2 = (7.3 \times 10^{-3}\text{ s})(0.69) = 5.0\text{ ms.}$$

(c) At this instant, $I = I_{max}/2 = V/2R = 120\text{ V}/60\ \Omega = 200$ mA, so the power being delivered by the battery is

$$P = IV = (0.20\text{ A})(12\text{ V}) = 2.4\text{ W.}$$

(d) From Eq. 31–8, the energy stored in an inductor L at any instant is

$$U = \frac{1}{2}LI^2$$

where I is the current in the inductor at that instant. The *rate* at which the energy changes is

$$\frac{dU}{dt} = LI\frac{dI}{dt}.$$

We must differentiate Eq. 31–11 to obtain dI/dt:

$$\frac{dU}{dt} = LI\left(\frac{V}{R}\right)\left(\frac{1}{\tau}\right)(e^{-t/\tau})$$

$$= (0.22\text{ H})(0.20\text{ A})\left(\frac{120\text{ V}}{30\ \Omega}\right)\left(\frac{1}{7.3 \times 10^{-3}\text{ s}}\right)(e^{-\ln 2})$$

$$= 1.2\text{ W.}$$

Since only part of the battery's power is feeding the inductor at this instant, where is the rest going?

31–5   *LC* Circuits and Electromagnetic Oscillations

In any electric circuit, there can be three basic components, resistance, capacitance, and inductance, in addition to a source of emf. (There can be more complex components, also, such as diodes or transistors.) We have previously discussed both *RC* and *LR* circuits. Now we look at an *LC* circuit, one that contains only a capacitance C and an inductance L, Fig. 31–5. (We are treating the ideal case in which we assume there is no resistance in the inductor; in the next section we include resistance.)

Suppose the capacitor in Fig. 31–5 is initially charged so that one plate has charge Q_0 and the other plate has charge $-Q_0$. Suppose that at $t = 0$, the switch is closed. The capacitor immediately begins to discharge. As it does so, the current I through the inductor increases. At every instant, the potential difference across the capacitor, $V = Q/C$ (where Q is the charge on the capacitor at that instant), must equal the potential difference across the inductor, which is equal to its emf, $-L(dI/dt)$. At the instant when the charge on the capacitor reaches zero ($Q = 0$), the current I in the inductor has reached its maximum value, but at this instant I is not changing ($-L\,dI/dt = Q/C = 0$). At this moment, the magnetic field B in the inductor is also a maximum. The current next begins to decrease as the flowing charge starts to accumulate on the opposite plate of the capacitor. When the current has dropped to zero, the capacitor has attained its maximum charge, which is equal to the negative of its initial value Q_0. (We shall show this in a

FIGURE 31–5

An *LC* circuit.

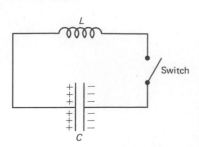

moment.) The capacitor then begins to discharge again, with the current now flowing in the opposite direction. This process of the charge flowing back and forth from one plate of the capacitor to the other, through the inductor, continues to repeat itself. This is called an *LC oscillation* or an *electromagnetic oscillation*. Not only does the charge oscillate back and forth, but so does the energy. The energy is initially stored in the electric field of the capacitor; when Q reaches zero, I is a maximum and all the energy is then stored in the magnetic field of the inductor; thus the energy oscillates between being stored in the electric field and in the magnetic field.

Now let us look at this *LC* oscillation analytically. At any instant of time, the potential difference across the capacitor must equal that across the inductor:

$$\frac{Q}{C} = -L\frac{dI}{dt},$$

where Q is the charge on one plate of the capacitor at a given instant and I is the current flowing in the inductor at the same instant. The current I is due solely to the flow of charge from the capacitor, and so $I = dQ/dt$. We can then rewrite the above relation as

$$\frac{d^2Q}{dt^2} + \frac{1}{LC}Q = 0. \qquad (31\text{–}13)$$

This is a familiar differential equation. It has the same form as that for simple harmonic motion, Eq. 14–3b, which is

$$\frac{d^2x}{dt^2} + \frac{k}{m}x = 0$$

where x represents the displacement from equilibrium of a mass m on the end of a spring whose stiffness constant is k. For an electromagnetic oscillation, the variable is the charge Q rather than x. We saw in Section 14–2 that the solution of the above equation for the simple harmonic oscillator is $x = A\cos(\omega t + \phi)$ where A is the amplitude, ω the angular frequency ($= 2\pi f$), and ϕ the phase angle. Since the two differential equations are identical except for the names of the variables and constants, it is clear that the solution for Q in Eq. 31–13 will also be sinusoidal:

$$Q = Q_0\cos(\omega t + \phi). \qquad (31\text{–}14)$$

For the simple harmonic oscillator, we saw (Eq. 14–4) that $\omega = \sqrt{k/m}$; since k/m multiplies the variable x in Eq. 14–3b, and $1/LC$ multiplies the variable Q in Eq. 31–13, we expect that for an *LC* oscillation

$$\omega = 2\pi f = \sqrt{\frac{1}{LC}}, \qquad (31\text{–}15)$$

where f is the frequency (in Hz). Just to confirm this, we insert Eq. 31–14 in Eq. 31–13, noting that $d^2Q/dt^2 = -\omega^2 Q_0\cos(\omega t + \phi)$; thus

$$-\omega^2 Q_0\cos(\omega t + \phi) + \frac{1}{LC}Q_0\cos(\omega t + \phi) = 0$$

or

$$\left(-\omega^2 + \frac{1}{LC}\right)\cos(\omega t + \phi) = 0.$$

This relation can be true for all times t only if $(-\omega^2 + 1/LC) = 0$, which gives us Eq. 31–15.

Equation 31–14 thus shows that the charge on the capacitor in an *LC* circuit oscillates sinusoidally. For the special case discussed above, $Q = Q_0$ at $t = 0$, so the phase angle $\phi = 0$.

The current in the inductor is

$$I = \frac{dQ}{dt} = -\omega Q_0 \sin(\omega t + \phi);$$ (31-16)

so it too is sinusoidal. The maximum value of I is $\omega Q_0 = Q_0/\sqrt{LC}$. Equations 31-15 and 31-16, for Q and I when $\phi = 0$, are plotted in Fig. 31-6; note that they confirm our qualitative description at the beginning of this section.

Now let us look at LC oscillations from the point of view of energy. The energy stored in the electric field of the capacitor at any time t is (see Eq. 25-5):

$$U_E = \frac{1}{2}\frac{Q^2}{C} = \frac{Q_0^2}{2C}\cos^2(\omega t + \phi),$$

whereas the energy stored in the magnetic field of the inductor at the same instant is (Eq. 31-8)

$$U_B = \frac{1}{2}LI^2 = \frac{L\omega^2 Q_0^2}{2}\sin^2(\omega t + \phi) = \frac{Q_0^2}{2C}\sin^2(\omega t + \phi)$$

where we used Eq. 31-15. If we take $\phi = 0$, then at times $t = 0$, $t = T$, $t = 2T$, and so on, (where T is the period $= 1/f = 2\pi/\omega$) we have $U_E = Q_0^2/2C$ and $U_B = 0$. That is, all the energy is stored in the electric field of the capacitor. But at $t = T/2$, $3T/2$, and so on, $U_E = 0$ and $U_B = Q_0^2/2C$, and so all the energy is stored in the magnetic field of the inductor. At any time t, the total energy is

$$U = U_E + U_B = \frac{Q_0^2}{2C}[\cos^2(\omega t + \phi) + \sin^2(\omega t + \phi)] = \frac{Q_0^2}{2C}.$$

Hence the total energy is constant, and energy is conserved.

EXAMPLE 31-6 A 1200-pF capacitor is charged by a 500-V battery. It is disconnected from the battery and is connected, at $t = 0$, to a 75-mH inductor. Determine (a) the initial charge on the capacitor; (b) the maximum current; (c) the frequency f and period T of oscillation; and (d) the total energy oscillating in the system.

SOLUTION (a) $Q_0 = CV = (1.2 \times 10^{-9} \text{ F})(500 \text{ V}) = 6.0 \times 10^{-7} \text{ C}$. (b) From Eq. 31-16, $I_{max} = \omega Q_0 = Q_0/\sqrt{LC} = (6.0 \times 10^{-7} \text{ C})/\sqrt{(1.2 \times 10^{-9} \text{ F})(0.075 \text{ H})} = 63 \text{ mA}$. (c) $f = \omega/2\pi = 1/(2\pi\sqrt{LC}) = 17 \text{ kHz}$, and $T = 1/f = 6.0 \times 10^{-5}$ s. (d) $U = Q_0^2/2C = (6.0 \times 10^{-7} \text{ C})^2/[2(1.2 \times 10^{-9} \text{ F})] = 1.5 \times 10^{-4}$ W.

FIGURE 31-6

Charge Q and current I in an LC circuit. The period

$$T = \frac{1}{f} = \frac{2\pi}{\omega} = 2\pi\sqrt{LC}.$$

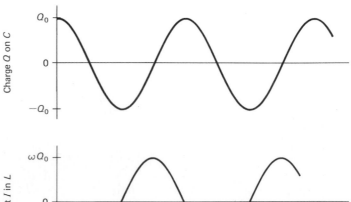

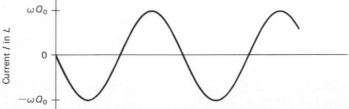

31–6 *LC* Oscillations with Resistance (*LRC* Circuit)

The *LC* circuit discussed in the previous section is an idealization. There is always some resistance *R* in any circuit, and so we now discuss such an *LRC* circuit, Fig. 31–7.

Suppose again that the capacitor is initially given a charge Q_0 and the switch is closed at $t = 0$. Since there is now a resistance in the circuit, we expect some of the energy to be converted to thermal energy, and so we don't expect undamped oscillations as in a pure *LC* circuit. Indeed, if we use Kirchhoff's loop rule around this circuit, we obtain

$$L\frac{dI}{dt} + IR + \frac{Q}{C} = 0$$

which is the same equation we had in Section 31–5 with the addition of the *IR* drop across the resistor. Since $I = dQ/dt$, this equation becomes

$$L\frac{d^2Q}{dt^2} + R\frac{dQ}{dt} + \frac{1}{C}Q = 0. \qquad (31\text{–}17)$$

This second-order differential equation in the variable *Q* has precisely the same form as that for the damped harmonic oscillator, Eq. 14–16:

$$m\frac{d^2x}{dt^2} + b\frac{dx}{dt} + kx = 0.$$

Hence we can analyze our *LRC* circuit in the same way as for damped harmonic motion, Section 14–7. Our system may undergo damped oscillations, as in Fig. 31–8a (underdamped system), or it may be critically damped or overdamped as in Figs. 31–8b and c, depending on the relative values of *R*, *L*, and *C*. Comparing Eq. 31–17 and the equation for damped harmonic motion just below it, we see that the constants have the following correspondence:

$$L \text{ plays the role of } m$$

$$R \text{ plays the role of } b$$

$$\frac{1}{C} \text{ plays the role of } k.$$

In the discussion in Section 14–7, we found that the system will be underdamped, and there will be oscillations, if

$$b^2 < 4mk.$$

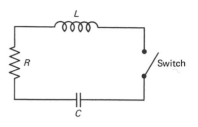

FIGURE 31–7
An *LRC* circuit.

FIGURE 31–8

Charge *Q* as a function of time in an *LRC* circuit: (a) underdamped, (b) critically damped, (c) overdamped.

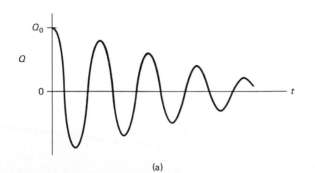

(a)

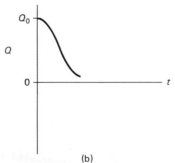

(b)

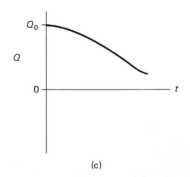

(c)

Then, in the electrical case, the system will be underdamped when

$$R^2 < \frac{4L}{C}.$$

If R is small compared to $\sqrt{4L/C}$, the angular frequency, ω', will be

$$\omega' = \sqrt{\frac{1}{LC} - \frac{R^2}{4L^2}} \qquad (31-18)$$

(compare to Eq. 14–21); and the charge Q as a function of time will be

$$Q = Q_0 e^{-\frac{R}{2L}t} \cos(\omega' t + \phi) \qquad (31-19)$$

where ϕ is a phase constant (compare Eq. 14–20).

Oscillators are an important element in many electronic devices: radios and television sets use them for tuning, tape recorders use them (the "bias frequency") when recording, and so on. Since some resistance is always present, electrical oscillators generally need a periodic input of power to compensate for the energy converted to thermal energy in the resistance.

EXAMPLE 31–7 At $t = 0$, a 40-mH inductor is placed in series with a resistance $R = 3.0 \ \Omega$ and a charged capacitor $C = 4.8 \ \mu F$. (a) Show that this circuit will oscillate. (b) Determine the frequency. (c) What is the time required for the charge amplitude to drop to half its starting value? (d) What is the current amplitude? (e) What value of R will make the circuit nonoscillating?

SOLUTION (a) To be underdamped, we must have $R^2 < 4L/C$. Since $R^2 = 9.0 \ \Omega^2$ and $4L/C = 4(0.040 \ \mathrm{H})/(4.8 \times 10^{-6} \ \mathrm{F}) = 3.3 \times 10^4 \ \Omega^2$, this relation is satisfied, so the circuit will oscillate.

(b) Since R is small compared to L/C, we can use Eq. 31–18:

$$f' = \frac{\omega'}{2\pi} = \frac{1}{2\pi} \sqrt{\frac{1}{LC} - \frac{R^2}{4L^2}} = 2.3 \times 10^3 \ \mathrm{Hz}.$$

(c) From Eq. 31–19, the amplitude will be half when

$$e^{-\frac{R}{2L}t} = \tfrac{1}{2}$$

or

$$t = \frac{2L}{R} \ln 2 = 18 \ \mathrm{ms}.$$

(d) We differentiate Eq. 31–19 to obtain I:

$$I = \frac{dQ}{dt} = Q_0 e^{-\frac{R}{2L}t} \left(-\frac{R}{2L} \cos \omega' t - \frac{1}{\sqrt{LC}} \sin \omega' t \right)$$

since $\omega' \approx \sqrt{1/LC}$. Because R is much less than $\sqrt{4L/C}$ (see part a above), we can ignore the $\cos \omega' t$ term, so

$$I \approx -\frac{Q_0}{\sqrt{LC}} e^{-\frac{R}{2L}t} \sin \omega' t.$$

Hence the initial current amplitude is $Q_0/\sqrt{LC}$. We cannot give a numerical value since we are not given Q_0.

(e) To make the circuit critically damped or overdamped, we must have

$$R^2 \geq 4L/C = 3.3 \times 10^4 \ \Omega^2.$$

Hence we must have $R > 180 \ \Omega$.

Summary

A changing current in a coil of wire will induce an emf in a second coil placed nearby. The mutual *inductance M* is defined as the proportionality constant between the induced emf $\mathscr{E}_2$ in the second coil and the time rate of change of current in the first: $\mathscr{E}_2 = -M dI_1/dt$. Within a single coil, a changing current induces a back emf, $\mathscr{E}$, so a coil has a *self-inductance L* defined by $\mathscr{E} = -L dI/dt$. This back emf acts as an *impedance* to the flow of an alternating current.

When the current in an inductance L is I, the energy stored in the inductance is given by $\frac{1}{2}LI^2$. This energy can be thought of as being stored in the magnetic field of the inductor. The energy density in any magnetic field B is given by $(\frac{1}{2})(B^2/\mu)$, where μ is the permeability in that region.

When an inductance L and resistor R are connected in series to a source of emf, the current rises according to an exponential of the form $(1 - e^{-t/\tau})$, where $\tau = L/R$ is the time constant—and eventually levels out at $I = V/R$. If the battery is suddenly switched out of the circuit (but the circuit remains complete), the current drops exponentially, $e^{-t/\tau}$, with the same time constant τ.

The current in a pure LC circuit (or charge on the capacitor) would oscillate indefinitely in simple harmonic motion. The current in a series LRC circuit (or charge on the capacitor) in which the capacitor at some instant is charged, can undergo damped oscillations, or undergo critically damped or overdamped oscillations.

Questions

1. In situations where a small signal must travel over a distance, a "shielded cable" is used in which the signal wire is surrounded by an insulator and then enclosed by a cylindrical conductor. Why is a "shield" necessary?

2. What is the advantage of placing the two electric wires carrying ac close together?

3. The primary of a transformer on a telephone pole has a resistance of $0.10\ \Omega$ and the input voltage is 2400 V ac. Can you estimate the current that will flow? Will it be 24,000 A? Explain.

4. A transformer designed for a 120-V ac input will often "burn out" if connected to a 120-V dc source. Explain. (Hint: the resistance of the primary coil is usually very low.)

5. How would you arrange two flat circular coils so that their mutual inductance was (a) greatest, (b) least (without separating them by a great distance)?

6. If the two coils in Fig. 31–1 were connected electrically, would there still be mutual inductance?

7. Suppose the second coil of N_2 turns in Fig. 31–2 were moved so it was near the end of the solenoid. How would this affect the mutual inductance?

8. If a solenoid has an iron core, is its self-inductance L a constant? Explain. What about the mutual inductance of an iron-core transformer?

9. A long narrow solenoid and a large-area, narrow-gap, parallel-plate capacitor are simple devices that represent ways of establishing uniform magnetic and electric fields. Compare and contrast these two devices in as many ways as possible.

10. Does a mutual inductance also have a self-inductance? Explain.

11. Is the self-inductance per unit length of a solenoid greater near its center or near its ends?

12. If you are given a fixed length of wire, how would you shape it to obtain the greatest self-inductance? The least?

13. In a battery, when the current is in the same direction as the emf, the energy of the battery decreases, whereas if the current is in the opposite direction, the energy of the battery increases (as in charging a battery). Is this also true for an inductor?

14. Is the energy density greatest near the ends of a solenoid or near its center?

15. Two solenoids have the same length and circular cross-sectional area. Both consist of tightly packed turns of wire, but one uses thicker wire than the other. Which has the greater inductance? Which has the greater inductive time constant?

16. Does the emf of the battery in Fig. 31–4a affect the time needed for the LR circuit to reach (a) a given fraction of its maximum possible current, (b) a given value of current?

17. A circuit with large inductive time constant carries a steady current. If a switch is opened, there can be a very large (and sometimes dangerous) spark or "arcing over." Explain.

18. Suppose you are looking along a line through the centers of two circular (but separate) wire loops, one behind the other. A battery is connected to the front loop, producing a counterclockwise current. (a) In what direction is the induced current in the rear loop? (b) How long does the induced current last? (c) If you doubled the diameter of the wire in each loop, how would this effect your answer to part (b)?

19. At the instant the battery is connected into the LR circuit of Fig. 31–4a, the emf in the inductor has its maximum value even though the current is zero. Explain.

20. Explain how a solenoid whose ends are connected together can by itself oscillate like an LC circuit. Where does the capacitance come from?

21. What keeps an *LC* circuit oscillating even after the capacitor has discharged completely?

22. Is the current in the inductor always the same as that in the resistor of the *LRC* circuit of Fig. 31–7?

23. Can an inductor ever have no resistance? Can an inductor ever have no capacitance? Can a capacitor ever have no inductance? Can a resistor ever have no inductance?

24. In the analogy between *LRC* oscillations and damped harmonic motion, discuss physically how *L* acts like *m*, *R* like *b*, and 1/*C* like *k*.

25. Suppose an inductor, capacitor, and (small) resistance were placed in *parallel*. At $t = 0$, $I = 0$ at all points in the circuit and the capacitor carries charge Q_0. Describe Q as a function of time.

 Problems ──

SECTION 31–1

1. (I) Suppose the second coil of Example 31–1 (Fig. 31–2) has twice the diameter of the solenoid, but is still concentric with it. What then will be the mutual-inductance? Assume the solenoid is very long.

2. (II) A 44-cm-long coil containing 500 loops is wound on an iron core (average $\mu = 2000\mu_0$) along with a second coil of 200 loops. The loops of each coil have a radius of 2.0 cm. If the current in the first coil drops uniformly from 4.0 A to zero in 80 ms, determine (*a*) the emf induced in the second coil; (*b*) the mutual inductance *M*.

3. (II) A small thin coil with N_2 loops, each of area A_2, is placed inside a long solenoid, near its center. The solenoid has N_1 loops in its length *L* and has area A_1. Determine the mutual inductance as a function of θ, the angle between the plane of the small coil and the axis of the solenoid.

4. (II) A long straight wire and a small rectangular wire loop lie in the same plane, Fig. 31–9. Determine the mutual inductance in terms of l_1, l_2, and *w*. Assume the wire is long compared to l_2 and *w*, and that the rest of its circuit is very far away compared to l_2 and *w*.

5. (II) Determine the mutual inductance per unit length between two long solenoids, one inside the other, whose radii are r_1 and r_2 ($r_2 < r_1$) and whose turns per unit length are n_1 and n_2.

6. (II) Estimate the mutual inductance *M* between two small coils, of radius R_1 and R_2 which are separated by a distance *l* that is large compared to R_1 and R_2, Fig. 31–10. Give *M* as a function of θ, the angle between the planes of the two coils. Assume the line joining their centers is perpendicular to the plane of coil 1. What if this assumption is not true?

SECTION 31–2

7. (I) If the current in a 130-mH coil changes steadily from 20.0 mA to 28.0 mA in 140 ms, what is the induced emf?

8. (I) What is the inductance *L* of a 1.2-m-long air-filled coil 5.7 cm in diameter containing 20,000 loops?

9. (I) What is the inductance of a coil if it produces an emf of 6.50 V when the current in it changes from -12.0 mA to $+23.0$ mA in 11.0 ms?

10. (I) What is the inductance of a coaxial cable 30 m long if its inner and outer conductors have diameters of 2.0 mm and 4.0 mm?

11. (I) A 35-V emf is induced in a 0.320-H coil by a current that rises uniformly from zero to I_0 in 2.0 ms. What is the value of I_0?

12. (II) How many turns does an air-filled coil have if it is 2.2 cm in diameter and 8.0 cm long and its inductance is 0.25 mH. How many turns are needed if it has an iron core and $\mu = 10^3\mu_0$?

13. (II) If the outer conductor of a coaxial cable has radius 3.0 mm, what should be the radius of the inner conductor so that the inductance per unit length does not exceed 40 nH per meter?

14. (II) (*a*) Show that if two circuits, such as the coils in Fig. 31–1, carry currents I_1 and I_2, the magnetic flux through each is $\Phi_1 = L_1 I_1 + M I_2$ and $\Phi_2 = L_2 I_2 + M I_1$. (*b*) Determine a formula for the emf induced in each coil in terms of the rate of change of current in the two coils.

FIGURE 31–9

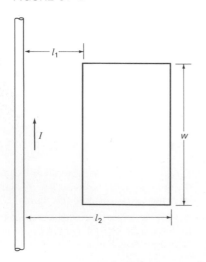

FIGURE 31–10

15. (II) The wire of a tightly wound solenoid is unwound and used to make another tightly wound solenoid of half the diameter. By what factor does the inductance change?

16. (II) A coil has 2.50-Ω resistance and 0.418-H inductance. If the current is 8.00 A and is increasing at a rate of 4.50 A/s, what is the potential difference across the coil?

17. (II) (a) Show that the self-inductance L of a torus (Fig. 29–10) of radius R containing N loops each of radius r is

$$L \approx \frac{\mu_0 N^2 r^2}{2R}$$

if $R \gg r$. Assume the field is uniform inside the toroid; is this actually true? Is this result consistent with L for a solenoid? Shoud it be? (b) Calculate the inductance L of a large toroid if the diameter of the coils is 2.0 cm and the diameter of the whole ring is 50 cm. Assume the field inside the toroid is uniform. There are a total of 300 loops of wire.

18. (II) (a) Ignoring any mutual inductance, what is the equivalent inductance of two inductors connected in series? (b) What if they are connected in parallel? (c) How does their mutual inductance (their geometrical relationship to each other) affect the results?

19. (II) A toroid has a rectangular cross-section as shown in Fig. 31–11. Show that the self-inductance is

$$L = \frac{\mu_0 N^2 h}{2\pi} \ln \frac{R_2}{R_1}$$

where N is the total number of turns and R_1, R_2, and h are the dimensions shown in the figure. (Hint: use Ampère's law to get B as a function of r inside the toroid, and integrate.)

20. (II) A pair of straight parallel thin wires, each of diameter r, are a distance l apart and carry current to a circuit some distance away. Ignoring the field within each wire, show that the inductance per unit length is $(\mu_0/\pi) \ln [(l - r)/r]$.

21. (II) The potential difference across a given coil is 15.5 V at an instant when the current is 360 mA and is changing at a rate of 240 mA/s. At a later instant, the potential difference is 6.2 V while the current is 300 mA and is decreasing at a rate of 180 mA/s. Determine the inductance and resistance of the coil.

22. (III) A long straight wire carries a current I that is uniform across its cross-sectional area A. (a) Determine the induct-

ance per unit length due only to the field inside the wire. (b) Can you calculate the self-inductance due to the field outside the wire as well? If so, do it. Discuss your answer.

23. (III) (a) Show that if two thin coils (inductance L_1 and L_2) are connected in series and placed close to one another, the net self-inductance is

$$L = L_1 + L_2 \pm M$$

where M is their mutual inductance. Explain the $\pm$ sign. (b) How can M be made zero (or nearly so)? (c) Determine the net inductance if the two coils are connected in parallel, assuming the mutual inductance is ignorable. If M is not ignorable, how would if affect your answer?

SECTION 31–3

24. (I) The magnetic field inside an air-filled solenoid 10 cm long and 2.0 cm in diameter is 0.50 T. Approximately how much energy is stored in this field?

25. (I) How much energy is stored in 400-mH inductor at an instant when the current is 2.0 A?

26. (I) Assuming the earth's magnetic field averages about 0.50×10^{-4} T near the surface of the earth, determine the total energy stored in this field in the first 10 km above the earth's surface.

27. (I) Typical large values for electric and magnetic fields attained in laboratories are about 1.0×10^4 V/m and 2.0 T. (a) Determine the energy density for each field and compare. (b) What value electric field would produce the same energy density as the 2.0-T magnetic field?

28. (I) In an electromagnetic plane wave (Chapter 33) the electric and magnetic fields are related by $E = cB$ where $c = 1/\sqrt{\varepsilon_0 \mu_0}$ is the speed of light. What is the ratio of the energy stored in the electric and the magnetic fields of such a wave?

29. (II) What is the energy density at the center of a circular loop of wire carrying a 30-A current if the radius of the loop is 8.0 cm?

30. (II) Calculate the magnetic and electric energy densities at the surface of a 3.0-mm-diameter copper wire carrying a 15-A current.

31. (II) For the toroid of Fig. 31–11, determine the energy density as a function of r ($R_1 < r < R_2$) and integrate this over the volume to obtain the total energy stored in the toroid, which carries a current I in each of its N loops.

32. (II) Determine the total energy stored per unit length in the magnetic field between the coaxial cylinders of a coaxial cable by using Eq. 31–9 for the energy density and integrating over the volume. Compare your answer to that obtained in Example 31–4.

SECTION 31–4

33. (I) How many time constants does it take for the potential difference across the resistor in an LR circuit (Fig. 31–4a) to drop to 1.0 percent of its original value?

34. (I) After how many time constants does the current in Fig. 31–4b reach within (a) 10 percent, (b) 1.0 percent, and (c) 0.1 percent of its maximum value?

35. (II) In Example 31–5, where is the power delivered by the battery going besides into the inductor? Do a calculation to show that energy is conserved.

FIGURE 31–11

A toroid of rectangular cross-section, with N turns carrying a current I.

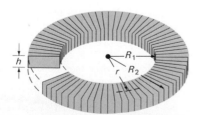

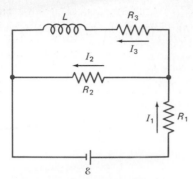

FIGURE 31–12

36. (II) Determine dI/dt at $t = 0$ (when the battery is connected) for the circuit of Fig. 31–4a and show that if I continued to increase at this rate, it would reach its maximum value in one time constant.

37. (II) It takes 1.56 ms for the current in an LR circuit to increase from zero to half its maximum value. Determine (a) the time constant of the circuit, (b) the resistance of the circuit if $L = 310$ H.

38. (II) Two tightly wound solenoids have the same length and circular cross-sectional area. But solenoid 1 uses wire that is half as thick as solenoid 2. (a) What is the ratio of their inductances? (b) What is the ratio of their inductive time constants (assuming no other resistance in the circuits)?

39. (II) (a) Determine the energy stored in the inductor L as a function of time for the LR circuit of Fig. 31–4a. (b) After how many time constants does the stored energy reach 99 percent of its maximum value?

40. (II) In the circuit of Fig. 31–12, determine the current in each resistor (I_1, I_2, I_3) (a) at the moment the switch is closed, (b) a long time after the switch is closed. After the switch has been closed a long time, and reopened, what is each current (c) just after it is opened, (d) after a long time?

SECTION 31–5

41. (I) The variable capacitor in the tuner of an AM radio has a capacitance of 1500 pF when the radio is tuned to a station at 550 kHz. (a) What must be the capacitance for a station at 1600 kHz? (b) What is the inductance (assumed constant)?

42. (I) (a) If the initial conditions of an LC circuit were $I = I_0$ and $Q = 0$ at $t = 0$, write Q as a function of time (b) Practically, how could you set up these initial conditions?

43. (I) Use the definitions of the farad and henry to show that $1/\sqrt{LC}$ has units of s^{-1}.

44. (II) A 660-pF capacitor is charged to 100 V and then quickly connected to a 75-mH inductor. Determine (a) the frequency of oscillation, (b) the peak value of the current, and (c) the maximum energy stored in the magnetic field of the inductor.

45. (II) At $t = 0$, $Q = Q_0$ and $I = 0$ in an LC circuit. At the first moment when the energy is shared equally by the inductor and the capacitor, (a) what is the charge on the capacitor? (b) How much time has elapsed (in terms of the period T)?

SECTION 31–6

46. (II) In an oscillating LRC circuit, how much time does it take for the energy stored in the fields of the capacitor and inductor to fall to half its initial value? (See Fig. 31–7; assume $R \ll \sqrt{4L/C}$.)

47. (II) How much resistance must be added to a pure LC circuit ($L = 200$ mH, $C = 1200$ pF) to change the oscillator's frequency by 0.10 percent? Will it be increased or decreased?

48. (II) A damped LC circuit loses 3.5 percent of its electromagnetic energy per cycle to thermal energy. If $L = 80$ mH and $C = 1.00$ μF, what is the value of R?

49. (II) (a) For an underdamped LRC circuit, determine a formula for the energy $U = U_E + U_B$ stored in the electric and magnetic fields as a function of time. Give in terms of the initial charge Q_0 on the capacitor. (b) Show how dU/dt is related to the rate energy is transformed in the resistor, $I^2 R$.

50. (II) (a) Obtain a differential equation for current I in the LRC circuit of Fig. 31–7, by differentiating Eq. 31–17. (b) Obtain a general solution for I, assuming the circuit is connected at $t = 0$ with charge Q_0 on the capacitor. Assume $R \ll \sqrt{4L/C}$. (c) Compare your result to part (d) of Example 31–7, and note any differences. (d) What complications would arise if R were not much less than $\sqrt{4L/C}$, say $R \approx \sqrt{L/C}$?

51. (II) Starting with Eq. 31–19, with $\phi = 0$, show that the current I in a lightly damped LRC circuit is given by

$$I \approx -\frac{Q_0}{\sqrt{LC}} e^{-\frac{R}{2L} t} \sin(\omega' t + \delta)$$

where

$$\delta = \tan^{-1} \frac{R}{2L\omega'}.$$

52. (II) Show that the phase constant ϕ in Eq. 31–19 is given by

$$\phi = \cot^{-1} \left(\frac{4L}{R^2 C} \right)^{1/2}$$

if $I = 0$ at $t = 0$ in the circuit of Fig. 31–7; assume that $R^2 \ll 4L/C$.

53. (III) Show that the fraction of electromagnetic energy lost (to thermal energy) per cycle in a lightly damped ($R^2 \ll 4L/C$) LRC circuit is approximately

$$\frac{\Delta U}{U} = \frac{2\pi R}{L\omega} = \frac{2\pi}{Q}.$$

The quantity Q, defined as $Q = L\omega/R$, is called the Q-value, or quality factor, of the circuit and is a measure of the damping present. A high Q-value means smaller damping and less energy input required to maintain oscillations.

54. (III) Suppose a battery of fixed voltage V is connected in the LRC circuit of Fig. 31–7. At $t = 0$, the switch is closed. Assuming this is a very overdamped circuit ($R^2 \gg 4L/C$) determine the current as a function of time. Plot I vs. t. Compare your result to a pure RC circuit, for which $L = 0$. Real RC circuits always have some inductance, so the result of this problem is more realistic than a pure RC circuit.

*55. (III) For the circuit described in Example 31–7, numerically integrate for the current $I (= dQ/dt)$ and the charge Q (on the capacitor) and plot each of these from $t = 0$ to $t = 3.0$ ms. Assume $Q = 3.50 \times 10^{-4}$ C at $t = 0$. Achieve an accuracy of 2 percent or better. (Hint: see footnote below.)

*56. (III) (a) What value must R have to make the circuit of

† See Section 2–10 and problems 64 and 67 in Chapter 14.

Example 31–7 critically damped? (b) Assume R has this value, and repeat problem 55 (see footnote below). (c) Assume R has half this value, and repeat problem 55.

*57. (III) Suppose that the capacitor of the circuit of Example 31–7 initially is uncharged and that at $t = 0$ a 12.0-V battery is connected in series into the circuit. Plot the current I as a function of time from $t = 0$ to $t = 5.0$ ms by numerically integrating.

AC Circuits

32

In earlier chapters (Sections 27–5, 31–4, 31–5 and 31–6), we discussed circuits that contain combinations of resistor, capacitor, and inductor (or all three), but only when they are connected to a dc source of emf or to no source (as in the discharge of a capacitor in an *RC* circuit or oscillation of an *LC* or *LRC* circuit). Now we discuss these circuit elements when connected to a source of alternating emf. Such *ac circuits* are important first of all because the output of most generators (Section 30–5) is sinusoidal and most electricity generated is of this type. Secondly, any voltage that varies in time, no matter how complex it is, can be written as a sum of sine and cosine terms of different frequencies in a Fourier series. Thus, the response of resistors, capacitors, and inductors to a sinusoidal source of emf is of basic importance.

 ## 32–1 Introduction: AC Circuits

We briefly discussed alternating currents in Section 26–7, and have seen that for sinusoidally varying currents and voltages, the rms and peak values are related by

$$V_{\mathrm{rms}} = \frac{V_0}{\sqrt{2}}, \qquad I_{\mathrm{rms}} = \frac{I_0}{\sqrt{2}}.$$

We will now assume that we have a source of alternating emf, represented by the symbol ——⊗——, which produces a sinusoidal emf of frequency f given by

$$\mathcal{E} = \mathcal{E}_0 \sin \omega t = \mathcal{E}_0 \sin 2\pi f t \tag{32–1}$$

where $\mathcal{E}_0$ is the peak emf (maximum value). We might expect that the current in the circuit will also be sinusoidal at the same frequency;[†] and indeed we will see

† See Section 14–8.

this is so, but it may be out of phase with the emf, so that

$$I = I_0 \sin(\omega t + \phi). \qquad (32\text{–}2)$$

For a given circuit, we will want to find the value of the peak current, I_0, and the phase angle ϕ, in terms of the parameters of the circuit such as $\mathcal{E}_0$, f, R, L, and/or C.

We first discuss, one at a time, how a resistor, a capacitor, and an inductor behave when connected to any source of alternating emf, whose output is Eq. 32–1.

32–2 AC Circuit Containing Only Resistance R

When a pure resistance R ($L = C = 0$) is connected to the ac source as in Fig. 32–1a, the current is found to increase and decrease in direct proportion to the emf as given by Ohm's law

$$I = \frac{V}{R} = \frac{\mathcal{E}}{R} = \frac{\mathcal{E}_0}{R} \sin \omega t, \qquad (32\text{–}3a)$$

where V is the voltage across the resistor at any moment and $V = \mathcal{E}$ by Kirchhoff's loop rule. ($V_0 = \mathcal{E}_0$ is the peak voltage.) This is graphed in Fig. 32–1b, where the voltage is the solid curve and the current is the dashed curve. The current and voltage are *in phase*, so $\phi = 0$ in Eq. 32–2. The peak current is

$$I_0 = \frac{V_0}{R} = \frac{\mathcal{E}_0}{R}. \qquad (32\text{–}3b)$$

Energy is transformed into heat, as discussed in Section 26–7, at an average rate

$$\bar{P} = \overline{IV} = I_{\text{rms}}^2 R = \frac{V_{\text{rms}}^2}{R}.$$

(a)

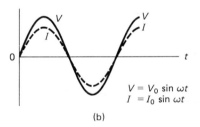

$$V = V_0 \sin \omega t$$
$$I = I_0 \sin \omega t$$

(b)

FIGURE 32–1

Resistor connected to ac source. Current is in phase with the voltage across resistor.

FIGURE 32–2

Capacitor connected to an ac source. Current leads voltage by a quarter cycle, or 90°. [Note: if we choose $I = I_0 \sin \omega t$, as we will in Section 32–5, then $V = V_0 \sin(\omega t - 90°)$].

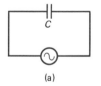

(a)

32–3 AC Circuit Containing Only Capacitance C

When a capacitor C is connected directly to a battery, the capacitor plates quickly acquire equal and opposite charges; but no steady current flows in the circuit. A capacitor prevents the flow of a dc current.[†] However, if a capacitor is connected to an alternating source of voltage, as in Fig. 32–2a (we assume $R = L = 0$ in this circuit), an alternating current will flow continuously. This can happen because when the ac voltage is first turned on, charge begins to flow so that one plate acquires a negative charge and the other a positive charge. But before the capacitor becomes fully charged, the voltage reverses itself, and the charges flow in the opposite direction. Thus for an alternating applied voltage, an ac current is present in the circuit continuously.

Let us look at this in more detail. By Kirchhoff's loop rule, the applied emf must equal the voltage V across the capacitor at every moment, which is given by $V = Q/C$ where Q is the charge on the capacitor plates. Thus

$$V = \frac{Q}{C} = \mathcal{E}_0 \sin \omega t$$

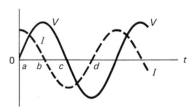

$$V = V_0 \sin \omega t$$
$$I = I_0 \cos \omega t = I_0 \sin(\omega t + 90°)$$

(b)

[†] Unless the capacitor is "leaky," in which case a leakage current would pass through the capacitor.

and the charge Q on the plates is in phase with the voltage. But what about the current in the circuit? At any instant, $I = dQ/dt$, so we have

$$I = \frac{dQ}{dt} = \omega C \mathscr{E}_0 \cos \omega t. \tag{32-4a}$$

By the trigonometric identity $\cos \theta = \sin (\theta + 90°)$, Eq. 32-4a becomes

$$I = I_0 \sin (\omega t + 90°), \tag{32-4b}$$

where we have written

$$I_0 = \omega C \mathscr{E}_0. \tag{32-4c}$$

Thus the peak current I_0 (see Eq. 32-2) is $I_0 = \omega C \mathscr{E}_0$ and the phase angle ϕ is $90°$. The current I and the voltage V ($= \mathscr{E}_0 \sin \omega t$) across the capacitor are graphed in Fig. 32-2b. It is clear from this graph, as well as from Eq. 32-4a or b, that the current and voltage are *not* in phase. In fact, they are out of phase by a quarter cycle, which is equivalent to $\pi/2$ radians or $90°$. It is clear from the graph that *current leads the voltage across a capacitor by $90°$*. That is, the current in the circuit reaches its peaks a quarter cycle before the voltage reaches its peaks. We could also say that the voltage lags the current by $90°$.

Why there is a phase difference between the current and the voltage across a capacitor can be seen intuitively as follows. At point a in Fig. 32-2b, when the voltage starts increasing, the charge on the plates is zero; thus charge flows readily toward the plates and the current is large. As the voltage approaches its maximum of V_0, the charge that has accumulated on the plates tends to prevent more charge from flowing, so the current drops to zero at point b. As V decreases, the charge that has accumulated now starts to flow off the plates and the current again increases (dashed curve), but in the opposite direction; it reaches a maximum (negatively) when the voltage is at point c. Thus the current follows the graph shown dashed in Fig. 32-2b, and leads the current by $90°$. Because the current and voltage are out of phase by $90°$, no power is taken from the battery and transformed to other forms of energy, on the average. This can be seen as follows. From point a to b in Fig. 32-2b the voltage is increasing from zero to a maximum. The current, however, is decreasing from a maximum and is approaching zero. The average power over this interval, VI, is positive. From b to c, however, V is positive whereas I is negative, so the product VI is negative; we can see from the symmetry that this contribution just balances out the positive contribution of the first quarter cycle. Similar considerations apply to the rest of the cycle. Thus, the average power transformed over one or many complete cycles is zero. We can see that energy from the source passes into the electric field of the capacitor where it is stored momentarily; then the field decreases and the energy is transferred back to the source. None is lost in this process. Compare this to a resistor, Fig. 32-1b, where the current is always in phase with the voltage and energy is transferred out of the source and never back into it. (The product VI is never negative.) This energy is not stored but is transformed to thermal energy in the resistor.

Just as a resistor impedes the flow of charge, so too a capacitor impedes the flow of charge in an alternating current, since charge builds up on the capacitor plates. For a resistor R, the peak current and peak voltage are related by Ohm's law:

$$V_0 = I_0 R.$$

We can write a similar relation for a capacitor:

$$V_0 = I_0 X_C \tag{32-5a}$$

where, from Eq. 32-4c (with $V_0 = \mathscr{E}_0$)

$$X_C = \frac{1}{\omega C}. \tag{32-5b}$$

The term X_C is called the *capacitive reactance*[†] and has units of ohms. Equation 32–5a, where X_C is given by Eq. 32–5b, relates the peak values of V and I. (It also relates the rms values: $V_{rms} = I_{rms}X_C$.) But because the peak current and voltage are not reached at the same time, Eq. 32–5a is *not* a valid relationship between V and I at a particular instant, as is the case for Ohm's law.

The dependence of X_C on C and ω, as given by Eq. 32–5b, makes sense as can be seen as follows. The larger the capacitance C, the more charge it can handle, so the less it would impede the flow of charge; hence we expect X_C to be inversely proportional to C. We also expect X_C to be inversely proportional to the frequency f ($= \omega/2\pi$) since, when the frequency is higher, there is less time per cycle for the charge to build up on the plates and impede the flow. Notice that if $f = 0$, which means a dc source, X_C becomes infinite; this is as it should be since a pure capacitor does not pass dc current.

EXAMPLE 32–1 What are the peak and rms currents in the circuit of Fig. 32–2a if $C = 1.0 \ \mu$F and $V_{rms} = 120$ V? Calculate (a) for $f = 60$ Hz, and then (b) for $f = 6.0 \times 10^5$ Hz.

SOLUTION (a) $V_0 = \sqrt{2} \ V_{rms} = 170$ V. Then

$$X_C = \frac{1}{\omega C} = \frac{1}{2\pi f C} = \frac{1}{(6.28)(60 \text{ s}^{-1})(1.0 \times 10^{-6} \text{ F})} = 2.7 \text{ k}\Omega.$$

Thus

$$I_0 = \frac{V_0}{X_C} = \frac{170 \text{ V}}{2.7 \times 10^3 \ \Omega} = 63 \text{ mA};$$

$$I_{rms} = \frac{V_{rms}}{X_C} = \frac{120 \text{ V}}{2.7 \times 10^3 \ \Omega} = 44 \text{ mA}.$$

(b) For $f = 6.0 \times 10^5$ Hz, X_C will be 0.27 Ω, $I_0 = 630$ A, and $I_{rms} = 440$ A. The dependence on f is dramatic.

Capacitors are used for a variety of purposes, some of which we have already described. Two other applications are illustrated in Fig. 32–3. In Fig. 32–3a, circuit A is said to be capacitively coupled to circuit B. The purpose of the capacitor is to prevent a dc voltage from passing from A to B but allowing an ac voltage to pass relatively unimpeded. If C is sufficiently large, the ac voltage will not be significantly attenuated. The capacitor in Fig. 32–3b also passes ac but not dc. In this case, a dc voltage can be maintained between circuits A and B; and if the capacitance C is large enough, the capacitor offers little impedance to an ac voltage leaving A. An ac voltage then passes to ground instead of into B. Thus the capacitor in Fig. 32–3b acts like a filter when a constant dc voltage is required; any sharp variation in voltage will pass to ground instead of into circuit B. Capacitors used in these two ways are very common in circuits.

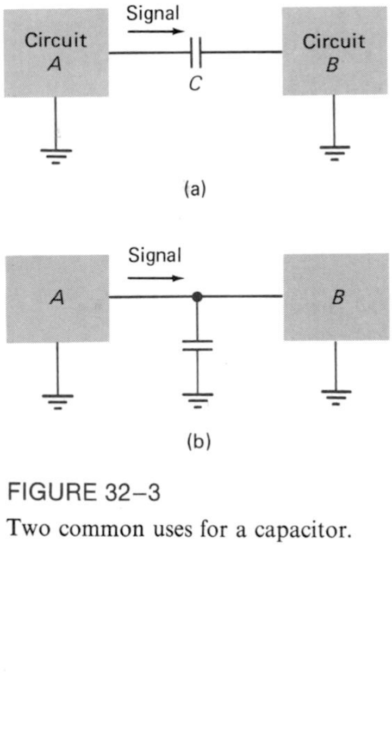

32–4 AC Circuit Containing Only Inductance L

In Fig. 32–4a, an inductor of inductance L is connected to the ac source. We ignore any resistance or capacitance it might have. The emf of the source will be equal to the "counter" emf generated in the inductor by the changing current as given

FIGURE 32–3
Two common uses for a capacitor.

[†] Sometimes it is also called the *impedance*, although impedance is usually reserved for use in a real circuit that contains more than a pure capacitor (see Section 32–5). To say it another way, capacitive reactance is one type of impedance, as is resistance.

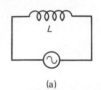

(a)

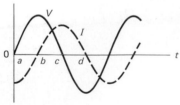

$V = V_0 \sin \omega t$
$I = -I_0 \cos \omega t = I_0 \sin (\omega t - 90°)$

(b)

FIGURE 32–4

Inductor connected to an ac source. Current lags voltage by a quarter cycle, or 90°. [Note: if we choose $I = I_0 \sin \omega t$, as we will in Section 32–5, then $V = V_0 \sin (\omega t + 90°)$].

by Eq. 31–7. Hence we have

$$\mathscr{E} - L\frac{dI}{dt} = 0$$

or

$$L\frac{dI}{dt} = \mathscr{E}_0 \sin \omega t.$$

To obtain the current I, we solve for dI in this last relation:

$$dI = (\mathscr{E}_0/L) \sin \omega t\, dt$$

and integrate

$$I = -\frac{\mathscr{E}_0}{\omega L} \cos \omega t. \tag{32–6a}$$

Since $\cos \theta = -\sin (\theta - 90°)$, this can be written

$$I = I_0 \sin (\omega t - 90°) \tag{32–6b}$$

where

$$I_0 = \frac{\mathscr{E}_0}{\omega L}. \tag{32–6c}$$

Thus the peak current is $I_0 = \mathscr{E}_0/\omega L$ and the phase angle $\phi = -90°$. The current I and the voltage $V\ (= \mathscr{E})$ across the inductor are plotted in Fig. 32–4b. It is clear that the current lags the voltage across an inductor by 90°. (Or, the voltage leads the current by 90°.)

Let us try to understand physically where this phase difference between V and I comes from. Since the voltage across the inductor is

$$V = L\frac{dI}{dt},$$

I is increasing most rapidly when V has its maximum value, $V = V_0$. And I will be decreasing most rapidly when $V = -V_0$. These two instants correspond to points b and d on the graph of voltage vs. time in Fig. 32–4b. At points a and c, $V = 0$; the above equation tells us that $dI/dt = 0$ at these instants, so I is not changing and therefore corresponds to the maximum and minimum values of the current I. These correspond to points a and c on the graph.

Because the current and voltage are out of phase by 90°, the average power dissipated in an inductor is zero, just as for a capacitor. Energy from the source is fed to the inductor, where it is stored in its magnetic field; when the field decreases, the energy returns to the source.

Thus, on the average, only a resistor will take energy from the source and dissipate it.

We can write a relationship between the peak voltage and peak current through an inductor just as we did for a capacitor:

$$V_0 = I_0 X_L. \tag{32–7a}$$

The term X_L is called the *inductive reactance* and, from Eq. 32–6c, is given by

$$X_L = \omega L. \tag{32–7b}$$

It, too, has units of ohms. Equation 31–7a relates the peak values of V and I, or rms values ($V_{rms} = I_{rms} X_L$); but it is not valid at a particular instant because I and V are not in phase.

We can see, physically, why X_L is proportional both to L and ω. From the fact that $V = -L\, dI/dt$, we see that the larger L is, the less will be the change in current dI during the time dt; hence I itself will be smaller at any instant for a given frequency. Thus $X_L \propto L$. Also the greater the frequency, the more rapidly the magnetic flux changes in the inductor. If the emf induced by this field is to remain equal to the applied emf, as it must, the magnitude of the current must then be less. Hence, the greater the frequency the greater the impedance, so $X_L \propto \omega L$. This is also consistent with the fact that if the frequency f is zero (so the current is dc) there is no back emf and no impedance to the flow of charge.

It is interesting to note that the reactance of an inductor increases with frequency, but that of a capacitor decreases with frequency.

 EXAMPLE 32–2 A coil has a resistance $R = 1.00\ \Omega$ and an inductance of 0.300 H. Determine the current in the coil if (*a*) 120-V dc is applied to it; (*b*) 120-V ac (rms) at 60.0 Hz is applied.

SOLUTION (*a*) There is no inductive impedance ($X_L = 0$ since $f = 0$), so we apply Ohm's law for the resistance

$$I = \frac{V}{R} = \frac{120\ \text{V}}{1.00\ \Omega} = 120\ \text{A}.$$

(*b*) The inductive impedance in this case is:

$$X_L = 2\pi f L = (6.28)(60.0\ \text{s}^{-1})(0.300\ \text{H}) = 113\ \Omega.$$

In comparison to this, the resistance can be ignored. Thus

$$I_{\text{rms}} = \frac{V_{\text{rms}}}{X_L} = \frac{120\ \text{V}}{113\ \Omega} = 1.06\ \text{A}.$$

(It might be tempting to say that the total impedance is $113\ \Omega + 1\ \Omega = 114\ \Omega$; this might imply that about 1 percent of the voltage drop is across the resistor, or about 1 V; and that across the inductance is 119 V. Although the 1 V(rms) across the resistor is accurate, the other statements are *not* true because of the alteration in phase in an inductor. This will be discussed in the next section.)

32–5 *LRC* Series AC Circuit

We now examine a circuit containing all three elements in series, a resistor R, an inductor L, and a capacitor C, Fig. 32–5. If a given circuit contains only two of these elements, we can still use the results of this section by setting $R = 0$, $L = 0$, or $C = \infty$ (infinity) as needed. We let V_R, V_L, and V_C represent the voltage across each element at a *given instant* in time; and V_{R0}, V_{L0}, and V_{C0} represent the *maximum* (peak) values of these voltages. The voltage across each of the elements will follow the phase relations we discussed in the previous sections. That is, V_R will be in phase with the current; V_L will lead the current by 90°; and V_C will lag behind the current by 90°. And at any instant the emf $\mathscr{E}$ (or V) supplied by the source will be

$$\mathscr{E} = V = V_R + V_L + V_C. \qquad (32\text{–}8)$$

FIGURE 32–5
An *LRC* circuit.

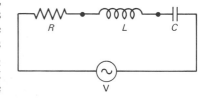

However, because the various voltages are not in phase, the rms voltages (which are what ac voltmeters usually measure) will not simply add up to give the rms voltage of the source. And the peak emf of the source, $\mathscr{E}_0$, will *not* equal $V_{R0} + V_{L0} + V_{C0}$. Let us now examine the circuit in detail. What we would like to find in particular is the impedance of the circuit as a whole, the rms current that flows, and the phase difference between the source voltage and the current. We will assume that the current at any instant is the same at all points in the circuit.[†]

[†] This is a good assumption if the frequency is not too high, so that the wavelength ($= c/f$ where c is the speed of light) is much larger than the dimensions of the circuit.

We could do our analysis by rewriting Eq. 32–8 as a differential equation (setting $V_C = Q/C$, $V_R = IR = (dQ/dt)R$, and $V_L = L\,dI/dt$) and trying to solve the differential equation. Instead we will use a simpler approach which is perhaps more physically illuminating. This approach uses the so-called *phasor diagram*. Arrows (acting like vectors) are drawn in an xy coordinate system to represent each voltage: the *length of each arrow represents the magnitude of the peak voltage across each element*:

$$V_{R0} = I_0 R, \qquad V_{L0} = I_0 X_L, \qquad V_{C0} = I_0 X_C.$$

The angle of each arrow represents the phase of each voltage, relative to the current (as we will explain in a moment). For convenience in this analysis, we choose

$$I = I_0 \sin \omega t.$$

Note that earlier in this chapter we chose $\mathscr{E} = \mathscr{E}_0 \sin \omega t$ and determined what I would be for each circuit element. But for our analysis now of the LRC series circuit it is clearer if we choose the form for the current I since it is the same for all points in the circuit at a given time t; that is, we choose $I = I_0 \sin \omega t$, and then establish V from this. (The physics, of course, is not affected by this choice.) At $t = 0$, the current $I = 0$ and we draw the vector representing I_0 along the positive x axis of our phasor diagram, Fig. 32–6a. Since the voltage across a resistor is always in phase with the current, the arrow representing V_{R0} is also drawn along the x axis at $t = 0$ as shown. Since the voltage across the inductor, V_L, leads the current by 90°, V_{L0} leads V_{R0} by 90° and is drawn for $t = 0$ along the positive y axis; V_C lags the current by 90°, so V_{C0} lags V_{R0} by 90°, and hence V_{C0} is drawn for $t = 0$ along the negative y axis (Fig. 32–6a). If we let this diagram rotate as a whole at angular frequency ω, then we obtain the diagram shown in Fig. 32–6b: after a time t, each arrow has rotated through an angle ωt. Then the *projections of each arrow on the* y *axis represent the voltages across each element at the instant* t. For example, the projection of V_{R0} on the y axis is $V_{R0} \sin \omega t$ ($= I_0 R \sin \omega t = IR$); and the projections of V_{L0} and V_{C0} on the y axis are $V_{L0} \cos \omega t = V_{L0} \sin (\omega t + 90°)$ and $-V_{C0} \cos \omega t = V_{C0} \sin (\omega t - 90°)$, consistent with our earlier results.[†] Maintaining the 90° angle between each vector ensures the correct phase relations. Although these facts show the validity of a phasor diagram, what we are really interested in is how to add the voltages.

The sum of the projections of the three vectors on the y axis is equal to the projection of their sum. But the sum of the projections represents the instantaneous voltage across the whole circuit, which is the source emf $\mathscr{E}$. Therefore the vector sum of these vectors will be the vector that represents the peak source emf, $\mathscr{E}_0$. This is shown in Fig. 32–7 where it is seen that $\mathscr{E}_0$ makes an angle ϕ with V_{R0} and I_0. As time passes, $\mathscr{E}_0$ rotates with the other vectors, so the instantaneous emf $\mathscr{E}$ (projection of $\mathscr{E}_0$ on y axis) is:

$$\mathscr{E} = \mathscr{E}_0 \sin (\omega t + \phi).$$

Thus we see that the voltage from the source is out of phase[‡] with the current by an angle ϕ.

From this analysis we can now draw some useful conclusions. First we determine the total **impedance** Z of the circuit which is defined by the relation

$$\mathscr{E}_{\text{rms}} = I_{\text{rms}} Z, \quad \text{or} \quad \mathscr{E}_0 = I_0 Z. \tag{32–9}$$

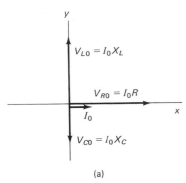

(a)

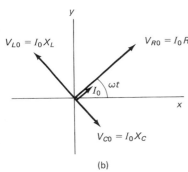

(b)

FIGURE 32–6

Phasor diagram for a series LRC circuit.

FIGURE 32–7

Phasor diagram for a series LRC circuit showing the sum vector, $\mathscr{E}_0$.

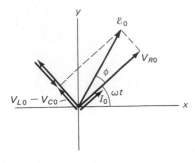

[†] Note that Eqs. 32–3a, 32–4b, and 32–6b gave the current when the source emf was chosen to be $\mathscr{E} = \mathscr{E}_0 \sin \omega t$. If instead we had chosen the current to be $I = I_0 \sin \omega t$, as we did in this section, then the voltages across each element would have been $V_R = IR = I_0 R \sin \omega t$, $V_L = L(dI/dt) = \omega L I_0 \cos \omega t = I_0 X_L \sin (\omega t + 90°)$ and $V_C = Q/C = (1/C) \int I\,dt = -(I_0/\omega C) \cos \omega t = I_0 X_C \sin (\omega t - 90°)$. These are the results we get from the phasor diagram above.

[‡] As a check, note that if $R = C = 0$, then $\phi = 90°$, and $\mathscr{E}_0$ would lead the current by 90°, as it must for an inductor alone. Similarly, if $R = L = 0$, $\phi = -90°$ and $\mathscr{E}_0$ would lag the current by 90° as it must for a capacitor alone.

$$\mathscr{E}_0 = \sqrt{V_{R0}^2 + (V_{L0} - V_{C0})^2}$$
$$= \sqrt{I_0^2 R^2 + (I_0 X_L - I_0 X_C)^2}$$
$$= I_0 \sqrt{R^2 + (X_L - X_C)^2}.$$

Thus, from Eq. 32–9,

$$Z = \sqrt{R^2 + (X_L - X_C)^2} \qquad (32\text{–}10a)$$

$$= \sqrt{R^2 + \left(\omega L - \frac{1}{\omega C}\right)^2}. \qquad (32\text{–}10b)$$

This gives the total impedance of the circuit. Also from Fig. 32–7 we can find the phase angle ϕ from

$$\tan \phi = \frac{V_{L0} - V_{C0}}{V_{R0}} = \frac{I_0(X_L - X_C)}{I_0 R} = \frac{X_L - X_C}{R} \qquad (32\text{–}11a)$$

or from

$$\cos \phi = \frac{V_{R0}}{\mathscr{E}_0} = \frac{I_0 R}{I_0 Z} = \frac{R}{Z}. \qquad (32\text{–}11b)$$

Note that Fig. 32–7 was drawn for the case $X_L > X_C$, and the current lags the source emf by ϕ; if the reverse is true, $X_L < X_C$, than ϕ in Eq. 32–11 is less than zero, and the current leads the source emf.

Finally, we can determine the power dissipated in the circuit. We saw earlier that power is dissipated only by a resistance; none is dissipated by inductance or capacitance. Therefore (see also Section 26–7) the average power $\bar{P} = I_{rms}^2 R$. But from Eq. 32–11b, $R = Z \cos \phi$. Therefore

$$\bar{P} = I_{rms}^2 Z \cos \phi$$
$$= I_{rms} \mathscr{E}_{rms} \cos \phi. \qquad (32\text{–}12)$$

The factor $\cos \phi$ is referred to as the *power factor* of the circuit. For a pure resistor, $\cos \phi = 1$ and $P = I_{rms} \mathscr{E}_{rms}$. For a capacitor or inductor alone, $\cos \phi = 0$ and no power is dissipated.

The test of this analysis is, of course, in experiment; and experiment is in full agreement with these results.

At the beginning of this section, we chose the phase of the current so that $I = I_0 \sin \omega t$. This choice is, of course, arbitrary. (What is physically important is the phase difference, ϕ, between current and voltage.) If we had chosen, instead,

$$\mathscr{E} = \mathscr{E}_0 \sin \omega t,$$

as we did in Sections 32–1 through 32–4, then the current I would be

$$I = I_0 \sin (\omega t - \phi)$$

where ϕ and I_0 have the same values as given by Eqs. 32–9, 32–10, and 32–11.

EXAMPLE 32–3 Suppose $R = 25 \, \Omega$, $L = 30 \, \text{mH}$ and $C = 12 \, \mu\text{F}$ in Fig. 32–5 and they are connected to a 90-V ac (rms), 500-Hz source. Calculate (*a*) the current in the circuit, (*b*) the voltmeter readings (rms) across each element, (*c*) the phase angle ϕ, and (*d*) the power dissipated in the circuit.

SOLUTION (*a*) First we find the individual impedances at $f = 500 \, \text{s}^{-1}$:

$$X_L = 2\pi f L = 94 \, \Omega,$$

$$X_C = \frac{1}{2\pi f C} = 27 \, \Omega.$$

Then

$$Z = \sqrt{R^2 + (X_L - X_C)^2} = \sqrt{(25\ \Omega)^2 + (94\ \Omega - 27\ \Omega)^2}$$
$$= 72\ \Omega.$$

From Eq. 32–9,

$$I_{rms} = \frac{\mathscr{E}_{rms}}{Z} = \frac{90\ V}{72\ \Omega} = 1.25\ A.$$

(b) The rms voltage across each element is

$$(V_R)_{rms} = I_{rms}R = (1.25\ A)(25\ \Omega) = 31\ V,$$
$$(V_L)_{rms} = I_{rms}X_L = 117\ V,$$
$$(V_C)_{rms} = I_{rms}X_C = 34\ V.$$

Notice that these do *not* add up to give the source voltage of 90 V_{rms}. Indeed, the rms voltage across the inductance *exceeds* the source voltage. This is due to the fact that the different voltages are out of phase with each other. The instantaneous voltages at any time do add up, of course; this can happen because at any instant one voltage can be negative to compensate for a large positive voltage of another. The rms voltages, however, are always positive by definition.

(c) Since $\cos \phi = R/Z = 25\ \Omega/72\ \Omega = 0.35$, then $\phi = 70°$.

(d) $P = I_{rms}V_{rms}\cos \phi = (1.25\ A)(90\ V)(0.35) = 39\ W.$

32–6 Resonance in AC Circuits

The rms current in an *LRC* circuit is given by (see Eq. 32–9 and 32–10b):

$$I_{rms} = \frac{\mathscr{E}_{rms}}{Z} = \frac{\mathscr{E}_{rms}}{\sqrt{R^2 + \left(\omega L - \dfrac{1}{\omega C}\right)^2}}. \qquad (32\text{–}13)$$

Because the impedance of inductors and capacitors depends on the frequency, f, of the source, the current in an *LRC* circuit will depend on frequency. From Eq. 32–13 we can see that the current will be maximum at a frequency such that

$$\left(\omega L - \frac{1}{\omega C}\right) = 0.$$

We solve this for ω, and call the solution ω_0:

$$\omega_0 = \sqrt{\frac{1}{LC}}. \qquad (32\text{–}14)$$

When $\omega = \omega_0$, the circuit is in *resonance*. At this frequency, $X_C = X_L$, so the impedance is purely resistive and $\cos \phi = 1$. A graph of I_{rms} vs. ω is shown in Fig. 32–8 for particular fixed values of $\mathscr{E}_0$, R, L, and C. The smaller R is compared to X_L and X_C, the higher and sharper will be the resonance peak. When R is small, the circuit approaches the pure *LC* circuit we discussed in Section 31–5.

This electrical resonance is analogous to mechanical resonance which we discussed in Chapter 14. The energy transferred to the system by the source is a maximum at resonance whether it is electrical resonance, the oscillation of a spring, or pushing a child on a swing (Section 14–8). That this is true in the electrical case can be seen from Eq. (32–12). At resonance, $\cos \phi = 1$, and I_{rms} is a maximum. For constant $\mathscr{E}_{rms}$, the power then is a maximum at resonance. A graph of power versus frequency peaks like that for the current, Fig. 32–8.

FIGURE 32–8

Current in an *LRC* circuit as a function of frequency, showing resonance peak at $\omega = \omega_0 = \sqrt{1/LC}$.

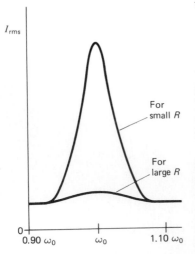

For small R

For large R

Electric resonance is used in many circuits. Radio and TV sets, for example, use resonant circuits for tuning in a station. Many frequencies reach the circuit, but a significant current flows only for those at or near the resonant frequency. Either L or C is variable so that different stations can be tuned in.

* 32–7   Impedance Matching

It is common to connect one electric circuit to a second circuit. For example, a TV antenna is connected to a TV set; an FM tuner is connected to an amplifier; the output of an amplifier is connected to a speaker; electrodes for an ECG or EEG are connected to an amplifier or a recorder. In many cases it is important that the maximum power be transferred from one to the other, with a minimum of loss. This can be achieved when the output impedance of the one device matches the input impedance of the second.

To show why this is true, we consider simple circuits that contain only resistance. In Fig. 32–9 the source in circuit 1 could represent a power supply, the output of an amplifier, or the signal from an antenna, a laboratory probe, or a set of electrodes. R_1 represents the resistance of this device and includes the internal resistance of the source. R_1 is called the output impedance (or resistance) of circuit 1. The output of circuit 1 is across the terminals x and y which are connected to the input of circuit 2. Circuit 2 may be very complicated. By combining the various resistors, we can find an equivalent resistance. This is represented by R_2, the "input resistance" of circuit 2.

The power delivered to circuit 2 is $P = I^2 R_2$ where $I = V/(R_1 + R_2)$. Thus

$$P = I^2 R_2 = \frac{V^2 R_2}{(R_1 + R_2)^2}.$$

We divide the top and bottom of the right side by R_1 and find

$$P = \frac{V^2}{R_1} \frac{\left(\dfrac{R_2}{R_1}\right)}{\left(1 + \dfrac{R_2}{R_1}\right)^2}.$$

The question is, if the resistance of the source is R_1, what value should R_2 have so that the maximum power is transferred to circuit B. To determine this, we take the derivative of P with respect to R_2 and set it equal to zero:

$$0 = \frac{dP}{dR_2} = \frac{V^2}{R_1^2} \frac{(1 - R_2/R_1)}{(1 + R_2/R_1)^3}.$$

This expression can be zero only if $(1 - R_2/R_1) = 0$, or

$$R_2 = R_1.$$

Thus, the maximum power is transmitted when the *output impedance* of one device *equals the input impedance* of the second. This is called *matching the impedances.*

In an ac circuit that contains capacitors and inductors, the different phases are important and the analysis is more complicated. However, the same result holds: to maximize power transfer it is important to match impedances ($Z_2 = Z_1$). In addition, one must be aware that it is possible to seriously distort a signal. For example, when a second circuit is connected, it may put the first circuit into resonance, or take it out of resonance for a certain frequency.

Without proper consideration of the impedances involved, one can make measurements that are completely meaningless. These considerations are normally examined by engineers when designing an integrated set of apparatus. However,

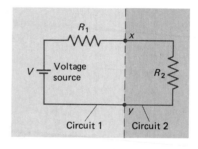

FIGURE 32–9

Output of the circuit on the left is input to the circuit on the right.

there have been cases when researchers connected several components to one another without regard for impedance matching, and made a "new discovery." Then, after their announcement of the "discovery," it was found that what they had observed was due to impedance mismatch rather than the natural phenomenon they had thought.

In some cases, a transformer is used to alter an impedance, so it can be matched to that of a second circuit. If Z_s is the secondary impedance and Z_p the primary impedance, then $V_s = I_s Z_s$ and $V_p = I_p Z_p$ (I and V are either peak or rms values of current and voltage). Hence

$$\frac{Z_p}{Z_s} = \frac{V_p I_s}{V_s I_p} = \left(\frac{N_p}{N_s}\right)^2$$

where we have used Eqs. 30–8 and 30–9 for a transformer. Thus the impedance can be changed with a transformer. A transformer is used for this purpose in some audio amplifiers; often it has several taps corresponding to 4 Ω, 8 Ω, and 16 Ω so the output can be matched to the impedance of any loudspeaker.

Some instruments, such as oscilloscopes, require only a signal voltage but very little power. Maximum power transfer is then not important and such instruments can have a high input impedance. This has the advantage that the instrument draws very little current and disturbs the original circuit as little as possible. This is often desirable in laboratory experiments.

* 32–8 Electric Hazards; Leakage Currents

An electric shock can cause damage to the body or may even be fatal. The severity of a shock depends on the magnitude of the current, how long it acts, and through what part of the body it passes. A current passing through vital organs such as the heart or brain is especially serious for it can interfere with their operation.

Most people can "feel" a current of about 1 mA. Currents of a few mA cause pain but rarely much damage in a healthy person. However, currents above 10 mA cause severe contraction of the muscles; in this case a person may not be able to release the source of the current (say a faulty appliance or wire). Death from paralysis of the respiratory system could then occur; artificial respiration, however, can often revive a victim. If a current above about 70 mA passes in the region of the heart for a second or more, the heart muscles will begin to contract irregularly and blood is not properly pumped. This condition is called "ventricular fibrillation." If it lasts for long, death results. Strangely enough, if the current is much larger, on the order of 1 A, the damage may be less and death less likely.[†]

The seriousness of a shock depends on the impedance of the body. Living tissue has quite low resistance since the fluid of cells contains ions that can conduct quite well. However, the outer layers of skin, when dry, offer much resistance. The resistance of a human body when the skin is dry is in the range of 10^4 to 10^6 Ω. However, when the skin is wet the resistance may be 10^3 Ω or less. A person in good contact with the ground who touches a 120-V dc line with wet hands can suffer a current

$$I = \frac{120 \text{ V}}{1000 \text{ }\Omega} = 120 \text{ mA}.$$

As we saw above, this could be lethal.

The human body acts as if it had capacitance in parallel with its resistance. A dc current can pass only through the resistance, and if this is large the current

---

[†] Apparently, larger currents bring the entire heart to a standstill. Upon release of the current the heart returns to its normal rhythm. This may not happen when fibrillation occurs since it is often hard to stop once it starts.

will be small. An ac current can pass through both the resistance and the capac-itance; since they are in parallel, the net impedance is less and the current, for a given V_{rms} will be greater than for the same dc voltage. At 60 Hz, the impedance may be less than half what it is for dc. Thus an ac voltage is more dangerous than an equal dc voltage.

Another danger is *leakage current*, by which we mean a current along an unintended path. Most commonly, leakage currents are capacitively or inductively coupled. For example, a wire in a lamp forms a capacitor with the metal case; charges moving in one conductor attract or repel charge in the other. Typical electrical codes limit leakage currents to 1 mA for any device. A 1-mA leakage current is usually harmless. It can be very dangerous, however, to a hospital patient with implanted electrodes connected to ground through the apparatus. This is because the current can pass directly through the heart as compared to the usual situation where it spreads out through the body. Although 70 mA may be needed to cause heart fibrillation (rapid irregular contractions of muscle fibers) when entering through the hands (very little of it actually passes through the heart), as little as 0.02 mA has been known to cause fibrillation when passing directly to the heart. Thus a "wired" patient is in considerable danger from leakage current even from as simple an act as touching a lamp.

Summary

Capacitance and inductance offer *impedance* to the flow of alternating current just as resistance does. This impedance is referred to as reactance, X. For capacitance and inductance the reactance is defined as the equivalent of R in Ohm's law: as the proportionality constant between voltage and current (either the rms or peak values); across a capacitor, $V_0 = I_0 X_C$, and across an inductor, $V_0 = I_0 X_L$. The reactance of a capacitor decreases with frequency: $X_C = 1/\omega C$; the reactance of an inductor increases with frequency: $X_L = \omega L$. Whereas the current through a resistor is always in phase with the voltage across it, this is not true for inductors and capacitors: in an inductor, the current lags the voltage by 90°, and in a capacitor the current leads the voltage by 90°.

In an LRC series circuit, the total impedance Z is defined by the equivalent of Ohm's law, $\mathscr{E}_0 = I_0 Z$ or $\mathscr{E}_{rms} = I_{rms} Z$; it is related to R, C, and L by $Z = \sqrt{R^2 + (X_L - X_C)^2}$. The current in the circuit lags (or leads) the source voltage by an angle ϕ given by $\cos\phi = R/Z$. Only the resistor in an LRC circuit dissipates energy, and at a rate $\bar{P} = I_{rms}^2 Z \cos\phi$.

An LRC series circuit resonates at an angular frequency $\omega_0 = \sqrt{LC}$. The rms current in the circuit is largest when the applied voltage has an angular frequency equal to ω_0. The lower the resistance R, the higher and sharper the resonance peak.

Questions

1. Under what conditions is the impedance in an LRC circuit a minimum?

2. Why can we assume that the current in an LRC circuit will have the same frequency as the applied emf?

3. How can we justify using phasor diagrams, with the potential differences as vectors, when we know that potential difference is not a vector?

4. In an LRC circuit, if $X_L > X_C$, the circuit is said to be predominantly "inductive." And if $X_C > X_L$, the circuit is said to be predominantly "capacitive." Discuss the reasons for these terms. In particular, do they say anything about the relative values of L and C at a given frequency?

5. Do the results of Section 32–5 approach the proper expected results when ω approaches zero? What are the expected results?

6. When an ac generator is connected to an LRC circuit, where does the energy come from ultimately? Where does it go? How do the values of L, C, and R affect the energy supplied by the generator?

7. Discuss the validity of both of Kirchhoff's rules (Section 27–3) when applied to ac circuits that contain several loops.

8. Is it possible for the instantaneous power output of an ac generator connected to an LRC circuit ever to be negative? Explain.

9. Can you tell whether the current in an LRC circuit leads or lags the applied voltage from a knowledge of the power factor, $\cos\phi$?

10. If $\cos\phi$ were less than zero, Eq. 32–12 tells us that $\bar{P} < 0$. Can this happen? Can $\cos\phi$ be made negative? Explain.

11. Does the power factor, $\cos\phi$, depend on frequency? Does the power dissipated in an *LRC* circuit depend on frequency?

12. Is ϕ the phase angle between (a) $\mathcal{E}_0$ and I_0, (b) $\mathcal{E}$ and I, (c) $\mathcal{E}_{rms}$ and I_{rms}?

13. What is the significance of the sign of $\phi(+$ or $-)$? Is this a convention, or is it a fixed rule?

14. Describe briefly how the frequency of the source emf affects the impedance of (a) a pure resistance, (b) a pure capacitance, (c) a pure inductance, (d) an *LRC* circuit near resonance (*R* small), (e) an *LRC* circuit far from resonance (*R* small).

15. Discuss the response of an *LRC* circuit as $R \rightarrow 0$ when the frequency is (a) at resonance, (b) near resonance, (c) far from resonance. Is there energy dissipation in each case?

16. Can you tell whether or not a circuit is in resonance if you are given the value of the power factor, $\cos\phi$?

17. An *LC* resonance circuit is often called an *oscillator* circuit. What is it that oscillates?

18. The response of an *LRC* circuit at resonance is the analog of damped harmonic motion; and the resonant angular frequency, $\omega_0 = 1/\sqrt{LC}$, for the current as given by Eq. 32–13 is the same as that for a pure (undamped, $R = 0$) *LC* circuit (Section 31–5). Yet we saw in Chapter 14 (and also in Section 31–6) that resonance in damped harmonic motion occurs at a slightly different frequency from that for undamped motion. Is this a contradiction? Explain. (Hint: consider the quantities that are "oscillating.")

Problems

SECTIONS 32–1 TO 32–4

1. (I) At what frequency will a 20.0-mH inductor have a reactance of 880 Ω?

2. (I) At what frequency will a 14.0-μF capacitor have a reactance of 28.2 kΩ?

3. (I) Plot a graph of the impedance of a 1.6-μF capacitor as a function of frequency from 10 Hz to 1000 Hz.

4. (I) Plot a graph of the impedance of a 2.0-mH inductor as a function of frequency from 100 Hz to 10,000 Hz.

5. (I) Calculate the impedance of, and rms current in, a 16.0-mH radio coil connected to a 400-V (rms) 33.3-kHz ac line. Ignore resistance.

6. (II) Show that if $I = I_0 \sin \omega t$, then the voltage (a) across a capacitor C is $V_C = -V_{C0} \cos \omega t = V_{C0} \sin (\omega t - 90°)$ where $V_{C0} = I_0/\omega C$, (b) across an inductor L is $V_L = V_{L0} \cos \omega t = V_{L0} \sin (\omega t + 90°)$ where $V_{L0} = I_0 \omega L$.

7. (II) (a) What is the impedance of a well-insulated 0.025-μF capacitor connected to a 2.1-kV (rms), 200-Hz line? (b) What will be the peak value of the current and its frequency?

8. (II) What is the inductance L of the primary of a transformer whose input is 110 V at 60 Hz and the current drawn is 2.2 A? Assume no current in the secondary.

9. (II) A current $I = 3.1 \cos 377 t$ (I in amps, t in seconds) flows in a series *LR* circuit in which $L = 0.85$ mH and $R = 160$ Ω. What is the average power dissipation?

10. (II) A capacitor is placed in parallel across a load as in Fig. 32–3b to filter out stray high-frequency signals, but to allow ordinary 60-Hz ac to pass through with little loss. Suppose circuit B in the figure is a resistance $R = 200$ Ω connected to ground, and that $C = 2.3$ μF. What percent of the incoming current will pass through C rather than R if (a) it is 60 Hz; (b) 60,000 Hz?

11. (II) Suppose circuit B in Fig. 32–3a is a resistance $R = 200$ Ω, and the capacitance $C = 2.0$ μF. Will this capacitor act to eliminate 60-Hz ac but pass a high-frequency signal of frequency 60,000 Hz? To check this, determine the voltage drop across R for a 30-mV signal of frequency (a) 60 Hz; (b) 60,000 Hz.

SECTION 32–5

12. (I) A 2.1-kΩ resistor and a 2.8-μF capacitor are connected in series to an ac source. Calculate the impedance of the circuit if the source frequency is (a) 60 Hz; (b) 60,000 Hz.

13. (I) A 4.0-kΩ resistor is in series with a 6.1-mH inductor and an ac source. Calculate the impedance of the circuit if the source frequency is (a) 60 Hz; (b) 30,000 Hz.

14. (I) For a 120-V, 60-Hz voltage, a current of 70 mA passing through the body for 1 s could be lethal. What must be the impedance of the body for this to occur?

15. (I) Construct a phasor diagram, like that of Fig. 32–7, for the case when $\mathcal{E} = \mathcal{E}_0 \sin \omega t$.

16. (II) (a) What is the rms current in an *RC* circuit if $R = 4.7$ kΩ, $C = 0.20$ μF, and the rms applied voltage is 120 V at 60 Hz? (b) What is the phase angle between voltage and current? (c) What is the power dissipated by the circuit? (d) What are the voltmeter readings across R and C?

17. (II) (a) What is the rms current in a series *LR* circuit when a 60.0-Hz, 120-V ac voltage is applied, where $R = 65.0$ Ω and $L = 50.0$ mH? (b) What is the phase angle between voltage and current? (c) How much power is dissipated? (d) What are the rms voltage readings across R and L?

18. (II) A voltage $V = 8.1 \sin 754t$ is applied to an *LRC* circuit (I is in amps, t is in seconds) which has $L = 12.0$ mH, $R = 2.0$ kΩ, and $C = 0.30$ μF. (a) What is the impedance and phase angle? (b) How much power is dissipated in the circuit? (c) What is the rms current and voltage across each element?

19. (II) A circuit contains two elements, but it is not known if they are L, R, or C. The current in this circuit when connected to a 120-V, 60-Hz source is 8.1 A and leads the voltage by 13°. What are the two elements and what are their values?

20. (II) A 35-mH inductor with 2.0-Ω resistance is connected in series to a 20-μF capacitor and a 60-Hz, 45-V source. Calculate (a) the rms current; (b) the phase angle; and (c) the energy dissipated in this circuit.

21. (II) A 23-mH coil whose resistance is 0.80 Ω is connected to a capacitor C and a 360-Hz source voltage. If the current and voltage are to be in phase, what value must C have?

22. (II) What is the resistance of a coil if its impedance is 225 Ω and its reactance is 20 Ω?

23. (II) Show that for the LRC circuit of Fig. 32–5 if $I = I_0 \cos \omega t$, then $V_R = I_0 R \cos \omega t$, $V_L = I_0 \omega L \cos (\omega t + \pi/2)$ and $V_C = (I_0/\omega C) \cos (\omega t - \pi/2)$ where $\omega = 2\pi f$.

24. (II) In the LRC circuit of Fig. 32–5, suppose $I = I_0 \sin \omega t$ and $\mathcal{E} = \mathcal{E}_0 \sin (\omega t + \phi)$. Determine the power dissipated in the circuit from $P = I\mathcal{E}$ using these equations and show that on the average, $\bar{P} = \frac{1}{2}\mathcal{E}_0 I_0 \cos \phi$, which confirms Eq. 32–12.

25. (II) If $\mathcal{E} = \mathcal{E}_0 \sin \omega t$, what is the average value of $\mathcal{E}$ over (a) a whole cycle, (b) a half cycle? How do these compare to $\mathcal{E}_{\text{rms}}$?

26. (II) An inductance coil draws 2.8 A dc when connected to a 45-V battery. When connected to a 60-Hz, 120-V source, the current drawn is 4.6 A rms. Determine the inductance and resistance of the coil.

27. (II) (a) Write Eq. 32–8 as a differential equation with the charge Q as variable, assuming $\mathcal{E} = \mathcal{E}_0 \sin \omega t$. (b) Show that a solution of the form $Q = Q_0 \sin (\omega t + \phi)$ satisfies this equation. (c) Solve for Q_0 and ϕ in terms of $\mathcal{E}_0$, L, R, C, and ω. (d) Differentiate the equation obtained in part (a) to obtain an equation for the current I. (e) Try a solution of the form $I = I_0 \sin (\omega t + \phi)$ and solve for I_0 and ϕ as above.

28. (III) Filter circuit. Figure 32–10 shows a simple filter circuit designed to pass dc voltages with minimal attenuation and to remove, as much as possible, any ac components (such as 60-Hz line voltage that could cause hum in a stereo receiver, for example). Assume $V_{\text{in}} = V_1 + V_2$ where V_1 is dc and $V_2 = V_{20} \sin \omega t$, and that any resistance is very small. (a) Determine the current through the capacitor; give amplitude and phase (assume $R = 0$ and $X_L > X_C$). (b) Show that the ac component of the output voltage, $V_{2\,\text{out}}$, equals Q/C where Q is the charge on the capacitor at any instant, and determine the amplitude and phase of $V_{2\,\text{out}}$. (c) Show that the attenuation of the ac voltage is greatest when $X_C \ll X_L$, and calculate the ratio of the output to input ac voltage in this case. (d) Compare the dc output to input voltage.

29. (III) Show that if the inductor L in the filter circuit of Fig. 32–10 (problem 28) is replaced by a large resistor R, there will still be significant attenuation of the ac voltage and little attenuation of the dc voltage if the input dc voltage is high and the current (and power) are low.

30. (III) A resistor R, capacitor C, and inductor L are connected in parallel across an ac generator as shown in Fig. 32–11. The source emf is $\mathcal{E} = \mathcal{E}_0 \sin \omega t$. Determine the

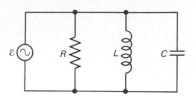

FIGURE 32–11

current as a function of time (including amplitude and phase) (a) in the resistor, (b) in the inductor, (c) in the capacitor. (d) What is the total current leaving the source? (Give amplitude I_0 and phase.) (e) Determine the impedance Z defined as $Z = \mathcal{E}_0/I_0$. (f) What is the power factor?

SECTION 32–6

31. (I) A 1200-pF capacitor is connected to a 16.0-μH coil of resistance 2.00 Ω. What is the resonant frequency of this circuit?

32. (I) What is the resonant frequency of the LRC circuit of Example 32–3? At what rate is energy taken from the generator, on the average, at this frequency?

33. (II) An LRC circuit has $L = 2.15$ mH and $R = 120$ Ω. (a) What value must C have to produce resonance at 33.0 kHz? (b) What will be the maximum current at resonance if the peak external voltage is 136 V? (c) At what frequencies will the current have half its maximum value?

34. (II) What will be the peak current in the previous problem if the capacitor is chosen so that the resonant frequency is twice the applied frequency of 33.0 kHz?

35. (II) (a) Show that oscillation of charge Q on the capacitor of an LRC circuit has amplitude

$$Q_0 = \frac{\mathcal{E}_0}{\sqrt{(\omega R)^2 + \left(\omega^2 L - \dfrac{1}{C}\right)^2}}.$$

(b) At what angular frequency, ω', will Q_0 be a maximum? (c) Compare to a damped harmonic oscillator, and discuss. (See also Question 18 in this chapter.)

36. (II) Show that the width of a sharp resonance peak, defined as the difference in (angular) frequency between the two frequencies where $I = \frac{1}{2}I_0$, is given by $\Delta\omega \approx \sqrt{3}R/L$.

37. (II) (a) Determine a formula for the average power P dissipated in an LRC circuit in terms of L, R, C, ω, and $\mathcal{E}_0$. (b) At what frequency is the power a maximum? (c) Find an approximate formula for the width of the resonance peak in average power, $\Delta\omega$, where $\Delta\omega$ is the difference in the two (angular) frequencies where $\bar{P}$ has half its maximum value. Assume a sharp peak.

38. (III) The Q value of a resonant circuit is defined as the ratio of the voltage across the capacitor (or inductor) to the voltage across the resistor, at resonance. The larger the Q value, the sharper the resonance curve. (a) Show that the Q value is given by the equation $Q = (1/R)\sqrt{L/C}$. (b) At a resonant frequency $f_0 = 2.0$ MHz, what must be the value of L and R to produce a Q value of 1000; assume $C = 0.018$ μF. (c) What is the Q value of the circuit in Example 32–3?

FIGURE 32–10

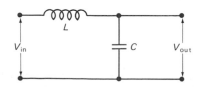

39. (II) A resonant circuit using a 120-pF capacitor is to resonate at 18.0 MHz. The air-core inductor is to be a solenoid with closely packed coils made from 12.0 m of insulated wire 1.1 mm in diameter. How many loops will the inductor contain?

SECTION 32–7

*40. (I) The output of an ECG amplifier has an impedance of 18,000 Ω. It is to be connected to an 8.0-Ω loudspeaker through a transformer. What should be the turns ratio of the transformer?

41. (I) An audio amplifier has output connections for 4 Ω, 8 Ω, and 16 Ω. If two 8-Ω speakers are to be connected in parallel, to which output terminals should they be connected?

SECTION 32–8

*42. (II) A current of 0.02 mA has been known to cause heart fibrillation in a patient with implanted electrodes. (a) What value of leakage capacitance would this correspond to for a 60-Hz, 120-V line? (b) If the conductors were, say, 2 mm apart, how large an area would each have to have. Is this unreasonable?

NUMERICAL/PROGRAMMABLE CALCULATOR†

*43. (III) A series LRC circuit, with $L = 110$ mH, $R = 150$ Ω, and $C = 200$ μF, is connected into a 150-V rms ac source of emf of variable frequency. Show that the differential equation describing such a circuit is

$$L \frac{d^2 I}{dt^2} + R \frac{dI}{dt} + \frac{1}{C} I = \frac{d\mathscr{E}}{dt}.$$

Design a program that will integrate this equation and will plot I for a source voltage $\mathscr{E} = \mathscr{E}_0 \sin \omega t$. Determine ω_0 for the circuit, and set $\omega = \omega_0/2$. Then plot I vs. t for about 10 cycles starting at $t = 0$; assume that at $t = 0$, $I = 0$ and $dI/dt = 0$. Note that there is a "transient behavior" before the current becomes a steady state pattern. Explain.

*44. (III) Repeat problem 43 for $\omega = \omega_0$. If you have done problem 43, compare the results and comment.

*45. (III) Repeat problem 43 for $\omega = 2\omega_0$. If you have done problem 43, compare the results and comment.

† See Section 2–10.

Maxwell's Equations and Electromagnetic Waves

33

The culmination of electromagnetic theory in the nineteenth century was the prediction, and the experimental verification, that waves of electromagnetic fields could travel through space. This achievement opened a whole new world of communication—first the wireless telegraph, then radio and television. And it yielded the spectacular prediction that light is an electromagnetic wave.

The theoretical prediction of electromagnetic waves was the work of the Scottish physicist James Clerk Maxwell (1831–1879), who unified, in one magnificent theory, all the phenomena of electricity and magnetism.

The development of electromagnetic theory in the early part of the nineteenth century by Oersted, Ampère, and others was not actually done in terms of electric and magnetic fields. The idea of the field was introduced somewhat later by Faraday. But the field concept was not generally used until Maxwell showed that all electric and magnetic phenomena could be described using only four equations involving electric and magnetic fields. These equations, known as *Maxwell's equations*, are as fundamental as Newton's laws. They are a magnificent summary of electromagnetic phenomena. Before proceeding to a discussion of Maxwell's equations and electromagnetic waves, we must discuss a major new prediction of Maxwell's and, in addition, Gauss's law for magnetism. These will be discussed in the first two sections.

33–1 Displacement Current; Changing Electric Fields Produce Magnetic Fields

That a magnetic field is produced by an electric current was discovered by Oersted and the mathematical relation is given by Ampère's law (Eq. 29–2):

$$\oint \mathbf{B} \cdot dl = \mu_0 I.$$

Magnetic fields might be produced in another way, Maxwell argued. For, if a changing magnetic field produces an electric field, as given by Faraday's law

(Chapter 30), then the reverse might be true as well: that *a changing electric field will produce a magnetic field*. Notice that this was a *hypothesis* by Maxwell, based on the idea of symmetry in the world. Indeed, the size of the effect in most cases is so small that Maxwell recognized it would be difficult to detect experimentally.

Maxwell used an indirect argument to back up his idea that a changing electric field would produce a magnetic field. The argument goes something like this. According to Ampère's law, you divide any closed path into short segments *dl*, multiply each segment by the parallel component of the magnetic field **B** at that segment, and then sum (integrate) all these products over the complete closed path; this sum will then equal μ_0 times the net current I that passes through the surface bounded by the path. When we applied Ampère's law to the field around a straight wire, we imagined the current as passing through the circle enclosed by our circular loop. This would be the flat surface 1 in Fig. 33–1. However, we could just as well use the sack-shaped surface 2 in the figure since the same current I passes through it. This naturally implies that the current passing through any surface enclosed by the closed path must be the same. Thus the current flowing into the volume enclosed by surfaces 1 and 2 together equals the current that flows out of this volume. This is basically Kirchhoff's point rule, and tells us that the rate at which charge enters the volume equals the rate at which it leaves.

Now consider a slightly different situation, Fig. 33–2a, in which a capacitor is being discharged. Ampère's law works for surface 1. But it does not work for surface 2 since no current passes through surface 2. There is a magnetic field around the wire, so the left side of Ampère's law is not zero; yet for surface 2, the right side *is* zero since $I = 0$. Similar things could be said for

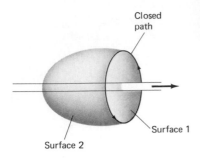

FIGURE 33–1

Ampère's law applied to two different surfaces bounded by the same closed path.

FIGURE 33–2

Capacitor discharging. No conduction current passes through surface 2 in either (a) or (b). An extra term is needed in Ampère's law.

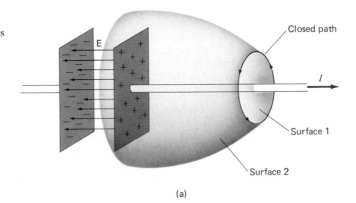

(a)

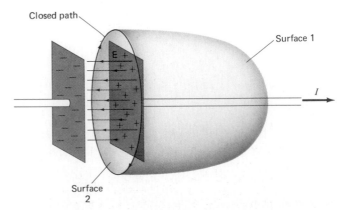

(b)

the closed path indicated in Fig. 33–2b that surrounds the region of the electric field of the plates; there is a current passing through surface 1 but not surface 2, and there is a magnetic field around the closed path. Actually, there is a magnetic field present only if charge is flowing to or away from the capacitor plates; in this case, the electric field between the plates is changing in time. Maxwell resolved the problem of no current through surface 2 (in either part a or b of the figure) by stating that the changing electric field between the plates is *equivalent* to an electric current. He called this a **displacement current**, I_D. (The name is not especially illuminating today.) An ordinary current, I, is then called a **conduction current**. Ampère's law can then be written

$$\oint \mathbf{B} \cdot d\mathbf{l} = \mu_0(I + I_D).$$ (33–1)

Ampère's law will then apply even for surface 2 in Fig. 33–2a or b where the displacement current I_D refers to the changing electric field.[†]

We can write the displacement current I_D in terms of the changing electric field between the capacitor plates in Fig. 33–2. The charge Q on a capacitor of capacitance C is $Q = CV$ where V is the potential difference between the plates. Also recall that $V = Ed$ where d is the (small) separation of the plates and E is the electric field strength between them, and we ignore any fringing of the field. Also, for a parallel-plate capacitor, $C = \varepsilon_0 A/d$, where A is the area of each plate (see Chapter 25). We combine these to obtain:

$$Q = CV = \left(\varepsilon_0 \frac{A}{d}\right)(Ed) = \varepsilon_0 AE.$$

Now if the charge on the plate changes at a rate dQ/dt, the electric field changes at a proportional rate. That is, by differentiating the expression above we have:

$$\frac{dQ}{dt} = \varepsilon_0 A \frac{dE}{dt}.$$

Now dQ/dt is the rate at which charge accumulates on or leaves the capacitor plates; it is therefore equal to the current flowing into or out of the capacitor. If the current flow into the capacitor, dQ/dt, is set equal to the displacement current I_D between the plates, then

$$I_D = \frac{dQ}{dt} = \varepsilon_0 A \frac{dE}{dt}.$$ (33–2)

This can be written

$$I_D = \varepsilon_0 \frac{d\Phi_E}{dt}$$

where $\Phi_E = EA$ is the *electric flux*. Then, Ampère's law becomes

$$\oint \mathbf{B} \cdot d\mathbf{l} = \mu_0 I + \mu_0 \varepsilon_0 \frac{d\Phi_E}{dt}.$$ (33–3)

This equation embodies Maxwell's idea that a magnetic field can be caused not only by an ordinary electric current, but also by a changing electric field or changing electric flux. Although we derived it for a special case, Eq. 33–3 has

---

[†] Notice that this interpretation of the changing electric field as a displacement current fits in well with our discussion in Chapter 32 where we saw that an alternating current passes through a capacitor. It also means that Kirchhoff's point rule will be valid even at a capacitor plate; for conduction current flows into the plate, but no conduction current flows out of the plate. Instead a "displacement current" flows out of the plate.

proved valid in general,[†] although normally the displacement current term (the second on the right) is very small as mentioned before.

◆ EXAMPLE 33–1 A 30-pF air-gap capacitor has circular plates of area $A = 100$ cm². It is charged by a 70-V battery through a 2.0-Ω resistor. At the instant the battery is connected, the electric field between the plates is changing most rapidly. At this instant, calculate (a) the displacement current between the plates, (b) the rate of change of electric field between the plates. (c) Determine the magnetic field induced between the plates. Assume **E** is uniform between the plates at any instant and is zero at all points beyond the edges of the plates.

SOLUTION (a) In Section 27–5 we discussed RC circuits and saw that the charge on the capacitor as a function of time is

$$Q = C\mathscr{E}(1 - e^{-t/RC}),$$

where $\mathscr{E}$ is the emf of the battery. To find the current at $t = 0$, we differentiate this and substitute the values $\mathscr{E} = 70$ V, $C = 30$ pF, $R = 2.0$ Ω:

$$\frac{dQ}{dt}\bigg|_{t=0} = \frac{C\mathscr{E}}{RC} e^{-t/RC}\bigg|_{t=0} = \frac{\mathscr{E}}{R} = \frac{70 \text{ V}}{2.0 \text{ }\Omega} = 35 \text{ A}.$$

This is the rate at which charge accumulates on the capacitor and equals the current flowing in the circuit. As discussed above, this also equals the displacement current, so $I_D = 35$ A.

(b) From Eq. 33–2,

$$\frac{dE}{dt} = \frac{I_D}{\varepsilon_0 A} = \frac{35 \text{ A}}{(8.85 \times 10^{-12} \text{ C}^2/\text{N·m}^2)(1.0 \times 10^{-2} \text{ m}^2)} = 4.0 \times 10^{14} \text{ V/m·s}.$$

(c) Although we will not prove it, we might expect the lines of **B** to be perpendicular to **E** and, because of symmetry, to be circles as shown in Fig. 33–3; this is the same symmetry we saw for the inverse situation of a changing magnetic field producing an electric field (Section 30–4, see Fig. 30–6). Now we apply Ampère's law, Eq. 33–3, between the plates, where the conduction current $I = 0$:

$$\oint \mathbf{B} \cdot d\mathbf{l} = \mu_0 \varepsilon_0 \frac{d\Phi_E}{dt}.$$

We choose our path to be a circle of radius r, centered at the center of the plate, and thus following a field line. For $r \le R$ (the radius of plate) we have

$$B(2\pi r) = \mu_0 \varepsilon_0 \frac{d}{dt} (\pi r^2 E)$$

$$= \mu_0 \varepsilon_0 \pi r^2 \frac{dE}{dt}.$$

Hence

$$B = \frac{\mu_0 \varepsilon_0}{2} r \frac{dE}{dt}. \qquad [r \le R]$$

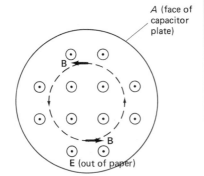

A (face of capacitor plate)

B

B

E (out of paper)

FIGURE 33–3

View of circular plate of parallel-plate capacitor. **E** between plates points out toward viewer; lines of **B** (dashed) are circles. (Example 33–1.)

[†] Actually, there is a third term on the right for the case when a magnetic field is produced by magnetized materials, the so-called magnetization current I_M discussed in Section 29–9; this can be accounted for by changing μ_0 to μ, but we will mainly be interested in cases where no magnetic material is present. In the presence of a dielectric ε_0 is replaced by $\varepsilon = K\varepsilon_0$ (see Section 25–5).

We assume $\mathbf{E} = 0$ for $r > R$, so for points beyond the edge of the plates

$$B(2\pi r) = \mu_0\varepsilon_0 \frac{d}{dt}(\pi R^2 E)$$

$$= \mu_0\varepsilon_0\pi R^2 \frac{dE}{dt}$$

or

$$B = \frac{\mu_0\varepsilon_0 R^2}{2r}\frac{dE}{dt}. \qquad [r \geq R]$$

B has its maximum value at $r = R$, which, from either relation above, is (since $R = \sqrt{A/\pi} = 5.6$ cm):

$$B = \frac{\mu_0\varepsilon_0 R}{2}\frac{dE}{dt}$$

$$= \tfrac{1}{2}(4\pi \times 10^{-7} \text{ T·m/A})(8.85 \times 10^{-12} \text{ C}^2/\text{N·m}^2)(5.6 \times 10^{-2} \text{ m})(4.0 \times 10^{14} \text{ V/m·s})$$

$$= 1.2 \times 10^{-4} \text{ T}.$$

This is a very small field and lasts only briefly (the time constant $RC = 6.0 \times 10^{-11}$ s) and so is very difficult to measure.

33–2 Gauss's Law for Magnetism

We are almost in a position to state Maxwell's equations, but first we need to discuss the magnetic equivalent of Gauss's law. As we saw in Chapter 30, for a magnetic field $\mathbf{B}$, the *magnetic flux* Φ_B through a surface is defined as

$$\Phi_B = \int \mathbf{B} \cdot d\mathbf{A}$$

where the integral is over the area of either an open or a closed surface. The magnetic flux through a closed surface—that is, a surface which completely encloses a volume—is written

$$\Phi_B = \oint \mathbf{B} \cdot d\mathbf{A}.$$

In the electric case, we saw in Section 23–2 that the electric flux Φ_E through a closed surface is equal to the total net charge Q enclosed by the surface, divided by ε_0 (Eq. 23–4):

$$\oint \mathbf{E} \cdot d\mathbf{A} = \frac{Q}{\varepsilon_0}.$$

This relation is Gauss's law for electricity.

We can write a similar relation for the magnetic flux. We have seen, however, that in spite of intense searches, no isolated magnetic poles, the magnetic equivalent of single electric charges, have ever been observed for certain. Hence, *Gauss's law for magnetism* is

$$\oint \mathbf{B} \cdot d\mathbf{A} = 0. \qquad (33-4)$$

In terms of magnetic lines of force, this relation tells us that as many lines enter the enclosed volume as leave it. If, indeed, magnetic poles do not exist, then there are no "sources" or "sinks" for magnetic field lines to start or stop on, as electric field lines start on positive charges and end on negative charges. Magnetic field lines must then be continuous. Even for a bar magnet, a magnetic field **B** exists inside as well as outside the magnetic material, and the lines of **B** are closed loops as shown in Fig. 33–4.

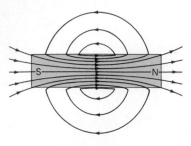

FIGURE 33–4

Magnetic field lines for a bar magnet.

 ### 33–3 Maxwell's Equations

With the extension of Ampère's law given by Eq. 33–3, plus Gauss's law for magnetism (Eq. 33–4), we are now ready to state all four of Maxwell's equations. We have seen them all before in the past dozen chapters. In the absence of dielectric or magnetic materials, **Maxwell's equations** are:

$$\oint \mathbf{E} \cdot d\mathbf{A} = \frac{Q}{\varepsilon_0} \tag{33–5a}$$

$$\oint \mathbf{B} \cdot d\mathbf{A} = 0 \tag{33–5b}$$

$$\oint \mathbf{E} \cdot d\mathbf{l} = -\frac{d\Phi_B}{dt} \tag{33–5c}$$

$$\oint \mathbf{B} \cdot d\mathbf{l} = \mu_0 I + \mu_0 \varepsilon_0 \frac{d\Phi_E}{dt}. \tag{33–5d}$$

The first two of Maxwell's equations are simply Gauss's law for electricity (Chapter 23, Eq. 23–4) and Gauss's law for magnetism (Section 33–2, Eq. 33–4). The third is Faraday's law (Chapter 30, Eq. 30–6) and the fourth is Ampère's law as modified by Maxwell (Section 33–1).

They can be summarized in words: (1) a generalized form of Coulomb's law relating electric field to its sources, electric charge; (2) the same for the magnetic field, except that if there are no magnetic monopoles, magnetic field lines are continuous—they do not begin or end (as electric field lines do on charges); (3) an electric field is produced by a changing magnetic field; (4) a magnetic field is produced by an electric current or by a changing electric field.

Maxwell's equations are the basic equations for all electromagnetism. They are fundamental in the sense that Newton's three laws of motion and the law of universal gravitation are for mechanics. In a sense they are even more fundamental, since they are valid even relativistically (Chapter 39) whereas Newton's laws are not. Because all of electromagnetism is contained in this set of four equations, Maxwell's equations are considered one of the great triumphs of the human mind.

In earlier chapters, we have seen that we can treat electric and magnetic fields separately if they do not vary in time. But we cannot treat them independently if they do change in time. For a changing magnetic field produces an electric field; and a changing electric field produces a magnetic field. An important outcome of this fact is the production of electromagnetic waves.

 ### 33–4 Production of Electromagnetic Waves

According to Maxwell, a magnetic field will be produced in empty space if there is a changing electric field. From this, Maxwell derived another startling conclusion. If a changing magnetic field produces an electric field, the electric field will itself be changing. This changing electric field will in turn produce a magnetic field. The latter will be changing and so will produce a changing electric field; and so on. When Maxwell manipulated his equations, he found that the net result of these interacting changing fields was a wave of electric and magnetic fields that can actually propagate through space! We now examine, in a qualitative way, how

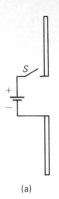
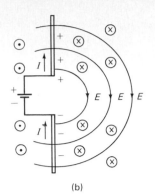

(a)

(b)

FIGURE 33–5

Fields produced by charge flowing into conductors. It takes time for the **E** and **B** fields to travel outward to distant points.

such **electromagnetic waves** are produced. (In Section 33–5 we will examine this more quantitatively.)

Consider two conducting rods that will serve as an "antenna," Fig. 33–5. Suppose these two rods are connected by a switch to the opposite terminals of a battery. As soon as the switch is closed, the upper rod quickly becomes positively charged and the lower one negatively charged. Electric field lines are formed as indicated by the lines in Fig. 33–5b. While the charges are flowing a current exists; the direction of conventional current is indicated by the arrows. A magnetic field is therefore produced; the magnetic field lines encircle the wires and therefore, in the plane of the paper, **B** points into the paper (⊗) on the right and out of the paper (⊙) on the left. Now we ask, how far out do these fields extend? In the static case, the fields extend outward indefinitely far. However, when the switch in Fig. 33–5 is closed, the fields quickly appear nearby but it takes time for them to reach distant points. Both electric and magnetic fields store energy, and this energy cannot be transferred to distant points with infinite speed.

This example illustrates that it takes time for electric and magnetic fields, once they are formed, to reach distant points. Now we look at a more interesting situation. Suppose our antenna is connected to an ac generator, Fig. 33–6. (The antenna in this case is called an "oscillating electric dipole antenna" since there is a separation of + and − charges at any instant, and its electric dipole moment is oscillating.) In Fig. 33–6a, the connection has just been completed. Charge starts building up and fields form just as in Fig. 33–5. The + and − signs indicate the net charge on each rod. The arrows indicate the direction of the current. The electric field is represented by lines in the plane of the paper; and the magnetic field, according to the right-hand rule, is into (⊗) or out of (⊙) the paper as shown. In Fig. 33–6b, the emf of the ac generator has reached its maximum and is starting to decrease; so the current is reversed and the new magnetic field is in the opposite direction. The old fields, however, don't suddenly disappear; they are on their way to distant points. But because the new fields have changed direction, the old lines fold back to connect up to some of the new lines and form closed loops as shown. In (c) the situation is shown a little later when the field lines have formed several loops (note change of scale). These field lines continue moving outward. The magnetic field lines also form closed loops; they are not shown since they are perpendicular to the paper.

The fields continue to move outward. Although the lines are shown only on the right of the source, fields also travel in other directions. (The field strengths are greatest in directions perpendicular to the oscillating charges; and they drop to zero along the direction of oscillation—above and below the antenna in Fig. 33–6.)

Several things can be noted from Fig. 33–6. First, *the electric and magnetic fields at any point are perpendicular to each other, and to the direction of motion.* Second, we can see that the fields alternate in direction (**B** is into the paper at some points and out of the paper at others; similarly for **E**). Thus the field strengths

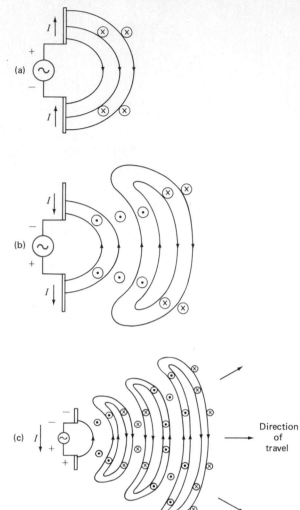

FIGURE 33-6

Sequence showing electric and magnetic fields that
spread outward from oscillating charges on two
conductors connected to an ac source (see the text).

vary from a maximum in one direction to zero to a maximum in the other direction. The electric and magnetic fields are "in phase"; that is, they each are zero at the same points and reach their maximum at the same points.

If the source emf varies sinusoidally, then the electric and magnetic field strengths will also vary sinusoidally. This is strictly true only for points far from the source. Indeed, the fields close to the antenna (the so-called "near field") is rather complex. We are mainly interested in the fields far from the antenna (the "radiation fields"), and, in fact, these are what are shown in Fig. 33-6c. The sinusoidal nature of the "far" or "radiation" fields cannot be seen in Fig. 33-6c which only shows the field lines. The sinusoidal character of the waves is diagrammed in Fig. 33-7 which shows the field *strengths* plotted as a function of distance. Notice that **B** and **E** are perpendicular to each other and to the direction of travel.

Electromagnetic (EM) waves are *transverse* waves and resemble other types of waves (Chapter 15). However, EM waves are waves of *fields*, not of matter as are waves on water or a rope.

We have seen in the above analysis that EM waves are produced by electric charges that are oscillating, and hence are undergoing acceleration. In fact, we can say in general that *accelerating electric charges give rise to electromagnetic waves.*

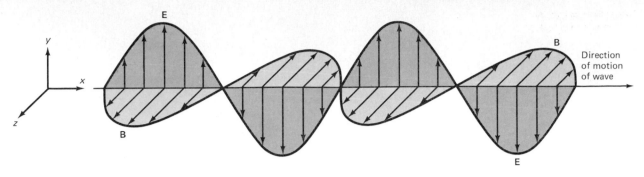

FIGURE 33–7

Electric and magnetic field strengths in an electromagnetic wave. E and B are at right angles to each other. The entire pattern moves in a direction perpendicular to both E and B.

33–5 Electromagnetic Waves, and Their Speed, from Maxwell's Equations

We now show how the existence of EM waves follows from Maxwell's equations. Also, we will see that not only was Maxwell's prediction of the existence of EM waves startling; equally startling was the speed at which they were predicted to travel.

We begin by considering a region of free space, where there are *no charges or conduction currents*—that is, far from the source so that the wave fronts (the field lines in Fig. 33–6c are essentially flat over a reasonable area. They are then called *plane waves*, meaning that at any instant, E and B are uniform over a plane perpendicular to the direction of propagation. We also assume, in a particular coordinate system, that the wave is traveling in the x direction with velocity $\mathbf{v} = v\mathbf{i}$, that E is parallel to the y axis, and that B is parallel to the z axis, as in Fig. 33–7.

If the wave is sinusoidal with wavelength λ and frequency f, then, as we saw in Chapter 15, such a traveling wave can be written as

$$E = E_y = E_0 \sin(kx - \omega t)$$
$$B = B_z = B_0 \sin(kx - \omega t) \tag{33–6}$$

where

$$k = \frac{2\pi}{\lambda}, \qquad \omega = 2\pi f, \quad \text{and} \quad f\lambda = \frac{\omega}{k} = v \tag{33–7}$$

with v being the speed of the wave. Although visualizing the wave as sinusoidal is helpful, we will not have to assume this in most of what follows.

Consider now a small rectangle in the plane of the electric field as shown in Fig. 33–8. This rectangle has a finite height Δy, and a very thin width which we take to be the infinitesimal distance dx. Let us apply Faraday's law, which is Maxwell's third equation (Eq. 33–5c),

$$\oint \mathbf{E} \cdot d\boldsymbol{l} = -\frac{d\Phi_B}{dt}$$

to the rectangle of height Δy and width dx shown in Fig. 33–8. First we consider[†] $\oint \mathbf{E} \cdot d\boldsymbol{l}$. Along the short top and bottom sections of length dx, E is perpendicular

---

[†] In order to show that E, B, and v are in the orientation shown, we apply Lenz's law to this rectangular loop. The changing magnetic flux through this loop is related to the electric field around the loop by Faraday's law (Maxwell's third equation, Eq. 33–5c). For the case shown, B through the loop is decreasing in time (the wave is moving to the right); so the electric field must be in a direction to oppose this change, meaning E must be greater on the right side of the loop than on the left, as shown (so it could produce a counterclockwise current whose magnetic field would act to oppose the change in Φ_B—but of course there is no current). This brief argument shows that the orientation of E, B, and v are in the correct relation as shown. That is, v is the direction of E × B.

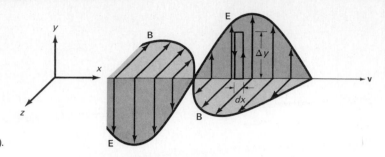

FIGURE 33–8

Applying Faraday's law to the rectangle $(\Delta y)(dx)$.

to dl, so $\mathbf{E} \cdot dl = 0$. Along the vertical sides, we let E be the electric field along the left side, and on the right side where it will be slightly larger, it is $E + dE$. Thus, if we take our loop counterclockwise,

$$\oint \mathbf{E} \cdot dl = (E + dE)\,\Delta y - E\,\Delta y = dE\,\Delta y.$$

The magnetic flux through this loop changes as

$$\frac{d\Phi_B}{dt} = \frac{dB}{dt}\,dx\,\Delta y$$

since the area of the loop, $(dx)(\Delta y)$, is not changing. Thus, Faraday's law gives us

$$dE\,\Delta y = -\frac{dB}{dt}\,dx\,\Delta y$$

or

$$\frac{dE}{dx} = -\frac{dB}{dt}.$$

Actually, both E and B are functions of position x and time t. We should therefore use partial derivatives:

$$\frac{\partial E}{\partial x} = -\frac{\partial B}{\partial t} \qquad (33\text{–}8)$$

where $\partial E/\partial x$ means the derivative of E with respect to x while t is held fixed, and $\partial B/\partial t$ is the derivative of B with respect to t while x is kept fixed.

We can obtain another important relation between E and B in addition to Eq. 33–8. To do so, we consider now a small rectangle in the plane of $\mathbf{B}$, whose length and width are Δz and dx as shown in Fig. 33–9. To this rectangular loop we apply Maxwell's fourth equation (the extension of Ampère's law):

$$\oint \mathbf{B} \cdot dl = \mu_0 \varepsilon_0 \frac{d\Phi_E}{dt}$$

where we have taken $I = 0$ since we assume the absence of conduction currents.

FIGURE 33–9

Applying Maxwell's fourth equation to the rectangle $(\Delta z)(dx)$.

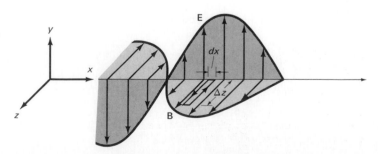

Along the short sides (dx), $\mathbf{B} \cdot d\mathbf{l}$ is zero since $\mathbf{B}$ is perpendicular to $d\mathbf{l}$. We next let B be the magnetic field along the left side of length Δz, and $B + dB$ be the field along the right side. We again integrate counterclockwise, so

$$\oint \mathbf{B} \cdot d\mathbf{l} = B\,\Delta z - (B + dB)\,\Delta z = -dB\,\Delta z.$$

Now we look at the right side of Maxwell's fourth equation, which is

$$\mu_0\varepsilon_0 \frac{d\Phi_E}{dt} = \mu_0\varepsilon_0 \frac{dE}{dt}\,dx\,\Delta z.$$

Equating the two expressions we obtain

$$-dB\,\Delta z = \mu_0\varepsilon_0 \frac{dE}{dt}\,dx\,\Delta z$$

or

$$\frac{\partial B}{\partial x} = -\mu_0\varepsilon_0 \frac{\partial E}{\partial t} \tag{33–9}$$

where we have replaced dB/dx and dE/dt by the proper partial derivatives as before.

We can use Eqs. 33–8 and 33–9 to obtain a relation between the magnitudes of $\mathbf{E}$ and $\mathbf{B}$, and the speed v. Let us assume, for the moment, that the wave is sinusoidal so that E and B are given by Eq. 33–6 as a function of x and t. When we apply Eq. 33–8, taking the derivatives of E and B as given by Eqs. 33–6, we obtain

$$kE_0 \cos(kx - \omega t) = \omega B_0 \cos(kx - \omega t)$$

or

$$\frac{E_0}{B_0} = \frac{\omega}{k} = v$$

since $v = \omega/k$ (see Eq. 33–7 or 15–12). Since E and B are in phase we see that E and B are related by

$$\frac{E}{B} = v \tag{33–10}$$

at any point in space, where v is the velocity of the wave.

Now we apply Eq. 33–9 to the sinusoidal fields (Eqs. 33–6) and we obtain

$$kB_0 \cos(kx - \omega t) = \mu_0\varepsilon_0\omega E_0 \cos(kx - \omega t)$$

or

$$\frac{B_0}{E_0} = \frac{\mu_0\varepsilon_0\omega}{k} = \mu_0\varepsilon_0 v.$$

But $B_0/E_0 = v$ from Eq. 33–10 so

$$\mu_0\varepsilon_0 v = \frac{1}{v}$$

or

$$v = \frac{1}{\sqrt{\varepsilon_0\mu_0}}. \tag{33–11}$$

Thus the speed of electromagnetic waves in free space is a constant, independent of the wavelength or frequency.

We can derive this important result without having to assume sinusoidal waves by combining Eqs. 33–8 and 33–9 as follows. We take the derivative with respect to t of Eq. 33–9

$$\frac{\partial^2 B}{\partial t\,\partial x} = -\mu_0\varepsilon_0\,\frac{\partial^2 E}{\partial t^2}.$$

We next take the derivative of Eq. 33–8 with respect to x:

$$\frac{\partial^2 E}{\partial x^2} = -\frac{\partial^2 B}{\partial t\,\partial x}.$$

Since $\partial^2 B/\partial t\,\partial x$ appears in both relations, we obtain

$$\frac{\partial^2 E}{\partial t^2} = \frac{1}{\mu_0\varepsilon_0}\,\frac{\partial^2 E}{\partial x^2}. \qquad (33\text{–}12a)$$

By taking other derivatives of Eqs. 33–8 and 33–9 we obtain the same relation for B:

$$\frac{\partial^2 B}{\partial t^2} = \frac{1}{\mu_0\varepsilon_0}\,\frac{\partial^2 B}{\partial t^2}. \qquad (33\text{–}12b)$$

Both of Eqs. 33–12 have the form of the *wave equation* for a plane wave traveling in the x direction,

$$\frac{\partial^2 y}{\partial t^2} = v^2\,\frac{\partial^2 y}{\partial x^2},$$

as discussed in Chapter 15, Section 15–4. We see that the velocity v is given by

$$v^2 = \frac{1}{\mu_0\varepsilon_0}$$

in agreement with Eq. 33–11. Thus we see that a natural outcome of Maxwell's equations is that E and B obey the wave equation for waves traveling with speed $v = 1/\sqrt{\mu_0\varepsilon_0}$. It was on this basis that Maxwell predicted the existence of electromagnetic waves.

Let us actually evaluate Eq. 33–11 for the speed of EM waves. We put in the values for ε_0 and μ_0 and we find

$$v = \frac{1}{\sqrt{\varepsilon_0\mu_0}} = \frac{1}{\sqrt{(8.85 \times 10^{-12}\ \text{C}^2/\text{N}\cdot\text{m}^2)(4\pi \times 10^{-7}\ \text{T}\cdot\text{m/A})}}$$
$$= 3.00 \times 10^8\ \text{m/s}.$$

This is a remarkable result. For this is precisely equal to the measured speed of light!

 ## 33–6   Light as an Electromagnetic Wave; the Electromagnetic Spectrum

The calculation at the end of the last section gives the result that Maxwell himself determined: that the speed of EM waves is 3.00×10^8 m/s, the same as the measured speed of light.

Light had been shown some 60 years previously to behave like a wave (we'll discuss this in Chapter 36). But nobody knew what kind of wave it was— that is, what is it that is oscillating in a light wave? Maxwell, on the basis of the calculated speed of EM waves, argued that light must be an electromagnetic wave. This soon came to be generally accepted by scientists, but not fully until after EM waves were experimentally detected. EM waves were first generated and detected experimentally by Heinrich Hertz (1857–1894) in 1887, 8 years after Maxwell's death. Hertz used a spark-gap apparatus in which charge was made to rush back and forth for a short time; this generated waves whose frequency was about 10^9 Hz. He detected them some distance away using a loop of wire in which an emf was produced when a changing magnetic field passed through. These waves were later shown to travel at the speed of light, 3.00×10^8 m/s, and to exhibit all the characteristics of light such as reflection, refraction, and interference. The only difference was that they were not visible. Hertz's experiment was a strong confirmation of Maxwell's theory.

The wavelengths of visible light were measured in the first decade of the nineteenth century, long before anyone imagined that light was an electromagnetic wave. The wavelengths were found to lie between 4.0×10^{-7} m and 7.5×10^{-7} m; or 400 nm to 750 nm. The frequencies of visible light rays can be found using Eq. 15–1 or 33–7, which we rewrite here:

$$f\lambda = c.$$

Here, c is the velocity of light, 3.00×10^8 m/s; it gets the special symbol c because of its universality for all EM waves in free space; f and λ are the frequency and wavelength of the wave. This equation tells us that the frequencies of visible light vary between 4.0×10^{14} Hz and 7.5×10^{14} Hz.

But visible light is only one kind of EM wave. As we have seen, Hertz produced EM waves of much lower frequency, about 10^9 Hz. These are called *radio waves*, since frequencies in this range are used today to transmit radio and TV signals. Electromagnetic waves, or EM radiation as we sometimes call it, have been produced or detected over a wide range of frequencies. They are usually categorized as shown in Fig. 33–10. This is known as the **electromagnetic spectrum**.

Radio waves and microwaves can be produced experimentally using electronic equipment, as discussed earlier; see Fig. 33–6. Higher-frequency waves are very difficult to produce electronically. These and other types of EM waves are produced in natural processes, as emission from atoms, molecules, and nuclei (more on this later). Generally, EM waves are produced by the acceleration of electrons or other charged particles, such as electrons accelerating in the antenna of Fig. 33–6. Another example is X rays which are produced (see Chapter 41) when fast-moving electrons are rapidly decelerated when striking a metal target. Even the visible light emitted by an ordinary incandescent light is due to electrons undergoing acceleration within the hot filament. We will meet various types of EM waves later. However, it is worth mentioning here that infrared (IR) radiation (EM waves whose frequency is just less than that of visible light) is mainly responsible for the heating effect of the sun. The sun emits not only visible light but substantial amounts of IR and UV (ultraviolet) as well. The molecules of our skin tend to "resonate" at infrared frequencies, so it is these that are preferentially absorbed and thus warm us up.

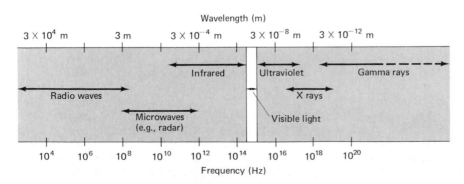

FIGURE 33–10
Electromagnetic spectrum.

EXAMPLE 33–2   Calculate the wavelength (a) of a 60-Hz EM wave; (b) of a 1240-kHz AM radio wave.

SOLUTION   (a) Since $c = \lambda f$,

$$\lambda = \frac{c}{f} = \frac{3.0 \times 10^8 \text{ m/s}}{60 \text{ s}^{-1}} = 5.0 \times 10^6 \text{ m},$$

or 5000 km. One wavelength stretches all the way across the USA!

(b) $$\lambda = \frac{3.0 \times 10^8 \text{ m/s}}{1.24 \times 10^6 \text{ s}^{-1}} = 240 \text{ m}.$$

33–7   Energy in EM Waves; the Poynting Vector

Electromagnetic waves carry energy from one region of space to another. This energy is associated with the moving electric and magnetic fields. In Chapter 25 (Eq. 25–6) we saw that the energy density (J/m³) stored in an electric field E is $u_E = \frac{1}{2}\varepsilon_0 E^2$, where u_E is the electric energy per unit volume. The energy stored in a magnetic field B, as we discussed in Chapter 31 (Eq. 31–9) is given by $u_B = \frac{1}{2}B^2/\mu_0$. Thus the total energy stored per unit volume in a region of space where there is an electromagnetic wave is

$$u = \frac{1}{2}\varepsilon_0 E^2 + \frac{1}{2}\frac{B^2}{\mu_0}. \tag{33–13}$$

In this equation, E and B represent the electric and magnetic field strengths of the wave at any instant at a given point in space. We can write Eq. 33–13 in terms of the E field only, since we know $\sqrt{\varepsilon_0\mu_0} = 1/c$, and $B = E/c$ (Eqs. 33–10 and 33–11.) We insert these into Eq. 33–13 to obtain

$$u = \frac{1}{2}\varepsilon_0 E^2 + \frac{1}{2}\frac{\varepsilon_0\mu_0 E^2}{\mu_0}$$

$$u = \varepsilon_0 E^2. \tag{33–14a}$$

Notice that the energy density associated with the B field is equal to that associated with the E field, so each contributes half to the total energy. We can also write the energy density in terms of the B field only, or in one term containing both: $u = \varepsilon_0 E^2 = \varepsilon_0 c^2 B^2 = \varepsilon_0 B^2/\varepsilon_0\mu_0$, or

$$u = \frac{B^2}{\mu_0}; \tag{33–14b}$$

also $u = \varepsilon_0 E^2 = \varepsilon_0 EcB = \varepsilon_0 EB/\sqrt{\varepsilon_0\mu_0}$, or

$$u = \sqrt{\frac{\varepsilon_0}{\mu_0}}\, EB. \tag{33–14c}$$

Equations 33–14 give the energy density in any region of space at any instant.

Now let us determine the energy the wave transports per unit time per unit area. This is given by a vector **S**, which is called the *Poynting vector*.[†] The units of **S** are W/m². The direction of **S** is the direction in which the energy is transported, which is the direction in which the wave is moving. Let us imagine the wave in passing through an area A perpendicular to the x axis as shown in

[†] After J. H. Poynting (1852–1914), a close friend and colleague of J. J. Thomson.

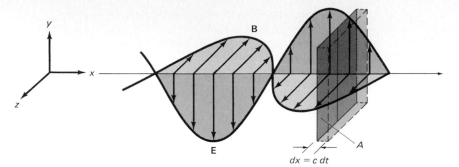

FIGURE 33–11
Electromagnetic wave carrying
energy through area A.

Fig. 33–11. In a short time dt, the wave moves to the right a distance $dx = c\,dt$ where c is the wave speed. The energy that has passed through A in the time dt is the energy that now occupies the volume $dV = A\,dx = Ac\,dt$. The energy density u is $u = \varepsilon_0 E^2$ where E is the electric field in this volume at the given instant. So the total energy dU contained in this volume dV is the energy density u times the volume: $dU = u\,dV = (\varepsilon_0 E^2)(Ac\,dt)$. Therefore the energy crossing the area A per time dt is

$$S = \frac{1}{A}\frac{dU}{dt} = \varepsilon_0 c E^2. \tag{33–15}$$

Since $E = cB$ and $c = 1/\sqrt{\varepsilon_0\mu_0}$, this can also be written:

$$S = \varepsilon_0 c E^2 = \frac{cB^2}{\mu_0} = \frac{EB}{\mu_0}.$$

The direction of $\mathbf{S}$ is along $\mathbf{v}$, perpendicular to $\mathbf{E}$ and $\mathbf{B}$, so the Poynting vector $\mathbf{S}$ can be written

$$\mathbf{S} = \frac{1}{\mu_0}(\mathbf{E} \times \mathbf{B}). \tag{33–16}$$

Equation 33–15 gives the energy transported per unit area per unit time at any *instant*. We often want to know the *average* over an extended period of time. If E and B are sinusoidal, then $\overline{E^2} = E_0^2/2$, just as for electric currents and voltages, where E_0 is the *maximum* value of E. Thus we can write for the magnitude of the Poynting vector, on the average,

$$\bar{S} = \frac{1}{2}\varepsilon_0 c E_0^2 = \frac{1}{2}\frac{c}{\mu_0}B_0^2 = \frac{E_0 B_0}{2\mu_0}, \tag{33–17}$$

where B_0 is the maximum value of B.

EXAMPLE 33–3 Radiation from the sun reaches the earth (above the atmosphere) at a rate of about 1350 J/s·m². Assume this is a single EM wave and calculate the maximum values of E and B.

SOLUTION Since $\bar{S} = 1350$ J/s·m² $= \varepsilon_0 c E_0^2/2$, then

$$E_0 = \sqrt{\frac{2\bar{S}}{\varepsilon_0 c}} = \sqrt{\frac{2(1350 \text{ J/s·m}^2)}{(8.85 \times 10^{-12} \text{ C}^2/\text{N·m}^2)(3.0 \times 10^8 \text{ m/s})}}$$

$$= 1.01 \times 10^3 \text{ V/m}.$$

From Eq. 33–10,

$$B_0 = \frac{E_0}{c} = \frac{1.01 \times 10^3 \text{ V/m}}{3.0 \times 10^8 \text{ m/s}} = 3.4 \times 10^{-6} \text{ T}.$$

This example illustrates the fact that B has a small numerical value compared to E; this is because of the different units for E and B and the way these units are defined; but, as we saw earlier, B contributes the same energy to the wave as E does.

* 33–8   Maxwell's Equations and Electromagnetic Waves in Matter

Maxwell's equations must be modified somewhat in the presence of dielectric materials and/or magnetic materials. We consider, briefly, only materials that are homogeneous and isotropic (the same properties in all directions), and for which the dielectric constant K and the magnetic permeability K_m are constants. Then in Maxwell's equations, Eqs. 33–5, we replace ε_0 by $\varepsilon = K\varepsilon_0$ and μ_0 by $\mu = K_m\mu_0$; also, Q in Eq. 33–5a must be only the free charge. (See Chapters 25 and 29.)

The speed of electromagnetic waves becomes

$$v = \frac{1}{\sqrt{\varepsilon\mu}} = \frac{c}{\sqrt{KK_m}} \tag{33–18}$$

where c is the speed of light in free space. Since $K_m \approx 1$ in most dielectrics, and K is always greater than 1.0, the speed of light in materials is always less than its speed in vacuum.

In a perfect conductor (resistivity $\rho = 0$), there can be no electric field. (An electric field would give rise to an infinite current). Hence an EM wave could not penetrate a perfect conductor and so must be totally reflected. In fact, polished metal surfaces are good reflectors, and are not transparent to EM waves. EM waves can penetrate real conductors ($\rho \neq 0$) somewhat and the electric currents produced dissipate some of the wave energy.

FIGURE 33–12

(a) Charge travels at speed v ($= c$ in vacuum and if $R = 0$) along transmission line after connection to constant source of emf, V_0. The dashed line indicates the wave "front". (b) V as a function of x along line at time t.

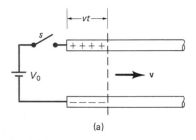

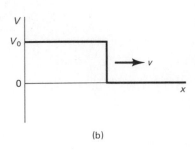

* 33–9   Transmission Lines

Consider a long transmission line—say a pair of parallel wires or a coaxial cable—which is connected at one end to a source of emf. The other end might be connected to some electrical or electronic device, or an antenna. Suppose, at $t = 0$, the two conductors of a transmission line are connected to the + and − terminals of a source of constant emf, V_0. The two conductors do not become charged instantly along their whole lengths. Nor does the current in the conductors, and the voltage across them, appear instantly at all points along the length. Rather, as indicated in Fig. 33–12a, the boundary of the charged portion of the wire moves outward with finite speed, v. If the conductors are resistanceless and the space around and between them is a vacuum (or air), then $v = c$, the speed of EM waves in free space. Similarly, the voltage across the two conductors as a function of distance x from the source is a "step voltage" as shown in Fig. 33–12b; its edge moves with speed c. What we have described is essentially a wave "pulse" traveling along the transmission lines. If the constant source of emf is replaced by a *sinusoidal* emf of frequency f, then there will be a sinusoidal wave traveling down the transmission line, whose wavelength is

$$\lambda = \frac{c}{f}$$

assuming zero resistance and a vacuum. If the space between the conductors of a coaxial cable is filled with a dielectric, the speed is reduced by the di-

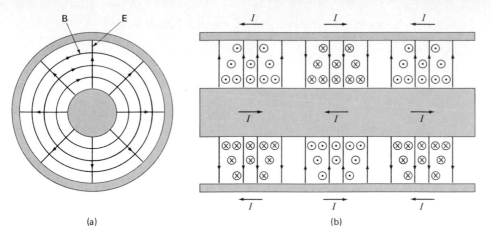

FIGURE 33–13

(a) End view, and (b) side view of coaxial cable transmission line, showing electric field lines (radial) and magnetic field lines (concentric circles). Also shown is the electric current (I) at points in the conductors.

(a)

(b)

electric constant K as given by Eq. 33–18. For ordinary 60-Hz ac lines, $\lambda = (3.0 \times 10^8 \text{ m/s})/(60 \text{ s}^{-1}) = 5000$ km, so the variation of voltage along a line would hardly be noticeable except on a cross-continental line. For higher frequencies, such as radio frequencies, the sinusoidal pattern is clear over much shorter distances.

Figure 33–13 shows the pattern of electric and magnetic fields between the inner and outer conductors of a resistanceless coaxial cable in a vacuum. Actually, more complicated E and B field patterns are possible and the one shown is the simplest; the electric field lines are radial and the magnetic field lines are concentric circles. This system can be analyzed either from a circuit point of view or from a traveling wave point of view. The analysis in either case can be very complicated. We note that when analyzed as a circuit, the cable has both capacitance and inductance; both the charge and current move along the cable at speed c. (The current at points in the conductor is indicated by the arrows labeled I.) In terms of waves, the electric and magnetic fields between the conductors travel with speed c much like a plane wave, but are restricted to the space between the conductors. When the wave reaches the end of the transmission line, where there may be a circuit or electrical device, part of the wave is reflected back unless the line terminates in a pure resistance of a certain value. That value of resistance that will prevent reflection is called the characteristic *impedance* of the line; cables with various impedances are in use, and a typical value is 50 Ω.

For very high frequencies, such as microwaves, *waveguides* are used to carry EM waves. A waveguide is a single hollow conductor, often rectangular in shape. The analysis of wave propagation is quite complicated, and even in vacuum the speed along a waveguide is not equal to c.

* 33–10 Differential Form of Maxwell's Equations

Maxwell's Equations can be written in another form that is often more convenient than Eqs. 33–5. This material is usually covered in more advanced courses, and is included here simply for completeness.

We quote here two theorems, without proof, that are derived in vector analysis textbooks. The first is called *Gauss's theorem* or the *divergence theorem*; it relates the integral over a surface of any vector function **F** to a volume integral over the volume enclosed by the surface:

$$\oint_{\text{Area } A} \mathbf{F} \cdot d\mathbf{A} = \int_{\text{Volume } V} \boldsymbol{\nabla} \cdot \mathbf{F} \, dV.$$

The operator $\mathbf{V}$ is the *gradient operator*, defined in Cartesian coordinates as

$$\mathbf{V} = \mathbf{i}\,\frac{\partial}{\partial x} + \mathbf{j}\,\frac{\partial}{\partial y} + \mathbf{k}\,\frac{\partial}{\partial z}.$$

The quantity

$$\mathbf{V} \cdot \mathbf{F} = \frac{\partial F_x}{\partial x} + \frac{\partial F_y}{\partial y} + \frac{\partial F_z}{\partial z}$$

is called the *divergence* of $\mathbf{F}$. The second theorem is *Stokes's theorem*, and relates a line integral around a closed path to a surface integral over any surface enclosed by that path:

$$\oint_{\text{Line}} \mathbf{F} \cdot d\mathbf{l} = \int_{\text{Area } A} \mathbf{V} \times \mathbf{F} \cdot d\mathbf{A}.$$

The quantity $\mathbf{V} \times \mathbf{F}$ is called the *curl* of $\mathbf{F}$. (See Section 10–1 on the vector product.)

We now use these two theorems to obtain the differential form of Maxwell's equations in free space. We apply Gauss's theorem to Eq. 33–5a (Gauss's law):

$$\oint_A \mathbf{E} \cdot d\mathbf{A} = \frac{Q}{\varepsilon_0} = \int \mathbf{V} \cdot \mathbf{E}\, dV.$$

Now the charge Q can be written as a volume integral over the charge density ρ: $Q = \int \rho\, dV$. Then

$$\int \mathbf{V} \cdot \mathbf{E}\, dV = \frac{1}{\varepsilon_0} \int \rho\, dV.$$

Both sides contain volume integrals over the same volume, and for this to be true over *any* volume, whatever its size or shape, the integrands must be equal:

$$\mathbf{V} \cdot \mathbf{E} = \frac{\rho}{\varepsilon_0}. \tag{33–19a}$$

This is the differential form of Gauss's law. The second of Maxwell's equations, $\oint \mathbf{B} \cdot d\mathbf{A} = 0$ is treated in the same way, and we obtain

$$\mathbf{V} \cdot \mathbf{B} = 0. \tag{33–19b}$$

Next, we apply Stokes's theorem to the third of Maxwell's equations,

$$\oint \mathbf{E} \cdot d\mathbf{l} = -\frac{d\Phi_B}{dt} = \int \mathbf{V} \times \mathbf{E} \cdot d\mathbf{A}.$$

Since the magnetic flux $\Phi_B = \int \mathbf{B} \cdot d\mathbf{A}$ we have

$$\int \mathbf{V} \times \mathbf{E} \cdot d\mathbf{A} = -\frac{\partial}{\partial t} \int \mathbf{B} \cdot d\mathbf{A}$$

where we use the partial derivative, $\partial \mathbf{B}/\partial t$, since B may also depend on position. These are surface integrals over the same area, and to be true over any area, even a very small one, we must have

$$\mathbf{V} \times \mathbf{E} = -\frac{\partial \mathbf{B}}{\partial t}. \tag{33–19c}$$

This is third of Maxwell's equations in differential form. Finally, to the last of Maxwell's equations,

$$\oint \mathbf{B} \cdot d\mathbf{l} = \mu_0 I + \mu_0 \varepsilon_0 \frac{d\Phi_E}{dt},$$

we apply Stokes's theorem and write $\Phi_E = \int \mathbf{E} \cdot d\mathbf{A}$:

$$\int \mathbf{\nabla} \times \mathbf{B} \cdot d\mathbf{A} = \mu_0 I + \mu_0\varepsilon_0 \frac{\partial}{\partial t} \int \mathbf{E} \cdot d\mathbf{A}.$$

The conduction current I can be written in terms of the current density $\mathbf{j}$, using Eq. 26–8:

$$I = \int \mathbf{j} \cdot d\mathbf{A}.$$

Then Maxwell's fourth equation becomes:

$$\int \mathbf{\nabla} \times \mathbf{B} \cdot d\mathbf{A} = \mu_0 \int \mathbf{j} \cdot d\mathbf{A} + \mu_0\varepsilon_0 \frac{\partial}{\partial t} \int \mathbf{E} \cdot d\mathbf{A}.$$

For this to be true over any area A, whatever its size or shape, the integrands on each side of the equation must be equal:

$$\mathbf{\nabla} \times \mathbf{B} = \mu_0 \mathbf{j} + \mu_0\varepsilon_0 \frac{\partial \mathbf{E}}{\partial t}. \qquad (33\text{–}19d)$$

Equations 33–19a, b, c, and d are Maxwell's equations in differential form for free space. They are summarized in Table 33–1.

TABLE 33–1
Maxwell's equations in free space

Integral Form	Differential Form
$\oint \mathbf{E} \cdot d\mathbf{A} = \dfrac{Q}{\varepsilon_0}$	$\mathbf{\nabla} \cdot \mathbf{E} = \dfrac{\rho}{\varepsilon_0}$
$\oint \mathbf{B} \cdot d\mathbf{A} = 0$	$\mathbf{\nabla} \cdot \mathbf{B} = 0$
$\oint \mathbf{E} \cdot dl = -\dfrac{d\Phi_B}{dt}$	$\mathbf{\nabla} \times \mathbf{E} = -\dfrac{\partial \mathbf{B}}{\partial t}$
$\oint \mathbf{B} \cdot dl = \mu_0 I + \mu_0\varepsilon_0 \dfrac{d\Phi_E}{dt}$	$\mathbf{\nabla} \times \mathbf{B} = \mu_0 \mathbf{j} + \mu_0\varepsilon_0 \dfrac{\partial \mathbf{E}}{\partial t}$

 Summary ——————————————————————————————————

James Clerk Maxwell synthesized an elegant theory in which all electric and magnetic phenomena could be described using four equations, now called *Maxwell's equations*. They are based on earlier ideas, but Maxwell added one more—that a changing electric field produces a magnetic field. Maxwell's equations are

$$\oint \mathbf{E} \cdot d\mathbf{A} = \frac{Q}{\varepsilon_0} \qquad \oint \mathbf{E} \cdot dl = -\frac{d\Phi_B}{dt}$$

$$\oint \mathbf{B} \cdot d\mathbf{A} = 0 \qquad \oint \mathbf{B} \cdot dl = \mu_0 I + \mu_0\varepsilon_0 \frac{d\Phi_E}{dt}.$$

The two on the left are Gauss's laws for electricity and for magnetism; the two on the right are Faraday's law and Ampère's law (as extended by Maxwell).

Maxwell's theory predicted that transverse *electromagnetic* (EM) *waves* would be produced by accelerating electric charges, and these waves would propagate through space at the speed of light, c. The oscillating electric and magnetic fields in an EM wave are perpendicular to each other and to the direction of propagation.

After EM waves were experimentally detected in the late 1800s, the idea that light is an EM wave (although of much higher frequency than those detected directly) became generally accepted. The *electromagnetic spectrum* includes EM waves of a wide variety of wavelengths, from microwaves and radio waves to visible light to X rays and gamma rays, all of which travel through space at a speed $c = 3.00 \times 10^8$ m/s.

The energy carried by EM waves can be described by the Poynting vector $\mathbf{S} = (1/\mu_0)(\mathbf{E} \times \mathbf{B})$ which gives the rate energy is carried across unit area per unit time when the electric and magnetic fields in an EM wave in free space are $\mathbf{E}$ and $\mathbf{B}$.

Questions

1. What is the direction of the displacement current in Fig. 33–2? (Note: the capacitor is discharging.)

2. Suppose you are looking along the same direction as an electric field E that is increasing. Will the induced magnetic field be clockwise or counterclockwise? What if E points toward you and is decreasing?

3. Why is it that the magnetic field of a displacement current in a capacitor is so much harder to detect than the magnetic field of a conduction current?

4. Are there any good reasons for calling the term $\mu_0 \varepsilon_0 \, d\Phi_E/dt$ in Eq. 33–3 a "current"?

5. The electric field in an EM wave traveling north oscillates in an east-west plane. Describe the direction of the magnetic field vector in this wave.

6. Is sound an electromagnetic wave? If not, what kind of wave is it?

7. Can EM waves travel through a perfect vacuum? Can sound waves?

8. How are light and sound alike? How are they different?

9. Are the wavelengths of radio and TV signals longer or shorter than those detectable by the human eye?

10. The carrier frequencies of FM broadcasts are much higher than for AM broadcasts. On the basis of what you learned about diffraction in Chapter 15, explain why AM signals can be detected more readily than FM signals behind low hills or buildings.

11. In the electromagnetic spectrum, what type of EM wave would have a wavelength of 10^3 km? 1 km? 1 m? 1 cm? 1 mm? 1 μm?

12. What does the result of Example 33–2 tell you about the phase of an ac current that starts at a power plant as compared to its phase at a house 200 km away?

13. When you connect up two loudspeakers to a stereo amplifier, should you be sure the lead wires are equal in length so there won't be a time lag between speakers? Explain.

14. When you flip a light switch, does the overhead light go on immediately? Explain.

*15. Show, on Fig. 33–13b, the sign of the charge along each conductor at the instant shown.

*16. Show the direction of the conduction current at various points on each conductor in Fig. 33–12. Explain your results.

Problems

SECTION 33–1

1. (I) Calculate the displacement current I_D between the square plates 1.0 cm on a side, of a capacitor if the electric field is changing at a rate of 3.0×10^6 V/m·s.

2. (I) At a given instant, a 2.8-A current flows in the wires connected to a parallel-plate capacitor. What is the rate at which the electric field is changing between the plates if the square plates are 1.00 cm on a side?

3. (II) Show that the displacement current through a parallel-plate capacitor can be written $I_D = C \, dV/dt$ where V is the voltage across the capacitor at any instant.

4. (II) A 1.20-μF capacitor with parallel circular plates accumulates charge at a rate of 12.0 mC/s. What will be the magnetic field strength 15.0 cm radially outward from the center of the plates whose radii are 0.600 cm?

5. (II) (a) Show that the magnetic field beyond the edges of the capacitor plates in Example 33–1 is given by

$$B = \frac{\mu_0 I_D}{2\pi r}$$

where I_D is the displacement current. (b) This is the same formula for the field outside a straight wire; explain the similarity. (c) Compare the magnetic field between the capacitor plates with that near the wire leading to the plates (assume its radius is 2.0 mm). Explain the large difference.

6. (II) Suppose an air-gap capacitor has circular plates of radius $R = 3.5$ cm and separation $d = 2.0$ mm. A 60-Hz emf, $\mathscr{E} = \mathscr{E}_0 \cos \omega t$, is applied to the capacitor. The maximum displacement current is 35 μA. Determine (a) the maximum conduction current I, (b) the value of

$\mathscr{E}_0$, (c) the maximum value of $d\Phi_E/dt$ between the plates. Neglect fringing.

SECTION 33–3

7. (II) If magnetic monopoles existed, which of Maxwell's equations would be altered, and what would be their new form. Let Q_m be the strength of a magnetic monopole, analogous to electric charge Q.

SECTIONS 33–5 AND 33–6

8. (I) In an EM wave traveling west, the B field oscillates vertically and has a frequency of 180 kHz and an rms strength of 8.65×10^{-9} T. What is the frequency and rms strength of the electric field and what is its direction?

9. (I) (a) An FM station broadcasts at 90.5 MHz; what is the wavelength of this wave? (b) What is the wavelength of an AM station at 1550 on the dial?

10. (II) Derive the wave equation for B, Eq. 33–12b.

11. (II) The electric field of a plane EM wave is given by $E_x = E_0 \cos(kz + \omega t)$, $E_y = E_z = 0$. Determine (a) the magnitude and direction of B, and (b) the direction of propagation.

SECTION 33–7

12. (I) The E field in an EM wave has a peak of 38 mV/m. What is the average rate at which this wave carries energy across unit area per unit time?

13. (I) The magnetic field in an EM wave has an rms strength of 2.50×10^{-9} T. How much energy does this wave transport per meter squared per second?

14. (I) How much energy is transported across a 1.0-cm^2 area per hour by an EM wave whose **E** field has an rms strength of 25 V/m?

15. (I) What is the energy contained per cubic meter near the earth's surface due to radiant energy from the sun?

16. (II) A 5.0 mW laser puts out a narrow beam 2.0 mm in diameter. What are the average (rms) values of E and B in the beam?

17. (II) Estimate the average power output of the sun using the fact that about 1350 W/m^2 reaches the upper atmosphere of the earth.

18. (II) A point source emits light energy uniformly in all directions at an average rate P_0 with a single frequency f. Show that the peak electric field in the wave is given by

$$E_0 = \sqrt{\frac{\mu_0 c P_0}{2\pi r^2}}.$$

19. (II) What are E_0 and B_0 10 m from a 100-W light source? Assume the bulb emits radiation of a single frequency uniformly in all directions.

20. (II) A 35-kW radio station emits EM waves uniformly in all directions from a vertical antenna. (a) What is the intensity of the signal (W/m^2) at a distance of 20.0 km from the transmitting antenna? (b) What is the rms magnitude of the **E** field at this point, assuming the station is operating at full power? (c) What is the voltage induced in a 1.0-m-long vertical car antenna? (d) What power output is needed for a station to produce an equal signal 100 km away? (This much power is illegal in the United States.)

21. (II) A 1.80-m-long FM antenna is oriented parallel to the electric field of an EM wave. How large must the **E** field strength be to produce a 1.0-mV rms voltage between the ends of the antenna? What is the rate of energy transport per unit area?

22. (II) How large an emf (rms) will be generated in an antenna that consists of a 600-loop circular coil of wire 0.50 cm in diameter if the EM wave has a frequency of 940 kHz and is transporting energy at an average rate of 2.0×10^{-4} W/m^2 at the antenna?

23. (II) (a) Show that the Poynting vector **S** points radially inward toward the center of a circular parallel-plate capacitor when it is being charged, as in Example 33–1. (b) Integrate **S** over the cylindrical boundary of the capacitor gap to show that the rate at which energy enters the capacitor

is equal to the rate at which electrostatic energy is being stored in the electric field of the capacitor (see Eq. 25–6). Ignore fringing of **E**.

24. (II) A cylindrical conductor of radius r and conductivity σ carries a steady current I distributed uniformly across its cross-section. (a) Determine **E** inside the conductor. (b) Determine **B** just outside the conductor. (c) Determine the Poynting vector **S** at the surface of the conductor and show that it is normal to the surface and points inward. (d) Integrate over **S** to show that the rate at which electromagnetic energy enters the sides of the conductor is equal to the rate at which energy is dissipated, $I^2 R$. Thus we can think of the energy entering the conductor in the form of electromagnetic fields from the sides rather than through the ends.

*SECTION 33–9

25. (I) Who will hear the voice of a singer first, a person in the balcony 50 m away from the stage, or a person 3000 km away at home whose ear is next to the radio? How much sooner? Assume the microphone is a few centimeters from the singer and the temperature is 20°C.

***26.** (II) Sketch the electric and magnetic field lines around the parallel wire transmission line of Fig. 33–12a.

***27.** (II) In Fig. 33–13b indicate the direction of the displacement current at various points for the instant shown.

28. (II) (a) In Fig. 33–13, apply Ampère's law to a circular path, concentric with the coaxial cables and between them, to show that the currents indicated are consistent with the magnetic field shown. (b) How is B related to the current as a function of the radial distance from the center?

***29.** (II) An emf $\mathscr{E} = \mathscr{E}_0 \sin 2\pi ft$, where $\mathscr{E}_0 = 20$ V and $f = 5.0 \times 10^7$ Hz, is applied to a long resistanceless transmission line. What is the potential difference V across the line (a) as a function of time at a point $x = 2.0$ m from the source, (b) as a function of distance x at time $t = 2.0$ s?

30. (III) For the coaxial cable of Fig. 33–13, the inner conductor has radius R_1 and the outer conductor has inner radius R_2. The terminals of a battery of emf V_0 is connected to the two conductors at one end of the cable, and a resistor R connects the two conductors at the other end. In the region between conductors ($R_1 < r < R_2$), determine (a) **E**, (b) **B**, (c) **S**. (d) Integrate **S** to show that energy flows through the empty space (between R_1 and R_2) at a rate V_0^2/R. Is this result surprising?

Light–Reflection and Refraction

34

The sense of sight is extremely important to us, for it provides us with a large part of our information about the world. How do we see? What is the something called *light* that enters our eyes and causes the sensation of sight? How does light behave so that we can see the great range of phenomena that we do? The subject of light will occupy us for the next several chapters.

We see an object in one of two ways: (1) the object may be a source, such as a light bulb, a flame, or a star, in which case we see the light emitted directly from the source; or (2), more commonly, we see an object by light reflected off of it. In the latter case the light may have originated from the sun, artificial lights, or some other source. An understanding of how bodies *emit* light was not achieved until the 1920s, and this will be discussed in Chapter 40. How light is *reflected* from objects was understood much earlier, and we will discuss this in Section 34–3.

34–1 The Ray Model of Light

A great deal of evidence suggests that light travels in straight lines under a wide variety of circumstances. For example, a point source of light like the sun casts distinct shadows; and the beam of a flashlight appears to be a straight line. In fact we infer the positions of objects in our environment by assuming that light moves from the object to our eyes in straight-line paths. Our whole orientation to the physical world is based on this assumption.

This reasonable assumption has led to the **ray model** of light. The straight-line paths that light follows are called light **rays**. Actually, a ray is an idealization: it is meant to represent an infinitely narrow beam of light. When we see an object, light reaches our eyes from each point on the object. Although light rays leave each point in many different directions, only a small bundle of these rays can enter an observer's eye as shown in Fig. 34–1. If the person's head moves to one side, a different bundle of rays will enter the eye from each point.

We saw in Chapter 33 that light can be considered as an electromagnetic wave. Although the ray model of light does not deal with this aspect of light

FIGURE 34–1

Light rays radiate from each single point on an object. A small bundle of rays leaving one point is shown entering a person's eye.

This bundle enters the eye

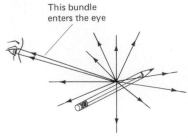

(we discuss the wave nature of light starting in Chapter 36), it has been very successful in dealing with many aspects of light such as reflection, refraction, and the formation of images by mirrors and lenses. Because these explanations involve straight-line rays at various angles, this subject is referred to as *geometrical optics*.[†]

34–2 The Speed of Light and Index of Refraction

Galileo attempted to measure the speed of light by trying to measure the time required for light to travel a known distance between two hilltops. He stationed an assistant on one hilltop and himself on another and ordered the assistant to lift the cover from a lamp the instant he saw a flash from Galileo's lamp. Galileo measured the time between the flash of his lamp and when he received the light from his assistant's lamp. The time was so short that Galileo concluded it merely represented human reaction time, and that the speed of light must be extremely high.

The first successful determination that the speed of light is finite was made by the Danish astronomer Ole Roemer (1644–1710). Roemer had noted that the carefully measured period of one of Jupiter's moons (Io, with an average period of 42.5 h) varied slightly, depending on the relative motion of Earth and Jupiter. When Earth was moving away from Jupiter, the period of the moon was slightly longer, and when Earth was moving toward Jupiter, the period was slightly shorter. He attributed this variation to the extra time needed for light to travel the increasing distance to Earth when Earth is receding, or to the shorter travel time for the decreasing distance when the two planets are approaching one another. Roemer concluded that the speed of light—though great—is finite.

Since then a number of techniques have been used to measure the speed of light. Among the most important were those carried out by the American Albert Michelson (1852–1931). Michelson used the rotating mirror apparatus diagrammed in Fig. 34–2 for a series of high-precision experiments carried out from 1880 to the 1920s. Light from a source was directed at one face of a rotating eight-sided mirror. The reflected light traveled to a stationary mirror a large distance away and back again as shown. If the rotating mirror was turning at just the right rate the returning beam of light would reflect from one face of the mirror into a small telescope through which the observer looked. At a different speed of rotation, the beam would be deflected to one side and would not be seen

[†] In geometrical optics we largely ignore the wave properties of light; this means that if light passes by objects or through apertures, these must be large compared to the wavelength of the light (so the wave phenomena of interference and diffraction, as discussed in Chapter 15, can be ignored), and we ignore what happens to the light at the edges of objects. But see Chapters 36 and 37.

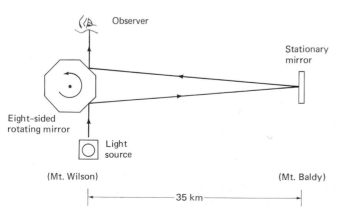

FIGURE 34–2

Michelson's speed-of-light apparatus (not to scale).

by the observer. From the required speed of the rotating mirror and the known distance to the stationary mirror, the speed of light could be calculated. In the 1920s, Michelson set up the rotating mirror on the top of Mt. Wilson in southern California and the stationary mirror on Mt. Baldy (Mt. San Antonio) 35 km away. He later measured the speed of light in vacuum using a long evacuated tube.

The accepted value today for the speed of light, c, in vacuum is

$$c = (2.99792458 \pm 0.00000001) \times 10^8 \text{ m/s}.$$

We usually round this off to

$$3.00 \times 10^8 \text{ m/s}$$

TABLE 34–1

Indices of refraction[†]

Material	$n = c/v$
Air (at STP)	1.0003
Water	1.33
Ethyl Alcohol	1.36
Glass	
Fused quartz	1.46
Crown glass	1.52
Light flint	1.58
Lucite or Plexiglas	1.51
Sodium chloride	1.53
Diamond	2.42

[†] $\lambda = 589$ nm

when extremely precise results are not required. In air, the speed is only slightly less. In other transparent materials such as glass and water, the speed is always less than that in vacuum. For example, in water it travels at about $\frac{3}{4}c$. The ratio of the speed of light in vacuum to the speed v in a given material is called the *index of refraction n* of that material:

$$n = \frac{c}{v}. \qquad (34\text{–}1)$$

The index of refraction for various materials is given in Table 34–1.[†] (As we shall see later, n varies somewhat with the wavelength of the light—except in vacuum—so a particular wavelength is specified, that of yellow light with wavelength $\lambda = 589$ nm.) For example, since $n = 2.42$ for diamond, the speed of light in diamond is

$$v = c/n = (3.00 \times 10^8 \text{ m/s})/2.42 = 1.24 \times 10^8 \text{ m/s}.$$

34–3 Reflection; Image Formation by a Plane Mirror

When light strikes the surface of an object, some of the light is reflected. The rest is either absorbed by the object (and transformed to heat) or, if the object is transparent like glass or water, part of it is transmitted through. For a very shiny object such as a silvered mirror, over 95 percent of the light may be reflected.

When a narrow beam of light strikes a flat surface we define the *angle of incidence*, θ_i, to be the angle an incident ray makes with the normal to the surface and the *angle of reflection*, θ_r, to be the angle the reflected ray makes with the normal. For flat surfaces, it is found that the incident and reflected rays lie in the same plane with the normal to the surface, and that

the angle of incidence equals the angle of reflection.

This is the **law of reflection** and is indicated in Fig. 34–3. It was known to the ancient Greeks, and you can confirm it yourself by shining a narrow flashlight beam at a mirror in a darkened room.

When light is incident upon a rough surface, even microscopically rough such as this page, it is reflected in many directions, Fig. 34–4. This is called *diffuse reflection*. The law of reflection still holds, however, at each small section of the surface. Because of diffuse reflection in all directions, an ordinary object can be seen from many different angles. When you move your head to the side, a

FIGURE 34–3

Law of reflection.

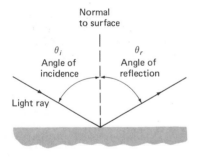

FIGURE 34–4

Diffuse reflection from a rough surface.

[†] From Eq. 33–18 it is clear that $n = \sqrt{KK_m}$ where K is the dielectric constant of a material and K_m is its magnetic permeability.

different bundle of reflected rays reach your eye from each point on the object, Fig. 34–5a. But when a narrow beam of light is shone on a mirror, the light will not reach your eye unless it is placed at just the right place where the law of reflection is satisfied, as shown in Fig. 34–5b. This is what gives rise to the unusual properties of mirrors. (Galileo, using similar arguments,[†] showed that the moon must have a rough surface rather than a highly polished surface like a mirror, as some people thought.)

When you look straight in a mirror your see what appears to be yourself as well as various objects around and behind you. Your face and the other objects look as if they are in front of you, beyond the mirror; but, of course, they are not. What you see in the mirror is an **image** of the objects.

Figure 34–6 shows how an image is formed by a plane (that is, flat) mirror. Rays from two different points on an object are shown. Rays leave each point on the object going in many directions, but only those that enclose the bundle of rays that reach the eye from the two points are shown. The diverging rays that enter the eye appear to come from behind the mirror as shown by the dashed lines.[‡] The point from which each bundle of rays seems to come is one point on the image. For each point on the object there is a corresponding image point. Let us concentrate on the two rays that leave the point A on the object and strike the mirror at points B and B'. The angles ADB and CDB are right angles. And angles ABD and CBD are equal because of the law of reflection. Therefore, the two triangles ABD and CDB are congruent and the length $AD = CD$. That is, the image is as far behind the mirror as the object is in front: the *image distance*, d_i (distance from mirror to image, Fig. 34–6), equals the *object distance*, d_o. From the geometry, we also see that the height of the image is the same as that of the object.

The light rays do not actually pass through the image itself. It merely *seems* like the light is coming from the image because our brains interpret any light entering our eyes as coming from in front of us. Since the rays do not actually pass through the image, a piece of white paper or film placed at the image would not detect the image. Therefore, it is called a *virtual image*. This is to distinguish it from a *real image* in which the light does pass through the image, and which therefore could appear on paper or film placed at the image position. We will see that curved mirrors and lenses can form real images.

[†] Galileo Galilei, *Dialogue Concerning the Two Chief World Systems*, trans. Stillman Drake (Berkeley: University of California Press, 1967). See Day 1.

[‡] Our eyes and brain interpret any rays that enter an eye as having traveled a straight-line path.

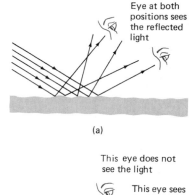

Eye at both
positions sees
the reflected
light

(a)

This eye does not
see the light

This eye sees
the light

(b)

FIGURE 34–5

A beam of light from a flashlight is shined on (a) white paper; (b) a small mirror. In part (a) you can see the white light reflected at various points because of diffuse reflection. But in part (b) you see the reflected light only when your eye is placed correctly ($\theta_r = \theta_i$).

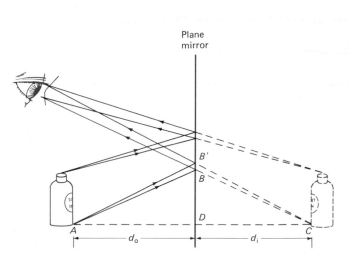

Plane
mirror

FIGURE 34–6

Formation of a virtual image by a plane mirror.

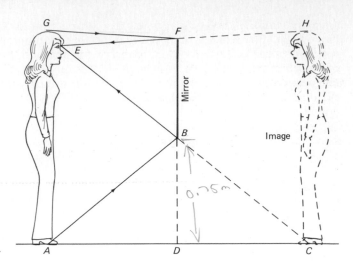

FIGURE 34–7

Seeing oneself in the mirror (Example 34–1).

EXAMPLE 34–1 A woman 1.60 m tall stands in front of a vertical plane mirror. What is the minimum height of the mirror and how high must its lower edge be above the floor if she is to be able to see her whole body? (Assume her eyes are 10 cm below the top of her head.)

SOLUTION The situation is diagrammed in Fig. 34–7. First consider the ray from the toe, *AB*, which upon reflection becomes *BE* and enters the eye *E*. Since light enters the eye from point *A* (the toes) after reflecting at *B*, the mirror needs to extend no lower than *B*. Because the angle of reflection equals the angle of incidence, the height *BD* is half of the height *AE*. Since *AE* = 1.60 m − 0.10 m = 1.50 m, *BD* = 0.75 m. Similarly, if the woman is to see the top of her head, the top edge of the mirror only needs to reach point *F*, which is 5 cm below the top of her head (half of *GE* = 10 cm). Thus *DF* = 1.55 m and the mirror need be only 1.55 m − 0.75 m = 0.80 m high; and its bottom edge must be 0.75 m above the floor. In general, a mirror need be only half as tall as a person for that person to see all of himself or herself. Does this result depend on the person's distance from the mirror?

34–4 Formation of Images by Spherical Mirrors

Reflecting surfaces do not have to be flat. The most common *curved* mirrors are *spherical*, which means they form a section of a sphere. A spherical mirror is called *convex* if the reflection takes place on the outer surface of the spherical shape so that the center of the mirror bulges out toward the viewer; and it is called *concave* if the reflecting surface is on the inner surface of the sphere so that the center of the mirror is farther from the viewer than the edges.

To see how spherical mirrors form images, we first consider an object that is very far from a concave mirror; in this case the rays from each point on the object that reach the mirror will be nearly parallel, as shown in Fig. 34–8. *For*

FIGURE 34–8

If the object's distance is large compared to the size of the mirror, the rays are nearly parallel.

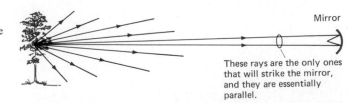

These rays are the only ones that will strike the mirror, and they are essentially parallel.

an object infinitely far away (the sun and stars approach this) *the rays would be precisely parallel.* Now consider such parallel rays falling on a concave mirror as in Fig. 34–9. The law of reflection holds for each of these rays at the point each strikes the mirror. As can be seen, they are not all brought to a single point. In order to form a sharp image, the rays must come to a point. Thus a spherical mirror will not make as sharp an image as a plane mirror will. However, if the mirror width is small compared to its radius of curvature so that the reflected rays make only a small angle upon reflection, then the rays will cross each other at nearly a single point, or *focus*, as shown in Fig. 34–10. In the case shown, the rays are parallel to the *principal axis*, which is defined as the straight line perpendicular to the curved surface at its center (line *CA* in the diagram). The point *F*, where rays parallel to the principal axis come to a focus, is called the **focal point** of the mirror; the distance between *F* and the center of the mirror, length *FA*, is called the **focal length**, *f*, of the mirror. Another way of defining the focal point is to say that it is the *image point for an object infinitely far away* along the principal axis. The image of the sun, for example, would be essentially at *F*.

Now we will show, for a mirror whose reflecting surface has a width that is small compared to the radius of curvature, that the rays do indeed meet at a common point, *F*, and we calculate the focal length *f*. We consider a ray that strikes the mirror at *B* in Fig. 34–10. The point *C* is the center of curvature of the mirror (the center of the sphere of which the mirror is a part). So the dashed line *CB* is equal to *r*, the radius of curvature, and *CB* is normal to the surface at *B*. From the law of reflection and the geometry, the three angles labeled θ are equal as indicated. The triangle *CBF* is isosceles because two of its angles are equal. Thus length *CF* = *BF*. We assume the mirror is small compared to its radius of curvature, so the angles are small and the length *FB* is nearly equal to length *FA*. In this approximation, *FA* = *FC*. But *FA* = *f*, the focal length, and *CA* = 2*FA* = *r*. Thus the focal length is half the radius of curvature:

$$f = r/2. \qquad (34\text{–}2)$$

This argument only assumed that the angle θ was small; so the same result applies for all the other rays; thus all the rays pass through the same point *F* in this approximation of a mirror small compared to its radius of curvature.

Since it is only approximately true that the rays come to a perfect focus at *F*, the larger the mirror, the worse the approximation (Fig. 34–9) and the more blurred the image. This "defect" of spherical mirrors is called *spherical aberration*; we will discuss it more with regard to lenses in Chapter 35. A *parabolic* reflector, on the other hand, will reflect the rays to a perfect focus. However, because parabolic shapes are much harder to make and thus much more expensive, spherical mirrors are used for most purposes. We consider here only spherical mirrors and we will assume that they are small compared to their radius of curvature so that the image is sharp and Eq. 34–2 holds.

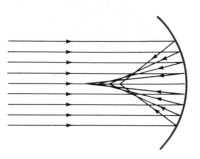

FIGURE 34–9
Parallel rays striking a concave spherical mirror do not focus at a single point.

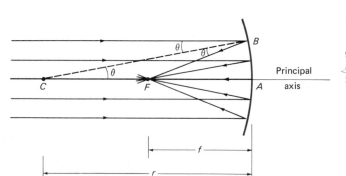

FIGURE 34–10
Rays parallel to the principal axis of a spherical mirror come to a focus at *F*, called the focal point, as long as the mirror is small in extent as compared to its radius of curvature *r*.

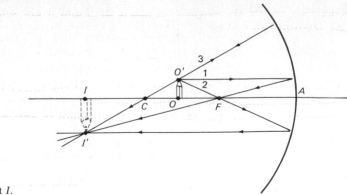

FIGURE 34–11

Rays from object at O form an image at I.

We saw that for an object at infinity the image is located at the focal point of a concave spherical mirror, where $f = r/2$. But where does the image lie for an object not at infinity? First consider the object shown in Fig. 34–11, which is placed at point O between F and C. Let us determine where the image will be for a given point O' on the object. To do this we can draw several rays and make sure these reflect from the mirror such that the reflection angle equals the incidence angle. This can involve much work and our task is simplified if we deal with three particularly simple rays. These are the rays labeled 1, 2, and 3 in the diagram. Ray 1 is drawn parallel to the axis; therefore it must pass through F (as in Fig. 34–10). Ray 2 is drawn through F; therefore it must reflect so it is parallel to the axis. Ray 3 is drawn so that it passes through C, the center of curvature, and thus is along a radius of the spherical surface; so it is perpendicular to the mirror and thus will be reflected back on itself. The point at which these rays cross is the image point I'. All other rays from the same object point will pass through this image point. We can show that this is true using Fig. 34–12 and at the same time obtain a formula for the position of the image for a given object. In Fig. 34–12, an arbitrary ray (OP) leaves a point O on the object (we have chosen a point on the principal axis only for convenience) and passes through point I, which will be its image point. Since $\gamma + \phi = 180°$, and the sum of the angles of a triangle is $180°$, we have

$$\gamma = \alpha + 2\theta.$$

Similarly

$$\beta = \alpha + \theta.$$

We eliminate θ between these relations and obtain

$$\alpha + \gamma = 2\beta.$$

FIGURE 34–12

Determining the image point I and deriving the mirror equation.

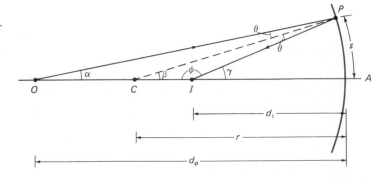

If we let s be the arc length PA, and if α, γ, and β are small angles, then this last relation can be written (in radian measure) to a good approximation as

$$\frac{s}{d_o} + \frac{s}{d_i} = \frac{2s}{r}$$

where d_o is the *object distance* (the distance of the object from the mirror), d_i is the *image distance*, and r is the radius of curvature of the mirror. We divide through by s and use Eq. 34–2 for the focal length to obtain

$$\frac{1}{d_o} + \frac{1}{d_i} = \frac{1}{f}. \qquad (34-3)$$

[handwritten: $\frac{1}{d_0} + \frac{1}{d_i} = \frac{1}{f}$]

This is called the *mirror equation*. The fact that d_i depends only on d_o and f, and not on the angle α of the particular ray OP, shows that all rays from point O are imaged at the same point I for a mirror that is small compared to its radius of curvature so that the angles involved are small. For larger mirrors, the approximation of small angles used above is not valid, and the image will be blurry, as was indicated in Fig. 34–9. We derived Eq. 34–3 for a point O on the principal axis. It will also be valid for points slightly above or below the axis since a spherical mirror is symmetric. Thus, all points of an object are imaged nearby to form a complete image of the object as indicated by the dashed lines in Fig. 34–11. Because the light actually passes through this image itself, this is a *real image*. Compare this to the virtual image formed by a plane mirror.

Equation 34–3 has shown that d_i, the image distance, is the same for all rays coming from one point on the object. This equation is also important because it gives us a way of determining the position of the image, given the object position and the focal length (or radius of curvature). Thus we have two ways of determining the position of an image: as in Fig. 34–11, we can draw rays 1, 2, and 3 (actually only two of these rays are needed, but the third serves as a check); or we can use Eq. 34–3. Of course the latter is analytic and relieves the difficulty of drawing a precise diagram; nonetheless, it is almost always helpful to draw at least a rough diagram to make sure the analytic answer is reasonable.

The *lateral magnification, m,* of a mirror is defined as the height of the image, h_i, divided by the height of the object, h_o. From the diagram of Fig. 34–13, the ray $O'AI'$ obeys the law of reflection so from similar triangles we have that the lateral magnification is

$$m = \frac{h_i}{h_o} = -\frac{d_i}{d_o}, \qquad (34-4)$$

[handwritten: $m = \frac{h_i}{h_o} = -\frac{d_i}{d_0}$]

where the minus sign is inserted as a convention; for consistency we must be careful about the signs of all quantities in Eq. 34–4. The conventions we use are: if h_o is considered positive, h_i is positive if the image is upright and

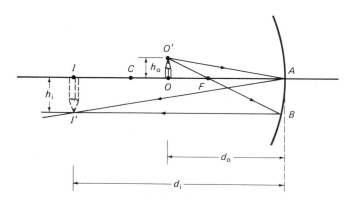

FIGURE 34–13

Diagram for deriving lateral magnification by a spherical mirror.

negative if inverted; d_i and d_o are positive if image and object are on the reflecting side of the mirror (as in Fig. 34–13), but if either image or object is behind the mirror the corresponding distance is negative (an example is shown in Fig. 34–14, Example 34–3). Thus the magnification (Eq. 34–4) is positive for an upright image and negative for an inverted image.

EXAMPLE 34–2 A 1.50-cm-high object is placed 20.0 cm from a concave mirror whose radius of curvature is 30.0 cm. Determine (a) the position of the image and (b) its size.

SOLUTION (a) The focal length $f = r/2 = 15.0$ cm. Then, since $d_o = 20.0$ cm, we have from Eq. 34–3 that

$$\frac{1}{d_i} = \frac{1}{f} - \frac{1}{d_o} = \frac{1}{15.0 \text{ cm}} - \frac{1}{20.0 \text{ cm}} = 0.0167 \text{ cm}^{-1}.$$

So $d_i = 1/0.0167 \text{ cm}^{-1} = 60.0$ cm; the image is 60.0 cm from the mirror on the same side as the object. (b) From Eq. 34–4, the lateral magnification is $m = -60.0 \text{ cm}/20.0 \text{ cm} = -3.00$. Therefore the image is $(-3.00)(1.5 \text{ cm}) = -4.5$ cm high, and is inverted.

EXAMPLE 34–3 A 1.00-cm-high object is placed 10.0 cm from a mirror whose radius of curvature is 30.0 cm. (a) Draw a ray diagram to locate the position of the image. (b) Determine the position of the image and the lateral magnification analytically.

SOLUTION (a) Since $f = 15.0$ cm, the object is between the mirror and the focal point. We draw the three rays as described earlier (for Fig. 34–11) and this is shown in Fig. 34–14. The rays reflected from the mirror diverge and so never meet at a point. They appear, however, to be coming from a point behind the mirror. The image is thus behind the mirror and is *virtual*. (Why?) (b) We use Eq. 34–3 to find d_i where $d_o = 10.0$ cm:

$$\frac{1}{d_i} = \frac{1}{15.0 \text{ cm}} - \frac{1}{10.0 \text{ cm}} = \frac{2 - 3}{30.0 \text{ cm}} = -\frac{1}{30.0 \text{ cm}}.$$

Therefore $d_i = -30.0$ cm. The minus sign means the image is behind the mirror. The lateral magnification is $m = -(-30.0 \text{ cm})/(10.0 \text{ cm}) = +3.00$. So the image is 3.00 times larger than the object; the plus sign indicates that the image is upright.

These examples show that a spherical mirror can produce a magnified image, one that is larger than the object. (There is an old story—maybe fable is a better word—that Julius Caesar spied on the British forces by setting up a curved mirror on the coast of Gaul. Is this reasonable?)

It is useful to compare Figs. 34–11 and 34–14. We can see that if the object is within the focal point, as in Fig. 34–14, the image is virtual, upright, and ·

FIGURE 34–14

Object placed within the focal point F (Example 34–3). The image is *behind* the mirror and is *virtual*.

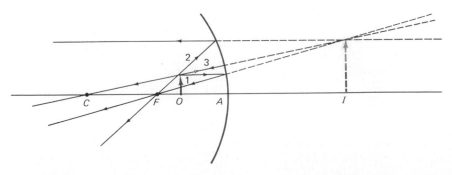

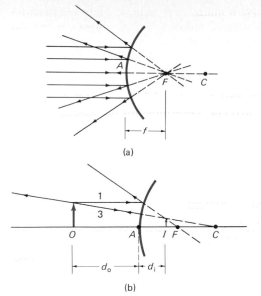

(a)

(b)

FIGURE 34–15

Convex mirror: (a) the focal point is at F, behind the mirror; (b) image I for the object at O is virtual, upright, and smaller than the object.

magnified. This is how a shaving or make-up mirror is used—you must place your head within the focal point if you are to see yourself right side up. If the object is *beyond* the focal point as in Fig. 34–11, the image is real and inverted. Whether the magnification is greater or less than 1.0 in this case depends on the position of the object relative to the center of curvature, point C.

The analysis used for concave mirrors can be applied to *convex* mirrors. Even the mirror equation (Eq. 34–3) holds for a convex mirror, although the quantities involved must be carefully defined. Fig. 34–15a shows parallel rays falling on a convex mirror. Again spherical aberration will be present, but we assume the mirror's size is small compared to its radius of curvature. The reflected rays diverge, but seem to come from point F behind the mirror. This is the *focal point*, and its distance from the center of the mirror is the focal length f. It is easy to show that again $f = r/2$. We see that an object at infinity produces a virtual image in a convex mirror. Indeed, no matter where the object is placed on the reflecting side of the mirror, the image will be virtual and erect as indicated in Fig. 34–15b. To find the image we draw rays 1 and 3 according to the rules used before on the concave mirror.

The mirror equation, Eq. 34–3, holds for convex mirrors but the focal length f must be considered negative, as must the radius of curvature. The proof is left as a problem (19). It is also left as a problem (17) to show that Eq. 34–4 for the magnification is also valid.

EXAMPLE 34–4 A convex rearview car mirror has a radius of curvature of 40.0 cm. Determine the location of the image and its magnification for an object 10.0 m from the mirror.

SOLUTION With $r = -40.0$ cm, $f = -20.0$ cm, the mirror equation gives:

$$\frac{1}{d_i} = \frac{1}{f} - \frac{1}{d_o} = -\frac{1}{0.200 \text{ m}} - \frac{1}{10.0 \text{ m}} = -\frac{51.0}{10.0 \text{ m}}.$$

So $d_i = -0.196$ m or 19.6 cm behind the mirror. The lateral magnification is $m = -d_i/d_o = -(-0.196 \text{ m})/(10.0 \text{ m}) = 0.0196$ or 1/51; so the upright image is reduced by a factor of 51.

We summarize the rules (the *sign conventions*) for applying Eqs. 34–3 and 34–4 to concave and convex mirrors: *When the object, image, or focal point is on the*

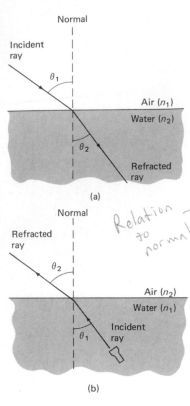

Incident
ray

θ_1

Air (n_1)

Water (n_2)

Refracted
ray

θ_2

(a)

Normal

Refracted
ray

θ_2

Air (n_2)

Water (n_1)

Incident
ray

θ_1

(b)

FIGURE 34–16
Refraction.

Relation to normal

FIGURE 34–17

Because of refraction, when a person
is standing in water, his or her legs
look shorter.

reflecting side of the mirror (on the left in all our drawings), *the corresponding distance is considered positive. If any of these points is behind the mirror* (on the right) *the corresponding distance must be considered negative.*[†] *Object and image heights, h_o and h_i, are considered positive or negative depending on whether the point is above or below the principal axis*. It is important to be consistent with these sign conventions.

34–5 Refraction; Snell's law

When light passes from one medium into another, part of the incident light is reflected at the boundary. The remainder passes into the new medium. If a ray of light is incident at an angle to the surface (other than perpendicular), the ray is bent as it enters the new medium. This bending is called **refraction**. Fig. 34–16a shows a ray passing from air into water. The angle θ_1 is the *angle of incidence* and θ_2 is the *angle of refraction*. Notice that the ray bends toward the normal when entering the water. This is always the case when the ray enters a medium where the speed of light is less. If light travels from one medium into a second where its speed is greater, the ray bends away from the normal; this is shown in Fig. 34–16b for a ray traveling from water to air.

Refraction is responsible for a number of common optical illusions. For example, a person standing in waist-deep water appears to have shortened legs, Fig. 34–17a. As shown in Fig. 34–17b, the rays leaving the person's foot are bent at the surface; the observer's eye (and brain) assumes the rays to have traveled a straight-line path, and so the feet appear to be higher than they really are. Similarly, when you put a stick in water, it appears to be bent.

The angle of refraction depends on the speed of light in the two media and on the incident angle. An analytical relation between θ_1 and θ_2 was arrived at experimentally about 1621 by Willebrord Snell (1591–1626). This is known as **Snell's law** and is written:

$$n_1 \sin \theta_1 = n_2 \sin \theta_2; \tag{34–5}$$

[†] We have seen examples where d_i and f are negative. The object distance d_o for any material object is, of course, always positive. However, if the mirror is used in conjunction with a lens or another mirror, the image formed by the first mirror or lens becomes the object for the second mirror; it is then possible for such an "object" to be behind the second mirror, in which case, d_o would be negative. These rules are also consistent with considering the focal length of a concave mirror positive and that of a convex mirror negative.

(a)

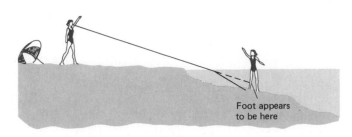

Foot appears
to be here

(b)

θ_1 is the angle of incidence and θ_2 is the angle of refraction; n_1 and n_2 are the respective indices of refraction in the materials (see Fig. 34–16). Snell's law[†] is the basic **law of refraction**.

It is clear from Snell's law that if $n_2 > n_1$, then $\theta_2 < \theta_1$; that is, if light enters a medium where n is greater (and its speed less) then the ray is bent toward the normal. And if $n_2 < n_1$, then $\theta_2 > \theta_1$, so the ray bends away from the normal. This is what we saw in Fig. 34–16.

 EXAMPLE 34–5 Light strikes a flat piece of glass at an incident angle of 60° as in Fig. 34–18. If the index of refraction of the glass is 1.50, (a) what is the angle of refraction θ_a in the glass; (b) what is the angle θ_b at which the ray emerges from the glass?

SOLUTION (a) We assume the incident ray is in air so $n_1 = 1.00$ and $n_2 = 1.50$. Then, from Eq. 34–5 we have

$$\sin \theta_a = \frac{1.00}{1.50} \sin 60° = 0.577,$$

so $\theta_a = 35.2°$. (b) Since the faces of the glass are parallel, the incident angle in this case is just θ_a, so $\sin \theta_a = 0.577$. This time $n_1 = 1.50$ and $n_2 = 1.00$. Thus $\theta_b (= \theta_2)$ is

$$\sin \theta_b = \frac{1.50}{1.00} \sin \theta_a = 0.866,$$

and $\theta_b = 60.0°$. The direction of the beam is thus unchanged by passing through a plane piece of glass. It should be clear that this is true for any angle of incidence. The ray is displaced slightly to one side however. You can observe this by looking through a piece of glass (near its edge) at some object and then moving your head to the side so that you see the object directly.

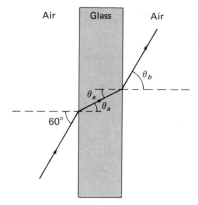

FIGURE 34–18

Light passing through a piece of glass (Example 34–5).

34–6 Total Internal Reflection; Fiber Optics

When light passes from one material into a second material where the index of refraction is less (say from water into air), the light bends away from the normal, as for ray A in Fig. 34–19. At a particular incident angle, the angle of refraction will be 90°, and the refracted ray would skim the surface (ray B) in this case. The

[†] Snell actually was not aware that the index of refraction is related to the speed of light in the particular medium. Only later was it found that the index of refraction could be written as the ratio of the speed of light in vacuum to that in the given material, Eq. 34–1. Snell's law can be derived from the wave theory of light (see next chapter), and in fact we already did so in Section 15–7, where Eq. 15–17 is just a combination of Eqs. 34–5 and 34–1.

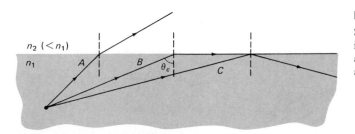

FIGURE 34–19

Since $n_2 < n_1$, light rays are totally internally reflected if $\theta > \theta_c$, as for ray C. If $\theta < \theta_c$, as for ray A, only a part of the light is reflected (this part not shown), and the rest is refracted.

incident angle at which this occurs is called the *critical angle* θ_c; from Snell's law, θ_c is given by

$$\sin \theta_c = \frac{n_2}{n_1} \sin 90° = \frac{n_2}{n_1}.$$ (34–6)

$\sin \theta_c = \frac{n_2}{n_1}$ but $n_1 > n_2$

For any incident angle less than θ_c there will be a refracted ray, although—as mentioned earlier—part of the light will also be reflected at the boundary. However, for incident angles greater than θ_c, Snell's law would tell us that $\sin \theta_2$ is greater than 1.00. Yet the sine of an angle can never be greater than 1.00. In this case there is no refracted ray at all, and *all of the light is reflected*, as for ray C in the diagram. This is called **total internal reflection**. But note that total internal reflection can occur only when light strikes a boundary where the medium beyond is optically less dense—that is, it has a lower index of refraction.

EXAMPLE 34–6   Describe what a person would see who looked up at the world from beneath the perfectly smooth surface of a lake or swimming pool.

SOLUTION   For an air-water interface, the critical angle is given by:

$$\sin \theta_c = \frac{1.00}{1.33} = 0.750$$

Therefore $\theta_c = 49°$. Thus the person would see the outside world compressed into a circle whose edge makes a 49° angle with the vertical. Beyond this angle the person will see reflections from the sides and bottom of the pool or lake.

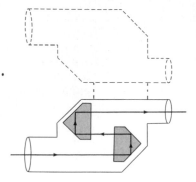

FIGURE 34–20

Prisms reflect light in binoculars.

FIGURE 34–21

Light reflected totally at the interior surface of a glass or transparent plastic fiber.

FIGURE 34–22

Fiber optic image.

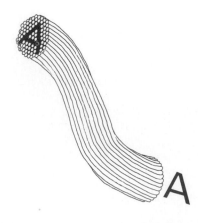

Many optical instruments such as binoculars use total internal reflection within a prism to reflect light. The advantage is that very nearly 100 percent of the light is reflected, whereas even the best mirrors reflect somewhat less than 100 percent. Thus the image is brighter. For glass with $n = 1.50$, $\theta_c = 41.8°$. Therefore, 45° prisms will reflect all the light internally as shown in the binoculars of Fig. 34–20.

Total internal reflection is the principle behind *fiber optics*. Very thin glass and plastic fibers can now be made as small as a few micrometers in diameter. A bundle of such tiny fibers is called a *light pipe* since light can be transmitted along it with almost no loss. Fig. 34–21 shows how light traveling down a thin fiber makes only glancing collisions with the walls so that total internal reflection occurs. Even if the light pipe is bent into a complicated shape, the critical angle won't (usually) be exceeded so light is transmitted practically undiminished to the other end. This effect is used in decorative lamps and to illuminate water streams in fountains. Light pipes can be used to illuminate difficult places to reach, such as inside the human body. They can be used to transmit telephone calls and other communication signals; the signal is a modulated light beam (variable intensity of the light beam) and is transmitted with less loss than an electrical signal in a copper wire. One sophisticated use of fiber optics, particularly in medicine, is to transmit a clear picture, Fig. 34–22. For example, a patient's stomach can be examined by inserting a light pipe through the mouth. Light is sent down one set of fibers to illuminate the stomach. The reflected light returns up another set of fibers. Light directly in front of each fiber travels up that fiber. At the opposite end, a viewer sees a series of bright and dark spots, much like a TV screen—that is, a picture of what lies at the opposite end.[†] The fibers must be optically insulated from one another, usually by a thin coating of material whose refractive index is less than that of the fiber. The fibers must be arranged precisely parallel to one another if the picture is to be clear. The more fibers there are, and the

---

[†] Lenses are used at each end: at the object end to bring the rays in parallel, and at the viewing end a telescope system for viewing.

smaller they are, the more detailed the picture. Such an "endoscope" is useful for observing the stomach or other hard to reach places for surgery or searching for lesions without surgery.

 Summary ———————————————————————————————————————

Light appears to travel in straight-line paths, called *rays*, at a speed v that depends on the index of refraction, n, of the material: $v = c/n$, where c is the speed of light in a vacuum.

When light reflects from a flat surface, the angle of reflection equals the angle of incidence. This *law of reflection* explains why mirrors can form *images*. In a plane mirror, the image is virtual, upright, the same size as the object, and as far behind the mirror as the object is in front.

Spherical mirrors, both convex and concave, can form real or virtual images. The *focal point* of a spherical mirror is the image point for an object at infinity (parallel incident rays) on the mirror's principal axis. The distance of the focal point from the center of the mirror is the *focal length*, f, and is equal to half the radius of curva-ture, r, of the mirror surface ($f = r/2$). The object distance d_o is related to the image distance d_i and the focal length by the *mirror equation*: $1/d_o + 1/d_i = 1/f$. The ratio of image height to object height is the lateral magnification, which is $m = (h_i/h_o) = -(d_i/d_o)$.

When light passes from one transparent medium into another, the rays bend or refract. The *law of refraction* (*Snell's law*) states that $n_1 \sin \theta_1 = n_2 \sin \theta_2$, where n_1 and θ_1 are the index of refraction and angle with the normal to the surface for the incident ray, and n_2 and θ_2 are for the refracted ray.

When light rays reach the boundary of a material where the index of refraction increases, the rays will be *totally internally reflected* if the incident angle, θ_1, is such that Snell's law would predict $\sin \theta_2 > 1$; this occurs if θ_1 exceeds the critical angle θ_c given by $\sin \theta_c = n_2/n_1$.

 Questions ———————————————————————————————————————

1. Give arguments to show why the moon must have a rough surface rather than a polished mirrorlike surface.

2. When you look at yourself in a tall plane mirror, you see the same amount of your body whether you are close to the mirror or far away. (Try it and see.) Use ray diagrams to show why this should be true.

3. When you look at the moon's reflection from a ripply sea it appears elongated. Explain.

4. If a perfect plane mirror reflected 100 percent of the light incident on it, could you see the surface of the mirror?

5. Although a plane mirror reverses left and right, it doesn't reverse up and down. Explain.

6. Archimedes is said to have burned the whole Roman fleet in the harbor of Syracuse by focusing the rays of the sun with a huge spherical mirror. Is this reasonable?

7. Show with diagrams that the magnification of a concave mirror is less than 1 if the object is beyond the center of curvature C and is greater than 1 if it is within this point.

8. If a concave mirror produces a real image, is the image necessarily inverted?

9. When you use a concave mirror, you cannot see an inverted image of yourself unless you place your head beyond the center of curvature C. Yet you can see an inverted image of another object placed between C and F as in Fig. 34–11. Explain. (Hint: you can see a real image only if your eye is behind the image, so that it can be formed.)

10. Using the rules for the three rays discussed with reference to Fig. 34–11 draw ray 2 for Fig. 34–15b.

11. What is the focal length of a plane mirror?

12. Does the mirror equation, Eq. 34–3, hold for a plane mirror? Explain.

13. What is the magnification of a plane mirror?

14. How might you determine the speed of light in a solid, rectangular, transparent object?

15. When a wide beam of parallel light enters water at an angle, the beam broadens. Explain.

16. Explain the origin of mirages, such as inverted images of cars seen at a distance on a hot highway. (Hint: consider the variation with temperature of the index of refraction of air.)

17. What is the angle of refraction when a light ray meets the boundary between two materials perpendicularly?

18. When you look down into a swimming pool or a lake, are you likely to underestimate or overestimate its depth? Explain. How does the apparent depth vary with the viewing angle? (Use ray diagrams.)

19. Draw a ray diagram to show why a stick looks bent when part of it is under water.

20. How are you able to "see" a round drop of water on a table even though the water is transparent and colorless?

21. Can a light ray traveling in air be totally reflected when it strikes a smooth water surface if the incident angle is right?

22. When you look up at an object in air from beneath the water in a pool, does the object appear to be the same size as when you see it directly in air? Explain.

Problems

SECTION 34–2

1. (I) What is the speed of light in (*a*) ethyl alcohol, (*b*) lucite?

2. (I) The speed of light in ice is 2.29×10^8 m/s. What is the index of refraction of ice?

3. (I) Light is emitted from an ordinary light bulb filament in wave-train bursts of about 10^{-8} s in duration. What is the length in space of such wave trains?

4. (I) How long does it take light to reach us from the sun, 1.49×10^8 km away?

5. (I) Our nearest star (other than the sun) is 4.2 light years away. That is, it takes 4.2 years for the light to reach earth. How far away is it in meters?

6. (II) (*a*) If the speed of light in a vacuum is to be measured with the uncertainty given in Section 34–2 for the best value, how accurately must the index of refraction of air be known if the measurement is actually done in air? (*b*) If $n = 1.00030 \pm 0.000010$ for air, what error will this introduce in the value for *c*?

7. (II) What is the minimum angular speed at which Michelson's eight-sided mirror would have had to rotate in order that light would be reflected into an observer's eye by succeeding mirror faces (Fig. 34–2)?

SECTION 34–3

8. (I) Suppose that you want to take a photograph of yourself as you look at your image in a mirror 2.2 m away. For what distance should the camera lens be focused?

9. (II) Stand up two plane mirrors so they form a right angle as in Fig. 34–23. When you look into this double mirror, you see yourself as others see you, instead of reversed as in a single mirror. Make a ray diagram to show how this occurs.

10. (II) A person whose eyes are 1.48 m above the floor stands 2.40 m in front of a vertical plane mirror whose bottom edge is 40 cm above the floor. What is the horizontal distance to the base of the wall supporting the mirror of the nearest point on the floor that can be seen reflected in the mirror?

FIGURE 34–23

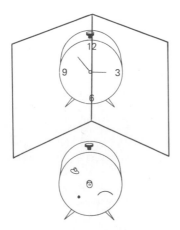

11. (II) Suppose you are 60 cm from a plane mirror. What area of the mirror is used to reflect the rays entering one eye from a point on the tip of your nose if your pupil diameter is 5.5 mm?

12. (II) Show that if two plane mirrors meet at an angle ϕ, a single ray reflected (successively) from both mirrors is deflected through an angle of 2ϕ independent of the incident angle. Assume $\phi < 90°$ and that only two reflections, one from each mirror, take place.

13. (III) Suppose a third mirror is placed beneath the two shown in Fig. 34–23, so that all three are perpendicular to each other. (*a*) Show that for such a "corner reflector," any incident ray will return in its original direction after three reflections. (*b*) What happens if it makes only two reflections?

SECTION 34–4

14. (I) What is the radius of a concave reflecting surface that brings parallel light to a focus 18.9 cm in front of it?

15. (I) How far from a convex mirror (radius 24.0 cm) must an object be placed if its image is to be at infinity?

16. (I) You try to look at yourself in a silvered ball of diameter 56.0 cm when you are 2.30 m away. Where is your image? Is it real or virtual? Can you see yourself clearly?

17. (II) Show, using a ray diagram, that the magnification *m* of a convex mirror is $m = -d_i/d_o$ just as for a concave mirror. (Hint: consider a ray from the top of the object that reflects at the center of the mirror.)

18. (II) A dentist wants a small mirror that, when 2.10 cm from a tooth, will produce a $5.5 \times$ upright image. What kind of mirror must be used and what must its radius of curvature be?

19. (II) Show that the mirror equation, Eq. 34–3, is valid for a convex mirror as long as *f* is considered negative.

20. (II) Use the mirror equation to show that the magnitude of the magnification of a concave mirror is less than 1 if the object is beyond the center of curvature C ($d_o > r$) and is greater than 1 if the object is within C ($d_o < r$).

21. (II) A 2.40-cm-tall object is placed 22.0 cm from a spherical mirror. It produces a virtual image 3.20 cm high. (*a*) What type of mirror is being used? (*b*) Where is the image located? (*c*) What is the radius of curvature of the mirror?

22. (II) The magnification of a convex mirror is $0.35 \times$ for objects 4.0 m away. What is the focal length of this mirror?

23. (II) (*a*) A plane mirror can be considered a limiting case of a spherical mirror; specify what this limit is. (*b*) Determine an equation that relates the image and object distances in this limit of a plane mirror. (*c*) Determine the magnification of a plane mirror in this same limit. (*d*) Are your results in parts (*b*) and (*c*) consistent with the discussion of Section 34–3 on plane mirrors?

24. (II) What is the radius of a concave mirror that gives a $1.6 \times$ magnification of a face 30 cm from it?

25. (II) A 1.58-m-tall person stands 4.50 m from a convex mirror and notices that he looks only half as tall as he does in a plane mirror placed at the same distance. What

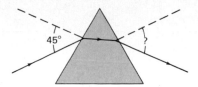

FIGURE 34–24

is the radius of curvature of the convex mirror? (Assume that $\sin \theta \approx \theta$.)

26. (III) A short thin object (like a short length of wire) of length l is placed on, and parallel to, the principal axis of a spherical mirror. Show that its image has length $l' = m^2 l$ so the *longitudinal magnification* is equal to $-m^2$ where m is the lateral magnification (Eq. 34–4). Why the minus sign?

SECTION 34–5

27. (I) A diver shines a flashlight upward from beneath the water at a 23.0° angle to the vertical. At what angle does the light leave the water?

28. (I) Rays of the sun are seen to make a 50° angle to the vertical beneath the water. At what angle above the horizon is the sun?

29. (II) Light is incident on an equilateral crown glass prism at a 45.0° angle to one face, Fig. 34–24. Calculate the angle at which light emerges from the opposite face. Assume that $n = 1.56$.

30. (II) A bright light is 2.50 m below the surface of a swimming pool and 1.45 m from one edge of the pool. At what angle does the light leave the water at the edge of the pool? Assume that the water reaches the top edge of the pool.

31. (II) Prove in general that for a light beam incident on a uniform layer of transparent material, as in Fig. 34–18, the direction of the emerging beam is parallel to the incident beam, independent of the incident angle θ and the thickness of the material.

32. (II) An aquarium filled with water has flat glass sides whose index of refraction is 1.58. A beam of light from outside the aquarium strikes the glass at a 43.7° angle to the perpendicular. What is the angle of this light ray when it enters the glass and then the water? What would it be if the ray entered the water directly?

FIGURE 34–25

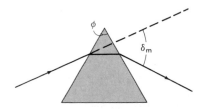

33. (II) When light passes through a prism, the angle that the refracted ray makes relative to the incident ray is called the deviation angle δ, Fig. 34–25. This angle is a minimum when the ray passes through the prism symmetrically— that is, for a prism shaped like an isosceles triangle, when the ray inside the prism is parallel to the base. Show that this minimum deviation angle, δ_m, is related to the prism's index of refraction n by

$$n = \frac{\sin \frac{1}{2}(\phi + \delta_m)}{\sin \phi/2}$$

where ϕ is the apex angle.

34. (II) A light ray is incident on a flat piece of glass as in Fig. 34–18. Show that if the incident angle θ is small, the ray is displaced a distance $d = t\theta(n - 1)/n$ where t is the thickness of the glass and θ is in radians.

35. (III) *Fermat's principle* states that "light travels between two points along that path which requires the least time, as compared to other nearby paths." From Fermat's principle derive (*a*) the law of reflection ($\theta_i = \theta_r$) and (*b*) the law of refraction (Snell's law). (Hint: choose two appropriate points so that a ray between them can undergo reflection or refraction; draw a rough path for a ray between these points, and write down an expression of the time required for light to travel the arbitrary path chosen. Then take the derivative.)

SECTION 34–6

36. (I) The critical angle for a certain liquid-air surface is 57°. What is the index of refraction of the liquid?

37. (II) A beam of light is emitted 6.0 cm beneath the surface of a liquid and strikes the surface 4.0 cm from the point directly above the source. If total internal reflection occurs, what can you say about the index of refraction of the liquid?

38. (II) The end faces of a cylindrical glass rod ($n = 1.54$) are perpendicular to the sides. Show that a light ray entering an end face at any angle will be totally internally reflected inside the rod when it strikes the sides. Assume the rod is in air. What if it were in water?

39. (II) Suppose a ray strikes the left face of the prism in Fig. 34–24 at 45° as shown, and is totally internally reflected at the opposite side. If the apex angle (at the top) is $\phi = 75°$, what can you say about the index of refraction of the prism?

40. (II) (*a*) What is the minimum index of refraction for a glass or plastic prism to be used in binoculars (Fig. 34–20) so that total internal reflection occurs at 45°? (*b*) Will binoculars work if its prisms are immersed in water? Assume $n = 1.50$. (*c*) What minimum n is needed if the prisms are immersed in water?

41. (II) If the apex angle of a prism is $\phi = 70°$ (see Fig. 34–25) what is the minimum incident angle for a ray if it is to emerge from the opposite side, if $n = 1.58$?

Lenses and Optical Instruments

35

The laws of reflection and refraction, particularly the latter, are the basis for explaining the operation of many optical instruments. In this chapter we discuss and analyze lenses using the model of ray optics discussed in the previous chapter. We also discuss a number of optical instruments, from the magnifying glass to telescopes, microscopes, and the human eye.

35–1 Refraction at a Spherical Surface

In preparation for our treatment of lenses in the next section, we first examine the refraction of rays at the spherical surface of a transparent material. Such a surface could be one face of a lens. To be general, let us consider an object which is located in a medium whose index of refraction is n_1, and rays from each point on the object can enter a medium whose index of refraction is n_2. The radius of curvature of the spherical boundary is R, and its center of curvature is at point C, Fig. 35–1. We now show that all rays leaving a point O on the object will be focused at a single point I, the image point, if we consider only rays that make a small angle with the axis and each other (such rays are called **paraxial rays**). To

FIGURE 35–1

Rays from a point O on an object will be focused at a single image point I by a spherical boundary between two transparent materials, as long as the rays make small angles with the axis ($n_2 > n_1$).

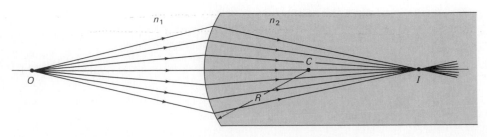

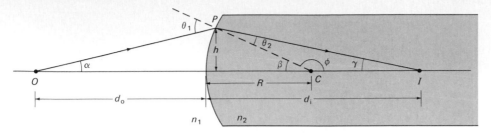

FIGURE 35-2

Diagram for proving all paraxial rays from O focus at the same point I $(n_2 > n_1)$.

do so, we consider a single ray that leaves point O as shown in Fig. 35-2. From Snell's law, Eq. 34-5, we have

$$n_1 \sin \theta_1 = n_2 \sin \theta_2.$$

We are assuming that angles θ_1, θ_2, α, β, and γ are small, so $\sin \theta \approx \theta$ (in radians), and Snell's law becomes, approximately,

$$n_1 \theta_1 = n_2 \theta_2.$$

Also, $\beta + \phi = 180°$ and $\theta_2 + \gamma + \phi = 180°$, so

$$\beta = \gamma + \theta_2.$$

Similarly for triangle OPC,

$$\theta_1 = \alpha + \beta.$$

These three relations can be combined to yield

$$n_1 \alpha + n_2 \gamma = (n_2 - n_1)\beta.$$

Since we are considering only the case of small angles, we can write, approximately,

$$\alpha = \frac{h}{d_o}, \quad \beta = \frac{h}{R}, \quad \gamma = \frac{h}{d_i},$$

where d_o and d_i are the object and image distances and h is the height as shown in Fig. 35-2. We substitute these into the previous equation, divide through by h, and obtain

$$\frac{n_1}{d_o} + \frac{n_2}{d_i} = \frac{n_2 - n_1}{R}. \tag{35-1}$$

For a given object distance d_o, it is clear that d_i, the image distance, does not depend on the angle of a ray. Hence all paraxial rays meet at the same point I. This is true, of course, only for rays that make small angles with the axis and with each other. This is equivalent to assuming that the height of the refracting spherical surface is small compared to its radius of curvature, so that only paraxial rays are refracted. If this assumption is not true, the rays will not converge to a point; there will be spherical aberration, just as for a mirror (see Fig. 34-9), and the image will be blurry. (Spherical aberration will be discussed further in Section 35-7.)

We derived Eq. 35-1 using Fig. 35-2 for which the spherical surface is convex (as viewed by the incoming ray). It is also valid for a concave surface—as can be seen using Fig. 35-3—if we make the following conventions:

1. If the surface is convex (so the center of curvature C is on the side of the surface opposite to that from which the light comes), R is positive; if the surface is concave (C on the same side from which the light comes) R is negative.

2. The image distance, d_i, follows the same convention: positive if on the opposite side from where the light comes, negative if on the same side.

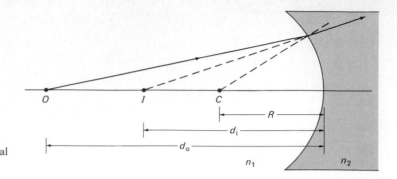

FIGURE 35–3

Rays from O refracted by concave surface form virtual image ($n_2 > n_1$).

3. The object distance is positive if on the same side from which the light comes (this is the normal case, although when several surfaces bend the light it may not be so), otherwise it is negative.

For the case shown in Fig. 35–3 with a concave surface, both R and d_i are negative when used in Eq. 35–1. Note, in this case, that the image is virtual.

■ **EXAMPLE 35–1** A person looks vertically down into a 4.0-m-deep lake. how deep does the lake appear to be?

SOLUTION A ray diagram is shown in Fig. 35–4. Point O represents a point on the lake bottom. The rays diverge and appear to come from point I, the image. We have $d_o = 4.0$ m and, for a flat surface $R = \infty$. Then Eq. 35–1 becomes

$$\frac{1.33}{4.0 \text{ m}} + \frac{1.00}{d_i} = \frac{(1.00 - 1.33)}{\infty} = 0$$

Hence $d_i = -(4.0 \text{ m})/(1.33) = -3.0$ m. So the lake appears to be only three-fourths as deep as it actually is. (At angles other than vertical, this conclusion must be modified.) The minus sign tells us the image point I is on the same side of the surface as O, and the image is virtual.

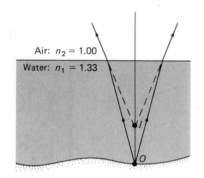

Air: $n_2 = 1.00$

Water: $n_1 = 1.33$

FIGURE 35–4

Example 35–1.

FIGURE 35–5

Types of lenses.

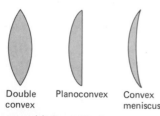

Double convex   Planoconvex   Convex meniscus

(a) Converging lenses

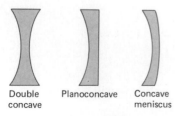

Double concave   Planoconcave   Concave meniscus

(b) Diverging lenses

▱ **35–2 Thin Lenses**

The most important simple optical device is no doubt the thin lens. The development of optical devices using lenses dates to the sixteenth and seventeenth centuries, although the earliest record of eyeglasses dates from the late thirteenth century. A thin lens is usually round, and its two faces are each a portion of a sphere. (Although cylindrical surfaces are also possible, we will concentrate on spherical ones.) The two faces can be concave, convex, or plane (see Fig. 35–5). The importance of lenses is that they form images of objects, and we can analyze a lens by applying the results of the previous section to the two lens surfaces. Before doing so, we look at some general properties of lenses.

Consider the rays parallel to the axis of the double convex lens which is shown in cross-section in Fig. 35–6a. We assume the lens is made of glass or transparent plastic so its index of refraction is greater than that of the air outside. From Snell's law we can see that each ray is bent toward the axis at both lens surfaces (note the dashed lines indicating the normals to each surface for the top ray). If rays parallel to the principal axis fall on a thin lens, they will be focused to a point called the *focal point*, F. This will not be precisely true for a lens with spherical surfaces. But it will be very nearly true—that is, parallel rays will be focused to a tiny region that is nearly a point—if the diameter of the lens is

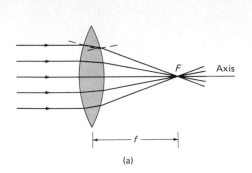
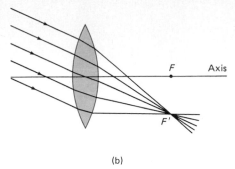

(a)   (b)

FIGURE 35-6

Parallel rays are brought to a focus by a converging lens.

small compared to the radii of curvature of the two lens surfaces. This criterion is satisfied by a **thin lens**, one that is very thin compared to its diameter; and it is only thin lenses that we consider here.

Since the rays from a distant object are essentially parallel (see Fig. 34-8), we can also say that *the focal point is the image point for an object on the principal axis at infinity.* The distance of the focal point from the center of the lens is called the *focal length, f.* Since the lens is assumed thin, f can be measured from the center or an edge of the lens without substantial error. A lens can be turned around so light can pass through it from the opposite side; the focal length is the same on both sides, as we shall see shortly, even if the curvatures of the two lens surfaces are different. If parallel rays fall on a lens at an angle, as in Fig. 35-6b, they focus at a point F'. The plane in which all points such as F and F' fall is called the *focal plane* of the lens.

Any lens which is thicker in the center than at the edges will make parallel rays converge to a point, and are called *converging lenses* (see Fig. 35-5a). Lenses which are thinner in the center than at the edges (Fig. 35-5b) are called *diverging lenses* because they make parallel light diverge, as shown in Fig. 35-7. The *focal point F* of a diverging lens is defined as that point from which refracted rays, originating from parallel incident rays, seem to emerge as shown in the figure. And the distance from F to the lens is called the focal length, just as for a converging lens. We can find the image formed by a lens for a given object by drawing rays for each point in the object; we need consider only three particular rays as indicated in Fig. 35-8. These three rays, emanating from a single point on the object, are drawn as if the lens were infinitely thin (thus the sharp bend within the lens). Ray 1 is drawn parallel to the axis; therefore it is refracted by the lens so it passes through the focal point F behind the lens. Ray 2 is drawn through the focal point F' on the same side of the lens as the object; it therefore emerges from the lens parallel to the axis. Ray 3 is directed toward the very center of the lens where the two surfaces are essentially parallel to each other; this ray therefore emerges from the lens at the same angle as it entered; as we saw in Example 34-5, the ray would be displaced slightly to one side, but since we assume the lens is thin, we draw ray 3 straight through as shown. Actually, any two of these rays will suffice to locate the image point, which is the point where they intersect. Drawing the third can serve as a check.

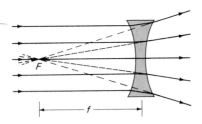

FIGURE 35-7

Diverging lens.

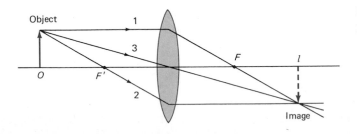

FIGURE 35-8

Finding the image by ray tracing—converging lens.

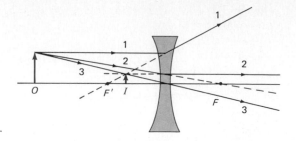

FIGURE 35–9

Finding image I by ray tracing—diverging lens.

In this way we can find the image point for one point of the object (the top of the arrow in Fig. 35–8). The image points for all other points on the object can be found similarly to determine the complete image of the object. Because the rays actually pass through the image for the case shown, it is a *real image.*

By drawing the same three rays we can determine the image position for a diverging lens, as shown in Fig. 35–9. Note that ray 1 is drawn parallel to the axis, but does not pass through the focal point F behind the lens; instead it seems to come from the focal point F' in front of the lens (dashed line). Ray 2 is directed toward F and is refracted parallel by the lens. The three refracted rays seem to emerge from a point on the left of the lens. This is the image, I. Since the rays do not pass through the image, it is a *virtual image.*

We now derive an equation that relates the image distance to the object distance and the properties of a thin lens. Consider the lens shown in Fig. 35–10 whose thickness at the center is t. Rays are shown leaving a point O on some object, passing through the lens, and coming to a focus at point I, the final image point. We assume the rays are paraxial, although for clarity in the diagram the ray angles shown are not actually small. At the front surface of the lens, Eq. 35–1 gives

$$\frac{1}{d_o} + \frac{n}{d_i'} = \frac{n-1}{R_1}$$

where R_1 is the radius of curvature of the front surface, and $n_1 = 1$ for air and $n_2 = n$ for the lens. Both d_o and d_i' (which is the image distance for the first surface alone) are measured from the front surface; for the case shown, d_i' is negative (the image is virtual, at point O'). Next we apply Eq. 35–1 to the second surface. The light rays striking the second surface have directions as if they were coming from point O'. Thus, the object distance for the second surface is $d_o' = -d_i' + t$, where the negative sign is necessary since $d_i' < O$. Thus Eq. 35–1 gives

$$\frac{n}{-d_i' + t} + \frac{1}{d_i} = \frac{1-n}{R_2}$$

where R_2 is the radius of curvature of the second surface (and is negative for the case shown). We now assume the thickness t of the lens is small compared to the

FIGURE 35–10

Deriving the lens equation.

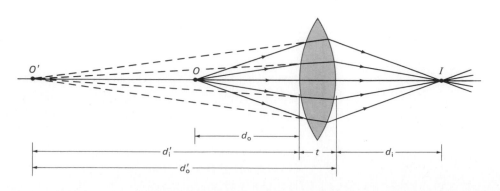

other distance involved, and we set $t = 0$. When we eliminate d_i' between these two equations we find:

$$\frac{1}{d_o} + \frac{1}{d_i} = (n - 1)\left(\frac{1}{R_1} - \frac{1}{R_2}\right). \qquad (35-2)$$

This is the relation we were looking for. It relates the object distance d_o to the image distance d_i (the distance from the lens of the final image formed by the lens) and to the properties R_1, R_2, and n of the lens. It is valid, of course, only for paraxial rays and only if the lens is very thin. Note that d_i does not depend on the angles of the rays, so all paraxial rays are imaged at the same point. For nonparaxial rays, and for nonthin lenses, the image may not be sharp (more on this later).

If we consider an object at infinity ($d_o = \infty$), the image distance is the focal length, $d_i = f$. Equation 35-2 then becomes

$$\frac{1}{f} = (n - 1)\left(\frac{1}{R_1} - \frac{1}{R_2}\right). \qquad (35-3)$$

This is called the **lens-maker's equation**, for it relates the focal length of any lens to the radii of curvature of its two surfaces and its index of refraction. Note (as discussed in Section 35-1) that a radius of curvature is positive if a surface is convex to the incoming light, and is negative if concave. Note also that if a lens is turned around, so light comes from the opposite direction, R_1 and R_2 exchange roles in Eq. 35-3 but the value of f is the same. So the position of the focal point F is the same on both sides of a lens, as we claimed earlier.

> **EXAMPLE 35-2** A Lucite planoconcave lens has one flat surface and the other has $R = 18.4$ cm. What is the focal length?
>
> **SOLUTION** From Table 34-1, n for Lucite is 1.51. A plane surface has infinite radius of curvature; we call this R_1, then $1/R_1 = 0$. Therefore
>
> $$\frac{1}{f} = (1.51 - 1.00)\left(-\frac{1}{18.4 \text{ cm}}\right).$$
>
> So $f = (-18.4 \text{ cm})/(0.51) = -36.0$ cm, and the lens is diverging.

When we combine Eqs. 35-3 and 35-2 we obtain

$$\frac{1}{d_o} + \frac{1}{d_i} = \frac{1}{f}. \qquad (35-4)$$

This is called the **lens equation**. If the focal length of a lens is known, the image distance for any object distance is readily obtained. It is one of the most useful equations in geometrical optics.

Graphical constructions are often useful, and we now use them briefly, as a check for consistency, to obtain the lens equation and to obtain the magnification of a lens. Consider the two rays shown in Fig. 35-11a for a converging lens (assumed to be very thin); h_o and h_i refer to the heights of the object and image, and d_o and d_i are their distances from the lens. The triangles FII' and FBA are similar; thus

$$\frac{-h_i}{h_o} = \frac{d_i - f}{f},$$

since length $AB = h_o$, and the minus sign on h_i is inserted because, by convention, we choose h_i (or h_o) to be negative if point I' (or O') is below the axis. (Thus $-h_i$ is positive in the case shown.) Triangles OAO' and IAI' are similar. Therefore,

$$-\frac{h_i}{h_o} = \frac{d_i}{d_o}.$$

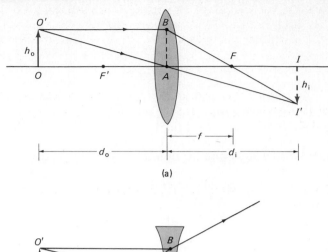

FIGURE 35–11

Deriving the lens equation for (a) converging and (b) diverging lenses.

The left sides of these two equations are the same; by equating the right sides, dividing by d_i, and rearranging, we obtain the lens equation:

$$\frac{1}{d_o} + \frac{1}{d_i} = \frac{1}{f}.$$

From the rays in Fig. 35–11b, we can derive the lens equation for a diverging lens. Triangles IAI' and OAO' are similar; and triangles IFI' and AFB are similar. Then we have (by our conventions of Section 35–1, $d_i < 0$ in Fig. 35–11b, so $-d_i$ is a positive quantity):

$$\frac{h_i}{h_o} = \frac{-d_i}{d_o}, \quad \text{and} \quad \frac{h_i}{h_o} = \frac{f - (-d_i)}{f}.$$

When these are equated and simplified we obtain

$$\frac{1}{d_o} + \frac{1}{d_i} = -\frac{1}{f}.$$

This equation becomes the same as Eq. 35–4 if we make f negative for a diverging lens. Thus, Eq. 35–4 will be valid for both converging and diverging lenses and for all object and image positions if we use the following conventions, which are consistent with those in Section 35–1 and our original derivation of Eqs. 35–2 to 35–4:

1. The focal length is positive for converging lenses and negative for diverging lenses. A radius of curvature is positive when light strikes a convex surface and negative when it strikes a concave surface.

2. The object distance is positive if it is on the side of the lens from which the light is coming (this is normally the case, although when lenses are used in combination, it might not be so), otherwise it is negative.

3. The image distance is positive if it is on the opposite side of the lens from where the light is coming; if it is on the same side, d_i is negative. Equivalently, the image distance is positive for a real image and negative for a virtual image.

4. Object and image heights, h_o and h_i, are positive for points above the axis, and negative for points below the axis.

The lateral *magnification* m of a lens is defined as the ratio of the image height to object height, $m = h_i/h_o$. From Fig. 35–11 and the conventions just stated, we have

$$m = \frac{h_i}{h_o} = -\frac{d_i}{d_o}.$$

(35–5)

For an upright image the magnification is positive, and for an inverted image m is negative.

$m = -\frac{d_i}{d_o}$

EXAMPLE 35–3 What is the position and size of the image of a 22.4-cm-high giant insect placed 1.50 m from a $+50.0$-mm-focal-length camera lens?

SOLUTION The camera lens is converging, with $f = +5.00$ cm, so Eq. 35–4 gives

$$\frac{1}{d_i} = \frac{1}{f} - \frac{1}{d_o} = \frac{1}{5.00\ \text{cm}} - \frac{1}{150\ \text{cm}} = \frac{30.0 - 1.00}{150\ \text{cm}},$$

and $d_i = 150\ \text{cm}/29.0 = 5.17$ cm or 51.7 mm behind the lens. The magnification $m = -d_i/d_o = -5.17\ \text{cm}/150\ \text{cm} = -0.0345$, so $h_i = (-0.0345)(22.4\ \text{cm}) = -0.773$ cm. The image is 7.73 mm high and is inverted as in Fig. 35–11a. Notice that the image is 1.7 mm farther from the lens than would be the image for an object at infinity. This is an example of the fact that when focusing a camera lens, the closer the object is to the camera, the farther the lens must be from the film.

$\frac{1}{f} = \frac{1}{d_o} + \frac{1}{d_i}$

$\frac{1}{10} = \frac{1}{15} + \frac{1}{d_i}$

$d_i = 30\text{cm}$

$-\frac{30}{15}$

EXAMPLE 35–4 An object is placed 10 cm from a 15-cm-focal-length converging lens. Determine the image position and size (a) analytically, (b) using a ray diagram.

SOLUTION (a) Since $f = 15$ cm and $d_o = 10$ cm,

$$\frac{1}{d_i} = \frac{1}{15\ \text{cm}} - \frac{1}{10\ \text{cm}} = -\frac{1}{30\ \text{cm}}$$

so $d_i = -30$ cm. Since d_i is negative, the image must be virtual and on the same side of the lens as the object. The magnification $m = -(-30\ \text{cm})/(10\ \text{cm}) = 3.0$; so the image is three times as large as the object and is upright. (b) The ray diagram is shown in Fig. 35–12 and confirms the result in part (a).

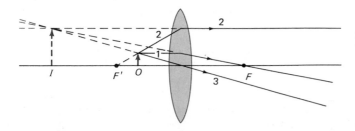

FIGURE 35–12

An object placed within the focal point of a converging lens produces a virtual image (Example 35–4).

EXAMPLE 35–5 Where must a small bearing be placed if a 25-cm-focal-length diverging lens is to form a virtual image 20 cm in front of the lens?

SOLUTION Since $f = -25$ cm and $d_i = -20$ cm, then

$$\frac{1}{d_o} = -\frac{1}{25 \text{ cm}} + \frac{1}{20 \text{ cm}} = \frac{1}{100 \text{ cm}}.$$

So the object must be 100 cm in front of the lens. The ray diagram is basically that of Fig. 35–11b.

EXAMPLE 35–6 To measure the focal length of a diverging lens, a converging lens is placed next to it as in Fig. 35–13. The sun's rays are focused by this combination at a point 28.5 cm behind them as shown. If the converging lens has a focal length f_C of 16.0 cm, what is the focal length f_D of the diverging lens?

SOLUTION Let $f_T = 28.5$ cm refer to the focal length of the total combination. If the diverging lens were absent, the converging lens would form the image at its focal point—that is, at a distance $f_C = 16.0$ cm behind it (dashed lines in Fig. 35–13). When the diverging lens is placed next to the converging lens (we assume both lenses are thin and the space between them is negligible) we treat the image formed by the first lens as the *object* for the second (diverging) lens. Since this object lies to the right of the diverging lens, this is a situation where d_o is negative. Thus, for the diverging lens, the object is virtual and $d_o = -16.0$ cm; and it forms the image a distance $d_i = 28.5$ cm away (this was given). Thus

$$\frac{1}{f_D} = \frac{1}{d_o} + \frac{1}{d_i} = -\frac{1}{16.0 \text{ cm}} + \frac{1}{28.5 \text{ cm}} = -0.0274 \text{ cm}^{-1}.$$

So $f_D = -1/(0.0274 \text{ cm}^{-1}) = -36.5$ cm. Note that the converging lens must have a focal length whose magnitude is less than that of the diverging lens if this technique is to work.

This last example is our first illustration of how to deal with lenses used in combination. In general, when light passes through several lenses, the image formed by one lens becomes the object for the next lens. The total magnification will be the product of the separate magnifications of each lens. We will see more examples later in this chapter.

EXAMPLE 35–7 A convex meniscus lens (see Fig. 35–5a) is made from glass with $n = 1.50$. The radius of curvature of the convex surface is 22.4 cm and that of the concave surface is 46.2 cm. (*a*) What is the focal length? (*b*) Where will it focus an object 2.00 m away?

FIGURE 35–13

Determining the focal length of a diverging lens (Example 35–6).

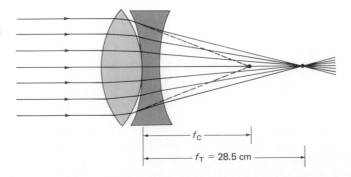

SOLUTION   (a) $R_1 = 22.4$ m and $R_2 = 46.2$ cm; both are positive since both are convex surfaces to the incoming light (assumed from the left.) Then

$$\frac{1}{f} = (1.50 - 1.00)\left(\frac{1}{22.4 \text{ cm}} - \frac{1}{46.2 \text{ cm}}\right) = 0.0114 \text{ cm}^{-1}.$$

So $f = 89$ cm and is converging. Notice that if we turn the lens around so that $R_1 = -46.2$ cm and $R_2 = -22.4$ cm we get the same result. (b) From the lens equation

$$\frac{1}{d_i} = \frac{1}{f} - \frac{1}{d_o} = \frac{1}{0.89 \text{ m}} - \frac{1}{2.00 \text{ m}} = 0.62 \text{ m}^{-1}$$

so $d_i = 1.60$ m.

Note a possible source of confusion involving the terms *convex* and *concave*: a so-called double convex lens, for example, as in Fig. 35–5a, has two convex surfaces as viewed from the outside; hence its name. But for purposes of assigning a sign to its radii of curvature, the second surface is concave to light passing through it.

35–3   The Magnifying Glass

This and the next few sections will deal with optical devices that are used to produce magnified images of objects. We first discuss the *simple magnifier* or *magnifying glass*, which is simply a converging lens.

How large an object appears, and how much detail we can see on it, depends on the angle subtended by the object at the eye. For example, a penny held 30 cm from the eye looks twice as high as one held 60 cm away because the angle it subtends is twice as great, Fig. 35–14. To examine detail on an object, we bring it up close to our eyes so it subtends a greater angle. However, our eyes can accommodate only up to a point. The nearest point at which an eye can focus clearly is called the *near point*; it varies from person to person (Section 35–6) although a good average is about 25 cm, and we take this as the standard or "normal" near point in what follows. The most distant point that can be focused clearly by the eye is called the *far point*, and occurs when the muscles of the eye are relaxed; for a normal eye the far point is quite large and we take it to be infinity.

A magnifying glass allows us to place an object closer to the eye so that it subtends a greater angle. As shown in Fig. 35–15a, the object is placed at the focal point or just within it. Then the converging lens produces a virtual image, which must be at least 25 cm from the eye if the eye is to focus on it. If the eye is relaxed, the image will be at infinity, and in this case the object is exactly at the focal point. (You make this slight adjustment yourself when you "focus" on the object by moving the lens.)

A comparison of part (a) of Fig. 35–15 with part (b), in which the same object is viewed at the near point with the unaided eye, reveals that the angle the object subtends at the eye is much larger when the magnifier is used. The *angular magnification* or *magnifying power*, M, of the lens is defined as the ratio of the angle subtended with the lens to that subtended by the unaided eye without the lens at a distance of 25 cm:

$$M = \frac{\theta'}{\theta}$$

where θ and θ' are shown in Fig. 35–15. This can be written in terms of the focal length f of the lens as follows. Suppose first that the image in Fig. 35–15a is at the near point N of the eye: $d_i = -N$, where $N = 25$ cm for the normal eye.

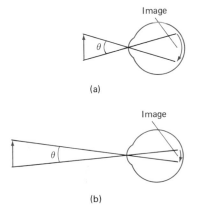

(a)

(b)

FIGURE 35–14

When the same object is viewed at a shorter distance, the image on the retina is greater; so the object appears larger and more detail can be seen. The angle θ the object subtends in (a) is greater than in (b).

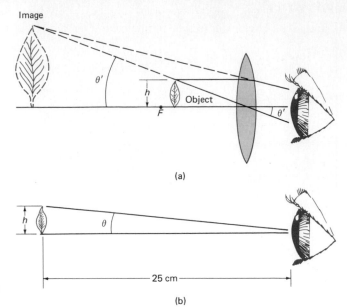

FIGURE 35–15

Leaf viewed (a) through a magnifying glass and (b) with the unaided eye focused at its near point.

Then the object distance d_o is given by

$$\frac{1}{d_o} = \frac{1}{f} - \frac{1}{d_i} = \frac{1}{f} + \frac{1}{N}$$

or $d_o = Nf/(f + N)$. (We see from this equation that $d_o < f$, as shown in Fig. 35–15a, since $N/(f + N)$ must be less than 1.) Let h be the height of an object, and we assume h is small so the angles θ and θ' are approximately equal to their sines and tangents; then $\theta' = h/d_o = (f + N)h/(Nf)$ and $\theta = h/N$. Thus

$$M = \frac{\theta'}{\theta} = \frac{(f + N)h}{Nf} \left(\frac{N}{h}\right)$$

or

$$M = 1 + \frac{N}{f}. \qquad \begin{bmatrix} \text{eye focused at near point, } N, \\ \text{which} = 25 \text{ cm for normal eye} \end{bmatrix} \quad (35\text{–}6a)$$

If the eye is relaxed when using the magnifying glass, the image is then at infinity, and the object is then precisely at the focal point. In this case $\theta' = h/f$ so

$$M = \frac{\theta'}{\theta} = \left(\frac{h}{f}\right)\left(\frac{N}{h}\right) = \frac{N}{f}. \qquad \begin{bmatrix} \text{eye focused at } \infty; N = \\ 25 \text{ cm for normal eye} \end{bmatrix} \quad (35\text{–}6b)$$

It is clear that the magnification is slightly greater when the eye is focused at its near point than when relaxed. And the shorter the focal length of the lens, the greater the magnification.

EXAMPLE 35–8 An 8-cm-focal-length converging lens is used as a magnifying glass by a person with normal eyes. Calculate (a) the maximum magnification; (b) the magnification when the eye is relaxed.

SOLUTION (a) The maximum magnification is obtained when the eye is focused at its near point:

$$M = 1 + \frac{N}{f} = 1 + \frac{25}{8} \approx 4X.$$

(b) With the eye focused at infinity, $M = 25 \text{ cm}/8 \text{ cm} \approx 3X$.

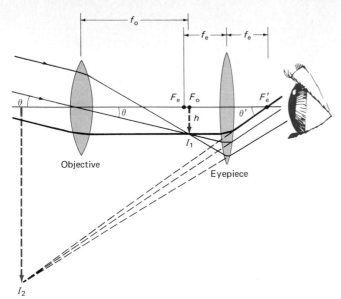

FIGURE 35–16

Astronomical telescope (refracting).

35–4   Telescopes

A telescope is used to magnify objects that are very far away; in most cases the object can be considered to be at infinity.

Galileo, although he did not invent it,[†] developed the telescope into a usable and important instrument; he was the first to train it on the heavens where he made a number of world-shaking discoveries (the moons of Jupiter, the phases of Venus, sunspots, the structure of the moon's surface, that the Milky Way is made up of a huge number of individual stars, among others).

Several types of astronomical telescope exist. The common *refracting* type, sometimes called "Keplerian," contains two converging lenses located at opposite ends of a long tube, Fig. 35–16. The lens closest to the object is called the *objective lens* and forms a real image I_1 of the object at its focal point F_o (or near it if the object is not at infinity). Although this image, I_1, is smaller than the original object, it is very close to the second lens, called the *eyepiece* which acts as a magnifier. That is, the eyepiece magnifies the image produced by the objective to produce a second, greatly magnified image, I_2, which is virtual. If the viewing eye is relaxed, the eyepiece is adjusted so the image I_2 is at infinity; then the real image I_1 is at the focal point F_e of the eyepiece, and the distance between the lenses is $f_o + f_e$ for an object at infinity.

To find the total magnification of this telescope, we note that the angle an object subtends as viewed by the unaided eye is just the angle θ subtended at the telescope objective. From Fig. 35–16 we can see that $\theta \approx h/f_o$ where h is the height of the image I_1 and we assume θ is small so $\tan \theta \approx \theta$. Note too that the darkest of the three rays drawn in the figure is parallel to the axis before it strikes the eyepiece and therefore passes through the focal point F'_e. Thus $\theta' \approx h/f_e$ and

---

[†] Galileo built his first telescope after having heard of such an instrument existing in Holland. The first telescopes magnified only 3 to 4 times, but Galileo soon made a 30-power instrument. The first Dutch telescope seems to date from about 1604, but there is a reference suggesting it may have been copied from an Italian one from as early as 1590. Kepler (see Chapter 5) gave a ray description (1611) of both the Keplerian and Galilean telescopes (that is, those with two lenses). The former is named for him because he first described it, although he did not build it.

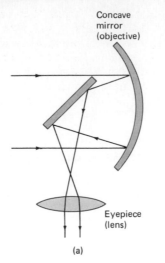

Concave
mirror
(objective)

Eyepiece
(lens)

(a)

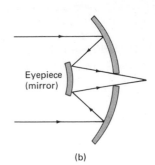

Eyepiece
(mirror)

(b)

FIGURE 35–17

A concave mirror can be used as the objective of an
astronomical telescope. Either a lens or a mirror can
be used as the eyepiece. Arrangement (a) is called
the Newtonian focus and (b) the Cassegrainian focus.
Other arrangements are also possible.

the total magnifying power of this telescope is

$$M = \frac{\theta'}{\theta} = -\frac{f_o}{f_e}, \tag{35–7}$$

where we have inserted a minus sign to indicate that the image is inverted. To
achieve a large magnification, the objective lens should have a long focal length
and the eyepiece a short one.

For an astronomical telescope to produce bright images of distant stars, the
objective lens must be large to allow in as much light as possible. The construction and grinding of large lenses is very difficult. Therefore, the largest telescopes
are *reflecting* telescopes using a curved mirror as the objective, Fig. 35–17, since
a mirror has only one surface to be ground and can be supported along its entire
surface. (A large lens, supported at its edges, would sag under its own weight.)[†]
Normally, the eyepiece lens or mirror (see Fig. 35–17) is removed so that the real
image formed by the objective can be recorded on film.

A *terrestrial telescope*, unlike its astronomical counterpart, must provide an
upright image. Two designs are shown in Fig. 35–18. The Galilean type shown
in part (a), which Galileo used for his great astronomical discoveries, has a
diverging lens as eyepiece which intercepts the converging rays from the objective
lens before they reach a focus, and acts to form a virtual upright image. This
design is often used in opera glasses. The tube is reasonably short, but the field

FIGURE 35–18

Terrestrial telescopes that produce an
upright image: (a) Galilean; (b) spyglass or field-lens type.

[†] Another advantage of mirrors is that they exhibit no chromatic aberration since the light doesn't
pass through them; and they can be ground in a parabolic shape to correct for spherical aberration (see
sec. 35–7). The reflecting telescope was first proposed by Newton.

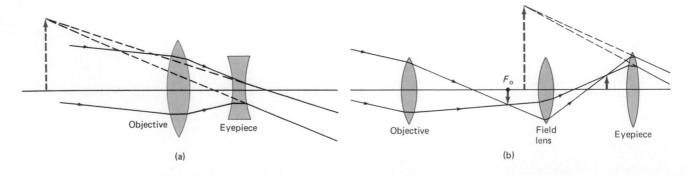

Objective   Eyepiece

(a)

Objective F_o Field   Eyepiece
                          lens

(b)

of view is small. The second type, shown in Fig. 35–18b is often called a *spyglass* and makes use of a third lens ("field lens") which acts to make the image upright as shown. A spyglass must be quite long. The most practical design today is the *prism binocular* which was shown in Fig. 34–20. The objective and eyepiece are converging lenses. The prisms reflect the rays by total internal reflection and shorten the physical size of the device, and they also act to produce an upright image; one prism reinverts the image in the vertical plane, the other in the horizontal plane.

 EXAMPLE 35–9 A Galilean telescope has an objective lens whose focal length is 28 cm and an eyepiece with focal length −8.0 cm. What is the magnification?

SOLUTION $M = -f_o/f_e = -(28 \text{ cm})/(-8.0 \text{ cm}) = 3.5 \times$.

35–5 Compound Microscope

The compound microscope, like the telescope, has both objective and eyepiece (or ocular) lenses, Fig. 35–19. The design is different from that for a telescope since a microscope is used to view objects that are very close, so the object distance is very small. The object is placed just beyond the objective's focal point as shown in Fig. 35–19. The image I_1 formed by the objective lens is real, quite far from the lens, and much enlarged. This image is magnified by the eyepiece into a very large virtual image, I_2, seen by the eye.

The total magnification of a microscope is the product of the magnifications produced by the two lenses. The image I_1 formed by the objective is a factor M_o greater than the object itself; from Fig. 35–19 and Eq. 35–5 for the magnification of a simple lens, we have

$$M_o = \frac{d_i}{d_o} = \frac{l - f_e}{d_o},$$

where l is the distance between the lenses (equal to the length of the barrel), and

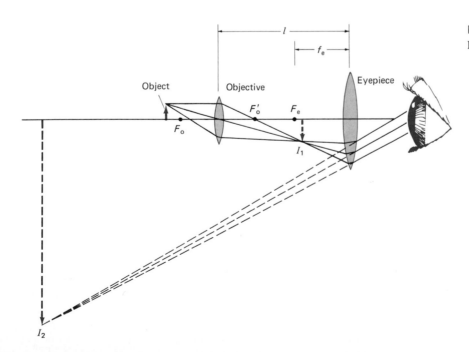

FIGURE 35–19
Ray diagram for a compound microscope.

we ignored the minus sign in Eq. 35–5 which only tells us that the image is inverted. The eyepiece acts like a simple magnifier; if we assume the eye is relaxed, its magnification M_e is (from Eq. 35–6b)

$$M_e = \frac{N}{f_e},$$

where the near point $N = 25$ cm for the normal eye. Since the eyepiece enlarges the image formed by the objective, the total magnification M is

$$M = M_e M_o = \left(\frac{N}{f_e}\right)\left(\frac{l - f_e}{d_o}\right) \qquad (35–8a)$$

$$\approx \frac{Nl}{f_e f_o}. \qquad (35–8b)$$

The approximation, Eq. 35–8b, is accurate when f_e and f_o are small compared to l so $l - f_e \approx l$ and $d_o \approx f_o$ (Fig. 35–19). This is a good approximation for large magnifications since these are obtained when f_e and f_o are very small (they are in the denominator of Eq. 35–8). (In order to make lenses of very short focal length, which can be done best for the objective lens, compound lenses involving several elements must be used to avoid serious aberrations as discussed in Section 35–7.)

EXAMPLE 35–10   A compound microscope consists of a $10\times$ eyepiece and a $50\times$ objective 18.0 cm apart. Determine (a) the total magnification; (b) the focal length of each lens; and (c) the position of the object when the final image is in focus with the eye relaxed. Assume a normal eye with $N = 25$ cm.

SOLUTION   (a) The total magnification is $10 \times 50 = 500\times$. (b) The eyepiece focal length is $f_e = N/M_e = 25$ cm$/10 = 2.5$ cm. It is easier to next find d_o (part c) before we find f_o since we can use the equation for M_o above; solving for d_o, we find $d_o = (l - f_e)/M_o = 15.5$ cm$/50 = 0.31$ cm. Then, from the lens equation:

$$\frac{1}{f_0} = \frac{1}{d_o} + \frac{1}{d_i} = \frac{1}{0.31\text{ cm}} + \frac{1}{15.5\text{ cm}} = \frac{51}{15.5\text{ cm}};$$

so $f_o = 0.30$ cm. (c) We just calculated $d_o = 0.31$ cm, which is very close to f_o.

* 35–6   The Human Eye; Corrective Lenses

The human eye, Fig. 35–20, is an enclosed volume into which light passes through a lens. There is a diaphragm called the *iris* (the colored part of your eye) which adjusts automatically to control the amount of light entering the eye. The hole in the iris through which light passes (the *pupil*) appears black because very little light is reflected back out from the interior of the eye. The *retina*, which plays the role of the film in a camera, is on the curved rear surface. It consists of a complex array of nerves and receptors known as rods and cones which act as transducers that change light energy into electrical signals that travel along the nerves. The construction of the image from all these tiny receptors is done mainly in the brain, although some analysis is apparently done in the complex interconnected nerve network at the retina itself. At the center of the retina is a small area called the *fovea*, about 0.25 mm in diameter, where the cones are very closely packed and the sharpest image and best color discrimination are found.

The lens of the eye does little of the bending of the light rays. Most of the refraction is done at the front surface of the *cornea*, which also acts as a protective

FIGURE 35–20

Diagram of a human eye.

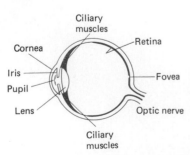

covering. The lens acts as a fine adjustment for focusing at different distances. This is accomplished by the ciliary muscles (Fig. 35–20) which change the curvature of the lens so that its focal length is changed. To focus on a distant object, the muscles are relaxed and the lens is thin, Fig. 35–21a. To focus on a nearby object, the muscles contract causing the center of the lens to be thicker, Fig. 35–21b, thus shortening the focal length. This adjustment is called *accommodation*. As discussed earlier, the closest distance at which the eye can focus clearly is called the *near point* of the eye. For young adults it is typically 25 cm, although younger children can often focus on objects as close as 10 cm; as people grow older, the ability to accommodate is reduced and the near point increases. A given person's *far point* is the farthest distance at which an object can be seen clearly. For some purposes it is useful to speak of a *normal eye* (a sort of average over the population), which is defined as one having a near point of 25 cm and a far point of infinity. Two common defects of the eye are nearsightedness and farsightedness. *Nearsightedness* or *myopia* refers to an eye that can focus only on nearby objects; the far point is not infinity but some shorter distance, so distant objects are not seen clearly. It is usually caused by an eyeball that is too long, although sometimes it is the curvature of the cornea that is too great. In either case, images of distant objects are focused in front of the retina. A diverging lens, because it causes parallel rays to diverge, allows the rays to be focused at the retina (Fig. 35–22a) and thus corrects this defect. *Farsightedness* or *hyperopia* refers to an eye which cannot focus on nearby objects. Although distant objects are usually seen clearly, the near point is somewhat greater than the "normal" 25 cm which makes reading difficult. This defect is caused by an eyeball that is too short, or (less often) by a cornea that is not sufficiently curved. It is corrected by a converging lens, Fig. 35–22b. Similar to hyperopia is *presbyopia* which refers to the lessening ability of the eye to accommodate as one ages, and the near point moves out. Converging lenses also compensate for this. *Astigmatism* is usually caused by an out-of-round cornea or lens so that point objects are focused as short lines. It is as if the cornea were spherical with a cylindrical section superimposed. As shown in Fig. 35–23, a cylindrical lens focuses a point into a line parallel to its axis. An astigmatic eye focuses rays in a vertical plane, say, at a shorter distance than it does for rays in a horizontal plane. Astigmatism is corrected with the use of a compensating cylindrical lens. Lenses for eyes that are nearsighted or farsighted as well as astigmatic are ground with superimposed spherical and cylindrical surfaces, so that the radius of curvature of the correcting lens is different in different planes.

Optometrists and ophthalmologists, instead of using the focal length, use the reciprocal of the focal length to specify the strength of lenses. This is called the *power*, P, of a lens:

$$P = \frac{1}{f}$$

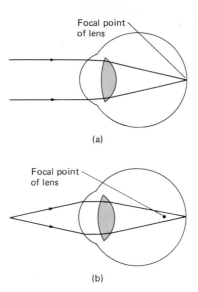

FIGURE 35–21

Accommodation by eye: (a) lens relaxed, focused on infinity; (b) lens thickened, focused on nearby object.

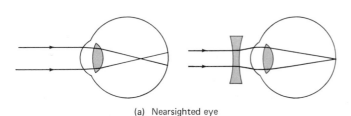

(a) Nearsighted eye

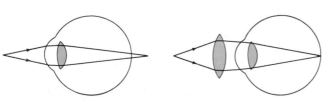

(b) Farsighted eye

FIGURE 35–22

(a) A nearsighted eye, which cannot focus clearly on distant objects, can be corrected by use of a diverging lens. (b) A farsighted eye, which cannot focus clearly on nearby objects, can be corrected by use of a converging lens.

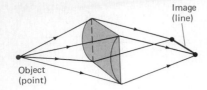

FIGURE 35-23

A cylindrical lens forms a line image of a point object because it is converging in one plane only.

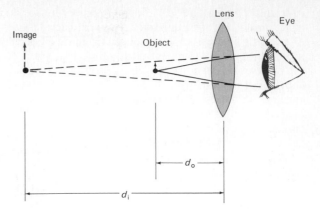

FIGURE 35-24

Lens of a reading glass (Example 35-11).

The unit for lens power is the diopter (D) which is an inverse meter: $1\ \text{D} = 1\ \text{m}^{-1}$. A 20-cm-focal-length lens, for example, has a power $P = 1/0.20\ \text{m} = 5.0\ \text{D}$. The power of a converging lens is positive, whereas that for a diverging lens is negative (since f is negative—see convention 1 in Section 35-2).

EXAMPLE 35-11 A particular farsighted person has a near point of 100 cm. Reading glasses must have what lens power so that this person can read a newspaper at a distance of 25 cm? Assume the lens is very close to the eye.

SOLUTION When the object is placed 25 cm from the lens, we want the image to be 100 cm away on the *same* side of the lens, and so it will be virtual, Fig. 35-24. Thus $d_o = 25$ cm, $d_i = -100$ cm and

$$\frac{1}{f} = \frac{1}{25\ \text{cm}} + \frac{1}{-100\ \text{cm}} = \frac{1}{33\ \text{cm}}.$$

So $f = 33$ cm $= 0.33$ m. The power P of the lens is $P = 1/f = +3.0$ D. The plus sign indicates it is a converging lens.

EXAMPLE 35-12 A nearsighted eye has near and far points of 12 cm and 17 cm, respectively. What lens power is needed for this person to see distant objects clearly, and what then will be the near point? Assume that each lens is 2.0 cm from the eye.

SOLUTION The lens must image distant objects ($d_o = \infty$) so they are 17 cm from the eye, or 15 cm in front of the lens ($d_i = -15$ cm):

$$\frac{1}{f} = -\frac{1}{15\ \text{cm}} + \frac{1}{\infty} = -\frac{1}{15\ \text{cm}}.$$

So $f = -15$ cm $= -0.15$ m or $P = 1/f = -6.7$ D; the minus sign indicates it must be a diverging lens. For the near point, the image must be 12 cm from the eye or 10 cm from the lens, so $d_i = -0.10$ m and

$$\frac{1}{d_o} = \frac{1}{f} - \frac{1}{d_i} = -\frac{1}{0.15\ \text{m}} + \frac{1}{0.10\ \text{m}} = \frac{1}{0.30\ \text{m}}.$$

So $d_o = 30$ cm which means the near point when the person is wearing glasses is 30 cm in front of the lens.

Contact lenses could be used to correct the eye in the last example. Since contacts are placed directly on the cornea, we would not subtract out the 2.0 cm. That is, for distant objects $d_i = -17$ cm so $P = 1/f = -5.9$ D. Thus we see that a contact lens and an eyeglass lens will require slightly different focal lengths for the same eye because of their different placement relative to the eye.

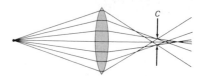

FIGURE 35–25

Spherical aberration (exaggerated). Circle of least confusion obtained at C.

⬡ * 35–7 Lens Aberrations

In Sections 35–1 and 2 we developed a theory of image formation by a thin lens. We found, for example, that all rays from each point on an object are brought to a single point as the image. This, and other results, were based on approximations such as that all rays make small angles with one another and we can use $\sin \theta \approx \theta$. Because of these approximations, we expect deviations from the simple theory and these are referred to as *lens aberrations*. There are several types of aberration; we will briefly discuss each of them separately, but all may be present at one time.

First consider a point object on the axis of a lens. Rays from this point that pass through the outer regions of the lens are brought to a focus at a different point than those that pass through the center of the lens; this is called *spherical aberration*, and is shown exaggerated in Fig. 35–25. Consequently, the image seen on a piece of film (say) will not be a point but a tiny circular patch of light If the film is placed at the point C as indicated, the circle will have its smallest diameter, which is referred to as the *circle of least confusion*. Spherical aberration is present whenever spherical surfaces are used. It can be corrected by using nonspherical lens surfaces, but to grind such lenses is very expensive. It can be minimized with spherical surfaces by choosing the curvatures so that equal amounts of bending occur at each lens surface; a lens can only be designed like this for one particular object distance. Spherical aberration is usually corrected (by which we mean reduced greatly) by the use of several lenses in combination.

For object points off the lens axis, additional aberrations occur. Rays passing through different parts of the lens cause spreading of the image that is noncircular. We won't go into the details but merely point out that there are two effects: *coma* (because the image is comet-shaped rather than a circle) and *off-axis astigmatism.*[†] Furthermore, the image points for objects off the axis but at the same distance from the lens do not fall on a flat plane but on a curved surface—that is, the focal plane is not flat. (We expect this since the points on a flat plane, such as the film in a camera, are not equidistant from the lens.) This aberration is known as *curvature of field* and is obviously a problem in cameras and other devices where the film is placed in a flat plane. In the eye, however, the retina is curved, which compensates for this effect. Another aberration, known as *distortion*, is a result of variation of magnification at different distances from the lens axis. Thus a straight line object not passing through the axis may form a curved image; a square grid of lines may be distorted to produce "pincushion distortion" or "barrel distortion," Fig. 35–26; the latter is common in extreme wide-angle lenses.

All the above aberrations occur for monochromatic light and hence are referred to as *monochromatic aberrations*. If the light is not monochromatic, there will also be *chromatic aberration*. This aberration arises because of dispersion—the variation of index of refraction of transparent materials with wavelength (Section 36–5). For example, blue light is bent more than red light by glass; so if white light is incident on a lens, the different colors are focused at different points, Fig. 35–27, and there will be colored fringes in the image. Chromatic aberration can be eliminated for any two colors (and reduced greatly for all others) by the use of two lenses made of different materials with different indices of refraction and dispersion. Normally one lens is converging and the other diverging and they

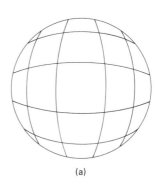

(a)

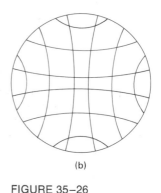

(b)

FIGURE 35–26

Distortion. Lenses may image a square grid of perpendicular lines to produce (a) barrel distortion or (b) pincushion distortion.

FIGURE 35–27

Chromatic aberration. Different colors are focused at different points.

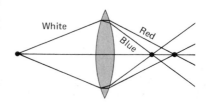

[†] Although the effect is the same as for astigmatism in the eye (Section 35–6) the cause is different. Off-axis astigmatism is no problem in the eye because objects are clearly seen only at the fovea which is on the lens axis.

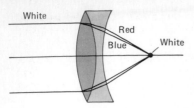

FIGURE 35–28
Achromatic doublet.

are often cemented together, Fig. 35–28. This is called an *achromatic doublet* (or "color corrected" lens).

It is not possible to fully correct all aberrations. Combining two or more lenses together can reduce them. High quality lenses used in cameras, microscopes and other devices are *compound lenses* consisting of many simple lenses (referred to as *elements*). A typical high quality camera lens may contain six to eight (or more) elements.

For simplicity we will normally indicate lenses in diagrams as if they were simple lenses; but it must be remembered that good quality lenses are compound.

 Summary

A lens uses refraction to produce a real or virtual image. The *focal point, F,* of a lens is the image point for an object at infinity (parallel incident rays). The distance of the focal point from the lens is called the *focal length, f,* of the lens. The *lens-maker's equation* relates the focal length of a lens to the index of refraction and radii of curvature of the lens: $1/f = (n - 1)(1/R_1 - 1/R_2)$. The position and size of the image formed by a lens of a given object can be found by ray tracing. Alternatively, the relation between image and object distances, d_i and d_o, and the focal length is given by the *lens equation*:

$$\frac{1}{d_o} + \frac{1}{d_i} = \frac{1}{f}.$$

The ratio of image height to object height, which equals the lateral magnification, is given by $m = (h_i/h_o) = -(d_i/d_o)$.

When using the various equations of geometrical optics, it is important to remember the sign conventions for all quantities involved.

A simple *magnifier* is a converging lens that forms a virtual image of an object placed at (or within) the focal point. The angular magnification, when viewed by a relaxed normal eye, is $M = N/f$, where f is the focal length of the lens and N is the near point of the eye (25 cm for a "normal" eye).

An *astronomical telescope* consists of an objective lens or mirror and an eyepiece that magnifies the real image formed by the objective. The magnification is equal to the ratio of the focal lengths of the objective and eyepiece, and the image is inverted. A terrestrial telescope uses extra lenses, prisms, or a diverging lens as eyepiece, so that the final image is right side up.

A compound *microscope* also uses objective and eyepiece lenses, and the final image is also inverted. The total magnification is the product of the magnifications of the two lenses and is approximately $(N/f_e)(l/f_o)$ where l is the distance between the lenses. N is the nearpoint of the eye (usually taken as 25 cm), and f_o and f_e are the focal lengths of objective and eyepiece.

 Questions

1. Where must the film be placed if a camera lens is to make a sharp image of an object very far away?

2. Can a diverging lens form a real image under any circumstances? Explain.

3. Why must the converging lens have a shorter focal length than the diverging lens if the latter's focal length is to be determined by combining them as in Example 35–6?

4. The thicker a double convex lens is in the center as compared to its edges, the shorter its focal length for a given lens diameter. Explain.

5. A lens is made of a material with an index of refraction $n = 1.30$. In air it is a converging lens. Will it still be a converging lens if placed in water? Explain, using a ray diagram.

6. Compare the mirror equation, 34–3, with the lens equation, 35–4. Discuss similarities and differences, and especially compare the sign conventions for the quantities involved.

7. Show that a real image formed by a thin lens is always

inverted, whereas a virtual image is always upright if the object is real.

8. Light rays are reversible. Is this consistent with the lens equation?

9. Can real images be projected on a screen? Can virtual images? Can either be photographed? Discuss carefully.

10. An unsymmetrical lens (say planoconvex) forms an image of a nearby object. Does the image point change if the lens is turned around?

11. A thin converging lens is moved closer to a nearby object. Does the real image formed change (*a*) in position, (*b*) in size? If yes, describe how.

12. Describe all the conditions for which a converging lens will form (*a*) an image larger than the object, (*b*) a virtual image, (*c*) an upright image, (*d*) a magnification of $+1$ or -1.

13. Describe all the conditions for which a diverging lens will form (*a*) an image larger than the object, (*b*) a virtual

image, (c) an upright image, (d) a magnification of $+1$ and -1.

14. Does the focal length of a lens depend on the material surrounding it? What about the focal length of a spherical mirror? Explain.

15. An underwater lens consists of a carefully shaped thin-walled plastic container filled with air. What shape should it have in order to be (a) converging (b) diverging? Use ray diagrams to support your answer.

16. Under what conditions might the double convex lens of Fig. 35–5a be diverging?

17. Complete the ray diagram of Fig. 35–18a by showing the intermediate image and the focal points.

18. Why must a camera lens be moved farther from the film to focus on a closer object?

19. A "pinhole" camera uses a tiny pinhole instead of a lens. Show, using ray diagrams, how reasonably sharp images can be formed using such a pinhole camera. In particular, consider two point objects 2.0 cm apart that are 1.0 m from a 1-mm-diameter pinhole. Show that on a piece of film 5 cm behind the pinhole that each object produces a tiny easily resolvable spot.

*20. Why are bifocals needed mainly by older persons and not generally by younger people?

*21. Explain why swimmers with good eyes see distant objects as blurry when they are under water. Use a diagram and also show why goggles correct this problem.

*22. Will a nearsighted person who wears corrective lenses be able to see clearly underwater when wearing glasses? Use a diagram to show why or why not.

*23. The human eye is much like a camera—yet when a camera shutter is left open and the camera moved, the image will be blurred; but when you move your head with your eyes open, you still see clearly. Explain.

*24. Reading glasses use converging lenses. A simple magnifier is also a converging lens. Are reading glasses therefore magnifiers? Discuss the similarities and differences between converging lenses as used for these two different purposes.

*25. Is the image formed on the retina of the human eye upright or inverted? Discuss the implications of this for our perception of objects.

*26. Inexpensive microscopes sold for children's use usually produce images that are colored at the edges. Why?

*27. Spherical aberration in a thin lens is minimized if rays are bent equally by the two surfaces. If a planoconvex lens is used to form a real image of an object at infinity which surface should face the object? Use ray diagrams to show why.

*28. Which aberrations present in a simple lens are not present (or are greatly reduced) in the human eye?

*29. Explain why chromatic aberration occurs for thin lenses but not for mirrors.

 Problems

SECTION 35–1

1. (II) (a) Suppose the radius of curvature of the surface in Fig. 35–1 is $R = 32$ cm. If $n_1 = 1.00$ and $n_2 = 1.50$, where will an object at infinity be focused? (b) Could you call this the focal point of the surface? If so, is there a second focal point (as for a lens)? If so, where is it? (c) Where is the image for an object 1.12 m from the surface?

2. (II) How deep would the lake of Example 35–1 seem to be if the viewer were looking at a 45° angle?

3. (II) A 20-cm-thick plane piece of glass ($n = 1.56$) lies on the surface of a 20-cm-deep pool of water. (a) How far below the top of the glass does the bottom of the pool seem, as viewed from directly above? (b) What is your answer if viewed at a 45° angle?

4. (III) Show that Eq. 35–1 is valid for both convex and concave spherical surfaces and for differently located objects and images as long as the conventions discussed in Section 35–1 are adhered to. Show this by using diagrams similar to Fig. 35–2 for all possible cases. Assume both $n_2 > n_1$ and then $n_2 < n_1$.

SECTION 35–2

5. (I) A certain lens focuses an object 32.0 cm away as an image 43.0 cm on the other side of the lens. What type of lens is it and what is its focal length? Is the image real or virtual?

6. (I) An object 34.0 cm in front of a certain lens is imaged 11.0 cm in front of that lens (on the same side as the object). What type of lens is this and what is its focal length? Is the image real or virtual?

7. (I) Both surfaces of a double convex lens have radii of 28.0 cm. If the focal length is 26.2 cm, what is the index of refraction of the lens material?

8. (I) A planoconvex lens is to have a focal length of 18.0 cm. If made from fused quartz, what must be the radius of curvature of the convex surface?

9. (I) How large is the image of the sun on the film used in a camera with a 50-mm-focal-length lens? The sun's diameter is 1.4×10^6 km and it is 1.5×10^8 km away.

10. (II) A 22.0-cm-focal-length converging lens is 15.0 cm behind a diverging lens. Parallel light strikes the diverging lens; after passing through the converging lens, the light is again parallel. What is the focal length of the diverging lens?

11. (II) (a) A 1.50-cm-high insect is 1.3 m from a $+135$-mm-focal-length lens; where is the image, how high is it, and what type is it? (b) Repeat for a -135-mm (diverging) lens.

12. (II) Show analytically that a diverging lens can never form a real image of a real object. Can you describe a situation in which a diverging lens can form a real image?

13. (II) A lens whose index of refraction is n is submerged in a material of index of refraction n' ($n' \neq 1$). Derive the equivalent of Eqs. 35–2, 3, 4 and 5 for this lens.

14. (II) How far from a 21.5-cm-focal-length converging lens should an object be placed if its image is to be magnified

3.00 times and be (a) virtual, (b) real. (c) Repeat for a diverging lens of the same focal length.

15. (II) A 35-mm slide (picture size is actually 24 mm × 36 mm) is to be projected on a screen 1.20 × 1.80 m placed 45.0 m from the projector. What focal-length lens should be used if the image is to cover the screen?

16. (II) Show analytically that the image formed by a converging lens is real and inverted if the object is beyond the focal point ($d_o > f$), and is virtual and upright if the object is within the focal point ($d_o < f$). Describe the image if the object is a virtual image (formed by another lens) for which $-d_o > f$, and for which $0 < -d_o < f$.

17. (II) A diverging lens is placed next to a converging lens of focal length f_C as in Fig. 35–13. If f_T represents the focal length of the combination, show that the focal length of the diverging lens, f_D, is given by

$$\frac{1}{f_D} = \frac{1}{f_T} - \frac{1}{f_C}.$$

18. (II) (a) Show that if two lenses of focal lengths f_1 and f_2, respectively, are placed next to each other, then the focal length of the combination f_T is given by $f_T = f_1 f_2/(f_1 + f_2)$. (b) Extend this formula to the case of three thin lenses in contact.

19. (II) How far apart are an object and an image formed by a 65-cm-focal-length converging lens if the image is 3.0 × larger than the object and is real?

20. (II) A crown glass double convex lens (see Fig. 35–5a) has radii of curvature of 21.5 cm and 17.2 cm. Where is the image, and its magnification, if the object is (a) 135 cm from the lens, (b) 46.5 cm from the lens?

21. (II) Show that the lens equation can be written in the *Newtonian form*:

$$xx' = f^2$$

where x is the distance of the object from the focal point on the front side of the lens, and x' is the distance of the image to the focal point on the other side of the lens.

22. (II) A diverging lens with $f = -32.5$ cm is placed 11.0 cm behind a converging lens with $f = 14.5$ cm. Where will an object at infinity be focused?

23. (II) Two 32.0-cm-focal-length converging lenses are placed 21.5 cm apart. An object is placed 55.0 cm in front of one. Where will the final image formed by the second lens be located? What is the total magnification?

24. (II) A bright object is placed on one side of a converging lens of focal length f and a white screen for viewing the image is on the opposite side. The distance $d_T = d_i + d_o$ between the object and the screen is kept fixed, but the lens can be moved. Show that (a) if $d_T > 4f$, there will be *two* positions where the lens can be placed and a sharp image will be produced on the screen; (b) if $d_T < 4f$, there will be no lens position where a sharp image is formed. (c) Determine the distance between the two images in part (a), and the ratio of their sizes.

SECTION 35–3

25. (I) A magnifier is rated at 3.0× for a normal eye focusing on an image at its near point. (a) What is its focal length? (b) What will it be if the 3.0× referred to a relaxed eye?

26. (I) A 6.0-cm-focal-length lens is used as a simple magnifier. To obtain maximum magnification where must the object be placed for a normal eye?

27. (II) A child has a near point of 10 cm. What is the maximum magnification the child can obtain using an 8.8-cm-focal-length magnifier? Compare to that for a normal eye.

28. (II) A 2.50-mm-wide stone is viewed with an 8.50-cm-focal-length lens. A normal eye views the image at its near point. Calculate (a) the angular magnification (b) the width of the image, (c) the object distance from the lens.

29. (II) A small insect is placed 3.80 cm from a +4.00-cm-focal-length lens. Calculate (a) the position of the image, (b) the angular magnification.

30. (II) A magnifying glass is rated at 3.3× for a normal eye that is relaxed. What would be the magnification for a relaxed eye whose near point is (a) 40 cm, (b) 15 cm?

SECTION 35–4

31. (I) What is the magnification of an astronomical telescope whose objective lens has a focal length of 50 cm and whose eyepiece has a focal length of 3.1 cm? What is the overall length of the telescope when adjusted for a relaxed eye?

32. (I) An astronomical telescope has an objective with focal length 80 cm and a +46-D eyepiece. What is the total magnification?

33. (I) What is the magnifying power of an astronomical telescope using a reflecting mirror whose radius of curvature is 4.80 m and an eyepiece whose focal length is 2.6 cm?

34. (II) A lawyer lost in the mountains tries to make a telescope using the lenses from his reading glasses. They have powers of +1.5 and +6.0 D, respectively. (a) What maximum magnification telescope is possible? (b) Which lens should be used as the eyepiece?

35. (II) A Galilean telescope adjusted for a relaxed eye is 33 cm long. If the objective lens has a focal length of 36 cm, what is the magnification?

36. (II) A 50× astronomical telescope is adjusted for a relaxed eye when the two lenses are 84 cm apart. What is the focal length of each lens?

37. (II) A 6.0× pair of binoculars has an objective focal length of 24 cm. If the binoculars are focused on an object 4.0 m away, what is the magnification? (The 6.0× refers to objects at infinity; Eq. 35–7 holds only for objects at infinity and not for nearby ones.)

SECTION 35–5

38. (I) A 900× microscope uses a 0.40-cm-focal-length objective lens. If the tube length is 16.8 cm, what is the focal length of the eyepiece? Assume a normal eye and that the final image is at infinity.

39. (II) The eyepiece of a compound microscope has focal length of 2.50 cm and the objective has $f = 0.800$ cm. If an object is placed 0.850 cm from the objective lens, calculate (a) the distance between the lenses when the microscope is adjusted for a relaxed eye, and (b) the total magnification.

40. (II) A microscope has a 15× eyepiece and a 40× objective 17.5 cm apart. Calculate (a) the total magnification, (b) the focal length of each lens, and (c) where the object must be for a normal relaxed eye to see it in focus.

41. (II) A microscope has a 2.0-cm-focal-length eyepiece and 1.0-cm-focal-length objective. Calculate (a) the position of

the object if the distance between the lenses is 18.0 cm, and (b) the total magnification, assuming a relaxed normal eye.

42. (II) Repeat problem 41, assuming the final image is located 25 cm from the eyepiece (near point of a normal eye).

SECTION 35-6

*43. (I) Reading glasses of what power are needed for a person whose near point is 120 cm so that he can read at 25 cm? Assume a lens-eye distance of 2.0 cm.

*44. (I) A person's left eye is corrected by a -6.5-D lens. (a) Is this person near- or farsighted? (b) What is this person's far point without glasses?

*45. (I) A man's left eye can see objects clearly only if they are between 18 cm and 36 cm away. (a) What power of contact lens is required so that objects far away are sharp? (b) What then will be his near point?

*46. (II) A 50-year-old man uses $+2.5$-D lenses to be able to read a newspaper 25 cm away. Ten years later, he finds he must hold the paper 45 cm away to see clearly with the same lenses. What power lenses does he need now?

*47. (II) A woman can see clearly with her right eye only when objects are between 35 cm and 210 cm away. Prescription bifocals should have what powers so that she can see distant objects clearly (upper part) and be able to read a book 25 cm away (lower part)? Assume that the glasses will be 2.0 cm from the eye.

*48. (II) A -3.0-D lens is held 20 cm from an ant 1.0 mm high. What is the position, type, and height of the image?

*49. (II) A glass lens ($n = 1.55$) in air has a power of $+3.0$ D. What would its power be if submerged in water?

*50. (II) (a) Show that if two lenses of powers P_1 and P_2 are put in contact, the power P of the combination will be $P = P_1 + P_2$. (b) An optometrist finds that a patient sees clearly when lenses of focal length 60 cm, -20 cm, and 100 cm are placed in contact. What power of lens should be prescribed?

*51. (II) One lens of a nearsighted person's eyeglasses has a focal length of -22.0 cm and the lens is 1.6 cm from the eye. If the person switches to contact lenses placed directly on the eye, what should be the focal length of the corresponding contact lens?

*52. (II) Approximately how much longer is the nearsighted eye of Example 35-11 than the 2.00 cm of a normal eye?

The Wave Nature of Light; Interference

36

That light carries energy is obvious to anyone who has focused the sun's rays with a magnifying glass on a piece of paper and burned a hole in it. But how does light travel and in what form is this energy carried? In our discussion of waves in Chapter 15, we noted that energy can be carried from place to place in basically two ways: by particles or by waves. In the first case, material bodies or particles can carry energy, such as a thrown baseball or rushing water. In the second case, water waves and sound waves, for example, can carry energy over long distances even though mass itself does not travel these distances. In view of this, what can we say about the nature of light: does light travel as a stream of particles away from its source; or does it travel in the form of waves that spread outward from the source? Historically, this question has turned out to be a difficult one. For one thing, light does not reveal itself in any obvious way as being made up of tiny particles nor do we see tiny light waves passing by as we do water waves. The evidence seemed to favor first one side and then the other until about 1830 when most physicists had accepted the wave theory. By the end of the nineteenth century, light was seen to be an *electromagnetic wave* (Chapter 33). Although modifications had to be made in the twentieth century (Chapter 40), the wave theory of light has proved very successful. We now investigate the evidence for the wave theory and how it has explained a wide range of phenomena.

 ## 36-1 Waves Versus Particles; Huygens' Principle and Diffraction

The Dutch scientist Christian Huygens (1629–1695), a contemporary of Newton, proposed a wave theory of light that had much merit. Still useful today is a technique he developed for predicting the future position of a wave front when an earlier position is known. This is known as **Huygens' principle** and can be stated as follows: *every point on a wave front can be considered as a source of tiny wavelets that spread out in the forward direction at the speed of the wave itself; the new wave front is the envelope of all the wavelets* (that is, *the tangent to all of them*).

As a simple example of the use of Huygens' principle, consider the wave front *AB* in Fig. 36–1 which is traveling away from a source *S*. We assume the medium is *isotropic*—that is, the speed *v* of the waves is the same in all directions. To find the wave front a short time *t* after it is at *AB*, tiny circles are drawn whose radius $r = vt$. The centers of these tiny circles are on the original wave front *AB* and the circles represent Huygens' wavelets. The tangent to all these wavelets, the line *CD*, is the new position of the wave front.

Huygens' principle is particularly useful when waves impinge on an obstacle and the wave fronts are partially interrupted. Huygens' principle predicts that waves bend in behind an obstacle as shown in Fig. 36–2. This is just what water waves do as we saw in Chapter 15 (Figs. 15–23 and 15–24). The bending of waves behind obstacles into the "shadow region" is known as **diffraction**. Since diffraction occurs for waves, but not for particles, it can serve as one means for distinguishing the nature of light.

Does light exhibit diffraction? In the mid-seventeenth century, a Jesuit priest, Francesco Grimaldi (1618–1663), had observed that when sunlight entered a darkened room through a tiny hole in a screen, the spot on the opposite wall was larger than would be expected from geometric rays; he also observed that the border of the image was not clear but was surrounded by colored fringes. Grimaldi attributed this to the diffraction of light. Newton, who favored a particle theory, was aware of Grimaldi's result. He felt that Grimaldi's result was due to the inter-action of light corpuscles ("little bodies") with the edges of the hole. If light were a wave, he said, the light waves should bend more than that observed. Newton's argument seems reasonable. Yet, as we saw in Chapter 15, diffraction is large only when the size of the obstacle or the hole is on the order of the wavelength of the wave (Fig. 15–24). Newton did not guess that the wavelengths of visible light might be incredibly tiny, and thus diffraction effects would be very small. (Indeed this is why geometrical optics using rays is so successful—normal openings and obstacles are much larger than the wavelength of the light, and so relatively little diffraction or bending occurs.)

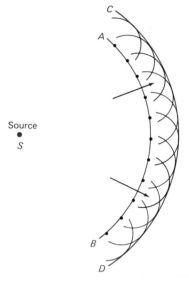

FIGURE 36–1

Huygens' principle used to determine wave front *CD* when *AB* is given.

 36–2 Huygens' Principle and the Law of Refraction

The laws of reflection and refraction were well known in Newton's time. The law of reflection could not distinguish the two theories. For when waves reflect from an obstacle, the angle of incidence equals the angle of reflection (Fig. 15–15); the same is true of particles—think of a tennis ball without spin striking a flat surface.

The law of refraction is another matter. Consider light entering a medium where it is bent toward the normal as when it travels from air into water. As shown in Fig. 36–3, this effect can be constructed using Huygens' principle if we assume the speed of light is less in the second medium. That is, in time *t*, the point *B* on wave front *AB* goes a distance $v_1 t$ to reach point *D*; point *A*, on the other hand, travels a distance $v_2 t$ to reach point *C*. Huygens' principle is applied to points

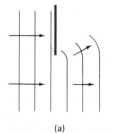

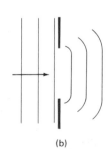

(a)　　　　(b)　　　　(c)

FIGURE 36–2

Huygens' principle is consistent with diffraction (a) around the edge of an obstacle, (b) through a large hole, (c) through a small hole whose size is on the order of the wavelength of the wave.

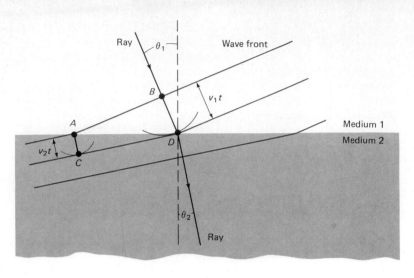

FIGURE 36–3

Refraction explained, using Huygens' principle.

A and *B* to obtain the curved wavelets shown at *C* and *D*. The wave front is tangent to these two wavelets so the new wave front is the line *CD*. Hence the rays (which are perpendicular to the wave fronts) bend toward the normal if $v_2 < v_1$ as drawn.[†] Newton's corpuscle theory predicted the opposite result. Newton argued that, if the path of light corpuscles entering a new medium changes direction, it must be because the medium exerts a force on the corpuscles at the boundary. This force was assumed to act perpendicular to the boundary and thus would affect only the perpendicular component of the corpuscles' velocity. When light enters a medium such as water where it is bent toward the normal, the force must accelerate the corpuscles so the perpendicular component of velocity is increased; only in this way will the refracted angle be less than the incident. In Newton's theory, then, the speed of light would be greater in the second medium ($v_2 > v_1$). Thus the wave theory predicts that the speed of light in water (say) is less than in air and Newton's corpuscle theory predicts the reverse. An experiment to actually measure the speed of light in water and confirm the wave theory prediction was not done until 1850 (by the French physicist Jean Foucault) and by then the wave theory was fully accepted as we shall see in the next section.

It is easy to show that Snell's law of refraction follows directly from Huygens' principle, given that the speed of light v in any medium is related to the speed in a vacuum, c, and the index of refraction, n, by Eq. 34–1: $v = c/n$. From the Huygens' construction of Fig. 36–3, angle *ADC* is equal to θ_2 and angle *BAD* is equal to θ_1. Then for the two triangles that have the common side *AD*, we have

$$\sin \theta_1 = \frac{v_1 t}{AD}, \quad \sin \theta_2 = \frac{v_2 t}{AD}.$$

We divide these two equations and obtain:

$$\frac{\sin \theta_1}{\sin \theta_2} = \frac{v_1}{v_2},$$

or, since $v_1 = c/n_1$ and $v_2 = c/n_2$,

$$n_1 \sin \theta_1 = n_2 \sin \theta_2$$

which is Snell's law of refraction. (The law of reflection can be derived from Huygens' principle in a similar way, and this is left as problem 1 at the end of the chapter.)

When a light wave travels from one medium to another, its frequency does not change but its wavelength does. This is clear from Fig. 36–3, where we assume

---

[†] This is basically the same as the discussion around Fig. 15–19.

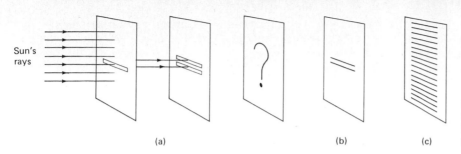

FIGURE 36-4

(a) Young's double-slit experiment. (b) If light consists of particles, we would expect to see two bright lines on the screen behind the slits. (c) Young observed many lines.

each of the lines representing a wave front corresponds to a crest (peak) of the wave. Then

$$\frac{\lambda_2}{\lambda_1} = \frac{v_2 t}{v_1 t} = \frac{v_2}{v_1} = \frac{n_1}{n_2}.$$

If medium 1 is a vacuum (or air), so $n_1 = 1$, $v_1 = c$, and we call λ_1 simply λ, then the wavelength in another medium of index of refraction n ($= n_2$) will be

$$\lambda_n = \frac{\lambda}{n}. \qquad (36-1)$$

This result is consistent with the frequency f being unchanged since $c = f\lambda$; combining this with $v = f\lambda_n$ in a medium where $v = c/n$ gives $\lambda_n = v/f = c/nf = f\lambda/nf = \lambda/n$.

 ## 36-3 Interference—Young's Double-Slit Experiment

In 1801, the Englishman Thomas Young (1773–1829) obtained convincing evidence for the wave nature of light and was even able to measure the wavelength. Figure 36–4 shows a diagram of Young's famous double-slit experiment. Light from a source (Young used the sun) passes through a slit S and then falls on a second screen containing two closely spaced slits S_1 and S_2. If light consists of particles, we would expect to see two bright lines on a screen placed behind the slits as in (b). But Young observed instead a series of bright lines as shown in (c). Young was able to explain this result as a wave-interference phenomenon. To see this, imagine plane waves of light of a single wavelength (called "monochromatic") falling on the two slits as shown in Fig. 36–5. Because of diffraction, the waves leaving the two small slits spread out as shown. This is equivalent to the interference pattern produced when two rocks are thrown into a lake (Fig. 15–21), or when sound from two loudspeakers interferes (Fig. 16–4).

　　To see how an interference pattern is produced on the screen, we make use of Fig. 36–6. Waves of wavelength λ are shown entering the slits S_1 and S_2 which

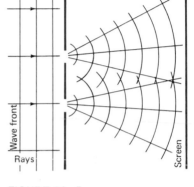

FIGURE 36-5

If light is a wave, light passing through one of two slits should interfere with light passing through the other slit.

FIGURE 36-6

How the wave theory explains the pattern of lines seen in the double-slit experiment.

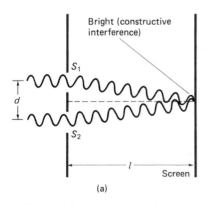

(a)

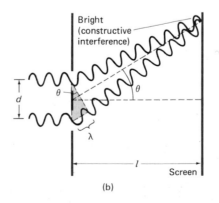

(b)

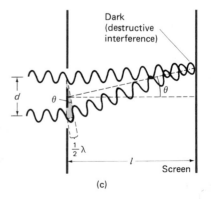

(c)

are a distance d apart. The waves spread out in all directions after passing through the slits, but they are shown only for three different angles θ. In (a) the waves reaching the center of the screen are shown ($\theta = 0$); the waves from the two slits travel the same distance, so they are in phase: constructive interference occurs and there is a bright spot at the center of the screen. There will also be constructive interference when the paths of the two rays differ by one wavelength (or any whole number of wavelengths) as shown in (b). But if one ray travels an extra distance of one-half a wavelength (or $\frac{3}{2}\lambda$, $\frac{5}{2}\lambda$, and so on) the two waves are exactly out of phase when they reach the screen, and so destructive interference occurs and the screen is dark. Thus there will be a series of bright and dark lines (or *fringes*).

To determine exactly where the bright lines fall, first note that Fig. 36–6 is somewhat exaggerated; in real situations, the distance d between the slits is very small compared to the distance l to the screen. The rays from each slit for each case will therefore be essentially parallel and θ is the angle they make with the horizontal. From the shaded right triangle in Fig. 36–6b, the extra distance traveled by the lower ray is $d \sin \theta$. Constructive interference will occur on the screen when $d \sin \theta$ equals a whole number of wavelengths.

$$d \sin \theta = m\lambda, \qquad m = 0, 1, 2, \ldots \qquad \begin{bmatrix} \text{constructive} \\ \text{interference} \end{bmatrix} \quad (36\text{–}2a)$$

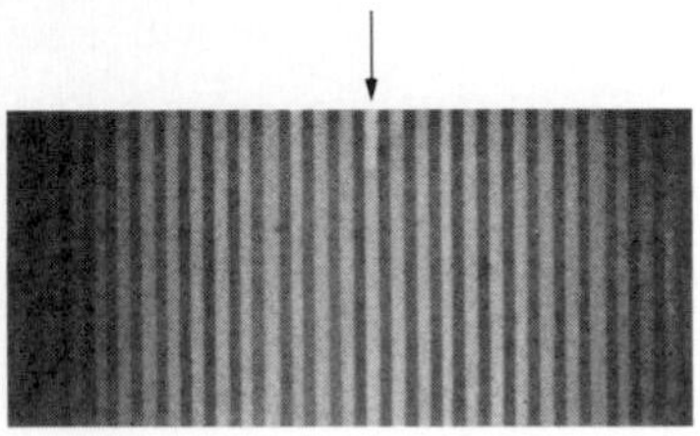

The value of m is called the *order* of the interference fringe. *Destructive interference* occurs when the extra distance $d \sin \theta$ is $\frac{1}{2}$, $\frac{3}{2}$, and so on, wavelengths:

$$d \sin \theta = (m + \tfrac{1}{2})\lambda, \qquad m = 0, 1, 2, \ldots \qquad \begin{bmatrix} \text{destructive} \\ \text{interference} \end{bmatrix} \quad (36\text{–}2b)$$

Figure 36–7 shows the interference pattern produced by a double-slit experiment as detected by photographic film placed on the viewing screen.

FIGURE 36–7

Interference fringes produced by double-slit experiment detected by photographic film placed on the viewing screen. The arrow marks the zeroth fringe. (From F. W. Sears, M. W. Zemansky, and Hugh D. Young, *College Physics*, 4th ed., © 1974, Addison-Wesley Publishing Co., Inc., Reading, Mass., Figure 42–3. Reprinted with permission.)

EXAMPLE 36–1 A screen containing two slits 0.100 mm apart is 1.20 m from the viewing screen. Light of wavelength $\lambda = 500$ nm falls on the slits from a distant source. Approximately how far apart will the bright interference fringes be on the screen?

SOLUTION Since $d = 0.100$ mm $= 1.00 \times 10^{-4}$ m and $l = 1.20$ m, the first-order fringe ($m = 1$) occurs at an angle θ given by:

$$\sin \theta_1 = \frac{m\lambda}{d} = \frac{(1)(500 \times 10^{-9} \text{ m})}{1.00 \times 10^{-4} \text{ m}} = 5.00 \times 10^{-3}.$$

This is a very small angle, so we can take $\sin \theta = \theta$. The first-order fringe will occur a distance x_1 above the center of the screen given by $x_1/l = \theta_1$, so

$$x_1 = l\theta_1 = (1.20 \text{ m})(5.00 \times 10^{-3}) = 6.00 \text{ mm}.$$

The second fringe ($m = 2$) will occur at

$$x_2 = l\theta_2 = l\frac{2\lambda}{d} = 12.0 \text{ mm}$$

above the center, and so on. Thus the fringes are 6.00 mm apart.

From Eq. 36–2 we can see that, except for the zeroth order fringe at the center, the position of the fringes depends on wavelength. Consequently, when white light falls on the two slits, as Young found in his experiments, the central fringe is white but the first (and higher) order fringes are like tiny rainbows; θ was found to be smallest for violet light and largest for red. By measuring the position of these fringes, Young became the first to determine the wavelengths of visible light (using Eq. 36–2). In doing so, he showed that what distinguishes different colors physically is their wavelength.

EXAMPLE 36–2 White light passes through two slits 0.50 mm apart and an interference pattern is observed on a screen 2.5 m away. The first-order fringe resembles a rainbow with violet and red light at either end. The violet light falls about 2.0 mm and the red 3.5 mm from the center of the central white fringe. Estimate the wavelengths of the violet and red lights.

SOLUTION We use Eq. 36–2a with $m = 1$ and $\sin \theta = \theta$. Then for violet light, $x = 2.0$ mm, so

$$\lambda = \frac{d\theta}{m} = \frac{d}{m}\frac{x}{l} = \frac{(5.0 \times 10^{-4} \text{ m})}{1}\frac{(2.0 \times 10^{-3} \text{ m})}{(2.5 \text{ m})} = 4.0 \times 10^{-7} \text{ m}$$

or 400 nm. For red light, $x = 3.5$ mm so λ is about 700 nm.

36–4 Coherence

The two slits in Fig. 36–6 act as if they were sources of radiation. They are called **coherent sources** because the waves leaving them bear the same phase relationship to each other at all times. An interference pattern is observed only when the sources are coherent. If two tiny light bulbs replaced the two slits (or separate light bulbs illuminated each slit) an interference pattern would not be seen; the light emitted by one light bulb would have a random phase with respect to the second bulb, and the screen would be more or less uniformly illuminated. Two such sources are called **incoherent sources**. The subject of coherence is rather complicated, and we discuss it here only briefly. Two light beams do not have to be in phase to be coherent and produce an interference pattern. For example, suppose a piece of glass were placed in front of the lower slit in Fig. 36–6 and suppose the glass is just thick enough to slow down the light so that it enters the lower slit a half wavelength behind the light entering the upper slit. The two beams would be a constant 180° out of phase, but there would still be an interference pattern on the screen. (Can you guess what it would look like? Hint: the central point would be dark instead of bright.) Two beams can be coherent whether they are in phase or out of phase; the important thing is that they have a *constant* phase relation to each other over time.

Coherent sources of water or sound waves are easier to obtain than are coherent sources of light—two loudspeakers sent the same pure frequency will be coherent sources. And two antennas connected to the same LC oscillator can be coherent sources of low-frequency electromagnetic waves. But LC oscillators at the high frequencies of visible light (10^{15} Hz) don't exist since L and C can't be made small enough. For sources of visible light we have to rely on the oscillations (or acceleration) of electric charge within atoms. In an incandescent light bulb, for example, the atoms in the filament are excited by heating, and they give off their excess energy in the form of "wave trains" of light, each of which lasts only about 10^{-8} s. The light we see is the sum of a great many such wave trains that bear a random phase relation to each other. Two light bulbs are thus not coherent, and an interference pattern would not be seen. The double-slit experiment can work with a single light bulb as source as long as the same part of the filament illuminates the two slits equally; that is, each spherical wave train should pass through both slits equally so that the two beams coming through the slits are essentially equal in amplitude and in phase at all times. But if the light passing through the two slits comes from different parts of the filament, there will be incoherence since the phases of the wave trains from different parts (that is, from different atoms—which serve as microscopic sources) will be random; then the interference pattern will be washed out to some extent; the two slits will not be perfectly coherent sources.

FIGURE 36–8

The spectrum of visible light, showing the range of wavelengths for the various colors.

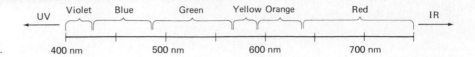

Since any macroscopic source consists of many microscopic sources (atoms), we can speak of the coherence of a single source. An incandescent light bulb is not a very coherent source of light since the light emitted by the different parts of the filament won't maintain a constant phase relation. It wasn't until the 1950s that a really coherent source of light was developed, the *laser*. How a laser produces its narrow coherent beam will be discussed in Chapter 41, along with some of its uses. Because of its coherence, laser light on a double slit produces a very "clean" interference pattern.

Coherence is a relative concept: perfect coherence of a beam would correspond to light that is perfectly sinusoidal (of one frequency) for all times, whereas complete incoherence occurs when there are waves whose phase relations are completely random over time. A measure of the "relative coherence" can be defined in terms of the sharpness of a two slit interference pattern.

The frequencies present in waves greatly affect the coherence. Consider, for example, two loudspeakers that emit the same single frequency tone. The phase between the two waves depends on your listening location (see Example 16–8) but at any given place, the phase remains constant in time, and so the waves are coherent. If the two frequencies are slightly different, the phase between them changes slowly (beats will be heard) and they will not be so coherent. Now suppose the speakers emit the same sound but it consists of many frequencies (such as music) which change in time. If you are precisely midway between the speakers, the two waves will be in phase. But at any other location, the two waves must travel different distances so different parts of the two waves will interfere; hence the phase between the two waves will be constantly changing, and they will be relatively incoherent.

FIGURE 36–9

Index of refraction as a function of wavelength.

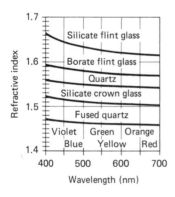

36–5 The Visible Spectrum and Dispersion

The two most obvious properties of light are readily describable in terms of the wave theory of light: intensity (or brightness) and color. The *intensity* of light is related to the square of the amplitude of the wave, just as for any wave (see Section 15–3, or Eqs. 33–15 and 33–17). The *color* of the light is related to the wavelength or frequency of the light. Visible light—that to which our eyes are sensitive—falls in the wavelength range of about 400 nm to 750 nm.[†] This is known as the *visible spectrum*, and within it lie the different colors from violet to red as shown in Fig. 36–8. Light with wavelengths shorter than 400 nm is called ultraviolet (UV) and that with wavelength greater than 750 nm is called infrared (IR).[‡] Although human eyes are not sensitive to UV or IR, some types of photographic film do respond to them.

It is a familiar fact that a prism separates white light into a rainbow of colors. This is due to the fact that the index of refraction of a material depends on the wavelength. This is shown for several materials in Fig. 36–9. White light is a mixture of all visible wavelengths; and when incident on a prism as in Fig. 36–10, the different wavelengths are bent to varying degrees. Since the index of refraction is greater for the shorter wavelengths, violet light is bent the most and red the

FIGURE 36–10

White light dispersed by a prism into the visible spectrum.

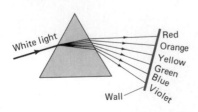

[†] Sometimes the Angstrom (Å) unit is used when referring to light: $1\text{Å} = 1 \times 10^{-10}$ m; then visible light falls in the wavelength range of 4000 Å to 7500 Å.

[‡] The complete electromagnetic spectrum is illustrated in Fig. 33–10.

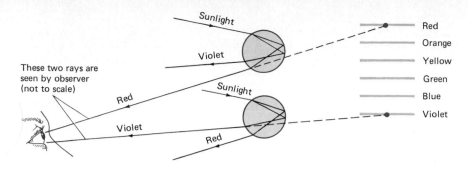

FIGURE 36-11

Formation of a rainbow.

least as indicated. This spreading of white light into the full spectrum is called *dispersion*.

Rainbows are a spectacular example of dispersion (by water in this case). You can see rainbows when you look at falling water with the sun at your back. Figure 36-11 shows how red and violet rays are bent by spherical water droplets and are reflected off the back surface. Red is bent the least and so reaches the observer's eyes from droplets higher in the sky. Thus the top of the rainbow is red.

Diamonds achieve their brilliance from a combination of dispersion and total internal reflection. Since diamonds have an incredibly high index of refraction of about 2.4, the critical angle for total internal reflection is only 25°. Incident light therefore strikes many of the internal surfaces before it strikes one at less than 25° and emerges. After many such reflections the light has traveled far enough that the colors have become sufficiently separated to be seen individually and brilliantly by the eye after leaving the crystal.

36-6 Intensity in the Double-Slit Interference Pattern

We saw in Section 36-3 that the interference pattern produced by the coherent light from two slits, S_1 and S_2 (Figs. 36-6 and 36-7), produces a series of bright and dark fringes. If the two monochromatic waves of wavelength λ are in phase at the slits, the maxima (brightest points) occur at angles θ given by

$$d \sin \theta = m\lambda,$$

and the minima (darkest points) when

$$d \sin \theta = (m + \tfrac{1}{2})\lambda,$$

where m is an integer ($m = 0, 1, 2, \ldots$).

We now determine the intensity of the light at all points in the pattern (that is, for all θ). For simplicity, let us assume that if either slit were covered, the light passing through the other would diffract sufficiently to illuminate a large portion of the screen uniformly. The intensity I of the light at any point is proportional to the square of its wave amplitude (Section 15-3). Treating light as an electromagnetic wave, I is proportional to the square of the electric field E or magnetic field B (Section 33-7). Since E and B are proportional to each other, it doesn't matter which we use, but it is conventional to use E and write the intensity as $I \propto E^2$. The electric field **E** at any point P (see Fig. 36-12) will be the sum of the electric field vectors of the waves coming from each of the two slits, $\mathbf{E}_1$ and $\mathbf{E}_2$. Since $\mathbf{E}_1$ and $\mathbf{E}_2$ are essentially parallel, the magnitude of the electric field at

FIGURE 36-12

Determining intensity in double-slit interference pattern.

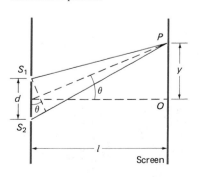

angle θ (that is, at point P) will be

$$E_\theta = E_1 + E_2.$$

Both E_1 and E_2 vary sinusoidally with frequency $f = c/\lambda$, but they differ in phase, depending on their different travel distances from the slits. The electric field at P can then be written for the light from each of the two slits:

$$E_1 = E_{10} \sin 2\pi ft$$
$$E_2 = E_{20} \sin (2\pi ft + \delta) \tag{36–3}$$

where E_{10} and E_{20} are their respective amplitudes and δ is the phase difference. The value of δ depends on the angle θ, so let us now determine δ as a function of θ. At the center of the screen (point O), $\delta = 0$. If the difference in path length from P to S_1 and S_2 is $d \sin \theta = \lambda/2$, the two waves are exactly out of phase so $\delta = \pi$ (or 180°). If $d \sin \theta = \lambda$, the two waves differ in phase by $\delta = 2\pi$. In general, then, δ is related to θ by

$$\frac{\delta}{2\pi} = \frac{d \sin \theta}{\lambda}$$

or

$$\delta = \frac{2\pi}{\lambda} d \sin \theta. \tag{36–4}$$

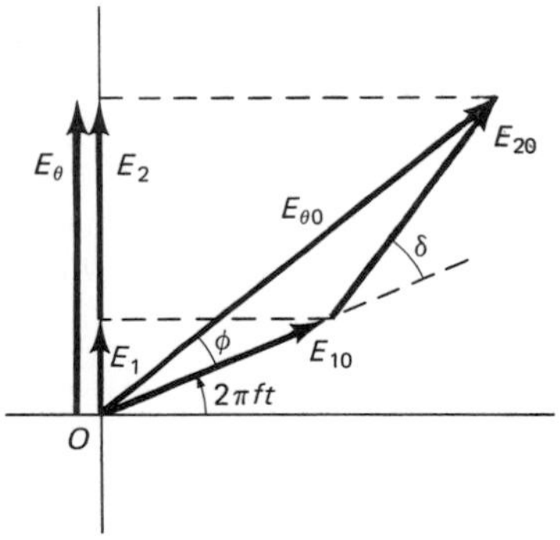

FIGURE 36–13

Phasor diagram for double-slit interference pattern.

To determine $E_\theta = E_1 + E_2$, we add the two scalars E_1 and E_2 which are sine functions differing by the phase δ. One way to determine the sum of E_1 and E_2 is to use a *phasor diagram*. (We used this technique before, in Chapter 32.) As shown in Fig. 36–13, we draw an arrow of length E_{10} to represent the amplitude of E_1 (Eq. 36–3); and the arrow of length E_{20}, which we draw to make a fixed angle δ with E_{10}, represents the amplitude of E_2. When the diagram rotates at frequency f about the origin, the projections of E_{10} and E_{20} on the vertical axis represent E_1 and E_2 as a function of time (see Eq. 36–3). We let $E_{\theta 0}$ be the "vector"† sum of E_{10} and E_{20}; it is the amplitude of the sum $E_\theta = E_1 + E_2$, and the projection of $E_{\theta 0}$ on the vertical axis is just $E_{\theta'}$. If the two slits provide equal illumination, so that $E_{10} = E_{20} = E_0$, then from the diagram the angle $\phi = \delta/2$, and we can write

$$E_\theta = E_{\theta 0} \sin \left(2\pi ft + \frac{\delta}{2} \right); \tag{36–5a}$$

from the diagram we can also see that

$$E_{\theta 0} = 2E_0 \cos \phi = 2E_0 \cos \frac{\delta}{2} \tag{36–5b}$$

and finally δ is given by Eq. 36–4.

We are not really interested in E_θ as a function of time, since for visible light the frequency (10^{14} to 10^{15} Hz) is much too high to be noticeable. We are interested in the average intensity, which is proportional to the amplitude squared, $E_{\theta 0}^2$. We now drop the word "average," and we let I_θ ($I_\theta \propto E_\theta^2$) be the intensity at any point P at an angle θ to the horizontal; and we let I_0 be the intensity at point O, the center of the screen, where $\theta = \delta = 0$, so $I_0 \propto (E_{10} + E_{20})^2 = (2E_0)^2$. Then the ratio I_θ/I_0 is equal to the ratio of the squares of the electric-field

---

† We are not adding the actual electric field vectors; instead we are adding the "phasors" in order to get the amplitude, taking into account the phase difference of the two waves.

Atomic hydrogen

Sodium

Helium

Neon

Mercury

Molecular hydrogen

Solar absorption spectrum

FIGURE 37–19

Line spectra for the elements indicated and spectrum of the sun, showing absorption lines. (Courtesy of Bausch and Lomb Incorporated)

amplitudes at these two points:

$$\frac{I_\theta}{I_0} = \frac{E_{\theta 0}^2}{(2E_0)^2} = \cos^2 \frac{\delta}{2}$$

where we used Eq. 36–5b. Thus the intensity I_θ at any point is related to the maximum intensity at the center of the screen by

$$I_\theta = I_0 \cos^2 \frac{\delta}{2} \qquad\qquad (36\text{--}6)$$

where δ is given by Eq. 36–4. This is the relation we sought. From Eq. 36–6 we see that maxima occur where $\cos \delta/2 = \pm 1$, which corresponds to $\delta = 0, 2\pi,$ $4\pi, \ldots$; from Eq. 36–4, δ has these values when

$$d \sin \theta = m\lambda. \qquad m = 0, 1, 2, \ldots$$

Minima occur where $\delta = \pi, 3\pi, 5\pi, \ldots$, which corresponds to

$$d \sin \theta = (m + \tfrac{1}{2})\lambda. \qquad m = 0, 1, 2, \ldots$$

These are the same results we obtained in Section 36–3. But now we know not only the position of maxima and minima, but from Eq. 36–6 we can determine the intensity at all points.

If the distance l to the screen from the slits is large compared to the slit separation d $(l \gg d)$, and we consider only points P whose distance y from the center (point O) is small compared to l $(y \ll l)$—see Fig. 36–12—then

$$\sin \theta = \frac{y}{l}.$$

From this it follows that

$$\delta = \frac{2\pi}{\lambda} \frac{d}{l} y.$$

Equation 36–6 then becomes

$$I = I_0 \left[\cos\left(\frac{\pi d}{\lambda l} y \right) \right]^2. \qquad [y \ll l, d \ll l] \quad (36\text{--}7)$$

The intensity I as a function of the phase difference δ is plotted in Fig. 36–14. In the approximation of Eq. 36–7, the horizontal axis could as well be y, the position on the screen.

The intensity pattern expressed in Eqs. 36–6 and 36–7, and plotted in Fig. 36–14, shows a series of maxima of equal height, and is based on the assumption that each slit (alone) would illuminate the screen uniformly. This is never quite true, as we shall see when we discuss diffraction in the next chapter; we will see that the center maximum is strongest and each succeeding maximum to each side is less strong.

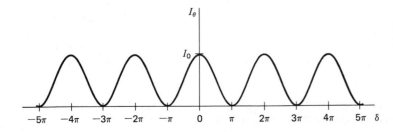

FIGURE 36–14

Intensity I as a function of phase difference δ and position on screen y (assuming $y \ll l$).

36–7 Intensity for Coherent and Incoherent Sources

Let us consider, briefly, the intensity pattern if the two slits were replaced by two *incoherent* sources of equal strength ($E_{10} = E_{20} = E_0$). Again assume uniform illumination of the screen. The phase difference δ between the two waves varies randomly at any point P; thus we must use the time average over $\cos^2 \delta/2$ in Eq. 36–6, which is $\frac{1}{2}$. The intensity, I_{inc}, of the two incoherent sources, will be

$$I_{\text{inc}} = \tfrac{1}{2} I_{\text{coh}}$$

where I_{coh} is the intensity due to two coherent sources at a maximum (I_{coh} is the I_0 of Eq. 36–6). This result, which is confirmed by experiment, can be obtained in another way: when the sources are coherent, we add their wave amplitudes and then square to get the intensity

$$I_{\text{coh}} \propto (E_{10} + E_{20})^2,$$

as in the previous section. But if the sources are incoherent, each wave produces an intensity unrelated to the other, and the two intensities add up:

$$I_{\text{inc}} \propto E_{10}^2 + E_{20}^2;$$

that is, we square the amplitudes and *then* add them. If $E_{10} = E_{20} = E_0$, then the two relations above become

$$I_{\text{coh}} \propto (2E_0)^2 = 4E_0^2$$
$$I_{\text{inc}} \propto E_0^2 + E_0^2 = 2E_0^2$$

so

$$I_{\text{inc}} = \tfrac{1}{2} I_{\text{coh}}$$

consistent with our result above.

It is important to notice that the total energy falling on the screen is the same for both the coherent and the noncoherent cases (assuming sources of the same intensity). The difference is that for incoherent sources the energy is distributed uniformly with intensity $\frac{1}{2}I_0$, whereas for two coherent sources it is concentrated in peaks but averages out to $\frac{1}{2}I_0$ since the average value of $\cos^2 \delta/2$ in Eq. 36–6 over many peaks is $\frac{1}{2}$. Thus there is no conflict with conservation of energy.

FIGURE 36–15

Light reflected from upper and lower surfaces of a thin film of oil lying on water.

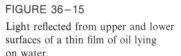

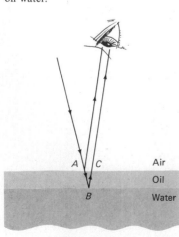

36–8 Interference by Thin Films

Interference of light gives rise to many everyday phenomena such as the bright colors reflected from soap bubbles and from thin oil films on water. In these and other cases, the colors are a result of constructive interference between light reflected from the two surfaces of the thin film. To see how this happens, consider a thin oil film lying on top of water, Fig. 36–15. Part of the incident light (say from the sun or street lights) is reflected at A on the top surface, and part of that transmitted is reflected at B on the lower surface. The part reflected at the lower surface must travel the extra distance ABC. If the distance ABC is equal to one or a whole number of wavelengths, the two waves (which are coherent since they originated from the same point on the source) will interfere constructively and the light will be bright.[†] But if ABC equals $\frac{1}{2}\lambda$, $\frac{3}{2}\lambda$, and so on, the two waves will

---

[†] As is discussed shortly, this is true if the refractive index of the oil is less than that of the water.

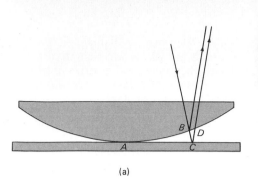

(a)

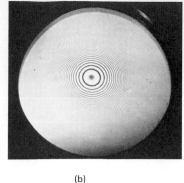

(b)

FIGURE 36–16

Newton's rings.
(The circular patches at various places on the photograph of the rings are an artifact (moiré pattern) of the printing process.) (Part b courtesy of Bausch & Lomb Incorporated.)

be out of phase and destructive interference occurs. The wavelength λ is that in the film (see Eq. 36–1).

When white light falls on such a film, the path ABC will equal λ (or $m\lambda$ with $m =$ an integer) for only one wavelength at a given viewing angle. This color will be seen as very bright. For light viewed at a slightly different angle, the path ABC will be longer or shorter and a different color will undergo constructive interference. Thus, for an extended (nonpoint) source emitting white light, a series of bright colors will be seen next to one another. Variations in thickness of the film will also alter the length ABC and therefore affect the color of light that is most strongly reflected.

When a curved glass surface is placed in contact with a flat glass surface, Fig. 36–16, a series of concentric rings is seen when illuminated from above by monochromatic light. These are called *Newton's rings*[†] and they are due to interference between rays reflected by the top and bottom surfaces of the *air gap* between the two pieces of glass. Because this gap (which is equivalent to a thin film) increases in width from the central contact point out to the edges, the extra path length for the lower ray (equal to BCD) varies; where it equals 0, $\frac{1}{2}\lambda$, λ, $\frac{3}{2}\lambda$, 2λ, and so on, it corresponds to constructive and destructive interference; and this gives rise to the series of bright and dark lines seen in Fig. 36–16b.

Note that the point of contact of the two glass surfaces (A in Fig. 36–16a) is dark in Fig. 36–16b. Since the path difference is zero here, we expect the rays reflected from each surface to be in phase and this point to be bright. But it is dark, which tells us the two rays must be 180° out of phase; this can happen only because one of the waves undergoes a change in phase of 180° upon reflection. Indeed, this and other experiments reveal that *a beam of light reflected by a material whose index of refraction is greater than that in which it is traveling changes phase by 180°*. If the index is less than that of the material in which it is traveling, no phase change occurs. (This can be derived from Maxwell's equations; it corresponds to the reflection of a wave traveling along a rope which consists of two sections of different thickness and density—if the second section is heavier than the first, the wave reflected at the boundary is inverted as we saw in Fig. 15–16, but if the second section is lighter, the wave does not change phase.) Thus the ray reflected by the curved surface above the air gap in Fig. 36–16a undergoes no change in phase. That reflected at the lower surface, where the beam in air strikes the glass, undergoes a 180° phase change. Thus the two rays reflected at the point of contact A of the two glass surfaces (where the air gap approaches zero thickness) will be 180°, or $\frac{1}{2}\lambda$, out of phase, and a dark spot occurs. Other dark bands will occur when the path difference BCD in Fig. 36–16a is equal to an integral number of wavelengths. Bright bands will occur when the path difference

---

[†] Although Newton gave an elaborate description of them, they had been first observed and described by his contemporary Robert Hooke. Newton did not realize their significance in support of a wave theory of light.

is $\frac{1}{2}\lambda$, $\frac{3}{2}\lambda$, and so on, since the phase change at one surface effectively adds another $\frac{1}{2}\lambda$.

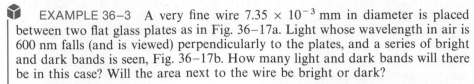

EXAMPLE 36–3 A very fine wire 7.35×10^{-3} mm in diameter is placed between two flat glass plates as in Fig. 36–17a. Light whose wavelength in air is 600 nm falls (and is viewed) perpendicularly to the plates, and a series of bright and dark bands is seen, Fig. 36–17b. How many light and dark bands will there be in this case? Will the area next to the wire be bright or dark?

SOLUTION The thin film is the wedge of air between the two glass plates. Because of the phase change at the lower surface, there will be a dark band when the path difference is 0, λ, 2λ, 3λ, and so on. Since the light rays are perpendicular to the plates, the extra path length equals $2t$ where t is the thickness of the air gap at any point; so dark bands occur where

$$2t = m\lambda, \quad m = 0, 1, 2, \ldots.$$

Bright bands occur when $2t = (m + \frac{1}{2})\lambda$, where m is an integer. At the position of the wire, $t = 7.35 \times 10^{-6}$ m. At this point there will be $(2)(7.35 \times 10^{-6}$ m)/ $(6.00 \times 10^{-7}$ m) $= 24.5$ wavelengths. Since this is a "half integer," the area next to the wire will be bright. There will be a total of 25 dark lines along the plates, corresponding to path lengths of 0λ, 1λ, 2λ, 3λ, $\ldots 24\lambda$, including the one at the point of contact A ($m = 0$). Between them there will be 24 bright lines plus the one at the end, or 25. The bright and dark bands will be smooth only if the glass plates are extremely smooth. If they are not, the pattern is not smooth, as in Fig. 36–17c. This is a very precise way of testing a glass surface for smoothness.

If the wedge between the two glass plates of Example 36–3 is filled with some transparent substance other than air—say water—the pattern shifts. This is because the wavelength of the light changes. In a material where the index of refraction is n, the wavelength is $\lambda_n = \lambda/n$ where λ is the wavelength in a vacuum. For instance, if the thin wedge of Example 36–3 were filled with water, $\lambda_n = 600$ nm/1.33 = 450 nm, and instead of 25 dark lines there would be 33.

When white light (rather than monochromatic light) is incident on the thin wedge of Figs. 36–16a or 36–17a a colorful series of fringes is seen. This is because constructive interference occurs in the reflected light at different locations along the wedge for different wavelengths. Such a difference in thickness is part of the reason bright colors appear when light is reflected from a soap bubble or a thin layer of oil on a puddle or lake; which wavelengths appear brightest also depends on the viewing angle as we saw earlier.

EXAMPLE 36–4 A soap bubble appears green ($\lambda = 540$ nm) at its point nearest the viewer. What is its minimum thickness? Assume $n = 1.35$.

SOLUTION The light is reflected perpendicularly from the point on a spherical surface nearest the viewer. Therefore the path difference is $2t$ where t is the thickness of the soap film. Light reflected from the outer surface undergoes a 180° phase change whereas that on the inner surface does not. Therefore, green light is bright when the minimum path difference equals $\frac{1}{2}\lambda$. Thus $2t = \lambda/2n$ so $t = 540$ nm/(4)(1.35) = 100 nm.

An important application of thin-film interference is in the coating of glass to make it "nonreflecting," particularly for lenses. Glass reflects about 4 percent of the light incident upon it. Good-quality cameras, microscopes, and other optical devices may contain six to ten thin lenses. Reflection from all these surfaces can reduce the light level considerably and multiple reflections produce a background haze that reduces the quality of the image. A very thin coating on the lens surfaces

FIGURE 36–17

(a) Light rays reflected from upper and lower surfaces of a thin wedge of air interfere to produce bright and dark bands. (b) Pattern observed when glass plates are optically flat (courtesy of The Van Keuren Co.); (c) pattern when plates are not so flat (courtesy of Bausch & Lomb Incorporated.)

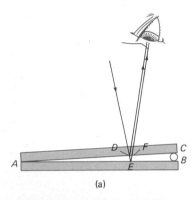

(a)

(b)

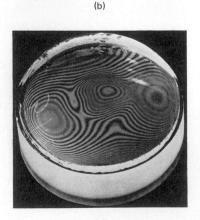

(c)

can reduce these problems considerably. The amount of reflection at a boundary depends on the difference of index of refraction between the two materials. Ideally, the coating material should have an index of refraction which is the geometric mean of those for air and glass, so that the amount of reflection at each surface is about equal; then destructive interference can occur nearly completely for one particular wavelength depending on the thickness of the coating. Nearby wavelengths will at least partially destructively interfere, but it is clear that a single coating cannot eliminate reflections for all wavelengths. Nonetheless, a single coating can reduce total reflection from 4 percent to 1 percent of the incident light. Often the coating is designed to eliminate the center of the spectrum (around 550 nm); the extremes of the spectrum—red and violet—will not be reduced as much. Since a mixture of red and violet produces purple, the light seen reflected from such coated lenses is purple. Lenses containing two or three separate coatings can more effectively reduce a wider range of reflecting wavelengths.

EXAMPLE 36-5 What is the thickness of an optical coating of MgF_2 whose index of refraction is $n = 1.38$ and is designed to eliminate wavelengths centered at 550 nm when incident normally on glass for which $n = 1.50$?

SOLUTION Light reflected at both the front and back surfaces of the coating undergoes a 180° phase shift. Thus, for normal incidence, the thickness of the coating to produce destructive interference must be a half integral number of wavelengths. That is $2t = (m + \frac{1}{2})\lambda$ where m is an integer. The minimum thickness ($m = 0$) is usually chosen because destructive interference will then occur over the widest angle. Then $t = \lambda_n/4 = \lambda/4n = (550 \text{ nm})/(4)(1.38) = 99.6 \text{ nm}$.

*36-9 Michelson Interferometer

Interference by thin films is the basis of the *Michelson interferometer* (Fig. 36–18),[†] invented by the American Albert A. Michelson (Section 34–2). Monochromatic light from a single point on an extended source is shown striking a half-silvered mirror M_s (called the *beam splitter*). Half of the beam passes through to a fixed mirror M_2, where it is reflected back. The other half is reflected by M_s up to a mirror M_1 which is movable (by a fine-thread screw), where it is also reflected back. Upon its return, part of beam 1 passes through M_s and reaches the eye; and part of beam 2, on its return, is reflected by M_s into the eye. (A compensator plate C of transparent glass, usually cut from the same plate as M_s, is placed in the path of beam 2 so both beams pass through the same thickness of glass, to

---

† There are other types of interferometer but Michelson's is the best known.

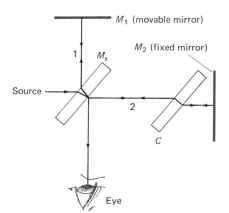

FIGURE 36–18
Michelson interferometer.

within a fraction of a wavelength.) If the two path lengths are identical, the two coherent beams entering the eye constructively interfere and brightness will be seen. If the movable mirror is moved a distance $\lambda/4$, one beam will travel an extra distance equal to $\lambda/2$ (because it travels back and forth over the distance $\lambda/4$); then the two beams will destructively interfere and darkness will be seen. As M_1 is moved farther, brightness will recur (when the path difference is λ), then darkness, and so on.

The eye sees the image of M_2 very close to the position of M_1; so in effect, there is a thin layer of air (a "thin film") between this image and M_1. If M_1 is tilted slightly, then the air gap is wedge shaped as in Fig. 36–17, and a series of bright and dark lines are seen; when mirror M_1 is then moved the bright and dark lines move to the left or right. When the path difference is many wavelengths, the observer no longer sees a nearly uniform brightness or darkness when the mirrors M_1 and M_2 are precisely aligned. Because the path difference is different for different angles of view, one sees a series of rings, much like the Newton ring pattern of Fig. 36–16b. When mirror M_1 is moved, dark and bright spots seem to emerge from (or disappear into) the center of the pattern.

Very precise length measurements can be made with an interferometer. The motion of mirror M_1 by only $\frac{1}{4}\lambda$ produces a clear difference between brightness and darkness. For $\lambda = 400$ nm, this means a precision of 100 nm or 10^{-4} mm! By observing the sideways motion of the fringes when the mirrors are not precisely aligned, even greater precision can be obtained. By counting the number of fringes, or fractions thereof, extremely precise length measurements can be made.

Michelson saw that the interferometer could be used to determine the length of the standard meter in terms of the wavelength of a particular light. Today, that standard is a particular orange-red line in the spectrum of krypton-86 (krypton atoms with atomic mass 86). Careful repeated measurements of the old standard meter (the distance between two marks on a platinum–iridium bar kept in Paris) were made to establish 1 meter as being 1,650,763.73 wavelengths of this light, which is now *defined* as the meter. Thus the primary standard of length can now be available in all laboratories of the world. And the accuracy of length measurements can be much higher.

 ## * 36–10   Luminous Intensity

Although the *intensity* of light (as for any electromagnetic wave) is measured by the Poynting vector in W/m^2, and the total power output of a source can be measured in watts (this is called the *radiant flux*), these are not adequate for measuring the visual sensation we call brightness. The reason is that we are really interested here only in the visible spectrum, whereas the two quantities just mentioned would take into account all wavelengths present. It is also important to take into account the eye's sensitivity to different wavelengths—the eye is most sensitive in the central, 550-nm (yellow), portion of the spectrum; so a yellow source would appear brighter than a red or blue source of the same power output.

These factors are taken into account in the quantity *luminous flux*, F, whose unit is the *lumen* (lm). One lumen is defined experimentally as the brightness of $\frac{1}{60}$ cm^2 of platinum surface at its melting temperature (1770°C). It is equivalent to $\frac{1}{621}$ watts of 550-nm light.

Since the luminous flux from a source may not be uniform, we define the *luminous intensity* I_l as the luminous flux per unit solid angle (steradian). Its unit is the candela (cd) where 1 cd = 1 lm/sr.

The *illuminance*, E, is the luminous flux incident on a surface per unit area of the surface: $E = F/A$. Its unit is the lumen per square meter (lm/m^2) and is a measure of the illumination falling on a surface.[†]

---

† The British unit is the foot-candle, or lumen per square foot.

We will not go into this subject in any more detail. We have introduced it here only for completeness since the *luminous intensity*, as measured by the candela in SI units, is one of the seven basic quantities in the SI. (See Section 1–5 and Table 1–2.) The other six (and their units) we have already met: length (m), time (s), mass (kg), electric current (A), temperature (K), and amount of substance (mol).

 EXAMPLE 36–6 The brightness of a particular type of 100-W light bulb is rated at 1700 lm. Determine (*a*) the luminous intensity and (*b*) the illuminance at a distance of 2.0 m. Assume the light output is uniform in all directions.

SOLUTION (*a*) A full sphere corresponds to 4π sr. Hence, $I_l = 1700$ lm/4π sr $= 135$ cd. It does not depend on distance. (*b*) At $d = 2.0$ m from the source, the luminous flux per unit area is

$$E = F/4\pi d^2 = 1700 \text{ lm}/(4\pi)(2.0 \text{ m})^2 = 34 \text{ lm/m}^2.$$

The illuminance decreases as the square of the distance.

Summary

The wave theory of light is strongly supported by the fact that light exhibits interference and diffraction; this theory also explains the refraction of light and the fact that light travels more slowly in transparent solids and liquids than it does in air. The wavelength of light in a medium with index of refraction n is $\lambda_n = \lambda/n$ where λ is the wavelength in a vacuum; the frequency is not changed.

Young's double-slit experiment clearly demonstrated the interference of light; the observed bright spots of the interference pattern were explained as constructive interference between the beams coming through the two slits, where they differ in path length by an integral number of wavelengths. The dark areas in between are due to destructive interference when the path lengths differ by $\frac{1}{2}\lambda$, $\frac{3}{2}\lambda$, and so on. The angles θ at which constructive interference occurs are given by $\sin\theta = m\lambda/d$, where λ is the wavelength of the light, d the separation of the slits, and m an integer (0, 1, 2, . . .). Complete destructive interference occurs at angles θ given by $\sin\theta = (m + \frac{1}{2})\lambda/d$. The light intensity I at any point in the interference pattern can be calculated using a phasor diagram, which predicts that $I = I_0 \cos^2 \delta/2$ where I_0 is the intensity at $\theta = 0$ and the phase angle $\delta = 2\pi d \sin\theta/\lambda$.

Two sources of light are perfectly *coherent* if the waves leaving them are sinusoidal of the same single frequency and maintain the same phase relationship at all times. If the light waves from the two sources have a random phase with respect to each other over time (as for two incandescent light bulbs) the two sources are *incoherent*.

The wavelength of light determines its color; the visible spectrum extends from 400 nm (violet) to about 750 nm (red). Glass prisms (and other transparent materials) can break white light down into its constituent colors because the index of refraction varies with wavelength, a phenomenon known as *dispersion*.

Light reflected from the front and rear surfaces of a thin film of transparent material can interfere. Such thin-film interference has many practical applications, such as lens coatings and Newton's rings.

Questions

1. Does Huygens' principle apply to sound waves? To water waves?
2. What is the evidence that light is energy?
3. Why is light sometimes described as rays and sometimes as waves?
4. We can hear sounds around corners but we cannot see around corners, yet both sound and light are waves. Explain the difference.
5. Can the wavelength of light be determined from reflection or refraction measurements?
6. Why was the observation of the double-slit interference pattern more convincing evidence for the wave theory of light than the observation of diffraction?
7. Suppose white light falls on the two slits of Fig. 36–6, but one slit is covered by a red filter (700 nm) and the other by a blue (450 nm) filter. Describe the interference pattern on the screen.
8. On what basis do we assume that the frequency of a light wave doesn't change when the light passes from one medium into another?

9. Compare a double-slit experiment for sound waves to that for light waves. Discuss the similarities and differences.

10. Two rays of light from the same source destructively interfere if their path lengths differ by how much?

11. If Young's double-slit experiment were submerged in water, how would the fringe pattern be changed?

12. How would you produce "beats" with light waves, a well-known phenomenon with sound waves? How might you detect light "beats"?

13. What is the purpose of the first screen, with a single slit, in Fig. 36–4? What happens if this slit is not very narrow? What if you used a laser source, rather than the sun?

14. Why doesn't the light from the two headlights of a distant car produce an interference pattern?

15. When white light passes through a flat piece of window glass, it is not broken down into colors as it is by a prism. Explain.

16. For both converging and diverging lenses, discuss how the focal length for red light differs from that for violet light.

17. Make a careful distinction between the angles θ, ϕ, and δ as discussed in Section 36–6.

18. Why are interference fringes noticeable only for a *thin* film like a soap bubble and not for a thick piece of glass, say?

19. Why are Newton's rings (Fig. 36–16) closer together farther from the center?

20. Some coated lenses appear greenish yellow when seen by reflected light. What wavelengths do you suppose they are designed to eliminate completely?

21. A drop of oil on a pond appears bright at its edges where its thickness is much less than the wavelengths of visible light. What can you say about the index of refraction of the oil?

22. Describe how a Michelson interferometer could be used to measure the index of refraction of air?

Problems

SECTION 36–2

1. (II) Derive the law of reflection—namely, that the angle of incidence equals the angle of reflection from a flat surface—using Huygens' principle for waves.

SECTION 36–3

2. (I) Monochromatic light falling on two slits 0.026 mm apart produces the fourth-order fringe at a 6.4° angle. What is the wavelength of the light used?

3. (I) Light of wavelength 680 nm falls on two slits and produces an interference pattern in which the fourth-order fringe is 28 mm from the central fringe on a screen 1.0 m away. What is the separation of the two slits?

4. (II) Television and radio waves can reflect from nearby mountains or from airplanes, and the reflections can interfere with the direct signal from the station. (a) Determine what kind of interference will occur when 75-MHz television signals arrive at a receiver directly from a distant station, and reflected from an airplane 118 m directly above the receiver. (Assume no change in phase of the signal upon reflection.) (b) What kind of interference will occur when the plane has moved 22 m farther toward the station?

5. (II) The shortest-wavelength visible light falls on two slits 2.80×10^{-2} mm apart. The slits are immersed in water, as is a viewing screen 18.5 cm away. How far apart are the fringes on the screen?

6. (II) If 520-nm and 660-nm light passes through two slits 0.50 mm apart, how far apart are the second-order fringes for these two wavelengths on a screen 1.5 m away?

7. (II) Suppose a thin piece of glass were placed in front of the lower slit in Fig. 36–6 so that the two waves enter the slits 180° out of phase. Describe in detail the interference pattern on the screen.

8. (II) When a charged particle, such as a proton, travels through a transparent medium at a speed v_p faster than the speed of light ($v = c/n$) in that medium, it emits electromagnetic radiation (light), called *Čerenkov radiation*. It is the electromagnetic equivalent of a shock wave (see Section 16–8), and is confined to a particular angle that depends on v_p and v. Determine this angle for a proton traveling 2.21×10^8 m/s in a plastic whose index of refraction is 1.52.

9. (II) A very thin sheet of plastic ($n = 1.60$) covers one slit of a double-slit apparatus illuminated by 510-nm light. The center point on the screen, instead of being a maximum, is dark. What is the (minimum) thickness of the plastic.

10. (II) Light of wavelength λ strikes a screen containing two slits a distance d apart at an angle θ_i to the normal. Determine the angle θ_m at which the mth order maximum occurs.

SECTION 36–4

11. (I) Estimate the values of L and C (inductance and capacitance) needed for an oscillator to produce visible light frequencies. Are these reasonable?

SECTION 36–5

12. (I) By what percent, approximately, does the speed of red light (700 nm) exceed that of violet light (400 nm) in silicate flint glass? (See Fig. 36–9.)

13. (II) A beam of light strikes a piece of glass at a 35.00° incident angle. The beam contains two wavelengths, 500.0 nm and 712.0 nm, for which the index of refraction of the glass is 1.4810 and 1.4742, respectively. What is the angle between the two refracted beams?

14. (II) A parallel beam of light containing two wavelengths, $\lambda_1 = 400$ nm and $\lambda_2 = 650$ nm, strikes a piece of silicate flint glass at an incident angle of 56.0°. Calculate the angle between the two color beams inside the glass. (See Fig. 36–9.)

15. (II) A double convex lens whose radii of curvature are both 18.0 cm is made of crown glass. Find the distance between the focal points for violet (400 nm) and red (700 nm) light.

16. (I) If one slit in Fig. 36–12 is covered, by what factor does the intensity at the center of the screen change?

17. (II) Show that the angular full width at half maximum of the central peak in a double-slit interference pattern is given by

$$\Delta\theta = \lambda/2d$$

if $\lambda \ll d$.

18. (II) Derive Eq. 36–5b analytically, without using a phasor diagram.

19. (II) Suppose that one slit of a double-slit apparatus is wider than the other so that the intensity of light passing through it is twice as great. Determine the intensity I as a function of position (θ) on the screen for coherent light.

20. (III) (a) Consider three equally spaced and equal-intensity coherent sources of light (such as adding a third slit to the two slits of Fig. 36–12). Use the phasor method to obtain the intensity as a function of phase difference δ (Eq. 36–4). (b) Determine the positions of maxima and minima.

21. (III) Extend the phasor method to use with four parallel slits separated by equal distances d. Assume coherent light and determine the intensity as a function of position on the screen and find the positions of maxima and minima.

22. (I) If a soap bubble is 120 nm thick, what color will appear at the center when illuminated by white light? Assume $n = 1.34$.

23. (I) How thick (minimum) should the air layer between two flat glass surfaces be if the glass is to appear bright when 640-nm light is incident normally? What if the glass is to appear dark?

24. (II) A glass lens appears greenish-yellow ($\lambda = 570$ nm is strongest) when white light reflects from it. What minimum thickness of coating ($n = 1.25$) do you think is used on such a lens, and why?

25. (II) Suppose you viewed the light *transmitted* through a thin film on a flat piece of glass. Draw a diagram, similar to Fig. 36–15, and describe the conditions required for maxima and minima; consider all possible values of index of refraction. Discuss the relative size of the minima compared to the maxima and to zero.

26. (II) *Lloyd's mirror* provides one way of obtaining a double slit interference pattern from a single source so the light is coherent; as shown in Fig. 36–19, the light that reflects from the plane mirror appears to come from the virtual image of the slit. Describe in detail the interference pattern on the screen.

27. (II) A total of 33 bright and 33 dark Newton's rings are observed when 450-nm light falls normally on a plano-convex lens resting on a flat glass surface (Fig. 36–16). How much thicker is the center than the edges?

28. (II) A fine metal foil separates one end of two pieces of optically flat glass as in Fig. 36–17. When light of wavelength 450 nm is incident normally, 52 dark lines are observed. How thick is the foil?

29. (II) When a Newton's ring apparatus (Fig. 36–16) is immersed in a liquid, the diameter of the eighth dark ring decreases from 2.92 cm to 2.48 cm. What is the refractive index of the liquid?

30. (II) Monochromatic light of variable wavelength is incident normally on a thin sheet of plastic film. The reflected light is a minimum only for $\lambda = 470$ nm and $\lambda = 650$ nm. What is the thickness of the film ($n = 1.60$)?

31. (II) A thin film of alcohol ($n = 1.36$) lies on a flat glass plate ($n = 1.58$). When monochromatic light, whose wavelength can be changed, is incident normally, the reflected light is a minimum for $\lambda = 520$ nm and a maximum for $\lambda = 640$ nm. What is the thickness of the film?

32. (II) Show that the radius r of the mth dark Newton's ring, as viewed from directly above (Fig. 36–16b), is given by $r = \sqrt{m\lambda R}$ where R is the radius of curvature of the curved glass surface and λ is the wavelength of light used. Assume that the thickness of the air gap is much less than R at all points and that $r \ll R$.

33. (II) Use the result of problem 32 to show that the distance between adjacent dark Newton's rings is

$$\Delta r \approx \sqrt{\lambda R/4\, m}$$

for the mth ring, assuming $m \gg 1$.

34. (II) A planoconvex Lucite lens 3.4 cm in diameter is placed on a flat piece of glass as in Fig. 36–16. When 580-nm light is incident normally, 88 bright rings are observed, the last one right at the edge. What is the radius of curvature of the lens surface, and the focal length of the lens?

35. (III) A single optical coating reduces reflection to zero for $\lambda = 550$ nm. By what factor is the intensity reduced by the coating for $\lambda = 450$ nm and $\lambda = 700$ nm as compared to no coating? Assume normal incidence.

*36. (II) What is the wavelength of the light entering an interferometer if 850 fringes are counted when the movable mirror moves 0.356 mm?

*37. (II) A micrometer is connected to the movable mirror of an interferometer. When the micrometer bears on a thin

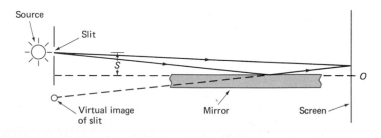

FIGURE 36–19
Lloyd's mirror.

metal foil, the net number of fringes that move, compared to the empty micrometer, is 328. What is the thickness of the foil? The wavelength of light used is 586 nm.

*38. (II) One of the beams of an interferometer passes through a small glass container containing a cavity 1.40 cm wide. When a gas is allowed to slowly fill the container, a total of 225 dark fringes move past a reference line. The light used has a wavelength of 546 nm. Calculate the index of refraction of the gas, assuming the interferometer is in a vacuum.

*39. (III) Suppose the mirrors in a Michelson interferometer are perfectly aligned and the path lengths to mirrors M_1 and M_2 are identical, so that an observer sees a bright maximum at the center of the viewing area. Determine a formula for the intensity at the center of the viewing area as a function of x, the distance the movable mirror is moved from the initial conditions.

*40. (III) The yellow "sodium-D" lines have wavelengths 589.0 nm and 589.6 nm. When they are used to illuminate a Michelson interferometer, it is noted that the interference fringes disappear and reappear periodically as the mirror M_1 is moved. Why does this happen? How far must the mirror move between one disappearance and the next?

*SECTION 36–10

*41. (I) The illuminance of direct sunlight on earth is about 10^5 lm/m^2. Estimate the luminous intensity and luminous flux of the sun.

*42. (II) The *luminous efficiency* of a light bulb is the ratio of luminous flux to electric power input. (a) What is the luminous efficiency of a 100-W, 1700-lm bulb? (b) How many 40-W, 60-lm/W fluorescent lamps would be needed to provide an illuminance of 250 lm/m^2 on a factory floor of area 25 m × 30 m? Assume the lights are 10 m above the floor and that half their flux reaches the floor.

Diffraction

37

Young's double-slit experiment put the wave theory of light on a firm footing. But full acceptance came only with studies on diffraction more than a decade later.

We have already discussed diffraction briefly with regard to water waves (Section 15–9) as well as for light (Section 36–1) and we have seen that it refers to the spreading or bending of waves around edges. Now we look at diffraction in more detail.

A part of the history of the wave theory of light belongs to Augustin Fresnel (1788–1827) who in 1819 presented to the French Academy a wave theory of light that predicted and explained interference and diffraction effects. Almost immediately Siméon Poisson (1781–1840) pointed out a counter-intuitive inference: that according to Fresnel's wave theory, if light from a point source were to fall on a solid disk, then light diffracted around the edges should constructively interfere at the center of the shadow (Fig. 37–1). That prediction seemed very unlikely. But when the experiment was actually carried out by François Arago, the bright spot was seen at the very center of the shadow! This was strong evidence for the wave theory.

Figure 37–2a is a photograph of the shadow cast by a coin using a (nearly) point source of light (a laser in this case). The bright spot is clearly present at the center. Note that there also are bright and dark fringes beyond the shadow. These resemble the interference fringes of a double slit. Indeed, they are due to interference of waves diffracted around different parts of the disk, and the whole is referred to as a *diffraction pattern*. A diffraction pattern exists around any sharp object illuminated by a point source, as shown in Fig. 37–2b and c. We are not always aware of them because most sources of light in everyday life are not points; so light from different parts of the source washes out the pattern.

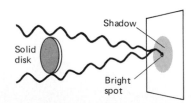

FIGURE 37–1

If a light is a wave, a bright spot will appear at the center of the shadow of a solid disk illuminated by a point source of monochromatic light.

FIGURE 37-2

Diffraction pattern of (a) a penny (courtesy of P. M. Rinard, from *Am. J. Phys.* 44, 1970, p. 70, (b) a razor blade (from F. W. Sears, *Optics,* © 1949, Addison-Wesley Publishing Co., Inc., Reading, Mass., Figure 9-8; reprinted with permission), (c) a single slit, illuminated by a nearly point source of monochromatic light (from M. Cagnet, M. Francon, and J. C. Thrierr, *Atlas of Optical Phenomena,* Springer-Verlag, Berlin, 1962).

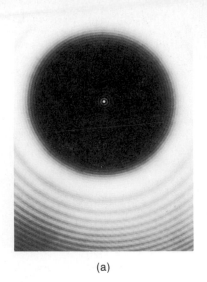

(a)

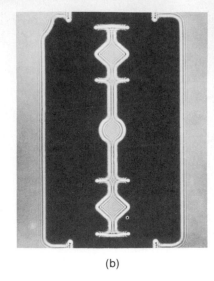

(b)

(c)

37-1   Diffraction by a Single Slit

To see how a diffraction pattern arises, we will analyze the important case of monochromatic light passing through a narrow slit. We will assume that parallel rays (or plane waves) of light fall on the slit of width D as shown in Fig. 37-3. If the viewing screen is infinitely far away, or a lens is placed behind the slit to focus parallel rays on the screen, the diffraction pattern is called *Fraunhofer diffraction.* If the screen is close and no lenses are used, it is called *Fresnel diffraction.* The analysis in the latter case is rather involved, so we consider only the limiting case of Fraunhofer diffraction. As we know from studying water waves and from Huygens' principle, the waves passing through the slit spread out in all directions. We will now examine how the waves passing through different parts of the slit interfere with each other.

Since the screen is assumed to be very far away, the rays heading for any point are essentially parallel. First we consider rays that pass straight through as in Fig. 37-3a. They are all in phase so there will be a central bright spot on the screen. In part (b), we consider rays moving at an angle θ such that the ray from the top of the slit travels exactly one wavelength farther than the ray from the bottom edge. The ray passing through the very center of the slit will travel one half wavelength farther than the ray at the bottom of the slit. These two rays will be exactly out of phase with one another and so will destructively interfere. Similarly, a ray slightly above the bottom one will cancel a ray that is the same distance above the central one. Indeed, each ray passing through the lower half of the slit will cancel with a corresponding ray passing through the upper half. Thus,

FIGURE 37-3

Analysis of diffraction patterns formed by light passing through a narrow slit.

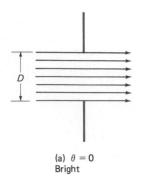

(a) $\theta = 0$
Bright

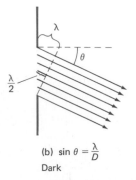

(b) $\sin \theta = \dfrac{\lambda}{D}$
Dark

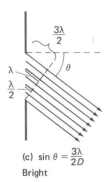

(c) $\sin \theta = \dfrac{3\lambda}{2D}$
Bright

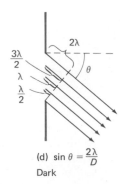

(d) $\sin \theta = \dfrac{2\lambda}{D}$
Dark

all the rays destructively interfere in pairs, and so no light will reach the viewing screen at this angle. The angle θ at which this occurs can be seen from the diagram to occur when $\lambda = D \sin \theta$, so

$$\sin \theta = \lambda/D. \qquad \text{[first minimum]} \quad (37\text{–}1)$$

The light intensity is a maximum at $\theta = 0°$ and decreases to a minimum (intensity = zero) at the angle θ given by Eq. 37–1.

Now consider a larger angle θ such that the top ray travels $\frac{3}{2}\lambda$ farther than the bottom ray as in Fig. 37–3c. In this case, the rays from the bottom third of the slit will cancel in pairs with those in the middle third since they will be $\lambda/2$ out of phase. However, light from the top third of the slit will still reach the screen, so there will be a bright spot, but not nearly as bright as the central spot at $\theta = 0°$. For an even larger angle θ such that the top ray travels 2λ farther than the bottom ray, Fig. 37–3d, rays from the bottom quarter of the slit will cancel with those in the quarter just above it since the path lengths differ by $\lambda/2$. And the rays through the quarter of the slit just above center will cancel with those through the top quarter. At this angle there will again be a minimum of zero intensity in the diffraction pattern. A plot of the intensity as a function of angle is shown in Fig. 37–4. This corresponds well with the photo of Fig. 37–2c. Notice that minima (zero intensity) occur at

$$D \sin \theta = m\lambda, \qquad m = 1, 2, 3, \ldots . \qquad (37\text{–}2)$$

but not at $m = 0$ where there is the strongest maximum. Between the minima, smaller brightness maxima occur.

EXAMPLE 37–1 Light of wavelength 750 nm passes through a slit 1.0×10^{-3} mm wide. How wide is the central maximum (a) in degrees, and (b) in centimeters, on a screen 20 cm away?

SOLUTION (a) The first minimum occurs at

$$\sin \theta = \frac{\lambda}{D} = \frac{7.5 \times 10^{-7} \text{ m}}{1 \times 10^{-6} \text{ m}} = 0.75.$$

So $\theta = 49°$. This is the angle between the center and the first minimum, Fig. 37–5. The angle subtended by the whole central maximum, between the minima above and below the center, is twice this or 98°. (b) The width of the central maximum is $2x$ where $\tan \theta = x/20$ cm. So $2x = 2(20 \text{ cm})(\tan 49°) = 46$ cm, so a large width of the screen will be illuminated. It will not be terribly bright, normally, since the amount of light that passes through such a small slit will be small and it is spread over a large area.

From Eq. 37–1 we can see that the smaller the aperture D, the larger the central diffraction maximum. This is consistent with our earlier study of waves in Chapter 15.

37–2 Intensity in Single-Slit Diffraction Pattern

We have determined the positions of the minima in the diffraction pattern produced by light passing through a single slit, Eq. 37–2. We now discuss a method for predicting the amplitude and intensity at any point in the pattern using the phasor technique already discussed in Section 36–6.

Let us consider the slit, assumed horizontal, divided into N very thin strips of (vertical) width Δy as indicated in Fig. 37–6. Each strip sends light in all directions toward a screen on the right. Again we take the case of Fraunhofer diffraction, so the rays heading for any particular point on the distant screen are

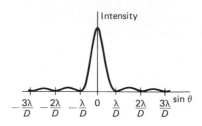

FIGURE 37–4

Intensity in diffraction pattern of a single slit as a function of $\sin \theta$.

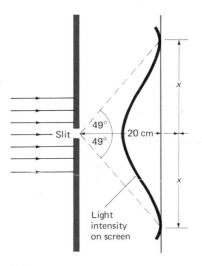

FIGURE 37–5

Example 37–1.

FIGURE 37–6

Slit of width D divided into N strips of width Δy.

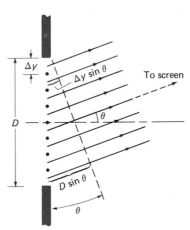

parallel, all making an angle θ with the horizontal as shown. We choose the strip width Δy to be much smaller than the wavelength λ of the monochromatic light falling on the slit, so all the light from a given strip is in phase. The strips are of equal size, and if the whole slit is uniformly illuminated, we can take the wave amplitudes ΔE_0 from each thin strip to be equal as long as θ is not too large. However, the separate amplitudes from the different strips will differ in phase. The phase difference in the light coming from adjacent strips will be (see Section 36–6, Eq. 36–4)

$$\Delta\beta = \frac{2\pi}{\lambda} \Delta y \sin \theta \qquad (37\text{–}3)$$

since the difference in path length is $\Delta y \sin \theta$.

The total amplitude on the screen at any angle θ will be the sum of the waves due to each strip; these wavelets have the same amplitude ΔE_0 but differ in phase. To obtain the total amplitude, we can use a phasor diagram as we did in

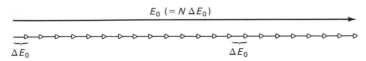

$E_0 \ (= N \Delta E_0)$

$\Delta E_0 \qquad\qquad \Delta E_0$

(a) At center, $\theta = 0$.

FIGURE 37–7

Phasor diagram for single-slit diffraction, giving total amplitude E_θ at various angles θ.

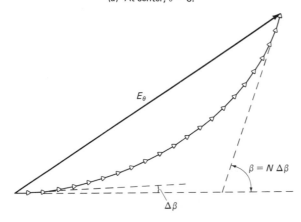

E_θ

$\beta = N \, \Delta\beta$

$\Delta\beta$

(b) Between center and first minimum.

(c) First minimum, $E_\theta = 0$ ($\beta = 360°$).

E_θ

(d) Near secondary maximum.

Section 36–6 (Fig. 36–13). The phasor diagrams for several angles θ are shown in Fig. 37–7. At the center of the screen, $\theta = 0$, the waves from each strip are all in phase ($\Delta\beta = 0$; see Eq. 37–3), so the arrows representing each ΔE_0 line up as shown in Fig. 37–7a. The total amplitude of the light arriving at the center of the screen is then $E_0 = N \Delta E_0$. At a small angle θ, not far from the center, Fig. 37–7b shows how the wavelets of amplitude ΔE_0 add up to give E_θ, the total amplitude on the screen at this angle θ. Note that each wavelet differs in phase from the adjacent one by $\Delta\beta$; the phase difference between the wavelets from the top and bottom edges of the slit is

$$\beta = N \Delta\beta = \frac{2\pi}{\lambda} N \Delta y \sin\theta = \frac{2\pi}{\lambda} D \sin\theta \qquad (37\text{–}4)$$

where $D = N \Delta y$ is the total width of the slit. Although the "arc" in Fig. 37–7b has length $N \Delta E_0$, and so equals E_0, the amplitude of the total wave E_θ is the *vector* sum of each wavelet amplitude and so is equal to the length of the chord as shown. Clearly, $E_\theta < E_0$.

For greater θ, we eventually come to the case, illustrated in Fig. 37–7c, where the chain of arrows closes on itself. In this case the vector sum is zero, so $E_\theta = 0$ for this angle θ. This corresponds to the first minimum. Since $\beta = N \Delta\beta$ is 360° or 2π in this case, we have from Eq. 37–3

$$2\pi = N \Delta\beta = N \left(\frac{2\pi}{\lambda} \Delta y \sin\theta \right)$$

or, since the slit width $D = N \Delta y$,

$$\sin\theta = \frac{\lambda}{D}.$$

Thus the first minimum ($E_\theta = 0$) occurs where $\sin\theta = \lambda/D$, which is the same result we obtained in the previous section, Eq. 37–1.

For even greater angles, θ, the chain of arrows spirals beyond 360°. Figure 37–7d shows the case near the secondary maximum next to the first minimum; here $\beta = N \Delta\beta \approx 360° + 180° = 540°$ or 3π. (Note that although $\beta \approx 540°$, θ may still be a very small angle, depending on the values of D and λ.) When $\beta = 4\pi$, we have a double circle and again a minimum ($\sin\theta = 2\lambda/D$; this corresponds to $m = 2$ in Eq. 37–2). When greater angles, θ, are considered, new maxima and minima occur. But since the total length of the coil remains constant, equal to $N \Delta E_0 \ (= E_0)$, each succeeding maximum is smaller and smaller as the coil winds in on itself.

To obtain a quantitative expression for the amplitude (and intensity) for any point on the screen (that is, for any angle θ), we now consider the limit $N \to \infty$ so Δy becomes the infinitesimal width dy. In this case, the diagrams of Fig. 37–7 become smooth curves, one of which is shown in Fig. 37–8. For any angle θ, the wave amplitude on the screen is E_θ, the chord in the figure; and the length of the arc is E_0, as before. If r is the radius of curvature of the arc, then

$$\frac{E_\theta}{2} = r \sin\frac{\beta}{2}.$$

And, using radian measure for $\beta/2$, we also have

$$\frac{E_0}{2} = r \frac{\beta}{2}.$$

We combine these to obtain

$$E_\theta = E_0 \frac{\sin\beta/2}{\beta/2}. \qquad (37\text{–}5)$$

FIGURE 37–8

Determining amplitude E_θ as a function of θ for single-slit diffraction.

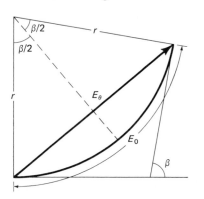

The angle β is the phase difference between the waves from the top and bottom edges of the slit; since the path difference for these two rays is $D \sin \theta$ (see Fig. 37–6 as well as Eq. 37–4):

$$\beta = \frac{2\pi}{\lambda} D \sin \theta. \tag{37–6}$$

Since the intensity is proportional to the square of the wave amplitude, the intensity I_θ at any angle θ is, from Eq. 37–5,

$$I_\theta = I_0 \left(\frac{\sin \beta/2}{\beta/2} \right)^2 \tag{37–7}$$

where I_0 ($\propto E_0^2$) is the intensity at $\theta = 0$ (the central maximum). We can combine Eqs. 37–7 and 37–6 (although it is often simpler to leave them as separate equations) to obtain

$$I_\theta = I_0 \left[\frac{\sin (\pi D \sin \theta / \lambda)}{\pi D \sin \theta / \lambda} \right]^2. \tag{37–8}$$

Note that the minima ($I_\theta = 0$) occur where $\sin (\pi D \sin \theta / \lambda) = 0$ which means $\pi D \sin \theta / \lambda$ must be π, 2π, 3π, and so on, or

$$D \sin \theta = m\lambda \qquad m = 1, 2, 3, \ldots$$

which is what we have obtained previously. Notice that m cannot be zero; for when $\beta/2 = \pi D \sin \theta / \lambda = 0$, the denominator as well as the numerator in Eq. 37–7 or 37–8 vanishes. We can evaluate the intensity in this case by taking the limit as $\theta \to 0$ (or $\beta \to 0$); for very small angles, $\sin \beta/2 \approx \beta/2$, so $(\sin \beta/2)/(\beta/2) \to 1$ and $I_\theta = I_0$, the maximum at the center of the pattern.

EXAMPLE 37–2   Determine the intensities of the secondary maxima ($m = 1, 2, \ldots$) approximately.

SOLUTION   The secondary maxima occur close to halfway between the minima, at

$$\frac{\beta}{2} = \frac{\pi D \sin \theta}{\lambda} \approx (m + \tfrac{1}{2})\pi, \qquad m = 1, 2, 3, \ldots.$$

(The actual maxima are not quite at this point, and can be determined by differentiating Eq. 37–7—see problem 7.) Thus

$$I_\theta = \frac{I_0}{(m + \tfrac{1}{2})^2 \pi^2}, \qquad m = 1, 2, 3, \ldots.$$

For $m = 1, 2$, and 3, we get

$$I_\theta = \frac{I_0}{22.2} = 0.045\, I_0 \qquad\qquad [m = 1]$$

$$I_\theta = \frac{I_0}{61.7} = 0.016\, I_0 \qquad\qquad [m = 2]$$

$$I_\theta = \frac{I_0}{121} = 0.0083\, I_0. \qquad\qquad [m = 3]$$

The first maximum to the side of the central peak has only 1/22 (or 4.5 percent) the intensity of the central intensity, and succeeding ones are smaller still. This corresponds to Fig. 37–4; but note that this relation is not clear in Fig. 37–2c which was overexposed so that the side maxima would be visible.

Diffraction by a circular opening produces a similar pattern (though circular rather than rectangular) and is of great practical importance, since lenses are essentially circular apertures through which light passes. We will discuss this in Section 37–4 and see how diffraction limits the resolution (or sharpness) of images.

37–3 Diffraction in the Double-Slit Experiment

When we analyzed Young's double-slit experiment in Section 36–6, we assumed that the central portion of the screen was uniformly illuminated. This is equivalent to assuming the slits are infinitesimally narrow, so that the central diffraction peak is spread out over the whole screen. This can never be the case for real slits, of course, and diffraction reduces the intensity of the bright interference fringes to the side of center so they are not all of the same height as they were shown in Fig. 36–14.

To calculate the intensity in a double-slit interference pattern, including diffraction, let us assume the slits have equal widths D and their centers are separated by a distance d. Since the distance to the screen is large compared to the slit separation d (we assume Fraunhofer diffraction), the diffraction patterns due separately to each slit essentially coincide. Then the wave amplitude at any angle θ will no longer be

$$E_{\theta 0} = 2E_0 \cos \delta/2$$

as was given by Eq. 36–5b; rather, it must be modified, because of diffraction, by Eq. 37–5, so that

$$E_{\theta 0} = 2E_0 \left(\frac{\sin \beta/2}{\beta/2}\right) \cos \frac{\delta}{2}.$$

(Note that in this chapter we have dealt only with the amplitude of E, and so we merely wrote E_θ rather than $E_{\theta 0}$ as was done in Chapter 36.) Thus the intensity will be given by

$$I_\theta = I_0 \left(\frac{\sin \beta/2}{\beta/2}\right)^2 \left(\cos \frac{\delta}{2}\right)^2 \tag{37–9}$$

where

$$\frac{\beta}{2} = \frac{\pi}{\lambda} D \sin \theta \quad \text{and} \quad \frac{\delta}{2} = \frac{\pi}{\lambda} d \sin \theta.$$

(These are from Eqs. 37–6 and 36–4.) The first term in parentheses in Eq. 37–9 is sometimes called the "diffraction factor" and the second one the "interference factor." These two factors are plotted in Fig. 37–9a and b for the case when $d = 6D$, and $D = 10\lambda$. (Figure 37–9b is essentially the same as Fig. 36–14, except for the factor I_0.) Figure 37–9c shows the product of these two curves, times I_0, which is the actual intensity as a function of θ (or as a function of position on the screen for θ not too large) as given by Eq. 37–9. As indicated by the dashed lines in Fig. 37–9c, the diffraction factor acts as a sort of envelope that limits the interference peaks. The decrease in intensity of the interference fringes away from the center can be seen in Fig. 36–7, but is clearer in Fig. 37–10 where the ratio d/D is smaller.

 EXAMPLE 37–3 Show why the central diffraction peak in Fig. 37–9 contains 11 interference fringes.

SOLUTION The first minimum in the diffraction pattern occurs where

$$\sin \theta = \lambda/D.$$

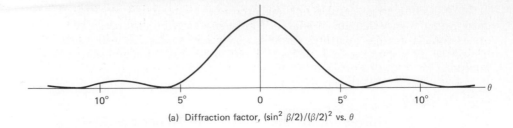

(a) Diffraction factor, $(\sin^2 \beta/2)/(\beta/2)^2$ vs. θ

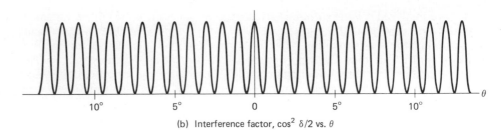

(b) Interference factor, $\cos^2 \delta/2$ vs. θ

FIGURE 37–9

(a) Diffraction factor, (b) interference factor, and (c) the resultant intensity I, plotted as a function of θ for $d = 6D = 60\lambda$.

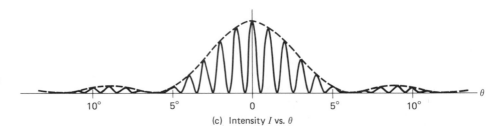

(c) Intensity I vs. θ

FIGURE 37–10

Photographs of double-slit interference pattern showing effects of diffraction; $d = 6D$ (from F. A. Jenkins and H. E. White, *Fundamental of Optics*, *4th ed.*, © 1976 by McGraw-Hill, Inc; Used with permission of McGraw-Hill Book Co.)

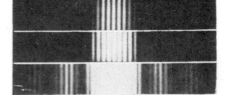

Since $d = 6D$,

$$d \sin \theta = 6D(\lambda/D) = 6\lambda.$$

From Eq. 36–2a, we see that interference peaks (maxima) occur for $d \sin \theta = m\lambda$ where m can be 0, 1, . . . or any integer. Thus the diffraction minimum coincides with $m = 6$ in the interference pattern, so the $m = 6$ peak won't appear. Hence the central diffraction peak encloses the central interference peak ($m = 0$) and five peaks ($m = 1$ to 5) on each side for a total of 11. Since the sixth order doesn't appear, it is said to be a "missing order."

Notice from this last example that the number of interference fringes in the central diffraction peak depends only on the ratio d/D. It does not depend on wavelength λ. The actual spacing (in angle, or in position on the screen) does depend on λ. For the case illustrated, $D = 10\lambda$, and so the first diffraction minimum occurs at $\sin \theta = \lambda/D = 0.10$ or about 6°.

The patterns due to interference and diffraction arise from the same phenomenon—the superposition of coherent waves of different phase. The distinction between them is thus not so much physical as for convenience of description, as in this section where we analyzed the two-slit pattern in terms of interference and diffraction separately. In general, we use the word "diffraction" when referring to an analysis by superposition of many infinitesimal and usually contiguous sources such as when we subdivide a source into infinitesimal parts. We use the term "interference" when we superpose the waves from a finite (and usually small) number of coherent sources.

 ## 37-4   Limits of Resolution; the Rayleigh Criterion

The ability of a lens to produce distinct images of two point objects very close together is called the **resolution** of the lens. The closer the two images can be and still be seen as distinct (rather than overlapping blobs), the higher the resolution. The resolution of a camera lens, for example, is often specified as so many lines per millimeter.[†] This is determined by photographing a standard set of parallel lines on fine grain film. The minimum spacing of lines distinguishable on film using the lens gives the resolution.

There are two principle factors that limit the resolution of a lens. The first is lens aberrations. As we saw, because of spherical and other aberrations, a point object is not a point on the image but a tiny blob. Careful design of compound lenses can reduce aberrations significantly, but they cannot be eliminated entirely. The second factor that limits resolution is *diffraction*, which cannot be corrected for because it is a natural result of the wave nature of light. We discuss it now.

In Section 37-1, we saw that because light travels as a wave, light from a point source passing through a slit is spread out into a diffraction pattern (Figs. 37-2 and 37-4). A lens, because it has edges, acts like a slit. When a lens forms the image of a point object, the image is actually a tiny diffraction pattern. Thus an image would be blurred even if aberrations were absent.

In the analysis that follows, we assume that the lens is free of aberrations, so that we can focus our attention on diffraction effects and how much they limit the resolution of a lens. In Fig. 37-4 we saw that the diffraction pattern produced by light passing through a rectangular slit has a central maximum in which most of the light falls. This central peak falls to a minimum on either side of its center at an angle $\theta \approx \sin \theta = \lambda/D$ (this is Eq. 37-1) where D is the diameter of the slit, λ is the wavelength of light used, and we assume θ is small. There are also low-intensity fringes beyond. For a lens, or any circular hole, the image of a point object will consist of a *circular* central peak (called the *diffraction spot* or *airy disk*) surrounded by faint circular fringes, as in Fig. 37-11a. The central maximum has an angular half width given by

$$\theta = \frac{1.22\lambda}{D}.$$

This differs from that for a slit (Eq. 37-1) by the factor 1.22. This factor comes from the fact that the width of a circular hole is not uniform (like a rectangular slit) but varies from its diameter D to zero. A careful analysis shows that the "average" width is $D/1.22$; hence we get the equation above rather than Eq. 37-1. The intensity of light in the diffraction pattern of a circular opening is shown in Fig. 37-12. Any image is thus a complex diffraction pattern. For most purposes we need consider only the central spot since the concentric rings are so much dimmer.

If two point objects are very close, the diffraction patterns of their images will overlap as shown in Fig. 37-11b. As the objects are moved closer, a point is reached where you can't tell if there are two overlapping images or a single image. Where this happens may be judged differently by different observers.

[†] This may be specified at the center of the field of view as well as at the edges, where it is usually less because of off-axis aberrations.

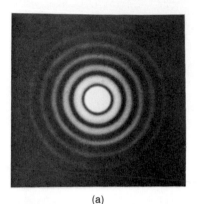

(a)

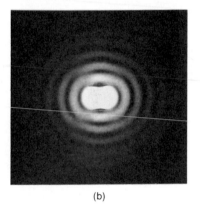

(b)

FIGURE 37-11

Photographs of images (greatly magnified) formed by a lens: (a) a single point object; (b) two point objects barely resolved. (From M. Cagnet, M. Françon, and J. C. Thrierr, *Atlas of Optical Phenomena*, Springer-Verlag, Berlin, 1962.)

FIGURE 37-12

Intensity of light across the diffraction pattern of a circular hole.

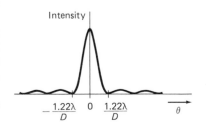

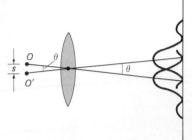

FIGURE 37–13

The *Rayleigh criterion*. Two images are just resolvable when the center of the diffraction peak of either is directly over the first minimum in the diffraction pattern of the other. The two point objects O and O' subtend an angle θ at the lens; one ray only is drawn for each point to indicate the center of the diffraction pattern of its image.

However, a generally accepted criterion is one proposed by Lord Rayleigh (1842–1919). This **Rayleigh criterion** states that *two images are just resolvable when the center of the diffraction disk of one is directly over the first minimum in the diffraction pattern of the other*. This is shown in Fig. 37–13. Since the first minimum is at an angle $\theta = 1.22\lambda/D$ from the central maximum, Fig. 37–13 shows us that two objects can be considered just resolvable if they are separated by this angle θ:

$$\theta = \frac{1.22\lambda}{D}. \tag{37–10}$$

This is the limit on resolution set by the wave nature of light due to diffraction.

37–5   Resolution of Telescopes and Microscopes

It might be thought that a microscope or telescope could be designed to produce any desired magnification depending on the choice of focal lengths. But this is not possible, because of diffraction. An increase in magnification above a certain point merely results in magnification of the diffraction pattern. This would be highly misleading since we might think we are seeing details of an object when we are really seeing details of the diffraction pattern. To examine this we apply the Rayleigh criterion, Eq. 37–10,

$$\theta = \frac{1.22\lambda}{D}$$

which is valid for either a microscope or a telescope where D is the diameter of the objective lens. For a telescope, the resolution is specified by stating θ as given by this equation.[†]

For a microscope, it is more convenient to specify the actual distance s between two points that are just barely resolvable, Fig. 37–13. Since objects are normally placed near the focal point of the objective, $\theta = s/f$ or $s = f\theta$. If we combine this with Eq. 37–10, we obtain for the *resolving power* (RP)

$$RP = s = f\theta = \frac{1.22\lambda f}{D}. \tag{37–11}$$

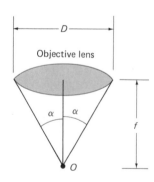

FIGURE 37–14

Objective lens of a microscope, showing the angle of acceptance, α.

This distance s is called the resolving power of the lens because it is the minimum separation of two object points that can just be resolved. The above equation is often written in terms of the angle of acceptance, α, of the objective lens as defined in Fig. 37–14. The derivation[‡] is left as problem 20 at the end of the chapter and the result is

$$RP = s = \frac{1.22\lambda}{2\sin\alpha} = \frac{0.61\lambda}{\sin\alpha}. \tag{37–12a}$$

The resolving power can be increased by placing a drop of oil that encloses the object and the front surface of the objective. This is called an *oil-immersion objective*. In the oil, the wavelength of the light is reduced to λ/n where n is the oil's index of refraction. Thus the resolving power becomes

$$RP = \frac{0.61\lambda}{n\sin\alpha}. \tag{37–12b}$$

[†] Telescopes with large-diameter objectives are usually limited not by diffraction but by other effects such as turbulence in the atmosphere. The resolution of a high-quality microscope, on the other hand, is normally limited by diffraction since microscope objectives are complex compound lenses containing many elements of small diameter (since f is small).

[‡] When the angle α is small, the derivation is easy since in this case $\sin\alpha = \tan\alpha = \alpha = D/2f$ (see Fig. 37–14) or $f/D = 1/(2\sin\alpha)$. Thus $s = 1.22\lambda f/D = 1.22\lambda/2\sin\alpha$. The general case is treated in problem 20.

The oil typically has $n \approx 1.5$ although n may be as great as 1.8. Thus oil immersion increases the resolution by 50 percent or more.

The quantity ($n \sin \alpha$) is called the *numerical aperture* (NA) of the lens:

$$NA = n \sin \alpha.$$

It is usually specified on the objective lens housing along with the magnification. The larger the value of the NA, the finer the resolving power.

EXAMPLE 37–4   What is the theoretical minimum angular separation of two stars that can just be resolved by the 200 inch telescope on Mt. Palomar? Assume the stars emit white light of average wavelength 550 nm.

SOLUTION   Since $D = 200$ in $= 5.1$ m, we have from Eq. 37–10 that $\theta = 1.22\lambda/D = 1.22(5.50 \times 10^{-7}\text{m})/5.1$ m $= 1.3 \times 10^{-7}$ rad or 0.75×10^{-5} deg. This is the limit set by diffraction. In actual fact, the resolution is not this good because of aberrations and, more importantly, turbulence in the atmosphere. In fact, large-diameter objectives are not justified by increased resolution, but by their greater light-gathering ability—they allow more light in, so dimmer objects can be seen.

EXAMPLE 37–5   Determine the NA and RP of the best oil-immersion microscopes where the index of refraction of the oil is $n = 1.8$ and $\sin \alpha \approx 0.90$. Assume that $\lambda = 550$ nm.

SOLUTION   The NA $= n \sin \alpha = 1.6$. The resolving power is RP $= 0.61\lambda/\text{NA} = (0.61)(5.50 \times 10^{-7}$ m$)/(1.6) \approx 2 \times 10^{-7}$ m $= 200$ nm. This is the best resolution that a visible-light microscope can attain.

Diffraction sets an ultimate limit on the detail that can be seen on any object. In Eq. 37–11 we note that the focal length of a lens cannot be made less than (approximately) the radius of the lens, and even that is very difficult—see the lens-maker's equation (Eq. 35–3); in this best case[†], Eq. 37–11 gives, with $f \approx D/2$

$$RP \approx \frac{\lambda}{2}. \tag{37–13}$$

Thus we can say, to within a factor of 2 or so, that

it is not possible to resolve detail of objects smaller than the wavelength of the radiation being used.

This is an important and useful rule of thumb.

Compound lenses are now designed so well that the actual limit on resolution is often set by diffraction—that is, by the wavelength of the light used. To obtain greater detail one must use radiation of shorter wavelength. The use of UV radiation can increase the resolution by a factor of perhaps 2. Far more important, however, was the discovery in the early twentieth century that electrons have wave properties (Chapter 40) and that their wavelengths can be very small; this fact is used in the electron microscope (Section 40–6), which can magnify 100 to 1000 times more than a visible-light microscope because of the much shorter wavelengths. X rays too have very short wavelengths and are often used to study objects in great detail using special techniques (Section 37–11).

[†] The same result can be obtained from Eq. 37–12a since $\sin \alpha$ can never exceed 1 and typically is 0.6 to 0.9 at most. With oil immersion, Eq. 37–12b gives, at best, RP $\approx \lambda/3$ which corresponds to the result of Example 37–5.

37−6 Resolution of the Human Eye and Useful Magnification

The resolution of the human eye is limited by several factors, all of roughly the same order of magnitude. The resolution is best at the fovea, where the cone spacing is smallest, about 3 μm ($=3000$ nm). The diameter of the pupil varies from about 0.1 cm to about 0.8 cm; so for $\lambda = 550$ nm (where the eye's sensitivity is greatest), the diffraction limit is about $\theta \approx 1.22\lambda/D \approx 8 \times 10^{-5}$ rad to 6×10^{-4} rad. Since the eye is about 2 cm long, this corresponds to a resolving power of $s \approx (8 \times 10^{-5}$ rad$)(2 \times 10^{-2}$ m$) \approx 2$ μm at best, to about 15 μm at worst (pupil small). Spherical and chromatic aberration also limit the resolution to about 10 μm. The net result is that the eye can resolve objects whose angular separation is about 5×10^{-4} rad at best. This corresponds to objects separated by 1 cm at a distance of about 20 m. The typical near point of a human eye is about 25 cm. At this distance, the eye can just resolve objects that are (25 cm)· $(5 \times 10^{-4}$ rad$) \approx 10^{-4}$ m apart. Since the best light microscopes can resolve objects no smaller than about 200 nm (see Example 37−5), the useful magnification is limited to about

$$\frac{10^{-4} \text{ m}}{200 \times 10^{-9} \text{ m}} = 500 \times.$$

In practice, magnifications of about 1000× are often used to minimize eyestrain. Any greater magnification would simply make visible the diffraction pattern produced by the microscope objective.

37−7 Multiple Slits—the Diffraction Grating

In the first three sections of this chapter, we examined the diffraction pattern produced when light passes through one or two slits. Now we look at a very useful device, the *diffraction grating*, which consists of a large number of equally spaced parallel slits. Gratings are often made by ruling very fine lines on glass with a diamond tip. The untouched spaces between the lines serve as the slits. Photographic transparencies of an original grating serve as inexpensive gratings. Photographic reduction can be used to make very fine gratings. Gratings containing 10,000 lines per centimeter are common today, and are very useful for measuring wavelengths very accurately (discussed in the next section). A diffraction grating containing slits is called a *transmission grating*. *Reflection gratings* are also possible; they can be made by ruling fine lines on a metallic or glass surface from which light is reflected and examined. The analysis is basically the same as for a transmission grating, and we consider only the latter here.

The analysis of a diffraction grating is much like that of Young's double-slit experiment. We assume parallel rays of light are incident on the grating as shown in Fig. 37−15. We also assume that the slits are narrow enough so that diffraction by each of them spreads light over a very wide angle on a distant screen behind the grating, and interference can occur with light from all the other slits. Light rays that pass through each slit without deviation ($\theta = 0°$) interfere constructively to produce a bright spot at the center. At an angle θ such that rays from adjacent slits travel an extra distance of $\Delta l = m\lambda$ where m is an integer, again constructive interference occurs. Thus, if d is the distance between slits,

$$\sin \theta = \frac{m\lambda}{d}, \qquad m = 0, 1, 2, \ldots \qquad \text{[principal maxima]} \quad (37\text{–}14)$$

FIGURE 37−15

Diffraction grating.

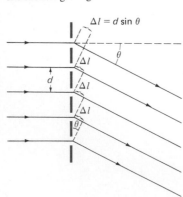

is the criterion to have a brightness maximum. This is the same equation as for the double-slit situation, and again m is called the *order* of the pattern.

There is an important difference between a double slit and a multiple-slit pattern, however. The bright maxima are much *sharper* and *narrower* for a grating. Why this happens can be seen as follows. Suppose that the angle θ is increased just slightly beyond that required for a maximum. In the case of only two slits, the two waves will be only slightly out of phase, so nearly full constructive interference occurs. This means the maxima are wide (see Fig. 36–14). For a grating, the waves from two adjacent slits will also not be significantly out of phase. But waves from one slit and those from a second one a few hundred slits away may be exactly out of phase; all or nearly all the light will cancel in pairs in this way. For example, suppose the angle θ is different from its first-order maximum so that the extra path length is not exactly λ but rather 1.0010λ. The wave through one slit and another one 500 slits below will be out of phase by 1.5000λ, or exactly $1\frac{1}{2}$ wavelengths, so the two will cancel. A pair of slits, one below each of these, will also cancel. That is, the light from slit 1 cancels with that from slit 501; light from slit 2 cancels with that from slit 502, and so on. Thus even for a tiny angle[†] corresponding to an extra path length of $\frac{1}{1000}\lambda$, there is much destructive interference, and so the maximum must be very narrow. The more lines there are in a grating the sharper will be the peaks (see Fig. 37–16). Because a grating produces much sharper (and brighter) lines than two slits alone, it is a far more precise device for measuring wavelengths.

Suppose the light striking a diffraction grating is not monochromatic, but rather consists of two or more distinct frequencies. Then for all orders other than $m = 0$, each wavelength will produce a maximum at a different angle (Fig. 37–17a). If white light strikes a grating, the central ($m = 0$) maximum will be a sharp white peak; but for all other orders, there will be a distinct rainbow spread out over a certain angular width, Fig. 37–17b. Because a diffraction grating spreads out light into its component wavelengths, the resulting pattern is called a *spectrum*.

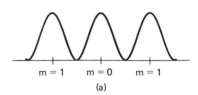

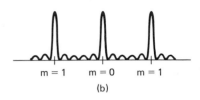

FIGURE 37–16

Intensity as a function of viewing angle θ (or position on screen) for (a) two slits, (b) six slits. For a diffraction grating, the number of slits is very large ($\sim 10^4$) and the peaks are narrower still.

[†] Depending on the number of slits, there may or may not be complete cancellation for such an angle, so there will be very tiny peaks between the main maxima (see Fig. 37–16b), but they are usually much too small to be seen. We discuss this in more detail in Section 37–9.

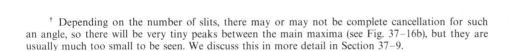

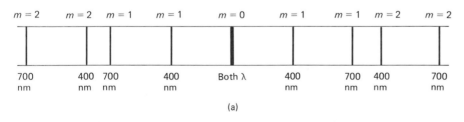

FIGURE 37–17

Spectra produced by a grating: (a) two wavelengths, 400 nm and 700 nm; (b) white light. The second order will be dimmer than the first order, normally; if grating spacing is small enough, second order may be missing. (Higher orders not shown.)

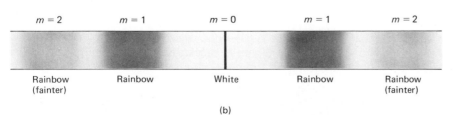

EXAMPLE 37–6 Calculate the first- and second-order angles for light of wavelength 400 nm and 700 nm if the grating contains 10,000 lines/cm.

SOLUTION Since the grating contains 10^4 lines/cm $= 10^6$ lines/m, the separation between slits is $d = 1/10^6$ m$^{-1} = 1.0 \times 10^{-6}$ m. In first order ($m = 1$), the angles are

$$\sin \theta_{400} = \frac{m\lambda}{d} = \frac{1(4.0 \times 10^{-7} \text{ m})}{1.0 \times 10^{-6} \text{ m}} = 0.400$$

$$\sin \theta_{700} = 0.700,$$

so $\theta_{400} = 23.6°$ and $\theta_{700} = 44.0°$. In second order

$$\sin \theta_{400} = \frac{2(4.0 \times 10^{-7} \text{ m})}{1.0 \times 10^{-6} \text{ m}} = 0.800$$

$$\sin \theta_{700} = 1.40,$$

so $\theta_{400} = 53.0°$ but the second order does not exist for $\lambda = 700$ nm since $\sin \theta$ cannot exceed 1. No higher orders will appear.

EXAMPLE 37–7 White light containing wavelengths from 400 nm to 750 nm strikes a grating containing 4000 lines/cm. Show that the violet of the third-order spectrum overlaps the red of the second order.

SOLUTION The grating spacing is $d = 1/(4000 \text{ cm}^{-1}) = 2.50 \times 10^{-6}$ m. The violet of the third order occurs at an angle θ given by

$$\sin \theta = \frac{(3)(4.00 \times 10^{-7} \text{ m})}{(2.50 \times 10^{-6} \text{ m})} = 0.480.$$

Red in second order occurs at

$$\sin \theta = \frac{(2)(7.50 \times 10^{-7} \text{ m})}{(2.50 \times 10^{-6} \text{ m})} = 0.600,$$

which is clearly a greater angle.

The diffraction grating is the essential component of a spectroscope, a device for precise measurement of wavelengths, and we discuss it next.

37–8 The Spectroscope and Spectroscopy

FIGURE 37–18

Spectroscope.

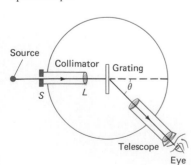

A *spectroscope*, Fig. 37–18, is a device to measure wavelengths accurately using a diffraction grating (discussed in Section 37–7), or a prism, to separate different wavelengths of light. Light from a source passes through a narrow slit S in the collimator. The slit is at the focal point of the lens L, so parallel light falls on the grating. The movable telescope can bring the rays to a focus. Nothing will be seen in the viewing telescope unless it is positioned at an angle θ that corresponds to a diffraction peak (first order is usually used) of a wavelength emitted by the source. The angle θ can be measured[†] to very high accuracy, so the wavelength of a line can be determined to high accuracy using Eq. 37–14:

[†] The angle θ for a given wavelength is usually measured on both sides of center because the grating cannot always be aligned precisely; the average of the two values is then taken.

$$\sin \theta = \frac{m\lambda}{d},$$

where m is an integer representing the order, and d is the distance between grating lines. (The line you see in a spectroscope corresponding to each wavelength is actually an image of the slit S; so the narrower the slit, the narrower—but dimmer—is the line, and the more precise its measurement. If the light contains a continuous range of wavelengths, you will then see a continuous spectrum in the spectroscope.)

In some spectroscopes, a reflection grating or a prism is used. A prism, because of dispersion (Section 36–5), bends light of different wavelengths into different angles. A prism has the disadvantage that it produces less sharp lines and is less able to separate closely spaced lines. But it has the advantage of deflecting more light (and so is more useful for dim sources) than a normal diffraction grating, since with the latter most of the light passes straight through to the central peak. However, reflection gratings can now be made that have shaped grooves, so that a large portion of the light is reflected at an appropriate angle to give a strong first-order peak.

If the spectrum of a source is recorded (say on film) rather than viewed by the eye, the device is called a *spectrometer*, as compared to a *spectroscope*, although these terms are sometimes used interchangeably. Devices which can also measure the intensity of light of a given wavelength are called *spectrophotometers*.

An important use of any of these devices is for the identification of atoms or molecules. When a gas is heated or a large electric current is passed through it, the gas emits a characteristic **line spectrum**. That is, only certain wavelengths of light are emitted, and these are different for different elements and compounds.[†] Figure 37–19 (color plate) shows the line spectra for a number of elements in the gas state. Line spectra occur only for gases at high temperatures and low pressure. The light from heated solids, such as a light bulb filament, and even from a dense gaseous object such as the sun, produces a *continuous spectrum* including a wide range of wavelengths.

As can be seen in Fig. 37–19, the sun's "continuous spectrum" contains a number of *dark* lines (only the most prominent are shown), called *absorption lines*. Atoms and molecules absorb light at the same wavelengths they emit. The sun's absorption lines are due to absorption by atoms and molecules in the cooler outer atmosphere of the sun, as well as by atoms and molecules in the earth's atmosphere. A careful analysis of all these thousands of lines reveals that at least two-thirds of all elements are present in the sun's atmosphere. The presence of elements in the atmosphere of other planets, in interstellar space, and in stars is also determined by spectroscopy.

Spectroscopy is useful for determining the presence of certain types of molecules in laboratory specimens where chemical analysis would be difficult. For example, biological DNA and different types of protein absorb light in particular regions of the spectrum (especially in the UV). The material to be examined, which is often in solution, is placed in a monochromatic light beam whose wavelength is chosen by placement of a prism or diffraction grating. The amount of absorption, as compared to a standard solution without the specimen, can reveal not only the presence of a particular type of molecule but also its concentration.

Light emission and absorption also occur outside the visible part of the spectrum, such as in the UV and IR regions. Since glass absorbs light in these regions, reflection gratings and mirrors (in place of lenses) are used. Special types of film, or photocell detectors, are used for detection.

[†] Why atoms and molecules emit line spectra was a great mystery for many years and played a central role in the development of modern quantum theory as we shall see in Chapter 40.

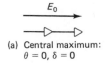

⬡ * 37–9 Phasor Diagrams for Multiple Slits; Peak Width

We now look at the pattern of maxima produced by a multiple-slit grating using phasor diagrams; we can determine a formula for the width of each peak, and we will see why there are tiny maxima between the principal maxima, as indicated in Fig. 37–16b. First of all, it should be noted that the two-slit and six-slit patterns shown in Fig. 37–16 were drawn assuming very narrow slits so that diffraction does not limit the height of the peaks. For real diffraction gratings, this is not normally the case; the slit width D is often not much smaller than the slit separation d, and diffraction thus limits the intensity of the peaks so the central peak ($m = 0$) is brighter than the side peaks. We won't worry about this effect on intensity except to note that if a diffraction minimum coincides with a particular order of the interference pattern, that order will not appear. (For example, if $d = 2D$, all the even orders, $m = 2, 4, \ldots$, will be missing. Can you see why? Hint: see Example 37–3.)

Figures 37–20 and 37–21 show phasor diagrams for a two-slit and a six-slit grating, respectively. Each short arrow represents the amplitude of a wave from a single slit, and their vector sum (as phasors) represents the total amplitude for a given viewing angle θ. Part (a) of each figure shows the phasor diagram at $\theta = 0°$, at the center of the pattern, which is the central maximum ($m = 0$). Part (b) of each figure shows the condition for the adjacent minimum: where the arrows first close on themselves (add to zero) so the amplitude E_θ is zero. For two slits, this occurs when the two separate amplitudes are 180° out of phase. For six slits, it occurs when each amplitude makes a 60° angle with its neighbor. For two slits, the minimum occurs when the phase between slits is $2\pi/2$ (in radians); for six slits it occurs when the phase δ is $2\pi/6$; and in the general case of N slits, the minimum occurs for a phase difference between adjacent slits of

$$\delta = \frac{2\pi}{N}. \tag{37–15}$$

What does this correspond to in θ? First note that δ is related to θ by

$$\frac{\delta}{2\pi} = \frac{d \sin \theta}{\lambda} \quad \text{or} \quad \delta = \frac{2\pi}{\lambda} d \sin \theta \tag{37–16}$$

just as in Eq. 36–4. Let us call $\Delta\theta_0$ the angular position of the minimum next to the peak at $\theta = 0°$. This corresponds to an extra path length between adjacent slits (see Fig. 37–15) of $\Delta l = d \sin \Delta\theta_0$, so

$$\frac{\delta}{2\pi} = \frac{\Delta l}{\lambda} = \frac{d \sin \Delta\theta_0}{\lambda}.$$

We insert Eq. 37–15 for δ and find

$$\sin \Delta\theta_0 = \frac{\lambda}{Nd}. \tag{37–17a}$$

Since $\Delta\theta_0$ is usually small (N is usually very large for a grating) $\sin \Delta\theta_0 \approx \Delta\theta_0$, so in the small angle limit we can write

$$\Delta\theta_0 = \frac{\lambda}{Nd}. \tag{37–17b}$$

It is clear from either of the last two relations that the larger N is, the narrower will be the central peak. (For $N = 2$, $\sin \Delta\theta_0 = \lambda/2d$, which is what we obtained earlier for the double slit, Eq. 36–2b with $m = 0$.)

Either of Eqs. 37–17 shows why the peaks become narrower for larger N. The origin of the small secondary maxima between the principal peaks (see

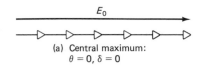

(a) Central maximum:
$\theta = 0$, $\delta = 0$

$E_\theta = 0$

$\delta = 180°$

(b) Minimum: $\delta = 180°$

FIGURE 37–20

Phasor diagram for two slits (a) at central maximum, (b) at minimum.

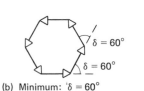

E_0

(a) Central maximum:
$\theta = 0$, $\delta = 0$

$\delta = 60°$

$\delta = 60°$

(b) Minimum: $\delta = 60°$

FIGURE 37–21

Phasor diagram for six slits (a) at central maximum, (b) at minimum.

Fig. 37–16b) can be deduced from the diagram of Fig. 37–22. This is just a continuation of Fig. 37–21b (where $\Delta\delta = 60°$); but now the phase $\Delta\delta$ has been increased to almost $90°$, where E_θ is a relative maximum; note that E_θ is much less than E_0 (Fig. 37–21a), so the intensity in this secondary maximum is much smaller than in a principal peak. As $\Delta\delta$ (and θ) is increased further, E_θ again decreases to zero (a "double circle"), then reaches another tiny maximum, and so on; eventually the diagram unfolds again and when $\Delta\delta = 360°$, all the amplitudes again lie in a straight line (as in Fig. 37–21a) corresponding to the next principal maximum ($m = 1$ in Eq. 37–14).

Equation 37–17b gives the half width of the central ($m = 0$) peak. To determine the half width of higher order peaks, $\Delta\theta_m$ for order m, we differentiate Eq. 37–16 so as to relate the change $\Delta\delta$ in δ, to the change $\Delta\theta$ in the angle θ:

$$\Delta\delta \approx \frac{d\delta}{d\theta}\Delta\theta = \frac{2\pi d}{\lambda}\cos\theta\,\Delta\theta.$$

If $\Delta\theta_m$ represents the half width of a peak of order m ($m = 1, 2, \ldots$)—that is, the angle between the peak maximum and the minimum to either side—then $\Delta\delta = 2\pi/N$ as given by Eq. 37–15. We insert this into the above relation and find

$$\Delta\theta_m = \frac{\lambda}{Nd\cos\theta_m}, \tag{37–18}$$

where θ_m is the angular position of the mth peak as given by Eq. 37–14. This derivation is valid, of course, only for small $\Delta\delta$ ($= 2\pi/N$) which is indeed the case for real gratings since N is on the order of 10^4 or more.

FIGURE 37–22

Phasor diagram for secondary peak.

 * 37–10   Resolving Power of a Diffraction Grating

An important property of any diffraction grating used in a spectroscope is its ability to resolve two very closely spaced wavelengths.

We define the *angular dispersion* of a grating as the rate of change in angle (for a given order) with wavelength, $d\theta/d\lambda$. From Eq. 37–14, the angle θ at which a wavelength λ is diffracted by a grating whose rulings are a distance d apart is

$$\sin\theta = \frac{m\lambda}{d},$$

where m is the order. We differentiate this to obtain

$$\cos\theta\,d\theta = \frac{m}{d}d\lambda$$

so that the angular dispersion is

$$\frac{d\theta}{d\lambda} = \frac{m}{d\cos\theta}. \tag{37–19}$$

If two monochromatic waves of wavelength λ_1 and λ_2 are to be distinguishable in a spectroscope, the angle $\Delta\theta$ between them (in a given order m) must be at least equal to the width of their diffraction peaks (the Rayleigh criterion). From Eq. 37–18, the angle $\Delta\theta$ (between a peak and the first minimum at its side) is given by

$$\Delta\theta = \frac{\lambda}{Nd\cos\theta},$$

where N is the total number of grating lines (and we dropped the subscript m).

Then the minimum separation in wavelength that can be resolved, $\Delta\lambda = \lambda_1 - \lambda_2$, is

$$\Delta\lambda \approx \frac{d\lambda}{d\theta}\,\Delta\theta = \left(\frac{d\cos\theta}{m}\right)\left(\frac{\lambda}{Nd\cos\theta}\right)$$

where we used Eq. 37–19 and took $\lambda_1 \approx \lambda_2 = \lambda$. This simplifies to

$$\Delta\lambda = \frac{\lambda}{Nm}. \tag{37–20}$$

The *resolving power* R of a grating is defined as

$$R = \frac{\lambda}{\Delta\lambda} \tag{37–21}$$

which, using Eq. 37–20, becomes

$$R = Nm. \tag{37–22}$$

The larger the value of R, the closer two wavelengths can be and still be resolvable. If R is given, the minimum separation $\Delta\lambda$ between two wavelengths near λ, is (see the previous two equations):

$$\Delta\lambda = \frac{\lambda}{R}.$$

EXAMPLE 37–8 Yellow sodium light, which consists of two wavelengths, $\lambda_1 = 589.00$ nm and $\lambda_2 = 589.59$ nm, falls on a 7500-line/cm diffraction grating. Determine (*a*) the maximum order m that will be present for sodium light, (*b*) the angular dispersion of the grating, (*c*) the width of grating necessary to resolve the two sodium lines, (*d*) the grating resolving power in this case, (*e*) the angular width of each sodium line.

SOLUTION   (*a*) Since $d = 1\ \text{cm}/7500 = 1.33 \times 10^{-6}$ m, then the maximum value of m for $\lambda = 589$ nm can be found from Eq. 37–14 with $\sin\theta \leq 1$:

$$m = \frac{d}{\lambda}\sin\theta \leq \frac{1.33 \times 10^{-6}\ \text{m}}{5.89 \times 10^{-7}\ \text{m}} = 2.25,$$

so $m = 2$ is the maximum order present.

(*b*) From Eq. 37–19 the angular dispersion is

$$\frac{d\theta}{d\lambda} = \frac{m}{d\cos\theta}.$$

The sodium lines in order 2 occur for $\sin\theta = 2\lambda/d = 2(589 \times 10^{-9}\ \text{m})/(1.33 \times 10^{-6}\ \text{m}) = 0.886$ so $\cos\theta = \sqrt{1 - \sin^2\theta} = 0.464$. Then

$$\frac{d\theta}{d\lambda} = \frac{2}{(1.33 \times 10^{-6}\ \text{m})(0.464)} = 3.24 \times 10^{-3}\ \text{rad/nm},$$

or 0.186 deg/nm.

(*c*) The resolving power needed is

$$R = \frac{\lambda}{\Delta\lambda} = \frac{589\ \text{nm}}{0.59\ \text{nm}} = 1000.$$

From Eq. 37–22, the total number N of lines needed is

$$N = \frac{R}{m} = \frac{1000}{2} = 500$$

so the grating need only be $500/7500\ \text{cm}^{-1} = 0.0667$ cm wide. A typical grating is a few centimeters wide, and so will easily resolve the two lines.

(d) The resolving power for $N = 500$, $m = 2$, is 1000 as calculated already. If the grating were 3.0 cm wide, it would contain 22,500 lines and the resolving power would be 45,000.

(e) For a grating with the minimum 500 lines, the angular width (Eq. 37–18) for $m = 2$ is

$$\Delta\theta = \frac{\lambda}{Nd\cos\theta} = \frac{589 \times 10^{-9}\text{ m}}{(500)(1.33 \times 10^{-6}\text{ m})(0.464)} = 0.0019\text{ rad}$$

or $0.11°$.

*37–11   X rays and X-ray Diffraction

In 1895, W. C. Roentgen (1845–1923) discovered that when electrons were accelerated by a high voltage in a vacuum tube and allowed to strike a glass (or metal) surface, fluorescent minerals some distance away would glow, and film would become exposed. Roentgen attributed these effects to a new type of radiation (different from cathode rays). They were given the name X rays after the algebraic symbol x meaning an unknown quantity. He soon found that X rays penetrated through some materials better than through others, and within a few weeks he presented the first X-ray photograph (of his wife's hand). The production of X rays today is done in a tube (Fig. 37–23) similar to Roentgen's, using voltages of typically 30 to 150 kV.

Investigations into the nature of X rays indicated they were not charged particles (such as electrons) since they could not be deflected by electric or magnetic fields. It was suggested that they might be a form of invisible light. However, they showed no diffraction or interference effects using ordinary gratings. Of course, if their wavelengths were much smaller than the typical grating spacing of 10^{-6} m ($= 10^3$ nm) no effects would be expected. Around 1912, it was suggested by Max von Laue (1879–1960) that if the atoms in a crystal were arranged in a regular array (see Fig. 17–1a), a theory generally held by scientists though not then fully tested, such a crystal might serve as a diffraction grating for very short wavelengths on the order of the spacing between atoms, estimated to be about 10^{-10} m ($= 10^{-1}$ nm). Experiments soon showed that X rays scattered from a crystal did indeed show the peaks and valleys of a diffraction pattern (Fig. 37–24). Thus it was shown, in a single blow, that X rays have a wave nature and that atoms are arranged in a regular way in crystals. Today X rays are recognized as electromagnetic radiation with wavelengths in the range of about 10^{-2} nm to 10 nm, the range readily produced in an X-ray tube.

We saw in Sections 37–4 and 37–5 that light of shorter wavelength provides greater resolution when we are examining an object microscopically. Since X rays have much shorter wavelengths than visible light, they should in principle offer much greater resolution. However, there seems to be no effective material to use as lenses for the very short wavelengths of X rays. Instead the clever but complicated technique of *X-ray diffraction* (or *crystallography*) has proved very effective for examining the microscopic world of atoms and molecules. In a simple crystal such as NaCl, the atoms are arranged in an orderly cubical fashion, Fig. 37–25, with atoms spaced a distance d apart. Suppose that a beam of X rays is incident on the crystal at an angle θ to the surface, and that the two rays shown are reflected from two subsequent planes of atoms as shown. The two rays will constructively interfere if the extra distance ray I travels is a whole number of wavelengths farther than what ray II travels. This extra distance is $2d\sin\theta$. Therefore, constructive interference will occur when

$$m\lambda = 2d\sin\theta, \quad m = 1, 2, 3, \ldots \quad (37\text{–}23)$$

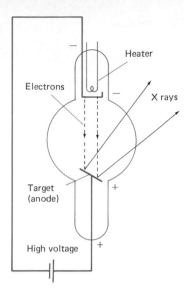

FIGURE 37–23

X-ray tube. Electrons emitted by a heated filament in a vacuum tube are accelerated by high voltage. When they strike the surface of the anode, the "target," X rays are emitted.

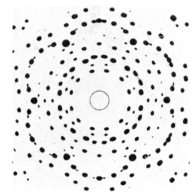

FIGURE 37–24

Diffraction pattern produced by X rays scattered from a beryl crystal. (Courtesy of Eastman Kodak.)

FIGURE 37–25

X-ray diffraction by a crystal.

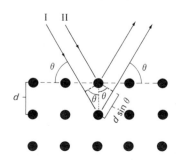

FIGURE 37–26

There are many possible planes existing within a crystal from which X rays can be diffracted.

where m can be any integer. (Notice that θ is *not* the angle with respect to the normal to the surface.) This is called the *Bragg equation* after W. L. Bragg (1890–1971), who derived it and who, together with his father W. H. Bragg (1862–1942), developed the theory and technique of X-ray diffraction by crystals in 1912–13. Thus, if the X-ray wavelength is known and the angle θ at which constructive interference occurs is measured, d can be obtained. This is the basis for X-ray crystallography.

Actual X-ray diffraction patterns are quite complicated. First of all, a crystal is a three-dimensional object, and X rays can be diffracted from different planes at different angles within the crystal, as shown in Fig. 37–26. Although the analysis is complex, a great deal can be learned about any substance that can be put in crystalline form. If the substance is not a single crystal but a mixture of many tiny crystals—as in a metal or a powder—then instead of a series of spots, as in Fig. 37–24, a series of circles is obtained, Fig. 37–27; each circle corresponds to diffraction of a certain order m (Eq. 37–23), from a particular set of parallel planes.

FIGURE 37–27

(a) Diffraction of X rays from a polycrystalline substance produces a set of circular rings as in (b) which is for polycrystalline sodium acetoacetate. (Part b Lester V. Bergman & Assoc., Inc.)

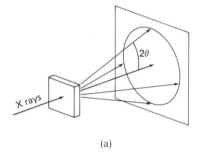

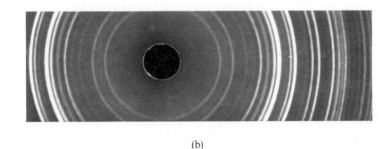

(a)

(b)

Summary

Diffraction refers to the fact that light, like other waves, bends around objects it passes and spreads out after passing through narrow slits. This bending gives rise to a diffraction pattern due to interference between rays of light that travel different distances. Light passing through a very narrow slit of width D will produce a pattern with a bright central maximum of half-width θ given by $\sin \theta = \lambda/D$, flanked by fainter lines to either side. The minima in the diffraction pattern occur at $D \sin \theta = m\lambda$ where $m = 1, 2, 3, \dots$, but not $m = 0$ (for which the pattern has its strongest maximum).

The intensity at any point in the single-slit diffraction pattern can be calculated using phasor diagrams. The same technique can be used to determine the intensity of the pattern produced by two slits; the pattern can be described as a series of maxima due to interference of light from the two slits, modified by an "envelope" due to diffraction at each slit.

The wave nature of light limits the sharpness or resolution of images. Because of diffraction, it is not possible to discern details smaller than the wavelength of the radiation being used.

A *diffraction grating* consists of many parallel slits or lines, each separated from its neighbors by a distance d. The peaks of constructive interference occur at angles θ given by $\sin \theta = m\lambda/d$, where $m = 0, 1, 2, \dots$. The peaks are much brighter and sharper for a diffraction grating than for the simple two-slit apparatus; the peak width is inversely proportional to the total number of lines in the grating.

A diffraction grating (or a prism) is used in a *spectroscope* to separate different colors or to observe *line spectra*, since for a given order m, θ depends on λ; precise determination of wavelength can be done with a spectroscope by careful measurement of θ.

1. Radio waves and light are both electromagnetic waves. Why can we hear a radio behind a hill when we cannot see the transmitting antenna?

2. Hold one hand close to your eye and focus on a distant light source through a narrow slit between two fingers. (Adjust your fingers to get best pattern.) Describe the pattern you see. Is this Fresnel or Fraunhofer diffraction?

3. A rectangular slit is twice as high as it is wide. Will the light spread more in the horizontal or in the vertical plane? Describe the pattern.

4. Explain why diffraction patterns are more difficult to observe with an extended light source than for a point source. Compare also a monochromatic source to white light.

5. For Fraunhofer diffraction by a single slit, what is the effect of increasing (a) the slit width, (b) the wavelength?

6. Describe the single-slit diffraction pattern when white light falls on a slit whose width is (a) 50 nm, (b) 50,000 nm.

7. What happens to the diffraction pattern of a single slit if the whole apparatus is immersed in (a) water, (b) a vacuum, instead of in air.

8. Discuss the similarities and differences between Fresnel and Fraunhofer diffraction by a single slit. Are the principles basically the same? How might the patterns differ?

9. Describe clearly the difference between the angles θ and β, as in Eq. 37–6.

10. In the single-slit Fraunhofer diffraction pattern, why does the first off-center maximum not occur at exactly $\sin \theta = \frac{3}{2}\lambda/D$?

11. Figure 37–9 shows a two-slit interference pattern for the case when d is larger than D. Can the reverse case occur, when d is less than D?

12. When both diffraction and interference are taken into account in the double slit experiment, discuss the effect of increasing (a) the wavelength, (b) the slit separation, (c) the slit width.

13. Discuss the similarities, and differences, of interference and diffraction.

14. Does diffraction limit the resolution of images formed by (a) spherical mirrors, (b) plane mirrors?

15. Do diffraction effects occur for virtual as well as real images?

16. What are the advantages (give at least two) for the use of large reflecting mirrors in astronomical telescopes.

17. Why can't a light microscope be used to observe molecules in a cell?

18. Atoms have diameters of about 10^{-8} cm. Can visible light be used to "see" an atom? Why or why not?

19. Which color of visible light would give the best resolution in a microscope?

20. If monochromatic light were used in a microscope, would the color affect the resolution? Explain.

21. Could a diffraction grating just as well be called an interference grating? Discuss.

22. For a diffraction grating, what is the advantage of (a) many slits, (b) closely spaced slits?

23. For light consisting of wavelengths between 400 nm and 700 nm, incident normally on a diffraction grating, for what orders (if any) would there be overlap? Does your answer depend on the slit width?

24. What is the difference in the interference patterns formed by two slits 10^{-4} cm apart and by a diffraction grating containing 10^4 lines/cm?

25. White light strikes (a) a diffraction grating and (b) a prism. A rainbow appears on a wall just below the direction of the incident beam in each case. What is the color of the top of the rainbow in each case?

26. Explain why there are tiny peaks between the main peaks produced by a diffraction grating illuminated with monochromatic light. Why are the peaks so tiny?

27. Why is a diffraction grating preferable to a prism for use in a spectroscope?

*28. Equation 37–22 tells us that the resolving power of a diffraction grating is greater the higher the order m. Yet Eq. 37–18 tells us that the width of the peaks is greater the higher the order. Aren't these contradictory? Explain.

 Problems

SECTION 37–1

1. (I) If 680-nm light falls on a slit 0.0245 mm wide, what is the angular width of the central diffraction peak?

2. (I) Monochromatic light of wavelength 589 nm falls on a slit. If the angle between the first bright fringes on either side of the central maximum is 33.0°, what is the slit width?

3. (I) For a given wavelength λ, what is the maximum slit width for which there will be no diffraction minima?

4. (II) How wide is the central diffraction peak on a screen 5.50 m behind a 0.101-mm-wide slit illuminated by 400-nm light?

5. (II) If parallel light falls on a single slit of width D at a 30° angle to the normal, describe the Fraunhofer diffraction pattern.

SECTION 37–2

6. (II) If you double the width of a single slit, the intensity of the light passing through the slit is doubled. (a) Show, however, that the intensity at the center of the screen increases by a factor of 4. (b) Explain why this does not violate conservation of energy.

7. (III) (a) Explain why the secondary maxima in the single-slit Fraunhofer diffraction pattern do not occur precisely

at $\beta/2 = (m + \frac{1}{2})\pi$ where $m = 1, 2, 3, \ldots$. (b) By differentiating Eq. 37–7, show that the secondary maxima occur when $\beta/2$ satisfies the relation $\tan(\beta/2) = \beta/2$. (c) Carefully and precisely plot the curves $y = \beta/2$ and $y = \tan \beta/2$. From their intersections, determine the values of β for the first and second secondary maxima. What is the percent difference from $\beta/2 = (m + \frac{1}{2})\pi$?

8. (III) Determine, approximately, the angular width at half maximum (where $I = \frac{1}{2}I_0$) of the central Fraunhofer diffraction peak for a single slit. (Hint: use graphical methods, or trial and error; the problem cannot be solved analytically.) To be concrete, assume $\lambda = 550$ nm and $D = 2.00 \times 10^{-3}$ mm.

SECTION 37–3

9. (II) Design a double-slit apparatus so that the central diffraction peak contains precisely fifteen fringes.

10. (II) If a double-slit pattern contains exactly eight fringes in the central diffraction peak, what can you say about the slit width and separation?

11. (II) Suppose $d = D$ in a double-slit apparatus, so that the two slits merge into one slit of width $2D$. Show that Eq. 37–9 reduces to the correct equation for single-slit diffraction.

12. (II) Two 0.010-mm-wide slits are 0.030 mm apart (center to center). Determine (a) the spacing between interference fringes for 550 nm light on a screen 1.0 m away and (b) the distance between the two diffraction minima on either side of the central maximum of the envelope.

13. (II) How many fringes are contained in the central diffraction peak for a double-slit pattern if (a) $d = 2.00D$, (b) $d = 12.0D$, (c) $d = 4.50D$, (d) $d = 7.20D$.

14. (III) Draw phasor diagrams (as in Fig. 37–7) for the double-slit experiment, including both interference and diffraction. Make a diagram for each of several crucial points in the pattern (see Fig. 37–9c) such as at the center, first minimum, next maximum, and at $\sin \theta = \lambda/D$.

SECTIONS 37–4 TO 37–6

15. (I) What is the angle of acceptance α of a microscope oil-immersion objective and its resolving power if $n = 1.60$ and the $NA = 1.35$? Use $\lambda = 500$ nm.

16. (I) What is the angular resolution limit set by diffraction for the 100-in Mt. Wilson telescope ($\lambda = 500$ nm)?

17. (II) A microscope objective is immersed in oil ($n = 1.60$) and accepts light scattered from the object up to $60°$ on either side of vertical. (a) What is the numerical aperture?

(b) What is the approximate resolution of the microscope if it uses 550-nm light?

18. (II) Two stars 10 light-years away are barely resolved by a 100-in (mirror diameter) telescope. How far apart are the stars? Assume $\lambda = 550$ nm.

19. (II) A certain sea organism has a pattern of dots on its surface with an average spacing of 0.40 μm. (a) If the specimen is viewed using 550-nm light, what minimum value must the numerical aperture be in order that the dots be resolved? (b) What minimum magnification is required to see the dots?

20. (II) (a) Can a normal human eye distinguish the two head-lights, 1.8 m apart, on a truck 10 km away? Consider only diffraction and assume an eye pupil diameter of 5.0 mm and a wavelength of 500 nm. (b) What is the maximum distance at which they can be resolved?

21. (II) Suppose that you wish to construct a telescope that can resolve features 10 km across on the moon, 384,000 km away. You have a 2.0-m-focal-length objective lens whose diameter is 10 cm. What focal-length eyepiece is needed if your eye can resolve objects 0.10 mm apart at a distance of 25 cm? What is the resolution limit set by the size of objective lens (that is, by diffraction)? Use $\lambda = 500$ nm.

22. (III) Derive Eq. 37–12a with the help of Fig. 37–28, which shows two object points, O and O', whose separation is s and whose image points I and I' are barely resolved. (O is on the axis of the lens.) (Hint: first show that the light waves diffracted from O' to the point I produce zero intensity at I and that therefore the path lengths $O'aI$ and $O'bI$ must differ by 1.22λ; then show, using a diagram, that $O'a$ is longer than Oa or Ob by ($s \sin \alpha$) and that $O'b$ is shorter than Oa or Ob by this amount.) Do not assume α is small. Note also that since the image distance is much greater than the object distance ($aI \gg Oa$), then $aI' \approx aI$.)

SECTIONS 37–7 AND 37–8

23. (I) At what angle will 840-nm light produce a third-order maximum when falling on a grating whose slits are 2.35×10^{-3} cm apart?

24. (I) How many lines per centimeter does a grating have if the third order occurs at a $12.0°$ angle for 650-nm light?

25. (I) A grating has 5000 lines/cm. How many spectral orders can be seen when it is illuminated by white light?

26. (I) The wings of a certain beetle have a series of parallel lines across them. When normally incident 560-nm light is

FIGURE 37–28

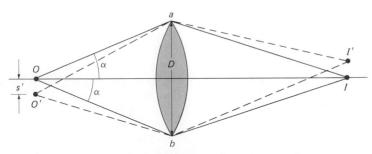

reflected from the wing, the wing appears bright when viewed at an angle of 48°. How far apart are the lines?

27. (I) A source produces first-order lines when incident normally on a 10,000-line/cm diffraction grating at angles 29.8°, 37.7°, 39.6°, and 48.9°. What are the wavelengths?

28. (II) White light containing wavelengths from 400 nm to 700 nm falls on a grating with 8000 lines/cm. How wide is the first-order spectrum on a screen 2.20 m away?

29. (II) How many lines per centimeter must a grating have if there is to be no second-order spectrum for any visible wavelength?

30. (II) Show that the second- and third-order spectra of white light produced by a diffraction grating always overlap.

31. (II) Two spectrum lines are measured by a 10,000 line/cm spectroscope at angles, on each side of center, of $+26°38'$, $+41°02'$ and $-26°18'$, $-40°27'$. What are the wavelengths?

32. (II) Suppose the angles measured in problem 31 were produced when the spectrometer (but not the source) was submerged in water. What then would be the wavelength?

33. (II) When yellow sodium light, $\lambda = 589$ nm, falls on a diffraction grating, its first-order peak on a screen 42.0 cm away falls 2.48 cm from the central peak. Another source produces a line 3.84 cm from the central peak. What is its wavelength?

34. (II) Light falling normally on a 10,000-line/cm grating is revealed to contain three lines in the first-order spectrum at angles of 31.2°, 36.4°, and 47.5°. What wavelengths are these?

35. (II) Monochromatic light falls on a transmission diffraction grating at an angle ϕ to the normal. Show that Eq. 37–14 for diffraction maxima must be replaced by

$$d(\sin \phi \pm \sin \theta) = m\lambda. \qquad m = 0, 1, 2, \ldots.$$

Explain the $\pm$ sign.

*SECTION 37–9

*36. (II) Missing orders occur for a diffraction grating when a diffraction minimum coincides with an interference maximum. Let D be the width of each slit and d the separation of slits and show (a) that if $d = 2D$, all even orders ($m = 2, 4, 6, \ldots$) are missing. (b) Show that there will be missing orders whenever

$$\frac{d}{D} = \frac{m_1}{m_2}$$

where m_1 and m_2 are integers. (c) Discuss the case $d = D$, the limit in which the space between slits becomes negligible.

*37. (II) Let 500-nm light be incident normally on a diffraction grating for which $d = 3.00D = 1200$ nm. (a) How many

orders (principal maxima) are present? (b) If the grating is 1.50 cm wide, what is the full angular width of each principal maximum?

*38. (II) For a diffraction grating, it is possible to show that the central maximum and the secondary peaks to either side (up to half the distance to the second principal peak) are at the same positions as those in the diffraction pattern of a single slit whose width equals the total width of the diffraction grating. Show this, with the aid of phasor diagrams.

*39. (III) (a) Derive an expression for the intensity in the interference pattern for three equally spaced slits. Express in terms of $\delta = 2\pi d \sin \theta/\lambda$ where d is the distance between adjacent slits and assume the slit width $D \approx \lambda$. (b) Show that there is only one secondary maximum between principal peaks.

*SECTION 37–10

*40. (II) How many lines per centimeter must a 4.0-cm-wide grating contain if it can just resolve two wavelengths, 4187.23 nm and 4187.41 nm in first order?

*41. (II) A 5500-line/cm diffraction grating is 3.6 cm wide. If 624-nm light falls on the grating, how close can two wavelengths be if they are to be resolved in any order? What order gives the best resolution?

*42. (II) A diffraction grating has 16,000 rulings in its 2.4 cm width. Determine (a) its angular dispersion in first and second orders, (b) its resolving power in first and second orders, and (c) the minimum wavelength resolution ($\Delta\lambda$) it can yield for $\lambda \approx 410$ nm.

*43. (II) For a fixed wavelength and diffraction angle, show that the resolving power of a diffraction grating depends on the total width, Nd, where N is the total number of slits each of width d.

*44. (II) Determine a formula for the minimum difference in frequency, Δf, that a diffraction grating can resolve when two frequencies, $f_1 \approx f_2 = f$, are incident on it.

*SECTION 37–11

*45. (II) X rays of wavelength 0.128 nm fall on a crystal whose atoms, lying in planes, are spaced 0.300 nm apart. At what angle must the X rays be directed if the first diffraction maximum is to be observed?

*46. (II) First-order Bragg diffraction is observed at 16.2° from a crystal with spacing between atoms of 0.24 nm. (a) At what angle will second order be observed? (b) What is the wavelength of the X rays?

*47. (II) If X-ray diffraction peaks corresponding to the first three orders ($m = 1, 2,$ and 3) are measured, can both the X-ray wavelength λ and lattice spacing d be determined? Prove your answer.

Polarization

38

In Chapters 36 and 37 we studied two aspects of light, interference and diffraction, that depend on its wave nature. In this chapter we examine another important and useful property of light: that it can be *polarized*. This phenomenon too can be explained if we assume light is a wave and, more precisely, a transverse wave.

 ## 38–1 Plane Polarization

To see what polarization of light means, let us first examine a simpler case, that of waves traveling on a rope. A rope may be set into vibration in a vertical plane as in Fig. 38–1a or in a horizontal plane as in part (b). In either case the wave is said to be **plane-polarized**—that is, the oscillations are in a plane.

If we now place an obstacle containing a vertical slit in the path of the wave, Fig. 38–2, a vertically polarized wave passes through, but a horizontally polarized wave will not. If a horizontal slit were used, the vertically polarized wave would be stopped. If both types of slit are used, neither wave could pass through. Note that polarization can exist *only* for *transverse waves*, and not for longitudinal waves. The latter vibrate only along the direction of motion and neither orientation of slit would stop them.

That light can be polarized was recognized only in the nineteenth century. However, even in Newton's time a phenomenon that depends on polarization was already known—namely that certain types of crystal, such as iceland spar, refract light into two rays; such crystals are called *doubly refracting* (Section 38–4). Certain other crystals, such as tourmaline, would not transmit one or the other of these two rays, depending on the orientation of the crystal. Today we recognize that the two rays in a doubly refracting crystal are plane-polarized in mutually perpendicular directions, and that the tourmaline acts as a "slit" to eliminate one or the other if properly oriented. However, it was not until after Young's and Fresnel's work in the early 1800s that this phenomenon was recognized as evidence that light is a transverse wave. A half century later, Maxwell's theory of light as electromagnetic (EM) waves was fully consistent with the facts of polarization

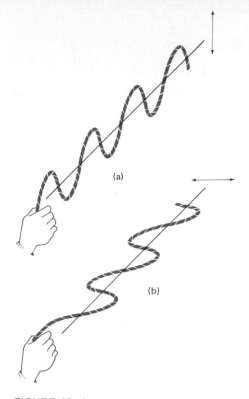

FIGURE 38-1

Transverse waves on a rope polarized (a) in a vertical plane and (b) in a horizontal plane.

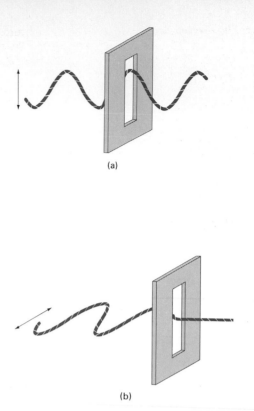

FIGURE 38-2

Vertically polarized wave passes through a vertical slit, but a horizontally polarized wave will not.

since an EM wave is a transverse wave. The plane of polarization in a plane-polarized EM wave is taken as the direction of the electric field vector (Fig. 33–7).

Light is not necessarily polarized. It can be *unpolarized*, which means that the source has vibrations in many planes at once, Fig. 38–3. An ordinary incandescent light bulb emits unpolarized light, as does the sun.

Plane-polarized light can be obtained from unpolarized light using certain crystals such as tourmaline. Or, more commonly today, we can use a *Polaroid sheet*. (Polaroid materials were invented in 1929 by Edwin Land.) A Polaroid sheet consists of complicated long molecules arranged with their axes parallel. Such a Polaroid acts as a series of parallel slits to allow one orientation of polarization to pass through nearly undiminished (this direction is called the *axis* of the Polaroid) whereas a perpendicular polarization is absorbed almost completely.[†] If a

FIGURE 38-3

Vibrations of the electric field vector in unpolarized light.

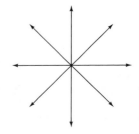

[†] How this occurs can be explained at the molecular level. An electric field **E** that oscillates parallel to the long molecules can set electrons into motion along the molecules, thus doing work on them and transferring energy; hence, if **E** is parallel to the "slits," it gets absorbed. An electric field **E** perpendicular to the long molecules (and thus perpendicular to the "slits") does not have this possibility of doing work and so passes through freely. Thus Fig. 38–2 does not apply for the **E** field of an EM wave (although it would for the **B** field). When we speak of the *axis* of a Polaroid, we mean the direction for which **E** is passed, so a Polaroid axis is perpendicular to the long molecules and the "slits" between them.

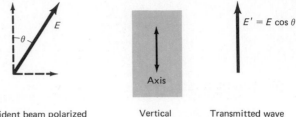

FIGURE 38–4

Vertical Polaroid transmits only the vertical component of a wave incident upon it.

Incident beam polarized at angle θ to the vertical and has amplitude E.

Vertical Polaroid

Transmitted wave

$E' = E \cos \theta$

Axis

beam of plane-polarized light strikes a Polaroid whose axis is at an angle θ to the incident polarization direction, the beam will emerge plane-polarized parallel to the Polaroid axis and its amplitude will be reduced by $\cos \theta$, Fig. 38–4. Thus a Polaroid passes only that component of polarization (the electric field vector, **E**) that is parallel to its axis. Since the intensity of a light beam is proportional to the square of the amplitude, we see that the intensity of a plane-polarized beam transmitted by a polarizer is

$$I = I_0 \cos^2 \theta, \qquad (38-1)$$

where θ is the angle between the polarizer axis and the plane of polarization of the incoming wave, and I_0 is the incoming intensity.[†]

A Polaroid can be used as a *polarizer* to produce plane-polarized light from unpolarized light, since only the component of light parallel to the axis is transmitted. A Polaroid can also be used as an *analyzer* to determine (1) if light is polarized and (2) what is the plane of polarization. A Polaroid acting as an analyzer will pass the same amount of light independent of the orientation of its axis if the light is unpolarized; try rotating one lens of a pair of Polaroid sunglasses while looking through it at a light bulb. If the light is polarized, however, when you rotate the Polaroid the transmitted light will be a maximum when the plane of polarization is parallel to the Polaroid's axis, and a minimum when perpendicular to it. If you do this while looking at the sky, preferably at right angles to the sun's direction, you will see that skylight is polarized. (Direct sunlight is unpolarized, but don't look directly at the sun, even through a polarizer, for damage to the eye may occur.) If the light transmitted by an analyzer Polaroid falls to zero at one orientation, then the light is 100 percent plane-polarized. If it merely reaches a minimum, the light is *partially polarized*.

Unpolarized light consists of light with random directions of polarization (electric field vector). Each of these polarization directions can be resolved into components along two mutually perpendicular directions. Thus, an unpolarized beam can be thought of as two plane-polarized beams of equal magnitude perpendicular to one another. When two Polaroids are *crossed*—that is, their axes are perpendicular to one another—unpolarized light can be entirely stopped (or nearly so—Polaroids are not quite perfect). As shown in Fig. 38–5, unpolarized light is made plane-polarized by the first Polaroid (the polarizer). The second Polaroid, the analyzer, then eliminates this component since its axis is perpendicular to the first. You can try this with Polaroid sunglasses. It should be clear that Polaroid sunglasses eliminate 50 percent of unpolarized light because of their polarizing property; they absorb even more because they are colored.

■ EXAMPLE 38–1 Unpolarized light passes through two Polaroids; the axis of one is vertical and that of the other is at 60° to the vertical. What is the orientation and intensity of the transmitted light?

[†] Equation 38–1 is often referred to as "Malus' law," after Etienne Malus, a contemporary of Fresnel.

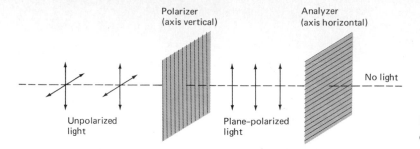

Polarizer
(axis vertical)

Analyzer
(axis horizontal)

No light

Unpolarized
light

Plane-polarized
light

FIGURE 38–5

Crossed Polaroids completely eliminate light.

SOLUTION The first Polaroid eliminates half the light so the intensity is reduced by half: $I_1 = \frac{1}{2}I_0$. The light reaching the second polarizer is vertically polarized and so is reduced in intensity (Eq. 38–1) by

$$I_2 = I_1 (\cos 60°)^2 = \tfrac{1}{4}I_1.$$

Thus $I_2 = \frac{1}{8}I_0$. The transmitted light has an intensity one-eighth that of the original and is plane-polarized at a 60° angle to the vertical.

 ## 38–2 Polarization by Reflection

Another means of producing polarized light from unpolarized light is by reflection. When light strikes a nonmetallic surface at any angle other than perpendicular, the reflected beam is polarized preferentially in the plane parallel to the surface, Fig. 38–6. In other words, the component with polarization in the plane perpendicular to the surface is preferentially transmitted or absorbed. You can check this fact by rotating Polaroid sunglasses while looking through them at a flat surface of a lake or road. Since most outdoor surfaces are horizontal, Polaroid sunglasses are made with their axis vertical to eliminate the stronger component, and thus reduce glare. This is well known by fishermen who wear Polaroids to eliminate reflected glare from the surface of a lake or stream and thus see beneath the water more clearly.

The amount of polarization in the reflected beam depends on the angle; it varies from no polarization at normal incidence to 100 percent polarization at an

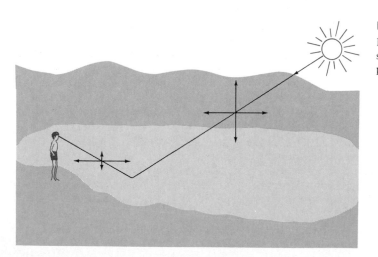

FIGURE 38–6

Light reflected from a nonmetallic surface, such as the smooth surface of water in this lake, is partially polarized parallel to the surface.

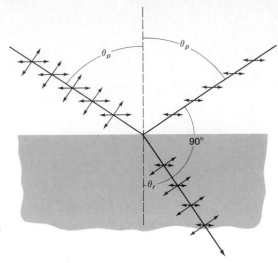

FIGURE 38–7

At θ_p the reflected light is plane-polarized parallel to the surface, and $\theta_p + \theta_r = 90°$, where θ_r is the refraction angle.

angle known as the *polarizing angle*, θ_p.[†] This angle is related to the index of refraction of the two materials on either side of the boundary by the equation

$$\tan \theta_p = \frac{n_2}{n_1}, \tag{38–2a}$$

where n_1 is the index of refraction of the material in which the beam is traveling, and n_2 is that of the medium beyond the reflecting boundary. If the beam is traveling in air, $n_1 = 1$ and

$$\tan \theta_p = n. \tag{38–2b}$$

The polarizing angle θ_p is also called *Brewster's angle*, and Eqs. 38–2 *Brewster's law*, after the Scottish physicist David Brewster (1781–1868) who worked it out experimentally in 1812. Equations 38–2 can be derived from the electromagnetic wave theory of light. It is interesting that at Brewster's angle, the reflected and transmitted rays make a 90° angle to each other; that is, $\theta_p + \theta_r = 90°$, Fig. 38–7. This can be seen by substituting Eq. 38–2a, $n_2 = n_1 \tan \theta_p = n_1 \sin \theta_p / \cos \theta_p$, into Snell's law, $n_1 \sin \theta_p = n_2 \sin \theta_r$, and get $\cos \theta_p = \sin \theta_r$ which can only hold if $\theta_p = 90° - \theta_r$.

  EXAMPLE 38–2 (*a*) At what incident angle is sunlight reflected from a lake plane-polarized? (*b*) What is the refraction angle?

SOLUTION (*a*) We use Eq. 38–2b with $n = 1.33$ so $\tan \theta_p = 1.33$ and $\theta_p = 53.1°$.
(*b*) $\theta_r = 90.0° - \theta_p = 36.9°$

✱ 38–3 Optical Activity

When a beam of plane-polarized light is passed through crystals and solutions, it is found that the plane of polarization is rotated through an angle. For example, Fig. 38–8 shows light passing through a polarizer and then through a sugar solution. The analyzer Polaroid behind the solution does not cut out all the light when

---

[†] Although all of one component is transmitted into a transparent medium at this angle, some of the other component is transmitted as well, so the transmitted beam is *not* 100 percent polarized.

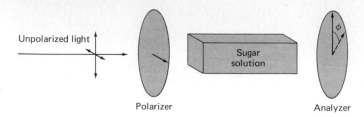

Unpolarized light

Polarizer

Sugar solution

Analyzer

FIGURE 38–8

Sugar solution rotates the plane of polarization of incident light so that it is no longer horizontal but at an angle ϕ. The analyzer thus extinguishes the light when rotated ϕ from the vertical (crossed) position.

placed at 90° to the polarizer. If the analyzer is rotated through an angle ϕ, however, there is no transmitted light; this indicates that the plane of polarization has been rotated through an angle ϕ by the intervening substance. Such substances are said to be *optically active*. Optical activity is due to asymmetry of the molecules, which may have a spiral shape as, for example, some proteins do. Substances which rotate the polarization plane to the right as viewed along the direction of the beam as in Fig. 38–8 are called *dextrorotatory* (also called *right-handed* or *positive*). Those that rotate it to the left are called *levorotatory* (*left-handed*, or *negative*). The most common sugar, dextrose or D-glucose, is dextrorotatory. Most naturally occurring amino acids and proteins are levorotatory.

The angle of rotation ϕ depends on the path length l (meters) through the substance and on the concentration c (kg/m^3) if it is in solution. For dilute solutions, this is a linear relationship and ϕ (in radians) is given by

$$\phi = \alpha l c.$$

The constant α is a property of the substance and is called the *specific rotation* or the *specific optical rotatory power* (it depends on the temperature and the wavelength of light used).

Since ϕ is proportional to the concentration, optical activity is a standard method to measure concentrations of solutions such as for sugars. It is also helpful in determining the three-dimensional shape of large molecules, such as proteins, or their change in shape when conditions are changed.

Glass and plastics become optically active when put under stress; the plane of polarization is rotated most where the stress is greatest. Models of bones or machine parts made of plastic can be observed between crossed Polaroids to determine where the points of greatest stress lie. This is called "optical stress analysis."

*38–4   Double Refraction: Birefringence and Dichroism

In many transparent materials, the speed of light is the same in all directions. Such materials are said to be *isotropic*. In some crystals and solutions, however, the speed of light is different in different directions, and these are referred to as being *anisotropic*. Such substances are also said to be *doubly refracting* or *birefringent* because they give rise to an unusual phenomenon.

There is one particular direction in a birefringent crystal, such as calcite, known as its *optic axis* (this is not a single line, but a direction in the crystal). If ordinary light enters a crystal along its optic axis nothing abnormal is noted. But if unpolarized light falls at an angle to the optic axis as in Fig. 38–9, a very unusual thing happens: there are two refracted rays. In the case shown, the incident ray is normal to the surface and the optic axis lies in the plane shown. One refracted ray, called the *ordinary ray* (o) passes straight through in a normal way. But the other ray, called the *extraordinary ray* (e), is refracted at an angle.

As can be seen, Snell's law does not hold in this case for the *e* ray. It does hold, however, for the *o* ray.

The *e* ray and *o* ray are found to be plane-polarized in mutually perpendicular directions. This is indicated in Fig. 38–9 by the dots on the *o* ray indicating

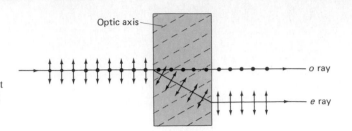

FIGURE 38–9

Unpolarized light, incident normally on a birefringent crystal such as calcite, is separated into two refracted beams. The dashed lines indicate the direction of the optic axis.

TABLE 38–1

Principal indices of refraction for some doubly refracting crystals ($\lambda = 589$ nm)

Crystal	n_o	n_e
Ice	1.309	1.313
Quartz	1.544	1.553
Calcite	1.658	1.486
Dolomite	1.681	1.500

the vibrations are perpendicular to the paper; and the polarization of the e ray is indicated by the short arrows.

The phenomenon of double refraction can be explained if we make the following assumption: that the speed of light is different depending on the orientation of the polarization vector with respect to the crystal's optic axis. As can be seen in Fig. 38–9 the polarization of the o ray is perpendicular to the optic axis; so its speed will be the same in all directions, as long as its polarization remains perpendicular to the optic axis. The e ray, on the other hand, has components of polarization both parallel and perpendicular to the optic axis, and it therefore travels with a different speed in different directions; it travels with the same speed as the o ray if its polarization vector is perpendicular to the optic axis; at other angles it has a higher speed (or, in certain crystals, a lower speed) which reaches a maximum (or minimum) when its polarization is parallel to the optic axis. (In what follows, we assume a crystal in which this speed is higher.) Thus the index of refraction n_o for the ordinary ray is the same for all directions whereas that for the extraordinary ray, n_e, depends on angle; usually n_e is specified for the e ray traveling perpendicular to the optic axis where its polarization direction is parallel to the optic axis. These principal values for n_o and n_e are given in Table 38–1 for several crystals.

Now we use the above assumptions, plus Huygens' principle, to explain double refraction. In Fig. 38–10a examine the ordinary wave. Huygens' wavelets are shown for two points on the wavefront just as it has entered the crystal. Since these wavelets travel at the same speed in all directions, they are spheres just as before; and the wave continues moving straight. For the e ray, shown in Fig. 38–10b, the Huygens' wavelets are elliptically shaped since the speed depends on direction. Ray 1 on this wavelet has polarization perpendicular to the optic axis, so is traveling at the same speed as the o ray; but rays 2, 3, and 4 are traveling at higher speeds, with 4 having the maximum. (This is the case for calcite; for a crystal like ice or quartz, Table 38–1, rays 2, 3, and 4 would be slower.) The advancing wavefront, which is tangent to the Huygens' wavelets, will thus be displaced to one side as shown. And this is the path taken by the e ray.

FIGURE 38–10

Huygens' principle used to explain double refraction.

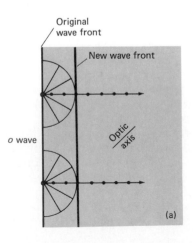

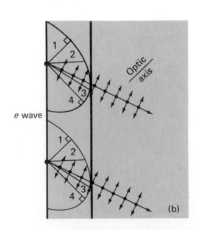

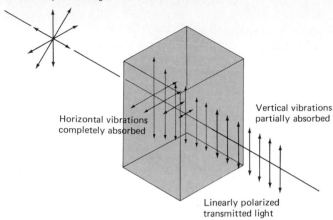

FIGURE 38–11
Dichroic crystal absorbs one polarized component
more than the other. If the crystal is thick enough,
the transmitted light is plane-polarized.

Some birefringent crystals, such as tourmaline, absorb one of the polarized
components more strongly than the other, Fig. 38–11. Such a crystal is said to
show *dichroism*. If the crystal is sufficiently thick, one component of unpolarized
light will be completely eliminated so the emerging light will be plane-polarized.
Dichroism is the basic principle behind Polaroid sheets.

* 38–5   Circular and Elliptic Polarization

Suppose a plane-polarized beam of light is incident on a birefringent crystal
(such as calcite) whose optic axis is parallel to the face of the crystal and makes
a 45° angle to the plane of polarization, as in Fig. 38–12. The incident electric
field vector **E** can be resolved into components parallel and perpendicular to
the optic axis, as shown, corresponding to *e* and *o* waves. These have equal
amplitudes, and they travel through the crystal along the same path. (Note that
the *e* wave is not refracted, as in Fig. 38–9, since its vibration is parallel to

FIGURE 38–12
(a) Wave polarized at 45° to optic axis
enters birefringent quarter-wave plate
normal to surface and to optic axis.
The *o* wave (E_o) is retarded by
one-quarter wavelength behind the e
wave (E_e) upon emergence. The
polarization vector (**E**) of the
emergent wave rotates in time as can
be seen by examining the components
for points 1, 2, 3, and 4 in (a) which
are redrawn in (b) looking along the
direction of wave travel. Thus the light
is circularly polarized (c).

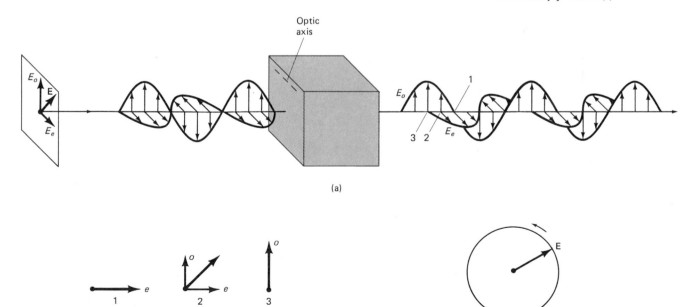

the optic axis.) However, the two waves travel at different speeds (since they travel perpendicular to the optic axis) and so emerge from the crystal out of phase. If the phase difference is 90°, the crystal is called a *quarter-wave plate* (since the two waves are shifted by one-quarter of a wavelength). This is the case shown in Fig. 38–12. The interesting result is that the polarization vector (the total electric field **E**, the sum of the two waves) is rotating in a circle of constant amplitude, as shown in parts (b) and (c) of the figure. Such light is called *circularly polarized*.

EXAMPLE 38–3 To make a quarter-wave plate of calcite for light of wavelength 589 nm, how thick should it be?

SOLUTION From Table 38–1, $n_e = 1.486$ and $n_o = 1.658$. In a thickness t of crystal, the number of wavelengths for the extraordinary ray will be

$$N_e = \frac{t}{\lambda_e} = \frac{tn_e}{\lambda}$$

where λ is the wavelength in air and $\lambda_e = \lambda/n_e$ is the wavelength in the crystal. For the ordinary ray, the number of wavelengths N_o is

$$N_o = \frac{tn_o}{\lambda}.$$

Since we want the waves to be one-quarter wavelength out of phase, we set $N_o - N_e = \frac{1}{4}$:

$$\frac{1}{4} = N_o - N_e = \frac{t}{\lambda}(n_o - n_e).$$

We solve for t and put in numbers:

$$t = \frac{\lambda}{4(n_o - n_e)} = \frac{589 \text{ nm}}{4(1.658 - 1.486)} = 856 \text{ nm}.$$

FIGURE 38–13

Unpolarized sunlight scattered by molecules of the air. An observer at right angles sees plane-polarized light, since the component of vibration along the line of sight emits no light along that line.

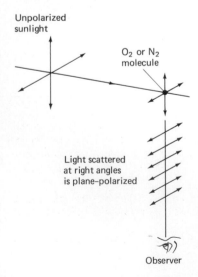

Unpolarized sunlight

O₂ or N₂ molecule

Light scattered at right angles is plane-polarized

Observer

If a birefringent crystal changes the phase by 180° (a *half-wave plate*), a diagram similar to Fig. 38–12 will show that the emerging wave will be plane-polarized. Indeed, if the crystal thickness is such that the phase difference is 0, π, 2π, or any integral multiple of π, the emerging beam will be plane-polarized. If the phase difference is $\pi/2$, $3\pi/2$, $5\pi/2$, and so on, the light will be circularly polarized. For all other phase differences, the emerging light will be *elliptically polarized*, meaning that the electric (polarization) vector rotates as in Fig. 38–12c, but its length (amplitude of **E**) varies so the path of its tip is an ellipse.

Up to now we assumed the incident wave has **E** making a 45° angle to the optic axis. But if the polarization vector of the incident beam makes an angle other than 45° to the optic axis (and not 0° or 90°) the emerging light for a quarter-wave plate (phase difference = 90°) will be elliptical; this is easy to see, since the two components (*e* and *o*) have unequal amplitudes.

* 38–6   Scattering of Light by the Atmosphere

Sunsets are red, the sky is blue, and skylight is polarized (at least partially). These phenomena can be explained on the basis of the *scattering* of light by the molecules of the atmosphere. In Fig. 38–13 we see unpolarized light from the sun impinging on a molecule of the earth's atmosphere. The electric field of the EM wave sets the electric charges within the molecule into motion, and the molecule absorbs some of the incident radiation. But it quickly reemits this light since the charges are oscillating. As discussed in Section 33–4, oscillating electric charges produce EM waves. The electric field of these waves is in the plane that includes the line of oscillation. The intensity is strongest along a line perpendicular to the oscillation, and drops to zero along the line of oscillation (Fig. 33–6). In Fig. 38–13 the motion of the charges is resolved into two components. An observer at right angles to the direction of the sunlight, as shown, will see plane-polarized light since no light is emitted along the line of the other component of the oscillation. (Another way to see this is to note that when viewing along the line of oscillation, one doesn't see the oscillation, and hence sees no waves made by it.) At other viewing angles, both components will be present; one will be stronger, however, so the light appears partially polarized. Thus, the process of scattering explains the polarization of skylight. (It can also explain complete polarization by reflection, Fig. 38–7. At Brewster's angle, as we saw, the angle between the reflected and refracted ray is 90°; if we think of the molecules of the medium oscillating perpendicular to the direction of the refracted ray, we can see that at 90° to this direction [the direction of the reflected ray] there will be only one component of polarization [Fig. 38–7] just as for scattering [Fig. 38–13].)

Scattering of light by the earth's atmosphere depends on λ. For particles much smaller than the wavelength of light (such as molecules of air), the particles will be less of an obstruction to long wavelengths than to short ones. The scattering decreases, in fact, as $1/\lambda^4$. Red and orange light is thus scattered much less than blue and violet, which is why the sky looks blue. At sunset, the sun's rays pass through a maximum length of atmosphere. Much of the blue has been taken out by scattering. The light that reaches the surface of the earth and reflects off clouds and haze is thus lacking in blue which is why sunsets appear reddish.

The dependence of scattering on $1/\lambda^4$ is valid only if the scattering objects are much smaller than the wavelength of the light. This is valid for oxygen and nitrogen molecules whose diameters are about 0.2 nm. Clouds, however, contain water droplets or crystals that are much larger than λ; they scatter all frequencies of light nearly uniformly. Hence clouds appear white (or gray, if shadowed).

 ## Summary

In *unpolarized light* the electric field vectors vibrate at all angles. If the electric vector vibrates only in one plane the light is said to be *plane-polarized*. Light can also be partially polarized. When an unpolarized light beam passes through a Polaroid sheet or a doubly refracting crystal, the emerging beams are plane-polarized. That a light beam is polarized can be determined experimentally by the fact that when it passes through a Polaroid, the intensity varies as the Polaroid is rotated. Thus a Polaroid can act as polarizer or as analyzer. Light can also be partially or fully polarized by reflection; if light traveling in air is reflected from a medium of index of refraction n, it will be completely plane-polarized if the incident angle θ_p is given by $\tan \theta_p = n$. That light can be polarized shows that it must be a transverse wave.

 ## Questions

1. What does polarization tell us about the nature of light?

2. What is the difference between a polarizer and an analyzer?

3. How can you tell if a pair of sunglasses is polarizing or not?

4. Sunlight will not pass through two Polaroids whose axes are at right angles. What happens if a third Polaroid, with axis at 45° to each of the other two, is placed between them?

*5. For which of the materials in Table 38–1 is the speed of the extraordinary ray less when its polarization vector is parallel to the optic axis than when it is perpendicular?

*6. What would be the color of the sky if the earth had no atmosphere?

*7. If the earth's atmosphere were 50 times denser than it is, would sunlight still be white, or would it be some other color?

*8. In a birefringent material, does the e ray always travel at the speed c/n_e?

*9. Of the materials listed in Table 38–1, which would make the thickest quarter-wave plate? Which the thinnest?

*10. How could you determine the direction of the optic axis for a quarter-wave plate?

*11. How could you determine if a given beam of light was circularly polarized or not?

*12. Could a sound wave be circularly polarized? Plane-polarized? What about a microwave beam?

 Problems _____

SECTION 38–1

1. (I) Two polarizers are oriented at 45° to one another. Unpolarized light falls on them. What fraction of the light intensity is transmitted?

2. (I) Two Polaroids are aligned so that the light passing through them is a maximum. At what angle should one of them be placed so the intensity is reduced by half?

3. (II) At what angle should the axes of two Polaroids be placed so as to reduce the intensity of the incident unpolarized light to (a) $\frac{1}{3}$, (b) $\frac{1}{10}$?

4. (II) Unpolarized light passes through five successive Polaroid sheets each of whose axis makes a 45° angle with the previous one. What is the intensity of the transmitted beam?

5. (II) Two polarizers are oriented at 34.0° to one another. Light polarized at a 17.0° angle to each polarizer passes through both. What reduction in intensity takes place?

6. (II) Show that if two equally intense sources of light produce light that is plane-polarized, but with their planes of polarization perpendicular to each other, then they cannot produce an interference pattern even if they are in phase at all moments.

7. (II) Unpolarized light falls on two polarizer sheets whose axes are at right angles. (a) What fraction of the incident light intensity is transmitted? (b) What fraction is transmitted if a third polarizer is placed between the first two so that its axis makes a 45° angle with each of their axes? (c) What if the third polarizer is in front of the other two?

8. (II) Describe how to rotate the plane of polarization of a plane-polarized beam of light by 90° and produce only a 10 percent loss in intensity using "perfect" polarizers.

9. (II) The percent polarization P of a partially polarized beam of light is defined as

$$P = \frac{I_{max} - I_{min}}{I_{max} + I_{min}} \times 100$$

where I_{max} and I_{min} are the maximum and minimum intensities that are obtained when the light passes through a polarizer that is slowly rotated. Such light can be considered as the sum of two unequal plane-polarized beams of intensities I_{max} and I_{min} perpendicular to each other. Show that the light transmitted by a polarizer, whose axis makes an angle ϕ to the direction in which I_{max} is obtained, has intensity

$$\frac{1 + p \cos 2\phi}{1 + p} I_{max}$$

where $p = P/100$ is the "fractional polarization."

SECTION 38–2

10. (I) What is Brewster's angle for an air-glass ($n = 1.56$) surface?

11. (I) What is Brewster's angle for a piece of glass ($n = 1.56$) submerged in water?

12. (I) At what angle above the horizon is the sun when light reflecting off a smooth lake is polarized most strongly?

13. (II) What would Brewster's angle be for reflections off the surface of water beneath the surface? Compare to the angle for total internal reflection, and to Brewster's angle from above the surface.

*SECTION 38–4

*14. (II) A region of a birefringent biological specimen is 1.65×10^{-3} cm thick with indices of refraction 1.322 and 1.331 for light whose wavelength in air is 580 nm. What phase difference arises between the two rays after passing through this region?

*15. (II) A narrow beam of unpolarized light strikes the face of a flat quartz crystal 2.00 cm thick at a 44.5° angle to the face. The optic axis is parallel to the face and perpendicular to the beam. (a) Calculate the distance between the ordinary and extraordinary ray when they emerge from the crystal. (Snell's law is valid for both rays, for beams traveling in the plane perpendicular to the optic axis.) (b) What is the polarization state of each emerging beam?

*SECTION 38–5

*16. (I) What minimum thickness of quartz will be a quarter-wave plate for a wavelength of 660 nm?

*17. (II) A beam of circularly polarized light, of intensity I, is incident on a Polaroid. Describe the nature of the transmitted light, and its intensity.

*18. (II) Describe the transmitted light when the light incident on a half-wave plate is (a) unpolarized, (b) plane-polarized, (c) circularly polarized. Give reasons for your answers.

*19. (II) What is the minimum thickness of a thin flat calcite crystal if 650-nm plane-polarized light, incident normally on one face of the crystal, is to be plane-polarized when it emerges? The optic axis is parallel to the faces of the crystal.

*20. (II) Show that a plane-polarized beam of light can be represented as the sum of two circularly polarized components.

*21. (II) Use a diagram to show that if the phase difference for the emerging waves in the setup of Fig. 38–12a is an even integer multiple of π (that is, 0, 2π, 4π, and so on), the plane of polarization is the same as that of the incident light; but if the phase difference is an odd multiple of π (π, 3π, 5π, and so on) the plane of polarization is perpendicular to the original.

*22. (III) (a) Use a diagram to show that if the phase difference between the waves emerging from the crystal of Fig. 38–12a is not an integral multiple of $\pi/2$, it will be elliptically polarized. (b) Describe the angle the principal axis makes with the incident polarization direction.

*23. (III) Suppose the polarization vector of the incident wave in Fig. 38–12a makes an angle with the optic axis that is not 45° (say it is 60°). Describe the emerging light for (a) a quarter-wave plate, (b) a half-wave plate.

The Special Theory of Relativity

39

Physics at the end of the nineteenth century looked back on a period of great progress. The theories developed over the preceding three centuries had been very successful in explaining a wide range of natural phenomena. Newtonian mechanics beautifully explained the motion of objects on earth and in the heavens; furthermore, it formed the basis for successful treatments of fluids, wave motion, and sound. Kinetic theory, on the other hand, explained the behavior of gases and other materials. And Maxwell's theory of electromagnetism not only brought together and explained electric and magnetic phenomena, but it predicted the existence of EM waves that would behave in every way just like light—so light came to be thought of as an electromagnetic wave. Indeed, it seemed that the natural world, as seen through the eyes of physicists, was very well explained; a few puzzles remained, and it was felt that these would soon be explained using already known principles.

But it did not turn out so simply. Instead, these puzzles were only to be solved by the introduction, in the early part of the twentieth century, of two revolutionary new theories that changed our whole conception of nature: the *theory of relativity* and *quantum theory*.

Physics as it was known at the end of the nineteenth century (what we've covered up to now in this book) is referred to as *classical physics*. The new physics that grew out of the great revolution at the turn of the twentieth century is now called *modern physics*. In this chapter we present the special theory of relativity, which was first proposed by Albert Einstein in 1905. In the following chapters we discuss the equally momentous quantum theory and then nuclear and elementary particle physics.

 ## 39–1 Galilean–Newtonian Relativity

Einstein's special theory of relativity deals with how we observe events, particularly how objects and events are observed from different frames of reference.[†]

[†] A reference frame is a set of coordinate axes fixed to some body (or group of bodies) such as the earth, a train, the moon, and so on. See Section 2–2.

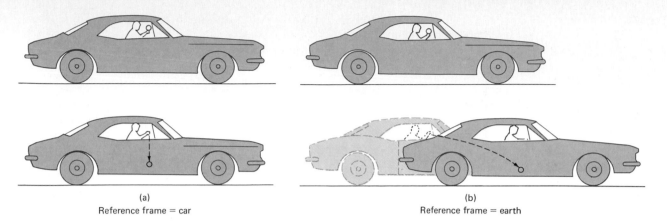

This had, of course, already been discussed by Galileo and Newton, and we first briefly discuss these earlier ideas.

We will mainly deal with so-called **inertial reference frames**. As discussed in Section 4–5, an inertial reference frame is one in which Newton's first law, the law of inertia, is valid. That is, if an object experiences no net force due to other bodies, the object either remains at rest or remains in motion with constant velocity in a straight line. Rotating, or otherwise accelerating frames of reference are noninertial frames,[†] and we won't be concerned with them here. The earth is not quite an inertial frame (it rotates). But it is close enough that for most purposes we can consider it an inertial frame.

A reference frame that moves with constant velocity with respect to an inertial frame is itself also an inertial frame.

Both Galileo and Newton were deeply aware of what we now call the *relativity principle*: *that the basic laws of physics are the same in all inertial reference frames.* You may have recognized its validity in everyday life: for example, objects move in the same way in a smoothly moving (constant velocity) train or airplane as they do on earth. (This assumes no vibrations or rocking— for they would make the reference frame noninertial.) When you walk, drink a cup of soup, play Ping-Pong, or drop a pencil on the floor, the bodies move just as they do on earth. Suppose you are in a car traveling rapidly along at constant velocity. If you release a coin from above your head inside the car, how will it fall? It falls straight downward and hits the floor directly below the point of release, Fig. 39–1a. (If you drop the object out the car's window, this won't happen because the moving air drags the object backward.) This is just how objects fall on the earth—straight down—and thus our experiment in the moving car is in accord with the relativity principle.

Note in this example, however, that to an observer on the earth the object follows a curved path, Fig. 39–1b. The actual path followed by the object is different as viewed from different frames of reference. This does not violate the relativity principle, because this principle states that the *laws* of physics are the same in all inertial frames. The same law of gravity, and the same laws of motion, apply in both reference frames. The difference in Figs. 39–1a and b is that in the earth's frame of reference, the coin has an initial velocity (equal to that of the car); the laws of physics therefore predict it will follow a parabolic path like any projectile. In the car's reference frame, there is no initial velocity and the laws of physics predict it will fall straight down. The laws are the same in both reference frames although the specific paths are different.[‡]

[†] On a rotating platform, for example, (say a merry-go-round) an object at rest starts moving outward even though no force is exerted on it. This is therefore not an inertial frame.

[‡] Galileo, in his great book *Dialogues on the Two Chief Systems of the World*, described a similar experiment and predicted the same results. Galileo's example involved a sailor dropping a knife from the top of the mast of a sailing vessel. If the vessel moves at constant speed, where will the knife hit the deck (ignoring earth's rotation)?

FIGURE 39–1

A coin is dropped by a person in moving car. (a) In the reference frame of the car, the coin falls straight down. (b) In a reference frame fixed on the earth, the coin follows a curved (parabolic) path. The upper views show the moment of the coin's release, and the lower views a short time later.

Galilean–Newtonian relativity involves certain unprovable assumptions that make sense from everyday experience. It is assumed that the lengths of objects are the same in one reference frame as in another, and that time passes at the same rate in different reference frames. In classical mechanics, space and time are considered to be *absolute*: their measurement doesn't change from one reference frame to another. The mass of an object, as well as all forces, are assumed to be unchanged by a change in inertial reference frame.

The position of an object is, of course, different when specified in different reference frames, and so is velocity. For example, a person may walk toward the front of a bus with a speed of 5 km/h; but if the bus moves 40 km/h with respect to the earth, the person's speed is then 45 km/h with respect to the earth. The acceleration of a body, however, is the same in any inertial reference frame according to classical mechanics. This is because the change in velocity, and the time, will be the same. For example, the person in the bus may accelerate from 0 to 5 km/h in 1.0 seconds, so $a = 5$ km/h/s in the reference frame of the bus. With respect to the earth, the acceleration is (45 km/h − 40 km/h)/ (1.0 s) = 5 km/h/s, which is the same.

Since neither F, m, nor a changes from one inertial frame to another, then Newton's second law, $F = ma$ does not change. Thus it satisfies the relativity principle. It is easily shown that the other laws of mechanics also satisfy the relativity principle.

That the laws of mechanics are the same in all inertial reference frames implies that no one inertial frame is special in any sense; we express this important conclusion by saying that *all inertial reference frames are equivalent* for the description of mechanical phenomena. No one inertial reference frame is any better than another. A reference frame fixed to a car or an aircraft traveling at constant velocity is as good as one fixed on the earth. When you travel smoothly at constant velocity in a car or airplane, it is just as valid to say you are at rest and the earth is moving as it is to say the reverse. There is no experiment one can do to tell which frame is "really" at rest and which is moving. Thus there is no way to single out one particular reference frame as being at absolute rest.

The situation changed somewhat in the last half of the nineteenth century. When Maxwell presented his comprehensive and very successful theory of electromagnetism (Chapter 33), he showed that light can be considered as an electromagnetic wave. Maxwell's equations predicted that the velocity of light c would be 3.00×10^8 m/s; and this is just what is measured within experimental error. The question then arose: in what reference frame does light have precisely the value predicted by Maxwell's theory? For it was assumed that light would have a different speed in different frames of reference. For example, if an observer were traveling on a rocket ship at a speed of 1.0×10^8 m/s toward a source of light, we might expect them to measure the speed of the light reaching them to be 3.0×10^8 m/s + 1.0×10^8 m/s = 4.0×10^8 m/s. But Maxwell's equations have no provision for relative velocity. They merely predicted the speed of light to be $c = 3.0 \times 10^8$ m/s. This seemed to imply there must be some special reference frame where c would have this value.

As we discussed in Chapters 15 and 16, waves travel on water and along ropes or strings, and sound waves travel in air and other materials. Since nineteenth-century physicists viewed the material world in terms of the laws of mechanics, it was natural for them to assume that light too must travel in some *medium*; they called this transparent medium the *ether* and assumed it permeated all space.[†]

[†] The medium for light waves could not be air, since light travels from the sun to earth through nearly empty space. Therefore another medium was postulated, the ether. The ether was not only transparent but, because of difficulty in detecting it, was assumed to have zero density. The properties of the ether were thus similar to caloric (the mysterious substance earlier assumed to be heat) whose existence had been rejected.

It was therefore presumed that the velocity of light given by Maxwell's equations must be with respect to this ether.

However, it appeared that Maxwell's equations (Chapter 33) did *not* satisfy the relativity principle. They were not the same in all inertial reference frames. They were simplest in the frame where $c = 3.00 \times 10^8$ m/s; that is, in a reference frame at rest in the ether. In any other reference frame, extra terms would have to be added to take into account the relative velocity. Thus, although most of the laws of physics obeyed the relativity principle, the laws of electricity and magnetism (or "electrodynamics") apparently did not. Instead, they seemed to single out one reference frame that was better than any other—a reference frame that could be considered absolutely at rest.

Scientists soon set out to determine the speed of the earth relative to this absolute frame whatever it might be. A number of clever experiments were designed to do this. The most direct were performed by A. A. Michelson and E. W. Morley. The details of their experiment are discussed in the next section. Briefly, what they did was measure the difference in the speed of light in different directions. They expected to find a difference depending on the orientation of their apparatus with respect to the ether. For just as a boat has different speeds relative to the ground when it moves upstream, downstream, or across the stream, so too light would be expected to have different speeds depending on the velocity of the ether past the earth.

Strange as it may seem, they detected no difference at all. This was a great puzzle. A number of explanations were put forth over a period of years; but they led to contradictions or were otherwise not generally accepted.

Then in 1905, Albert Einstein proposed a radical new theory that reconciled these many problems in a simple way. But at the same time, as we shall soon see, it completely changed our ideas of space and time.

* 39–2 The Michelson–Morley Experiment[†]

The Michelson–Morley experiment was designed to measure the speed of the *ether*—the medium in which light was assumed to travel—with respect to the earth. The experimenters thus hoped to find an absolute reference frame, one that could be considered to be at rest.

One of the possibilities nineteenth-century scientists considered was that the ether is fixed relative to the sun, for even Newton had taken the sun as the center of the universe. If this were the case (there was no guarantee, of course), the earth's speed of about 3×10^4 m/s in its orbit around the sun would produce a change of 1 part in 10^4 in the speed of light (3.0×10^8 m/s). Direct measurement of the speed of light to this accuracy was not possible. But A. A. Michelson, later with the help of E. W. Morley, was able to use his interferometer (Section 36–9) to measure the difference in the speed of light in different directions to this accuracy. This famous experiment is based on the principle shown in Fig. 39–2. Part (a) is a simplified diagram of the Michelson interferometer, and for simplicity it is assumed that the ether "wind" is moving with speed v to the right. (Alternately, the earth is assumed to move to the left with respect to the ether at speed v.)

Whether constructive or destructive interference occurs at the center of the interference pattern (Section 36–9) depends on the relative phases of the two beams after they have traveled their separate paths. To examine this we will use an analogy of a boat traveling up and down, and across, a river whose current moves with speed v as shown in Fig. 39–2b. In still water the boat can travel with speed c (not the speed of light in this case).

[†] This section depends on material covered in Section 36–9.

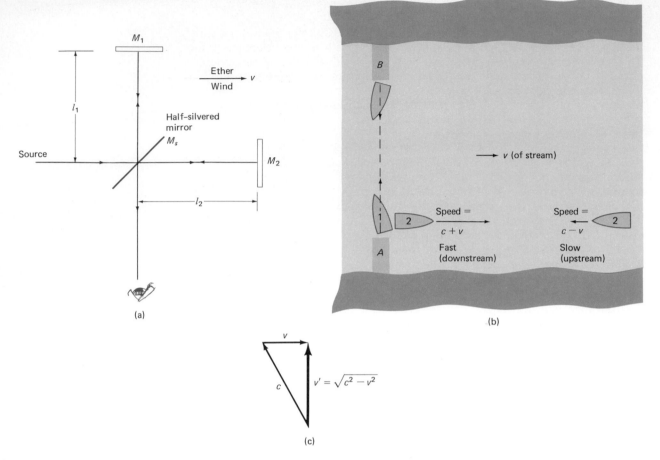

FIGURE 39–2

The Michelson-Morley experiment.
(a) Michelson interferometer. (b) Boat
analogy: boat 1 goes across the stream
and back; boat 2 goes downstream
and back upstream. (c) Calculation of
velocity of boat (or light beam)
traveling perpendicular to the current
(or ether wind).

First we consider beam 2 in Fig. 39–2a which travels parallel to the "ether wind." In its journey from M_s to M_2, we expect the light to travel with speed $c + v$, just as a boat traveling downstream (see Fig. 39–2b) acquires the speed of the river current. Since the beam travels a distance l_2, the time it takes to go from M_s to M_2 is $t = l_2/(c + v)$. To make the return trip from M_2 to M_s, the light is moving against the ether wind (like the boat going upstream) so its relative speed is expected to be $c - v$; the time for the return trip is $l_2/(c - v)$. The total time required for beam 2 to travel from M_s to M_2 and back to M_s is

$$t_2 = \frac{l_2}{c + v} + \frac{l_2}{c - v}$$

$$= \frac{2l_2}{c(1 - v^2/c^2)}.$$

Now let us consider beam 1, which travels crosswise to the ether wind. Here the boat analogy, part (b), is especially helpful. The boat wants to get from wharf A to wharf B directly across the stream. If it heads directly across, the stream's current will drag it downstream. In order to reach wharf B, the boat must head at an angle upstream. The precise angle depends on the magnitudes of c and v, but is of no interest to us in itself. Part (c) of Fig. 39–2 shows how to calculate the velocity v' of the boat relative to earth as it crosses the stream. Since c, v, and v' form a right triangle, we have that $v' = \sqrt{c^2 - v^2}$. The boat has this same velocity when it returns. If we now apply these same principles to light beam 1 in Fig. 39–2a, we see that the beam travels with a speed $\sqrt{c^2 - v^2}$

in going from M_s to M_1 and back again. The total distance traveled is $2l_1$, so the time required for beam 1 to make the round trip is $2l_1/\sqrt{c^2 - v^2}$, or

$$t_1 = \frac{2l_1}{c\sqrt{1 - v^2/c^2}}.$$

If $l_1 = l_2 = l$, then we see that beam 1 will lag behind beam 2 by an amount

$$\Delta t = t_2 - t_1 = \frac{2l}{c}\left(\frac{1}{1 - v^2/c^2} - \frac{1}{\sqrt{1 - v^2/c^2}}\right).$$

If $v = 0$, then $\Delta t = 0$ and the two beams will return in phase since they were initially in phase. But if $v \neq 0$, then $\Delta t \neq 0$, and the 2 beams will return out of phase. If this change of phase from the condition $v = 0$ to that for $v = v$ could be measured, then v could be determined. But the earth cannot be stopped. Furthermore, it is not possible to independently assume $l_1 = l_2$.

Michelson and Morley realized that they could detect the difference in phase (assuming $v \neq 0$) if they rotated their apparatus by 90°, for then the interference pattern between the two beams should change. In the rotated position, beam 1 would now move parallel to the ether and beam 2 perpendicular to it. Thus the roles would be reversed and in the rotated positions the times (designated by primes) would be

$$t_1' = \frac{2l_1}{c(1 - v^2/c^2)} \quad \text{and} \quad t_2' = \frac{2l_2}{c\sqrt{1 - v^2/c^2}}.$$

The time lag between the two beams in the nonrotated position (unprimed) would be

$$\Delta t = t_2 - t_1 = \frac{2l_2}{c(1 - v^2/c^2)} - \frac{2l_1}{c\sqrt{1 - v^2/c^2}}.$$

In the rotated position the time difference would be

$$\Delta t' = t_2' - t_1' = \frac{2l_2}{c\sqrt{1 - v^2/c^2}} - \frac{2l_1}{c(1 - v^2/c^2)}.$$

When the rotation is made, the fringe pattern will shift an amount determined by the difference in these:

$$\Delta t - \Delta t' = \frac{2}{c}(l_1 + l_2)\left(\frac{1}{1 - v^2/c^2} - \frac{1}{\sqrt{1 - v^2/c^2}}\right).$$

This can be considerably simplified if we assume $v/c \ll 1$. For in this case we can use the binomial expansion so

$$\frac{1}{(1 - v^2/c^2)} \approx 1 + \frac{v^2}{c^2} \quad \text{and} \quad \frac{1}{\sqrt{1 - v^2/c^2}} \approx 1 + \frac{1}{2}\frac{v^2}{c^2}.$$

Then

$$\Delta t - \Delta t' \approx \frac{2}{c}(l_1 + l_2)\left(1 + \frac{v^2}{c^2} - 1 - \frac{1}{2}\frac{v^2}{c^2}\right)$$

$$\approx (l_1 + l_2)\frac{v^2}{c^3}.$$

Now we take $v = 3.0 \times 10^4$ m/s, the speed of the earth in its orbit around the sun. In Michelson and Morley's experiments, the arms l_1 and l_2 were about 11 m long. The time difference would then be about $(22 \text{ m})(3.0 \times 10^4 \text{ m/s})^2/(3.0 \times 10^8 \text{ m/s})^3 \approx 7.0 \times 10^{-16}$ s. For visible light of wavelength $\lambda = 5.5 \times 10^{-7}$ m,

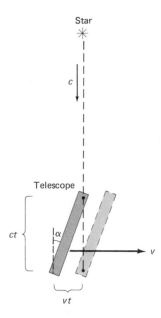

FIGURE 39–3

Aberration of starlight. If a telescope moves with the earth at speed v with respect to the ether, the telescope must be aimed slightly away from the star so that light entering the telescope will travel down the tube (since the tube moves to the right with speed v), rather than hitting the sides. Thus an overhead star will appear to be at an angle $\alpha = v/c$ to the vertical and, as the earth revolves, will seem to trace a circular path.

say, the frequency would be $f = c/\lambda = (3.0 \times 10^8 \text{ m/s})/(5.5 \times 10^{-7} \text{ m}) = 5.5 \times 10^{14}$ Hz, which means wavecrests pass by a point every $1/(5.5 \times 10^{14} \text{ Hz}) = 1.8 \times 10^{-15}$ s. Thus, with a time difference of 7×10^{-16} s, Michelson and Morley should have noted a movement in the interference pattern of $(7 \times 10^{-16} \text{ s})/(1.8 \times 10^{-15} \text{ s}) \approx 0.4$ fringe. They could easily have detected this since their apparatus was capable of observing a fringe shift as small as 0.01 fringe.

But they found *no significant fringe shift*. They set their apparatus at various orientations. They made observations day and night, so they would be at various orientations with respect to the sun. They tried at different seasons of the year. Never did they observe a fringe shift.

This "null" result was one of the great puzzles of physics at the end of the nineteenth century. To explain it was a difficult challenge. One possibility was that the ether is not at rest with respect to the sun or other stars, but instead is at rest with respect to the earth; in this case v would be zero and no fringe shift would be seen. But this implies the earth is somehow a preferred object; only with respect to the earth would the speed of light be c as predicted by Maxwell's equations. This is tantamount to assuming the earth is the central body of the universe, an ancient idea that had been rejected centuries earlier. Another possibility was that the ether was dragged along by the earth and other bodies, so that its speed at the earth's surface would be zero. However, experiments in high-flying balloons, where at least some ether movement might be detected, also gave a null result. Both possibilities were negated by the observation of "aberration of starlight." Over an observation period of a year, the stars appear to move in tiny ellipses; this is explained, if the earth is moving relative to the ether, by assuming a telescope must be pointed not exactly at the star to be observed but rather at an angle as explained in Fig. 39–3. If the ether were at rest with respect to the earth, you could point directly at the star and no "aberration angle" would be present. But the aberration angle is there, so the ether could not be at rest with respect to the earth.

Another theory to explain the null result was put forth independently by G. F. Fitzgerald and H. A. Lorentz (in the 1890s) in which they proposed that any length (including the arm of an interferometer) contracts by a factor $\sqrt{1 - v^2/c^2}$ in the direction of motion through the ether. According to Lorentz, this could be due to the ether affecting the forces between the molecules of a substance which were assumed to be electrical in nature. This theory, as useful as it was, was eventually replaced by the far more comprehensive theory proposed by Albert Einstein in 1905—the special theory of relativity.

39–3 Postulates of the Special Theory of Relativity

The problems that existed at the turn of the century with regard to electromagnetic theory and Newtonian mechanics were beautifully resolved by Einstein's introduction of the theory of relativity in 1905. Einstein, however, was apparently not influenced directly by the null result of the Michelson–Morley experiment. He had studied and admired the theoretical work of Lorentz. What motivated Einstein were certain questions regarding electromagnetic theory and light waves. For example, he asked himself: "What would I see if I rode a light beam?" The answer was that instead of a traveling electromagnetic wave, he would see electric and magnetic fields at rest whose magnitude changed in space, but did not change in time. Such fields, he realized, had never been detected and indeed were not consistent with Maxwell's electromagnetic theory.[†] He argued, therefore, that it was unreasonable to think that the speed of light relative to any observer could be reduced to zero, or in fact reduced at all; this became the second postulate of his theory of relativity.

[†] We saw in Section 33–5 that sinusoidally varying electric and magnetic fields are solutions of Maxwell's equations only if they are traveling at speed $c = 1/\sqrt{\varepsilon_0 \mu_0}$. Sinusoidal fields that are static speed $= 0$) or with any speed $\neq c$ could not be solutions.

Einstein concluded that the inconsistencies he found in electromagnetic theory were due to the assumption that an absolute space exists. In his famous 1905 paper, he proposed doing away completely with the idea of the ether and the accompanying assumption of an absolute reference frame at rest. This proposal was embodied in two postulates. The first postulate was an extension of the Newtonian relativity principle to include not only the laws of mechanics but also those of the rest of physics, including electricity and magnetism:

> **First postulate** (*the relativity principle*): **The laws of physics have the same form in all inertial reference frames.**

The second postulate is consistent with the first:

> **Second postulate** (*constancy of the speed of light*): **Light propagates through empty space with a definite speed c independent of the speed of the source or observer.**

These two postulates form the foundation of Einstein's **special theory of relativity**. It is called "special" to distinguish it from his later "general theory of relativity," which deals with noninertial (accelerating) reference frames. The special theory, which is what we will discuss here, deals only with inertial frames.

The second postulate seems to be the hardest to accept, for it violates commonsense notions. First of all, we have to think of light traveling through empty space. Giving up the ether is not too hard, however, for after all it could never be detected. But the second postulate also tells us that the speed of light in vacuum is always the same, 3.00×10^8 m/s, no matter what the speed of the observer or the source. Thus, a person traveling toward or away from a source of light will measure the same speed for that light as someone at rest with respect to the source. This conflicts with our everyday notions, for we would expect to have to add in the velocity of the observer. Part of the problem is that in our everyday experience we do not measure velocities anywhere near the speed of light; thus we can't expect our everyday experience to be helpful when dealing with such a high velocity. On the other hand, the Michelson–Morley experiment is fully consistent with the second postulate[†].

The beauty of Einstein's proposal is evident. For by doing away with the idea of an absolute reference frame, it was possible to reconcile Maxwell's electromagnetic theory with mechanics. The speed of light predicted by Maxwell's equations *is* the speed of light in vacuum in *any* reference frame.

Einstein's theory required giving up commonsense intuitive notions of space and time, and in the following sections we will examine some strange but interesting consequences of Einstein's theory. Our arguments for the most part will be simple ones. We will use a technique that Einstein himself did: we will imagine very simple experimental situations in which little mathematics is needed. In this way we can see what the consequences of relativity theory are without getting involved in detailed calculations. Einstein called these "gedanken" experiments, which is German for "thought" experiments. Starting in Section 39–8 we will look more fully into the mathematics of relativity.

 ## 39–4 Simultaneity

One of the important consequences of the theory of relativity is that we can no longer regard time as an absolute quantity. No one doubts that time flows onward and never turns back. But, as we shall see in this section and the next, the time between two events and even whether two events are simultaneous depends on the observer's reference frame.

[†] The Michelson–Morley experiment can also be considered as evidence for the first postulate, for it was intended to measure the motion of the earth relative to an absolute reference frame; its failure to do so implies the absence of any such preferred frame.

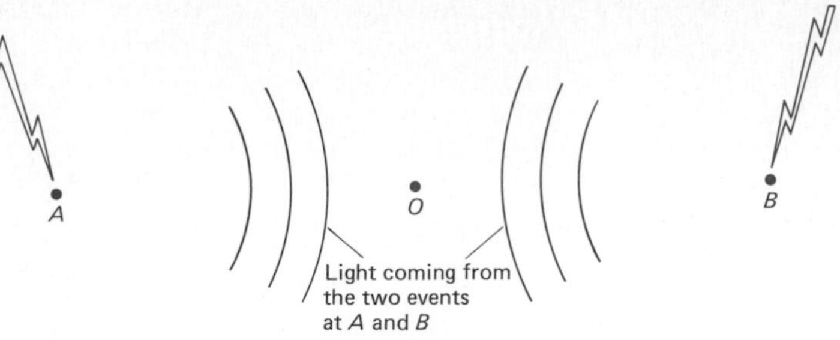

FIGURE 39-4

A moment after lightning strikes points A and B, the pulses of light are traveling toward O, but O "sees" the lightning only when the light reaches him or her.

A

O

B

Light coming from the two events at A and B

Two events are said to occur simultaneously if they occur at exactly the same time. But how do we tell if two events occur precisely at the same time? If they occur at the same point in space—such as two apples falling on your head at the same time—it is easy. But if the two events occur at widely separated places, it is more difficult to know since we have to take into account the time it takes for the light to reach us. Because light travels at finite speed, a person who sees two events must calculate back to find out when they actually occurred. For example, if two events are *observed* to occur at the same time, but one actually took place farther from the observer than the other, then the former must have occurred earlier, and the two events were not simultaneous.

To avoid making calculations, we will now make use of a simple thought experiment. We assume an observer, called O, is located exactly halfway between points A and B where two events occur, Fig. 39-4. The two events may be lightning that strikes the points A and B, as shown, or any other type of events. For brief events like lightning, only short pulses of light will travel outward from A and B and reach O. O "sees" the events when the pulses of light reach him. If the two pulses reach O at the same time, then the two events had to be simultaneous. This is because the two light pulses travel at the same speed; and since the distance OA equals OB, the time for the light to travel from A to O and B to O must be the same. Observer O can then definitely state the two events occurred simultaneously. On the other hand, if O sees the light from one event before that from the other, then it is certain the former event occurred first.

The question we really want to examine is this: if two events are simultaneous to an observer in one reference frame, are they also simultaneous to another observer moving with respect to the first? Let us call the observers O_1 and O_2 and we assume they are fixed in reference frames 1 and 2 that move with speed v relative to one another. These two reference frames can be thought of as railroad cars (Fig. 39-5). O_2 says that O_1 is moving to the right with speed v, as in (a); and O_1 says O_2 is moving to the left, as in (b). Both viewpoints are legitimate according to the relativity principle. (There is, of course, no third point of view which will tell us which one is "really" moving.)

FIGURE 39-5

Observers O_1 and O_2, on two different trains (two different reference frames), are moving with relative velocity v. O_2 says that O_1 is moving to the right (a); O_1 says that O_2 is moving to the left (b). Both viewpoints are legitimate—it all depends on your reference frame.

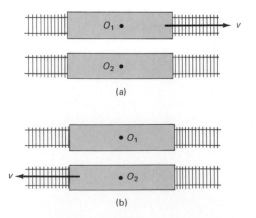

O_1

v

O_2

(a)

O_1

v

O_2

(b)

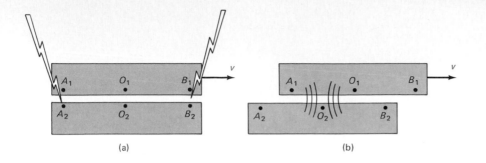

FIGURE 39-6

Thought experiment on simultaneity. To observer O_2, the reference frame of O_1 is moving to the right. In (a), one lightning bolt strikes the two reference frames at A_1 and A_2, and a second lightning bolt strikes at B_1 and B_2. According to observer O_2, the two bolts of lightning strike simultaneously. (b) A moment later the light from the two events reaches O_2 at the same time (simultaneously). But in O_1's reference frame, the light from B_1 has already reached O_1, whereas the light from A_1 has not yet reached O_1. So in O_1's reference frame the event at B_1 must have preceded the event at A_1. Time is not absolute!

Now suppose two events occur that are seen by both observers. Let us assume again that the two events are the striking of lightning and that the lightning marks both trains where it struck: at A_1 and B_1 on O_1's train, and at A_2 and B_2 on O_2's train. For simplicity, we assume O_1 happens to be exactly halfway between A_1 and B_1, and that O_2 is halfway between A_2 and B_2. We must now put ourselves in one reference frame or the other. Let us put ourselves in O_2's reference frame, so we see O_1 moving to the right with speed v. Let us also assume that the two events occur *simultaneously* in O_2's frame, and just at the time when O_1 and O_2 are opposite each other, Fig. 39-6a. A short time later, Fig. 39-6b, the light reaches O_2 from A_2 and B_2 at the same time. Since O_2 knows (or measures) the distances O_2A_2 and O_2B_2 as equal, he knows the two events are simultaneous.

But what does observer O_1 see? From our (O_2) reference frame, we can see that O_1 moves to the right during the time the light is traveling to O_1 from A_1 and B_1. As can be seen in Fig. 39-6b, the light from B_1 has already passed O_1, whereas that from A_1 has not yet reached O_1. Therefore, it is clear that O_1 sees the light from B_1 before that from A_1. Now O_1's frame is as good as O_2's. Light travels at the same speed for O_1 as for O_2, and is the same from A_1 to O_1 as from B_1 to O_1. Since the distance O_1A_1 equals O_1B_1, observer O_1 must conclude that the event at B_1 occurred before the event at A_1.

We thus find that *two events which are simultaneous to one observer are not necessarily simultaneous to a second observer.*

It may be tempting to ask: "Which observer is right, O_1 or O_2?" The answer, according to relativity, is that they are *both* right. There is no "best" reference frame we can choose to determine which observer is right. Both frames are equally good. We can only conclude that simultaneity is not an absolute concept, but is relative. We are not aware of it in everyday life, however, since the effect is noticeable only when the relative speed of the two reference frames is very large (near c), or the distances involved are very large.

Because of the principle of relativity, the argument we gave for the thought experiment of Fig. 39-6 can be done from O_1's reference frame as well. In this case, O_1 will be at rest and will see event B_1 occur before A_1. But O_1 will recognize (by drawing a diagram equivalent to Fig. 39-6) that O_2, who is moving with speed v to the left, will see the two events as simultaneous. The analysis is left as an exercise (Question 8).

 ## 39-5 Time Dilation and the Twin Paradox

The fact that two events simultaneous to one observer may not be simultaneous to a second observer suggests that time itself is not absolute; could it be that time passes differently in one reference frame than another? This is, indeed, just what Einstein's theory of relativity predicts, as the following thought-experiment shows.

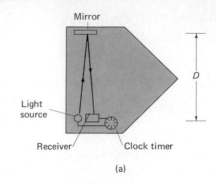

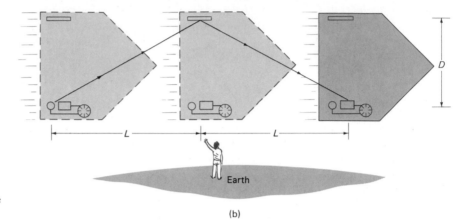

FIGURE 39–7

Time dilation can be shown by a thought experiment: the time it takes for light to travel over and back on a spaceship is longer for the earth observer (b) than for the observer on the spaceship (a).

Figure 39–7 shows a spaceship traveling past earth at high speed. The point of view of an observer on the spaceship is shown in part (a), and that of an observer on earth in part (b). Both observers have accurate clocks. The person on the spaceship (a) flashes a light and measures the time it takes the light to travel across the spaceship and return after reflecting from a mirror. The light travels a distance $2D$ at speed c so the time required, which we call Δt_0, is

$$\Delta t_0 = \frac{2D}{c}.$$

The observer on earth, Fig. 39–7b, observes the same process. But to this observer, the spaceship is moving; so the light travels the diagonal path shown in going across the spaceship, reflecting off the mirror, and returning to the sender. Although the light travels at the same speed to this observer (the second postulate), it travels a greater distance. Hence the time required, as measured by the earth observer, will be greater than that measured by the observer on the spaceship. The time interval, Δt, as observed by the earth observer can be calculated as follows. In the time Δt, the spaceship travels a distance $2L = v \, \Delta t$ where v is the speed of the spaceship (Fig. 39–7b). Thus, the light travels a total distance on its diagonal path of $2\sqrt{D^2 + L^2}$ and therefore

$$c = \frac{2\sqrt{D^2 + L^2}}{\Delta t} = \frac{2\sqrt{D^2 + v^2(\Delta t)^2/4}}{\Delta t}.$$

We square both sides and solve for Δt to find

$$c^2 = \frac{4D^2}{(\Delta t)^2} + v^2,$$

$$\Delta t = \frac{2D}{c\sqrt{1 - v^2/c^2}}.$$

We combine this with the formula above for Δt_0 and find:

$$\Delta t = \frac{\Delta t_0}{\sqrt{1 - v^2/c^2}}. \tag{39-1}$$

Since $\sqrt{1 - v^2/c^2}$ is always less than 1, we see that $\Delta t > \Delta t_0$. That is, the time between the two events (the sending of the light, and its reception at the receiver) is *greater* for the earth observer than for the traveling observer. This is a general result of the theory of relativity, and is known as **time dilation**. Stated simply, the time dilation effect says that *moving clocks are measured to run slowly*. However, we should not think that the clocks are somehow at fault. To the contrary, we assume the clocks are good ones. Time is actually measured to pass more slowly in any moving reference frame as compared to your own. This remarkable result is an inevitable outcome of the two postulates of the theory of relativity.

The concept of time dilation may be hard to accept, for it violates our commonsense understanding. We can see from Eq. 39-1 that the time-dilation effect is negligible unless v is reasonably close to c. The speeds we experience in everyday life are very much smaller than c, so it is little wonder we haven't noticed time dilation as a real effect. Experiments have been done to test the time-dilation effect, and have confirmed Einstein's predictions. In 1971, for example, extremely precise atomic clocks were flown around the world in jet planes. Since the speed of the planes (10^3 km/h) is much less than c, the clocks had to be accurate to nanoseconds (10^{-9} s) in order to detect the time-dilation effect. They were this accurate and confirmed Eq. 39-1 to within experimental error. Time dilation had been confirmed decades earlier, however, by observation on "elementary particles" (see Chapter 43) which have very small masses (typically) 10^{-30} to 10^{-27} kg) and so require little energy to be accelerated to speeds close to c. Many of these elementary particles are not stable and decay after a time into simpler particles. One example is the muon whose mean lifetime is 2.2 μs when at rest. Careful experiments showed that when a muon is traveling at high speeds, its lifetime increases just as predicted by the time-dilation formula.

EXAMPLE 39-1 What will be the mean lifetime of a muon as measured in the laboratory if it is traveling $0.60c = 1.8 \times 10^8$ m/s with respect to the laboratory? Its mean life at rest is 2.2×10^{-6} s.

SOLUTION If an observer were to move along with the muon (the muon would be at rest to this observer) the muon would have a mean life of 2.2×10^{-6} s. To an observer in the lab, the muon lives longer because of time dilation. From Eq. 39-1 with $v = 0.60c$, we have

$$t = \frac{t_0}{\sqrt{1 - v^2/c^2}} = \frac{2.2 \times 10^{-6} \text{ s}}{\sqrt{1 - \frac{0.36c^2}{c^2}}} = \frac{2.2 \times 10^{-6} \text{ s}}{\sqrt{0.64}} = 2.8 \times 10^{-6} \text{ s}.$$

We need to make a comment about the use of Eq. 39-1 and the meaning of Δt and Δt_0. The equation is true only when Δt_0 represents the time interval between the two events in a reference frame where the two events occur *at the same point in space* (as in Fig. 39-7a). This time interval, Δt_0, is often called the **proper time**. Then Δt in Eq. 39-1 represents the time interval between the two events as measured in a reference frame moving with speed v with respect to the first. In Example 39-1 above, Δt_0 (and not Δt) was set equal to 2.2×10^{-6} s because it is only in the rest frame of the muon that the two events ("birth" and "decay") occur at the same point in space.

Time dilation has aroused interesting speculation about space travel. Under the old time regime, to reach a star 100 light-years away would not be possible by ordinary mortals (1 light-year is the distance light can travel in 1 year = 3.0×10^8 m/s $\times 3.15 \times 10^7$ s $= 9.5 \times 10^{15}$ m). Even if a spaceship could travel at close to the speed of light, it would take over 100 years to reach such a star. But time dilation tells us that the time involved would be less for an astronaut. In a spaceship traveling at $v = 0.999c$, the time for such a trip would be only about $\Delta t_0 = \Delta t \sqrt{1 - v^2/c^2} = (100 \text{ yr})\sqrt{1 - (0.999)^2} = 4.5 \text{ yr}$. Thus a person could make such a trip. Time dilation allows such a trip, but the enormous practical problems of achieving such speeds will not be overcome in the foreseeable future.

Notice, in this example, that whereas 100 years would pass on earth, only 4.5 years would pass for the astronaut on the trip. Is it just the clocks that would slow down for the astronaut? The answer is no. All processes, including life processes, run more slowly for the astronaut according to the earth observer. But to the astronaut, time would pass in a normal way. The astronaut would experience 4.5 years of normal sleeping, eating, reading, and so on. And people on earth would experience 100 years of ordinary activity.

Not long after Einstein proposed the special theory of relativity, an apparent paradox was pointed out. According to this "twin paradox," suppose one of a pair of 20-year-old twins takes off in a spaceship traveling at very high speed to a distant star and back again, while the other twin remains on earth. According to the earth twin, the traveling twin will age less. Whereas 20 years might pass for the earth twin, perhaps only 1 year (depending on the spacecraft's speed) would pass for the traveler. Thus, when the traveler returns, the earthbound twin could expect to be 40 years old whereas his twin would be only 21.

This is the viewpoint of the twin on the earth. But what about the traveling twin? Since "everything is relative," all inertial reference frames are equally good. Won't the traveling twin make all the claims the earth twin does, only in reverse? Can't the astronaut twin claim that since the earth is moving away at high speed, time passes more slowly on earth and the twin on earth will age less? This is the opposite of what the earth twin predicts. They cannot both be right, for after all the spacecraft returns to earth and a direct comparison of ages and clocks can be made.

This is, however, not a paradox at all. The consequences of the special theory of relativity—in this case, time dilation—can be applied only by observers in inertial reference frames. The earth is such a frame (or nearly so) whereas the spacecraft is not. The spacecraft accelerates at the start and end of its trip and, more importantly, when it turns around at the far point of its journey. During these acceleration periods, the spacecraft's predictions based on special relativity are not valid. The twin on earth is in an inertial frame and can make valid predictions. Thus, there is no paradox. The traveling twin's point of view expressed above is not correct. The predictions of the earth twin are valid, and the prediction that the traveling twin returns having aged less is the proper one.[†]

39–6 Length Contraction

Not only time intervals are different in different reference frames; space intervals—lengths and distances—are different as well, according to the special theory of relativity, and we illustrate this with a thought experiment.

An observer on Earth watches a spacecraft traveling at speed v from Earth to, say, Neptune, Fig. 39–8a. The distance between the planets, as measured by an Earth observer, is L_0. The time required for the trip, measured from Earth, is $\Delta t = L_0/v$. In Fig. 39–8b we see the point of view of an observer on the spacecraft. In this frame of reference, the spaceship is at rest; Earth and Neptune move with speed v. (We assume v is much greater than the relative speed of Neptune and Earth, so the latter can be ignored.) But the time between the departure of

---

[†] Einstein's general theory of relativity, which deals with accelerating reference frames, confirms this result.

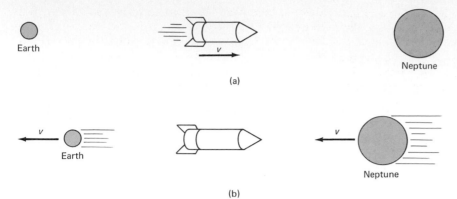

(a)

(b)

FIGURE 39-8

(a) A spaceship traveling at very high speed from Earth to Neptune, as seen from Earth's frame of reference. (b) As viewed by an observer on the spaceship, Earth and Neptune are moving at the very high velocity v. Note in (b) that each planet does not look shortened since at high speeds we see the trailing edge (as in Fig. 39-9), and the net effect is to leave its appearance as a circle.

Earth and arrival of Neptune is less for the spacecraft observer than the earth observer, because of time dilation. From Eq. 39-1, the time for the trip as viewed by the spacecraft is $\Delta t_0 = \Delta t \sqrt{1 - v/c^2}$. Since the spacecraft observer measures the same speed but less time between these two events, he must also measure the distance as less. If we let L be the distance between the planets as viewed by the spacecraft observer, then $L = v \Delta t_0$. We have already seen that $\Delta t_0 = \Delta t \sqrt{1 - v^2/c^2}$ and $\Delta t = L_0/v$, so we have $L = v \Delta t_0 = v \Delta t \sqrt{1 - v^2/c^2} = L_0 \sqrt{1 - v^2/c^2}$. That is

$$L = L_0 \sqrt{1 - v^2/c^2}. \tag{39-2}$$

This is a general result of the special theory of relativity and applies to lengths of objects as well as to distance. The result can be stated most simply in words as: *the length of an object is measured to be shorter when it is moving than when it is at rest.* This is called **length contraction**. The length L_0 in Eq. 39-2 is called the **proper length**. It is the length of the object (or distance between two points) as measured by an observer at rest with respect to it. Equation 39-2 gives the length that will be measured when the object travels past an observer at speed v. It is important to note, however, that the length contraction occurs *only along the direction of motion.* For example, the moving spaceship in Fig. 39-8a is shortened in length but its height is the same as when it is at rest.

Length contraction, like time dilation, is not noticeable in everyday life because the factor $\sqrt{1 - v^2/c^2}$ in Eq. 39-2 differs from 1.00 significantly only when v is very large.

EXAMPLE 39-2 A spaceship passes the earth at speed $v = 0.80c$. Describe the changes in length of a meter stick as it is slowly rotated from vertical to horizontal by a person inside as viewed (a) by another person in the spaceship, (b) by an observer on earth.

SOLUTION (a) The meter looks 1.0 m long in all orientations since it is at rest. (b) The meter stick varies in length from 1.0 m (vertical) to $L = (1.0 \text{ m}) \times \sqrt{1 - (0.80)^2} = 0.60$ m long in the horizontal direction (assuming that is the direction of motion).

Equation 39-2 tells us what the length of an object will be *measured* to be when traveling at speed v. The *appearance* of the object is another matter. Suppose, for example, you are traveling to the left past a tall building at speed $v = 0.85c$. This is equivalent to the building moving past you to the right at speed v. The building will look narrower (and the same height), but you will also be able to see the side of the building even if you are directly in front of it. This is shown in

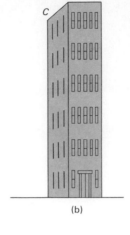

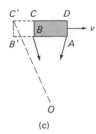

FIGURE 39–9

Building seen (a) at rest, (b) moving at high speed. (c) Diagram explains why the side of the building is seen (see the text).

(a)   (b)   (c)

Fig. 39–9b—part (a) shows the building at rest. The fact that you see the side is not really a relativistic effect, but is due to the finite speed of light. To see how this occurs, we look at Fig. 39–9c which is a top view of the building, looking down. At the instant shown, the observer O is directly in front of the building. Light from points A and B reach O at the same time. If the building were at rest, light from point C could never reach O. But the building is moving at very high speed and does "get out of the way" so that light from C can reach O. Indeed, at the instant shown, light from point C when it was at an earlier location (C' on the diagram) can reach O because the building has moved. In order to reach the observer at the same time as light from A and B, light from C had to leave at an earlier time since it must travel a greater distance. Thus it is light from C' that reaches the observer at the same time as light from A and B. This, then, is how an observer might see both the front and side of an object at the same time even when directly in front of it.[†] It can be shown, by the same reasoning, that spherical objects will actually still have a circular outline even at high speeds. That is why the planets in Fig. 39–8b are drawn round rather than contracted.

39–7 Four-Dimensional Space–Time

Let us suppose a person is on a train moving at a very high speed, say $0.65c$, Fig. 39–10. This person begins a meal at 7:00 and finishes at 7:15, according to a clock on the train. The two events, beginning and ending the meal, take place at the same point on the train. So the proper time between these two events is 15 min. To observers on earth, the meal will take longer—20 min according to Eq. 39–1. Let us assume that the meal was served on a 20-cm-diameter plate. To observers on the earth, the plate is only 15 cm wide (length contraction). Thus to observers on the earth the food looks smaller but lasts longer.

In a sense these two effects, time dilation and length contraction, balance each other. When viewed from the ground, what the food seems to lose in size it gains in length of time it lasts. Space, or length, is exchanged for time.

Considerations like this led to the idea of *four-dimensional space–time*: space takes up three dimensions and time is a fourth dimension. Space and time are intimately connected. Just as when we squeeze a balloon we make one dimension

---

[†] It would be an error to think that the building in Fig. 39–9b would look rotated. This is not correct since in that case side A would look shorter than side B. In fact, if the observer is directly in front these sides appear equal in height. Thus the building looks contracted in its front face, but we also see the side, as described above.

(a)   (b)

FIGURE 39–10

According to an accurate clock on a fast-moving train, a person begins dinner at 7:00 (a) and finishes at 7:15 (b). At the beginning of the meal, observers on the earth set their watches to correspond with the clock on the train. These observers measure the eating time as 20 min.

larger and another smaller, so when we examine objects and events from different reference frames a certain amount of space is exchanged for time, or vice versa.

Although the idea of four dimensions may seem strange, it refers to the fact that any object or event is specified by four quantities—three to describe where in space, and one to describe when in time. The really unusual aspect of four-dimensional space–time is that space and time can intermix: a little of one can be exchanged for a little of the other when the reference frame is changed.

It is difficult for most of us to understand the idea of four-dimensional space–time. Somehow we feel, just as physicists did before the advent of relativity, that space and time are completely separate entities. Yet we have found in our thought experiments that they are not completely separate. Our difficulty in accepting this is reminiscent of the situation in the seventeenth century at the time of Galileo and Newton. Before Galileo, the vertical direction, that in which objects fall, was considered to be entirely different from the two horizontal dimensions. Galileo showed that the vertical dimension differs only in that it happens to be the direction in which gravity acts. Otherwise, all three dimensions are equivalent, a fact that we all accept today. Now we are asked to accept one more dimension, time, which we had previously thought of as being somehow different. This is not to say that there is no distinction whatsoever between space and time. What relativity has shown is that space and time determinations are not independent of one another.

 ## 39–8   Galilean and Lorentz Transformations

We now examine in detail the mathematics of relating quantities in one inertial reference frame to the equivalent quantities in another. In particular, we will see how positions and velocities *transform* (that is, change) from one frame to the other.

We begin with the classical or Galilean viewpoint. Consider two reference frames S and S' which are each characterized by a set of coordinate axes, Fig. 39–11. The axes x and y (z is not shown) refer to S and x' and y' to S'. The x' and

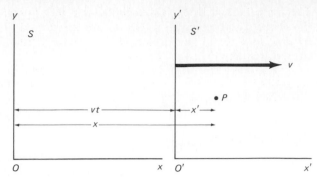

FIGURE 39–11

Inertial reference frame S' moves to the right at speed v with respect to frame S.

x axes overlap one another, and we assume that frame S' moves to the right (in the x direction) at speed v with respect to S. And for simplicity let us assume the origins O and O' of the two reference frames are superimposed at time $t = 0$.

Now consider an event that occurs at some point P (Fig. 39–11) represented by the coordinates x', y', z' in reference frame S' at the time t'. What will be the coordinates of P in S? Since S and S' overlap precisely initially, after a time t, S' will have moved a distance vt'. Therefore, at time t', $x = x' + vt'$. The y and z coordinates, on the other hand, are not altered by motion along the x axis; thus $y = y'$ and $z = z'$. Finally, since time is assumed to be absolute in Galilean–Newtonian physics, clocks in the two frames will agree with each other; so $t = t'$. We summarize these in the following **Galilean transformation equations**:

$$x = x' + vt'$$
$$y = y'$$
$$z = z' \qquad (39\text{--}3)$$
$$t = t'.$$

These equations give the coordinate of an event in the S frame when those in the S' frame are known. If those in the S system are known, then the S' coordinates are obtained from

$$x' = x - vt, \quad y' = y, \quad z' = z, \quad t' = t.$$

These four equations are the "inverse" transformation and are very easily obtained from Eqs. 39–3. Notice that the effect is merely to exchange primed and unprimed quantities and replace v by $-v$. This makes sense because from the S' frame, S moves to the left (negative x direction) with speed v.

Now suppose the point P in Fig. 39–11 represents a particle that is moving. Let the components of its velocity vector in S' be u'_x, u'_y, u'_z. (We use u to distinguish it from the relative velocity of the two frames, v.) Now $u'_x = dx'/dt'$, $u'_y = dy'/dt'$ and $u'_z = dz'/dt'$. The velocity of P as seen from S will have components u_x, u_y, and u_z. We can show how these are related to the velocity components in S' by differentiating Eq. 39–3; for u_x we get

$$u_x = \frac{dx}{dt} = \frac{d(x' + vt')}{dt'} = u'_x + v$$

since v is assumed constant. For the other components, $u'_y = u_y$ and $u'_z = u_z$, so we have

$$u_x = u'_x + v$$
$$u_y = u'_y \qquad (39\text{--}4)$$
$$u_z = u'_z.$$

These are known as the *Galilean velocity transformation equations.* We see that the y and z components of velocity are unchanged, but the x components differ by $v: u_x = u_x' + v$. This is just what we have used before when dealing with relative velocity.

The Galilean transformations, Eqs. 39–3 and 39–4, are valid only when the velocities involved are much less than c. We can see, for example, that the first of Eqs. 39–4 will not work for the speed of light; for light traveling in S' with speed $u_x' = c$ will have speed $c + v$ in S, whereas the theory of relativity insists it must be c in S. Clearly, then, a new set of transformation equations are needed to deal with relativistic velocities.

We will derive the required equations in a simple way, again looking at Fig. 39–11. We assume the transformation is linear and of the form

$$x = \gamma(x' + vt'), \qquad y = y', \qquad z = z'.$$

That is, we modify the first of Eqs. 39–3 by multiplying by a constant γ which is yet to be determined; but we assume the y and z equations are unchanged since there is no length contraction in these directions. We won't assume a form for t, but will derive it. The inverse equations must have the same form with v replaced by $-v$. (The principle of relativity demands it, since S' moving to the right with respect to S is equivalent to S moving to the left with respect to S'.) Therefore

$$x' = \gamma(x - vt).$$

Now if a light pulse leaves the common origin of S and S' at time $t = t' = 0$, after a time t it will have traveled a distance $x = ct$ or $x' = ct'$ along the x axis. Therefore, from the equations for x and x' above,

$$ct = \gamma(ct' + vt') = \gamma(c + v)t',$$

$$ct' = \gamma(ct - vt) = \gamma(c - v)t.$$

We substitute t' from the second equation into the first and find $ct = \gamma(c + v)\gamma(c - v)(t/c) = \gamma^2(c^2 - v^2)t/c$. We cancel out the t on each side and solve for γ to find

$$\gamma = \frac{1}{\sqrt{1 - v^2/c^2}}.$$

Now that we have found γ, we need only find the relation between t and t'. To do so, we combine $x' = \gamma(x - vt)$ with $x = \gamma(x' + vt')$:

$$x' = \gamma(x - vt) = \gamma(\gamma[x' + vt'] - vt).$$

We solve for t and find $t = \gamma(t' + vx'/c^2)$. In summary

$$x = \frac{1}{\sqrt{1 - v^2/c^2}}(x' + vt')$$

$$y = y'$$
$$z = z'$$

(39–5)

$$t = \frac{1}{\sqrt{1 - v^2/c^2}}(t' + vx'/c^2).$$

These are called the **Lorentz transformation equations.** They were first proposed, in a slightly different form, by Lorentz in 1904 to explain the null result of the Michelson–Morley experiment and to make Maxwell's equations take the same form in all inertial systems. A year later Einstein derived them independently based on his theory of relativity. Notice that not only is the x equation modified as compared to the Galilean transformation, but so is the t equation; indeed, we see directly in this last equation how the space and time coordinates mix.

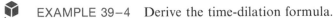 EXAMPLE 39–3   Derive the length contraction formula, Eq. 39–2, from the Lorentz transformation equations.

SOLUTION   Let an object of length L_0 be at rest on the x axis in S. The coordinates of its two end points are x_1 and x_2, so that $x_2 - x_1 = L_0$. At any instant in S', the end points will be at x_1' and x_2' as given by the Lorentz transformation equations. The length measured in S' is $L = x_2' - x_1'$. An observer in S' measures this length by measuring x_2' and x_1' at the same time, so $t_2' = t_1'$. Thus, from the first of Eqs. 39–5,

$$L_0 = x_2 - x_1 = \frac{1}{\sqrt{1 - v^2/c^2}} (x_2' + vt_2' - x_1' - vt_1').$$

Since $t_2' = t_1'$, we have

$$L_0 = \frac{1}{\sqrt{1 - v^2/c^2}} (x_2' - x_1') = \frac{L}{\sqrt{1 - v^2/c^2}},$$

or

$$L = L_0 \sqrt{1 - v^2/c^2}$$

which is Eq. 39–2.

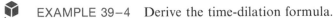 EXAMPLE 39–4   Derive the time-dilation formula.

SOLUTION   The time Δt_0 between two events that occur at the same place $(x_2' = x_1')$ in S' is measured to be $\Delta t_0 = t_2' - t_1'$. Since $x_2' = x_1'$, then from the last of Eqs. 39–5, the time Δt between the events as measured in S is

$$\Delta t = t_2 - t_1 = \frac{1}{\sqrt{1 - v^2/c^2}} \left(t_2' + \frac{vx_2'}{c^2} - t_1' - \frac{vx_1'}{c^2} \right)$$

$$= \frac{1}{\sqrt{1 - v^2/c^2}} (t_2' - t_1') = \frac{\Delta t_0}{\sqrt{1 - v^2/c^2}}$$

which is Eq. 39–1. Notice that we chose S' to be the frame where the two events occur at the same place, so that $x_1' = x_2'$ and these terms would cancel out.

The relativistically correct velocity equations are readily obtained by differentiating Eqs. 39–5 with respect to time. For example (we let $\gamma = 1/\sqrt{1 - v^2/c^2}$ and use the chain rule for derivatives):

$$u_x = \frac{dx}{dt} = \frac{d}{dt} [\gamma(x' + vt')]$$

$$= \frac{d}{dt'} [\gamma(x' + vt')] \frac{dt'}{dt} = \gamma \left[\frac{dx'}{dt'} + v \right] \frac{dt'}{dt}.$$

But $dx'/dt' = u_x'$ and $dt'/dt = 1/(dt/dt') = 1/[\gamma(1 + vu_x'/c^2)]$ where we have differentiated the last of Eqs. 39–5 with respect to time. Therefore

$$u_x = [\gamma(u_x' + v)]/[\gamma(1 + vu_x'/c^2)] = \frac{u_x' + v}{1 + vu_x'/c^2}.$$

The others are obtained in the same way and we collect them here:

$$u_x = \frac{u'_x + v}{1 + vu'_x/c^2} \qquad (39\text{–}6a)$$

$$u_y = \frac{u'_y\sqrt{1 - v^2/c^2}}{1 + vu'_x/c^2} \qquad (39\text{–}6b)$$

$$u_z = \frac{u'_z\sqrt{1 - v^2/c^2}}{1 + vu'_x/c^2}. \qquad (39\text{–}6c)$$

We will have more to say about Eqs. 39–6 in the next section. Note that even though the relative velocity **v** is in the x direction, the transformation of all the components of a particle's velocity are affected by v and the x component of the particle's velocity; this was not true for the Galilean transformation, Eqs. 39–4.

 EXAMPLE 39–5   Calculate the speed of rocket 2 in Fig. 39–12 with respect to earth.

SOLUTION   Rocket 2 moves with speed $u' = 0.60c$ with respect to rocket 1. Rocket 1 has speed $v = 0.60c$ with respect to earth. The velocities are along the same straight line which we take to be the x (and x') axis. We need use only the first of Eqs. 39–6. Then the speed of rocket 2 with respect to earth is

$$u = \frac{u' + v}{1 + vu'/c^2}$$

$$= \frac{0.60c + 0.60c}{1 + \dfrac{(0.60c)(0.60c)}{c^2}} = \frac{1.20c}{1.36c} = 0.88c.$$

(The Galilean transformation would have given $u = 1.20c$.)

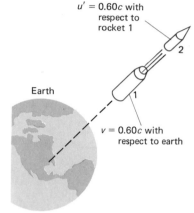

FIGURE 39–12
Rocket 2 is fired from rocket 1 with speed $u' = 0.60c$. What is the speed of rocket 2 with respect to the earth?

Notice that Eqs. 39–6 reduce to the classical (Galilean) forms for velocities small compared to the speed of light.

39–9   Relativistic Mass and Momentum

The three basic mechanical quantities are length, time intervals, and mass. The first two have been shown to be relative—their value depends on the reference frame from which they are measured. We might expect that mass too is a relative quantity. Indeed, Einstein showed that *the mass of an object increases as its speed increases* according to the formula

$$m = \frac{m_0}{\sqrt{1 - v^2/c^2}}. \qquad (39\text{–}7)$$

In this mass-increase formula, m_0 is called the **rest mass** of the object—the mass it has as measured in a reference frame in which it is at rest; and m is the mass it will be measured to have in a reference frame in which it moves at speed v.

Relativistic mass increase has been tested countless times on tiny elementary particles (such as muons) and the mass has been found to increase in accord with Eq. 39–7.

Using a thought experiment, we now show that Eq. 39–7 follows from the theory of relativity and the law of conservation of momentum. We assume that the momentum of a particle is given by

$$\mathbf{p} = m\mathbf{v},$$

which is just like the classical formula for momentum, except that we will let the mass m of the object be a function of its speed v [written as $m(v)$]. As we shall see, this is necessary if the law of conservation of momentum is to be valid in the relativistic domain.

Our thought experiment involves the elastic collision of two *identical* balls. If the two balls travel at the same speed, v, they will have the same mass $m(v)$ and this may be different from the rest mass which is the same m_0 for each. If they travel at different speeds, their masses may be different.

The collision in this thought experiment takes place as follows. We consider two inertial reference frames, A and B, moving with a velocity v with respect to each other. In reference frame A, a ball (call it ball A) is thrown with speed u along the y_A axis, perpendicular to v; in reference frame B, a second ball (B) is thrown with speed u along the negative y_B axis. The two balls are thrown at just the right time so that they collide. We assume they rebound elastically, and each moves with the same speed u back along the y axis in its thrower's reference frame. Figure 39–13a shows the collision as seen by the person in reference frame A; and Fig. 39–13b shows the collision as seen from reference frame B. In reference frame A, ball A has velocity $+u$ along the y axis before the collision and $-u$ along the y axis after the collision; in frame B, ball A has, both before and after the collision, an x component of velocity equal to v, and a y component (see Eq. 39–6b with $u'_x = 0$) of magnitude

$$u\sqrt{1 - v^2/c^2}.$$

The same holds true for ball B, except in reverse, and these velocity components are indicated in the figure. Let us further assume that $u \ll v$, so that the mass of

FIGURE 39–13

Deriving the mass formula.
Collision as seen (a) from reference frame A, (b) from reference frame B.

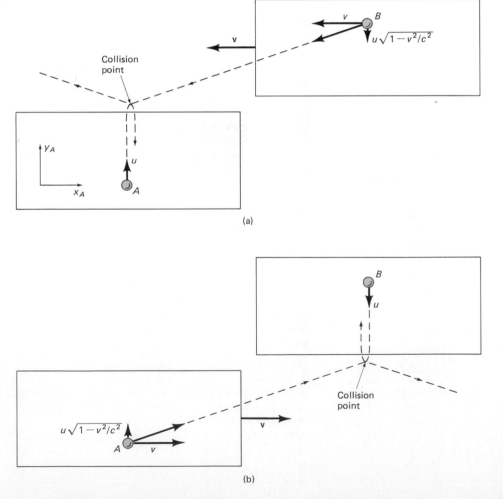

(a)

(b)

A as seen in reference frame *B* can be written $m(v)$, and the mass of *B* in reference frame *A* is $m(v)$. We now apply the law of conservation of momentum (which we hope remains valid in relativity, even if it has to be redefined): we therefore assume that the total momentum before the collision is equal to the total after the collision. We apply this to the *y* component of momentum in reference frame *A* (Fig. 39–13a):

$$m(u)u - m(v)u\sqrt{1 - v^2/c^2} = -m(u)u + m(v)u\sqrt{1 - v^2/c^2}.$$

We solve this for $m(v)$ and obtain

$$m(v) = \frac{m(u)}{\sqrt{1 - v^2/c^2}}.$$

If we now let *u* become very small so that it approaches zero (this corresponds to a glancing collision with one of the balls essentially at rest and the other moving with speed *v*), then $m(u)$ becomes the rest mass m_0. So the above equation becomes

$$m(v) = \frac{m_0}{\sqrt{1 - v^2/c^2}}.$$

Although $m(v)$ referred to particle *A*, and m_0 referred to particle *B*, m_0 is the same for both particles. Hence the above equation is valid for ball *A*. (Using Fig. 39–13b we can derive the same relation for ball *B*.) This is just Eq. 39–7, which we set out to prove.

We define the relativistic momentum of a particle as

$$\mathbf{p} = m\mathbf{v} = \frac{m_0\mathbf{v}}{\sqrt{1 - v^2/c^2}} \tag{39–8}$$

since then, as we saw above, the law of conservation of momentum will remain valid even in the relativistic realm. Newton's second law, stated in its most general form

$$F = \frac{d\mathbf{p}}{dt} = \frac{d}{dt}(m\mathbf{v}) \tag{39–9}$$

also remains valid relativistically. But since the mass of a particle cannot be considered constant, relativistically, Newton's second law written as $\mathbf{F} = m\mathbf{a}$ is not valid.

39–10   The Ultimate Speed

A basic result of the special theory of relativity is that the speed of an object cannot equal or exceed the speed of light. That the speed of light is a natural speed limit in the universe can be seen from any of Eqs. 39–1, 39–2, 39–7, or the addition of velocities formula. It is perhaps easiest to see it from the mass-increase formula, $m = m_0/\sqrt{1 - v^2/c^2}$. As an object is accelerated to greater and greater speeds, its mass becomes larger and larger. Indeed, if *v* were to equal *c*, the denominator in this equation would be zero, and the mass *m* would be infinite. To accelerate an object up to *v* = *c* would thus require infinite energy, and so is not possible.

If *v* were to exceed *c*, then the factor $\sqrt{1 - v^2/c^2}$ would be the square root of a negative number, which is imaginary: so lengths, time intervals, and mass would not be real. We conclude, then, that ordinary objects cannot equal or exceed the speed of light. However, as was pointed out in the late 1960s, Einstein's equations do not rule out the possibility that objects exist whose speed is *always* greater than *c*. If such particles exist (the name "tachyon"—meaning "fast"—was proposed) the rest mass m_0 would have to be imaginary; in this way the mass *m* would be the ratio of two imaginary numbers for *v* > *c*, which is real. For such hypothetical particles, c would be a *lower* limit on their speed. In spite of extensive searches for tachyons, none has been found. It seems that the speed of light *is* the ultimate speed in the universe.

39–11 $E = mc^2$; Mass and Energy

When a steady net force is applied to an object of rest mass m_0, the object increases in speed. Since the force is acting through a distance, work is done on the object and its energy increases. As the speed of the object approaches c, the speed cannot increase indefinitely since it cannot exceed c. On the other hand, the mass of the object increases with increasing speed. That is, the work done on an object not only increases its speed but also contributes to increasing its *mass*. Normally, the work done on an object increases its energy. This new twist from the theory of relativity leads to the idea that mass is a form of energy, a crucial part of Einstein's theory of relativity.

To find the mathematical relationship between mass and energy, we assume that the work-energy theorem is still valid in relativity, and we take the motion to be along the x axis. The work done to increase a particle's speed from zero to v is

$$W = \int_i^f F \, dx = \int_i^f \frac{dp}{dt} \, dx = \int_i^f \frac{dp}{dt} v \, dt = \int_i^f v \, dp$$

where i and f refer to the initial ($v = 0$) and final ($v = v$) states. Since $d(pv) = p \, dv + v \, dp$ we can write

$$v \, dp = d(pv) - p \, dv$$

so

$$W = \int_i^f d(pv) - \int_i^f p \, dv.$$

Since integration is the exact inverse of differentiation, the first term on the right becomes

$$\int_i^f d(pv) = pv \Big|_i^f = mv^2 \Big|_0^v = mv^2.$$

where m is a function of v. Therefore

$$W = mv^2 - \int_0^v mv \, dv$$

$$= mv^2 - \int_0^v \frac{m_0 v}{\sqrt{1 - v^2/c^2}} \, dv.$$

The second term is easily integrated since

$$\frac{d}{dv}\left(\sqrt{1 - v^2/c^2}\right) = -(v/c^2)/\sqrt{1 - v^2/c^2}.$$

Then

$$W = mv^2 + (m_0 c^2 \sqrt{1 - v^2/c^2})\Big|_0^v$$

$$= mv^2 + m_0 c^2 \sqrt{1 - v^2/c^2} - m_0 c^2$$

$$= mv^2 + mc^2(1 - v^2/c^2) - m_0 c^2$$

where we have used Eq. 39–7. Since $mc^2(1 - v^2/c^2) = mc^2 - mv^2$, we have finally

$$W = mc^2 - m_0 c^2.$$

By the work-energy theorem, the work done must equal the final kinetic energy
since the particle started from rest. Therefore

39–11 $E = mc^2$; Mass
and Energy   **767**

$$\text{KE} = mc^2 - m_0c^2 \qquad (39\text{--}10)$$

where m is the mass of the particle traveling at speed v and m_0 is its <u>rest mass</u>.
Clearly, then, $\text{KE} \neq \frac{1}{2}mv^2$ at high speeds; nor can we substitute $m = m_0/\sqrt{1 - v^2/c^2}$
into the classical expression and get a correct result: $\text{KE} \neq \frac{1}{2}m_0v^2/\sqrt{1 - v^2/c^2}$. The
correct expression is Eq. 39–10.

Equation 39–10 requires some interpretation. First of all, what does the
second term in Eq. 39–10 mean, the m_0c^2? Consistent with the idea that mass is
a form of energy, Einstein called m_0c^2 the *rest energy* of the object. We can rearrange Eq. 39–10 to get $mc^2 = m_0c^2 + \text{KE}$. We call mc^2 the *total energy E* of the
particle (assuming no potential energy), and we see that the total energy equals
the rest energy plus the kinetic energy:

$$\begin{aligned} E &= mc^2 \\ &= m_0c^2 + \text{KE}. \end{aligned} \qquad (39\text{--}11)$$

Here we have Einstein's famous formula $E = mc^2$.

For a particle at rest, its total energy is $E_0 = m_0c^2$ which we have called its
rest energy. This formula mathematically relates the concepts of energy and mass.
But if this idea is to have any meaning from a practical point of view, then mass
ought to be convertible to energy and vice versa. That is, if mass is just one form
of energy then it should be convertible to other forms of energy just as other types
of energy are interconvertible. Einstein suggested that this might be possible, and
indeed this has been experimentally confirmed countless times and forms the basis
for many important processes.

The interconversion of mass and energy is most easily detected in nuclear
and elementary particle physics. For example, the neutral pion (π^0) of rest mass
2.4×10^{-28} kg is observed to decay into pure electromagnetic radiation (photons); the π^0 completely disappears in the process. The amount of electromagnetic
energy produced is found to be exactly equal to that predicted by Einstein's formula, $E = m_0c^2$. The reverse process is also commonly observed in the laboratory:
electromagnetic radiation under certain conditions can be converted into material
particles such as electrons. On a larger scale, the energy produced in nuclear power
plants is a result of the loss in mass of the uranium fuel as it undergoes the process
called fission (Chapter 42). Even the radiant energy we receive from the sun is an
instance of $E = mc^2$; the sun's mass is continually decreasing as it radiates energy
outward.

EXAMPLE 39–6   A π^0 meson ($m_0 = 2.4 \times 10^{-28}$ kg) travels at a speed $v = 0.80c = 2.4 \times 10^8$ m/s. What is its kinetic energy? Compare to a classical calculation.

SOLUTION   The mass of the π^0 at $v = 0.80c$ is

$$m = \frac{m_0}{\sqrt{1 - v^2/c^2}} = \frac{2.4 \times 10^{-28}\text{ kg}}{\sqrt{1 - (0.80)^2}} = 4.0 \times 10^{-28}\text{ kg}.$$

Thus its KE is

$$\begin{aligned} \text{KE} &= (m - m_0)c^2 = (4.0 \times 10^{-28}\text{ kg} - 2.4 \times 10^{-28}\text{ kg})(3.0 \times 10^8\text{ m/s})^2 \\ &= 1.4 \times 10^{-11}\text{ J}. \end{aligned}$$

Notice that the units of mc^2 are kg·m²/s² which is the joule. A classical calculation
would give $\text{KE} = \frac{1}{2}m_0v^2 = \frac{1}{2}(2.4 \times 10^{-28}\text{ kg})(2.4 \times 10^8\text{ m/s})^2 = 6.9 \times 10^{-12}$ J,
but this is not a correct result.

EXAMPLE 39–7 How much energy would be released if the π^0 meson in the above example is transformed by decay completely into electromagnetic radiation?

SOLUTION   The rest energy of the π^0 is

$$E_0 = m_0 c^2 = (2.4 \times 10^{-28} \text{ kg})(3.0 \times 10^8 \text{ m/s})^2 = 2.2 \times 10^{-11} \text{ J}.$$

This is how much energy would be released if the π^0 decayed at rest. If it has $\text{KE} = 1.4 \times 10^{-11}$ J, the total energy released would be 3.6×10^{-11} J.

EXAMPLE 39–8 The energy required or released in nuclear reactions and decays comes from a change in mass between the initial and final particles. In one type of radioactive decay (Section 42–4) an atom of uranium ($m = 232.03714$ u) decays to an atom of thorium ($m = 228.02873$ u) plus an atom of helium ($m = 4.00260$ u) where the masses given are in atomic mass units (Section 4–3; 1 u $= 1.6606 \times 10^{-27}$ kg). Calculate the energy released in this decay.

SOLUTION   The initial mass is 232.03714 u and after the decay it is 228.02873 u $+ 4.00260$ u $= 232.03133$ u, so there is a decrease in mass of 0.00581 u. This mass, which equals $(0.00581 \text{ u})(1.66 \times 10^{-27} \text{ kg}) = 9.64 \times 10^{-30}$ kg, is changed into energy. By $E = mc^2$ we have

$$E = (9.64 \times 10^{-30} \text{ kg})(3.0 \times 10^8 \text{ m/s})^2 = 8.68 \times 10^{-13} \text{ J}.$$

Since 1 MeV $= 1.60 \times 10^{-13}$ J, the energy released is 5.4 MeV.

Equation 39–10 for the kinetic energy can be written in terms of the speed v of the object with the help of Eq. 39–7:

$$\text{KE} = m_0 c^2 \left(\frac{1}{\sqrt{1 - v^2/c^2}} - 1 \right). \tag{39–12}$$

At low speeds, where $v \ll c$, we can expand the square root in Eq. 39–12 using the binomial expansion $(1 + x)^n = 1 + nx + n(n - 1)x^2/2! + \dots$. Then we get

$$\text{KE} \approx m_0 c^2 \left(1 + \frac{1}{2}\frac{v^2}{c^2} + \dots - 1 \right)$$

$$\approx \tfrac{1}{2} m_0 v^2$$

where the dots in the first expression represent very small terms in the expansion which we have neglected since we assumed $v \ll c$. Thus at low speeds, the relativistic form for kinetic energy reduces to the classical form $\text{KE} = \frac{1}{2}m_0 v^2$. This is, of course, what must be. Relativity would not be a valuable theory if it didn't predict accurate results at low speeds as well as at high. Indeed, the other equations of special relativity also reduce to their classical equivalents at ordinary speeds: length contraction, time dilation, and mass increase all disappear for $v \ll c$ since $\sqrt{1 - v^2/c^2} \approx 1$.

A useful relation between the total energy E of a particle and its momentum p can also be derived. Since $E = mc^2$ and $p = mv$ where $m = m_0/\sqrt{1 - v^2/c^2}$ we have

$$E^2 = m^2 c^4 = m^2 c^2 (c^2 + v^2 - v^2)$$
$$= m^2 c^2 v^2 + m^2 c^2 (c^2 - v^2)$$
$$= p^2 c^2 + \frac{m_0^2 c^4 (1 - v^2/c^2)}{1 - v^2/c^2}$$

$$E^2 = p^2c^2 + m_0^2c^4. \qquad (39-13)$$

Thus the total energy can be written in terms of the momentum p or in terms of the kinetic energy (Eq. 39–11).

39–12   The Impact of Special Relativity

A great many experiments have been performed to test the predictions of the special theory of relativity. Within experimental error, no contradictions have been found. The vast majority of scientists have therefore accepted relativity as an accurate description of nature.

At speeds much less than the speed of light, the relativistic formulas reduce to the old classical ones, as we have discussed. We would, of course, hope—or rather, insist—that this be true since Newtonian mechanics works so well for objects moving with speeds $v \ll c$. This insistence that a more general theory (such as relativity) give the same results as a more restricted theory (such as classical mechanics which works for $v \ll c$) is called the **correspondence principle**. The two theories must correspond where their realms of validity overlap. Relativity thus does not contradict classical mechanics. Rather, it is a more general theory, of which classical mechanics is now considered to be a special case.

The importance of relativity is not simply that it gives more accurate results, especially at very high speeds. Much more than that, it has changed the way we view the world. The concepts of space and time are now seen to be relative, and intertwined with one another, whereas before they were considered absolute and separate. Even our concepts of matter and energy have changed: either can be converted to the other. The impact of relativity extends far beyond physics. It has influenced the other sciences, and even the world of art and literature; it has, indeed, entered the general culture.

From a practical point of view, we do not have much opportunity in our daily lives to use the mathematics of relativity. For example, the factor $\sqrt{1 - v^2/c^2}$ which appears in many relativistic formulas has a value of 0.995 when $v = 0.1c$. Thus, for speeds even as high as $0.1c = 3 \times 10^7$ m/s, the factor $\sqrt{1 - v^2/c^2}$ in relativistic formulas gives a numerical correction of less than 1 percent. For speeds less than $0.1c$, or unless mass and energy get interchanged, we thus don't need to use the more complicated relativistic formulas, and can use the simpler classical formulas.

Summary

An *inertial reference frame* is one in which Newton's law of inertia holds. Inertial reference frames can move at constant velocity relative to each other; accelerating reference frames are noninertial.

Einstein's *special theory of relativity* is based on two principles: the *relativity principle*, which states that the laws of physics are the same in all inertial reference frames, and the principle of the *constancy of the speed of light*, which states that the speed of light in empty space has the same value in all inertial reference frames.

One consequence of relativity theory is that two events that are simultaneous in one reference frame may not be simultaneous in another. Other effects are *time dilation*—moving clocks are measured to run slowly; *length contraction*—the length of an object is measured to be shorter when it is moving than when it is at rest; *mass increase*—the mass of a body increases with speed; furthermore, velocity addition must be done in a special way. All these effects are significant only at high speeds, close to the speed of light, which can be thought of as the utimate speed in the universe.

The *Lorentz transformations* relate the positions and times of events in one inertial frame to their positions and times in a second inertial frame.

The theory of relativity has changed our notions of space and time, and of mass and energy. Space and time are seen to be intimately connected, with time being a fourth dimension in addition to space's three dimensions.

Mass and energy are interconvertible; the equation

$$E = mc^2$$

tells how much energy E is needed to create mass m, or vice versa. The law of conservation of energy must include mass as a form of energy. The total energy E and momentum p of a particle are related by $E^2 = p^2c^2 + m_0^2 c^4$.

 Questions _____

1. Give some examples of noninertial reference frames.

2. A woman stands on top of a moving railroad car. She throws a heavy ball straight up (it seems to her) in the air. Ignoring air resistance, will the ball land on the car or behind it?

3. According to the principle of relativity, it is just as legitimate for a person riding in a uniformly moving automobile to say that the car is at rest and the earth is moving beneath it, as it is for a person on the ground to say that the car is moving and the earth is at rest. Do you agree, or are you reluctant to accept this? Discuss the reasons for your response.

4. Does the earth really go around the sun? Or is it also valid to say that the sun goes around the earth? Discuss in view of the first principle of relativity (that there is no best reference frame).

5. If you were on a spaceship traveling at $0.5c$ away from a star, at what speed would the starlight pass you?

6. Will two events that occur at the same place and same time for one observer be simultaneous to a second observer moving with respect to the first?

7. (a) Explain why two events will be simultaneous to each of two observers moving with respect to each other only if the two events occur at the same point in each reference frame. (b) Under what other conditions will two events be simultaneous to each observer?

8. Analyze the thought experiment of Section 39–4 from O_1's point of view. (Make a diagram analogous to Fig. 39–6.)

9. The time-dilation effect says that "moving clocks run slowly." Actually, this effect has nothing to do with motion affecting the functioning of clocks. What then does it deal with?

10. Does time dilation means that time actually passes more slowly in moving reference frames or that it only *seems* to pass more slowly?

11. Today's subways are said to age people prematurely. Suppose that in the future, subway trains which traveled very close to the speed of light could be designed. How do you think this would affect the aging process?

12. A young-looking woman astronaut has just arrived home from a long trip. She rushes up to an old gray-haired man and in the ensuing conversation refers to him as her son. How is this possible?

13. If you were traveling away from earth at a speed of $0.5c$, would you notice a change in your heartbeat? Would your mass, height, or waistline change? What would observers on earth using telescopes say about these things?

14. The predictions of the theory of relativity seem to contradict some of our everyday notions about the world and therefore don't seem to make sense. Examine these notions in detail and determine if any measurable contradiction exists.

15. Discuss how our everyday lives would be different if the speed of light were only 25 m/s.

16. Do mass increase, time dilation, and length contraction occur at ordinary speeds, say 90 km/h?

17. Suppose the speed of light were infinite. What would happen to the relativistic predictions of length contraction, time dilation, and mass increase?

18. Explain how the length-contraction and time-dilation formulas might be used to indicate that c is the limiting speed in the universe.

19. Consider an object of mass m to which is applied a constant force for an indefinite period of time. Discuss how its velocity and mass change with time.

20. A white-hot iron bar is cooled to room temperature. Does its mass change?

21. Does the equation $E = mc^2$ conflict with the conservation of energy principle? Explain.

22. Does $E = mc^2$ apply to particles that travel at the speed of light? Does it apply only to them?

23. An electron is limited to travel at speeds less than c. Does this put an upper limit on the momentum of an electron? If so, what is this upper limit?

24. If mass is a form of energy, does this mean that a spring has more mass when compressed than when relaxed?

25. It is not correct to say that "matter can neither be created nor destroyed." What must we say instead?

26. A neutrino is an elementary particle with zero rest mass that travels at the speed of light. Could you ever catch up to a neutrino that passed you?

 Problems _____

SECTIONS 39–5 AND 39–6

1. (I) A beam of a certain type of elementary particle travels at a speed of 2.85×10^8 m/s. At this speed, the average lifetime is measured to be 2.50×10^{-8} s. What is the particle's lifetime at rest?

2. (I) What is the speed of a beam of pions if their average lifetime is measured to be 3.5×10^{-8} s? At rest, their lifetime is 2.6×10^{-8} s.

3. (I) A spaceship passes you at a speed of $0.80c$. You measure its length to be 90 m. How long would it be when at rest?

4. (I) You are sitting in your sports car when a very fast

sports car passes you at a speed of $0.18c$. A person in the car says his car is 6.00 m long and yours is 6.15 m long. What do you measure for these two lengths?

5. (I) Suppose you decide to travel to a star 65 light-years away. How fast would you have to travel so the distance would be only 20 light-years?

6. (II) A certain star is 36 light-years away. How long would it take a spacecraft traveling $0.98c$ to reach that star from earth as measured by observers (a) on earth; (b) on the spacecraft? (c) What is the distance traveled according to observers on the spacecraft? (d) What will the spacecraft occupants compute their speed to be from the results of (b) and (c)?

7. (II) A friend of yours travels by you in her fast sports car at a speed of $0.760c$. It appears to be 5.80 m long and 1.45 m high. (a) What will be its length and height at rest? (b) How many seconds did you see elapse on your friend's watch when 20.0 s passed on yours? (c) How fast did you appear to be traveling to your friend? (d) How many seconds did she see elapse on your watch when she saw 20.0 s pass on hers?

8. (II) The nearest star to earth is Alpha Centauri, 4.0 light-years away. (a) At what constant velocity must a spacecraft travel from earth if it is to reach the star in 3.0 years time, as measured by travelers on the spacecraft? (b) How long does the trip take according to earth observers?

9. (II) How fast must a pion be traveling to travel 20 m before it decays? The average lifetime, at rest, is 2.6×10^{-8} s.

SECTION 39–8

10. (I) Suppose in Fig. 39–11 that the origins of S and S' overlap at $t = t' = 0$ and that S' moves at speed $v = 30$ m/s with respect to S. In S', a person is resting at a point whose coordinates are $x' = 25$ m, $y' = 20$ m, and $z' = 0$. Calculate this person's coordinates in S (x, y, z) at (a) $t = 2.5$ s, (b) $t = 10.0$ s. Use the Galilean transformation.

11. (I) Repeat problem 10 using the Lorentz transformations and a relative speed $v = 1.50 \times 10^8$ m/s, but choose the time to be (a) 2.5 μs and (b) 10.0 μs.

12. (I) A person on a rocket traveling at $0.50c$ (with respect to the earth) observes a meteor come from behind and pass him at a speed he measures as $0.50c$. How fast is the meteor moving with respect to the earth?

13. (II) Two spaceships leave the earth in opposite directions, each with a speed of $0.50c$ with respect to the earth. (a) What is the velocity of spaceship 1 relative to spaceship 2? (b) What is the velocity of spaceship 2 relative to spaceship 1?

14. (II) A spaceship leaves earth traveling $0.68c$. A second spaceship leaves the first at a speed of $0.86c$ with respect to the first. Calculate the speed of the second ship with respect to earth if it is fired (a) in the same direction the first spaceship is already moving, (b) directly backward toward earth.

15. (II) In problem 10, suppose that the person moves with a velocity whose components are $u'_x = u'_y = 25.0$ m/s. What will be his velocity with respect to S? (Give magnitude and direction.)

16. (II) In problem 11, suppose that the person moves with a velocity (in a rocket) whose components are $u'_x = u'_y = $

2.3×10^8 m/s. What will be his velocity (magnitude and direction) with respect to S?

17. (II) A spaceship traveling $0.66c$ away from earth fires a module with a speed of $0.82c$ at right angles to its own direction of travel (as seen by the first spaceship). What is the speed of this second spaceship, and its direction of travel (relative to the first spaceship's direction), as seen by an observer on earth?

18. (II) If a particle moves in the xy plane of system S (Fig. 39–11) in a direction that makes an angle θ with the x axis, show that it makes an angle θ' in S' given by $\tan \theta' = (\sin \theta)\sqrt{1 - v^2/c^2}/(\cos \theta - v/u)$.

19. (II) A stick of length L_0, at rest in reference frame S, makes an angle θ with the x axis. In reference frame S', which moves to the right with speed $\mathbf{v} = v\mathbf{i}$ with respect to S, determine (a) the length L of the stick, and (b) the angle θ' it makes with the x' axis.

20. (II) In the old West, a marshal riding on a train traveling 50 m/s sees a duel between two men standing on the earth 50 m apart parallel to the train. The marshal's instruments indicate that in his reference frame the two men fired simultaneously. (a) Which of the two men, the first one the train passes (A) or the second (B) should be arrested for firing the first shot? That is, in the gunfighters' frame of reference, who fired first? (b) How much earlier did he fire? (c) Who was struck first?

SECTION 39–9

21. (I) What is the mass of a proton traveling at $v = 0.75c$?

22. (I) At what speed will an object's mass be twice its rest mass?

23. (II) At what speed v will the mass of an object be 1 percent greater than its rest mass?

24. (II) Escape velocity from the earth is 40,000 km/h. What would be the percent increase in mass of a 3.8×10^5-kg spacecraft traveling at that speed?

25. (II) (a) What is the speed of an electron whose mass is 10,000 times its rest mass? Such speeds are reached in the Stanford Linear Accelerator, SLAC. (b) If the electrons travel in the lab through a tube 3.0 km long (as at SLAC), how long is this tube in the electrons' reference frame?

26. (II) Derive a formula showing how the density of an object changes with speed v.

SECTION 39–11

27. (I) How much energy can be obtained from conversion of 1.0 mg of mass? How much mass could this energy raise to a height of 100 m?

28. (I) What is the kinetic energy of an electron whose mass is 5.0 times its rest mass?

29. (I) A certain chemical reaction requires 2.56×10^4 J of energy input for it to go. What is the increase in mass of the products over the reactants?

30. (I) Calculate the rest energy of an electron in joules and in MeV (1 MeV $= 1.60 \times 10^{-13}$ J).

31. (II) Suppose a spacecraft of rest mass 20,000 kg is accelerated to $0.25c$. (a) How much kinetic energy would it have? (b) If you used the classical formula for KE, by what percentage would you be in error?

32. (II) Calculate the mass of a proton ($m_0 = 1.67 \times 10^{-27}$ kg) whose kinetic energy is half its total energy. How fast is it traveling?

33. (II) Calculate the kinetic energy and momentum of a proton ($m_0 = 1.67 \times 10^{-27}$ kg) traveling 8.3×10^7 m/s. By what percentages would your calculations have been in error if you had used classical formulas?

34. (II) What is the speed and momentum of a proton ($m_0 = 1.67 \times 10^{-27}$ kg) whose kinetic energy is half its rest energy?

35. (II) An electron ($m_0 = 9.11 \times 10^{-31}$ kg) is accelerated from rest to speed v by a conservative force. In this process its potential energy decreases by 4.20×10^{-14} J. Determine the electron's speed, v.

36. (II) How many grams of matter would have to be totally destroyed to run a 100-W light bulb for 1 year?

37. (II) How much energy would be required to break a helium nucleus into its constituents, two protons and two neutrons? The rest masses of proton (including an electron), neutron, and helium are 1.00783 u, 1.00867 u, and 4.00260 u respectively. (This is called the *total binding energy* of the $_2^4$He nucleus.)

38. (II) Make a graph of the kinetic energy versus momentum for (*a*) a particle of nonzero rest mass, (*b*) a particle with zero rest mass.

39. (II) Determine the change in mass of a spherical conductor of radius $R = 1.0$ m when it is given a charge $Q = +85\ \mu$C. Consider the mass change due to (*a*) loss of electrons, (*b*) energy increase ($E = mc^2$).

*40. (II) What magnetic field intensity is needed at the 1.0 km radius of the Fermilab synchrotron for 400 GeV protons? Use the relativistic mass. The proton's rest mass is 0.938 GeV/c^2. (See Section 28–8.)

*41. (II) Show that the energy of a particle of charge e in a synchrotron (Section 28–8) in the relativistic limit ($v \approx c$) is given by E (in eV) = Brc where B is magnetic field strength and r is the radius of the orbit (SI units).

42. (II) Show that the kinetic energy (KE) of a particle of rest mass m_0 is related to its momentum p by the equation $p = \sqrt{(\text{KE})^2 + (\text{KE})(m_0c^2)}/c$.

43. (II) A pi meson of rest mass m_π decays at rest into a mu meson (rest mass m_μ) and a neutrino of zero rest mass. Show that the kinetic energy of the muon is $\text{KE}_\mu = (m_\pi - m_\mu)^2 c^2 / 2m_\pi$.

44. (III) (*a*) In reference frame S, a particle has momentum $\mathbf{p} = p_x\mathbf{i}$ along the positive x axis. Show that in frame S', which moves with speed v as in Fig 39–11, the momentum has components

$$p'_x = \frac{p_x - vE/c^2}{\sqrt{1 - v^2/c^2}}$$

$$p'_y = p_y$$

$$p'_z = p_z$$

$$E' = \frac{E - p_x v}{\sqrt{1 - v^2/c^2}}.$$

(These transformation equations hold, actually, for any direction of **p**.) (*b*) Show that p_x, p_y, p_z, E/c transform according to the Lorentz transformation in the same way as x, y, z, ct.

Early Quantum Theory and Models of the Atom

40

The second aspect of the revolution that shook the world of physics in the early part of the twentieth century (the first half was Einstein's theory of relativity) is the quantum theory. Unlike the special theory of relativity—whose basic tenets were put forth mainly by one person in a single year—the revolution of quantum theory required almost three decades to unfold, and many scientists contributed to its development. It began in 1900 with Planck's quantum hypothesis and culminated in the mid-1920s with the theory of quantum mechanics of Schrödinger and Heisenberg which has been so effective in explaining the structure of matter.

 40–1 Planck's Quantum Hypothesis

One of the observations that was unexplained at the end of the nineteenth century was the spectrum of light emitted by hot objects. We saw in Chapter 19 that all objects emit radiation whose total intensity is proportional to the fourth power of the Kelvin temperature ($\propto T^4$). At normal temperatures, we are not aware of this electromagnetic radiation because of its low intensity. At higher temperatures, there is sufficient infrared radiation that we can feel heat if we are close to the object. At still higher temperatures (on the order of 1000 K) objects actually glow, such as a red-hot electric stove burner or element in a toaster. At temperatures above 2000 K objects glow with a yellow or whitish color, such as white-hot iron and the filament of a light bulb. As the temperature increases, the electromagnetic radiation emitted by bodies is strongest at higher and higher frequencies.

The spectrum of light emitted by a hot dense object is shown in Fig. 40–1 for an idealized *blackbody*; such a body would absorb all the radiation falling on it and the radiation it emits, called **blackbody radiation**, is the easiest to deal with. As can be seen, the spectrum contains a continuous range of frequencies; such a continuous spectrum is emitted by any heated solid or liquid, and even by dense gases. The 6000-K curve in Fig. 40–1, corresponding to the temperature of the sun (not quite a blackbody), peaks in the visible part of the spectrum.

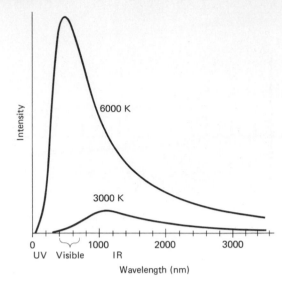

FIGURE 40–1

Spectrum of frequencies emitted by a blackbody at two
different temperatures.

For lower temperatures the total radiation drops considerably and the peak occurs
at higher wavelengths; hence the blue end of the visible spectrum (and the UV)
is relatively weaker. It is found that the wavelength at the peak of the spectrum,
λ_p, is related to the Kelvin temperature T by:

$$\lambda_p T = 2.90 \times 10^{-3} \text{ m·k}.$$

This is known as *Wien's displacement law* and for the sun's temperature gives
us $\lambda_p = (2.90 \times 10^{-3} \text{ m·k})/(6.0 \times 10^3 \text{ k}) = 500$ nm, which is in the visible spectrum (Fig. 40–1).

A major problem facing scientists in the 1890s was to explain blackbody
radiation. Maxwell's electromagnetic theory had predicted that oscillating electric
charges produce electromagnetic waves, and the radiation emitted by a hot object
could be due to the oscillations of electric charges in the molecules of the material.
Although this would explain where the radiation came from, it did not correctly
predict the observed spectrum of emitted light. Two important theoretical curves
based on classical ideas were those proposed by Wien (1896) and by Rayleigh
(1900); the latter was slightly modified later by Jeans and since then has been
known as the Rayleigh-Jeans theory. As experimental data came in, it became
clear that neither Wien's nor the Rayleigh-Jeans formulations were in accord with
experiment: Wien's was accurate at short wavelength but deviated from experiment at longer wavelengths whereas the reverse was true for the Rayleigh-Jeans
theory (see Fig. 40–2).

The break came in late 1900 when Max Planck (1858–1947) proposed an
empirical formula that nicely fit the data. He then sought a theoretical basis
for the formula and within 2 months found that he could obtain the formula by
making a new and radical (though not so recognized at the time) assumption:
that the energy distributed among the molecular oscillators is not continuous
but instead consists of a finite number of very small discrete amounts, each
related to the frequency of oscillation by

$$E_{\min} = hf.$$

Here h is a constant, now called *Planck's constant*, whose value was estimated
by Planck by fitting his formula[†] for the blackbody radiation curve to experiment.
The value accepted today is

$$h = (6.626176 \pm 0.000036) \times 10^{-34} \text{ J·s}$$

[†] Planck's blackbody radiation formula is

$$\text{Intensity} = f(\lambda, T) = \frac{8\pi hc\lambda^{-5}}{e^{-hc/\lambda T} - 1}.$$

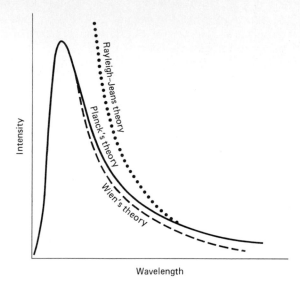

FIGURE 40–2

Comparison of the Wien and the Rayleigh-Jeans theories
to that of Planck which closely follows experiment.

or, for most computations,

$$h = 6.626 \times 10^{-34} \text{ J} \cdot \text{s}.$$

Planck's assumption suggests that the energy of any molecular vibration could
be only some whole number multiple of hf:

$$E = nhf, \qquad n = 1, 2, 3, \ldots . \tag{40–1}$$

This is often called *Planck's quantum hypothesis* ("quantum" means "fixed
amount") although little attention was brought to this point at the time. In fact,
it appears that Planck considered it more as a mathematical device to get the
"right answer" rather than as a discovery comparable to those of Newton.
Planck himself continued to seek a classical explanation for the introduction
of h. The recognition that this was an important and radical innovation did not
come until later, after about 1905 when others, particularly Einstein, entered the
field.

The quantum hypothesis, Eq. 40–1, as we accept it today, says that the
energy can be $E = hf$, or $2hf$, or $3hf$, and so on, but there cannot be vibrations whose energy lies between these values. That is, energy would not be a
continuous quantity as had been believed for centuries; rather it is *quantized*—
it exists only in discrete amounts. The smallest amount of energy possible (hf)
is called the *quantum of energy*. Another way of expressing the quantum hypothesis is that not just any amplitude of vibration is possible. The possible values for
the amplitude are related to the frequency f.

 ## 40–2 Photon Theory of Light and
the Photoelectric Effect

In 1905, the same year as he introduced the special theory of relativity, Einstein
made a bold extension of the quantum idea by proposing a new theory of light.
Planck's work suggested that the vibrational energy of molecules in a radiating
object is quantized with energy $E = nhf$. Einstein reasoned that if the energy
of the molecular oscillators is thus quantized, then to conserve energy, the light
ought to be emitted in packets or quanta, each with an energy

$$E = hf. \tag{40–2}$$

Again h is Planck's constant. Since all light ultimately comes from a radiating
source, this suggests that perhaps *light is transmitted as tiny particles*, or **photons**,

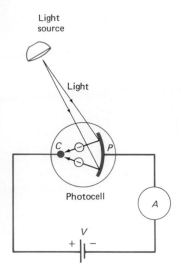

Light source

Light

C   P

Photocell

A

V
+ | | −

FIGURE 40–3

Photoelectric effect.

as they are now called, rather than as waves. This, too, was a radical departure from classical ideas. Einstein proposed a simple test of the quantum theory of light: quantitative measurements on the photoelectric effect.

The **photoelectric effect** refers to the fact that when light shines on a metal surface, electrons are emitted from the surface. (The photoelectric effect occurs in other materials but is most easily observed with metals.) This effect can be observed using the apparatus shown in Fig. 40–3. A metal plate P along with a smaller electrode C are placed inside an evacuated glass tube, called a *photocell*. The two electrodes are connected to an ammeter and a source of emf as shown. When the photocell is in the dark, the ammeter reads zero. But when light of sufficiently high frequency is shone on the plate, the ammeter indicates a current flowing in the circuit. To explain completion of the circuit we can imagine electrons flowing across the tube from the plate to the "collector" C as shown in the diagram.

That electrons should be emitted when light shines on a metal is consistent with the electromagnetic (EM) wave theory of light, since the electric field of an EM wave could exert a force on electrons in the metal and thrust some of them out. Einstein pointed out, however, that the wave theory and the photon theory of light give very different predictions on the details of this effect. For example, one thing that can be measured with the apparatus of Fig. 40–3 is the maximum kinetic energy ($\mathrm{KE_{max}}$) of the emitted electrons. This can be done by using a variable voltage source and reversing the terminals so that electrode C is negative and P is positive. The electrons emitted from P will be repelled by the negative electrode, but if this reverse voltage is small enough, the fastest electrons will still reach C and there will be a current in the circuit. If the reversed voltage is increased, a point is reached where the current reaches zero—no electrons have sufficient kinetic energy to reach C. This is called the "stopping potential" V_0, and from its measurement $\mathrm{KE_{max}}$ can be determined:

$$\mathrm{KE_{max}} = eV_0.$$

Now let us examine the details of the photoelectric effect in view of the wave theory versus Einstein's particle theory. First the wave theory, assuming monochromatic light. The two important properties of a light wave are its intensity and its frequency (or wavelength). When these two quantities are varied, the wave theory makes the following predictions.

1. If the light intensity is increased, the number of electrons ejected and their maximum KE should be increased because the higher intensity means a greater electric field amplitude, and the greater electric field should thrust electrons out with higher speed.

2. The frequency of the light should not affect the KE of the ejected electrons. Only the intensity should affect $\mathrm{KE_{max}}$.

The photon theory makes completely different predictions. First we note that in a monochromatic beam, all photons have the same energy ($= hf$). Increasing the intensity of the light beam means increasing the number of photons in the beam but does not affect the energy of each as long as the frequency is not changed. According to Einstein's theory, an electron is ejected from the metal by a collision with a single photon. In the process, all the photon energy is transferred to the electron and the photon ceases to exist. Since electrons are held in the metal by attractive forces, some minimum energy W_0 (called the *work function*, which is on the order of a few electron volts for most metals) is required just to get an electron out of the surface. If the frequency f of the incoming light is so low that hf is less than W_0, then the photons will not have enough energy to eject any electrons at all. If $hf > W_0$, then electrons will be ejected and energy will be conserved in the process. That is, the input energy (of the photon), hf, will equal the outgoing KE of the electron plus the energy required to get it out of the metal W:

$$hf = \mathrm{KE} + W. \qquad (40\text{–}3a)$$

For the least tightly held electrons, W is the work function W_0, and KE in this equation becomes KE_{max}:

$$hf = \text{KE}_{max} + W_0. \qquad (40\text{–}3b)$$

Many electrons will require more energy than the bare minimum (W_0) to just get out of the metal, and thus the KE of such electrons will be less than the maximum.

From these considerations, the photon theory makes the following predictions.

1. An increase in intensity of the light beam means more photons are incident, so more electrons will be ejected; but since the energy of each photon is not changed, the maximum KE of electrons is not changed.

2. If the frequency of the light is increased, the maximum KE of the electrons increases linearly, according to Eq. 40–3b. That is,

$$\text{KE}_{max} = hf - W_0.$$

This is plotted in Fig. 40–4.

3. If the frequency f is less than the "cutoff" frequency f_0, where $hf_0 = W_0$, no electrons will be ejected at all.

These predictions of the photon theory are clearly very different from the predictions of the wave theory. In 1913–1914 careful experiments were carried out by R. A. Millikan. The results were fully in agreement with Einstein's photon theory.

A number of other experiments were carried out in the early twentieth century which also supported the photon theory. One of these was the **Compton effect** (1923), named after its discoverer, A. H. Compton (1892–1962). Compton scattered short-wavelength light (actually X rays) from various materials. He found the scattered light had a slightly lower frequency than did the incident light, indicating a loss of energy. This, he showed, could be explained on the basis of the photon theory as incident photons colliding with electrons of the material, Fig. 40–5; he applied the laws of conservation of energy and momentum to such collisions and found that the predicted energies of scattered photons was in accord with experimental results. (See problem 18.) Thus the photon theory of light rests on a firm experimental foundation.

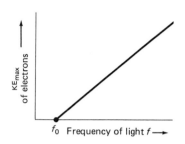

FIGURE 40–4

Photoelectric effect: maximum kinetic energy of ejected electrons increases linearly with frequency of incident light. No electrons are emitted if $f < f_0$.

EXAMPLE 40–1 Calculate the energy of a photon of blue light, $\lambda = 450$ nm.

SOLUTION Since $f = c/\lambda$, we have

$$E = hf = \frac{hc}{\lambda} = \frac{(6.63 \times 10^{-34}\ \text{J·s})(3.0 \times 10^8\ \text{m/s})}{(4.5 \times 10^{-7}\ \text{m})} = 4.4 \times 10^{-19}\ \text{J},$$

or $(4.4 \times 10^{-19}\ \text{J})/(1.6 \times 10^{-19}\ \text{J/eV}) = 2.7$ eV.

FIGURE 40–5

Compton effect.

EXAMPLE 40–2 What is the maximum kinetic energy and speed of an electron ejected from a sodium surface whose work function is $W_0 = 2.28$ eV when illuminated by light of wavelength (a) 410 nm; (b) 550 nm?

SOLUTION (a) For $\lambda = 410$ nm, $hf = hc/\lambda = 4.85 \times 10^{-19}$ J or 3.03 eV. From Eq. 40–3b, $\text{KE}_{max} = 3.03$ eV $- 2.28$ eV $= 0.75$ eV or 1.2×10^{-19} J. Since $\text{KE} = \frac{1}{2}mv^2$ where $m = 9.1 \times 10^{-31}$ kg,

$$v = \sqrt{2\,\text{KE}/m} = 5.1 \times 10^5\ \text{m/s}.$$

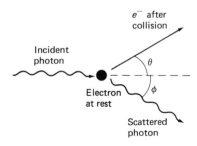

Notice that we used the nonrelativistic equation for KE. If v had turned out to be more than about $0.1c$, our calculation would have been inaccurate by more than a percent or so, and we would probably prefer to redo it using the relativistic form (Eq. 39–12).

(b) For $\lambda = 550$ nm, $hf = 3.60 \times 10^{-19}$ J $= 2.25$ eV. Since this photon energy is less than the work function, no electrons are ejected.

  EXAMPLE 40–3   In *photosynthesis*, which is the process by which pigments such as chlorophyll in plants use the energy of sunlight to change CO_2 to useful carbohydrate, it is believed that about nine photons are needed to change one molecule of CO_2 to carbohydrate and O_2. Assuming light of wavelength $\lambda = 670$ nm (chlorophyll absorbs most strongly in the range 650 nm to 700 nm), how efficient is the photosynthetic process? The reverse chemical reaction has a heat of combustion of 4.9 eV/molecule of CO_2.

SOLUTION   The energy of nine photons, each of energy $hf = hc/\lambda$ is $(9)(6.6 \times 10^{-34}$ J·s$)(3.0 \times 10^8$ m/s$)/(6.7 \times 10^{-7}$ m$) = 2.7 \times 10^{-18}$ J or 17 eV. Thus the process is (4.9 eV/17 eV) = 29 percent efficient.

40–3   Photons and Pair Production

The photon is truly a relativistic particle—it travels at the speed of light. Thus we must use relativistic formulas for dealing with its mass, energy, and momentum. The mass m of any particle is given by $m = m_0/\sqrt{1 - v^2/c^2}$. Since $v = c$ for a photon, the denominator is zero. So the rest mass, m_0, of a photon must also be zero, or its energy $E = mc^2$ would be infinite. Of course a photon is never at rest. The momentum of a photon, from Eq. 39–13 with $m_0 = 0$, is

$$p = \frac{E}{c}.$$

Since $E = hf$, the momentum of a photon is related to its wavelength by

$$p = \frac{hf}{c} = \frac{h}{\lambda}. \tag{40–4}$$

A photon has energy, and it can be changed into matter. The most common process is the production of an electron and a positron, Fig. 40–6. (A positron has the same mass as an electron, but the opposite charge, $+e$.)[†] This is called **pair production** and the photon disappears in the process. This is an example of rest mass being created from pure energy, and it occurs in accord with Einstein's equation $E = mc^2$. Notice that a photon cannot create a single electron since electric charge would not then be conserved.

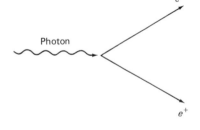

FIGURE 40–6

Pair production: a photon disappears and produces an electron and a positron.

  EXAMPLE 40–4   What is the minimum energy of a photon, and its wavelength, that can produce an electron-positron pair?

SOLUTION   Because $E = mc^2$, the photon must have energy $E = 2(9.1 \times 10^{-31}$ kg$)(3.0 \times 10^8$ m/s$)^2 = 1.64 \times 10^{-13}$ J or 1.02 MeV. A photon with less energy cannot undergo pair production. Since $E = hf = hc/\lambda$, the wavelength of a 1.02-MeV photon is

$$\lambda = \frac{hc}{E} = \frac{(6.6 \times 10^{-34} \text{ J·s})(3.0 \times 10^8 \text{m/s})}{(1.64 \times 10^{-13} \text{ J})} = 1.2 \times 10^{-12} \text{ m},$$

which is 0.0012 nm. Thus the wavelength must be very short. Such photons are in the gamma-ray (or very short X-ray) region of the electromagnetic spectrum.

Pair production cannot occur in empty space for energy and momentum could not simultaneously be conserved. In Example 40–4, for instance, energy is con-

---

[†] Positrons do not last long in nature because, when they collide with an electron, they annihilate each other and produce two or more photons. This pair annihilation process is also an example of $E = mc^2$.

served, but the electron–positron pair has no momentum to carry away the initial momentum of the photon. Indeed, it can be shown that at any energy an additional massive object, such as an atomic nucleus, must take part in the interaction to carry off some of the momentum.

40–4 Wave–Particle Duality; the Principle of Complementarity

The photoelectric effect, the Compton effect, and other experiments (see, for example, Section 41–8 on X rays) have placed the particle theory of light on a firm experimental basis. But what about the classic experiments of Young and others (Chapters 36 and 37) on interference and diffraction which showed that the wave theory of light also rests on a firm experimental basis?

We seem to be in a dilemma. Some experiments indicate that light behaves like a wave; and others indicate that it behaves like a stream of particles. These two theories seem to be incompatible but both have been shown to have validity. Physicists have finally come to the conclusion that this duality of light must be accepted as a fact of life. It is referred to as the **wave-particle duality**. Apparently, light is a more complex phenomenon than just a simple wave or a simple beam of particles.

To clarify the situation, the great Danish physicist Niels Bohr (1885–1962) proposed his famous **principle of complementarity**. It states that to understand any given experiment, we must use either the wave or the photon theory, but not both. Yet we must be aware of both the wave and particle aspects of light if we are to have a full understanding of light. Therefore these two aspects of light complement one another.

It is not possible to "visualize" this duality. We cannot picture a combination of wave and particle. Instead, we must recognize that the two aspects of light are different "faces" that light shows to experimenters.

Part of the difficulty stems from how we think. Visual pictures (or models[†]) in our minds are based on what we see in the everyday world. We apply the concepts of waves and particles to light because in the macroscopic world we see that energy is transferred from place to place by these two methods. We cannot see directly whether light is a wave or particle—so we do indirect experiments. To explain the experiments we apply the models of waves or of particles to the nature of light. But these are abstractions of the human mind. When we try to conceive of what light really "is," we insist on a visual picture. Yet there is no reason why light should conform to these models (or visual images) taken from the macroscopic world. The "true" nature of light—if that means anything—is not possible to visualize. The best we can do is recognize that our knowledge is limited to the indirect experiments, and that in terms of everyday language and images, light reveals both wave and particle properties.

It is worth noting that Einstein's equation $E = hf$ itself links the particle and wave properties of a light beam. In this equation E refers to the energy of a particle; and on the other side of the equation we have the frequency f of the corresponding wave.

40–5 Wave Nature of Matter

In 1923 Louis de Broglie (1892–) extended the idea of the wave-particle duality. He sensed deeply the symmetry in nature and argued that if light sometimes behaves like a wave and sometimes like a particle, then perhaps those things in

---

[†] As we discussed in Chapter 1, a model is a kind of analogy or mental picture using something familiar for the purpose of describing a certain set of phenomena at a deeper level than we can see directly.

nature thought to be particles—such as electrons and other material objects—might also have wave properties. De Broglie proposed that the wavelength of a material particle would be related to its momentum in the same way as for a photon,[†] Eq. 40–4, $p = h/\lambda$. That is, for a particle of mass m traveling with speed v, the wavelength λ is given by

$$\lambda = \frac{h}{mv}. \tag{40–5}$$

This is sometimes called the *de Broglie wavelength* of a particle.

EXAMPLE 40–5   Calculate the de Broglie wavelength of a 0.20-kg ball moving with a speed of 15 m/s.

SOLUTION $\lambda = h/mv = (6.6 \times 10^{-34} \text{ J·s})/(0.20 \text{ kg})(15 \text{ m/s})$
$$= 2.2 \times 10^{-34} \text{ m.}$$

This is an incredibly small wavelength. Even if the speed were extremely small, say 10^{-4} m/s, the wavelength would be about 10^{-29} m. Indeed, the wavelength of any ordinary object is much too small to be measured and detected. The problem is that the properties of waves, such as interference and diffraction, are significant only when the size of objects or slits is not much larger than the wavelength. And there are no known objects or slits to diffract waves only 10^{-30} m long, so the wave properties of ordinary objects go undetected.

But tiny elementary particles, such as electrons are another matter. Since the mass m appears in the denominator in Eq. 40–5, a very small mass should give a much larger wavelength.

EXAMPLE 40–6   Determine the wavelength of an electron that has been accelerated through a potential difference of 100 V.

SOLUTION   We assume that the speed of the electron will be much less than c so we use nonrelativistic mechanics. (If this assumption were to come out wrong, we would have to recalculate using relativistic formulas.) The gain in KE will equal the loss in PE, so $\frac{1}{2}mv^2 = eV$ and

$$v = \sqrt{2\,eV/m} = \sqrt{(2)(1.6 \times 10^{-19}\text{C})(100 \text{ V})/(9.1 \times 10^{-31} \text{ kg})} = 5.9 \times 10^6 \text{ m/s.}$$

Then

$$\lambda = \frac{h}{mv} = \frac{(6.6 \times 10^{-34} \text{ J·s})}{(9.1 \times 10^{-31} \text{ kg})(5.9 \times 10^6 \text{ m/s})} = 1.2 \times 10^{-10} \text{ m}$$

or 0.12 nm.

FIGURE 40–7

Diffraction pattern of electrons scattered from aluminum foil, as recorded on film. (Courtesy of Education Development Center, Newton, Mass. 02160.)

From this example, we see that electrons can have wavelengths on the order of 10^{-10} m. Although small, this wavelength can be detected: the spacing of atoms in a crystal is on the order of 10^{-10} m and the orderly array of atoms in a crystal could be used as a type of diffraction grating, as was done earlier for X rays (see Section 37–11). C. J. Davisson and L. H. Germer performed the crucial experiment; they scattered electrons from the surface of a metal crystal and, in early 1927, observed that the electrons came off in regular peaks. When they interpreted these peaks as a diffraction pattern, the wavelength of the diffracted electron wave was found to be just that predicted by de Broglie, Eq. 40–5. In the same year, G. P. Thomson (son of J. J. Thomson), using a different experimental arrangement, also detected diffraction of electrons (see Fig. 40–7). Later experiments showed that protons, neutrons, and other particles also have wave properties.

[†] de Broglie chose this formula (rather than, say, $E = hf$, which is not consistent with $p = h/\lambda$ for a particle with rest mass), because it allowed him to explain, or give a reason for, Bohr's model of the atom. This will be discussed in Section 40–10.

Thus the wave–particle duality applies to material objects as well as to light. The principle of complementarity applies to matter as well. That is, we must be aware of both the particle and wave aspects in order to have an understanding of matter, including electrons. But again we must recognize that a visual picture of a "wave–particle" is not possible.

We might ask ourselves: "What is an electron?" The early experiments of J. J. Thomson (see Section 28–5) indicated a glow in a tube that moved when a magnetic field was applied. The results of these and other experiments were best interpreted as being caused by tiny negatively charged particles which we now call electrons. No one, however, has actually seen an electron directly. The drawings we sometimes make of electrons as tiny spheres with a negative charge on them are merely convenient pictures (now recognized to be inaccurate). Again we must rely on experimental results, some of which are best interpreted using the particle model and others using the wave model. These again are mere pictures that we use to extrapolate from the macroscopic world to the tiny microscopic world of the atom. And there is no reason to expect that these models somehow reflect the reality of an electron. We thus use a wave or a particle model (whichever works best in a situation) so that we can talk about what is happening. But we shouldn't be led to believe that an electron *is* a wave or a particle. Instead we could say that an electron is the set of its properties that we can measure. Bertrand Russell said it well when he wrote that an electron is "a logical construction."

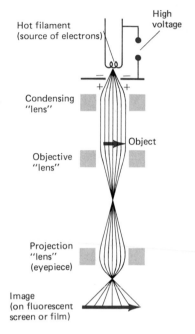

FIGURE 40–8

Transmission electron microscope. The squares represent magnetic field coils for the "magnetic lenses," which bend the electron paths and bring them to a focus as shown.

 *40–6 Electron Microscopes

The idea that electrons have wave properties led to the development of the *electron microscope*, which can produce images of much greater magnification than a light microscope. Figures 40–8 and 40–9 are diagrams of two types, the *transmission electron microscope*, which produces a two-dimensional image, and the more recently developed *scanning electron microscope*, which produces images with a three-dimensional quality. In each design, the objective and eyepiece lenses are actually magnetic fields produced by current-carrying coils of wire. Photographs using each type are shown in Fig. 40–10.

Electrons accelerated by voltages on the order of 10^5 eV have wavelengths on the order of 0.004 nm. The maximum resolution obtainable would be on this order (see Sections 37–4 and 37–5) but in practice aberrations in the magnetic lenses limit the resolution in transmission electron microscopes to at best about

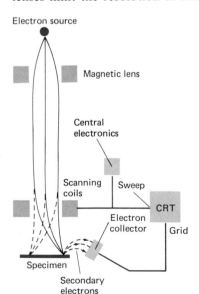

FIGURE 40–9

Scanning electron microscope. Scanning coils move an electron beam back and forth across the specimen. Secondary electrons produced when the beam strikes the specimen are collected and modulate intensity of the beam in the CRT to produce a picture.

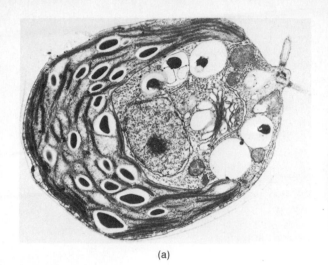

(a)

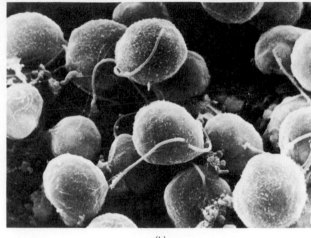

(b)

FIGURE 40–10

Micrographs of the green alga *Tetraspora* sp.: (a) made by the transmission electron microscope and (b) made by the scanning electron microscope. (Courtesy of J. D. Pickett–Heaps.)

0.2 to 0.5 nm. This is still 10^3 times finer than that attainable with a light microscope, and corresponds to a useful magnification of about a million. Such magnifications are difficult to attain, and more common magnifications are 10^4 to 10^5. The maximum resolution attainable with a scanning electron microscope is somewhat less, about 5 to 10 nm at best.

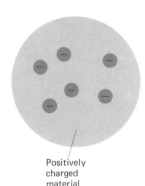

Positively
charged
material

FIGURE 40–11

Plum-pudding model of the atom.

  40–7 Early Models of the Atom

The idea that matter is made up of atoms was accepted by most scientists by 1900. With the discovery of the electron in the 1890s, scientists began to think of the atom itself as having a structure and that electrons were part of that structure. We now trace, in the remainder of this chapter and the next, the development of our modern understanding of the atom, and of the quantum theory with which it is intertwined.[†]

A typical model of the atom in the 1890s visualized the atom as a homogeneous sphere of positive charge inside of which there were the negatively charged electrons, a little like plums in a pudding, Fig. 40–11. J. J. Thomson, soon after his discovery of the electron in 1897, argued that the electrons in this model should be moving.

Around 1911, Ernest Rutherford (1871–1937) and his colleagues performed experiments whose results contradicted Thomson's model of the atom. In these experiments a beam of positively charged "alpha (α) particles" was directed at a thin sheet of metal foil such as gold, Fig. 40–12a. (These newly discovered α particles were emitted by certain radioactive materials and were soon shown to be ionized helium atoms—see Chapter 42.) It was expected from Thomson's model that the alpha particles would not be deflected significantly since electrons are so much lighter than alpha particles, and the alpha particles didn't approach any massive concentration of positive charge to strongly repel them. The experimental results completely contradicted these predictions. It was found that most of the alpha particles passed through the foil unaffected, as if the foil were mostly empty space. And of those deflected, a few were deflected at very large angles—some even in nearly the direction from which they had come. This could happen, Rutherford reasoned, only if the positively charged alpha particles were being repelled by a

---

[†] Some readers may say: "Tell us the facts as we know them today, and don't bother us with the historical background and its outmoded theories." Such an approach would not only ignore the creative aspect of science and thus give a false impression of how science develops, but it would not really be possible to understand today's view of the atom and the quantum without discussing the concepts that led to it.

massive positive charge concentrated in a very small region of space (see Fig. 40–12b). He theorized that the atom must consist of a tiny but massive positively charged nucleus, containing over 99.9 percent of the mass of the atom, surrounded by electrons some distance away. The electrons would be moving in orbits about the nucleus—much like the planets move around the sun—because if they were at rest they would fall into the nucleus due to electrical attraction, Fig. 40–13. Rutherford's experiments suggested that the nucleus must have a radius of about 10^{-15} to 10^{-14} m. From kinetic theory, and especially Einstein's analysis of Brownian movement (see Section 17–1), the radius of atoms was estimated to be about 10^{-10} m. Thus the electrons would seem to be at a distance from the nucleus of about 10,000 to 100,000 times the radius of the nucleus itself; so an atom would be mostly empty space.

Rutherford's "planetary" model of the atom was a major step toward how we view the atom today. It was, however, not a complete model and presented some major problems as we shall see.

 ## 40–8 Atomic Spectra: Key to the Structure of the Atom

At the beginning of this chapter we saw that heated solids (as well as liquids and dense gases) emit light with a continuous spectrum of wavelengths. This radiation is assumed to be due to oscillations of atoms and molecules which are largely governed by the interaction of each atom or molecule with its neighbors.

Rarefied gases can also be excited to emit light. This is done by intense heating or more commonly by applying a high voltage to a "discharge tube" containing the gas at low pressure, Fig. 40–14. The radiation from excited gases had been observed early in the nineteenth century, and it was found that the spectrum was not continuous, but *discrete*. Since excited gases emit light of only certain wavelengths, when this light is analyzed through the slit of a spectroscope or spectrometer, a *line spectrum* is seen rather than a continuous spectrum. The line spectrum emitted by a number of elements in the visible region was shown in Fig. 37–19 (color plate). The *emission spectrum* is characteristic of the material and can serve as a type of "fingerprint" for identification of the gas. As we also saw (Chapter 37), if a continuous spectrum passes through a gas, dark lines are observed in the spectrum corresponding to lines normally emitted by the gas. This is called an *absorption spectrum*, and it became clear that gases absorb light at the same frequencies at which they emit. With the use of film sensitive to ultraviolet and to infrared light, it was found that gases emit and absorb discrete frequencies in these regions as well as in the visible.

For our purposes here, the importance of the line spectra is that they are emitted (or absorbed) by gases that are not dense. In such thin gases, the atoms are far apart on the average and hence the light emitted or absorbed must be by individual atoms rather than through interactions between atoms as in a solid, liquid, or dense gas. Thus the line spectra serve as a key to the structure of the atom: any theory of atomic structure must be able to explain why atoms emit light only of discrete wavelengths, and it should be able to predict what these frequencies are.

Hydrogen is the simplest atom—it has only one electron orbiting its nucleus. It also has the simplest spectrum. The spectrum of most atoms shows little apparent regularity; but the spacing between lines in the hydrogen spectrum decreases in a regular way (Fig. 40–15). Indeed, in 1885, J. J. Balmer (1825–1898) showed that the four visible lines in the hydrogen spectrum (measured to be 656 nm, 486 nm, 434 nm, and 410 nm) would fit the following formula

$$\frac{1}{\lambda} = R\left(\frac{1}{2^2} - \frac{1}{n^2}\right), \qquad n = 3, 4, \ldots \tag{40–6}$$

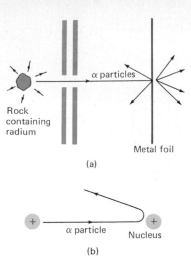

FIGURE 40–12

(a) Experimental setup for Rutherford's experiment: α particles emitted by radium strike metallic foil and some rebound backward; (b) backward rebound of α particles explained as repulsion from heavy positively charged nucleus.

FIGURE 40–13

Rutherford's model of the atom, in which electrons orbit a tiny positive nucleus (not to scale). The atom is visualized as mostly empty space.

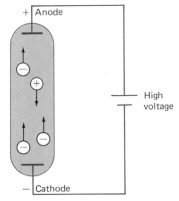

FIGURE 40–14

Gas discharge tube.

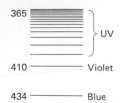

365 ⟩ UV

410 ——————— Violet

434 ——————— Blue

486 ——————— Blue-green

λ
(nm)

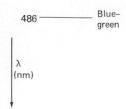

656 ——————— Red

FIGURE 40–15

Balmer series of lines for
hydrogen.

where n takes on the values 3, 4, 5, 6 for the four lines, and R, called the *Rydberg constant,*[†] has the value $R = 1.097 \times 10^7 \, \text{m}^{-1}$. Later it was found that this **Balmer series** of lines extended into the UV region, ending at $\lambda = 365$ nm, as shown in Fig. 40–15. Balmer's formula, Eq. 40–6, also worked for these lines with higher integer values of n. The lines near 365 nm became too close together to distinguish but the limit of the series at 365 nm corresponds to $n = \infty$ (so $1/n^2 = 0$ in the formula).

Later experiments on hydrogen showed that there were other series of lines in the UV and IR and these had a pattern just like the Balmer series, but at different wavelengths. Each of these series was found to fit a formula resembling Eq. 40–6 but with the $1/2^2$ replaced by $1/1^2$, $1/3^2$, $1/4^2$, and so on. For example, the so-called *Lyman series* contains lines with wavelengths from 91 nm to 122 nm and fits the formula

$$\frac{1}{\lambda} = R\left(\frac{1}{1^2} - \frac{1}{n^2}\right), \qquad n = 2, 3, \ldots.$$

And the wavelengths of the *Paschen series* fit

$$\frac{1}{\lambda} = R\left(\frac{1}{3^2} - \frac{1}{n^2}\right), \qquad n = 4, 5, \ldots.$$

The Rutherford model, as it stood, was unable to explain why atoms emit line spectra. It had other difficulties as well. According to the Rutherford model, electrons orbit the nucleus and since their paths are curved the electrons are accelerating. Hence they should give off light like any other accelerating electric charge (Chapter 33). Then, since energy is conserved, the electron's own energy must decrease to compensate. Hence electrons would be expected to spiral into the nucleus. As they spiraled inward, their frequency would increase gradually and so too would the frequency of the light emitted. Thus the two main difficulties of the Rutherford model are these: (1) it predicts that light of a continuous range of frequencies will be emitted, whereas experiment shows line spectra; (2) it predicts that atoms are unstable—electrons quickly spiral into the nucleus—but we know that atoms in general are stable, since the matter around us is stable.

Clearly Rutherford's model was not sufficient. Some sort of modification was needed, and it was Niels Bohr who provided it by adding an essential idea—the quantum hypothesis.

 40–9 The Bohr Model

Bohr had studied in Rutherford's laboratory for several months in 1912 and was convinced that Rutherford's planetary model of the atom had validity. But in order to make it work he felt that the newly developing quantum theory would somehow have to be incorporated in it. The work of Planck and Einstein had shown that in heated solids, the energy of oscillating electric charges must change discontinuously—from one discrete energy state to another with the emission of a quantum of light. Perhaps, Bohr argued, the electrons in an atom also cannot lose energy continuously but must do so in quantum "jumps." In working out his theory during the next year, Bohr postulated that electrons move about the nucleus in circular orbits, but that only certain orbits are allowed. He further postulated that an electron in each orbit would have a definite energy and would move in the orbit *without radiating energy* (even though this violated classical ideas). He thus called the possible orbits **stationary states**. Light is emitted, he hypothesized, only when an electron jumps from one stationary state to another of lower energy. When such a jump occurs, a single photon of light would be

[†] After J. R. Rydberg who generalized Balmer's formula to elements beyond hydrogen.

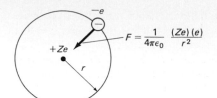

FIGURE 40–16

Electric force (Coulomb's law) keeps negative electron in orbit around a positively charged nucleus.

emitted whose energy, since energy is conserved, is given by

$$hf = E_u - E_l \qquad (40\text{–}7)$$

where E_u refers to the energy of the upper state and E_l the energy of the lower state.

He next set out to determine what energies these orbits would have, since then the spectrum of light emitted could be predicted from Eq. 40–7. When he became aware of the Balmer formula in early 1913, he had the key he was looking for. Bohr quickly found that his theory would be in accord with the Balmer formula if he assumed the electron's angular momentum L was equal to an integer n times $h/2\pi$. As we saw in Chapter 10, the angular momentum of a particle of mass m moving in a circle of radius r with speed v is $L = mvr$. Bohr's *quantum condition* then is

$$L = mvr_n = n\frac{h}{2\pi}, \qquad n = 1, 2, 3, \ldots. \qquad (40\text{–}8)$$

Here n is an integer and r_n is the radius of the nth possible orbit. The allowed orbits are numbered 1, 2, 3, . . . , according to the value of n, which is called the **quantum number** of the orbit.

Equation 40–8 did not have a firm theoretical foundation. Bohr had searched for some "quantum condition," and such tries as $E = hf$ (where E represents the energy of the electron in an orbit) did not give results in accord with experiment. Bohr's reason for using Eq. 40–8 was simply that it worked; and we now look at how.

An electron in a circular orbit of radius r_n would have a centripetal acceleration v^2/r_n produced by the electrical force of attraction between the negative electron and the positive nucleus according to Coulomb's law,

$$F = \frac{1}{4\pi\varepsilon_0}\frac{(Ze)(e)}{r^2},$$

Fig. 40–16. The charge on the electron is $q_1 = -e$, and that on the nucleus is $q_2 = +Ze$, where Z is the number of positive charges[†] (or protons). For the hydrogen atom, $Z = +1$.

From Newton's second law, $F = ma$, we substitute $a = v^2/r_n$ and Coulomb's law for F and obtain

$$\frac{1}{4\pi\varepsilon_0}\frac{Ze^2}{r_n^2} = \frac{mv^2}{r_n}.$$

We solve this for r_n and substitute for v from Eq. 40–8 (which says $v = nh/2\pi mr_n$):

$$r_n = \frac{Ze^2}{4\pi\varepsilon_0 mv^2} = \frac{Ze^2 4\pi^2 mr_n^2}{4\pi\varepsilon_0 n^2 h^2}.$$

We solve for r_n (it appears on both sides) and find:

$$r_n = \frac{n^2 h^2 \varepsilon_0}{\pi m Z e^2}. \qquad (40\text{–}9)$$

[†] We include Z in our derivation so we can treat other single-electron ("hydrogenlike") atoms such as the ions He$^+$ ($Z = 2$) and Li^{2+} ($Z = 3$). Helium in the neutral state has two electrons; if one electron is missing the remaining He$^+$ ion consists of one electron revolving around a nucleus of charge $+2e$; similarly, doubly ionized lithium, Li^{2+}, also has a single electron, and in this case $Z = 3$.

This equation gives the radii of the possible orbits in the Bohr model of hydrogen. The smallest orbit has $n = 1$ and for hydrogen ($Z = 1$) it has the value

$$r_1 = \frac{(1)^2(6.626 \times 10^{-34}\,\text{J·s})^2(8.85 \times 10^{-12}\,\text{C}^2/\text{N·m}^2)}{(3.14)(9.11 \times 10^{-31}\,\text{kg})(1)(1.602 \times 10^{-19}\,\text{C})^2}$$

or

$$r_1 = 0.529 \times 10^{-10}\,\text{m}. \qquad (40\text{–}10)$$

This smallest radius, r_1, is sometimes called the **Bohr radius**. From Eq. 40–10, we can see that the radii of the larger orbits[†] increase as n^2, so

$$r_2 = 4r_1 = 2.11 \times 10^{-10}\,\text{m}$$

$$r_3 = 9r_1 = 4.74 \times 10^{-10}\,\text{m}$$

and so on. The first four are shown in Fig. 40–17.

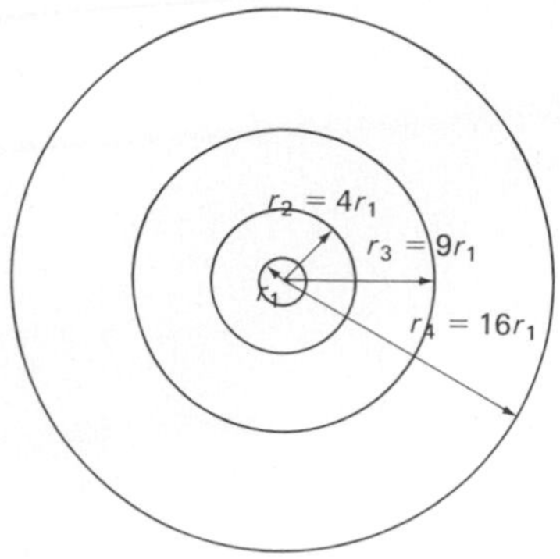

In each of its possible orbits, the electron would have a definite energy, as the following calculation shows. The total energy equals the sum of the kinetic and potential energies. The potential energy of the electron is given by $U = qV = -eV$, where V is the potential due to a point charge $+Ze$ as given by Eq. 24–5: $V = (1/4\pi\varepsilon_0)(Q/r) = (1/4\pi\varepsilon_0)(Ze/r)$. So

$$U = -eV = -\frac{1}{4\pi\varepsilon_0}\left(\frac{Ze^2}{r}\right).$$

The total energy E_n for an electron in the nth orbit of radius r_n is the sum of its kinetic and potential energies:

$$E_n = \tfrac{1}{2}mv^2 - \frac{1}{4\pi\varepsilon_0}\frac{Ze^2}{r_n}.$$

When we substitute v from Eq. 40–8 and r_n from Eq. 40–9 into this equation, we obtain

$$E_n = -\left(\frac{Z^2e^4m}{8\varepsilon_0^2 h^2}\right)\left(\frac{1}{n^2}\right), \qquad n = 1, 2, 3, \ldots. \qquad (40\text{–}11)$$

For hydrogen ($Z = 1$), the lowest energy level has $n = 1$ and when numbers are substituted into Eq. 40–11, we have

$$E_1 = -2.17 \times 10^{-18}\,\text{J} = -13.6\,\text{eV}$$

where we have converted joules to electron volts ($1\,\text{eV} = 1.6 \times 10^{-19}\,\text{J}$) as is customary in atomic physics. Since n^2 appears in the denominator of Eq. 40–11, the energies of the larger orbits are given by

$$E_n = -\frac{13.6\,\text{eV}}{n^2}.$$

For example,

$$E_2 = \frac{-13.6\,\text{eV}}{4} = -3.40\,\text{eV}$$

$$E_3 = \frac{-13.6\,\text{eV}}{9} = -1.51\,\text{eV}.$$

We see from Eq. 40–11 that not only are the orbit radii quantized in the Bohr model, but so is the energy. Notice that although the energy for the larger orbits

FIGURE 40–17

Possible orbits in the Bohr model of hydrogen; $r_1 = 0.527 \times 10^{-10}$ m.

[†] Be careful not to believe that these well-defined orbits actually exist. The Bohr model is only a model, not reality; and, as we shall see in the next chapter, the idea of electron orbits has been rejected. Instead, today electrons are better thought of as forming "clouds," as discussed in Chapter 41.

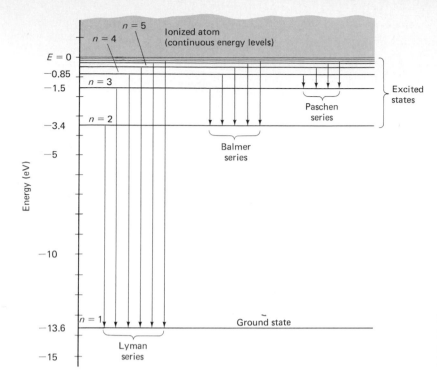

FIGURE 40–18

Energy-level diagram for the hydrogen atom, showing origin of spectral lines for the Lyman, Balmer, and Paschen series.

has a smaller numerical value, all of the energies are less than zero. Thus -3.4 eV is a greater energy than -13.6 eV. Hence the orbit closest to the nucleus (r_1) has the lowest energy. The reason the energies have negative values has to do with the way we defined the zero for potential energy; for two point charges, $U = (1/4\pi\varepsilon_0)(q_1 q_2/r)$ corresponds to zero PE when the two are infinitely far apart. Thus, an electron without KE that is free from the atom would have $E = 0$, corresponding to $n = \infty$. Electrons that are part of an atom would have $E < 0$. Also, it is clear that energy would be required to remove an electron from an atom—this is called the **binding energy** or **ionization energy**—so if $E = 0$ for a free electron, an electron bound to an atom must have $E < 0$. The ionization energy for hydrogen has been measured to be 13.6 eV, and this corresponds precisely to removing an electron from the lowest state, $E_1 = -13.6$ eV, up to $E = 0$ where it will be free.

It is useful to show the various possible energy values as horizontal lines on an energy-level diagram.[†] This is shown for hydrogen in Fig. 40–18. The quantum number n that labels the orbit radii also labels the energy levels. The lowest *energy level* or *energy state* has energy E_1, and is called the **ground state**. The higher states, E_2, E_3, and so on, are called **excited states**. The electron in a hydrogen atom can be in any one of these levels according to Bohr theory. But it could never be in between, say at -9.0 eV. At room temperature, nearly all H atoms will be in the ground state. At higher temperatures, or during an electric discharge when there are many collisions between free electrons and atoms, many electrons can be in excited states. Once in an excited state, an electron can jump down to a lower state, and give off a photon in the process. This is, according to the Bohr model, the origin of the emission spectra of excited gases. The vertical arrows in Fig. 40–18 represent the transitions or jumps that correspond to the

[†] Note that an electron can have any energy above $E = 0$, for there it is free; thus there is a continuum of energy states above $E = 0$, as indicated in the energy-level diagram of Fig. 40–18.

various observed spectral lines. For example, an electron jumping from the level $n = 3$ to $n = 2$ would give rise to the 656-nm line in the Balmer series, and the jump from $n = 4$ to $n = 2$ would give rise to the 486-nm line (see Fig. 40–15). We can predict wavelengths of the spectral lines emitted by combining Eq. 40–7 with Eq. 40–11. Since $hf = hc/\lambda$, we have

$$\frac{1}{\lambda} = \frac{hf}{hc} = \frac{1}{hc}(E_n - E_{n'})$$

or

$$\frac{1}{\lambda} = \frac{Z^2 e^4 m}{8\varepsilon_0^2 h^3 c}\left(\frac{1}{(n')^2} - \frac{1}{(n)^2}\right) \qquad (40–12)$$

where n refers to the upper state and n' to the lower state. This theoretical formula has the same form as the experimental Balmer formula, Eq. 40–6, with $n' = 2$. Thus we see that the Balmer series of lines corresponds in the Bohr model to transitions or "jumps" that bring electrons down to the second energy level. Similarly, $n' = 1$ corresponds to the Lyman series and $n' = 3$ to the Paschen series (see Fig. 40–18). When the constant $(Z^2 e^4 m/8\varepsilon_0^2 h^3 c)$ in Eq. 40–12 is evaluated with $Z = 1$, it is found to have the measured value of the Rydberg constant, $R = 1.0974 \times 10^7$ m^{-1}, in accord with experiment.

The great success of Bohr's theory is that it gives a model for why atoms emit line spectra and accurately predicts, for hydrogen, the wavelengths of emitted light. The Bohr theory also offers an explanation for absorption spectra: photons of just the right wavelength can knock an electron from one energy level to a higher one. To conserve energy, the photon must have just the right energy. This explains why a continuous spectrum passing through a gas will have dark (absorption) lines at the same frequencies as the emission lines.

The Bohr theory also ensures the stability of atoms. It establishes stability by fiat: the ground state is the lowest state for an electron and there is no lower energy level to which it can go and emit more energy. Finally, as we saw above, the Bohr theory accurately predicts the ionization energy of 13.6 eV for hydrogen.

EXAMPLE 40–7 Use Fig. 40–18 to determine the wavelength of the first Lyman line, the transition from $n = 2$ to $n = 1$.

SOLUTION In this case, $hf = E_2 - E_1 = 13.6$ eV $- 3.4$ eV $= 10.2$ eV $= 1.63 \times 10^{-18}$ J. Since $\lambda = c/f$, we have

$$\lambda = \frac{c}{f} = \frac{hc}{E_2 - E_1} = \frac{(6.63 \times 10^{-34} \text{ J·s})(3.00 \times 10^8 \text{ m/s})}{1.63 \times 10^{-18} \text{ J}} = 1.22 \times 10^{-7} \text{ m}$$

or 122 nm, which is in the UV.

EXAMPLE 40–8 Determine the wavelength of light emitted when a hydrogen atom makes a transition from the $n = 6$ to the $n = 2$ energy level of the Bohr model.

SOLUTION We can use Eq. 40–12 or its equivalent, Eq. 40–6, with $R = 1.097 \times 10^7$ m^{-1}. Thus

$$\frac{1}{\lambda} = (1.097 \times 10^7 \text{ m}^{-1})(\tfrac{1}{4} - \tfrac{1}{36}) = 2.44 \times 10^6 \text{ m}^{-1}.$$

So $\lambda = 4.10 \times 10^{-7}$ m or 410 nm. This is the fourth line in the Balmer series, Fig. 40–15, and is violet in color.

EXAMPLE 40–9 Use the Bohr model to determine the ionization energy of the ion He$^+$ which has a single electron. Also calculate the minimum wavelength a photon must have to cause ionization.

SOLUTION We want to determine the energy required to lift the electron from its ground state to the free state, $E = 0$. The ground state energy of He$^+$ is given by Eq. 40–11 with $n = 1$ and $Z = 2$. Since all the symbols in Eq. 40–11 are the same as for the calculation for hydrogen, except Z is 2 instead of 1, we see that E_1 will be $Z^2 = 2^2 = 4$ times larger than for hydrogen. That is

$$E_1 = 4(-13.6 \text{ eV}) = -54.4 \text{ eV}.$$

Thus, to ionize the He$^+$ ion should require 54.4 eV, and this agrees with experiment. The minimum wavelength photon that can cause ionization will have energy $hf = 54.4$ eV and wavelength $\lambda = c/f = hc/hf = (6.63 \times 10^{-34} \text{ J·s}) \times (3.00 \times 10^8 \text{ m/s})/(54.4 \text{ eV})(1.60 \times 10^{-19} \text{ J/eV}) = 22.8$ nm. If the atom absorbed a photon of greater energy (wavelength shorter than 22.8 nm), the atom could still be ionized and the freed electron would have kinetic energy of its own.

In the last example, we saw that E_1 for the He$^+$ ion is four times lower than that for hydrogen. Indeed, the energy-level diagram for He$^+$ looks just like that for hydrogen, Fig. 40–18, except that the numerical values for each energy level are four times larger. It is important to note, however, that we are talking here about the He$^+$ *ion*. Normal (neutral) helium has two electrons and its spectrum is entirely different.

We should note that Bohr made some radical assumptions that were at variance with classical ideas. He assumed that electrons in fixed orbits do not radiate light even though they are accelerating, and he assumed that angular momentum is quantized. Furthermore, he was not able to say how an electron moved when it made a transition from one energy level to another. On the other hand, there is no real reason to expect that in the tiny world of the atom electrons would behave as ordinary-sized objects do. Nonetheless, he felt that where quantum theory overlaps with the macroscopic world, it should predict classical results. This is the *correspondence principle* already mentioned in regard to relativity (Section 39–12). This principle does seem to work for Bohr's theory of the hydrogen atom. The orbit sizes and energies are quite different for $n = 1$ and $n = 2$, say. But orbits with $n = 100,000,000$ and $100,000,001$ would be very close in size and energy (see Fig. 40–18). Indeed, jumps between such large orbits, which would approach everyday sizes, would be imperceptible; such orbits would thus appear to be continuous, which is what we expect in the everyday world.

40–10 de Broglie's Hypothesis

Bohr's theory was largely of an *ad hoc* nature. Assumptions were made so that theory would agree with experiment. But Bohr could give no reason why the orbits were quantized. Finally, ten years later, a reason was found by de Broglie.

We saw in Section 40–5 that in 1923, Louis de Broglie proposed that material particles, such as electrons, have a wave nature; and that this hypothesis was confirmed by experiment several years later.

One of de Broglie's original arguments in favor of the wave nature of electrons was that it provided an explanation for Bohr's theory of the hydrogen atom.

According to de Broglie, a particle of mass m moving with speed v would have a wavelength (Eq. 40–5) of

$$\lambda = \frac{h}{mv}.$$

Each electron orbit in an atom, he proposed, is actually a standing wave. As we saw in Chapter 15, when a violin or guitar string is plucked, a vast number of wavelengths are excited. But only certain ones—those that have nodes at the ends—are sustained. These are the *resonant* modes of the string. All other wavelengths interfere with themselves upon reflection and their amplitudes quickly drop to zero. Since electrons move in circles, according to Bohr's theory, de Broglie argued that the electron wave must be a *circular* standing wave that closes on itself, Fig. 40–19. If the wavelength of a wave does not close on itself, as in Fig. 40–20, destructive interference takes place as the wave travels around the loop and it quickly dies out. Thus, the only waves that persist are those for which the circumference of the circular orbit contains a whole number of wavelengths, Fig. 40–21. The circumference of a Bohr orbit of radius r_n is $2\pi r_n$ so we must have

$$2\pi r_n = n\lambda, \qquad n = 1, 2, 3, \ldots.$$

When we substitute $\lambda = h/mv$, we get

$$2\pi r_n = \frac{nh}{mv},$$

or

$$mvr_n = \frac{nh}{2\pi}.$$

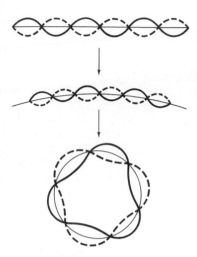

FIGURE 40–19

An ordinary standing wave compared to a circular standing wave.

FIGURE 40–20

When a wave does not close (and hence interferes with itself), it rapidly dies out.

This is just the *quantum condition* proposed by Bohr on an *ad hoc* basis, Eq. 40–8. And it is from this equation that the discrete orbits and energy levels were derived. Thus we have an explanation for the quantized orbits and energy states in the Bohr model: they are due to the wave nature of the electron and the fact that only resonant "standing" waves can persist. This implies that the *wave–particle duality* is at the root of atomic structure.

It should be noted in viewing the circular electron waves of Fig. 40–21 that the electron is not to be thought of as following the oscillating wave pattern. In the Bohr model of hydrogen, the electron, considered as a particle, moves in a circle. The circular wave, on the other hand, represents the *amplitude* of the electron "matter wave", and in Fig. 40–21 the wave amplitude is shown superimposed on the circular path of the particle orbit for convenience.

Bohr's theory worked well for hydrogen and for one-electron ions. It did not prove as successful for multielectron atoms. We will discuss this and other problems with the Bohr theory in the next chapter, and we will see how a new and radical theory, quantum mechanics, finally solved the problem of atomic structure and gave us a very different view of the atom: the idea of electrons in well-defined orbits was replaced with a picture of electron "clouds." And this new (and now generally accepted) theory of quantum mechanics has given us a wholly different view of the basic mechanisms underlying physical processes.

FIGURE 40–21

Standing circular waves for two, three, and five wavelengths on the circumference; n, the number of wavelengths, is also the quantum number.

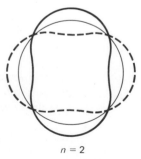

$n = 2$

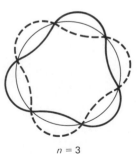

$n = 3$

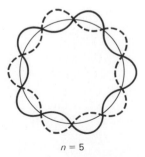

$n = 5$

Summary

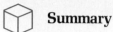

Quantum theory has its origins in the *quantum hypothesis* that molecular oscillations are quantized: their energy E can only be integer (n) multiples of hf, where h is Planck's constant and f is the natural frequency of oscillation: $E = nhf$. This hypothesis explained the spectrum of radiation emitted by (black) bodies at high temperature.

Einstein proposed that for some purposes light might be pictured as being emitted and absorbed as quanta (particles) we now call *photons*, each with energy

$$E = hf.$$

He proposed the photoelectric effect as a test for the photon theory of light. In this effect, the photon theory says that each incident photon can strike an electron in a material and eject it if it has sufficient energy; the maximum energy of ejected electrons is then linearly related to the frequency of the incident light. The photon theory is also supported by the Compton effect and the observation of electron-positron *pair production*.

The wave–particle duality refers to the idea that light and matter (such as electrons) have both wave and particle properties. The wavelength of a material object is

$$\lambda = \frac{h}{mv},$$

where mv is the momentum of the object. The *principle of complementarity* states that we must be aware of both the particle and wave properties of light and of matter for a complete understanding of them.

Early models of the atom include Thomson's modified plum-pudding model and Rutherford's planetary (or nuclear) model. Rutherford's model, which was created to explain the back-scattering of alpha particles from thin metal foils, assumes that an atom consists of a massive positively charged nucleus surrounded (at a relatively great distance) by electrons.

To explain the line spectra emitted by atoms, as well as their stability, Bohr proposed a theory which postulated (1) that electrons bound in an atom can only occupy orbits for which the angular momentum is quantized, which results in discrete values for the radius and energy; (2) that an electron in such a *stationary state* emits no radiation; (3) that, if an electron jumps to a lower state, it emits a photon whose energy equals the difference in energy between the two states; (4) that the angular momentum L of atomic electrons is quantized by the rule $L = nh/2\pi$, where n is an integer called a *quantum number*. The $n = 1$ state in hydrogen is the *ground state*, which has an energy $E_1 = -13.6$ eV; higher values of n correspond to *excited states* and their energies are $E_n = -13.6$ eV/n^2. Atoms are excited to these higher states by collisions with other atoms or electrons or by absorption of a photon of just the right frequency.

De Broglie's hypothesis that electrons (and other matter) have a wavelength $\lambda = h/mv$ gave an explanation for Bohr's quantized orbitals by bringing in the wave–particle duality: the orbits correspond to circular standing waves in which the circumference of the orbit equals a whole number of wavelengths.

Questions

1. What can be said about the relative temperature of yellowish, reddish, and bluish stars?

2. If all objects radiate energy, why can't we see them in the dark?

3. Does a light bulb at a temperature of 2500 K produce as white a light as the sun at 6000 K? Explain.

4. An ideal blackbody can be approximated by a small hole in an otherwise enclosed cavity. Explain. (Hint: the pupil of your eye is an approximate case.)

5. Show that the units of Planck's constant are those for angular momentum.

6. Why do jewelers often examine diamonds in daylight rather than with artificial indoor light?

7. "Orthochromatic" film is not sensitive to red light. Give an explanation based on the photon theory of light.

8. If the threshold wavelength in the photoelectric effect increases when the emitting metal is changed, what can you say about the work functions of the two metals?

9. UV light causes sunburn whereas visible light does not. Explain.

10. Explain why the existence of a cutoff frequency in the photoelectric effect more strongly favors a particle theory rather than a wave theory of light.

11. Consider a point source of light. How would the intensity of light vary with distance from the source according to (a) wave theory, (b) particle (photon) theory? Would this help to distinguish the two theories?

12. If an X-ray photon is scattered by an electron, does its wavelength change? If so, does it increase or decrease?

13. Explain how the photoelectric circuit of Fig. 40–3 could be used in (a) a burglar alarm, (b) a smoke detector, (c) a photographic light meter, (d) a spectrophotometer (see Section 37–8), (e) a film sound track.

14. Why do we say that light has wave properties? Why do we say that light has particle properties?

15. Why do we say that electrons have wave properties? Why do we say that electrons have particle properties?

16. What is the difference between a photon and an electron? Be specific.

17. If an electron and a proton travel at the same speed, which has the shorter wavelength?

18. Is it possible for the de Broglie wavelength of a "particle" to be greater than the dimensions of the particle? To be smaller? Is there any direct connection?

19. Approximately how large would h have to be so that quantum effects would be noticeable in the macroscopic world of everyday life?

20. In Rutherford's planetary model of the atom, what keeps the electrons from flying off into space?

21. Which of the following can emit a line spectrum: (a) gases, (b) liquids, (c) solids. Which can emit a continuous spectrum?

22. Why doesn't the O_2 gas in the air around us give off light?

23. How can you tell if there is oxygen on the sun?

24. When a wide spectrum of light passes through hydrogen gas at room temperature, absorption lines are observed that correspond only to the Lyman series. Why don't we observe the other series?

25. Explain how the closely spaced energy levels for hydrogen near the top of Fig. 40–18 correspond to the closely spaced spectral lines at the left of Fig. 40–15.

26. Discuss the differences between Rutherford's and Bohr's theory of the atom.

27. In a helium atom, which contains two electrons, do you think that on the average the electrons are closer to the nucleus or farther away than in a hydrogen atom? Why?

28. How can the spectrum of hydrogen contain so many lines when hydrogen contains only one electron?

29. Why was the Balmer series of lines from hydrogen the first to be observed and analyzed rather than, say, the Lyman or Paschen series?

30. Use conservation of momentum to explain why photons emitted by hydrogen atoms have slightly less energy than that predicted by Eq. 40–7.

31. Does it make sense that the potential energy of the electron in a hydrogen atom is negative and greater in magnitude than the kinetic energy, as given by the Bohr theory? What is the significance of this?

 Problems _____

SECTION 40–1

1. (I) An HCl molecule vibrates with a natural frequency of 8.1×10^{13} Hz. What is the difference in energy (in joules and electron volts) between possible values of the oscillation energy?

2. (I) (a) At what temperature will the peak of a blackbody spectrum be at 1.0 nm? (b) What is the wavelength at the peak of a blackbody spectrum if the body is at a temperature of 800 K?

3. (II) A child's swing has a natural frequency of 0.40 Hz. (a) What is the separation between possible energy values (in joules)? (b) If the swing reaches a height of 30 cm above its lowest point and has a mass of 20 kg (including the child), what is the value of the quantum number n? (c) What is the fractional change in energy between levels whose quantum numbers are n (as just calculated) and $n + 1$? Would quantization be measurable in this case?

4. (III) Planck's radiation law is given by:

$$f(\lambda, T) = \frac{8\pi hc\lambda^{-5}}{e^{hc/\lambda kT} - 1}$$

where $f(\lambda, T)$ is the rate energy is radiated per unit surface area per unit wavelength interval at wavelength λ and Kelvin temperature T. (This is essentially what is plotted in Figs. 40–1 and 40–2.) (a) Show that Wien's displacement law follows from this relationship. (b) Determine the value of h from the experimental value of $\lambda_p T$ given in the text. (c) Derive the Stefan-Boltzmann law (the T^4 dependence of the rate at which energy is radiated—Eq. 19–3), by integrating Planck's formula over all wavelengths; that is, show that

$$\int f(\lambda, T)\, d\lambda \propto T^4.$$

SECTION 40–2

5. (I) What is the energy (in electron volts) of a photon of wavelength (a) 400 nm, (b) 700 nm, and (c) of photons emitted by a 100-MHz FM radio station?

6. (I) What is the longest wavelength of light that will eject electrons from a metal whose work function is 2.3 eV?

7. (I) Barium has a work function of 2.48 eV. What is the maximum kinetic energy of electrons if the metal is illuminated by light of wavelength 480 nm? What is their speed?

8. (I) When UV light of wavelength 280 nm falls on a metal surface, the maximum kinetic energy of emitted electrons is 0.85 eV. What is the work function of the metal?

9. (II) Show that the energy E (in electron volts) of a photon whose wavelength λ (meters) is given by

$$E = 1.24 \times 10^{-6}/\lambda.$$

10. (II) Sunlight reaching the earth has an intensity of 1300 W/m². How many photons per square meter per second does this represent? Take the average wavelength to be 550 nm.

11. (II) The threshold wavelength for emission of electrons from a given surface is 380 nm. What will be the maximum kinetic energy of ejected electrons when the wavelength is changed to (a) 480 nm, (b) 280 nm?

12. (II) A certain type of film is sensitive only to light whose wavelength is less than 640 nm. What is the energy (kcal/mol) needed for the chemical reaction to occur which causes the film to change?

13. (II) If a 100-W light bulb emits 3.0 percent of the input energy as visible light (average wavelength 550 nm) uniformly in all directions, approximately how many photons per second of visible light will strike the pupil (4.0 mm diameter) of the eye of an observer 10 km away?

14. (II) When 230-nm light falls on a metal, the current through a photoelectric circuit (Fig. 40–3) is brought to zero at a reverse voltage of 1.64 V. What is the work function of the metal?

15. (II) In an X-ray tube (see Fig. 37–23 and discussion in Section 37–11), the high voltage between filament and target is V. After being accelerated through this voltage, an electron strikes the target where it is decelerated (by positively charged nuclei) and in the process one or more X-ray photons are emitted. (a) Show that the shortest wavelength photon will have

$$\lambda_0 = \frac{hc}{eV}.$$

(b) What is the shortest wavelength X ray emitted when accelerated electrons strike the face of a 30-kV television picture tube?

16. (II) (a) Show that if the two electrodes in the photocell of Fig. 40–3 are made of the same material, the maximum kinetic energy of electrons is related to the stopping voltage V_0 by

$$\mathrm{KE}_{max} = eV_0.$$

(b) If the two electrodes are made of different materials, show that

$$\mathrm{KE}_{max} = eV_0 - (W_{0e} - W_{0c})$$

where W_{0e} and W_{0c} are the work functions for the emitting electrode (P in Fig. 40–3) and the collecting electrode (C) respectively.

17. (III) In the Compton effect, a 0.100-nm photon strikes a free electron in a head-on collision and knocks it into the forward direction. The rebounding photon recoils directly backward. Use conservation of energy and momentum to determine (a) the kinetic energy of the electron, (b) the wavelength of the recoiling photon. Assume that the electron's kinetic energy is given by the nonrelativistic formula.

18. (III) In the *Compton effect* (see Fig. 40–5), use conservation of energy and of linear momentum to show that the wavelength λ' of the scattered photon is given by

$$\lambda' = \lambda + \frac{h}{m_0 c}(1 - \cos \phi)$$

where ϕ is the angle between the scattered photon's direction and that of the incident photon whose wavelength is λ, and m_0 is the rest mass of the electron. Use relativistic formulas. Note that $\Delta\lambda = \lambda' - \lambda$ does not depend on wavelength.

SECTION 40–3

19. (I) How much total kinetic energy will an electron-positron pair have if produced by a 3.6-MeV photon?

20. (II) What is the momentum and effective mass of a 0.10-nm X-ray photon?

21. (II) What is the minimum photon energy needed to produce a $\mu^+ - \mu^-$ pair? The mass of each muon is 207 times the mass of the electron. What is the wavelength of such a photon?

22. (II) An electron-positron pair, each with a kinetic energy of 435 keV, is produced from a photon. What was the energy and wavelength of the photon?

SECTION 40–5

23. (I) What is the wavelength of a neutron ($m_0 = 1.67 \times 10^{-27}$ kg) traveling 2.5×10^4 m/s?

24. (II) Show that if an electron and a proton have the same kinetic energy, the proton has the shorter wavelength.

25. (II) Calculate the ratio of the KE of an electron to that of a proton whose wavelengths are equal. Assume that the speeds are much less than the speed of light.

26. (II) A beam of 65 eV electrons is scattered from a crystal, as in X-ray diffraction, and a first-order peak is observed at $\theta = 45°$. What is the spacing between planes in the diffracting crystal? (See Section 37–11.)

27. (II) What is the wavelength of a proton whose KE is 1.0 GeV?

28. (II) Show that the wavelength of a particle of mass m_0 with kinetic energy KE is given by the relativistic formula $\lambda = hc/\sqrt{(\mathrm{KE})^2 + 2m_0 c^2 (\mathrm{KE})}$.

29. (II) What is the kinetic energy and wavelength of a "thermal" neutron (one that is in equilibrium at room temperature—see Chapter 18).

30. (III) Show that for a particle of rest mass m_0, if $\lambda = h/mv$ then it cannot be true that $E = hf$ where E is (a) kinetic energy, or (b) KE plus rest mass energy, and $v = f\lambda$ is the speed of the particle.

*SECTION 40–6

*31. (II) What is the theoretical limit of resolution for an electron microscope whose electrons are accelerated through 50 kV? (Relativistic formulas should be used.)

*32. (II) Electrons are accelerated by 2200 V in an electron microscope. To achieve a resolution of 4.0 nm, what numerical aperture is required?

SECTION 40–7

33. (II) In certain of Rutherford's experiments, the alpha particles (mass = 6.68×10^{-27} kg) had a kinetic energy of 4.8 MeV. How close could they get to a gold nucleus (charge = $+79e$)? Ignore the recoil motion of the nucleus.

SECTION 40–9

34. (I) For what maximum kinetic energy is a collision between an electron and a hydrogen atom in its ground state definitely elastic?

35. (I) (a) Determine the wavelength of the second Balmer line ($n = 4$ to $n = 2$ transition) using Fig. 40–18. Determine likewise (b) the wavelength of the first Lyman line and (c) the wavelength of the fourth Balmer line.

36. (I) Calculate the ionization energy of doubly ionized lithium, Li^{2+}, which has $Z = 3$.

37. (I) What is the longest wavelength light capable of ionizing a hydrogen atom in the ground state?

38. (I) At low temperatures, nearly all the atoms in hydrogen gas will be in the ground state. What minimum frequency photon is needed if the photoelectric effect is to be observed?

39. (II) Construct the energy-level diagram for the He$^+$ ion (see Fig. 40–18).

40. (II) Construct the energy-level diagram for doubly ionized lithium, Li^{2+}.

41. (II) By what fraction does the mass of an H atom decrease when it makes an $n = 2$ to $n = 1$ transition?

42. (II) What is the potential energy and kinetic energy of an electron in the ground state of the hydrogen atom?

43. (II) An excited hydrogen atom could, in principle, have a radius of 1.0 mm. What would be the value of n for a Bohr orbit of this size? What would its energy be?

44. (II) Calculate the ratio of the gravitational to electric force for the electron in a hydrogen atom. Can the gravitational force be safely ignored?

45. (II) Are nonrelativistic formulas justified in the Bohr atom? To check, calculate the electron's velocity, v, in terms of c for the ground state of hydrogen.

46. (II) Suppose an electron was bound to a proton, as in the hydrogen atom, by the gravitational force rather than by the electric force. What would be the radius, and energy, of the first Bohr orbit?

47. (II) Electrons accelerated by a potential difference of 12.3 V pass through a gas of hydrogen atoms at room temperature. What wavelengths of light will be emitted?

48. (II) Atoms can be formed in which a mu meson (mass = 207 times the mass of an electron) replaces one of the electrons in an atom. Calculate, using Bohr theory, the energy of the photon emitted when a mu meson makes a transition from $n = 2$ to $n = 1$ in a mu-mesic $^{208}_{82}$Pb (lead whose nucleus has a mass 208 times the proton mass and charge $+82e$).

49. (II) *Correspondence principle*: Show that for large values of n, the difference in radius Δr between two adjacent orbits (with quantum numbers n and $n - 1$) is given by

$$\Delta r = r_n - r_{n-1} \approx \frac{2r_n}{n}$$

so $\Delta r / r_n \to 0$ as $n \to \infty$ in accordance with the correspondence principle. [Note that we can check the correspondence principle by either considering large values of n ($n \to \infty$) or by letting $h \to 0$. Are these equivalent?]

50. (III) (a) For very large values of n, show that when an electron jumps from the level n to the level $n - 1$, the frequency of the light emitted is equal to

$$f = \frac{v}{2\pi r_n}.$$

(b) Show that this is also just the frequency predicted by classical theory for an electron revolving in a circular orbit of radius r_n with speed v. (c) Explain why this is consistent with the correspondence principle.

SECTION 40–10

51. (III) Suppose a particle of mass m is confined to a one-dimensional box of width L. According to quantum theory, the particle's wave (with $\lambda = h/mv$) is a standing wave with nodes at the edges of the box. (a) Show the possible modes of vibration on a diagram. (b) Show that the kinetic energy of the particle has quantized energies given by KE = $n^2h^2/8mL^2$ where n is an integer. (c) Calculate the ground-state energy ($n = 1$) for an electron confined to a box of width 0.50×10^{-10} m. (d) What is the ground-state energy for a baseball ($m = 140$ g) in a box 0.50 m wide? (e) An electron confined to a box has a ground-state energy of 10 eV. What is the width of the box?

Quantum Mechanics of Atoms and Molecules

41

Bohr's model of the atom gave us a first (though rough) picture of what an atom is like. It proposed an explanation for why there should be emission and absorption of light by atoms at discrete wavelengths, as well as the stability of atoms. The wavelengths of the line spectra and ionization energy for hydrogen and one-electron ions are calculated to be in excellent agreement with experiment. But the Bohr theory had important limitations. It was not able to predict the line spectra for more complex atoms—not even for the neutral helium atom, which has only two electrons. Nor could it explain why the emission lines, when viewed with great precision, actually consist of two or more very closely spaced lines (referred to as *fine structure*). The Bohr theory also didn't explain why some spectral lines were brighter than others. And it couldn't explain the bonding of atoms in molecules or in solids and liquids.

From a theoretical point of view, too, the Bohr theory was not really satisfactory. For it was a strange mixture of classical and quantum ideas. And the wave–particle duality was still not really resolved.

We mention these limitations of the Bohr theory not to disparage it—for it was a landmark in the history of science. Rather, we had to mention them to show why, in the early 1920s, it became increasingly evident that a new, more comprehensive theory was needed. It was not long in coming. Less than 2 years after de Broglie gave us his matter-wave hypothesis, Erwin Schrödinger (1887–1961) and Werner Heisenberg (1901–1976) independently developed a new comprehensive theory. Their separate approaches were quite different but were soon shown to be fully compatible.

41–1 Quantum Mechanics—A New Theory

The new theory, called **quantum mechanics**, unifies the wave–particle duality into a single consistent theory. As a theory, quantum mechanics has been extremely successful. It has successfully dealt with the spectra emitted by complex atoms, even the fine details. It explains the relative brightness of spectral lines and how atoms form molecules. It is also a much more general theory that covers all quantum phenomena from blackbody radiation to atoms and molecules. It has explained a wide range of natural phenomena and from its predictions many new practical devices have become possible. Indeed, it has been so successful that it is accepted today by nearly all physicists as the fundamental theory underlying physical processes.

Quantum mechanics deals mainly with the microscopic world of atoms and light; but in our everyday macroscopic world we do perceive light and we accept that ordinary objects are made up of atoms. This new theory must therefore also account for the verified results of classical physics; that is, when it is applied to macroscopic phenomena, quantum mechanics must be able to produce the old classical laws. This, the *principle of correspondence* (already mentioned in Section 40–9), is met fully by quantum mechanics. This doesn't mean we throw away classical theories such as Newton's laws. In the everyday world the latter are far easier to apply and give an accurate description. But when we deal with high speeds, close to the speed of light, we must use the theory of relativity; and when we deal with the tiny world of the atom, we use quantum mechanics.

Although we won't go into the detailed mathematics of quantum mechanics, we will discuss the main ideas and how they involve the wave and particle properties of matter to explain atomic structure and other applications.

 ## 41–2 The Wave Function and Its Interpretation; The Double-slit Experiment

The important properties of any wave are its wavelength, frequency, and amplitude. For an electromagnetic wave, the wavelength determines whether the light is visible or not, and if so, what color it is. We also have seen that the wavelength (or frequency) is a measure of the energy of the corresponding photon ($E = hf$). The amplitude of an electromagnetic wave at any point is the strength of the electric (or magnetic) field at that point, and is related to the intensity of the wave (the brightness of the light).

For material particles such as electrons, quantum mechanics relates the wavelength to momentum according to de Broglie's formula, $\lambda = h/mv$. But what about the amplitude of a matter wave? In quantum mechanics the amplitude of, say, an electron wave is called the **wave function** and is given the symbol Ψ (the Greek letter psi). Thus Ψ represents the amplitude, as a function of time and position, of a new kind of field which we might call a "matter" field or a matter wave.

To calculate the wave function Ψ in a given situation (say, for an electron in an atom) is one of the basic tasks of quantum mechanics. Indeed, the development of an equation to do so was Schrödinger's great contribution. The *Schrödinger wave equation*, as it is called, is considered to be the basic equation for the description of nonrelativistic material particles.[†] Since our treatment of quantum mechanics is necessarily brief (whole books are needed for the subject), the actual form of the Schrödinger equation, and how it is solved, will not concern us here. What will be useful are the solutions obtained and the meaning of the wave function, Ψ, itself. One way to interpret Ψ is simply as the amplitude at any point in space and time of a "matter wave", just as **E** (the electric field vector) represents the amplitude of an electromagnetic wave. Another interpreta-

---

[†] The Schrödinger equation differs only slightly from the ordinary wave equation (Eq. 15–16) and for the one-dimensional case can be written

$$-\frac{\hbar^2}{2m}\frac{\partial^2 \Psi(x, t)}{\partial x^2} + U(x)\Psi(x, t) = i\hbar\frac{\partial \Psi(x, t)}{\partial t}$$

where $U(x)$ is the potential energy of the "particle" of mass m described by the wave function $\Psi(x, t)$ which is a function of position x and time t; $\hbar = h/2\pi$ and $i = \sqrt{-1}$.

tion is possible, however, based on the wave–particle duality. To understand this, we make an analogy with light.

We saw in Chapter 15 that the intensity I of any wave is proportional to the square of the amplitude. This holds true for light waves as well, as we saw in Chapter 33; that is,

$$I \propto E^2$$

where E is the electric field strength. From the *particle* point of view, the intensity of a light beam is proportional to the number of photons, N, that pass through a given area per unit time. The more photons (of a given wavelength) there are, the greater the intensity. Thus

$$I \propto E^2 \propto N.$$

This proportion can be turned around so we have

$$N \propto E^2.$$

That is, the number of photons (striking a page of this book, say) is proportional to the square of the electric field strength.

If the light beam is very weak, only a few photons will be involved. Indeed, it is possible to "build up" a photograph in a camera using very weak light so the effect of individual photons can be seen. If we are dealing with only one photon, the relationship above ($N \propto E^2$) can be interpreted in a slightly different way. At any point the square of the electric field strength, E^2, is a measure of the *probability* that a photon will be at that location. At points where E^2 is large, there is a high probability the photon will be there; where E^2 is small, the probability is low.

We can interpret matter waves in the same way. The wave function Ψ may vary in magnitude from point to point in space and time. If Ψ describes a collection of many electrons, then Ψ^2 at any point will be proportional to the number of electrons expected to be found at that point. When dealing with small numbers of electrons we can't make very exact predictions, so Ψ^2 takes on the character of a probability. If Ψ, which depends on time and position, represents a single electron (say in an atom) then Ψ^2 is interpreted as follows: Ψ^2 *at a certain point in space and time represents the probability of finding the electron at the given position and time.*

To understand this better, let's examine the familiar double-slit experiment, and consider it both for light and for electrons.

Consider two slits whose size and separation are on the order of the wavelength of whatever we direct at them, either light or electrons, Fig. 41–1. We know very well what happens in this case for light, since this is just Young's double-slit experiment: an interference pattern will be seen on the screen behind. If light is replaced by electrons with wavelength comparable to the slit size, they too will produce an interference pattern. (Recall the Davisson–Germer experiment, Fig. 40–7.) In the case of light, the pattern will be visible to the eye or can be recorded on film. For electrons, a fluorescent screen can be used (it glows where an electron strikes).

Now, if we reduce the flow of electrons (or photons) so that they pass through the slits one at a time, we will see a flash each time one strikes the screen. At first, the flashes will seem random. Indeed, there is no way to predict just where any one electron will hit the screen. If we let the experiment run for a long time, however, and keep track of where each electron hits the screen, we will soon see a pattern emerging—namely the interference pattern predicted by the wave theory. Thus, although we cannot predict where a given electron will strike the screen, we can predict probabilities. (The same can be said for photons.) The probability, as mentioned before, is proportional to Ψ^2. Where Ψ is zero, we get a minimum in the interference pattern. And where Ψ is a maximum, we get a peak in the interference pattern.

FIGURE 41–1

Parallel beam of light or electrons falls on two slits whose sizes are comparable to the wavelength. An interference pattern is observed.

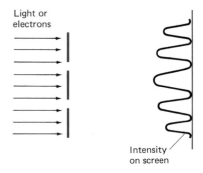

Light or electrons

Intensity on screen

Since the interference pattern occurs even when electrons (or photons) pass through the slit one at a time, it is clear that the interference pattern does not arise from the interaction of one electron with another. It is as if the electron passed through both slits at the same time! This is possible, for remember that an electron is not precisely a particle. It is as much a wave as it is a particle, and a wave can certainly travel through both slits at once. But what happens if we cover one of the slits so we know that the electron passed through the other one, and after a time we cover the second slit so the electron has to have passed through the first? The result is that no interference pattern is seen. We merely see two bright areas on the screen behind each slit. This confirms our idea that when both slits are open, each electron passes through both, as if it were a wave. Yet each makes a tiny spot on the screen as if it were a particle.

The main point of this discussion is this. If we treat electrons (and other matter) as if they were waves, then Ψ represents the wave amplitude. If we treat them as particles, then we must treat them on a *probabilistic* basis. The square of the wave function, Ψ^2, is the probability of finding a given electron at a given point. We cannot predict—or even follow—the path of a single electron precisely through space and time.

 ## 41–3   The Heisenberg Uncertainty Principle

Whenever a measurement is made, some uncertainty or error is always involved. For example, you cannot make an absolutely exact measurement of the length of a table. Even with a measuring stick that has markings 1 mm apart there will be an inaccuracy of about $\frac{1}{2}$ mm or so. More precise instruments will produce more precise measurements; but there is always some uncertainty involved in a measurement no matter how good the measuring device. We expect that by using more precise instruments the uncertainty in a measurement can be made indefinitely small.

But according to quantum mechanics there is actually a limit to the accuracy of certain measurements. This limit is not a restriction on how well instruments can be made; rather, it is inherent in nature. It is the result of two factors: the wave–particle duality; and the unavoidable interaction between the thing observed and the observing instrument. Let us look at this in more detail.

To make a measurement on an object without somehow disturbing it, at least a little, is not possible. Consider trying to locate a Ping-Pong ball in a completely dark room. You grope about trying to find its position; and just when you touch it with your finger, it bounces away. Whenever we measure the position of an object, whether it's a Ping-Pong ball or an electron, we always touch it with something else which gives us the information about its position. To locate a lost Ping-Pong ball in a dark room you could probe about with your hand or a stick; or you could shine a light and detect the light reflecting off the ball. When you search with your hand or a stick, you find the ball's position when you touch it. But when you touch the ball you unavoidably bump it, and give it some momentum; thus you won't know its *future* position. The same would be true, but to a much lesser extent, if you observe the Ping-Pong ball using light: in order to "see" the ball, at least one photon must scatter from it, and the reflected photon must enter your eye or some other detector. When a photon strikes an ordinary-sized object it does not appreciably alter the motion or position of the object. But when a photon strikes a very tiny object like an electron, it can transfer much of its momentum to the electron, and thus greatly change the object's motion and position in an unpredictable way. The mere act of measuring the position of an object at one time makes our knowledge of its future position inaccurate.

Now let us see where the wave–particle duality comes in. Imagine a thought experiment in which we are trying to measure the position of an object (say an electron) with photons, although the arguments would be similar if we were using an electron microscope. As we saw in Chapter 37, objects can be seen to an accuracy no greater than the wavelength of the radiation used. If we want an accurate position measurement, we must use a short wavelength. But a short wavelength corresponds to high frequency and high energy (since $E = hf$); and the more energy the photons have, the more momentum they can give the object when they strike it. If photons of longer wavelength, and correspondingly lower energy are used, the object's motion when struck by the photons will not be affected as much; but its position will be less accurately known. Thus the act of observing produces a significant uncertainty in either the *position* or the *momentum* of the electron. This is the essence of the *uncertainty principle* first enunciated by Heisenberg in 1927.

Quantitatively, we can make an approximate calculation of the magnitude of this effect. If we use light of wavelength λ, the position can be measured at best to an accuracy of about λ. That is, the uncertainty in the position measurement, Δx, is approximately

$$\Delta x \approx \lambda.$$

Suppose that the object can be detected by a single photon. The photon has a momentum $p = h/\lambda$; and when it strikes our object it will give some or all of this momentum to the object. Therefore, the final momentum of our object will be uncertain in the amount

$$\Delta p \approx h/\lambda$$

since we can't tell beforehand how much momentum will be transferred. The product of these uncertainties is

$$(\Delta x)(\Delta p) \approx h.$$

Of course, the uncertainties could be worse than this, depending on the apparatus and the number of photons needed for detection. In Heisenberg's more careful calculation,[†] he found that at the very best

$$(\Delta x)(\Delta p) \gtrsim \frac{h}{2\pi}. \tag{41–1}$$

This is a mathematical statement of **Heisenberg's uncertainty principle**. It tells us that we cannot measure both the position and momentum of an object precisely at the same time. The more accurately we try to measure the position, so that Δx is small, the greater will be the uncertainty in momentum, Δp. If we try to measure the momentum very precisely, then the uncertainty in the position becomes large. The uncertainty principle does not forbid single exact measurements, however. For example, in principle we could measure the position of an object exactly. But then its momentum would be completely unknown. Thus, although we might know the position of the object exactly at one instant, we could have no idea at all where it would be a moment later.

Another useful form of the uncertainty principle relates energy and time, and we examine this as follows. The object to be detected has an uncertainty in position $\Delta x \approx \lambda$. Now the photon (or electron) used to detect it travels with speed $v \, (= c$ for a photon); and it takes a time $\Delta x/v \approx \lambda/v$ to pass through the distance of uncertainty. Hence, the measured time when our object is at a given position is uncertain by about

$$\Delta t \approx \frac{\lambda}{v}.$$

[†] In the language of statistics and probability, the uncertainties Δx and Δp refer to the *standard deviations* (or *variance*) in measurements of x and p.

Since the photon can transfer some or all of its energy ($= hf = hv/\lambda$) to our object, the uncertainty in energy of our object as a result is

$$\Delta E \approx \frac{hv}{\lambda}.$$

The product of these two uncertainties is

$$(\Delta E)(\Delta t) \approx h.$$

Heisenberg's more careful calculation gives

$$(\Delta E)(\Delta t) \geq \frac{h}{2\pi}. \qquad (41-2)$$

This form of the uncertainty principle tells us that the energy of an object can be uncertain, or may even be nonconserved, by an amount ΔE for a time $\Delta t \approx h/(2\pi \, \Delta E)$.

We have been discussing the position and velocity of an electron as if it were a particle. But it isn't a particle. Indeed, we have the uncertainty principle because an electron—and matter in general—is not purely particulate. What the uncertainty principle really tells us is that if we insist on thinking of the electron as a particle, then there are certain limitations on this simplified view—namely that the position and velocity cannot both be known precisely at the same time; and that the energy can be uncertain (or nonconserved) in the amount ΔE for a time $\Delta t \approx h/(2\pi \, \Delta E)$.

Because Planck's constant, h, is so small, the uncertainties expressed in the uncertainty principle are usually negligible on the macroscopic level. But at the level of the atom, the uncertainties are significant. Because ordinary sized objects are made up of atoms containing nuclei and electrons, the uncertainty principle is relevant to our understanding of all of nature. The uncertainty principle expresses, perhaps most clearly, the probabilistic nature of quantum mechanics; it thus is often used as a basis for philosophic discussion.

 ## 41–4   Philosophic Implications; Probability versus Determinism

The classical Newtonian view of the world is a deterministic one (see Section 5–7). One of its basic ideas is that once the position and velocity of an object are known at a particular time, its future position can be predicted if the forces on it are known. For example, if a stone is thrown a number of times with the same initial velocity and angle, and the forces on it remain the same, the path of the projectile will always be the same. If the forces are known (gravity and air resistance, if any) the stone's path, and where it will travel, can be precisely predicted. This mechanistic view implies that the future unfolding of the universe, assumed to be made up of particulate bodies, is completely determined.

This classical deterministic view of the physical world has been radically altered by quantum mechanics. As we saw in the analysis of the double-slit experiment (Section 41–2), electrons all prepared in the same way will not all end up in the same place; according to quantum mechanics, certain probabilities exist that an electron will arrive at different points. This is very different from the classical view, in which the path of a particle is precisely predictable from the initial position and velocity and the forces exerted on it. According to quantum mechanics, the position and velocity of an object cannot even be known accurately at the same time. This is expressed in the uncertainty principle, and is a result of the fact that basic entities, such as electrons, are not considered simply as particles since they have wave properties as well. Quantum mechanics only allows us to calculate the probability that, say, an electron (when thought of as a particle) will be observed at various different places. Only approximate predictions are possible. Quantum mechanics says there is an inherent unpredictability in nature.

Since matter is considered to be made up of atoms, even ordinary-sized objects are expected to be governed by chance and probability, rather than by strict determinism. For example, there is a finite (but very small) probability that when you throw a stone its path will suddenly curve upward instead of following the downward-curved parabola of normal projectile motion. Quantum mechanics predicts with very high probability that ordinary objects will behave just as the classical laws of physics predict; but these predictions are probabilities, not certainties. The reason that macroscopic objects behave in accordance with classical laws with very high probability is due to the large number of molecules involved: when large numbers of objects are present in a statistical situation, deviations from the average are negligible. It is the average configuration of vast numbers of molecules that follows the so-called fixed laws of classical physics with such high probability, and gives rise to an apparent "determinism." Deviations from classical laws are readily observed when small numbers of molecules are dealt with. We can say, then, that although there are no precise deterministic laws in quantum mechanics, there are statistical laws based on probability.

It is important to note that there is a difference in the probability imposed by quantum mechanics and that used in the nineteenth century to understand thermodynamics and the behavior of gases in terms of molecules (Chapters 18 and 21). In thermodynamics, probability is used because there are far too many molecules to be kept track of. But the molecules were still assumed to move and interact in a deterministic way according to Newton's laws. Probability in quantum mechanics is quite different; it is seen as *inherent* in nature, and not as a limitation on our abilities to calculate.

Although a few physicists have not given up the deterministic view of nature and have refused to accept quantum mechanics as a complete theory—one was Einstein—nonetheless, the vast majority of physicists do accept quantum mechanics and the probabilistic view of nature. This view, which as presented here is the generally accepted one, is called the *Copenhagen interpretation* of quantum mechanics in honor of Niels Bohr's home, since it was largely developed there through discussions between Bohr and other prominent physicists.

Because electrons are not simply particles, they cannot be thought of as following particular paths in space and time. This suggests that a description of matter in space and time may not be completely correct. This deep and far-reaching conclusion has been a lively topic of discussion among philosophers. Perhaps the most important and influential philosopher of quantum mechanics was Bohr. He argued that a space–time description of actual atoms and electrons is not possible; but a description of experiments on atoms or electrons must be given in terms of space and time and other concepts familiar to ordinary experience, such as waves and particles. Yet we must not let our *descriptions* of experiments lead us into believing that atoms or electrons themselves actually exist in space and time as particles. This distinction between our interpretations of experiments and what is "really" happening in nature is crucial.

41–5 Quantum Mechanical View of Atoms

At the beginning of this chapter, we discussed the limitations of the Bohr theory of atomic structure. Now we examine the quantum mechanical theory of atoms, which is far more complete than the old Bohr theory. It retains certain aspects of the older theory, such as that electrons in an atom exist only in discrete states of definite energy; and that a photon of light is emitted (or absorbed) when an electron makes a transition from one state to another. But quantum mechanics is not merely an extension of the Bohr theory. It is a much deeper theory, and has provided us with a very different view of the atom. According to quantum mechanics, electrons do not exist in well-defined circular orbits as in the Bohr theory. Rather, the electron (because of its wave nature) is spread out in space, as a "cloud" of negative charge. The size and shape of the electron cloud can be calculated for a given state of an atom. For the ground state in the hydrogen atom, the solution of the Schrödinger equation (see Section 41–2) gives

$$\psi(r) = \sqrt{\frac{1}{\pi r_1^3}}\, e^{-r/r_1}$$

where $\psi(r)$ is the wave function[†] as a function of position and it depends only on the radial distance r from the center, and not on angle θ or ϕ. (The constant r_1 has a value that happens, by chance, to come out equal to the first Bohr radius.) Thus the electron cloud for the ground state of hydrogen is spherically symmetric as shown in Fig. 41–2. The electron cloud roughly indicates the "size" of an atom, but just as a cloud may not have a distinct border, atoms do not have a precise boundary or a well-defined size. Not all electron clouds have a spherical shape as we shall see later in this chapter. But note that $\psi(r)$, while becoming extremely small for large r (see above equation), does not equal zero in any finite region. So quantum mechanics says that an atom is not mostly empty space; indeed, since $\psi \to 0$ only for $r \to \infty$ we conclude that there is no truly empty space in the universe.

The electron cloud can be interpreted using either the particle or the wave viewpoint. Remember that by a particle we mean something that is localized in space—it has a definite position at any given instant. But a wave is spread out in space. Thus the electron cloud, spread out in space as in Fig. 41–2, is a result of the wave nature of electrons. Electron clouds can also be interpreted as *probability distributions* for a particle. If you were to make 500 different measurements of the position of an electron (considering it as a particle) the majority of the results would show the electron at points where the probability is high (dark area in Fig. 41–2); only occasionally would the electron be found where the probability is low. We cannot predict the path an electron will follow. As we saw in Section 41–3, after one measurement of its position we cannot predict exactly where it will be at a later time. We can only calculate the probability that it will be found at different points. This is clearly different from classical Newtonian physics. Indeed, as Bohr later pointed out, it is meaningless even to ask how an electron gets from one state to another when the atom emits a photon of light.

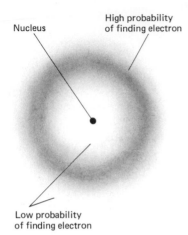

Nucleus

High probability of finding electron

Low probability of finding electron

FIGURE 41–2

Electron cloud or "probability distribution" for the ground state of the hydrogen atom. The cloud is densest—indicating the highest probability—at a distance from the nucleus of 0.53×10^{-10} m, which is just what the Bohr model predicts for the radius of the first orbit. But unlike the Bohr model, quantum mechanics tells us the electron can be within or beyond that distance at any given time.

41–6 Quantum Mechanics of the Hydrogen Atom; Quantum Numbers

We now look more closely at what quantum mechanics tells us about the hydrogen atom. Much of what we say here also applies to more complex atoms, which are discussed in the next section.

Quantum mechanics predicts exactly the same energy levels (Fig. 40–18) for the hydrogen atom as does the Bohr theory (a fortunate accident). That is,

$$E_n = -\frac{13.6\,\text{eV}}{n^2} \qquad n = 1, 2, 3, \ldots$$

where n is an integer. In the simple Bohr theory, there was only one quantum number, n. In quantum mechanics, it turns out that four different quantum numbers are needed to specify each state in the atom.

The quantum number n from Bohr theory is retained in quantum mechanics and is called the **principal quantum number**. It can have any integer value from 1

[†] We use the lowercase ψ to represent the wave function written as a function of position only. When the time dependence is included we use the capital letter $\Psi(\mathbf{r}, t)$ as before.

to ∞. The total energy of a state in the hydrogen atom depends on n, as we saw above.

41–6 Quantum Mechanics of the Hydrogen Atom; Quantum Numbers **803**

The **orbital quantum number**, l, is related to the angular momentum of the electron; l can take on integer values from 0 to $(n - 1)$. For the ground state, $n = 1$, l can only be zero; but for $n = 3$, say, l can be 0, 1, or 2. The actual magnitude of the angular momentum L is related to the quantum number l by

$$L = \sqrt{l(l + 1)} \, \frac{h}{2\pi}.$$

The value of l does not affect the total energy in the hydrogen atom;[†] only n does. But in atoms with two or more electrons, the energy does depend on l as well as n, as we shall see in the next section.

The **magnetic quantum number**, m_l, is related to the direction of the electron's angular momentum and it can take on integer values ranging from $-l$ to $+l$; for example, if $l = 2$, then m_l can be -2, -1, 0, $+1$, or $+2$. Since angular momentum is a vector, it is not surprising that both its magnitude and its direction would be quantized. For $l = 2$, the five different directions allowed can be represented by the diagram of Fig. 41–3. In quantum mechanics the direction of the angular momentum is usually specified by giving its component along the z axis. Then L_z is related to m_l by the equation

$$L_z = m_l \frac{h}{2\pi}.$$

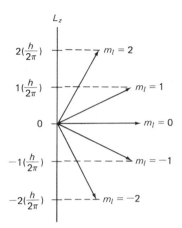

FIGURE 41–3

Quantization of angular momentum direction for $l = 2$.

The name for m_l derives not from theory (which relates it to L_z) but from experiment: it was found that when a gas discharge tube was placed in a magnetic field, the spectral lines were split into several very closely spaced lines. (This is called the Zeeman effect.) This implies that the energy levels must be split (Fig. 41–4), and thus the energy of a state depends not only on n but also on m_l when a magnetic field is applied—hence the name "magnetic quantum number." (Why the energy should depend on the direction of **L** can be seen from a semiclassical view of a moving electron as an electric current which interacts with the magnetic field [see Chapter 28].)

Finally, there is the **spin quantum number**, m_s, and it can have only two values, $m_s = +\frac{1}{2}$ and $m_s = -\frac{1}{2}$. The existence of this quantum number did not come out of Schrödinger's original theory, as did n, l, and m_l. Instead, a subsequent modification due to P. A. M. Dirac showed that it was present as a relativistic effect. The fact that m_s was needed, however, first came from experiment. A careful study of the spectral lines of hydrogen showed that each actually consisted of two (or more) very close lines. This was called **fine structure**. It was at first hypothesized that this tiny splitting of energy levels might be due to angular momentum associated with a spinning of the electron; that is, the electron might spin on its axis as well as orbit the nucleus just as the earth spins on its axis as it orbits the sun. The interaction between the tiny current of the spinning electron could then interact with the magnetic field due to the orbiting charge and cause the small observed splitting of energy levels. (So the energy depends slightly on m_l and m_s.) This picture of the electron as spinning is, however, wholly discredited today. We cannot even view an electron as a localized object, much less a spinning one. What is important is that the electron can have two different states due to some intrinsic property and (perhaps unfortunately) we still call this property "spin." The two possible values of $m_s (+\frac{1}{2}$ and $-\frac{1}{2})$ are often said to be "spin up" and "spin down," referring to the two possible directions of the spin angular momentum.

The possible values of the four quantum numbers for an electron in the hydrogen atom are summarized in Table 41–1.

FIGURE 41–4

When a magnetic field is applied, an $n = 3$, $l = 2$ energy level is split into five separate levels, corresponding to the five values of m_l (2, 1, 0, -1, -2). An $n = 2$, $l = 1$ level is split into three levels ($m_l = 1$, 0, -1). Transitions can occur between levels (not all are shown), with photons being given off with several frequencies, which are only slightly different.

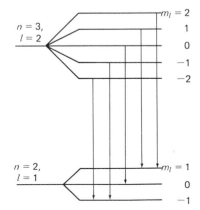

† Not quite precisely true due to normally very small interactions within the atom; see discussion of *fine structure* later in this section.

TABLE 41–1

Quantum numbers for an electron

Name	Symbol	Possible Values
Principal	n	$1, 2, 3, \ldots, \infty$.
Orbital	l	For a given n, l can be $0, 1, 2, \ldots, n-1$.
Magnetic	m_l	For given n and l, m_l can be l, $l-1, \ldots, 0, \ldots, -l$.
Spin	m_s	For each set of n, l, and m_l, m_s can be $+\frac{1}{2}$ or $-\frac{1}{2}$.

EXAMPLE 41–1 How many different states are possible for an electron whose principal quantum number is $n = 3$?

SOLUTION For $n = 3$, l can have the values $l = 2, 1, 0$. For $l = 2$, m_l can be $2, 1, 0, -1, -2$ which is five different possibilities; for each of these, m_s can be either up or down ($+\frac{1}{2}$ or $-\frac{1}{2}$); so for $l = 2$ there are $2 \times 5 = 10$ states. For $l = 1$, m_l can be $1, 0, -1$, and since m_s can be $+\frac{1}{2}$ or $-\frac{1}{2}$ for each of these, we have 6 more possible states. Finally, for $l = 0$, m_l can only be 0, and there are only two states corresponding to $m_s = +\frac{1}{2}$ and $-\frac{1}{2}$. The total number of states is $10 + 6 + 2 = 18$.

Although l and m_l do not significantly affect the energy levels in hydrogen they do affect the electron probability distribution in space. For $n = 1$, l and m_l can only be zero and the electron distribution is as shown in Fig. 41–2. For $n = 2$, l can be 0 or 1. The distribution for $n = 2$, $l = 0$ is shown in Fig. 41–5a, and it is seen to differ from that for the ground state, although it is still spherically symmetric. For $n = 2$, $l = 1$, the distribution is not spherically symmetric, but is more barbell-shaped as shown in Fig. 41–5b. The distribution shown is for one value of m_l, say $m_l = 1$; for $m_l = 0$ and $m_l = -1$, the distribution is also barbell-shaped but is symmetric around the other axes.

FIGURE 41–5

Electron cloud (or probability distribution) for $n = 2$ states in hydrogen.

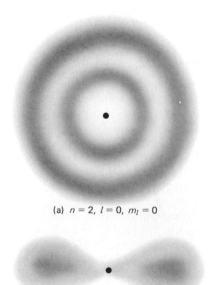

(a) $n = 2$, $l = 0$, $m_l = 0$

(b) $n = 2$, $l = 1$, $m_l = 1$

Although the spatial distribution of the electron can be calculated for the various states, it is difficult to measure them experimentally. Indeed, most of the experimental information about the atom has come from a careful examination of the emission spectra under various conditions.

 ## 41–7 Complex Atoms; the Exclusion Principle

We have discussed the hydrogen atom in detail because it is the simplest to deal with. Now we briefly discuss more complex atoms, those that contain more than one electron, and whose energy levels can be determined experimentally from an analysis of the emission spectra. The number of electrons in a neutral atom is called its *atomic number*, Z; Z is also the number of positive charges (protons) in the nucleus, and determines what kind of atom it is.

Although modifications to the Bohr theory had been attempted in order to deal with complex atoms, the development of quantum mechanics in the years after 1925 proved far more successful. The mathematics becomes very difficult, however, since in multielectron atoms, each electron is not only attracted to the nucleus but is repelled by the other electrons.

The simplest approach has been to treat each electron in an atom as occupying a particular state characterized by the quantum numbers n, l, m_l, and m_s. But to understand the possible arrangements of electrons in an atom, a new principle was needed. It was introduced by Wolfgang Pauli (1900–1958) and is called the **Pauli exclusion principle**. It states:

No two electrons in an atom can occupy the same quantum state.

That is, no two electrons can have exactly the same set of the quantum numbers n, l, m_l, and m_s. The Pauli exclusion principle[†] forms the basis not only for understanding complex atoms, but also for understanding molecules and bonding, and other phenomena as well.

Let us now look at the structure of some of the simpler atoms when they are in the ground state. After hydrogen, the next simplest atom is helium with two electrons. Both electrons can have $n = 1$, since one can have spin up ($m_s = +\frac{1}{2}$) and the other spin down ($m_s = -\frac{1}{2}$), thus satisfying the exclusion principle. Of course, since $n = 1$, l and m_l must be zero (Table 41–1). Thus the two electrons have the quantum numbers indicated in the table in the margin.

Lithium has three electrons, two of which can have $n = 1$. But the third cannot have $n = 1$ without violating the exclusion principle. Hence the third electron must have $n = 2$. Since it happens that the $n = 2$, $l = 0$ level has a lower energy than $n = 2$, $l = 1$, the electrons in the ground state have the quantum numbers indicated in the table in the margin. Of course, the quantum numbers of the third electron could also be $(3, 1, -1, \frac{1}{2})$, say; but the atom in this case would be in an excited state since it would have greater energy. It would not be long before it jumped to the ground state with the emission of a photon. At room temperature, unless energy is put in (as in a discharge tube), the vast majority of atoms are in the ground state.

We can continue in this way to describe the quantum numbers of each electron in the ground state of larger and larger atoms. That for sodium, with its eleven electrons, is shown in the table in the margin.

The ground-state configuration for all atoms is given in the *periodic table* (Table 41–2). Each square contains the atomic number Z, the symbol for the element, and the atomic mass (in atomic mass units). Finally, in the lower left corner

[†] It applies to all particles whose spin quantum number is a half-integer ($\frac{1}{2}, \frac{3}{2}$, and so on), including electrons, protons, and neutrons, but not to particles with integer spin (0, 1, 2, and so on), such as the photon and π meson.

Helium, $Z = 2$

n	l	m_l	m_s
1	0	0	$\frac{1}{2}$
1	0	0	$-\frac{1}{2}$

Lithium, $Z = 3$

n	l	m_l	m_s
1	0	0	$\frac{1}{2}$
1	0	0	$-\frac{1}{2}$
2	0	0	$\frac{1}{2}$

Sodium, $Z = 11$

n	l	m_l	m_s
1	0	0	$\frac{1}{2}$
1	0	0	$-\frac{1}{2}$
2	0	0	$\frac{1}{2}$
2	0	0	$-\frac{1}{2}$
2	1	1	$\frac{1}{2}$
2	1	1	$-\frac{1}{2}$
2	1	0	$\frac{1}{2}$
2	1	0	$-\frac{1}{2}$
2	1	-1	$\frac{1}{2}$
2	1	-1	$-\frac{1}{2}$
3	0	0	$\frac{1}{2}$

TABLE 41–2
Periodic table of the elements[†]

Legend:
Symbol — **Cl** 17 — Atomic number
Atomic mass[‡] — 35.453
$3p^5$ — Electron configuration

Transition elements

Element	Z	Atomic mass	Electron configuration
H	1	1.0079	$1s^1$
He	2	4.0026	$1s^2$
Li	3	6.94	$2s^1$
Be	4	9.01218	$2s^2$
B	5	10.81	$2p^1$
C	6	12.011	$2p^2$
N	7	14.0067	$2p^3$
O	8	15.9994	$2p^4$
F	9	18.9984	$2p^5$
Ne	10	20.17	$2p^6$
Na	11	22.9898	$3s^1$
Mg	12	24.305	$3s^2$
Al	13	26.9815	$3p^1$
Si	14	28.0855	$3p^2$
P	15	30.9738	$3p^3$
S	16	32.06	$3p^4$
Cl	17	35.453	$3p^5$
Ar	18	39.948	$3p^6$
K	19	39.0983	$4s^1$
Ca	20	40.08	$4s^2$
Sc	21	44.9559	$3d^14s^2$
Ti	22	47.90	$3d^24s^2$
V	23	50.9415	$3d^34s^2$
Cr	24	51.996	$3d^54s^1$
Mn	25	54.9380	$3d^54s^2$
Fe	26	55.847	$3d^64s^2$
Co	27	58.9332	$3d^74s^2$
Ni	28	58.71	$3d^84s^2$
Cu	29	63.546	$3d^{10}4s^1$
Zn	30	65.38	$3d^{10}4s^2$
Ga	31	69.735	$4p^1$
Ge	32	72.59	$4p^2$
As	33	74.9216	$4p^3$
Se	34	78.96	$4p^4$
Br	35	79.904	$4p^5$
Kr	36	83.80	$4p^6$
Rb	37	85.467	$5s^1$
Sr	38	87.62	$5s^2$
Y	39	88.9059	$4d^15s^2$
Zr	40	91.22	$4d^25s^2$
Nb	41	92.9064	$4d^45s^1$
Mo	42	95.94	$4d^55s^1$
Tc	43	98.9062	$4d^55s^2$
Ru	44	101.07	$4d^75s^1$
Rh	45	102.906	$4d^85s^1$
Pd	46	106.4	$4d^{10}5s^0$
Ag	47	107.868	$4d^{10}5s^1$
Cd	48	112.41	$4d^{10}5s^2$
In	49	114.82	$5p^1$
Sn	50	118.69	$5p^2$
Sb	51	121.75	$5p^3$
Te	52	127.60	$5p^4$
I	53	126.90	$5p^5$
Xe	54	131.30	$5p^6$
Cs	55	132.905	$6s^1$
Ba	56	137.33	$6s^2$
(57–71)	57–71[§]		
Hf	72	178.49	$5d^26s^2$
Ta	73	180.947	$5d^36s^2$
W	74	183.85	$5d^46s^2$
Re	75	186.207	$5d^56s^2$
Os	76	190.2	$5d^66s^2$
Ir	77	192.22	$5d^76s^2$
Pt	78	195.09	$5d^96s^1$
Au	79	196.966	$5d^{10}6s^1$
Hg	80	200.59	$5d^{10}6s^2$
Tl	81	204.37	$6p^1$
Pb	82	207.2	$6p^2$
Bi	83	208.980	$6p^3$
Po	84	(209)	$6p^4$
At	85	(210)	$6p^5$
Rn	86	(222)	$6p^6$
Fr	87	(223)	$7s^1$
Ra	88	226.025	$7s^2$
(89–103)	89–103[‖]		
Rf	104	(260)	$6d^27s^2$
Ha	105	(260)	$6d^37s^2$
	106	(263)	

[§]Lanthanide series

Element	Z	Atomic mass	Electron configuration
La	57	139.906	$5d^16s^2$
Ce	58	140.12	$5d^14f^16s^2$
Pr	59	140.908	$4f^36s^2$
Nd	60	144.24	$4f^46s^2$
Pm	61	(145)	$4f^56s^2$
Sm	62	150.4	$4f^66s^2$
Eu	63	151.96	$4f^76s^2$
Gd	64	157.25	$5d^14f^76s^2$
Tb	65	158.925	$5d^14f^86s^2$
Dy	66	162.50	$4f^{10}6s^2$
Ho	67	164.930	$4f^{11}6s^2$
Er	68	167.26	$4f^{12}6s^2$
Tm	69	168.934	$4f^{13}6s^2$
Yb	70	173.04	$4f^{14}6s^2$
Lu	71	174.967	$5d^14f^{14}6s^2$

[‖]Actinide series

Element	Z	Atomic mass	Electron configuration
Ac	89	(227)	$6d^17s^2$
Th	90	232.038	$6d^27s^2$
Pa	91	231.036	$5f^26d^17s^2$
U	92	238.029	$5f^36d^17s^2$
Np	93	237.048	$5f^56d^17s^2$
Pu	94	(244)	$5f^66d^07s^2$
Am	95	(243)	$5f^76d^07s^2$
Cm	96	(247)	$5f^76d^17s^2$
Bk	97	(247)	$5f^86d^17s^2$
Cf	98	(251)	$5f^{10}6d^07s^2$
Es	99	(254)	$5f^{11}6d^07s^2$
Fm	100	(257)	$5f^{12}6d^07s^2$
Md	101	(258)	$5f^{13}6d^07s^2$
No	102	(259)	$6d^07s^2$
Lr	103	(260)	$6d^17s^2$

[†] Atomic mass values averaged over isotopes in percentages they occur on earth's surface.

[‡] For many unstable elements, mass number of the most stable known isotope is given in parentheses.

the configuration of the ground state of the atom is given. This requires some explanation. Electrons with the same value of n are referred to as being in the same *shell*. Electrons with $n = 1$ are in one shell (called the K shell); those with $n = 2$ are in a second shell (the L shell), and so on. Electrons with the same value of l are referred to as being in the same *subshell*. Letters are often used to specify the value of l as follows:

$$l = 0 \quad 1 \quad 2 \quad 3 \quad 4 \ldots$$
$$\text{Symbol} = s \quad p \quad d \quad f \quad g \ldots$$

That is, $l = 0$ is the s subshell; $l = 1$ is the p subshell; $l = 2$ is the d subshell; beginning with $l = 3$, the letters follow the alphabet, f, g, h, i, and so on. (The first letters s, p, and d were originally abbreviations of "sharp," "principle," and "diffuse," and were experimental terms referring to the spectra.) The Pauli exclusion principle limits the number of electrons possible in each shell and subshell. Notice, for example, that for sodium (just discussed) the $n = 1$ and $n = 2$ shells are filled (as are their subshells) and the eleventh electron goes into the $n = 3$ shell.

Since the energy levels depend almost entirely on the values of n and l, it is customary to specify the electron configuration simply by giving the n value and the appropriate letter for l; and the number of electrons in each subshell is given as a superscript. The ground state configuration of sodium, for example, is written as $1s^2 2s^2 2p^6 3s^1$. In the periodic table (Table 41–2) this is simplified by specifying the configuration only of the outermost electrons and any other nonfilled subshells. The grouping of atoms in the periodic table is according to increasing mass; there is also a regularity according to chemical properties, and this is discussed in chemistry textbooks.

* 41–8 X rays and Atomic Number

The line spectra of atoms in the visible, UV, and IR regions of the EM spectrum are mainly due to transitions between states of the outer electrons. For these electrons, much of the charge of the nucleus is shielded from them by the negative charge on the inner electrons. But the innermost electrons in the $n = 1$ shell see the full charge of the nucleus. Since the energy of a level is proportional to Z^2 (see Eq. 40–11), for an atom with $Z = 50$, we would expect wavelengths about $50^2 = 2500$ times shorter than those found in the Lyman series of hydrogen (around 100 nm), or 10^{-2} to 10^{-1} nm. These are in the X-ray region of the spectrum!

As we saw in Section 37–11, X rays are produced when electrons accelerated by a high voltage strike the metal target inside the X-ray tube. If we look at the spectrum of wavelengths emitted by an X-ray tube we see that the spectrum consists of two parts: a continuous spectrum with a cutoff at some λ_0 which depends only on the voltage across the tube; and peaks which are superimposed on top. A typical example is shown in Fig. 41–6. The smooth curve and the cutoff wavelength λ_0 move to the left as the voltage across the tube increases. But the peaks (labeled K_α and K_β in Fig. 41–6) remain at the same wavelength when the voltage is changed, but are located at different places when different target materials are used. This suggests that the peaks are characteristic of the material used. Indeed, we can explain them by imagining that the electrons accelerated by the high voltage of the tube can reach sufficient energies that when they collide with the atoms of the target, they can knock out one of the very tightly held inner electrons. These *characteristic X rays* (the peaks in Fig. 41–6) are photons emitted when an electron in an upper state drops down to fill the vacated lower state.

FIGURE 41–6

Spectrum of X rays emitted from a molybdenum target in an X-ray tube operated at 50 kV.

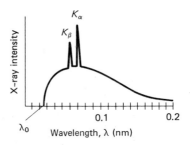

Measurement of the characteristic X-ray spectra has allowed a determination of the inner energy levels of atoms. It has also allowed the determination of Z values for many atoms, since (as we have seen) the wavelength of the shortest X rays emitted will be proportional to Z^2. Actually, for an electron jumping from say the $n = 2$ to the $n = 1$ level, the wavelength is inversely proportional to $(Z - 1)^2$ because the nucleus is shielded by the one electron that still remains in the $1s$ level. In 1914, H. G. J. Moseley found that a plot of $\sqrt{1/\lambda}$ versus Z produced a straight line, Fig. 41–7. The Z values of a number of elements were determined by fitting them to such a "Moseley plot." The work of Moseley put the concept of atomic number on a firm experimental basis.

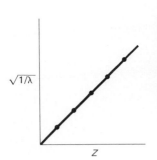

FIGURE 41–7

Plot of $\sqrt{1/\lambda}$ versus Z for K_α X-ray lines.

◆ **EXAMPLE 41–2** Estimate the wavelength for an $n = 2$ to $n = 1$ transition in molybdenum ($Z = 42$).

SOLUTION We use the Bohr formula, Eq. 40–12, with Z^2 replaced by $(Z - 1)^2 = (41)^2$. Or, more simply, we can use the result of Example 40–7 for the $n = 2$ to $n = 1$ transition in hydrogen ($Z = 1$). Since $\lambda \propto 1/(Z - 1)^2$, we will have

$$\lambda = (1.22 \times 10^{-7} \text{ m})/(41)^2 = 0.073 \text{ nm}.$$

This is close to the measured value (Fig. 41–6) of 0.071 nm.

Now we briefly analyze the continuous part of an X-ray spectrum (Fig. 41–6) based on the photon theory of light. When electrons strike the target, they collide with atoms of the material and give up most of their energy as heat (about 99 percent, so X-ray tubes must be cooled—usually with water). However, electrons can also give up energy by emitting a photon of light. As shown in Fig. 41–8, an electron can be decelerated by the positive nucleus of an atom; but an accelerating charge can emit radiation (Chapter 33), and in this case it is called *bremsstrahlung* (German for "braking radiation"). Because energy is conserved, the energy of the emitted photon, hf, must equal the loss of kinetic energy of the electron, $\Delta KE = KE - KE'$, so

$$hf = \Delta KE.$$

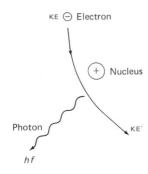

FIGURE 41–8

Bremsstrahlung photon produced by an electron that is decelerated by a positive nucleus.

An electron may lose all or a part of its energy in such a collision. The continuous X-ray spectrum (Fig. 41–6) is explained as being due to such bremsstrahlung collisions in which varying amounts of energy are lost by the electrons. The shortest wavelength X ray produced (the highest frequency) must be due to an electron that gives up all its kinetic energy to one photon in a single collision. Since the initial kinetic energy of an electron is equal to the energy given it by the accelerating voltage, V, then $KE = eV$; in a single collision in which the electron is brought to rest we have

$$hf_0 = eV$$

or

$$\lambda_0 = \frac{hc}{eV} \tag{41-3}$$

where $\lambda_0 = h/f_0$ is the cutoff wavelength, Fig. 41–6. This prediction for λ_0 corresponds precisely with that observed experimentally. This result is further evidence that X rays are a form of light[†] and that the photon theory of light is valid.

[†] If X rays were not photons but rather a neutral particle with rest mass m_0, Eq. 41–3 would not hold.

EXAMPLE 41-3 What is the shortest wavelength X-ray photon emitted in an X-ray tube subjected to 50 kV?

SOLUTION From Eq. 41-3,

$$\lambda_0 = \frac{(6.6 \times 10^{-34}\ \text{J·s})(3.0 \times 10^8\ \text{m/s})}{(1.6 \times 10^{-19}\ \text{C})(5.0 \times 10^4\ \text{V})} = 2.5 \times 10^{-11}\ \text{m}$$

or 0.025 nm. This agrees well with experiment, Fig. 41-6.

*41-9 Lasers and Holography

A laser is a device that can produce a very narrow intense beam of monochromatic coherent light. The emitted beam is a nearly perfect plane wave. An ordinary light source, on the other hand, emits light in all directions (so the intensity decreases rapidly with distance), and the emitted light is incoherent. It is incoherent (the different parts of the beam are not in phase with each other) since the excited atoms which emit the light act independently so each photon emitted can be considered as a short wave train, typically 30 cm long and lasting 10^{-8} s; these wave trains bear no phase relation to one another. (See Section 36-4.)

The action of a laser is based on quantum theory. We have seen that a photon can be absorbed by an atom if (and only if) its energy hf corresponds to the energy difference between an occupied energy level of the atom and an available excited state, Fig. 41-9a. This is, in a sense, a resonant situation. If the atom is already in the excited state, it may of course jump spontaneously to the lower state with the emission of a photon. However, if a photon with this same energy strikes the excited atom, it can stimulate the atom to make the transition sooner to the lower state, Fig. 41-9b. This is called **stimulated emission**, and it can be seen that not only do we still have the original photon, but also a second one of the same frequency as a result of the atom's transition. And these two photons are exactly *in phase*. This is how coherent light is produced in a laser; hence the name "laser," which is an acronym for **l**ight **a**mplification by **s**timulated **e**mission of **r**adiation.

Normally, most atoms are in the lower state, so incident photons will mostly be absorbed. In order to obtain the coherent light from stimulated emission, two conditions must be satisfied. First, the atoms must be excited to the higher state. That is, an *inverted population* is needed, one in which more atoms are in the upper state than in the lower one, so that *emission* of photons will dominate over absorption. And secondly, the higher state must be a *metastable state*—a state in which the electrons remain longer than usual so that the transition to the lower state occurs by stimulated emission rather than spontaneously. How these conditions are achieved for different lasers will be discussed shortly. For now, we assume that the atoms have been excited to an upper state. Figure 41-10 is a schematic diagram of a laser: the "lasing" material is placed in a long narrow tube at the ends of which are two mirrors, one of which is partially transparent (perhaps 1 or 2 percent). Some of the excited atoms drop down fairly soon after being excited. One

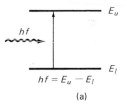

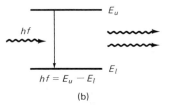

FIGURE 41-9

(a) Absorption of a photon.
(b) Stimulated emission.

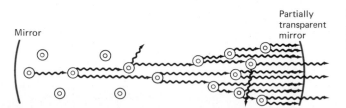

FIGURE 41-10

Laser diagram, showing excited atoms stimulated to emit light.

of these is the atom shown on the far left in Fig. 41–10. If the emitted photon strikes another atom in the excited state, it stimulates this atom to emit a photon of the same frequency and in phase with it. These two photons then move on to strike other atoms causing more stimulated emission. As the process continues, the number of photons multiplies. When the photons strike the end mirrors, most are reflected back and as they move in the opposite direction, they continue to stimulate more atoms to emit photons. As the photons move back and forth between the mirrors, a small percentage pass through the partially transparent mirror at one end. These photons make up the narrow coherent external laser beam.

Inside the tube, some photons will not be emitted parallel to the axis, and these will merely go out the side of the tube and not contribute to the main beam. Thus the beam can be very narrow. In a well-designed laser, the spreading of the beam is limited only by diffraction, so the angular spread is $\approx \lambda/D$ (see Eq. 37–1) where D is the diameter of the end mirror; the diffraction spreading can be incredibly small. The light energy, instead of spreading out in space as it does for an ordinary light source, is directed in a pencil-thin beam.

The excitation of the atoms in a laser can be done in several ways to produce the necessary inverted population. In a ruby laser, the lasing material is a ruby rod consisting of Al_2O_3 with a small percentage of aluminum (Al) atoms replaced by chromium (Cr) atoms. The Cr atoms are the ones involved in lasing. The atoms are excited by strong flashes of light of wavelength 550 nm, which corresponds to a photon energy of 2.2 eV. As shown in Fig. 41–11 the atoms are excited from state E_0 to state E_2. This process is called *optical pumping*. The atoms quickly decay either back to E_0 or to the intermediate state E_1 which is metastable with a lifetime of about 3×10^{-3} s (compared to 10^{-8} s for ordinary levels). With strong pumping action, more atoms can be forced into the E_1 state than are in the E_0 state. Thus we have the inverted population needed for lasing. As soon as a few atoms in the E_1 state jump down to E_0, they produce stimulated emission of the other atoms and the lasing action begins. A ruby laser thus emits a beam whose photons have energy 1.8 eV and a wavelength of 694.3 nm (or "ruby-red" light).

In a helium-neon (He-Ne) laser, the lasing material is a gas, a mixture of about 15 percent He and 85 percent Ne. This combination works as a lasing material because of certain compatible properties of the two gases. In this laser, the atoms are excited by applying a high voltage to the tube so that an electric discharge takes place within the gas. In the process, some of the He atoms are raised to the metastable state E_1 shown in Fig. 41–12 which corresponds to a jump of 20.61 eV. Now Ne atoms have an excited state that is almost exactly the same energy above the ground state, 20.66 eV. The He atoms do not quickly return to the ground state by spontaneous emission, but instead often give their excess energy to a Ne atom when they collide; in such a collision, the He drops to the ground state and the Ne atom is excited to the state E_3' (the prime refers

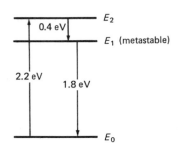

FIGURE 41–11

Energy levels of chromium in ruby crystal. Photons of energy 2.2 eV "pump" atoms from E_0 to E_2, which then decay to metastable state E_1. Lasing action occurs by stimulated emission of photons in transition from E_1 to E_0.

FIGURE 41–12

Energy levels for He and Ne. He is excited in electric discharge to the E_1 state. This energy is transferred to the E_3' level of Ne by collision. E_3' is metastable and decays to E_2' by stimulated emission.

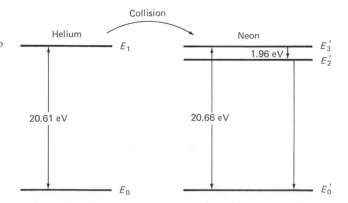

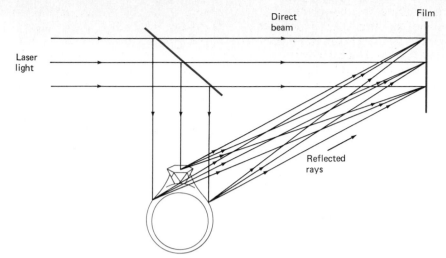

FIGURE 41–13

Making a hologram.

to neon states). The slight difference in energy (0.05 eV) is supplied by the kinetic energy of the moving molecules. In this manner, the E'_3 state in Ne—which is metastable—becomes more populated than the E'_2 level. Hence we have the inverted population needed for lasing.

Another type of laser makes use of a p–n junction in which the transitions occur between the bottom of the conduction band and the upper part of the valence band (Section 41–11).

One of the many interesting applications of laser light is the production of three-dimensional images called *holograms*. In an ordinary photograph, the film simply records the intensity of light reaching it at each point; when the photograph or transparency is viewed, light reflecting from it or passing through it gives us a two-dimensional picture. In holography, the images are formed by interference, without lenses. When a laser hologram is made on film, a broadened laser beam is split into two parts by a half-silvered mirror, Fig. 41–13. One part goes directly to the film; the rest passes to the object to be photographed, from which it is reflected to the film. Light from every point on the object reaches each point on the film, and the interference of the two beams allows the film to record both the intensity and relative phase of the light at each point. After the film is developed, it is placed again in a laser beam and a three-dimensional image of the object is seen. You can walk around such an image and see it from different sides as if it were the original object. Yet, if you try to touch it with your hand, there will be nothing material there.

The details of how the image is formed are quite complicated. But we can get the basic idea by considering one single point on the object. In Fig. 41–14a the rays OA and OB have reflected from one point on our object. The rays CA and DB come directly from the source and interfere with OA and OB at points A and B on the film. A set of interference fringes is produced as shown in Fig. 41–14b. The spacing between the fringes changes from top to bottom as shown. Why this happens is explained in Fig. 41–15. Thus the hologram of a single point object would have the pattern shown in Fig. 41–14b. The film in this case looks like a diffraction grating with variable spacing. Hence, when coherent laser light is passed back through the developed film, the diffracted rays in the first order maxima occur at slightly different angles because the spacing changes (remember Eq. 37–1, $\sin \theta = \lambda/D$; so where the spacing D is greater, the angle θ is less). Hence, the rays diffracted upward (in first order) seem to diverge from a single point, Fig. 41–16. This is a virtual image of the original object, which can be seen with the eye. Rays diffracted in first order *downward* converge to make a real image, which can be seen and also photographed. (Note that the

FIGURE 41–14

Light from point O on the object interferes with light of direct beam (rays CA and DB).

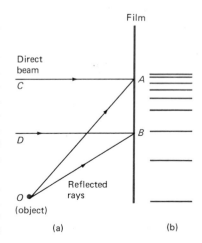

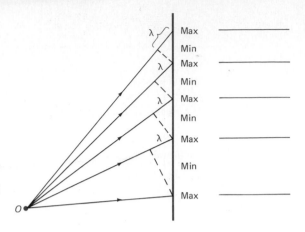

FIGURE 41-15

Each of the rays shown leaving point O is one wavelength shorter than the one above it. If the top ray is in phase with the direct beam (not shown), which has the same phase at all points on the screen, all the rays shown produce constructive interference. From this diagram it can be seen that the fringe spacing increases toward the bottom.

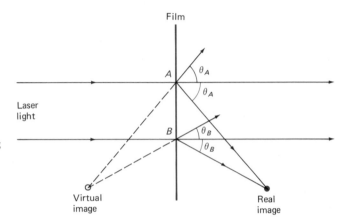

FIGURE 41-16

Laser beam strikes film that is like a diffraction grating of variable spacing. Rays corresponding to the first diffraction maxima are shown emerging. The angle $\theta_A > \theta_B$ because the spacing at B is greater than at A ($\sin \theta = \lambda/D$). Hence real and virtual images of the point are reproduced as shown.

straight-through undiffracted rays are of no interest.) Of course real objects consist of many points, so a hologram will be a complex interference pattern which, when laser light is incident on it, will reproduce an image of the object. Each image point will be at the correct (three-dimensional) position with respect to other points, so the image accurately represents the original object; and it can be viewed from different angles as if viewing the original object. Holograms can be made in which a viewer can walk entirely around the image (360°) and see all sides of it.

*41-10   Molecular Spectra

In this and the next section we discuss some great successes of the quantum theory involving aggregates of atoms. But we can do so only very briefly.

When atoms combine to form molecules, the energy levels of the outer electrons are altered because they now interact with each other. Additional energy levels also become possible because the atoms can vibrate with respect to each other, and the molecule as a whole can rotate. The energy levels for both vibrational and rotational motion are quantized. However, they are very close together, particularly for the rotational levels (typically 10^{-3} eV apart). Each atomic energy level thus becomes a set of closely spaced levels corresponding to the vibrational and rotational motions, Fig. 41-17. Transitions from one level to another appear as many *very* closely spaced lines. In fact, the lines are not always distinguishable, and these spectra are called *band spectra*. Each type

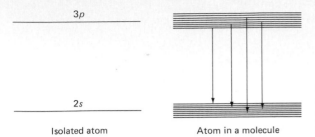

3p

2s

Isolated atom

Atom in a molecule

FIGURE 41–17

The individual energy levels of an isolated atom become bands of closely spaced levels in molecules as well as in solids and liquids.

of molecule has its own characteristic spectrum, which can be used for identification and for determination of structure.

When atoms or molecules are in the solid (or liquid) state, their outer electrons also overlap. Hence their energy levels are changed somewhat, and because of the interaction, the energy levels are spread out into *energy bands* as in Fig. 41–17. Because there are so many atoms present, the energy levels in such a band are extremely close together. They thus seem continuous, which is why the spectrum of heated solids, such as red-hot iron or a light bulb filament, appears continuous. The energy bands in semiconductors play an important role for transistors and diodes as we discuss in the next section.

* 41–11 Semiconductors

Quantum mechanics has been a great tool for understanding the structure of solids. This active field of research today is called *solid-state physics* (or *condensed-matter physics*, so as to include liquids as well). We now examine, very briefly, one aspect of this field—semiconductors—and some of its applications in modern electronics. First we look at some properties of the most commonly used semiconductors, germanium and silicon.

An atom of silicon or germanium has four outer electrons which act to hold the atoms in the regular lattice structure of the crystal, Fig. 41–18a. Germanium and silicon acquire useful properties for use in electronics only when a tiny amount of impurity is introduced into the crystal structure (about 1 part in 10^6). This is called "doping" the semiconductor. Two kinds of semiconductors can be made, depending on the type of impurity used. If the impurity is a material whose atoms have five outer electrons, such as arsenic, we have the situation shown in Fig. 41–18b. Only four of arsenic's electrons fit into the crystal structure. The fifth does not fit in and can move relatively freely, much like the electrons in a conductor. Because of this small number of extra electrons, a doped semiconductor becomes slightly conducting. An arsenic-doped germanium crystal is called an *n-type semiconductor* because electrons (negative charge) carry the electric current.

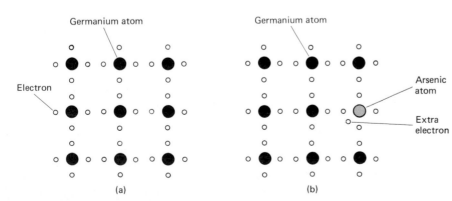

FIGURE 41–18

A germanium crystal. (a) Four (outer) electrons surround each germanium atom. (b) Germanium crystal doped with a few arsenic atoms; the extra electron doesn't fit into the crystal lattice and so is free to move about. This is an *n*-type semiconductor.

Germanium atom

Electron

(a)

Germanium atom

Arsenic atom

Extra electron

(b)

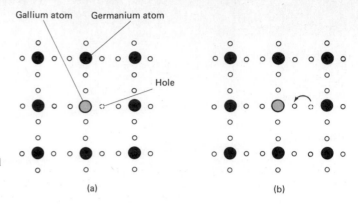

FIGURE 41–19

A *p*-type semiconductor, gallium-doped germanium.
(a) Gallium has only three outer electrons, so there is
an empty spot, or *hole*, in the structure. (b) Electrons
from germanium atoms can jump into the hole and fill
it. As a result, the hole moves to a new location (to
the right in this figure).

In a *p-type semiconductor*, a small amount of impurity with three outer
electrons—such as gallium—is added to the semiconductor. As shown in Fig.
41–19a, there is a "hole" in the lattice structure next to a gallium atom since
it has only three outer electrons. Electrons from nearby germanium atoms can
jump into this hole and fill it. But this leaves a "hole" where that electron had
previously been, Fig. 41–19b. The vast majority of atoms are germanium, so
the hole is almost always next to a germanium atom. Since germanium atoms
require four outer electrons to be neutral, this means there is a net positive
charge at a hole. Whenever an electron moves to fill a hole, the positive hole
is then at the previous position of that electron. Another electron can then fill
this hole, and the hole thus moves to a new location; and so on. This type of
semiconductor is called *p-type* because it is the *p*ositive holes that seem to carry
the electric current. Note, however, that *p*-type and *n*-type semiconductors have
no net charge on them.

Semiconductor behavior can be explained at the atomic level using the
so-called *electron band theory* of solids which is based on quantum mechanics.
Electrons in a single atom can occupy only certain energy levels, as we have
seen. The lowest possible energy level is called the ground state and higher ones
are called excited states. In a solid, these energy levels spread out into wide
"bands" because of the interaction among the atoms, as described in the previous
section. The outer electrons can be considered to be in either of two bands:[†]
the lower *valence band*, which corresponds to the ground state; or the upper
conduction band. No electron can have an energy in the "forbidden" energy gap
between the two bands. Normally the electrons reside in the valence band where
they are held rather tightly to individual atoms. In an insulator, the valence
band is full, the conduction band is essentially empty, and the forbidden gap
between them is fairly large; so there are no free states for the electrons to
move into, unless given sufficient energy (say by collision with another electron)
to jump up to the higher conduction band. Once there, an electron would no
longer be bound to a particular atom but could move about freely in the lattice.
If the gap between the bands is large, the material will be an insulator since
very few electrons would have sufficient energy to reach the conduction band;
see Fig. 41–20a. In a good conductor, on the other hand, there is no gap
(Fig. 41–20b); the two bands may overlap, or there is simply one band that
is not filled and the electrons are free to move easily to other states; they can
thus move about freely and carry an electric current. In a pure (or *intrinsic*)
semiconductor such as carbon, germanium or silicon, the forbidden energy gap
between valence and conduction bands is small, Fig. 41–20c. A small percentage
of electrons can have enough energy to jump the gap, so there will be a very
slight amount of conduction. (At room temperature, the average KE of electrons
is on the order of $kT \approx \frac{1}{40}$ eV.) If the temperature is raised, more electrons

---

[†] Additional bands, of lower energy, normally have all states occupied.

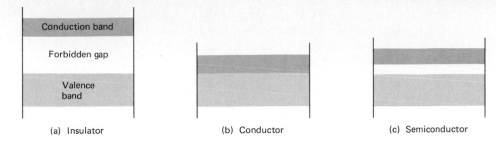

(a) Insulator   (b) Conductor   (c) Semiconductor

FIGURE 41–20
Energy bands in solids, showing (idealized) only bands for outer electrons. The bands for other electrons are of lower energy.

will have enough energy to jump the gap; this effect can often more than offset the effects of reduced mean free path due to increased disorder at increased temperature, so that the resistivity of semiconductors can *decrease* with temperature (see Table 26–1).

In a doped semiconductor, the impurity provides additional energy states between the bands as shown in Fig. 41–21. In an *n*-type semiconductor, the impurity energy level lies just below the conduction band. Electrons in this energy level can easily acquire sufficient energy to reach the conduction band. Since this level supplies electrons to the conduction band, it is called a *donor* level. In *p*-type semiconductors, the impurity energy level is just above the valence band. It is called an *acceptor* level because electrons from the valence band can easily jump into it. A positive hole is left behind; and as other electrons move into this hole, the hole moves about as discussed earlier.

Semiconductor diodes and transistors are essential components of modern electronic devices. The miniaturization achieved today allows many thousands of diodes, transistors, resistors, and so on, to be placed on a single *chip* only a centimeter on a side. We now discuss, briefly and simply, the operation of diodes and transistors.

When an *n*-type semiconductor is joined to a *p*-type, a *p-n junction diode* is formed. Separately, the two semiconductors are electrically neutral. When joined, a few electrons near the junction diffuse from the *n*-type into the *p*-type semiconductor, where they fill a few of the holes. The *n*-type is left with a positive charge, and the *p*-type acquires a net negative charge. Thus a potential difference is established, with the *n* side positive relative to the *p* side, and this prevents further diffusion of electrons.

If a battery is connected to a diode with the positive terminal to the *p* side and the negative terminal to the *n* side as in Fig. 41–22a, such a voltage opposes the internal potential difference and the diode is said to be *forward biased*. If the voltage is great enough (about 0.3 V for Ge, 0.6 V for Si at room temperature) a current will flow. The positive holes in the *p*-type semiconductor are repelled by the positive terminal of the battery and the electrons in the *n*-type are repelled by the negative terminal of the battery. The holes and electrons meet at the junction, and the electrons cross over and fill the holes. In the meantime, the positive terminal of the battery is continually pulling electrons off the *p* end, forming new holes, and electrons are being supplied by the negative terminal at the *n* end. Consequently a large current flows through the diode.

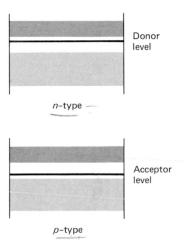

n-type

Donor level

Acceptor level

p-type

FIGURE 41–21
Impurity levels in doped semiconductors.

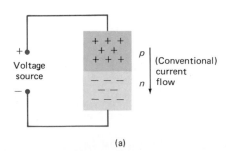

(a)

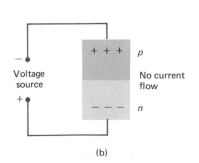

(b)

FIGURE 41–22
Schematic diagram showing how a semiconductor diode operates. Current flows when the voltage is connected in forward bias, as in (a), but not when in reverse bias, as in (b).

When the diode is *reverse biased*, as in Fig. 41–22b, the holes in the *p* end are attracted to the battery's negative terminal and the electrons in the *n* end are attracted to the positive terminal. The current carriers do not meet near the junction and, ideally, no current flows.

Since a *p-n* junction diode allows current to flow only in one direction, it can serve as a *rectifier*—to change ac into dc. A simple rectifier circuit is shown in Fig. 41–23a where the arrow inside the symbol for a diode indicates the direction in which a diode conducts conventional (+) current. The ac source applies a voltage across the diode alternately positive and negative. Only during half of each cycle will a current pass through the diode; so only then is there a current through the resistor R. Hence the voltage V_{ab} across R as a function of time looks like that shown in the graph in Fig. 41–23b. This *half-wave rectification* is not exactly dc, but it is unidirectional. More useful is a *full–wave rectifier* circuit which uses two diodes (or sometimes four) as shown in Fig. 41–24a. At any given instant, either one diode or the other will conduct current to the right. Therefore the output across the load resistor R will be as shown in Fig. 41–24b; actually this is the voltage if the capacitor C were not in the circuit. The capacitor tends to store charge, and thus helps to smooth out the current as shown in Fig. 41–24c.

Rectifier circuits are important because most line voltage is ac, and most electronic devices require a dc voltage for their operation. Hence, diodes are found in nearly all electronic devices including radio and TV sets.

A simple *junction transistor* consists of a crystal of one type of semiconductor sandwiched between two crystals of the opposite type. Both *pnp* and *npn* transistors are made and they are shown schematically in Fig. 41–25a. The three semiconductors are given the names collector, base, and emitter. The symbols for *npn* and *pnp* transistors are shown in Fig. 41–25b. The arrow is always placed on the emitter and indicates the direction of (conventional) current flow in normal operation.

The operation of a transistor can be analyzed as follows. Consider an *npn* transistor connected as shown in Fig. 41–26. A voltage V_{CE} is maintained between the collector and emitter by the battery $\mathscr{E}_C$. The voltage applied to the base is called the *base bias voltage*, V_{BE}; if it is positive, electrons in the emitter are attracted into the base. Since the base region is very thin (perhaps 1 μm), most of the electrons flow right across into the collector, which is maintained at a positive voltage. A large current, I_C, flows between collector and emitter and a much smaller current, I_B, through the base. A small variation in the base voltage due to an input signal causes a large change in the collector current and therefore a large change in the voltage drop across the output resistor R_C. Hence a transistor can amplify a small signal into a larger one. In fact transistors are the basic elements in modern electronic amplifiers of all sorts.

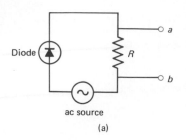

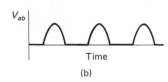

FIGURE 41–23

(a) Simple rectifier circuit, using a semiconductor diode. (b) Output voltage across R as a function of time.

FIGURE 41–24

(a) Full-wave rectifier circuit. (b) Output voltage in the absence of capacitor C. (c) Output voltage with capacitor in the circuit.

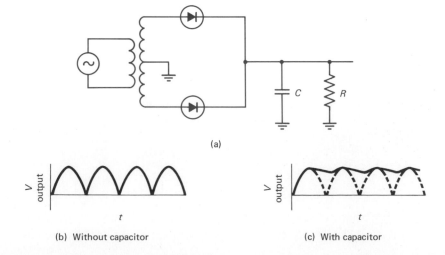

(a)

(b) Without capacitor

(c) With capacitor

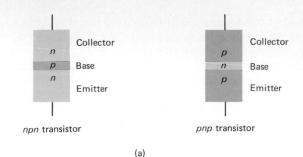

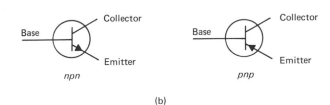

FIGURE 41–25

(a) Schematic diagram of *npn* and *pnp* transistors.
(b) Symbols for *npn* and *pnp* transistors.

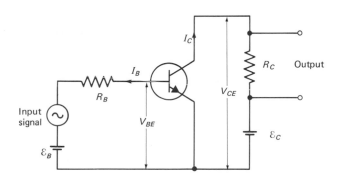

FIGURE 41–26

An *npn* transistor used as an amplifier.

A *pnp* transistor operates in the same fashion, except that holes move instead of electrons. The collector voltage is negative, and so is the base voltage in normal operations.

 Summary ⎯⎯⎯⎯⎯⎯⎯⎯⎯⎯⎯⎯⎯⎯⎯⎯⎯⎯⎯⎯⎯⎯⎯⎯⎯⎯⎯⎯⎯⎯⎯

In 1925, Schrödinger and Heisenberg separately worked out a new theory, *quantum mechanics*, which is now considered to be the basic theory at the atomic level. It is a statistical theory rather than a deterministic one. According to quantum mechanics, the electrons in an atom do not have well-defined orbits but instead exist as a "cloud." Electron clouds can be interpreted as an electron wave spread out in space, or as a probability distribution for electrons as particles.

An important aspect of quantum mechanics is the Heisenberg *uncertainty principle*. It results from the wave–particle duality and the unavoidable interaction between the observed object and the observer. One form of the uncertainty principle states that the position x and momentum p of an object cannot both be measured precisely at the same time; the products of the uncertainties, $(\Delta x)(\Delta p)$, can be no less than $h/2\pi$:

$$(\Delta p)(\Delta x) \geq \frac{h}{2\pi}.$$

Another form states that the energy can be uncertain, or nonconserved, by an amount ΔE for a time Δt given by:

$$(\Delta E)(\Delta t) \geq \frac{h}{2\pi}.$$

According to quantum mechanics, the state of an electron in an atom is specified by four quantum numbers: n, l, m_l, and m_s. The principal quantum number, n, can take on any integer values (1, 2, 3, . . .) and corresponds to the quantum number of the old Bohr theory; l can take on values from 0 up to $n - 1$; m_l can take on integer values from $-l$ to $+l$; and m_s can be $+\frac{1}{2}$ or $-\frac{1}{2}$. The energy levels in the hydrogen atom depend on n, whereas in other atoms they depend on n and l; when an external magnetic field is applied, the spectral lines are split (the Zeeman effect), indicating that the energy depends also on m_l in this case. Even in the absence of a magnetic field, precise measurements of spectral lines show a tiny splitting of the lines called fine structure, whose explanation is that the energy depends very slightly on m_l and m_s.

The arrangement of electrons in multielectron atoms is governed by the *Pauli exclusion principle*, which states that no two electrons can occupy the same quantum state—that is, they cannot have the same set of quantum numbers n, l, m_l, and m_s. Electrons, as a result, are grouped into shells (according to the value of n) and subshells (according to l). This shell structure gives rise to a periodicity in the properties of the elements.

 ## Questions

1. Compare a matter wave ψ to (a) a wave on a string, (b) an EM wave. Discuss similarities and differences.

2. Explain why Bohr's theory of the atom is not compatible with quantum mechanics, particularly the uncertainty principle.

3. Explain why it is that the more massive an object is the easier it becomes to predict its future position.

4. In view of the uncertainty principle, why does a baseball seem to have a well-defined position and speed whereas an electron does not?

5. Discuss whether something analogous to the uncertainty principle operates when taking a public opinion survey. That is, do we alter what we are trying to measure when we take such a survey?

6. A cold thermometer is placed in a hot bowl of soup. Will the temperature reading of the thermometer be the same as the temperature of the hot soup before the measurement was made?

7. If Planck's constant were much larger than it is, how would this affect our everyday life?

8. In what ways is Newtonian mechanics contradicted by quantum mechanics?

9. Discuss the differences between Bohr's view of the atom and the quantum-mechanical view.

10. The 589-nm yellow line in sodium is actually two very closely spaced lines. It is due to an "internal" Zeeman effect. Can you explain this? (Hint: put yourself in the reference frame of the electron.)

11. Which of the following electron configurations are forbidden? (a) $1s^2 2s^2 2p^6 3s^3$; (b) $1s^2 2s^2 2p^4 3s^2 4p^2$; (c) $1s^2 2s^2 2p^8 3s^1$.

12. Give the complete electron configuration for a uranium atom.

*13. Compare spontaneous emission to stimulated emission.

*14. How does laser light differ from ordinary light? How is it the same?

*15. Explain how a 0.0005-W laser beam, photographed at a distance, can seem much stronger than a 1000-W street lamp.

*16. A silicon semiconductor is doped with phosphorus. Will these atoms be donors or acceptors? What type of semiconductor will this be?

*17. Explain how a transistor could be used as a switch.

*18. Describe how a *pnp* transistor can operate as an amplifier.

 ## Problems

SECTION 41–3

1. (I) A proton is traveling with a speed of $(8.880 \pm 0.012) \times 10^5$ m/s. With what maximum accuracy can its position be ascertained?

2. (I) An electron remains in an excited state of an atom for typically 10^{-8} s. What is the minimum uncertainty in the energy of the state (in eV)? What is this uncertainty, in percent, of the first excited state in hydrogen?

3. (I) If an electron's position can be measured to an accuracy of 1.6×10^{-8} m, how accurately can its velocity be known?

4. (II) Estimate the least energy of a neutron contained in a typical nucleus of radius 10^{-15} m.

5. (II) A 12-g bullet leaves a rifle at a speed of 450 m/s. (a) What is the wavelength of this bullet? (b) If the position of the bullet is known to an accuracy of 0.55 cm (radius of the barrel), what is the minimum uncertainty in its momentum? (c) If the accuracy of the bullet were determined only by the uncertainty principle (an unreasonable assumption), by how much might the bullet miss a pinpoint target 300 m away?

6. (II) An electron and a 150-g baseball are each traveling 220 m/s measured to an accuracy of 0.065 percent. Calculate and compare the uncertainty in position of each.

7. (II) Use the uncertainty principle to show that if an electron were present in the nucleus ($r \approx 10^{-15}$ m), its energy would be hundreds of MeV. (Since such electron energies are not observed, we conclude that electrons are not present in the nucleus.)

8. (II) The uncertainty principle can be stated in terms of angular qualities as follows:

$$\Delta L \, \Delta \phi \geq \frac{h}{2\pi}.$$

Here, L stands for angular momentum along a given axis, and ϕ for the angular position measured in a plane per-

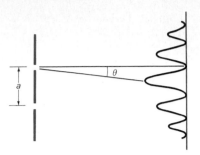

FIGURE 41–27

pendicular to that axis. (a) Make a plausibility argument for this relation. (b) Electrons in atoms have well-defined, quantized, values of angular momentum, with no uncertainty. What does this say about the uncertainty in angular position and the concept of electron orbits?

9. (II) Use the uncertainty principle to estimate the position uncertainty for the electron in the ground state of the hydrogen atom. (Hint: determine the momentum using the Bohr model of Section 40–9.) How does this compare to the Bohr radius?

10. (II) In a double-slit experiment on electrons (or photons), suppose that we use indicators to determine which slit each electron went through (Section 41–2). These indicators must tell us the y coordinate to within $a/2$ where a is the distance between slits. Use the uncertainty principle to show that the interference pattern will be destroyed. (Note: first show that the angle θ between maxima and minima of the interference pattern is given by $\lambda/2a$, Fig. 41–27.)

11. (II) How accurately can the position of a 1.50-keV electron be measured assuming its energy is known to 1.00 percent?

12. (III) A 20-cm-long pencil is balanced on its point. Classically, this is a configuration of (unstable) equilibrium, so the pencil could remain there forever if it were perfectly placed. A quantum mechanical analysis shows that the pencil must fall. (a) Why is this the case? (b) Estimate (within a factor of 2) how long it will take the pencil to hit the table if it is positioned initially as well as possible?

SECTIONS 41–6 AND 41–7

13. (I) How many different states are possible for an electron whose principal quantum number is $n = 5$?

14. (I) List the quantum numbers for each electron in the ground state of magnesium ($Z = 12$).

15. (I) How many electrons can be in the $n = 6$, $l = 3$ subshell?

16. (I) For $n = 4$, $l = 3$, what are the possible values of m_l and m_s?

17. (I) List the quantum numbers for each electron in the ground state of carbon ($Z = 6$).

18. (I) Calculate the magnitude of the angular momentum of an electron in the $n = 3$, $l = 2$ state of hydrogen.

19. (II) Using the Bohr formula for the radius of an electron orbit, estimate the average distance from the nucleus for an electron in the innermost ($n = 1$) orbit in uranium ($Z = 92$). Approximately how much energy would be required to remove this innermost electron?

20. (II) Estimate the binding energy of the third electron in lithium ($Z = 3$) using Bohr theory. The measured value is 5.36 eV.

21. (II) The ionization (binding) energy of the outermost electron in boron is 8.26 eV. (a) Determine the "effective charge," Z_{eff}, seen by this electron. (Hint: use the Bohr model.) (b) Estimate the average orbital radius.

22. (II) Show that the maximum number of electrons allowed in any subshell is equal to $2(2l + 1)$ where l is the angular momentum quantum number of the subshell.

*SECTION 41–8

*23. (I) What are the shortest-wavelength X rays emitted by electrons striking the face of a 30-kV TV picture tube? What are the longest wavelengths?

*24. (II) Use the result of Example 41–2 to estimate the X-ray wavelength emitted when a Ti ($Z = 22$) atom jumps from $n = 2$ to $n = 1$.

*25. (II) Estimate the wavelength for an $n = 2$ to $n = 1$ transition in iron ($Z = 26$).

*26. (II) Use Bohr theory to estimate the wavelength for an $n = 3$ to $n = 1$ transition in molybdenum. The measured value is 0.063 nm. Why do we not expect perfect agreement?

*27. (II) An X-ray tube operates at 100 kV with a current of 20 mA and nearly all the electron energy goes into heat. If the specific heat capacity of the 78-g plate is 0.11 kcal/kg·C°, what will be the temperature rise per minute if no cooling water is used?

*SECTION 41–9

*28. (II) A laser used to weld detached retinas puts out 25-ms-long pulses of 640-nm light which average 0.50-W output during a pulse. How much energy can be deposited per pulse and how many photons does each pulse contain?

*29. (II) Estimate the angular spread of a laser beam due to diffraction if the beam emerges through a 5.0-mm-diameter mirror. Assume that $\lambda = 694$ nm. How large would be the diameter of this beam if it struck a satellite 1000 km above the earth?

Nuclear Physics

42

In the early part of the twentieth century Rutherford's experiments led to the idea that at the center of an atom there is a tiny but massive nucleus. At the same time that the quantum theory was being developed and scientists were attempting to understand the structure of the atom and its electrons, investigations into the nucleus itself had also begun. In this chapter we take a brief look at *nuclear physics*.

 42–1   Structure of the Nucleus

An important question to physicists in the early part of this century was whether the nucleus had a structure, and what this structure might be. It turns out that the nucleus is a complicated entity and is not fully understood even today. However, by the early 1930s a model of the nucleus had been developed that is still useful. According to this model, a nucleus is considered as an aggregate of two types of particles: protons and neutrons. (Of course, we must remember that these "particles" also have wave properties, but for ease of visualization and language we often refer to them simply as "particles".) A **proton** is the nucleus of the simplest atom, hydrogen. It has a positive charge ($= +e = +1.6 \times 10^{-19}$ C) and a mass

$$m_p = (1.6726485 \pm 0.0000086) \times 10^{-27} \text{ kg}$$

or, rounded off,

$$m_p = 1.6726 \times 10^{-27} \text{ kg}.$$

The **neutron**, whose existence was ascertained only in 1932 by the Englishman James Chadwick (1891–1974), is electrically neutral ($q = 0$), as its name implies; its mass, which is almost identical to that of the proton, is

$$m_n = (1.674954 \pm 0.000009) \times 10^{-27} \text{ kg}$$

$$m_n = 1.6750 \times 10^{-27} \text{ kg.}$$

These two constituents of a nucleus, neutrons and protons, are referred to collectively as **nucleons**.

Although the hydrogen nucleus consists of a single proton alone, the nuclei of other elements consist of both neutrons and protons. The different types of nuclei are often referred to as **nuclides**. The number of protons in a nucleus (or nuclide) is called the **atomic number** and is designated by the symbol Z. The total number of nucleons, neutrons plus protons, is designated by the symbol A and is called the **atomic mass number**. This name is used since the mass of a nucleus is very closely A times the mass of one nucleon. A nuclide with 7 protons and 8 neutrons thus has $Z = 7$ and $A = 15$. The *neutron number N* is $N = A - Z$.

To specify a given nuclide, we need give only A and Z. A special symbol is commonly used which takes the form

$$^A_Z X$$

where X is the chemical symbol for the element (see Table 41–2), A is the atomic mass number, and Z is the atomic number. For example, $^{15}_7 N$ means a nitrogen nucleus containing 7 protons and 8 neutrons for a total of 15 nucleons. In a neutral atom, the number of electrons orbiting the nucleus is equal to the atomic number Z (since the charge on an electron has the same magnitude but opposite sign to that of a proton). The main properties of an atom are determined by the number of electrons. Hence Z determines what kind of atom it is: carbon, oxygen, gold, or whatever. It is redundant to specify both the symbol of a nucleus and its atomic number Z as described above. If the nucleus is nitrogen, for example, we know immediately that $Z = 7$. The subscript Z is thus sometimes dropped and $^{15}_7 N$ is then written simply ^{15}N; in words we say "nitrogen fifteen." The subscript Z is often kept, nonetheless, for convenience.

For a particular type of atom (say carbon), nuclei are found to contain different numbers of neutrons, although they all have the same number of protons. For example carbon nuclei always have 6 protons but they may have 5, 6, 7, 8, 9, or 10 neutrons. Nuclei that contain the same number of protons but different numbers of neutrons are called **isotopes**. Thus $^{11}_6 C$, $^{12}_6 C$, $^{13}_6 C$, $^{14}_6 C$, $^{15}_6 C$, and $^{16}_6 C$ are all isotopes of carbon. Of course, the isotopes of a given element are not all equally common. For example, 98.9 percent of naturally occurring carbon (on earth) is the isotope $^{12}_6 C$ and about 1.1 percent is $^{13}_6 C$. These percentages are referred to as the **natural abundances**.[†] Many isotopes that do not occur naturally can be produced in the laboratory by means of nuclear reactions (more on this later). Indeed, all elements beyond uranium ($Z > 92$) do not occur naturally and are only produced artificially.

The approximate size of nuclei was determined originally by Rutherford by scattering of charged particles. Of course, we cannot speak about a definite size for nuclei because of the wave–particle duality: their spatial extent must remain somewhat fuzzy. Nonetheless a rough "size" can be measured, and the most accurate recent measurements have been done by scattering of high-speed electrons off nuclei. It is found that nuclei have a roughly spherical shape with a radius that increases with A according to the approximate formula

$$r \approx (1.2 \times 10^{-15} \text{ m})(A^{1/3}). \tag{42–1}$$

Since the volume of a sphere is $V = \frac{4}{3}\pi r^3$, we see that the volume is proportional to the number of nucleons, $V \propto A$. This is what we would expect if nucleons were like impenetrable billiard balls: if you double the number of balls, you double the total volume. Hence, all nuclei have nearly the same density.

[†] The mass values for the elements as given in the periodic table (Table 41–2) are averaged over the natural abundances.

TABLE 42–1

Rest masses in kilograms, unified atomic mass units, and MeV/c^2

Object	Mass		
	kg	**u**	**MeV/c^2**
Electron	9.1095×10^{-31}	0.00054858	0.51100
Proton	1.67265×10^{-27}	1.007276	938.28
$^1_1$H atom	1.67356×10^{-27}	1.007825	938.79
Neutron	1.67500×10^{-27}	1.008665	939.57

The masses of nuclei can be determined, in one method, by measuring the radius of curvature of fast-moving nuclei in a magnetic field using a mass spectrometer, as discussed in Section 28–7. Indeed, as mentioned there, the existence of different isotopes of the same element was discovered using this device. Nuclear masses are specified in *unified atomic mass units* (u). On this scale, a neutral $^{12}_6$C atom is given the precise value 12.000000 u. A neutron then has a measured mass of 1.008665 u, a proton 1.007276 u, and a neutral hydrogen atom, $^1_1$H (proton plus electron) 1.007825 u. The masses of many nuclides are given in Appendix D. (It should be noted that the masses in this table, as is customary, are for the *neutral atom*, and not for a bare nucleus.)

Masses are often specified using the electron-volt energy unit. This can be done because mass and energy are related, and the precise relationship is given by Einstein's equation $E = mc^2$ (Chapter 39). Since the mass of a neutral $^1_1$H atom is 1.6736×10^{-27} kg or 1.007825 u, then 1.0000 u $= (1.0000/1.007825) \cdot (1.6736 \times 10^{-27}$ kg$) = 1.6606 \times 10^{-27}$ kg; this is equivalent to an energy $E = mc^2 = (1.6606 \times 10^{-27}$ kg$)(2.998 \times 10^8$ m/s$)^2/(1.602 \times 10^{-19}$ J/eV$) = 931.5$ MeV. The current precise values are

$$1 \text{ u} = (1.6605655 \pm 0.0000086) \times 10^{-27} \text{ kg} = (931.5016 \pm 0.0026) \text{ MeV}/c^2;$$

which can be rounded off to

$$1 \text{ u} = 1.6606 \times 10^{-27} \text{ kg} = 931.5 \text{ MeV}/c^2.$$

The rest masses of some of the basic particles are given in Table 42–1.

42–2 Binding Energy and Nuclear Forces

The total mass of a nucleus is always less than the sum of the masses of its constituent protons and neutrons as the following example shows.

 EXAMPLE 42–1 Compare the mass of a $^4_2$He nucleus to that of its constituent nucleons.

SOLUTION The mass of a neutral $^4_2$He atom, from Appendix D, is 4.002603 u. The mass of two neutrons and two protons (including the two electrons) is

$$\begin{aligned} 2m_n &= 2.017330 \text{ u} \\ 2m(^1_1\text{H}) &= \underline{2.015650 \text{ u}} \\ &\quad\ 4.032980 \text{ u} \end{aligned}$$

(We almost always deal with masses of neutral atoms—that is, nuclei with Z electrons—since this is how masses are measured. We must therefore be sure to balance out the electrons when we compare masses, which is why we used the mass of $^1_1$H in this example rather than that of the proton alone.)

Thus the mass of $_2^4$He is measured to be 4.032980 u − 4.002603 u = 0.030377 u less than the masses of its constituents. How can this be? Where has this mass gone?

It has, in fact, gone into energy of another kind (such as radiation, kinetic energy, and so on). The mass (or energy) difference in the case of $_2^4$He, is (0.030377 u)(931.5 MeV/u) = 28.30 MeV. This difference is referred to as the **total binding energy** of the nucleus. The total binding energy represents the amount of energy that must be put into a nucleus in order to break it apart into its constituent protons and neutrons. If the mass of, say, a $_2^4$He nucleus were exactly equal to the mass of two neutrons plus two protons, the nucleus could fall apart without any input of energy. To be stable the mass *must* be less than that of its constituents. Note that the binding energy is not something a nucleus has—it is energy it "lacks" relative to the total mass of its separate constituents.

This situation can be compared to the binding energy of electrons in an atom. We saw in Chapter 40 that the binding energy of the one electron in the hydrogen atom, for example, is 13.6 eV. The mass of a $_1^1$H atom is less than that of a single proton plus a single electron by 13.6 eV. Compared to the total mass (938 MeV) this is incredibly small (1 part in 10^8), and for practical purposes the mass difference can be ignored. The binding energies of nuclei are on the order of 10^6 times greater than the binding energies of atoms and are therefore far more important.

The **average binding energy per nucleon** is defined as the total binding energy of a nucleus divided by A, the total number of nucleons. For $_2^4$He it is 28.3 MeV/4 = 7.1 MeV. Figure 42–1 shows the average binding energy per nucleon as a function of A for stable nuclei. The curve rises as A increases and reaches a plateau at about 8 MeV per nucleon above $A \approx 15$. Beyond about $A = 60$, the curve decreases slowly, indicating that larger nuclei are held together a little less tightly than those in the middle of the periodic table. (We will see later that these facts allow the release of nuclear energy in the processes of fission and fusion.)

We can analyze nuclei not only from the point of view of energy but also from the point of view of the forces that hold them together. We would not expect a collection of protons and neutrons to come together spontaneously since protons are all positively charged and thus exert repulsive forces on each

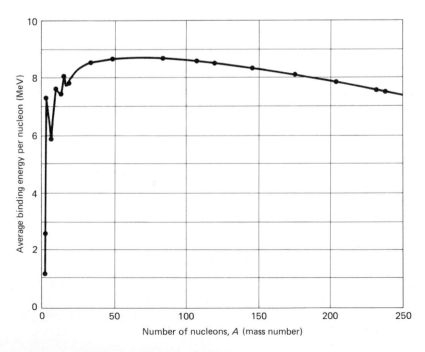

FIGURE 42–1

Average binding energy per nucleon as a function of mass number A for stable nuclei.

other. Indeed, the question arises as to how a nucleus stays together at all in view of the fact that the electric force between protons would tend to break it apart. Since stable nuclei *do* stay together, it is clear that another force must be acting. Because this new force is stronger than the electric force (which in turn is much stronger than gravity at the nuclear level) it is called the **strong nuclear force**. The strong nuclear force is an attractive force that acts between all nucleons—protons and neutrons alike. Thus protons attract each other via the nuclear force at the same time they repel each other via the electric force. Neutrons, since they are electrically neutral, only attract other neutrons or protons via the nuclear force.

The nuclear force turns out to be far more complicated than the gravitational and electromagnetic forces. A precise mathematical description is not yet possible. Nonetheless, a great deal of work has been done to try to understand the nuclear force. One important aspect of the strong nuclear force is that it is a *short-range* force: it acts only over a very short distance. It is very strong between two nucleons if they are less than about 10^{-15} m apart; but it is essentially zero if they are separated by a distance greater than this. Compare this to electric and gravitational forces, which can act over great distances and are therefore called *long-range* forces. The strong nuclear force has some strange quirks. For example, if a nuclide contains too many or too few neutrons relative to the number of protons, the nuclear force is weakened; nuclides that are too unbalanced in this regard are unstable. As shown in Fig. 42–2, stable nuclei tend to have the same number of protons as neutrons ($N = Z$) up to about $A \approx 30$ or 40. Beyond this, stable nuclei contain more neutrons than protons. This reflects the fact that as Z increases, the electrical repulsion increases, so a greater number of neutrons—which exert only the attractive nuclear force—are required to maintain stability. For very large Z, no number of neutrons can overcome the greatly increased electric repulsion; indeed, there are no completely stable nuclides above $Z = 82$.

What we mean by a stable nucleus is one that stays together indefinitely. What then is an unstable nucleus? It is one that comes apart; and this results in radioactive decay. Before we discuss the important subject of radioactivity (the next section), we note that there is a second type of nuclear force that is much weaker than the strong nuclear force. It is called the **weak nuclear force**, and we are aware of its existence only because it shows itself in certain types of radioactive decay. These two nuclear forces, the strong and the weak, together with the gravitational and electromagnetic forces, comprise the four known types of force in nature (more on this in Chapter 43).

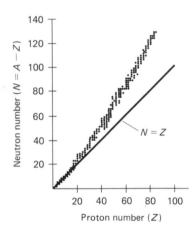

FIGURE 42–2

Number of neutrons versus number of protons for stable nuclides, which are represented by dots.

 42–3   Radioactivity

Nuclear physics had its beginnings in 1896. In that year, Henri Becquerel (1852–1908) made an important discovery: in his studies of phosphorescence, he found that a certain mineral (which happened to contain uranium) would darken a photographic plate even when the plate was wrapped to exclude light. It was clear that the mineral emitted some new kind of radiation which, unlike X rays, occurred without any external stimulus. This new phenomenon eventually came to be called **radioactivity**.

Soon after Becquerel's discovery, Marie Curie (1867–1934) and her husband, Pierre Curie (1859–1906), isolated two previously unknown elements which were very highly radioactive. These were named polonium and radium. Other radioactive elements were soon discovered as well. The radioactivity was found in every case to be unaffected by the strongest physical and chemical treatments including strong heating or cooling and the action of strong chemical reagents. It soon became clear that the source of radioactivity must be deep within the atom, that it must emanate from the nucleus. And it became apparent that radioactivity is the result of the *disintegration* or *decay* of an unstable nucleus. Certain isotopes are not stable under the action of the nuclear force, and they decay with the emission of some type of radiation or "rays."

Many unstable isotopes occur in nature, and such radioactivity is called "natural radioactivity." Other unstable isotopes can be produced in the laboratory by nuclear reactions (Section 42–11); these are said to be produced "artificially" and they are said to have "artificial radioactivity."

Rutherford and others began studying the nature of the rays emitted in radioactivity about 1898. They found that the rays could be classified into three distinct types according to their penetrating power. One type of radiation could barely penetrate a piece of paper. The second type could pass through as much as 3 mm of aluminum. The third was extremely penetrating: it could pass through several centimeters of lead and still be detected on the other side. They named these three types of radiation alpha (α), beta (β), and gamma (γ), respectively, after the first three letters of the Greek alphabet.

Each type of ray was found to have a different charge and hence is bent differently in a magnetic field, Fig. 42–3; α rays are positively charged, β rays are negatively charged, and γ rays are neutral. It was soon found that all three types of radiation consisted of familiar kinds of particles. Gamma rays are very high energy photons whose energy is even higher than that of X rays. Beta rays are electrons, identical to those that orbit the nucleus (but they are created within the nucleus itself). Alpha rays (or α particles) are simply the nuclei of helium atoms, ^4_2He; that is, an α ray consists of two protons and two neutrons bound together.

We now discuss each of these three types of radioactivity, or decay, in more detail.

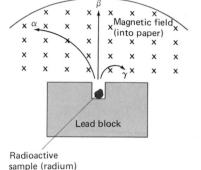

FIGURE 42–3
Alpha and beta rays are bent in opposite directions by a magnetic field, whereas gamma rays are not bent at all.

 ## 42–4 Alpha Decay

When a nucleus emits an α particle (^4_2He), it is clear that the remaining nucleus will be different from the original: for it has lost two protons and two neutrons. Radium 226 ($^{226}_{88}\text{Ra}$), for example, is an α emitter. It decays to a nucleus with $Z = 88 - 2 = 86$ and $A = 226 - 4 = 222$. The nucleus with $Z = 86$ is radon (Rn)—see Table 41–2. Thus the radium decays to radon with the emission of an α particle. This is written

$$^{226}_{88}\text{Ra} \rightarrow {}^{222}_{86}\text{Rn} + {}^4_2\text{He}.$$

It is clear that when α decay occurs, a new element is formed. The *daughter* nucleus ($^{222}_{86}\text{Rn}$ in this case) is different from the *parent* nucleus ($^{226}_{88}\text{Ra}$ in this case). This changing of one element into another is called **transmutation**.

Alpha decay occurs because the strong nuclear force is unable to hold very large nuclei together. Because the nuclear force is a short-range force, it acts only between neighboring nucleons. But the electric force can act clear across the nucleus. For very large nuclei, the large Z means the repulsive electric force becomes very large (Coulomb's law) and acts between all protons; the strong nuclear force, since it acts only between neighboring nucleons, is overpowered and is unable to hold the nucleus together.

We can express the instability in terms of binding energy: the binding energy of a radioactive nucleus is too low for it to remain stable. That is, the mass of the parent nucleus is greater than the mass of the daughter nucleus plus the mass of the α particle. The mass difference appears as kinetic energy, which is carried away mainly by the α particle. If the parent has *less* mass than the daughter plus the α particle, the decay could not occur, for the conservation of energy law would be violated.

EXAMPLE 42–2   Calculate the kinetic energy of the α particle emitted when a $^{232}_{92}U$ nucleus (mass = 232.03714 u) decays to $^{228}_{90}Th$ (228.02873 u).

SOLUTION   Since the mass of a ^4_2He nucleus is 4.002603 u the total mass in the final state is 228.02873 u + 4.002603 u = 232.03133 u. The mass lost when the $^{232}_{92}U$ nucleus decays is 232.03714 u − 232.03133 u = 0.00581 u. This mass appears as kinetic energy; since 1 u = 931.5 MeV, the KE released is (0.00581 u) · (931.5 MeV/u) ≈ 5.4 MeV. (Using conservation of momentum, it can be shown that the α particle in this decay has a KE of about 5.3 MeV. Thus the daughter nucleus—which recoils in the opposite direction from the emitted α particle—has about 0.1 MeV of kinetic energy.)

Why, you may wonder, do nuclei emit this combination of four nucleons called an α particle? Why not just four nucleons, or even one? The answer lies in the fact that the α particle is very strongly bound and its mass is so much less than four separate nucleons. As we saw in Example 42–1, two protons and two neutrons separately have a total mass of about 4.03298 u. A $^{228}_{90}Th$ nucleus plus four nucleons has a total mass of 232.06171 u which is greater than the mass of the parent nucleus. Such a decay could not occur because it would violate conservation of energy. Similarly, it is almost always true that the emission of a single nucleon is energetically not possible.

42–5   Beta Decay

Transmutation of elements also occurs when a nucleus decays by β decay—that is, with the emission of an electron or β particle. The nucleus $^{14}_6C$, for example, decays as follows:

$$^{14}_6C \rightarrow ^{14}_7N + ^{\ 0}_{-1}e.$$

(The symbol $^{\ 0}_{-1}e$ stands for an electron whose charge corresponds to $Z = -1$ and, since it is not a nucleon and has very small mass, has $A = 0$.) No nucleons are lost when an electron is emitted, and the total number of nucleons, A, is the same in the daughter as in the parent. But because an electron has been emitted, the charge on the daughter is different from the parent. The parent had $Z = +6$. In the decay, the nucleus loses a charge of -1, so the nucleus remaining behind (from charge conservation) must have an extra + charge for a total of 7. So the daughter has $Z = 7$, which is a nitrogen nucleus.

It must be carefully noted that the electron emitted in β decay is *not* an orbital electron. Instead, the electron is created *within the nucleus itself*. It is as if one of the neutrons changes to a proton and in the process (to conserve charge) throws off an electron. Indeed, free neutrons actually do decay in this fashion: $n \rightarrow p + e^-$. Because of their origin in the nucleus, the electrons emitted in β decay are often referred to as "β particles," rather than as electrons, to remind us of their origin; they are, nonetheless, indistinguishable from orbital electrons.

EXAMPLE 42–3   How much energy is released when $^{14}_6C$ decays to $^{14}_7N$ by β emission? Use Appendix D.

SOLUTION   Because the masses given in Appendix D are those of the neutral atom, we have to keep track of the electrons involved. Assume the parent nucleus has six orbiting electrons so it is neutral and its mass is 14.003242 u. The daughter, in this decay $^{14}_7N$, is not neutral however since it has the same six electrons circling it but the nucleus has a charge of $+7e$. However, the mass of this daughter with its six electrons, plus the mass of the emitted electron (which makes a total of seven electrons), is just the mass of a neutral nitrogen atom. That is, the mass in the final state is

$$\text{(mass of } {}^{14}_{7}\text{N nucleus} + 6 \text{ electrons)} + \text{(mass of 1 electron)}$$
$$= \text{mass of neutral } {}^{14}_{7}\text{N (includes 7 electrons)}$$
$$= 14.003074 \text{ u.}$$

Hence the mass before decay is 14.003242 u and after decay is 14.003074 u, so the mass difference is 0.000168 u which corresponds to 0.156 MeV or 156 keV.

According to this example, we would expect the emitted electron to have a kinetic energy of 156 keV. (The daughter nucleus, because its mass is very much larger than that of the electron, recoils with very low velocity and hence gets very little of the kinetic energy.) Indeed, very careful measurements indicate that a few emitted β particles do have kinetic energy close to this calculated value; but the vast majority of emitted electrons have somewhat less energy. In fact, the energy of the emitted electron can be anywhere from zero up to the maximum value as calculated above. This was found to be true for any β decay. It was as if the law conservation of energy was being violated! Careful experiments also indicated that linear momentum and angular momentum also did not seem to be conserved. Physicists were troubled at the prospect of having to give up these laws, which had worked so well in all previous situations. In 1930, Wolfgang Pauli proposed an alternate solution: perhaps a new particle that was very difficult to detect was emitted during β decay in addition to the electron. This hypothesized particle could be carrying off the energy, momentum, and angular momentum required to maintain the conservation laws. This new particle was named the **neutrino**—meaning "little neutral one"—by the great Italian physicist Enrico Fermi (1901–1954) who in 1934 worked out a detailed theory of β decay. (It was Fermi, in this theory, who postulated the existence of the fourth force in nature, which we call the weak nuclear force.) The neutrino has zero charge and seems to have zero rest mass, although there are recent suggestions that it may have a very tiny rest mass. If its rest mass is zero, it is much like a photon in that it is neutral and travels at the speed of light (or near it in the case of nonzero mass); but it is far more difficult to detect. In 1956, complex experiments produced further evidence for the existence of the neutrino; but by then most physicists had already accepted its existence.

The symbol for the neutrino is the Greek letter nu (ν). The correct way of writing the decay of ${}^{14}_{6}\text{C}$ is then

$${}^{14}_{6}\text{C} \rightarrow {}^{14}_{7}\text{N} + {}^{0}_{-1}e + \bar{\nu}.$$

The bar (‾) over the neutrino symbol is to indicate that it is an "antineutrino." (More on antiparticles later; why this is called an antineutrino rather than simply a neutrino need not concern us now; it is discussed in Chapter 43.)

Many isotopes decay by electron emission. They are always isotopes that have too many neutrons compared to the number of protons. That is, they are isotopes that lie above the stable isotopes plotted in Fig. 42–2. But what about unstable isotopes that have too few neutrons compared to their number of protons—those that fall below the stable isotopes of Fig. 42–2? These, it turns out, decay by emitting a **positron** instead of an electron. A positron (sometimes called an e^+ or β^+ particle) has the same mass as the electron, but it has a positive charge of $+1e$. Because it is so like an electron, except for its charge, it is called the *antiparticle*[†] to the electron. An example of a β^+ decay is that of ${}^{19}_{10}\text{Ne}$:

$${}^{19}_{10}\text{Ne} \rightarrow {}^{19}_{9}\text{F} + {}^{0}_{1}e + \nu$$

[†] Discussed in Chapter 43. Briefly, an antiparticle has the same mass as its corresponding particle, but opposite charge.

where $_1^0e$ stands for a positron (e^+). (Note that the ν emitted here is a neutrino, whereas that emitted in β^- decay is called an antineutrino. Thus an antielectron (= positron) is emitted with a neutrino, whereas an antineutrino is emitted with an electron; this is discussed in Chapter 43.)

Besides β^- and β^+ emission, there is a third related process. This is *electron capture* and occurs when a nucleus absorbs one of its orbiting electrons. An example is $_4^7$Be, which as a result becomes $_3^7$Li; the process is written

$$_4^7\text{Be} + {_{-1}^0}e \rightarrow {_3^7}\text{Li} + \nu.$$

Usually it is an electron in the innermost (K) shell that is captured, and then it is called "K-capture." The electron disappears in the process and a proton in the nucleus becomes a neutron; a neutrino is emitted as a result. This process is inferred experimentally by detection of emitted X rays (due to electrons jumping down to fill the empty state) of just the proper wavelength.

In β decay it is the weak nuclear force that plays the crucial role. The neutrino is unique in that it interacts with matter only via the weak force, which is why it is so hard to detect.

 ## 42–6   Gamma Decay

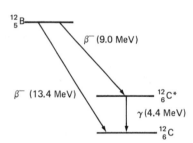

FIGURE 42–4

Energy-level diagram, showing how $_5^{12}$B can decay to the ground state of $_6^{12}$C by β decay (total energy released = 13.4MeV) or can β-decay to an excited state of $_6^{12}$C (indicated by *), which subsequently decays to its ground state by emitting a 4.4-MeV γ ray.

Gamma rays are photons having very high energy. And the decay of a nucleus by emission of a γ ray is much like emission of photons by excited atoms. Like an atom, a nucleus itself can be in an excited state. When it jumps down to a lower energy state, or to the ground state, it emits a photon. The possible energy levels of a nucleus are much farther apart in energy than those of an atom: on the order of keV or MeV, as compared to a few eV for electrons in an atom. Hence the emitted photons have energies that can range from a few keV to several MeV. For a given decay, the γ ray always has the same energy. Since a γ ray carries no charge, there is no change in the element as a result of a γ decay.

How does a nucleus get into an excited state? It may occur because of a violent collision with another particle; or more commonly the nucleus remaining after a previous radioactive decay may be in an excited state. A typical example is shown in the energy-level diagram of Fig. 42–4. $_5^{12}$B can decay by β decay directly to the ground state of $_6^{12}$C; or it can go by β decay to an excited state of $_6^{12}$C which then decays by emission of a 4.4 MeV γ ray to the ground state.[†]

In some cases a nucleus may remain in an excited state for some time before it emits a γ ray as in Fig. 42–4. The nucleus is then said to be in a *metastable state* and is called an *isomer*.

42–7   Conservation of Nucleon Number and Other Conservation Laws

In all three types of radioactive decay the classical conservation laws hold. Energy, linear momentum, angular momentum, and electric charge are all conserved; these quantities are the same before the decay as after. But a new conservation law is also revealed. This is the **law of conservation of nucleon number**. According to this law, the total number of nucleons (A) remains constant in any process,

---

[†] An excited nucleus can sometimes return to the ground state by another process known as *internal conversion*. In this process, the excited nucleus interacts with one of the orbital electrons and ejects this electron from the atom with the same KE that an emitted γ ray would have had.

Table 42–2

42–8 Half-life and Rate of Decay   **829**

The three types of radioactive decay[†]

α decay
$$N(A, Z) \rightarrow N(A - 4, Z - 2) + {}_{2}^{4}\text{He}$$

β decay
$$\beta^{-}: N(A, Z) \rightarrow N(A, Z + 1) + {}_{-1}^{0}e + \bar{\nu}$$
$$\beta^{+}: N(A, Z) \rightarrow N(A, Z - 1) + {}_{1}^{0}e + \nu$$
$$\text{Electron capture: } N(A, Z) + {}_{-1}^{0}e \rightarrow N(A, Z - 1) + \nu$$

γ decay
$$N^{*}(A, Z) \rightarrow N(A, Z) + \gamma$$

[†] $N(A, Z)$ means a nucleus with atomic number Z and mass number A; * means excited state of a nucleus.

although one type can change into the other type (protons into neutrons or vice versa). This law holds in all three types of decay. See Table 42–2 which gives a summary of α, β, γ decay.

42–8 Half-life and Rate of Decay

A macroscopic sample of any radioactive isotope consists of a vast number of radioactive nuclei. These nuclei do not all decay at one time. Rather, they decay one by one over a period of time. This is a random process: we can't predict exactly when a given nucleus will decay. But we can determine, on a probabilistic basis, approximately how many nuclei in a sample will decay over a given time period.

The number of decays ΔN that occur in a very short time interval Δt is found to be proportional to Δt and to the total number N of radioactive nuclei present:

$$\Delta N = -\lambda N \, \Delta t. \tag{42–2}$$

In this equation, λ is a constant of proportionality called the **decay constant**, which is different for different isotopes. The greater λ is, the greater the rate of decay and the more radioactive that isotope is said to be. The quantity ΔN in Eq. 42–2 represents the number of decays that occur in the short time interval Δt; we call this ΔN since each decay that occurs corresponds to a decrease by one in the number N of nuclei present. That is, radioactive decay is a "one-shot" process, Fig. 42–5. Once a particular parent nucleus decays into its daughter, it cannot do it again. The minus sign in Eq. 42–2 is needed to indicate that N is decreasing.

If we take the limit $\Delta t \rightarrow 0$ in Eq. 42–2, ΔN will be small compared to N, and we can write the equation in infinitesimal form as

$$dN = -\lambda N \, dt. \tag{42–3}$$

We can determine N as a function of t by rearranging this equation to

$$\frac{dN}{N} = -\lambda \, dt$$

and then integrating from $t = 0$ to $t = t$:

$$\int_{N_0}^{N} \frac{dN}{N} = -\int_{0}^{t} \lambda \, dt,$$

FIGURE 42–5
Radioactive nuclei decay one by one. Hence the number of parent nuclei in a sample is continually decreasing. When a ${}_{6}^{14}\text{C}$ nucleus emits the electron, it becomes a ${}_{7}^{14}\text{N}$ nucleus.

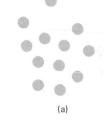

(a)

(b)

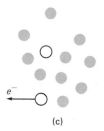

(c)

Legend

● ${}_{6}^{14}\text{C}$ atom (parent)

○ ${}_{7}^{14}\text{N}$ atom (daughter)

where N_0 is the number of parent nuclei present at $t = 0$ and N is the number remaining at time t. The integration gives

$$\ln \frac{N}{N_0} = -\lambda t$$

or

$$N = N_0 e^{-\lambda t}. \tag{42-4}$$

Equation 42–4 is called the **radioactive decay law**. It tells us that the number of radioactive nuclei in a given sample decreases exponentially in time. This is shown in Fig. 42–6a for the case of $^{14}_{6}C$ whose decay constant is $\lambda = 3.8 \times 10^{-12} \text{ s}^{-1}$.

The rate of decay, or number of decays per second, in a pure sample is

$$\frac{dN}{dt}$$

which is also called the *activity* of a given sample. From Eqs. 42–3 and 42–4,

$$\frac{dN}{dt} = -\lambda N = -\lambda N_0 e^{-\lambda t}.$$

At $t = 0$, the activity is

$$\left(\frac{dN}{dt} \right)_0 = -\lambda N_0.$$

Hence

$$\frac{dN}{dt} = \left(\frac{dN}{dt} \right)_0 e^{-\lambda t} \tag{42-5}$$

so the activity also decreases exponentially at the same rate N does (see Fig. 42–6b).

The rate of decay of any isotope is often specified by giving its half-life rather than the decay constant λ. The **half-life** of an isotope is defined as the time it takes for half the original amount of isotope in a given sample to decay. For example, the half-life of $^{14}_{6}C$ is about 5730 yr. If at some time a piece of petrified wood contains, say, 1.00×10^{22} $^{14}_{6}C$ nuclei, then 5700 yr later it will contain only 0.50×10^{22} nuclei. After another 5700 yr it will contain 0.25×10^{22} nuclei, and so on. This is shown in Fig. 42–6a. Since the rate of decay dN/dt is proportional to N, it too decreases by a factor of two every half-life, Fig. 42–6b.

The half-lives of known radioactive isotopes vary from as short as 10^{-22} s to 10^{28} s (about 10^{21} yr). The half-lives of many isotopes are given in Appendix D. It should be clear that the half-life (which we designate $T_{\frac{1}{2}}$) bears an inverse relationship to the decay constant. The longer the half-life of an isotope,

FIGURE 42–6

(a) The number N of parent nuclei in a given sample of $^{14}_{6}C$ decreases exponentially. (b) The number of decays per second also decreases exponentially. The half-life of $^{14}_{6}C$ is about 5700 yr, which means that the number of parent nuclei, N, and the rate of decay, dN/dt, decrease by half every 5700 yr.

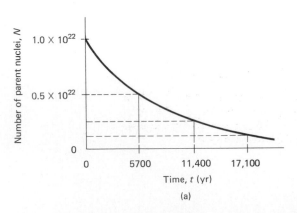

(a)

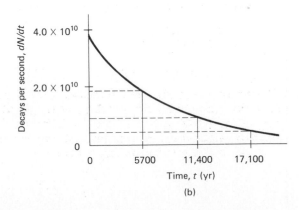

(b)

from Eq. 42–4 by setting $N = N_0/2$ at $t = T_{\frac{1}{2}}$:

$$\frac{N_0}{2} = N_0 e^{-\lambda T_{\frac{1}{2}}}$$

$$T_{\frac{1}{2}} = \frac{\ln 2}{\lambda} = \frac{0.693}{\lambda}. \tag{42–6}$$

EXAMPLE 42–4 The isotope $^{14}_{6}C$ has a half-life of 5730 yr. If at some time a sample contains 1.0×10^{22} carbon 14 nuclei, what is the activity of the sample?

SOLUTION First we calculate the decay constant λ from Eq. 42–6 and obtain

$$\lambda = \frac{0.693}{T_{\frac{1}{2}}} = \frac{0.693}{(5730 \text{ yr})(3.15 \times 10^7 \text{ s/yr})} = 3.84 \times 10^{-12} \text{ s}^{-1}$$

since there are $(60)(60)(24)(365) = 3.15 \times 10^7$ s in a year. From Eq. 42–3, the activity or rate of decay is (we ignore the minus sign):

$$\frac{dN}{dt} = \lambda N = (3.84 \times 10^{-12} \text{ s}^{-1})(1.0 \times 10^{22})$$

$$= 3.8 \times 10^{10} \text{ decays/s}.$$

Note that the graph of Fig. 42–6b starts at this value, corresponding to the original value of $N = 1.0 \times 10^{22}$ nuclei in Fig. 42–6a.

EXAMPLE 42–5 A laboratory has 1.49 μg of pure $^{13}_{7}N$, which has a half-life of 10.0 min. (a) How many nuclei are present initially? (b) What is the activity initially? (c) What is the activity after 1.00 h? (d) After approximately how long will the activity drop to less than one per second?

SOLUTION (a) Since the atomic mass is 13.0 then 13.0 g will contain 6.02×10^{23} nuclei (Avogadro's number). Since we have only 1.49×10^{-6} g, we have N nuclei as given by the ratio

$$\frac{N}{6.02 \times 10^{23}} = \frac{1.49 \times 10^{-6} \text{ g}}{13.0 \text{ g}}$$

so $N = 6.90 \times 10^{16}$ nuclei.

(b) From Eq. 42–6, $\lambda = (0.693)/(600 \text{ s}) = 1.16 \times 10^{-3} \text{ s}^{-1}$. Then, at $t = 0$,

$$\left(\frac{dN}{dt}\right)_0 = \lambda N = 8.00 \times 10^{13} \text{ decays/s}.$$

(c) After 1.00 h = 3600 s, the activity will be

$$\frac{dN}{dt} = \left(\frac{dN}{dt}\right)_0 e^{-\lambda t}$$

$$= (8.00 \times 10^{13} \text{ s}^{-1}) e^{-(1.16 \times 10^{-3} \text{s}^{-1})(3600 \text{s})} = 1.23 \times 10^{12} \text{ s}^{-1}.$$

This result can be obtained in another way: since 1.00 h represents six half-lives $(6 \times 10.0 \text{ min})$, the activity will decrease by $(\frac{1}{2})(\frac{1}{2})(\frac{1}{2})(\frac{1}{2})(\frac{1}{2})(\frac{1}{2}) = (\frac{1}{2})^6 = \frac{1}{64}$ of its original value, or $(800 \times 10^{13} \text{ s}^{-1})/64 = 1.25 \times 10^{12} \text{ s}^{-1}$. (The discrepancy between the two values arises because we kept only three significant figures.)

(d) We determine the time t when

$$\frac{dN}{dt} = 1.00 \text{ s}^{-1}.$$

From Eq. 42–5, we have

$$e^{-\lambda t} = \frac{1.00 \text{ s}^{-1}}{8.00 \times 10^{13} \text{ s}^{-1}} = 1.25 \times 10^{-14}.$$

Thus

$$t = -\frac{\ln(1.25 \times 10^{-14})}{\lambda} = 2.76 \times 10^4 \text{ s}$$

or 7.66 h.

 ## 42–9 Stability and Tunneling

We have seen that radioactive decay occurs only when the mass of the parent nucleus is greater than the sum of the masses of the daughter nucleus and all particles emitted. Since systems tend to go in the direction that reduces their internal or potential energy (a ball rolls downhill, a positive charge moves toward a negative charge), you may wonder why an unstable nucleus doesn't fall apart immediately. In other words, why do $^{238}_{92}\text{U}$ ($T_{\frac{1}{2}} = 4.5 \times 10^9$ yr) and other isotopes have such long half-lives?

The answer has to do with quantum theory and the nature of the forces involved. One way to view the situation is with the aid of a potential-energy diagram, as in Fig. 42–7. Here we consider the particular case of the decay $^{238}_{92}\text{U} \rightarrow ^{234}_{90}\text{Th} + ^{4}_{2}\text{He}$. The curved line represents the potential energy, including rest mass. Actually, it is simplest to imagine the α particle as a separate entity in the $^{238}_{92}\text{U}$ nucleus. Then the curve of Fig. 42–7 can represent the energy of the α particle when it is within the uranium nucleus (point A on the diagram) and when it is free of the uranium nucleus (point C). In order to get to point C, the α particle has to get by the barrier shown. In classical physics, this could only be done by putting in an energy equal to the height of the barrier above point A.

According to quantum mechanics, nuclei decay spontaneously, without any input of energy, by actually passing through the barrier in a process known as **tunneling**. Classically, this couldn't happen because an α particle at point B (within the barrier) in the diagram would be violating the conservation of energy principle. The uncertainty principle, however, tells us that energy conservation can be violated by an amount ΔE for a length of time Δt given by

$$(\Delta E)(\Delta t) \approx h/2\pi.$$

Thus, quantum mechanics allows conservation of energy to be violated for brief periods that may be long enough for an α particle to "tunnel" through the barrier. The higher and wider the barrier, the less time the α particle has to escape and the less likely it is to do so. It is therefore the height and width of this barrier that controls the rate of decay and half-life of an isotope.

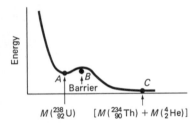

FIGURE 42–7

Potential-energy diagram for the decay $^{238}_{92}\text{U} \rightarrow ^{234}_{90}\text{Th} + ^{4}_{2}\text{He}$.

 ## * 42–10 Radioactive Dating

Radioactive decay has many interesting applications. One is the technique of *radioactive dating* in which the age of ancient materials can be determined.

The age of any object made from once living matter, such as wood, can be determined using the natural radioactivity of $^{14}_{6}\text{C}$. All living plants absorb carbon dioxide (CO_2) from the air, utilizing the carbon and expelling oxygen. The vast majority of these carbon atoms are $^{12}_{6}\text{C}$, but a small fraction, about 1.3×10^{-12}, is the radioactive isotope $^{14}_{6}\text{C}$. The ratio of $^{14}_{6}\text{C}$ to $^{12}_{6}\text{C}$ in the atmosphere has remained roughly constant over many thousands of years, in spite of the fact that $^{14}_{6}C$ decays with a half-life of about 5700 yr. This is because neutrons in the cosmic radiation that impinges on the earth from outer space collide with atoms of the atmosphere. In particular, collisions with nitrogen nuclei produce the following nuclear transformation: $n + {}^{14}_{7}N \rightarrow {}^{14}_{6}C + p$. That is, a neutron strikes and is absorbed by a $^{14}_{7}N$ nucleus, and a proton is knocked out in the process. The remaining nucleus is $^{14}_{6}C$. This continual production of $^{14}_{6}C$ in the atmosphere roughly balances the loss of $^{14}_{6}C$ by radioactive decay. As long as a plant or tree is alive, it continually uses the carbon from carbon dioxide in the air to build new tissue and to replace old. Animals eat plants, so they too are continually receiving a fresh supply of carbon for their tissues. Organisms cannot distinguish[†] $^{14}_{6}C$ from $^{12}_{6}C$ and since the ratio of $^{14}_{6}C$ to $^{12}_{6}C$ in the atmosphere remains nearly constant, the ratio of the two isotopes within the living organism remains nearly constant as well. But when the organism dies, carbon dioxide is no longer absorbed and utilized; and because the $^{14}_{6}C$ decays radioactively, the ratio of $^{14}_{6}C$ to $^{12}_{6}C$ in the dead organism decreases in time. Since the half-life of $^{14}_{6}C$ is about 5700 yr, the $^{14}_{6}C/^{12}_{6}C$ ratio decreases by half every 5700 yr. If, for example, the $^{14}_{6}C/^{12}_{6}C$ ratio of an ancient wooden tool is half of what it is in living trees, then the object must have been made from a tree that was felled about 5700 yr ago. Actually, corrections must be made for the fact that the $^{14}_{6}C/^{12}_{6}C$ ratio in the atmosphere has not remained precisely constant over time. The determination of what this ratio has been over the centuries has been done using techniques such as comparing the expected ratio to actual ratio for objects whose age is known, such as very old trees whose annual rings can be counted.

Carbon dating is useful only for determining the age of objects less than about 40,000 yr old. The amount of $^{14}_{6}C$ remaining in older objects is usually too small to measure accurately. However, radioactive isotopes with longer half-lives can be used in certain circumstances to obtain the age of older objects. For example, the decay of $^{238}_{92}U$, because of its long half-life of 4.5×10^9 years, is useful to determine the age of rocks on a geologic time scale.

42–11 Nuclear Reactions and the Transmutation of Elements

When a nucleus undergoes α or β decay, the daughter nucleus is that of a different element from the parent. The transformation of one element into another, called *transmutation*, also occurs by means of nuclear reactions. A **nuclear reaction** is said to occur when a given nucleus is struck by another nucleus, or by a simpler particle such as a γ ray or neutron, so that an interaction takes place. Ernest Rutherford was the first to report seeing a nuclear reaction. In 1919 he observed that some of the α particles passing through nitrogen gas were absorbed and protons emitted. He concluded that nitrogen nuclei had been transformed into oxygen nuclei via the reaction

$$^{4}_{2}He + {}^{14}_{7}N \rightarrow {}^{17}_{8}O + {}^{1}_{1}H$$

where $^{4}_{2}He$ is an α particle and $^{1}_{1}H$ is a proton. Since then, a great many nuclear reactions have been observed, both naturally occurring and produced in the laboratory.

Nuclear reactions are sometimes written in a shortened form; for example, the reaction $^{1}_{0}n + {}^{14}_{7}N \rightarrow {}^{14}_{6}C + {}^{1}_{1}H$ is written $^{14}_{7}N(n, p)^{14}_{6}C$. The symbols outside the parentheses on left and right represent the initial and final nuclei, respectively. The symbols inside the parentheses (*n* for neutron, *p* for proton) represent the bombarding particle (first) and the emitted small particle (second).

[†] Organisms operate almost exclusively via chemical reactions—which involve only the electrons of the atom; extra neutrons have almost no effect.

In any nuclear reaction, both electric charge and nucleon number are conserved. These conservation laws are often useful as the following example shows.

EXAMPLE 42–8 A neutron is observed to strike an $^{16}_{8}O$ nucleus and a deuteron is given off. (A *deuteron*, or *deuterium*, is the isotope of hydrogen containing one proton and one neutron, $^{2}_{1}H$.) What is the nucleus that results?

SOLUTION We have the reaction $^{1}_{0}n + ^{16}_{8}O \rightarrow ? + ^{2}_{1}H$. The total number of nucleons initially is $16 + 1 = 17$, and the total charge is $8 + 0 = 8$; the same totals apply to the right side of the reaction. Hence the product nucleus must have $Z = 7$ and $A = 15$. From the periodic table, we find that it is nitrogen that has $Z = 7$, so the nucleus produced is $^{15}_{7}N$. The reaction can be written $^{16}_{8}O(n, d)^{15}_{7}N$ where d represents deuterium, $^{2}_{1}H$.

Energy (as well as momentum) is conserved in nuclear reactions. This fact can be used to determine whether a given reaction can occur or not. For example, if the total mass of the products is less than the total mass of the initial particles, then energy will be released by the reaction—it will appear as kinetic energy of the outgoing particles. But if the total mass of the products is greater than the total mass of the initial reactants, the reaction requires energy; the reaction will not occur unless the bombarding particle has sufficient kinetic energy. This is demonstrated in the next example.

EXAMPLE 42–9 Can the reaction $^{13}_{6}C(p, n)^{13}_{7}N$ occur when $^{13}_{6}C$ is bombarded by 2.0 MeV protons?

SOLUTION We look up the masses of the nuclei in Appendix D. The total masses before and after the reaction are:

Before	After
$m(^{13}_{6}C) = 13.003355$	$m(^{13}_{7}N) = 13.005739$
$m(^{1}_{1}H) = 1.007825$	$m(^{1}_{0}n) = 1.008665$
14.011180	14.014404

(We must use the mass of the $^{1}_{1}H$ atom rather than that of the bare proton because the masses of $^{13}_{6}C$ and $^{13}_{7}N$ include the electrons and we must include an equal number of electrons on each side of the equation since none are created or destroyed.) The products have an excess mass of 0.003224 amu × 931.5 MeV/amu = 3.00 MeV. This reaction requires energy, and the 2.0 MeV protons do not have enough to make it go. Hence this reaction won't occur. The proton would have to have somewhat more than 3.00 MeV of KE; 3.00 MeV would be enough to conserve energy, but a proton of this energy would produce the $^{13}_{7}N$ and $^{1}_{0}n$ with no KE and hence no momentum; since the incident proton has momentum, conservation of momentum would be violated. A more complicated calculation shows that to conserve both energy and momentum the minimum proton energy required (called the *threshold energy*) is 3.23 MeV in this case. See problem 40.

The artificial transmutation of elements took a great leap forward in the 1930s when Enrico Fermi realized that neutrons would be the most effective projectiles for causing nuclear reactions and in particular for producing new elements. Because neutrons have no net electric charge, they are not repelled by positively charged nuclei as protons or alpha particles are. (The latter must overcome the so-called "coulomb barrier.") Hence the probability of a neutron reaching the nucleus and causing a reaction is much greater than for charged projectiles,[†] particularly at low energies. Between 1934 and 1936, Fermi and his co-workers in Rome produced many previously unknown isotopes by bombarding different elements with neutrons. Fermi realized that if the heaviest known element, uranium, were bombarded with neutrons, it might be possible to produce new elements whose atomic numbers were greater than that of uranium. After several years of hard work, it was suspected that two new elements had been produced, neptunium ($Z = 93$) and plutonium ($Z = 94$). The full confirmation that such "transuranic" elements could be produced came several years later at the University of California, Berkeley. The reactions are shown in Fig. 42–8.

It was soon shown that what Fermi actually had observed when he bombarded uranium was an even stranger process—one that was destined to play an extraordinary role in the world at large.

(a) $n + {}^{238}_{92}U \rightarrow {}^{239}_{92}U$

Neutron captured by ${}^{238}_{92}U$

(b) ${}^{239}_{92}U \rightarrow {}^{239}_{93}Np + e^- + \bar{\nu}$

${}^{239}_{92}U$ decays by β decay to neptunium 239.

(c) ${}^{239}_{93}Np \rightarrow {}^{239}_{94}Pu + e^- + \bar{\nu}$

${}^{239}_{93}Np$ itself decays by β decay to produce plutonium 239.

FIGURE 42–8

Neptunium and plutonium are produced in the series of reactions, after bombardment of ${}^{238}_{92}U$ by neutrons.

FIGURE 42–9

Fission of a ${}^{235}_{92}U$ nucleus after capture of a neutron.

42–12 Nuclear Fission

In 1938, the German scientists Otto Hahn and Fritz Strassmann made an amazing discovery. Following up on Fermi's work, they found that uranium bombarded by neutrons sometimes produced smaller nuclei which were roughly half the size of the original uranium nucleus. Lise Meitner and Otto Frisch, two refugees from Nazi Germany working in Scandinavia, quickly realized what had happened: the uranium nucleus, after absorbing a neutron, actually had split into two roughly equal pieces. This was startling, for until then the known nuclear reactions involved knocking out only a tiny fragment (for example, n, p, or α) from a nucleus.

This new phenomenon was named **nuclear fission** because of its resemblance to biological fission (cell division). It occurs much more readily for ${}^{235}_{92}U$ than for the more common ${}^{238}_{92}U$. The process can be visualized by imagining the uranium nucleus to be like a liquid drop. According to this **liquid-drop model**, the neutron absorbed by the ${}^{235}_{92}U$ nucleus gives the nucleus extra internal energy (like heating a drop of water). This intermediate state, or **compound nucleus**, is ${}^{236}_{92}U$ (because of the absorbed neutron). The extra energy of this nucleus—it is in an excited state—appears as increased motion of the individual nucleons which causes the nucleus to take on abnormal elongated shapes, Fig. 42–9. When the nucleus elongates into the shape shown in Fig. 42–9c, the attraction of the two ends via the short-range nuclear force is greatly weakened by the excess separation distance, but the electric repulsive force is weakened only slightly and becomes predominant; so the nucleus splits in two. The two resulting nuclei, N_1 and N_2, are called *fission fragments*, and in the process a number of neutrons (typically two or three) are also given off. The reaction can be written

$$ {}^{1}_{0}n + {}^{235}_{92}U \rightarrow {}^{236}_{92}U \rightarrow N_1 + N_2 + \text{neutrons.} \qquad (42\text{–}7) $$

The compound nucleus, ${}^{236}_{92}U$, exists for less than 10^{-12} s, so the process occurs very quickly. The two fission fragments have roughly half the mass of the uranium, although rarely are they exactly equal in mass. A typical fission reaction is

$$ {}^{1}_{0}n + {}^{235}_{92}U \rightarrow {}^{141}_{56}Ba + {}^{92}_{36}Kr + 3{}^{1}_{0}n, \qquad (42\text{–}8) $$

although many others also occur.

A tremendous amount of energy is released in a fission reaction because the mass of ${}^{235}_{92}U$ is considerably greater than that of the fission fragments. This can be seen from the binding-energy-per-nucleon curve of Fig. 42–1; the binding energy per nucleon for uranium is about 7.6 MeV/nucleon, but for fission fragments which have intermediate mass (in the center portion of the graph, $Z \approx 100$),

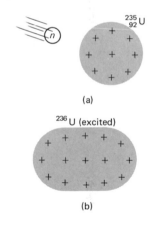

${}^{235}_{92}U$

(a)

${}^{236}U$ (excited)

(b)

(c)

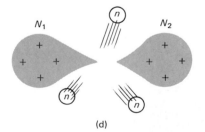

N_1 N_2

(d)

[†] That is, positively charged particles. Electrons rarely cause nuclear reactions because they do not partake of the strong nuclear force.

the average binding energy per nucleon is about 8.5 MeV/nucleon. Since the fission fragments are more tightly bound, they have less mass. The difference in mass (or energy) between the original uranium nucleus and the fission fragments is about $8.5 - 7.6 = 0.9$ MeV per nucleon. Since there are 236 nucleons involved in each fission, the total energy released per fission is

$$(0.9 \text{ MeV/nucleon})(236 \text{ nucleons}) \approx 200 \text{ MeV.}$$

This is an enormous amount of energy on the nuclear scale. At a practical level, the energy from one fission is, of course, tiny. However, a great deal of energy at the macroscopic level would be available if many such fissions could occur at once. A number of physicists, including Fermi, recognized that the neutrons released in each fission (Eq. 42–7 or 42–8) could be used to create a **chain reaction**: one neutron initially causes one fission of a uranium nucleus; the two or three neutrons released can go on to cause additional fissions, so the process multiplies as shown schematically in Fig. 42–10. If a *self-sustaining chain reaction* was actually possible in practice, the enormous energy available in fission could be released. Fermi and his co-workers (at the University of Chicago) showed it was possible by constructing the first **nuclear reactor** in 1942.

Several problems have to be overcome in any nuclear reactor. First, the probability that a $^{235}_{92}$U nucleus will absorb a neutron is large only for slow neutrons, but the neutrons emitted during a fission, and which are needed to sustain a chain reaction, are moving very fast. A substance known as a **moderator** must be used to slow down the neutrons. The most effective moderator will consist of atoms whose mass is as close as possible to that of the neutrons so there will be maximum energy loss per collision. (See Chapter 8, especially Example 8–10.) The best moderator would thus contain $^1_1$H atoms; unfortunately, $^1_1$H tends to absorb neutrons; but *deuterium*, $^2_1$H, does not absorb many neutrons and is thus an ideal moderator. Either $^1_1$H or $^2_1$H can be used in the form of water; in the latter case, it is *heavy water*, where the hydrogen atoms have been replaced by deuterium. Another common moderator is *graphite*, which consists of $^{12}_6$C atoms.

A second problem is that the neutrons produced in one fission may be absorbed and produce other nuclear reactions with other nuclei in the reactor, rather than produce further fissions. In a "light water" reactor the $^1_1$H nuclei absorb neutrons, as does $^{238}_{92}$U to form $^{239}_{92}$U in the reaction $n + {}^{238}_{92}\text{U} \rightarrow {}^{239}_{92}\text{U} + \gamma$. Naturally occurring uranium contains 99.3 percent $^{238}_{92}$U and only 0.7 percent fissionable $^{235}_{92}$U. To increase the probability of fission of $^{235}_{92}$U nuclei, natural

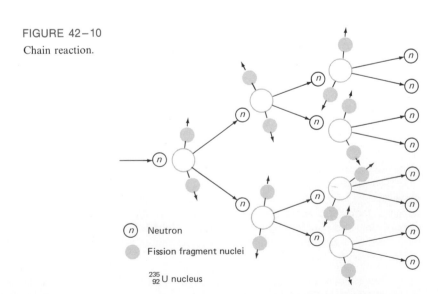

FIGURE 42–10

Chain reaction.

$\bigcirc\!\!n$ Neutron

● Fission fragment nuclei

$^{235}_{92}$U nucleus

uranium is often **enriched**[†] to increase the percentage of $^{235}_{92}$U using processes such as diffusion or centrifugation.

The third problem is that some neutrons will escape through the surface of the reactor core before they cause further fissions (Fig. 42–11). Thus the mass of fuel must be sufficiently large for a self-sustaining chain reaction to take place. The minimum mass of uranium needed is called the **critical mass**. The value of the critical mass depends on the moderator, the fuel ($^{239}_{94}$Pu may be used[‡] instead of $^{235}_{92}$U), and how much the fuel is enriched, if at all. Typical values are on the order of a few kilograms (that is, not grams nor thousands of kilograms).

To have a self-sustaining chain reaction, it is clear that on the average at least one neutron produced in each fission must go on to produce another fission. The average number of neutrons per each fission that do go on to produce further fissions is called the *multiplication factor*, f. For a self-sustaining chain reaction we must have $f \geq 1$. If $f < 1$, the reactor is "subcritical"; if $f > 1$, it is "supercritical." Reactors are equipped with movable *control rods* (usually of cadmium), whose function is to absorb neutrons and maintain the reactor[§] at just barely "critical," $f = 1$. A diagram of a typical nuclear-power reactor is shown in Fig. 42–12. The energy of fission appears as heat which is used to heat water to steam to drive a turbine connected to an electric generator.

The difference between a nuclear reactor and a fission bomb ("atomic bomb") is the rate of release of energy: in a bomb, the neutron multiplication factor, instead of being kept near 1 as for a reactor, is somewhat greater than 1 so the chain reaction occurs swiftly and an explosion occurs. This can be done by bringing two subcritical masses together to form a single supercritical mass at the moment of detonation. A nuclear bomb not only releases immense energy, but also intense and very damaging radiation. It comes from the fission fragments

---

[†] Enrichment is not usually necessary for reactors using heavy water as moderator.

[‡] $^{238}_{92}$U will fission, but only with fast neutrons ($^{238}_{92}$U is more stable than $^{235}_{92}$U). The probability of absorbing a fast neutron and producing a fission is too low to produce a self-sustaining chain reaction.

[§] The release of neutrons and subsequent fissions caused by them occurs so quickly that manipulation of the control rods to maintain $f = 1$ would not be possible if it weren't for the fortunate fact that a small percentage of neutrons released are "delayed." They come from the decay of neutron-rich fission fragments (or their daughters) having lifetimes on the order of seconds—sufficient to allow enough reaction time to maintain $f = 1$.

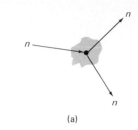

(a)

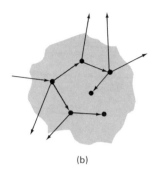

(b)

FIGURE 42–11

If the amount of uranium exceeds the critical mass, as in (b), a sustained chain reaction is possible. If the mass is less than critical, as in (a), most neutrons escape before additional fissions occur, and the chain reaction is not sustained.

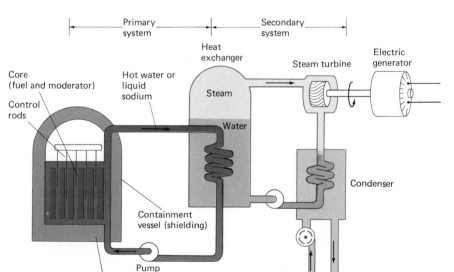

FIGURE 42–12

Nuclear reactor. The heat generated by the fission process in the fuel rods is carried off by hot water or liquid sodium and is used to boil water to steam in the heat exchanger. The steam drives a turbine to generate electricity and is then cooled in the condenser.

which, like their uranium or plutonium parents, have about 50 percent more neutrons than protons. Nuclei with atomic number in the typical range for fission fragments ($Z \approx 30$ to 60) are stable only if they have more nearly equal numbers of protons and neutrons (see Fig. 42–2). Hence the highly neutron-rich fission fragments are very unstable and decay radioactively. The radioactivity released in a nuclear explosion in the atmosphere is known as *nuclear fallout*. The same radioactive fission fragments are produced in a nuclear-power reactor, and it is this "nuclear waste" that must be disposed of.

 ## 42–13   Fusion

The mass of every stable nucleus is less than the sum of the masses of its constituent protons and neutrons. For example, the mass of the helium isotope ^4_2He is less than the mass of two protons plus the mass of two neutrons. Thus, if two protons and two neutrons were to come together to form a helium nucleus there would be a loss of mass. This mass loss is manifested in the release of a large amount of energy. The process of building up nuclei by bringing together individual protons and neutrons, or building larger nuclei by combining small nuclei, is called **nuclear fusion**. A glance at Fig. 42–1 shows how small nuclei can combine to form larger ones with the release of energy: it is because the binding energy per nucleon is smaller for light nuclei than it is for those of intermediate size ($A = 50$ to 100). It is believed that all the elements in the universe were originally formed through the process of fusion. Today, fusion is continually taking place within the stars, including our sun, producing the intense light energy they emit.

EXAMPLE 42–10   One of the simplest fusion reactions involves the production of deuterium ^2_1H from a neutron and a proton: $^1_1\text{H} + ^1_0 n \rightarrow ^2_1\text{H} + \gamma$. How much energy is released in this reaction?

SOLUTION   From Appendix D the initial rest mass is $1.007825\ \text{u} + 1.008665\ \text{u} = 2.016490\ \text{u}$ and after the reaction the mass is that of the ^2_1H, namely $2.014102\ \text{u}$. The energy released is thus $(0.002388\ \text{u})(931.5\ \text{MeV/u}) = 2.22$ MeV, and is carried off by the ^2_1H nucleus and the γ ray.

The possibility of utilizing the energy released in fusion to make a power reactor is very attractive, but so far a successful reactor has not been achieved. The fusion reactions most likely to succeed in a reactor involve the isotopes of hydrogen, ^2_1H (deuterium) and ^3_1H (tritium), and are as follows, with the energy released given in parentheses:

$$^2_1\text{H} + ^2_1\text{H} \rightarrow ^3_1\text{H} + ^1_1\text{H} \qquad (4.03\ \text{MeV}) \qquad\qquad (42\text{–}9\text{a})$$

$$^2_1\text{H} + ^2_1\text{H} \rightarrow ^3_2\text{He} + ^1_0 n \qquad (3.27\ \text{MeV}) \qquad\qquad (42\text{–}9\text{b})$$

$$^2_1\text{H} + ^3_1\text{H} \rightarrow ^4_2\text{He} + ^1_0 n \qquad (17.59\ \text{MeV}) \qquad\qquad (42\text{–}9\text{c})$$

$$^3_1\text{H} + ^3_1\text{H} \rightarrow ^4_2\text{He} + 2^1_0 n \qquad (11.33\ \text{MeV}). \qquad\qquad (42\text{–}9\text{d})$$

The energy released in fusion reactions is greater for a given mass of fuel than in fission. Furthermore, fusion presents less of a radioactive-waste problem. For fuel, a fusion reactor could use deuterium, which is very plentiful in the water of the oceans (the natural abundance of ^2_1H is 0.015 percent, or about 1 g of deuterium per 60 L of water).

Unfortunately, considerable difficulties still exist for making a usable fusion reactor. The problems are associated with the fact that all nuclei have a positive charge and thus repel each other. However, if they can be brought close enough together so that the short-range attractive nuclear force can come into play, the latter can pull the nuclei together and fusion will occur. In order for the nuclei to get close enough together they must have very high speeds. Since high speed of atoms corresponds to high temperature, very high temperatures are required for fusion to occur; hence fusion devices are often referred to as *thermonuclear devices*. The sun and other stars are very hot, many millions of degrees, so the nuclei are moving fast enough for fusion to take place and the energy released keeps the temperature high so that further fusion reactions can occur. The sun and the stars represent self-sustaining thermonuclear reactors, but on earth such high temperatures are not easily attained in a controlled manner.

It was realized after World War II that the temperature produced within a fission (or "atomic") bomb was close to 10^8 K; this suggested that a fission bomb could be used to ignite a fusion bomb (popularly known as a thermonuclear or hydrogen bomb) to release the vast energy of fusion. The uncontrollable release of fusion energy in an H-bomb was fairly easy to obtain. But to realize usable energy from fusion at a slow and controlled rate turned out to be very difficult. The high temperatures needed can now be produced by an infusion of concentrated energy, such as from a high-powered laser. The real difficulty is to contain the nuclei long enough for sufficient reactions to occur that a usable amount of energy is obtained. At the temperatures needed for fusion (approximately 10^8 K) the atoms are ionized, and this collection of nuclei and electrons is referred to as a **plasma**. Ordinary materials vaporize at a few thousand degrees at best, and hence could not be used to contain a high temperature plasma. One technique is to confine a plasma with a magnetic field; unfortunately, all attempted configurations of magnetic field confinement have developed "leaks" and the charged particles leak out before sufficient fusion takes place. One promising configuration of magnetic fields is the so-called Tokamak, and confinement times are being increased. The plasma must also be sufficiently dense so that a significant number of interactions takes place during the confinement time. In 1957, J. D. Lawson showed that for the output energy of a fusion reactor to exceed the input energy, the product of the ion density n (number/cm^3) and confinement time τ must satisfy (approximately)

$$n\tau \gtrsim 3 \times 10^{14} \text{ s/cm}^3.$$

This is known as the *Lawson criterion*, and over the years it has been more closely approached, but not yet reached.

Another containment technique is to form solid pellets of fuel which are quickly heated by an intense laser or electron beam. This *inertial confinement* technique too shows promise, but controlled fusion remains elusive.

* 42-14   Measurement of Radiation-Dosimetry

The passage through matter (including the human body) of charged particles, such as α and β rays, as well as γ and X rays, can ionize atoms and molecules and cause considerable damage. Radiation can also be used to treat certain diseases, particularly cancer. It is therefore important to be able to quantify the amount, or *dose*, of radiation that passes through a material. This is the subject of *dosimetry*. We now discuss the more important ways of measuring dosage.

The strength of a source can be specified at a given time by stating the activity or how many disintegrations occur per second. A special unit is used, called the *curie* (Ci), defined as

$$1 \text{ Ci} = 3.70 \times 10^{10} \text{ disintegrations per second.}$$

Of course commercial suppliers of *radioisotopes* (radioactive isotopes) specify the activity at a given time; since the activity decreases in time, particularly for short-lived isotopes, it is important to take this into account.

▼ **EXAMPLE 42–6** In a certain experiment, 0.016 μCi of $^{32}_{15}$P is injected into a medium containing a culture of bacteria. After 1 h, the cells are washed and a detector that is 70 percent efficient (counts 70 percent of emitted β rays) records 720 counts per minute from all the cells. What percentage of the original $^{32}_{15}$P was taken up by the cells?

SOLUTION The total number of disintegrations per second originally was $(0.016 \times 10^{-6})(3.7 \times 10^{10}) = 590$. The counter could be expected to count 70 percent of this or 410 per second. Since it counted $720/60 = 12$ per second, then $12/410 = 0.029$ or 2.9 percent was incorporated into the cells.

The earliest unit of dosage was the *roentgen* (R) which was defined in terms of the amount of ionization produced by the radiation. Today 1 R is defined as the amount of X or γ radiation that deposits 0.878×10^{-2} J of energy per kilogram of air.

The roentgen has been largely superseded by another unit of dosage applicable to any type of radiation, the *rad: 1 rad is that amount of radiation which deposits 1.00×10^{-2} J/kg of any absorbing material.*[†] (This is quite close to the roentgen for X and γ rays.) The radiation dose in terms of rads depends not only on the strength of a given radiation beam (number of particles per second) and the energy per particle, but also on the type of material that is absorbing the radiation. Since bone, for example, is denser than flesh and absorbs more of the radiation normally used, the same beam passing through the body deposits a greater dose (in rads) in bone than in the flesh.

The rad is a physical unit of dose—it is the energy deposited per gram of material. It is, however, not the most meaningful unit for measuring the biological damage produced by radiation. This is because equal doses of different types of radiation cause differing amounts of damage. For example, 1 rad of α radiation does 10 to 20 times the amount of damage as does 1 rad of β or γ rays. This is largely due to the fact that α rays (and other heavy particles such as protons and neutrons) move much more slowly than equal energy β and γ rays because of their greater mass. Hence ionizing collisions occur closer together so more irreparable damage is done. The *relative biological effectiveness* (RBE) or *quality factor* (QF) of a given type of radiation is defined as the number of rads of X or γ radiation that produces the same biological damage as 1 rad of the given radiation. The QF for several types of radiation is given in Table 42–3. The numbers are approximate since they depend somewhat on the energy of the particles and on the type of damage that is used as the criterion.

The product of the dose in rads and the QF gives a unit known as the *rem* (which stands for *rad equivalent man*):

$$\text{rem} = \text{rad} \times \text{QF}.$$

By this definition, 1 rem of any type of radiation does approximately the same amount of biological damage. For example, 50 rem of fast neutrons does the same damage as 50 rem of γ rays. But note that 50 rem of fast neutrons is only 5 rads whereas 50 rem of γ rays is 50 rads.

We are constantly exposed to low-level radiation from natural sources: cosmic rays, natural radioactivity in rocks and soil, and naturally occurring radioactive isotopes that occur in our food such as $^{40}_{19}$K. The natural radioactive background supplies about 0.13 rem per year per person on the average. From medical X rays the average person receives about 0.07 rem per year. The U.S. government specifies the recommended upper limit of allowed radiation for an individual in the general populace at about 0.5 rem, exclusive of natural sources. However, since even low doses of radiation are believed to increase the chances of

TABLE 42–3

Quality factor (QF) of different kinds of radiation

Type	QF
X and γ rays	≈ 1
β (electrons)	≈ 1
Fast protons	1
Slow neutrons	≈ 3
Fast neutrons	Up to 10
α particles and heavy ions	Up to 20

[†] The gray (Gy), equal to 1 J/kg, is the new official SI unit.

cancer or genetic defects, the attitude today is to keep the radiation dose as low as possible.

People who work around radiation—in hospitals, in power plants, in research—often are subjected to much higher doses than 0.5 rem/yr. The upper limit for such occupational exposures has been set somewhat higher, on the order of 5 rem/yr whole-body dose (presumably because such people know what they are getting into).

 EXAMPLE 42–7 What whole-body dose is received by a 70-kg patient exposed to a 1000-Ci $^{60}_{27}$Co source if 2.0 percent of the γ rays reach the patient? $^{60}_{27}$Co emits γ rays of energy 1.33 MeV and 1.17 MeV in equal amounts. Approximately 50 percent of the γ rays interact in the body and deposit all their energy. (The rest pass through.)

SOLUTION The average γ-ray energy is 1.25 MeV so the total energy passing through the body is $(1000 \text{ Ci})(3.7 \times 10^{10} \text{ decays/Ci·s})(1.25 \text{ MeV})(0.020) = 9.3 \times 10^{11}$ MeV/s. (The factor 0.020 represents the 2.0 percent that reach the body.) Only half this energy is deposited in the body, so the dose rate is $(\frac{1}{2})(9.3 \times 10^{11} \text{ MeV/s})/(1.6 \times 10^{-13} \text{ J/MeV}) = 7.4 \times 10^{-2}$ J/s. Since 1 rad $= 10^{-2}$ J/kg, the whole-body dose rate is $(7.4 \times 10^{-2} \text{ J/s})/(10^{-2} \text{ J/kg·rad})(70 \text{ kg}) = 0.11$ rad/s.

Summary

Nuclear physics is the study of atomic nuclei. Nuclei contain protons and neutrons, which are collectively known as *nucleons*. The total number of nucleons, A, is the *atomic mass number*. The number of protons, Z, is the *atomic number*. The number of neutrons equals $A - Z$. *Isotopes* are nuclei with the same Z, but with different numbers of neutrons. For an element X, an isotope of given Z and A is represented by $^A_Z X$. The nuclear radius is proportional to $A^{1/3}$, indicating that all nuclei have about the same density. Nuclear masses are specified in atomic mass units (u), where the mass of $^{12}_6$C is defined as exactly 12.000 u, or in terms of their energy equivalent (because $E = mc^2$), where 1 u = 931.5 MeV/c^2.

The mass of a nucleus is less than the sum of the masses of its constituent nucleons. The difference in mass (times c^2) is equal to the *total binding energy*; it represents the energy needed to break the nucleus into its constituent nucleons. The *binding energy per nucleon* averages about 8 MeV per nucleon, and is lowest for very light and very heavy nuclei.

Unstable nuclei undergo *radioactive decay*; they change into other nuclei with the emission of an α, β, or γ particle. An α particle is a $^4_2$He nucleus; a β particle is an electron or positron; and a γ ray is a high-energy photon. In β decay, a *neutrino* is also emitted.

Nuclei are held together by the *strong nuclear force* which has a short range. The *weak nuclear force* makes itself apparent in β decay. These two forces, plus the gravitational and electromagnetic forces, are the four known types of force.

Radioactive decay is a statistical process. The number that decay (ΔN) in a time Δt is proportional to the number N of parent nuclei present: $\Delta N = -\lambda N \Delta t$; the proportionality constant, λ, is called the decay constant and is characteristic of the given nucleus. The *half-life*, $T_{\frac{1}{2}}$, is the time required for half the nuclei of a radioactive sample to decay. It is related to the decay constant by $T_{\frac{1}{2}} = \ln 2/\lambda$. Electric charge, linear and angular momentum, mass-energy, and nucleon number are conserved in all decays. Radioactive decay occurs spontaneously only when the total mass of the products is less than the mass of the parent nucleus; the loss in mass appears as kinetic energy of the products.

A *nuclear reaction* occurs when two nuclei collide and two or more other nuclei (or particles) are produced. In this process, as in radioactivity, *transmutation* (change) of elements occurs.

In *fission* a heavy nucleus such as uranium splits into two intermediate-sized nuclei after being struck by a neutron. $^{235}_{92}$U is fissionable by slow neutrons, whereas some fissionable nuclei require fast neutrons. Much energy is released in fission because the binding energy per nucleon is lower for heavy nuclei than it is for intermediate-sized nuclei, so the mass of a heavy nucleus is greater than the total mass of its fission products. The fission process releases neutrons so that a *chain reaction* is possible. The *critical mass* is the minimum mass of fuel needed to sustain a chain reaction. In a *nuclear reactor* or nuclear bomb, a moderator is needed to slow down the released neutrons.

The *fusion* process, in which small nuclei combine to form larger ones, also releases energy. It has not yet been possible to build a fusion reactor for power generation, because of the difficulty in containing the fuel long enough at the high temperature required.

1. How do we know there is such a thing as the strong nuclear force?

2. What do different isotopes of a given element have in common? How are they different?

3. What are the elements represented by the X in the following: (a) $^{232}_{92}X$; (b) $^{18}_{7}X$; (c) $^{1}_{1}X$; (d)$^{82}_{38}X$; (e)$^{247}_{97}X$?

4. How many protons and how many neutrons do each of the isotopes in question 3 have?

5. Why are the atomic masses of many elements (Table 41–2) not close to whole numbers?

6. What is the experimental evidence in favor of radioactivity being a nuclear process?

7. The isotope $^{64}_{29}Cu$ is unusual in that it can decay by γ, β^-, and β^+ emission. What is the resulting nuclide for each case?

8. A $^{238}_{92}U$ nucleus decays to a nucleus containing how many neutrons?

9. Describe, in as many ways as possible, the differences between α, β, and γ rays.

10. What element is formed by the radioactive decay of (a) $^{24}_{11}Na(\beta^-)$; (b) $^{22}_{11}Na(\beta^+)$; (c) $^{210}_{84}Po(\alpha)$?

11. What element is formed by the decay of (a) $^{32}_{15}P(\beta^-)$; (b) $^{35}_{16}S(\beta^-)$; (c) $^{211}_{83}Bi(\alpha)$?

12. Fill in the missing particle or nucleus:
    (a) $^{45}_{20}Ca \rightarrow ? + e^- + \bar{\nu}$
    (b) $^{58}_{29}Cu \rightarrow ? + \gamma$
    (c) $^{46}_{24}Cr \rightarrow ^{46}_{23}V + ?$
    (d) $^{234}_{94}Pu \rightarrow ? + \alpha$
    (e) $^{239}_{93}Np \rightarrow ^{239}_{92}U + ?$

13. Immediately after a $^{238}_{92}U$ nucleus decays to $^{234}_{90}Th + ^{4}_{2}He$, the daughter thorium nucleus still has 92 electrons circling it. Since thorium normally holds only 90 electrons, what do you suppose happens to the two extra ones?

14. Do isotopes that undergo electron capture generally lie above or below the line of stability in Fig. 42–2?

15. Why are many artificially produced radioactive isotopes rare in nature?

16. Describe how the potential energy curve for an α particle in an α-emitting nucleus differs from that for a stable nucleus.

17. Fill in the missing particles or nuclei: (a) $^{137}_{56}Ba\ (n, \gamma)$?; (b) $^{137}_{56}Ba\ (n, ?)\ ^{137}_{55}Cs$; (c) $^{2}_{1}H\ (d, ?)\ ^{4}_{2}He$; (d) $^{197}_{79}Au\ (\alpha, d)$?

18. The isotope $^{32}_{15}P$ is produced by an (n, p) reaction. What must be the target nucleus?

19. When $^{22}_{11}Na$ is bombarded by deuterons ($^{2}_{1}H$), an α particle is emitted. What is the resulting nuclide?

20. Are fission fragments β^+ or β^- emitters?

21. $^{235}_{92}U$ releases an average of 2.5 neutrons per fission compared to 2.7 for $^{239}_{94}Pu$. Pure samples of which of these two nuclei do you think would have the smaller critical mass?

22. If $^{235}_{92}U$ released only 1.5 neutrons per fission on the average, would a chain reaction be possible? What would be different?

23. Discuss the relative merits and disadvantages, including pollution and safety, of power generation by fossil fuels, nuclear fission, and nuclear fusion.

24. What is the reason for the "secondary system" in Fig. 42–12? That is, why is the water heated by the fuel in a nuclear reactor not used directly to drive the turbines?

25. The energy from nuclear fission appears in the form of thermal energy—but the thermal energy of what?

26. Why would a porous block of uranium be more likely to explode if kept under water rather than in air?

27. Does $E = mc^2$ apply in (a) fission, (b) fusion, (c) nuclear reactions?

28. Light energy emitted by the sun and stars comes from the fusion process. What conditions in the interior of stars makes this possible?

 Problems

SECTION 42–1

1. (I) What is the rest energy of an α particle in MeV/c^2?

2. (I) A pi meson has a mass of 139 MeV/c^2. What is this in atomic mass units?

3. (I) (a) What is the approximate radius of a $^{64}_{29}Cu$ nucleus? (b) Approximately what is the value of A for a nucleus whose radius is 3.6×10^{-15} m?

4. (II) How much energy must an α particle have to just "touch" the surface of a $^{238}_{92}U$ nucleus?

5. (II) (a) Determine the density of nuclear matter in kg/m^3. (b) What would be the radius of the earth if it had its actual mass but had the density of nuclei? (c) What would be the radius of a $^{238}_{92}U$ nucleus if it had the density of the earth?

6. (II) What stable nucleus (approximately) has half the radius of a uranium nucleus?

SECTION 42–2

7. (I) Estimate the total binding energy for $^{40}_{20}Ca$?

8. (I) Use Fig. 42–1 to estimate the total binding energy of (a) $^{238}_{92}U$, (b) $^{107}_{47}Ag$.

9. (I) Use Appendix D to calculate the binding energy of $^{2}_{1}H$.

10. (I) Calculate the binding energy, and binding energy per nucleon for $^{6}_{3}Li$. Use Appendix D.

11. (II) Calculate the binding energy of the last neutron in a $^{12}_{6}C$ nucleus.

12. (II) (a) Show that the nucleus $^{8}_{4}Be$ (mass = 8.005308 u) is unstable to decay into two α particles. (b) Is $^{12}_{6}C$ stable

against decay into three α particles? Show why or why not.

SECTIONS 42–3 TO 42–7

13. (I) $^{60}_{27}\text{Co}$ in an excited state emits a 1.33 MeV γ ray as it jumps to the ground state. What is the mass of the excited cobalt atom?

14. (II) Show that the decay $^{11}_{6}\text{C} \rightarrow ^{10}_{5}\text{B} + p$ is not possible because energy would not be conserved.

15. (II) A $^{232}_{92}\text{U}$ nucleus emits an α particle with KE = 5.32 MeV. What is the final nucleus and what is the approximate mass (in u) of the final atom?

16. (II) When $^{23}_{10}\text{Ne}$ (mass = 22.9945 u) decays to $^{23}_{11}\text{Na}$ (mass = 22.9898 u), what is the maximum kinetic energy of the emitted electron? What is its minimum energy? What is the energy of the neutrino in each case?

17. (II) The nuclide $^{32}_{15}\text{P}$ decays by emitting an electron whose maximum kinetic energy can be 1.71 MeV. (a) What is the daughter? (b) Calculate its atomic mass (in u).

18. (II) The isotope $^{218}_{84}\text{Po}$ (mass = 218.008969 u) can decay by either α or β^- emission. What is the energy release in each case?

19. (II) How much energy is released in electron capture by beryllium: $^{7}_{4}\text{Be} + _{-1}^{0}e \rightarrow ^{7}_{3}\text{Li} + \nu$?

20. (II) The nuclide $^{191}_{76}\text{Os}$ decays with β^- energy of 0.14 MeV accompanied by γ rays of energy 0.042 MeV and 0.129 MeV. (a) What is the daughter nucleus? (b) Draw an energy-level diagram showing the ground states of the parent and daughter and excited states of the daughter. To which of the daughter states does β decay of $^{191}_{76}\text{Os}$ occur?

21. (II) (a) Show that when a nucleus decays by β^+ decay, the total energy released is equal to $(M_P - M_D - 2m_e)c^2$ where M_P and M_D are the masses of the parent and daughter atoms (neutral) and m_e is the mass of an electron or positron. (b) Determine the maximum kinetic energy of β^+ particles released when $^{11}_{6}\text{C}$ decays to $^{11}_{5}\text{B}$. What is the maximum energy the neutrino can have? What is its minimum energy?

22. (III) In α decay of, say, a $^{226}_{88}\text{Ra}$ nucleus, show that the nucleus carries away a fraction $1/(1 + A_D/4)$ of the total energy available, where A_D is the mass number of the daughter nucleus. (Hint: use conservation of momentum as well as conservation of energy.) Approximately what percentage of the energy available is thus carried off by the α particle for the case cited?

SECTIONS 42–8 TO 42–10

23. (I) (a) What is the decay constant of $^{238}_{92}\text{U}$ whose half-life is 4.5×10^9 yr? (b) The decay constant of a given nucleus is 6.2×10^{-5} s^{-1}. What is its half-life?

24. (I) What fraction of a sample of $^{68}_{32}\text{Ge}$, whose half-life is about 9 months, will remain after 4.5 yr?

25. (II) In a series of decays, the nuclide $^{235}_{92}\text{U}$ becomes $^{207}_{82}\text{Pb}$. How many α and β^- particles are emitted in this series?

26. (II) $^{124}_{55}\text{Cs}$ has a half-life of 30.8 s. (a) If we have 6.2 μg initially, how many nuclei are present? (b) How many are present 1.20 min later? (c) What is the activity at this time? (d) After how much time will the activity drop to less than about 1 per second?

27. (II) The activity of a sample of $^{35}_{16}\text{S}$ is 8.8×10^6 decays per second. What is the mass of sample present?

28. (II) A radioactive nuclide registers 2880 counts per minute on a Geiger counter at one time, and 1.6 h later registers 820 counts per minute. What is the half-life of the nuclide?

29. (II) The $^{3}_{1}\text{H}$ isotope of hydrogen, tritium, has a half-life of 12.33 yr. It can be used to measure the age of objects up to about 100 yr. It is produced in the upper atmosphere by cosmic rays and is brought to earth by rain. As an application, determine the age of a bottle of wine whose $^{3}_{1}\text{H}$ radiation is about $\frac{1}{10}$ that present in new wine.

30. (II) The rubidium isotope $^{87}_{37}\text{Rb}$, a β emitter with a half-life of 4.9×10^{10} yr, is used to determine the age of rocks and fossils. Rocks containing fossils of early animals contain a ratio of $^{87}_{38}\text{Sr}$ to $^{87}_{37}\text{Rb}$ of 0.018. Assuming there was no $^{87}_{38}\text{Sr}$ present when the rocks were formed, calculate the age of these fossils.

31. (II) At $t = 0$, a pure sample of radioactive nuclei contains N_0 nuclei whose decay constant is λ. Determine a formula for the number of daughter nuclei, N_D, as a function of time; assume $N_D = 0$ at $t = 0$.

32. (II) (a) Show that the *mean life*, or *average lifetime* of a radioactive nuclide, defined as

$$\tau = \frac{\int_0^\infty N(t)t\, dt}{\int_0^\infty N(t)\, dt}$$

is $\tau = 1/\lambda$. (b) What fraction of the original number of nuclei remains after one mean life?

33. (II) An ancient club is found that contains 240 g of carbon and has an activity of 5.0 decays per second. Determine the age of the club using the fact that in living trees the ratio $^{14}\text{C}/^{12}\text{C}$ is about 1.3×10^{-12}.

34. (III) At $t = 0$, a pure sample of a radioactive nuclide (the parent) contains N_{P0} nuclei whose half-life is T_P. The daughter nuclide is also radioactive, with half-life T_D. (a) Determine the number of daughter nuclei, N_D, as a function of time, assuming $N_D = 0$ at $t = 0$. (b) Plot N_D versus t for the cases $T_P = T_D$, $T_P = 3T_D$, $T_P = \frac{1}{3}T_D$. (Hint: Eq. 42–3 must be modified.)

SECTION 42–11

35. (I) Determine whether the reaction $^{2}_{1}\text{H}(d, n)^{3}_{2}\text{He}$ requires a threshold energy.

36. (I) Is the reaction $^{238}_{92}\text{U}(n, \gamma)^{239}_{92}\text{U}$ possible with slow neutrons? Explain.

37. (II) Does the reaction $^{7}_{3}\text{Li}(p, \alpha)^{4}_{2}\text{He}$ require energy or does it release energy? How much energy?

38. (II) (a) Can the reaction $^{24}_{12}\text{Mg}(n, d)^{23}_{11}\text{Na}$ occur if the bombarding particles have 10.0 MeV of KE? (b) If so, how much energy is released?

39. (II) In the reaction $^{14}_{7}\text{N}(\alpha, p)^{17}_{8}\text{O}$, the incident α particles have 7.68 MeV of kinetic energy. (a) Can this reaction occur? (b) If so, what is the total kinetic energy of the products? The mass of $^{17}_{8}\text{O}$ is 16.999131 u.

40. (III) Use conservation of energy and momentum to show that a bombarding proton must have an energy of 3.23 MeV in order to make the reaction $^{13}_{6}\text{C}(p, n)^{13}_{7}\text{N}$ occur. (See Example 42–9.)

41. (III) Show, using the laws of conservation of energy and momentum, that for a nuclear reaction requiring energy the minimum kinetic energy of the bombarding particle (the *threshold energy*) is equal to $[Qm_{pr}/(m_{pr} - m_b)]$, where Q is the energy required (difference in total mass between products and reactants), m_b is the rest mass of the bombarding particle and m_{pr} the total rest mass of the products. Assume the target nucleus is at rest before an interaction takes place.

SECTION 42–12

42. (I) Calculate the energy released in the fission reaction $^1_0n + ^{235}_{92}U \rightarrow ^{88}_{38}Sr + ^{136}_{54}Xe + 12^1_0n$. Use Appendix D, and assume the initial KE of the neutron is very small.

43. (I) What is the energy released in the fission reaction of Eq. 42–8? (The masses of $^{141}_{56}Ba$ and $^{92}_{36}Kr$ are 140.9141 and 91.9250, respectively.)

44. (I) How many fissions take place per second in a 25-MW reactor? Assume that 200 MeV is released per fission.

45. (II) One means of enriching uranium is by diffusion of the gas UF_6. Calculate the ratio of the speeds of molecules of this gas containing $^{235}_{92}U$ and $^{238}_{92}U$, on which this process depends.

46. (II) Suppose that the average power consumption, day and night, of an average house is 300 W. What mass of $^{235}_{92}U$ would have to undergo fission to supply the electrical needs of such a house for a year? (Assume 200 MeV is released per fission.) What total mass of $^{235}_{92}U$ would be needed ($T_{\frac{1}{2}} = 7.0 \times 10^8$ yr)?

47. (II) Suppose that the neutron multiplication factor is 1.0004. If the average time between successive fissions in a chain of reactions is 1.0 ms, by what factor will the reaction rate increase in 1.0 s?

48. (III) Consider a system of nuclear power plants that produce 4000 MW. (a) What total mass of $^{235}_{92}U$ fuel would be required to operate these plants for 1 yr, assuming that 200 MeV is released per fission? (b) Typically 3 percent of the $^{235}_{92}U$ mass that fissions is converted to $^{90}_{38}Sr$, a β^- emitter with a half-life of 29 yr. What is the total radioactivity of the $^{90}_{38}Sr$, in curies, after its production? (Neglect the fact that some of it decays during the 1-yr period.)

SECTION 42–13

49. (I) What is the average kinetic energy of protons at the center of a star where the temperature is 10^7 K?

50. (II) If an average house requires 300 W of electric power on average, how much deuterium would have to be used in a year to supply these electrical needs? Assume the reaction of Eq. 42–9b.

51. (II) Calculate the energy release per gram of fuel for the reactions of Eqs. 42–9a and 42–9c. Compare to the energy release per gram of uranium in the fission process.

52. (II) In the so-called carbon cycle that occurs in the sun, ^4_2He is built from four protons starting with $^{12}_6C$. First $^{12}_6C$ absorbs a proton to form nucleus X_1. X_1 decays by β^+ emission to X_2. X_2 absorbs a proton to become X_3, which itself absorbs a proton to become X_4. X_4 decays to X_5 by β^+ decay and X_5 reacts via $X_5(p, \alpha)X_6$. (a) Determine the intermediate nuclei and write out each step in detail, and show that X_6 is again $^{12}_6C$, which is thus not used up in the process. (b) Determine the energy release for each step.

53. (II) How much energy (J) is contained in 1.00 kg of water if the natural deuterium is used in the fusion reaction of Eq. 42–9a? Compare this to the energy obtained from the burning of 1.00 kg of gasoline, about 5×10^7 J.

*SECTION 42–14

***54. (I)** A 0.018-μCi sample of $^{32}_{15}P$ is injected into an animal for tracer studies. If a Geiger counter intercepts 20 percent of the emitted β particles and is 90 percent efficient in counting them, what will be the counting rate?

***55. (I)** An average adult body has about 0.10 μCi of $^{40}_{19}K$, which comes from food. How many decays occur per second?

***56. (I)** (a) A dose of 500 rem of γ rays would be lethal to about half the people subjected to it. How many rads is this? (b) 50 rad of α-particle radiation is equivalent to how many rads of X rays, in terms of biological damage? (c) How many rads of slow neutrons will do as much biological damage as 50 rads of fast neutrons?

***57. (II)** A 1.0-mCi source of $^{32}_{15}P$ (in NaHPO$_4$), a β^- emitter, is implanted in an organ where it is to administer 5000 rad. The half-life of $^{32}_{15}P$ is 14.3 days and 1 mCi delivers about 1 rad/min. Approximately how long should the source remain implanted?

***58. (II)** $^{57}_{27}Co$ emits 122-keV γ rays. If a 70-kg person swallowed 2.0 μCi of $^{57}_{27}Co$, what would be the dose rate (rad/day) averaged over the whole body? Assume 50 percent of the γ-ray energy is deposited in the body.

***59 (II)** A shielded γ-ray source yields a dose rate of 0.055 rad/h at a distance of 1.0 m for an average-size person. If workers are allowed a maximum dose rate of 5.0 rem/yr, how close to the source may they operate assuming a 35-h work week? Assume the intensity of radiation falls off as the square of the distance. (It actually falls off more rapidly than $1/r^2$ because of absorption in the air, so the answer above will give a better-than-permissible value.)

Elementary
Particles

43

In this, the final chapter, we discuss the exciting subject of *elementary particle* physics, which represents the human endeavor to understand the basic building blocks of all matter.

In the years after World War II, it was found that if the incoming particle in a nuclear reaction has sufficient energy, new types of particles can be produced. In order to produce high-energy particles, physicists have constructed various types of particle accelerators; see, for example, Section 28–8. Most commonly they accelerate protons or electrons, although heavy ions can also be accelerated depending on the design. These high-energy accelerators have been used to probe the nucleus more deeply, to produce and study new particles, and to give us information about the basic forces and constituents of nature.

 43–1   High-Energy Projectiles

Particles accelerated to high speeds are projectiles that can probe the interior of nuclei they strike. An important factor is that faster moving projectiles can reveal more detail. The wavelength of projectile particles is given by de Broglie's wavelength formula

$$\lambda = \frac{h}{mv}, \qquad (43-1)$$

from which we see that the greater the momentum of the bombarding particle, the shorter the wavelength and the more detail that can be obtained; as discussed in Chapter 37, resolution of details is limited by the wavelength, so we see one reason why particle accelerators of higher and higher energy have been built in recent years.

EXAMPLE 43–1   To explore the distribution of charge within nuclei, very-high-energy electrons are used. (Electrons are used rather than protons because they do not partake in the strong nuclear force, so only the electric force is involved.) Experiments at the Stanford linear accelerator using electrons with KE of 1.3 GeV obtained the charge distribution for the bismuth nucleus shown in Fig. 43–1. What is the expected resolution—that is, the size of the smallest details that can be detected?

SOLUTION   1.3 GeV = 1300 MeV which is about 2500 times the mass of the electron. We are clearly dealing with relativistic speeds here, and it is easily shown that the speed of the electron is nearly $c = 3.0 \times 10^8$ m/s. (From Eq. 39–10, KE $= mc^2 - m_0c^2 \approx mc^2$; and from Eq. 39–13, $E^2 = p^2c^2 + m_0^2c^4 \approx p^2c^2$ since m_0c^2 is small compared to pc; hence $p^2c^2 = m^2v^2c^2 \approx m^2c^4$, so $v \approx c$.) Therefore

$$\lambda = \frac{h}{mv} \approx \frac{h}{mc} = \frac{hc}{mc^2}$$

where $mc^2 = 1.3$ GeV. Hence

$$\lambda = \frac{(6.6 \times 10^{-34}\ \text{J·s})(3.0 \times 10^8\ \text{m/s})}{(1.3 \times 10^9\ \text{eV})(1.6 \times 10^{-19}\ \text{J/eV})} = 0.96 \times 10^{-15}\ \text{m},$$

or 0.96 fm (1 fm $= 10^{-15}$ m = 1 femtometer or 1 fermi—in honor of Enrico Fermi). This is somewhat less than the size of nuclei. Notice in Fig. 43–1 that nuclei have a nearly uniform charge distribution on the interior, but their boundary is not distinct. Studies with even higher-energy electrons bombarding protons and neutrons show structural details that may indicate nucleons consist of still smaller particles.

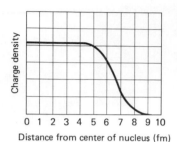

FIGURE 43–1

Distribution of electric charge for bismuth nucleus, determined using 1.3-GeV electrons.

High energy particles produced in accelerators are used not only to investigate nuclear structure, but also to produce many new types of particles as we shall see.

 ## 43–2   Beginnings of Elementary Particle Physics—the Yukawa Particle

By the mid 1930s, it was recognized that all atoms can be considered to be made up of neutrons, protons, and electrons. The basic constituents of the universe were no longer considered to be atoms but rather the proton, neutron, and electron. Besides these three *elementary particles*, as they could be called, there were several others also known: the positron (a negative electron), the neutrino, and the γ particle (or photon), for a total of six elementary particles.

Looking back, things seemed fairly simple in 1935. But in the decades that followed, hundreds of other elementary particles were discovered. The properties and interactions of these particles, and which ones should be considered as fundamental or "elementary," became the substance of research in the field of **elementary particle physics.**

Elementary particle physics, as it exists today, can be said to have begun in 1935 when the Japanese physicist Hideki Yukawa (1907–1981) predicted the existence of a new particle that would in some way mediate the strong nuclear force. To understand Yukawa's idea, we first look at the electromagnetic force. When we first discussed electricity we saw that the electric force acts over a distance, without contact. To better perceive how a force can act over a distance, we saw that Faraday introduced the idea of a *field*. The force that one charged particle exerts on a second can be said to be due to the electric field set up by the first. Similarly, the magnetic field can be said to carry the magnetic force. Later (Chapter 33) we saw that electromagnetic fields can travel through space as waves. Finally, in Chapter 40, we saw that electromagnetic radiation (light) can be considered as either a wave or as a collection of particles called photons. Because of this wave–particle duality, it is possible to imagine that the electromagnetic force

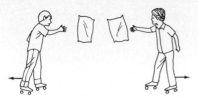

(a) Repulsive force (children
throwing pillows)

(b) Attractive force (children grabbing
pillows from each other's hand)

FIGURE 43-2

Forces equivalent to particle exchange. (a) Repulsive
force (children throwing pillows). (b) Attractive force
(children grabbing pillows from each other's hand).

between charged particles is due (1) to the EM field set up by one and felt by the
other, or (2) to an exchange of photons or γ-particles between them. It is (2) that
we want to concentrate on here, and an example of how an exchange of particles
could give rise to a force is illustrated in the crude picture of Fig. 43–2. In part (a),
two children start throwing pillows at each other; each catch results in the child
being moved backward by the impulse. This is the equivalent of a repulsive force.
On the other hand, if the two children exchange pillows by grabbing them out
of the other person's hand, they will be pulled toward each other, as when an
attractive force acts. For the electromagnetic force between two charged particles,
it is photons that are exchanged between the two particles that give rise to the
force. A simple diagram describing this is shown in Fig. 43–3. Such a diagram is
called a *Feynman diagram* and the theory on which it is based, *quantum electrodynamics* (QED). The case shown is the simplest in which a single photon is
exchanged. One of the charged particles emits the photon and recoils somewhat
as a result; and the second particle absorbs the photon. In any such collision or
interaction, energy and momentum are transferred from one particle to the other,
and it is carried by the photon. Because the photon is absorbed by the second
particle very shortly after it is emitted by the first, it is not observable, and it is
referred to as being a *virtual* photon, as compared to one that is free and can be
detected by instruments.

Now to Yukawa's prediction. In analogy to photon exchange to mediate the
electromagnetic force, Yukawa argued that there ought to be a particle that me-
diates the strong nuclear force—the force that holds nucleons together in the
nucleus. Just as the photon is called the quantum of the electromagnetic field or
force, so the Yukawa particle would represent the quantum of the strong nuclear
force.

Yukawa predicted that this new particle would have a mass intermediate
between that of the electron and the proton. Hence it was called a *meson*, meaning
"in the middle," and Fig. 43–4 is a Feynman diagram of meson exchange. We
can make a rough approximation of the mass of the meson as follows. Suppose
the proton on the left in Fig. 43–4 is at rest. For it to emit a meson would
require energy (to make the mass) which would have to come from nowhere; such
a process would violate conservation of energy. But the uncertainty principle
allows nonconservation of energy of an amount ΔE if it occurs only for a time
Δt given by $(\Delta E)(\Delta t) \approx h/2\pi$. We set ΔE equal to the energy needed to create
the mass m of the meson: $\Delta E = mc^2$. Now conservation of energy is violated only
as long as the meson exists, which is the time Δt required for the meson to pass
from one nucleon to the other. If we assume the meson travels at relativistic
speed, close to the speed of light c, then Δt will be at most about $\Delta t = d/c$,

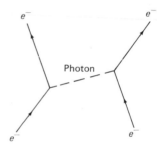

FIGURE 43-3

Feynman diagram, showing how
a photon acts as carrier of
electromagnetic force between
two electrons.

FIGURE 43-4

Meson exchange when proton
and neutron interact via strong
nuclear force.

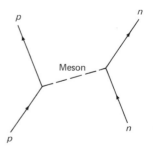

where d is the maximum distance that can separate the interacting nucleons. Thus we have

$$\Delta E \, \Delta t \approx \frac{h}{2\pi}$$

$$mc^2 \left(\frac{d}{c}\right) \approx \frac{h}{2\pi}$$

or

$$mc^2 \approx \frac{hc}{2\pi d}. \tag{43-2}$$

The range of the strong nuclear force is about $d \approx 1.5 \times 10^{-15}$ m (this is the maximum distance away it can be felt) so

$$mc^2 \approx \frac{(6.6 \times 10^{-34} \text{ J} \cdot \text{s})(3.0 \times 10^8 \text{ m/s})}{(6.28)(1.5 \times 10^{-15} \text{ m})} \approx 2.2 \times 10^{-11} \text{ J} = 130 \text{ MeV}.$$

The mass of the predicted meson is thus very roughly 130 MeV/c^2 or about 250 times the electron mass of 0.51 MeV/c^2. (Note, incidentally, that since the electromagnetic force has infinite range, $d = \infty$, Eq. 43–2 tells us that the exchanged particle for the electromagnetic force, the photon, will have zero rest mass.)

Just as photons can be observed as free particles, as well as acting in an exchange, so it was expected that mesons might be observed directly. Such a meson was searched for in the cosmic radiation that enters the earth's atmosphere from the sun and other sources in the universe. In 1937 a new particle was discovered whose mass was 106 MeV (207 times the electron mass). This is quite close to the mass predicted. But it turned out that this new particle, called the *muon* (or *mu meson*), did not interact strongly with matter. It could hardly mediate the strong nuclear force if it didn't interact via the strong nuclear force. Thus the muon, which can have either a + or a − charge and seems to be nothing more than a very massive electron, is not the Yukawa particle.

The particle predicted by Yukawa was finally found in 1947. It is called the "π" or pi meson, or simply the *pion*. It comes in three charge states, +, −, or 0. The π^+ and π^- have mass of 139.6 MeV/c^2 and the π^0 a mass of 135.0 MeV/c^2. All three interact strongly with matter. Soon after their discovery in cosmic rays, pions were produced in the laboratory using a particle accelerator. Reactions observed included

$$p + p \rightarrow p + p + \pi^0$$

$$p + p \rightarrow p + n + \pi^+. \tag{43-3}$$

The incident proton from the accelerator must have sufficient energy to produce the additional mass of the pion. A number of other mesons were discovered in subsequent years which were also considered to mediate the strong nuclear force. (The recent theory of quantum chromodynamics, involving quarks, has replaced mesons with gluons as exchange particles for the strong force; see Section 43–9.)

So far we have discussed the particles that mediate the electromagnetic and strong nuclear forces. But there are four known types of force—or interaction—in nature. What about the other two: the weak nuclear force, and gravity? Theorists believe that these are also mediated by particles. The particles presumed to transmit the weak force are referred to as the W^+, W^-, and Z^0. Early in 1983, after extensive searches, the long-awaited discovery of the W particle was announced by Carlo Rubbia, the scientist behind the building of the proton-antiproton collider at the very high energy accelerator at CERN[†] in Geneva, Switzerland, that produced the results, and leader of a large group of scientists (over 100) who

[†] CERN is an acronym for European Center for Nuclear Research (in French).

TABLE 43–1

The four forces in nature

Type	Relative Strength (approx.)	Field Particle
Strong nuclear	1	Mesons/gluons[†]
Electromagnetic	10^{-2}	Photon
Weak nuclear	10^{-13}	$W^{\pm}$ and Z^0
Gravitational	10^{-40}	Graviton (?)

[†] See Section 43–9.

worked on the project. The quantum of the gravitational force, called the *graviton*, has not yet been identified. A comparison of the four forces is given in Table 43–1, where they are listed according to their (approximate) relative strengths. Notice that although gravity may be the most obvious force in daily life (because of the huge mass of the earth), on a nuclear scale it is much the weakest of the four forces and its effect at the nuclear level can nearly always be ignored.

 ## 43–3 Particles and Antiparticles

The positron, as we have seen, is basically a positive electron. That is, many of its properties are the same as for the electron, such as mass, but it has the opposite charge. The positron is said to be the **antiparticle** to the electron. After the positron was discovered in 1932, it was predicted that other particles also ought to have antiparticles. In 1955 the antiparticle to the proton was found, the *antiproton* ($\bar{p}$), see Fig. 43–5. (The bar over the *p* is used to indicate antiparticle.) A large amount of energy was needed to produce this massive particle (mass = proton's mass). Its discovery (by E. Segré and O. Chamberlain) was made only after the completion of the large accelerator (the Bevatron) at the University of California at Berkeley. Soon after, the antineutron ($\bar{n}$) was found. Most other particles also have antiparticles. But the photon, the π^0, and a few other particles do not have distinct antiparticles—or we say that they are their own antiparticles.

Antiparticles are produced in nuclear reactions when there is sufficient energy available, and they do not live very long in the presence of matter. For example, when a positron encounters an electron, the two annihilate each other. The energy of their vanished mass, plus any kinetic energy they possess, is converted to the energy of γ rays or of other particles. Annihilation occurs for all other particle–antiparticle pairs.

 ## 43–4 Particle Interactions and Conservation Laws

One of the important uses of high-energy accelerators is to study the interactions of elementary particles with each other. As a means of ordering this subnuclear world, the conservation laws are indispensable. The laws of conservation of energy, of momentum, of angular momentum, and of electric charge are found to hold precisely in all particle interactions. (Although the uncertainty principle allows, for example, nonconservation of energy for times $\Delta t \approx h/2\pi\,\Delta E$, as we have seen, the times involved are much too short for experimental observation and violation has never been detected.)

A study of particle interactions has revealed a number of new conservation laws, some of which we now discuss. These new conservation laws (just like the old ones) are ordering principles: they help to explain why some reactions occur

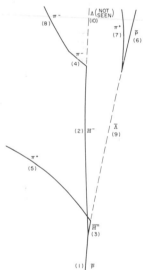

FIGURE 43–5

Liquid-hydrogen bubble-chamber photograph of an antiproton ($\bar{p}$) colliding with a proton, producing a hyperon pair ($\bar{p} + p \rightarrow \Xi^- + \overline{\Xi^-}$), which subsequently decay into other particles. The drawing indicates the assignment of a particle to each track. Neutral-particle paths are shown by dashed lines since neutral particles produce no bubbles. (Courtesy of Brookhaven National Laboratory.)

849

and others do not. For example, the following reaction has never been found to occur:

$$p + n \not\rightarrow p + p + \bar{p}$$

even though charge, energy, and so on, are conserved ($\bar{p}$ means an antiproton and $\not\rightarrow$ means the reaction does not occur). To understand why it doesn't occur, physicists hypothesized a new conservation law, the conservation of **baryon number**. (Baryon number is the same as nucleon number, which we saw earlier is conserved in nuclear reactions.) An important addition to this law is the proposal that whereas all nucleons have baryon number $B = +1$, all antinucleons (antiprotons, antineutrons) have $B = -1$. The reaction above does not conserve baryon number since on the left side we have $B = (+1) + (+1) = +2$ and on the right $B = (+1) + (+1) + (-1) = +1$. On the other hand, the following reaction does conserve B and *does* occur if the incoming proton has sufficient energy:

$$p + n \rightarrow \quad p + n + \bar{p} + p$$
$$B = +1 + 1 = +1 + 1 - 1 + 1.$$

As indicated, $B = +2$ on both sides of this equation. From these and other reactions, the conservation of baryon number has been established as a basic law of physics.

Other useful "number" laws are associated with weak interactions, mainly decays. In ordinary β decay an electron or positron is emitted along with a neutrino or antineutrino; in a similar type of decay a muon is emitted instead of an electron. The neutrino that accompanies an emitted electron (v_e) is found to be different from the neutrino that accompanies an emitted muon (v_μ). Each of these neutrinos has an antiparticle: $\bar{v}_e$ and $\bar{v}_\mu$. In ordinary β decay we have, for example

$$n \rightarrow p + e^- + \bar{v}_e$$

but never $n \rightarrow p + e^- + \bar{v}_\mu$ or $n \rightarrow p + e^- + \bar{v}_e + v_e$. To explain why these do not occur, the concept of electron lepton number, L_e, was invented. If the electron (e^-) and the electron neutrino (v_e) are given $L_e = +1$, and e^+ and $\bar{v}_e$ are given $L_e = -1$, whereas all other particles have $L_e = 0$, then all observed decays conserve L_e. For example, in $n \rightarrow p + e^- + \bar{v}_e$, $L_e = 0$ initially, and $L_e = 0 + (+1) + (-1) = 0$ after the decay. Decays that do not conserve L_e yet would obey the other conservation laws are not observed to occur; hence it is believed that L_e is conserved in all interactions.

In a decay involving muons, such as

$$\pi^+ \rightarrow \mu^+ + v_\mu,$$

a second quantum number, muon lepton number (L_μ) is conserved. The μ^- and v_μ are assigned $L_\mu = +1$ and μ^+ and $\bar{v}_\mu$ have $L_\mu = -1$, whereas other particles have $L_\mu = 0$. It is believed that L_μ is also conserved in all interactions or decays. Similar assignments can be made for a third lepton number, L_τ, associated with the newly discovered τ lepton and its neutrino, v_τ.

Keep in mind that antiparticles have not only opposite electric charge from their particles, but also opposite B, L_e, and L_μ.

EXAMPLE 43–2 Which of the following decay schemes is possible for muon decay: (a) $\mu^- \rightarrow e^- + \bar{v}_e$; (b) $\mu^- \rightarrow e^- + \bar{v}_e + v_\mu$; (c) $\mu^- \rightarrow e^- + v_e$?

SOLUTION A μ^- has $L_\mu = +1$, and $L_e = 0$. This is the initial state, and the final state (after decay) must also have $L_\mu = +1$, $L_e = 0$. In (a), the final state has $L_\mu = 0 + 0 = 0$, and $L_e = +1 - 1 = 0$; L_μ would not be conserved and indeed this decay is not observed to occur. The final state of (b) has $L_\mu = 0 + 0 + 1 = +1$ and $L_e = +1 - 1 + 0 = 0$, so both L_μ and L_e are conserved. This is in fact the most common decay mode of the μ^-. Finally (c) does not occur because $L_e (= +2$ in final state) is not conserved.

Recent theoretical work suggests that the conservation laws of baryon and lepton numbers may be only approximate, rather than exact, as we shall discuss in Section 43–10.

 ## 43–5 Particle Classification

In the decades following the discovery of the π meson in the late 1940s, a great many other subnuclear particles were discovered; today they number in the hundreds. Much theoretical and experimental work has been done to try to understand this multitude of particles. One important aid to understanding is to arrange the particles in categories according to their properties. One way of doing this is according to their interactions. Since not all particles take part in all four of the forces known in nature, this fact is used as a classification scheme. Table 43–2 lists many of the stable (and "long-lived") particles classified in this way along with many of their properties. The *photon* takes part only in the electromagnetic force, and it is in a class by itself. The **leptons** are those particles that do not interact via the strong force but do interact via the weak nuclear force (as well as the much weaker gravitational force); those that carry electric charge also interact via the electromagnetic force. The four well-known leptons are the electron, the muon, and two types of neutrino: the electron neutrino, ν_e, and the muon neutrino,

TABLE 43–2
Elementary particles (stable or "long-lived")[†]

Category	Particle Name	Symbol	Antiparticle	Rest Mass (MeV/c^2)	B	L_e	L_μ	L_τ	S	Lifetime (s)	Principal Decay Modes
Photon	Photon	γ	Self	0	0	0	0	0	0	Stable	
Leptons	Electron	e^-	e^+	0.511	0	+1	0	0	0	Stable	
	Neutrino (e)	ν_e	$\bar{\nu}_e$	0 (?)	0	+1	0	0	0	Stable	
	Muon	μ^-	μ^+	105.7	0	0	+1	0	0	2.20×10^{-6}	$e^- \bar{\nu}_e \nu_\mu$
	Neutrino (μ)	ν_μ	$\bar{\nu}_\mu$	0 (?)	0	0	+1	0	0	Stable	
	Tau	τ^-	τ^+	1784.	0	0	0	−1	0	$<4 \times 10^{-13}$	$\mu^- \bar{\nu}_\mu \nu_\tau, e^- \bar{\nu}_e \nu_\tau$, hadrons
	Neutrino (τ)	ν_τ	$\bar{\nu}_\tau$	0 (?)	0	0	0	−1	0	Stable	
Hadrons											
Mesons	Pion	π^+	π^-	139.6	0	0	0	0	0	2.60×10^{-8}	$\mu^+ \nu_\mu$
		π^0	Self	135.0	0	0	0	0	0	0.83×10^{-16}	2γ
	Kaon	K^+	K^-	493.7	0	0	0	0	+1	1.24×10^{-8}	$\mu^+ \nu_\mu, \pi^+ \pi^0$
		K_S^0	$\bar{K}_S^0$	497.7	0	0	0	0	+1	0.89×10^{-10}	$\pi^+ \pi^-, 2\pi^0$
		K_L^0	$\bar{K}_L^0$	497.7	0	0	0	0	+1	5.2×10^{-8}	$\pi^\pm e^\mp \overset{(-)}{\nu}_e$ $\pi^+ \mu^\mp \overset{(-)}{\nu}_\mu$ $3\pi^0$
	Eta	η^0	Self	548.8	0	0	0	0	0	$<10^{-18}$	$2\gamma, 3\pi$
Baryons	Proton	p	$\bar{p}$	938.3	+1	0	0	0	0	Stable[‡]	
	Neutron	n	$\bar{n}$	939.6	+1	0	0	0	0	920	$pe^- \bar{\nu}_e$
	Lambda	Λ^0	$\overline{\Lambda^0}$	1115.6	+1	0	0	0	−1	2.6×10^{-10}	$p\pi^-, n\pi^0$
	Sigma	Σ^+	$\overline{\Sigma^-}$	1189.4	+1	0	0	0	−1	0.80×10^{-10}	$p\pi^0, n\pi^+$
		Σ^0	$\overline{\Sigma^0}$	1192.5	+1	0	0	0	−1	6×10^{-20}	$\Lambda^0 \gamma$
		Σ^-	$\overline{\Sigma^+}$	1197.3	+1	0	0	0	−1	1.5×10^{-10}	$n\pi^-$
	Xi	Ξ^0	$\overline{\Xi^0}$	1315	+1	0	0	0	−2	2.9×10^{-10}	$\Lambda^0 \pi^0$
		Ξ^-	Ξ^+	1321	+1	0	0	0	−2	1.64×10^{-10}	$\Lambda^0 \pi^-$
	Omega	Ω^-	Ω^+	1672	+1	0	0	0	−3	0.82×10^{-10}	$\Xi^0 \pi^0, \Lambda^0 K^-$

[†] See also Table 43–4.
[‡] $>6 \times 10^{31}$/yr.

v_μ. They each have antiparticles as indicated in Table 43–2. The recent evidence for another lepton, the τ, and its neutrino, v_τ, brings the total number of leptons today to six.

The third category of particle is the **hadron**. Hadrons are those particles that can interact via the strong nuclear force. Hence they are said to be **strongly interacting particles**. They also interact via the other forces, but the strong force predominates at short distances. The hadrons include nucleons, pions, and a large number of other particles. They are divided into two subgroups:[†] **baryons**, which are those particles that have baryon number $+1$ (or -1 in the case of their antiparticles); and **mesons**, which have baryon number $= 0$.

Notice that the baryons Λ, Σ, Ξ, and Ω all decay to lighter-mass baryons, and eventually to a proton or neutron. All these processes conserve baryon number. Since there is no lighter particle than the proton with $B = +1$, if baryon number is strictly conserved, the proton itself cannot decay and is stable (but see Section 43–10).

43–6 Particle Stability and Resonances

The particles listed in Table 43–2 are those that are either stable or are rather long-lived (lifetime $\gtrsim 10^{-19}$ s). The lifetime of an unstable particle depends on which force is most active in causing the decay. When we say the strong nuclear force is stronger than the electromagnetic, we mean that two particles will interact more quickly and more frequently if this force is acting. When a stronger force influences a decay, that decay occurs more quickly. Decays caused by the weak force have lifetimes of 10^{-10} s or longer. Particles that decay via the electromagnetic force have much shorter lifetimes, typically about 10^{-16} to 10^{-19} s. The unstable particles listed in Table 43–2 decay either via the weak or the electromagnetic interaction; those that involve a γ (photon) are electromagnetic, and the others shown decay via the weak interaction (note the lifetimes).

A great many particles have been found that can decay via the strong interaction, and these are not listed in Table 43–2. Such particles decay into other strongly interacting particles (say, n, p, π, but not involving γ, e, and so on) and their lifetimes are very short, typically 10^{-23} s. In fact their lifetimes are so short that they do not go far enough to be detected before decaying. Their decay products can be detected, however, and it is from them that the existence of such short-lived particles is inferred. To see how this is done, let us consider the first such particle discovered (by Fermi). Fermi used a beam of π^+ directed through a hydrogen target (protons) with varying amounts of energy. A graph of the number of interactions (π^+ scattered) versus the pion's kinetic energy is shown in Fig. 43–6. The large peak around 200 MeV was much higher than expected and certainly much higher than the number of interactions at neighboring energies. This led Fermi to conclude that the π^+ and proton combined momentarily to form a short-lived particle before coming apart again, or at least they resonated back and forth for a short time. Indeed, the large peak in Fig. 43–6 resembles a resonance curve (see Figs. 14–13 and 32–8) and this new "particle"—now called the Δ—is referred to as a **resonance**. Hundreds of other resonances have been found in a similar way. Many resonances are regarded as excited states of other particles such as the nucleon.

The width of the resonances—in Fig. 43–6 the width of the Δ peak is about 100 MeV—is an interesting application of the uncertainty principle. If a

---

[†] Originally, particles were divided according to their mass into leptons (meaning light particles), baryons (meaning "heavy"), and those of intermediate mass, the mesons (meaning "middle"). The newer classification according to their interactions is almost consistent with this. The muon, however, is now called a lepton (it doesn't interact strongly) although it was once called the mu meson because of its mass. Other exceptions are the ψ particles which are very heavy but have $B = 0$, and so are classified as mesons (Section 43–8).

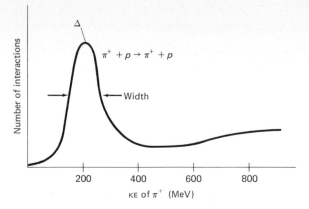

FIGURE 43-6

Number of π^+ being scattered by a proton target as a function of the incident π^+ kinetic energy.

particle lives only 10^{-23} s, then its mass (that is, its rest energy) will be uncertain by an amount $\Delta E \approx h/2\pi\,\Delta t \approx (6.6 \times 10^{-34}\,\text{J}\cdot\text{s})/(6)(10^{-23}\,\text{s}) \approx 10^{-11}\,\text{J} \approx 100\,\text{MeV}$, which is what is observed. Actually, the lifetimes of $\approx 10^{-23}$ s for such resonances are inferred by the reverse process: from the measured width being ≈ 100 MeV.

43-7 Strange Particles

In the early 1950s, certain of the newly found particles, namely the K, Λ, and Σ, were found to behave rather strangely in two ways. First, they were always produced in pairs. For example, the reaction

$$\pi^- + p \to K^0 + \Lambda^0$$

occurred with high probability, but the reaction $\pi^- + p \to K^0 + n$ was never observed to occur; this seemed strange since no known conservation law would have been violated and there was plenty of energy available. The second feature of these *strange particles* (as they came to be called) was that although they were clearly produced via the strong interaction (that is, at a high rate), they did not decay at a rate characteristic of the strong interaction; and they decayed only into strongly interacting particles (for example $K \to 2\pi$, $\Sigma^+ \to p + \pi^0$). Instead of lifetimes of 10^{-23} s as expected for strongly interacting particles, strange particles have lifetimes of 10^{-10} to 10^{-8} s which are characteristic of the weak interaction.

To explain these observations, a new quantum number, **strangeness**, and a new conservation law, conservation of strangeness, were introduced. By assigning the strangeness numbers (S) indicated in Table 43-2, the production of strange particles in pairs was readily explained; antiparticles were assigned opposite strangeness from their particles: one of each pair was assigned $S = +1$ and the other $S = -1$ (see Table 43-2). For example, in the reaction $\pi^- + p \to K^0 + \Lambda^0$, the initial state has strangeness $S = 0 + 0 = 0$, and the final state has $S = +1 - 1 = 0$, so strangeness is conserved. But for $\pi^- + p \not\to K^0 + n$, the initial state has $S = 0$ and the final state has $S = +1 + 0 = +1$, so strangeness would not be conserved; and the reaction isn't observed.

To explain the decay of strange particles, it is assumed that strangeness is conserved in the strong interaction but is *not* conserved in the weak interaction. Thus, although the strange particles were forbidden by strangeness conservation to decay to lower mass nonstrange particles via the strong interaction, they could undergo such decay by means of the weak interaction. This would occur much more slowly, of course, which accounts for their longer lifetimes of 10^{-10} to 10^{-8} s.

The conservation of strangeness was the first example of a "partially conserved" quantity. In this case, the quantity strangeness is conserved by strong interactions but not by weak.

 ## 43-8   Quarks and Charm

Nearly all observed particles fall into one of two groups: leptons or hadrons. The principal difference between these two groups is that the hadrons interact via the strong interaction whereas the leptons do not. Another important difference that physicists had to deal with in the 1960s was that there were only four known leptons (e^-, μ^-, v_e, v_μ) but there are well over a hundred hadrons.

The leptons are considered to be truly elementary particles since they do not seem to break down into smaller entities, do not show any internal structure, and have no measurable size. (Attempts to determine the size of leptons have put an upper limit of about 10^{-18} m.)

The hadrons, on the other hand, are more complex. Experiments indicate they do have an internal structure. And the fact that there are so many of them suggests that they can't all be elementary. To deal with this problem, M. Gell-Mann and G. Zweig in 1963 independently proposed that none of the hadrons so far observed is elementary. Instead, they proposed that the hadrons are made up of combinations of three, more fundamental, pointlike entities called **quarks**.[†] Quarks, then, would be considered truly elementary particles, like leptons. The three quarks were labeled u, d, s, and given the names *up*, *down*, and *sideways* (or, more commonly now, *strange*). They were assumed to have fractional charge ($\frac{1}{3}$ or $\frac{2}{3}$ the charge on the electron—that is, less than the previously thought smallest charge); other properties of quarks and antiquarks are indicated in Table 43-3. All hadrons known at the time could be constructed in theory from these three types of quark. Mesons would consist of a quark–antiquark pair. For example, a π^+ meson is considered a $u\bar{d}$ pair (note that

[†] Gell-Mann chose the word from the phrase "Three quarks for Muster Mark" in James Joyce's *Finnegan's Wake*.

TABLE 43-3
Properties of quarks and antiquarks

Quarks

Name	Symbol	Spin	Charge	Baryon Number	Strangeness	Charm	Bottomness	Topness
Up	u	$\frac{1}{2}$	$+\frac{2}{3}e$	$\frac{1}{3}$	0	0	0	0
Down	d	$\frac{1}{2}$	$-\frac{1}{3}e$	$\frac{1}{3}$	0	0	0	0
Strange	s	$\frac{1}{2}$	$-\frac{1}{3}e$	$\frac{1}{3}$	-1	0	0	0
Charmed	c	$\frac{1}{2}$	$+\frac{2}{3}e$	$\frac{1}{3}$	0	$+1$	0	0
Bottom	b	$\frac{1}{2}$	$-\frac{1}{3}e$	$\frac{1}{3}$	0	0	$+1$	0
Top (?)	t	$\frac{1}{2}$	$+\frac{2}{3}e$	$\frac{1}{3}$	0	0	0	$+1$

Antiquarks

Name	Symbol	Spin	Charge	Baryon Number	Strangeness	Charm	Bottomness	Topness
Up	$\bar{u}$	$\frac{1}{2}$	$-\frac{2}{3}e$	$-\frac{1}{3}$	0	0	0	0
Down	$\bar{d}$	$\frac{1}{2}$	$+\frac{1}{3}e$	$-\frac{1}{3}$	0	0	0	0
Strange	$\bar{s}$	$\frac{1}{2}$	$+\frac{1}{3}e$	$-\frac{1}{3}$	$+1$	0	0	0
Charmed	$\bar{c}$	$\frac{1}{2}$	$-\frac{2}{3}e$	$-\frac{1}{3}$	0	-1	0	0
Bottom	$\bar{b}$	$\frac{1}{2}$	$+\frac{1}{3}e$	$-\frac{1}{3}$	0	0	-1	0
Top (?)	$\bar{t}$	$\frac{1}{2}$	$-\frac{2}{3}e$	$-\frac{1}{3}$	0	0	0	-1

for the $u\bar{d}$ pair, $Q = \frac{2}{3}e + \frac{1}{3}e = +1e$, $B = \frac{1}{3} - \frac{1}{3} = 0$, $S = 0 + 0 = 0$, as it must for a π^+). On the other hand a $K^+ = u\bar{s}$ with $Q = +1$, $B = 0$, $S = +1$. Baryons, on the other hand, would consist of three quarks; for example a neutron is $n = ddu$ whereas an antiproton is $\bar{p} = \bar{u}\bar{u}\bar{d}$.

Soon after the quark theory was proposed, physicists began looking for these fractionally charged particles. Although there is indirect experimental evidence in favor of their existence, direct detection of them remains elusive. Indeed, there are suggestions that quarks are so tightly bound together that they may not ever exist in the free state.

In 1964, several physicists proposed that there ought to be a fourth quark. Their argument was based on the expectation that there exists a deep symmetry in nature, including a connection between quarks and leptons. If there are four leptons (as was thought in the 1960s) then symmetry in nature would suggest there should also be four quarks. The fourth quark was said to be *charmed*; its charge would be $+\frac{2}{3}e$ and it would have another property to distinguish it from the other three quarks. This new property, or quantum number, was called **charm** (see Table 43–3). Charm was assumed to be like strangeness: it would be conserved in strong and electromagnetic interactions, but would not be conserved by the weak. The new charmed quark would have charm $C = +1$ and its antiquark $C = -1$.

Experimentally, however, there seemed to be no need for a charmed quark. Before 1974, all known hadrons could be explained as combinations of the three original quarks. In fact, hadrons corresponding to all three quark, and quark–antiquark, combinations had been found. Furthermore the Ω^- baryon had been predicted by the three-quark theory ($\Omega^- = sss$) and was discovered soon after, Fig. 43–7. But in 1974, a new heavy meson was discovered simultaneously by two different groups of experimenters. This new meson, called the J/ψ (often simply the ψ), did not fit the old three-quark scheme. The J/ψ, whose mass is 3100 MeV/c^2, far higher than other known mesons, could also not be an excited state of a smaller-mass meson (into which it would decay) because its lifetime would have to be about 10^{-23} s. In fact, its lifetime was found to be 1000 times greater than this, about 10^{-20} s. It soon became clear that the existence of the J/ψ could be accounted for on the basis of the charmed quark: a J/ψ would be a combination of a charmed quark and its antiquark ($c\bar{c}$). The charm of the J/ψ itself is zero ($C = +1 - 1 = 0$), so it can decay strongly into hadrons (such as several pions, $p\bar{p}$, $\Lambda\bar{\Lambda}$, etc), which is observed. The question of why it lives 1000 times longer than other strongly decaying particles is theorized as being due to the fact that the c and $\bar{c}$ quarks that make it up must each be converted into noncharmed quarks that become the hadrons seen in the decay. This process would inhibit the rate at which the decay could occur, so the charmed-quark model yielded a useful explanation.

Soon after the J/ψ was discovered, a similar meson, called the ψ', was also found whose mass is about 3685 MeV/c^2. Many other ψ-like mesons were also found, and these are all believed to be bound states of a $c\bar{c}$ pair.

Although the J/ψ meson and its relatives have no net charm themselves, it stands to reason that in the decay of one of the larger-mass $c\bar{c}$ combinations there ought to be mesons that do have charm. That is, the c and $\bar{c}$ quarks ought to appear in separate particles and lend a charm to them of $+1$ or -1. In such a decay process, other quarks would also be produced. For example the decay

$$\psi \to D^+ + D^-$$

might be written as

$$\psi = c\bar{c} \to c\bar{c}d\bar{d} \to c\bar{d} + \bar{c}d.$$

The $d\bar{d}$ quarks in the third step are produced from energy; since the d and $\bar{d}$ quarks have opposite quantum numbers and charge, no conservation laws are violated—only energy is required to produce their mass. The $c\bar{d}$ combination

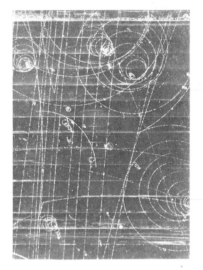

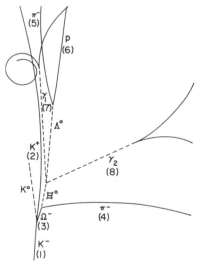

FIGURE 43–7

Liquid-hydrogen bubble-chamber photograph showing the production of a negatively charged omega baryon (Ω^-) by the interaction of a negative K meson (K^-) with a proton (a hydrogen nucleus in the bubble chamber). (Courtesy of Brookhaven National Laboratory.)

which has $Q = +1$, $B = 0$, $S = 0$, $C = +1$, has been dubbed the D^+ meson, and $\bar{c}d$ is its antiparticle, D^-. This meson, along with its neutral sister, D^0, was searched for and believed found in 1977 with a mass around 1870 MeV/c^2. More recent experiments indicate that charmed baryons also exist. Some of the new particles are listed in Table 43–4.

Another important recent development is the experimental evidence for the tau (τ) lepton, with a mass of 1784 MeV/c^2. This lepton, like the electron and muon, presumably has a neutrino associated with it. Thus the family of leptons is at present believed to have six members. This would upset the balance between leptons and quarks, the presumed basic building blocks of matter, unless two new quarks also exist. Indeed, theoretical physicists have postulated the existence of a fifth and sixth quark. These have been named *top* and *bottom* quarks, since they resemble the "up" and "down" quarks. (Some physicists prefer the names *truth* and *beauty* for these t and b quarks.) The names apply also to the new properties (quantum numbers) that distinguish the new quarks from the old quarks. (These are included in Table 43–3.) Indeed, another new meson, Υ, has been detected at about 9400 MeV/c^2, which is considered to be a $b\bar{b}$ combination. There is evidence also for a "bottom" meson (consisting of only one b quark plus a non-b quark). But for the t quark there is no experimental evidence as yet. It is believed that particles containing the t quark will have very large mass, requiring higher-energy accelerators to produce.

43–9   The "Standard Model": Quantum Chromodynamics (QCD) and the Electroweak Theory

Not long after the quark theory was proposed, it was suggested that quarks have another property (or quality) called **color**. The distinction between the five or six quarks [u, d, s, c, b, t (?)] was referred to as **flavor**. According to theory, each of the flavors of quark can have three colors, usually designated red, green, and blue. [Note that the names "color" and "flavor" have nothing to do with our senses, but are purely whimsical—as are other names (like charm) in this new field.] The antiquarks are colored antired, antigreen, and antiblue. Baryons are made up of three quarks, one of each color; mesons consist of a quark–antiquark pair of a particular color and its anticolor; thus baryons and mesons are white or colorless.

Originally, the idea of quark color was proposed to preserve the Pauli exclusion principle (Section 41–7) which applies to particles of spin $\frac{1}{2}$ (or any half-integral spin, like $\frac{3}{2}$, $\frac{5}{2}$, and so on), such as electrons and nucleons. Since quarks have spin $\frac{1}{2}$, they ought to obey the exclusion principle; yet for three particular baryons (uuu, ddd, sss) all three quarks would have the same quantum numbers, and at least two of them have their spin in the same direction

TABLE 43–4

Partial list of hadrons associated with charm and bottomness ($L_e = L_\mu = L_\tau = 0$)

Category	Particle	Antiparticle	Rest Mass (MeV/c^2)	Baryon Number	Strangeness	Charm	Lifetime(s)	Principal Decay Modes
Mesons	D^+	D^-	1869	0	0	+1	9×10^{-13}	K + others, e + others
	D^0	$\overline{D^0}$	1865	0	0	+1	5×10^{-13}	K + others
	F^+	F^-	2021	0	+1	+1	$\sim 2 \times 10^{-13}$	$n\pi$, $n3\pi$
	J/ψ (3100)	Self	3097	0	0	0	$\sim 1 \times 10^{-20}$	Hadrons
	ψ' (3685)	Self	3686	0	0	0	$\sim 3 \times 10^{-20}$	$J/\psi\,\pi$, hadrons
	Υ	Self	9460	0	0	0	$\sim 1 \times 10^{-20}$	$\mu^+\mu^-$, e^+e^-
Baryon	Λ_C^+	Λ_C^-	2280	+1	0	+1	1×10^{-13}	Hadrons

(since there are only two choices, spin up $[m_s = +\frac{1}{2}]$ or spin down $[m_s = -\frac{1}{2}]$). This would seem to violate the exclusion principle, but if quarks have an additional quantum number (color), which could be different for each quark, it would serve to distinguish them and the exclusion principle would hold. Although quark color, and the resulting increase in number of quarks (three-fold), was thus originally an *ad hoc* idea, it also served to bring the theory into better agreement with experiment, such as predicting the correct lifetime of the π^0 meson. The idea of color soon became, in addition, a central feature of the theory as determining the force binding quarks together in a hadron. Each quark is assumed to carry a *color charge*, analogous to electric charge, and the strong force between quarks is often referred to as the **color force**. This new theory of the strong force is called **quantum chromodynamics** (*chrome* = color in Greek), or QCD, to indicate that the force acts between color charges (and not between, say, electric charges). The strong force between hadrons[†] is considered to be a force between the quarks that make them up, as suggested in Fig. 43–8. The particles that transmit the force (analogous to photons for the EM force) are called **gluons** (a play on "glue"). There are eight gluons, according to the theory, all massless, and six of them have color charge.[‡] Thus gluons have replaced mesons (Table 43–1) as the particles mediating the strong (color) force.

The weak force, as we have seen, is thought to be mediated by the W^+, W^-, and Z^0 particles. It acts between the "weak charge" that each particle has. Each elementary particle thus has electric charge, weak charge, color charge, and gravitational mass, although one or more of these could be zero. For example, all leptons have color charge of zero, so they do not interact via the strong force.

To summarize, the latest theories consider the truly elementary particles to be the photon, leptons, quarks, gluons, and $W^\pm$ and Z^0. The photon and the leptons are observed in experiments, and finally so were the W^+ and W^-. But so far only combinations of quarks (baryons, mesons) have been observed, and it seems likely that free quarks are unobservable. On the other hand, some physicists believe that leptons and quarks are not fundamental, but are composites of still more fundamental objects. Only the future can tell.

One important aspect of new theoretical work is the attempt to find a unified basis for the different forces in nature. This was a long-held hope of Einstein, which he was never able to fulfill. A so-called "gauge" theory that unified the weak and electromagnetic interactions was put forward in the early 1960s by Weinberg, Glashow, and Salam. In this **electroweak theory**, the weak and electromagnetic forces are seen as two different manifestations of a single, more fundamental, electroweak interaction. The electroweak theory has had many successes, including the prediction of the $W^\pm$ particles, as intermediates for the weak force, having masses of $82 \pm 2 \ \text{GeV}/c^2$ in excellent agreement with the measured (1983) values of $81 \pm 5 \ \text{GeV}/c^2$. The electroweak theory and QCD for the strong interaction are often referred to today as the "standard model."

43–10 Grand Unified Theories

With the success of the unified electroweak theory, attempts have recently been made also to incorporate it and QCD for the strong (color) force into a so-called **grand unified theory** (GUT). One type of such a grand unified theory of the electromagnetic, weak, and strong forces has been worked out in which there is only one

---

[†] The strong force between hadrons appears feeble, however, in comparison to the force directly between quarks within a hadron.

[‡] Compare to the EM interaction where the photon has no electric charge. Because gluons have color charge, they could attract each other and form composite particles (photons cannot). Such "glueballs" are being searched for and may have been observed.

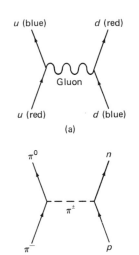

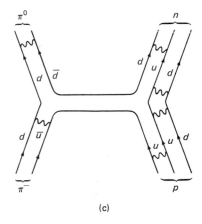

FIGURE 43–8

(a) Force between two quarks holding them together as a proton, say, is carried by a gluon which in this case involves a change in color. (b) Strong interaction $\pi^- \, p \rightarrow \pi^0 \, n$ with the exchange of a charged π meson (+ or − depending on whether it is considered moving to the left or to the right). (c) Quark representation of the same interaction $\pi^- \, p \rightarrow \pi^0 \, n$; the intermediate pion ($d\bar{u}$ or $\bar{d}u$) can be interpreted as the annihilation of a $u\bar{u}$ pair with production of a $d\bar{d}$ pair. The wavy lines between quarks represent gluon exchanges holding the hadrons together.

class of particle—leptons and quarks belonging to the same family and able to change freely from one type to the other—and the three forces are different aspects of a single underlying force. The unity is predicted to occur, however, only on a scale of less than about 10^{-31} m. If two elementary particles (leptons or quarks) approach each other to within this *unification scale*, the apparently fundamental distinction between them would not exist at this level, and a quark could readily change to a lepton, or vice versa. Baryon and lepton numbers would not be conserved. Leptons and quarks would belong to a single family. The weak, electromagnetic, and strong (color) force would blend to a force of a single strength.

How could a lepton become a quark, or vice versa? The theory predicts the existence of particles that can be exchanged between a quark and a lepton that allows one to change into the other, just as the charged pion exchanged between the π^- and p in Fig. 43–8b allows the proton to become a neutron. The mass of the new exchange particles, consistent with the uncertainty principle as applied earlier in this chapter (see Eq. 43–2 with $d \approx 10^{-31}$ m), would be about 10^{15} GeV/c^2, or 10^{15} times the proton mass. With such an incredibly large mass, there is little hope of seeing them in the laboratory. It is also this huge mass which keeps baryon and lepton numbers conserved in observed reactions, since the likelihood of producing such a massive particle, even as a virtual exchange particle, is extremely small at even the highest laboratory energies.

What happens between the unification distance of 10^{-31} m and more normal (larger) distances is referred to as *symmetry breaking*. As an analogy, consider an atom in a crystal. Deep within the atom, there is much symmetry—in the innermost regions the electron cloud is spherically symmetric. Further out, this symmetry breaks down—the electron clouds are distributed preferentially along the lines joining the atoms in the crystal. In a similar way, at 10^{-31} m the force between elementary particles appears as one—it is symmetrical and does not single out one type of "charge" over another. But at larger distances, that symmetry is broken and we see three distinct forces. (In the "standard model" of electroweak interactions, Section 43–9, the symmetry breaking between the electromagnetic and the weak interactions occurs at about 10^{-18} m.)

Since unification occurs at such tiny distances and huge energies, the theory is difficult to test experimentally. But it is not completely impossible. One possibly testable prediction is the basis for the idea, suggested at the end of Section 43–4, that the proton might decay (via, for example, $p \rightarrow \pi^0 e^+$) and violate conservation of baryon number. This could happen if two quarks approached to within 10^{-31} m of each other; but it is very unlikely at normal temperature and energy, so the decay of a proton can only be an unlikely process. In the simplest form of GUT, the theoretical estimate of the proton lifetime is $\approx 10^{31}$ yr, and this has just come within the realm of testability. Proton decays have still not been seen and experiments in 1983 put the lower limit on the proton lifetime to be 6.5×10^{31} yr, almost an order of magnitude greater than this prediction. This may seem a disappointment, but on the other hand, it presents a challenge; indeed more complex GUTs are not affected by this result.

Another interesting prediction of unified theories relates to cosmology. It is thought that during the first 10^{-40} s after the theorized big bang that created the universe, the temperature was so extremely high that particles had energies corresponding to the unification scale; then baryon number would not have been conserved. This could account for the observed predominance of matter ($B > 0$) over antimatter ($B < 0$) in the universe.

This last example is interesting, for it illustrates a deep connection between investigations at either end of the size scale: theories about the tiniest objects (elementary particles) have a strong bearing on the understanding of the universe as a whole.

Even more ambitious than grand unified theories are attempts to incorporate also gravity, and thus unify all four forces in nature into a single theory. The world of elementary particles is opening new vistas. What happens in the near future is bound to be exciting.

Summary

Particle accelerators are used to accelerate charged particles such as electrons and protons to very high energy. High-energy particles have short wavelength and so can be used to probe in great detail; high kinetic energy also allows the creation of new particles through collision (via $E = mc^2$).

Just as the electromagnetic force can be said to be due to an exchange of photons, the strong nuclear force is thought to be carried by *mesons* that have rest mass, or, according to more recent theory, by massless *gluons*.

An *antiparticle* has the same mass as a particle but opposite charge. Certain other properties may also be opposite: for example, the antiproton has *baryon number* (nucleon number) opposite to that for the proton. In all nuclear and particle reactions, the following conservation laws hold: momentum, mass-energy, angular momentum, electric charge, baryon number, and the three lepton numbers. Certain particles have a property, called *strangeness*, which is conserved by the strong force but not by the weak force.

Particles can be classified as *leptons* and *hadrons*, plus the photon. Leptons participate in the weak and electromagnetic interactions. Hadrons participate in the strong interaction as well. The hadrons can be classified as *mesons*, with baryon number zero, and *baryons* with nonzero baryon number.

All particles, except for the photon, electron, neutrinos, and (so far at least) proton, decay with measurable half-lives varying from 10^{-23} s to 10^3 s. The half-life depends on which force is predominant in the decay. Weak decays have half-lives greater than about 10^{-10} s. Electromagnetic decays have half-lives on the order of 10^{-16} to 10^{-19} s. The shortest-lived particles, called *resonances*, decay via the strong interaction and live typically for only about 10^{-23} s.

The latest theories of elementary-particle physics postulate the existence of *quarks* as the basic building blocks of the hadrons. Initially, three quarks were proposed. Recent evidence suggests that a fourth, *charmed*, quark is needed and also a fifth and perhaps a sixth. It is expected that there are the same number of quarks as leptons, and that quarks and leptons are the truly elementary particles. Quarks are said to have *color*, and, according to *quantum chromodynamics*, the strong color force acts between their color charges and is transmitted by *gluons*. A unified theory of forces suggests that at very short distances (10^{-31} m) and very high energy, the weak, electromagnetic, and strong forces appear as a single force, and the fundamental difference between quarks and leptons disappears.

Questions

1. Give a reaction between two nucleons that could produce a π^-, similar to Eq. 43–3.

2. Draw a Feynman diagram for $n + p \rightarrow n + p + \pi^0$.

3. Draw a Feynman diagram for $p\bar{p}$ annihilation with the production of two pions.

4. Draw a Feynman diagram for the photoelectric effect.

5. What would an "antiatom," made up of antiparticles to constituents of normal atoms, consist of? What might happen if *antimatter*, made of such antiatoms, came in contact with our normal world of matter?

6. Why is it that a neutron decays via the weak interaction even though the neutron and one of its decay products (proton) are strongly interacting?

7. Which of the four interactions (strong, electromagnetic, weak, gravitational) does an electron take part in? A neutrino? A proton?

8. Check that charge and baryon number are conserved in each of the decays in Table 43–2.

9. Which of the particle decays in Table 43–2 occur via the electromagnetic interaction?

10. Which of the particle decays in Table 43–2 occur by the weak interaction?

11. Which of the following decays are possible? For those that are forbidden, explain which laws are violated.
    (a) $\Xi^0 \rightarrow \Sigma^+ + \pi^-$
    (b) $\Omega^- \rightarrow \Sigma^0 + \pi^- + \nu$
    (c) $\Sigma \rightarrow \Lambda + \gamma + \gamma$

12. Which of the following reactions are possible, and by what interaction could they occur? For those forbidden, explain why.
    (a) $\pi^- p \rightarrow K^+ \Sigma^-$
    (b) $\pi^+ p \rightarrow K^+ \Sigma^+$
    (c) $\pi^- p \rightarrow \Lambda^0 K^0 \pi^0$
    (d) $\pi^+ p \rightarrow \Sigma^0 \pi^0$
    (e) $\pi^- p \rightarrow p e^- \bar{\nu}_e$
    (f) $\pi^- p \rightarrow K^0 p \pi^0$
    (g) $K^- p \rightarrow \Lambda^0 \pi^0$
    (h) $K^+ n \rightarrow \Sigma^+ \pi^0 \gamma$
    (i) $K^+ \rightarrow \pi^0 \pi^0 \pi^+$
    (j) $\pi^+ \rightarrow e^+ \nu_e$

13. What are the quark combinations that can form a (a) neutron, (b) antineutron, (c) Λ^0, (d) $\overline{\Xi^0}$?

14. What particles do the following quark combinations produce? (a) uud, (b) $\bar{u}\bar{u}\bar{s}$, (c) $\bar{u}s$, (d) $d\bar{u}$, (e) $\bar{c}s$.

15. What is the quark combination needed to produce a D^0 meson ($Q = B = S = 0$, $C = +1$)?

16. The F^+ meson has $Q = S = C = +1$, $B = 0$. What quark combination would produce it?

17. Draw a possible quark Feynman diagram (see Fig. 43–8c) for the reaction $K^- p \rightarrow K^- p$.

Problems

SECTION 43–1

1. (I) Calculate the wavelength of 30 GeV electrons.

2. (I) What is the total energy of a proton whose kinetic energy is 25 GeV? What is its wavelength?

3. (I) What is the wavelength, and maximum resolving power attainable, using 400-GeV protons at Fermilab?

SECTION 43–2

4. (II) What minimum kinetic energy must a neutron and proton each have if they are traveling at the same speed toward each other, collide, and produce a K^+K^- pair in addition to themselves? (See Table 43–2.)

5. (II) The mass of a π^0 can be measured by observing the reaction $\pi^- + p \to \pi^0 + n$ at very low incident π^- kinetic energy (assume it is zero). The neutron is observed to be emitted with a KE of 0.60 MeV. Use conservation of energy and momentum to determine the π^0 mass.

6. (III) Could a π^+ meson be produced if a 100-MeV proton struck a proton at rest? What minimum KE must it have?

SECTION 43–3

7. (I) (a) How much energy is released when an electron and a positron annihilate each other? (b) How much energy is released when a proton and an antiproton annihilate each other?

8. (I) How much energy is required to produce a neutron–antineutron pair?

9. (II) What are the wavelengths of the two photons when a proton–antiproton pair at rest annihilate?

SECTION 43–5

10. (I) How much energy is released in the decay $\pi^+ \to \mu^+ + \nu_\mu$?

11. (I) About how much energy is released when a Λ^0 decays? (See Table 43–2.)

12. (II) What would be the wavelengths of the two photons produced when an electron–positron pair, each with 300 keV of KE, annihilate?

13. (II) In the rare decay $\pi^+ \to e^+ + \nu_e$, what is the kinetic energy of the positron?

SECTIONS 43–6 TO 43–10

14. (I) The measured width of the ψ meson is 63 keV. Estimate its lifetime.

15. (I) The measured width of the ψ' meson is 215 keV. Estimate its lifetime.

16. (II) (a) Show that the so-called unification distance of 10^{-31} m in recent unified theory is equivalent to an energy of about 10^{15} GeV. Use either the uncertainty principle or de Broglie's wavelength formula and explain how they apply. (b) Calculate what temperature this corresponds to.

17. (II) Symmetry breaking occurs in the electroweak theory at about 10^{-18} m. Show that this corresponds to an energy which is on the order of the mass of the $W^\pm$.

18. (II) Draw possible Feynman diagrams using quarks (as in Fig. 43–8c) for the reactions (a) $pn \to pn$, (b) $\bar{p}p \to \pi^+\pi^-$.

Summary

Particle accelerators are used to accelerate charged particles such as electrons and protons to very high energy. High-energy particles have short wavelength and so can be used to probe in great detail; high kinetic energy also allows the creation of new particles through collision (via $E = mc^2$).

Just as the electromagnetic force can be said to be due to an exchange of photons, the strong nuclear force is thought to be carried by *mesons* that have rest mass, or, according to more recent theory, by massless *gluons*.

An *antiparticle* has the same mass as a particle but opposite charge. Certain other properties may also be opposite: for example, the antiproton has *baryon number* (nucleon number) opposite to that for the proton. In all nuclear and particle reactions, the following conservation laws hold: momentum, mass-energy, angular momentum, electric charge, baryon number, and the three lepton numbers. Certain particles have a property, called *strangeness*, which is conserved by the strong force but not by the weak force.

Particles can be classified as *leptons* and *hadrons*, plus the photon. Leptons participate in the weak and electromagnetic interactions. Hadrons participate in the strong interaction as well. The hadrons can be classified as *mesons*, with baryon number zero, and *baryons* with nonzero baryon number.

All particles, except for the photon, electron, neutrinos, and (so far at least) proton, decay with measurable half-lives varying from 10^{-23} s to 10^3 s. The half-life depends on which force is predominant in the decay. Weak decays have half-lives greater than about 10^{-10} s. Electromagnetic decays have half-lives on the order of 10^{-16} to 10^{-19} s. The shortest-lived particles, called *resonances*, decay via the strong interaction and live typically for only about 10^{-23} s.

The latest theories of elementary-particle physics postulate the existence of *quarks* as the basic building blocks of the hadrons. Initially, three quarks were proposed. Recent evidence suggests that a fourth, *charmed*, quark is needed and also a fifth and perhaps a sixth. It is expected that there are the same number of quarks as leptons, and that quarks and leptons are the truly elementary particles. Quarks are said to have *color*, and, according to *quantum chromodynamics*, the strong color force acts between their color charges and is transmitted by *gluons*. A unified theory of forces suggests that at very short distances (10^{-31} m) and very high energy, the weak, electromagnetic, and strong forces appear as a single force, and the fundamental difference between quarks and leptons disappears.

 Questions

1. Give a reaction between two nucleons that could produce a π^-, similar to Eq. 43–3.

2. Draw a Feynman diagram for $n + p \to n + p + \pi^0$.

3. Draw a Feynman diagram for $p\bar{p}$ annihilation with the production of two pions.

4. Draw a Feynman diagram for the photoelectric effect.

5. What would an "antiatom," made up of antiparticles to constituents of normal atoms, consist of? What might happen if *antimatter*, made of such antiatoms, came in contact with our normal world of matter?

6. Why is it that a neutron decays via the weak interaction even though the neutron and one of its decay products (proton) are strongly interacting?

7. Which of the four interactions (strong, electromagnetic, weak, gravitational) does an electron take part in? A neutrino? A proton?

8. Check that charge and baryon number are conserved in each of the decays in Table 43–2.

9. Which of the particle decays in Table 43–2 occur via the electromagnetic interaction?

10. Which of the particle decays in Table 43–2 occur by the weak interaction?

11. Which of the following decays are possible? For those that are forbidden, explain which laws are violated.
   (a) $\Xi^0 \to \Sigma^+ + \pi^-$
   (b) $\Omega^- \to \Sigma^0 + \pi^- + v$
   (c) $\Sigma \to \Lambda + \gamma + \gamma$

12. Which of the following reactions are possible, and by what interaction could they occur? For those forbidden, explain why.
   (a) $\pi^- p \to K^+ \Sigma^-$
   (b) $\pi^+ p \to K^+ \Sigma^+$
   (c) $\pi^- p \to \Lambda^0 K^0 \pi^0$
   (d) $\pi^+ p \to \Sigma^0 \pi^0$
   (e) $\pi^- p \to p e^- \bar{v}_e$
   (f) $\pi^- p \to K^0 p \pi^0$
   (g) $K^- p \to \Lambda^0 \pi^0$
   (h) $K^+ n \to \Sigma^+ \pi^0 \gamma$
   (i) $K^+ \to \pi^0 \pi^0 \pi^+$
   (j) $\pi^+ \to e^+ v_e$

13. What are the quark combinations that can form a (a) neutron, (b) antineutron, (c) Λ^0, (d) $\overline{\Xi^0}$?

14. What particles do the following quark combinations produce? (a) uud, (b) $\bar{u}\bar{u}\bar{s}$, (c) $\bar{u}s$, (d) $d\bar{u}$, (e) $\bar{c}s$.

15. What is the quark combination needed to produce a D^0 meson ($Q = B = S = 0$, $C = +1$)?

16. The F^+ meson has $Q = S = C = +1$, $B = 0$. What quark combination would produce it?

17. Draw a possible quark Feynman diagram (see Fig. 43–8c) for the reaction $K^- p \to K^- p$.

Problems _____

SECTION 43–1

1. (I) Calculate the wavelength of 30 GeV electrons.

2. (I) What is the total energy of a proton whose kinetic energy is 25 GeV? What is its wavelength?

3. (I) What is the wavelength, and maximum resolving power attainable, using 400-GeV protons at Fermilab?

SECTION 43–2

4. (II) What minimum kinetic energy must a neutron and proton each have if they are traveling at the same speed toward each other, collide, and produce a K^+K^- pair in addition to themselves? (See Table 43–2.)

5. (II) The mass of a π^0 can be measured by observing the reaction $\pi^- + p \rightarrow \pi^0 + n$ at very low incident π^- kinetic energy (assume it is zero). The neutron is observed to be emitted with a KE of 0.60 MeV. Use conservation of energy and momentum to determine the π^0 mass.

6. (III) Could a π^+ meson be produced if a 100-MeV proton struck a proton at rest? What minimum KE must it have?

SECTION 43–3

7. (I) (*a*) How much energy is released when an electron and a positron annihilate each other? (*b*) How much energy is released when a proton and an antiproton annihilate each other?

8. (I) How much energy is required to produce a neutron–antineutron pair?

9. (II) What are the wavelengths of the two photons when a proton–antiproton pair at rest annihilate?

SECTION 43–5

10. (I) How much energy is released in the decay $\pi^+ \rightarrow \mu^+ + \nu_\mu$?

11. (I) About how much energy is released when a Λ^0 decays? (See Table 43–2.)

12. (II) What would be the wavelengths of the two photons produced when an electron–positron pair, each with 300 keV of KE, annihilate?

13. (II) In the rare decay $\pi^+ \rightarrow e^+ + \nu_e$, what is the kinetic energy of the positron?

SECTIONS 43–6 TO 43–10

14. (I) The measured width of the ψ meson is 63 keV. Estimate its lifetime.

15. (I) The measured width of the ψ' meson is 215 keV. Estimate its lifetime.

16. (II) (*a*) Show that the so-called unification distance of 10^{-31} m in recent unified theory is equivalent to an energy of about 10^{15} GeV. Use either the uncertainty principle or de Broglie's wavelength formula and explain how they apply. (*b*) Calculate what temperature this corresponds to.

17. (II) Symmetry breaking occurs in the electroweak theory at about 10^{-18} m. Show that this corresponds to an energy which is on the order of the mass of the $W^\pm$.

18. (II) Draw possible Feynman diagrams using quarks (as in Fig. 43–8c) for the reactions (*a*) $pn \rightarrow pn$, (*b*) $\bar{p}p \rightarrow \pi^+\pi^-$.

Appendix A
Mathematical Formulas

a QUADRATIC FORMULA

If

$$ax^2 + bx + c = 0$$

then

$$x = \frac{-b \pm \sqrt{b^2 - 4ac}}{2a}$$

b BINOMIAL EXPANSION

$$(1 + x)^n = 1 + nx + \frac{n(n-1)}{2!}x^2 + \frac{n(n-1)(n-2)}{3!}x^3 + \cdots$$

$$(x + y)^n = x^n\left(1 + \frac{y}{x}\right)^n = x^n\left[1 + n\frac{y}{x} + \frac{n(n-1)}{2!}\frac{y^2}{x^2} + \cdots\right]$$

c OTHER EXPANSIONS

$$e^x = 1 + x + \frac{x^2}{2!} + \frac{x^3}{3!} + \cdots$$

$$\ln(1 + x) = x - \frac{x^2}{2} + \frac{x^3}{3} - \frac{x^4}{4} + \cdots$$

$$\sin\theta = \theta - \frac{\theta^3}{3!} + \frac{\theta^5}{5!} - \cdots$$

$$\cos\theta = 1 - \frac{\theta^2}{2!} + \frac{\theta^4}{4!} - \cdots$$

In general:

$$f(x) = f(0) + \left(\frac{df}{dx}\right)_0 x + \left(\frac{d^2f}{dx^2}\right)_0 \frac{x^2}{2!} + \cdots$$

d AREAS AND VOLUMES

Object	Surface Area	Volume
Circle, radius r	πr^2	—
Sphere, radius r	$4\pi r^2$	$\frac{4}{3}\pi r^3$
Right circular cylinder, radius r, height h	$2\pi r^2 + 2\pi rh$	$\pi r^2 h$

e TRIGONOMETRIC FUNCTIONS

(See diagram)

$$\sin\theta = \frac{o}{h} \qquad\qquad \csc\theta = \frac{1}{\sin\theta} = \frac{h}{o}$$

$$\cos\theta = \frac{a}{h} \qquad\qquad \sec\theta = \frac{1}{\cos\theta} = \frac{h}{a}$$

$$\tan\theta = \frac{o}{a} = \frac{\sin\theta}{\cos\theta} \qquad \cot\theta = \frac{1}{\tan\theta} = \frac{a}{o}$$

$$a^2 + o^2 = h^2 \qquad \text{(Pythagorean theorem)}$$

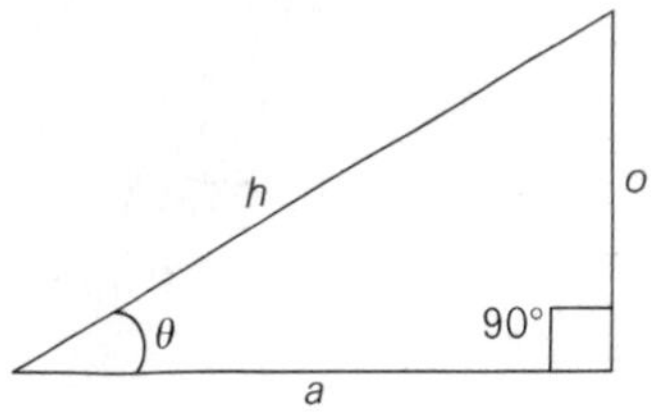

f TRIGONOMETRIC IDENTITIES

$$\sin^2\theta + \cos^2\theta = 1 \qquad \sec^2\theta - \tan^2\theta = 1 \qquad \csc^2\theta - \cot^2\theta = 1$$

$$\sin 2\theta = 2\sin\theta\cos\theta$$

$$\cos 2\theta = \cos^2\theta - \sin^2\theta = 2\cos^2\theta - 1 = 1 - 2\sin^2\theta$$

$$\tan 2\theta = \frac{2\tan\theta}{1 - \tan^2\theta}$$

$$\sin(A \pm B) = \sin A\cos B \pm \cos A\sin B$$

$$\cos(A \pm B) = \cos A\cos B \mp \sin A\sin B$$

$$\tan(A \pm B) = \frac{\tan A \pm \tan B}{1 \mp \tan A\tan B}$$

$$\sin\tfrac{1}{2}\theta = \sqrt{\frac{1 - \cos\theta}{2}} \qquad \cos\tfrac{1}{2}\theta = \sqrt{\frac{1 + \cos\theta}{2}} \qquad \tan\tfrac{1}{2}\theta = \sqrt{\frac{1 - \cos\theta}{1 + \cos\theta}}$$

$$\sin A \pm \sin B = 2\sin\left(\frac{A \pm B}{2}\right)\cos\left(\frac{A \mp B}{2}\right)$$

Appendix B
Derivatives and Integrals

---

a DERIVATIVES: GENERAL RULES

$$\frac{dx}{dx} = 1$$

$$\frac{d}{dx}\left[af(x)\right] = a\frac{df}{dx} \qquad (a = \text{constant})$$

$$\frac{d}{dx}\left[f(x) + g(x)\right] = \frac{df}{dx} + \frac{dg}{dx}$$

$$\frac{d}{dx}\left[f(x)g(x)\right] = \frac{df}{dx}g + f\frac{dg}{dx}$$

$$\frac{d}{dx}\left[f(y)\right] = \frac{df}{dy}\frac{dy}{dx} \qquad [\text{chain rule}]$$

$$\frac{dx}{dy} = \frac{1}{\left(\dfrac{dy}{dx}\right)} \qquad \text{if} \quad \frac{dy}{dx} \neq 0.$$

b DERIVATIVES: PARTICULAR FUNCTIONS

$$\frac{da}{dx} = 0 \qquad (a = \text{constant})$$

$$\frac{d}{dx}\,x^n = nx^{n-1}$$

$$\frac{d}{dx}\sin ax = a\cos ax$$

$$\frac{d}{dx}\cos ax = -a\sin ax$$

$$\frac{d}{dx}\tan ax = \sec^2 ax$$

$$\frac{d}{dx}\ln ax = \frac{1}{x}$$

$$\frac{d}{dx}\,e^{ax} = ae^{ax}$$

c INDEFINITE INTEGRALS: GENERAL RULES

$$\int dx = x$$

$$\int af(x)\,dx = a \int f(x)\,dx \quad (a = \text{constant})$$

$$\int [f(x) + g(x)]\,dx = \int f(x)\,dx + \int g(x)\,dx$$

$$\int u\,dv = uv - \int v\,du \quad (\text{integration by parts})$$

d INDEFINITE INTEGRALS: PARTICULAR FUNCTIONS

(An arbitrary constant can be added to the right side of each equation.)

$$\int a\,dx = ax \quad (a = \text{constant})$$

$$\int x^m\,dx = \frac{1}{m+1} x^{m+1} \quad (m \neq -1)$$

$$\int \sin ax\,dx = -\frac{1}{a} \cos ax$$

$$\int \cos ax\,dx = \frac{1}{a} \sin ax$$

$$\int \tan ax\,dx = \frac{1}{a} \ln |\sec ax|$$

$$\int \frac{1}{x}\,dx = \ln x$$

$$\int e^{ax}\,dx = \frac{1}{a} e^{ax}$$

$$\int \frac{dx}{x^2 + a^2} = \frac{1}{a} \tan^{-1} \frac{x}{a}$$

$$\int \frac{dx}{x^2 - a^2} = \frac{1}{2a} \ln \left(\frac{x-a}{x+a} \right) \quad (x^2 > a^2)$$

$$= -\frac{1}{2a} \ln \left(\frac{a+x}{a-x} \right) \quad (x^2 < a^2)$$

Appendix C
Polar Coordinates

It is often convenient to use polar coordinates, r and θ, particularly for circular motion: r is the distance of a point from the origin (equal to the magnitude of the position vector, **r**), and θ is the angle the position vector makes with the x axis, Fig. C–1. The position of a particle in a plane can be specified at any moment by giving either its x and y coordinates, or by giving r and θ. These coordinates are related (see Fig. C–1) by

$$x = r \cos \theta, \qquad y = r \sin \theta,$$

or, in reverse, by

$$r = \sqrt{x^2 + y^2}, \qquad \tan \theta = y/x.$$

Just as we found the unit vectors **i** and **j** useful for rectangular coordinates, so too in polar coordinates we define two unit vectors $\hat{\mathbf{r}}$ and $\hat{\boldsymbol{\theta}}$. The unit vector $\hat{\mathbf{r}}$ at any point has magnitude equal to one, and points in the direction of increasing **r** at that point; $\hat{\boldsymbol{\theta}}$ has magnitude 1 and points in the direction of increasing θ; see Fig. C–1. Notice that although **i** and **j** are truly constant, $\hat{\mathbf{r}}$ and $\hat{\boldsymbol{\theta}}$ are constant in magnitude only ($=$ one unit) but their directions vary from point to point.

As an example, we treat the case of a particle moving in a circle (Sections 3–9 and 3–10). We can write the velocity and acceleration vectors for the particle in terms of these unit vectors $\hat{\mathbf{r}}$ and $\hat{\boldsymbol{\theta}}$. Since **v** is always tangent to the circle,

$$\mathbf{v} = v\hat{\boldsymbol{\theta}}.$$

From our discussion in Section 3–10 (see Eqs. 3–25 and 3–26, and Figs. 3–20 and C–1) the acceleration is

$$\mathbf{a} = a_c\hat{\mathbf{r}} + a_t\hat{\boldsymbol{\theta}}$$

$$= -\frac{v^2}{r}\hat{\mathbf{r}} + \frac{dv}{dt}\hat{\boldsymbol{\theta}}. \qquad (C-1)$$

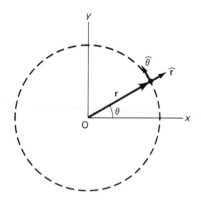

FIGURE C–1
Polar coordinates related to Cartesian coordinates.

The first term is the centripetal acceleration and the second term is the tangential acceleration; the former has a minus sign because it points toward the center of the circle, opposite to the direction of $\hat{\mathbf{r}}$.

Notice that if motion occurs at constant speed, $a_t = dv/dt = 0$, and we obtain the result of Section 3–9.

For motion in a curve which is not a circle, the formulas for a_t and a_c (Eq. 3–26 or C–1) can still be used if for r we use the radius of curvature of the path at each particular point. In this case a_c is the component of the acceleration perpendicular to the path at a given moment, and a_t is the component tangent to the path at that moment.

Appendix D
Selected Isotopes[†]

(1) Atomic Number Z	(2) Element	(3) Symbol	(4) Mass Number, A	(5) Atomic Mass[‡]	(6) Percent Abundance, Or Decay Mode If Radioactive	(7) Half-life (If Radioactive)
0	(Neutron)	n	1	1.008665	β^-	10.6 min
1	Hydrogen	H	1	1.007825	99.985	
	Deuterium	D	2	2.014102	0.015	
	Tritium	T	3	3.016049	β^-	12.33 yr
2	Helium	He	3	3.016029	0.00014	
			4	4.002603	≈ 100	
3	Lithium	Li	6	6.015123	7.5	
			7	7.016005	92.5	
4	Beryllium	Be	7	7.016930	EC, γ	53.3 days
			9	9.012183	100	
5	Boron	B	10	10.012938	19.8	
			11	11.009305	80.2	
6	Carbon	C	11	11.011433	β^+, EC	20.4 min
			12	12.000000	98.89	
			13	13.003355	1.11	
			14	14.003242	β^-	5730 yr
7	Nitrogen	N	13	13.005739	β^+	9.96 min
			14	14.003074	99.63	
			15	15.000109	0.37	
8	Oxygen	O	15	15.003065	β^+, EC	122 s
			16	15.994915	99.76	
			18	17.999159	0.204	
9	Fluorine	F	19	18.998403	100	
10	Neon	Ne	20	19.992439	90.51	
			22	21.991384	9.22	
11	Sodium	Na	22	21.994435	β^+, EC, γ	2.602 yr
			23	22.989770	100	
			24	23.990964	β^-, γ	15.0 h
12	Magnesium	Mg	24	23.985045	78.99	
13	Aluminum	Al	27	26.981541	100	

[†] Data are taken from *Chart of the Nuclides,* 12th ed., 1977, and from C. M. Lederer and V. S. Shirley, eds., *Table of Isotopes,* 7th ed., John Wiley & Sons, Inc., New York, 1978.
[‡] The masses given in column (5) are those for the neutral atom, including the Z electrons.

(1) Atomic Number Z	(2) Element	(3) Symbol	(4) Mass Number, A	(5) Atomic Mass‡	(6) Percent Abundance, Or Decay Mode If Radioactive	(7) Half-life (If Radioactive)
14	Silicon	Si	28	27.976928	92.23	
			31	30.975364	β^-, γ	2.62 h
15	Phosphorus	P	31	30.973763	100	
			32	31.973908	β^-	14.28 days
16	Sulfur	S	32	31.972072	95.0	
			35	34.969033	β^-	87.4 days
17	Chlorine	Cl	35	34.968853	75.77	
			37	36.965903	24.23	
18	Argon	Ar	40	39.962383	99.60	
19	Potassium	K	39	38.963708	93.26	
			40	39.964000	$\beta^-, EC, \gamma, \beta^+$	1.28×10^9 yr
20	Calcium	Ca	40	39.962591	96.94	
21	Scandium	Sc	45	44.955914	100	
22	Titanium	Ti	48	47.947947	73.7	
23	Vanadium	V	51	50.943963	99.75	
24	Chromium	Cr	52	51.940510	83.79	
25	Manganese	Mn	55	54.938046	100	
26	Iron	Fe	56	55.934939	91.8	
27	Cobalt	Co	59	58.933198	100	
			60	59.933820	β^-, γ	5.271 yr
28	Nickel	Ni	58	57.935347	68.3	
			60	59.930789	26.1	
29	Copper	Cu	63	62.929599	69.2	
			65	64.927792	30.8	
30	Zinc	Zn	64	63.929145	48.6	
			66	65.926035	27.9	
31	Gallium	Ga	69	68.925581	60.1	
32	Germanium	Ge	72	71.922080	27.4	
			74	73.921179	36.5	
33	Arsenic	As	75	74.921596	100	
34	Selenium	Se	80	79.916521	49.8	
35	Bromine	Br	79	78.918336	50.69	
36	Krypton	Kr	84	83.911506	57.0	
37	Rubidium	Rb	85	84.911800	72.17	
38	Strontium	Sr	86	85.909273	9.8	
			88	87.905625	82.6	
			90	89.907746	β^-	28.8 yr
39	Yttrium	Y	89	88.905856	100	
40	Zirconium	Zr	90	89.904708	51.5	
41	Niobium	Nb	93	92.906378	100	
42	Molybdenum	Mo	98	97.905405	24.1	
43	Technetium	Tc	98	97.907210	β^-, γ	4.2×10^6 yr
44	Ruthenium	Ru	102	101.904348	31.6	
45	Rhodium	Rh	103	102.90550	100	
46	Palladium	Pd	106	105.90348	27.3	
47	Silver	Ag	107	106.905095	51.83	
			109	108.904754	48.17	
48	Cadmium	Cd	114	113.903361	28.7	
49	Indium	In	115	114.90388	95.7; β^-	5.1×10^{14} yr
50	Tin	Sn	120	119.902199	32.4	

(1) Atomic Number Z	(2) Element	(3) Symbol	(4) Mass Number, A	(5) Atomic Mass‡	(6) Percent Abundance, Or Decay Mode If Radioactive	(7) Half-life (If Radioactive)
51	Antimony	Sb	121	120.903824	57.3	
52	Tellurium	Te	130	129.90623	34.5; β^-	2×10^{21} yr
53	Iodine	I	127	126.904477	100	
			131	130.906118	β^-, γ	8.04 days
54	Xenon	Xe	132	131.90415	26.9	
			136	135.90722	8.9	
55	Cesium	Cs	133	132.90543	100	
56	Barium	Ba	137	136.90582	11.2	
			138	137.90524	71.7	
57	Lanthanum	La	139	138.90636	99.911	
58	Cerium	Ce	140	139.90544	88.5	
59	Praseodymium	Pr	141	140.90766	100	
60	Neodymium	Nd	142	141.90773	27.2	
61	Promethium	Pm	145	144.91275	EC, α, γ	17.7 yr
62	Samarium	Sm	152	151.91974	26.6	
63	Europium	Eu	153	152.92124	52.1	
64	Gadolinium	Gd	158	157.92411	24.8	
65	Terbium	Tb	159	158.92535	100	
66	Dysprosium	Dy	164	163.92918	28.1	
67	Holmium	Ho	165	164.93033	100	
68	Erbium	Er	166	165.93031	33.4	
69	Thulium	Tm	169	168.93423	100	
70	Ytterbium	Yb	174	173.93887	31.6	
71	Lutecium	Lu	175	174.94079	97.39	
72	Hafnium	Hf	180	179.94656	35.2	
73	Tantalum	Ta	181	180.94801	99.988	
74	Tungsten (wolfram)	W	184	183.95095	30.7	
75	Rhenium	Re	187	186.95577	62.60, β^-	4×10^{10} yr
76	Osmium	Os	191	190.96094	β^-, γ	15.4 days
			192	191.96149	41.0	
77	Iridium	Ir	191	190.96060	37.3	
			193	192.96294	62.7	
78	Platinum	Pt	195	194.96479	33.8	
79	Gold	Au	197	196.96656	100	
80	Mercury	Hg	202	201.97063	29.8	
81	Thallium	Tl	205	204.97441	70.5	
82	Lead	Pb	206	205.97446	24.1	
			207	206.97589	22.1	
			208	207.97664	52.3	
			210	209.98418	α, β^-, γ	22.3 yr
			211	210.98874	β^-, γ	36.1 min
			212	211.99188	β^-, γ	10.64 h
			214	213.99980	β^-, γ	26.8 min
83	Bismuth	Bi	209	208.98039	100	
			211	210.98726	α, β^-, γ	2.15 min
84	Polonium	Po	210	209.98286	α, γ	138.38 days
			214	213.99519	α, γ	164 μs
85	Astatine	At	218	218.00870	α, β^-	≈ 2 s
86	Radon	Rn	222	222.017574	α, γ	3.8235 days
87	Francium	Fr	223	223.019734	α, β^-, γ	21.8 min

(1) Atomic Number Z	(2) Element	(3) Symbol	(4) Mass Number, A	(5) Atomic Mass‡	(6) Percent Abundance, Or Decay Mode If Radioactive	(7) Half-life (If Radioactive)
88	Radium	Ra	226	226.025406	α, γ	1.60×10^3 yr
89	Actinium	Ac	227	227.027751	α, β^-, γ	21.773 yr
90	Thorium	Th	228	228.02873	α, γ	1.9131 yr
			232	232.038054	100, α, γ	1.41×10^{10} yr
91	Protactinium	Pa	231	231.035881	α, γ	3.28×10^4 yr
92	Uranium	U	232	232.03714	α, γ	72 yr
			233	233.039629	α, γ	1.592×10^5 yr
			235	235.043925	0.72; α, γ	7.038×10^8 yr
			236	236.045563	α, γ	2.342×10^7 yr
			238	238.050786	99.275; α, γ	4.468×10^9 yr
			239	239.054291	β^-, γ	23.5 min
93	Neptunium	Np	239	239.052932	β^-, γ	2.35 days
94	Plutonium	Pu	239	239.052158	α, γ	2.41×10^4 yr
95	Americium	Am	243	243.061374	α, γ	7.37×10^3 yr
96	Curium	Cm	245	245.065487	α, γ	8.5×10^3 yr
97	Berkelium	Bk	247	247.07003	α, γ	1.4×10^3 yr
98	Californium	Cf	249	249.074849	α, γ	351 yr
99	Einsteinium	Es	254	254.08802	α, γ, β^-	276 days
100	Fermium	Fm	253	253.08518	EC, α, γ	3.0 days
101	Mendelevium	Md	255	255.0911	EC, α	27 min
102	Nobelium	No	255	255.0933	EC, α	3.1 min
103	Lawrencium	Lr	257	257.0998	α	≈ 35 s
104	Rutherfordium (?)	Rf	261	261.1087	α	1.1 min
105	Hahnium (?)	Ha	262	262.1138	α	0.7 min
106			263	263.1184	α	0.9 s

Answers to Odd-Numbered Problems

CHAPTER 1

1 1%
3 $(1.4 \pm 0.4) \times 10^{10}$ cm²
5 (a) 1 MV, (b) 1 μm, (c) 40 megadays, (d) 2 kilobucks, (e) 2 nanopieces
7 1 km = 0.621 mi, 1 km/h = 0.621 mi/h.
9 3.3×10^{-10} ft
11 9%
13 (a) 9.5×10^{15} m, (b) 6.3×10^4, (c) 7.2 AU/h
15 m/s⁴, m/s²
17 $T \propto \sqrt{m/k}$

CHAPTER 2

1 50 m
3 (a) 89 km/h, (b) 25 m/s, (c) 81 ft/s
5 (a) 0.19 mi/min, (b) 0
7 (a) $t = 0$ to $t = 20$s, (b) $t \approx 28$s, (c) $t \approx 37$s, (d) both
9 920 km/h
11 3500 cars/h
13 (a)13 m/s, (b) 4.5 m/s away from master
15 (a) 7.5 m, 10.8 m, 11.9 m, (b) 2.2 m/s, (c) 2.2 m/s, 0 m/s
17 3.3 min, 5.0 km; 25 s, 0.62 km
19 4.2 m/s
21 (a) 24 Bt, (b) $A + 300 B$, 120 B, (c) $A - 3 Bt^{-4}$
23 (a) at $t(s) = .12, .37, .62, .87, 1.25, 1.75$, etc., $v(m/s) = 0.44, 1.40, 2.40, 3.52, 5.36, 7.86, 10.48, 13.14, 15.90, 18.68, 21.44, 23.86, 25.92, 27.80$; (b) at $t(s) = .06, .25, .50, .75, 1.06, 1.50, 2.00$, etc., $a(m/s^2) = 3.52, 3.84, 4.00, 4.48, 4.84, 5.00, 5.24, 5.32, 5.52, 5.56, 5.52, 4.84, 4.12, 3.76$
25 -2.6 m/s²
27 391 m
29 (a) 170 m, (b) 14 s, (c) 24 m, 21 m
31 14 g
33 (a) 48 m, (b) 33 m
35 (b) 3.27 s

CHAPTER 3

1 Resultant ≈ 6.3 m, 5° S of E
5 1.5
7 (a) 843 km/h N, 537 km/h W; (b) 2530 km N, 1610 km E
9 (7,4,5), 9.5
11 (25 m, 18 m, 36 m), $|\mathbf{D}| = 47$ m
13 (a) 6.7, 27° (b) 4.7, 122°, (c) 7.8, 63°, (d) 8.6, 173°
15 10.5 km/h
17 $v_T/\tan \theta$
19 16.0 km/h
21 (a) 93 m, (b) 110 s
23 32 km/h, 48° N of W
25 43.5° N of E
27 head boat 25° upstream; 37 min
29 Parabola in xz plane at $y = 6.05$
31 (a) 20.0 m/s, 30° N of E, (b) 4.44 m/s², 30° S of E, (c) 27.3 m/s; must assume $\mathbf{a}$ = constant.
33 (a) $(-\sin 3.0 \, t\mathbf{i} + \cos 3.0 \, t\mathbf{j})(18 \text{ m/s})$, (b) $(-\cos 3.0 \, t\mathbf{i} - \sin 3.0 \, t\mathbf{j})(54 \text{ m/s}^2)$, (c) circle of radius 6.0 m, (d) $r = -a/9.0, 180°$
35 13 m
37 10.0 m/s
39 12.9 m
41 7.1 s

CHAPTER 4

1 975 N
3 0.020 N
5 3.9×10^2 N
7 2.1×10^4 N; 0.82 m
9 2.0×10^2 N
11 accelerate downward at 1.8 m/s²
13 (a) 9.6 m/s, (b) 3.6×10^3 N upward
15 6.5 N upward
17 (a) 7.4×10^2 N, (b) 1.3×10^2 N, (c) 6.5×10^2 N, (d) 0
19 (a) 78 N, (b) 2.2×10^2 N, (c) 86 N
21 (a) 4.4 m/s², (b) 17 N
23 46 N in top rope, 23 N in bottom rope
25 3.9 m
27 (a) $g(2y - L)/L$, (b) $\sqrt{2gy_0(1 - (y_0/L))}$, (c) $\frac{2}{3}\sqrt{gL}$
29 0.69
31 37 N, 0.54
33 -7.84 m/s²
35 10.0 kg
37 1.5 m
39 26 m/s
41 (b) 37 m, (c) 220 m
43 12 m/s
45 5.4 m/s²
47 No

39 108 km/h
41 (a) 3.6 s, (b) 36 m/s
43 6.4 m/s
45 5.8 s
47 -14 m/s²
51 (a) 4.8 s, (b) 37 m/s
53 (a) 8.43 m/s, (b) 0.93 s or 2.65 s, (c) first is on the way up, second on the way down
55 56 m
57 9.1 m/s
61 140 m
63 (a) $3.2 (e^{2t} - 1)$, (b) $1.6 (e^{2t} - 1)$, (c) $a = 2v + 6.4$, (d) 2.1×10^3 m, 4.3×10^3 m/s
65 (a) $v = (1 - e^{-kt}) g/k$, (b) g/k

43 22 m
47 $(v_0 \cos \theta_0/g)(v_0 \sin \theta \pm [v_0^2 \sin^2 \theta_0 - 2gh]^{1/2})$
49 (a) 60 m, (b) 56 m, (c) 34 m/s, $-75°$
51 (a) 3.4 m/s, 48°, (b) 0.32 m above board, (c) 10.5 m/s, 77°
53 $\theta = \phi/2 + \pi/4$
55 8.5 m/s, 118°
57 0.54 m/s²
59 5.2×10^{-3} m/s²
61 3.36×10^{-2} m/s², $3.4 \times 10^{-3} g$
63 (a) 1.86 m/s, (b) 5.50 m/s²
65 (a_t, a_c) (a) (7.0 m/s², 0), (b) (7.0 m/s², 25 m/s²), (c) (7.0 m/s², 98 m/s²)

49 1.6 m/s^2
51 (a) 26 N·s/m^2, (b) 120 N
53 (a) This force opposes the motion

CHAPTER 5

1 (a) 5.90×10^{-3} m/s^2, (b) 3.53×10^{22} N gravitational force of sun
3 0.39
5 (a) 0.29 m/s^2, (b) 7.2 N
9 0.10
11 0.16 N at top, 6.73 N at bottom
13 (a) on the part farthest from the axis of rotation, (b) 1.1×10^3 rev/day
15 6.1×10^3 N down embankment
17 920 N
19 2.64×10^3 km
21 3.3×10^{-8} N toward center of square
25 2.0×10^{30} kg
27 (b) g decreases as r increases, (c) 9.49 m/s^2
31 6.5×10^3 m/s
33 (a) 540 N, (b) 540 N, (c) 720 N, (d) 360 N, (e) 0
35 (a) 18 N toward moon, (b) 2.2×10^2 N away from moon
37 1.4 h
39 (b) 9.6×10^{26} kg
41 85 min
43 225 days
45 2.96×10^4 m/s, 3.06×10^4 m/s
47 6.0×10^{-3} m/s^2

CHAPTER 6

1 2.1×10^3 J
3 46 N
5 (a) 1.8×10^3 J, (b) 6.9×10^3 J
7 (a) 3.29×10^5 J, (b) 4.5×10^5 J
9 (0.1) mgh
13 840
17 mgy_0
25 31 J
29 4×10^3 J
31 6.83×10^3 m/s
33 2.0×10^{-18} J
35 2.5×10^4 N/m
37 18 m/s
39 4.2 m/s, 2.1 m/s
41 (a) 1.8×10^5 J, (b) 22 m/s, (c) 3.2 m
43 0.56 m
45 (a) 4.85 J, (b) 0.79 J
47 (a) $\frac{1}{2} k(x^2 - x_0^2)$, (b) 0
49 (a) $9.3°$, (b) $14.8°$

CHAPTER 7

1 $\mathbf{F} = -(2x + 2y)\mathbf{i} - (2x + 8yz)\mathbf{j} - (4y^2)\mathbf{k}$
3 (a) yes, since total work over closed path is zero, (b) $\frac{1}{2} kx^2 - \frac{1}{4} ax^4 - \frac{1}{5} bx^5$
5 No
7 7.4 m/s

9 3.3 m
11 (a) 21.1 m/s, (b) 23.1 m
13 24.4 m/s at B, 10.2 m/s at C, 18.9 m/s at D
15 (a) mg/k, (b) 2 mg/k
17 (a) 7.8×10^4 J, (b) 6.1×10^2 N
19 $(C/2r_0)(1 - 1/n^2)$
23 (a) 21 m/s, (b) 1.2 N
25 (a) 12 m/s, (b) 87 m
27 (a) 0.15 m, (b) 1.1 m, (c) 0.90 J
33 (a) 1.78×10^{32}, (b) 1.78×10^{32}, (c) 0
35 (a) $r_e^2 v_0^2/(2GMe - r_e v_0^2)$, (b) 7.5×10^6 m
37 (a) 3.56×10^3 m/s, (b) 3.65×10^9 J, (c) 1.12×10^9 N, (d) 3.65×10^9 J
43 $a^2/4b$
47 (a) 1.0×10^3 J, (b) 1.0×10^3 W
49 4.1×10^2 W
51 (a) 8.7×10^5 J, (b) 61 W, 8.1×10^{-2} hp, (c) 0.54 hp
53 3.4 W
55 300 W
57 (a) 50 m/s, (b) 8.1×10^5 W
59 (a) -490 W, (b) 7.4 kW, (c) 2.4 kW

CHAPTER 8

1 4.84×10^{-11} m
3 1.02 m
5 9.4% of body height
7 $x_{cm} = 0$, $y_{cm} = 2r/\pi$
9 $x_{cm} = 0$, $y_{cm} = 4r/3\pi$
11 (a) 4.62×10^6 m from center of earth
13 (a) $7 D/3$, (b) $5 D$
15 $mv/(m + M)$; the balloon stops
17 $8.0\,\mathbf{i} - 3.9\,\mathbf{k}$
19 (a) 0.75 kg·m/s, (b) 0.51 kg·m/s
21 $(-4.4\,\mathbf{i} + 4.4\,\mathbf{j})$ kg·m/s
23 1.1 m/s
25 14.8 m/s
27 3.5×10^3 m/s
29 $\frac{3}{2} v_0\mathbf{i} - v_0\mathbf{j}$
31 (a) no, (b) $-m_2/m_1$, (c) m_2/m_1, (d) it does not move, (e) friction introduces an external force which influences the motion
33 (a) 7.4 km/s and 5.0 km/s in the original direction, (b) 6.8×10^5 J
35 130 N, no
37 (a) 0.17 m/s, (b) 9.4×10^2 N, (c) 470 J for astronaut and 32 J for capsule
39 1.9×10^3 N
41 (a) 4.5 N·s, (b) 75 m/s
43 52 m
45 3.0 m/s and 2.0 m/s, respectively
47 2 m
49 (a) -4.2 m/s, 4.2 m/s, (b) height $= 0.90$ m
51 (a) 1.00, (b) 0.890, (c) 0.286, (d) 0.019
53 (a) 0.43 m, (b) -2.1 m/s, 5.9 m/s, (c) yes
55 positive x axis, 4.8 m/s, 3.0 m/s
61 4.1×10^5 m/s, $60°$
63 6.4×10^{-10} J
65 (a) 2.5×10^{-16} m/s, (b) 1.7×10^{-17}, (c) 1.9×10^{-7} J
67 1800 J, 2700 J
69 $141°$
71 0.89 hp

73 (a) $[6.00 \times 10^4/(4.00 \times 10^3 + t)]$ m/s, where t is in minutes, (b) 14.8 m/s, yes
75 86 kg/s
77 7.0 s

CHAPTER 9

1 3.4 km
3 (a) 210 rad/s, (b) 21 m/s
5 $L/2$, $L/2$
7 460 m/s, 300 m/s
9 (a) -280 rad/s, (b) -340 rad/s^2, (c) $5.0 + 6.0t - 18.0t^3$, (d) $6.0 - 54.0t^2$, (e) -463 rad/s, -480 rad/s^2
11 (a) 6.4 rad/s^2, (b) 90 m/s^2, 1.3 m/s^2
15 (a) 6.5 s, (b) 0.53 rad/s^2
17 (a) -1.5 rad/s, (b) 20 s
19 (a) left along axis, up, (b) 75 rad/s, $37°$ above horizontal, (c) $\omega_1\omega_2 = 2700$ rad/s^2, perpendicular to ω_1 and ω_2
21 367 $\mathbf{k}$
23 4 Ms^2
25 0.16 kg/m^2
27 (a) 3.5 kg·m^2, (b) 0.024 N·m
29 0.19 m
31 8.7×10^4 N·m
33 $Mg\sqrt{h(2R_0 - h)}/(R_0 - h)$
35 9.2×10^3 rev, 110 s
37 (a) 0.26 m/s^2, (b) 0.066 N·m
39 (a) 150 rad/s^2, (b) 1100 N
41 0.15 kg·m^2
45 (a) $\frac{1}{2} MR_0^2$, (b) $\frac{3}{2} MR_0^2$
47 $\frac{1}{2} M (R_1^4 + R_2^4)/(R_1^2 + R_2^2)$
51 $Ms^2/6$
53 (a) 2.6 kg·m^2/s, (b) 0.37 N·m
55 (a) $\Delta L = 0$, $\Delta I > 0$, so $\Delta\omega < 0$, (b) 1.5
57 0.040 rad/s
59 $M/2$
61 2.2×10^4 J
63 156 hp
65 54 hp
67 12 m/s
69 (b) 2.2×10^3 rad/s, (c) 23 min
71 (a) 3.4 m, (b) 4.0 s
73 (a) $\frac{1}{2} mv_1^2 [(R_1/R_2)^2 - 1]$, (b) yes
75 0.056
77 (a) 5.7 m/s, (b) 15.8 J, (c) 0.9 J, (d) 5.3 m/s, 16.7 J, 0

CHAPTER 10

9 (a) 4.5 $\mathbf{j} + 7.8$ $\mathbf{k}$, (b) no, only the net torque (which includes torque exerted by the bearings) is proportional to $\boldsymbol{\alpha}$
13 (a) mgd, (b) $mgdt$
15 (a) $L^2 \omega^2(\frac{7}{9} m + \frac{1}{6} M)$, (b) $\omega L^2(\frac{14}{9} m + \frac{1}{3} M)$
17 8.2×10^{-6}
19 angular velocity 80 rad/s about the center of mass, which moves with linear velocity 21 m/s
21 (b) $36°$, (c) 4.2 m
23 $\omega^2 \sin\theta \cos\theta (m_1r_1^2 + m_2r_2^2)/d$

25 (a) 5.1×10^2 N on each, (b) on the diameter passing through the axis of rotation and 18 cm on the other side of the center
27 3.3×10^{-4} kg·m²
29 44 rad/s

CHAPTER 11

1 0.88 N
3 72 kg
5 (a) 2.0×10^2 N toward arm and 27 N down, (b) 1.7×10^3 N toward arm and 340 N down
7 top: 43 N toward hinge and 64 N up; bottom: 43 N toward door and 64 N up
9 (a) -1800 N, 2400 N, (b) -2100 N, 3000 N
11 390 N
13 0.84 m from the top of head
15 $T_1 = 4.55$ mg, $T_2 = 4.97$ mg, $h = 158$ m
17 (a) 160 N up, 29 N toward wall, (b) 0.26
21 (a) yes, (b) 140 km/h
23 72 N
25 (a) tip over, (b) 0.50 m
27 (a) $F_m = 1.5 \times 10^3$ N, $F_{Jx} = 0.51 \times 10^3$ N, $F_{Jy} = 2.4 \times 10^3$ N, (b) $F_m = 1.72 \times 10^3$ N, $F_{Jx} = 0.59 \times 10^3$ N, $F_{Jy} = 2.42 \times 10^3$ N
29 (a) 1.5×10^5 N/m², (b) 2.9×10^{-6}, (c) 3.1×10^{-5} m
31 9.9×10^7 N/m², or 990 atm
33 1.0×10^7 N/m²
35 (a) 93 N·m
37 2.7×10^{-5} m²
39 (a) 1.1×10^9 N/m², (b) will break, (c) 2.3×10^7 N/m², will not break
41 15 total
43 2.0 cm

CHAPTER 12

1 2.7×10^{11} kg
3 0.89
5 7.4×10^7 N/m², 740 atm
7 0.052
11 3.3×10^9 N
13 5.3×10^{18} kg
15 (a) $P_0 + \rho(g + a)h$, (b) $P_0 + \rho(g - a)h$, (c) P_0
17 1.15×10^3 kg/m²
21 (a) 3.9×10^5 N down, (b) 3.9×10^5 N up
23 1.4×10^2 N
25 (a) 1.41×10^5 N/m², (b) 0.98×10^5 N/m²
27 140 cm²
29 13 m
31 (a) 1.2×10^{-3} N, (b) 0.19 N
33 1.01×10^3 kg/m³
35 5.94×10^4 N

37 0.83
39 16 N; no
41 1.9×10^7 kg
43 0.95×10^3 kg/m³
45 (a) 585 kg/m³, (b) $\rho_l = \rho\left(1 - \dfrac{m'}{m}\right)$
47 0.57 kg
49 1.2×10^3 N·m
51 (a) $F/4\pi r$, (b) 2.1×10^{-2} N/m
53 $3.0 \times$ (ignoring the bottom surfaces)
55 0.12 m
57 1.7 mm
59 1.5×10^{-7} m (20°C)

CHAPTER 13

1 47 cm/s
3 380 h
7 2.4 atm
9 1.0×10^5 N up
11 3.6×10^{-3} m³/s
15 190 m/s
19 (a) $\sqrt{2gh\,\rho_m/\rho_h}$, (b) 2.3 m/s
23 (a) $2\sqrt{h_1(h_2 - h_1)}$, (b) $h_2 - h_1$
25 7.0×10^{-2} Pa · s
27 5.7×10^{-4} m³/s
29 0.96 Pa/cm
31 0.93 m
33 16 cm
39 (a) 5.0×10^{-3} m³/s, (b) 10 m
41 10^{-3}
43 9.8 mm/s
45 4.6 yr
47 $2(\rho - \rho_f)gr^2/9v_T$

CHAPTER 14

1 20 N/m
3 (a) $(0.20$ m$)\cos(4.2\,t)$, (b) 6.2 cm
5 3.0 Hz
7 (d) C^4/D^3, (e) $2\pi\sqrt{mD^3/C^4}$
11 $\dfrac{1}{2\pi}\sqrt{\dfrac{2k}{m}}$
13 (a) 1.60 s, 0.62 Hz, (b) 2.1 m and -4.7 m/s, (c) 9.1 m/s and 9.6 m/s²
15 1.3 Hz
19 (a) $(0.0800$ m$)\sin(28.9\,(t - 0.060))$, (b) $t = (0.114 + 0.217\,n)$ s for max, $(0.005 + 0.217\,n)$ s for min, where $n = 0, 1, 2, \ldots$; (c) 19.7 N, (d) -7.89 cm, (e) 2.31 m/s at 0.060 s
21 (a) $A/\sqrt{2}$, (b) $^1/_4$ is potential, $^3/_4$ is kinetic
23 5.5 m/s
25 420 m/s
27 4.4 cm, 1.7 m
31 0.248 m
33 (a) 0.83 Hz, (b) 0.33 m/s
35 (a) $-0.4°$, (b) $-10°$, (c) 12°
37 $^1/_3$
39 (a) 23°, (b) 7°

41 $\dfrac{1}{2\pi}\sqrt{\dfrac{2g}{3R}}$
43 (a) $d^2\theta/dt^2 = -K\theta/I$, (b) $2\pi\sqrt{I/K}$
45 (b) 1.7 s (depends on assumptions)
51 (a) $\pi/2$, (b) 0, max
53 1.65×10^8
59 (a) $x = 2A\cos(\omega t - \pi/4)$, $y = A\cos\omega t$; (b) $x = A\cos(\omega t - \pi/4)$, $y = A\cos 2\omega t$
61 (a) $x = A\cos(\omega t + \phi_0)$, $y = A\cos(\omega t + \phi_0 \pm \dfrac{\pi}{2})$, where $A = 5.2$ m; (b) $f = \omega/2\pi = 0.74$ Hz, (c) $\phi_0 = 54°$

CHAPTER 15

1 3.0 m/s
3 0.114 s
5 176 Hz
7 (a) 1400 m/s, (b) 4100 m/s
9 1500 m
11 (b) $2\sqrt{L/g}$
13 (a) 1 : 4, (b) 1 : 2
21 (a) 41 m/s, (b) 6.4×10^4 m/s², (c) 40 m/s, 8.2×10^3 m/s²
23 D = $(0.020$ cm$)\cos(4.77\,x \pm 1650\,t + \pi)$
27 (c) $2k_2/(k_1 + k_2)$
29 8.6°
31 (c) The energy is entirely kinetic (string is moving).
33 261 Hz
35 2.79 m
39 (a) 4.8 cm, (b) 2.8 cm, 8.4 Hz, 80 cm/s, (c) 80 cm/s
41 0.85 m/s
43 308 Hz

CHAPTER 16

1 (a) 5.76 mm, (b) 1.37 mm
3 70 m
5 (a) 6.0 m, (b) 850 Hz, (c) 5100 m/s, (d) 3.0×10^{-11} m
9 (a) 49 dB, (b) 3.2×10^{-9} W/m²
11 140 dB
13 1.12
15 92 dB
17 1.6×10^{-9} W
19 16 W
21 45 dB
23 18.1 W
25 25 N
27 6.7 cm
29 0.0176
31 0.44, 0.20, 0.088
33 (a) 65.0 cm, (b) 773 Hz
35 (a) 115 overtones, (b) 116 overtones
37 (a) 0.12 and 0.015, (b) 9 dB and 20 dB
39 780 Hz
41 (a) (132.0 ± 0.5) Hz, (b) 0.76%
43 (a) 0.69 m, (b) any separation

45 (a) 12 Hz, (b) 28.6 m

47 $\dfrac{\lambda}{8d} \sqrt{l^2 + 4d^2}$

49 $f' = f(v \pm v_0)/(v \mp v_s)$

51 1.6 kHz

53 0.25 m/s

55 (a) 606 Hz, (b) 697 Hz, (c) 648 Hz, (d) 648 Hz, (e) 761 Hz, (f) 708 Hz

59 Pressure decreases to 0.60 times the original pressure, compared to 0.25 times the original pressure if it obeyed an inverse-square law

59 0.60 (compared to 0.25 for $1/h^2$)

CHAPTER 17

1 3.53 kg

3 75°

5 −40°

7 (a) low, (b) 0.024%

9 −85°C

13 5.7 minutes fast

15 (a) $L_2 = L_1(1 + \alpha(T_2 - T_1))$,

(b) $L_2 = L_1 \left(1 + \int_{T_1}^{T_2} \alpha(T)\,dT\right)$,

(c) $L_2 = L_1(1 + \alpha_0(T_2 - T_1) + \dfrac{b}{2}(T_2^2 - T_1^2))$

19 1.6 mL

21 (a) lower, (b) 0.43%

23 (a) 3.9 cm

25 (a) 9.4×10^7 N/m², (b) no, (c) 9.4×10^6 N/m², yes

27 (a) 310 K, (b) 300 K, (c) 77 K

29 (a) 4.3×10^3 K, 1.5×10^7 K, (b) 6.4%, 1.8×10^{-3}%

31 $^1/_5$

33 1.43 kg/m³

35 4.21 kg

37 (a) 770 kg, (b) 65 kg leaves the house

39 8.0%

41 12 L

43 $^1/_T$

45 2.9 cm³

47 2.69×10^{25} molecules

49 270 molecules/cm³

51 1.8×10^3 N outward

53 (a) 71.2 torr, (b) 157°C

55 (a) 0.15 K, (b) 0.04%

CHAPTER 18

1 (a) 5.65×10^{-21} J, (b) 3.65×10^3 J

3 1.17

7 370 m/s

9 1.4×10^5 K

11 (a) 460 m/s, (b) 27

13 0.996

15 (a) 22.5 m/s, (b) 24.2 m/s, (c) 15 m/s

19 0.31 atm

21 120°C

23 1.74×10^3 Pa

25 4.8 kg

27 29%

29 (a) 52 atm, (b) 56 atm

31 (c) 3.2×10^{-10} m

33 (a) 3.9×10^{-10} m, (b) 1.8×10^{-10} m

35 ≈0.003

37 1.8×10^{-3} Pa

41 (a) 1.52×10^{-10} m, (b) 1.28×10^{-10} m

43 (b) 1.6×10^{-5} m²/s

45 (c) N_2, by 7%

CHAPTER 19

1 1.67×10^6 J

3 188 kg

5 0.36 kcal/kg · C°

7 1.00

9 430 Cal

11 39.3°C

13 15.3°C

15 2.8°C

17 .0593 kcal/kg · C°

19 0.334 kg

21 0.81 kg (25 moles)

23 2.7 kcal/kg

25 3.0 g

27 (a) 18.4 W, (b) 5.5 W

29 250 W

31 10°C

33 $(k_1 A_1/l_1 + k_2 A_2/l_2)\,\Delta T$

39 (a) 2.3 K/s, (b) 125°C

CHAPTER 20

1 (a) 13 kJ, (b) 18 kJ

5 $RT \ln\left(\dfrac{V_2 - b}{V_1 - b}\right) + \left(\dfrac{a}{V_2} - \dfrac{a}{V_1}\right)$

7 (a) 1.6 kJ, (b) 1.6 kJ

9 -5.5×10^5 J

11 52 kJ

13 30 C°

17 0.06%

19 4.25×10^6 J

21 (a) 780 kcal, (b) 220 kcal, (c) 560 kcal

23 158 J

25 (a) 696 cal, (b) −324 cal, (c) 372 cal

31 (a) 295 K initial and 142 K final, (b) −15.9 kJ, (c) −15.9 kJ, (d) 0

35 (b) 280 m/s

CHAPTER 21

1 23.7%

3 11 km³/day, yes, 57 km²

5 (a) b to c compression, (c) 0.51, (d) no, no

9 540°C

11 1.77×10^{10} J/h

13 720°C

15 1.4×10^8 kg/hr

17 (b) $\left(\dfrac{1}{\varepsilon} - 1\right)$, (c) 6.8

19 0.025 J/K

21 0.25 cal/s · K

23 50 J/K

25 330 J/K

27 (a) 91°C, (b) 2.5 J/K

29 (a) 11.5 J/K, (b) 0, (c) 12.5 J/K

31 (b) net heat flow into the system

33 (a) 7.5 K, (b) 110 J/kg · K

35 (a) 20/64, (b) 1/64

37 (a) 2.5×10^{-23} J/K, (b) -9.2×10^{-22} J/K

CHAPTER 22

1 6.25×10^{14}

3 6.0×10^{-3} N

5 5.1 m

7 53 N, 16 N, 37 N

9 1.6×10^5 N, toward center of square

11 10.6 cm beyond the smaller charge

13 $|Q_1| = |Q_2| = |Q_T|$, Q_1 or $Q_2 = 0$

15 4.8 μC

17 2.96×10^{10} Q^2/l^2

19 2.57×10^8 N/C

21 1.02×10^{-7} N/C

23 2.6×10^6 N/C, 90°; 6.7×10^6 N/C, 56°

25 4.9×10^7 N/C, 45°

27 $a/\sqrt{2}$

29 (a) $(Q/2\pi\varepsilon_0 a^2)\left(1 - \dfrac{x}{(x^2 + a^2)^{1/2}}\right)$

31 $(\sigma/\pi\varepsilon_0)\,\sin^{-1}[L^2/(L^2 + 4z^2)]$

33 2.39×10^4 N/C, 7.4×10^2 N/C; .005, .031

39 (a) 2.0 mm, (b) 3.3 ns

41 16°

43 (a) 3.4×10^{-20}C, (c) 8.5×10^{-26} N · m, (d) 2.5×10^{-26} N · m

45 (b) $(1/2\pi)\sqrt{pE/I}$

CHAPTER 23

1 (a) 25 N · m²/C, (b) 18 N · m²/C, (c) 0

3 $Q/6\varepsilon_0$

5 1.28×10^{-8} C

9 2.1×10^{-11} C

11 (a) $\rho r/3\varepsilon_0$, (b) $\rho r_0^3/3\varepsilon_0 r^2$

13 (a) -6.8×10^5 C, (b) 8.3×10^9

15 (a) 0, (b) σ/ε_0

17 $\varepsilon_0 bl^3$

19 (a) always, (b) $Q = 0$

21 (a) $Q/[4\pi\varepsilon_0 r_0^2(1 + t^2/900)^2]$ up to $t = 30$ s, (b) $Q/64\pi\varepsilon_0 r_0^2$

23 (a) 0, (b) $(Q/4\pi\varepsilon_0)(1 - r_1^2/r^2)/(r_0^2 - r_1^2)$, (c) $Q/(4\pi\varepsilon_0 r^2)$

25 (a) $(\sigma/\varepsilon_0)(R_0/r)$, (b) 0

27 (a) 4.9×10^{10} N/C, (b) 1.4×10^6 N/C

29 (a) $Q/\pi r_0^4$, (b) $(Q/\varepsilon_0)(r^4/r_0^4)$, (c) $Q/4\pi\varepsilon_0 r^2$

CHAPTER 24

1 -4.8 mJ
3 (a) 1.1×10^9 J, (b) 2500 kg
7 2.2×10^4 V/m
9 $1.7°$
11 24 kV
13 6.2×10^7 m/s
15 (a) -2.0×10^8 V, (b) 1.3×10^9 V/m, 26° N of E
17 (a) 8.9 cm beyond -2.0 μC charge, (b) 4.0 cm beyond -2.0 μC charge; 0.80 cm from -2.0 μC charge, between the two charges
19 (a) 0.043 V, (b) 0.031 V, (c) -0.031 V
21 (a) $(Q/4\pi\varepsilon_0)(2r/(r^2 + l^2) - 2/r)$
25 (a) -9.6×10^8 V, (b) 9.6×10^8
27 1.0 cm, 3.3×10^{-8} C
29 9.4 MV
31 1.6×10^{12}
33 (a) $V_0 - (\sigma R_0/\varepsilon_0) \ln (r/R_0)$, (b) V_0, (c) No
35 (a) $Q/4\pi\varepsilon_0 r$, (b) $(Q/8\pi\varepsilon_0 R)(3 - r^2/R^3)$
37 (a) $Q/4\pi\varepsilon_0 r$, (b) $(Q/4\pi\varepsilon_0 R)(5/4 - r^4/5R^4)$
41 (b) $2 p \cos \theta/4\pi\varepsilon_0 r^3$, $p \sin \theta/4\pi\varepsilon_0 r^3$
43 (b) 96 keV, (b) 190 keV
45 43 eV
47 (a) -3.1×10^{-20} V, (b) -2.0×10^{-20} V, (c) 0, (d) $+3.1 \times 10^{20}$ V
49 (a) $Q^2(4 + \sqrt{2})/4\pi\varepsilon_0 b$, (b) $\sqrt{2} \, Q^2/\pi\varepsilon_0 b$
51 0.529×10^{-10} m
53 $3 \, Q^2/20\pi\varepsilon_0 R$

CHAPTER 25

1 300 μC
3 $Q_0 C_1/(C_1 + C_2)$, $Q_0 C_2/(C_1 + C_2)$; $Q_0/(C_1 + C_2)$
5 4.8 C
7 1.0×10^9 m^2
9 7.1×10^{-4} F
13 (a) $4\pi\varepsilon_0 R_1 R_2/(R_2 - R_1)$
15 9.0 μF, 0.25 μF
17 1400 pF in series
19 (a) $C_1 + C_2 C_3/(C_2 + C_3)$, (b) 200 μC, 67 μC, 67 μC
21 0.017 μF, series; 0.0014 μF
23 (a) 3.7 μF (b) 21 V, 29 V, 50 V
25 $(\varepsilon_0 \sqrt{A}/\tan\theta) \ln (1 + \sqrt{A} \tan\theta/d)$
27 4.0×10^{-6} J
29 840 J
31 9.0×10^{-6} J
33 (a) $\ln (2R_2/R_1)/\ln (R_2/R_1)$, (b) $\ln (R_2/R_1)/\ln (2R_2/R_1)$
35 (a) $-\frac{1}{2}\varepsilon_0 A V_0^2 \, l/d(d - l)$, (b) $Q_0^2 l/2\varepsilon_0 A$
39 33 pF
41 $\varepsilon_0 A V^2 (1 - K)/2d$
43 (c) 1.84×10^{-8} C, (d) 1.17×10^5 V/m, (e) 3.33×10^4, (f) 150 V, (h) 2.58×10^{-8} C; others the same
45 $2\varepsilon_0 A K_1 K_2/d(K_1 + K_2)$
47 22.2%

CHAPTER 26

1 1.2×10^5 C
3 4.0×10^{-11} C
5 1.9×10^{18} per min
7 6.7×10^{-2} mho
9 1.09 Ω
11 1/4
13 (a) 0.38×10^{-3} Ω, (b) 1.5×10^{-3} Ω, (c) 6.0×10^{-3} Ω
17 9.0 Ω
19 $(r_2 - r_1)/4\pi\sigma r_1 r_2$
21 1.73×10^{-4} m, 53.4 m
23 6.25 mol/m^3
27 3.6 W
29 $0.10
31 $18
33 3.2×10^{-4} m
35 71%
37 4.4 mm
39 0.0473 kg/s
41 840 W
43 0.94 A, 1.33 A
45 0.15 A
47 5.8 kW, 12 kW, 0

CHAPTER 27

1 150 Ω, 6.0 Ω
3 ratio of resistors in series : 2.1 : 3.9
5 3.4 kΩ
7 37 Ω
9 1.8 kΩ
11 130 Ω
13 42 V
15 3.5 mA, 2.2 mA, 1.32 mA, 0.88 mA, 0.44 mA, 0.44 mA
17 0.58 A
19 0.050 Ω
21 0.16 A
23 17.4 V, 13.3 V
25 77 V, 43 V
27 0.45 A
29 1.3 A
31 (a) 8.0 V, (b) 16V, (c) 8.0 V (at equilibrium), (d) -5.76 μC, (e) 2.1 μs
33 11 ms
35 (a) $R_1 R_2 C/(R_1 + R_2)$, (b) $C\mathscr{E}R_2/(R_1 + R_2)$
37 2.1 μs
39 1.5 MΩ
41 1.0×10^{-3} Ω in parallel
43 (a) 1.2×10^{-4} Ω in parallel, (b) 5.0×10^7 Ω in series
45 1.78 A
47 7.9 V
51 3.0×10^4 Ω
53 0.301 V
55 4.472 V
57 13.6 Ω

CHAPTER 28

1 (a) 17.9 N/m, (b) 12.6 N/m
3 1.6 N
5 $2 \pi \, RIBd/(d^2 + R^2)^{1/2}$
9 Counterclockwise circle of radius 3.8 mm
13 13 T, north
15 1.5×10^{-8} m
17 $-(2.3 \, \mathbf{i} + 2.8 \, \mathbf{j}) \times 10^6$ m/s
23 2.6×10^6 V/m
25 6
27 1.2×10^5 m/s, 3.4×10^5 m/s
29 $\pm 3.4 \times 10^5$ V/m
31 70, 72, 73, 74 u
33 1.2 m
35 1.4 T
37 34 MeV, 4.1×10^7 m/s
39 17 MeV, 44 MeV, 13 MHz, 11 MHz
41 (b) IBl/neA, (c) $IBl/e\mathscr{E}A$
43 1.8 T
47 0.51 mm
51 39.6 μA

CHAPTER 29

1 1.5×10^3 A
3 1.2×10^{-4} g
5 48° N of W
7 (a) $4.0 \times 10^{-6} \, (I - 10)$ T, (b) $4.0 \times 10^{-6} \, (I + 10)$ T
9 $\mu_0 (d - 2x)/2\pi x (d - x)$
11 1.27×10^{-17} N
13 7.2 A
17 $\mu_0 \, jt/2$
19 1.4×10^4
21 0.23 N, attractive
23 $A : 2.6 \times 10^{-5}$ N/m; $B, C : 1.5 \times 10^{-5}$ N/m
25 (a) 9.7×10^3 A, parallel, (b) unstable, (c) 1.08×10^4 A, opposite, vertical stability only
27 $\mu_0 I\theta (R_2 - R_1)/4\pi R_1 R_2$
31 1.2×10^{-8} N · m
35 (b) 94, (c) 1.2×10^{-3} T, (d) no
37 (a) $\frac{1}{2} \mu_0 NIR^2 \, [1/(R^2 + x^2)^{3/2} + 1/(R^2 + (R - x)^2)^{3/2}]$, (c) 5.5×10^{-2} T
39 $\mu_0 Il/2\pi (x^2 + l^2/4)(x^2 + l^2/2)^{1/2}$
43 1.5×10^6 A/m

CHAPTER 30

1 0.023 V
3 36 V
5 1.9×10^{-4} J
7 (a) $(0.75 \, t^2 - 1.26)$ V; (b) -0.51 V, 17.5 V
9 (a) 0.0765 A, (b) 4.9×10^{-4} W

59 68°C or -18°C
61 (a) $a = 4.99 \times 10^{-5}$ V/C°, $b = -5.83 \times 10^{-8}$ V/(C°)^2, (b) 113°C

11 2.7 mV

15 (a) 0.20 V, (b) 8.0 mA

19 $-\frac{1}{2} BR^2 \omega$

21 (b) $v = v_0 e^{-(B^2 l^2/mR)t}$

23 0.403 V/m

25 (b) $(l^2/R)(dB/dt)$, (c) $(l^2/8)(dB/dt)$

27 (b) clockwise, (c) increase

29 (a) IR/l, (b) $(\mathcal{E}_0/l)e^{-(B^2 l^2/mR)t}$

33 8.5 Hz

37 (b) 28.1 A, (c) 4.72 A

39 $B^2 l^2 A \omega d/\rho$

41 0.44

43 8.0 V, 0.53 A

45 (a)17.5 kV, (b) 1.04×10^7 W, (c) 3.1×10^5 W, (c) 1.01×10^7 W

47 46 kW

CHAPTER 31

1 unchanged

3 $\mu_0 N_1 N_2 A_2 \cos \theta/L$

5 $\mu_0 n_1 n_2 \pi r_2^2$

7 7.43 mV

9 2.04 H

11 0.22 A

13 2.46 mm

15 $\frac{1}{2}$

17 (b) 23 mH

21 31 Ω, 18 H

25 0.80 J

27 (a) 4.4×10^{-4} J/m^3, 1.6×10^6 J/m^3, (b) 6.8×10^8 V/m

29 2.2×10^{-2} J/m^3

31 $\mu_0 N^2 I^2/8\pi^2 r^2$; $(\mu_0 N^2 I^2 h/4\pi) \ln (R_2/R_1)$

33 4.61

37 (a) 2.25 ms, (b) 1.38×10^5 Ω

39 (a) $(LV^2/2R^2)(1 - e^{-Rt/L})^2$, (b) 5.30

41 (a) 177 pF, (b) 56 mH

45 (a) $Q_0/\sqrt{2}$, (b) $T/8$

47 1150 Ω

49 (a) $(Q_0^2/2C)e^{-Rt/L}$

CHAPTER 32

1 7.0 kHz

5 3.35×10^3 Ω, 0.120 A

7 (a) 3.2×10^4 Ω, (b) 93 mA, 200 Hz

9 770 W

11 (a) 4.5 mV, (b) 30.0 mV

13 (a) 4.0 kΩ, (b) 4.2 kΩ

17 (a) 1.77 A, (b) 16°, (b) 204 W, (d) 115 V, 33 V

19 $R = 14.4$ Ω, $C = 800$ μF

21 8.5 μF

25 (a) 0, (b) $2\mathcal{E}_0/\pi$

27 (c) $Q_0 = \mathcal{E}_0/\sqrt{(\omega R)^2 + (\omega^2 L - 1/C)^2}$, $\tan\phi = R\omega/(\omega^2 L - 1/C)$, (e) $I_0 = \mathcal{E}_0/\sqrt{R^2 + (\omega L - 1/\omega C)^2}$, $\tan \phi = (1/C - \omega^2 L)\omega R$

31 1.15 MHz

33 (a) 1.08×10^{-8} F, (b) 1.13 A, (c) 26 kHz, 42kHz

35 (b) $\sqrt{1/LC - R^2/2L^2}$

37 (a) $\mathcal{E}_0^2 R/2[R^2 + (\omega L - 1/\omega C)^2]$, (b) $1/\sqrt{LC}$, (c) R/L

39 2.0×10^4

41 4 Ω

CHAPTER 33

1 2.7×10^{-9} A

7 $\oint \mathbf{B} \cdot d\mathbf{A} = Q_m$; $\oint \mathbf{E} \cdot dl = -d\Phi_B/dt + \mu_0 dQ_m/dt$

9 (a) 3.31 m, (b) 194 m

11 (a) (E_0/c) **j**

13 1.49×10^{-3} J/s · m^2

15 4.5×10^{-6} J/m^3

17 3.82×10^{26} W

19 7.75 V/m, 2.58×10^{-8} T

21 $E_0 = 7.9 \times 10^{-4}$ V/m; 8.2×10^{-10} W/m^2

25 at home; 0.14 s

29 (a) $20 \sin [3.14 \times 10^8$ s^{-1} $(0.67 \times 10^{-8}$ s $- t)]$, (b) $20 \sin [1.05$ m^{-1} $(x - 6.00 \times 10^8$m)]

CHAPTER 34

1 (a) 2.21×10^8 m/s, (b) 1.99×10^8 m/s

3 3 m

5 4.0×10^{16} m

7 536 rev/s

11 circle of diameter 2.75 mm

15 12.0 cm

21 (a) concave, (b) -29.3 cm, (c) 177 cm

23 (a) $r = \infty$, (b) $d_i = -d_o$, (c) 1, (d) yes

25 -4.50 m

27 31.3°

29 54.3°

37 >1.80

39 >1.4

41 53.8°

CHAPTER 35

1 (a) 96 cm, (c) 2.2 m

3 (a) 28 cm, (b) 23 cm

5 converging, 18.3 cm, real

7 1.53

9 0.47 mm

11 (a) 151 mm, -1.74 mm, real, (b) -122 mm, 1.41 mm, virtual

13 $1/d_0 + 1/d_i = (n/n' - 1)(1/R_1 - 1/R_2)$; $1/f = (n/n' - 1)(1/R_1 - 1/R_2)$; same; same

15 88 cm

19 3.5 m

23 20.2 cm behind second lense, m = -0.511

25 (a) 12.5 cm, (b) 8.3 cm

27 2.1 $\times$, vs. 3.8 $\times$, but note that child sees more detail without magnifier than does a normal eye

29 (a) -76 cm, (b) 6.6$\times$

31 $-16\times$, 53 cm

33 92$\times$

35 12$\times$

37 6.4$\times$

39 (a) 16.1 cm, (b) 160$\times$

41 (a) 1.07 cm, (b) 190$\times$

43 3.5 D

45 (a) -2.8 D, (b) 36 cm

47 -0.48 D, 1.3 D

49 0.90 D

51 -23.6 cm

CHAPTER 36

3 0.097 mm

5 1.99 mm

9 425 nm

11 $LC \approx 10^{-31}$ HF

13 0.111°

15 0.51 cm

19 $I_0 (3 + 2 \sqrt{2} \cos \delta)/(3 + 2\sqrt{2})$

21 $I_0 \cos^2 \delta \cos^2 \frac{\delta}{2}$; maximum at $\sin \theta = m\lambda/d$ for $m = 0, 1, 2, \ldots$; minimum at $\sin \theta = m\lambda/4d$ for $m = 1, 2, 3, 5, 6, \ldots$ (not for m divisible by 4)

23 160 nm; 0 or 320 nm

27 7.4 μm

29 1.39

31 510 nm

35 reduced to 0.117, 0.109

37 96.1 μm

39 $I_0 \cos^2 (2\pi x/\lambda)$

41 2.3×10^{27} lm/sr; 2.8×10^{28} lm

CHAPTER 37

1 3.18°

3 $D = \lambda$

5 $D (\sin 30° - \sin \theta) = m\lambda$, $m = \pm 1, \pm 2, \ldots$ for minima, etc.

7 $\beta/2 \approx \pm 1.4303 \pi$, $\pm 2.4590 \pi$; 4.65%, 1.64%

9 $d = 8D$

13 (a) 3, (b) 23, (c) 9, (d) 15

15 57.5°, 230 nm

17 (a) 1.4, (b) 240 nm

19 (a) 0.84, (b) 250 $\times$

21 0.13 m; 2.3 km on moon

23 6.16°

25 all of $m = 2$, parts of $m = 3$ and 4

27 497 nm, 612 nm, 637 nm, 754 nm

29 12,500/cm

31 446 nm, 653 nm

33 910 nm

37 (a) 2 on each side of center, (b) 6.67×10^{-5} rad, 7.34×10^{-5} rad, 1.21×10^{-4} rad

39 (a) $\frac{1}{9} I_0 (1 + 2 \cos \delta)^2 (\sin^2 \beta/2)/(\beta/2)^2$, where $\beta = \delta D/d$
41 0.032 nm, $m = 2$
45 12.3°

CHAPTER 38

1 $\frac{1}{4}$
3 (a) 35°, (b) 63°
5 reduced by 37.1%
7 (a) 0, (b) $\frac{1}{8}$, (c) 0
11 49.6°
13 37°, 49°, 53°
15 (a) 7.4×10^{-3} cm
17 plane-polarized, $I = I_0/2$
19 1.89 μm
23 (a) elliptically polarized, principal axes parallel and perpendicular to optic axis, (b) plane-polarized, 60° to original

CHAPTER 39

1 7.81×10^{-9} s
3 150 m
5 2.85×10^8 m/s
7 (a) 8.92 m, 1.45 m, (b) 13.0 s, (c) $-0.760\,c$, (d) 13.0 s
9 $0.93\,c$
11 (a) (397 m, 20 m, 0), (b) (1522 m, 20 m, 0)
13 (a) $0.80\,c$, (b) $-0.80\,c$
15 60 m/s, 24°
17 $0.90\,c$, 43°
19 (a) $L_0 \sqrt{(1/\gamma^2 - 1) \cos^2 \theta + 1}$, (b) $\tan^{-1}(\gamma \tan\theta)$
21 2.52×10^{-27} kg
23 $0.14\,c$
25 (a) $(c - v)/c = 5.0 \times 10^{-9}$, (b) 30 cm
27 9.0×10^{10} J, 9.2×10^7 kg
29 2.84×10^{-13} kg
31 (a) 5.9×10^{19} J, (b) 4.7%
33 6.10×10^{-12} J, 1.44×10^{-19} kg · m/s; 5.8%, 3.9%
35 2.25×10^8 m/s
37 28.3 MeV
39 (a) 4.8×10^{-16} kg, (b) 3.2×10^{-16} kg

CHAPTER 40

1 5.4×10^{-20} J, 0.34 eV
3 (a) 2.7×10^{-34} J, (b) 2.2×10^{35}, (c) 4.5×10^{-36}, no
5 (a) 3.10 eV, (b) 1.77 eV, (c) 4.14×10^{-7} eV
7 0.11 eV, 1.9×10^5 m/s
11 (a) none, (b) 1.17 eV
13 8.3×10^4 s^{-1}
15 (b) 0.041 nm
17 (a) 570 eV, (b) 0.105 nm
19 2.6 MeV
21 212 MeV, 5.85 fm
23 1.6×10^{-11} m
25 1836
27 7.3×10^{-16} m
29 0.15 nm, 0.038 eV
31 5.4×10^{-12} m
33 4.7×10^{-14} m
35 (a) 487 nm, (b) 122 nm, (c) 411 nm
37 91.2 nm
41 1.1×10^{-8}
43 ≈ 4300, -7×10^{-7} eV
45 $7.3 \times 10^{-3}\,c$
47 103 nm, 122 nm, 656 nm
51 (c) 150 eV, (d) 10^{-47} eV, (e) 0.19 nm

CHAPTER 41

1 5.3×10^{-11} m
3 7.2×10^3 m/s
5 (a) 1.2×10^{-34} m, (b) 1.9×10^{-32} kg · m/s, (c) 1.1×10^{-30} m
9 0.53×10^{-10} m
11 1.0 nm
13 50
15 14
19 5.8×10^{-13} m; 0.12 MeV
21 (a) 1.56 e, (b) 1.36×10^{-10} m
23 4.1×10^{-11} m; ∞
25 1.95×10^{-10} m
27 3.3×10^3 C°/min
29 1.7×10^{-4} rad; 340 m

CHAPTER 42

1 3727 MeV/c^2
3 4.8×10^{-15} m, (b) 27
5 (a) 2.3×10^{17} kg/m^3, (b) 180 m, (c) 2.6×10^{-10} m
7 350 MeV
9 2.224 MeV
11 18.72 MeV
13 59.93525
15 $^{228}_{90}$Th; 228.0288
17 (a) $^{32}_{16}$S, (b) 31.97207
19 0.862 MeV
21 (b) 0.960 MeV; 0.960 MeV, 0
23 4.9×10^{-18} s^{-1}, (b) 1.1×10^4 s
25 7, 4
27 5.6×10^{-12} kg
29 41 yr
31 $N_0 (1 - e^{-\lambda t})$
33 19,000 yr
35 no
37 releases 17.35 MeV
39 (a) yes, (b) 6.49 MeV
43 174.7 MeV
45 1.0043
47 1.49
49 1.3 keV
51 9.7×10^{10} J/g, 3.4×10^{11} J/g; 6×10^8 J/g (unenriched uranium) or 8×10^{10} J/g (pure $^{235}_{92}$U)
53 1.6×10^9 J/kg, 30×
55 3700 decays/s
57 90 h
59 4.5 m

CHAPTER 43

1 0.041 fm
3 3×10^{-18} m
5 133.5 MeV
7 (a) 1.022 MeV, (b) 1876.6 MeV
9 1.32 fm
11 37.7 MeV, 41.0 MeV
13 69.3 MeV
15 3.1×10^{-21} s

Index

Electric power, 508–10
  in ac circuits, 510–11, 623
  and impedance matching, 625–26
  transmission of, 592–93
Electric quadrupole, 449 *p*, 476 *p*
Electrode, 499
Electrodynamics *(defn)*, 747
Electromagnet, 542, 573
Electromagnetic force, 91, 428 *fn*,
    846–49, 851, 852, 857 (*see also*
    Electric force, Magnetic force)
Electromagnetic induction, 582 *ff*
Electromagnetic oscillations, 607–10
Electromagnetic spectrum, 643, 696
Electromagnetic (EM) waves, 636–46,
    690 (*see also* Light)
Electrometer, 431, 482–83, 498, 529
Electromotive force, (emf), 498, 499,
    518, 581–85
  back, 589–90
  counter, 589–90
  of generator, 588–89
  Hall, 551–52
  induced, 581–85
  series and parallel, 522
  sources of, 499, 518, 522–23, 581–85
Electron:
  in atoms (*see* Atomic structure), 429
  charge on, 432, 546–47
  as current, 500
  discovery of, 545–46
  as elementary particle, 850–51
  mass of, 547
  measurement of charge on, 546–47
  measurement of *e/m*, 545–46
  in pair production, 778–79
  wavelength of, 780
  wave-particle nature of, 781
  "what is", 781
Electron band theory of solids, 814–15
Electron capture, 828
Electron "cloud," 802, 804 (*see also*
    Probability distributions of
    electrons)
Electron gun, 548
Electronic devices, 815–17
Electron lepton number, 850–51
Electron microscope, 781–82
Electron neutrino, 850–51
Electron-positron pair production,
    778–79
Electron spin, 571, 803
Electron volt (unit), 466
Electroscope, 431, 482–83
Electrostatic unit, 432 *fn*
Electrostatics, 428–92
Electroweak theory, 857–88
Elementary charge, 432

Elementary particles, 845–58
  classification of, 851–52
  and conservation laws, 849–51
  grand unified theories of, 857–58
  lifetimes, 851–52
  short-lived (resonances), 852–53
  stability of, 852–53
  and standard model, 857
  strange, 853–54
  table of, 851, 856
  wavelength of, 845
  Yukawa, 846–48
Elements:
  of lens, 686
  periodic table of, 805–7
EM waves (*see* Electromagnetic waves)
Emf (*see* Electromotive force)
Emission spectra, 783–84, 787–89, 795,
    803, 805
Emissivity, 384
Emitter, 816–17
Energy, 97, 104–30, 183–85
  activation, 355
  binding, 134 *p*, 787, 822–25
  conservation of, 116–22, 392
  degradation of, 420
  in EM waves, 644–65
  equipartition of, 398
  equivalence to mass, 766–68
  and first law of thermodynamics,
    392
  interaction, 474
  internal, 121, 375–76, 393, 419
  ionization, 787
  kinetic, 104–6, 108 *fn*, 183–85,
    767–68
  mechanical, 116–20
  molecular rotational and vibrational,
    397–98
  potential, 107–10, 115–17, 121,
    123–28
  relation to work, 104–10
  in relativity theory, 766–69
  in simple harmonic motion, 265–67
  thermal, 120, 375–76
  threshold, 834
  transformation of, 110, 120, 766–68
  unavailability of, 420
  vibrational, 265–67
  of waves, 290–91
Energy bands, 813–15
Energy density:
  in electric field, 486
  in magnetic field, 604
Energy level (*see* Energy states)
Energy level diagram:
  atomic, 787
  for lasers, 809–11

Energy level diagram (*cont.*)
  in molecules, liquids, and solids,
    812–15
  nuclear, 828
Energy states:
  in atoms, 787–88, 801, 803, 807
  in molecules, 812–13
  in nuclei, 828
Engine, heat (*see* Heat engine)
Enrichment of uranium, 836–37
Entropy, 412–22
  in life processes, 419
Equation of continuity, 243–44
Equation of motion, 261, 272, 274
Equation of state, 338, 340, 362–63
  Clausius, 362
  ideal gas, 340
  van der Waals, 362–63
Equilibrium, 127, 208–17
  thermal, 333–34
Equilibrium position (vibrational motion),
    259
Equipartition of energy, 398
Equipotential lines, 465
Equipotential surface, 465
Erg (unit), 98, 106
Estimating, 9, 73
Ether, 746–51
Evaporation, 355–56, 381
Evolution and entropy, 419
Exchange coupling, 571
Exchange of particles, as carriers of
    force, 846–48
Excited state:
  of atom, 787
  of nucleus, 828
Exclusion principle, 805, 807
Expansion, thermal (*see* Thermal
    expansion)
Extended body, motion of, 139–43 (*see
    also* Rigid body, rotational motion
    of)
Extraordinary ray, 737
Eye, 682–85
  defects of, 683
  far and near points of, 677, 683
  normal *(defn)*, 683
  resolution of, 720
  structure and function of, 682–83
Eyeglass lenses, 683–85
Eyepiece, 679, 681

F

Fahrenheit temperature scale, 331–32
Falling bodies, 24–26

Motion (*cont.*)
  uniform rotational, 168
  vibrational, 259–77
  of waves (*see* Waves)
Motor, electric, 554–55, 589–90
  counter emf in, 589–90
Multimeter, 528
Multiplication factor, in nuclear reactor,
    837
Mu meson (*see* Muon)
Muon, 848, 850–51
Muon lepton number, 850–51
Muon neutrino, 850–51
Muscles and joints, forces in, 212,
    218–21 *p*
Musical instruments, 314–18
Mutual inductance, 599–601
Myopic eye, 683

N

Natural abundances, 821
Natural frequency, 273 (*see also*
    Resonant frequency)
Near point, of eye, 677, 683
Nearsighted eye, 683
Negative electric charge, 429
Negative pressure, 236–37
Neptune, 91
Neutrino, 827–28, 846, 850
Neutron, 820, 822, 851
  delayed, 837 *fn*
  role in fission, 835–37
  used to produce new elements, 834–35
Neutron number, 821
Newton (unit), 61
Newton, Isaac, 1, 12, 59–63, 82–83,
    90–91, 141, 680 *fn*, 691–92, 701
    fn, 745
Newtonian focus, 680
Newton's law of cooling, 388 *p*
Newton's law of universal gravitation,
    82–85
Newton's laws of motion, 58–72, 84, 97,
    139–41, 765
  for rotational motion, 171, 173–74,
    181, 196–98
Newton's rings, 701
Newton's synthesis, 90–91
Nodes, 302
Nonconductors, 430, 442
Noninductive winding, 601
Nonohmic device (*defn*), 501
Normal force, 65, 68
npn transistor, 816–17
n-type semiconductor, 813
Nuclear binding energy, 822–24

Nuclear energy:
  released in fission, 835–36
  released in fusion, 838–39
Nuclear fallout, 838
Nuclear fission, 835–38
  and bomb, 837–38
  and chain reaction, 836
  energy release in, 835–36
  in nuclear reactors, 836–38
Nuclear forces, 91, 823–24, 825
  and binding energy, 823–24
  grand unified theories of, 857–58
  role in radioactive decay, 825, 827
  short range of, 824, 835, 838–39, 848
  standard model of, 857
  strong, 824, 835, 838–39, 847–49,
    851–52, 857–58
  weak, 824, 827, 848–49, 851–52,
    857–58
Nuclear fusion, 838–39
Nuclear masses, 822, 866–69
Nuclear physics, 820–41, (*see also*
    Elementary particles)
Nuclear radiation, dosimetry and
    measurement of, 739–41
Nuclear reactions, 833–35
Nuclear reactors, 836–68
Nuclear transmutation, 825, 833–35
Nuclei
  compound, 835
  daughter, 825
  half-lives of, 830–31
  masses of, 822, 866–69
  parent, 825
  radioactive decay of, 824–33
  size of, 821
  stability of, 824, 825, 832
  structure, 825 *ff*
  table of, 866–69
Nucleon (*defn*), 821
Nuclide (*defn*), 821
  masses of, table of, 866–69
Numerical aperture, 719
Numerical integration, 30–32

O

Object distance, 655, 659, 662, 669,
    670, 673, 674
Objective lens, 679, 681
  oil-immersion, 718–19
Ocular lens (*see* Eyepiece)
Oersted, H. C., 540, 631
Ohm (unit of resistance), 501
Ohm, G. S., 500
Ohmmeter, 528, 537 *p*

Ohm's law, 501, 507
Oil-drop experiment, 546–47
Oil-immersion objective lens, 717–18
Omega particle, 851, 855
Onnes, H. K., 504
Operational definitions, 7
Optical activity, 736–37
Optical illusions, 662
Optical instruments and devices, 670,
    677–85, 703–4, 722–23
Optical pumping, 810
Optical rotatory power, 737
Optic axis, 737
Optics:
  geometrical, 652–86
  physical, 690–741
Orbital quantum number, 803
Order and disorder, 418–20, 422
Order of interference or diffraction
    pattern, 694, 720–21, 728
Order of magnitude and rapid estimating,
    9, 73
Ordinary ray, 737
Oscillations (*see* Vibrations)
Oscillator, electronic, 610
Oscilloscope, 549
Otto cycle, 425 *p*
Overtone, 303

P

Pair production, 778–79
Parabolic reflector, 657
Paramagnetism, 573–74
Parallel axis theorem, 178–79
Paraxial ray, 668
Parent nucleus, 825
Partial pressure, 342–43, 357–58
  Dalton's law of, 343
Particle (*defn*), 12
Particle exchange as carrier of force,
    846–49
Pascal (unit of pressure), 225, 229 (*table*)
Pascal, B., 230
Pascal's principle, 230–31
Paschen series, 784, 787
Pauli, W., 805, 827
Pauli exclusion principle, 805, 807
Pendulum:
  physical, 269–71
  simple, 8–9, 102–3, 267–69
Period:
  of simple pendulum, 8, 9, 268
  of vibration, 260, 263
Periodic motion (*defn*), 259
Periodic table, 805–7

Periodic wave, 284
Permeability:
  magnetic, 561, 566, 572
  relative, 572
Permittivity, 434, 486, 567
Perpendicular-axis theorem, 178–79
Perturbations, 91
Phase:
  in ac circuits, 610, 616–24
  changes of, 358–61, 379–81
  in light beams, 695, 698–700, 712–14,
    724–25, 809
  of matter, 223, 330–31
  of waves, 292, 300
Phase angle, 262, 607
Phase diagram, 359–61
Phase velocity, 292
Phasor diagram:
  ac circuits, 622
  interference and diffraction of light,
    698, 711–15, 724–25
Phonograph cartridge, 593
Photocell, 776
Photoelectric effect, 776–77
Photon as elementary particle, 846, 849,
    851, 857
  virtual, 847
Photon exchange, 846–47
Photon theory of light, 775–79, 797,
    808 ●
  and atoms, 784–85, 788, 801
  (*see also* Gamma rays, X rays)
Photosynthesis, 778
Physical pendulum, 269–71
Piezoelectric effect, 531
Pi meson, (*see* Pion)
Pion, 848, 851
Pipe, vibrating air columns in, 315–18
Pitch of a sound, 310
Planck, Max, 774–75, 784
Planck's constant, 774–75
Planck's quantum hypothesis, 773–75
Plane-polarized light, 732–39, 740, 741
Planetary motion, 89–91
Plane waves, 639
Plasma, 223, 839
Plum-pudding model of the atom, 782
Pluto, 91
Plutonium, 835, 837–38
pn junction diode, 815–16
pnp transistor, 816–17
Point charge, 432
  electric field due to, 436
  electric potential due to, 466–68
Poise (unit), 249
Poiseuille, J. L., 250
Poiseuille's equation, 249–52
Poisson, S., 709

Polar coordinates, 52, 865
Polarization of light, 732–41
  circular, 739–40
  elliptical, 740
  plane, 732–39, 740, 741
  by reflection, 735–36, 741
  of skylight, 740–41
Polarization vector, electric, 491–92
Polarized light (*see* Polarization of light)
Polarizer, 734
Polarizing angle, 736
Polar molecules, 430, 468
Polaroid, 733–35
Poles, magnetic, 538–40
  of Earth, 539–40
Positive electric charge, 429
Positive holes, 814
Positivism, 62
Positron, 827–28, 846, 849
Potential difference, electric, 462, 463
  (*see also* Electric potential)
Potential energy, 107–10
  elastic, 109
  electric, 461–63, 473–74, 485–86
  gravitational, 107–9
  vibrational, 266, 398
  (*see also* Nuclear energy)
Potential-energy diagrams, 125–28
Potentiometer, 529–30
Pound (unit), 61
Power, 128–30, 184
  in forced vibrations, 281
  (*see also* Electric power, Nuclear
    power)
Power factor (ac circuit), 623
Power of a lens, 683–84
Poynting vector, 644–45
Precession, 203–4
Pressure, 225–31
  atmospheric, 227, 228
  of gas, 337–43, 358–63
  gauge, 228
  measurement of, 228–30
  negative, 236–37
  partial, 342–43, 357–58
  units for and conversions, 225,
    228–29
  vapor, 356–58
Pressure amplitude, 311
Pressure transducers, 531
Pressure waves, 311
Prestressed concrete, 217
Primary, of transformer, 591
Principal axis, 657
Principal quantum number, 802
Principle of complementarity, 779, 781
Principle of correspondence, 769, 789,
    796

Prism, 723
Prism binoculars, 664, 681
Probability:
  and determinism, 800–801
  in nuclear decay, 829
  in quantum mechanics, 797–98
  and second law of thermodynamics,
    420–22, 801
  and wave function, 797–98
Probability distributions of electrons,
    802, 804
Problem solving techniques, 22, 72–73
Processes, reversible and irreversible
    (defn), 409
Projectile motion, 47–50, 51, 67
Proper length, 757
Proper time, 755
Proton, 820, 822, 851, 858
  decay of, 858
Pseudovector, 170 *fn*
PT diagram, 359–61
Ptolemy, 2
p-type semiconductor, 814
PV diagram, 358–59, 363, 395

Q

Quadratic formula, 26 *fn*, 861
Quadrupole, electric, 449 *p*, 476 *p*
Quality factor (QF) of nuclear radiation,
    840
Quality of a sound, 318–19
Quality value (Q-value) of resonant
    system, 275, 281 *p*, 614 *p*, 629 *p*
Quantities, base and derived, 6–7
Quantization:
  of angular momentum, 785, 803
  of electric charge, 432, 546
  of energy, 775, 784, 786, 801
Quantum, 775
Quantum chromodynamics (QCD), 857
Quantum condition, Bohr's, 785, 790
Quantum electrodynamics (QED), 847
Quantum hypothesis, Planck's, 773–75
Quantum mechanics, 795–815, 832
  of atoms, 801–7
  Copenhagen interpretation of, 801
  of molecules, 812–13
Quantum number, 785, 802–4
  magnetic, 803
  orbital, 803
  principal, 802
  spin, 803
Quantum theory, 773–817
  of atoms, 784–90, 795, 801–7
  of blackbody radiation, 773–75

Rutherford's model of the atom, 782–84
Rydberg constant, 784, 788

S

Safety factor, 216
Sailboat and Bernoulli's principle, 246–47
Satellites, 87–89
Saturated vapor pressure, 356–58
Scalar, 37, 38
Scalar product, 99–100
Scattering of light, 740–41
Schrödinger, Erwin, 773, 795, 796
Schrödinger wave equation, 796
Search coil, 596 p
Second (unit), 5, 6
Secondary of transformer, 591
Second law of thermodynamics, 404–22
　general statements of, 417, 418
　statistical interpretation of, 420–22
Sedimentation, 80, 254
Sedimentation velocity, 254
Segrè, E., 849
Self-inductance, 601–3
Semiconductor, 430, 503–4, 813–17
　in diodes, transistors, 815–17
　doping of, 813
　energy bands in, 814–15
　intrinsic, (defn), 814
　n and p types, 813–14
　resistivity of, 503–4, 815
Semiconductor circuits and devices, 815–17
Sensitivity of meters, 526, 528
Shear modulus, 214–15
Shear strength, 216
Shear stress, 215
Shells, atomic, 807
SHM (see Simple harmonic motion)
Shock waves, 323–24
Short circuit, 509
Shunt resistor, 526–27
Sign conventions (geometric optics), 662, 669–70, 674–75
Significant figures, 4–5
Silicon, 813
Simple harmonic motion (SHM), 260–67, 276–77
　applied to pendulum, 267–69
Simple harmonic oscillator (SHO), 260–67
Simple pendulum, 8–9, 102–3, 267–69
Simultaneity, 751–53

Single-slit diffraction, 710–15
SI units (Système International), 5–6
Sky, color of, 740–41
Slope, of a curve, 16–17, 19–20, 26–27
Slug (unit), 61
Snell, W., 662
Snell's law, 662–63, 692
Solar constant, 385
Solenoid, 563–65
Solids, 223, 331, 398, 813–15
　electron band theory of, 814–15
　energy levels in, 813–15
　(see also Phase, changes of)
Solid-state devices, 815–17
Solid-state physics (defn), 813
Sonic boom, 324
Sound "barrier," 324
Sounding board, 315
Sound waves, 309–24
　adiabatic character of, 400–1
　Doppler shift of, 321–23
　infrasonic, 310
　intensity of (and dB), 312–14
　interference of, 319–21
　sources of, 314–18
　speed of, 309–10, 400–401
　supersonic, 310
　ultrasonic, 310
Source of emf, 499, 518, 522–23
Space:
　absolute, 746
　relativity of, 756–59
Space-time, 758–59
Special theory of relativity (see Relativity, special theory of)
Specific gravity, 225
Specific heat, 376–78, 396
Spectrograph, mass, 549
Spectrometer:
　light, 723
　mass, 549
Spectrophotometer, 723
Spectroscope and spectroscopy, 722–23, 725
Spectrum:
　absorption, 783, 788
　atomic emission, 783–84, 787–89, 795, 803, 805
　band, 812–13
　continuous, 723, 773
　electromagnetic, 643, 696
　emitted by hot object, 773–75
　of hydrogen, 783–84, 787–88, 803
　line, 723, 783
　molecular, 812–13
　visible light, 696
　x-ray, 807–8

Speed, 13, 15, 17
　of EM waves, 639–43, 646
　molecular, 349–55
　most probable, 354
　rms, 352
　(see also Velocity)
Speed of light, 642–43, 653–54, 746–51
　constancy of, 751
　measurement of, 653–54
　as ultimate speed, 765–66
Speed of sound, 309–10, 400
　supersonic, 310, 323
Spherical aberration, 657, 669, 685
Spin, electron, 571, 803
Spinning top, 203–4
Spin quantum number, 803
Spring, PE of, 109
Spring constant, 102, 260
Spring, vibration of (see Vibrations)
Spyglass, 681
Stable and unstable equilibrium, 127
Standard conditions (STP), 340
Standard model, 857–58
Standards and units, 5–6
Standing waves, 302–4
　circular, 790
State:
　changes of, 358–61, 379–81
　energy (see Energy states)
　equation of, 338, 340, 362–63
　of matter, 223, 330–31
　metastable, 809
　as physical condition of system, 329
State variable, 329, 393, 414, 415
Static electricity, 428–92
Static equilibrium, 208 (defn), 208–17
Static friction, 68–69, 72
Statics, 208–17
Stationary states in atom, 784 (see also Energy states)
Statistics and entropy, 420–22
Steam engine, 405–6
Stefan-Boltzmann constant, 384
Stimulated emission, 809
Stokes's equation, 254
Stokes's theorem, 648
Stopping distances for car, 23–24, 35 p
Strain, 214–15
Strain gauge, 531, 537 p
Strangeness, 853–54
Strange particles, 853–54
Strassmann, F., 835
Streamline (defn), 243
Streamline flow, 242–52
Strength of materials, 213, 216
Stress, 213–17
　compressive, 215
　shear, 215

Weightlessness, 88–89
Wheatstone bridge, 530–31
Wien's displacement law, 774
Wien's radiation theory, 774–75
Wing of airplane, lift on, 246–47
Work, 97–99, 100–106, 110, 114–16,
    184
  compared to heat, 390
  done by a gas, 390–92
  in first law of thermodynamics,
    392–95
  from heat engines, 405 *ff*
  relation to energy, 104–10, 116, 184
Work-energy theorem, 104–6, 116, 119
    fn, 184

Work function, 776
Working fluid, 406

X

X-ray crystallography, 727–28
X-ray diffraction, 727–28
X rays, 643, 727–28, 807–9
  and atomic number, 808
  characteristic, 807–8
  continuous spectrum of, 807–9
  in electromagnetic spectrum, 643

Y

Young, Thomas, 693, 694
Young's double-slit experiment (*see*
    Double slit experiment)
Young's modulus, 213–14
Yukawa, H., 846–48

Z

Z particle, 848–49, 857
Zeeman effect, 557 *p*, 803
Zero, absolute, 339, 423
Zweig, G., 854

CONVERSION FACTORS

Length

1 in = 2.54 cm
1 cm = 0.394 in
1 ft = 30.5 cm
1 m = 39.4 in = 3.28 ft
1 mi = 5280 ft = 1.61 km
1 km = 0.621 mi
1 nautical mile = 6080 ft = 1.85 km
1 fermi = 1 femtometer (fm) = 10^{-15} m
1 angstrom (Å) = 10^{-10} m
1 light-year = 9.46×10^{15} m

Time

1 day = 8.64×10^4 s
1 year = 3.156×10^7 s

Speed

1 mi/h = 1.47 ft/s = 1.61 km/h = 0.447 m/s
1 km/h = 0.278 m/s = 0.621 mi/h
1 ft/s = 0.305 m/s = 0.682 mi/h
1 m/s = 3.28 ft/s = 3.60 km/h
1 knot = 1.151 mi/h = 0.5144 m/s

Angle

1 radian (rad) = 57.30° = 57°18′
1° = 0.01745 rad
1 rev/min (rpm) = 0.1047 rad/s

Mass

1 atomic mass unit (u) = 1.6606×10^{-27} kg
1 slug = 14.6 kg
1 kg = 0.0685 slug
[1 kg has a weight of 2.21 lb where g = 9.80 m/s².]

Force

1 lb = 4.45 N
1 N = 0.225 lb = 10^5 dyne

Energy and Work

1 J = 0.738 ft · lb = 10^7 ergs
1 ft · lb = 1.36 J = 1.29×10^{-3} Btu = 3.25×10^{-4} kcal
1 kcal = 4.18×10^3 J = 3.97 Btu
1 Btu = 252 cal = 778 ft · lb = 1054 J
1 eV = 1.602×10^{-19} J
1 kWh = 3.80×10^6 J = 860 kcal

Power

1 W = 1 J/s = 0.738 ft · lb/s
1 hp (U.S.) = 550 ft · lb/s = 746 W
1 hp (metric) = 750 W

Pressure

1 atm = 1.013 bar = 1.013×10^5 N/m²
        = 14.7 lb/in² = 760 torr
1 lb/in² = 6.90×10^3 N/m²
1 Pa = 1 N/m² = 1.45×10^{-4} lb/in²